BIOCHEMICAL INTERACTIONS

Fundamentals of Biochemistry is bundled with *Biochemical Interactions*, a CD-ROM that expands upon the information presented in the textbook through the use of a variety of interactive three-dimensional molecular graphics displays and animations. These take the form of:

1. **Interactive Exercises**, 60 Chime™-based molecular graphics displays of proteins and nucleic acids that can be interactively rotated and otherwise manipulated. These are keyed to figures in the textbook as is indicated by a disk icon. ⊙

2. **Kinemages**, alternative types of molecular graphics displays. These are presented in the form of 21 Exercises comprising 54 kinemages that amplify specific aspects of protein and nucleic acid structures. They are also keyed to figures in the textbook as is indicated by the disk icon.

3. **Guided Explorations**, more complex interactive computer graphics displays and computerized animations, dealing with specific subjects in the textbook. These are indicated by a disk icon in the margin of the page where each subject is initially discussed.

The Interactive Exercises and Guided Explorations were produced by ScienceMedia Inc in collaboration with Donald Voet and Judith G. Voet. The Kinemages were produced by Donald Voet and Judith G. Voet.

For the student, the CD-ROM extends the learning process from the textbook to the multimedia environment by drawing upon motion, color, and three-dimensionality to illustrate aspects of molecular form and function that would otherwise be difficult to envision.

For the instructor, the CD-ROM is designed to be used as a teaching tool in computer presentation-equipped classrooms.

The Tables of Contents for these exercises (with text references in parentheses) may be found on the preceding pages.

FUNDAMENTALS OF

BIOCHEMISTRY

UPGRADE EDITION

FUNDAMENTALS OF
BIOCHEMISTRY
UPGRADE EDITION

Donald Voet
University of Pennsylvania

Judith G. Voet
Swarthmore College

Charlotte W. Pratt
Seattle, Washington

John Wiley & Sons, Inc.

EXECUTIVE EDITOR   *David Harris*
DEVELOPMENTAL EDITOR   *Barbara Heaney*
NEW MEDIA EDITOR   *Linda Muriello*
MARKETING MANAGER   *Robert Smith*
PRODUCTION MANAGER   *Jeanine Furino*
PRODUCTION EDITOR   *Sandra Russell*
OUTSIDE PRODUCTION MANAGEMENT   *Ingrao Associates*
TEXT DESIGNER   *Madelyn Lesure*
COVER DESIGNER   *Judy Allan, TopDesk Publishers' Group*
PHOTO EDITOR   *Hilary Newman*
ILLUSTRATION EDITOR   *Sigmund Malinowski*
ILLUSTRATORS   *Precision Graphics/ J/B Woolsey Associates*
INDEXER   *Kevin Mulrooney*

The front cover shows some of the molecular assemblies that form the circle of life: ***DNA makes RNA makes protein makes DNA.***

The images are (*clockwise from the top*):

1. B-DNA, *based on an X-ray structure by Richard Dickerson and Horace Drew.*
2. The nucleosome, *courtesy of Timothy Richmond.*
3. Model of the *lac* repressor in complex with DNA and CAP protein, *courtesy of Ponzy Lu and Mitchell Lewis.*
4. Ribozyme RNA, *based on an X-ray structure by Jennifer Doudna.*
5. The ribosome in complex with tRNAs, *courtesy of Joachim Frank.*
6. DNA polymerase in complex with DNA, *courtesy of Tom Ellenberger.*

The central image is based on Leonardo da Vinci's drawing *Study of Proportions.* It represents for us the never ending human quest for understanding. (© G. Bartholomew/ Westlight)

This book was set in 10/12 Times Ten by York Graphic Services, Inc. and printed and bound by Von Hoffmann Press, Inc. The cover was printed by Lehigh Press, Inc.

This book is printed on acid-free paper. ∞

To order books or for customer service please, call 1(800)-CALL-WILEY (225-5945).

Library of Congress Cataloging-in-Publication Data
Voet, Donald.
    Fundamentals of biochemistry upgrade / Donald Voet, Judith G. Voet, Charlotte W. Pratt.—[Rev. ed]

        p.   cm.
    Rev. ed. of Fundamentals of Biochemistry, 1999.
    Includes index.
    ISBN 0-471-41759-9 (cloth : alk. paper)
    1. Biochemistry.   I. Voet, Judith G.   II. Pratt, Charlotte W.   III. Title.

QD415.V63   2001
572 — dc21   CIP   2001017815

Printed in the United States of America

10 9 8 7 6 5

DEDICATION

In memory of Irving Geis,
Artist, teacher, friend

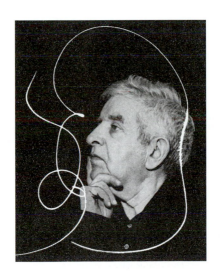

[Courtesy of Sandy Geis]

PREFACE

The pleasure of mastering a difficult subject is matched only by the excitement of successfully conveying our knowledge to others. Our objective in writing this textbook therefore was to approach biochemistry as instructors attuned to students' needs. Accordingly, we have constructed a text that is carefully organized, clearly written, and generously illustrated. We sought not to be encyclopedic but instead to present a broad and lucid survey of biochemistry. There is no substitute for a well-crafted text to guide students through an ever-expanding body of knowledge whose relevance and utility is beyond question.

Readers familiar with *Biochemistry* by Donald Voet and Judith G. Voet will be pleased to find that *Fundamentals of Biochemistry Upgrade Edition* retains the overall philosophy of the larger book but dispenses with the level of detail that some students find overly burdensome. However, *Fundamentals of Biochemistry Upgrade Edition* is by no means an abridgment of *Biochemistry;* rather, it is an entirely new work with its own organization and style. *Fundamentals of Biochemistry Upgrade Edition,* like its parent, presents biochemistry with chemical rigor, focusing on the structures of biomolecules, chemical mechanisms, and evolutionary relationships. It is written to impart a sense of the intellectual history of biochemistry, an understanding of the tools and approaches used to solve biochemical puzzles, and a hint of the excitement that accompanies new discoveries. We have also been attentive to the need for up-to-date coverage, particularly regarding human health and disease, since many students of biochemistry subsequently pursue careers in this area. Ultimately, we hope to convey an appreciation for the awe-inspiring beauty of the structure and chemistry of life.

Organization

Fundamentals of Biochemistry Upgrade Edition is organized into five parts:

1. Two introductory chapters covering the origin of life, evolution, an introduction to thermodynamics, the properties of water, and acid–base chemistry.
2. Eight chapters on biomolecular structure. These cover nucleotides and nucleic acids, amino acids, proteins, carbohydrates, lipids, and biological membranes.
3. Two chapters on enzymes.
4. Ten chapters on metabolism, including an introductory chapter to provide an overview of metabolic pathways, the thermodynamics of "high-energy"

compounds, and redox chemistry. A chapter on the integration of metabolism highlights organ specialization and metabolic regulation in mammals.

5. Five chapters to describe the biochemistry of nucleic acids. An introductory chapter includes a unique section on DNA binding proteins to set the stage for understanding many of the proteins that catalyze and regulate replication, transcription, and translation.

We have organized the material in *Fundamentals of Biochemistry Upgrade Edition* according to the way we would teach it. Yet we recognize that many instructors adhere to different syllabi. The chapters of *Fundamentals of Biochemistry Upgrade Edition* are therefore divided into sections and subsections that make it easy for instructors and students alike to locate particular subjects and to discern the thematic links among them. We hope that this format allays the anxiety of the reader who "skips around" yet fears missing critical information.

Thermodynamics is introduced in Chapter 1 since it is needed to understand the hydrophobic effect (Chapter 2) and protein structure (Chapter 6). The thermodynamics of metabolic reactions is revisited in Chapter 13, the introduction to metabolism. Here, oxidation–reduction reactions are discussed, although the material on electrochemistry could be deferred until Chapter 17 (Electron Transport and Oxidative Phosphorylation).

Early coverage of nucleotides and nucleic acids (Chapter 3) pays homage to the central role these substances play in biochemistry. Virtually every area of protein chemistry depends on cloning, sequencing, expression, and mutagenesis of genes that encode proteins. Therefore, to comprehend how proteins are studied and how proteins can reveal evolutionary history, students should understand how nucleic acids underlie all of biochemistry. This chapter also serves as a review for students who are already familiar with the biological roles of DNA and RNA and allows better coverage of protein evolution before nucleic acids are covered in detail in the last part of the book. It also provides an introduction to nucleotides such as ATP that play an important role in metabolism. Chapter 3 is designed so that it can also be covered later in the course, after other macromolecules have been presented, or in conjunction with Chapter 23, which takes up the finer points of nucleic acid structure and nucleic acid–protein interactions.

The two chapters on enzymes (Chapters 11 and 12) appear between chapters on molecular structure and chapters on metabolism but could just as easily be covered immediately after protein structure (Chapter 6) or the chapter on protein function (Chapter 7). A discussion of enzyme mechanisms (Chapter 11) precedes the treatment of enzyme kinetics (Chapter 12) because it is easier for students to see how enzymes work before being presented with the more abstract reaction kinetics.

Similarly, chapters on carbohydrates (Chapter 8), lipids (Chapter 9), and membranes (Chapter 10) can be covered along with other macromolecules (Chapters 3–7), following enzymes (Chapters 11 and 12), or in conjunction with the corresponding chapters in the metabolism section.

Central metabolic pathways are presented in detail (e.g., glycolysis and the citric acid cycle) so that students can appreciate how individual enzymes catalyze reactions and understand how enzymes work in concert to perform complicated biochemical tasks. The regulation of pathways is also a central feature. Not all pathways are described in full detail, particularly some lipid and amino acid biosynthetic pathways. Instead, key enzymatic reactions are highlighted for their interesting chemistry or regulatory properties. The focus is on mammalian metabolism, with mention of interesting variations in other types of organisms. Thus, students can focus on human health and disease.

Virtually all of carbohydrate metabolism is covered in Chapters 14 and 15. The absence of intervening material permits students to develop a better appreciation of the features of opposing metabolic pathways (e.g., glucose or glycogen synthesis and degradation). A chapter on the integration of mammalian metabolism (Chapter 21) highlights such interorgan metabolic processes as the Cori cycle and the development of diabetes, which can be fully comprehended only in this context.

Chapters 24–26 discuss DNA replication, transcription, and translation in somewhat parallel fashion so that students can more easily spot similarities in the initiation, elongation, and termination phases of these processes.

Chapter 27 deals with a variety of regulatory mechanisms that have been collected here so that the chapters on transcription and translation will not contain material that might be considered optional, thereby making it easier to focus on the primary pathways. Material from Chapter 27, of course, can be interspersed with material from Chapters 25 and 26 according to individual instructors' preferences.

Pedagogical Features

We have built several features into the text to guide students, to help them discern the fundamental principles, and to help them study. To begin, each chapter opens with a figure that illustrates a principle that is covered in that chapter. Material within chapters is arranged in **outline form** to help the student understand the relationships among various topics.

The names of biochemical processes, compounds, enzymes, and diseases are highlighted in boldface at their first appearance. A list of **key terms** at the end of each chapter prompts students for definitions or explanations of the most important biochemical terms. Definitions for these and other terms are included in a **glossary** at the end of the book for easy reference. **Key sentences** emphasizing experimental conclusions and major biochemical principles are italicized.

Key figures and tables, focusing on structure, function, and metabolism, are identified for more careful study. Examples are the mechanism of force generation in muscle (Fig. 7-29), the catalytic mechanism of serine proteases (Fig. 11-26), the reactions of the citric acid cycle Fig. 16-2), and the elongation cycle in *E. coli* ribosomes (Fig. 26-28). **Overview figures** at various points in Chapters 13–21 help students follow complicated metabolic processes.

The **illustration program** includes a variety of types of figures, on the premise that students benefit from seeing biomolecules and processes depicted in different ways, in many cases as presented by the investigators who first described them. Accordingly, the text is illustrated with reproductions of figures from the research literature, computer-generated molecular models, electron micrographs, line drawings, tables, and schematic diagrams.

Those figures that are highlighted with a disk icon (◉) are presented as interactive molecular graphics diagrams on the **CD-ROM** that is placed in the back of this textbook. These are in the form of **Interactive Exercises** (Chime™-based images) and **Kinemages** that students can rotate, animate, and otherwise manipulate. The CD-ROM additionally contains a series of computer graphics-animated **Guided Explorations** that deal with a variety of topics.

Optional enrichment material is placed in **boxes** so that the main text contains fewer digressions. Three types of boxes, which in all cases are clearly linked to the text, offer additional information and food for thought. **Biochemistry in Focus** boxes include descriptions of techniques and approaches to biochemical problems as well as upper-level information that might otherwise be beyond the scope of the text (e.g., Box 3-2, Uses of PCR; Box 11-2, Catalytic Antibodies; Box 23-3, Packaging Viral Nucleic Acids). **Biochemistry in Context** boxes are devoted to topics that are of a more theoretical nature and are intended to prompt students to link their biochemical knowledge to other areas of study (e.g., Box 15-1, Optimizing Glycogen Structure; Box 24-5, Why Doesn't DNA Contain Uracil?). **Biochemistry in Health and Disease** boxes include descriptions of diseases resulting from biochemical defects (e.g., Box 6-4, Diseases Related to Protein Folding; Box 20-2, The Porphyrias).

Chapter summaries repeat the chapter's main points for quick review. A set of **study exercises** allows students to identify major themes of each chapter and to check their mastery of the facts. For example, we ask students to describe the hydrogen bonding pattern of an α helix (Chapter 6), what the metabolic advantage of a substrate cycle is (Chapter 14), and what the functions of the three eukaryotic RNA polymerases are (Chapter 25). Answers to these questions are not provided explicitly but can be found in the text.

Each chapter contains at least 10 thought-provoking **problems.** These are not simple regurgitative exercises but require application of newly mastered principles. **Detailed solutions** to all of these problems are provided at the end of the book.

Sample calculations are included for problems in thermodynamics, pH determination, enzyme kinetics, and redox chemistry.

A few **references,** which are predominantly review articles, are listed at the end of each chapter to provide students with additional information, not to serve as a comprehensive bibliography. Some primary sources of particular importance or historical interest are included. The text also indicates (with URLs provided) how the Internet can be used to access databases on protein and nucleic acid sequences, molecular structures, enzyme classification, and metabolic pathways.

A glossary containing the definitons of many of the biochemical terms used in this textbook is provided at the end of the book.

Supplements

The following supplements to *Fundamentals of Biochemistry Upgrade Edition* are available:

- The CD-ROM that accompanies this textbook is produced by ScienceMedia, Inc. It contains an extensive series of computer graphics-animated Interactive Exercises and Guided Explorations, all keyed to the textbook as indicated by a disk icon (🔵). In addition, the CD contains a set of Kinemages by Donald Voet and Judith G. Voet. These are computer-animated color images of selected proteins and nucleic acids that students can manipulate and which are also keyed to the textbook.

- *Student's Companion to Fundamentals of Biochemistry Upgrade Edition* by Akif Uzman, Joseph Eichberg, William Widger, Donald Voet, Judith G. Voet, and Charlotte W. Pratt. It contains learning objectives, numerous new problems and their detailed answers, key terms and concepts, and chapter summaries.

- An art notebook, *Take Note,* containing selected figures from the textbook, reproduced in black and white and designed to facilitate student note-taking during lectures.

- A CD-ROM containing nearly all of the illustrations from *Fundamentals of Biochemistry Upgrade Edition* to be used for computerized classroom projection or from which to print transparencies.

- A Web site (http://www.wiley.com/college/voetfob) containing a suite of computer-graded quizzes and animated figures. Here the student can find practice quizzes to take on their own to test their knowledge and graded quizzes that an instructor can assign with the grades returned by email. Many of the questions in the practice quizzes are accompanied by animations of figures in the textbook. Animated figures, which can be accessed independently of the quizzes, are indicated by a web icon (✳) following the caption to the corresponding figure in the textbook.

ACKNOWLEDGMENTS

This textbook is the result of the dedicated effort of many individuals, several of whom deserve special mention:

Judith Allan designed the book's cover. Linda Muriello & Ellen Bari guided us through and oversaw the development of the CD-ROM and Web site. Irving Geis provided us with his extraordinary molecular art and gave freely of his wise counsel. Laura Ierardi cleverly combined text, figures, and tables in designing each of the textbook's pages. Suzanne Ingrao, our Production Manager, skillfully managed the production of the textbook. Barbara Heaney, our Developmental Editor, coordinated both the art and writing programs and kept our noses to the grindstone. Madelyn Lesure designed the book's typography and provided valuable artistic advice. Cliff Mills, our Acquisitions Editor, skillfully organized and managed the entire project. Hilary Newman and Ramon Rivera-Moret acquired many of the photographs in the textbook and kept track of all of them. Connie Parks, our copy editor, put the final polish on the manuscript and eliminated large numbers of grammatical and typographical errors. Sandra Russell, Jeanine Furino, and Pamela Kennedy Oborski were our in-house production managers at Wiley. Edward Starr, Sigmund Malinowski, and Ishaya Monokoff coordinated the illustration program.

The atomic coordinates of many of the proteins and nucleic acids that we have drawn for use in this textbook were obtained from the Protein Data Bank at Brookhaven National Laboratory. We created these drawings using the molecular graphics programs RIBBONS by Mike Carson; GRASP by Anthony Nicholls, Kim Sharp, and Barry Honig; and INSIGHT II from BIOSYM Technologies. Many of the drawings generously contributed by others were made using either these programs or MIDAS by Thomas Ferrin, Conrad Huang, Laurie Jarvis, and Robert Langridge; MOLSCRIPT by Per Kraulis; and O by Alwyn Jones.

The interactive computer graphics diagrams that are presented in the CD-ROM that accompanies this textbook are either Chime™ images or Kinemages. Chemscape Chime™, which is based on the program RasMol by Roger Sayle, was developed and generously made publically available by MDL Information Systems, Inc. Kinemages are displayed by the program MAGE, which was written and generously provided by David C. Richardson who also wrote and provided the program PREKIN, which DV and JGV used to help generate the Kinemages.

We wish especially to thank those colleagues who reviewed this text:

Marjorie A. Bates
University of California at Los Angeles

Charles E. Bowen
California Polytechnic University

Caroline Breitenberger
The Ohio State University

Scott Champney
East Tennessee State University

Kathleen Cornely
Providence College

Bonnie Diehl
The Johns Hopkins University

Jacquelyn Fetrow
University of Albany

Jeffrey A. Frick
Illinois Wesleyan University

Michael E. Friedman
Auburn University

Arno L. Greenleaf
Duke University

Michael D. Griswold
Washington State University

James Hageman
New Mexico State University

Lowell P. Hager
University of Illinois at Urbana-Champaign

LaRhee Henderson
Drake University

Diane W. Husic
East Stroudsburg University

Larry L. Jackson
Montana State University

Jason D. Kahn
University of Maryland at College Park

Barrie Kitto
University of Texas

Anita S. Klein
University of New Hampshire

Paul C. Kline
Middle Tennessee State University

W. E. Kurtin
Trinity University

Robley J. Light
Florida State University

Robert D. Lynch
University of Massachusetts-Lowell

Dave Mascotti
John Carroll University

Gary E. Means
The Ohio State University

Laura Mitchell
Saint Joseph's University

Tim Osborne
University of California at Irvine

G. R. Parslow
University of Melbourne

Allen T. Phillips
Pennsylvania State University

Leigh Plesniak
University of San Diego

Stephan Quirk
Georgia Institute of Technology

Raghu Sarma
State University of New York at Stony Brook

Bryan Spangelo
University of Nevada at Las Vegas

Gary Spedding
Butler University

Pam Stacks
San Jose State University

Scott Taylor
University of Toronto

David C. Teller
University of Washington

Steven B. Vik
Southern Methodist University

Jubran M. Wakim
Middle Tennessee State University

Joseph T. Warden
Rensselaer Polytechnic Institute

William Widger
University of Houston

Bruce Wightman
Muhlenberg College

Kenneth O. Willeford
Mississippi State University

Robert P. Wilson
Mississippi State University

Adele Wolfson
Wellesley College

Cathy Yang
Rowan University

Leon Yengoyan
San Jose State University

Ryland F. Young
Texas A&M University

Finally, DV and JGV wish to thank Joel Sussman, Michal Harel, and their colleagues at the Weizmann Institute of Science, Israel, for their sumptuous hospitality and stimulating conversations while we wrote much of this textbook, and Yeda Computers for their generous loan of a Macintosh computer for our use at the Weizmann Institute.

BRIEF CONTENTS

CONTENTS

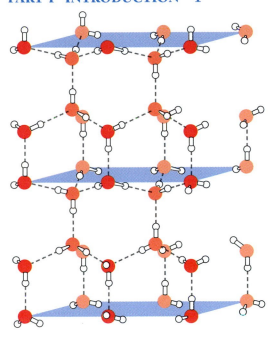
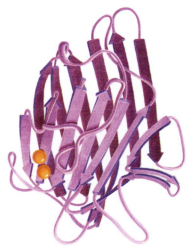

PART III   ENZYMES   279

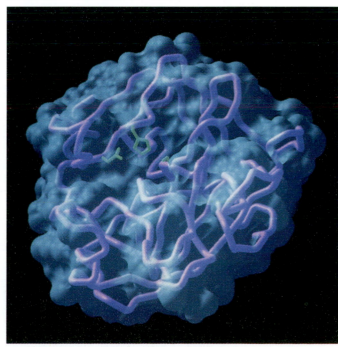

CHAPTER 11   ENZYMATIC CATALYSIS   281

PART IV   METABOLISM   351

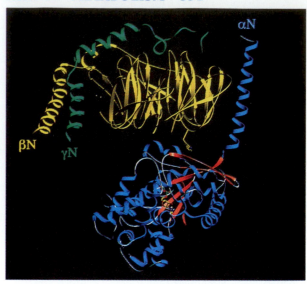

PART V   GENE EXPRESSION AND REPLICATION   723

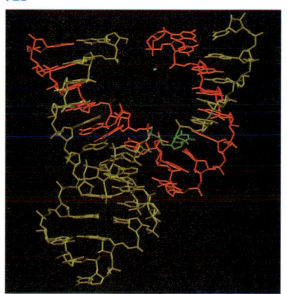

CHAPTER 24   DNA REPLICATION, REPAIR, AND RECOMBINATION   772

CHAPTER 25   TRANSCRIPTION AND RNA PROCESSING   813

CHAPTER 26   TRANSLATION   844

A NOTE TO STUDENTS

You are about to embark on a voyage of discovery, that of the chemistry of life. This body of knowledge, compiled by the skilled and dedicated efforts of many tens of thousands of researchers over more than a century, has had an enormous impact on medicine, agriculture, and the way we view ourselves and our world. It forms an epic and awe-inspiring tale that has provided a lifetime of fascination for many of those who study it.

Before you begin your voyage, we have a few words of advice. The vocabulary of biochemistry, to which you will be introduced here, is nearly as rich as that of a foreign language, and many chapters build on their predecessors. Thus, it is important to keep your studies current with the course lectures. The end-of-chapter materials, including Study Exercises and Problems, are designed to aid you in mastering and applying principles. Try to complete the Problems before you look at the Solutions. Avail yourselves of additional information in the supplements, on the Internet, and at your library, using the References as a guide. Biochemistry is a challenging subject that yields to hard work and diligence. The more effort you put into learning the material, the more rewarding these efforts will be. Enjoy meeting the challenge, and bon voyage!

I

INTRODUCTION

Early earth, a tiny speck in a galaxy, contained simple inorganic molecules that gave rise to the first biological macromolecules. These, in turn, gained the ability to self-organize and self-replicate, eventually forming cellular life forms.
[© Lynette Cook/Photo Researchers.]

LIFE

Biochemistry is, literally, the study of the chemistry of life. Although it overlaps other disciplines, including cell biology, genetics, immunology, microbiology, pharmacology, and physiology, biochemistry is largely concerned with a limited number of issues. These are

1. What are the chemical and three-dimensional structures of biological molecules?
2. How do biological molecules interact with each other?
3. How does the cell synthesize and degrade biological molecules?
4. How is energy conserved and used by the cell?
5. What are the mechanisms for organizing biological molecules and coordinating their activities?
6. How is genetic information stored, transmitted, and expressed?

Biochemistry, like other modern sciences, relies on sophisticated instruments to dissect the architecture and operation of systems that are inaccessible to the human senses. In addition to the chemist's tools for separating, quantifying, and otherwise analyzing biological materials, biochemists take advantage of the uniquely biological aspects of their subject by examining the evolutionary histories of organisms, metabolic systems, and individual molecules. In addition to its obvious implications for human health, biochemistry reveals the workings of the natural world, allowing us to understand and appreciate the unique and mysterious condition that we call life.

1. THE ORIGIN OF LIFE

Certain biochemical features are common to all organisms: for example, the way hereditary information is encoded and expressed, and the way biological molecules are built and broken down for energy. The underlying genetic and biochemical unity of modern organisms suggests they are descended from a single ancestor. All known cultures, past and present, have some sort of creation myth that rationalizes how life first arose. Only in the modern era, however, has it been possible to consider the origin of life in scientific terms.

A. The Prebiotic World

Living matter consists of a relatively small number of elements (Table 1-1). C, N, O, H, Ca, P, K, and S account for ~98% of the dry weight of living things (most organisms are ~70% water). The balance consists of elements that are present in only trace quantities. With the exceptions of oxygen and calcium, the biologically most abundant elements are only minor constituents of the earth's crust (which contains 47% O, 28% Si, 7.9% Al, 4.5% Fe, and 3.5% Ca). The mechanism by which life evolved from this somewhat limited set of elements is not known with certainty, but a plausible scenario based on paleontological and laboratory evidence is put forth below.

The earliest known fossil evidence of life is ~3.5 billion years old (Fig. 1-1). The preceding **prebiotic era,** which began with the formation of the earth ~4.6 billion years ago, left no direct record, but scientists can experimentally duplicate the sorts of chemical reactions that might have given rise to living organisms during that billion-year period.

The atmosphere of the early earth probably consisted of H_2O, N_2, CO_2, and smaller amounts of CH_4, NH_3, SO_2, and possibly H_2. In the 1930s,

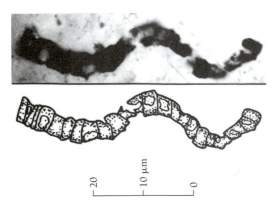

Figure 1-1. **Microfossil of filamentous bacterial cells.** This fossil (shown with an interpretive drawing) is from ~3400 million-year-old rock from Western Australia. [Courtesy of J. William Schopf, UCLA.]

Table 1-1.   Elemental Composition of the Human Body[a]

Element	Dry Weight (%)	Elements Present in Trace Amounts
C	61.7	B
N	11.0	F
O	9.3	Si
H	5.7	V
Ca	5.0	Cr
P	3.3	Mn
K	1.3	Fe
S	1.0	Co
Cl	0.7	Cu
Na	0.7	Zn
Mg	0.3	Se
		Mo
		Sn
		I

[a]Calculated from Frieden, E., *Sci. Am.* **227**(1), 54–55 (1972).

Alexander Oparin and J. B. S. Haldane independently suggested that ultraviolet radiation from the sun or lightning discharges caused the molecules of the primordial atmosphere to react to form simple **organic** (carbon-containing) compounds. This reaction process was replicated in 1953 by Stanley Miller and Harold Urey, who subjected a mixture of H_2O, CH_4, NH_3, and H_2 to an electric discharge for about a week. The resulting solution contained water-soluble organic compounds, the most abundant of which are listed in Table 1-2. Several of the soluble compounds are amino acid components of proteins and many of the others, as we shall see, are also biochemically significant.

Table 1-2.   Yields from Sparking a Mixture of CH_4, NH_3, H_2O, and H_2

Compound	Yield (%)
Formic acid	4.0
Glycine[a]	2.1
Glycolic acid	1.9
Alanine[a]	1.7
Lactic acid	1.6
β-Alanine	0.76
Propionic acid	0.66
Acetic acid	0.51
Iminodiacetic acid	0.37
α-Amino-*n*-butyric acid	0.34
α-Hydroxybutyric acid	0.34
Succinic acid	0.27
Sarcosine	0.25
Iminoaceticpropionic acid	0.13
N-Methylalanine	0.07
Glutamic acid[a]	0.051
N-Methylurea	0.051
Urea	0.034
Aspartic acid[a]	0.024
α-Aminoisobutyric acid	0.007

[a]Amino acid constituent of proteins.
Source: Miller, S.J. and Orgel, L.E., *The Origins of Life on Earth*, p. 85, Prentice–Hall (1974).

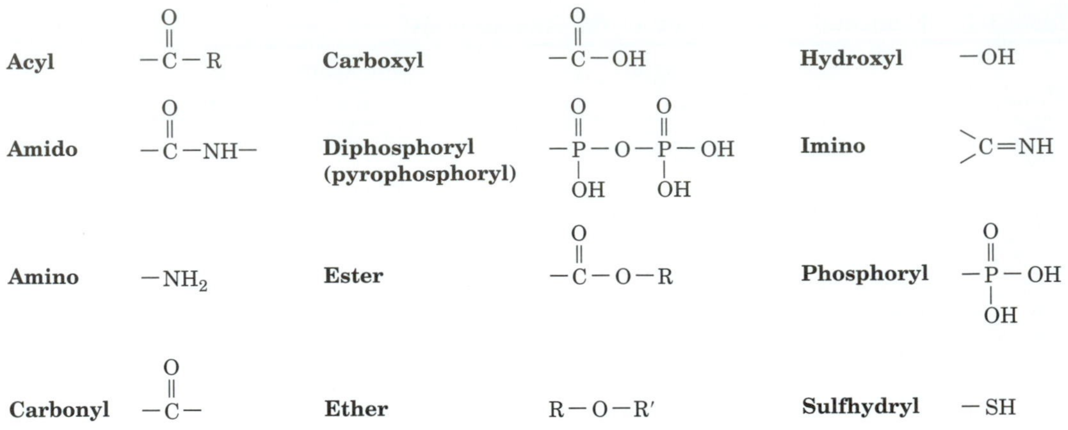

Acyl	$-\overset{\overset{\displaystyle O}{\|\|}}{C}-R$	**Carboxyl**	$-\overset{\overset{\displaystyle O}{\|\|}}{C}-OH$	**Hydroxyl**	$-OH$
Amido	$-\overset{\overset{\displaystyle O}{\|\|}}{C}-NH-$	**Diphosphoryl** (pyrophosphoryl)	$-\overset{\overset{\displaystyle O}{\|\|}}{\underset{\underset{\displaystyle OH}{\|}}{P}}-O-\overset{\overset{\displaystyle O}{\|\|}}{\underset{\underset{\displaystyle OH}{\|}}{P}}-OH$	**Imino**	$>C=NH$
Amino	$-NH_2$	**Ester**	$-\overset{\overset{\displaystyle O}{\|\|}}{C}-O-R$	**Phosphoryl**	$-\overset{\overset{\displaystyle O}{\|\|}}{\underset{\underset{\displaystyle OH}{\|}}{P}}-OH$
Carbonyl	$-\overset{\overset{\displaystyle O}{\|\|}}{C}-$	**Ether**	$R-O-R'$	**Sulfhydryl**	$-SH$

Figure 1-2. **Key to Structure. Common functional groups in biochemistry.** Amino, carboxyl, and phosphoryl groups are frequently ionized under physiological conditions.

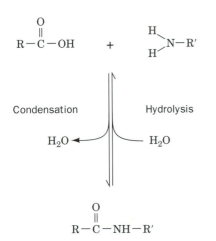

Figure 1-3. **Reaction of a carboxylic acid with an amine.** The elements of water are released during condensation. In the reverse process—hydrolysis—water is added to cleave the amide bond. In living systems, condensation reactions are not freely reversible.

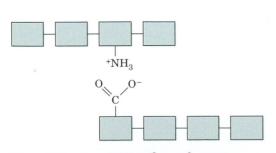

Figure 1-4. **Association of complementary molecules.**

Other experiments have shown that HCN and formaldehyde (CH_2O) can give rise to nucleic acid bases and sugars. It is probably no accident that these compounds are the basic components of biological molecules: They were apparently among the most common organic substances in prebiotic times. The so-called **functional groups** in these substances are critical for their modern-day biological activities. Some of the most common functional groups in biochemistry are shown in Fig. 1-2.

B. Chemical Evolution

Even with the raw materials of life present, and possibly abundant where they had accumulated in tidal pools or shallow lakes, life did not immediately arise. During a period of **chemical evolution,** simple molecules condensed to form more complex molecules or combined end to end as **polymers** of repeating units. In a **condensation reaction,** the elements of water are lost. The rate of condensation of simple compounds to form a stable polymer must therefore be greater than the rate of **hydrolysis** (splitting by adding the elements of water; Fig. 1-3). In the prebiotic environment, minerals such as clay may have catalyzed polymerization reactions and sequestered the reaction products from water. The size and composition of prebiotic macromolecules would have been limited by the availability of small molecular starting materials, the efficiency with which they could be joined, and their resistance to degradation.

Obviously, *combining different functional groups into a single large molecule increases the chemical versatility of that molecule,* allowing it to perform chemical feats beyond the reach of simpler molecules. (This principle of emergent properties can be expressed as "the whole is greater than the sum of its parts.") Separate macromolecules with complementary arrangements of functional groups can associate with each other (Fig. 1-4), giving rise to more complex molecular assemblies with an even greater range of functional possibilities.

Specific pairing between complementary functional groups means that one member of a pair can determine the identity and orientation of the other member. *Such* **complementarity** *makes it possible for a macromolecule to* **replicate,** *or copy itself, by directing the assembly of a new molecule from smaller complementary units.* Replication of a simple polymer with

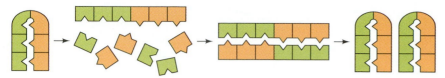

Figure 1-5.   **Replication through complementarity.** In this simple case, a polymer serves as a template for the assembly of a complementary molecule, which, because of intramolecular complementarity, is an exact copy of the original.

intramolecular complementarity is illustrated in Fig. 1-5. A similar phenomenon is central to the function of DNA, where the sequence of bases on one strand (e.g., A-C-G-T) absolutely specifies the sequence of bases on the strand to which it is paired (T-G-C-A). When DNA replicates, the two strands separate and direct the synthesis of complementary daughter strands. Complementarity is also the basis for transcribing DNA into RNA and for translating RNA into protein.

A critical moment in chemical evolution was the transition from systems of randomly generated molecules to systems in which molecules were organized and specifically replicated. Once macromolecules gained the ability to self-perpetuate, the primordial environment would have become enriched in molecules that were best able to survive and multiply. The first replicating systems were no doubt somewhat sloppy, with progeny molecules being imperfectly complementary to their parents. Over time, **natural selection** would have favored molecules that made more accurate copies of themselves.

2. CELLULAR ARCHITECTURE

The types of systems described so far would have had to compete with all the other components of the primordial "pond" for the available resources. Changing environmental conditions might also have influenced the survival of a self-replicating system. A selective advantage would have accrued to a system that was sequestered and protected by boundaries of some sort. How these boundaries first arose, or even what they were made from, is obscure. One theory is that membranous **vesicles** (fluid-filled sacs) first attached to and then enclosed self-replicating systems. These vesicles would have become the first cells.

A. The Evolution of Cells

The advantages of **compartmentation** are several. In addition to receiving some protection from adverse environmental effects, an enclosed system can maintain high local concentrations of components that would otherwise diffuse away. More concentrated substances can react more readily, leading to increased efficiency in polymerization and other types of chemical reactions.

A membrane-bounded compartment that protected its contents would gradually become quite different in composition from its surroundings. Modern cells contain high concentrations of ions, small molecules, and large molecular aggregates that are found in only traces—if at all—outside the cell. For example, the *Escherichia coli* cell contains millions of molecules representing some 3000 to 6000 different compounds (Fig. 1-6). A typical animal cell may contain 100,000 different types of molecules.

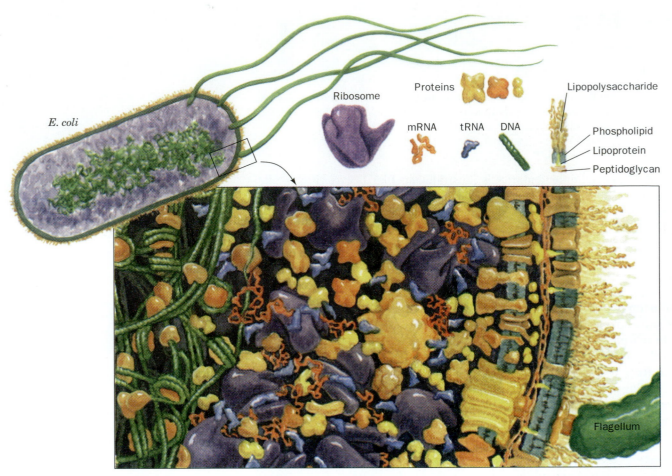

Proteins
Ribosome
mRNA   tRNA   DNA
Lipopolysaccharide
Phospholipid
Lipoprotein
Peptidoglycan
E. coli
Flagellum

Figure 1-6.   Cross section of an *E. coli* cell. The right side of the drawing shows the multilayered cell wall and membrane. The cytoplasm in the middle region of the drawing is filled with ribosomes engaged in protein synthesis. The left side of the drawing contains a dense tangle of DNA. This drawing corresponds to a millionfold magnification. Only the largest macromolecules and molecular assemblies are shown. In a living cell, the remaining space in the cytoplasm would be crowded with smaller molecules and water (the water molecules would be about the size of the period at the end of this sentence). [After a drawing by David Goodsell, UCLA.]

Early cells depended on the environment to supply building materials. As some of the essential components in the prebiotic soup became scarce, natural selection favored organisms that developed mechanisms for synthesizing the required compounds from simpler but more abundant **precursors.** The first metabolic reactions may have used metal or clay catalysts co-opted from the inorganic surroundings (a catalyst is a substance that promotes a chemical reaction without itself being changed). In fact, metal ions are still at the heart of many chemical reactions in modern cells. Some catalysts may also have arisen from polymeric molecules that had the appropriate functional groups.

In general, biosynthetic reactions require energy; hence, the first cellular reactions also needed an energy source. The eventual depletion of pre-existing energy-rich substances in the prebiotic environment would have stimulated the development of energy-producing metabolic pathways. For example, photosynthesis evolved relatively early to take advantage of a practically inexhaustible energy supply, the sun. However, the accumulation of O_2 generated from H_2O by photosynthesis (the modern atmosphere is 21% O_2) presented an additional challenge to organisms adapted to life in an oxygen-poor atmosphere. Metabolic refinements eventually permitted organisms not only to avoid oxidative damage but to use O_2 for oxidative metabolism, a much more efficient form of energy metabolism than anaerobic metabolism. Vestiges of ancient life can be seen in the anaerobic metabolism of certain modern organisms.

Early organisms that developed metabolic strategies to synthesize biological molecules, conserve and utilize energy in a controlled fashion, and replicate within a protective compartment were able to propagate in an ever-widening range of habitats. Adaptation of cells to different external conditions ultimately led to the present diversity of species. Specialization of individual cells also made it possible for groups of differentiated cells to work together in multicellular organisms.

B. Prokaryotes and Eukaryotes

All modern organisms are based on the same morphological unit, the cell. There are two major classifications of cells: the **eukaryotes** (Greek: *eu*, good or true + *karyon*, kernel or nut), which have a membrane-enclosed **nucleus** encapsulating their DNA; and the **prokaryotes** (Greek: *pro*, before), which lack a nucleus. *Prokaryotes, comprising the various types of bacteria, have relatively simple structures and are invariably unicellular* (although they may form filaments or colonies of independent cells). *Eukaryotes, which are multicellular as well as unicellular, are vastly more complex than prokaryotes.* (**Viruses** are much simpler entities than cells and are not classified as living because they lack the metabolic apparatus to reproduce outside their host cells.)

Prokaryotes are the most numerous and widespread organisms on earth. This is because their varied and often highly adaptable metabolisms suit them to an enormous variety of habitats. Prokaryotes range in size from 1 to 10 μm and have one of three basic shapes (Fig. 1-7): spheroidal (cocci), rodlike (bacilli), and helically coiled (spirilla). Except for an outer cell membrane, which in most cases is surrounded by a protective cell wall, prokaryotes lack cellular membranes. However, the prokaryotic **cytoplasm** (cell contents) is by no means a homogeneous soup. Different metabolic functions are believed to be carried out in different regions of the cytoplasm (see Fig. 1-6). The best-characterized prokaryote is *Escherichia coli*, a 2 μm by 1 μm rodlike bacterium that inhabits the mammalian colon.

Eukaryotic cells are generally 10 to 100 μm in diameter and thus have a thousand to a million times the volume of typical prokaryotes. It is not size, however, but a profusion of membrane-enclosed **organelles** that best characterizes eukaryotic cells (Fig. 1-8). In addition to a nucleus, eukaryotes have an **endoplasmic reticulum,** the site of synthesis of many cellular components, some of which are subsequently modified in the **Golgi apparatus.** Aerobic metabolism takes place in **mitochondria** in almost all eukaryotes, and photosynthetic cells contain **chloroplasts.** Other organelles, such as **lysosomes** and **peroxisomes,** perform specialized functions. **Vacuoles,** which are more prominent in plant cells, usually function as storage depots. The **cytosol** (the cytoplasm minus its membrane-bounded organelles) is organized by the **cytoskeleton,** an extensive array of filaments that also gives the cell its shape and the ability to move.

The various organelles that compartmentalize eukaryotic cells represent a level of complexity that is largely lacking in prokaryotic cells. Nevertheless, prokaryotes are actually more efficient than eukaryotes in many respects. Prokaryotes have exploited the advantages of simplicity and miniaturization. Their rapid growth rates permit them to occupy ecological niches in which there may be drastic fluctuations of the available nutrients. In con-

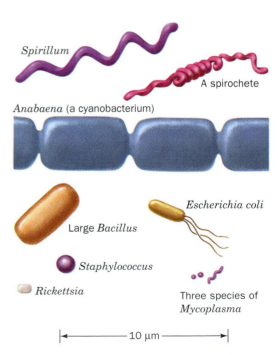

Figure 1-7.   **Scale drawings of some prokaryotic cells.**

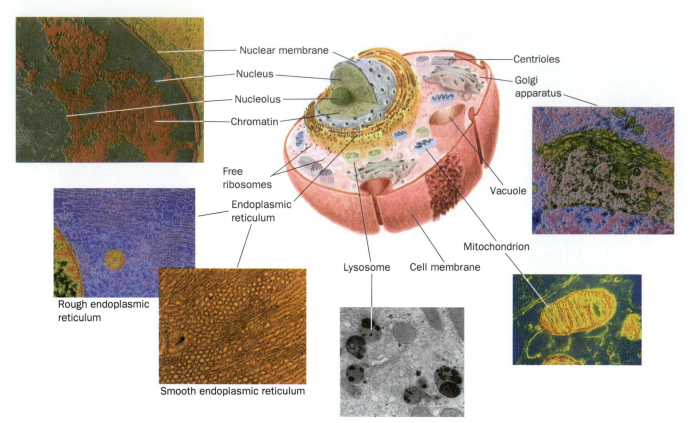

Figure 1-8.   **Diagram of a typical animal cell accompanied by electron micrographs of its organelles.** Membrane-bounded organelles include the nucleus, endoplasmic reticulum, lysosome, peroxisome (not pictured), mitochondrion, vacuole, and Golgi apparatus. The nucleus contains chromatin (a complex of DNA and protein) and the nucleolus (the site of ribosome synthesis). The rough endoplasmic reticulum is studded with ribosomes; the smooth endoplasmic reticulum is not. A pair of centrioles help organize cytoskeletal elements. A typical plant cell differs mainly by the presence of an outer cell wall and chloroplasts in the cytosol. [Nucleus: Tektoff-RM, CNRI/Photo Researchers; rough endoplasmic reticulum and Golgi apparatus: Secchi-Lecaque/ Roussel-UCLAF/CNRI/Photo Researchers; smooth endoplasmic reticulum: David M. Phillips/Visuals Unlimited; mitochondrion: CNRI/Photo Researchers; lysosome: Biophoto Associates/Photo Researchers.]

trast, the complexity of eukaryotes, which renders them larger and more slowly growing than prokaryotes, gives them the competitive advantage in stable environments with limited resources. It is therefore erroneous to consider prokaryotes as evolutionarily primitive compared to eukaryotes. Both types of organisms are well adapted to their respective lifestyles.

3. ORGANISMAL EVOLUTION

Tracing the origins of different species, that is, defining their probable evolutionary history, is valuable because *the biological purpose of a particular biochemical adaptation is often best appreciated by examining how it evolved.* Such evolutionary information is often as useful as information about structure and chemistry for understanding how life operates at the molecular level.

A. Taxonomy and Phylogeny

The practice of lumping all prokaryotes in a single category based on what they lack—a nucleus—obscures their metabolic diversity and evolutionary history. Conversely, the remarkable morphological diversity of eukaryotic organisms (consider the anatomical differences among, say, an amoeba, an oak tree, and a human being) masks their fundamental similarity at the cellular level. Traditional taxonomic schemes (**taxonomy** is the science of biological classification), which are based on gross morphology, have proved inadequate to describe the actual relationships between organisms as revealed by their evolutionary history (**phylogeny**). In fact, the nomenclature devised by Carolus Linnaeus in the mid-1700s, with its hierarchy of kingdom, phylum, class, order, family, genus, and species, imposes artificial distinctions on more than a few groups of organisms and is all but inapplicable to prokaryotes. (However, the convenience of the Linnaean system for identifying unique organisms by a genus and species name is not likely to be abandoned any time soon.)

Biological classification schemes based on reproductive or developmental strategies more accurately reflect evolutionary history than those based solely on adult morphology. But *phylogenetic relationships are best deduced by comparing polymeric molecules—RNA, DNA, or protein—from different organisms*. For example, analysis of RNA led Carl Woese to group all organisms into three domains (Fig. 1-9). The **archaea** (also known as **archaebacteria**) are a group of prokaryotes that are as distantly related to other prokaryotes (the **bacteria,** sometimes called **eubacteria**) as both groups are to eukaryotes (**eukarya**). The archaea include some unusual organisms: the **methanogens** (which produce CH_4), the **halobacteria** (which thrive in concentrated brine solutions), and certain **thermophiles** (which inhabit hot springs). Evidence that the archaeotes are not just unusual bacteria lies in their genetic material, some of which is more closely similar to that of eukaryotes than bacteria.

The pattern of branches in Woese's diagram indicates the divergence of different types of organisms (each branch point represents a common ancestor). The three-domain scheme also shows that animals, plants, and fungi constitute only a small portion of all life forms. Such phylogenetic trees supplement the fossil record, which provides a patchy record of life prior to about 600 million years before the present (multicellular organisms arose about 700–900 million years ago).

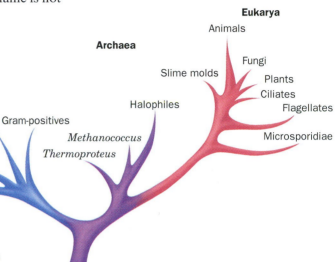

Figure 1-9.　Phylogenetic tree showing three domains of organisms. The branches indicate the pattern of divergence from a common ancestor. The archaea are prokaryotes, like bacteria, but share some features with eukaryotes. [After Wheelis, M.L., Kandler, O., and Woese, C.R., *Proc. Natl. Acad. Sci.* **89**, 2931 (1992).]

B. The Origins of Complexity

The last common ancestor of bacteria, archaea, and eukaryotes was no doubt a relatively complex organism, which accounts for the shared characteristics of all present-day organisms. *It is unlikely that eukaryotes are descended from a highly developed prokaryote because the differences between bacteria and eukaryotes are so profound*. The first eukaryote instead appears to have evolved from a primordial life form that was relatively rare, according to fossil evidence. Only after it developed complex membrane-bounded organelles did it become successful enough to generate significant fossil remains.

Among the significant evolutionary developments that produced the present-day variety of bacteria, archaea, and eukaryotes is the emergence of mechanisms for sexual reproduction. The exchange of genetic material between organisms increased their adaptability to changing conditions. A related development was the appearance of multiple chromosomes (prokaryotes typically have only one chromosome), which allowed eukaryotes to efficiently store and replicate greater amounts of genetic material.

Additional clues to the origin of eukaryotic cellular complexity lie in the mitochondria and chloroplasts. Both organelles resemble bacteria in size and shape, and both contain their own genetic material and protein synthetic machinery. These observations led Lynn Margulis to hypothesize that mitochondria and chloroplasts evolved from free-living aerobic bacteria that formed **symbiotic** relationships with primordial eukaryotes. Presumably, the eukaryotic host provided nutrients and a protected environment to the prokaryotic symbiont and was repaid severalfold by the highly efficient aerobic metabolic feats of the prokaryote. This hypothesis is corroborated by the observation that certain eukaryotes that lack mitochondria or chloroplasts permanently harbor symbiotic bacteria.

At some point in evolutionary history, individual eukaryotic cells acting in a mutually beneficial manner gave rise to multicellular organisms for whom the division of labor provided a competitive advantage. Similar principles no doubt characterized the development of higher order systems, such as societies of individuals and interacting species within an ecosystem.

C. How Do Organisms Evolve?

The natural selection that guided prebiotic evolution continues to direct the evolution of organisms. Richard Dawkins has likened evolution to a blind watchmaker capable of producing intricacy by accident, although such an image fails to convey the vast expanse of time and the incremental, trial-and-error manner in which complex organisms emerge. Small **mutations** (changes in an individual's genetic material) arise at random as the result of physical damage or inherent errors in the replication process. *A mutation that increases the chances of survival of the individual increases the likelihood that the mutation will be passed on to the next generation.* Beneficial mutations tend to spread rapidly through a population; deleterious changes tend to die along with the organisms that harbor them.

Although the theory of evolution by natural selection was first articulated by Charles Darwin in the 1860s, it has only recently been tested experimentally. Bacteria are particularly useful for studies of evolution, since they rapidly reproduce under laboratory conditions, with generation times as short as 20 min. Richard Lenski, for example, has documented changes in fitness (adaptation to certain conditions) over thousands of bacterial generations. The concurrence of experimental evidence, molecular information, and the fossil record highlight some important—and often misunderstood—principles of evolution:

1. **Evolution is not directed toward a particular goal.** It proceeds by random changes that may affect the ability of an organism to reproduce under the prevailing conditions. An organism that is well adapted to its environment may fare better or worse when conditions change.

2. **Evolution requires some built-in sloppiness,** which allows organisms to adapt to unexpected changes. This is one reason why genetically homogeneous populations (e.g., some crop species) are so susceptible to a single challenge (e.g., an insect). A more heterogeneous population has greater means to resist the adversity and recover.

3. **Evolution is constrained by its past.** New structures and metabolic functions emerge from preexisting elements. For example, insect wings did not erupt spontaneously but probably developed gradually from small heat-exchange structures.

4. **Evolution is ongoing,** although it does not proceed exclusively toward complexity. An anthropocentric view places human beings at the pinnacle of an evolutionary scheme, but a quick survey of life's diversity reveals that simpler species have not died out or stopped evolving.

The near universal acceptance of evolutionary theory lies in its explanatory power. Yet no matter how consistently experimental observations conform to the theory of evolution by natural selection, the complexity of evolutionary history makes it impossible to say exactly what happened and when. Nor is it possible, at our current level of comprehension, to predict where evolution is headed.

4. THERMODYNAMICS

The normal activities of living organisms—moving, growing, reproducing—demand an almost constant input of energy. Even at rest, organisms devote a considerable portion of their biochemical apparatus to the acquisition and utilization of energy. The study of energy and its effects on matter fall under the purview of **thermodynamics** (Greek: *therme,* heat + *dynamis,* power). Although living systems present some practical challenges to thermodynamic analysis, *life obeys the laws of thermodynamics.*

A. *The First Law of Thermodynamics: Energy Is Conserved*

In thermodynamics, a **system** is defined as the part of the universe that is of interest, such as a reaction vessel or an organism; the rest of the universe is known as the **surroundings.** *The first law of thermodynamics states that energy (U) is conserved;* it can be neither created nor destroyed. The energy change of a system is defined as the difference between the **heat** (**q**) absorbed by the system from the surroundings, and the **work** (**w**) done by the system on the surroundings.

$$\Delta U = U_{\text{final}} - U_{\text{initial}} = q - w \qquad [1\text{-}1]$$

Heat is a reflection of random molecular motion, whereas work, which is defined as force times the distance moved under its influence, is associated with organized motion. Force may assume many different forms, including the gravitational force exerted by one mass on another, the expansional force exerted by a gas, the tensional force exerted by a spring or muscle fiber, the electrical force of one charge on another, and the dissipative forces of friction and viscosity.

Most biological processes take place at constant pressure. Under such conditions, the work done by the expansion of a gas (pressure–volume work) is $P\Delta V$. Consequently, it is convenient to define a new thermodynamic quantity, the **enthalpy** (Greek: *enthalpein,* to warm in), abbreviated **H**:

$$H = U + PV \qquad [1\text{-}2]$$

Then at constant pressure,

$$\Delta H = \Delta U + P\Delta V = q_P - w + P\Delta V \qquad [1\text{-}3]$$

where q_P is defined as the heat at constant pressure. Now, admitting only pressure–volume work (other types of work in chemical reactions are usually negligible),

$$\Delta H = q_P - P\Delta V + P\Delta V = q_P \qquad [1\text{-}4]$$

Box 1-1

BIOCHEMISTRY IN FOCUS

Biochemical Conventions

Modern biochemistry generally uses Système International (SI) units, including meters (m), kilograms (kg), and seconds (s) and their derived units, for various thermodynamic and other measurements. The following table lists the commonly used biochemical units, some useful biochemical constants, and a few conversion factors.

Units

Energy, heat	joule (J)	$kg \cdot m^2 \cdot s^{-2}$ or $C \cdot V$
Electric potential	volt (V)	$J \cdot C^{-1}$

Prefixes for units

mega (M)	10^6	nano (n)	10^{-9}
kilo (k)	10^3	pico (p)	10^{-12}
milli (m)	10^{-3}	femto (f)	10^{-15}
micro (μ)	10^{-6}	atto (a)	10^{-18}

Constants

Avogadro's number (N)	6.0221×10^{23} molecules·mol^{-1}

Coulomb (C)	6.241×10^{18} electron charges
Faraday ($\mathscr{F}$)	96,485 $C \cdot mol^{-1}$ or 96,485 $J \cdot V^{-1} \cdot mol^{-1}$
Gas constant (R)	8.3145 $J \cdot K^{-1} \cdot mol^{-1}$
Boltzmann constant (k_B)	1.3807×10^{-23} $J \cdot K^{-1}$ (R/N)
Planck's constant (h)	6.6261×10^{-34} $J \cdot s$

Conversions

angstrom (Å)	10^{-10} m
calorie (cal)	4.184 J
Kelvin (K)	degrees Celsius (°C) + 273.15

Throughout this text, molecular masses of particles are expressed in units of **daltons (D)**, which are defined as 1/12th the mass of a ^{12}C atom (1000 D = 1 **kilodalton, kD**). Biochemists also use **molecular weight**, a dimensionless quantity defined as the ratio of the particle mass to 1/12th the mass of a ^{12}C atom, which is symbolized M_r (for relative molecular mass).

Moreover, the volume changes in most biochemical reactions are insignificant, so the differences between their ΔU and ΔH values are negligible. Enthalpy, like energy, heat, and work, is given units of joules (some commonly used units and biochemical constants and other conventions are given in Box 1-1).

Thermodynamics is useful for indicating the spontaneity of a process. A **spontaneous process** occurs without the input of additional energy from outside the system. (Thermodynamic spontaneity has nothing to do with how quickly a process occurs.) The first law of thermodynamics, however, cannot by itself determine whether a process is spontaneous. Consider two objects of different temperatures that are brought together. Heat spontaneously flows from the warmer object to the cooler one, never vice versa. Yet either process would be consistent with the first law of thermodynamics since the aggregate energy of the two objects does not change. Therefore, an additional criterion of spontaneity is needed.

B. The Second Law of Thermodynamics: Entropy Tends to Increase

According to the second law of thermodynamics, spontaneous processes are characterized by the conversion of order to disorder. In this context, disorder is defined as the number of energetically equivalent ways, **W**, of arranging the components of a system. To make this concept concrete, consider a system consisting of two bulbs of equal volume, one of which contains molecules of an ideal gas (Fig. 1-10). When the stopcock connecting the bulbs is open, the molecules become randomly but equally distributed between the two bulbs. The equal number of gas molecules in each bulb is not the result of any law of motion; it is because the probabilities

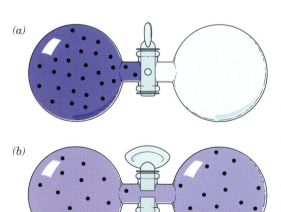

(a)

(b)

Figure 1-10.   Random distribution of gas molecules. In (a), a gas occupies the leftmost of two equal-sized bulbs. When the stopcock is opened (b), the gas molecules diffuse back and forth between the bulbs and eventually become distributed evenly, half in each bulb.

of all other distributions of the molecules are so overwhelmingly small. Thus, the probability of all the molecules in the system spontaneously rushing into the left bulb (the initial condition) is nil, even though the energy and enthalpy of this arrangement is exactly the same as that of the evenly distributed molecules.

The degree of randomness of a system is indicated by its **entropy** (Greek: *en*, in + *trope*, turning), abbreviated **S:**

$$S = k_B \ln W \qquad [1\text{-}5]$$

where k_B is the **Boltzmann constant.** The units of S are $J \cdot K^{-1}$ (absolute temperature, in units of Kelvin, is a factor because entropy varies with temperature; e.g., a system becomes more disordered as its temperature rises). The most probable arrangement of a system is the one that maximizes W and hence S. Thus, if a spontaneous process, such as the one shown in Fig. 1-10, has overall energy and enthalpy changes (ΔU and ΔH) of zero, its entropy change (ΔS) must be greater than zero; that is, the number of equivalent ways of arranging the final state must be greater than the number of ways of arranging the initial state. Furthermore, because

$$\Delta S_{\text{system}} + \Delta S_{\text{surroundings}} = \Delta S_{\text{universe}} > 0 \qquad [1\text{-}6]$$

all processes increase the entropy—that is, the disorder—of the universe.

In chemical and biological systems, it is impractical, if not impossible, to determine the entropy of a system by counting all the equivalent arrangements of its components (W). However, there is an entirely equivalent expression for entropy that applies to the constant temperature conditions typical of biological systems: for a spontaneous process,

$$\Delta S \geq \frac{q}{T} \qquad [1\text{-}7]$$

Thus, the entropy change in a process can be experimentally determined from measurements of heat.

C. Free Energy

The spontaneity of a process cannot be predicted from a knowledge of the system's entropy change alone. For example, 2 mol of H_2 and 1 mol of O_2, when sparked, react to form 2 mol of H_2O. Yet two water molecules, each of whose three atoms are constrained to stay together, are more ordered than are the three diatomic molecules from which they formed.

What, then, is the thermodynamic criterion for a spontaneous process? Equations 1-4 and 1-7 indicate that at constant temperature and pressure

$$\Delta S \geq \frac{q_P}{T} = \frac{\Delta H}{T} \qquad [1\text{-}8]$$

Thus,

$$\Delta H - T\Delta S \leq 0 \qquad [1\text{-}9]$$

This is the true criterion for spontaneity as formulated, in 1878, by J. Willard Gibbs. He defined the **Gibbs free energy** (**G,** usually called just **free energy**) as

$$G = H - TS \qquad [1\text{-}10]$$

Consequently, spontaneous processes at constant temperature and pressure have

$$\boxed{\Delta G = \Delta H - T\Delta S < 0} \qquad [1\text{-}11]$$

***Table 1-3.*   Variation of Reaction Spontaneity (Sign of ΔG) with the Signs of ΔH and ΔS**

ΔH	ΔS	$\Delta G = \Delta H - T\Delta S$
−	+	The reaction is both enthalpically favored (exothermic) and entropically favored. It is spontaneous (exergonic) at all temperatures.
−	−	The reaction is enthalpically favored but entropically opposed. It is spontaneous only at temperatures *below* $T = \Delta H/\Delta S$.
+	+	The reaction is enthalpically opposed (endothermic) but entropically favored. It is spontaneous only at temperatures *above* $T = \Delta H/\Delta S$.
+	−	The reaction is both enthalpically and entropically opposed. It is *un*spontaneous (endergonic) at all temperatures.

Such processes are said to be **exergonic** (Greek: *ergon*, work). Processes that are not spontaneous have positive ΔG values ($\Delta G > 0$) and are said to be **endergonic;** they must be driven by the input of free energy. If a process is exergonic, the reverse of that process is endergonic and vice versa. Processes at **equilibrium,** those in which the forward and reverse reactions are exactly balanced, are characterized by $\Delta G = 0$. For the most part, only changes in free energy, enthalpy, and entropy (ΔG, ΔH, and ΔS) can be measured, not their absolute values.

A process that is accompanied by an increase in enthalpy ($\Delta H > 0$), which opposes the process, can nevertheless proceed spontaneously if the entropy change is sufficiently positive ($\Delta S > 0$; Table 1-3). Conversely, a process that is accompanied by a decrease in entropy ($\Delta S < 0$) can proceed if its enthalpy change is sufficiently negative ($\Delta H < 0$). It is important to emphasize that *a large negative value of ΔG does not ensure that a process such as a chemical reaction will proceed at a measurable rate. The rate depends on the detailed mechanism of the reaction, which is independent of ΔG.*

Free energy as well as energy, enthalpy, and entropy are **state functions.** In other words, their values depend only on the current state or properties of the system, not on how the system reached that state. Therefore, *thermodynamic measurements can be made by considering only the initial and final states of the system and ignoring all the stepwise changes in enthalpy and entropy that occur in between.* For example, it is impossible to directly measure the energy change for the reaction of glucose with O_2 in a living organism because of the numerous other simultaneously occurring chemical reactions. But since ΔG depends on only the initial and final states, the combustion of glucose can be analyzed in any convenient apparatus, using the same starting materials (glucose and O_2) and end products (CO_2 and H_2O) that would be obtained *in vivo.*

D. Chemical Equilibria and the Standard State

The entropy (disorder) of a substance increases with its volume. For example, a collection of gas molecules, in occupying all of the volume available to it, maximizes its entropy. Similarly, dissolved molecules become uniformly distributed throughout their solution volume. Entropy is therefore a function of concentration.

If entropy varies with concentration, so must free energy. Thus, *the free energy change of a chemical reaction depends on the concentrations of both its reactants and its products.* This phenomenon has great significance because many biochemical reactions operate in either direction depending on the relative concentrations of their reactants and products.

Equilibrium Constants Are Related to ΔG

The relationship between the concentration and the free energy of a substance A is approximately

$$\overline{G}_A - \overline{G}_A^\circ = RT \ln [A] \qquad [1\text{-}12]$$

where $\overline{G}_A$ is known as the **partial molar free energy** or the **chemical potential** of A (the bar indicates the quantity per mole), $\overline{G}_A^\circ$ is the partial molar free energy of A in its **standard state,** R is the gas constant, and [A] is the molar concentration of A. Thus, for the general reaction

$$a\text{A} + b\text{B} \rightleftharpoons c\text{C} + d\text{D}$$

the free energy change is

$$\Delta G = c\overline{G}_C + d\overline{G}_D - a\overline{G}_A - b\overline{G}_B \qquad [1\text{-}13]$$

and

$$\Delta G^\circ = c\overline{G}_C^\circ + d\overline{G}_D^\circ - a\overline{G}_A^\circ - b\overline{G}_B^\circ \qquad [1\text{-}14]$$

because free energies are additive and the free energy change of a reaction is the sum of the free energies of the products less those of the reactants. Substituting these relationships into Eq. 1-12 yields

$$\Delta G = \Delta G^\circ + RT \ln \left(\frac{[C]^c[D]^d}{[A]^a[B]^b} \right) \qquad [1\text{-}15]$$

where ΔG° is the free energy change of the reaction when all of its reactants and products are in their standard states (see below). Thus, the expression for the free energy change of a reaction consists of two parts: (1) a constant term whose value depends only on the reaction taking place and (2) a variable term that depends on the concentrations of the reactants and the products, the stoichiometry of the reaction, and the temperature.

For a reaction at equilibrium, there is no *net* change because the free energy of the forward reaction exactly balances that of the reverse reaction. Consequently, $\Delta G = 0$, so Eq. 1-15 becomes

$$\boxed{\Delta G^\circ = -RT \ln K_{eq}} \qquad [1\text{-}16]$$

where K_{eq} is the familiar **equilibrium constant** of the reaction:

$$K_{eq} = \frac{[C]_{eq}^c[D]_{eq}^d}{[A]_{eq}^a[B]_{eq}^b} = e^{-\Delta G^\circ/RT} \qquad [1\text{-}17]$$

The subscript "eq" denotes reactant and product concentrations at equilibrium (the equilibrium condition is usually clear from the context of the situation, so equilibrium concentrations are usually expressed without this subscript). *The equilibrium constant of a reaction can therefore be calculated from standard free energy data and vice versa.*

K Depends on Temperature

The manner in which the equilibrium constant varies with temperature can be seen by substituting Eq. 1-11 into Eq. 1-16 and rearranging:

$$\ln K_{eq} = \frac{-\Delta H^\circ}{R} \left(\frac{1}{T} \right) + \frac{\Delta S^\circ}{R} \qquad [1\text{-}18]$$

where H° and S° represent enthalpy and entropy in the standard state.

Equation 1-18 has the form $y = mx + b$, the equation for a straight line. A plot of $\ln K_{eq}$ versus $1/T$, known as a **van't Hoff plot,** permits the values of $\Delta H°$ and $\Delta S°$ (and hence $\Delta G°$) to be determined from measurements of K_{eq} at two (or more) different temperatures. This method is often more practical than directly measuring ΔH and ΔS by calorimetry (which measures the heat, q_P, of a process).

Standard State Conventions in Biochemistry

In order to compare free energy changes for different reactions, it is necessary to express ΔG values relative to some standard state (likewise, we refer the elevations of geographic locations to sea level, which is arbitrarily assigned the height of zero). According to the convention used in physical chemistry, a solute is in its standard state when the temperature is 25°C, the pressure is 1 atm, and the solute has an **activity** of 1 (the activity of a substance is its concentration corrected for its nonideal behavior at concentrations higher than infinite dilution).

The concentrations of reactants and products in most biochemical reactions are usually so low (on the order of millimolar or less) that their activities are closely approximated by their molar concentrations. Furthermore, because biochemical reactions occur near neutral pH, biochemists have adopted a somewhat different standard state convention:

1. The activity of pure water is assigned a value of 1, even though its concentration is 55.5 M. This practice simplifies the free energy expressions for reactions in dilute solutions involving water as a reactant, because the $[H_2O]$ term can then be ignored.

2. The hydrogen ion activity is assigned a value of 1 at the physiologically relevant pH of 7. Thus, the biochemical standard state is pH 7.0 (neutral pH, where $[H^+] = 10^{-7}$ M) rather than pH 0 ($[H^+] = 1$ M), the physical chemical standard state, where many biological substances are unstable.

3. The standard state of a substance that can undergo an acid–base reaction is defined in terms of the total concentration of its naturally occurring ion mixture at pH 7. In contrast, the physical chemistry convention refers to a pure species whether or not it actually exists at pH 0. The advantage of the biochemistry convention is that the total concentration of a substance with multiple ionization states, such as most biological molecules, is usually easier to measure than the concentration of one of its ionic species. Since the ionic composition of an acid or base varies with pH, however, the standard free energies calculated according to the biochemical convention are valid only at pH 7.

Under the biochemistry convention, the standard free energy changes of reactions are customarily symbolized by $\Delta G°'$ to distinguish them from physical chemistry standard free energy changes, $\Delta G°$. If a reaction includes neither H_2O, H^+, nor an ionizable species, then $\Delta G°' = \Delta G°$.

E. Life Obeys the Laws of Thermodynamics

At one time, biologists believed that life, with its inherent complexity and order, somehow evaded the laws of thermodynamics. However, elaborate measurements on living animals are consistent with the conservation of energy predicted by the first law. Unfortunately, experimental verification of the second law is not practicable, since it requires dismantling an organism to its component molecules, which would result in its irreversible death.

Consequently, it is possible to assert only that the entropy of living matter is less than that of the products to which it decays. *Life persists, however, because a system can be ordered at the expense of disordering its surroundings to an even greater extent* (Eq. 1-6). Living organisms achieve order by disordering (breaking down) the nutrients they consume. Thus, the entropy content of food is as important as its energy content.

Living Organisms Are Open Systems

Classical thermodynamics applies primarily to reversible processes in **isolated** or **closed systems,** which, respectively, cannot exchange matter or energy or can only exchange energy with their surroundings. An isolated system inevitably reaches equilibrium. For example, if its reactants are in excess, the forward reaction will proceed faster than the reverse reaction until equilibrium is attained ($\Delta G = 0$), at which point the forward and reverse reactions exactly balance each other. In contrast, **open systems,** which exchange both matter and energy with their surroundings, can never be at equilibrium.

Living organisms, which take up nutrients, release waste products, and generate work and heat, are open systems and therefore can never be at equilibrium. They continuously ingest high-enthalpy, low-entropy nutrients, which they convert to low-enthalpy, high-entropy waste products. The free energy released in this process powers the cellular activities that produce the high degree of organization characteristic of life. If this process is interrupted, the system ultimately reaches equilibrium, which for living things is synonymous with death. An example of energy flow in an open system is illustrated in Fig. 1-11. Through photosynthesis, plants convert radiant energy from the sun, the primary energy source for life on earth, to the chemical energy of carbohydrates and other organic substances. The plants, or the animals that eat them, then metabolize these substances to power such functions as the synthesis of biomolecules, the maintenance of intracellular ion concentrations, and cellular movements.

Living Things Maintain a Steady State

Even in a system that is not at equilibrium, matter and energy flow according to the laws of thermodynamics. For example, materials tend to move from areas of high concentration to areas of low concentration. This is why blood takes up O_2 in the lungs, where O_2 is abundant, and releases it to the tissues, where O_2 is scarce.

Living systems are characterized by being in a **steady state.** This means that all flows in the system are constant so that the system does not change with time. Energy flow in the biosphere (Fig. 1-11) is an example of a system in a steady state. Slight perturbations from the steady state give rise to changes in flows that restore the system to the steady state. In all living systems, energy flow is exclusively "downhill" ($\Delta G < 0$). In addition, nature is inherently dissipative, so the recovery of free energy from a biochemical process is never total and some energy is always lost to the surroundings.

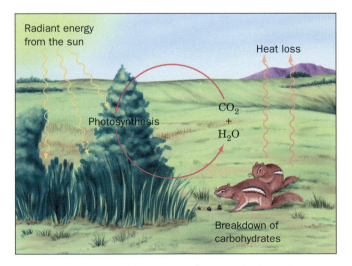

Figure 1-11. Energy flow in the biosphere. Plants use the sun's radiant energy to synthesize carbohydrates from CO_2 and H_2O. Plants or the animals that eat them eventually metabolize the carbohydrates to release their stored free energy and thereby return CO_2 and H_2O to the environment.

Enzymes Catalyze Biochemical Reactions

Nearly all the molecular components of an organism can potentially react with each other, and many of these reactions are thermodynamically favored (spontaneous). Yet only a subset of all possible reactions actually occur to a significant extent in a living organism. The rate of a particular reaction depends not on the free energy difference between the initial and final states but on the actual path through which the reactants are transformed to products. Living organisms take advantage of **catalysts,** substances that increase the rate at which a reaction approaches equilibrium without affecting the reaction's ΔG. Biological catalysts are referred to as **enzymes,** most of which are proteins.

Enzymes accelerate biochemical reactions by physically interacting with the reactants and products to provide a more favorable pathway for the transformation of one to the other. Enzymes increase the rates of reactions by increasing the likelihood that the reactants can interact productively. However, enzymes cannot promote reactions whose ΔG values are positive.

A multitude of enzymes mediate the flow of energy in every cell. As free energy is harvested, stored, or used to perform cellular work, it may be transferred to other molecules. And although it is tempting to think of free energy as something that is stored in chemical bonds, chemical energy can be transformed into heat, electrical work, or mechanical work, according to the needs of the organism and the biochemical machinery with which it has been equipped through evolution.

SUMMARY

1. A model for the origin of life proposes that organisms ultimately arose from simple organic molecules that polymerized to form more complex molecules capable of replicating themselves.

2. Compartmentation gave rise to cells that developed metabolic reactions for synthesizing biological molecules and generating energy.

3. All cells are either prokaryotic or eukaryotic. Eukaryotic cells contain a variety of membrane-bounded organelles.

4. Phylogenetic evidence groups organisms into three domains: archaea, bacteria, and eukarya.

5. Natural selection directs the evolution of species.

6. The first law of thermodynamics (energy is conserved) and the second law (spontaneous processes increase the disorder of the universe) apply to biochemical processes. The spontaneity of a process is determined by its free energy change ($\Delta G = \Delta H - T\Delta S$): spontaneous reactions have $\Delta G < 0$ and nonspontaneous reactions have $\Delta G > 0$.

7. The equilibrium constant for a process is related to the standard free energy change for that process.

8. Living organisms are open systems that maintain a steady state.

REFERENCES

Origin and Evolution of Life

de Duve, C., *Blueprint for a Cell. The Nature and Origin of Life,* Carolina Biological Supply Co. (1991).

Knoll, A.H., The early evolution of eukaryotes: A geological perspective, *Science* **256,** 622–627 (1992).

Orgel, L.E., Molecular replication, *Nature* **358,** 203–209 (1992).

Schopf, J.W., Microfossils of the early Archaean Apex chert: New evidence of the antiquity of life, *Science* **260,** 640–646 (1993).

Szathmáry, E. and Smith, J.M., The major evolutionary transitions, *Nature* **374,** 227–232 (1995). [A summary of the evolution of complexity, from the origin of prebiotic systems to the emergence of language.]

Volkenstein, M.V., *Physical Approaches to Biological Evolution,* Springer-Verlag (1994).

Cells

Alberts, B., Bray, D., Lewis, J., Raff, M., Roberts, K., and Watson, J.D., *Molecular Biology of the Cell* (3rd ed.), Chapters 1 and 2, Garland Publishing (1994). [This and other cell biology textbooks offer thorough reviews of cellular structure.]

Attenborough, D., *Life on Earth,* Little, Brown (1980). [A beautifully illustrated exposition of evolutionary development.]

Campbell, N.A., *Biology* (3rd ed.), Chapters 7 and 8, Benjamin/Cummings (1993).

Goodsell, D.S., A look inside the living cell, *Am. Scientist* **80,** 457–465 (1992); *and* Inside a living cell, *Trends Biochem. Sci.* **16,** 203–206 (1991).

Lodish, H., Baltimore, D., Berk, A., Zipursky, S.L., Matsudaria, P., and Darnell, J., *Molecular Cell Biology* (3rd ed.), Chapter 5, Scientific American Books (1995).

Margulis, L. and Sagan, D., *What Is Life,* Simon & Schuster (1995). [A nontechnical review of the origin and development of life on earth.]

Thermodynamics

Tinoco, I., Jr., Sauer, K., and Wang, J.C., *Physical Chemistry. Principles and Applications in Biological Sciences* (3rd ed.), Chapters 2–5, Prentice–Hall (1996). [Most physical chemistry texts treat thermodynamics in some detail.]

van Holde, K.E., Johnson, W.C., and Hu, P.S., *Principles of Physical Biochemistry,* Chapters 2 and 3, Prentice–Hall (1998).

KEY TERMS

organic compound	Golgi apparatus	thermophiles	G
polymer	mitochondrion	symbiosis	exergonic
condensation reaction	chloroplast	mutation	endergonic
hydrolysis	lysosome	thermodynamics	equilibrium
replication	peroxisome	system	state function
vesicle	vacuole	surroundings	$\overline{G}_A$
compartmentation	cytosol	U	$\overline{G}_A^\circ$
catalyst	cytoskeleton	q	standard state
eukaryote	taxonomy	q_P	equilibrium constant
prokaryote	phylogeny	w	isolated system
virus	archaea	H	closed system
nucleus	bacteria	W	open system
organelle	eukarya	S	steady state
cytoplasm	methanogens	k_B	enzyme
endoplasmic reticulum	halobacteria	spontaneous process	

STUDY EXERCISES

1. Summarize the major stages of chemical and organismal evolution.

2. Describe the process of evolution by natural selection.

3. What kinds of organisms are found in each of the three major evolutionary domains?

4. Explain the first and second laws of thermodynamics.

5. How does the free energy change in a process depend on its enthalpy and entropy changes?

6. What is the biochemistry standard state?

7. How does life persist despite the second law of thermodynamics?

PROBLEMS

1. Identify the functional groups in coenzyme A (Fig. 3-5).

2. Why is the cell membrane not an absolute barrier between the cytoplasm and the external environment?

3. A spheroidal bacterium with a diameter of 1 μm contains two molecules of a particular protein. What is the molar concentration of the protein?

4. How many glucose molecules does the cell in Problem 3 contain when its internal glucose concentration is 1.0 mM?

5. (a) Which has greater entropy, liquid water at 0°C or ice at 0°C? (b) How does the entropy of ice at −5°C differ, if at all, from its entropy at −50°C?

6. Does entropy increase or decrease in the following processes?

(a) $N_2 + 3\,H_2 \longrightarrow 2\,NH_3$

(b) $H_2N - \overset{\overset{\displaystyle O}{\|}}{C} - NH_2 \;+\; H_2O \longrightarrow CO_2 \;+\; 2\,NH_3$

Urea

(c)

1 M NaCl ⟶ 0.5 M NaCl

(d)

$$
\begin{array}{c}
COO^- \\
| \\
HC - OH \\
| \\
H_2C - OPO_3^{2-}
\end{array}
\longrightarrow
\begin{array}{c}
COO^- \\
| \\
HC - OPO_3^{2-} \\
| \\
H_2C - OH
\end{array}
$$

3-Phosphoglycerate   **2-Phosphoglycerate**

7. Consider a reaction with $\Delta H = 15$ kJ and $\Delta S = 50\ \mathrm{J \cdot K^{-1}}$. Is the reaction spontaneous (a) at 10°C, (b) at 80°C?

8. Calculate the equilibrium constant for the reaction glucose-1-phosphate + $H_2O \longrightarrow$ glucose + $H_2PO_4^-$ at pH 7.0 and 25°C ($\Delta G^{\circ\prime} = -20.9\ \mathrm{kJ \cdot mol^{-1}}$).

9. Calculate $\Delta G^{\circ\prime}$ for the reaction A + B $\rightleftharpoons$ C + D at 25°C when the equilibrium concentrations are [A] = 10 μM, [B] = 15 μM, [C] = 3 μM, and [D] = 5 μM. Is the reaction exergonic or endergonic under standard conditions?

10. $\Delta G^{\circ\prime}$ for the isomerization reaction

Glucose-1-phosphate (G1P) $\rightleftharpoons$ glucose-6-phosphate (G6P)

is $-7.1\ \mathrm{kJ \cdot mol^{-1}}$. Calculate the equilibrium ratio of [G1P] to [G6P] at 25°C.

Coral reefs support a variety of vertebrates and invertebrates. The water that surrounds them is critical for their existence, acting as a solvent for biochemical reactions and, to a large extent, determining the structures of the macromolecules that carry out these reactions. [© Jeff Hunter/The Image Bank.]

WATER

Any study of the chemistry of life must include a study of water. Biological molecules and the reactions they undergo can be best understood in the context of their aqueous environment. Not only are organisms made mostly of water (~70% of the mass of the human body is water), they are surrounded by water on this, the "blue planet." Aside from its sheer abundance, water is central to biochemistry for the following reasons:

1. Nearly all biological molecules assume their shapes (and therefore their functions) in response to the physical and chemical properties of the surrounding water.

2. The medium for the majority of biochemical reactions is water. Reactants and products of metabolic reactions, nutrients as well as waste products, depend on water for transport within and between cells.

3. Water itself actively participates in many chemical reactions that support life. Frequently, the ionic components of water, the H^+ and OH^- ions, are the true reactants. In fact, the reactivity of many functional groups on biological molecules depends on the relative concentrations of H^+ and OH^- in the surrounding medium.

4. The oxidation of water to produce molecular oxygen, O_2, is a fundamental reaction of photosynthesis, the process that converts the sun's energy to a usable, chemical form. Expenditure of that energy ultimately leads to the reduction of O_2 back to H_2O.

All organisms require water, from the marine creatures who spend their entire lives in an aqueous environment to terrestrial organisms who must guard their watery interiors with protective skins. Not surprisingly, living organisms can be found wherever there is liquid water—in springs as hot as 105°C and in the cracks and crevices between rocks hundreds of meters beneath the surface of the earth. Organisms that survive desiccation do so only by becoming dormant, as seeds or spores.

An examination of water from a biochemical point of view requires a look at the physical properties of water, its properties as a solvent, and its chemical behavior—that is, the nature of aqueous acids and bases.

1. PHYSICAL PROPERTIES OF WATER

The colorless, odorless, and tasteless nature of water belies its fundamental importance to living organisms. Despite its bland appearance to our senses, water is anything but inert. Its physical properties—unique among molecules of similar size—give it unparalleled strength as a solvent. And yet, its limitations as a solvent also have important implications for the structures and functions of biological molecules.

A. Structure of Water

A water molecule consists of two hydrogen atoms bonded to an oxygen atom. The O—H bond distance is 0.958 Å (1 Å = 10^{-10} m), and the angle formed by the three atoms is 104.5° (Fig. 2-1). The hydrogen atoms are not arranged linearly, because the oxygen atom's four sp^3 hybrid orbitals extend roughly toward the corners of a tetrahedron. Hydrogen atoms occupy two corners of this tetrahedron, and the nonbonding electron pairs of the oxygen atom occupy its other two corners (in a perfectly tetrahedral molecule, such as methane, CH_4, the bond angles are 109.5°).

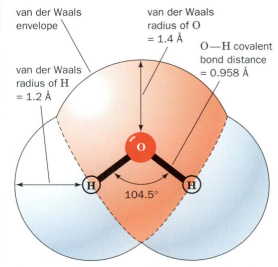

Figure 2-1. **Structure of the water molecule.** The shaded outline represents the van der Waals envelope, the effective "surface" of the molecule.

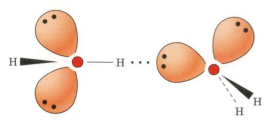

Figure 2-2.   A hydrogen bond in water. The strength of the interaction is maximal when the O—H covalent bond of one molecule points directly toward the lone-pair electron cloud of the other.

Water Molecules Form Hydrogen Bonds

The angular geometry of the water molecule has enormous implications for living systems. Water is a **polar** molecule: The oxygen atom with its unshared electrons carries a partial negative charge ($\delta-$) of $-0.66e$, and the hydrogen atoms each carry a partial positive charge ($\delta+$) of $+0.33e$, where e is the charge of the electron. Electrostatic attractions between the dipoles of water molecules are crucial to the properties of water itself and to its role as a biochemical solvent. Neighboring water molecules tend to orient themselves so that the O—H bond of one water molecule (the positive end) points toward one of the electron pairs of the other water molecule (the negative end). The resulting directional intermolecular association is known as a **hydrogen bond** (Fig. 2-2).

In general, a hydrogen bond can be represented as D—H···A, where D—H is a weakly acidic "donor" group such as O—H, N—H, or sometimes S—H, and A is a weakly basic "acceptor" atom such as O, N, or occasionally S. Hydrogen bonds are structurally characterized by an H···A distance that is at least 0.5 Å shorter than the calculated **van der Waals distance** (the distance of closest approach between two nonbonded atoms). In water, for example, the O···H hydrogen bond distance is ~1.8 Å versus 2.6 Å for the corresponding van der Waals distance.

A single water molecule contains two hydrogen atoms that can be "donated" and two unshared electron pairs that can act as "acceptors," so each molecule can participate in a maximum of four hydrogen bonds with other water molecules. Although the energy of an individual hydrogen bond (~20 $kJ \cdot mol^{-1}$) is relatively small (e.g., the energy of an O—H covalent bond is 460 $kJ \cdot mol^{-1}$), the sheer number of hydrogen bonds in a sample of water is the key to its remarkable properties.

Ice Is a Crystal of Hydrogen-Bonded Water Molecules

The structure of ice provides a striking example of the cumulative strength of many hydrogen bonds. X-Ray and neutron diffraction studies have established that water molecules in ice are arranged in an unusually open structure. Each water molecule is tetrahedrally surrounded by four nearest neighbors to which it is hydrogen bonded (Fig. 2-3). As a consequence of its open structure, water is one of the very few substances that expands on freezing (at 0°C, liquid water has a density of 1.00 $g \cdot mL^{-1}$, whereas ice has a density of 0.92 $g \cdot mL^{-1}$).

The expansion of water on freezing has overwhelming consequences for life on earth. Suppose that water contracted on freezing, that is, became more dense rather than less dense. Ice would then sink to the bottoms of lakes and oceans rather than float. This ice would be insulated from the sun so that oceans, with the exception of a thin surface layer of liquid in warm weather, would be permanently frozen solid (the water at great depths, even in tropical oceans, is close to 4°C, its temperature of maximum density). Thus, the earth would be locked in a permanent ice age and life might never have arisen.

The melting of ice represents the collapse of the strictly tetrahedral orientation of hydrogen-bonded water molecules, although hydrogen bonds between water molecules persist in the liquid state. In fact, liquid water is only ~15% less hydrogen bonded than ice at 0°C. Indeed, the boiling point of water is 264°C higher than that of methane, a substance with nearly the same molecular mass as H_2O but which is incapable of hydrogen bonding (substances with similar intermolecular associations and equal molecular masses should have similar boiling points). This difference reflects the

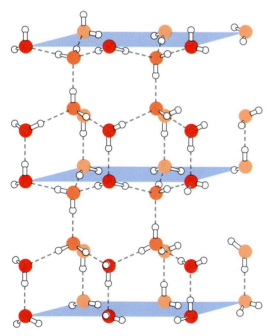

Figure 2-3.   The structure of ice. Each water molecule interacts tetrahedrally with four other water molecules. Oxygen atoms are red and hydrogen atoms are white. Hydrogen bonds are represented by dashed lines. [After Pauling, L., *The Nature of the Chemical Bond* (3rd ed.), p. 465, Cornell University Press (1960).]

extraordinary internal cohesiveness of liquid water resulting from its intermolecular hydrogen bonding.

The Structure of Liquid Water Is Irregular

Because each molecule of liquid water reorients about once every 10^{-12} s, very few experimental techniques can explore the instantaneous arrangement of these water molecules. Theoretical considerations and spectroscopic evidence suggest that molecules in liquid water are each hydrogen bonded to four nearest neighbors, as they are in ice. These hydrogen bonds are distorted, however, so the networks of linked molecules are irregular and varied. For example, three- to seven-membered rings of hydrogen-bonded molecules commonly occur in liquid water (Fig. 2-4), in contrast to the six-membered rings characteristic of ice (Fig. 2-3). Moreover, the networks are continually breaking up and re-forming over times on the order of 2×10^{-11} s. *Liquid water therefore consists of a rapidly fluctuating, three-dimensional network of hydrogen-bonded H_2O molecules.*

Hydrogen Bonds and Other Weak Interactions in Biological Molecules

Biochemists are concerned not just with the strong covalent bonds that define chemical structure but with the weak forces that act under relatively mild physical conditions. The structures of most biological molecules are determined by the collective influence of many individually weak interactions. The weak electrostatic forces that interest biochemists include ionic interactions, hydrogen bonds, and van der Waals forces.

The strength of association of ionic groups of opposite charge depends on the identity of the ions, the distance between them, and the polarity of the medium. In general, the strength of the interaction between two charged groups (i.e., the energy required to completely separate them in the medium of interest) is less than the energy of a covalent bond but greater than the energy of a hydrogen bond (Table 2-1).

The noncovalent associations between neutral molecules, collectively known as **van der Waals forces,** arise from electrostatic interactions among permanent or induced dipoles (the hydrogen bond is a special kind of dipolar interaction). Interactions among permanent dipoles such as carbonyl

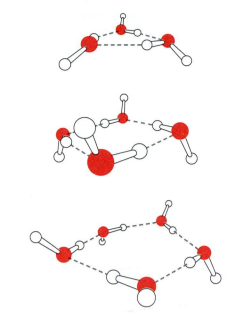

Figure 2-4. **Structure of a water trimer, tetramer, and pentamer.** These models are based on theoretical predictions and spectroscopic data. [After Liu, K., Cruzan, J.D., and Saykally, R.J., *Science* **271**, 929 (1996).]

Table 2-1. **Bond Energies in Biomolecules**

Type of Bond	Example	Bond Strength $(kJ \cdot mol^{-1})$
Covalent	O—H	460
	C—H	414
	C—C	348
Noncovalent		
Ionic interaction	$-COO^- ---{}^+H_3N-$	86
van der Waals forces		
Hydrogen bond	$-O-H---O\langle$	20
Dipole–dipole interaction	$\rangle C{=}O ---\rangle C{=}O$	9.3
London dispersion forces	$-\overset{\text{H}}{\underset{\text{H}}{C}}-H --- H-\overset{\text{H}}{\underset{\text{H}}{C}}-$	0.3

Figure 2-5. **Dipole–dipole interactions.** The strength of each dipole is indicated by the thickness of the accompanying arrow. (*a*) Interaction between permanent dipoles. (*b*) Dipole–induced dipole interaction. (*c*) London dispersion forces.

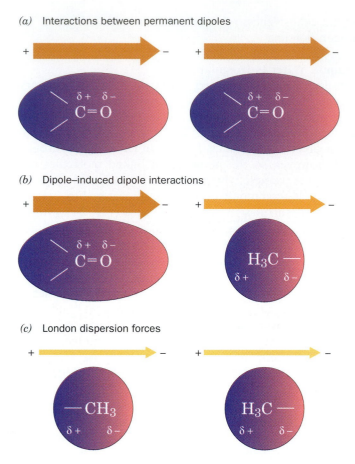

(*a*) Interactions between permanent dipoles

(*b*) Dipole–induced dipole interactions

(*c*) London dispersion forces

groups (Fig. 2-5*a*) are much weaker than ionic interactions. A permanent dipole also induces a dipole moment in a neighboring group by electrostatically distorting its electron distribution (Fig. 2-5*b*). Such dipole–induced dipole interactions are generally much weaker than dipole–dipole interactions.

At any instant, nonpolar molecules have a small, randomly oriented dipole moment resulting from the rapid fluctuating motion of their electrons. This transient dipole moment polarizes the electrons in a neighboring group (Fig. 2-5*c*), so that the groups are attracted to each other. These so-called **London dispersion forces** are extremely weak and fall off so rapidly with distance that they are significant only for groups in close contact. They are, nevertheless, extremely important in determining the structures of biological molecules, whose interiors contain many closely packed groups.

B. *Water as a Solvent*

Solubility depends on the ability of a solvent to interact with a solute more strongly than solute particles interact with each other. Water is said to be the "universal solvent." Although this statement cannot literally be true, water certainly dissolves more types of substances and in greater amounts than any other solvent. In particular, the polar character of water makes it an excellent solvent for polar and ionic materials, which are said to be **hydrophilic** (Greek: *hydro*, water + *philos*, loving). On the other hand, nonpolar substances are virtually insoluble in water ("oil and water don't mix") and are consequently described as **hydrophobic** (Greek: *phobos*, fear). Nonpolar substances, however, are soluble in nonpolar solvents such as CCl_4

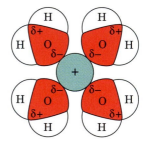

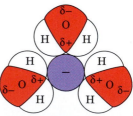

Figure 2-6. **Solvation of ions.** The dipoles of the surrounding water molecules are oriented according to the charge of the ion. Only one layer of solvent molecules is shown.

and hexane. This information is summarized by another maxim, "like dissolves like."

Why do salts dissolve in water? Polar solvents, such as water, weaken the attractive forces between oppositely charged ions and can therefore hold the ions apart (in nonpolar solvents, ions of opposite charge attract each other so strongly that they coalesce to form a solid). An ion immersed in a polar solvent such as water attracts the oppositely charged ends of the solvent dipoles (Fig. 2-6). The ion is thereby surrounded by several concentric shells of oriented solvent molecules. Such ions are said to be **solvated** or, when water is the solvent, to be **hydrated.**

The bond dipoles of uncharged polar molecules make them soluble in aqueous solutions for the same reasons that ionic substances are water soluble. The solubilities of polar and ionic substances are enhanced when they carry functional groups, such as hydroxyl (OH), carbonyl (C=O), carboxyl (COOH), and amino (NH_2) groups, that can form hydrogen bonds with water as illustrated in Fig. 2-7. Indeed, water-soluble biomolecules such as proteins, nucleic acids, and carbohydrates bristle with just such groups. Nonpolar substances, in contrast, lack hydrogen bonding donor and acceptor groups.

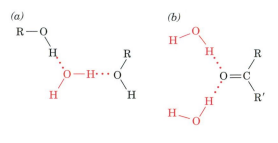

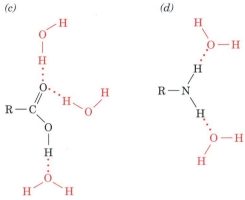

Figure 2-7.   Hydrogen bonding by functional groups. Water forms hydrogen bonds with (*a*) hydroxyl groups, (*b*) keto groups, (*c*) carboxyl groups, and (*d*) amino groups.

C. The Hydrophobic Effect

When a nonpolar substance is added to an aqueous solution, it does not dissolve but instead is excluded by the water. *The tendency of water to minimize its contacts with hydrophobic molecules is termed the* **hydrophobic effect.** Many large molecules and molecular aggregates, such as proteins and cellular membranes, assume their shapes in response to the hydrophobic effect.

Consider the thermodynamics of transferring a nonpolar molecule from an aqueous solution to a nonpolar solvent. In all cases, the free energy change is negative, which indicates that such transfers are spontaneous processes (Table 2-2). Interestingly, these transfer processes are either endothermic (positive ΔH) or athermic ($\Delta H = 0$); that is, it is enthalpically more or less equally favorable for nonpolar molecules to dissolve in water than in nonpolar media. In contrast, the entropy change (expressed as $-T\Delta S$) is large and negative in all cases. Clearly, the transfer of a hydrocarbon from an aqueous medium to a nonpolar medium is entropically driven (i.e., the free energy change is mostly due to an entropy change).

Entropy, or "randomness," is a measure of the order of a system (Section 1-4B). If entropy increases when a nonpolar molecule leaves an aqueous solution, entropy must decrease when the molecule enters water. This

Table 2-2.   Thermodynamic Changes for Transferring Hydrocarbons from Water to Nonpolar Solvents at 25°C

Process	ΔH (kJ·mol^{-1})	$-T\Delta S$ (kJ·mol^{-1})	ΔG (kJ·mol^{-1})
CH_4 in H_2O ⇌ CH_4 in C_6H_6	11.7	−22.6	−10.9
CH_4 in H_2O ⇌ CH_4 in CCl_4	10.5	−22.6	−12.1
C_2H_6 in H_2O ⇌ C_2H_6 in benzene	9.2	−25.1	−15.9
C_2H_4 in H_2O ⇌ C_2H_4 in benzene	6.7	−18.8	−12.1
C_2H_2 in H_2O ⇌ C_2H_2 in benzene	0.8	−8.8	−8.0
Benzene in H_2O ⇌ liquid benzene[a]	0.0	−17.2	−17.2
Toluene in H_2O ⇌ liquid toluene[a]	0.0	−20.0	−20.0

[a]Data measured at 18°C.
Source: Kauzmann, W., *Adv. Protein Chem.* **14,** 39 (1959).

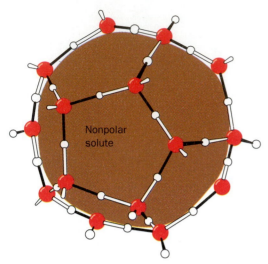

Figure 2-8.   Orientation of water molecules around a nonpolar solute. In order to maximize their number of hydrogen bonds, water molecules form a "cage" around the solute. Black lines represent hydrogen bonds.

decrease in entropy when a nonpolar molecule is solvated by water is an experimental observation, not a theoretical conclusion. Yet, the entropy changes are too large to reflect only the changes in the conformations of the hydrocarbons. Thus the entropy changes must arise mainly from some sort of ordering of the water itself. What is the nature of this ordering?

The extensive hydrogen-bonding network of liquid water molecules is disrupted when a nonpolar group intrudes. A nonpolar group can neither accept nor donate hydrogen bonds, so the water molecules at the surface of the cavity occupied by the nonpolar group cannot hydrogen bond to other molecules in their usual fashion. In order to recover their lost hydrogen-bonding energy, these surface water molecules orient themselves to form a hydrogen-bonded network enclosing the cavity (Fig. 2-8). This orientation constitutes an ordering of the water structure since the number of ways that water molecules can form hydrogen bonds around the surface of a nonpolar group is less than the number of ways they can form hydrogen bonds in bulk water.

Unfortunately, the complexity of liquid water's basic structure has not yet allowed a detailed description of this ordering process. One model proposes that water forms icelike hydrogen-bonded "cages" around the nonpolar groups. The water molecules of the cages are tetrahedrally hydrogen bonded to other water molecules, and the ordering of water molecules extends several layers beyond the first hydration shell of the nonpolar solute.

The unfavorable free energy of hydration of a nonpolar substance caused by its ordering of the surrounding water molecules has the net result that the nonpolar substance tends to be excluded from the aqueous phase. This is because the surface area of a cavity containing an aggregate of nonpolar molecules is less than the sum of the surface areas of the cavities that each of these molecules would individually occupy. *The aggregation of the nonpolar groups thereby minimizes the surface area of the cavity and therefore the entropy loss of the entire system.* In a sense, the nonpolar groups are squeezed out of the aqueous phase.

Amphiphiles Form Micelles and Bilayers

Most biological molecules have both polar (or charged) and nonpolar segments and are therefore simultaneously hydrophilic and hydrophobic. Such molecules, for example, fatty acid ions (soaps; Fig. 2-9), are said to be **amphiphilic** or **amphipathic** (Greek: *amphi,* both; *pathos,* passion). How do amphiphiles interact with an aqueous solvent? Water tends to hydrate the hydrophilic portion of an amphiphile, but it also tends to exclude the hydrophobic portion. Amphiphiles consequently tend to form structurally ordered aggregates. For example, **micelles** are globules of up to several thousand amphiphilic molecules arranged so that the hydrophilic

$$CH_3CH_2CH_2CH_2CH_2CH_2CH_2CH_2CH_2CH_2CH_2CH_2CH_2CH_2 \overset{\overset{\displaystyle O}{\|}}{-C} - O^-$$

Palmitate ($C_{15}H_{31}COO^-$)

$$CH_3CH_2CH_2CH_2CH_2CH_2CH_2CH_2 \overset{\overset{\displaystyle H \quad H}{|\quad |}}{-C = C} - CH_2CH_2CH_2CH_2CH_2CH_2CH_2 \overset{\overset{\displaystyle O}{\|}}{-C} - O^-$$

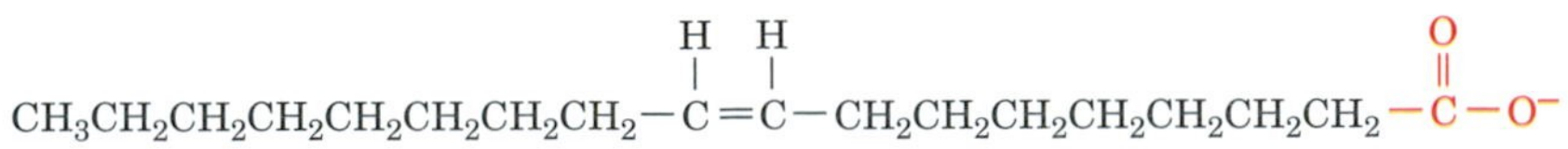

Oleate ($C_{17}H_{33}COO^-$)

Figure 2-9.   Fatty acid anions (soaps). Palmitate and oleate are amphiphilic compounds; each has a polar carboxylate group and a long nonpolar hydrocarbon chain.

groups at the globule surface can interact with the aqueous solvent while the hydrophobic groups associate at the center, away from the solvent (Fig. 2-10a). Alternatively, the amphiphiles may arrange themselves to form **bilayered** sheets or vesicles in which the polar groups face the aqueous phase (Fig. 2-10b). In either case, the aggregate is stabilized by the hydrophobic effect, the tendency of water to exclude hydrophobic groups.

The consequences of the hydrophobic effect are often called hydrophobic forces or hydrophobic "bonds." However, the term *bond* implies a discrete directional relationship between two entities. The hydrophobic effect acts indirectly on nonpolar groups and lacks directionality. Despite the temptation to attribute some mutual attraction to a collection of nonpolar groups excluded from water, their exclusion is largely a function of the entropy of the surrounding water molecules, not some "hydrophobic force" among them.

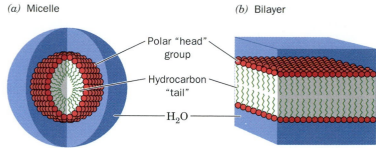

(a) Micelle   *(b)* Bilayer

Polar "head" group

Hydrocarbon "tail"

H_2O

Figure 2-10. Structures of micelles and bilayers. In aqueous solution, the polar head groups of amphipathic molecules are hydrated while the nonpolar tails aggregate by exclusion from water. (*a*) A micelle is a spheroidal aggregate. (*b*) A bilayer is an extended planar aggregate. The bilayer may form a closed spheroidal shell, known as a **vesicle,** that encloses a small amount of the aqueous solution.

D. Osmosis and Diffusion

Paleobiologists believe that the first organisms arose from pools of nutrient-rich water, the so-called primordial soup. In fact, the composition of the fluid that bathes the cells in our bodies bears some resemblance to seawater (Table 2-3). Thus our watery interiors are quite unlike pure water: The fluid in the cells and surrounding the cells in multicellular organisms is full of dissolved substances ranging from small inorganic ions to huge molecular aggregates. Furthermore, the amount of "free" water in which these substances move and collide is only a portion of the total amount of water, because *each dissolved molecule is surrounded by a shell of relatively immobile water molecules, called* **water of hydration.**

In addition, the solute concentration affects water's **colligative properties,** the physical properties that depend on the concentration of dissolved substances rather than on their chemical features. For example, solutes depress the freezing point and elevate the boiling point of water by making it more difficult for water molecules to crystallize as ice or to escape from solution into the gas phase.

Osmotic pressure also depends on the solute concentration. When a solution is separated from pure water by a semipermeable membrane that permits the passage of water molecules but not solutes, water tends to move into the solution in order to equalize its concentration on both sides of the membrane. **Osmosis** is the movement of solvent from a region of high concentration (here, pure water) to a region of relatively low concentration

Table 2-3. Ionic Composition of Seawater and Extracellular Fluid

Ion	Seawater (mM)	Extracellular Fluid (mM)
Na^+	468	145
Mg^{2+}	53	1.5
Ca^{2+}	10	2.5
K^+	10	4
Cl^-	550	115
HPO_4^{2-}	0.001	2
SO_4^{2-}	28	1
HCO_3^-	2.3	30

***Figure 2-11.* Osmotic pressure.** (*a*) A water-permeable membrane separates a tube of concentrated solution from pure water. (*b*) As water moves into the solution by osmosis, the height of the solution in the tube increases. (*c*) The pressure that prevents the influx of water is the osmotic pressure (22.4 atm for a 1 M solution).

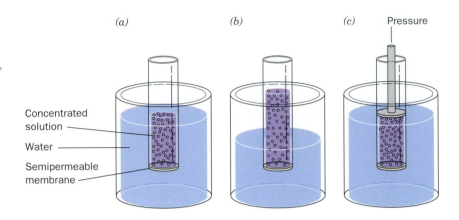

(water containing dissolved solute). The **osmotic pressure** of a solution is the pressure that must be applied to the solution to prevent the inward flow of water; it is proportional to the concentration of the solute (Fig. 2-11). For a 1 M solution, the osmotic pressure is 22.4 atm. Consider the implications of osmotic pressure for living cells, which are essentially semipermeable sacs of aqueous solution. One strategy used by many animal cells to minimize osmotic influx of water (which would burst the relatively weak cell membrane) is to surround the cell with a solution of similar osmotic pressure (so there is no net flow of water). Another strategy, used by most plants and bacteria, is to enclose the cell with a rigid cell wall that can withstand the osmotic pressure generated within.

When an aqueous solution is separated from pure water by a membrane that is permeable to both water and solutes, solutes move out of the solution even as water moves in. The molecules move randomly, or **diffuse,** until the concentration of the solute is the same on both sides of the membrane. At this point, equilibrium is established; that is, there is no further *net* flow of water or solute (although molecules continue to move and collide).

Diffusion of solutes is the basis for the laboratory technique of **dialysis.** In this process, solutes smaller than the pore size of the dialysis membrane freely exchange between the sample and the bulk solution until equilibrium is reached (Fig. 2-12). Larger substances cannot cross the membrane and remain where they are. Dialysis is particularly useful for separating large molecules, such as proteins and nucleic acids, from smaller molecules. And because small soluble particles move freely between the sample and the surrounding medium, dialysis can be repeated several times to replace the sample medium with another solution.

The tendency for solutes to diffuse from an area of high concentration to an area of low concentration (i.e., down a concentration gradient) is thermodynamically favored because it is accompanied by an increase in entropy. However, diffusion, the result of the random wandering of particles, does have constraints (see Box 2-1). For example, increasing the viscosity of the medium decreases the rate of diffusion. Furthermore, since diffusion is a random process, the rate of diffusion varies with the square of the distance diffused. Thus, if a particle, on average, diffuses 1 cm in 1 s, the same particle would require 100 s to diffuse 10 cm.

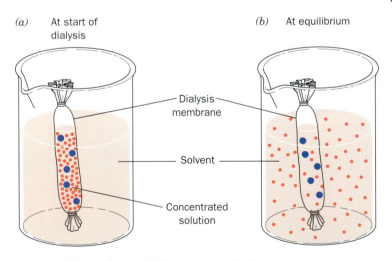

***Figure 2-12.* Dialysis.** (*a*) A concentrated solution is separated from a large volume of solvent by a dialysis membrane (shown here as a tube knotted at both ends). Only small molecules can diffuse through the pores in the membrane. (*b*) At equilibrium, the concentrations of small molecules are nearly the same on either side of the membrane, whereas the macromolecules remain inside the dialysis bag.

Diffusion Rates and the Sizes of Organisms

Small organisms depend on diffusion to acquire nutrients and dispose of wastes. But diffusion is efficient only when the ratio of surface area (where exchange with the surroundings takes place) to internal volume is relatively large. For any solid object, surface area is proportional to the square of the length, but volume is proportional to the cube of length.

Solid	Surface Area	Volume	Surface Area/ Volume
Cube	$6l^2$	l^3	$6/l$
Cylinder	$2\pi rh + 2\pi r^2$	$\pi r^2 h$	$2/r + 2/h$
Sphere	$4\pi r^2$	$\frac{4}{3}\pi r^3$	$3/r$

Therefore, a large object has less surface area relative to its volume than a small object of the same shape. It is hardly surprising that although the sizes of organisms range over some eight orders of magnitude, the diameter of the basic unit of life—the cell—varies by only about 1000-fold. Clearly, a certain level of organization has been retained through evolutionary history.

In multicellular organisms, specialization of form and function helps overcome the surface-to-volume ratio problem. Structures where gas exchange and nutrient absorption take place, such as gills, lungs, and intestines, are characterized microscopically by tremendous surface areas. For example, the human lungs, with an air capacity of ~6 L divided among ~30 million alveoli (small sacs), have a surface area of 50–100 m². Nevertheless, the rates of diffusion of nutrients and wastes are high enough to support life only in organisms that are no larger than about 1 mm in their smallest dimension. Larger organisms require a circulatory system such as the bloodstream to actively transport substances from sites where they are produced or absorbed to sites where they are utilized or eliminated.

2. CHEMICAL PROPERTIES OF WATER

Water is not just a passive component of the cell or extracellular environment. By virtue of its physical properties, water defines the solubilities of other substances. Similarly, water's chemical properties determine the behavior of other molecules in solution.

A. Ionization of Water

Water is a neutral molecule with a slight tendency to ionize. We usually express this ionization as

$$H_2O \rightleftharpoons H^+ + OH^-$$

There is actually no such thing as a free proton (H^+) in solution. Rather, the proton is associated with a water molecule as a **hydronium ion, H_3O^+**. The association of a proton with a cluster of water molecules also gives rise to structures with the formulas $H_5O_2^+$, $H_7O_3^+$, and so on. For simplicity, however, we collectively represent these ions by H^+.

The proton of a given hydronium ion can jump rapidly from one water molecule to another (Fig. 2-13). For this reason, the ionic mobilities of H^+ and OH^- are much higher than for other ions. *Proton jumping is also responsible for the observation that acid–base reactions are among the fastest reactions that take place in aqueous solution.*

The ionization of water is described by an equilibrium expression in which the concentration of the parent substance is in the denominator and the concentrations of the dissociated products are in the numerator:

$$K = \frac{[H^+][OH^-]}{[H_2O]} \qquad [2\text{-}1]$$

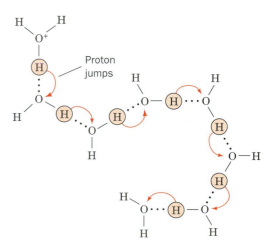

Figure 2-13. Proton jumping. Proton jumps occur more rapidly than direct molecular migration, accounting for the observed high ionic mobilities of hydronium ions (and hydroxide ions) in aqueous solutions.

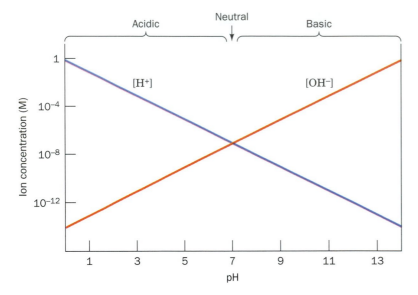

Figure 2-14. Relationship of pH and the concentrations of H$^+$ and OH$^-$ in water. Because the product of [H$^+$] and [OH$^-$] is a constant (10^{-14} M^2), [H$^+$] and [OH$^-$] are reciprocally related. Solutions with relatively more H$^+$ are acidic (pH < 7), solutions with relatively more OH$^-$ are basic (pH > 7), and solutions in which [H$^+$] = [OH$^-$] = 10^{-7} M are neutral (pH = 7). Note the logarithmic scale for ion concentration.

Table 2-4. pH Values of Some Common Substances

Substance	pH
1 M NaOH	14
Household ammonia	12
Seawater	8
Blood	7.4
Milk	7
Saliva	6.6
Tomato juice	4.4
Vinegar	3
Gastric juice	1.5
1 M HCl	0

K is the **dissociation constant** (here and throughout the text, quantities in square brackets symbolize the molar concentrations of the indicated substances). Since the concentration of the undissociated H$_2$O ([H$_2$O]) is so much larger than the concentrations of its component ions, it can be considered constant and incorporated into K to yield an expression for the ionization of water,

$$K_w = [H^+][OH^-] \qquad [2\text{-}2]$$

The value of K_w, the ionization constant of water, is 10^{-14} M^2 at 25°C.

Pure water must contain equimolar amounts of H$^+$ and OH$^-$, so [H$^+$] = [OH$^-$] = $(K_w)^{1/2}$ = 10^{-7} M. Since [H$^+$] and [OH$^-$] are reciprocally related by Eq. 2-2, when [H$^+$] is greater than 10^{-7} M, [OH$^-$] must be correspondingly less and vice versa. Solutions with [H$^+$] = 10^{-7} M are said to be **neutral,** those with [H$^+$] > 10^{-7} M are said to be **acidic,** and those with [H$^+$] < 10^{-7} M are said to be **basic.** Most physiological solutions have hydrogen ion concentrations near neutrality. For example, human blood is normally slightly basic with [H$^+$] = 4.0×10^{-8} M.

The values of [H$^+$] for most solutions are inconveniently small and thus impractical to compare. A more practical quantity, which was devised in 1909 by Søren Sørenson, is known as the **pH:**

$$pH = -\log[H^+] \qquad [2\text{-}3]$$

The higher the pH, the lower the H$^+$ concentration; the lower the pH, the higher the H$^+$ concentration (Fig. 2-14). The pH of pure water is 7.0, whereas acidic solutions have pH < 7.0 and basic solutions have pH > 7.0. Note that solutions that differ by one pH unit differ in [H$^+$] by a factor of 10. The pH values of some common substances are given in Table 2-4.

B. Acid–Base Chemistry

H$^+$ and OH$^-$ ions derived from water are fundamental to the biochemical reactions we shall encounter later in this book. Biological molecules, such as proteins and nucleic acids, have numerous functional groups that act as acids or bases, for example, carboxyl and amino groups. These molecules influence the pH of the surrounding aqueous medium, and their structures and reactivities are in turn influenced by the ambient pH. An appreciation of acid–base chemistry is therefore essential for understanding the biological roles of many molecules.

An Acid Can Donate a Proton

According to a definition coined in the 1880s by Svante Arrhenius, an **acid** is a substance that can donate a proton, and a **base** is a substance that can donate a hydroxide ion. This definition is rather limited. For example, it does not account for the observation that NH$_3$, which lacks an OH group, exhibits basic properties. In a more general definition, which was formulated in 1923 by Johannes Brønsted and Thomas Lowry, *an acid is a substance that can donate a proton* (as in the Arrhenius definition), *and a base is a substance that can accept a proton.* Under the Brønsted–Lowry definition, an acid–base reaction can be written as

$$HA + H_2O \rightleftharpoons H_3O^+ + A^-$$

An acid (HA) reacts with a base (H_2O) to form the **conjugate base** of the acid (A^-) and the **conjugate acid** of the base (H_3O^+). Accordingly, the acetate ion (CH_3COO^-) is the conjugate base of acetic acid (CH_3COOH), and the ammonium ion (NH_4^+) is the conjugate acid of ammonia (NH_3). The acid–base reaction is frequently abbreviated $HA \rightleftharpoons H^+ + A^-$ with the participation of H_2O implied.

The Strength of an Acid Is Specified by Its Dissociation Constant

The equilibrium constant for an acid–base reaction is expressed as a dissociation constant with the concentrations of the "reactants" in the denominator and the concentrations of the "products" in the numerator:

$$K = \frac{[H_3O^+][A^-]}{[HA][H_2O]} \qquad [2\text{-}4]$$

In dilute solutions, the water concentration is essentially constant, 55.5 M ($1000 \text{ g} \cdot L^{-1}/18.015 \text{ g} \cdot mol^{-1} = 55.5$ M). Therefore, the term $[H_2O]$ is customarily combined with the dissociation constant, which then takes the form

$$K_a = K[H_2O] = \frac{[H^+][A^-]}{[HA]} \qquad [2\text{-}5]$$

For brevity, however, we shall henceforth omit the subscript "a."

The dissociation constants of some common acids are listed in Table 2-5. Because acid dissociation constants, like $[H^+]$ values, are sometimes cumbersome to work with, they are transformed to **pK** values by the formula

$$pK = -\log K \qquad [2\text{-}6]$$

which is analogous to Eq. 2-3.

Acids can be classified according to their relative strengths, that is, their abilities to transfer a proton to water. The acids listed in Table 2-5 are known

Table 2-5.   **Dissociation Constants and pK Values at 25°C of Some Acids**

Acid	K	pK
Oxalic acid	5.37×10^{-2}	1.27 (pK_1)
H_3PO_4	7.08×10^{-3}	2.15 (pK_1)
Formic acid	1.78×10^{-4}	3.75
Succinic acid	6.17×10^{-5}	4.21 (pK_1)
Oxalate$^-$	5.37×10^{-5}	4.27 (pK_2)
Acetic acid	1.74×10^{-5}	4.76
Succinate$^-$	2.29×10^{-6}	5.64 (pK_2)
2-(N-Morpholino)ethanesulfonic acid (MES)	8.13×10^{-7}	6.09
H_2CO_3	4.47×10^{-7}	6.35 (pK_1)[a]
Piperazine-N,N'-bis(2-ethanesulfonic acid) (PIPES)	1.74×10^{-7}	6.76
$H_2PO_4^-$	1.51×10^{-7}	6.82 (pK_2)
3-(N-Morpholino)propanesulfonic acid (MOPS)	7.08×10^{-8}	7.15
N-2-Hydroxyethylpiperazine-N'-2-ethanesulfonic acid (HEPES)	3.39×10^{-8}	7.47
Tris(hydroxymethyl)aminomethane (Tris)	8.32×10^{-9}	8.08
NH_4^+	5.62×10^{-10}	9.25
Glycine	1.66×10^{-10}	9.78
HCO_3^-	4.68×10^{-11}	10.33 (pK_2)
Piperidine	7.58×10^{-12}	11.12
HPO_4^{2-}	4.17×10^{-13}	12.38 (pK_3)

Source: Dawson, R.M.C., Elliott, D.C., Elliott, W.H., and Jones, K.M., *Data for Biochemical Research* (3rd ed.), pp. 424–425, Oxford Science Publications (1986) *and* Good, N.E., Winget, G.D., Winter, W., Connolly, T.N., Izawa, S., and Singh, R.M.M., *Biochemistry* **5**, 467 (1966).

[a]The pK for the overall reaction $CO_2 + H_2O \rightleftharpoons H_2CO_3 \rightleftharpoons H^+ + HCO_3^-$; see Box 2-2.

S A M P L E   C A L C U L A T I O N

Calculate the pH of a 2 L solution containing 10 mL of 5 M acetic acid and 10 mL of 1 M sodium acetate.

First, calculate the concentrations of the acid and conjugate base, expressing all concentrations in units of moles per L.

Acetic acid: $(0.01\ L)(5\ M)/(2\ L) = 0.025\ M$
Sodium
    acetate: $(0.01\ L)(1\ M)/(2\ L) = 0.005\ M$

Substitute the concentrations of the acid and conjugate base into the Henderson–Hasselbalch equation. Find the pK for acetic acid in Table 2-5.

pH = pK + log ([acetate]/[acetic acid])
pH = 4.76 + log (0.005/0.025)
pH = 4.76 − 0.70
pH = 4.06

as **weak acids** because they are only partially ionized in aqueous solution ($K < 1$). Many of the so-called mineral acids, such as $HClO_4$, HNO_3, and HCl, are **strong acids** ($K \gg 1$). Since strong acids rapidly transfer all their protons to H_2O, *the strongest acid that can stably exist in aqueous solutions is H_3O^+.* Likewise, *there can be no stronger base in aqueous solutions than OH^-.* Virtually all the acid–base reactions that occur in biological systems involve H_3O^+ (and OH^-) and weak acids (and their conjugate bases).

The pH of a Solution Is Determined by the Relative Concentrations of Acids and Bases

The relationship between the pH of a solution and the concentrations of an acid and its conjugate base is easily derived. Equation 2-5 can be re-arranged to

$$[H^+] = K\frac{[HA]}{[A^-]} \qquad [2\text{-}7]$$

Taking the negative log of each term (and letting pH = $-\log[H^+]$; Eq. 2-3) gives

$$pH = -\log K + \log\frac{[A^-]}{[HA]} \qquad [2\text{-}8]$$

Substituting pK for $-\log K$ (Eq. 2-6) yields

$$\boxed{pH = pK + \log\frac{[A^-]}{[HA]}} \qquad [2\text{-}9]$$

This relationship is known as the **Henderson–Hasselbalch equation.** *When the molar concentrations of an acid (HA) and its conjugate base (A^-) are equal, log ([A^-]/[HA]) = 0, and the pH of the solution is numerically equivalent to the pK of the acid.* The Henderson–Hasselbalch equation is invaluable for calculating, for example, the pH of a solution containing a known quantity of a weak acid and its conjugate base. However, since the Henderson–Hasselbalch equation does not account for the ionization of water itself, it is not useful for calculating the pH of solutions of strong acids or bases. For example, a 1 M solution of a strong acid has a pH of 0, and a 1 M solution of a strong base has a pH of 14.

C. Buffers

Adding a 0.01-mL droplet of 1 M HCl to 1 L of pure water changes the pH of the water from 7 to 5, which represents a 100-fold increase in [H^+]. Such a huge change in pH would be intolerable to most biological systems, since even small changes in pH can dramatically affect the structures and functions of biological molecules. Maintaining a relatively constant pH is therefore of paramount importance for living systems. To understand how this is possible, consider the titration of a weak acid with a strong base.

Figure 2-15 shows how the pH values of solutions of acetic acid, $H_2PO_4^-$, and ammonium ion (NH_4^+) vary as OH^- is added. **Titration curves** such as these can be constructed from experimental observation or by using the Henderson–Hasselbalch equation. When OH^- reacts with HA, the products are A^- and water.

Several details about the titration curves in Fig. 2-15 should be noted:

1. The curves have similar shapes but are shifted vertically along the pH axis.

2. The pH at the midpoint of each titration is numerically equivalent to the pK of its corresponding acid; at this point, [HA] = [A$^-$].

3. The slope of each titration curve is much lower near its midpoint than near its wings. This indicates that *when [HA] ≈ [A$^-$], the pH of the solution is relatively insensitive to the addition of strong base or strong acid.* Such a solution, which is known as an acid–base **buffer,** resists pH changes because small amounts of added H$^+$ or OH$^-$ react with A$^-$ or HA, respectively, without greatly changing the value of log([A$^-$]/[HA]).

Substances that can undergo more than one ionization, such as H$_3$PO$_4$ and H$_2$CO$_3$, are known as **polyprotic acids.** The titration curves of such molecules, as illustrated in Fig. 2-16 for H$_3$PO$_4$, are more complicated than the titration curves of monoprotic acids such as acetic acid. A polyprotic acid has multiple pK's, one for each ionization step. H$_3$PO$_4$, for example, has three dissociation constants because the ionic charge resulting from one proton dissociation electrostatically inhibits further proton dissociation, thereby increasing the corresponding pK's. Similarly, a molecule with more than one ionizable group has a discrete pK for each group. In a biomolecule that contains numerous ionizable groups with different pK's, the many dissociation events may yield a titration curve without any clear "plateaus."

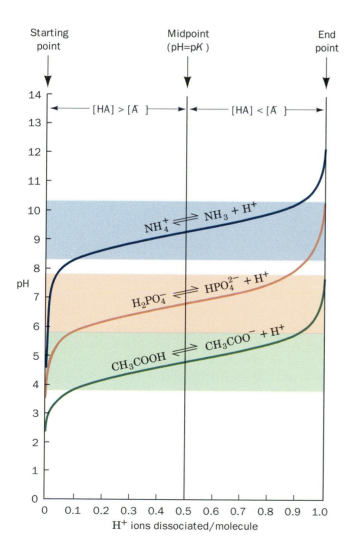

Figure 2-15. **Titration curves for acetic acid, phosphate, and ammonia.** At the starting point, the acid form predominates. As strong base (e.g., NaOH) is added, the acid is converted to its conjugate base. At the midpoint of the titration, where pH = pK, the concentrations of the acid and the conjugate base are equal. At the end point (equivalence point), the conjugate base predominates, and the total amount of OH$^-$ that has been added is equivalent to the amount of acid that was present at the starting point. The shaded bands indicate the pH ranges over which the corresponding solution can function as a buffer.
✳ See the Animated Figures.

***Figure 2-16.* Titration of a polyprotic acid.**
The first and second equivalence points for titration of H_3PO_4 occur at the steepest parts of the curve. The pH at the midpoint of each stage provides the pK value of the corresponding ionization. ✳ **See the Animated Figures.**

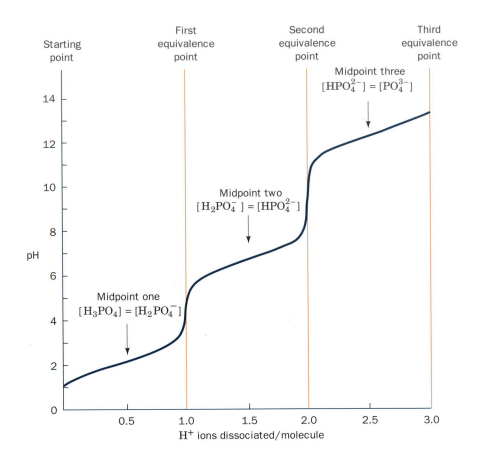

Biological fluids, both intracellular and extracellular, are heavily buffered. For example, the pH of the blood in healthy individuals is closely controlled at pH 7.4 (see Box 2-2). The phosphate and carbonate ions in most biological fluids are important buffering agents because they have pK's in this range (Table 2-5). Moreover, many biological molecules, such as proteins, nucleic acids, and lipids, as well as numerous small organic molecules, bear multiple acid–base groups that are effective buffer components in the physiological pH range.

The concept that the properties of biological molecules vary with the acidity of the solution in which they are dissolved was not fully appreciated before the twentieth century. Many early biochemical experiments were undertaken without controlling the acidity of the sample, so the results were often poorly reproducible. Nowadays, biochemical preparations are routinely buffered to simulate the properties of naturally occurring biological fluids. A number of synthetic compounds have been developed for use as buffers; some of these are included in Table 2-5. The **buffering capacity** of these weak acids (their ability to resist pH changes on addition of acid or base) is maximal when pH = pK. It is helpful to remember that a weak acid is in its useful buffer range within one pH unit of its pK (e.g., the shaded regions of Fig. 2-15). Above this range, where the ratio $[A^-]/[HA] > 10$, the pH of the solution changes rapidly with added strong base. A buffer is similarly impotent with the addition of strong acid when its pK exceeds the pH by more than one unit.

In the laboratory, the desired pH of the buffered solution determines which buffering compound is selected. Typically, the acid form of the compound and one of its soluble salts are dissolved in the (nearly equal) molar ratio necessary to provide the desired pH, and, with the aid of a pH meter, the resulting solution is fine-tuned by titration with strong acid or base.

The Blood Buffering System

Bicarbonate is the most significant buffer compound in human blood; other buffering agents, including proteins and organic acids, are present at much lower concentrations. The buffering capacity of blood depends primarily on two equilibria: (1) between gaseous CO_2 dissolved in the blood and carbonic acid formed by the reaction

$$CO_2 + H_2O \rightleftharpoons H_2CO_3$$

and (2) between carbonic acid and bicarbonate formed by the dissociation of H^+

$$H_2CO_3 \rightleftharpoons H^+ + HCO_3^-$$

The overall pK for these two sequential reactions is 6.35. (The further dissociation of HCO_3^- to CO_3^{2-}, pK = 10.33, is not significant at physiological pH.)

When the pH of the blood falls due to metabolic production of H^+, the bicarbonate–carbonic acid equilibrium shifts toward more carbonic acid. At the same time, carbonic acid loses water to become CO_2, which is then expired in the lungs as gaseous CO_2. Conversely, when the blood pH rises, relatively more HCO_3^- forms. Breathing is adjusted so that increased amounts of CO_2 in the lungs can be reintroduced into the blood for conversion to carbonic acid. In this manner, a near-constant hydrogen ion concentration can be maintained.

Disturbances in the blood buffer system can lead to conditions known as **acidosis**, with a pH as low as 7.1, or **alkalosis**, with a pH as high as 7.6. (Deviations of less than 0.05 pH unit from the "normal" value of 7.4 are not significant.) For example, obstructive lung diseases that prevent efficient expiration of CO_2 can cause respiratory acidosis. Hyperventilation accelerates the loss of CO_2 and causes respiratory alkalosis. Overproduction of organic acids from dietary precursors or sudden surges in lactic acid levels during exercise can lead to metabolic acidosis.

Acid–base imbalances are best alleviated by correcting the underlying physiological problem. In the short term, acidosis is commonly treated by administering $NaHCO_3$ intravenously. Alkalosis is more difficult to treat. Metabolic alkalosis sometimes responds to KCl or NaCl (the additional Cl^- helps minimize the secretion of H^+ by the kidneys), and respiratory alkalosis can be ameliorated by breathing an atmosphere enriched in CO_2.

SUMMARY

1. Water is essential for all living organisms.
2. Water molecules can form hydrogen bonds with other molecules because they have two H atoms that can be donated and two unshared electron pairs that can act as acceptors.
3. Liquid water is an irregular network of water molecules that each form up to four hydrogen bonds with neighboring water molecules.
4. Hydrophilic substances such as ions and polar molecules dissolve readily in water.
5. The hydrophobic effect is the tendency of water to minimize its contacts with nonpolar substances.
6. Water molecules move from regions of high concentration to regions of low concentration by osmosis; solutes move from regions of high concentration to regions of low concentration by diffusion.
7. Water ionizes to H^+ (which represents the hydronium ion, H_3O^+) and OH^-.
8. The concentration of H^+ in solutions is expressed as a pH value; in acidic solutions pH < 7, in basic solutions pH > 7, and in neutral solutions pH = 7.
9. Acids can donate protons and bases can accept protons. The strength of an acid is expressed as its pK.
10. The Henderson–Hasselbalch equation relates the pH of a solution to the concentrations of an acid and its conjugate base.
11. Buffered solutions resist changes in pH within about one pH unit of the pK of the buffering species.

REFERENCES

Cooke, R. and Kuntz, I.D., The properties of water in biological systems, *Annu. Rev. Biophys. Bioeng.* **3**, 95–126 (1974).

Franks, F., *Water,* The Royal Society of Chemistry (1993).

Good, N.E., Winget, G.D., Winter, W., Connolly, T.N., Izawa, S., and Singh, R.M.M., Hydrogen ion buffers for biological research, *Biochemistry* **5**, 467–477 (1966).

Jeffrey, G.A. and Saenger, W., *Hydrogen Bonding in Biological Structures,* Chapters 1, 2, and 21, Springer-Verlag (1994). [Reviews hydrogen bond chemistry and its importance in small molecules and macromolecules.]

Segel, I.H., *Biochemical Calculations* (2nd ed.), Chapter 1, Wiley (1976). [An intermediate level discussion of acid–base equilibria with worked-out problems.]

Stillinger, F.H., Water revisited, *Science* **209**, 451–457 (1980). [An excellent outline of water structure on an elementary level.]

Tanford, C., *The Hydrophobic Effect: Formation of Micelles and Biological Membranes* (2nd ed.), Chapters 5 and 6, Wiley–Interscience (1980). [Discusses the structures of water and micelles.]

KEY TERMS

polar	amphipathic	dissociation constant	weak acid
hydrogen bond	micelle	K_w	strong acid
van der Waals distance	bilayer	neutral solution	Henderson–Hasselbalch
van der Waals forces	water of hydration	acidic solution	equation
London dispersion forces	colligative properties	basic solution	titration curve
hydrophilic	osmosis	pH	buffer
hydrophobic	osmotic pressure	acid	polyprotic acid
solvation	diffusion	base	buffering capacity
hydration	dialysis	conjugate base	
hydrophobic effect	hydronium ion	conjugate acid	
amphiphilic	proton jumping	pK	

STUDY EXERCISES

1. Compare hydrogen bonding in ice and hydrogen bonding in liquid water.

2. Explain why polar substances dissolve in water while nonpolar substances do not.

3. Describe the contribution of entropy to the hydrophobic effect.

4. Why do amphiphiles form micelles in water?

5. How does osmosis differ from diffusion?

6. Compare the Arrhenius and Brønsted–Lowry definitions of acids and bases.

7. Explain why a 1 M solution of HCl has a pH of 0.

PROBLEMS

1. Identify the potential hydrogen bond donors and acceptors in the following molecules:

   (a)
   (b)
   (c)
   $$H-\underset{\underset{NH_3^+}{|}}{\overset{\overset{COO^-}{|}}{C}}-CH_2-OH$$

2. Where would the following substances partition in water containing palmitic acid micelles? (a) $^+H_3N-CH_2-COO^-$, (b) $^+H_3N-(CH_2)_{11}-COO^-$, (c) $H_3C-(CH_2)_{11}-COO^-$.

3. Describe what happens when a dialysis bag containing pure water is suspended in a beaker of seawater. What would happen if the dialysis membrane were permeable to water but not solutes?

4. Compare the surface-to-volume ratios for a bacterium (length 3 μm, diameter 0.5 μm) and a fish (length 30 cm, diameter 5 cm). Assume each organism is shaped like a cylinder.

5. Draw the structures of the conjugate bases of the following acids:

   (a)
   $$\underset{\underset{\underset{|}{COOH}}{HC}}{\overset{\overset{\overset{|}{CH}}{COO^-}}{\|}}$$

   (b)
   $$H-\underset{\underset{NH_3^+}{|}}{\overset{\overset{COOH}{|}}{C}}-H$$

(c)
$$H-\underset{\underset{NH_3^+}{|}}{\overset{\overset{COO^-}{|}}{C}}-H$$

(d)
$$H-\underset{\underset{NH_3^+}{|}}{\overset{\overset{COO^-}{|}}{C}}-CH_2-COOH$$

6. Indicate the ionic species that predominates at pH 4, 8, and 11 for (a) ammonia, and (b) phosphoric acid.

7. Calculate the pH of a 1 L solution containing (a) 10 mL of 5 M NaOH, (b) 10 mL of 100 mM glycine and 20 mL of 5 M HCl, and (c) 10 mL of 2 M acetic acid and 5 g of sodium acetate (formula weight 82 g·mol⁻¹).

8. How many grams of sodium succinate (formula weight 140 g·mol⁻¹) and disodium succinate (formula weight 162 g·mol⁻¹) must be added to 1 L of water to produce a solution with pH 6.0 and a total solute concentration of 50 mM?

9. Estimate the volume of a solution of 5 M NaOH that must be added to adjust the pH from 4 to 9 in 100 mL of a 100 mM solution of phosphoric acid.

10. (a) Would phosphoric acid or succinic acid be a better buffer at pH 5? (b) Would ammonia or piperidine be a better buffer at pH 9? (c) Would HEPES or Tris be a better buffer at pH 7.5?

II

BIOMOLECULES

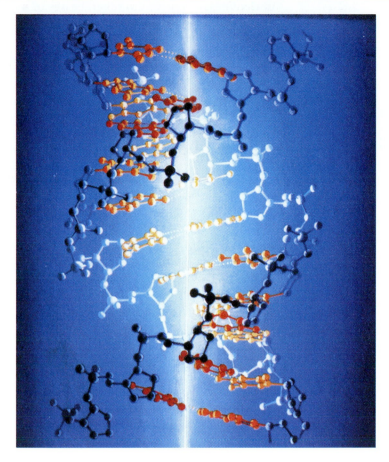

A DNA molecule consists of two strands that wind around a central axis, shown here as a glowing wire. A complete set of genetic instructions contains just four types of monomeric units. The sequence in which these monomers are linked constitutes a form of biological information that can be efficiently decoded and faithfully copied. [Figure copyrighted © by Irving Geis.]

NUCLEOTIDES AND NUCLEIC ACIDS

If the major molecular constituents of living cells are classified according to their physical and chemical properties, none are more versatile than the **nucleotides.** Compared to other classes of molecules—amino acids, carbohydrates, and lipids—nucleotides are notable for their involvement in the reactions that are central to the maintenance and propagation of life. Specifically, nucleotides participate in energy transfer, and their polymeric forms, the **nucleic acids,** are the primary players in the storage and decoding of genetic information. Nucleotides and nucleic acids also perform structural and catalytic roles in cells. No other class of molecules undertakes such varied functions or so many functions that are essential for life.

Evolutionists postulate that the appearance of nucleotides permitted the evolution of organisms that could harvest and store energy from their surroundings and, most importantly, could make copies of themselves. Although the chemical and biological details of early life forms are the subject of speculation, it is incontrovertible that life as we know it is inextricably linked to the chemistry of nucleotides and nucleic acids.

In this chapter, we examine the structures of nucleotides and the nucleic acids RNA and DNA. We also discuss how information, as a sequence of nucleotides, is contained in DNA molecules and how that information can be manipulated *in vitro* by recombinant DNA technology. In later chapters, we examine in greater detail the participation of nucleotides and nucleic acids in metabolism and the storage and expression of genetic information.

1. NUCLEOTIDE STRUCTURE AND FUNCTION

Nucleotides are ubiquitous molecules with considerable structural diversity. *There are eight common varieties of nucleotides, each composed of a nitrogenous base linked to a sugar to which at least one phosphate group is also attached.* The bases of nucleotides are planar, aromatic, heterocyclic molecules that are structural derivatives of either **purine** or **pyrimidine** (although they are not synthesized *in vivo* from either of these organic compounds).

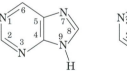

Purine   **Pyrimidine**

The most common purines are **adenine (A)** and **guanine (G),** and the major pyrimidines are **cytosine (C), uracil (U),** and **thymine (T).** The purines form bonds to a five-carbon sugar (a pentose) via their N9 atoms, whereas pyrimidines do so through their N1 atoms (Table 3-1).

In **ribonucleotides** (Fig. 3-1*a*), the pentose is **ribose,** while in **deoxyribonucleotides** (or just **deoxynucleotides;** Fig. 3-1*b*), the sugar is **2′-deoxyribose** (i.e., the carbon at position 2′ lacks a hydroxyl group). Note that the "primed" numbers refer to the atoms of the ribose; "unprimed" numbers refer to the atoms of the nitrogenous base. The phosphate group may be bonded to C3′ or C5′ of a pentose to form its 3′-nucleotide or its 5′-nucleotide, respectively. When the phosphate group is absent, the compound is known as a **nucleoside.** A 5′-nucleotide can therefore be called a nucleoside-5′-phosphate.

The structures, names, and abbreviations of the common bases, nucleosides, and nucleotides are given in Table 3-1. Ribonucleotides are found in

Table 3-1.   **Names and Abbreviations of Nucleic Acid Bases, Nucleosides, and Nucleotides**

Base Formula	Base (X = H)	Nucleoside (X = ribose[a])	Nucleotide[b] (X = ribose phosphate[a])
	Adenine	Adenosine	Adenylic acid
	Ade	Ado	Adenosine monophosphate
	A	A	AMP
	Guanine	Guanosine	Guanylic acid
	Gua	Guo	Guanosine monophosphate
	G	G	GMP
	Cytosine	Cytidine	Cytidylic acid
	Cyt	Cyd	Cytidine monophosphate
	C	C	CMP
	Uracil	Uridine	Uridylic acid
	Ura	Urd	Uridine monophosphate
	U	U	UMP
	Thymine	Deoxythymidine	Deoxythymidylic acid
	Thy	dThd	Deoxythymidine monophosphate
	T	dT	dTMP

[a]The presence of a 2′-deoxyribose unit in place of ribose, as occurs in DNA, is implied by the prefixes "deoxy" or "d." For example, the deoxynucleoside of adenine is deoxyadenosine or dA. However, for thymine-containing residues, which rarely occur in RNA, the prefix is redundant and may be dropped. The presence of a ribose unit may be explicitly implied by the prefix "ribo" or "r." Thus the ribonucleotide of thymine is ribothymidine or rT.

[b]The position of the phosphate group in a nucleotide may be explicitly specified as in, for example, 3′-AMP and 5′-GMP.

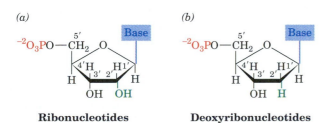

Figure 3-1.   **Chemical structures of (a) ribonucleotides and (b) deoxyribonucleotides.** The purine or pyrimidine base is linked to C1′ of the pentose and at least one phosphate (*red*) is attached.

RNA (ribonucleic acid), whereas deoxynucleotides are found in **DNA (deoxyribonucleic acid).** Adenine, guanine, and cytosine are found as both ribonucleotides and deoxynucleotides (accounting for six of the eight common nucleotides), but uracil is found primarily as a ribonucleotide, and thymine as a deoxynucleotide. Free nucleotides, which are anionic, are usually associated with the counterion Mg^{2+} in cells.

Functions of Nucleotides and Nucleotide Derivatives

The bulk of the nucleotides in any cell are found in polymeric forms, as either RNA or DNA, whose primary functions are information storage and transfer. However, free nucleotides and nucleotide derivatives perform an enormous variety of metabolic functions not related to the management of genetic information.

Perhaps the best known nucleotide is **adenosine triphosphate (ATP),** a nucleotide containing adenine, ribose, and a triphosphate group. ATP is often mistakenly referred to as an energy-storage molecule, but it is more accurately termed an energy carrier or energy transmitter. The process of photosynthesis or the breakdown of metabolic fuels such as carbohydrates and fatty acids leads to the formation of ATP from **adenosine diphosphate (ADP):**

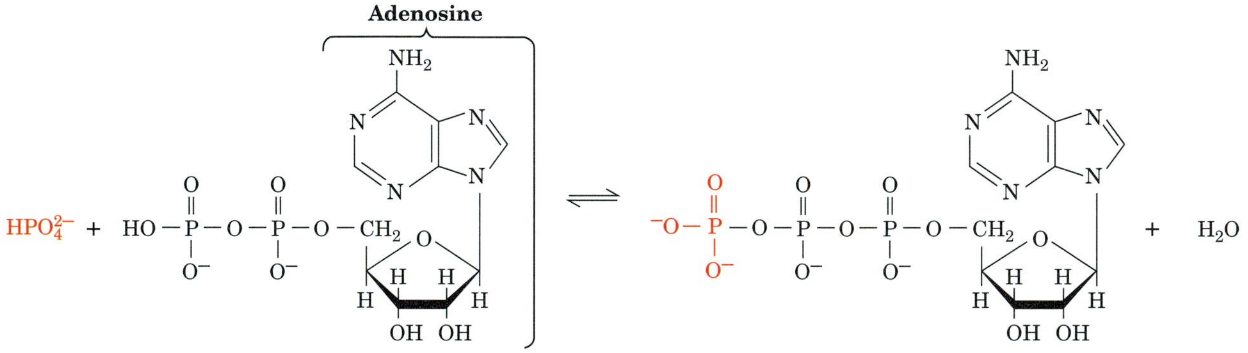

Adenosine diphosphate (ADP)   Adenosine triphosphate (ATP)

ATP diffuses throughout the cell to provide energy for other cellular work, such as biosynthetic reactions, ion transport, and cell movement. The energy of ATP is made available when it transfers one (or two) of its phosphate groups to another molecule. This process can be represented by the reverse of the above reaction, namely, the hydrolysis of ATP to ADP. (As we shall see in later chapters, the interconversion of ATP and ADP in the cell is not freely reversible, and free phosphate groups are seldom released directly from ATP.) The degree to which ATP participates in routine cellular activities is illustrated by calculations indicating that while the concentration of cellular ATP is relatively moderate (~5 mM), the average human turns over his or her own weight of ATP each day.

In many metabolic reactions, a nucleotide transfers not a phosphoryl group but another constituent that has taken its place. For example, starch synthesis in plants proceeds by repeated addition of glucose units donated by ADP–glucose (Fig. 3-2). Other examples of groups donated by nucleotide derivatives can be found in the synthetic pathways for other complex carbohydrates (Sections 15-2 and 15-5) and membrane lipids (Section 19-6).

The observation that all known organisms use ATP as an energy carrier suggests that the role of ATP was established very early in evolution. Other evidence for the dominance of nucleotides in ancient metabolism can be

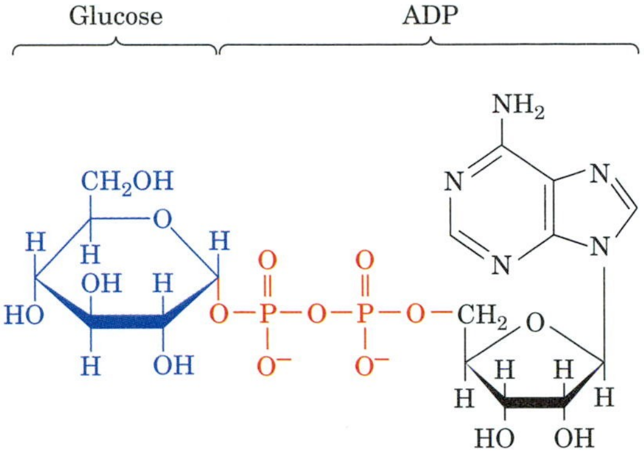

Figure 3-2. **ADP–glucose.** In this nucleotide derivative, glucose (*blue*) is attached to adenosine (*black*) by two phosphate groups (*red*).

found in the metabolic pathways of modern-day organisms. *All the basic processes by which energy is recovered—whether from the sun, inorganic (mineral) components, or organic compounds—rely on series of nucleotide derivatives that transfer energy in discrete quantities.* Despite the enormous variations in energy metabolism in different organisms, several nucleotide derivatives stand out in their ubiquity. Among them are flavin adenine dinucleotide, nicotinamide adenine dinucleotide, and coenzyme A.

Flavin adenine dinucleotide (FAD) contains adenosine linked via two phosphate groups to **riboflavin** (Fig. 3-3). The riboflavin portion of FAD is synthesized by many organisms but not by humans, who must obtain riboflavin (also called **vitamin B$_2$**) from their diet. The heterocyclic ring system of the riboflavin portion of FAD can be reversibly reduced (Section 13-3A). For this reason, FAD participates in many biological oxidation–reduction reactions.

Nicotinamide adenine dinucleotide (NAD$^+$), like FAD, participates in oxidation–reduction reactions (Section 13-3A). In NAD$^+$ and the related compound **nicotinamide adenine dinucleotide phosphate (NADP$^+$),** adenosine is linked via two phosphate groups to ribose and **nicotinamide**

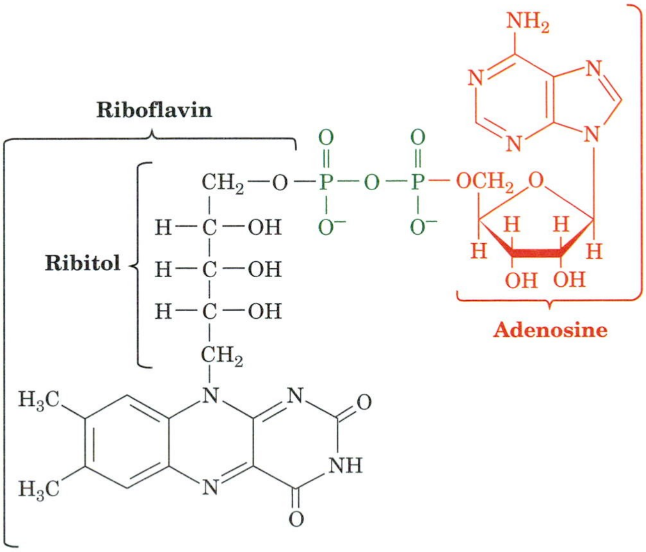

Figure 3-3. **Flavin adenine dinucleotide (FAD).** Adenosine (*red*) is linked to riboflavin (*black*) by two phosphate groups.

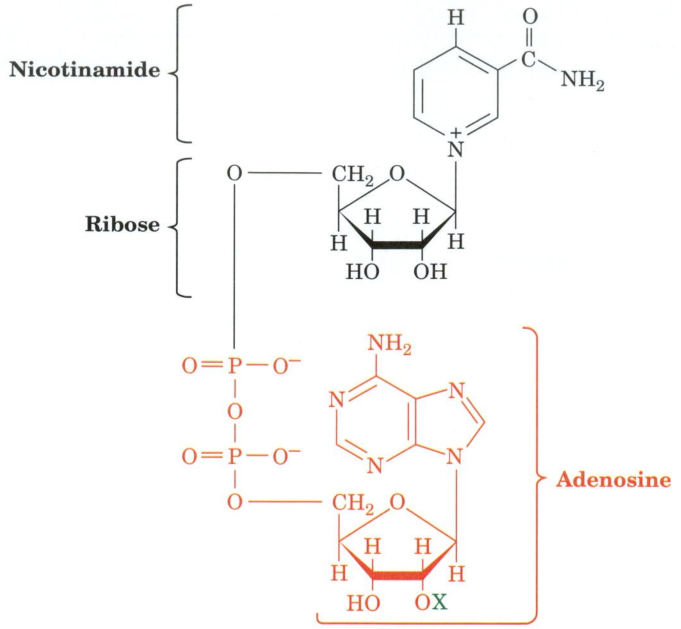

Figure 3-4. **Nicotinamide adenine dinucleotide (NAD$^+$) and nicotinamide adenine dinucleotide phosphate (NADP$^+$).** These dinucleotides contain adenosine linked to a nicotinamide nucleotide. NADP$^+$ contains an additional phosphate group at C2$'$ of its adenosine residue.

X = H   Nicotinamide adenine dinucleotide (NAD$^+$)
X = PO$_3^{2-}$ Nicotinamide adenine dinucleotide phosphate (NADP$^+$)

(Fig. 3-4). In NADP$^+$, a third phosphate group is attached to the ribose of adenosine at its 2$'$ position. The nicotinamide portion of NAD$^+$ and NADP$^+$, which is derived from the vitamin **niacin** (Section 11-1C), is the site of reversible reduction. Note that the nicotinamide base of NAD$^+$ is joined to a phosphorylated ribose, thereby forming one of the two nucleotides in this dinucleotide.

Coenzyme A (CoA; Fig. 3-5), another nucleotide derivative, plays a central role in metabolism although it does not undergo oxidation or reduction.

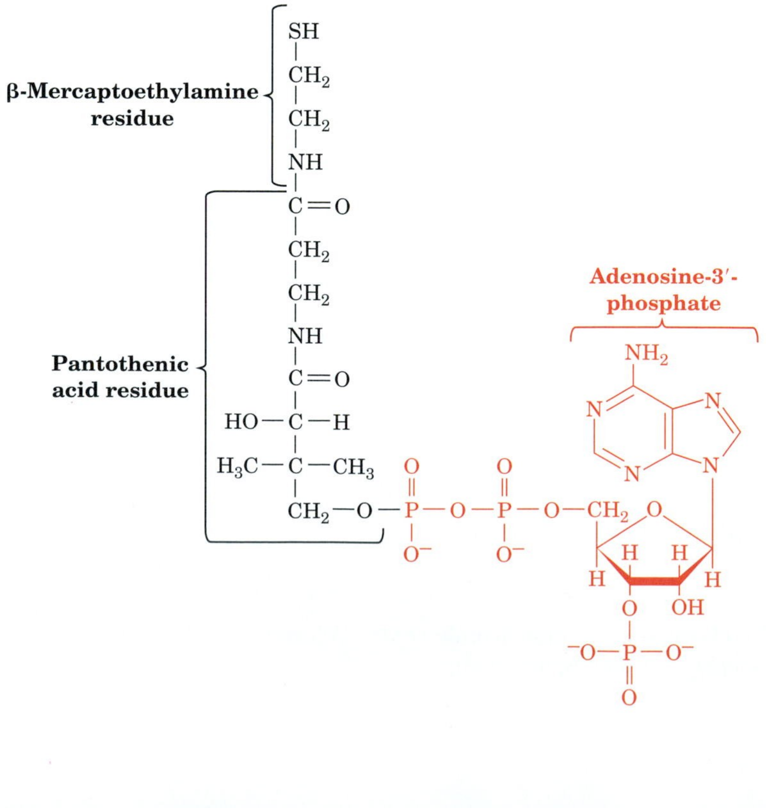

Figure 3-5. **Coenzyme A (CoA).** This adenosine derivative carries acyl groups that are covalently linked as thioesters to the sulfhydryl group.

CoA is a carrier of acyl groups [$CH_3(CH_2)_nCO—$] rather than electrons. The acyl group (often an acetyl group, $CH_3CO—$) is linked to the sulfhydryl group at the end of the mercaptoethylamine portion of the molecule. Coenzyme A is derived from **pantothenic acid** (also known as **vitamin B₃**).

The essential nature of compounds such as FAD, NAD^+, and CoA makes it difficult to imagine what sort of chemical species might have preceded them in prebiotic evolution. Once nucleotides evolved, they gradually assumed varied roles. They have remained key players in the metabolic field even while other structures and processes have grown up around them.

2. NUCLEIC ACID STRUCTURE

Nucleotides can be joined to each other to form the polymers that are familiar to us as RNA and DNA. The nucleic acids are chains of nucleotides whose phosphates bridge the 3′ and 5′ positions of neighboring ribose units (Fig. 3-6). The phosphates of these **polynucleotides** are acidic, so at physiological pH, nucleic acids are polyanions.

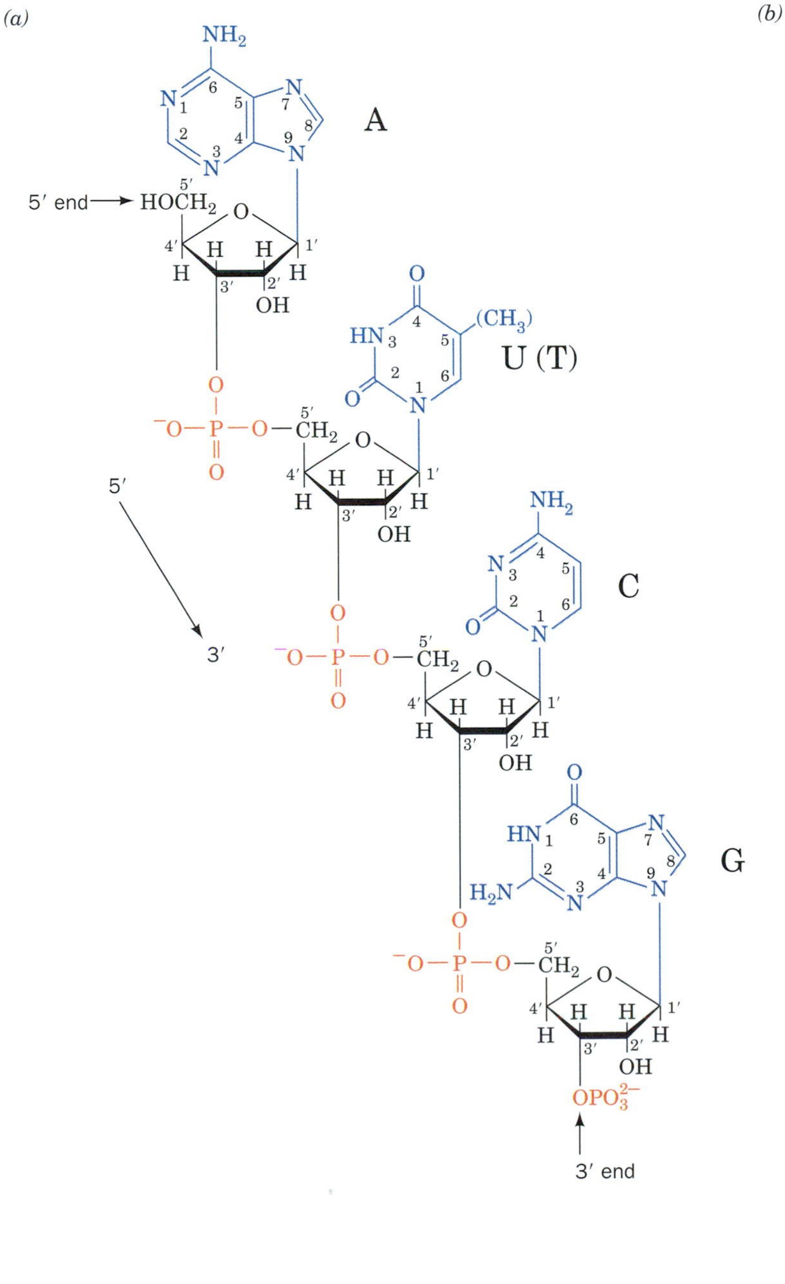

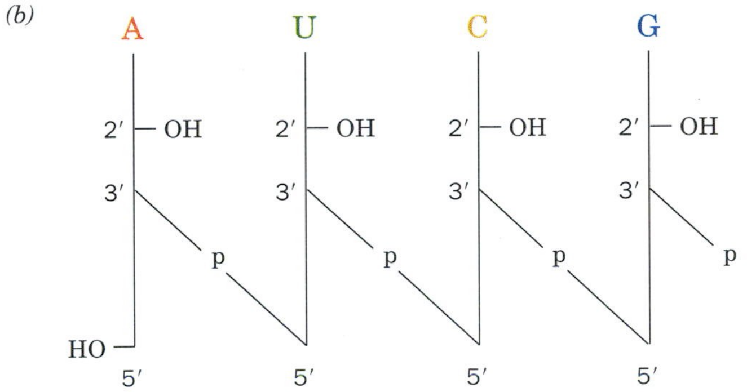

Figure 3-6.   Key to Structure. Chemical structure of a nucleic acid. (*a*) The tetraribonucleotide adenylyl-3′,5′-uridylyl-3′,5′-cytidylyl-3′,5′-guanylyl-3′-phosphate. The sugar atoms are primed to distinguish them from the atoms of the bases. By convention, a polynucleotide sequence is written with the 5′ end at the left and the 3′ end at the right. Thus, reading left to right, the phosphodiester bond links neighboring ribose residues in the 5′ → 3′ direction. The sequence shown here can be abbreviated ApUpCpGp or just AUCGp (the "p" to the right of a nucleoside symbol indicates a 3′ phosphoryl group). The corresponding deoxytetranucleotide, in which the 2′-OH groups are replaced by H and the uracil (U) is replaced by thymine (T), is abbreviated d(ApTpCpGp) or d(AUCGp). (*b*) Schematic representation of AUCGp. A vertical line denotes a ribose residue, the attached base is indicated by a single letter, and a diagonal line flanking an optional "p" represents a phosphodiester bond. The atom numbers for the ribose residue may be omitted. The equivalent representation of d(ATCGp) differs only by the absence of the 2′-OH group and the replacement of U by T.

The linkage between individual nucleotides is known as a **phosphodiester bond,** so named because the phosphate is esterified to two ribose units. Each nucleotide that has been incorporated into the polynucleotide is known as a **nucleotide residue.** The terminal residue whose C5′ is not linked to another nucleotide is called the **5′ end,** and the terminal residue whose C3′ is not linked to another nucleotide is called the **3′ end.**

The properties of a polymer such as a nucleic acid may be very different from the properties of the individual units, or **monomers,** before polymerization. As the size of the polymer increases from **dimer, trimer, tetramer,** and so on through **oligomer** (Greek: *oligo,* few), physical properties such as charge and solubility may change. In addition, *a polymer of nonidentical residues has a property that its component monomers do not have—namely, it contains information in the form of its sequence of residues.*

A. The Base Composition of DNA

Although there appear to be no rules governing the nucleotide composition of typical RNA molecules, DNA has equal numbers of adenine and thymine residues (A = T) and equal numbers of guanine and cytosine residues (G = C). These relationships, known as **Chargaff's rules,** were discovered in the late 1940s by Erwin Chargaff, who devised the first reliable quantitative methods for the compositional analysis of DNA.

DNA's base composition varies widely among different organisms. It ranges from ~25 to 75% G + C in different species of bacteria. However, it is more or less constant among related species; for example, in mammals G + C ranges from 39 to 46%. The significance of Chargaff's rules was not immediately appreciated, but we now know that the structural basis for the rules derives from DNA's double-stranded nature.

B. The Double Helix

The determination of the structure of DNA by James Watson and Francis Crick in 1953 is often said to mark the birth of modern molecular biology. The **Watson–Crick structure** of DNA not only provided a model of what is arguably the central molecule of life, it also suggested the molecular mechanism of heredity. Watson and Crick's accomplishment, which is ranked as one of science's major intellectual achievements, was based in part on two pieces of evidence in addition to Chargaff's rules: the correct tautomeric forms of the bases and indications that DNA is a helical molecule.

The purine and pyrimidine bases of nucleic acids can assume different tautomeric forms (**tautomers** are easily converted isomers that differ only in hydrogen positions; Fig. 3-7). X-Ray, nuclear magnetic resonance (NMR), and spectroscopic investigations have firmly established that the nucleic acid bases are overwhelmingly in the keto tautomeric forms shown in Fig. 3-6. In 1953, however, this was not generally appreciated. Information about the dominant tautomeric forms was provided by Jerry Donohue, an office mate of Watson and Crick and an expert on the X-ray structures of small organic molecules.

Evidence that DNA is a helical molecule was provided by an X-ray diffraction photograph of a DNA fiber taken by Rosalind Franklin (Fig. 3-8). A description of the photograph enabled Crick, an X-ray crystallographer by training, to deduce (a) that DNA is a helical molecule and (b) that its planar aromatic bases form a stack that is parallel to the fiber axis.

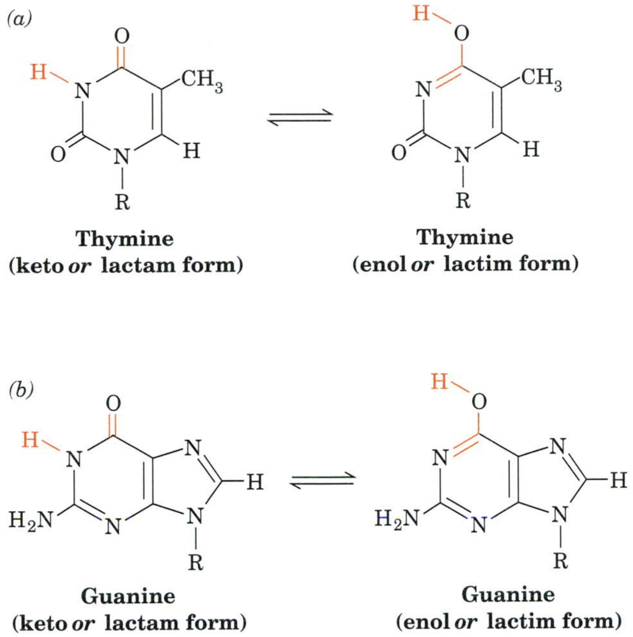

(a)

Thymine
(keto *or* lactam form)

Thymine
(enol *or* lactim form)

(b)

Guanine
(keto *or* lactam form)

Guanine
(enol *or* lactim form)

Figure 3-7. **Tautomeric forms of bases.** Some of the possible tautomeric forms of (*a*) thymine and (*b*) guanine are shown. Cytosine and adenine can undergo similar proton shifts.

The limited structural information, along with Chargaff's rules, provided but few clues to the structure of DNA; Watson and Crick's model sprang mostly from their imaginations and model-building studies. Once the Watson–Crick model had been published, however, its basic simplicity combined with its obvious biological relevance led to its rapid acceptance. Later investigations have confirmed the general accuracy of the Watson–Crick model, although its details have been modified.

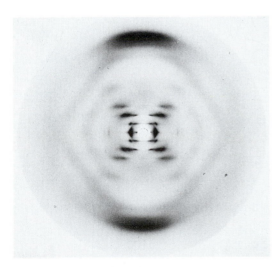

Figure 3-8. **An X-ray diffraction photograph of a vertically oriented DNA fiber.** This photograph, taken by Rosalind Franklin, provided key evidence for the elucidation of the Watson–Crick structure. The central X-shaped pattern indicates a helix, whereas the heavy black arcs at the top and bottom of the diffraction pattern reveal the spacing of the stacked bases (3.4 Å). [Courtesy of Maurice Wilkins, King's College, London.]

Figure 3-9.   **Three-dimensional structure of DNA.** The repeating helix is based on the structure of the self-complementary dodecamer d(CGCGAATTCGCG) determined by Richard Dickerson and Horace Drew. The view in this ball-and-stick model is perpendicular to the helix axis. The sugar–phosphate backbones (*blue, with green ribbon outlines*) wind around the periphery of the molecule. The bases (*red*) form hydrogen-bonded pairs that occupy the core. H atoms have been omitted for clarity. The two strands run in opposite directions. [Figure copyrighted © by Irving Geis.] ● **See the Interactive Exercises and Kinemage Exercise 2-1.**

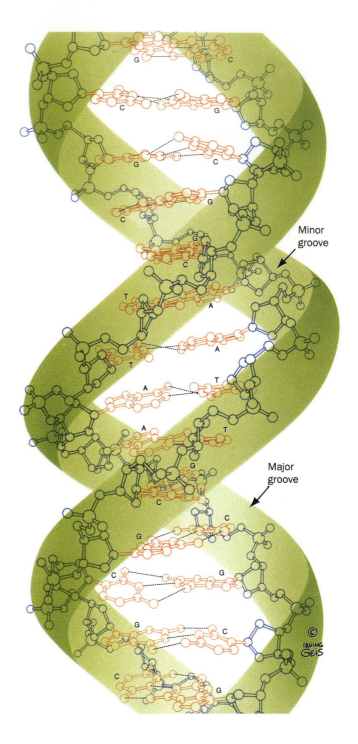

Figure 3-10.   **Diagrams of left- and right-handed helices.** In each case, the fingers curl in the direction the helix turns when the thumb points in the direction the helix rises. Note that the handedness is retained when the helices are turned upside down.

The Watson–Crick model of DNA has the following major features:

1. Two polynucleotide chains wind around a common axis to form a **double helix** (Fig. 3-9).

2. The two strands of DNA are **antiparallel** (run in opposite directions), but each forms a right-handed helix. (The difference between a right-handed and a left-handed helix is shown in Fig. 3-10.)

3. The bases occupy the core of the helix and sugar–phosphate chains run along the periphery, thereby minimizing the repulsions between charged phosphate groups. The surface of the double helix contains two grooves of unequal width: the **major** and **minor grooves.**

4. Each base is hydrogen bonded to a base in the opposite strand to form a planar **base pair.** The Watson–Crick structure can accommodate only two types of base pairs. Each adenine residue must pair with a thymine residue and vice versa, and each guanine residue must pair with a cytosine residue and vice versa (Fig. 3-11). These hydrogen-bonding interactions, a phenomenon known as **complementary base pairing,** result in the specific association of the two chains of the double helix.

The Watson–Crick structure can accommodate any sequence of bases on one polynucleotide strand if the opposite strand has the complementary base sequence. This immediately accounts for Chargaff's rules. More importantly, it suggests that *each DNA strand can act as a* **template** *for the*

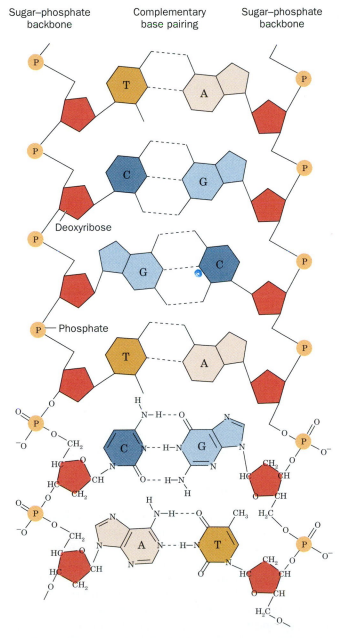

Figure 3-11. **Complementary strands of DNA.** Two polynucleotide chains associate by base pairing to form double-stranded DNA. A pairs with T, and G pairs with C by forming specific hydrogen bonds. ● See Kinemage Exercise 2-2.

Table 3-2.   **Sizes of Some DNA Molecules**

Organism	Number of Base Pairs (kb)[a]
Viruses	
Polyoma, SV40	5.1
λ bacteriophage	48.6
T2, T4, T6 bacteriophage	166
Fowlpox	280
Bacteria	
Mycoplasma hominis	760
Escherichia coli	4,600
Eukaryotes	
Yeast (in 16 chromosomes)	12,600
Drosophila (in 4 chromosomes)	165,000
Human (in 23 chromosomes)	2,900,000

[a]kb = kilobase pair = 1000 base pairs (bp).

Source: Kornberg, A. and Baker, T.A., *DNA Replication* (2nd ed.), p. 20, Freeman (1992).

synthesis of its complementary strand and hence that hereditary information is encoded in the sequence of bases on either strand.

Most DNA molecules are extremely large, in keeping with their role as the depository of a cell's genetic information. With few exceptions, more complex organisms contain more DNA (Table 3-2). Of course, an organism's **genome,** its unique DNA content, may be allocated among several **chromosomes** (Greek: *chromos,* color + *soma,* body), each of which contains a separate DNA molecule. Note that many organisms are **diploid;** that is, they contain two equivalent sets of chromosomes, one from each parent. Their content of unique **(haploid)** DNA is half their total DNA. For example, humans are diploid organisms that carry 46 chromosomes per cell; their haploid number is therefore 23.

Because of their great lengths, DNA molecules are described in terms of the number of base pairs **(bp)** or thousands of base pairs **(kilobase pairs,** or **kb**). Although individual DNA molecules are long and relatively stiff, they are not completely rigid. We shall see later that the DNA double helix forms coils and loops when it is packaged inside the cell. Furthermore, depending on the nucleotide sequence, DNA may adopt slightly different helical conformations. Finally, in the presence of other cellular components, the DNA may bend sharply or the two strands may partially unwind. (We consider the structure of DNA in greater detail in Chapter 23.)

C. Single-Stranded Nucleic Acids

Single-stranded DNA is rare, occurring mainly as the hereditary material of certain viruses. In contrast, RNA occurs primarily as single strands, which usually form compact structures rather than loose extended chains (double-stranded RNA is the hereditary material of certain viruses). An RNA strand—which is identical to a DNA strand except for the presence of 2′-OH groups and the substitution of uracil for thymine—can base-pair with a complementary strand of RNA or DNA. As expected, A pairs with U (or T in DNA), and G with C. Base pairing often occurs intramolecularly, giving rise to **stem-and-loop** structures (Fig. 3-12) or, when loops interact with each other, to more complex structures.

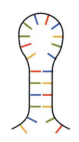

***Figure 3-12.*   Formation of a stem-and-loop structure.** Base pairing between complementary sequences within an RNA strand allows the polynucleotide to fold back on itself.

The intricate structures that can potentially be adopted by single-stranded RNA molecules provide additional evidence that RNA can do more than just store and transmit genetic information. Numerous investigations have found that certain RNA molecules can specifically bind small organic molecules and can catalyze reactions involving those molecules. These findings provide substantial support for theories that *many of the processes essential for life began through the chemical versatility of small polynucleotides.*

3. OVERVIEW OF NUCLEIC ACID FUNCTION

DNA is the carrier of genetic information in all cells and in many viruses. Yet a period of over 75 years passed from the time the laws of inheritance were discovered by Gregor Mendel until the biological role of DNA was elucidated. Even now, many details of how genetic information is expressed and transmitted to future generations are still unclear.

Mendel's work with garden peas led him to postulate that an individual plant contains a pair of factors (which we now call **genes**), one inherited from each parent. But Mendel's theory of inheritance, reported in 1866, was almost universally ignored by his contemporaries, whose knowledge of anatomy and physiology provided no basis for its understanding. Eventually, genes were hypothesized to be part of chromosomes, and the pace of genetic research greatly accelerated.

A. DNA Carries Genetic Information

Until the 1940s, it was generally assumed that genes were made of protein, since proteins were the only biochemical entities that, at the time, seemed complex enough to serve as agents of inheritance. Nucleic acids, which had first been isolated in 1869 by Friedrich Miescher, were believed to have monotonously repeating nucleotide sequences and were therefore unlikely candidates for transmitting genetic information.

It took the efforts of Oswald Avery, Colin MacLeod, and Maclyn McCarty to demonstrate that DNA carries genetic information. Their experiments, completed in 1944, showed that DNA—not protein—extracted from a virulent (pathogenic) strain of the bacterium *Diplococcus pneumoniae* was the substance that **transformed** (permanently changed) a nonpathogenic strain of the organism to the virulent strain (Fig. 3-13). Avery's discovery was initially greeted with skepticism, but it influenced Erwin

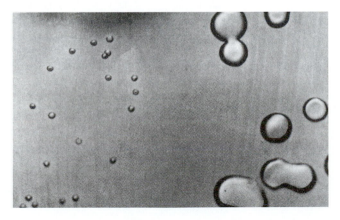

Figure 3-13. Transformed pneumococci. The large colonies are virulent pneumococci that resulted from the transformation of nonpathogenic pneumococci (smaller colonies) by DNA extracted from the virulent strain. We now know that this DNA contained a gene that was defective in the nonpathogenic strain. [From Avery, O.T., MacLeod, C.M., and McCarty, M., *J. Exp. Med.* **79**, 153 (1944). Copyright © 1944 by Rockefeller University Press.]

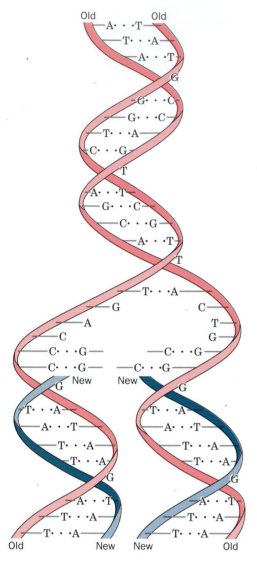

Figure 3-14. DNA replication. Each strand of parental DNA (*red*) acts as a template for the synthesis of a complementary daughter strand (*green*). Thus, the resulting double-stranded molecules are identical.

Chargaff, whose rules (Section 3-2A) led to subsequent models of the structure and function of DNA.

The double-stranded, or duplex, nature of DNA facilitates its **replication.** When a cell divides, each DNA strand acts as a template for the assembly of its complementary strand (Fig. 3-14). Consequently, every progeny cell contains a complete DNA molecule (or a set of DNA molecules in organisms whose genomes contain more than one chromosome). Each DNA molecule consists of one parental strand and one daughter strand. Daughter strands are synthesized by the stepwise polymerization of nucleotides that specifically pair with bases on the parental strands. The mechanism of replication, while straightforward in principle, is exceedingly complex in the cell, requiring a multitude of cellular factors to proceed with fidelity and efficiency, as we shall see in Chapter 24.

B. Genes Direct Protein Synthesis

The question of how sequences of nucleotides control the characteristics of organisms took some time to be answered. In experiments with the mold *Neurospora crassa* in the 1940s, George Beadle and Edward Tatum found that *there is a specific connection between genes and enzymes, the one gene–one enzyme theory.* Beadle and Tatum showed that mutant varieties of *Neurospora* that were generated by irradiation with X-rays required additional nutrients in order to grow. Presumably, the offspring of the radiation-damaged cells lacked specific enzymes necessary to synthesize those nutrients.

The link between DNA and enzymes (nearly all of which are proteins) is RNA. According to the so-called **central dogma of molecular biology,** formulated by Crick in 1958, *DNA directs its own replication as well as its* **transcription** *to form an RNA of complementary sequence. The sequence of bases in the RNA is then* **translated** *into the corresponding sequence of amino acids to form a protein* (Fig. 3-15). Just as the daughter strands of DNA are synthesized from free deoxynucleotides that pair with bases in the parent DNA strand, RNA strands are synthesized from free ribonucleotides that pair with the complementary bases in one DNA strand of a gene (transcription is described in greater detail in Chapter 25). The RNA

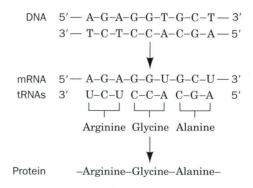

Figure 3-15.   Transcription and translation. One strand of DNA directs the synthesis of messenger RNA (mRNA). The base sequence of the transcribed RNA is complementary to that of the DNA strand. The message is translated when transfer RNA (tRNA) molecules align with the mRNA by complementary base pairing between three-nucleotide segments. Each tRNA carries a specific amino acid. These amino acids are covalently joined to form a protein. Thus, the sequence of bases in DNA specifies the sequence of amino acids in a protein.

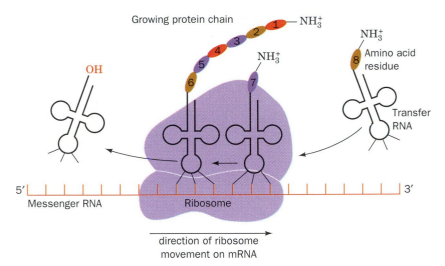

Figure 3-16.   Translation. tRNA molecules with their attached amino acids bind to complementary three-nucleotide sequences on mRNA. The ribosome facilitates the alignment of the tRNA and the mRNA and catalyzes the joining of amino acids to produce a protein chain. When a new amino acid is added, the preceding tRNA is ejected, and the ribosome proceeds along the mRNA.

that corresponds to a protein-coding gene (called **messenger RNA,** or **mRNA**) makes its way to a **ribosome,** an organelle that is itself composed largely of RNA (**ribosomal RNA, or rRNA**). At the ribosome, each set of three nucleotides in the mRNA pairs with three complementary nucleotides in a small RNA molecule—a **transfer RNA, or tRNA** (Fig. 3-16). Attached to each tRNA molecule is its corresponding amino acid. The ribosome catalyzes the joining of amino acids, which are the monomeric units of proteins (protein synthesis is described in detail in Chapter 26). Amino acids are added to the growing protein chain according to the order in which the tRNA molecules bind to the mRNA. Since the nucleotide sequence of the mRNA in turn reflects the sequences of nucleotides in the gene, DNA directs the synthesis of proteins. It follows that alterations to the genetic material of an organism **(mutations)** may manifest themselves as proteins with altered structures and functions.

C. The RNA World

The chemical reaction that covalently joins amino acids appears to be catalyzed by rRNA (Section 26-4B), which is an example of an RNA catalyst, one that has persisted for billions of years of evolution. Laboratory experiments have also produced RNA molecules that can carry out reactions that are chemically related to those required for replicating DNA, transcribing it to RNA, and attaching amino acids to transfer RNA. These results are consistent with a precellular world in which RNA molecules enjoyed a more exalted position as the catalytic workhorses of biochemistry. The extant RNA catalysts are presumably vestiges of this earlier "RNA world."

Proteins have largely eclipsed RNA as cellular catalysts, presumably because of the greater chemical versatility of proteins. Whereas nucleic acids are polymers of four types of monomeric units, proteins have at their disposal 20 types of amino acids, some of which possess functional groups such as hydroxyl, sulfhydryl, amido, and carboxyl groups (Section 4-1C), which are not present in nucleic acids. The expanded repertoire of functional groups of proteins relative to nucleic acids apparently gives proteins a competitive edge in evolutionary terms.

4. NUCLEIC ACID SEQUENCING

Much of our current understanding of protein structure and function rests squarely on information gleaned not from the proteins themselves, but indirectly from their genes. *The ability to determine the sequence of nucleotides in nucleic acids has made it possible to deduce the amino acid sequences of their encoded proteins and, to some extent, the structures and functions of those proteins. Nucleic acid sequencing has also revealed information about the regulation of genes.* Portions of genes that are not actually transcribed into RNA nevertheless may influence how often a gene is transcribed and translated, that is, **expressed.** Moreover, efforts to elucidate the sequences in hitherto unmapped regions of DNA have led to the discovery of new genes and new regulatory elements. *Once in hand, a nucleic acid sequence can be duplicated, modified, and expressed, making it possible to study proteins that could not otherwise be obtained in useful quantities.* In this section, we describe how nucleic acids are sequenced and what information the sequences may reveal. In the following section, we discuss the manipulation of purified nucleic acid sequences for various purposes.

The overall strategy for sequencing any polymer of nonidentical units is

1. Cleave the polymer into specific fragments that are small enough to be fully sequenced.
2. Determine the sequence of residues in each fragment.
3. Determine the order of the fragments in the original polymer by repeating the preceding steps using a degradation procedure that yields a set of fragments that overlap the cleavage points in the first step.

The first efforts to sequence RNA used nonspecific enzymes to generate relatively small fragments whose nucleotide composition was then determined by partial digestion with an enzyme that selectively removed nucleotides from one end or the other (Fig. 3-17). Sequencing RNA in this manner was tedious and time-consuming. Using such methods, it took Robert Holley 7 years to determine the sequence of a 76-residue tRNA molecule.

After 1975, dramatic progress was made in nucleic acid sequencing technology. The advances were made possible by the discovery of enzymes that could cleave DNA at specific sites and by the development of rapid sequencing techniques for DNA. The advent of modern molecular cloning techniques (Section 3-5) also made it possible to produce sufficient quantities of specific DNA to be sequenced. These cloning techniques are necessary because most specific DNA sequences are normally present in a genome in only a single copy.

A. Restriction Endonucleases

Many bacteria are able to resist infection by **bacteriophages** (viruses that are specific for bacteria) by virtue of a **restriction–modification system.** The bacterium modifies certain nucleotides in specific sequences of its own DNA by adding a methyl ($-CH_3$) group in a reaction catalyzed by a **modification methylase. A restriction endonuclease,** which recognizes the same nucleotide sequence as does the methylase, cleaves any DNA that has not been modified on at least one of its two strands. (An **endonuclease** cleaves

```
G C A C U U G A
          | snake venom
          | phosphodiesterase
          ↓
G C A C U U G A
G C A C U U G
G C A C U U
G C A C U
G C A C
G C A
G C    + Mononucleotides
```

Figure 3-17.   Determining the sequence of an oligonucleotide. The oligonucleotide is partially digested with snake venom phosphodiesterase, which breaks the phosphodiester bonds between nucleotide residues, starting at the 3′ end of the oligonucleotide. The result is a mixture of fragments of all lengths, which are then separated. Comparing the base composition of a pair of fragments that differ in length by one nucleotide establishes the identity of the 3′-terminal nucleotide in the larger fragment. Analysis of each pair of fragments reveals the sequence of the original oligonucleotide.

Table 3-3. **Recognition and Cleavage Sites of Some Type II Restriction Enzymes**

Enzyme	Recognition Sequence[a]	Microorganism
*Alu*I	AG↓C*T	*Arthrobacter luteus*
*Bam*HI	G↓GATC*C	*Bacillus amyloliquefaciens* H
*Bgl*I	GCCNNNNN↓NGGC	*Bacillus globigii*
*Bgl*II	A↓GATCT	*Bacillus globigii*
*Eco*RI	G↓AA*TTC	*Escherichia coli* RY13
*Eco*RII	↓CC*(^A_T)GG	*Escherichia coli* R245
*Eco*RV	GA*T↓ATC	*Escherichia coli* J62 pLG74
*Hae*II	RGCGC↓Y	*Haemophilus aegyptius*
*Hae*III	GG↓C*C	*Haemophilus aegyptius*
*Hind*III	A*↓AGCTT	*Haemophilus influenzae* R$_d$
*Hpa*II	C↓C*GG	*Haemophilus parainfluenzae*
*Msp*I	C*↓CGG	*Moraxella* species
*Pst*I	CTGCA*↓G	*Providencia stuartii* 164
*Pvu*II	CAG↓C*TG	*Proteus vulgaris*
*Sal*I	G↓TCGAC	*Streptomyces albus* G
*Taq*I	T↓CGA*	*Thermus aquaticus*
*Xho*I	C↓TCGAG	*Xanthomonas holcicola*

[a]The recognition sequence is abbreviated so that only one strand, reading 5′ to 3′, is given. The cleavage site is represented by an arrow (↓) and the modified base, where it is known, is indicated by an asterisk (A* is N^6-methyladenine and C* is 5-methylcytosine). R, Y, and N represent a purine nucleotide, a pyrimidine nucleotide, and any nucleotide, respectively.

[*Source*: Roberts, R.J. and Macelis, D., REBASE—the restriction enzyme database, http://www.neb.com/rebase.]

a nucleic acid within the polynucleotide strand; an **exonuclease** cleaves a nucleic acid by removing one of its terminal residues.) This system destroys foreign (phage) DNA containing a recognition site that has not been modified by methylation. The host DNA is always at least half methylated, because although the daughter strand is not methylated until shortly after it is synthesized, the parental strand to which it is paired is already modified (and thus protects both strands of the DNA from cleavage by the restriction enzyme).

Type II restriction endonucleases are particularly useful in the laboratory. These enzymes cleave DNA within the four- to eight-base sequence that is recognized by their corresponding modification methylase. (Type I and Type III restriction endonucleases cleave DNA at sites other than their recognition sequences.) About 2500 Type II restriction enzymes with nearly 200 different recognition sequences have been characterized. Some of the more widely used restriction enzymes are listed in Table 3-3. A restriction enzyme is named by the first letter of the genus and the first two letters of the species of the bacterium that produced it, followed by its serotype or strain designation, if any, and a roman numeral if the bacterium contains more than one type of restriction enzyme. For example, *Eco*RI is produced by *E. coli* strain RY13.

Interestingly, most Type II restriction endonucleases recognize and cleave palindromic DNA sequences. A **palindrome** is a word or phrase that reads the same forward or backward. Two examples are "refer" and "Madam, I'm Adam." In a palindromic DNA segment, the sequence of nucleotides is the same in each strand, and the segment is said to have twofold symmetry (Fig. 3-18). Most restriction enzymes cleave the two strands of

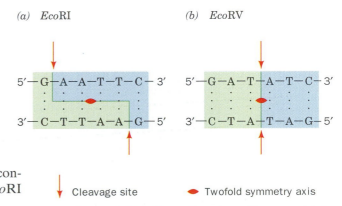

Figure 3-18. **Restriction sites.** The recognition sequences for Type II restriction endonucleases are palindromes, sequences with a twofold axis of symmetry. (*a*) Recognition site for *Eco*RI, which generates DNA fragments with sticky ends. (*b*) Recognition site for *Eco*RV, which generates blunt-ended fragments.

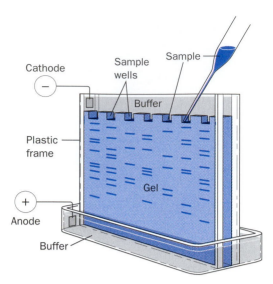

Figure 3-19. **Apparatus for gel electrophoresis.** Samples are applied in slots at the top of the gel and electrophoresed in parallel lanes. Negatively charged molecules such as DNA migrate through the gel matrix toward the anode in response to an applied electric field. Because smaller molecules move faster, the molecules in each lane are separated according to size. Following electrophoresis, the separated molecules may be visualized by staining, fluorescence, or a radiographic technique.

DNA at positions that are staggered, producing DNA fragments with complementary single-strand extensions. Restriction fragments with such **sticky ends** can associate by base pairing with other restriction fragments generated by the same restriction enzyme. Some restriction endonucleases cleave the two strands of DNA at the symmetry axis to yield restriction fragments with fully base-paired **blunt ends.**

B. Electrophoresis and Restriction Mapping

Treating a DNA molecule with a restriction endonuclease produces a series of precisely defined fragments that can be separated according to size. **Gel electrophoresis** is commonly used for the separation. In principle, a charged molecule moves in an electric field with a velocity proportional to its overall charge density, size, and shape. For molecules with a relatively homogeneous composition (such as nucleic acids), shape and charge density are constant, so the velocity depends primarily on size. Electrophoresis is carried out in a gel-like matrix, usually made from **agarose** (carbohydrate polymers that form a loose mesh) or **polyacrylamide** (a more rigid cross-linked synthetic polymer). The gel is typically held between two glass or plastic plates (Fig. 3-19), or a thin layer of the gel covers the surface of a flat support. The molecules to be separated are applied to one end of the gel, and the molecules move through the pores in the matrix under the influence of the electric field. Smaller molecules move more rapidly though the gel and therefore migrate farther in a given time.

Following electrophoresis, the separated molecules may be visualized in the gel by an appropriate technique, such as addition of a stain that binds tightly to the DNA or by radioactive labeling. Depending on the dimensions of the gel and the visualization technique used, samples containing less than a nanogram of material can be separated and detected by gel electrophoresis. Several samples can be electrophoresed simultaneously. For example, the fragments obtained by digesting a DNA sample with different restriction endonucleases can be visualized side by side (Fig. 3-20). The sizes of the various fragments can be determined by comparing their electrophoretic mobilities to the mobilities of fragments of known size.

The results of gels such as the one in Fig. 3-20 can be used to construct a diagram called a **restriction map.** Consider as an example a 4-kb linear DNA molecule cleaved by *Bam*HI, *Hin*dIII, or both and subjected to gel electrophoresis (Fig. 3-21*a*). The sizes of the restriction fragments are used to deduce the positions of the restriction sites in the intact DNA and to construct the restriction map diagrammed in Fig. 3-21*b*. Restriction maps are useful laboratory tools because restriction sites are physical reference points on a DNA molecule. Restriction maps are therefore a convenient framework for locating particular base sequences or genes on a chromosome and for comparing different chromosomes (see Box 3-1).

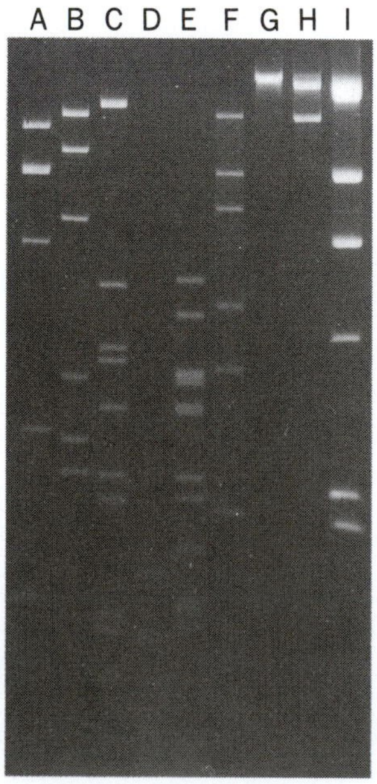

A B C D E F G H I

Figure 3-20. **Electrophoretogram of restriction digests.** The plasmid pAgK84 has been digested with (A) *Bam*HI, (B) *Pst*I, (C) *Bgl*II, (D) *Hae*III, (E) *Hinc*II, (F) *Sac*I, (G) *Xba*I, and (H) *Hpa*I. Lane I contains bacteriophage λ DNA digested with *Hin*dIII as a standard since these fragments have known sizes. The restriction fragments in each lane are made visible by fluorescence against a black background. [From Slota, J.E. and Farrand, S.F., *Plasmid* **8**, 180 (1982). Copyright © 1982 by Academic Press.]

BIOCHEMISTRY IN FOCUS

Restriction Fragment Length Polymorphisms

Individuality in humans and other species derives from their high degree of genetic polymorphism. Homologous human chromosomes (e.g., the pairs of maternally and paternally inherited chromosomes) differ in sequence, on average, every 200 to 500 bp. These genetic differences create or eliminate restriction sites. Restriction enzyme digests of homologous chromosomes therefore contain fragments with different lengths; that is, these DNAs exhibit **restriction fragment length polymorphisms (RFLPs)**.

The two homologous chromosomal segments shown here differ in the number of restriction sites. An individual with two copies of chromosome I would yield fragments A and B in RFLP analysis; an individual with two copies of chromosome II would yield fragment C. An individual with one of each chromosome would yield fragments A, B, and C.

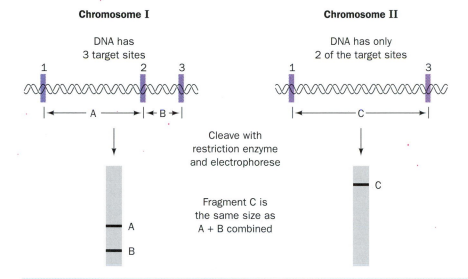

RFLPs are particularly valuable for diagnosing inherited diseases for which the molecular defect is unknown. If a particular RFLP is closely linked to a defective gene, detecting that RFLP in an individual indicates that there is a high probability that the individual has also inherited the defective gene. For example, Huntington's disease, a fatal neurological disorder whose symptoms first appear around age 40, is caused by a dominant genetic defect (Box 27-1). The identification of an RFLP that is closely linked to the defective Huntington's gene has permitted the children of Huntington's disease victims (50% of whom inherit this devastating condition) to make informed decisions in ordering their lives.

Figure 3-21. **Construction of a restriction map.** (*a*) The gel electrophoretic pattern of digests of a hypothetical 4-kb DNA molecule with *Hind*III, *Bam*HI, and their mixture. The sizes of the various fragments are indicated. (*b*) A restriction map of the DNA resulting from the information in *a*. The distance in kb between each restriction site corresponds to the size of the corresponding restriction fragment.

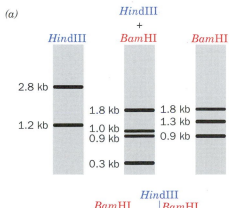

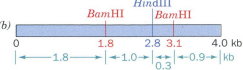

C. The Chain-Terminator Method of Sequencing

The first specific method for sequencing long stretches of DNA was the chemical cleavage method devised by Allan Maxam and Walter Gilbert. However, this technique has been largely superseded by the **chain-terminator procedure,** devised by Frederick Sanger, which we discuss here. The first step in DNA sequencing is obtaining single polynucleotide strands. Complementary DNA strands can be separated by heating, which breaks the hydrogen bonds between bases. Next, polynucleotide fragments that terminate at positions corresponding to each of the four nucleotides are generated. Finally, the fragments are separated and detected.

The Chain-Terminator Method Uses DNA Polymerase

The chain-terminator method (also called the **dideoxy method**) uses an *E. coli* enzyme to make complementary copies of the single-stranded DNA

> ● See Guided Exploration 2:
>
> DNA Sequence Determination by the Chain-Terminator Method.

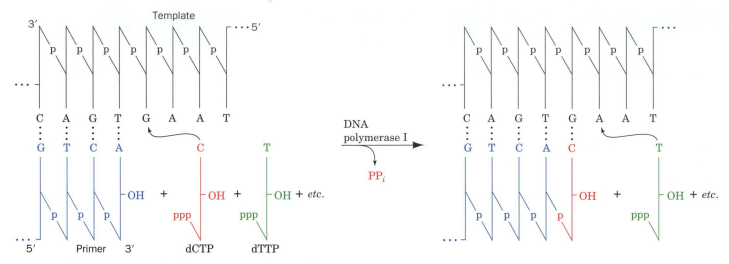

***Figure* 3-22.   Action of DNA polymerase I.** Using a single DNA strand as a template, the enzyme elongates the primer by stepwise addition of complementary nucleotides. Incoming nucleotides pair with bases on the template strand and are joined to the growing polynucleotide strand in the $5' \rightarrow 3'$ direction. The polymerase-catalyzed reaction requires a free $3'$-OH group on the growing strand. **Pyrophosphate** ($P_2O_7^{4-}$; PP_i) is released with each nucleotide addition.

being sequenced. The enzyme is a fragment of **DNA polymerase I,** one of the enzymes that participates in replication of bacterial DNA (Section 24-2A). Using the single DNA strand as a template, DNA polymerase I assembles the four nucleoside triphosphates **(dNTPs),** dATP, dCTP, dGTP, and dTTP, into a complementary polynucleotide chain that it elongates in the $5' \rightarrow 3'$ direction (Fig. 3-22).

DNA polymerase I can sequentially add deoxynucleotides only to the $3'$ end of a polynucleotide. Hence, replication is initiated in the presence of a short polynucleotide (a **primer**) that is complementary to the $3'$ end of the template DNA and thus becomes the $5'$ end of the new strand. The primer base-pairs with the template strand, and nucleotides are sequentially added to the $3'$ end of the primer. If the DNA being sequenced is a restriction fragment, as it usually is, it begins and ends with a restriction site. The primer can therefore be a short DNA segment with the sequence of this restriction site.

DNA Synthesis Terminates after Specific Bases

In the chain-terminator technique (Fig. 3-23), the DNA to be sequenced is incubated with DNA polymerase I, a suitable primer, and the four dNTP substrates for the polymerization reaction. The reaction mixture also includes a "tagged" compound, either one of the dNTPs or the primer. The tag, which may be a radioactive isotope (e.g., ^{32}P) or a fluorescent label, permits the products of the polymerase reaction to be easily detected.

The key component of the reaction mixture is a small amount of a **2′,3′-dideoxynucleoside triphosphate (ddNTP),**

**2′,3′-Dideoxynucleoside
triphosphate**

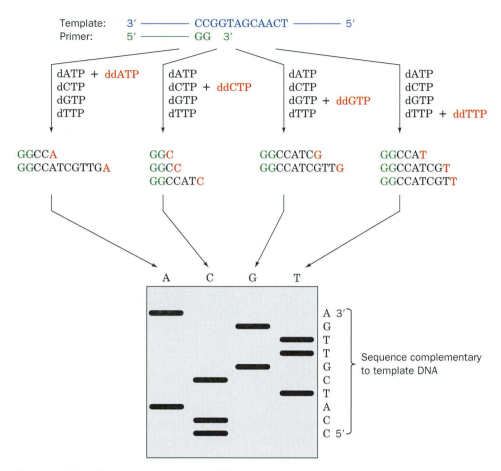

Figure 3-23. The chain-terminator (dideoxy) method of DNA sequencing. Each of the four reaction mixtures includes the single-stranded DNA to be sequenced (the template), a primer, the four deoxynucleoside triphosphates (represented as dATP, etc.), and one of the four dideoxynucleoside triphosphates (ddATP, etc.). Extension of the primer by the action of DNA polymerase generates stretches of DNA terminating with a dideoxynucleotide. Gel electrophoresis in parallel lanes of the fragments from the four reaction mixtures yields a set of polynucleotides whose 3′ terminal residues are known. The sequence obtained by "reading" from the smallest fragment to the largest (i.e., from the bottom to the top of the gel) is complementary to the sequence of the template DNA.

which lacks the 3′-OH group of deoxynucleotides. *When the dideoxy analog is incorporated into the growing polynucleotide in place of the corresponding normal nucleotide, chain growth is terminated because addition of the next nucleotide requires a free 3′-OH.* By using only a small amount of the ddNTP, a series of truncated chains is generated, each of which ends with the dideoxy analog at one of the positions occupied by the corresponding base.

Relatively modest sequencing tasks use four reaction mixtures, each with a different ddNTP, and the reaction products are electrophoresed in parallel lanes. The lengths of the truncated chains indicate the positions where the dideoxynucleotide was incorporated. Thus, the sequence of the repli-

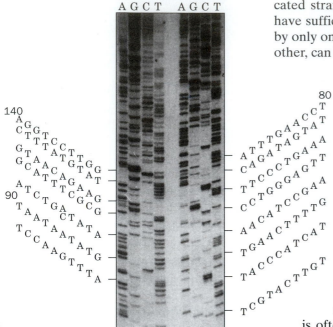

Figure 3-24.   An autoradiogram of a sequencing gel. The positions of radioactive DNA fragments produced by the chain-terminator method were visualized by laying X-ray film over the gel after electrophoresis. A second loading of the gel (four lanes at right) was made 90 min after the initial loading in order to obtain the sequences of the smaller fragments. The deduced sequence of 140 nucleotides is written along the side. [From Hindley, J., DNA sequencing, *in* Work, T.S. and Burdon, R.H. (Eds.), *Laboratory Techniques in Biochemistry and Molecular Biology,* Vol. 10, p. 82, Elsevier (1983). Used by permission.]

cated strand can be directly read from the gel (Fig. 3-24). The gel must have sufficient resolving power to separate fragments that differ in length by only one nucleotide. Two sets of gels, one run for a longer time than the other, can be used to obtain the sequence of up to 800 bases of DNA. Note that the sequence obtained by the chain-terminator method is complementary to the DNA strand being sequenced.

Automated Sequencing

Large-scale sequencing operations are accelerated by automation. In one variation of the chain-terminator method, the primers used in the four chain-extension reactions are each linked to a different fluorescent dye. The separately reacted mixtures are combined and subjected to gel electrophoresis in a single lane. The terminal base on each fragment is identified by its characteristic fluorescence (Fig. 3-25). Using computer-controlled fluorescence detectors, automated systems can identify ~10,000 bases per day, in contrast to ~50,000 bases per year obtained by manual methods.

Nucleic acid sequencing has become so routine that directly determining a protein's amino acid sequence (Section 5-3) is often far more difficult than determining the base sequence of its corresponding gene. In fact, nucleic acid sequencing is invaluable for studying genes whose products have not yet been identified. If the gene can be sequenced, the probable function of its protein product may be deduced by comparing the base sequence to those of genes whose products are already characterized. An important advance in human genetics was the isolation of the gene linked to cystic fibrosis, a common inherited disease. In 1989, when Francis Collins identified this gene, nothing was known about the product of the normal gene or how a defective gene product contributed to the disease. By examining the nucleotide sequence and inferring its amino acid sequence, Collins was able to postulate that the gene coded for a transmembrane protein whose absence led to abnormal secretion of chloride ions in the respiratory and gastrointestinal tracts.

The advent of nucleic acid sequencing techniques brought with it the dream of ultimately sequencing the entire human genome. In fact, this gargantuan project was completed years faster than had first been anticipated.

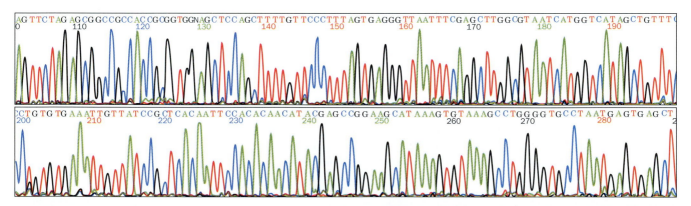

Figure 3-25.   Automated DNA sequencing. In this variant of the technique, a different fluorescent dye is attached to the primer in each of the four reaction mixtures in the chain-terminator procedure. The four reaction mixtures are combined for electrophoresis. Each of the four colored curves therefore represents the electrophoretic pattern of fragments containing one of the dideoxynucleotides: Green, red, black, and blue correspond to fragments ending in ddATP, ddTTP, ddGTP, and ddCTP, respectively. The 3'-terminal base of each oligonucleotide is identified by the fluorescence of its gel band. This portion of the readout corresponds to nucleotides 100–290 of the DNA segment being sequenced. [Courtesy of Mark Adams, The Institute for Genomic Research, Rockville, Maryland.]

Table 3-4.   **Some Genome Sequencing Projects**

Organism	Genome Size (kb)	Number of Chromosomes
Mycoplasma genitalium (human parasite)	580	1
Rickettsia prowazekii (putative relative of mitochondria)	1,100	1
Methanococcus jannaschii (thermophilic methanogen)	1,700	1
Haemophilus influenzae (human pathogen)	1,830	1
Synechocystis sp. (cyanobacterium)	3,570	1
Escherichia coli (human symbiont)	4,600	1
Saccharomyces cerevisiae (baker's yeast)	12,100	16
Plasmodium falciparum (protozoan that causes malaria)	27,000	14
Caenorhabditis elegans (nematode)	100,000	6
Arabidopsis thaliana (dicotyledonous plant)	117,000	5
Drosophila melanogaster (fruit fly)	180,000	4
Danio rerio (zebrafish)	1,700,000	25
Homo sapiens	3,200,000	23

However, the determination of the human genome sequence is only a first step: understanding its full meaning will require perhaps decades of additional research. This undertaking will be greatly aided by comparisons with the genome sequences of other organisms (Table 3-4).

D. Sequences, Mutation, and Evolution

Perhaps the richest reward of nucleic acid sequencing technology is the information it provides about the mechanisms of evolution. The chemical and physical properties of DNA, such as its regular three-dimensional shape and the elegant process of replication, may leave the impression that genetic information is relatively static. In fact, *DNA is a dynamic molecule, subject to changes that alter genetic information.* For example, the mispairing of bases during DNA replication introduces errors known as **point mutations** in the daughter strands. Mutations also result from DNA damage by chemicals or radiation. More extensive alterations in genetic information are caused by faulty **recombination** (exchange of DNA between chromosomes) and the **transposition** of genes from one chromosome to another and, in some cases, from one organism to another. All these alterations to DNA provide the raw material for natural selection. When a mutated gene is transcribed and the messenger RNA is subsequently translated, the resulting protein may have properties that confer some advantage to the individual. As a beneficial change is passed from generation to generation, it becomes part of the standard genetic makeup of the species. Of course, many changes occur as a species evolves, not all of them simple and not all of them gradual.

Figure 3-26. **Maize and teosinte.** Despite the large differences in phenotype—maize (*bottom*) has hundreds of easily chewed kernels whereas teosinte (*top*) has only a few hard, inedible kernels—the plants differ in only a few genes. The ancestor of maize is believed to be a mutant form of teosinte in which the kernels were more exposed. [John Doebley/Visuals Unlimited.]

Phylogenetic relationships can be revealed by comparing the sequences of similar genes in different organisms. The number of nucleotide differences in a gene roughly corresponds to the degree to which the organisms have diverged by evolution. The regrouping of prokaryotes into archaea and bacteria (Section 1-3A) according to rRNA sequences present in all organisms illustrates the impact of sequence analysis.

Nucleic acid sequencing also reveals that species differing in **phenotype** (physical characteristics) are nonetheless remarkably similar at the molecular level. For example, humans and chimpanzees share 98–99% of their DNA. Studies of corn (maize) and its putative ancestor, teosinte, suggest that the plants differ in only a handful of genes governing kernel development (teosinte kernels are encased by an inedible shell; Fig. 3-26).

Small mutations in DNA are apparently responsible for relatively large evolutionary leaps. This is perhaps not so surprising when the nature of genetic information is considered. A mutation in a gene segment that does not encode protein might interfere with the binding of cellular factors that influence the timing of transcription. A mutation in a gene encoding an RNA might interfere with the binding of factors that affect the efficiency of translation. Even a minor rearrangement of genes could disrupt an entire developmental process, resulting in the appearance of a novel species. Notwithstanding the high probability that most sudden changes would lead to diminished individual fitness or the inability to reproduce, the capacity for sudden changes in genetic information is consistent with the fossil record. Ironically, the discontinuities in the fossil record that are probably caused in part by sudden genetic changes once fueled the adversaries of Charles Darwin's theory of evolution by natural selection.

5. RECOMBINANT DNA TECHNOLOGY

Along with nucleic acid sequencing, techniques for manipulating DNA *in vitro* and *in vivo* (in the test tube and in living systems) have produced dramatic advances in biochemistry, cell biology, and genetics. In many cases,

recombinant DNA technology has made it possible to purify specific DNA sequences and to prepare them in quantities sufficient for study. Consider the problem of isolating a 1000-bp length of chromosomal DNA from *E. coli*. A 10-L culture of cells grown at a density of $\sim 10^{10}$ cells $\cdot$ mL^{-1} contains only ~ 0.1 mg of the desired DNA, which would be all but impossible to separate from the rest of the DNA using classical separation techniques (Sections 5-2 and 23-3). *Recombinant DNA technology, also called **molecular cloning** or **genetic engineering**, makes it possible to isolate, amplify, and modify specific DNA sequences.*

A. Cloning Techniques

The following approach is used to obtain and amplify a segment of DNA:

1. A fragment of DNA of the appropriate size is generated using restriction endonucleases and then isolated.

2. The fragment is incorporated into another DNA molecule known as a **vector,** which contains the sequences necessary to direct DNA replication.

3. The vector—with the DNA of interest—is introduced into cells, where it is replicated.

4. Cells containing the desired DNA are identified, or **selected.**

Cloning refers to the production of multiple identical organisms derived from a single ancestor. In this case, the **clone** is the collection of cells that contain the vector carrying the DNA of interest. In a suitable host organism, such as *E. coli* or yeast, large amounts of the inserted DNA can be produced.

Cloned DNA can be purified and sequenced (Section 3-4). Alternatively, if a cloned gene is flanked by the properly positioned regulatory sequences for RNA and protein synthesis, the host may also produce large quantities of the RNA and protein specified by that gene. Thus, cloning provides materials (nucleic acids and proteins) for other studies and also provides a means for studying gene expression under controlled conditions.

Cloning Vectors

A variety of small, autonomously replicating DNA molecules are used as cloning vectors. **Plasmids** are circular DNA molecules of 1 to 200 kb found in bacteria or yeast cells. Plasmids can be considered molecular parasites, but in many instances they benefit their host by providing functions, such as resistance to antibiotics, that the host lacks.

Some types of plasmids are present in one or a few copies per cell and replicate only when the bacterial chromosome replicates. However, the plasmids used for cloning are typically present in hundreds of copies per cell and can be induced to replicate until the cell contains two or three thousand copies (representing about half of the cell's total DNA). The plasmids that have been constructed for laboratory use are relatively small, replicate easily, carry genes specifying resistance to one or more antibiotics, and contain a number of conveniently located restriction endonuclease sites into which foreign DNA can be inserted. Plasmid vectors can be used to clone DNA segments of no more than ~ 10 kb. The *E. coli* plasmid designated pBR322 (Fig. 3-27) is a representative cloning vector.

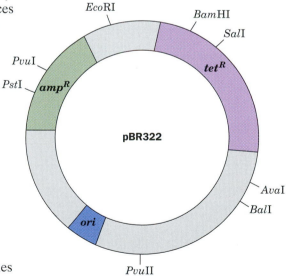

Figure 3-27. **The plasmid pBR322.** Several restriction sites, where foreign DNA can be easily inserted, are indicated. The *amp*^{R} gene confers resistance to the antibiotic ampicillin; *tet*^{R} confers resistance to tetracycline. *ori* represents the point where plasmid replication begins. The plasmid contains 4362 bp, with each of its two strands forming a covalently closed circle.

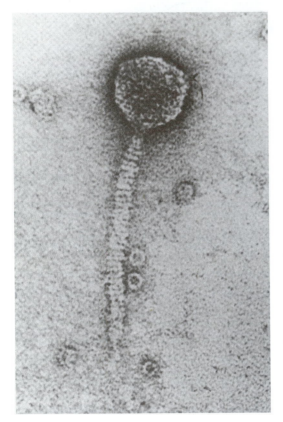

Bacteriophage λ (Fig. 3-28) is an alternative cloning vector that can accommodate DNA inserts up to 16 kb. The central third of the 48.5-kb phage genome is not required for infection and can therefore be replaced by foreign DNAs of similar size. The resulting **recombinant,** or **chimera** (named after the mythological monster with a lion's head, goat's body, and serpent's tail), is packaged into phage particles that can then be introduced into the host cells. One advantage of using phage vectors is that the recombinant DNA is produced in large amounts in easily purified form.

Much larger DNA segments up to several hundred kilobase pairs can be cloned in **yeast artificial chromosomes (YACs).** YACs are linear DNA molecules that contain all the chromosomal structures required for normal replication and segregation during yeast cell division.

Ligation

A DNA segment to be cloned is often obtained through the action of restriction endonucleases. Most restriction enzymes cleave duplex DNA to yield sticky ends (Section 3-4A). Therefore, as Janet Mertz and Ron Davis first demonstrated in 1972, *a restriction fragment can be inserted into a cut made in a cloning vector by the same restriction enzyme* (Fig. 3-29). The complementary ends of the two DNAs form base pairs **(anneal)** and the sugar–phosphate backbones are covalently **ligated,** or spliced together, through the action of an enzyme named **DNA ligase** (a ligase produced by a bacteriophage is used to join blunt-ended restriction fragments). A great advantage of using a restriction enzyme to construct a chimera is that the DNA insert can later be precisely excised from the cloned vector by cleaving it with the same restriction enzyme.

Transformation and Selection

The expression of a chimeric plasmid in a bacterial host was first demonstrated in 1973 by Herbert Boyer and Stanley Cohen. A host bacterium can take up a plasmid when the two are mixed together, but the vector becomes permanently established in its bacterial host (transformation) with an efficiency of only ~0.1%. However, a single transformed cell can multiply without limit, producing large quantities of recombinant DNA. Bacterial cells are typically plated on a semisolid growth medium at a low enough density that discrete colonies, each arising from a single cell, are visible.

It is essential to select only those host organisms that have been transformed and that contain a properly constructed vector. In the case of plasmid transformation, selection can be accomplished through the use of an

Figure 3-28. **Bacteriophage λ.** During phage infection, DNA contained in the "head" of the phage particle enters the bacterial cell, where it is replicated ~100 times and packaged to form progeny phage. [Electron micrograph courtesy of A.F. Howatson. From Lewin, B., *Gene Expression,* Vol. 3, Fig. 5.23, Wiley (1977).]

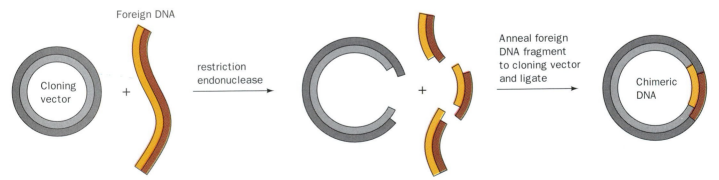

Figure 3-29. **Construction of a recombinant DNA molecule.** The cloning vector and the foreign DNA are cut by the same restriction endonuclease. The sticky ends of the vector and the foreign DNA fragments anneal and are covalently joined by DNA ligase. The result is a chimeric DNA containing a portion of the foreign DNA inserted into the vector.
✳ **See the Animated Figures.**

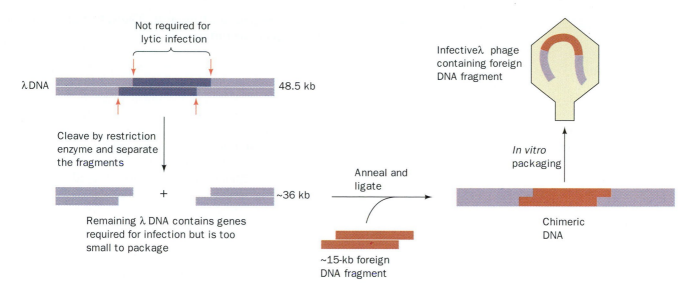

Figure 3-30.   Cloning with bacteriophage λ. Removal of a nonessential portion of the phage genome allows a segment of foreign DNA to be inserted. The chimeric DNA can be packaged into an infectious phage particle only if the insert DNA has the appropriate size. ✷ **See the Animated Figures.**

tibiotics. For example, *E. coli* cells transformed by plasmid pBR322 (Fig. 3-27) containing a foreign DNA insert at its single *Bam*HI site are ampicillin resistant but not tetracycline resistant, because the insert interrupts the protein-coding sequence of the tet^R (tetracycline resistance) gene. In contrast, bacteria that have taken up pBR322 that lacks a foreign DNA insert are resistant to both ampicillin and tetracycline, whereas bacteria that have not taken up the plasmid lack both the amp^R and the tet^R genes and are sensitive to both ampicillin and tetracycline. Thus, successfully transformed bacterial cells can grow in a medium containing ampicillin but not in a medium that contains ampicillin and tetracycline. Genes such as amp^R and tet^R are therefore known as **selectable markers.**

Genetically engineered λ bacteriophage variants contain restriction sites that flank the dispensable central third of the phage genome. This segment can be replaced by foreign DNA, but the chimeric DNA is packaged in phage particles only if its length is from 75 to 105% of the 48.5-kb wild-type λ genome (Fig. 3-30). Consequently, λ phage vectors that have failed to acquire a foreign DNA insert are unable to propagate because they are too short to form infectious phage particles. Of course, the production of infectious phage particles results not in a growing bacterial colony but in a **plaque,** a region of lysed bacterial cells, on a culture plate containing a "lawn" of the host bacteria.

B. Genomic Libraries

In order to clone a particular DNA fragment, it must first be obtained in relatively pure form. The magnitude of this task can be appreciated by considering that, for example, a 1-kb fragment of human DNA represents only 0.000035% of the 2.9 billion-bp human genome. Of course, identifying a particular DNA fragment requires knowing something about its nucleotide sequence or its protein product. In practice, it is usually more difficult to identify a particular gene from an organism and then clone it than it is to clone the organism's entire genome as DNA fragments and then identify the clones containing the sequence of interest. The set of all the cloned fragments is known as a **genomic library.**

Shotgun Cloning

Genomic libraries are generated by a procedure known as **shotgun cloning.** The chromosomal DNA of the organism is isolated, cleaved to fragments of clonable size, and inserted into a cloning vector. The DNA is usually fragmented by partial rather than exhaustive restriction digestion so that the genomic library contains intact representatives of all the organism's genes, including those that contain restriction sites. DNA in solution can also be mechanically fragmented **(sheared)** by rapid stirring.

Given the large size of the genome relative to a gene, the shotgun cloning method is subject to the laws of probability. The number of randomly generated fragments that must be cloned to ensure a high probability that a desired sequence is represented at least once in the genomic library is calculated as follows: The probability P that a set of N clones contains a fragment that constitutes a fraction f, in bp, of the organism's genome is

$$P = 1 - (1 - f)^N \qquad [3\text{-}1]$$

Consequently,

$$N = \log (1 - P) \, / \log (1 - f) \qquad [3\text{-}2]$$

Thus, in order for $P = 0.99$ for fragments averaging 10 kb in length, $N = 2116$ for the 4600-kb *E. coli* chromosome and 76,000 for the 165,000-kb *Drosophila* genome. The use of YAC-based genomic libraries with their large fragment lengths therefore greatly reduces the effort necessary to obtain a given gene segment from a large genome.

Screening

Once the requisite number of clones is obtained, the genomic library must be **screened** for the presence of the desired gene. This can be done by a process known as **colony** or *in situ* hybridization (Latin: *in situ,* in position; Fig. 3-31). The cloned yeast colonies, bacterial colonies, or phage plaques to be tested are transferred, by **replica plating,** from a master plate

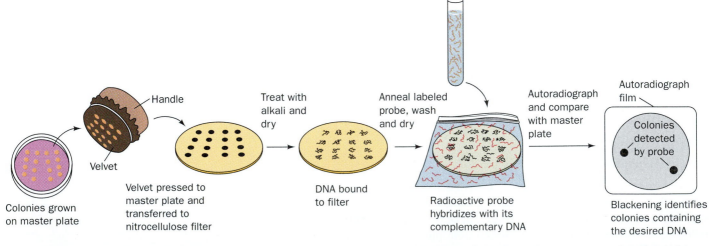

Figure 3-31.   Colony (*in situ*) hybridization. Colonies are transferred from a "master" culture plate by replica plating. Clones containing the DNA of interest are identified by the ability to bind a specific probe. Here, binding is detected by laying X-ray film over the dried filter. Since the colonies on the master plate and on the filter have the same spatial distribution, positive colonies are easily retrieved.

to a nitrocellulose filter (replica plating is also used to transfer colonies to plates containing different growth media). Next, the filter is treated with NaOH, which lyses the cells or phages and separates the DNA into single strands, which preferentially bind to the nitrocellulose. The filter is then dried to fix the DNA in place and incubated with a labeled **probe.** The probe is a short segment of DNA or RNA whose sequence is complementary to a portion of the DNA of interest. After washing away unbound probe, the presence of the probe on the nitrocellulose is detected by a technique appropriate for the label used (e.g., exposure to X-ray film for a radioactive probe, a process known as **autoradiography,** or illumination with an appropriate wavelength for a fluorescent probe). Only those colonies or plaques containing the desired gene bind the probe and are thereby detected. The corresponding clones can then be retrieved from the master plate. Using this technique, a human genomic library of ~1 million clones can be readily screened for the presence of one particular DNA segment.

Choosing a probe for a gene whose sequence is not known requires some artistry. The corresponding mRNA can be used to do so if it is produced in sufficient quantities to be isolated. Alternatively, if the amino acid sequence of the protein encoded by the gene is known, the probe may be a mixture of the various synthetic oligonucleotides that are complementary to a segment of the gene's inferred base sequence

C. DNA Amplification by the Polymerase Chain Reaction

Although molecular cloning techniques are indispensable to modern biochemical research, the **polymerase chain reaction (PCR)** is often a faster and more convenient method for amplifying a specific DNA. Segments of up to 6 kb can be amplified by this technique, which was devised by Kary Mullis in 1985. *In PCR, a DNA sample is separated into single strands and incubated with DNA polymerase, dNTPs, and two oligonucleotide primers whose sequences flank the DNA segment of interest. The primers direct the DNA polymerase to synthesize complementary strands of the target DNA* (Fig. 3-32). Multiple cycles of this process, each doubling the amount of the target DNA, geometrically amplify the DNA starting from as little as a single gene copy. In each cycle, the two strands of the duplex DNA are separated by heating, the primers are annealed to their complementary segments on the DNA, and the DNA polymerase directs the synthesis of the complementary strands. The use of a heat-stable DNA polymerase, such as **Taq polymerase** isolated from *Thermus aquaticus,* a bacterium that thrives at 75°C, eliminates the need to add fresh enzyme after each round of heating (heat inactivates most enzymes). Hence, in the presence of sufficient quantities of primers and dNTPs, PCR is carried out simply by cyclically varying the temperature.

Twenty cycles of PCR increase the amount of the target sequence around a millionfold ($\sim2^{20}$) with high specificity. Indeed, PCR can amplify a target DNA present only once in a sample of 10^5 cells, so this method can be used without prior DNA purification (see Box 3-2). The amplified DNA can then be sequenced or cloned.

D. Applications of Recombinant DNA Technology

Aside from its utility in the laboratory, molecular cloning techniques have been developed and perfected for a variety of medical and commercial purposes. The ability to manipulate genes *in vitro* allows the genes to be altered, leading to the production of proteins with improved functional properties or to the correction of genetic defects.

> ◉ See Guided Exploration 3:
> PCR and Site-Directed Mutagenesis.

***Figure 3-32.*   The polymerase chain reaction (PCR).** In each cycle of the reaction, the strands of the duplex DNA are separated by heating, the reaction mixture is cooled to allow primers to anneal to complementary sequences on each strand, and DNA polymerase extends the primers. The number of "unit-length" strands doubles with every cycle after the second cycle. By choosing primers specific for each end of a gene, the gene can be amplified over a millionfold.

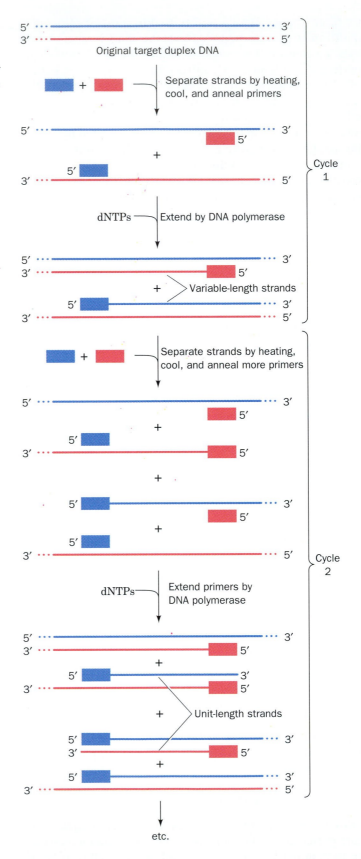

Uses of PCR

PCR amplification has become an indispensable tool. Clinically, it is used to diagnose infectious diseases and to detect rare pathological events such as mutations leading to cancer. Forensically, the DNA from a single hair or sperm can be amplified by PCR so that its RFLPs (Box 3-1) can be used to identify the donor. Traditional ABO blood-type analysis requires a coin-sized drop of blood; PCR is effective on pinhead-sized samples of biological fluids. Most courts now consider DNA sequences as unambiguous identifiers of individuals, as are fingerprints, because the chance of two individuals sharing extended sequences of DNA is typically one in a million or more. In a few cases, PCR has dramatically restored justice to convicts who were released from prison on the basis of PCR results that proved their innocence—even many years after the crime-scene evidence had been collected.

PCR is also largely responsible for the new science of molecular paleontology (paleontology is the study of ancient life forms from their fossil remains). For example, DNA has been extracted from a 120 to 135 million-year-old amber-entombed insect such as the one shown here (amber is fossilized pine resin). The DNA was amplified by PCR and sequenced. DNA fragments from several insects in amber, from fossilized plant leaves, and from frozen mammoths have been subjected to PCR. Comparing the sequences of DNA fragments from ancient organisms and from contemporary species provides information on evolutionary relationships. Amber is an ideal medium for preserving DNA, since it excludes water, which causes DNA degradation over many years. Less well preserved specimens are more likely to contain DNA that is too degraded to use for PCR. Of course, the

enormous amplifying power of PCR requires that particular care must be taken to avoid contamination by modern-day DNA, particularly DNA from the human operators carrying out the PCR analysis. Indeed, many DNAs that were reported to have been isolated from fossils were later shown to be of modern origin.

Population geneticists use PCR to probe the relatedness of extant species and populations, including humans. PCR has also been used to trace the origins of ancient human populations who buried their dead in peat bogs (where lack of oxygen inhibits decomposition) or in areas where the bodies remain permanently frozen. Even when phenotypic and linguistic information is available, DNA sequences obtained by PCR provide more valuable quantitative data for tracing population movements.

[Figure © John D. Cunningham/Visuals Unlimited.]

Protein Production

The production of large quantities of scarce or novel proteins is relatively straightforward only for bacterial proteins: A cloned gene must be inserted into an **expression vector,** a plasmid that contains properly positioned transcriptional and translational control sequences. The production of a protein of interest may reach 30% of the host's total cellular protein. Such genetically engineered organisms are called **overproducers.** Bacterial cells often sequester large amounts of useless and possibly toxic (to the bacterium) protein as insoluble inclusions, which require more elaborate purification methods than soluble or secreted proteins.

Bacteria can produce eukaryotic proteins only if the recombinant DNA that carries the protein-coding sequence also includes bacterial transcriptional and translational control sequences. Synthesis of eukaryotic proteins in bacteria also presents other problems. For example, many eukaryotic genes are large and contain stretches of nucleotides **(introns)** that are transcribed and excised before translation (Section 25-3A); bacteria lack the machinery to excise the introns. In addition, many eukaryotic proteins are

Table 3-5. **Some Proteins Produced by Genetic Engineering**

Protein	Use
Human insulin	Treatment of diabetes
Human growth hormone	Treatment of some endocrine disorders
Erythropoietin	Stimulation of red blood cell production
Colony-stimulating factors	Production and activation of white blood cells
Coagulation factors IX and X	Treatment of blood clotting disorders (hemophilia)
Tissue-type plasminogen activator	Lysis of blood clots in heart attack and stroke
Bovine growth hormone	Production of milk in cows

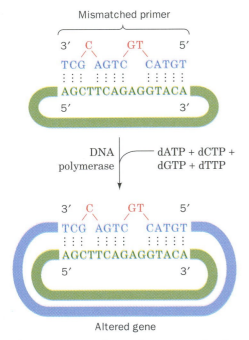

Mismatched primer

3' C GT 5'
TCG AGTC CATGT
AGCTTCAGAGGTACA
5' 3'

DNA polymerase | dATP + dCTP + dGTP + dTTP

3' C GT 5'
TCG AGTC CATGT
AGCTTCAGAGGTACA
5' 3'

Altered gene

Figure 3-33. **Site-directed mutagenesis.** A chemically synthesized oligonucleotide incorporating the desired base changes anneals to the DNA encoding the gene to be altered (*green strand*). The mismatched primer is then extended by DNA polymerase, generating the mutated gene (*blue strand*). The altered gene can be inserted into a suitable cloning vector to be amplified, expressed, or used to generate a mutant organism. ✳ See the Animated Figures.

posttranslationally modified by the addition of carbohydrates or by other reactions. These problems can be overcome by using expression vectors that propagate in eukaryotic hosts, such as yeast or cultured insect or animal cells.

Table 3-5 lists some recombinant proteins produced for medical and agricultural use. In many cases, purification of these proteins directly from human or animal tissues is unfeasible on ethical or practical grounds. Expression systems permit large-scale, efficient preparation of the proteins while minimizing the risk of contamination by viruses or other pathogens from tissue samples.

Site-Directed Mutagenesis

After isolating a gene, it is possible to modify the nucleotide sequence to alter the amino acid sequence of the encoded protein. **Site-directed mutagenesis,** a technique pioneered by Michael Smith, *mimics the natural process of evolution and allows predictions about the structural and functional roles of particular amino acids in a protein to be rigorously tested in the laboratory.*

Synthetic oligonucleotides are required to specifically alter genes through site-directed mutagenesis. An oligonucleotide whose sequence is identical to a portion of the gene of interest except for the desired base changes is used as a primer for DNA polymerase I replication of the gene. The primer hybridizes to the corresponding **wild-type** (naturally occurring) sequence if there are only a few mismatched base pairs. Extension of the primer by DNA polymerase I yields the desired altered gene (Fig. 3-33). The altered gene can then be inserted into an appropriate vector. A mutagenized primer can also be used to generate altered genes by PCR.

Transgenic Organisms

For many purposes it is preferable to tailor an intact organism rather than just a protein—true genetic engineering. Multicellular organisms expressing a gene from another organism are said to be **transgenic,** and the transplanted foreign gene is called a **transgene.**

For the change to be permanent, that is, heritable, a transgene must be stably integrated into the organism's germ cells. For mice, this is accomplished by microinjecting cloned DNA encoding the desired altered characteristics into a fertilized egg and implanting it into the uterus of a foster mother. A well-known example of a transgenic mouse contains extra copies of a growth hormone gene (Fig. 3-34).

Transgenic farm animals have also been developed. Ideally, the genes of such animals could be tailored to allow the animals to grow faster on less food or to be resistant to particular diseases. Some transgenic farm animals have been engineered to secrete medically useful proteins into their milk. Harvesting such a substance from milk is much more cost-effective than producing the same substance in bacterial cultures. Transgenic plants are also available, including freeze-tolerant strawberries, slow-ripening tomatoes, and crops that are more resistant to insect infestation.

Transgenic organisms have greatly enhanced our understanding of gene expression. Animals that have been engineered to contain a defective gene or that lack a gene entirely (a so-called **gene knockout**) also serve as experimental models for human diseases.

Gene Therapy

Gene therapy is the transfer of new genetic material to the cells of an individual in order to produce a therapeutic effect. Although the potential benefits of this as yet rudimentary technology are enormous, there are numerous practical obstacles to overcome. The longest running experiment in gene therapy was begun in 1990, when W. French Anderson and Michael Blaese introduced the gene for **adenosine deaminase (ADA)** into several children. Individuals with an inherited ADA deficiency have a condition called **severe combined immunodeficiency** (**SCID;** Box 22-2) and must be isolated to avoid potentially fatal infections. Among the ethical concerns of gene therapy (see Box 3-3) are the difficulties of abandoning successful conventional treatment protocols in favor of experimental therapies. In the

Figure 3-34. **Transgenic mouse.** The gigantic mouse on the left was grown from a fertilized ovum that had been microinjected with DNA containing the rat growth hormone gene. He is nearly twice the weight of his normal littermate on the right. [Courtesy of Ralph Brinster, University of Pennsylvania.]

Box 3-3

BIOCHEMISTRY IN CONTEXT

Ethical Aspects of Recombinant DNA Technology

In the early 1970s, when genetic engineering was first discussed, little was known about the safety of the proposed experiments. After considerable debate, during which there was a moratorium on such experiments, regulations for recombinant DNA research were drawn up. The rules prohibit obviously dangerous experiments (e.g., introducing the gene for diphtheria toxin into *E. coli*, which would convert this human symbiont into a deadly pathogen). Other precautions limit the risk of accidentally releasing potentially harmful organisms into the environment. For example, many vectors must be cloned in host organisms with special nutrient requirements. These organisms are unlikely to survive outside the laboratory.

The proven value of recombinant DNA technology has silenced nearly all its early opponents. Certainly, it would not have been possible to study some pathogens, such as the virus that causes AIDS, without cloning. The lack of recombinant-induced genetic catastrophes so far does not guarantee that recombinant organisms won't ever adversely affect the environment. Nevertheless, the techniques used by genetic engineers mimic those used in nature—that is, mutation and selection—so natural and man-made organisms are fundamentally similar. In any case, mankind has been breeding plants and animals for several millennia already, and for many of the same purposes that guide experiments with recombinant DNA.

There are other ethical considerations to be faced as new genetic engineering techniques become available. Bacterially produced human growth hormone is now routinely prescribed to increase the stature of abnormally short children. However, should athletes be permitted to use this protein, as some reportedly have, to increase their size and strength? Few would dispute the use of gene therapy, if it can be developed, to cure such genetic defects as sickle-cell anemia (Section 7-2D) and Lesch–Nyhan syndrome (Section 22-1D). If, however, it becomes possible to alter complex (i.e., multigene) traits such as athletic ability and intelligence, which changes would be considered desirable and who would decide whether to make them? Should gene therapy be used only to correct an individual's defects, or should it also be used to alter genes in the individual's germ cells so that succeeding generations would not inherit the defect? If it becomes easy to determine an individual's genetic makeup, should this information be used in evaluating applicants for educational and employment opportunities or health insurance?

As new genes are discovered and existing genes sequenced, the potential commercial value of genetic information increases. Should nucleotide sequences be patented? To what extent would such proprietary rights impede the free exchange of ideas and information that have hitherto characterized the pursuit of scientific knowledge?

case of the SCID children, their continued treatment with exogenous (externally administered) ADA renders the success of the recombinant ADA treatment difficult to quantify. Nevertheless, gene therapy retains considerable promise for treating the approximately 4000 known genetic diseases, particularly those for which there is no alternative treatment.

SUMMARY

1. Nucleotides consist of a purine or pyrimidine base linked to ribose to which at least one phosphate group is attached. RNA is made of ribonucleotides; DNA is made of deoxynucleotides (which contain 2′-deoxyribose).

2. Nucleotides such as ATP and nucleotide derivatives such as FAD, NAD$^+$, and CoA play central roles in energy metabolism.

3. In DNA, two antiparallel chains of nucleotides linked by phosphodiester bonds form a double helix. Bases in opposite strands pair: A with T, and G with C.

4. Single-stranded nucleic acids, such as RNA, can adopt stem-and-loop structures.

5. DNA carries genetic information in its sequence of nucleotides. When DNA is replicated, each strand acts as a template for the synthesis of a complementary strand.

6. According to the central dogma of molecular biology, one strand of the DNA of a gene is transcribed into mRNA. The RNA is then translated into protein by the ordered addition of amino acids that are bound to tRNA molecules that base-pair with the mRNA at the ribosome.

7. Restriction endonucleases that recognize certain sequences of DNA are used to specifically cleave DNA molecules.

8. Gel electrophoresis is used to separate and measure the sizes of DNA fragments.

9. In the chain-terminator method of DNA sequencing, the sequence of nucleotides in a DNA strand is determined by enzymatically synthesizing complementary polynucleotides that terminate with a dideoxy analog of each of the four nucleotides. Polynucleotide fragments of increasing size are separated by electrophoresis to reconstruct the original sequence.

10. Mutations and other changes to DNA are the basis for the evolution of organisms.

11. In molecular cloning, a fragment of foreign DNA is inserted into a vector for amplification in a host cell. Transformed cells can be identified by selectable markers.

12. Genomic libraries contain all the DNA of an organism. Clones harboring particular DNA sequences are identified by screening procedures.

13. The polymerase chain reaction amplifies selected sequences of DNA.

14. Recombinant DNA methods are used to produce wild-type or selectively mutagenized proteins in cells or entire organisms.

REFERENCES

General

Hartl, D.L., *Essential Genetics,* Jones and Bartlett (1996). [Includes chapters on DNA structure and function and recombinant DNA.]
McCarty, M., *The Transforming Principle,* Norton (1985). [A chronicle of the discovery that genes are DNA.]

Nucleic Acid Structure and Function

Gesteland, R.F. and Atkins, J.F. (Eds.), *The RNA World,* Cold Spring Harbor Laboratory Press (1993). [A collection of papers; Chapters 1, 2, and 4 provide an overview of the importance of RNA in the prebiotic world.]
Neidle, S., *DNA Structure and Recognition,* IRL Press (1994).
Watson, J.D. and Crick, F.H.C., Molecular structure of nucleic acids, *Nature* **171,** 737–738 (1953); *and* Genetical implications of the structure of deoxyribonucleic acid, *Nature* **171,** 964–967 (1953). [The seminal papers that are widely held to mark the origin of modern molecular biology.]

DNA Sequencing

Lander, E.S., The new genomics: Global views of biology, *Science* **274,** 536–539 (1996). [Lists some specific aims of genome sequencing efforts.]
Lipschutz, R.J. and Fodor, S.P.A., Advanced DNA sequencing technologies, *Curr. Opin. Struct. Biol.* **4,** 376–380 (1994).
Sanger, F., Sequences, sequences, sequences, *Annu. Rev. Biochem.* **57,** 1–28 (1988). [A scientific memoir.]
Venter, J.C., et al. The sequence of the human genome, *Science* **291,** 1304–1351 (2001); *and* International Human Genome Sequencing Consortium, Initial sequencing and analysis of the human genome, *Nature* **409,** 860–921 (2001).

Recombinant DNA

Erlich, H.A. and Arnheim, N., Genetic analysis using the polymerase chain reaction, *Annu. Rev. Genet.* **26,** 479–506 (1992).
Fersht, A. and Winter, G., Protein engineering, *Trends Biochem. Sci.* **17,** 292–294 (1992).

Glover, D.M. and Hames, B.D. (Eds.), *DNA Cloning: A Practical Approach* (2nd ed.), Vols. 1 and 2, IRL Press (1995).

Mullis, K.B., The unusual origin of the polymerase chain reaction. *Sci. Am.* **262**(4), 56–65 (1990).

Pääbo, S., Ancient DNA, *Sci. Am.* **269**(5), 86–92 (1993).

Rosenberg, J.M., Structure and function of restriction endonucleases, *Curr. Opin. Struct. Biol.* **1**, 104–113 (1991).

Sambrook, J., Fritsch, E.F., and Maniatis, T., *Molecular Cloning*

(2nd ed.), Cold Spring Harbor Laboratory (1989). [A three-volume "bible" of laboratory protocols with accompanying background explanations.]

Watson, J.D., Gilman, M., Witkowski, J., and Zoller, M., *Recombinant DNA* (2nd ed.), Freeman (1992). [A detailed exposition of the methods, findings, and results of recombinant DNA technology and research.]

KEY TERMS

nucleic acid
nucleotide
nucleoside
RNA
DNA
CoA
polynucleotide
phosphodiester bond
nucleotide residue
5′ end
3′ end
monomer
dimer
trimer
tetramer
oligomer
Chargaff's rules
tautomer
double helix
antiparallel
major groove
minor groove

complementary base pairing
genome
chromosome
diploid
haploid
bp
kb
stem-and-loop
gene
transformation
replication
transcription
translation
mRNA
rRNA
tRNA
ribosome
gene expression
bacteriophage
modification methylase
endonuclease
restriction endonuclease

exonuclease
palindrome
sticky ends
blunt ends
gel electrophoresis
restriction map
dNTP
primer
ddNTP
point mutation
recombination
transposition
phenotype
cloning
clone
vector
plasmid
recombinant
YAC
anneal
ligation
selectable marker

plaque
genomic library
shotgun cloning
colony (*in situ*) hybridization
replica plating
autoradiography
PCR
expression vector
overproducer
site-directed mutagenesis
wild type
transgenic organism
transgene
gene knockout
gene therapy

STUDY EXERCISES

1. Draw the structures of adenine, guanine, cytosine, uracil, and thymine.

2. Draw general structures for a ribonucleotide and a deoxynucleotide.

3. List the chemical and biological differences between DNA and RNA.

4. Summarize the central dogma of molecular biology.

5. Describe the Watson–Crick model of DNA.

6. How does the restriction–modification system operate?

7. Explain the chain-terminator (dideoxy) procedure for sequencing DNA.

PROBLEMS

1. Kinases are enzymes that transfer a phosphoryl group from a nucleoside triphosphate. Which of the following are valid kinase-catalyzed reactions?
   (a) ATP + GDP → ADP + GTP
   (b) ATP + GMP → AMP + GTP
   (c) ADP + CMP → AMP + CDP
   (d) AMP + ATP → 2 ADP

2. A diploid organism with a 45,000-kb haploid genome con-

tains 21% G residues. Calculate the number of A, C, G, and T residues in the DNA of each cell in this organism.

3. Draw the tautomeric forms of (a) adenine and (b) cytosine.

4. How many different amino acids could theoretically be encoded by nucleic acids containing four different nucleotides if (a) each nucleotide coded for one amino acid; (b) consecutive sequences of two nucleotides coded for one amino acid;

(c) consecutive sequences of three nucleotides coded for one amino acid; (d) consecutive sequences of four nucleotides coded for one amino acid?

5. Using the data in Table 3-3, identify restriction enzymes that (a) produce blunt ends; (b) recognize and cleave the same sequence (called **isoschizomers**); (c) produce identical sticky ends.

6. Construct a restriction map for a circular plasmid from the following data:

Restriction enzyme	Fragment sizes (kb)
*Eco*RI	4.0
*Hae*II	1.6, 2.4
*Pst*I	1.9, 2.1
*Eco*RI and *Hae*II	0.7, 1.6, 1.7
*Eco*RI and *Pst*I	0.8, 1.3, 1.9
*Hae*II and *Pst*I	0.6, 0.9, 1.0, 1.5

7. Describe the outcome of a chain-terminator sequencing procedure in which (a) too little ddNTP is added; (b) too much

ddNTP is added; (c) too few primers are present; (d) too many primers are present.

8. Calculate the number of clones required to obtain with a probability of 0.99 a specific 5-kb fragment from *C. elegans* (Table 3-4).

9. Describe how to select recombinant clones if a foreign DNA is inserted into the *Pst*I site of pBR322 and then introduced into *E. coli* cells.

10. Describe the possible outcome of a PCR experiment in which (a) one of the primers is inadvertently omitted from the reaction mixture; (b) one of the primers is complementary to several sites in the starting DNA sample; (c) there is a single-stranded break in the target DNA sequence, which is present in only one copy in the starting sample; (d) there is a double-stranded break in the target DNA sequence, which is present in only one copy in the starting sample.

11. Write the sequences of the two 12-residue primers that could be used to amplify the following DNA segment by PCR.

ATAGGCATAGGCCCATATGGCATAAGGCTT-
TATAATATGCGATAGGCGCTGGTCAG

[Problem provided by Bruce Wightman, Muhlenberg College.]

Each amino acid is a small molecule with unique biochemical properties determined by its functional groups. Glutamic acid, for example, contains an amino group and two carboxylic acid groups, all of which are ionized at neutral pH. It is shown here as a translucent space-filling model containing the corresponding stick model.

AMINO ACIDS

1. AMINO ACID STRUCTURE
   A. General Properties
   B. Peptide Bonds
   C. Classification and Characteristics
   D. Acid–Base Properties
   E. A Few Words on Nomenclature

2. STEREOCHEMISTRY

3. NONSTANDARD AMINO ACIDS
   A. Amino Acid Derivatives in Proteins
   B. D–Amino Acids
   C. Biologically Active Amino Acids

When scientists first turned their attention to nutrition, early in the nineteenth century, they quickly discovered that natural products containing nitrogen were essential for the survival of animals. In 1839, the Dutch chemist G.J. Mulder coined the term **protein** (Greek: *proteios,* primary) for this class of compounds. The physiological chemists of that time did not realize that proteins were actually composed of smaller components, amino acids, although the first amino acids had been isolated in 1830. In fact, for many years, it was believed that substances from plants—including proteins—were incorporated whole into animal tissues. This misconception was laid to rest when the process of digestion came to light. After it became clear that ingested proteins were broken down to smaller compounds containing amino acids, scientists turned their attention to the nutritive qualities of those compounds.

Experiments with animal diets showed that the amino acid content of a protein determined the nutritional adequacy of the protein. For example, certain cereal proteins are deficient in the amino acid lysine, and animals fed a diet lacking lysine do not thrive. By the time the structure of the twentieth common amino acid was determined, in 1925, the relative nutritive importance of different amino acids in proteins had been thoroughly documented.

Modern studies of proteins and amino acids owe a great deal to nineteenth century experiments. We now understand that nitrogen-containing amino acids are essential for life and that they are the building blocks of proteins. The central role of amino acids in biochemistry is perhaps not surprising: Several amino acids are among the organic compounds believed to have appeared early in the earth's history (Section 1-1A). Amino acids, as ancient and ubiquitous molecules, have been co-opted by evolution for a variety of purposes in living systems. We begin this chapter by discussing the structures and chemical properties of the common amino acids, including their stereochemistry, and end with a brief summary of the structures and functions of some unusual amino acids.

1. AMINO ACID STRUCTURE

The analyses of a vast number of proteins from almost every conceivable source have shown that *all proteins are composed of 20 "standard" amino acids.* Not every protein contains all 20 types of amino acids, but most proteins contain most if not all of the 20 types.

The common amino acids are known as **α-amino acids** because they have a primary amino group (—NH$_2$) and a carboxylic acid group (—COOH) as substituents of the same carbon atom (the **α carbon;** Fig. 4-1). The sole exception is proline, which has a secondary amino group (—NH—), although for uniformity we shall refer to proline as an α-amino acid. The 20 standard amino acids differ in the structures of their side chains **(R groups).**

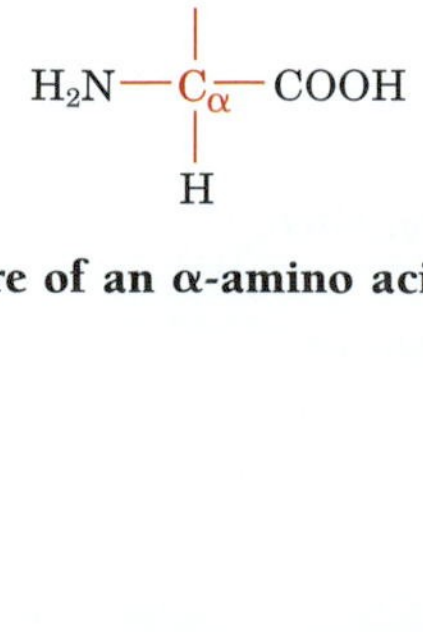

$$H_2N—\overset{\displaystyle R}{\underset{\displaystyle H}{\overset{|}{\underset{|}{C_\alpha}}}}—COOH$$

Figure 4-1.　General structure of an α-amino acid. The R groups differentiate the 20 standard amino acids.

Table 4-1 displays the names and complete chemical structures of the 20 standard amino acids.

A. General Properties

The amino and carboxylic acid groups of amino acids readily ionize. The pK values of the α-carboxylic acid groups (represented by pK_1 in Table 4-1) lie in a small range around 2.2, while the pK values of the α-amino groups (pK_2) are all near 9.4. *At physiological pH (~7.4), the amino groups are protonated and the carboxylic acid groups are in their conjugate base (carboxylate) form* (Fig. 4-2). An amino acid can therefore act as both an acid and a base. Table 4-1 also lists the pK values for the side chains that contain ionizable groups (pK_R).

Molecules such as amino acids, which bear charged groups of opposite polarity, are known as **zwitterions** or **dipolar ions.** The zwitterionic character of the α-amino acids has been established by several methods including spectroscopic measurements and X-ray crystal structure determinations. Amino acids, like other ionic compounds, are more soluble in polar solvents than in nonpolar solvents. As we shall see, the ionic properties of the side chains influence the physical and chemical properties of free amino acids and amino acids in proteins.

B. Peptide Bonds

Amino acids can be polymerized to form chains. This process can be represented as a **condensation** reaction (elimination of a water molecule), as shown in Fig. 4-3. The resulting CO—NH linkage, an amide linkage, is known as a **peptide bond.**

Polymers composed of two, three, a few (3–10), and many amino acid units are known, respectively, as **dipeptides, tripeptides, oligopeptides,** and **polypeptides.** These substances, however, are often referred to simply as "peptides." After they are incorporated into a peptide, the individual amino acids (the monomeric units) are referred to as **amino acid residues.**

Polypeptides are linear polymers; that is, each amino acid residue participates in two peptide bonds and is linked to its neighbors in a head-to-tail fashion rather than forming branched chains. The residues at the two ends of the polypeptide each participate in just one peptide bond. The residue with a free amino group (by convention, the leftmost residue, as shown in Fig. 4-3) is called the **amino terminus** or **N-terminus.** The residue with a free carboxylate group (at the right) is called the **carboxyl terminus** or **C-terminus.**

Proteins are molecules that contain one or more polypeptide chains. *Variations in the length and the amino acid sequence of polypeptides contribute to the diversity in the shape and biological functions of proteins,* as we shall see in succeeding chapters.

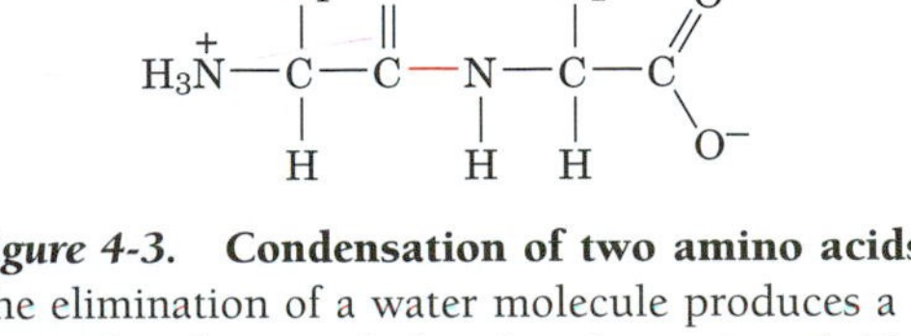

C. Classification and Characteristics

The most useful way to classify the 20 standard amino acids is by the polarities of their side chains. According to the most common classification scheme, there are three major types of amino acids: (1) those with nonpolar R groups, (2) those with uncharged polar R groups, and (3) those with charged polar R groups.

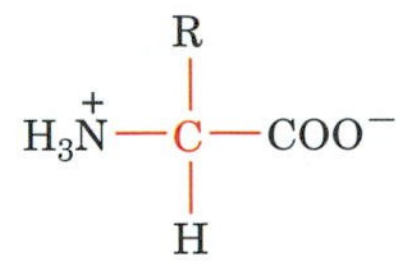

Figure 4-2. A zwitterionic amino acid. At physiological pH, the amino group is protonated and the carboxylic acid group is unprotonated.

Figure 4-3. Condensation of two amino acids. The elimination of a water molecule produces a dipeptide. The peptide bond is shown in red. The residue with a free amino group is the N-terminus of the peptide, and the residue with a free carboxylate group is the C-terminus.

Table 4-1.   Key to Structure. Covalent Structures and Abbreviations of the "Standard" Amino Acids of Proteins, Their Occurrence, and the pK Values of Their Ionizable Groups

Name, Three-letter Symbol, and One-letter Symbol	Structural Formula[a]	Residue Mass (D)[b]	Average Occurrence in Proteins (%)[c]	pK_1 α-COOH[d]	pK_2 α-NH$_3^+$[d]	pK_R Side Chain[d]
Amino acids with nonpolar side chains						
Glycine / Gly / G	H–C(H)(NH$_3^+$)COO$^-$	57.0	7.2	2.35	9.78	
Alanine / Ala / A	H–C(CH$_3$)(NH$_3^+$)COO$^-$	71.1	7.8	2.35	9.87	
Valine / Val / V	H–C(CH(CH$_3$)$_2$)(NH$_3^+$)COO$^-$	99.1	6.6	2.29	9.74	
Leucine / Leu / L	H–C(CH$_2$–CH(CH$_3$)$_2$)(NH$_3^+$)COO$^-$	113.2	9.1	2.33	9.74	
Isoleucine / Ile / I	H–C(C*(CH$_3$)(H)–CH$_2$–CH$_3$)(NH$_3^+$)COO$^-$	113.2	5.3	2.32	9.76	
Methionine / Met / M	H–C(CH$_2$–CH$_2$–S–CH$_3$)(NH$_3^+$)COO$^-$	131.2	2.2	2.13	9.28	
Proline / Pro / P	(pyrrolidine ring: C^2(COO$^-$)(H)–C^{3}H$_2$–C^{4}H$_2$–C^{5}H$_2$–N$^+$H$_2$)	97.1	5.2	1.95	10.64	
Phenylalanine / Phe / F	H–C(CH$_2$–C$_6$H$_5$)(NH$_3^+$)COO$^-$	147.2	3.9	2.20	9.31	
Tryptophan / Trp / W	H–C(CH$_2$–indole)(NH$_3^+$)COO$^-$	186.2	1.4	2.46	9.41	

[a]The ionic forms shown are those predominating at pH 7.0 (except for that of histidine[f]) although residue mass is given for the neutral compound. The C$_\alpha$ atoms, as well as those atoms marked with an asterisk, are chiral centers with configurations as indicated according to Fischer projection formulas. The standard organic numbering system is provided for heterocycles.

[b]The residue masses are given for the neutral residues. For the molecular masses of the parent amino acids, add 18.0 D, the molecular mass of H$_2$O, to the residue masses. For side chain masses, subtract 56.0 D, the formula mass of a peptide group, from the residue masses.

[c]Calculated from a database of nonredundant proteins containing 300,688 residues as compiled by Doolittle, R.F. in Fasman, G.D. (Ed.), *Predictions of Protein Structure and the Principles of Protein Conformation*, Plenum Press (1989).

[d]Data from Dawson, R.M.C., Elliott, D.C., Elliott, W.H., and Jones, K.M., *Data for Biochemical Research* (3rd ed.), pp. 1–31, Oxford Science Publications (1986).

[e]The three- and one-letter symbols for asparagine *or* aspartic acid are Asx and B, whereas for glutamine *or* glutamic acid they are Glx and Z. The one-letter symbol for an undetermined or "nonstandard" amino acid is X.

[f]Both neutral and protonated forms of histidine are present at pH 7.0, since its pK$_R$ is close to 7.0.

Table 4-1. **(continued)**

Name, Three-letter Symbol, and One-letter Symbol	Structural Formula[a]	Residue Mass (D)[b]	Average Occurrence in Proteins (%)[c]	pK$_1$ α-COOH[d]	pK$_2$ α-NH$_3^+$[d]	pK$_R$ Side Chain[d]
Amino acids with uncharged polar side chains						
Serine Ser S	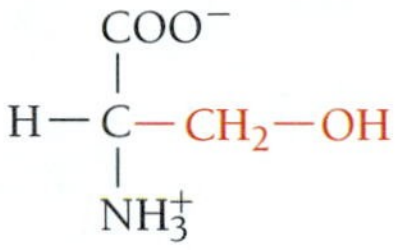	87.1	6.8	2.19	9.21	
Threonine Thr T	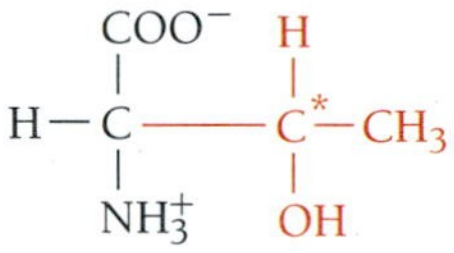	101.1	5.9	2.09	9.10	
Asparagine[e] Asn N	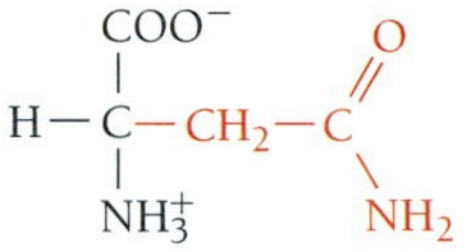	114.1	4.3	2.14	8.72	
Glutamine[e] Gln Q	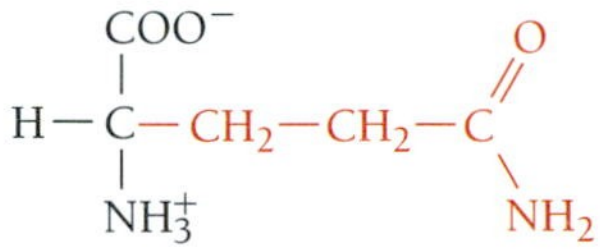	128.1	4.3	2.17	9.13	
Tyrosine Tyr Y	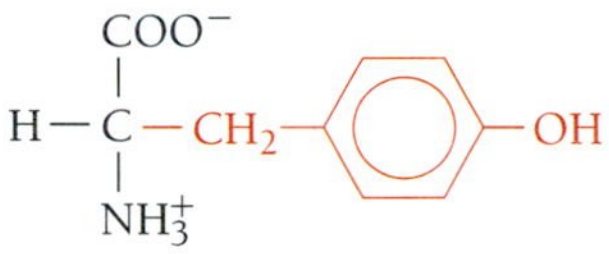	163.2	3.2	2.20	9.21	10.46 (phenol)
Cysteine Cys C	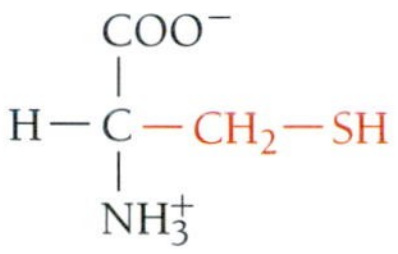	103.1	1.9	1.92	10.70	8.37 (sulfhydryl)
Amino acids with charged polar side chains						
Lysine Lys K	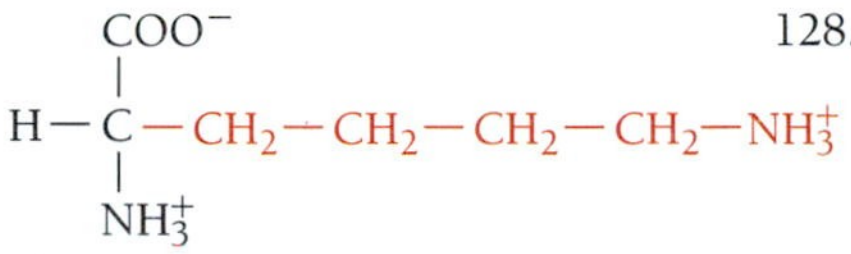	128.2	5.9	2.16	9.06	10.54 (ε-NH$_3^+$)
Arginine Arg R	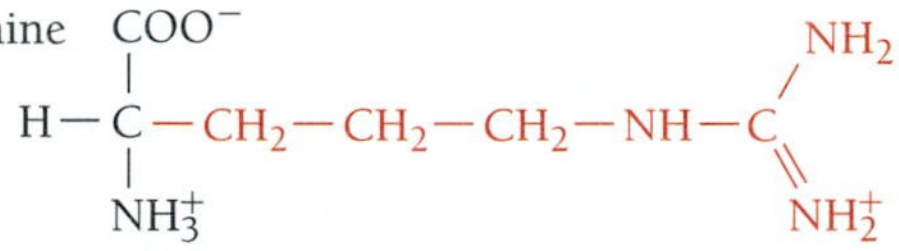	156.2	5.1	1.82	8.99	12.48 (guanidino)
Histidine[f] His H	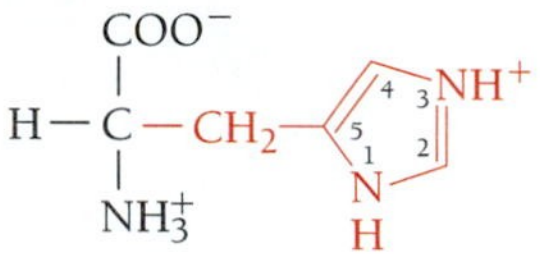	137.1	2.3	1.80	9.33	6.04 (imidazole)
Aspartic acid[e] Asp D	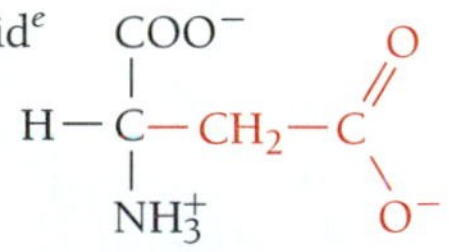	115.1	5.3	1.99	9.90	3.90 (β-COOH)
Glutamic acid[e] Glu E	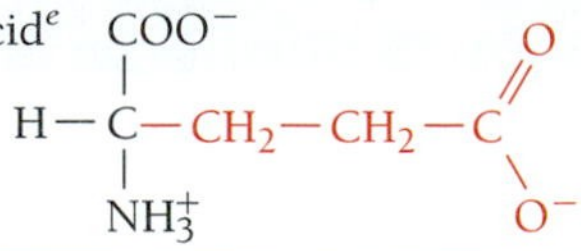	129.1	6.3	2.10	9.47	4.07 (γ-COOH)

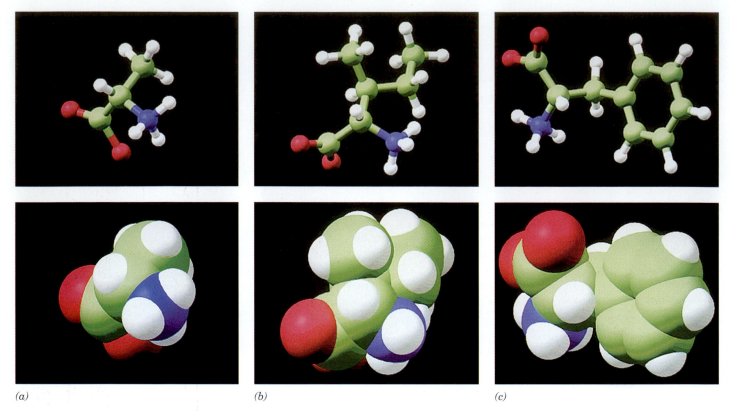

(a)

(b)

(c)

Figure 4-4.   **Models of some amino acids with nonpolar side chains.** Ball-and-stick structures (*top*) and the corresponding space-filling models (*bottom*) are shown for three amino acids: (*a*) alanine, (*b*) isoleucine, and (*c*) phenylalanine. All models are drawn to the same scale with C green, O red, N blue, and H white.

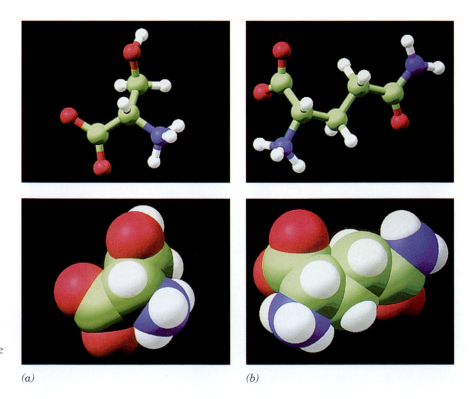

Figure 4-5.   **Models of some amino acids with uncharged polar side chains.** (*a*) Serine. (*b*) Glutamine. Atoms are colored as in Fig. 4-4. Note the presence of electronegative atoms on the side chains.

(a)

(b)

The Nonpolar Amino Acid Side Chains Have a Variety of Shapes and Sizes

Nine amino acids are classified as having nonpolar side chains. The three-dimensional shapes of some of these amino acids are shown in Fig. 4-4. **Glycine** has the smallest possible side chain, an H atom. **Alanine, valine, leucine,** and **isoleucine** have aliphatic hydrocarbon side chains ranging in size from a methyl group for alanine to isomeric butyl groups for leucine and isoleucine. **Methionine** has a thiol ether side chain that resembles an *n*-butyl group in many of its physical properties (C and S have nearly equal electronegativities, and S is about the size of a methylene group). **Proline** has a cyclic pyrrolidine side group. **Phenylalanine** (with its phenyl moiety) and **tryptophan** (with its indole group) contain aromatic side groups, which are characterized by bulk as well as nonpolarity.

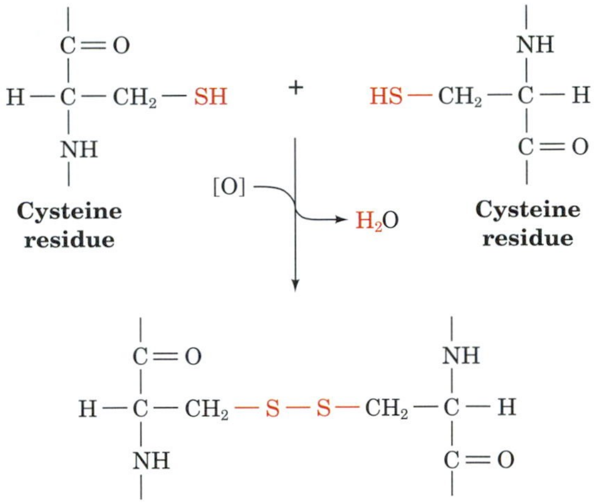

Uncharged Polar Side Chains Have Hydroxyl, Amide, or Thiol Groups

Six amino acids are commonly classified as having uncharged polar side chains (Table 4-1 and Fig. 4-5). **Serine** and **threonine** bear hydroxylic R groups of different sizes. **Asparagine** and **glutamine** have amide-bearing side chains of different sizes. **Tyrosine** has a phenolic group (and, like phenylalanine and tryptophan, is aromatic). **Cysteine** is unique among the 20 amino acids in that it has a thiol group that can form a disulfide bond with another cysteine (Fig. 4-6) through the oxidation of the two thiol groups. This dimeric compound was referred to in the older biochemical literature as the amino acid **cystine,** and cysteine was occasionally called a half-cystine residue.

Charged Polar Side Chains Are Positively or Negatively Charged

Five amino acids have charged side chains (Table 4-1 and Fig. 4-7). The side chains of the basic amino acids are positively charged at physiological pH values; they are **lysine,** which has a butylammonium side chain, **arginine,** which bears a guanidino group, and **histidine,** which carries an imid-

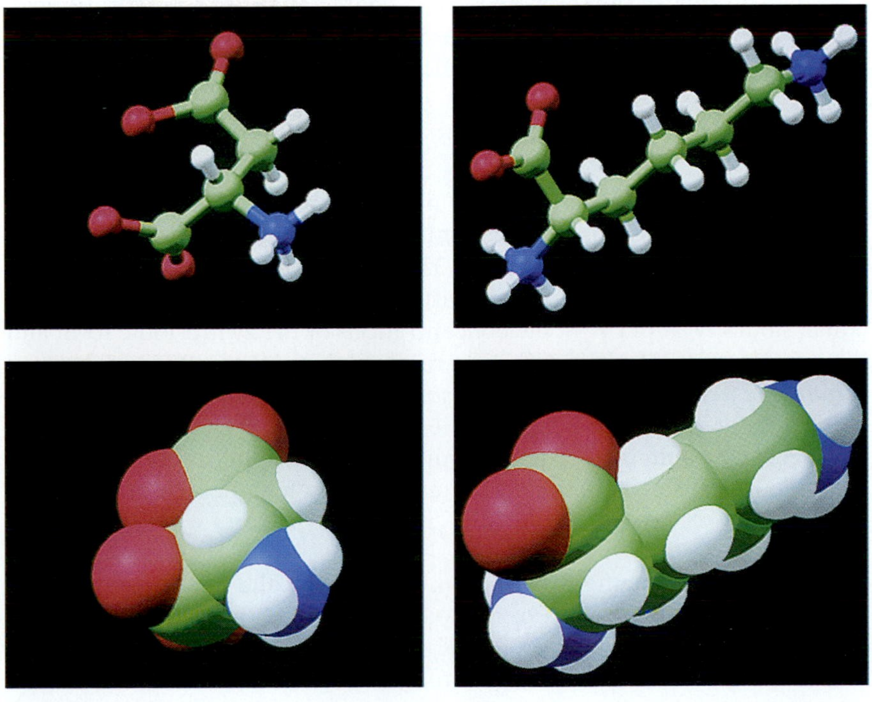

Figure 4-6.   Disulfide-bonded cysteine residues. The disulfide bond forms when the two thiol groups are oxidized.

(a)   *(b)*

*Figure 4-7.   **Models of some amino acids with charged polar side chains.** (a)* Aspartate. *(b)* Lysine. Atoms are colored as in Fig. 4-4.

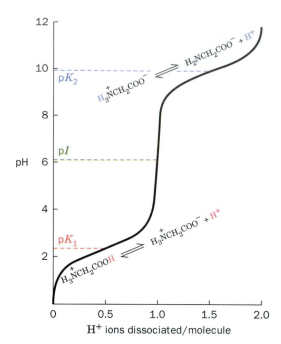

Figure 4-8.   Titration of glycine. [After Meister, A., *Biochemistry of Amino Acids* (2nd ed.), Vol. 1, p. 30, Academic Press (1965).]
✳ **See the Animated Figures.**

azolium moiety. Of the 20 α-amino acids, only histidine, with $pK_R = 6.0$, ionizes within the physiological pH range.

The side chains of the acidic amino acids, **aspartic acid** and **glutamic acid,** are negatively charged above pH 3; in their ionized state, they are often referred to as **aspartate** and **glutamate.** Asparagine and glutamine are, respectively, the amides of aspartic acid and glutamic acid.

The allocation of the 20 amino acids among the three different groups is somewhat arbitrary. For example, glycine and alanine, the smallest of the amino acids, and tryptophan, with its heterocyclic ring, might just as well be classified as uncharged polar amino acids. Similarly, tyrosine and cysteine, with their ionizable side chains, might also be thought of as charged polar amino acids, particularly at higher pH values.

Inclusion of a particular amino acid in one group or another reflects not just the properties of the isolated amino acid, but its behavior when it is part of a polypeptide. The structures of most polypeptides depend on a tendency for polar and ionic side chains to be solvated and for nonpolar side chains to associate with each other rather than with water. This property of polypeptides is the hydrophobic effect (Section 2-1C) in action. As we shall see, the chemical and physical properties of amino acid side chains also govern the chemical reactivity of the polypeptide. It is worthwhile studying the structures of the 20 standard amino acids in order to appreciate how they vary in polarity, acidity, aromaticity, bulk, conformational flexibility, ability to cross-link, ability to hydrogen bond, and reactivity toward other groups.

D. Acid–Base Properties

The α-amino acids have two or, for those with ionizable side chains, three acid–base groups. The titration curve of glycine, the simplest amino acid, is shown in Fig. 4-8. At low pH values, both acid–base groups of glycine are fully protonated, so that the cationic form ($^+H_3NCH_2COOH$) predominates. In the course of titration with a strong base such as NaOH, glycine loses two protons in the stepwise fashion characteristic of a polyprotic acid.

The pK values of glycine's two ionizable groups are sufficiently different so that the Henderson–Hasselbalch equation (Section 2-2B)

$$pH = pK + \log \frac{[A^-]}{[HA]} \qquad [4\text{-}1]$$

adequately describes each leg of the titration curve. Consequently, the pK for each ionization step is the midpoint of the corresponding leg of the titration curve (Section 2-2C). At pH 2.35, the concentrations of the cationic form ($^+H_3NCH_2COOH$) and the zwitterionic form ($^+H_3NCH_2COO^-$) are equal; similarly, at pH 9.78, the concentrations of the zwitterionic form and the anionic form ($H_2NCH_2COO^-$) are equal. Note that *amino acids never assume the neutral form in aqueous solution.*

The pH at which a molecule carries no net electric charge is known as its **isoelectric point, pI.** For the α-amino acids, the application of the Henderson–Hasselbalch equation indicates that, to a high degree of precision,

$$pI = \tfrac{1}{2}(pK_i + pK_j) \qquad [4\text{-}2]$$

where K_i and K_j are the dissociation constants of the two ionizations involving the neutral species. For monoamino, monocarboxylic acids such as glycine, K_i and K_j represent K_1 and K_2. However, for aspartic and glutamic acids, K_i and K_j are K_1 and K_R, whereas for arginine, histidine, and lysine, these quantities are K_R and K_2.

The pK Values of Ionizable Groups Depend on Nearby Groups

The value of pK_1 of glycine (2.35) is much lower than the pK of a simple monocarboxylic acid such as acetic acid (CH_3COOH, $pK = 4.76$). The large difference in pK values for the same functional group is caused by the electrostatic influence of glycine's positively charged ammonium group. The NH_3^+ group electrostatically stabilizes the COO^- group more than the COOH group. However, the NH_3^+ group of glycine ($pK_2 = 9.78$) is significantly more acidic than are aliphatic amines ($pK \approx 10.7$) because of the electron-withdrawing character of glycine's carboxylate group, but it is less acidic than glycine ethyl ester ($pK = 7.75$), whose carboxyl group is uncharged. Thus, both electronic and electrostatic effects influence the pK of the NH_3^+ group.

The electronic influence of one functional group on another depends on the distance between the groups. For example, the ionization constant of lysine's side chain amino group (separated from the α carbon by four methylene groups) is indistinguishable from that of an aliphatic amine.

Of course, amino acid residues in the interior of a polypeptide chain do not have free amino and carboxyl groups that can ionize (these groups are joined in peptide bonds; Fig. 4-3). Furthermore, the pK values of all ionizable groups, including the N- and C-termini, may differ from the pK values listed in Table 4-1 for free amino acids. The pK values of α-carboxyl groups in unfolded proteins range from 3.5 to 4.0, while the pK values for α-amino groups range from 8.0 to 9.0. The three-dimensional structure of a folded polypeptide chain may bring polar side chains and the N- and C-termini close together. The resulting electrostatic interactions between these groups may shift their pK values up to several pH units from the values in the corresponding free amino acids.

E. A Few Words on Nomenclature

The three-letter abbreviations for the 20 standard amino acids given in Table 4-1 are widely used in the biochemical literature. Most of these abbreviations are taken from the first three letters of the name of the corresponding amino acid and are pronounced as written. The symbol **Glx** indicates Glu or Gln, and, similarly, **Asx** means Asp or Asn. This ambiguous notation stems from laboratory experience: Asn and Gln are easily hydrolyzed to Asp and Glu, respectively, under the acidic or basic conditions often used to recover them from proteins (Section 5-3A). Without special precautions, it is impossible to tell whether a detected Glu was originally Glu or Gln, and likewise for Asp and Asn.

The one-letter symbols for the amino acids are also given in Table 4-1. This more compact code is often used when comparing the amino acid sequences of several similar proteins. Note that the one-letter symbol is usually the first letter of the amino acid's name. However, for sets of residues that have the same first letter, this is true only of the most abundant residue of the set.

Amino acid residues in polypeptides are named by dropping the suffix **-ine** in the name of the amino acid and replacing it by **-yl.** Polypeptide chains are described by starting at the N-terminus and proceeding to the C-terminus. The amino acid at the C-terminus is given the name of its parent amino acid. Thus, the compound in the margin is called alanyltyrosyl-aspartylglycine. Obviously, such names for polypeptide chains of more than a few residues are extremely cumbersome. This tetrapeptide (*at right*) can also be written as Ala-Tyr-Asp-Gly using the three-letter abbreviations, or AYDG using the one-letter symbols.

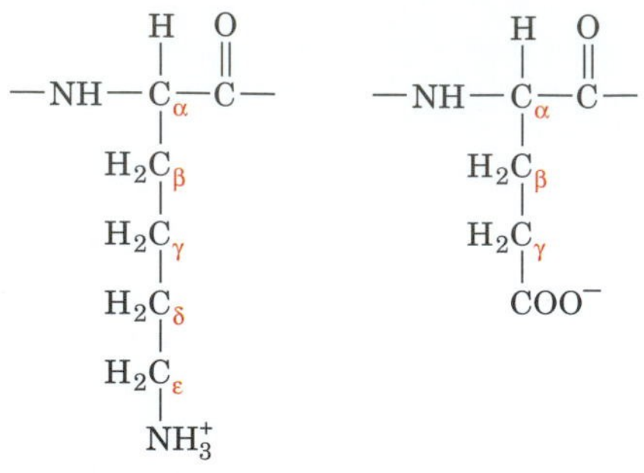

H O
| ||
—NH—C$_\alpha$—C—
|
H$_2$C$_\beta$
|
H$_2$C$_\gamma$
|
H$_2$C$_\delta$
|
H$_2$C$_\epsilon$
|
NH$_3^+$

Lys

H O
| ||
—NH—C$_\alpha$—C—
|
H$_2$C$_\beta$
|
H$_2$C$_\gamma$
|
COO$^-$

Glu

Figure 4-9. **Greek nomenclature for amino acids.** The carbon atoms are assigned sequential letters in the Greek alphabet, beginning with the carbon next to the carboxyl group.

The various atoms of the amino acid side chains are often named in sequence with the Greek alphabet, starting at the carbon atom adjacent to the peptide carbonyl group. Therefore, as Fig. 4-9 indicates, the Lys residue is said to have an ε-amino group and Glu has a γ-carboxyl group. Unfortunately, this labeling system is ambiguous for several amino acids. Consequently, standard numbering schemes for organic molecules are also employed (and are indicated in Table 4-1 for the heterocyclic side chains).

2. STEREOCHEMISTRY

With the exception of glycine, all the amino acids recovered from polypeptides are **optically active;** that is, they rotate the plane of polarized light. The direction and angle of rotation can be measured using an instrument known as a **polarimeter** (Fig. 4-10).

Optically active molecules are asymmetric; that is, they are not superimposable on their mirror image in the same way that a left hand is not superimposable on its mirror image, a right hand. This situation is characteristic of substances containing tetrahedral carbon atoms that have four different substituents. The two such molecules depicted in Fig. 4-11 are not superimposable since they are mirror images. The central atoms in such molecules are known as **asymmetric centers** or **chiral centers** and are said to have the property of **chirality** (Greek: *cheir,* hand). The C$_\alpha$ atoms of the amino acids (except glycine) are asymmetric centers. Glycine, which has two H atoms attached to its C$_\alpha$ atom, is superimposable on its mirror image and is therefore not optically active. Many biological molecules in addition to amino acids contain one or more chiral centers.

Chiral Centers Give Rise to Enantiomers

Molecules that are nonsuperimposable mirror images are known as **enantiomers** of one another. Enantiomeric molecules are physically and chemically indistinguishable by most techniques. *Only when probed asymmetrically, for example, by plane-polarized light or by reactants that also contain chiral centers, can they be distinguished or differentially manipulated.*

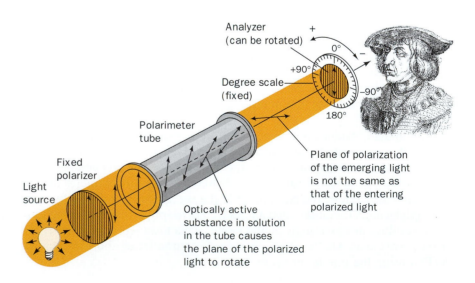

Figure 4-10. **Diagram of a polarimeter.** This device is used to measure optical rotation.

Unfortunately, there is no clear relationship between the structure of a molecule and the degree or direction to which it rotates the plane of polarized light. For example, leucine isolated from proteins rotates polarized light 10.4° to the left, whereas arginine rotates polarized light 12.5° to the right. (The enantiomers of these compounds rotate polarized light to the same degree but in the opposite direction.) It is not yet possible to predict optical rotation from the structure of a molecule, or to derive the **absolute configuration** (spatial arrangement) of chemical groups around a chiral center from optical rotation measurements.

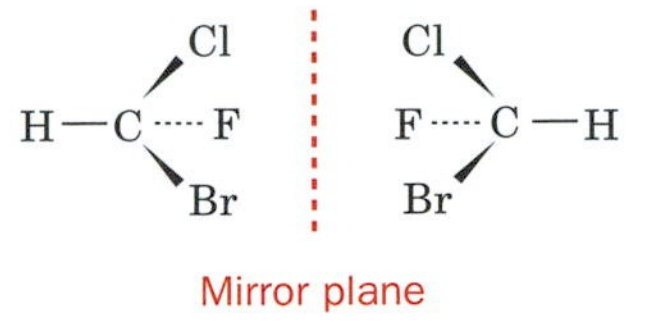

The Fischer Convention Describes the Configuration of Asymmetric Centers

Biochemists commonly use the **Fischer convention** to describe different forms of chiral molecules. In this system, the configuration of the groups around an asymmetric center is compared to that of **glyceraldehyde,** a molecule with one asymmetric center. In 1891, Emil Fischer proposed that the spatial isomers, or **stereoisomers,** of glyceraldehyde be designated D-glyceraldehyde and L-glyceraldehyde (Fig. 4-12). The prefix L (note the use of a small uppercase letter) signified rotation of polarized light to the left (Greek: *levo,* left), and the prefix D indicated rotation to the right (Greek: *dextro,* right) by the two forms of glyceraldehyde. Fischer assigned the prefixes to the structures shown in Fig. 4-12 without knowing whether the structure on the left and the structure on the right were actually **levorotatory** and **dextrorotatory,** respectively. Only in 1949 did experiments confirm that Fischer's guess was indeed correct.

Fischer also proposed a shorthand notation for molecular configurations, known as **Fischer projections,** which are also given in Fig. 4-12. In the Fischer convention, horizontal bonds extend above the plane of the paper and vertical bonds extend below the plane of the paper.

The configuration of groups around any chiral center can be related to that of glyceraldehyde by chemically converting the groups to those of glyceraldehyde. For α-amino acids, the amino, carboxyl, R, and H groups around the C_α atom correspond to the hydroxyl, aldehyde, CH_2OH, and H groups, respectively, of glyceraldehyde.

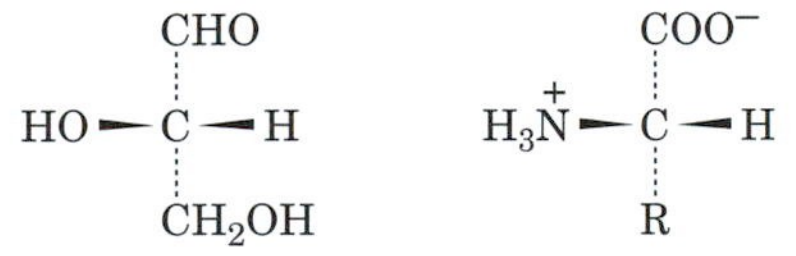

L-Glyceraldehyde   **L-α-Amino acid**

Therefore, L-glyceraldehyde and L-α-amino acids are said to have the same **relative configuration.** *All amino acids derived from proteins have the* L *stereochemical configuration;* that is, they all have the same relative configuration around their C_α atoms. Of course, the L or D designation of an amino acid does not indicate its ability to rotate the plane of polarized light: Many L-amino acids are dextrorotatory.

The Fischer system has some shortcomings, particularly for molecules with multiple asymmetric centers. Each asymmetric center can have two possible configurations, so a molecule with n chiral centers has 2^n different possible stereoisomers. Threonine and isoleucine, for example, each have two chiral carbon atoms, and therefore each has four stereoisomers, or two pairs of enantiomers [the enantiomers (mirror images) of the L forms are the D forms]. For most purposes, the Fischer system provides an adequate description of biological molecules. A more precise nomenclature system is also occasionally used by biochemists (see Box 4-1).

Figure 4-11. **The two enantiomers of fluorochlorobromomethane.** The four substituents are tetrahedrally arranged around the central carbon atom. A dotted line indicates that a substituent lies behind the plane of the paper, a wedged line indicates that it lies above the plane of the paper, and a thin line indicates that it lies in the plane of the paper. The mirror plane relating the enantiomers is represented by a vertical dashed line.

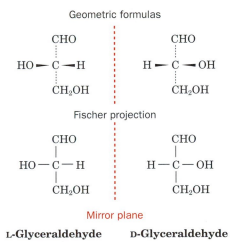

Geometric formulas

Fischer projection

Mirror plane

L-Glyceraldehyde   **D-Glyceraldehyde**

Figure 4-12. **The Fischer convention.** The enantiomers of glyceraldehyde are shown as geometric formulas (*top*) and as Fischer projections (*bottom*). In a Fischer projection, horizontal lines represent bonds that extend above the page and vertical lines represent bonds that extend below the page (in some Fischer projections, the central chiral carbon atom is not shown explicitly).

Box 4-1

BIOCHEMISTRY IN FOCUS

The RS System

A system to unambiguously describe the configurations of molecules with more than one asymmetric center was devised in 1956 by Robert Cahn, Christopher Ingold, and Vladimir Prelog. In the **Cahn–Ingold–Prelog** or *RS* system, the four groups surrounding a chiral center are ranked according to a specific although arbitrary priority scheme: Atoms of higher atomic number rank above those of lower atomic number (e.g., —OH ranks above —CH₃). If the first substituent atoms are identical, the priority is established by the next atom outward from the chiral center (e.g., —CH₂OH takes precedence over —CH₃). The order of priority of some common functional groups is

$$SH > OH > NH_2 > COOH > CHO > CH_2OH >$$
$$C_6H_5 > CH_3 > H$$

The prioritized groups are assigned the letters W, X, Y, Z such that their order of priority ranking is W > X > Y > Z. To establish the configuration of the chiral center, it is viewed from the asymmetric center toward the Z group (lowest priority). If the order of the groups W → X → Y is clockwise, the configuration is designated R (Latin: *rectus*, right). If the order

of W → X → Y is counterclockwise, the configuration is designated S (Latin: *sinistrus*, left).

L-Glyceraldehyde is (S)-glyceraldehyde because the three highest priority groups are arranged counterclockwise when the H atom (*dashed lines*) is positioned behind the chiral C atom (*large circle*).

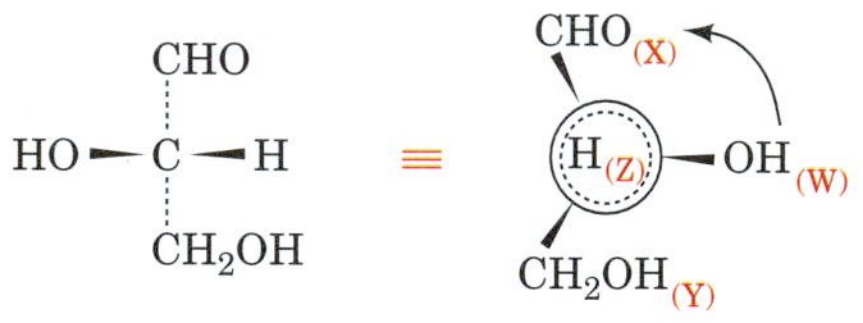

L-Glyceraldehyde   **(S)-Glyceraldehyde**

All the L-amino acids in proteins are (S)-amino acids except cysteine, which is (R)-cysteine because the S in its side chain increases its priority. Other closely related compounds with the same designation under the Fischer DL convention may have different representations under the RS system. The RS system is particularly useful for describing the chiralities of compounds with multiple asymmetric centers. Thus, L-threonine can also be called (2S,3R)-threonine.

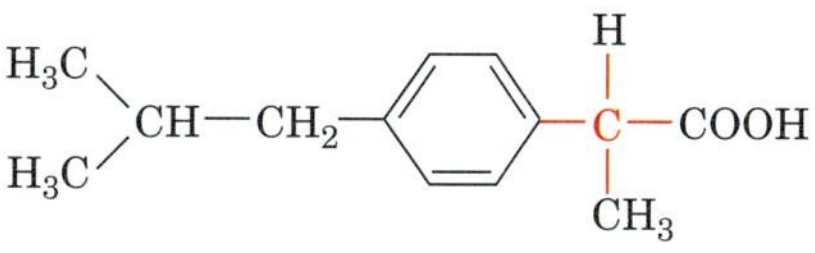

Ibuprofen

Figure 4-13. **Ibuprofen.** Only the enantiomer shown has anti-inflammatory action. The chiral carbon is red.

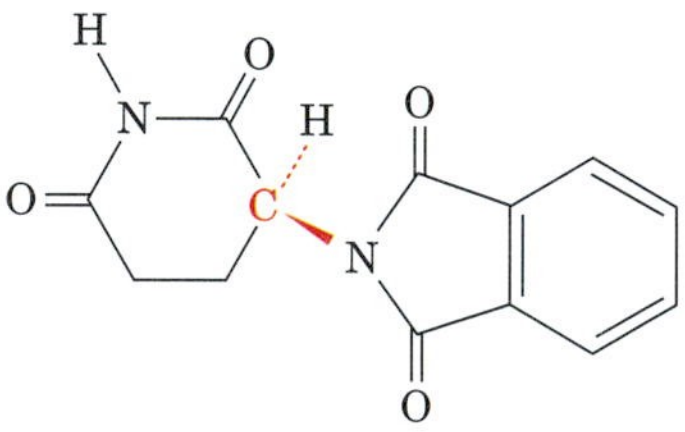

Thalidomide

Figure 4-14. **Thalidomide.** This drug was widely used in Europe as a mild sedative in the early 1960s. Its inactive enantiomer, which was present in equal amounts in the formulations used, causes severe birth defects in humans when taken during the first trimester of pregnancy. Thalidomide was commonly prescribed to alleviate the nausea (morning sickness) that is common during that period.

Life Is Based on Chiral Molecules

Consider the ordinary chemical synthesis of a chiral molecule, which produces a **racemic** mixture (containing equal amounts of each enantiomer). In order to obtain a product with net asymmetry, a chiral process must be employed. One of the most striking characteristics of life is its production of optically active molecules. *Biosynthetic processes almost invariably produce pure stereoisomers.* The fact that the amino acid residues of proteins all have the L configuration is just one example of this phenomenon (see Box 4-2). Furthermore, because most biological molecules are chiral, a given molecule—present in a single enantiomeric form—will bind to or react with only a single enantiomer of another compound. For example, a protein made of L-amino acid residues that reacts with a particular L-amino acid does not readily react with the D form of that amino acid. An otherwise identical synthetic protein made of D-amino acid residues, however, readily reacts only with the corresponding D-amino acid.

The importance of stereochemistry in living systems is also a concern of the pharmaceutical industry. *Many drugs are chemically synthesized as racemic mixtures, although only one enantiomer has biological activity.* In most cases, the opposite enantiomer is biologically inert and is therefore packaged along with its active counterpart. This is true, for example, of the anti-inflammatory agent **ibuprofen,** only one enantiomer of which is physiologically active (Fig. 4-13). Occasionally, the enantiomer of a useful drug produces harmful effects and must therefore be eliminated from the racemic mixture. The most striking example of this is the drug **thalidomide** (Fig. 4-14), a mild sedative whose inactive enantiomer causes severe birth defects.

Why Are Proteins Made of 20 L-Amino Acids?

The almost exclusive use of L-amino acids rather than D-amino acids in proteins prompts the question, Why did nature "choose" L-amino acids over D-amino acids? And why just 20 amino acids and none of the dozens of others that might have been incorporated into proteins?

There are no compelling chemical or physical reasons why L-amino acids should be favored over D-amino acids. The D and L forms of amino acids are equally stable, and a protein made of D-amino acids is physically identical in all other respects to its L counterpart. A quick glance at the hundreds of known amino acids reveals that the few that are used to build proteins do not represent the most naturally abundant or the most easily synthesized or even the most chemically versatile.

The reasons for the limited makeup of modern proteins lie buried in the earliest days of evolution and are accessible only through inference. The first amino acids that appeared on earth were probably racemic mixtures, because they were generated by purely chemical events. It is possible that some early forms of life used D-amino acids but were overtaken by more efficient life forms that, by chance, used L-amino acids. Other chance occurrences decreed that 20 particular amino acids take their places as the raw materials for building proteins. The specificity (and stereospecificity) of present-day biological processes is so complete that investigators in the early twentieth century were able to take advantage of biology to purify certain D-amino acids: When animals were fed a racemic mixture of the amino acid, they metabolized the L but not the D isomer so that the excreted, unused amino acid was almost entirely of the D form.

Just as humans are unlikely to suddenly develop gills or sprout wings, useful though they might be, proteins containing D-amino acids are unlikely to appear. At this point in evolution, L-amino acids are so deeply enmeshed in the biochemical fabric of cells that D-amino acids, however worthy from a strictly chemical point of view, cannot participate in any of the processes that have evolved along with the L-amino acids. In essence, modern proteins are made of 20 L-amino acids because there is no longer any alternative.

3. NONSTANDARD AMINO ACIDS

The 20 common amino acids are by no means the only amino acids that occur in biological systems. "Nonstandard" amino acid residues are often important constituents of proteins and biologically active peptides. In addition, many amino acids are not constituents of polypeptides at all but independently play a variety of biological roles.

A. Amino Acid Derivatives in Proteins

The "universal" genetic code, which is nearly identical in all known life forms (Section 26-1C), specifies only the 20 standard amino acids of Table 4-1. Nevertheless, many other amino acids, some of which are shown in Fig. 4-15, are components of certain proteins. *In almost all cases, these unusual amino acids result from the specific modification of an amino acid residue after the polypeptide chain has been synthesized.*

Amino acid modifications include the simple addition of small chemical groups to certain amino acid side chains: hydroxylation, methylation, acetylation, carboxylation, and phosphorylation. Larger groups, including lipids and carbohydrate polymers, are attached to particular amino acid residues of certain proteins. More elaborate chemical modifications are also found in some amino acid residues. The free amino and carboxyl groups at the N- and C-termini of a polypeptide can also be chemically modified. In many cases, the modifications are important, if not essential, for the function of the protein.

B. D-Amino Acids

D-Amino acid residues are components of some relatively short (<20 residues) bacterial polypeptides. These polypeptides are perhaps most

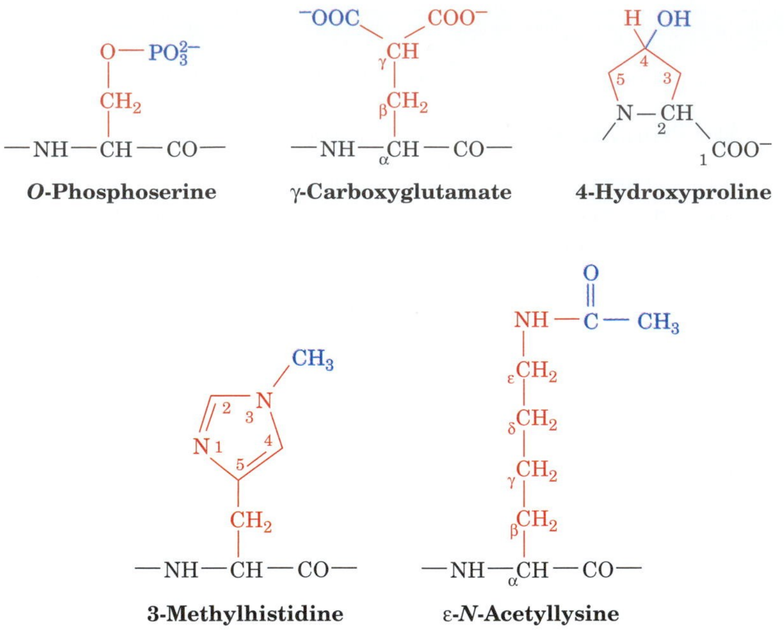

Figure 4-15.   Some modified amino acid residues in proteins. The side chains of these residues are derived from one of the 20 standard amino acids after the polypeptide has been synthesized. The standard R groups are red, and the modifying groups are blue.

widely distributed as constituents of bacterial cell walls (Section 8-3B). The presence of the D-amino acids renders bacterial cell walls less susceptible to attack by the **peptidases** (enzymes that hydrolyze peptide bonds) that are produced by other organisms to digest bacteria. Likewise, D-amino acids are components of many bacterially produced peptide antibiotics including **valinomycin** and **gramicidin A** (Section 10-4B) and **actinomycin D** (Box 25-2).

Most peptides containing D-amino acids are not synthesized by the standard protein synthetic machinery, in which messenger RNA is translated at the ribosome by transfer RNA molecules with attached L-amino acids. Instead, the D-amino acids are directly joined together by the action of specific bacterial enzymes. In a few cases, D-amino acids are components of ribosomally synthesized proteins in prokaryotes and eukaryotes. These D-amino acid residues are posttranslationally formed, probably by enzymatic inversion of the pre-existing L-amino acid residues.

C. Biologically Active Amino Acids

The 20 standard amino acids undergo a bewildering number of chemical transformations to other amino acids and related compounds as part of their normal cellular synthesis and degradation. In a few cases, the intermediates of amino acid metabolism have functions beyond their immediate use as precursors or degradation products of the 20 standard amino acids. Moreover, many amino acids are synthesized not to be residues of polypeptides but to function independently. We shall see that many organisms use certain amino acids to transport nitrogen in the form of amino groups (Section 21-2B). Amino acids may also be oxidized as metabolic fuels to provide energy (Section 20-4). In addition, amino acids and their de-

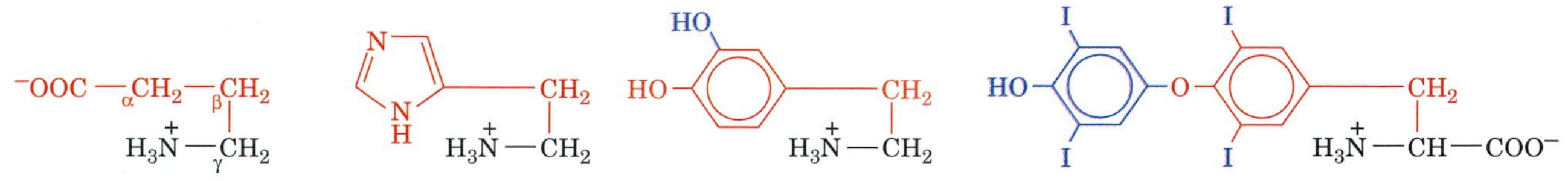

Figure 4-16. Some biologically active amino acid derivatives. The remaining portions of the parent amino acids are black and red, and additional groups are blue.

rivatives often function as chemical messengers for communication between cells (Fig. 4-16). For example, glycine, **γ-aminobutyric acid (GABA;** a glutamine decarboxylation product), and **dopamine** (a tyrosine derivative) are **neurotransmitters,** substances released by nerve cells to alter the behavior of their neighbors. **Histamine** (the decarboxylation product of histidine) is a potent local mediator of allergic reactions. **Thyroxine** (another tyrosine derivative) is an iodine-containing thyroid hormone that generally stimulates vertebrate metabolism.

The use of amino acids for functions other than protein synthesis is an example of nature adapting the materials at hand for new purposes. About 250 different amino acids have been found in various plants and fungi. For the most part, their biological roles are obscure, but the fact that many are toxic to other organisms suggests that they have a protective function.

SUMMARY

1. At neutral pH, the amino group of an amino acid is protonated and its carboxylic acid group is ionized.

2. Proteins are polymers of amino acids joined by peptide bonds.

3. The 20 standard amino acids can be classified as nonpolar (Gly, Ala, Val, Leu, Ile, Met, Pro, Phe, Trp), uncharged polar (Ser, Thr, Asn, Gln, Tyr, Cys), and charged (Lys, Arg, His, Asp, Glu).

4. The pK values of the ionizable groups of amino acids are influenced by neighboring groups and may be altered when the amino acid is part of a polypeptide.

5. Amino acids are chiral molecules. Only L-amino acids are found in proteins (some bacterial peptides contain D-amino acids).

6. Amino acids may be covalently modified after they have been incorporated into a polypeptide.

7. Individual amino acids and their derivatives have diverse physiological functions.

REFERENCES

Barrett, G.C. (Ed.), *Chemistry and Biochemistry of Amino Acids,* Chapman & Hall (1985). [Includes structures of the common amino acids along with a discussion of their chemical reactivities and information on analytical properties.]

Davies, J.S. (Ed.), *Amino Acids and Peptides,* Chapman & Hall (1985). [A sourcebook on amino acids.]

Jakubke, H.-D. and Jeschkeit, H., *Amino Acids, Peptides and Proteins,* translated into English by Cotterrell, G.P., Wiley (1977).

Lamzin, V.S., Dauter, Z., and Wilson, K.S., How nature deals with stereoisomers, *Curr. Opin. Struct. Biol.* **5,** 830–836 (1995). [Discusses proteins synthesized from D-amino acids.]

Solomons, T.W.G., *Organic Chemistry* (6th ed.), Chapter 5, Wiley (1996). [A discussion of chirality. Most other organic chemistry textbooks contain similar material.]

KEY TERMS

protein	tripeptide	optical activity	dextrorotatory
α-amino acid	oligopeptide	chiral center	Fischer projection
α carbon	polypeptide	enantiomers	racemic mixture
R group	residue	absolute configuration	peptidase
zwitterion	N-terminus	Fischer convention	neurotransmitter
peptide bond	C-terminus	stereoisomers	
dipeptide	pI	levorotatory	

STUDY EXERCISES

1. Draw the structures of the 20 standard amino acids and give their one- and three-letter abbreviations.

2. Classify the 20 standard amino acids by polarity, structure, type of functional group, and acid–base properties.

3. Describe the ionization states of amino acids that have ionizable side chains.

4. Explain how the Fischer convention describes the absolute configuration of a chiral molecule.

5. List some covalent modifications of amino acids in proteins.

PROBLEMS

1. Identify the amino acids that differ from each other by a single methylene group.

2. Identify the hydrogen bond donor and acceptor groups in asparagine.

3. Draw the dipeptide Asp-His at pH 7.0.

4. Calculate the number of possible pentapeptides that contain one residue each of Ala, Gly, His, Lys, and Val.

5. Determine the net charge of the predominant form of Asp at (a) pH 1.0, (b) pH 3.0, (c) pH 6.0, and (d) pH 11.0.

6. Calculate the pI of (a) Ala, (b) His, and (c) Glu.

7. Draw the four stereoisomers of threonine.

8. The two C_α H atoms of Gly are said to be prochiral, because when one of them is replaced by another group, C_α becomes chiral. Draw a Fischer projection of Gly and indicate which H must be replaced with CH_3 to yield D-Ala.

9. Describe isoleucine (as shown in Table 4-1) using the *RS* system.

10. Identify the amino acid from which the following compounds are synthesized:

(a)

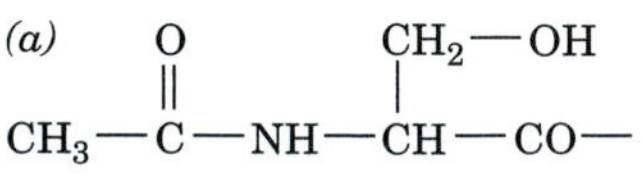

$$CH_3-\overset{\overset{O}{\|}}{C}-NH-\underset{\underset{|}{CH_2-OH}}{CH}-CO-$$

(b)

$$\overset{\overset{NH_3^+}{|}}{\underset{6}{C}H_2}$$
$$\underset{5}{CH}-OH$$
$$\underset{4}{CH_2}$$
$$\underset{3}{CH_2}$$
$$-NH-\underset{2}{CH}-\underset{1}{CO}-$$

(c)

$$\overset{S-CH_3}{|}$$
$$CH_2$$
$$CH_2$$
$$HC-NH-\underset{\underset{|}{}}{CH}-CO-$$
with O double-bonded to HC

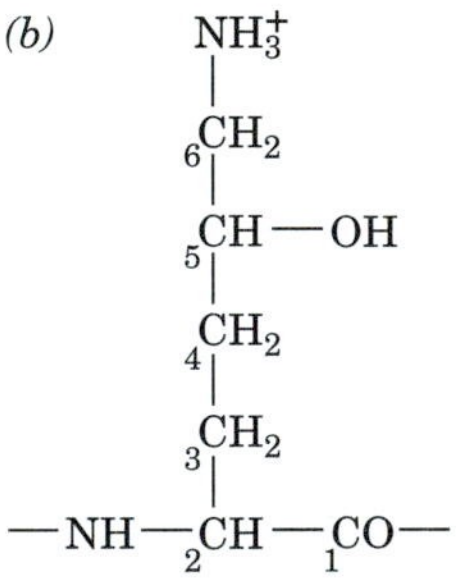

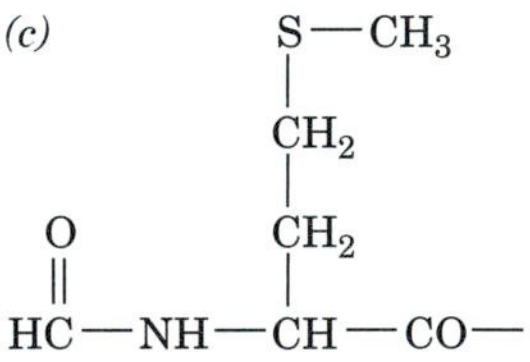

11. Draw the peptide ATLDAK. (a) Calculate its approximate pI. (b) What is its net charge at pH 7.0? [Problem provided by Kathleen Cornely, Providence College.]

12. The protein insulin consists of two polypeptides termed the A and B chains. Insulins from different organisms have been isolated and sequenced. Human and duck insulins have the same amino acid sequence with the exception of six amino acid residues, as shown below. Is the pI of human insulin lower than or higher than that of duck insulin?

Amino acid residue

	A8	A9	A10	B1	B2	B27
human	Thr	Ser	Ile	Phe	Val	Thr
duck	Glu	Asn	Pro	Ala	Ala	Ser

[Problem provided by Kathleen Cornely, Providence College.]

The great variation in structure and function among proteins reflects the astronomical variation in the sequences of their component amino acids—there are far more possible amino acid sequences than there are stars in the sky. How can such diversity be assessed in the laboratory and what does it reveal about the evolutionary relationships among proteins? [© John Chumack/Photo Researchers.]

PROTEINS: PRIMARY STRUCTURE

Proteins are at the center of action in biological processes. Nearly all the molecular transformations that define cellular metabolism are mediated by protein catalysts. Proteins also perform regulatory roles, monitoring extracellular and intracellular conditions and relaying information to other cellular components. In addition, proteins are essential structural components of cells. A complete list of known protein functions would contain many thousands of entries, including proteins that transport other molecules and proteins that generate mechanical and electrochemical forces. And such a list would not account for the thousands of proteins whose functions are not yet fully characterized or, in many cases, are completely unknown.

One of the keys to deciphering the function of a given protein is to understand its structure. Like the other major biological macromolecules, the nucleic acids (Section 3-2) and the polysaccharides (Section 8-2), proteins are polymers of smaller units. But unlike many nucleic acids, proteins do not have uniform, regular structures. This is, in part, because the 20 kinds of amino acid residues from which proteins are made have widely differing chemical and physical properties (Section 4-1C). By examining how these amino acids are strung together, we can endeavor to understand the chemical and physical properties of proteins and, ultimately, their mechanisms of action in living organisms.

The sheer variety of proteins demands a sophisticated arsenal of preparative and analytical techniques for studying them. In this chapter, we describe methods for purifying proteins. We also discuss amino acid sequencing and the information it provides about protein function and evolution. In Chapter 6, we explore the three-dimensional structures of proteins.

1. POLYPEPTIDE DIVERSITY

Like all polymeric molecules, proteins can be described in terms of levels of organization, in this case, their primary, secondary, tertiary, and quaternary structures. *A protein's **primary structure** is the amino acid sequence of its polypeptide chain or chains,* if the protein consists of more than one polypeptide. An example of an amino acid sequence is given in Fig. 5-1. Each residue is linked to the next via a peptide bond (Fig. 4-3). Higher levels of protein structure—secondary, tertiary, and quaternary—refer to the three-dimensional shapes of folded polypeptide chains and will be described in the following chapter.

Proteins are synthesized *in vivo* by the stepwise polymerization of amino acids in the order specified by the sequence of nucleotides in a gene. The direct correspondence between one linear polymer (DNA) and another (a polypeptide) illustrates the elegant simplicity of living systems and allows us to extract information from one polymer and apply it to the other.

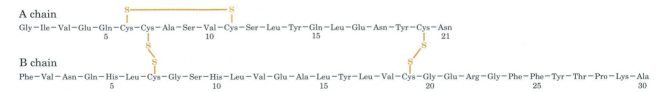

Figure 5-1. **The primary structure of bovine insulin.** Note the intrachain and interchain disulfide bond linkages.

The Theoretical Possibilities for Polypeptides Are Unlimited

With 20 different choices available for each amino acid residue in a polypeptide chain, it is easy to see that a huge number of different protein molecules are possible. For a protein of n residues, there are 20^n possible sequences. A relatively small protein molecule consists of a single polypeptide chain of 100 residues. There are $20^{100} \approx 1.27 \times 10^{130}$ possible unique polypeptide chains of this length, a quantity vastly greater than the estimated number of atoms in the universe (9×10^{78}). Clearly, evolution has produced only a tiny fraction of the theoretical possibilities—a fraction that nevertheless represents an astronomical number of different polypeptides (it is possible to generate even more in the laboratory; see Box 5-1).

Actual Polypeptides Are Somewhat Limited in Size and Composition

In general, proteins contain at least 40 residues or so; polypeptides smaller than that are simply called **peptides.** The largest known polypeptide chain belongs to the 26,926-residue **titin,** a giant (2,990 kD) protein that helps arrange the repeating structures of muscle fibers. However, *the vast majority of polypeptides contain between 100 and 1000 residues* (Table 5-1). **Multisubunit** proteins contain several identical and/or nonidentical chains called **subunits.** Some proteins are synthesized as single polypeptides that are later cleaved into two or more chains that remain associated; **insulin** is such a protein (Fig. 5-1).

The size range in which most polypeptides fall probably reflects the optimization of several biochemical processes:

1. Forty residues appear to be near the minimum for a polypeptide chain to fold into a discrete and stable shape that allows it to carry out a particular function.

2. Polypeptides with many hundreds of residues may approach the limits of efficiency of the protein synthetic machinery. The longer the polypeptide (and the longer its corresponding mRNA and its gene), the greater is the likelihood of introducing errors during transcription and translation.

In addition to these mild constraints on size, polypeptides are subject to more severe limitations on amino acid composition. The 20 standard amino acids do not appear with equal frequencies in proteins (Table 4-1 lists the

Table 5-1. **Compositions of Some Proteins**

Protein	Amino Acid Residues	Subunits	Molecular Mass (D)
Proteinase inhibitor III (melon)	30	1	3,409
Cytochrome *c* (human)	104	1	13,000
Ribonuclease H (*E. coli*)	155	1	17,600
Interferon-γ (rabbit)	288	2	34,200
Chorismate mutase (*Bacillus subtilis*)	381	3	43,500
Triose phosphate isomerase (*E. coli*)	510	2	56,400
Hemoglobin (human)	574	4	64,500
RNA polymerase (bacteriophage T7)	883	1	98,000
Nucleoside diphosphate kinase (*Dictyostelium discoideum*)	930	6	102,000
Pyruvate decarboxylase (yeast)	2252	4	250,000
Glutamine synthetase (*E. coli*)	5616	12	600,000
Titin (human)	26,926	1	2,990,000

Box 5-1
BIOCHEMISTRY IN FOCUS

Combinatorial Peptide Libraries

Where nature leaves off, the laboratory picks up, at least as far as peptide diversity is concerned. In pioneering work in the 1980s, Mario Geysen showed that it is possible to chemically synthesize large numbers of oligopeptides whose residues vary systematically at selected positions. For example, a synthetic hexapeptide might have the sequence X-X-A-B-X-X, where X represents any one of the 20 randomly incorporated amino acids and A and B are known. Geysen's approach to hexapeptide synthesis converts an impossible task (synthesizing all possible sequences, which would amount to 20^6, or 64 million, peptides) to a manageable one (synthesizing and testing 20^2, or 400, peptides). The so-called **peptide libraries** are therefore of great value in developing new drugs and probes of protein function.

The library of 400 hexapeptides can be rapidly screened for their ability to bind other molecules (antibodies, in Geysen's original experiments) or to produce biological effects. A "positive" X-X-A-B-X-X sequence can then become the starting point for synthesizing a new set of 400 hexapeptides with the sequence X-A-Y-Z-B-X, where Y and Z now represent the two chosen, invariant residues (this method is not limited to hexapeptides by any means).

An extended sequence with high biological activity can be constructed by repeated synthesis and screening. In a sense, the chemist exploits the same principles that guide evolution, but on a much smaller time scale. Peptide libraries may include novel peptides—those that don't appear in nature—or those that would be all but impossible to locate and purify from natural sources.

Using similar principles, oligonucleotide libraries have been synthesized *in vitro* and tested for their ability to bind other molecules or catalyze chemical reactions. If the oligonucleotides are incorporated into an expression vector, the corresponding oligopeptides can be synthesized in a suitable host organism and evaluated for the desired activity in that *in vivo* system. Oligonucleotide libraries have the added advantage that a single copy of an oligonucleotide can later be amplified by PCR (Section 3-5C) and sequenced.

---

average occurrence of each amino acid residue). For example, the most abundant amino acids in proteins are Leu, Ala, Gly, Ser, Val, and Glu; the rarest are Trp, Cys, Met, and His. Because each amino acid residue has characteristic chemical and physical properties, its presence at a particular position in a protein influences the properties of that protein. In particular, as we shall see, the three-dimensional shape of a folded polypeptide chain is a consequence of the intramolecular forces among its various residues. In general, a protein's hydrophobic residues cluster in its interior, out of contact with water, whereas its hydrophilic side chains tend to occupy the protein's surface.

The characteristics of an individual protein depend more on its amino acid sequence than on its amino acid composition per se, for the same reason that "kitchen" and its anagram "thicken" are quite different words. In addition, many proteins consist of more than just amino acid residues. They may form complexes with metal ions such as Zn^{2+} and Ca^{2+}, they may covalently or noncovalently bind certain small organic molecules, and they may be covalently modified by the post-translational attachment of groups such as phosphates and carbohydrates.

2. PROTEIN PURIFICATION

Fortunately, variations in polypeptide size and chemical composition make it easier to devise methods to separate proteins from each other and from other biological molecules. Purification is an all but mandatory step in studying macromolecules, but it is not necessarily easy. Typically, a substance that makes up <0.1% of a tissue's dry weight must be brought to ~98% purity. Purification problems of this magnitude would be considered unreasonably difficult by most synthetic chemists! The following sections outline some of the most common techniques for purifying and, to some extent, characterizing proteins. Most of these techniques can be used, sometimes in slightly modified form, for nucleic acids and other types of biological molecules.

A. *General Approach to Purifying Proteins*

The task of purifying a protein present in only trace amounts was once so arduous that many of the earliest proteins to be characterized were studied in part because they are abundant and easily isolated. For example, **hemoglobin,** which accounts for about one-third the weight of red blood cells, has historically been among the most extensively studied proteins. Most of the enzymes that mediate basic metabolic processes or that are involved in the expression and transmission of genetic information are common to all species. For this reason, a given protein is frequently obtained from a source chosen primarily for convenience, for example, tissues from domesticated animals or easily obtained microorganisms such as *E. coli* and *Saccharomyces cerevisiae* (baker's yeast).

The development of molecular cloning techniques (Section 3-5) allows almost any protein-encoding gene to be isolated from its parent organism, specifically altered (genetically engineered) if desired, and expressed at high levels in a microorganism. Indeed, the cloned protein may constitute up to 40% of the microorganism's total cell protein (Fig. 5-2). This high level of protein production generally renders the cloned protein far easier to isolate than it would be from its parent organism (in which it may occur in vanishingly small amounts).

The first step in the isolation of a protein or other biological molecule is to get it out of the cell and into solution. Many cells require some sort of mechanical disruption to release their contents. Most of the procedures for lysing cells use some variation of crushing or grinding followed by filtration or centrifugation to remove large insoluble particles. If the target protein is tightly associated with a lipid membrane, a detergent or organic solvent may be used to solubilize the lipids and recover the protein.

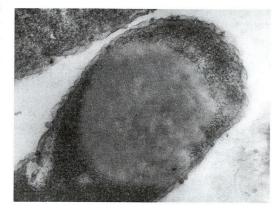

Figure 5-2.   Inclusion body. A genetically engineered organism that produces large amounts of a foreign protein often sequesters it in **inclusion bodies.** This electron micrograph shows an inclusion body of the protein prochymosin in an *E. coli* cell. [Courtesy of Teruhiko Beppu, Nikon University, Japan.]

Stabilizing Proteins

Once a protein has been removed from its natural environment, it becomes exposed to many agents that can irreversibly damage it. These influences must be carefully controlled at all stages of a purification process. The following factors should be considered:

1. **pH.** Biological materials are routinely dissolved in buffer solutions effective in the pH range over which the materials are stable (buffers are described in Section 2-2C). Failure to do so could cause their **denaturation** (structural disruption), if not their chemical degradation.

2. **Temperature.** The thermal stability of proteins varies. Although some proteins denature at low temperatures, most proteins denature at high temperatures, some only a few degrees higher than their native environment. Protein purification is normally carried out at temperatures near 0°C.

3. **Presence of degradative enzymes.** When tissues are destroyed to liberate the molecule of interest, degradative enzymes are also released. These include **nucleases** (enzymes that degrade nucleic acids) and **proteases** (enzymes that cleave the peptide bonds of proteins). Degradative enzymes can be inhibited by adjusting the pH or temperature to values that inactivate them (provided this does not adversely affect the protein of interest) or by adding compounds that specifically block their action.

4. **Adsorption to surfaces.** Many proteins are denatured by contact with the air–water interface or with glass or plastic surfaces. Hence, protein solutions are handled so as to minimize foaming and are kept relatively concentrated.

5. **Long-term storage.** All the factors listed above must be considered when a purified protein sample is to be kept stable. In addition, processes such as slow oxidation and microbial contamination must be prevented. Protein solutions are sometimes stored under nitrogen or argon gas (rather than under air containing ~21% O_2) and frozen at −70°C or at −196°C (the temperature of liquid nitrogen).

Assaying Proteins

Purifying a substance requires some means for quantitatively detecting it. Accordingly, an **assay** must be devised that is specific for the target protein, highly sensitive, and convenient to use (especially if it must be repeated at every stage of the purification process).

Among the most straightforward of protein assays are those for enzymes that catalyze reactions with readily detected products, because *the rate of product formation is proportional to the amount of enzyme present.* Substances with colored or fluorescent products have been developed for just this purpose. If no such substance is available for the enzyme being assayed, the product of the enzymatic reaction may be converted, by the action of another enzyme, to an easily quantified substance. This is known as a **coupled enzymatic reaction.** Proteins that are not enzymes can be detected by their ability to bind specific substances or to produce observable biological effects.

Immunochemical procedures are among the most sensitive of assay techniques. **Immunoassays** use **antibodies,** proteins produced by an animal's immune system in response to the introduction of a foreign substance (an **antigen**). Antibodies recovered from the blood serum of an immunized animal or from cultures of immortalized antibody-producing cells bind specifically to the original protein antigen (antibodies are discussed in more detail in Section 7-4).

A protein in a complex mixture can be detected by its binding to its corresponding antibodies. In one technique, known as a **radioimmunoassay** (**RIA**), the protein is indirectly detected by determining the degree to which it competes with a radioactively labeled standard for binding to the antibody. Another technique, the **enzyme-linked immunosorbent assay** (**ELISA**), has many variations, one of which is diagrammed in Fig. 5-3.

Separation Techniques

Proteins are purified by **fractionation procedures.** In a series of independent steps, the various physicochemical properties of the protein of interest are used to separate it progressively from other substances. The idea is not necessarily to minimize the loss of the desired protein, but to *eliminate selectively the other components of the mixture so that only the required substance remains.*

It may not be philosophically possible to prove that a substance is pure. In practice, *purity is established by showing that the sample of interest consists of only one component, using all available analytical methods.* Standards of purity are continually revised as new separation methods are developed. Experience has shown that a "pure" substance often reveals its heterogeneity when subjected to a new separation technique.

Protein purification is considered as much an art as a science, with many options available at each step. While a trial-and-error approach can work,

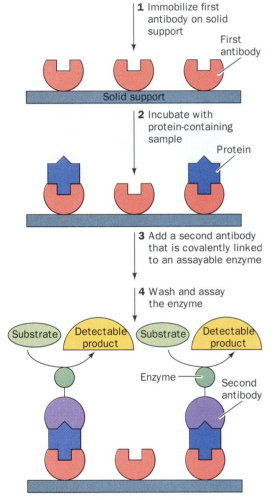

1 Immobilize first antibody on solid support

First antibody

Solid support

2 Incubate with protein-containing sample

Protein

3 Add a second antibody that is covalently linked to an assayable enzyme

4 Wash and assay the enzyme

Substrate   Detectable product   Substrate   Detectable product

Enzyme   Second antibody

Figure 5-3. **Enzyme-linked immunosorbent assay (ELISA).** (1) An antibody against the protein of interest is immobilized on an inert solid such as polystyrene. (2) The solution to be assayed is applied to the antibody-coated surface. The antibody binds the protein of interest, and other proteins are washed away. (3) The protein–antibody complex is reacted with a second protein-specific antibody to which an enzyme is attached. (4) Binding of the second antibody–enzyme complex is measured by assaying the activity of the enzyme. The amount of substrate converted to product indicates the amount of protein present. ✳ **See the Animated Figures.**

knowing something about the target protein (or the proteins it is to be separated from) simplifies the selection of separation procedures. The characteristics of proteins and other biomolecules that are used in the various separation procedures are solubility, ionic charge, molecular size, and binding specificity for other biological molecules. Some of the procedures we discuss and the protein characteristics they depend on are as follows:

Characteristic	Procedure
Charge	Ion exchange chromatography
	Electrophoresis
Polarity	Hydrophobic interaction chromatography
Size	Gel filtration chromatography
	SDS-PAGE
	Ultracentrifugation
Binding specificity	Affinity chromatography

B. Protein Solubility

Because a protein contains multiple charged groups, its solubility depends on the concentrations of dissolved salts, the polarity of the solvent, the pH, and the temperature. Some or all of these variables can be manipulated to selectively precipitate certain proteins while others remain soluble.

The solubility of a protein at low ion concentrations increases as salt is added, a phenomenon called **salting in.** The additional ions shield the protein's multiple ionic charges, thereby weakening the attractive forces between individual protein molecules (such forces can lead to aggregation and precipitation). However, as more salt is added, the solubility of the protein again decreases. This **salting out** effect is primarily a result of the competition between the added salt ions and the other dissolved solutes for molecules of solvent. At very high salt concentrations, so many of the added ions are solvated that there is significantly less bulk solvent available to dissolve other substances, including proteins.

Since different proteins precipitate at different salt concentrations, salting out is the basis of one of the most commonly used protein purification procedures. Adjusting the salt concentration in a solution containing a mixture of proteins to just below the precipitation point of the protein to be purified eliminates many unwanted proteins from the solution (Fig. 5-4). Then, after removing the precipitated proteins by filtration or centrifugation, the salt concentration of the remaining solution is increased to precipitate the desired protein. This procedure results in a significant purification and concentration of large quantities of protein.

Ammonium sulfate, $(NH_4)_2SO_4$, is the most commonly used reagent for salting out proteins, because its high solubility (3.9 M in water at 0°C) allows the preparation of solutions with high ionic strength. The pH may be adjusted to approximate the isoelectric point (pI) of the desired protein since a protein is least soluble when its net charge is zero. The pI's of some proteins are listed in Table 5-2.

C. Chromatography

The process of **chromatography** (Greek: *chroma*, color + *graphein*, to write) was discovered in 1903 by Mikhail Tswett, who separated solubilized plant pigments using solid adsorbents. In most modern chromatographic procedures, a mixture of substances to be fractionated is dissolved in a liquid (the **mobile phase**) and percolated through a column containing a

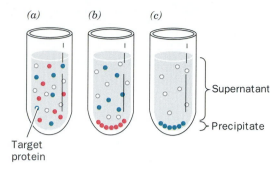

Figure 5-4. **Fractionation by salting out.** (*a*) The salt of choice, often ammonium sulfate, is added to a solution of macromolecules to a concentration just below the precipitation point of the protein of interest. (*b*) After centrifugation, the unwanted precipitated proteins (*red spheres*) are discarded and more salt is added to the supernatant to a concentration sufficient to salt out the desired protein (*green spheres*). (*c*) After a second centrifugation, the protein is recovered as a precipitate, and the supernatant is discarded.

Table 5-2. **Isoelectric Points of Several Common Proteins**

Protein	pI
Pepsin	<1.0
Ovalbumin (hen)	4.6
Serum albumin (human)	4.9
Tropomyosin	5.1
Insulin (bovine)	5.4
Fibrinogen (human)	5.8
γ-Globulin (human)	6.6
Collagen	6.6
Myoglobin (horse)	7.0
Hemoglobin (human)	7.1
Ribonuclease A (bovine)	9.4
Cytochrome *c* (horse)	10.6
Histone (bovine)	10.8
Lysozyme (hen)	11.0
Salmine (salmon)	12.1

porous solid matrix (the **stationary phase**). As solutes flow through the column, they interact with the stationary phase and are retarded. The retarding force depends on the properties of each solute. If the column is long enough, substances with different rates of migration will be separated. The chromatographic procedures that are most useful for purifying proteins are classified according to the nature of the interaction between the protein and the stationary phase.

One of the earliest chromatographic techniques used strips of filter paper as the stationary phase, a process called **paper chromatography.** Modern column chromatography uses derivatives of cellulose, agarose, or dextran (all carbohydrate polymers) or synthetic substances such as cross-linked polyacrylamide or silica. **High-performance liquid chromatography (HPLC)** employs automated systems with precisely applied samples, controlled flow rates at high pressures (up to 5000 psi), a chromatographic matrix of specially fabricated 3- to 300-μm-diameter glass or plastic beads coated with a uniform layer of chromatographic material, and on-line sample detection. This greatly improves the speed, resolution, and reproducibility of the separation—features that are particularly desirable when chromatographic separations are repeated many times or when they are used for analytical rather than preparative purposes.

Ion Exchange Chromatography

In **ion exchange chromatography,** charged molecules bind to oppositely charged groups that have been immobilized on the matrix. Anions bind to cationic groups on **anion exchangers,** and cations bind to anionic groups on **cation exchangers.** Perhaps the most frequently used anion exchanger is a matrix with attached **diethylaminoethyl (DEAE)** groups, and the most frequently used cation exchanger is a matrix bearing **carboxymethyl (CM)** groups

$$\text{DEAE: Matrix}-CH_2-CH_2-NH(CH_2CH_3)_2^+$$

$$\text{CM: Matrix}-CH_2-COO^-$$

Cellulose- and agarose-based resins are among the most frequently used matrix materials in the ion exchange chromatography of proteins.

Proteins and other **polyelectrolytes** (polyionic polymers) that bear both positive and negative charges can bind to both cation and anion exchangers, depending on their net charge. *The binding affinity of a particular protein depends on the presence of other ions that compete with the protein for binding to the ion exchanger and on the pH of the solution, which influences the net charge of the protein.*

The proteins to be separated are dissolved in a buffer of an appropriate pH and salt concentration and are applied to a column containing the ion exchanger. The column is then washed with the buffer (Fig. 5-5). As the column is washed, proteins with relatively low affinities for the ion exchanger move through the column faster than proteins that bind with higher affinities. The column effluent is collected in a series of fractions. Proteins that bind tightly to the ion exchanger can be **eluted** (washed through the column) by applying a buffer, called the **eluant,** that has a higher salt concentration or a pH that reduces the affinity with which the matrix binds the protein. The eluted fractions can then be assayed for the protein of interest.

The concentration of proteins in the column effluent is frequently monitored by absorbance spectroscopy. Measurements are often made at a wavelength near 280 nm, where most proteins absorb light. The three aromatic residues (especially Trp) account for most of this absorbance. Of

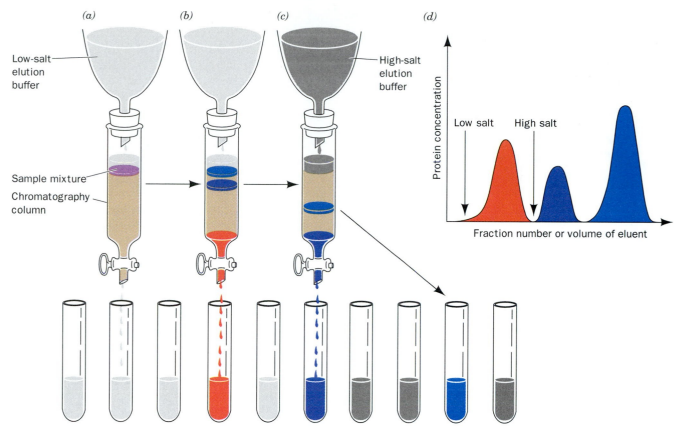

Figure 5-5.   Ion exchange chromatography. The tan region of the column represents the ion exchanger and the colored bands represent proteins. (*a*) A mixture of proteins dissolved in a small volume of buffer is applied to the top of the matrix in the column. (*b*) As elution progresses, the proteins separate into discrete bands as a result of their different affinities for the exchanger. In this diagram, the first protein (*red*) has passed through the column and has been isolated as a separate fraction. The other proteins remain near the top of the column. (*c*) The salt concentration in the eluant is increased to elute the remaining proteins. (*d*) The elution diagram of the protein mixture from the column. ✳ **See the Animated Figures.**

course, absorbance at 280 nm provides only a rough measure of protein concentration, since a protein's absorbance depends on its content of aromatic residues and other light-absorbing groups.

Hydrophobic Interaction Chromatography

Hydrophobic interactions between proteins and the chromatographic matrix can be exploited to purify the proteins. In **hydrophobic interaction chromatography,** the matrix material is lightly substituted with octyl or phenyl groups. Nonpolar groups on the surface of proteins "interact" with the hydrophobic groups; that is, both types of groups are excluded by the polar solvent. The eluant is typically an aqueous buffer with decreasing salt concentration (hydrophobic effects are augmented by increased ionic strength), increasing concentrations of detergent (which disrupts hydrophobic interactions), or changes in pH.

Gel Filtration Chromatography

In **gel filtration chromatography** (also called **size exclusion** or **molecular sieve chromatography**), molecules are separated according to their size and shape. The stationary phase consists of gel beads containing pores that span a relatively narrow size range. The pore size is typically determined

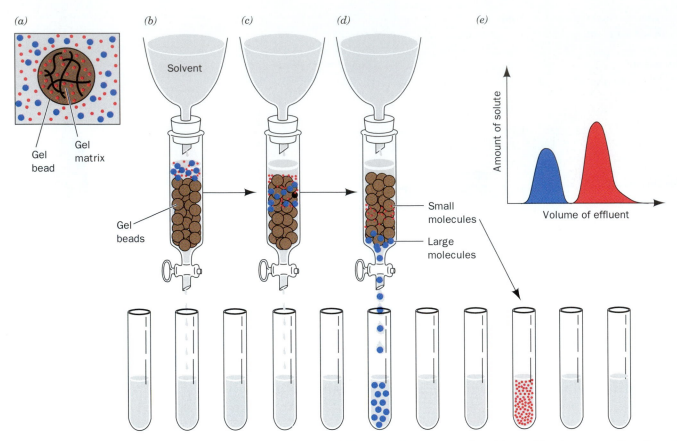

Figure 5-6. **Gel filtration chromatography.** (*a*) A gel bead consists of a gel matrix (*wavy solid lines*) that encloses an internal solvent space. Small molecules (*red dots*) can freely enter the internal space of the gel bead. Large molecules (*blue dots*) cannot penetrate the gel pores. (*b*) The sample solution is applied to the top of the column (the gel beads are represented as brown spheres). (*c*) The small molecules can penetrate the gel and consequently migrate through the column more slowly than the large molecules that are excluded from the gel. (*d*) The large molecules elute first and are collected as fractions. Small molecules require a larger volume of solvent to elute. (*e*) The elution diagram, or chromatogram, indicating the complete separation of the two components. ❋ **See the Animated Figures.**

by the extent of cross-linking between the polymers of the gel material. If an aqueous solution of molecules of various sizes is passed through a column containing such "molecular sieves," the molecules that are too large to pass through the pores are excluded from the solvent volume inside the gel beads. *These large molecules therefore traverse the column more rapidly than small molecules that pass through the pores* (Fig. 5-6). Because the pore size in any gel varies to some degree, gel filtration can be used to separate a range of molecules; larger molecules with access to fewer pores elute sooner (i.e., in a smaller volume of eluant) than smaller molecules that have access to more of the gel's interior volume.

Within the size range of molecules separated by a particular pore size, there is a linear relationship between the relative elution volume of a substance and the logarithm of its molecular mass (assuming the molecules have similar shapes). If a given gel filtration column is calibrated with several proteins of known molecular mass, the mass of an unknown protein can be conveniently estimated by its elution position.

Affinity Chromatography

A striking characteristic of many proteins is their ability to bind specific molecules tightly but noncovalently. This property can be used to purify such proteins by **affinity chromatography** (Fig. 5-7). In this technique, a molecule (a **ligand**) that specifically binds to the protein of interest is covalently attached to an inert matrix. *When an impure protein solution is passed through this chromatographic material, the desired protein binds to the immobilized ligand, whereas other substances are washed through the column with the buffer.* The desired protein can then be recovered in highly purified form by changing the elution conditions to release the protein from the matrix. The great advantage of affinity chromatography is its ability to exploit the desired protein's unique biochemical properties rather than the small differences in physicochemical properties between proteins used by other chromatographic methods.

Affinity chromatography columns can be constructed by chemically attaching small molecules or proteins to a chromatographic matrix. In **immunoaffinity chromatography,** an antibody is attached to the matrix in order to purify the protein against which the antibody was raised. In all cases, the ligand must have an affinity high enough to capture the protein of interest but not high enough to prevent the protein's subsequent release without denaturing it. The bound protein can be eluted by washing the column with a solution containing a high concentration of free ligand or a solution of different pH or ionic strength. An example of protein purification by affinity chromatography is shown in Fig. 5-8. The separation

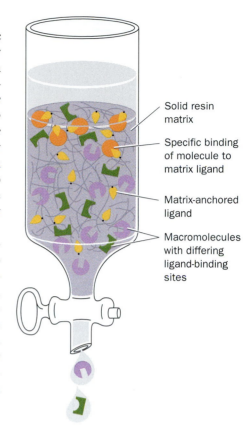

Figure 5-7.   Affinity chromatography. A ligand (shown here in yellow) is immobilized by covalently binding it to the chromatographic matrix. The cutout squares, semicircles, and triangles represent ligand-binding sites on macromolecules. Only certain molecules (represented by orange circles) specifically bind to the ligand. The other components are washed through the column.

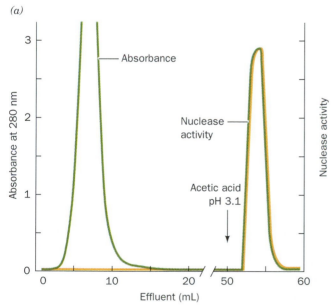

(b)

Figure 5-8.   Purification of an enzyme by affinity chromatography. (*a*) Staphylococcal nuclease (a DNA-hydrolyzing enzyme) was purified by passing a crude preparation through a column containing an immobilized diphosphothymidine derivative (shown in *b*). Approximately 40 mg of material was applied to the column. Only the nuclease binds to the immobilized ligand. After washing the column with 50 mL of a buffer with pH 8.0, 8.2 mg of pure enzyme was eluted with 0.1 M acetic acid. Note that essentially all the enzymatic activity (*yellow trace*) elutes with the nuclease. [After Cuatrecasas, P., Wilchek, M., and Anfinsen, C.B., *Proc. Natl. Acad. Sci.* **61**, 636 (1968).]

Table 5-3.   Purification of Rat Liver Glucokinase

Stage	Specific Activity $(nkat \cdot g^{-1})^a$	Yield (%)	Fold[b] Purification
Scheme A: A "traditional" chromatographic procedure			
1. Liver supernatant	0.17	100	1
2. $(NH_4)_2SO_4$ precipitate	*c*	*c*	*c*
3. DEAE chromatography I	4.9	52	29
4. DEAE chromatography II	23	45	140
5. DEAE chromatography III	44	33	260
6. DEAE chromatography IV	80	15	480
7. Gel filtration chromatography	130	15	780
Scheme B: An affinity chromatography procedure			
1. Liver supernatant	0.092	100	1
2. DEAE chromatography	20.1	104	220
3. Affinity chromatography[d]	420	83	4500

[a]A katal (abbreviation kat) is the amount of enzyme that catalyzes the transformation of 1 mol of substrate per second under standard conditions. One nanokatal (nkat) is 10^{-9} kat.

[b]Calculated from specific activity; the first step is arbitrarily assigned unity.

[c]The activity could not be accurately measured as this stage because of uncertainty in correcting for contamination by other enzymes.

[d]The ligand for affinity chromatography was glucosamine, an inhibitor of glucokinase.

Source: Cornish-Bowden, A., *Fundamentals of Enzyme Kinetics*, p. 48, Butterworths (1979), as adapted from Parry, M.J. and Walker, D.G., *Biochem. J.* **99**, 266 (1966), for Scheme A and from Holroyde, M.J., Allen, B.M., Storer, A.C., Warsey, A.S., Chesher, J.M.E., Trayer, I.P., Cornish-Bowden, A., and Walker, D.G., *Biochem. J.* **153**, 363 (1976), for Scheme B.

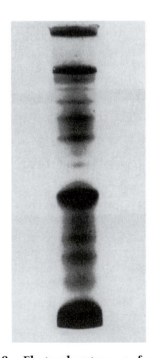

Figure 5-9.   Electrophoretogram of proteins in human serum. A sample of serum was applied to the top of a 0.5×4.0-cm glass tube containing polyacrylamide gel. Following electrophoresis, the proteins were stained with **amido black**. [Courtesy of Robert W. Hartley, NIH.]

power of affinity chromatography for a specific protein is often greater than that of other chromatographic techniques (e.g., Table 5-3).

D. Electrophoresis of Proteins

Electrophoresis, the migration of ions in an electric field, is described in Section 3-4B. Here, we describe its application to protein purification and analysis.

Polyacrylamide Gel Electrophoresis

Electrophoresis of proteins is typically carried out in agarose or polyacrylamide gels with a characteristic pore size. *The molecular separations are therefore based on sieving effects as well as electrophoretic mobility.* However, electrophoresis differs from gel filtration in that the electrophoretic mobility of small molecules is greater than the mobility of large molecules with the same charge density. The pH of the gel is high enough (usually about pH 9) so that nearly all proteins have net negative charges and move toward the anode when the current is switched on. Molecules of similar size and charge move as a band through the gel.

Following electrophoresis, the separated bands can be visualized by an appropriate method, such as staining (Fig. 5-9). In preparative gel electrophoresis, the proteins are eluted from gel slices, usually without staining. If the proteins in a sample are radioactive, the gel can be dried and then clamped over a sheet of X-ray film. After a time (ranging from a few minutes to many weeks, depending on the radiation intensity), the film is developed and the resulting **autoradiograph** shows the positions of the radioactive components by a blackening of the film. Alternatively, a position-sensitive radiation detector (electronic film) can be used to reveal the locations of the radioactive components within even a few seconds. If an antibody to the protein of interest is available, it can be used to specifically

Figure 5-10. **SDS-PAGE.** Samples of supernatants (*left*) and membrane fractions (*right*) from a preparation of the bacterium *Salmonella typhimurium* were electrophoresed in parallel lanes on a 35-cm-long by 0.8-mm-thick polyacrylamide slab. The lane marked MW contains molecular weight standards. [Courtesy of Giovanna F. Ames, University of California at Berkeley.]

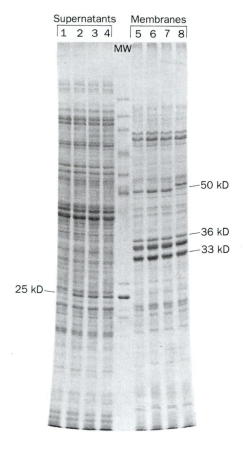

detect the protein on a gel in the presence of many other proteins, a process called **immunoblotting** or **Western blotting.**

SDS-PAGE

In one form of **polyacrylamide gel electrophoresis (PAGE)**, the detergent sodium dodecyl sulfate (SDS)

$$[CH_3—(CH_2)_{10}—CH_2—O—SO_3^-]Na^+$$

is used to denature proteins. Amphiphilic molecules (Section 2-1C) such as SDS interfere with the hydrophobic interactions that normally stabilize proteins. Proteins assume a rodlike shape in the presence of SDS. Furthermore, most proteins bind SDS in a ratio of about 1.4 g SDS per g protein (about one SDS molecule for every two amino acid residues). The large negative charge that the SDS imparts masks the proteins' intrinsic charge. The net result is that SDS-treated proteins have similar shapes and charge-to-mass ratios. *SDS-PAGE therefore separates proteins by gel filtration effects,* that is, according to molecular mass. Figure 5-10 is an example of the resolving power and the reproducibility of SDS-PAGE.

The molecular masses of proteins are routinely determined to an accuracy of 5 to 10% through SDS-PAGE. The relative mobilities of proteins on such gels vary linearly with the logarithm of their molecular masses (Fig. 5-11). In practice, a protein's molecular mass is determined by electrophoresing it together with several "marker" proteins of known molecular masses that bracket that of the protein of interest. Because SDS disrupts noncovalent interactions between polypeptides, SDS-PAGE yields the molecular masses of the subunits of multisubunit proteins. The possibility that subunits are linked by disulfide bonds can be tested by preparing samples for SDS-PAGE in the presence and absence of a reducing agent, such as **2-mercaptoethanol** ($HSCH_2CH_2OH$), that breaks these bonds.

Capillary Electrophoresis

Although gel electrophoresis in its various forms is highly effective at separating charged molecules, it can require up to several hours and is difficult to quantitate and automate. These disadvantages are largely overcome through the use of **capillary electrophoresis (CE)**, a technique in which electrophoresis is carried out in very thin capillary tubes (20 to 75 μm inner diameter). Such narrow capillaries rapidly dissipate heat and hence permit the use of very high electric fields, which reduces separation times to a few minutes. The CE techniques have extremely high resolution and can be automated in much the same way as is HPLC, that is, with automatic sample loading and on-line sample detection. Since CE can separate only small amounts of material, it is largely limited to use as an analytical tool.

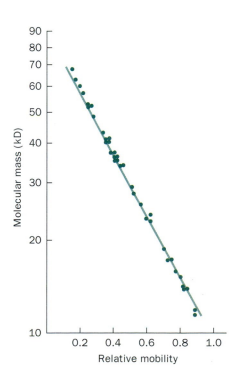

Figure 5-11. **Logarithmic relationship between the molecular mass of a protein and its electrophoretic mobility in SDS-PAGE.** The masses of 37 proteins ranging from 11 to 70 kD are plotted. [After Weber, K. and Osborn, M., *J. Biol. Chem.* **244,** 4406 (1969).]

E. Ultracentrifugation

If a container of sand and water is shaken and then allowed to stand quietly, the sand rapidly sediments to the bottom of the container due to the influence of the earth's gravity (an acceleration g of 9.821 m·s^{-2}). Yet macromolecules in solution, which experience the same gravitational field, do not exhibit any perceptible sedimentation because their random thermal (Brownian) motion keeps them uniformly distributed throughout the solution. *Only when subjected to enormous accelerations do macromolecules begin to sediment as do sand grains.*

The **ultracentrifuge,** which was developed around 1923 by the Swedish biochemist The Svedberg, can attain rotational speeds as high as 80,000 rpm so as to generate centrifugal fields in excess of 600,000g. Using this instrument, Svedberg first demonstrated that proteins are macromolecules with homogeneous compositions and that many proteins contain subunits.

The rate at which a particle sediments in the ultracentrifuge is related to its mass (the density of the solution and the shape of the particle also affect the sedimentation rate). A protein's sedimentation coefficient (s, sedimentation velocity per unit of centrifugal force) is usually expressed in units of 10^{-13} s, which are known as **Svedbergs (S).** The sedimentation coefficient is customarily corrected to the value that would be obtained at 20°C in pure water: $s_{20,w}$. Table 5-4 lists the values of $s_{20,w}$ for some proteins. Note that the relationship between molecular mass and sedimentation coefficient is not linear; therefore, s values are not additive. (**Fibrinogen's** anomalous sedimentation is due to its elongated fibrous shape, which slows its sedimentation relative to that of other, more nearly spherical proteins of the same mass.) The sedimentation coefficients of proteins range from about 1S to about 50S; viruses have sedimentation coefficients in the range of 40S to 1000S. Subcellular particles such as mitochondria have sedimentation coefficients of tens of thousands.

Table 5-4. **Sedimentation Coefficients of Some Proteins**

Protein	Molecular Mass (kD)	Sedimentation Coefficient, $s_{20,w}$ (S)
Lipase (milk)	6.7	1.14
Ribonuclease A (bovine pancreas)	12.6	2.00
Cytochrome c (bovine heart)	13.4	1.71
Myoglobin (horse heart)	16.9	2.04
α-Chymotrypsin (bovine pancreas)	21.6	2.40
Crotoxin (rattlesnake)	29.9	3.14
Concanavalin B (jack bean)	42.5	3.50
Diphtheria toxin	70.4	4.60
Cytochrome oxidase (*P. aeruginosa*)	89.8	5.80
Lactate dehydrogenase H (chicken)	150	7.31
Catalase (horse liver)	222	11.20
Fibrinogen (human)	340	7.63
Hemocyanin (squid)	612	19.50
Glutamate dehydrogenase (bovine liver)	1015	26.60
Turnip yellow mosaic virus protein	3013	48.80

Source: Smith, M.H., *in* Sober, H.A. (Ed.), *Handbook of Biochemistry and Molecular Biology* (2nd ed.), p. C-10, CRC Press (1970).

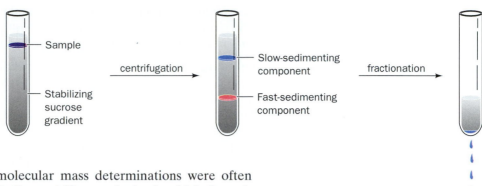

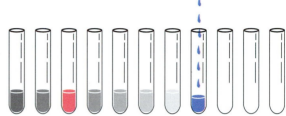

Before about 1970, molecular mass determinations were often made using an **analytical ultracentrifuge,** a device in which the sedimentation of macromolecules can be optically observed. More recently, gel filtration chromatography and SDS-PAGE have proved to be more convenient methods. The analytical ultracentrifuge is still used, however, for characterizing systems of noncovalently associating molecules, including subunits of proteins and other molecules that form macromolecular complexes.

In **preparative ultracentrifugation,** sedimentation is carried out in a solution of an inert substance in which the concentration, and therefore the density, of the solution increases from the top to the bottom of the centrifuge tube. The use of such **density gradients** enhances the resolving power of the ultracentrifuge. In **zonal ultracentrifugation,** a macromolecular solution is layered on top of a preformed density gradient, usually made with sucrose. During centrifugation, each species of macromolecule moves through the gradient at a rate largely determined by its sedimentation coefficient and therefore travels as a zone that can be separated from other such zones (Fig. 5-12). After centrifugation, the tube is punctured to collect fractions containing the separated macromolecules.

In **equilibrium density gradient centrifugation,** the sample is dissolved in a relatively concentrated solution of a dense, fast-diffusing substance such as CsCl. Under the high gravitational field produced at high spin rates, the CsCl forms a density gradient. The sample components form bands at positions where their densities are equal to that of the solution. The individual bands can then be separated as in zonal ultracentrifugation (Fig. 5-12).

Figure 5-12. **Zonal ultracentrifugation.** The sample is layered onto a preformed sucrose density gradient (*left*). During centrifugation (*middle*), each particle sediments at a rate that depends largely on its mass. After centrifugation, the tube is punctured and the separated particles (zones) are collected (*right*). In equilibrium density gradient centrifugation, the centrifuge tube is filled with a sample solution containing a substance, usually CsCl, that forms a density gradient as the tube spins.

3. PROTEIN SEQUENCING

The first protein whose sequence was determined was the bovine hormone insulin. Its complete sequence was reported in 1953 by Frederick Sanger (who later devised the chain-terminator method of DNA sequencing; Section 3-4C), thereby definitively establishing that proteins have unique covalent structures. The amino acid sequences of hundreds of thousands of polypeptides are now known. Such information is valuable for the following reasons:

1. Knowledge of a protein's amino acid sequence is prerequisite for determining its three-dimensional structure and is essential for understanding its molecular mechanism of action.

2. Sequence comparisons among analogous proteins from different species yield insights into protein function and reveal evolutionary relationships among the proteins and the organisms that produce them.

● See Guided Exploration 4:

Protein Sequence Determination.

3. Many inherited diseases are caused by mutations leading to an amino acid change in a protein. Amino acid sequence analysis can assist in the development of diagnostic tests and effective therapies.

Determining the sequence of insulin's 51 residues (Fig. 5-1) took about 10 years and required ~100 g of protein. Procedures for primary structure determination have since been so refined and automated that most proteins can be automatically sequenced in a few days using only a few micrograms of material. Polypeptides of up to ~25 residues can be sequenced by mass spectrometry, a technique in which the masses of ionized peptide fragments are measured directly. However, the basic approach to sequencing proteins is similar to the procedure developed by Sanger. Briefly, *the protein must be broken down into fragments small enough to be individually sequenced. The primary structure of the intact protein is then reconstructed from the sequences of overlapping fragments.*

A. Preliminary Steps

The complete amino acid sequence of a protein includes the sequence of each of its subunits, if any, so the subunits must be identified and isolated before sequencing begins.

End Group Analysis Reveals the Number of Different Types of Subunits

Each polypeptide chain (if it is not chemically blocked) has an N-terminal and a C-terminal residue. *Identifying these **end groups** can establish the number of chemically distinct polypeptides in a protein.* For example, insulin has equal amounts of the N-terminal residues Gly and Phe, which indicates that it has equal numbers of two nonidentical polypeptide chains.

The N-terminus of a polypeptide can be determined by several methods. The fluorescent compound **1-dimethylaminonaphthalene-5-sulfonyl chloride (dansyl chloride)** reacts with primary amines to yield dansylated polypeptides (Fig. 5-13). Acid hydrolysis liberates the modified N-terminal residue, which is separated chromatographically and identified by its intense yellow fluorescence. The N-terminal residue can also be identified by performing the first step of Edman degradation (Section 5-3C), a procedure that liberates amino acids one at a time from the N-terminus of a polypeptide.

There is no reliable chemical procedure for identifying the C-terminal residue of a polypeptide. However, this can often be done by using **carboxypeptidases,** enzymes that catalyze the hydrolytic excision of a polypeptide's C-terminal residue:

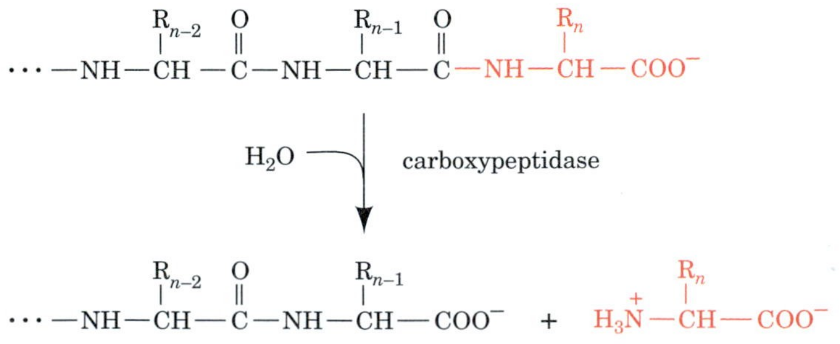

$$\cdots -NH-\underset{\underset{R_{n-2}}{|}}{CH}-\underset{\underset{O}{\parallel}}{C}-NH-\underset{\underset{R_{n-1}}{|}}{CH}-\underset{\underset{O}{\parallel}}{C}-NH-\underset{\underset{R_n}{|}}{CH}-COO^-$$

$$H_2O \downarrow \text{carboxypeptidase}$$

$$\cdots -NH-\underset{\underset{R_{n-2}}{|}}{CH}-\underset{\underset{O}{\parallel}}{C}-NH-\underset{\underset{R_{n-1}}{|}}{CH}-COO^- \quad + \quad H_3\overset{+}{N}-\underset{\underset{R_n}{|}}{CH}-COO^-$$

The liberated amino acid can then be isolated and identified. **Aminopeptidases** similarly cleave the N-terminal residues of polypeptides; aminopeptidases and carboxypeptidases are collectively known as **exopeptidases.**

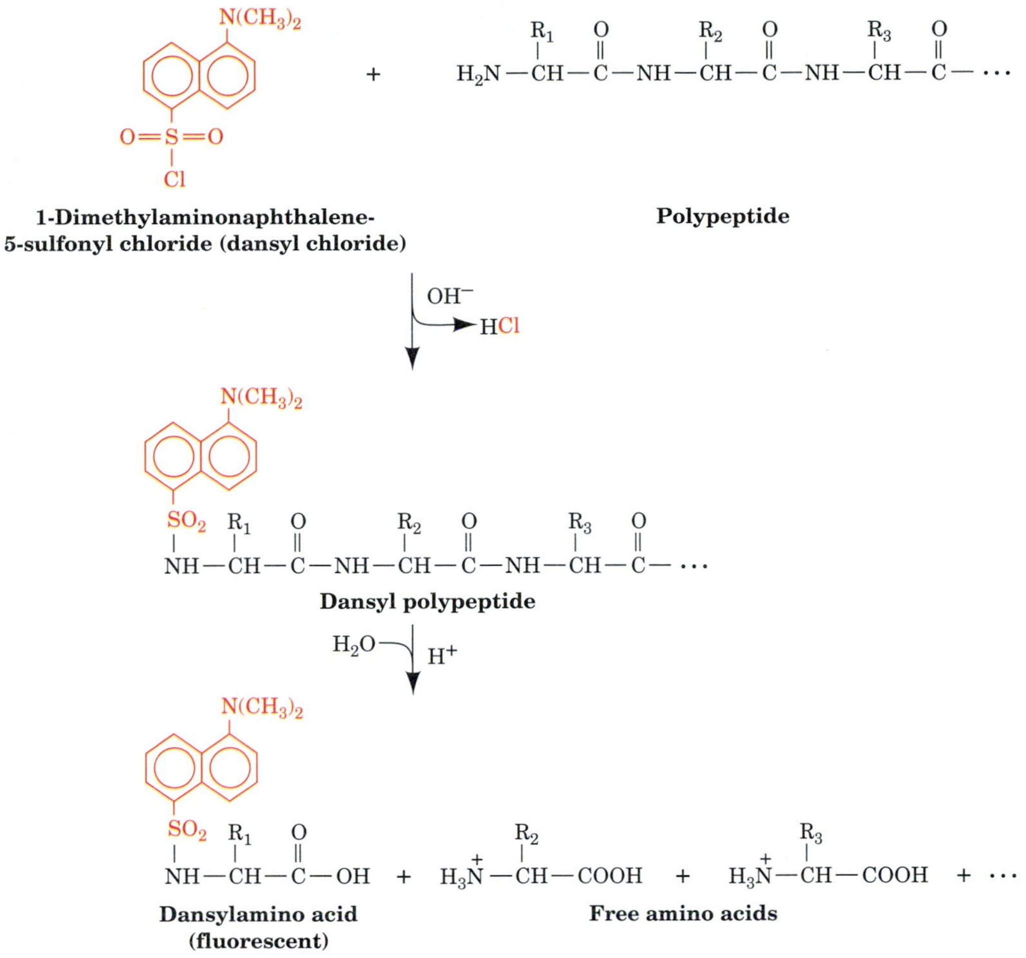

Figure 5-13. **The dansyl chloride reaction.** The reaction of dansyl chloride with primary amino groups is used for end group analysis.

Carboxypeptidases, like all enzymes, are highly specific (selective) for the chemical identities of their targets. For example, **carboxypeptidase A,** an intestinal digestive enzyme, does not cleave C-terminal Arg or Lys residues or residues that are next to Pro. **Carboxypeptidase B,** on the other hand, hydrolyzes only C-terminal Arg and Lys residues, but only if they are not preceded by Pro. Such specificity requires that the results of enzymatic end group analysis be treated with caution. Thus, if a carboxypeptidase cleaves the first residue slowly and the second residue quickly, its use may yield two amino acids, suggesting the presence of two different polypeptide chains.

Disulfide Bonds between and within Polypeptides Are Cleaved

Disulfide bonds between Cys residues must be cleaved to separate polypeptide chains—if they are disulfide-linked—and to ensure that polypeptide chains are fully linear (residues in polypeptides that are "knotted" with disulfide bonds may not be accessible to all the enzymes and reagents used for sequencing). Disulfide bonds can be cleaved oxidatively by **performic acid** or reduced by **mercaptans,** compounds containing —SH groups. Performic acid oxidation, which was pioneered by Sanger, converts all Cys residues, whether linked by S—S bridges or not, to **cysteic acid** residues:

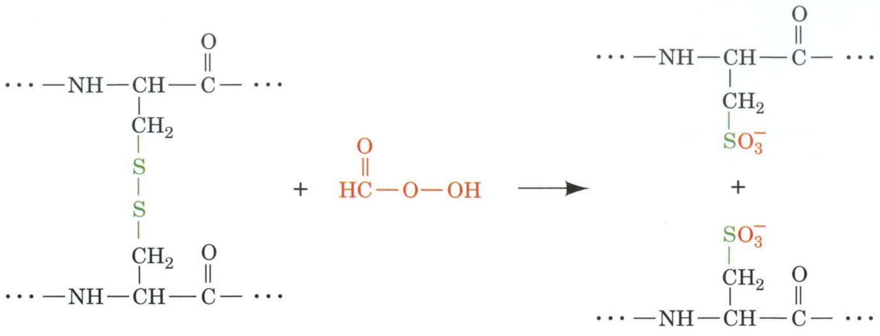

$$
\text{Cystine} \quad + \quad \text{Performic acid} \quad \longrightarrow \quad \text{Cysteic acid}
$$

A major disadvantage of performic acid treatment is that it also oxidizes Met residues and partially destroys the indole side chain of Trp.

Reductive cleavage of disulfide bonds is usually achieved by treatment with 2-mercaptoethanol or another mercaptan:

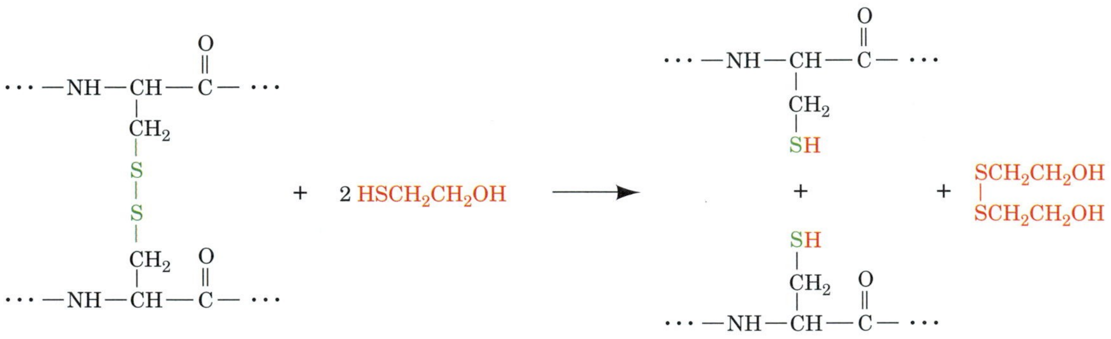

$$
\text{Cystine} \quad + \quad 2\ \text{HSCH}_2\text{CH}_2\text{OH} \quad \longrightarrow \quad \text{Cysteine}
$$

The resulting free sulfhydryl groups are then alkylated, usually by treatment with **iodoacetic acid,** to prevent the re-formation of disulfide bonds through oxidation by O_2:

$$
\underset{\text{Cysteine}}{\text{Cys}-\text{CH}_2-\text{SH}} \ + \ \underset{\text{Iodoacetate}}{\text{ICH}_2\text{COO}^-} \ \longrightarrow \ \underset{\textit{S}\text{-Carboxymethylcysteine}}{\text{Cys}-\text{CH}_2-\text{S}-\text{CH}_2\text{COO}^-} \ + \ \text{HI}
$$

The Amino Acid Composition of a Polypeptide May Be Determined

In some cases, it is desirable to know the **amino acid composition** of a polypeptide, that is, the number of each type of amino acid residue present. This information may provide clues to the protein's structure, but it is not required for determining its amino acid sequence.

The amino acid composition of a polypeptide is determined by its complete hydrolysis followed by the analysis of the liberated amino acids. Polypeptide hydrolysis can be accomplished by either chemical (acid or base) or enzymatic means, although none of these methods alone is fully satisfactory. For example, acid hydrolysis degrades Ser, Thr, Tyr, and Trp and converts Asn and Gln to Asp and Glu. Base hydrolysis destroys Cys, Ser, Thr, and Arg. Enzymatic hydrolysis of polypeptides is often incomplete and is complicated by the fact that the peptidases, being proteins themselves, are subject to proteolytic degradation, thereby contributing to the total amino acid content of the reaction mixture.

Figure 5-14. Amino acid analysis. The amino acids were derivatized with a fluorescent tag before their separation by HPLC. [After Hunkapiller, M.W., Strickler, J.E., and Wilson, K.J., *Science* **226**, 309 (1984).]

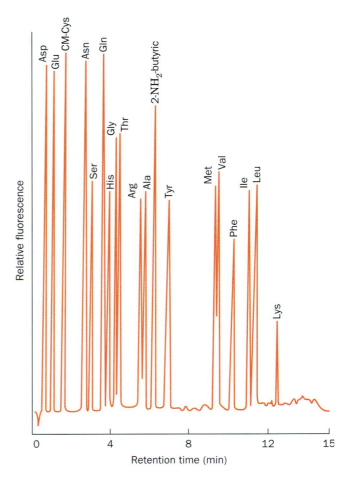

The quantitative analysis of the polypeptide hydrolysate is performed by an instrument that separates amino acids by chromatography and derivatizes them (either before or after chromatography) with an easily detected tag. The amino acids are then identified by their characteristic elution volumes (retention times on HPLC; Fig. 5-14) and quantified by their absorbance or fluorescence intensities. Modern amino acid analyzers can completely analyze a protein digest containing as little as 1 pmol of each amino acid in <1 h.

B. Polypeptide Cleavage

Polypeptides that are longer than 40 to 100 residues cannot be directly sequenced by the Edman degradation procedure and must therefore be cleaved, either enzymatically or chemically, to specific fragments that are small enough to be sequenced.

Various **endopeptidases** (enzymes that catalyze the hydrolysis of internal peptide bonds) can be used to fragment polypeptides. These enzymes, like exopeptidases, have side chain requirements for the residues flanking the scissile (to be cleaved) peptide bond (Table 5-5). The digestive enzyme **trypsin** has the greatest specificity and is therefore the most valuable member of the arsenal of peptidases used to fragment polypeptides. It cleaves peptide bonds on the C side (toward the carboxyl termi-

Table 5-5.   Specificities of Various Endopeptidases

$$-NH-CH-C \longrightarrow NH-CH-C-$$

with R_{n-1}, O above the first $CH-C$ and R_n, O above the second, pointing to the **Scissile peptide bond**

Enzyme	Source	Specificity	Comments
Trypsin	Bovine pancreas	R_{n-1} = positively charged residues: Arg, Lys; $R_n \neq$ Pro	Highly specific
Chymotrypsin	Bovine pancreas	R_{n-1} = bulky hydrophobic residues: Phe, Trp, Tyr; $R_n \neq$ Pro	Cleaves more slowly for R_{n-1} = Asn, His, Met, Leu
Elastase	Bovine pancreas	R_{n-1} = small neutral residues: Ala, Gly, Ser, Val; $R_n \neq$ Pro	
Thermolysin	*Bacillus thermoproteolyticus*	R_n = Ile, Met, Phe, Trp, Tyr, Val; $R_{n-1} \neq$ Pro	Occasionally cleaves at R_n = Ala, Asp, His, Thr; heat stable
Pepsin	Bovine gastric mucosa	R_n = Leu, Phe, Trp, Tyr; $R_{n-1} \neq$ Pro	Also others; quite nonspecific; pH optimum = 2
Endopeptidase V8	*Staphylococcus aureus*	R_{n-1} = Glu	

nus) of the positively charged residues Arg and Lys if the next residue is not Pro.

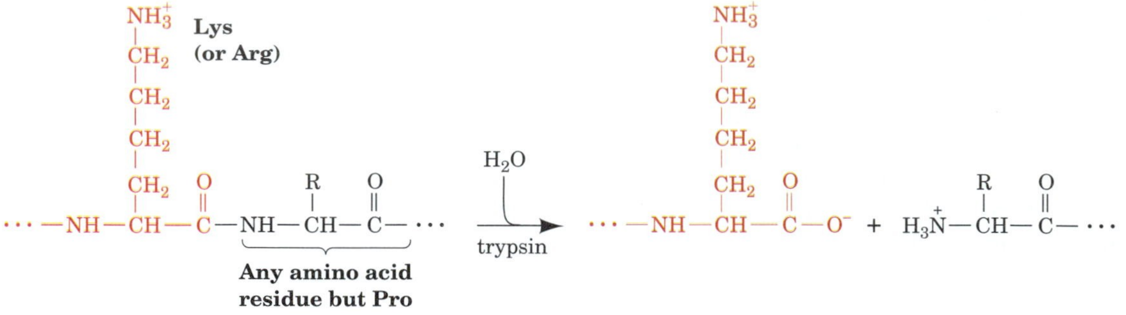

The other endopeptidases listed in Table 5-5 exhibit broader side chain specificities than trypsin and often yield a series of peptide fragments with overlapping sequences. However, through **limited proteolysis,** that is, by adjusting reaction conditions and limiting reaction times, these less specific endopeptidases can yield a set of discrete, nonoverlapping fragments.

Several chemical reagents promote peptide bond cleavage at specific residues. The most useful of these, **cyanogen bromide** (CNBr), cleaves on the C side of Met residues.

A peptide fragment generated by a specific cleavage process may still be too large to sequence. In that case, it is purified and then subjected to a second round of fragmentation using a different cleavage technique.

C. Edman Degradation

Once the peptide fragments formed through specific cleavage reactions have been isolated, their amino acid sequences can be determined. This is usually accomplished through repeated cycles of **Edman degradation.** In this process (named after its inventor, Pehr Edman), **phenylisothiocyanate** (**PITC**) reacts with the N-terminal amino group of a polypeptide under mildly alkaline conditions to form a **phenylthiocarbamyl** (**PTC**) adduct (Fig. 5-15). This product is treated with anhydrous **trifluoroacetic acid,** which cleaves the N-terminal residue as a **thiazolinone** derivative but does not hydrolyze other peptide bonds. Edman degradation therefore releases the N-terminal amino acid residue but leaves intact the rest of the polypeptide chain. The thiazolinone-amino acid is selectively extracted into an organic solvent and is converted to the more stable **phenylthiohydantoin** (**PTH**) derivative by treatment with aqueous acid. This PTH-amino acid can later be identified by chromatography. Thus, *it is possible to determine the amino acid sequence of a polypeptide chain from the N-terminus inward by subjecting the polypeptide to repeated cycles of Edman degradation and, after every cycle, identifying the newly liberated PTH-amino acid.*

The Edman degradation technique has been automated and refined, resulting in great savings of time and material. The first automated device, called a sequenator, was developed by Edman and Geoffrey Begg. In modern instruments, the peptide sample is dried onto a disk of glass fiber paper, and accurately measured quantities of reagents are delivered as vapors in a stream of argon at programmed intervals. Up to 100 residues can be identified before the cumulative effects of incomplete reactions, side reactions, and peptide loss make further amino acid identification unreliable. Since less than a picomole of a PTH-amino acid can be detected and identified, sequence analysis can be carried out on as little as 5 to 10 pmol of a peptide (<0.1 μg—an invisibly small amount).

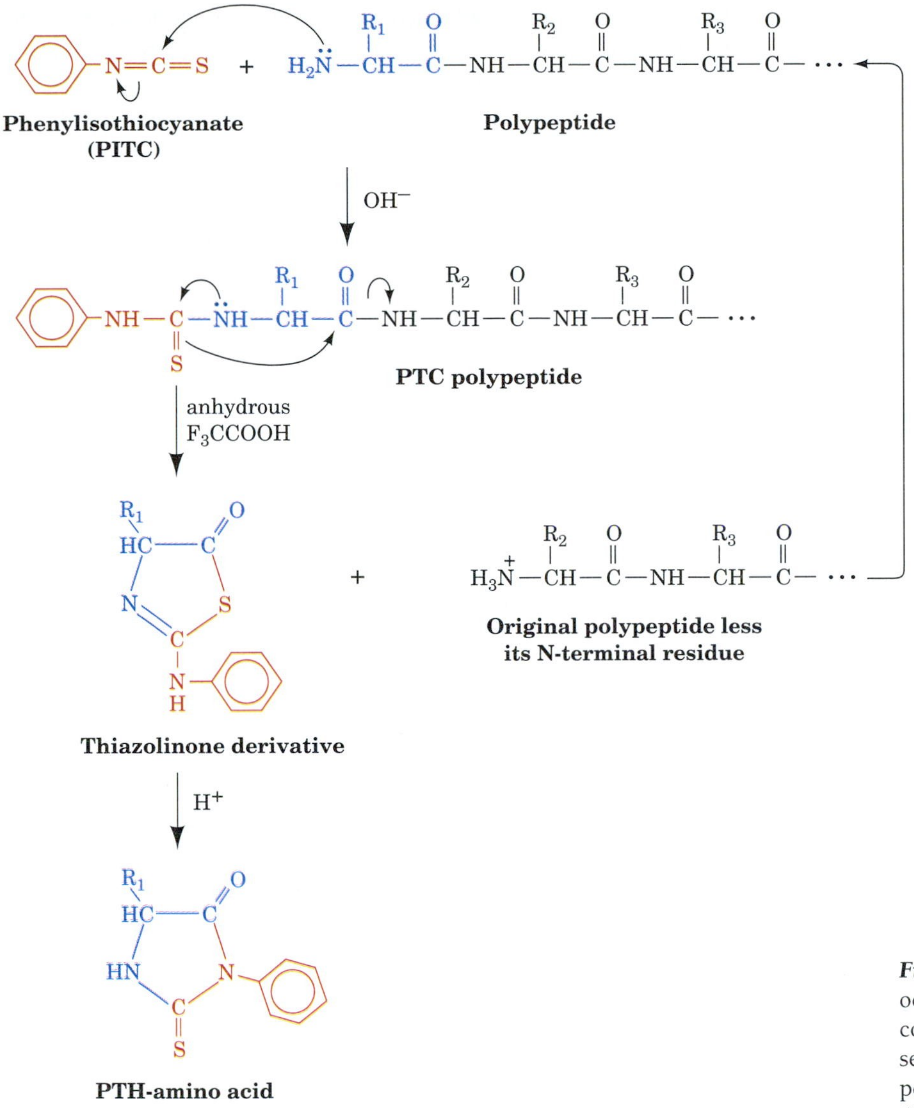

Figure 5-15. Edman degradation. The reaction occurs in three stages, each requiring different conditions. Amino acid residues can therefore be sequentially removed from the N-terminus of a polypeptide in a controlled stepwise fashion. ✳ See the Animated Figures.

D. Reconstructing the Protein's Sequence

After individual peptide fragments have been sequenced, their order in the original polypeptide must be elucidated. This is accomplished by conducting a second round of protein cleavage with a reagent of different specificity and then comparing the amino acid sequences of the overlapping sets of peptide fragments (Fig. 5-16).

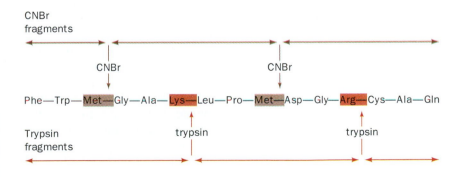

Figure 5-16. Generating overlapping fragments to determine the amino acid sequence of a polypeptide. In this example, two sets of overlapping peptide fragments are made by using trypsin to cleave the polypeptide after all its Arg and Lys residues and, in a separate reaction, using CNBr to cleave it after all its Met residues. ✳ See the Animated Figures.

The final step in an amino acid sequence analysis is to determine the positions (if any) of the disulfide bonds. This can be done by cleaving a sample of the protein, with its disulfide bonds intact, to yield pairs of peptide fragments, each containing a single Cys, that are linked by a disulfide bond. After isolating a disulfide-linked polypeptide fragment, the disulfide bond is cleaved and alkylated (Section 5-3A), and the sequences of the two peptides are determined (Fig. 5-17). The various pairs of such polypeptide fragments are identified by comparing their sequences with that of the protein, thereby establishing the locations of the disulfide bonds.

After a protein's amino acid sequence has been determined, the information is customarily deposited in a public database such as Swiss-Prot. Databases for nucleotide sequences, from which amino acid sequences can be inferred, include GenBank. These databases, as well as others with more specialized contents, are accessible via the World Wide Web (http://expasy.hcuge.ch for Swiss-Prot and http://www.ncbi.nlm.nih.gov for GenBank). Electronic links between databases allow rapid updates and cross-checking of sequence information.

Amino acid sequence information is no less valuable when the nucleotide sequence of the corresponding gene is also known, because the protein sequence sometimes provides information about protein structure that is not revealed by nucleic acid sequencing (see Box 5-2). Armed with the appropriate software (which is often publically available at the database sites), a researcher can search a database to find proteins with similar sequences in various organisms. The sequence of even a short peptide fragment may be sufficient to "fish out" the parent protein or its homolog from another species.

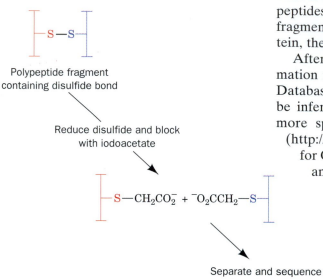

Figure 5-17.   Determining the positions of disulfide bonds. In this method, disulfide-linked peptide fragments from a protein are reduced and separately sequenced to identify the positions of the disulfide bonds in the intact protein.

4. PROTEIN EVOLUTION

An organism's genetic material specifies the amino acid sequences of all its proteins. Changes in genes due to random mutation often alter a protein's primary structure. A mutation in a protein is propagated only if it somehow increases, or at least does not decrease, the probability that its owner will survive to reproduce. Many mutations are deleterious or produce lethal effects and therefore rapidly die out. On rare occasions, however, a mutation arises that improves the fitness of its host. This is the essence of **Darwinian evolution.**

> ● See Guided Exploration 5:
>
> Protein Evolution.

A. Protein Sequence Evolution

The primary structures of a given protein from related species closely resemble one another. Consider **cytochrome *c*,** a protein found in nearly all eukaryotes. Cytochrome *c* is a component of the mitochondrial electron-transport system (Section 17-2), which is believed to have taken its present form between 1.5 and 2 billion years ago, when organisms evolved the ability to respire. Emanuel Margoliash, Emil Smith, and others have elucidated the amino acid sequences of the cytochromes *c* from over 100 eukaryotic species ranging in complexity from yeast to humans. The cytochromes *c* from different species are single polypeptides of 104 to 112 residues. The sequences of 38 of these proteins are arranged in Table 5-6 to show the similarities between vertically aligned residues (the residues have been color-coded according to their physical properties). A survey of the aligned sequences (bottom line of Table 5-6) shows that at 38 positions (23 posi-

Box 5-2
BIOCHEMISTRY IN CONTEXT

Protein versus Nucleic Acid Sequencing

Although it is often easier to deduce a protein's sequence by identifying and sequencing the corresponding gene, direct protein sequencing is an indispensable tool for several reasons. For example, only direct protein sequencing can reveal the locations of disulfide bonds in proteins. In addition, many proteins are modified after they are synthesized. For example, certain residues may be excised to produce the "mature" protein (the 51-residue insulin, shown in Fig. 5-1, is actually synthesized as an 84-residue polypeptide that is proteolytically processed to its smaller two-chain form). Amino acid side chains may also be modified by the addition of carbohydrates, phosphate groups, or acetyl groups, to name only a few. Although some of these modifications occur at characteristic amino acid sequences and are therefore identifiable in nucleotide sequences, only the actual protein sequence can confirm whether and where they actually occur.

Any biochemical procedure, no matter how carefully executed, occasionally produces errors. Nucleotide sequencing errors may have greater repercussions than errors in protein sequencing: Because the genetic code consists of consecutive triplets of nucleotides, each specifying a single amino acid residue, the inadvertent insertion or deletion of a single nucleotide, a common error, shifts the "reading frame" of the gene and thereby alters the predicted amino acid sequence from that point on. Thus, the correspondence between protein and nucleic acid sequences makes it possible to use one to verify the accuracy of the other. However, anomalies sometimes indicate exceptions to the "universal" genetic code; for example, mitochondria and certain protozoa translate DNA sequences into a somewhat different sequence of amino acids than do most organisms (Section 26-1C). In some cases, discrepancies between nucleic acid and protein sequences simply reflect the natural heterogeneity of a protein in a population.

---

tions in the complete set of >100 sequences), the same amino acid appears in all species. Most of the remaining positions are occupied by chemically similar residues in different organisms. In only eight positions does the sequence accommodate six or more different residues.

According to evolutionary theory, *related species have evolved from a common ancestor, so it follows that the genes specifying each of their proteins must likewise have evolved from the corresponding gene in that ancestor.* The sequence of the ancestral cytochrome *c* is accessible only indirectly, by examining the sequences of extant proteins. Cytochrome *c* is an **evolutionarily conservative** protein; that is, its sequence has undergone only modest evolutionary changes.

Sequence Comparisons Provide Information on Protein Structure and Function

*In general, comparisons of the primary structures of **homologous proteins** (evolutionarily related proteins) indicate which of the protein's residues are essential to its function, which are less significant, and which have little specific function.* For example, finding the same residue at a particular position in the amino acid sequence of a series of related proteins suggests that the chemical or structural properties of that so-called **invariant residue** uniquely suit it to some essential function of the protein. Other amino acid positions may have less stringent side chain requirements and can therefore accommodate residues with similar characteristics (e.g., Asp or Glu, Ser or Thr, etc.); such positions are said to be **conservatively substituted.** On the other hand, a particular amino acid position may tolerate many different amino acid residues, indicating that the functional requirements of that position are rather nonspecific. Such a position is said to be **hypervariable.**

Why is cytochrome *c*—an ancient and essential protein—not identical in all species? Even a protein that is well adapted to its function, that is,

Table 5-6. **Amino Acid Sequences of Cytochromes *c* from 38 Species**[a]

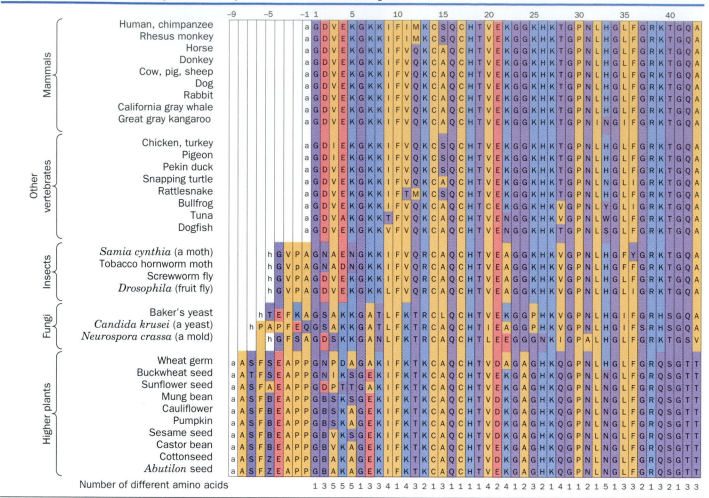

[a]The amino acid side chains have been shaded according to their polarity characteristics so that an invariant or conservatively substituted residue is identified by a vertical band of a single color. The letter a at the beginning of the chain indicates that the N-terminal amino group is acetylated; an h indicates that the acetyl group is absent.

Source: After Dickerson, R.E., *Sci. Am.* **226**(4); 58–72 (1972), with corrections from Dickerson, R.E., and Timkovich, R., *in* Boyer, P.D. (Ed.), *The Enzymes* (3rd ed.), Vol. 11, pp. 421–422, Academic Press (1975). Table copyrighted © by Irving Geis.

one that is not subject to physiological improvement, nevertheless continues evolving. *The random nature of mutational processes will, in time, change such a protein in ways that do not significantly affect its function, a process called* **neutral drift** (deleterious mutations are, of course, rapidly rejected through natural selection). Hypervariable residues are apparently particularly subject to neutral drift.

Constructing Phylogenetic Trees

Far-reaching conclusions about evolutionary relationships can be drawn by comparing the amino acid sequences of homologous proteins. The simplest way to assess evolutionary differences is to count the amino acid differences between proteins. For example, the data in Table 5-6 show that primate cytochromes *c* more nearly resemble those of other mammals than

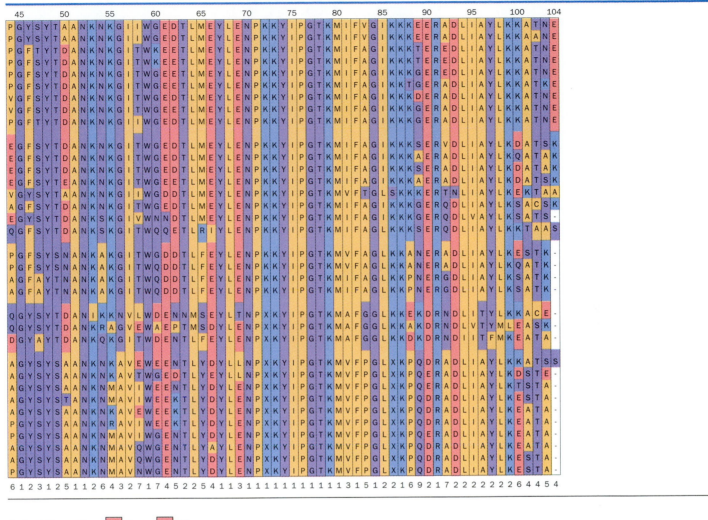

Hydrophilic, acidic: D Asp   E Glu

Hydrophilic, basic: H His   K Lys   R Arg   X TrimethylLys

Polar, uncharged: B Asn or Asp   G Gly   N Asn   Q Gln
                     S Ser   T Thr   W Trp   Y Tyr   Z Gln or Glu

Hydrophobic: A Ala   C Cys   F Phe   I Ile   L Leu
               M Met   P Pro   V Val

they do those of insects (8–12 differences among mammals versus 26–31 differences between mammals and insects). Similarly, the cytochromes *c* of fungi differ as much from those of mammals (45–51 differences) as they do from those of insects (41–47) or higher plants (47–54). The order of these differences largely parallels that expected from classical taxonomy.

The sequences of homologous proteins can be analyzed by computer to construct a **phylogenetic tree,** a diagram that indicates the ancestral relationships among organisms that produce the protein. The phylogenetic tree

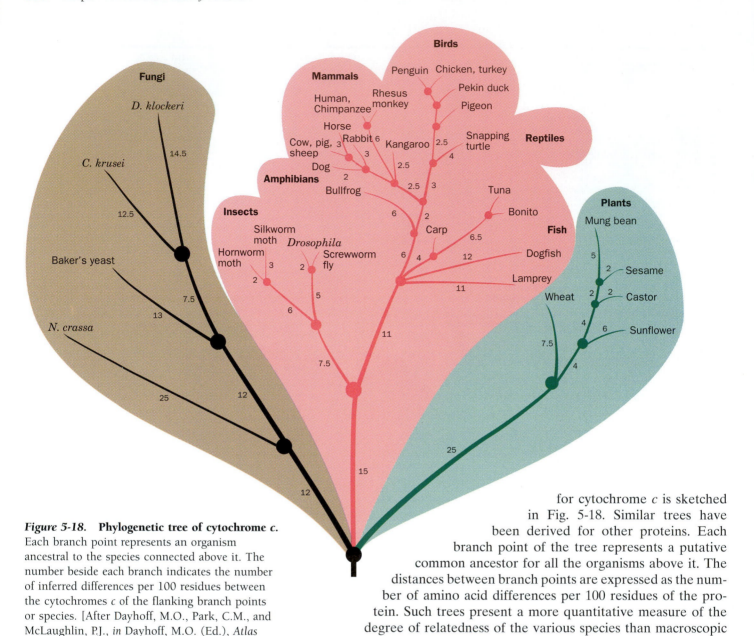

Figure 5-18.   Phylogenetic tree of cytochrome c.
Each branch point represents an organism
ancestral to the species connected above it. The
number beside each branch indicates the number
of inferred differences per 100 residues between
the cytochromes c of the flanking branch points
or species. [After Dayhoff, M.O., Park, C.M., and
McLaughlin, P.J., *in* Dayhoff, M.O. (Ed.), *Atlas
of Protein Sequence and Structure,* p. 8, National
Biomedical Research Foundation (1972).]

for cytochrome *c* is sketched
in Fig. 5-18. Similar trees have
been derived for other proteins. Each
branch point of the tree represents a putative
common ancestor for all the organisms above it. The
distances between branch points are expressed as the number of amino acid differences per 100 residues of the protein. Such trees present a more quantitative measure of the
degree of relatedness of the various species than macroscopic
taxonomy can provide.

Note that the evolutionary distances from all modern cytochromes *c* to
the lowest point, the earliest common ancestor producing this protein, are
approximately the same. Thus, "lower" organisms do not represent life
forms that appeared early in history and ceased to evolve further. The cytochromes *c* of all the species included in Fig. 5-18—whether called "primitive" or "advanced"—have evolved to about the same extent.

Proteins Evolve at Characteristic Rates

The protein sequence differences between various species can be plotted against the time when, according to the fossil record, the species diverged. The plot for a given protein is essentially linear, indicating that its
mutations accumulate at a constant rate over a geological time scale. However, rates of evolution vary among proteins (Fig. 5-19). This does not imply that the rates of mutation of the DNAs specifying those proteins differ, but rather that *the rate at which mutations are accepted into a protein
depends on the extent to which amino acid changes affect the protein's func-*

tion. For example, **histone H4,** a protein that binds to DNA in eukaryotes (Section 23-5A), is among the most highly conserved proteins (the histones H4 from peas and cows, species that diverged 1.2 billion years ago, differ by only two conservative changes in their 102 residues). Evidently, histone H4 is so important for packaging DNA in cells that it is extremely intolerant of any mutations. Cytochrome *c* is only slightly more tolerant. It is a relatively small protein that binds to several other proteins. Hence, any changes in its amino acid sequence must be compatible with all its binding partners. Hemoglobin, which functions as a free-floating molecule, is subject to less selective pressure than histone H4 or cytochrome *c*, so its surface residues are more easily substituted by other amino acids. The **fibrinopeptides** are ~20-residue fragments that are cleaved from the vertebrate blood protein **fibrinogen** to induce blood clotting. Once they have been removed, the fibrinopeptides are discarded, so there is little selective pressure to maintain their amino acid sequences.

Of course, protein sequences alone cannot reveal the complete story of evolution. Slight differences in protein sequence are not enough to account for the sometimes dramatic morphological features that differentiate even closely related species. Thus, the proteins of humans and chimpanzees are >99% identical on average (e.g., their cytochromes *c* are identical), but their anatomical and behavioral differences are so great that they are classified in separate families. Large evolutionary steps that mark the divergence of species are probably accomplished by mutations in the segments of DNA that control gene expression, that is, how much of each protein is made, where, and when. Other changes in DNA, such as rearrangements and duplication of genes, give rise to new proteins that are then subject to natural selection.

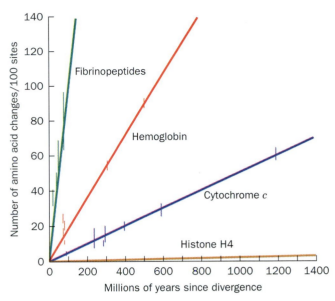

Figure 5-19. Rates of evolution of four proteins. The graph was constructed by plotting the number of different amino acid residues in the proteins on two sides of a branch point of a phylogenetic tree versus the time, according to the fossil record, since the corresponding species diverged from their common ancestor. [Figure copyrighted © by Irving Geis.]

B. Gene Duplication and Protein Families

It is not surprising that proteins with similar functions have similar sequences; such proteins presumably evolved from a common ancestor. Interestingly, protein sequence analysis has revealed that some proteins with widely different physiological functions also have similar sequences of amino acids. In fact, most proteins have extensive sequence similarities with other proteins from the same organism (we will see in the next chapter that three-dimensional protein structures are likewise highly conserved). Such proteins arose through **gene duplication,** an aberrant genetic recombination event in which one chromosome acquired both copies of the primordial gene (genetic recombination is discussed in Section 24-6). *Gene duplication is a particularly efficient mode of evolution because one copy of the gene can evolve a new function through natural selection while its counterpart continues to direct the synthesis of the original protein.*

The **globin family** of proteins provides an excellent example of evolution through gene duplication. Hemoglobin, which transports O_2 from the lungs (or gills or skin) to the tissues, is a tetramer with the subunit composition $\alpha_2\beta_2$ (i.e., two α polypeptides and two β polypeptides). The sequences of the α and β subunits are similar to each other and to the sequence of the protein **myoglobin,** which facilitates oxygen diffusion through muscle tissue (hemoglobin and myoglobin are discussed in more detail in Chapter 7). The primordial globin probably functioned simply as an oxygen-storage protein. Duplication allowed one globin to evolve into a

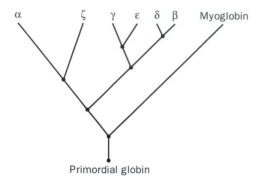

Figure 5-20.   Genealogy of the globin family.
Each branch point represents a gene duplication
event. Myoglobin is a single-chain protein. The
globins identified by Greek letters are subunits of
hemoglobins.

monomeric hemoglobin α chain. Duplication of the α chain gave rise to the
β chain. Other members of the globin family include the β-like γ chain that
is present in fetal hemoglobin, an $\alpha_2\gamma_2$ tetramer, and the β-like ε and α-
like ζ chains that appear together early in embryogenesis as $\zeta_2\varepsilon_2$ hemoglobin. Primates contain a relatively recently duplicated globin, the β-like δ
chain, which appears as a minor component ($\sim$1%) of adult hemoglobin.
Although the $\alpha_2\delta_2$ hemoglobin has no known unique function, perhaps it
may eventually evolve one. The genealogy of the members of the globin
family is diagrammed in Fig. 5-20. The human genome also contains the
relics of globin genes that are not expressed (Section 27-1B). These **pseudogenes** can be considered the dead ends of protein evolution.

C. Protein Modules

Gene duplication is not the only mechanism for generating new proteins.
Analysis of protein sequences has revealed that many proteins are mosaics
of sequence motifs, or **modules,** of about 40–100 amino acid residues. In
some large proteins, a single module is repeated many times (Fig. 5-21*a*);
other proteins contain a variety of modules. Some of the proteins involved
in blood clotting, for example, are built from a set of smaller modules (Fig.
5-21*b*). The sequence homology between modules is imperfect, since the
sequence of a given protein evolves independently of other sequences. The
functions of individual modules are not always known: Some appear to have
discrete activities, such as catalyzing a certain type of chemical reaction or
binding a particular molecule, but others may merely be spacers or scaffolding for other modules.

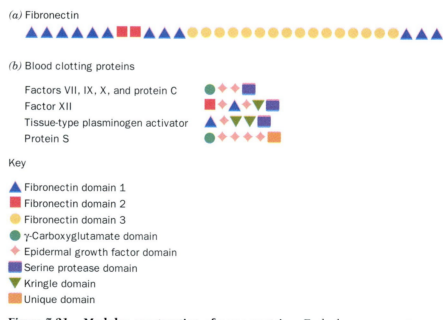

Figure 5-21.   Modular construction of some proteins. Each shape represents a
segment of $\sim$40–100 residues that appears, with some sequence variation, several
times in the same protein or in a number of related proteins. (*a*) **Fibronectin,** a
protein of the extracellular matrix, is composed mostly of repeated modules of three
types. (*b*) Some of the proteins that participate in blood clotting are built from a
small set of modules (the epidermal growth factor domain is so named because it
was first observed as a component of **epidermal growth factor**). [After Baron, M.,
Norman, D.G., and Campbell, I.D., *Trends Biochem. Sci.* **16,** 14 (1991).]

Generating new genes by shuffling modules is a much faster process than duplicating an entire gene and allowing it to mutate over time. Nevertheless, both mechanisms have played important roles in the evolution of species. The appearance of a protein with a novel sequence is a rare event in biology; most proteins are variations on other proteins or parts thereof.

SUMMARY

1. The properties of proteins depend largely on the sizes and sequences of their component polypeptides.

2. Fractionation procedures are used to purify proteins on the basis of solubility, charge, size, and binding specificity.

3. A purified protein must be stabilized.

4. Differences in solubility permit proteins to be concentrated and purified by salting out.

5. Chromatography, the separation of soluble substances by their rates of movement through an insoluble matrix, is a technique for purifying molecules by charge (ion exchange chromatography), hydrophobicity (hydrophobic interaction chromatography), size (gel filtration chromatography), and binding properties (affinity chromatography). Binding and elution often depend on the salt concentration and the pH.

6. Electrophoresis separates molecules by charge and size; SDS-PAGE separates them primarily by size.

7. Proteins can be separated by mass in an ultracentrifuge.

8. Analysis of a protein's sequence begins with end group analysis, to determine the number of different subunits, and the cleavage of disulfide bonds. The amino acid composition may also be determined.

9. Polypeptides are cleaved into fragments suitable for sequencing by the Edman degradation, in which residues are removed, one at a time, from the N-terminus.

10. A protein's sequence is reconstructed from the sequences of overlapping peptide fragments and from information about the locations of disulfide bonds.

11. Proteins evolve through changes in primary structure. Comparisons of proteins in different species may reveal which residues are most essential for a protein's structure and function.

12. Comparisons of polypeptide sequences also reveal the evolutionary relationships between species and between different proteins within a species.

REFERENCES

Protein Purification

Arakawa, T. and Timasheff, S.N., Theory of protein solubility, *Methods Enzymol.* **114,** 49–77 (1985).

Creighton, T.E. (Ed.), *Protein Structure. A Practical Approach,* IRL Press (1989). [Chapters 1–3 describe various electrophoretic methods. Chapters 6 and 7 discuss amino acid composition and identifying disulfide bonds in proteins.]

Freifelder, D., *Physical Biochemistry: Applications to Biochemistry and Molecular Biology* (2nd ed.), Freeman (1982). [Chapters 6, 8, 9, 11, and 14 include the theoretical underpinnings and practical applications of autoradiography, chromatography, electrophoresis, sedimentation, and spectroscopy, respectively.]

Hames, B.D. and Rickwood, D. (Eds.), *Gel Electrophoresis of Proteins. A Practical Approach* (2nd ed.), IRL Press (1990).

Scopes, R., *Protein Purification: Principles and Practice* (3rd ed.), Springer-Verlag (1994).

Wilson, K. and Walker, J.M. (Eds.), *Principles and Techniques of Practical Biochemistry,* Cambridge University Press (1994). [Includes reviews of centrifugation, spectroscopy, electrophoresis, and chromatography.]

Protein Sequencing

Benson, D.A., Boguski, M.S., Lipman, D.J., Ostell, J., and Ouellette, B.F., GenBank, *Nucleic Acids Res.* **26,** 1–7 (1998).

[Describes how protein and nucleic acid sequences can be retrieved from databases.]

Blackman, D.S., *The Logic of Biochemical Sequencing,* CRC Press (1994). [Question-and-answer format that provides an overview of strategies and techniques for sequencing proteins and nucleic acids.]

Findlay, J.B.C. and Geisow, M.J. (Eds.), *Protein Sequencing. A Practical Approach,* IRL Press (1989).

Gallop, M.A., Barrett, R.W., Dower, W.J., Fodor, S.P.A., and Gordon, E.M., Application of combinatorial technologies to drug discovery. 1. Background and peptide combinatorial libraries, *J. Med. Chem.* **37,** 1233–1251 (1994).

Sanger, F., Sequences, sequences, sequences, *Annu. Rev. Biochem.* **57,** 1–28 (1988). [A scientific autobiography that provides a glimpse of the early difficulties in sequencing proteins.]

Protein Evolution

Doolittle, R.F., The multiplicity of domains in proteins, *Annu. Rev. Biochem.* **64,** 287–314 (1995). [Summarizes the theory of domain shuffling.]

Doolittle, R.F., Feng, D.-F., Tsang, S., Cho, G., and Little, E., Determining divergence times of the major kingdoms of living organisms with a protein clock, *Science* **271,** 470–477 (1996). [Demonstrates how protein sequences can be used to draw phylogenetic trees.]

KEY TERMS

primary structure	mobile phase	autoradiograph	limited proteolysis
peptide	stationary phase	SDS-PAGE	Edman degradation
subunit	HPLC	capillary electrophoresis	Darwinian evolution
multisubunit protein	ion exchange chroma-	ultracentrifugation	homologous proteins
denaturation	tography	Svedberg (S)	conservative substitution
nuclease	anion exchanger	analytical ultracentrifuge	invariant residue
protease	cation exchanger	preparative ultracentrifuge	hypervariable residue
assay	polyelectrolyte	density gradient	neutral drift
coupled enzymatic reaction	elution	zonal ultracentrifugation	phylogenetic tree
immunoassay	eluant	equilibrium density gradient	gene duplication
antibody	hydrophobic interaction	centrifugation	pseudogene
antigen	chromatography	end group analysis	module
RIA	gel filtration	endopeptidase	
ELISA	affinity chromatography	carboxypeptidase	
fractionation procedure	ligand	aminopeptidase	
salting in	immunoaffinity chroma-	exopeptidase	
salting out	tography	mercaptan	
chromatography	immunoblot (Western blot)	amino acid composition	

STUDY EXERCISES

1. List the 20 amino acids in order of their abundance in proteins.

2. List some factors that influence the stability of purified proteins.

3. List the separation techniques that exploit the following molecular properties: charge, polarity, size, and specificity.

4. How does ion exchange chromatography differ from affinity chromatography?

5. What kinds of information can be gathered by ultracentrifugation?

6. How can the N- and the C-terminal residues of proteins be identified?

7. Why should disulfide bonds be cleaved before protein sequencing?

8. Why might it be helpful to know a protein's amino acid composition before trying to sequence it?

9. Describe the steps of Edman degradation.

10. List some advantages of automating laboratory procedures.

11. What information is provided by comparing the sequences of proteins from different organisms?

12. How is a phylogenetic tree constructed?

PROBLEMS

1. (a) In what order would the amino acids Arg, His, and Leu be eluted from a carboxymethyl column at pH 6? (b) In what order would Glu, Lys, and Val be eluted from a diethylaminoethyl column at pH 8?

2. Which peptide has greater absorbance at 280 nm?

    A. Gln-Leu-Glu-Phe-Thr-Leu-Asp-Gly-Tyr

    B. Ser-Val-Trp-Asp-Phe-Gly-Tyr-Trp-Ala

3. Explain why a certain protein has an apparent molecular mass of 90 kD when determined by gel filtration and 60 kD when determined by SDS-PAGE in the presence or absence of 2-mercaptoethanol. Which molecular mass determination is more accurate?

4. Determine the subunit composition of a protein from the following information:

Molecular mass by gel filtration: 200 kD

Molecular mass by SDS-PAGE: 100 kD

Molecular mass by SDS-PAGE with 2-mercaptoethanol: 40 kD and 60 kD

5. Explain why a protein, which has a sedimentation coefficient of 2.6S when ultracentrifuged in a solution containing 0.1 M NaCl, has a sedimentation coefficient of 4.3S in a solution containing 1 M NaCl.

6. Explain why the dansyl chloride treatment of a single polypeptide chain followed by its complete acid hydrolysis yields several dansylated amino acids.

7. Identify the first residue obtained by Edman degradation of cytochrome *c* from (a) *Drosophila,* (b) baker's yeast, and (c) wheat germ.

8. Cleavage of a polypeptide by CNBr and chymotrypsin yields fragments with the following amino acid sequences. What is the sequence of the intact polypeptide?

CNBr treatment	*Chymotrypsin*
1. Arg-Ala-Tyr-Gly-Asn	4. Met-Arg-Ala-Tyr
2. Leu-Phe-Met	5. Asp-Met-Leu-Phe
3. Asp-Met	6. Gly-Asn

9. Treatment of a polypeptide with 2-mercaptoethanol yields two polypeptides:

   1. Ala-Val-Cys-Arg-Thr-Gly-Cys-Lys-Asn-Phe-Leu

   2. Tyr-Lys-Cys-Phe-Arg-His-Thr-Lys-Cys-Ser

   Treatment of the intact polypeptide with trypsin yields fragments with the following amino acid compositions:

   3. (Ala, Arg, Cys$_2$, Ser, Val)

   4. (Arg, Cys$_2$, Gly, Lys, Thr, Phe)

   5. (Asn, Leu, Phe)

   6. (His, Lys, Thr)

   7. (Lys, Tyr)

   Indicate the positions of the disulfide bonds in the intact polypeptide

10. Why has the oxidative cleavage of disulfide bonds preparatory to amino acid sequencing fallen from use?

11. What fractionation procedure could be used to purify protein 1 from a mixture of three proteins whose amino acid compositions are:

    1. 25% Ala, 20% Gly, 20% Ser, 10% Ile, 10% Val, 5% Asn, 5% Gln, 5% Pro

    2. 30% Gln, 25% Glu, 20% Lys, 15% Ser, 10% Cys

    3. 25% Asn, 20% Gly, 20% Asp, 20% Ser, 10% Lys, 5% Tyr

    All three proteins are similar in size and p*I*, and there is no antibody available for protein 1. [Problem provided by Bruce Wightman, Muhlenberg College.]

12. You wish to sequence the light chain of a protease inhibitor from the *Brassica nigra* plant. Cleavage of the light chain by trypsin and chymotrypsin yields the following fragments. What is the sequence of the light chain?

Chymotrypsin

   1. Leu-His-Lys-Gln-Ala-Asn-Gln-Ser-Gly-Gly-Gly-Pro-Ser

   2. Gln-Gln-Ala-Gln-His-Leu-Arg-Ala-Cys-Gln-Gln-Trp

   3. Arg-Ile-Pro-Lys-Cys-Arg-Lys-Phe

Trypsin

   4. Arg

   5. Ala-Cys-Gln-Gln-Trp-Leu-His-Lys

   6. Cys-Arg

   7. Gln-Ala-Asn-Gln-Ser-Gly-Gly-Gly-Pro-Ser

   8. Phe-Gln-Gln-Ala-Gln-His-Leu-Arg

   9. Ile-Pro-Lys

   10. Lys

[Problem provided by Kathleen Cornely, Providence College.]

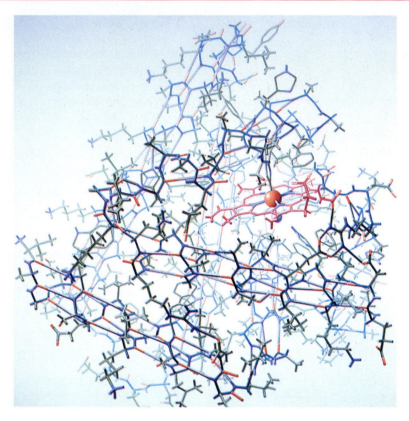

The atomic structure of myoglobin, an oxygen binding protein, is drawn here as a stick model. The overall conformation of a protein such as myoglobin is a function of its amino acid sequence. How do noncovalent forces act on a polypeptide chain to stabilize its unique three-dimensional arrangement of atoms? [Figure copyrighted © by Irving Geis.]

PROTEINS: THREE-DIMENSIONAL STRUCTURE

For many years, it was thought that proteins were colloids of random structure and that the enzymatic activities of certain crystallized proteins were due to unknown entities associated with an inert protein carrier. In 1934, J.D. Bernal and Dorothy Crowfoot Hodgkin showed that a crystal of the protein **pepsin** yielded a discrete diffraction pattern when placed in an X-ray beam. This result provided the first evidence that pepsin was not a random colloid but an ordered array of atoms organized into a large yet uniquely structured molecule.

Even relatively small proteins contain thousands of atoms, almost all of which occupy definite positions in space. The first X-ray structure of a protein, that of sperm whale myoglobin, was reported in 1958 by John Kendrew and co-workers. At the time—only 5 years after James Watson and Francis Crick had elucidated the simple and elegant structure of DNA (Section 3-2B)—protein chemists were chagrined by the complexity and apparent lack of regularity in the structure of myoglobin. In retrospect, such irregularity seems essential for proteins to fulfill their diverse biological roles. However, comparisons of the ~7000 protein structures now known have revealed that proteins actually exhibit a remarkable degree of structural regularity.

As we saw in Section 5-1, the primary structure of a protein is its linear sequence of amino acids. In discussing protein structure, three further levels of structural complexity are customarily invoked:

- **Secondary structure** is the local spatial arrangement of a polypeptide's backbone atoms without regard to the conformations of its side chains.
- **Tertiary structure** refers to the three-dimensional structure of an entire polypeptide.
- Many proteins are composed of two or more polypeptide chains, loosely referred to as subunits. A protein's **quaternary structure** refers to the spatial arrangement of its subunits.

The four levels of protein structure are summarized in Fig. 6-1.

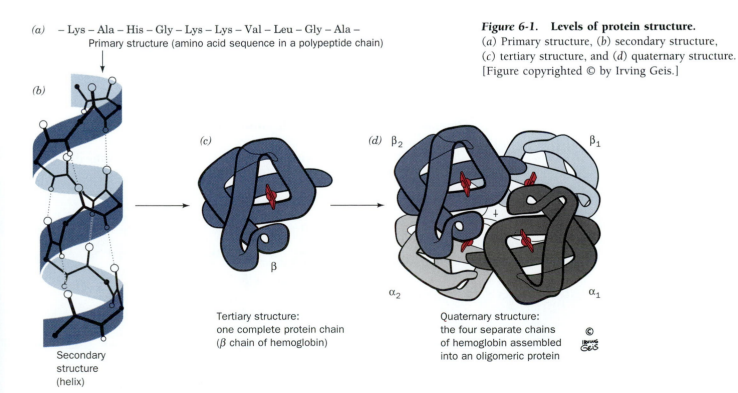

(a)   – Lys – Ala – His – Gly – Lys – Lys – Val – Leu – Gly – Ala –
Primary structure (amino acid sequence in a polypeptide chain)

(b)

(c)

(d) β₂   β₁

β

α₂   α₁

Secondary
structure
(helix)

Tertiary structure:
one complete protein chain
(β chain of hemoglobin)

Quaternary structure:
the four separate chains
of hemoglobin assembled
into an oligomeric protein

Figure 6-1.   **Levels of protein structure.**
(*a*) Primary structure, (*b*) secondary structure, (*c*) tertiary structure, and (*d*) quaternary structure. [Figure copyrighted © by Irving Geis.]

In this chapter, we explore secondary through quaternary structure, including examples of proteins that illustrate each of these levels. We also introduce methods for determining three-dimensional molecular structure and discuss the forces that stabilize folded proteins.

1. SECONDARY STRUCTURE

Protein secondary structure includes the regular polypeptide folding patterns such as helices, sheets, and turns. However, before we discuss these basic structural elements, we must consider the geometric properties of peptide groups, which underlie all higher order structures.

A. The Peptide Group

In the 1930s and 1940s, Linus Pauling and Robert Corey determined the X-ray structures of several amino acids and dipeptides in an effort to elucidate the conformational constraints on a polypeptide chain. These studies indicated that *the peptide group has a rigid, planar structure as a consequence of resonance interactions that give the peptide bond ~40% double-bond character:*

This explanation is supported by the observations that a peptide group's C—N bond is 0.13 Å shorter than its N—C_α single bond and that its C=O bond is 0.02 Å longer than that of aldehydes and ketones. The planar conformation maximizes π-bonding overlap, which accounts for the peptide group's rigidity.

Peptide groups, with few exceptions, assume the **trans conformation,** in which successive C_α atoms are on opposite sides of the peptide bond joining them (Fig. 6-2). The **cis conformation,** in which successive C_α atoms are on the same side of the peptide bond, is ~8 kJ·mol^{-1} less stable than the trans conformation because of steric interference between neighboring side chains. However, this steric interference is reduced in peptide bonds to Pro residues, so *~10% of the Pro residues in proteins follow a cis peptide bond.*

Torsion Angles between Peptide Groups Describe Polypeptide Chain Conformations

The **backbone** or **main chain** of a protein refers to the atoms that participate in peptide bonds, ignoring the side chains of the amino acid

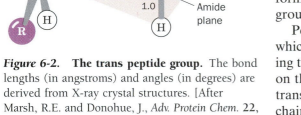

Figure 6-2.   The trans peptide group. The bond lengths (in angstroms) and angles (in degrees) are derived from X-ray crystal structures. [After Marsh, R.E. and Donohue, J., *Adv. Protein Chem.* **22,** 249 (1967).] ● See Kinemage Exercise 3-1.

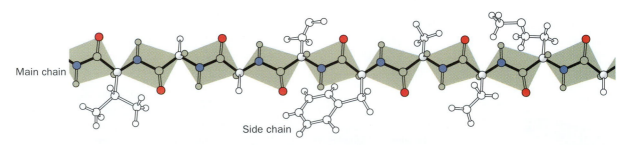

Figure 6-3.   Extended conformation of a polypeptide. The backbone is shown as a series of planar peptide groups. [Figure copyrighted © by Irving Geis.]

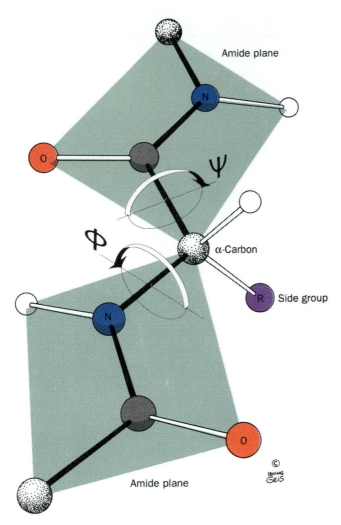

Figure 6-4. Torsion angles of the polypeptide backbone. Two planar peptide groups are shown. The only reasonably free movements are rotations around the C_α—N bond (measured as ϕ) and the C_α—C bond (measured as ψ). By convention, both ϕ and ψ are 180° in the conformation shown and increase, as indicated, in the clockwise direction when viewed from C_α. [Figure copyrighted © by Irving Geis.]
● See Kinemage Exercise 3-1.

residues. The backbone can be drawn as a linked sequence of rigid planar peptide groups (Fig. 6-3). *The conformation of the backbone can therefore be described by the* **torsion angles** *(also called* **dihedral angles** *or rotation angles) around the C_α—N bond (ϕ) and the C_α—C bond (ψ) of each residue* (Fig. 6-4). These angles, ϕ and ψ, are both defined as 180° when the polypeptide chain is in its fully extended conformation and increase clockwise when viewed from C_α.

The conformational freedom and therefore the torsion angles of a polypeptide backbone are sterically constrained. Rotation around the C_α—N and C_α—C bonds to form certain combinations of ϕ and ψ angles may cause the amide hydrogen, the carbonyl oxygen, or the substituents of C_α of adjacent residues to collide (e.g., Fig. 6-5). Certain conformations of longer polypeptides can similarly produce collisions between residues that are far apart in sequence.

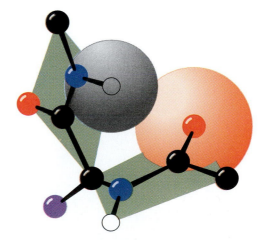

Figure 6-5. Steric interference between adjacent peptide groups. Rotation can result in a conformation in which the amide hydrogen of one residue and the carbonyl oxygen of the next are closer than their van der Waals distance. [Figure copyrighted © by Irving Geis.]

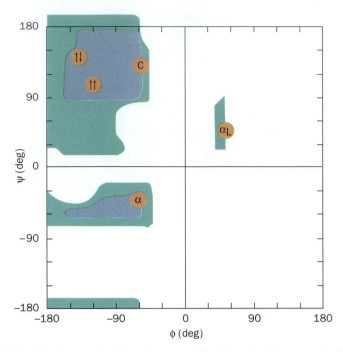

Figure 6-6. **The Ramachandran diagram.** The blue-shaded regions indicate the sterically allowed ϕ and ψ angles for all residues except Gly and Pro. The green-shaded regions indicate the more crowded (outer limit) ϕ and ψ angles. The orange circles represent conformational angles of several secondary structures: α, right-handed α helix; $\uparrow\uparrow$, parallel β sheet; $\uparrow\downarrow$, antiparallel β sheet; C, collagen helix; α_L, left-handed α helix.

The Ramachandran Diagram Indicates Allowed Conformations of Polypeptides

The sterically allowed values of ϕ and ψ can be calculated. Sterically forbidden conformations, such as the one shown in Fig. 6-5, have ϕ and ψ values that would bring atoms closer than the corresponding van der Waals distance (the distance of closest contact between nonbonded atoms). Such information is summarized in a **Ramachandran diagram** (Fig. 6-6), which is named after its inventor, G. N. Ramachandran.

Most areas of the Ramachandran diagram (most combinations of ϕ and ψ) represent forbidden conformations of a polypeptide chain. Only three small regions of the diagram are physically accessible to most residues. The observed ϕ and ψ values of accurately determined structures nearly always fall within these allowed regions of the Ramachandran plot. There are, however, some notable exceptions:

1. The cyclic side chain of Pro limits its range of ϕ values to angles of around $-60°$, making it, not surprisingly, the most conformationally restricted amino acid residue.

2. Gly, the only residue without a C_β atom, is much less sterically hindered than the other amino acid residues. Hence, its permissible range of ϕ and ψ covers a larger area of the Ramachandran diagram. At Gly residues, polypeptide chains often assume conformations that are forbidden to other residues.

B. Regular Secondary Structure: The α Helix and the β Sheet

A few elements of protein secondary structure are so widespread that they are immediately recognizable in proteins with widely differing amino acid sequences. Both the **α helix** and the **β sheet** are such elements; they are

● See Guided Exploration 6:

Stable Helices in Proteins: The α helix.

Figure 6-7.   *Key to Structure*. The α helix. This right-handed helical conformation has 3.6 residues per turn. Dashed lines indicate hydrogen bonds between C=O groups and N—H groups that are four residues farther along the polypeptide chain. [Figure copyrighted © by Irving Geis.] ⬤ See Kinemage Exercise 3-2. ✳ See the Animated Figures.

called **regular secondary structures** because they are composed of sequences of residues with repeating φ and ψ values.

The α Helix

Only one polypeptide helix has both a favorable hydrogen bonding pattern and φ and ψ values that fall within the fully allowed regions of the Ramachandran diagram: the α helix. Its discovery by Linus Pauling in 1951, through model building, ranks as one of the landmarks of structural biochemistry.

The α helix (Fig. 6-7) is right-handed; that is, it turns in the direction that the fingers of a right hand curl when its thumb points in the direction that the helix rises (see Fig. 3-10). The α helix has 3.6 residues per turn and a **pitch** (the distance the helix rises along its axis per turn) of 5.4 Å. The α helices of proteins have an average length of ~12 residues, which corresponds to over three helical turns, and a length of ~18 Å.

In the α helix, the backbone hydrogen bonds are arranged such that the peptide C=O bond of the nth residue points along the helix axis toward the peptide N—H group of the (n + 4)th residue. This results in a strong hydrogen bond that has the nearly optimum N···O distance of 2.8 Å. Amino acid side chains project outward and downward from the helix (Fig. 6-8), thereby avoiding steric interference with the polypeptide backbone and with each other. The core of the helix is tightly packed; that is, its atoms are in van der Waals contact.

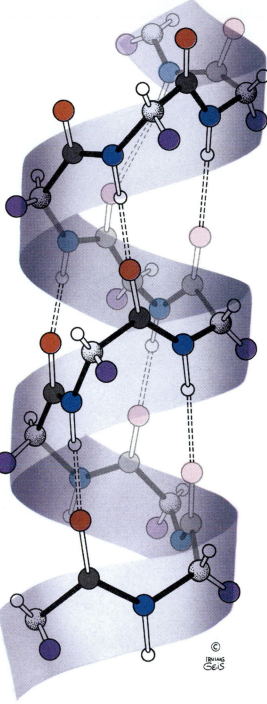

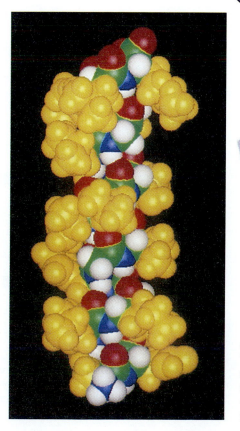

Figure 6-8.   Space-filling model of an α helix. The backbone atoms are colored with carbon atoms green, nitrogen atoms blue, oxygen atoms red, and hydrogen atoms white. The side chains (*yellow*) project away from the helix. This α helix is a segment of sperm whale myoglobin.

● See Guided Exploration 7:

Hydrogen Bonding in β Sheets.

β Sheets

In 1951, the same year Pauling proposed the α helix, Pauling and Corey postulated the existence of a different polypeptide secondary structure, the β sheet. Like the α helix, the β sheet uses the full hydrogen-bonding capacity of the polypeptide backbone. *In β sheets, however, hydrogen bonding occurs between neighboring polypeptide chains rather than within one as in an α helix.*

Sheets come in two varieties:

1. The **antiparallel β sheet,** in which neighboring hydrogen-bonded polypeptide chains run in opposite directions (Fig. 6-9*a*).
2. The **parallel β sheet,** in which the hydrogen-bonded chains extend in the same direction (Fig. 6-9*b*).

The conformations in which these β structures are optimally hydrogen bonded vary somewhat from that of the fully extended polypeptide shown in Fig. 6-3. They therefore have a rippled or pleated edge-on appearance (Fig. 6-10) and for that reason are sometimes called "pleated sheets." Successive side chains of a polypeptide chain in a β sheet extend to opposite sides of the sheet with a two-residue repeat distance of 7.0 Å.

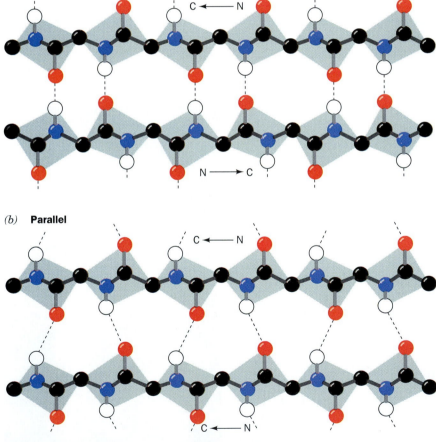

(a) **Antiparallel**

C ← N

N → C

(b) **Parallel**

C ← N

C ← N

Figure 6-9. *Key to Structure.* **β Sheets.** Dashed lines indicate hydrogen bonds between polypeptide strands. Side chains are omitted for clarity. (*a*) An antiparallel β sheet. (*b*) A parallel β sheet. [Figure copyrighted © by Irving Geis.] ● **See Kinemage Exercise 3-3.** ✳ **See the Animated Figures.**

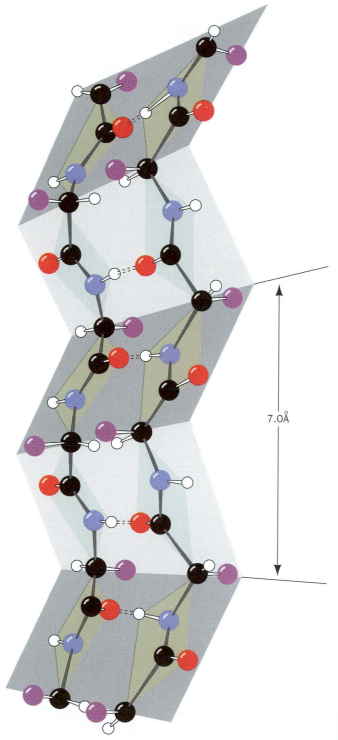

7.0Å

Figure 6-10.   **Pleated appearance of a β sheet.** Dashed lines indicate hydrogen bonds. The R groups (*purple*) on each polypeptide chain alternately extend to opposite sides of the sheet and are in register on adjacent chains. [Figure copyrighted © by Irving Geis.] ● See Kinemage Exercise 3-3.

β Sheets in proteins contain 2 to >12 polypeptide strands, with an average of 6 strands. Each strand may contain up to 15 residues, the average being 6 residues. A six-stranded antiparallel β sheet is shown in Fig. 6-11.

Figure 6-11.   Space-filling model of a β sheet. The backbone atoms are colored with carbon atoms green, nitrogen atoms blue, oxygen atoms red, and hydrogen atoms white. The R groups are represented by large purple spheres. This six-stranded β sheet is from the jack bean protein **concanavalin A.**

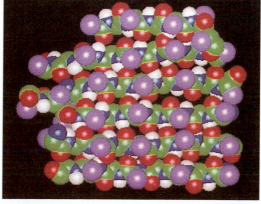

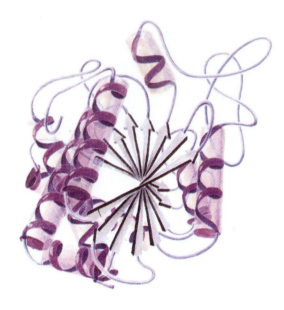

Figure 6-12.   Diagram of a β sheet in bovine carboxypeptidase A. The polypeptide backbone is represented by a ribbon with α helices drawn as coils and strands of the β sheet drawn as arrows pointing toward the C-terminus. Side chains are not shown. The eight-stranded β sheet forms a saddle-shaped curved surface with a right-handed twist. [After a drawing by Jane Richardson, Duke University.]

Parallel β sheets containing fewer than five strands are rare. This observation suggests that parallel β sheets are less stable than antiparallel β sheets, possibly because the hydrogen bonds of parallel sheets are distorted compared to those of the antiparallel sheets (Fig. 6-9). β Sheets containing mixtures of parallel and antiparallel strands frequently occur.

β Sheets almost invariably exhibit a pronounced right-handed twist when viewed along their polypeptide strands (Fig. 6-12). Conformational energy calculations indicate that the twist is a consequence of interactions between chiral L-amino acid residues in the extended polypeptide chains. The twist actually distorts and weakens the β sheet's interchain hydrogen bonds. The geometry of a particular β sheet is thus a compromise between optimizing the conformational energies of its polypeptide chains and preserving its hydrogen bonding.

The **topology** (connectivity) of the polypeptide strands in a β sheet can be quite complex. The connection between two antiparallel strands may be just a small loop (Fig. 6-13*a*), but the link between tandem parallel strands must be a crossover connection that is out of the plane of the β sheet (Fig. 6-13*b*). The connecting link in either case can be extensive, often containing helices (e.g., Fig. 6-12).

C. Fibrous Proteins

Proteins have historically been classified as either **fibrous** or **globular,** depending on their overall morphology. This dichotomy predates methods for determining protein structure on an atomic scale and does not do justice to proteins that contain both stiff, elongated, fibrous regions as well as more compact, highly folded, globular regions. Nevertheless, the division helps emphasize the properties of fibrous proteins, which often have a protective, connective, or supportive role in living organisms. The three well-characterized fibrous proteins we discuss here—keratin, silk fibroin, and collagen—are highly elongated molecules whose shapes are dominated by a single type of secondary structure. They are therefore useful examples of these structural elements.

(a)   *(b)*

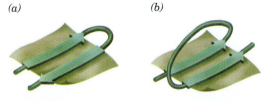

Figure 6-13.   Connections between adjacent strands in β sheets. (*a*) Antiparallel strands may be connected by a small loop. (*b*) Parallel strands require a more extensive cross-over connection. [After Richardson, J.S., *Adv. Protein Chem.* **34,** 196 (1981).]

α Keratin—A Coiled Coil

Keratin is a mechanically durable and chemically unreactive protein that occurs in all higher vertebrates. It is the principal component of their horny outer epidermal layer and its related appendages such as hair, horn, nails, and feathers. Keratins have been classified as either α keratins, which occur in mammals, or β keratins, which occur in birds and reptiles. Mammals each have about 30 keratin variants that are expressed in a tissue-specific manner.

The X-ray diffraction pattern of α keratin resembles that expected for an α helix (hence the name α keratin). However, α keratin exhibits a 5.1-Å spacing rather than the 5.4-Å distance corresponding to the pitch of the α helix. This discrepancy is the result of *two α keratin polypeptides, each of which forms an α helix, twisting around each other to form a left-handed coil.* The normal 5.4-Å repeat distance of each α helix in the pair is thereby tilted relative to the axis of this assembly, yielding the observed 5.1-Å spacing. The assembly is said to have a **coiled coil** structure because each α helix itself follows a helical path.

The conformation of α keratin's coiled coil is a consequence of its primary structure: The central ~310-residue segment of each polypeptide chain has a 7-residue pseudorepeat, *a-b-c-d-e-f-g,* with nonpolar residues predominating at positions *a* and *d*. Since an α helix has 3.6 residues per turn, α keratin's *a* and *d* residues line up along one side of each α helix (Fig. 6-14*a*). The hydrophobic strip along one helix associates with the hydrophobic strip on another helix. Because the 3.5-residue repeat in α keratin is slightly smaller than the 3.6 residues per turn of a standard α helix, the two keratin helices are inclined about 18° relative to one another, resulting in the coiled coil arrangement. This conformation allows the contacting side chains to interdigitate (Fig. 6-14*b*).

The higher order structure of α keratin is not well understood. The N- and C-terminal domains of each polypeptide facilitate the assembly of coiled coils (dimers) into protofilaments, two of which constitute a protofibril (Fig. 6-15). Four protofibrils constitute a microfibril, which associates with other microfibrils to form a macrofibril. A single mammalian hair consists of layers of dead cells, each of which is packed with parallel macrofibrils.

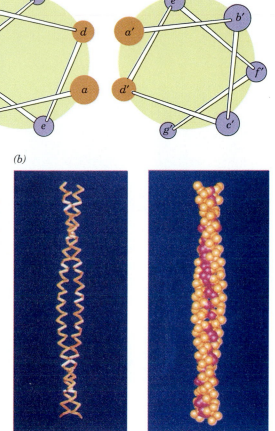

Figure 6-14.   The coiled coil of α keratin. (*a*) View down the coil axis showing the alignment of nonpolar residues along one side of each α helix. The helices have the pseudorepeating sequence *a-b-c-d-e-f-g* in which residues *a* and *d* are predominately nonpolar. [After McLachlan, A.D. and Stewart, M., *J. Mol. Biol.* **98**, 295 (1975).] (*b*) Side view of the polypeptide backbone in skeletal (*left*) and space-filling (*right*) forms. Note that the contacting side chains (red spheres in the space-filling model) interlock. [Courtesy of Carolyn Cohen, Brandeis University.] ● **See Kinemage Exercises 4-1 and 4-2.**

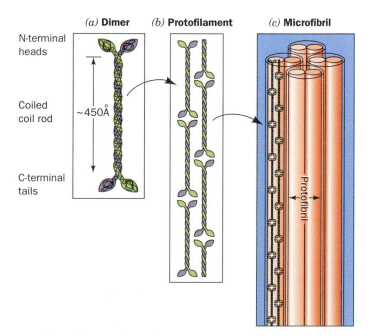

Figure 6-15.   Higher order α keratin structure. (*a*) Two keratin polypeptides form a dimeric coiled coil. (*b*) Protofilaments are formed from two staggered rows of head-to-tail associated coiled coils. (*c*) Protofilaments dimerize to form a protofibril, four of which form a microfibril. The structures of the latter assemblies are poorly characterized.

α Keratin is rich in Cys residues, which form disulfide bonds that cross-link adjacent polypeptide chains. The α keratins are classified as "hard" or "soft" according to whether they have a high or low sulfur content. Hard keratins, such as those of hair, horn, and nail, are less pliable than soft keratins, such as those of skin and callus, because the disulfide bonds resist deformation. The disulfide bonds can be reductively cleaved with mercaptans (Section 5-3A). Hair so treated can be curled and set in a "permanent wave" by applying an oxidizing agent that reestablishes the disulfide bonds in the new "curled" conformation. Conversely, curly hair can be straightened by the same process.

The springiness of hair and wool fibers is a consequence of the coiled coil's tendency to recover its original conformation after being untwisted by stretching. If some of its disulfide bonds have been cleaved, however, an α keratin fiber can be stretched to over twice its original length. At this point, the polypeptide chains assume a β sheet conformation. β Keratin, such as that in feathers, exhibits a β-like pattern in its native state.

Silk Fibroin—A β Sheet

Insects and arachnids (spiders) produce various silks to fabricate structures such as cocoons, webs, nests, and egg stalks. **Silk fibroin,** the fibrous protein from the cultivated larvae (silkworms) of the moth *Bombyx mori,* consists of antiparallel β sheets whose chains extend parallel to the fiber axis. Sequence studies have shown that long stretches of silk fibroin contain a six-residue repeat:

$$(\text{-Gly-Ser-Gly-Ala-Gly-Ala-})_n$$

Since the side chains from successive residues of a β strand extend to opposite sides of the β sheet (Fig. 6-10), silk's Gly side chains project from one surface of a β sheet and its Ser and Ala side chains project from the opposite surface. *The β sheets stack to form a microcrystalline array in which layers of contacting Gly side chains from neighboring sheets alternate with layers of contacting Ser and Ala side chains* (Fig. 6-16).

The β sheet structure of silk accounts for its mechanical properties. Silk, which is among the strongest of fibers, is only slightly extensible because appreciable stretching would require breaking the covalent bonds of its nearly fully extended polypeptide chains. Yet silk is flexible because neighboring β sheets associate only through relatively weak van der Waals forces.

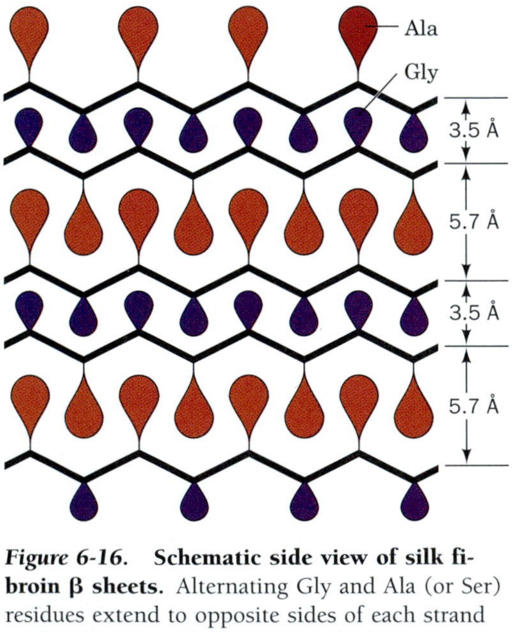

Collagen—A Triple Helix

Collagen, which occurs in all multicellular animals, is the most abundant vertebrate protein. Its strong, insoluble fibers are the major stress-bearing components of connective tissues such as bone, teeth, cartilage, tendon, and the fibrous matrices of skin and blood vessels. A single collagen molecule consists of three polypeptide chains. Mammals have about 30 genetically distinct chains that are assembled into at least 19 collagen varieties found in different tissues in the same individual. One of the most common collagens, called Type I, consists of two $\alpha_1(I)$ chains and one $\alpha_2(I)$ chain. It has a molecular mass of ~285 kD, a width of ~14 Å, and a length of ~3000 Å.

Collagen has a distinctive amino acid composition: Nearly one-third of its residues are Gly; another 15 to 30% of its residues are Pro and **4-hydroxyprolyl (Hyp)**. **3-Hydroxyprolyl** and **5-hydroxylysyl (Hyl)** residues also occur in collagen, but in smaller amounts.

Ala
Gly

3.5 Å

5.7 Å

3.5 Å

5.7 Å

Figure 6-16. **Schematic side view of silk fibroin β sheets.** Alternating Gly and Ala (or Ser) residues extend to opposite sides of each strand so that the Gly side chains (*purple*) from one sheet nestle efficiently between those of the neighboring sheet and likewise for the Ser and Ala side chains (*brown*). The intersheet spacings consequently have the alternating values of 3.5 and 5.7 Å. [Figure copyrighted © by Irving Geis.]

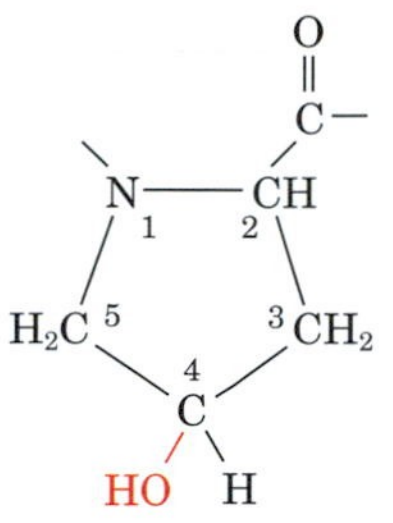

4-Hydroxyprolyl residue (Hyp) 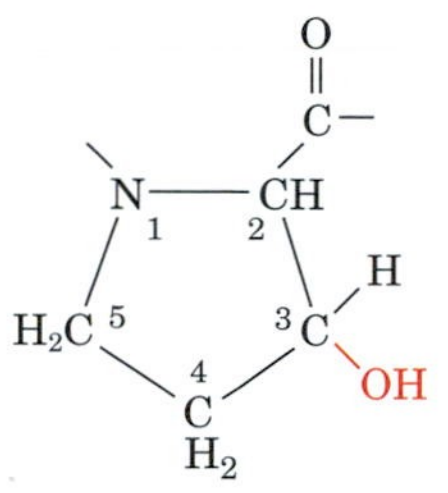 **3-Hydroxyprolyl residue** 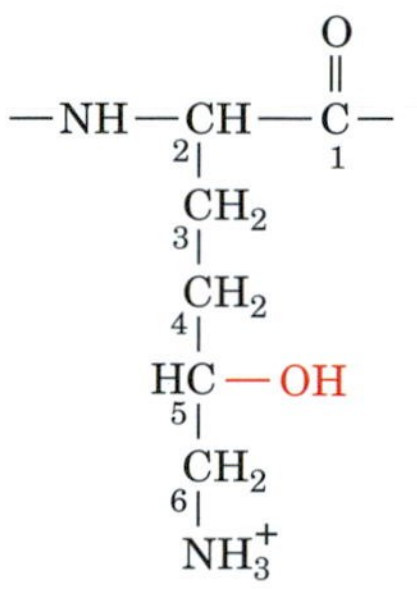 **5-Hydroxylysyl residue (Hyl)**

These nonstandard residues are formed after the collagen polypeptides are synthesized. For example, Pro residues are converted to Hyp in a reaction catalyzed by **prolyl hydroxylase.** This enzyme requires **ascorbic acid (vitamin C)** to maintain its activity.

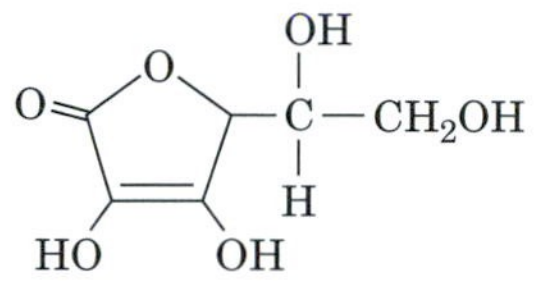

Ascorbic acid (vitamin C)

The disease **scurvy** results from the dietary deficiency of vitamin C (see Box 6-1).

BIOCHEMISTRY IN HEALTH AND DISEASE

Collagen Diseases

Some collagen diseases have dietary causes. In scurvy (caused by vitamin C deficiency), Hyp production decreases because prolyl hydroxylase requires vitamin C. Thus, in the absence of vitamin C, newly synthesized collagen cannot form fibers properly, resulting in skin lesions, fragile blood vessels, poor wound healing, and, ultimately, death. Scurvy was common in sailors on long voyages whose diets were devoid of fresh foods. The introduction of limes to the diet of the British navy by the renowned explorer Capt. James Cook alleviated scurvy and led to the nickname "limey" for the British sailor.

The disease **lathyrism** is caused by regular ingestion of the seeds from the sweet pea *Lathyrus odoratus,* which contain a compound that specifically inactivates lysyl oxidase. The resulting reduced cross-linking of collagen fibers produces serious abnormalities of the bones, joints, and large blood vessels.

Several rare heritable disorders of collagen are known. Mutations of Type I collagen, which constitutes the major structural protein in most human tissues, usually result in **osteogenesis imperfecta** (brittle bone disease). The severity of this disease varies with the nature and position of the mutation:

Even a single amino acid change can have lethal consequences. For example, the central Gly → Ala substitution in the model polypeptide shown in Fig. 6-18b locally distorts the already internally crowded collagen helix. This ruptures the hydrogen bond from the backbone N—H of each Ala (normally Gly) to the carbonyl group of the adjacent Pro in a neighboring chain, thereby reducing the stability of the collagen structure.

Mutations may affect the structure of the collagen molecule or how it forms fibrils. These mutations tend to be dominant because they affect either the folding of the triple helix or fibril formation even when normal chains are also involved.

Many collagen disorders are characterized by deficiencies in the amount of a particular collagen type synthesized, or by abnormal activities of collagen-processing enzymes such as lysyl hydroxylase and lysyl oxidase. One group of at least 10 different collagen deficiency diseases, the **Ehlers–Danlos syndromes,** are all characterized by the hyperextensibility of the joints and skin. The "India-rubber man" of circus fame had an Ehlers–Danlos syndrome.

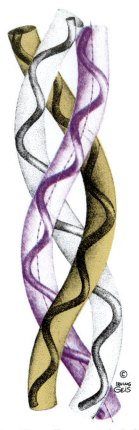

Figure 6-17.   The collagen triple helix. Left-handed polypeptide helices are twisted together to form a right-handed superhelical structure. [Figure copyrighted © by Irving Geis.]

The amino acid sequence of a typical collagen polypeptide consists of monotonously repeating triplets of sequence Gly-X-Y over a segment of ~1000 residues, where X is often Pro, and Y is often Hyp. Hyl sometimes appears at the Y position. Collagen's Pro residues prevent it from forming an α helix (Pro residues cannot assume the α-helical backbone conformation and lack the backbone N—H groups that form the intrahelical hydrogen bonds shown in Fig. 6-7). Instead, *the collagen polypeptide assumes a left-handed helical conformation with about three residues per turn. Three parallel chains wind around each other with a gentle, right-handed, ropelike twist to form the triple-helical structure of a collagen molecule* (Fig. 6-17).

Every third residue of each polypeptide chain passes through the center of the triple helix, which is so crowded that only a Gly side chain can fit there. This crowding explains the absolute requirement for a Gly at every third position of a collagen polypeptide chain. The three polypeptide chains are staggered so that Gly, X, and Y residues from each of the three chains occur at the same level along the helix axis. The peptide groups are oriented such that the N—H of each Gly makes a strong hydrogen bond with the carbonyl oxygen of an X residue on a neighboring chain (Fig. 6-18*a*). The bulky and relatively inflexible Pro and Hyp residues confer rigidity on the entire assembly.

This model of the collagen structure has been confirmed by Barbara Brodsky and Helen Berman, who determined the X-ray crystal structure of the collagenlike polypeptide (Pro-Hyp-Gly)₄-(Pro-Hyp-Ala)-(Pro-Hyp-Gly)₅. Three of these polypeptides associate to form a triple-helical structure that closely resembles the above model (Fig. 6-18*b*). The X-ray structure further reveals that the 87-Å-long cylindrical molecule is surrounded by a sheath of ordered water molecules that apparently stabilizes the collagen structure. These water mol-

Figure 6-18.   Molecular interactions in collagen. (*a*) Hydrogen bonding in the collagen triple helix. This view down the helix axis shows one Gly and two Pro residues (X and Y) in each chain. The residues are staggered so that one Gly, X, and Y occur at every level along the axis. The dashed lines represent hydrogen bonds between each Gly N—H group and the oxygen of the succeeding Pro residue on a neighboring chain. Every third residue on each chain must be Gly because no other residue can fit near the helix axis. The bulky Pro side chains are on the periphery of the helix, where they are sterically unhindered. [After Yonath, A. and Traub, W., *J. Mol. Biol.* **43**, 461 (1969).] (*b*) Space-filling model of a collagenlike peptide. The three parallel polypeptide chains (*blue, purple,* and *green*) are staggered by one residue. Ala residues (*yellow*), which replace the normally occurring Gly residue in each chain cause a significant distortion of the normal collagen structure. [Courtesy of Helen Berman, Rutgers University.] ⬤ See Kinemage Exercises 4-3 and 4-4.

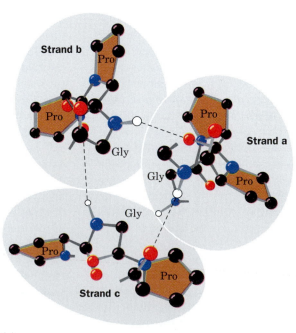

(*a*)

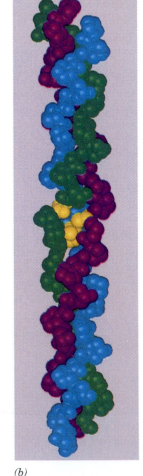

(*b*)

ecules form a hydrogen-bonded network that is anchored to the polypeptides in large part through hydrogen bonds to the 4-OH group of Hyp.

Collagen's well-packed, rigid, triple-helical structure is responsible for its characteristic tensile strength. The twist in the helix cannot be pulled out under tension because its component polypeptide chains are twisted in the opposite direction (Fig. 6-17). Successive levels of fiber bundles in high-quality ropes and cables, as well as in other proteins such as keratin (Fig. 6-14), are likewise oppositely twisted.

Several types of collagen molecules assemble to form loose networks or thick fibrils arranged in bundles or sheets, depending on the tissue. The collagen molecules in fibrils are organized in staggered arrays that are stabilized by hydrophobic interactions resulting from the close packing of triple-helical units. Collagen is also covalently cross-linked, which accounts for its poor solubility. The cross-links cannot be disulfide bonds, as in keratin, because collagen is almost devoid of Cys residues. Instead, the cross-links are derived from Lys and His side chains in reactions such as those shown in Fig. 6-19. **Lysyl oxidase,** the enzyme that converts Lys residues to those of the aldehyde **allysine,** is the only enzyme implicated in this cross-linking process. Up to four side chains can be covalently bonded to each other. The cross-links do not form at random but tend to occur near the N- and C-termini of the collagen molecules. The degree of cross-linking in a particular tissue increases with age. This is why meat from older animals is tougher than meat from younger animals.

D. Nonrepetitive Protein Structure

The majority of proteins are globular proteins that, unlike the fibrous proteins discussed in the preceding section, may contain several types of regular secondary structure, including α helices, β sheets, and other recognizable elements. A significant portion of a protein's structure may also be irregular or unique.

Irregular Structures

Segments of polypeptide chains whose successive residues do not have similar ϕ and ψ values are sometimes called coils. However, you should not confuse this term with the appellation **random coil,** which refers to the totally disordered and rapidly fluctuating conformations assumed by **denatured** (fully unfolded) proteins in solution. In **native** (folded) proteins, *nonrepetitive structures are no less ordered than are helices or β sheets; they are simply irregular and hence more difficult to describe.*

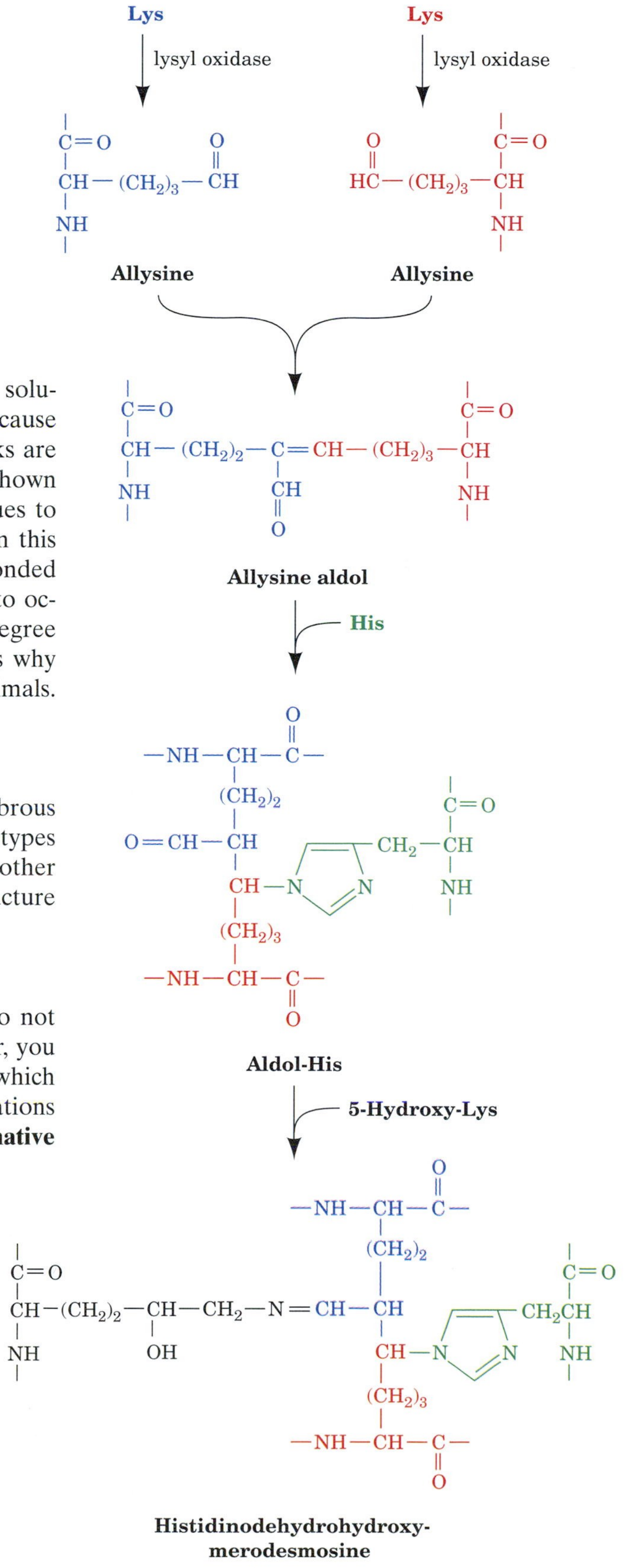

Figure 6-19. A reaction pathway for cross-linking side chains in collagen. The first step is the lysyl oxidase–catalyzed oxidative deamination of Lys to form the aldehyde allysine. Two allysines then undergo an aldol condensation to form allysine aldol. This product can react with His to form aldol histidine, which can in turn react with 5-hydroxylysine to form a Schiff base (an imine bond), thereby cross-linking four side chains.

Box 6-2
BIOCHEMISTRY IN CONTEXT

Protein Structure Prediction and Protein Design

Hundreds of thousands of protein sequences are known either through direct protein sequencing (Section 5-3) or, more commonly, through nucleic acid sequencing (Section 3-4). Yet the structures of only ~7000 of these proteins have as yet been determined by X-ray crystallography or NMR techniques. Determining the function of a newly discovered protein often requires knowledge of its three-dimensional structure.

There are currently several major approaches to protein structure prediction. The simplest and most reliable approach, **homology modeling**, aligns the sequence of interest with the sequence of a homologous protein of known structure—compensating for amino acid substitutions, insertions, and deletions—through modeling and energy minimization calculations. This method yields reliable models for proteins that have as little as 25% sequence identity with a protein of known structure, although, of course, the accuracy of the model increases with the degree of sequence identity.

Distantly related proteins may be structurally similar even though they have diverged to such an extent that their sequences show no obvious resemblance. **Threading** is a computational technique that attempts to determine the unknown structure of a protein by ascertaining whether it is consistent with a known protein structure. It does so by placing (threading) the residues of the unknown protein along the backbone of a known protein structure and then determining whether the amino acid side chains of the unknown protein are stable in that arrangement. This method is not yet reliable, although it has yielded encouraging results.

Empirical methods based on experimentally determined statistical information such as the α helix and β sheet propensities deduced by Chou and Fasman (Table 6-1) have been moderately successful in predicting the secondary structures of proteins. Their advantage is their simplicity (they don't require a computer).

Since the native structure of a protein depends only on its amino acid sequence, it should be possible, in principle, to predict the structure of a protein based only on its chemical and physical properties (e.g., the hydrophobicity, size, hydrogen-bonding propensity, and charge of each of its amino acid residues). Such *ab initio* (from the beginning) methods are still in their infancy. They are moderately successful in predicting simple structures such as a single α helix but have failed miserably when tested with larger polypeptides whose structures have been experimentally determined. Nevertheless, a recently developed algorithm that simulates the hierarchical protein folding pathway has yielded structural models that are surprisingly similar to those of the corresponding observed protein structures.

Protein design, the experimental inverse of protein structure prediction, has provided insights into protein folding and stability. Protein design begins with a target structure such as a simple sandwich of β sheets or a bundle of four α helices. It attempts to construct an amino acid sequence that will form that structure. The designed polypeptide is then chemically or biologically synthesized, and its structure is determined. Fortunately, protein folding seems to be governed more by extended sequences of amino acids than by individual residues, which allows some room for error in designing polypeptides. Experimental results suggest that the greatest challenge of protein design may lie not in getting the polypeptide to fold to the desired conformation but in preventing it from folding into other, unwanted conformations. In this respect, science lags far behind nature. However, the recent successful design, using computationally based techniques, of a 28-residue polypeptide that stably folds to the desired structure (the smallest known polypeptide that is capable of folding into a unique structure without the aid of disulfide bonds, metal ions, or other subunits) indicates that significant progress in our understanding of protein folding has been made.

Variations in Standard Secondary Structure

Variations in amino acid sequence as well as the overall structure of the folded protein can distort the regular conformations of secondary structural elements. For example, the α helix frequently deviates from its ideal conformation in its initial and final turns of the helix. Similarly, a strand of polypeptide in a β sheet may contain an "extra" residue that is not hydrogen bonded to a neighboring strand, producing a distortion known as a **β bulge.**

Many of the limits on amino acid composition and sequence (Section 5-1) may be due in part to conformational constraints in the three-dimensional structure of proteins. For example, a Pro residue produces a kink in an α helix or β sheet. Similarly, steric clashes between several sequential amino acid residues with large branched side chains (e.g., Ile and Tyr) can destabilize α helices.

Analysis of known protein structures by Peter Chou and Gerald Fasman revealed the propensity P of a residue to occur in an α helix or a β sheet (Table 6-1). Chou and Fasman also discovered that certain residues not only have a high propensity for a particular secondary structure but they tend to disrupt or break other secondary structures. Such data are useful for predicting the secondary structures of proteins with known amino acid sequences (see Box 6-2).

The presence of certain residues outside of α helices or β sheets may also be nonrandom. For example, α helices are often flanked by residues such as Asn and Gln, whose side chains can fold back to form hydrogen bonds with one of the four terminal residues of the helix, a phenomenon termed **helix capping.** Recall that the four residues at each end of an α helix are not fully hydrogen bonded to neighboring backbone segments (Fig. 6-7).

Turns and Loops

Segments with regular secondary structure such as α helices or strands of β sheets are typically joined by stretches of polypeptide that abruptly change direction. Such **reverse turns** or **β bends** (so named because they often connect successive strands of antiparallel β sheets) almost always occur at protein surfaces. Most reverse turns involve four successive amino acid residues more or less arranged in one of two ways, Type I and Type II, that differ by a 180° flip of the peptide unit linking residues 2 and 3 (Fig. 6-20). Both types of turns are stabilized by a hydrogen bond, although deviations from these ideal conformations often disrupt this hydrogen bond. In Type II turns, the oxygen atom of residue 2 crowds the $C_β$ atom of residue 3, which is therefore usually Gly. Residue 2 of either type of turn is often Pro since it can assume the required conformation.

Almost all proteins with more than 60 residues contain one or more loops of 6 to 16 residues, called **Ω loops.** These loops, which have the

Table 6-1. **Propensities of Amino Acid Residues for α Helical and β Sheet Conformations**

Residue	$P_α$	$P_β$
Ala	1.42	0.83
Arg	0.98	0.93
Asn	0.67	0.89
Asp	1.01	0.54
Cys	0.70	1.19
Gln	1.11	1.10
Glu	1.51	0.37
Gly	0.57	0.75
His	1.00	0.87
Ile	1.08	1.60
Leu	1.21	1.30
Lys	1.16	0.74
Met	1.45	1.05
Phe	1.13	1.38
Pro	0.57	0.55
Ser	0.77	0.75
Thr	0.83	1.19
Trp	1.08	1.37
Tyr	0.69	1.47
Val	1.06	1.70

Source: Chou, P.Y. and Fasman, G.D., *Annu. Rev. Biochem.* 47, 258 (1978).

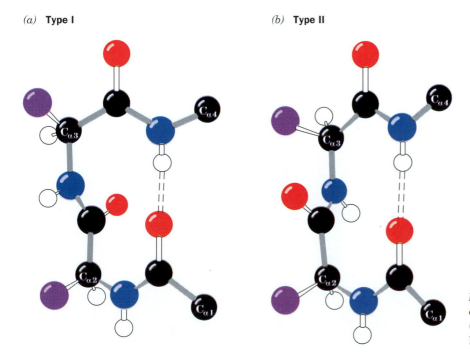

(a) **Type I**   (b) **Type II**

Figure 6-20. **Reverse turns in polypeptide chains.** Dashed lines represent hydrogen bonds. (a) Type I. (b) Type II. [Figure copyrighted © by Irving Geis.] ● **See Kinemage Exercise 3-4.**

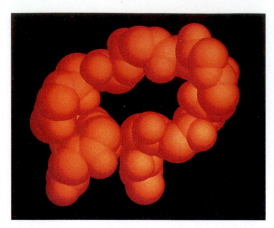

Figure 6-21. Space-filling model of an Ω loop. Only backbone atoms are shown; the side chains would fill the loop. This structure is residues 40 to 54 from cytochrome *c*. [Courtesy of George Rose, The Johns Hopkins University.]

necked-in shape of the Greek uppercase letter omega (Fig. 6-21), are compact globular entities because their side chains tend to fill in their internal cavities. Since Ω loops are almost invariably located on the protein surface, they may have important roles in biological recognition processes.

2. TERTIARY STRUCTURE

The tertiary structure of a protein describes the folding of its secondary structural elements and specifies the positions of each atom in the protein, including those of its side chains. The known protein structures have come to light through **X-ray crystallographic** or **nuclear magnetic resonance (NMR)** studies. The atomic coordinates of most of these structures are deposited in a database known as the Protein Data Bank (PDB). These data are readily available via the Internet (http://www.rcsb.org), which allows the tertiary structures of a variety of proteins to be analyzed and compared. The common features of protein tertiary structure reveal much about the biological functions of the proteins and their evolutionary origins.

A. Determining Protein Structure

X-Ray crystallography is one of the most powerful methods for studying macromolecular structure. According to optical principles, the uncertainty in locating an object is approximately equal to the wavelength of the radiation used to observe it. X-Rays can directly image a molecule because X-ray wavelengths are comparable to covalent bond distances (~1.5 Å; individual molecules cannot be seen in a light microscope because visible light has a minimum wavelength of 4000 Å).

When a crystal of the molecule to be visualized is exposed to a collimated (parallel) beam of X-rays, the atoms in the molecule scatter the X-rays, with the scattered rays canceling or reinforcing each other in a process known as diffraction. The resulting **diffraction pattern** is recorded on photographic film (Fig. 6-22) or by a radiation counter. The intensities of the

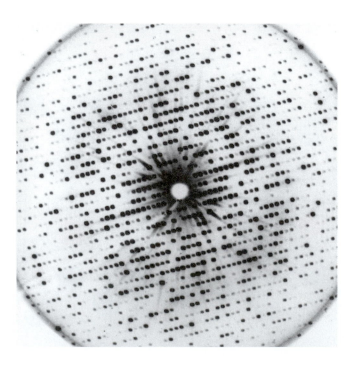

Figure 6-22. An X-ray diffraction photograph of a crystal of sperm whale myoglobin. The intensity of each diffraction maximum (the darkness of each spot) is a function of the crystal's electron density. [Courtesy of John Kendrew, Cambridge University, U.K.]

diffraction maxima (darkness of the spots on the film) are then used to mathematically construct the three-dimensional image of the crystal structure. The photograph in Fig. 6-22 represents only a small portion of the total diffraction information available from a crystal of myoglobin, a small globular protein. In contrast, fibrous proteins do not crystallize but, instead, can be drawn into fibers whose X-ray diffraction patterns contain only a few spots and thus contain comparatively little structural information. Likewise, the diffraction pattern of a DNA fiber (Fig. 3-8) is relatively simple.

X-Rays interact almost exclusively with the electrons in matter, not the atomic nuclei. An X-ray structure is therefore an image of the electron density of the object under study. This information can be shown as a three-dimensional **contour map** (Fig. 6-23). Hydrogen atoms, which have only one electron, are not visible in macromolecular X-ray structures.

The X-ray structures of small organic molecules can be determined with a resolution on the order of ~1 Å. Few protein crystals have this degree of organization. Furthermore, not all proteins can be coaxed to crystallize, that is, to precipitate in ordered three-dimensional arrays. The protein crystals that do form (Fig. 6-24) differ from those of most small organic molecules in being highly hydrated; protein crystals are typically 40 to 60% water by volume. The large solvent content gives protein crystals a soft, jellylike consistency so that the molecules are typically disordered by a few angstroms. This limits their resolution to about 2 to 3.5 Å, although a few protein crystals are better ordered (have higher resolution).

A resolution of a few angstroms is too coarse to clearly reveal the positions of individual atoms, but the distinctive shape of the polypeptide backbone can usually be traced. The positions and orientations of its side chains can therefore be deduced. However, since many side chains have similar sizes and shapes, *knowledge of the protein's primary structure is required to fit the sequence of amino acids to its electron density map.* Mathematical techniques can then refine the atomic positions to within ~0.1 Å in high-resolution structures.

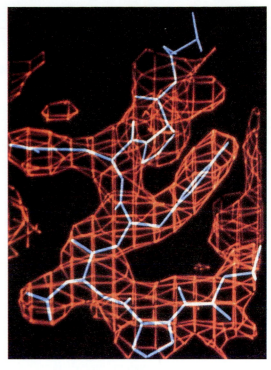

Figure 6-23.   An electron density map. The three-dimensional outline of the electron density (*orange*) is shown with a superimposed atomic model of the corresponding polypeptide segment (*white*). This structure is a portion of human rhinovirus (the cause of the common cold). [Courtesy of Michael Rossmann, Purdue University.]

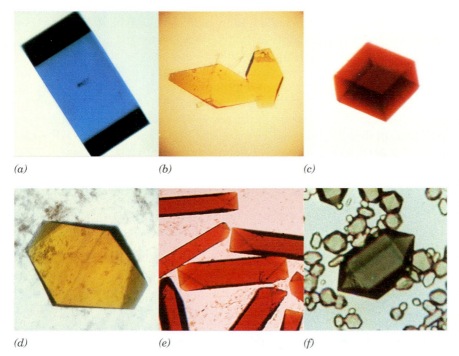

(a)   (b)   (c)

(d)   (e)   (f)

Figure 6-24.   Protein crystals. (*a*) Azurin from *Pseudomonas aeruginosa*, (*b*) flavodoxin from *Desulfovibrio vulgaris*, (*c*) rubredoxin from *Clostridium pasteurianum*, (*d*) azidomet myohemerythrin from the marine worm *Siphonosoma funafuti*, (*e*) lamprey hemoglobin, and (*f*) bacteriochlorophyll *a* protein from *Prosthecochloris aestuarii*. These crystals are colored because the proteins contain light-absorbing groups; proteins are colorless in the absence of such groups. [Parts *a–c* courtesy of Larry Siecker, University of Washington; parts *d* and *e* courtesy of Wayne Hendrikson, Columbia University; and part *f* courtesy of John Olsen, Brookhaven National Laboratories, and Brian Matthews, University of Oregon.]

Box 6-3

BIOCHEMISTRY IN FOCUS

Protein Structure Determination by NMR

The determination of the three-dimensional structures of small proteins (<250 residues) in aqueous solution has become possible since the mid-1980s, through the development of two-dimensional (2D) NMR spectroscopy (and, more recently, of 3D and 4D techniques), in large part by Kurt Wüthrich. A sample is placed in a magnetic field so that the spins of its protons are aligned. When radiofrequency pulses are applied, the protons are excited and then emit signals whose frequency depends on the molecular environment of the protons. Proteins contain so many atoms that their standard (one-dimensional) NMR spectra consist almost entirely of overlapping signals that are impossible to interpret.

In 2D NMR, the characteristics of the applied signal are varied in order to obtain additional information from interactions between protons that are <5 Å apart in space or are covalently connected by only one or two other atoms. The resulting set of distances, together with known geometric constraints such as covalent bond distances and angles, group planarity, chirality, and van der Waals radii, are used to compute the protein's three-dimensional structure. However, since interproton distance measurements are imprecise, they are insufficient to imply a unique structure. For this reason, the NMR structure of a protein (or any other macromolecule

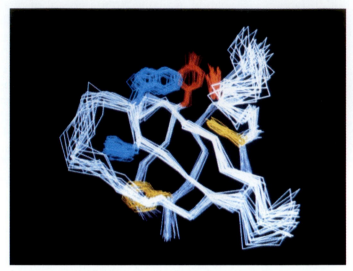

[Figure courtesy of Stuart Schreiber, Harvard University.]

with a well-defined structure) is often presented as an ensemble of closely related structures. The above NMR structure of a 64-residue polypeptide is presented as 20 superimposed C_α traces (*white*), each of which is consistent with the NMR data and geometric constraints. A few side chains (*red, yellow,* and *blue*) are also shown.

Another consequence of the large solvent content of protein crystals is that *crystalline proteins maintain their native conformations and therefore their functions.* Indeed, the degree of hydration of proteins in crystals is similar to that in cells. Thus, the X-ray crystal structures of proteins often provide a basis for understanding their biological activities. Crystal structures have been used as starting points for designing drugs that can interact specifically with target proteins under physiological conditions.

Recent advances in NMR spectroscopy have permitted the determination of the structures of proteins and nucleic acids in solution (but limited to a size of <30 kD). Thus, NMR techniques can be used to elucidate the structures of proteins and other macromolecules that fail to crystallize (see Box 6-3). In the several instances in which both the X-ray and NMR structures of a particular protein were determined, the two structures had few, if any, significant differences.

Visualizing Proteins

The huge number of atoms in proteins makes it difficult to visualize them using the same sorts of models employed for small organic molecules. Ball-and-stick representations showing all or most atoms in a protein (as in Figs. 6-7 and 6-10) are exceedingly cluttered, and space-filling models (as in Figs. 6-8 and 6-11) obscure the internal details of the protein. Accordingly, computer-generated or artistic renditions (e.g., Fig. 6-12) are often more useful for representing protein structures. The course of the polypeptide chain can be followed by tracing the positions of its C_α atoms or by representing helices as helical ribbons or cylinders, and β sheets as sets of flat arrows pointing from the N- to the C-termini.

B. Motifs (Supersecondary Structures) and Domains

In the years since Kendrew solved the structure of myoglobin, nearly 7000 protein structures have been reported. No two are exactly alike, but they exhibit remarkable consistencies.

Side Chain Location Varies with Polarity

The primary structures of globular proteins generally lack the repeating sequences that support the regular conformations seen in fibrous proteins. However, *the amino acid side chains in globular proteins are spatially distributed according to their polarities:*

1. The nonpolar residues Val, Leu, Ile, Met, and Phe occur mostly in the interior of a protein, out of contact with the aqueous solvent. The hydrophobic effects that promote this distribution are largely responsible for the three-dimensional structure of native proteins.

2. The charged polar residues Arg, His, Lys, Asp, and Glu are usually located on the surface of a protein in contact with the aqueous solvent. This is because immersing an ion in the virtually anhydrous interior of a protein is energetically unfavorable.

3. The uncharged polar groups Ser, Thr, Asn, Gln, and Tyr are usually on the protein surface but also occur in the interior of the molecule. When buried in the protein, these residues are almost always hydrogen bonded to other groups; in a sense, the formation of a hydrogen bond "neutralizes" their polarity. This is also the case with the polypeptide backbone.

These general principles of side chain distribution are evident in individual elements of secondary structure (Fig. 6-25) as well as in whole proteins

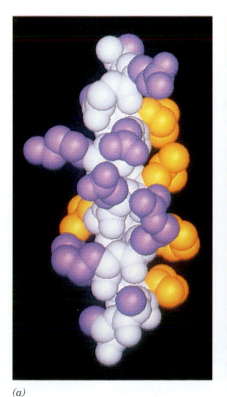
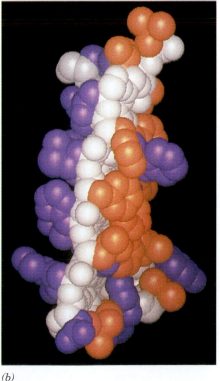

(a)   *(b)*

Figure 6-25. Side chain locations in an α helix and a β sheet. In these space-filling models, the main chain is white, nonpolar side chains are yellow or brown, and polar side chains are purple. (*a*) An α helix from sperm whale myoglobin. Note that the nonpolar residues are primarily on one side of the helix. (*b*) An antiparallel β sheet from concanavalin A (*side view*). The protein interior is to the right and the exterior is to the left.

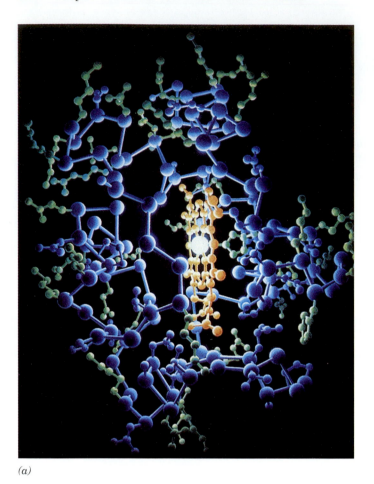

(a)

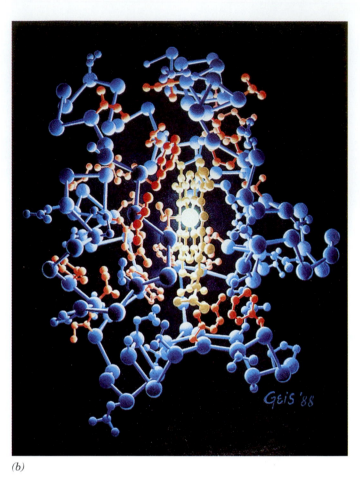

(b)

Figure 6-26. **Side chain distribution in horse heart cyto-chrome *c*.** In these paintings, based on the X-ray structure determined by Richard Dickerson, the protein is illuminated by its single iron atom centered in a heme group. Hydrogen atoms are not shown. In (*a*) the hydrophilic side chains are green, and in (*b*) the hydrophobic side chains are orange. [Figures copyrighted © by Irving Geis.] ⊙ **See Kinemage Exercise 5.**

(Fig. 6-26). Polar side chains tend to extend toward—and thereby help form—the protein's surface, whereas nonpolar side chains largely extend toward—and thereby occupy—its interior.

Most proteins are quite compact, with their interior atoms packed together even more efficiently than the atoms in a crystal of small organic molecules. Nevertheless, the atoms of protein side chains almost invariably have low-energy arrangements. Evidently, interior side chains adopt relaxed conformations despite the profusion of intramolecular interactions. Closely packed protein interiors generally exclude water. When water molecules are present, they often occupy specific positions where they can form hydrogen bonds, sometimes acting as a bridge between two hydrogen-bonding protein groups.

Helices and Sheets Can Be Combined in Various Ways

The major types of secondary structural elements, α helices and β sheets, occur in globular proteins in varying proportions and combinations. Some proteins, such as hemoglobin subunits, consist only of α helices spanned by short connecting links (Fig. 6-27*a*). Others, such as **concanavalin A,** have a large proportion of β sheets and are devoid of α helices (Fig. 6-27*b*). Most proteins, such as **triose phosphate isomerase** (Fig. 6-27*c*) and carboxypeptidase A (Fig. 6-12), have significant amounts of both types of secondary structure (on average, ~31% α helix and ~28% β sheet).

⊙ **See Guided Exploration 8:**

Secondary Structures in Proteins (Myoglobin, Concanavalin A, and TIM).

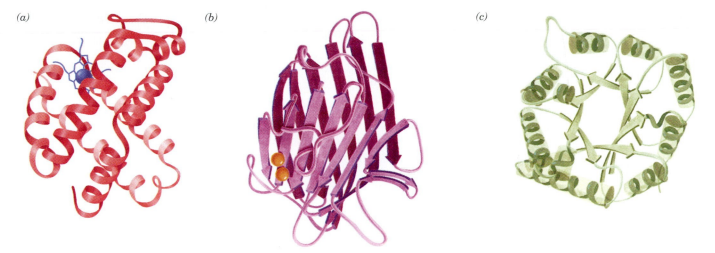

Figure 6-27. **Examples of globular proteins.** Different proteins contain different proportions and arrangements of secondary structural elements. In these models, α helices are drawn as helical ribbons, and strands of β sheets are drawn as flat arrows pointing toward the C-terminus. (*a*) A hemoglobin subunit, with its heme group shown as a skeletal model. ● See Kinemage Exercise 6-2. (*b*) Jack bean concanavalin A. The spheres represent metal ions. (*c*) **Triose phosphate isomerase** from chicken muscle. [After drawings by Jane Richardson, Duke University.] ● See Kinemage Exercise 12-1.

Certain groupings of secondary structural elements, called **supersecondary structures** or **motifs,** occur in many unrelated globular proteins:

1. The most common form of supersecondary structure is the **βαβ motif,** in which an α helix connects two parallel strands of a β sheet (Fig. 6-28*a*).

2. Another common supersecondary structure, the **β hairpin** motif, consists of antiparallel strands connected by relatively tight reverse turns (Fig. 6-28*b*).

3. In an **αα motif,** two successive antiparallel α helices pack against each other with their axes inclined. This permits energetically favorable

Figure 6-28. **Protein motifs.** (*a*) A βαβ motif, (*b*) a β hairpin, (*c*) an αα motif, and (*d*) β barrels. The β barrel composed of overlapping βαβ units (*far right*) is known as an α/β barrel. It is shown in top view in Fig. 6-27*c*.

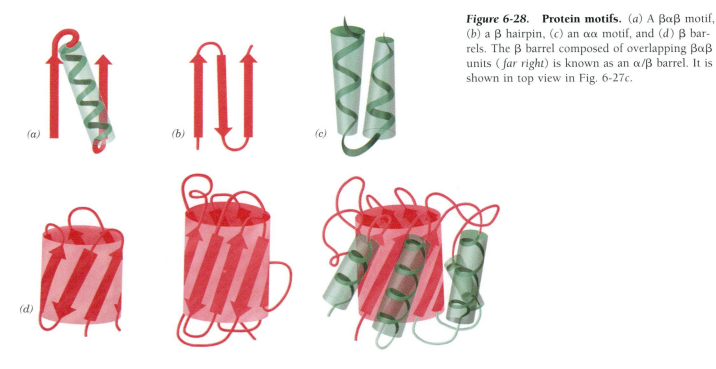

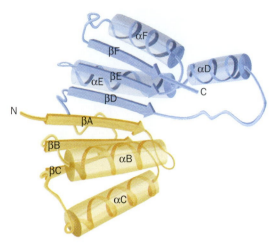

Figure 6-29.   An idealized dinucleotide-binding (Rossmann) fold. The two structurally similar βαβαβ units (*yellow* and *blue*) can each bind a nucleotide portion of the dinucleotide NAD$^+$ (not shown). [After Rossmann, M.G., Liljas, A., Bränden, C.-I., and Banaszak, L.J., *in* Boyer, P.D. (Ed.), *The Enzymes,* Vol. 11 (3rd ed.), p. 68, Academic Press (1975).] ● See Kinemage Exercise 21-1.

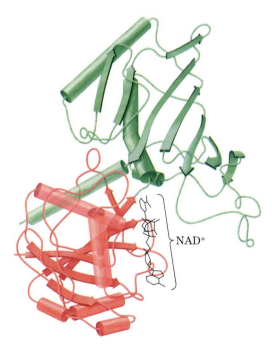

Figure 6-30.   The two-domain protein glyceraldehyde-3-phosphate dehydrogenase. The first domain (*red*) binds NAD$^+$ (*black*), and the second domain (*green*) binds glyceraldehyde-3-phosphate (not shown). [After Biesecker, G., Harris, J.I., Thierry, J.C., Walker, J.E., and Wonacott, A., *Nature* **266**, 331 (1977).] ● See the Interactive Exercises.

intermeshing of their contacting side chains (Fig. 6-28c). Such associations stabilize the coiled coil conformation of α keratin (Section 6-1C).

4. Extended β sheets often roll up to form **β barrels.** Three different types of β barrels are shown in Fig. 6-28d.

Motifs may have functional as well as structural significance. For example, Michael Rossmann showed that a βαβαβ unit, in which the β strands form a parallel sheet with α helical connections, often acts as a nucleotide-binding site. In most proteins that bind dinucleotides (such as nicotinamide adenine dinucleotide, NAD$^+$; Section 3-1), two such βαβαβ units combine to form a motif known as a **dinucleotide-binding fold,** or **Rossmann fold** (Fig. 6-29).

Large Polypeptides Form Domains

Polypeptide chains containing more than ~200 residues usually fold into two or more globular clusters known as **domains,** which give these proteins a bi- or multilobal appearance. Most domains consist of 100 to 200 amino acid residues and have an average diameter of ~25 Å. Each subunit of the enzyme **glyceraldehyde-3-phosphate dehydrogenase,** for example, has two distinct domains (Fig. 6-30). A polypeptide chain wanders back and forth within a domain, but neighboring domains are usually connected by only one or two polypeptide segments. *Consequently, many domains are structurally independent units that have the characteristics of small globular proteins.* Nevertheless, the domain structure of a protein is not necessarily obvious since its domains may make such extensive contacts with each other that the protein appears to be a single globular entity.

An inspection of the various protein structures diagrammed in this chapter reveals that domains consist of two or more layers of secondary structural elements. The reason for this is clear: At least two such layers are required to seal off a domain's hydrophobic core from its aqueous environment.

Domains often have a specific function such as the binding of a small molecule. In Fig. 6-30, for example, NAD$^+$ binds to the first domain of glyceraldehyde-3-phosphate dehydrogenase (note its dinucleotide-binding fold). In multidomain proteins, binding sites often occupy the clefts between domains; that is, the small molecules are bound by groups from two domains. In such cases, the relatively pliant covalent connection between the domains allows flexible interactions between the protein and the small molecule.

C. Protein Families

The thousands of known protein structures, comprising an even greater number of separate domains, can be grouped into families by examining the overall paths followed by their polypeptide chains. When folding patterns are compared without regard to the amino acid sequence or the presence of surface loops, the number of unique structural domains drops to only a few hundred. (Although not all protein structures are known, estimates place an upper limit of about 1000 on the total number of unique protein domains in nature.) Surprisingly, a few dozen folding patterns account for about half of all known protein structures.

There are several possible reasons for the limited number of known domain structures. The numbers may reflect database bias; that is, the collection of known protein structures may not be a representative sample of all protein structures. However, the rapidly increasing number of proteins whose structures have been determined makes this possibility less and less

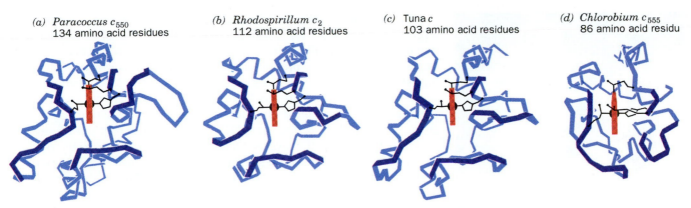

(a) *Paracoccus* c_{550}
134 amino acid residues

(b) *Rhodospirillum* c_2
112 amino acid residues

(c) Tuna c
103 amino acid residues

(d) *Chlorobium* c_{555}
86 amino acid residu

Figure 6-31.   Three-dimensional structures of c-type cytochromes. The polypeptide backbones (*blue*) are shown in analogous orientations such that their heme groups (*red*) are viewed edge-on. The Cys, Met, and His side chains that covalently link the heme to the protein are also shown. (*a*) Cytochrome c_{550} from *Paracoccus denitrificans* (134 residues), (*b*) cytochrome c_2 from *Rhodospirillum rubrum* (112 residues), (*c*) cytochrome c from tuna (103 residues), and (*d*) cytochrome c_{555} from *Chlorobium thiosulfatophilum* (86 residues). [Figures copyrighted © by Irving Geis.] ● See Kinemage Exercise 5.

plausible. More likely, the common protein structures may be evolutionary sinks—domains that arose and persisted because of their ability (1) to form stable folding patterns; (2) to tolerate amino acid deletions, substitutions, and insertions, thereby making them more likely to survive evolutionary changes; and (3) to support essential biological functions.

Polypeptides with similar sequences tend to adopt similar backbone conformations. This is certainly true for evolutionarily related proteins that carry out similar functions. For example, the cytochromes c of different species are highly conserved proteins with closely similar sequences (see Table 5-6) and three-dimensional structures.

Cytochrome c occurs only in eukaryotes, but prokaryotes contain proteins known as **c-type cytochromes,** which perform the same general function (that of an electron carrier). The c-type cytochromes from different species exhibit only low degrees of sequence similarity to each other and to eukaryotic cytochromes c. Yet their X-ray structures are clearly similar, particularly in polypeptide chain folding and side chain packing in the protein interior (Fig. 6-31). The major structural differences among c-type cytochromes lie in the various polypeptide loops on their surfaces. The sequences of the c-type cytochromes have diverged so far from one another that, in the absence of their X-ray structures, they can be properly aligned only through the use of recently developed and mathematically sophisticated computer programs. Thus, *it appears that the essential structural and functional elements of proteins, rather than their amino acid residues, are conserved during evolution.*

Structural similarities in proteins with only distantly related functions are commonly observed. For example, many NAD^+-binding enzymes that participate in widely different metabolic pathways contain similar dinucleotide-binding folds (see Fig. 6-30) coupled to diverse domains that carry out specific enzymatic reactions.

3. QUATERNARY STRUCTURE AND SYMMETRY

Most proteins, particularly those with molecular masses >100 kD, consist of more than one polypeptide chain. These polypeptide subunits associate with a specific geometry. The spatial arrangement of these subunits is known as a protein's quaternary structure.

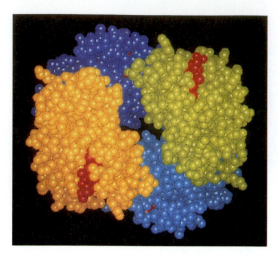

***Figure 6-32.*   Quaternary structure of hemoglobin.** In this space-filling model, the α_1, α_2, β_1, and β_2 subunits are colored yellow, green, cyan, and blue, respectively. Heme groups are red.

There are several reasons why multisubunit proteins are so common. In large assemblies of proteins, such as collagen fibrils, the advantages of subunit construction over the synthesis of one huge polypeptide chain are analogous to those of using prefabricated components in constructing a building: Defects can be repaired by simply replacing the flawed subunit; the site of subunit manufacture can be different from the site of assembly into the final product; and the only genetic information necessary to specify the entire edifice is the information specifying its few different self-assembling subunits. In the case of enzymes, increasing a protein's size tends to better fix the three-dimensional positions of its reacting groups. *Increasing the size of an enzyme through the association of identical subunits is more efficient than increasing the length of its polypeptide chain since each subunit has an active site. More importantly, the subunit construction of many enzymes provides the structural basis for the regulation of their activities* (Sections 7-2E and 12-3).

Subunits Usually Associate Noncovalently

A multisubunit protein may consist of identical or nonidentical polypeptide chains. Hemoglobin, for example, has the subunit composition $\alpha_2\beta_2$ (Fig. 6-32). Proteins with more than one subunit are called **oligomers,** and their identical units are called **protomers.** A protomer may therefore consist of one polypeptide chain or several unlike polypeptide chains. In this sense, hemoglobin is a dimer of $\alpha\beta$ protomers.

The contact regions between subunits closely resemble the interior of a single-subunit protein. They contain closely packed nonpolar side chains, hydrogen bonds involving the polypeptide backbones and their side chains, and, in some cases, interchain disulfide bonds.

Subunits Are Symmetrically Arranged

In the vast majority of oligomeric proteins, the protomers are symmetrically arranged; that is, each protomer occupies a geometrically equivalent position in the oligomer. Proteins cannot have inversion or mirror symmetry, however, because bringing the protomers into coincidence would require converting chiral L residues to D residues. Thus, *proteins can have only* **rotational symmetry.**

In the simplest type of rotational symmetry, **cyclic symmetry,** protomers are related by a single axis of rotation (Fig. 6-33*a*). Objects with two-, three-, or *n*-fold rotational axes are said to have C_2, C_3, or C_n symmetry, respectively. C_2 symmetry is the most common; higher cyclic symmetries are relatively rare.

Dihedral symmetry (D_n), a more complicated type of rotational symmetry, is generated when an *n*-fold rotation axis intersects a two-fold rotation axis at right angles (Fig. 6-33*b*). An oligomer with D_n symmetry consists of 2*n* protomers. D_2 symmetry is the most common type of dihedral symmetry in proteins.

Other possible types of rotational symmetry are those of a tetrahedron, cube, and icosahedron (Fig. 6-33*c*). Some multienzyme complexes and spherical viruses are built on these geometric plans.

4. PROTEIN FOLDING AND STABILITY

Incredible as it may seem, thermodynamic measurements indicate that *native proteins are only marginally stable under physiological conditions.* The free energy required to denature them is ~ 0.4 kJ·mol^{-1} per amino acid residue, so a fully folded 100-residue protein is only about 40 kJ·mol^{-1}

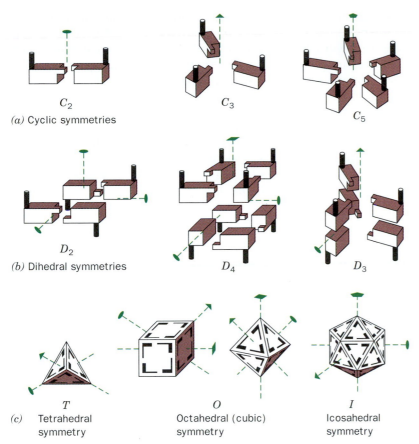

C_2
(a) Cyclic symmetries

C_3

C_5

D_2
(b) Dihedral symmetries

D_4

D_3

(c) T
Tetrahedral
symmetry

O
Octahedral (cubic)
symmetry

I
Icosahedral
symmetry

Figure 6-33.   **Some symmetries for oligomeric proteins.** The oval, the triangle, the square, and the pentagon at the ends of the dashed green lines indicate, respectively, the unique two-fold, three-fold, four-fold, and five-fold rotational axes of the objects shown. (*a*) Assemblies with cyclic (*C*) symmetry. (*b*) Assemblies with dihedral (*D*) symmetry. In these objects, a 2-fold axis is perpendicular to another rotational axis. (*c*) Assemblies with the rotational symmetries of a tetrahedron (*T*), a cube or octahedron (*O*), and an icosahedron (*I*). [Figure copyrighted © by Irving Geis.] ✳ See the Animated Figures.

more stable than its unfolded form (for comparison, the energy required to break a typical hydrogen bond is $\sim$20 kJ · mol^{-1}). The various noncovalent influences on proteins—hydrophobic effects, electrostatic interactions, and hydrogen bonding—each have energies that may total thousands of kilojoules per mole over an entire protein molecule. Consequently, a protein structure is the result of a delicate balance among powerful countervailing forces. In this section, we discuss the forces that stabilize proteins and the processes by which proteins achieve their most stable folded state.

A. Forces That Stabilize Protein Structure

Protein structures are governed primarily by hydrophobic effects and, to a lesser extent, by interactions between polar residues and other types of bonds.

The Hydrophobic Effect

The hydrophobic effect, which causes nonpolar substances to minimize their contacts with water (Section 2-1C), is the major determinant of native protein structure. The aggregation of nonpolar side chains in the interior of

Table 6-2.   Hydropathy Scale for Amino Acid Side Chains

Side Chain	Hydropathy
Ile	4.5
Val	4.2
Leu	3.8
Phe	2.8
Cys	2.5
Met	1.9
Ala	1.8
Gly	−0.4
Thr	−0.7
Ser	−0.8
Trp	−0.9
Tyr	−1.3
Pro	−1.6
His	−3.2
Glu	−3.5
Gln	−3.5
Asp	−3.5
Asn	−3.5
Lys	−3.9
Arg	−4.5

Source: Kyte, J. and Doolittle, R.F., *J. Mol. Biol.* **157**, 110 (1982).

a protein is favored by the increase in entropy of the water molecules that would otherwise form ordered "cages" around the hydrophobic groups. The combined hydrophobic and hydrophilic tendencies of individual amino acid residues in proteins can be expressed as **hydropathies** (Table 6-2). The greater a side chain's hydropathy, the more likely it is to occupy the interior of a protein and vice versa. Hydropathies are good predictors of which portions of a polypeptide chain are inside a protein, out of contact with the aqueous solvent, and which portions are outside (Fig. 6-34).

Site-directed mutagenesis experiments in which individual interior residues have been replaced by a number of others suggest that the factors that affect stability are, in order, the hydrophobicity of the substituted residue, its steric compatibility, and, last, the volume of its side chain.

Electrostatic Interactions

In the closely packed interiors of native proteins, van der Waals forces, which are relatively weak (Section 2-1A), are nevertheless an important stabilizing influence. This is because these forces act over only short distances and hence are lost when the protein is unfolded.

Perhaps surprisingly, *hydrogen bonds, which are central features of protein structures, make only minor contributions to protein stability.* This is because hydrogen-bonding groups in an unfolded protein form energetically equivalent hydrogen bonds with water molecules. Nevertheless, hydrogen bonds are important determinants of native protein structures, because if a protein folded in a way that prevented a hydrogen bond from forming, the stabilizing energy of that hydrogen bond would be lost. Hydrogen bonding therefore fine-tunes tertiary structure by "selecting" the unique native structure of a protein from among a relatively small number of hydrophobically stabilized conformations.

The association of two ionic protein groups of opposite charge (e.g., Lys and Asp) is known as an **ion pair** or **salt bridge.** About 75% of the charged residues in proteins are members of ion pairs that are located mostly on the protein surface. Despite the strong electrostatic attraction between the

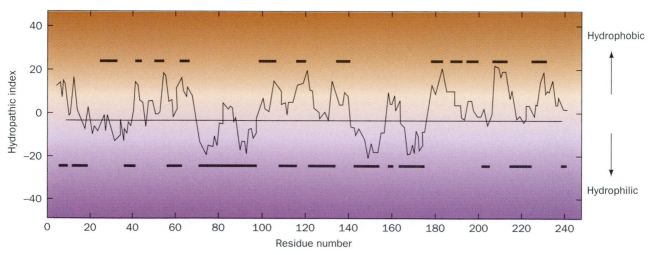

Figure 6-34.   A hydropathic index plot for bovine chymotrypsinogen. The sum of the hydropathies of nine consecutive residues is plotted versus residue sequence number. A large positive hydropathic index indicates a hydrophobic region of the polypeptide, whereas a large negative value indicates a hydrophilic region. The upper bars denote the protein's interior regions, as determined by X-ray crystallography, and the lower bars denote the protein's exterior regions. [After Kyte, J. and Doolittle, R.F., *J. Mol. Biol.* **157**, 111 (1982).]

oppositely charged members of an ion pair, these interactions contribute little to the stability of a native protein. This is because the free energy of an ion pair's charge–charge interactions usually fails to compensate for the loss of entropy of the side chains and the loss of solvation free energy when the charged groups form an ion pair. This accounts for the observation that ion pairs are poorly conserved among homologous proteins.

Chemical Cross-links

Disulfide bonds (Fig. 4-6) within and between polypeptide chains form as a protein folds to its native conformation. Some polypeptides whose Cys residues have been derivatized to prevent disulfide bond formation can still assume their fully active conformations, suggesting that disulfide bonds are not essential stabilizing forces. They may, however, be important for "locking in" a particular backbone folding pattern as the protein proceeds from its fully extended state to its mature form.

Disulfide bonds are rare in intracellular proteins because the cytoplasm is a reducing environment. Most disulfide bonds occur in proteins that are secreted from the cell into the more oxidizing extracellular environment. The relatively hostile extracellular world (e.g., uncontrolled temperature and pH) apparently requires the additional structural constraints conferred by disulfide bonds.

Metal ions may also function to internally cross-link proteins. For example, at least ten motifs collectively known as **zinc fingers** have been described in nucleic acid–binding proteins. These structures contain about 25–60 residues arranged around one or two Zn^{2+} ions that are tetrahedrally coordinated by the side chains of Cys, His, and occasionally Asp or Glu (Fig. 6-35). The Zn^{2+} allows relatively short stretches of polypeptide chain to fold into stable units that can interact with nucleic acids. Zinc fingers are too small to be stable in the absence of Zn^{2+}. Zinc is ideally suited to its structural role in intracellular proteins: Its filled *d* electron shell permits it to interact strongly with a variety of ligands (e.g., sulfur, nitrogen, or oxygen) from different amino acid residues. In addition, zinc has only one stable oxidation state (unlike, for example, copper and iron), so it does not undergo oxidation–reduction reactions in the cell.

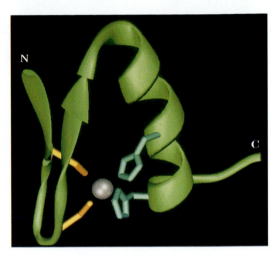

Figure 6-35. **A zinc finger motif.** This structure, from the DNA-binding protein **Zif268**, is known as a Cys_2–His_2 zinc finger because the zinc atom (*silver*) is coordinated by two Cys residues (*yellow*) and two His residues (*cyan*). [Based on an X-ray structure by Carl Pabo, MIT.]

B. Protein Denaturation and Renaturation

The low conformational stabilities of native proteins make them easily susceptible to denaturation by altering the balance of the weak nonbonding forces that maintain the native conformation. Proteins can be denatured by a variety of conditions and substances:

1. Heating causes a protein's conformationally sensitive properties, such as optical rotation (Section 4-2), viscosity, and UV absorption, to change abruptly over a narrow temperature range. Such a sharp transition indicates that the entire polypeptide unfolds or "melts" **cooperatively,** that is, nearly simultaneously. Most proteins have melting points well below 100°C. Among the exceptions are the proteins of thermophilic bacteria, organisms that inhabit hot springs or submarine volcanic vents with temperatures near 100°C. Amazingly, the X-ray structures of these heat-stable proteins are only subtly different from those of their low-temperature homologs.

2. pH variations alter the ionization states of amino acid side chains, thereby changing protein charge distributions and hydrogen bonding requirements.

3. Detergents associate with the nonpolar residues of a protein, thereby interfering with the hydrophobic interactions responsible for the protein's native structure.

4. The **chaotropic agents** guanidinium ion and urea,

$$H_2N-\overset{\overset{\displaystyle NH_2^+}{\|}}{C}-NH_2 \qquad\qquad H_2N-\overset{\overset{\displaystyle O}{\|}}{C}-NH_2$$

Guanidinium ion   **Urea**

in concentrations in the range 5 to 10 M, are the most commonly used protein denaturants. Chaotropic agents are ions or small organic molecules that increase the solubility of nonpolar substances in water. Their effectiveness as denaturants stems from their ability to disrupt hydrophobic interactions, although their mechanism of action is not well understood.

Denatured Proteins Can Be Renatured

In 1957, the elegant experiments of Christian Anfinsen on **ribonuclease A (RNase A)** showed that proteins can be denatured reversibly. RNase A, a 124-residue single-chain protein, is completely unfolded and its four disulfide bonds reductively cleaved in an 8 M urea solution containing 2-mercaptoethanol (Fig. 6-36). Dialyzing away the urea and reductant and exposing the resulting solution to O_2 at pH 8 (which oxidizes the SH groups to form disulfides) yields a protein that is virtually 100% enzymatically active and physically indistinguishable from native RNase A. The protein must therefore **renature** spontaneously.

The renaturation of RNase A demands that its four disulfide bonds reform. The probability of one of the eight Cys residues randomly forming a disulfide bond with its proper mate among the other seven Cys residues is 1/7; that one of the remaining six Cys residues then randomly forming its proper disulfide bond is 1/5; etc. The overall probability of RNase A reforming its four native disulfide links at random is

$$\frac{1}{7} \times \frac{1}{5} \times \frac{1}{3} \times \frac{1}{1} = \frac{1}{105}$$

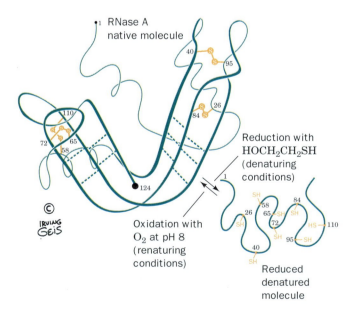

Figure 6-36. The reductive denaturation and oxidative renaturation of RNase A. [Figure copyrighted © by Irving Geis.]

Clearly, the disulfide bonds do not randomly re-form under renaturing conditions, since, if they did, only 1% of the refolded protein would be catalytically active. Indeed, if the RNase A is reoxidized in 8 M urea so that its disulfide bonds re-form while the polypeptide chain is a random coil, then after removal of the urea, the RNase A is, as expected, only ~1% active. This "scrambled" protein can be made fully active by exposing it to a trace of 2-mercaptoethanol, which breaks the improper disulfide bonds and allows the proper bonds to form. *Anfinsen's work demonstrated that proteins can fold spontaneously into their native conformations under physiological conditions. This implies that a protein's primary structure dictates its three-dimensional structure.*

C. Protein Folding Pathways

Studies of protein stability and renaturation suggest that protein folding is directed largely by the residues that occupy the interior of the folded protein. But *how* does a protein fold to its native conformation? One might guess that this process occurs through the protein's random exploration of all the conformations available to it until it eventually stumbles onto the correct one. A simple calculation first made by Cyrus Levinthal, however, convincingly demonstrates that this cannot possibly be the case: Assume that an *n*-residue protein's 2^n torsion angles, ϕ and ψ, each have three stable conformations. This yields $3^{2n} \approx 10^n$ possible conformations for the protein (a gross underestimate because we have completely neglected its side chains). Then, if the protein could explore a new conformation every 10^{-13} s (the rate at which single bonds reorient), the time *t*, in seconds, required for the protein to explore all the conformations available to it is

$$t = \frac{10^n}{10^{13}}$$

For a small protein of 100 residues, $t = 10^{87}$ s, which is immensely greater than the apparent age of the universe (20 billion years, or 6×10^{17} s).

In fact, many proteins fold to their native conformations in less than a few seconds. This is because *proteins fold to their native conformations via directed pathways rather than stumbling on them through random conformational searches.* Thus, as a protein folds, its conformational stability increases sharply (i.e., its free energy decreases sharply), which makes folding a one-way process. A hypothetical folding pathway is diagrammed in Fig. 6-37.

Experimental observations indicate that protein folding begins with the formation of local segments of secondary structure (α helices and β sheets). This early stage of protein folding is extremely rapid, with much of the native secondary structure in small proteins appearing within 5 ms of the initiation of folding. Since native proteins contain compact hydrophobic cores, it is likely that the driving force in protein folding is what has been termed a **hydrophobic collapse.** The collapsed state is known as a **molten globule,** a species that has much of the secondary structure of the native protein but little of its tertiary structure. Over the next 5 to 1000 ms, the secondary structure becomes stabilized and tertiary structure begins to form. During this intermediate stage, the nativelike elements are thought to take the form of subdomains that are not yet properly docked to form domains. In the final stage of folding, which for small single-domain proteins occurs over the next few seconds, the protein undergoes a series of complex motions in which it attains its relatively rigid internal side chain packing and hydrogen bonding while it expels the remaining water molecules from its hydrophobic core.

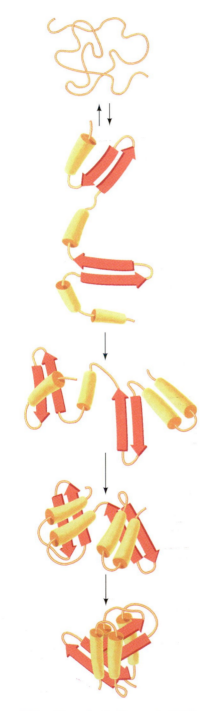

Figure 6-37. **Hypothetical protein folding pathway.** This example shows a linear pathway for folding a two-domain protein. [After Goldberg, M.E., *Trends Biochem. Sci.* **10,** 389 (1985).]

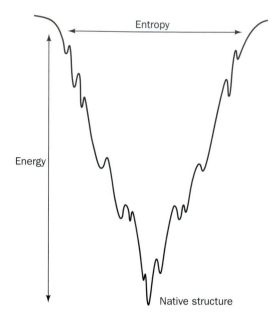

Figure 6-38.   Energy–entropy diagram for protein folding. The width of the diagram represents entropy, and the depth, the energy. The unfolded polypeptide proceeds from a high-entropy, disordered state (*wide*) to a single low-entropy (*narrow*), low-energy native conformation. [After Onuchic, J.N., Wolynes, P.G., Luthey-Schulten, Z., and Socci, N.D., *Proc. Natl. Acad. Sci.* **92**, 3626 (1995).]

In multidomain and multisubunit proteins, the respective units then assemble in a similar manner, with a few slight conformational adjustments required to produce the protein's native tertiary or quaternary structure. Thus, *proteins appear to fold in a hierarchical manner, with small local elements of structure forming and then coalescing to yield larger elements, which coalesce with other such elements to form yet larger elements, etc.*

Folding, like denaturation, appears to be a cooperative process, with small elements of structure accelerating the formation of additional structures. A folding protein must proceed from a high-energy, high-entropy state to a low-energy, low-entropy state. This energy–entropy relationship is diagrammed in Fig. 6-38. An unfolded polypeptide has many possible conformations (high entropy). As it folds into an ever-decreasing number of possible conformations, its entropy and free energy decrease. The energy–entropy diagram is not a smooth valley but a jagged landscape. Minor clefts and gullies represent conformations that are temporarily trapped until, through random thermal activation, they overcome a slight "uphill" free energy barrier and can then proceed to a lower energy conformation. Evidently, *proteins have evolved to have efficient folding pathways as well as stable native conformations.* Nevertheless, misfolded proteins do occur in nature, and their accumulation is believed to be the cause of a variety of neurological diseases (see Box 6-4).

Protein Disulfide Isomerase

Even under optimal experimental conditions, proteins often fold more slowly *in vitro* than they fold *in vivo*. One reason is that folding proteins often form disulfide bonds not present in the native proteins, which then slowly form native disulfide bonds through the process of disulfide interchange. **Protein disulfide isomerase (PDI)** catalyzes this process. Indeed, the observation that RNase A folds so much faster *in vivo* than *in vitro* led Anfinsen to discover this enzyme.

PDI binds to a wide variety of unfolded polypeptides via a hydrophobic patch on its surface. A Cys —SH group on PDI reacts with a disulfide group on the polypeptide to form a mixed disulfide and a Cys —SH group on the polypeptide (Fig. 6-39a). Another disulfide group on the polypeptide, brought into proximity by the spontaneous folding of the polypeptide, is attacked by this Cys —SH group. The newly liberated Cys —SH group then repeats this process with another disulfide bond, and so on, ultimately yielding the polypeptide containing only native disulfide bonds, along with regenerated PDI.

Oxidized (disulfide-containing) PDI also catalyzes the initial formation of a polypeptide's disulfide bonds by a similar mechanism (Fig. 6-39b). In this case, the reduced PDI reaction product must be reoxidized by cellular oxidizing agents in order to repeat the process.

Molecular Chaperones

Proteins begin to fold as they are being synthesized, so the renaturation of a denatured protein *in vitro* may not mimic the folding of a protein *in vivo*. In addition, proteins fold *in vivo* in the presence of extremely high concentrations of other proteins with which they can potentially interact. ***Molecular chaperones*** *are essential proteins that bind to unfolded and partially folded polypeptide chains to prevent the improper association of exposed hydrophobic segments that might lead to non-native folding as well as polypeptide aggregation and precipitation.* This is especially important for multidomain and multisubunit proteins, whose components must fold fully before they can properly associate with each other. Molecular chap-

(a)

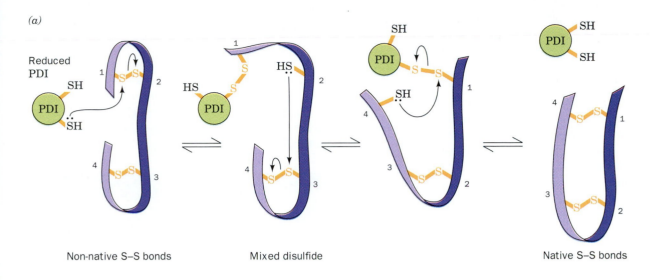

(b)

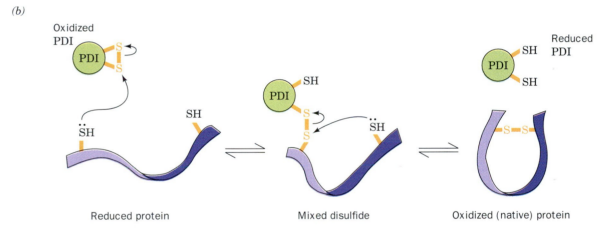

Figure 6-39.   Mechanism of protein disulfide isomerase.
(*a*) Reduced (SH-containing) PDI catalyzes the rearrangement of a polypeptide's non-native disulfide bonds via disulfide inter-change reactions to yield native disulfide bonds. (*b*) Oxidized (disulfide-containing) PDI catalyzes the initial formation of a polypeptide's disulfide bonds through the formation of a mixed disulfide. Reduced PDI can then react with a cellular oxidizing agent to regenerate oxidized PDI. ✳ **See the Animated Figures.**

erones also allow misfolded proteins to refold into their native conformations.

Many molecular chaperones were first described as **heat shock proteins (Hsp)** because their rate of synthesis is increased at elevated temperatures. Presumably, the additional chaperones are required to recover heat-denatured proteins or to prevent misfolding under conditions of environmental stress. There are two major classes of molecular chaperones in both prokaryotes and eukaryotes: the **Hsp70** family of 70-kD proteins and the **chaperonins,** which are large multisubunit proteins.

An Hsp70 protein binds to a newly synthesized polypeptide, possibly as soon as the first 30 amino acids have been polymerized at the ribosome. The Hsp70 chaperone probably helps prevent premature folding.

Chaperonins consist of two types of proteins:

1. The **Hsp60 proteins (GroEL** in *E. coli*), which are composed of 14 identical ~60-kD subunits arranged in two stacked rings of 7 subunits

Box 6-4
BIOCHEMISTRY IN HEALTH AND DISEASE

Diseases Related to Protein Folding

A number of neurological diseases are characterized by insoluble protein aggregates, called **amyloid deposits**, in brain and other tissues. These diseases, including the most common, **Alzheimer's disease**, result from the precipitation of a single kind of protein. Protein chemists are familiar with partially folded polypeptides that precipitate indiscriminately *in vitro*, probably due to exposure of hydrophobic patches. It is somewhat surprising that *in vivo*, in the presence of high concentrations of many other proteins, partially folded protein molecules aggregate only with other molecules of the same protein. What is even more surprising is that the molecules that precipitate are apparently misfolded versions of molecules that are normally present in the same tissues.

The **β amyloid protein** that forms the fibrous deposits, or **plaques**, in the brains of Alzheimer's patients is a 40-residue segment that is cleaved from a larger precursor protein. Its normal function is not known. Some mutations in the precursor increase β amyloid production. The increased concentration of the peptide apparently increases the likelihood that a misfolded peptide will initiate the aggregation that eventually produces an amyloid plaque. Plaques appear to be the primary cause, not a side effect, of the neurological deterioration in the disease.

Abnormal protein folding is almost certainly a factor in a group of infectious disorders that include **bovine spongiform encephalopathy** ("mad cow disease"), **scrapie** in sheep, and several rare fatal human diseases including **Creutzfeldt–Jakob disease**. According to a hypothesis advanced by Stan-

ley Prusiner, these diseases are caused by a protein known as a **prion**. The prion protein is present in normal brain tissue, where it apparently has a mostly α helical conformation (the structural model on the left, which was derived using structure-prediction techniques).

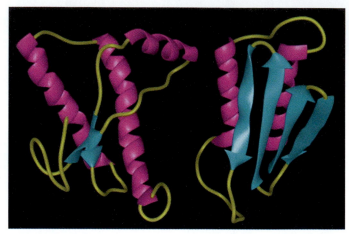

[Figure adapted from Fred Cohen, University of California at San Francisco.]

In diseased brains, the same protein—now apparently a mixture of β sheet and α helices (model on the right)—forms insoluble fibrous aggregates that damage brain cells. The infectious nature of prions is believed to result from the ability of an abnormally folded prion protein to catalyze the misfolding of normal prion proteins that then aggregate.

---

each, thereby forming a hollow cylinder with D_7 symmetry (Section 6-3).

2. The **Hsp10 proteins** (**GroES** in *E. coli*), which consist of 7 identical ~10-kD subunits arranged with 7-fold rotational (C_7) symmetry to form a dome-shaped complex.

The X-ray structure of the GroEL–GroES complex determined by Paul Sigler shows, in agreement with previous electron microscopy studies, that one open end of the GroEL cylinder is capped by a GroES complex (Fig. 6-40). The interior of the cylinder provides a protected environment in which a protein can fold without aggregating with other partially folded proteins.

The interior of the GroEL cylinder contains hydrophobic patches that bind the exposed groups of its enclosed and improperly folded protein. The GroEL subunits bind ATP and catalyze its hydrolysis to ADP and inorganic phosphate (P_i), a process that motivates a conformational change that masks the hydrophobic patches. The bound protein is thereby released and stimulated to continue folding. The binding and release, in effect, frees the partially folded protein from its entrapment in a local free energy minimum, such as those in Fig. 6-38, which permits the folding protein to con-

(a)

(b)

(c)

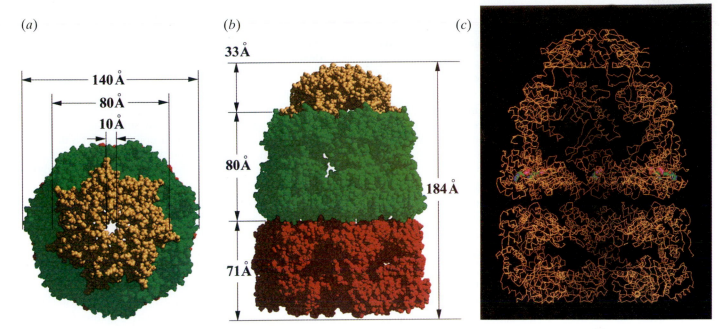

Figure 6-40.   The X-ray structure of the GroEL–GroES–(ADP)₇ chaperonin complex. Two seven-membered rings of GroEL subunits stack to form a hollow cylinder that is capped at one end by a seven-membered ring of GroES subunits. (*a*) A space-filling representation of the complex as viewed down its 7-fold axis of rotation. The GroES ring is gold, its adjacent GroEL ring is green, and the adjoining GroEL ring is red. (*b*) As in *a* but viewed perpendicularly to the 7-fold axis. Note the different con-formations of the two GroEL rings. (*c*) The C_α backbones of the complex as viewed perpendicularly to the 7-fold axis and cut away in the plane containing this axis. The ADPs, which are bound to the lower portion of each subunit in the upper GroEL ring, are shown in space-filling form. Note the large cavity formed by the GroES cap and the upper GroEL ring in which a polypeptide can fold in isolation. [Courtesy of Paul Sigler, Yale University.]

tinue its descent down its funnel toward the native state. In the protected environment inside the hollow GroEL–GroES (Hsp60–Hsp10) barrel, a >70-kD protein can fold out of contact with other proteins with which it otherwise might aggregate. *This cycle of ATP-driven binding, release, and refolding is repeated until the protein achieves its native conformation.* The Hsp70 proteins are thought to follow a similar pathway of binding and ATP-driven release of a folding protein.

D. Protein Dynamics

The precision with which protein structures are determined may leave the false impression that proteins have fixed and rigid structures. In fact, *proteins are flexible and rapidly fluctuating molecules whose structural mobilities are functionally significant.* Groups ranging in size from individual side chains to entire domains or subunits may be displaced by up to several angstroms through random intramolecular movements or in response to a trigger such as the binding of a small molecule. Extended side chains, such as Lys, and the N- and C-termini of polypeptide chains are especially prone to wave around in solution because there are few forces holding them in place.

Theoretical calculations by Martin Karplus indicate that a protein's native structure probably consists of a large collection of rapidly intercon-verting conformations that have essentially equal stabilities (Fig. 6-41). Conformational flexibility, or **breathing,** with structural displacement of up to ~2 Å, allows small molecules to diffuse in and out of the interior of certain proteins.

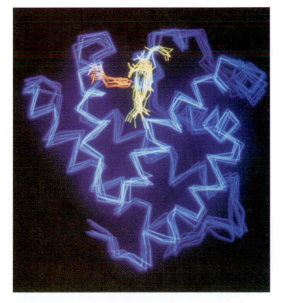

Figure 6-41.   Molecular dynamics of myoglobin. Several "snapshots" of the protein calculated at intervals of 5×10^{-12} s are superimposed. The backbone is blue, the heme group is yellow, and the His side chain linking the heme to the protein is orange. [Courtesy of Martin Karplus, Harvard University.]

SUMMARY

1. Four levels of structural complexity are used to describe the three-dimensional shapes of proteins.

2. The conformational flexibility of the peptide group is described by its ϕ and ψ torsion angles.

3. The α helix is a regular secondary structure in which hydrogen bonds form between backbone groups four residues apart. In the β sheet, hydrogen bonds form between the backbones of separate polypeptide strands.

4. Fibrous proteins are characterized by a single type of secondary structure: α keratin is a left-handed coil of two α helices; silk fibroin is an array of stacked β sheets; and collagen is a left-handed triple helix with three residues per turn.

5. Nonrepetitive structures include variations in regular secondary structures, turns, and loops.

6. The tertiary structures of proteins, which can be determined by X-ray crystallography or NMR techniques, may contain motifs (supersecondary structures) and domains.

7. The nonpolar side chains of a globular protein tend to occupy the protein's interior, whereas the polar side chains tend to define its surface.

8. Protein structures can be grouped into families according to their folding patterns. Structural elements are more likely to be evolutionarily conserved than are amino acid sequences.

9. The individual units of multisubunit proteins are usually symmetrically arranged.

10. Native protein structures are only slightly more stable than their denatured forms. The hydrophobic effect is the primary determinant of protein stability. Hydrogen bonding and ion pairing contribute relatively little to a protein's stability.

11. Studies of protein denaturation and renaturation indicate that the primary structure of a protein determines its three-dimensional structure.

12. Proteins fold to their native conformations via directed pathways in which small elements of structure coalesce into larger structures.

13. Protein disulfide isomerase and molecular chaperones facilitate protein folding *in vivo*.

14. Proteins have some conformational flexibility that results in small molecular motions.

REFERENCES

General

Branden, C. and Tooze, J., *Introduction to Protein Structure*, Garland Publishing (1991).

Chothia, C. and Finkelstein, A.V., The classification and origins of protein folding patterns, *Annu. Rev. Biochem.* **59**, 1007–1039 (1990).

Creighton, T.E., *Proteins* (2nd ed.), Chapters 4–6, Freeman (1993).

Darby, N.J. and Creighton, T.E., *Protein Structure*, IRL Press (1993).

Holm, L. and Sander, C., Mapping the protein universe, *Science* **273**, 595–602 (1996). [Discusses the limited numbers of structural domains.]

Kyte, J., *Structure in Protein Chemistry*, Garland Publishing (1995).

Specific Proteins

Baldwin, M.A., Cohen, F.E., and Prusiner, S.B., Prion protein isoforms, a convergence of biological and structural investigations, *J. Biol. Chem.* **270**, 19197–19200 (1995).

Bella, J., Eaton, M., Brodsky, B., and Berman, H.M., Crystal and molecular structure of a collagen-like peptide at 1.9 Å resolution, *Science* **266**, 75–81 (1994).

Kaplan, D., Adams, W.W., Farmer, B., and Viney, C., *Silk Polymers*, American Chemical Society (1994).

Prokop, D.J. and Kivirikko, K.I., Collagens: biology, disease, and potentials for therapy, *Annu. Rev. Biochem.* **64**, 403–434 (1995).

Techniques

Karplus, M. and Petsko, G.A., Molecular dynamics simulations in biology, *Nature* **347**, 631–639 (1990).

Rhodes, G., *Crystallography Made Crystal Clear*, Academic Press (1993).

Wagner, G., Hyberts, S.G., and Havel, T.F., NMR structure determination in solution. A critique and comparison with X-ray crystallography, *Annu. Rev. Biophys. Biomol. Struct.* **21**, 167–198 (1992).

Folding and Stabilization

Aurora, R. and Rose, G.D., Helix capping, *Protein Sci.* **7**, 21–38 (1998).

Bardwell, J.C.A. and Beckwith, J., The bonds that tie: catalyzed disulfide bond formation, *Cell* **74**, 769–771 (1993).

Bukau, B. and Horwich, A.L., The Hsp70 and Hsp60 chaperone machines, *Cell* **92**, 351–356 (1998).

Cordes, M.H.J., Davidson, A.R., and Sauer, R.T., Sequence space, folding and protein design, *Curr. Opin. Struct. Biol.* **6**, 3–10 (1996).

Dahiyat, B. and Mayo, S.L., De novo protein design: Fully automated sequence selection, *Science* **278**, 82–87 (1997).

Dill, K.A. and Chan, H.S., From Levinthal pathways to funnels, *Nature Struct. Biol.* **4**, 10–19 (1997). [A highly readable review of modern theories of protein folding.]

Frydman, J. and Hartl, F.U., Principles of chaperone-assisted protein folding: differences between in vitro and in vivo mechanisms, *Science* **272**, 1497–1502 (1996).

Honig, B. and Yang, A.-S., Free energy balance in protein fold-ing, *Adv. Prot. Chem.* **46,** 27–58 (1995).

Matthews, B.W., Structural and genetic analysis of protein sta-bility, *Annu. Rev. Biochem.* **62,** 139–160 (1993).

Rost, B. and Sander, C., Bridging the protein-sequence–structure gap by structure predictions, *Annu. Rev. Biophys. Biomol. Struct.* **25,** 113–136 (1996).

Schwabe, J.W.R. and Klug, A., Zinc mining for protein domains, *Nature Struct. Biol.* **1,** 345–349 (1994).

Stickle, D.F., Presta, L.G., Dill, K.A., and Rose, G.D., Hydrogen bonding in globular proteins, *J. Mol. Biol.* **226,** 1143–1159 (1992).

KEY TERMS

secondary structure
tertiary structure
quaternary structure
trans conformation
cis conformation
backbone
torsion (dihedral) angle
ϕ
ψ
Ramachandran diagram
α helix
pitch
parallel β sheet
antiparallel β sheet

topology
fibrous protein
globular protein
coiled coil
denatured
native
β bulge
helix capping
reverse turn (β bend)
Ω loop
X-ray crystallography
NMR
diffraction pattern
contour map

supersecondary structure
(motif)
$\beta\alpha\beta$ motif
β hairpin
$\alpha\alpha$ motif
β barrel
dinucleotide-binding
(Rossmann) fold
domain
oligomer
protomer
rotational symmetry
cyclic symmetry
dihedral symmetry

hydropathy
ion pair (salt bridge)
zinc finger
renaturation
cooperativity
chaotropic agent
hydrophobic collapse
molten globule
molecular chaperone
heat shock protein
breathing

STUDY EXERCISES

1. Explain why the conformational freedom of peptide bonds is limited.

2. What distinguishes regular and irregular secondary struc-tures?

3. Describe the hydrogen bonding pattern of an α helix.

4. Why are β sheets pleated?

5. What properties do fibrous proteins confer on substances such as hair, horns, bones, and tendons?

6. Why do turns and loops most often occur on the protein sur-face?

7. Which side chains usually occur in a protein's interior? On its surface?

8. Give some reasons why the number of possible protein struc-tures is much less than the number of amino acid sequences.

9. List the advantages of multiple subunits in proteins.

10. Why can't proteins have mirror symmetry?

11. Describe the forces that stabilize proteins.

12. Describe the energy and entropy changes that occur during protein folding.

13. How does protein renaturation *in vitro* differ from protein folding *in vivo*?

PROBLEMS

1. Draw a cis peptide bond and identify the groups that expe-rience steric interference.

2. Helices can be described by the notation n_m where n is the number of residues per helical turn and m is the number of atoms, including H, in the ring that is closed by the hydro-gen bond. (a) What is this notation for the α helix? (b) Is the 3_{10} helix steeper or shallower than the α helix?

3. Calculate the length in angstroms of a 100-residue segment of the α keratin coiled coil.

4. Is it possible for a native protein to be entirely irregular, that is, without α helices, β sheets, or other repetitive secondary structure?

5. (a) Is Trp or Gln more likely to be on a protein's surface? (b) Is Ser or Val less likely to be in the protein's interior? (c) Is Leu or Ile less likely to be found in the middle of an α helix? (d) Is Cys or Ser more likely to be in a β sheet?

6. Describe the structure of glyceraldehyde-3-phosphate dehy-drogenase (Fig. 6-30) as a linear sequence of α helix ("α") and strands of β sheet ("β") starting from the N-terminus. The N-terminal domain is red.

7. What types of rotational symmetry are possible for a protein with (a) four or (b) six identical subunits?

8. Given enough time, can all denatured proteins spontaneously renature?

9. Describe the intra- and intermolecular bonds/interactions that are broken or retained when collagen is heated to produce gelatin.

10. Under physiological conditions, polylysine assumes a random coil conformation. Under what conditions might it form an α helix?

11. It is often stated that proteins are quite large compared to the molecules they bind. However, what constitutes a large number depends on your point of view. Calculate the ratio of the volume of a hemoglobin molecule (65 kD) to that of the four O_2 molecules that it binds and the ratio of the volume of a typical office ($4 \times 4 \times 3$ m) to that of the typical (70 kg) office worker that occupies it. Assume that the molecular volumes of hemoglobin and O_2 are in equal proportions to their molecular masses and that the office worker has a density of 1.0 g/cm^3. Compare these ratios. Is this the result you expected?

12. Which of the following polypeptides is most likely to form an α helix? Which is least likely to form a β strand?

    (a) CRAGNRKIVLETY

    (b) SEDNFGAPKSILW

    (c) QKASVEMAVRNSG

[Problem by Bruce Wightman, Muhlenberg College.]

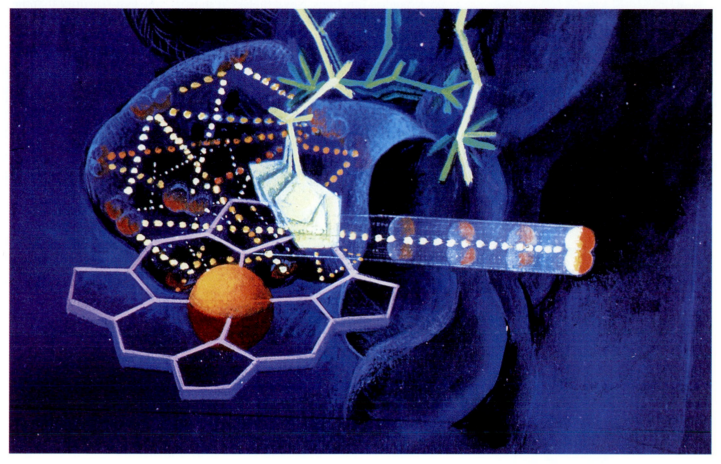

The structure of a protein determines its biological role, whether it is binding a small molecule or interacting with another large molecule. The oxygen binding site of myoglobin, for example, is structured so that O_2 can readily bind to it or, as pictured above, escape from the protein altogether. [Figure copyrighted © by Irving Geis.]

PROTEIN FUNCTION

The preceding two chapters have painted a broad picture of the chemical and physical properties of proteins but have not delved deeply into their physiological functions. Nevertheless, it should come as no surprise that the structural complexity and variety of proteins allow them to carry out an enormous array of specialized biological tasks. For example, the enzyme catalysts of virtually all metabolic reactions are proteins (we consider enzymes in detail in Chapters 11 and 12). Genetic information would remain locked in DNA were it not for the proteins that participate in decoding and transmitting that information. Remarkably, the thousands of proteins that participate in building, supporting, recognizing, transporting, and transforming cellular components act with incredible speed and accuracy and in many cases are subject to multiple regulatory mechanisms.

The specialized functions of proteins, from the fibrous proteins we examined in Section 6-1C to the precisely regulated metabolic enzymes we discuss in later chapters, can all be understood in terms of how proteins bind to and interact with other components of living systems. We shall encounter proteins that bind small molecules, proteins that interact with other proteins, and proteins that bind to much larger entities such as nucleic acids. In this chapter, we focus on just four of the many possible types of proteins whose functions are relatively well understood. These four examples—myoglobin, hemoglobin, muscle myosin and actin, and antibodies—are highlighted for several reasons. Studies of these proteins have been the source of some of the most significant advances in biochemistry; their proper functioning is vital for human health; and they are valuable models for many of the proteins we shall examine later when we discuss metabolism and the management of genetic information.

1. MYOGLOBIN

We begin our study of protein function with **myoglobin,** the first protein whose structure was determined by X-ray crystallography. In addition to myoglobin's importance as an oxygen-binding protein, its structure and function provide insights into the structure and function of hemoglobin, which is a tetramer of myoglobinlike polypeptides.

A. Myoglobin Structure

Myoglobin is a small intracellular protein in vertebrate muscle. Its X-ray structure, determined by John Kendrew in 1959, revealed that most of myoglobin's 153 residues are members of eight α helices (traditionally labeled A through H) that are arranged to form a globular protein with approximate dimensions $44 \times 44 \times 25$ Å (Fig. 7-1).

Myoglobin, other members of the globin family of proteins (Section 5-4B), and some other proteins such as cytochrome *c* (Sections 5-4A and 6-2C) contain a single **heme** group (Fig. 7-2). The heme is tightly wedged in a hydrophobic pocket between the E and F helices in myoglobin. The heterocyclic ring system of heme is a **porphyrin** derivative containing four **pyrrole** groups (labeled A–D) linked by methene bridges (other porphyrins vary in the substituents attached to rings A–D). The Fe(II) atom at the center of heme is coordinated by four porphyrin N atoms and one N from a His side chain (called, in a nomenclature peculiar to myoglobin and hemoglobin, His F8 because it is the eighth residue of the F helix). A molecule of oxygen (O_2) can act as a sixth ligand to the iron atom. His E7 (the seventh residue of helix E) hydrogen bonds to

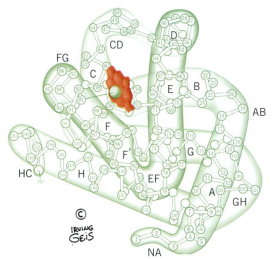

Figure 7-1.   Structure of sperm whale myoglobin. This 153-residue monomeric protein consists of eight α helices, labeled A through H, that are connected by short polypeptide links (the last half of what was originally thought to be the EF corner has been shown to form a short helix that is designated the F′ helix). The heme group is shown in red. [Figure copyrighted © by Irving Geis.] ● See Kinemage Exercise 6-1.

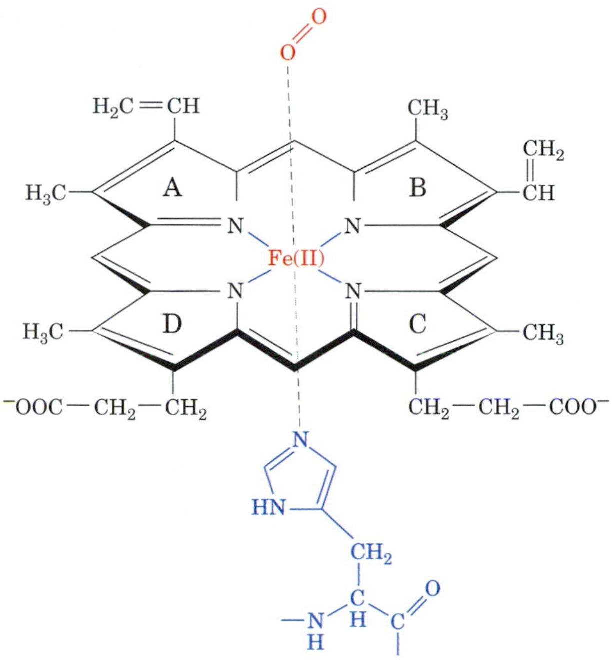

Figure 7-2.   The heme group. The central Fe(II) atom is shown liganded to four N atoms of the porphyrin ring, whose pyrrole groups are labeled A–D. The heme is a conjugated system, so all the Fe—N bonds are equivalent. The Fe(II) is also liganded to a His side chain and, when it is present, to O_2.

the O_2 with the geometry shown in Fig. 7-3. Two hydrophobic side chains on the O_2-binding side of the heme, Val E11 and Phe CD1 (the first residue in the segment between helices C and D), help hold the heme in place. These side chains presumably swing aside as the protein "breathes" (Section 6-4D), allowing O_2 to enter and exit.

When exposed to oxygen, the Fe(II) atom of isolated heme is irreversibly oxidized to Fe(III), a form that cannot bind O_2. The protein portion of myoglobin (and of hemoglobin, which contains four heme groups in four globin chains) prevents this oxidation and makes it possible for O_2 to bind reversibly to the heme group. **Oxygenation** alters the electronic state of the Fe(II)–heme complex, as indicated by its color change from dark purple (the color of hemoglobin in venous blood) to brilliant scarlet (the color of hemoglobin in arterial blood). Under some conditions, the Fe(II) of myoglobin or hemoglobin becomes oxidized to Fe(III) to form **metmyoglobin** or **methemoglobin,** respectively; these proteins are responsible for the brown color of old meat and dried blood.

In addition to O_2, certain other small molecules such as CO, NO, and H_2S can bind to heme groups in proteins. These other compounds bind with much higher affinity than O_2, which accounts for their toxicity. CO, for example, has 200-fold greater affinity for hemoglobin than does O_2.

B. Myoglobin Function

Although myoglobin was originally thought to be only an oxygen-storage protein, it is now apparent that *its major physiological role is to facilitate oxygen transport in muscle* (the most rapidly respiring tissue under conditions of high exertion). The rate at which O_2 can diffuse from the capillaries to the tissues is limited by its low solubility in aqueous solution ($\sim 10^{-4}$ M in blood). Myoglobin increases the effective solubility of O_2 in muscle, act-

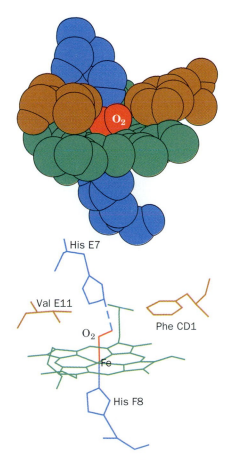

Figure 7-3.   The heme complex in myoglobin. In the upper drawing, atoms are represented in space-filling form (H atoms are not shown). The lower drawing shows the corresponding skeletal model with a dashed line representing the hydrogen bond between His E7 and the bound O_2.
● **See Kinemage Exercise 6-1.**

ing as a kind of molecular bucket brigade to boost the O_2 diffusion rate. The oxygen storage function of myoglobin is probably significant only in aquatic mammals such as seals and whales, whose muscle myoglobin concentrations are around 10-fold greater than that in terrestrial mammals (which is one reason why Kendrew chose the sperm whale as a source of myoglobin for his X-ray crystallographic studies).

The reversible binding of O_2 to myoglobin **(Mb)** is described by a simple equilibrium reaction:

$$Mb + O_2 \rightleftharpoons MbO_2$$

The dissociation constant, K, for the reaction is

$$K = \frac{[Mb][O_2]}{[MbO_2]} \qquad [7\text{-}1]$$

Note that biochemists usually express equilibria in terms of dissociation constants, the reciprocal of the association constants favored by chemists. The O_2 dissociation of myoglobin can be characterized by its **fractional saturation, Y_{O_2},** which is defined as the fraction of O_2-binding sites occupied by O_2:

$$Y_{O_2} = \frac{[MbO_2]}{[Mb] + [MbO_2]} = \frac{[O_2]}{K + [O_2]} \qquad [7\text{-}2]$$

Since O_2 is a gas, its concentration is conveniently expressed by its **partial pressure, pO_2** (also called the oxygen tension). Equation 7-2 can therefore be expressed as

$$Y_{O_2} = \frac{pO_2}{K + pO_2} \qquad [7\text{-}3]$$

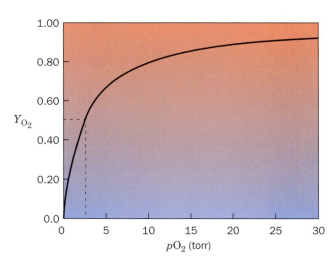

Figure 7-4.   **Oxygen binding curve of myoglobin.** Myoglobin is half-saturated with O_2 ($Y_{O_2} = 0.5$) at an oxygen pressure (pO_2) of 2.8 torr. The hyperbolic shape of myoglobin's binding curve is typical of the simple binding of a small molecule to a protein.

*This equation describes a rectangular **hyperbola** and is identical in form to the equations that describe a hormone binding to its cell-surface receptor or a small molecular substrate binding to the active site of an enzyme.* This hyperbolic function can be represented graphically as shown in Fig. 7-4. At low pO_2, very little O_2 binds to myoglobin (Y_{O_2} is very small). As the pO_2 increases, more O_2 binds to myoglobin. At very high pO_2, virtually all the O_2-binding sites are occupied and myoglobin is said to be **saturated** with O_2.

The steepness of the hyperbola for a simple binding event, such as O_2 binding to myoglobin, increases as the value of K decreases. When $pO_2 = K$, myoglobin is half-saturated with oxygen. This can be shown algebraically by substituting K for pO_2 in Eq. 7-3. Thus, K can be operationally defined as the value of pO_2 at which $Y = 0.5$ (Fig. 7-4).

It is convenient to define K as **p_{50}**, that is, the oxygen pressure at which myoglobin is 50% saturated. The p_{50} for myoglobin is 2.8 torr (760 torr = 1 atm). Over the physiological range of pO_2 in the blood (100 torr in arterial blood and 30 torr in venous blood), myoglobin is almost fully saturated with oxygen; for example, $Y_{O_2} = 0.97$ at $pO_2 = 100$ torr, and 0.91 at 30 torr. Consequently, *myoglobin efficiently relays oxygen from the capillaries to muscle cells.*

Myoglobin, a single polypeptide chain with one heme group and hence one oxygen-binding site, is a useful model for other binding proteins. Even proteins with multiple binding sites for the same small molecule, or **ligand,** generate hyperbolic binding curves like myoglobin's when the ligands interact with each binding site independently. In practice, the affinity of a ligand for its binding protein may not be known. Constructing a binding curve such as the one shown in Fig. 7-4 may provide this information.

2. HEMOGLOBIN

Hemoglobin, the intracellular protein that gives red blood cells their color, is one of the best-characterized proteins and was one of the first proteins to be associated with a specific physiological function (oxygen transport). However, hemoglobin is not just a simple oxygen tank; it is a sophisticated delivery system that provides the proper amount of oxygen to the tissues under a wide variety of circumstances. Animals that are too large (>1 mm thick) for simple diffusion to deliver sufficient oxygen to their tissues have circulatory systems containing hemoglobin or a protein of similar function (see Box 7-1) that does so.

The efficiency with which hemoglobin binds and releases O_2 is reminiscent of the specificity and efficiency of metabolic enzymes. Because many of the theories formulated to explain O_2 binding to hemoglobin also explain the control of enzyme activity, hemoglobin has been dubbed an "honorary enzyme." This section includes an examination of hemoglobin's structure, its cooperative oxygen binding behavior, and diseases resulting from hemoglobin defects.

A. Hemoglobin Structure

Mammalian hemoglobin, as we saw in Fig. 6-32, is a tetrameric protein with the quaternary structure $\alpha_2\beta_2$ (a dimer of $\alpha\beta$ protomers). The α and β subunits are structurally and evolutionarily related to each other and to myoglobin (the genealogy of the globin polypeptides is discussed in Section 5-4B).

Box 7-1

BIOCHEMISTRY IN CONTEXT

Other Oxygen-Transport Proteins

The presence of O_2 in the earth's atmosphere and its use in the combustion of metabolic fuels have driven the evolution of various mechanisms for storing and transporting oxygen. Small organisms rely on diffusion to supply their respiratory oxygen needs. However, since the rate at which a substance diffuses varies inversely with the square of the distance it must diffuse, organisms of >1 mm thickness overcome the constraints of diffusion with circulatory systems and boost the limited solubility of O_2 in water with specific O_2-transport proteins.

Many invertebrates, and even some plants and bacteria, contain heme-based O_2-binding proteins. Single-subunit and multimeric hemoglobins are found both as intracellular proteins and as extracellular components of blood and other body fluids. The sporadic appearance of hemoglobinlike proteins in bacteria raises the possibility of gene transfer from animals to bacteria at one or more points during evolution. In some leguminous plants, the so-called **leghemoglobins** bind O_2 that would otherwise interfere with bacterial nitrogen fixation. The **chlorocruorins**, which occur in some annelids (e.g., earthworms), contain a somewhat differently derivatized porphyrin than that in hemoglobin, which accounts for the green color of chlorocruorins.

The two other types of O_2-binding proteins, **hemocyanin** and **hemerythrin** (neither of which contains heme groups), occur only in invertebrate animals. Hemocyanin, which consists of ~75-kD subunits that each contain two Cu atoms ligated by His residues, is blue when oxygenated and colorless when deoxygenated. Oxygenated hemerythrin (subunit mass ~13 kD) contains two Fe atoms ligated by His and acidic residues and is purple (colorless when deoxygenated). Hemerythrin is exclusively intracellular, whereas hemocyanin is exclusively extracellular. Some invertebrates have myoglobin in their muscles and chlorocruorin or hemocyanin in their blood.

In general, most extracellular O_2-transport molecules are large multimeric proteins with masses up to several million daltons and over 100 O_2-binding sites. These proteins are often the predominant extracellular protein and may therefore have additional functions as buffers against pH changes and osmotic fluctuations. In some invertebrate species, the O_2-binding proteins may serve as a nutritional reserve, for example, during metamorphosis or molting.

The structure of hemoglobin was determined by Max Perutz, the father of protein X-ray crystallography, who worked with enormous optimism and tenacity for about 30 years to obtain a high-resolution X-ray structure of horse methemoglobin in 1968. Only about 18% of the residues are identi-

(a)

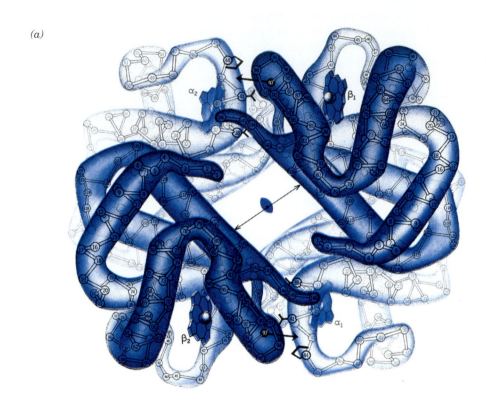

(b)

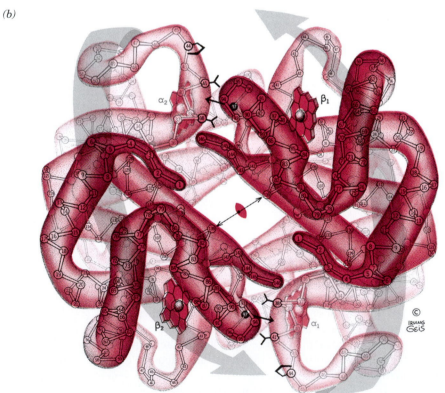

Figure 7-5.   Hemoglobin structure. (a) Deoxyhemoglobin and (b) oxyhemoglobin. The $\alpha_1\beta_1$ protomer is related to the $\alpha_2\beta_2$ protomer by a 2-fold axis of symmetry (*lenticular symbol*), which is perpendicular to the page. Oxygenation brings the β chains closer together (compare the lengths of the double-headed arrows) and shifts the contacts between subunits at the α_1–β_2 and α_2–β_1 interfaces (some of the relevant side chains are shown in black). The large gray arrows in *b* indicate the molecular movements that accompany oxygenation. [Figure copyrighted © by Irving Geis.] ● **See Kinemage Exercises 6-2 and 6-3.**

Figure 7-6.   The major structural differences between the quaternary conformations of (*a*) deoxyhemoglobin and (*b*) oxyhemoglobin. On oxygenation, the $\alpha_1\beta_1$ (*shaded*) and $\alpha_2\beta_2$ (*outlined*) protomers move, as indicated on the right, as rigid units such that there is an ~15° off-center rotation of one protomer relative to the other that preserves the molecule's exact 2-fold symmetry. Note how the position of His FG4β (*pentagons*) changes with respect to Thr C3α, Thr C6α, and Pro CD2α (*yellow dots*) at the α_1–β_2 and α_2–β_1 interfaces. The view is from the right side relative to that in Fig. 7-5. [Figure copyrighted © by Irving Geis.]

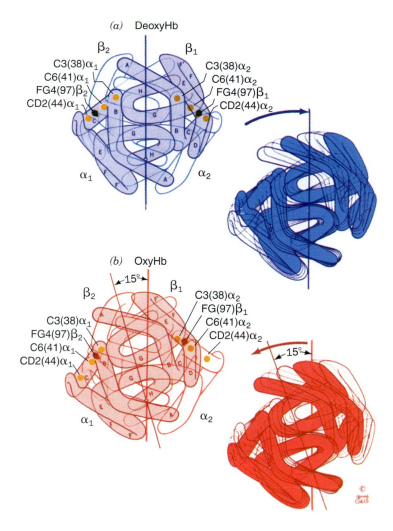

cal in myoglobin and in the α and β subunits of hemoglobin, but the three polypeptides have remarkably similar tertiary structures (hemoglobin subunits follow the myoglobin helix-labeling system, although the α chain has no D helix). The αβ protomers of hemoglobin are symmetrically related by a 2-fold rotation (i.e., a rotation of 180° brings the protomers into coincidence). In addition, hemoglobin's structurally similar α and β subunits are related by an approximate 2-fold rotation (pseudosymmetry) whose axis is perpendicular to that of the exact 2-fold rotation. Thus, hemoglobin has exact C_2 symmetry and pseudo-D_2 symmetry (Section 6-3; objects with D_2 symmetry have the rotational symmetry of a tetrahedron). The hemoglobin molecule has overall dimensions of about $64 \times 55 \times 50$ Å.

Oxygen binding alters the structure of the entire hemoglobin tetramer, so the structures of **deoxyhemoglobin** (Fig. 7-5*a*) and **oxyhemoglobin** (Fig. 7-5*b*) are noticeably different. In both forms of hemoglobin, the α and β subunits form extensive contacts: Those at the α_1–β_1 interface (and its α_2–β_2 symmetry equivalent) involve 35 residues, and those at the α_1–β_2 (and α_2–β_1) interface involve 19 residues. These associations are predominantly hydrophobic, although numerous hydrogen bonds and several ion pairs are also involved. Note, however, that the α_1–α_2 and β_1–β_2 interactions are tenuous at best because these subunit pairs are separated by an ~20-Å-diameter solvent-filled channel that parallels the 50-Å length of hemoglobin's exact 2-fold axis (Fig. 7-5).

When oxygen binds, the α_1–β_2 (and α_2–β_1) contacts shift, producing a change in quaternary structure. Oxygenation rotates one αβ dimer ~15° with respect to the other αβ dimer (Fig. 7-6), which brings the β subunits closer together and narrows the solvent-filled central channel of hemoglobin (Fig. 7-5). Some atoms in the α_1–β_2 and α_2–β_1 interfaces shift by as much as 6 Å (oxygenation causes such extensive quaternary structural changes that crystals of deoxyhemoglobin shatter on exposure to O_2). This structural rearrangement is a crucial element of hemoglobin's oxygen-binding behavior.

B. Oxygen Binding to Hemoglobin

Hemoglobin has an overall p_{50} of 26 torr (i.e., hemoglobin is half-saturated with O_2 at an oxygen pressure of 26 torr), which is nearly 10 times greater than the p_{50} of myoglobin. Moreover, hemoglobin does not exhibit a myoglobinlike hyperbolic oxygen binding curve. Instead, O_2 binding to hemo-

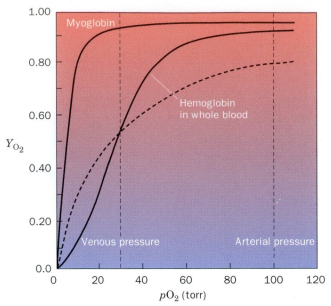

Figure 7-7. **Key to Function.** **Oxygen binding curve of hemoglobin.** In whole blood, hemoglobin is half-saturated at an oxygen pressure of 26 torr. The normal sea level values of human arterial and venous pO_2 are indicated (atmospheric pO_2 is 150–160 torr at sea level). The O_2 binding curve for myoglobin is included for comparison. The dashed line is a hyperbolic O_2 binding curve with the same p_{50} as hemoglobin. ✴ See the Animated Figures.

globin is described by a **sigmoidal** (S-shaped) curve (Fig. 7-7). *This permits the blood to deliver much more O_2 to the tissues than if hemoglobin had a hyperbolic curve with the same p_{50}* (dashed line in Fig. 7-7). For example, hemoglobin is nearly fully saturated with O_2 at arterial oxygen pressures ($Y_{O_2} = 0.95$ at 100 torr) but only about half-saturated at venous oxygen pressures ($Y_{O_2} = 0.55$ at 30 torr). This 0.40 difference in oxygen saturation, a measure of hemoglobin's ability to deliver O_2 from the lungs to the tissues, would be only 0.25 if hemoglobin exhibited hyperbolic binding behavior.

In any binding system, a sigmoidal curve is diagnostic of a **cooperative** *interaction between binding sites.* This means that the binding of a ligand to one site affects the binding of additional ligands to the other sites. In the case of hemoglobin, O_2 binding to one subunit increases the O_2 affinity of the remaining subunits. The initial slope of the oxygen binding curve (Fig. 7-7) is low, as hemoglobin subunits independently compete for the first O_2. However, an O_2 molecule bound to one of hemoglobin's subunits increases the O_2 binding affinity of its other subunits, thereby accounting for the increasing slope of the middle portion of the sigmoidal curve.

The Hill Equation Describes Hemoglobin's O_2 Binding Curve

The earliest attempt to analyze hemoglobin's sigmoidal O_2 dissociation curve was formulated by Archibald Hill in 1910. Hill assumed that hemoglobin (Hb) bound n molecules of O_2 in a single step,

$$Hb + n\,O_2 \longrightarrow Hb(O_2)_n$$

that is, with infinite cooperativity. Thus, in analogy with the derivation of Eq. 7-3,

$$Y_{O_2} = \frac{(pO_2)^n}{(p_{50})^n + (pO_2)^n} \qquad [7\text{-}4]$$

which is known as the **Hill equation.** It describes the degree of saturation of hemoglobin as a function of pO_2.

Infinite O_2 binding cooperativity, as Hill assumed, is a physical impossibility. Nevertheless, n may be taken to be a nonintegral parameter related to the degree of cooperativity among interacting hemoglobin subunits rather than the number of subunits that bind O_2 in one step. The Hill equation can then be taken as a useful empirical curve-fitting relationship rather than as an indicator of a particular model of ligand binding.

The quantity n, the **Hill constant,** *increases with the degree of cooperativity of a reaction and therefore provides a convenient although simplistic characterization of a ligand-binding reaction.* If $n = 1$, Eq. 7-4 describes a hyperbola as does Eq. 7-3 for myoglobin, and the O_2-binding reaction is said to be **noncooperative.** If $n > 1$, the reaction is described as being **positively cooperative,** because O_2 binding increases the affinity of hemoglobin for further O_2 binding (cooperativity is infinite in the limit that $n = 4$, the number of O_2-binding sites in hemoglobin). Conversely, if $n < 1$, the reaction is said to be **negatively cooperative,** because O_2 binding would then reduce the affinity of hemoglobin for subsequent O_2 binding.

The Hill constant, n, and the value of p_{50} that best describe hemoglobin's saturation curve can be graphically determined by rearranging Eq. 7-4 as follows:

$$\frac{Y_{O_2}}{1 - Y_{O_2}} = \frac{(pO_2)^n}{(p_{50})^n} \qquad [7\text{-}5]$$

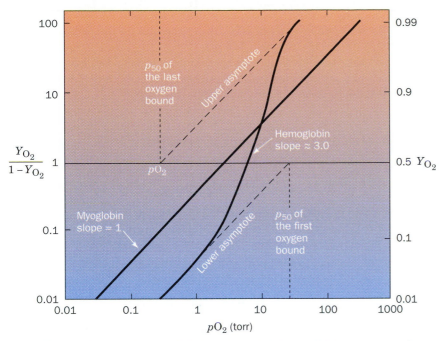

Figure 7-8.   Hill plots for myoglobin and purified hemoglobin. Note that this is a log–log plot. At $pO_2 = p_{50}$, $Y_{O_2}/(1 - Y_{O_2}) = 1$.

Taking the log of both sides yields a linear equation:

$$\log\left(\frac{Y_{O_2}}{1 - Y_{O_2}}\right) = n \log pO_2 - n \log p_{50} \qquad [7\text{-}6]$$

The linear plot of $\log[Y_{O_2}/(1 - Y_{O_2})]$ versus $\log pO_2$, the **Hill plot,** has a slope of n and an intercept on the $\log pO_2$ axis of $\log p_{50}$ (recall that the linear equation $y = mx + b$ describes a line with a slope of m and an x in-tercept of $-b/m$).

Figure 7-8 shows the Hill plots for myoglobin and hemoglobin. For myo-globin, the plot is linear with a slope of 1, as expected. Although hemo-globin does not bind O_2 in a single step as was assumed in deriving the Hill equation, its Hill plot is essentially linear for values of Y_{O_2} between 0.1 and 0.9. Its maximum slope, which occurs near $pO_2 = p_{50}$ $[Y_{O_2}/(1 - Y_{O_2}) = 1]$, is customarily taken to be the Hill constant. For normal human hemoglo-bin, the Hill constant is between 2.8 and 3.0; that is, hemoglobin's oxygen binding is highly, but not infinitely, cooperative. Many abnormal hemoglo-bins exhibit smaller Hill constants (Section 7-2D), indicating that they have a less than normal degree of cooperativity.

At Y_{O_2} values near zero, when few hemoglobin molecules have bound even one O_2 molecule, the Hill plot for hemoglobin assumes a slope of 1 (Fig. 7-8, lower asymptote) because the hemoglobin subunits independently compete for O_2 as do molecules of myoglobin. At Y_{O_2} values near 1, when at least three of each of hemoglobin's four O_2-binding sites are occupied, the Hill plot also assumes a slope of 1 (Fig. 7-8, upper asymptote) because the few remaining unoccupied sites are on different molecules and there-fore bind O_2 independently.

Extrapolating the lower asymptote in Fig. 7-8 to the horizontal axis in-dicates, according to Eq. 7-6, that $p_{50} = 30$ torr for binding the first O_2 to hemoglobin. Likewise, extrapolating the upper asymptote yields $p_{50} = 0.3$ torr for binding hemoglobin's fourth O_2. Thus, *the fourth O_2 to bind to* *hemoglobin does so with 100-fold greater affinity than the first.* This difference, as we shall see in Section 7-2C, is entirely due to the influence of the globin chain on the O₂ affinity of heme.

More realistic models than that used in deriving the Hill equation have been developed for analyzing the cooperative binding of O_2 to hemoglobin. We discuss them in Section 7-2E.

C. Mechanism of Oxygen Binding Cooperativity

The cooperativity of oxygen binding to hemoglobin arises from the effect of the ligand-binding state of one heme group on the ligand-binding affinity of another. Yet the hemes are 25 to 37 Å apart—too far to interact electronically. Instead, information about the O_2-binding status of a heme group is mechanically transmitted to the other heme groups by motions of the protein. These movements are responsible for the different quaternary structures of oxy- and deoxyhemoglobin depicted in Fig. 7-5.

Hemoglobin Has Two Conformational States

On the basis of the X-ray structures of oxy- and deoxyhemoglobin, Perutz formulated a mechanism for hemoglobin oxygenation. *In the Perutz mechanism, hemoglobin has two stable conformational states, the T state (the conformation of deoxyhemoglobin) and the R state (the conformation of oxyhemoglobin).* The conformations of all four subunits in T-state hemoglobin differ from those in the R state. Oxygen binding initiates a series of coordinated movements that result in a shift from the T state to the R state within a few microseconds:

1. In the T state, the Fe(II) in each of the four hemes is situated ~0.6 Å out of the heme plane because of a pyramidal doming of the porphyrin group toward His F8 (Fig. 7-9). O_2 binding changes the heme's electronic state, which shortens the Fe—N$_{porphyrin}$ bonds by ~0.1 Å and causes the porphyrin doming to subside. Consequently, during the T → R transition, the Fe(II) moves into the center of the heme plane.

2. The Fe(II) drags the covalently linked His F8 along with it. However, the direct movement of His F8 by 0.6 Å toward the heme plane would cause it to collide with the heme. To avoid this steric clash, the attached F helix tilts and translates by ~1 Å across the heme plane.

3. The changes in tertiary structure are coupled to a shift in the arrangement of hemoglobin's four subunits. The largest change produced by the T → R transition is the result of movements of residues at the α_1–β_2 and α_2–β_1 interfaces. In the T state, His FG4(97) (the fourth residue in the polypeptide segment that connects helices F and G, as well as the 97th residue overall) in the β chains contacts Thr C6(41) (the sixth residue of the C helix, or residue 41) in the α chains (Fig. 7-10a). In the R state, however, His FG4 contacts Thr C3(38), one turn back along the C helix (Fig. 7-10b). In both conformations, the "knobs" on one subunit mesh nicely with the "grooves" on the other. An intermediate position would be severely strained because it would bring His FG4 and Thr C6 too close together (i.e., knobs on knobs).

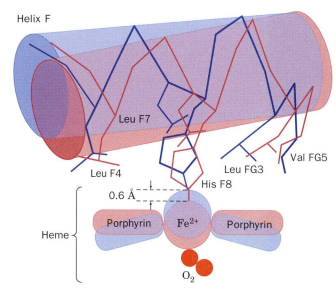

Figure 7-9. Movements of the heme and the F helix during the T → R transition in hemoglobin. In the T form *(blue)*, the Fe is 0.6 Å above the center of the domed porphyrin ring. On assuming the R form *(red)*, the Fe moves into the plane of the now undomed porphyrin, where it can more tightly bind O_2, and, in doing so, pulls His F8 and its attached F helix with it. ● See Kinemage Exercise 6-4. ✳ See the Animated Figures.

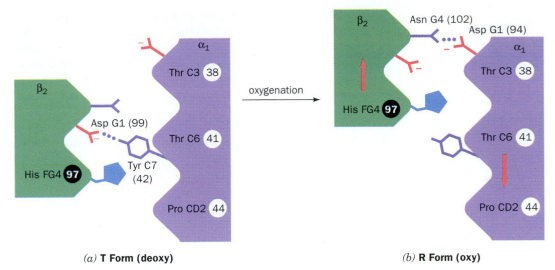

(a) **T Form (deoxy)**

(b) **R Form (oxy)**

Figure 7-10.   Changes at the α₁–β₂ interface during the T → R transition in hemoglobin. (a) In the T form, His FG4 (His 97) in the β₂ chain fits into the helical groove next to Thr C6 (Thr 41) in the α₁ chain. (b) In the R form, His FG4 has shifted position by one turn of the C helix, to lie next to Thr C3

(Thr 38) of α₁. These interactions are also shown in Figs. 7-5 and 7-6. Note that in the two conformations, the subunits are linked by different but equivalent hydrogen bonds. During the T → R transition, the subunit contacts snap from one position to the other with no stable intermediate. ● **See Kinemage Exercise 6-5.**

4. The C-terminal residues of each subunit (Arg 141α and His 146β) in T-state hemoglobin each participate in a network of intra- and inter-subunit ion pairs (Fig. 7-11) that stabilize the T state. However, the conformational shift in the T → R transition tears away these ion pairs in a process that is driven by the energy of formation of the Fe—O₂ bonds.

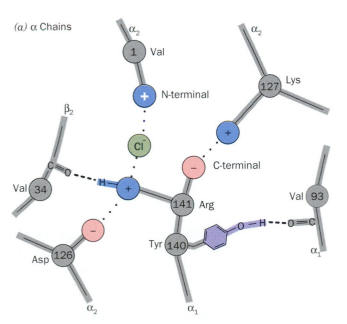

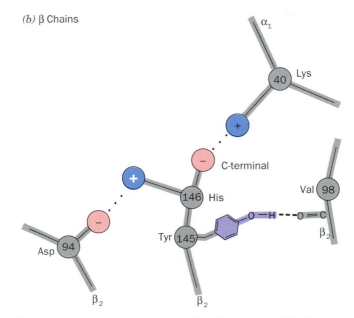

Figure 7-11.   Networks of ion pairs and hydrogen bonds in deoxyhemoglobin. These bonds, which involve the last two residues of (a) the α chains and (b) the β chains, are ruptured in

the T → R transition. Two groups that become partially deprotonated in the R state (part of the Bohr effect) are indicated by white plus signs. [Figure copyrighted © by Irving Geis.]

The essential feature of hemoglobin's T → R transition is that *its subunits are so tightly coupled that large tertiary structural changes within one subunit cannot occur without quaternary structural changes in the entire tetrameric protein*. Hemoglobin is limited to only two quaternary forms, T and R, because the intersubunit contacts shown in Fig. 7-10 act as a binary switch that permits only two stable positions of the subunits relative to each other. The inflexibility of the α_1–β_1 and α_2–β_2 interfaces requires that the T → R shift occur simultaneously at both the α_1–β_2 and α_2–β_1 interfaces. No one subunit or dimer can greatly change its conformation independently of the others.

We are now in a position to structurally rationalize the cooperativity of oxygen binding to hemoglobin. The T state of hemoglobin has low O_2 affinity, mostly because of the 0.1-Å greater length of its Fe—O_2 bond relative to that of the R state (e.g., the blue structure shown in Fig. 7-9). Experimental evidence indicates that when at least one O_2 has bound to each $\alpha\beta$ dimer, the strain in the T-state hemoglobin molecule is sufficient to tear away the C-terminal ion pairs, thereby snapping the protein into the R state. All the subunits are thereby simultaneously converted to the R-state conformation whether or not they have bound O_2. Unliganded subunits in the R-state conformation have increased oxygen affinity because they are already in the O_2-binding conformation. This accounts for the high O_2 affinity of nearly saturated hemoglobin.

Carbon Dioxide Transport and the Bohr Effect

The conformational changes in hemoglobin that occur on oxygen binding decrease the pK's of several groups, including the N-terminal amino group of the α subunits and the C-terminal His of the β subunits (see Fig. 7-11). In T-state hemoglobin, these positively charged acidic groups participate in ion pairs, which increases their pK's (makes them more basic), whereas those interactions are absent in R-state hemoglobin. Consequently, under physiological conditions, hemoglobin releases ~0.6 protons for each O_2 it binds. Conversely, increasing the pH, that is, removing protons, stimulates hemoglobin to bind more O_2 at lower oxygen pressures (Fig. 7-12). This phenomenon is known as the **Bohr effect** after Christian Bohr (father of the physicist Niels Bohr), who first reported it in 1904.

The Bohr effect has important physiological functions in transporting O_2 from the lungs to respiring tissue and in transporting the CO_2 produced by respiration back to the lungs (Fig. 7-13). The CO_2 produced by respiring tissues diffuses from the tissues to the capillaries. This dissolved CO_2 forms bicarbonate (HCO_3^-) only very slowly, by the reaction

$$CO_2 + H_2O \rightleftharpoons H^+ + HCO_3^-$$

However, in the **erythrocyte** (red blood cell; Greek: *erythrose*, red + *kytos*, a hollow vessel), the enzyme **carbonic anhydrase** greatly accelerates this reaction. Accordingly, most of the CO_2 in the blood is carried in the form of bicarbonate.

In the capillaries, where pO_2 is low, the H^+ generated by bicarbonate formation is taken up by hemoglobin in forming the ion pairs of the T state, thereby inducing hemoglobin to unload its bound O_2. This H^+ uptake, moreover, facilitates CO_2 transport by stimulating bicarbonate formation. Conversely, in the lungs, where pO_2 is high, O_2 binding by hemoglobin dis-

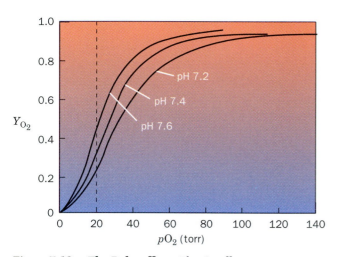

Figure 7-12. The Bohr effect. The O_2 affinity of hemoglobin increases with increasing pH. The dashed line indicates the pO_2 in actively respiring muscle. [After Benesch, R.E. and Benesch, R., *Adv. Protein Chem.* **28**, 212 (1974).] ✳ See the Animated Figures.

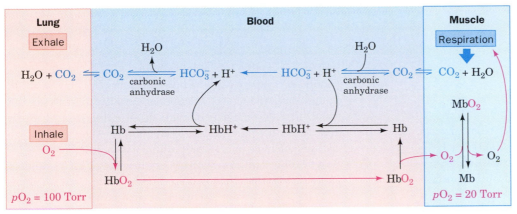

Figure 7-13. *Key to Function.* The roles of hemoglobin and myoglobin in transporting O_2 from the lungs to respiring tissues and CO_2 (as HCO_3^-) from the tissues to the lungs. Oxygen is inhaled into the lungs at high pO_2, where it binds to hemoglobin in the blood. The O_2 is then transported to respiring tissue, where the pO_2 is low. The O_2 therefore dissociates from the Hb and diffuses into the tissues, where it is used to oxidize metabolic fuels to CO_2 and H_2O. In rapidly respiring muscle tissue, the O_2 first binds to myoglobin (whose oxygen affinity is higher than that of hemoglobin). This increases the rate that O_2 can diffuse from the capillaries to the tissues by, in effect, increasing its solubility. The Hb and CO_2 (mostly as HCO_3^-) are then returned to the lungs, where the CO_2 is exhaled.

rupts the T-state ion pairs to form the R state, thereby releasing the Bohr protons, which recombine with bicarbonate to drive off CO_2. These reactions are closely matched, so they cause very little change in blood pH (see Box 2-2).

The Bohr effect provides a mechanism whereby additional oxygen can be supplied to highly active muscles, where the pO_2 may be <20 torr. Such muscles generate lactic acid (Section 14-3A) so fast that they lower the pH of the blood passing through them from 7.4 to 7.2. At a pO_2 of 20 torr, hemoglobin releases ~10% more O_2 at pH 7.2 than it does at pH 7.4 (Fig. 7-12).

CO_2 also modulates O_2 binding to hemoglobin by combining reversibly with the N-terminal amino groups of blood proteins to form **carbamates:**

$$R—NH_2 + CO_2 \rightleftharpoons R—NH—COO^- + H^+$$

The T (deoxy) form of hemoglobin binds more CO_2 as carbamates than does the R (oxy) form. When the CO_2 concentration is high, as it is in the capillaries, the T state is favored, stimulating hemoglobin to release its bound O_2. The protons released by carbamate formation further promote O_2 release through the Bohr effect. Although the difference in CO_2 binding between the oxy and deoxy states of hemoglobin accounts for only ~5% of the total blood CO_2, it is nevertheless responsible for around half the CO_2 transported by the blood. This is because only ~10% of the total blood CO_2 is lost through the lungs in each circulatory cycle.

Bisphosphoglycerate Binds to Deoxyhemoglobin

Highly purified ("stripped") hemoglobin has a much greater oxygen affinity than hemoglobin in whole blood (Fig. 7-14). This observation led Joseph Barcroft, in 1921, to speculate that blood contains some other sub-

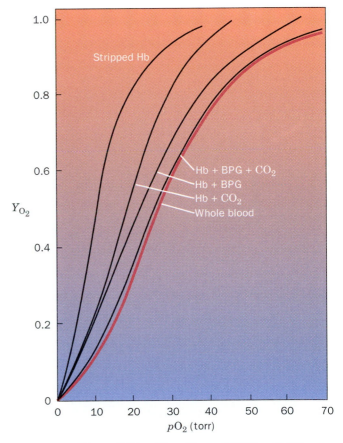

Figure 7-14. **The effects of BPG and CO_2 on hemoglobin's O_2 dissociation curve.** Stripped hemoglobin (*left*) has higher O_2 affinity than whole blood (*red curve*). Adding BPG or CO_2 or both to hemoglobin shifts the dissociation curve back to the right (lowers hemoglobin's O_2 affinity). [After Kilmartin, J.V. and Rossi-Bernardi, L., *Physiol. Rev.* **53**, 884 (1973).] ✳ See the Animated Figures.

Box 7-2

BIOCHEMISTRY IN HEALTH AND DISEASE

High-Altitude Adaptation

Atmospheric pressure decreases with altitude, so that the oxygen pressure at 10,000 feet is only ~110 torr, 70% of its sea-level pressure. A variety of physiological responses are required to maintain normal arterial pO_2 levels (pO_2 levels of 85 torr or less result in mental impairment).

High-altitude adaptation is a complex process that involves increases in the amount of hemoglobin per erythrocyte and in the number of erythrocytes. It normally requires several weeks to complete. Yet, as is clear to anyone who has climbed to high altitude, even a 1-day stay there results in a noticeable degree of adaptation. This effect results from a rapid increase in the amount of BPG synthesized in erythrocytes. The consequent decrease in hemoglobin's O_2-binding affinity, as indicated by its elevated p_{50}, increases the amount of O_2 that hemoglobin unloads in the capillaries (ingesting or injecting BPG would have no effect because BPG cannot cross the red cell membrane). Similar increases in BPG concentration occur in individuals suffering from disorders that limit the oxygenation of the blood (**hypoxia**), such as various anemias and cardiopulmonary insufficiency.

The BPG concentration in erythrocytes can be adjusted more rapidly than hemoglobin can be synthesized. An altered BPG level is also a more sensitive regulator of arterial pO_2 than an altered respiratory rate. Hyperventilation, another early response to high altitude, may lead to respiratory alkalosis (see Box 2-2). Interestingly, individuals in long-established Andean and Himalayan populations exhibit high lung capacity, along with high hemoglobin levels and, often, enlarged right ventricles (reflecting increased cardiac output), compared to individuals from low-altitude populations.

---

stance besides CO_2 that affects oxygen binding to hemoglobin. This compound is **D-2,3-bisphosphoglycerate (BPG).**

D-2,3-Bisphosphoglycerate (BPG)

BPG binds tightly to deoxyhemoglobin but only weakly to oxyhemoglobin. *The presence of BPG in mammalian erythrocytes therefore decreases hemoglobin's oxygen affinity by keeping it in the deoxy conformation.* In other vertebrates, different phosphorylated compounds elicit the same effect.

BPG has an indispensable physiological function: In arterial blood, where pO_2 is ~100 torr, hemoglobin is ~95% saturated with O_2, but in venous blood, where pO_2 is ~30 torr, it is only 55% saturated (Fig. 7-7). Consequently, in passing through the capillaries, hemoglobin unloads ~40% of its bound O_2. In the absence of BPG, little of this bound O_2 would be released since hemoglobin's O_2 affinity is increased, thus shifting its O_2 dissociation curve significantly toward lower pO_2 (Fig. 7-14, *left*). BPG also plays an important role in adaptation to high altitudes (see Box 7-2).

The X-ray structure of a BPG–deoxyhemoglobin complex shows that BPG binds in the central cavity of deoxyhemoglobin (Fig. 7-15). The anionic groups of BPG are within hydrogen bonding and ion pairing distances of the N-terminal amino groups of both β subunits. The T → R transformation brings the two β H helices together, which narrows the central cavity (compare Figs. 7-5a and 7-5b) and expels the BPG. It also widens the distance between the β N-terminal amino groups from 16 to 20 Å, which

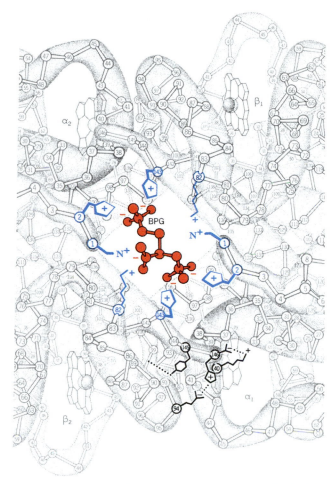

Figure 7-15. **Binding of BPG to deoxyhemoglobin.** BPG (*red*) binds in hemoglobin's central cavity. The BPG, which has a charge of −5 under physiological conditions, is surrounded by eight cationic groups (*blue*) extending from the two β subunits. In the R state, the central cavity is too narrow to contain BPG. The ion pairs and hydrogen bonds that help stabilize the T state (Fig. 7-11*b*) are indicated at the lower right. [Figure copyrighted © by Irving Geis.] ● **See Kinemage Exercise 6-3.**

prevents their simultaneous hydrogen bonding with BPG's phosphate groups. BPG therefore stabilizes the T conformation of hemoglobin by cross-linking its β subunits. This shifts the T ⇌ R equilibrium toward the T state, which lowers hemoglobin's O_2 affinity.

Fetal Hemoglobin Has Low BPG Affinity

The effects of BPG also help supply the fetus with oxygen. A fetus obtains its O_2 from the maternal circulation via the placenta. The concentration of BPG is the same in adult and fetal erythrocytes, but BPG binds more tightly to adult hemoglobin than to fetal hemoglobin. The higher oxygen affinity of fetal hemoglobin facilitates the transfer of O_2 to the fetus.

Fetal hemoglobin has the subunit composition $\alpha_2\gamma_2$ in which the γ subunit is a variant of the β chain (Section 5-4B). Residue 143 of the β chain of adult hemoglobin has a cationic His residue, whereas the γ chain has an uncharged Ser residue. The absence of this His eliminates a pair of interactions that stabilize the BPG–deoxyhemoglobin complex (Fig. 7-15).

Table 7-1.   **Some Hemoglobin Variants**

Name[a]	Mutation	Effect
Hammersmith	Phe CD1(42)$\beta \rightarrow$ Ser	Weakens heme binding
Bristol	Val E11(67)$\beta \rightarrow$ Asp	Weakens heme binding
Bibba	Leu H19(136)$\alpha \rightarrow$ Pro	Disrupts the H helix
Savannah	Gly B6(24)$\beta \rightarrow$ Val	Disrupts the B–E helix interface
Philly	Tyr C1(35)$\alpha \rightarrow$ Phe	Disrupts hydrogen bonding at the α_1–β_1 interface
Boston	His E7(58)$\alpha \rightarrow$ Tyr	Promotes methemoglobin formation
Milwaukee	Val E11(67)$\beta \rightarrow$ Glu	Promotes methemoglobin formation
Iwate	His F8(87)$\alpha \rightarrow$ Tyr	Promotes methemoglobin formation
Yakima	Asp G1(99)$\beta \rightarrow$ His	Disrupts a hydrogen bond that stabilizes the T conformation
Kansas	Asn G4(102)$\beta \rightarrow$ Thr	Distrupts a hydrogen bond that stabilizes the R conformation

[a]Hemoglobin variants are usually named after the place where they were discovered (e.g., hemoglobin Hammersmith).

(a)

(b)

Figure 7-16.   Scanning electron micrographs of human erythrocytes. (*a*) Normal erythrocytes are flexible biconcave disks that can tolerate slight distortions as they pass through the capillaries (many of which have smaller diameters than erythrocytes). [David M. Phillips/Visuals Unlimited.] (*b*) Sickled erythrocytes from an individual with sickle-cell anemia are elongated and rigid and cannot easily pass through capillaries. [Bill Longcore/Photo Researchers, Inc.]

D. Abnormal Hemoglobins

Before the advent of recombinant DNA techniques, mutant hemoglobins provided what was essentially a unique opportunity to study structure–function relationships in proteins. This is because, for many years, hemoglobin was the only protein of known structure that had a large number of well-characterized naturally occurring **variants.** The examination of individuals with physiological disabilities, together with the routine electrophoretic screening of human blood samples, has led to the discovery of nearly 500 variant hemoglobins, ~95% of which result from single amino acid substitutions in a globin polypeptide chain. Indeed, about 5% of the world's population are carriers of an inherited variant hemoglobin.

Not all hemoglobin variants produce clinical symptoms, but some abnormal hemoglobin molecules do cause debilitating diseases (naturally occurring hemoglobin variants that are lethal are, of course, never observed). Table 7-1 lists several of these hemoglobin variants. Mutations that destabilize hemoglobin's tertiary or quaternary structure alter hemoglobin's oxygen binding affinity (p_{50}) and reduce its cooperativity (Hill constant). Moreover, the unstable hemoglobins are degraded by the erythrocytes, and their degradation products often cause the erythrocytes to **lyse** (break open). The resulting **hemolytic anemia** (anemia is the loss of red blood cells) compromises O_2 delivery to tissues.

Certain mutations at the O_2-binding site favor the oxidation of Fe(II) to Fe(III). Individuals carrying the resulting methemoglobin exhibit **cyanosis,** a bluish skin color, due to the presence of methemoglobin in their arterial blood. These hemoglobins have reduced cooperativity (Hill constant ~1.2 compared to a maximum value of 2, since only two subunits in each of these methemoglobins can bind oxygen).

Mutations that increase hemoglobin's oxygen affinity lead to increased numbers of erythrocytes in order to compensate for the less than normal amount of oxygen released in the tissues. Individuals with this condition, which is named **polycythemia,** often have a ruddy complexion.

Sickle-Cell Anemia

Most harmful hemoglobin variants occur in only a few individuals, in many of whom the mutation apparently originated. However, ~10% of American blacks and as many as 25% of black Africans carry a single copy of (are **heterozygous** for) the gene for **sickle-cell hemoglobin (hemoglobin S).**

Figure 7-17. **Structure of a deoxyhemoglobin S fiber.** (*a*) The arrangement of deoxyhemoglobin S molecules in the fiber. Only three subunits of each deoxyhemoglobin S molecule are shown. (*b*) The side chain of the mutant Val 6 in the β_2 chain of one hemoglobin S molecule (yellow knob in *a*) binds to a hydrophobic pocket on the β_1 subunit of a neighboring deoxyhemoglobin S molecule. [Figure copyrighted © by Irving Geis.]

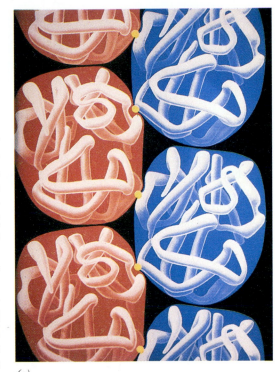

(*a*)

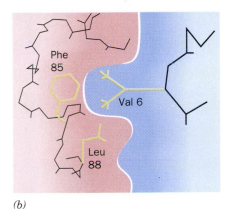

(*b*)

Individuals who carry two copies of (are **homozygous** for) the gene for hemoglobin S suffer from **sickle-cell anemia,** in which deoxyhemoglobin S forms insoluble filaments that deform erythrocytes (Fig. 7-16). In this painful, debilitating, and often fatal disease, the rigid, sickle-shaped cells cannot easily pass through the capillaries. Consequently, in a sickle-cell "crisis," the blood flow to some tissues may be completely blocked, resulting in tissue death. In addition, the mechanical fragility of the misshapen cells results in hemolytic anemia. Heterozygotes, whose hemoglobin is ~40% hemoglobin S, usually lead a normal life, although their erythrocytes have a shorter than normal lifetime.

In 1945, Linus Pauling hypothesized that sickle-cell anemia was the result of a mutant hemoglobin, but the molecular defect was not identified until 1956, when Vernon Ingram showed that hemoglobin S contains Val rather than Glu at the sixth position of each β chain. This was the first time an inherited disease was shown to arise from a specific amino acid change in a protein.

The X-ray structure of deoxyhemoglobin S has revealed that one mutant Val side chain in each hemoglobin S molecule nestles into a hydrophobic pocket on the surface of a β subunit in another hemoglobin molecule (Fig. 7-17). This intermolecular contact allows hemoglobin S molecules to form linear polymers. Aggregates of 14 strands that wind around each other form fibers with a diameter of ~220 Å. The fibers extend throughout the length of the erythrocyte (Fig. 7-18). The hydrophobic pocket on the β subunit cannot accommodate the normally occurring Glu side chain, and the pocket is absent in oxyhemoglobin. Consequently, neither normal hemoglobin nor oxyhemoglobin S can polymerize. In fact, hemoglobin S fibers dissolve essentially instantaneously on oxygenation, so none are present in arterial blood. The danger of sickling is greatest when erythrocytes pass through the capillaries, where deoxygenation occurs. The polymerization of hemoglobin S molecules is time and concentration dependent, which explains why blood flow blockage occurs only sporadically (in a sickle-cell "crisis").

Interestingly, many hemoglobin S homozygotes have only a mild form of sickle-cell anemia because they express relatively high levels of fetal hemoglobin, which contains γ chains rather than the defective β chains. The administration of **hydroxyurea**

$$H_2N-\overset{\overset{\displaystyle O}{\|}}{C}-NH-OH$$

Hydroxyurea

can also ameliorate the symptoms of sickle-cell anemia by increasing the fraction of cells containing fetal hemoglobin (the mechanism whereby hy-

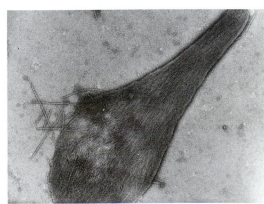

Figure 7-18. **Electron micrograph of deoxyhemoglobin S fibers spilling out of a ruptured erythrocyte.** [Courtesy of Robert Josephs, University of Chicago.]

Malaria and Hemoglobin S

Malaria is the most lethal infectious disease that presently affects humanity: Of the 2.5 billion people living within malaria-endemic areas, 100 million are clinically ill with the disease at any given time and around 1 million, mostly very young children, die from it each year. Malaria is caused by the mosquito-borne protozoan *Plasmodium falciparum,* which resides within an erythrocyte during much of its 48-h life cycle. Infected erythrocytes adhere to capillary walls, causing death when cells impede blood flow to a vital organ.

The regions of equatorial Africa where malaria is a major cause of death *(blue areas on map below)* coincide closely with those areas where the sickle-cell gene is prevalent *(pink areas),* thereby suggesting that the sickle-cell gene confers resistance to malaria.

How does it do so? Plasmodia increase the acidity of infected erythrocytes by ~0.4 pH units. The lower pH favors the formation of deoxyhemoglobin via the Bohr effect, thereby increasing the likelihood of sickling in erythrocytes that contain hemoglobin S. Erythrocytes damaged by sickling are normally removed from the circulation by the spleen. During the early stages of a malarial infection, parasite-enhanced sickling probably allows the spleen to preferentially remove infected erythrocytes. In the later stages of infection, when the parasitized erythrocytes attach to the capillary walls (presumably to prevent the spleen from removing them from the circulation), sickling may mechanically disrupt the parasite. Consequently, heterozygous carriers of hemoglobin S in a malarial region have an adaptive advantage: They are more likely to survive to maturity than individuals who are homozygous for normal hemoglobin. Thus, in malarial regions, the fraction of the population who are heterozygotes for the sickle-cell gene increases until their reproductive advantage is balanced by the correspondingly increased proportion of homozygotes (who, without modern medical treatment, die in childhood).

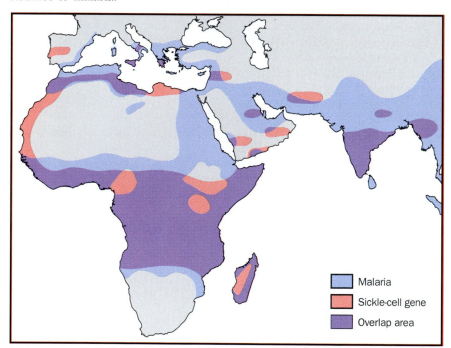

	Malaria
	Sickle-cell gene
	Overlap area

droxyurea acts is unknown). The increased fetal hemoglobin dilutes the hemoglobin S, making it more difficult for hemoglobin S to aggregate during the 10–20 s it takes for an erythrocyte to travel from the tissues to the lungs for reoxygenation.

Before the advent of modern palliative therapies, individuals with sickle-cell anemia rarely survived to maturity. Natural selection has not minimized the prevalence of the hemoglobin S variant, however, because heterozygotes are more resistant to **malaria** (see Box 7-3).

E. Allosteric Proteins

The cooperativity of oxygen binding to hemoglobin is a classic model for the behavior of many other multisubunit proteins (including certain enzymes) that bind small molecules. In some cases, binding of a ligand to one site increases the affinity of other binding sites on the same protein (as in O_2 binding to hemoglobin). In other cases, a ligand decreases the affinity of other binding sites (as when BPG binding decreases the O_2 affinity of hemoglobin). All these effects are the result of **allosteric interactions** (Greek: *allos*, other + *stereos*, solid or space). *Allosteric effects, in which the binding of a ligand at one site affects the binding of another ligand at another site, generally require interactions among subunits of oligomeric proteins.* The T → R transition in hemoglobin subunits explains the difference in the oxygen affinities of oxy- and deoxyhemoglobin. Other proteins exhibit similar conformational shifts, although the molecular mechanisms that underlie these phenomena are not completely understood.

Two models that account for cooperative ligand binding have received the most attention. One of them, the **symmetry model** of allosterism, formulated in 1965 by Jacques Monod, Jeffries Wyman, and Jean-Pierre Changeux, is defined by the following rules:

1. An allosteric protein is an oligomer of symmetrically related subunits (although the α and β subunits of hemoglobin are only pseudosymmetrically related).

2. Each oligomer can exist in two conformational states, designated R and T; these states are in equilibrium.

3. The ligand can bind to a subunit in either conformation. *Only the conformational change alters the affinity for the ligand.*

4. *The molecular symmetry of the protein is conserved during the conformational change.* The subunits must therefore change conformation in a concerted manner; in other words, there are no oligomers that simultaneously contain R- and T-state subunits.

The symmetry model is diagrammed for a tetrameric binding protein in Fig. 7-19. If a ligand binds more tightly to the R state than to the T state, ligand binding will promote the T → R shift, thereby increasing the affinity of the unliganded subunits for the ligand.

One major objection to the symmetry model is that it is difficult to believe that oligomeric symmetry is perfectly preserved in all proteins, that is, that the T → R shift occurs simultaneously in all subunits regardless of the number of ligands bound. In addition, the symmetry model can account only for positive cooperativity, although some proteins exhibit negative cooperativity.

An alternative to the symmetry model is the **sequential model** of allosterism, proposed by Daniel Koshland. According to this model, ligand binding induces a conformational change in the subunit to which it binds, and cooperative interactions arise through the influence of those conformational changes on neighboring subunits. The conformational changes occur sequentially as more ligand-binding sites are occupied (Fig. 7-20). The

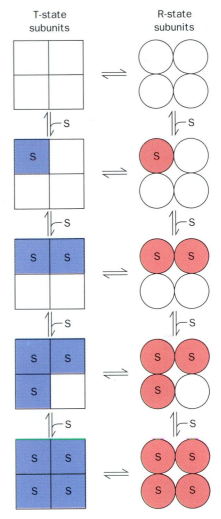

T-state subunits   **R-state subunits**

Figure 7-19. **The symmetry model of allosterism.** Squares and circles represent T- and R-state subunits, respectively, of a tetrameric protein. The T and R states are in equilibrium regardless of the number of ligands (represented by S) that have bound to the protein. All the subunits must be in either the T or the R form; the model does not allow combinations of T- and R-state subunits in the same protein.

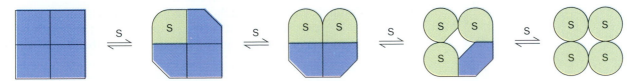

Figure 7-20. **The sequential model of allosterism.** Ligand binding progressively induces conformational changes in the subunits, with the greatest changes occurring in those subunits that have bound ligand. The symmetry of the oligomeric protein is not preserved in this process as it is in the symmetry model.

ligand-binding affinity of a subunit varies with its conformation and may be higher or lower than that of the subunits in the ligand-free protein. Thus, proteins that follow the sequential model of allosterism may be positively or negatively cooperative.

If the mechanical coupling between subunits in the sequential model is particularly strong, the conformational changes occur simultaneously and the oligomer retains its symmetry, as in the symmetry model. Thus, the symmetry model of allosterism may be considered to be an extreme case of the more general sequential model.

Oxygen binding to hemoglobin exhibits features of both models. The quaternary T → R conformational change is concerted, as the symmetry model requires. Yet ligand binding to the T state does cause small tertiary structural changes, as the sequential model predicts. These minor conformational shifts are undoubtedly responsible for the buildup of strain that eventually triggers the T → R transition. It therefore appears that the complexity of ligand–protein interactions in hemoglobin and other proteins allows binding processes to be fine-tuned to the needs of the organism under changing internal and external conditions. We shall revisit allosteric effects when we discuss enzymes in Chapter 12.

3. MYOSIN AND ACTIN

One of the most striking characteristics of living things is their capacity for organized movement. Such phenomena occur at all structural levels and include such diverse vectorial processes as the separation of replicated chromosomes during cell division, the beating of flagella and cilia, and muscle contraction. In this section, we consider the structural and chemical basis of movement in **striated muscle,** one of the best understood motility systems.

A. *Structure of Striated Muscle*

The voluntary muscles, which include the skeletal muscles, have a striated (striped) appearance when viewed by light microscopy (Fig. 7-21). Such muscles consist of long multinucleated cells (the muscle fibers) that contain parallel bundles of **myofibrils** (Greek: *myos,* muscle; Fig. 7-22). Electron micrographs show that muscle striations arise from the banded structure of multiple in-register myofibrils. The bands are formed by alternating regions of greater and lesser electron density called **A bands** and **I bands,** respectively (Fig. 7-23). The myofibril's repeating unit, the **sarcomere** (Greek: *sarkos,* flesh), is bounded by **Z disks** at the center of each I band. The A band is centered on the **H zone,** which in turn is centered on the **M disk.** The A band contains 150-Å-diameter **thick filaments,** and the I band contains 70-Å-diameter **thin filaments.** The two sets of filaments are linked by cross-bridges where they overlap.

A contracted muscle can be as much as one-third shorter than its fully extended length. The contraction results from a decrease in the length of the sarcomere, caused by reductions in the lengths of the I band and the

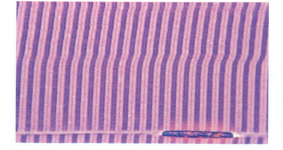

Figure 7-21. **Photomicrograph of a muscle fiber.** The longitudinal axis of the fiber is horizontal (perpendicular to the striations). The alternating pattern of dark A bands and light I bands from multiple in-register myofibrils is clearly visible. [J.C. Revy, CNRI/Photo Researchers.]

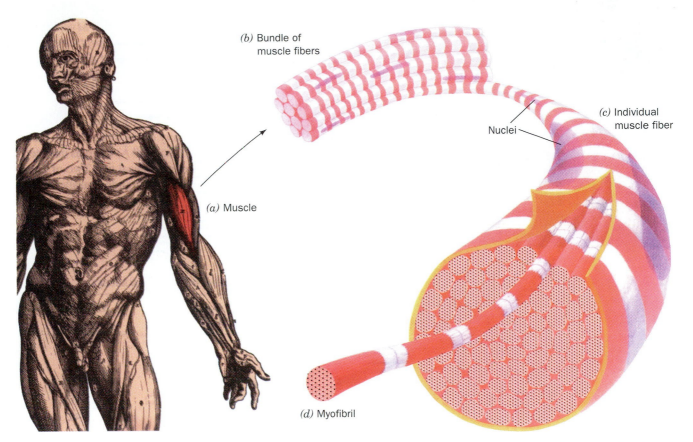

(b) Bundle of
muscle fibers

(c) Individual
muscle fiber

Nuclei

(a) Muscle

(d) Myofibril

Figure 7-22.   Skeletal muscle organization. A muscle (a) consists of bundles of muscle fibers (b), each of which is a long, thin, multinucleated cell (c) that may run the length of the muscle. Muscle fibers contain bundles of laterally aligned myofibrils (d), which in turn consist of bundles of alternating thick and thin filaments.

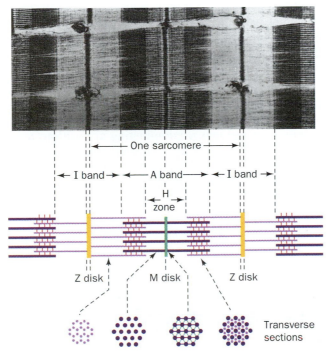

One sarcomere

I band   A band   I band

H
zone

Z disk   M disk   Z disk

Transverse
sections

Figure 7-23.   Anatomy of the myofibril. The electron micrograph shows parts of three myofibrils, which are separated by horizontal gaps. The accompanying interpretive drawing shows the major features of the myofibril: the light I band, which contains only thin filaments; the A band, whose dark H zone contains only thick filaments and whose darker outer segments contain overlapping thick and thin filaments; the Z disk, to which the thin filaments are anchored; and the M disk, which arises from a bulge at the center of each thick filament. The myofibril's functional unit, the sarcomere, is the region between two successive Z disks. [Courtesy of Hugh Huxley, Brandeis University.]

(a)

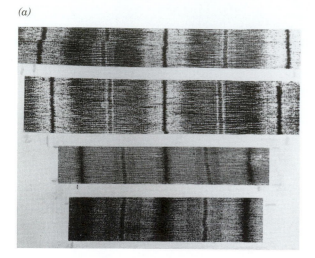

(b)

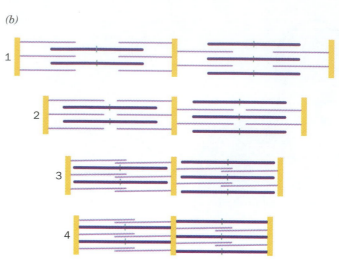

Figure 7-24.   Myofibril contraction. (*a*) Electron micrographs showing myofibrils in progressively more contracted states. The lengths of the I band and H zone decrease on contraction, whereas the lengths of the thick and thin filaments remain constant. (*b*) Interpretive drawings showing interpenetrating sets of thick and thin filaments sliding past each other. [Courtesy of Hugh Huxley, Brandeis University.]

H zone (Fig. 7-24*a*). These observations are explained by the **sliding filament model** in which interdigitated thick and thin filaments slide past each other (Fig. 7-24*b*).

B. *Structures of Thick and Thin Filaments*

Vertebrate thick filaments are composed almost entirely of a single type of protein, **myosin,** which consists of six polypeptide chains: two 220-kD **heavy chains** and two pairs of different **light chains,** the so-called **essential** and **regulatory light chains** (**ELC** and **RLC**) that vary in size between 15 and 22 kD, depending on their source. As is diagrammed in Fig. 7-25, the N-terminal half of each heavy chain forms an elongated globular head (55 × 165 Å) to which one subunit each of ELC and RLC bind. The C-terminal half of the heavy chain forms a long fibrous α-helical tail, two of which associate to form a left-handed coiled coil. Thus, myosin consists of a 1600-Å-long rodlike segment with two globular heads. The amino acid sequence of myosin's α-helical tail is characteristic of coils such as those in keratin (Section 6-1C): It has a seven-residue pseudorepeat, *a-b-c-d-e-f-g*, with nonpolar residues predominating at positions *a* and *d*.

Figure 7-25.   The myosin molecule. It contains two identical myosin heavy chains (*green and orange*), which have N-terminal globular heads and α helical tails. Each head associates with an essential light chain (*pink*) and a regulatory light chain (*blue*). The tails wind around each other to form a 1600-Å-long parallel coiled coil. ● **See Kinemage Exercise 4-1.**

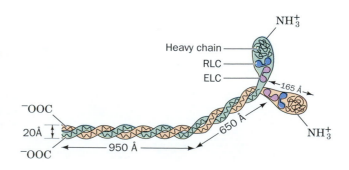

(a)

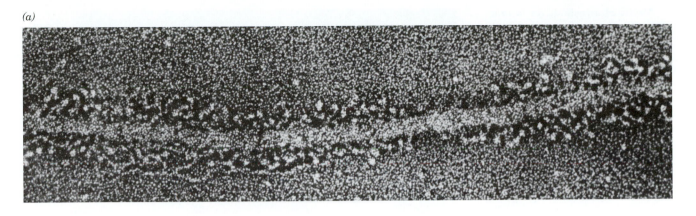

(b)

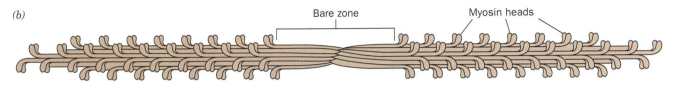

Figure 7-26. Structure of the thick filament. (*a*) Electron micrograph showing the myosin heads projecting from the thick filament. [From Trinick, J. and Elliott, A., *J. Mol. Biol.* **131**, 135 (1977).] (*b*) Drawing of a thick filament, in which several hundred myosin molecules form a staggered array with their globular heads pointing away from the filament.

Under physiological conditions, several hundred myosin molecules aggregate to form a thick filament. The rodlike tails pack end to end in a regular staggered array, leaving the globular heads projecting to the sides on both ends (Fig. 7-26). These myosin heads form the cross-bridges to thin filaments in intact myofibrils. The myosin head, which is an **ATPase** (ATP-hydrolyzing enzyme), has its ATP-binding site located in a 13-Å-deep V-shaped pocket.

Thin filaments consist of three proteins, actin, tropomyosin, and troponin, that combine as diagrammed in Fig. 7-27. **Actin** is a 375-residue bilobal globular protein that polymerizes to form a fiber at the core of the thin filament. **Tropomyosin** is a two-chain coiled coil of α helices. This 400-Å-long molecule associates in a head-to-tail fashion with other tropomyosin molecules to form a cable that winds in the helical groove of the actin fiber such that each tropomyosin molecule contacts seven actin monomers. Each tropomyosin molecule binds a single heterotrimeric **troponin** molecule, which binds Ca^{2+} ions.

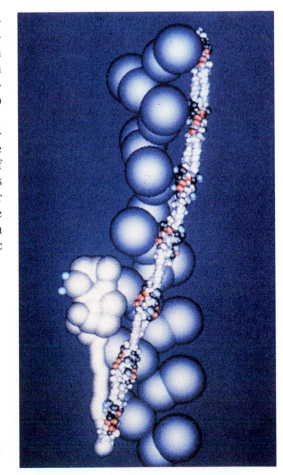

Figure 7-27. Structure of the thin filament. This model is based on electron micrographs of actin fibers and the X-ray structure of tropomyosin. Actin (*large blue spheres*) polymerizes to form the core of the thin filament. Tropomyosin, a parallel coiled coil of two α helical subunits, wraps in the groove of the helical actin fiber such that each tropomyosin molecule contacts seven consecutive bilobal actin monomer units (only one tropomyosin molecule is shown). The seven red and blue regions of tropomyosin identify the seven actin monomer units to which it binds. Each tropomyosin molecule binds a single troponin molecule (*large white spheres and stalk*). The small blue spheres represent bound Ca^{2+} ions. The tropomyosin molecule that winds around the opposite face of the actin filament has been omitted for clarity. [Courtesy of George N. Phillips, Jr., Rice University.]

Figure 7-28. **Model of myosin–actin interaction.** This space-filling model was constructed from the X-ray structures of actin and the myosin head and electron micrographs of their complex. The actin filament is at the top. A myosin head (*brightly colored structure* with ELC *yellow* and RLC *magenta*) is attached to the actin polymer at an angle of ~45°. The myosin coiled-coil tail is not shown. An ATP-binding site is located in a cleft in the red domain of the myosin head. In a myofibril, every actin monomer has the potential to bind a myosin head, and the thick filament has many myosin heads projecting from it. [Courtesy of Ivan Rayment and Hazel Holden, University of Wisconsin.]

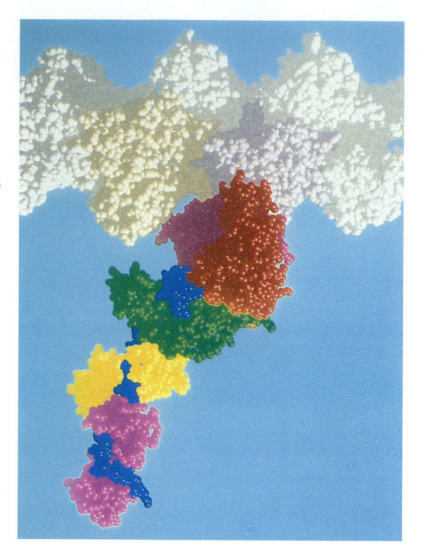

Each actin monomer can bind a myosin head (Fig. 7-28), probably by ion pairing and by the association of hydrophobic patches on each protein. The tropomyosin–troponin complex regulates muscle contraction by controlling the access of myosin heads on thick filaments to their binding sites on the actin of thin filaments.

Myosin and actin together account for 60 to 95% of total muscle protein. Other proteins form the Z disk and M disk and organize the arrays of thick and thin filaments.

C. Mechanism of Muscle Contraction

In order for the sliding filament model to explain muscle contraction, each myosin cross-bridge to actin must repeatedly detach and reattach itself at a new site further along the thin filament toward the Z disk. The actual contractile force is provided by ATP hydrolysis. Edwin Taylor formulated a model for myosin-mediated ATP hydrolysis, which has been refined by the structural studies of Ivan Rayment, Hazel Holden, and Ronald Milligan as follows (Fig. 7-29):

1. ATP binds to a myosin head in a manner that causes the actin-binding site to open up and release its bound actin.

2. The active site closes around the ATP. The resulting hydrolysis of the

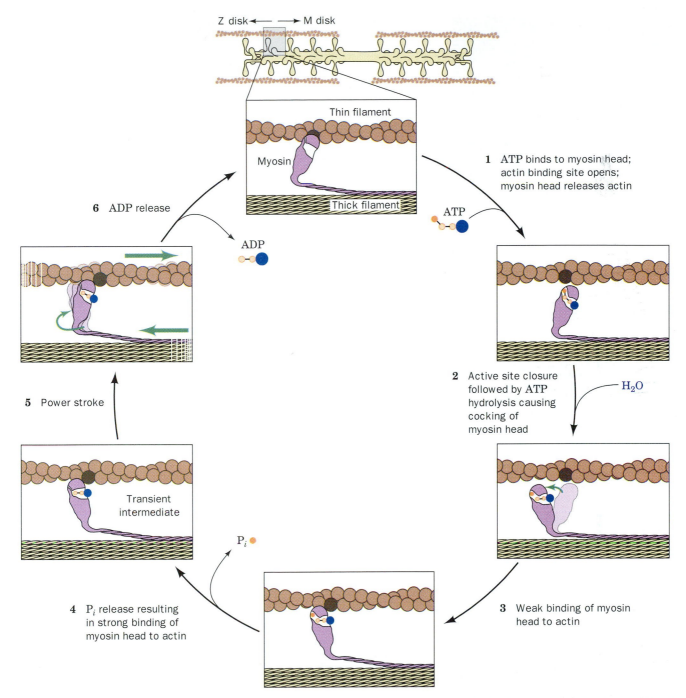

Figure 7-29. *Key to Function.* **Mechanism of force generation in muscle.** The myosin head "walks" up the actin thin filament through a unidirectional cyclic process that is driven by ATP hydrolysis to ADP and P_i. Only one myosin head is shown. The actin monomer to which the myosin head is bound at the beginning of the cycle is more darkly colored for reference. [After Rayment, I. and Holden, H.M., *Curr. Opin. Struct. Biol.* **3**, 949 (1993).] ✳ **See the Animated Figures.**

ATP "cocks" the myosin head, that is, puts it into its "high-energy" conformation in which it is approximately perpendicular to the thick filament.

3. The myosin head binds weakly to an actin monomer that is closer to the Z disk than the one to which it had been bound previously.

4. Myosin releases P_i, which causes a conformational shift that increases myosin's affinity for actin.

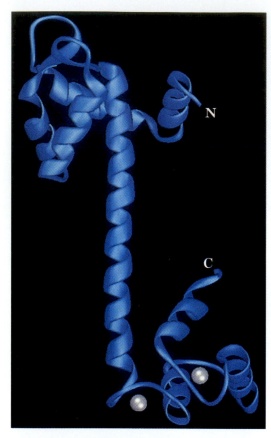

Figure 7-30.   The X-ray structure of troponin C from chicken skeletal muscle. The two structurally similar globular domains, which are connected by a nine-turn helix, can each bind two Ca^{2+} ions. Two Ca^{2+} ions (*silver spheres*) remain bound at low cellular [Ca^{2+}]. At higher [Ca^{2+}], additional Ca^{2+} ions bind to the upper domain. [Based on an X-ray structure by Muttaiya Sundaralingam, University of Wisconsin.]

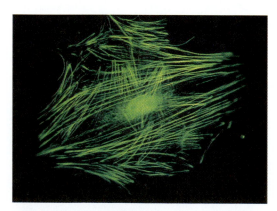

Figure 7-31.   Actin microfilaments. The microfilaments in a fibroblast resting on the surface of a culture dish are revealed by immunofluorescence microscopy using a fluorescently labeled antibody to actin. When the cell begins to move, the filaments disassemble to form a diffuse mesh. [Courtesy of Elias Lazarides, Tanabe Research Laboratories, Inc.]

5. The resulting transient state is immediately followed by the power stroke, a further conformational shift that sweeps the myosin head's C-terminal tail by an estimated 60 Å toward the Z disk relative to the actin-binding site on its head, thus translating the attached thin filament by this distance toward the M disk.

6. ADP is released, thereby completing the cycle.

The ~500 myosin heads on every thick filament asynchronously cycle through this reaction sequence about five times each per second during a strong muscular contraction. *The myosin heads thereby "walk" or "row" up adjacent thin filaments toward the Z disk with the concomitant contraction of the muscle.*

Control of Muscle Contraction

Crude muscle extracts contract only in the presence of Ca^{2+}, but highly purified actin and myosin can contract regardless of the Ca^{2+} concentration. Sensitivity to Ca^{2+} in intact muscle is provided by **troponin C** (Fig. 7-30), a subunit of the thin filament protein troponin (Section 7-3B). Stimulation of a myofibril by a nerve impulse results in an influx of Ca^{2+} from the sarcoplasmic reticulum (a system of flattened vesicles derived from the endoplasmic reticulum). As a result, the intracellular [Ca^{2+}] increases from ~10^{-7} to ~10^{-5} M. At the higher concentration, additional Ca^{2+} ions bind to troponin C, causing a conformational change in the troponin–tropomyosin complex. This movement exposes the site on actin where the myosin head binds. When the myofibril [Ca^{2+}] is low (Ca^{2+} is specifically ejected from the myofibril by ATP-driven protein pumps), the troponin–tropomyosin complex assumes its resting conformation, blocking myosin binding to actin and causing the muscle to relax.

D. Actin in Nonmuscle Cells

Although actin and myosin are most prominent in muscle, they also occur in other tissues. In fact, actin is ubiquitous and is usually the most abundant cytoplasmic protein in eukaryotic cells, typically accounting for 5 to 10% of their total protein.

Nonmuscle actin forms ~70-Å-diameter fibers known as **microfilaments** that can be visualized by **immunofluorescence microscopy** (in which a fluorescent-tagged antibody is used to "stain" the actin to which it binds; Fig. 7-31). The microfilaments, along with other protein fibers, are elements of the cytoskeleton. The assembly of actin monomers into fibers requires ATP. Typically, actin subunits are added and removed in a dynamic fashion. **Treadmilling,** the removal of monomeric units from one end of the fiber and their addition at the other end, can occur even though the net length of the fiber remains constant. The assembly and disassembly of actin filaments are the basis of such cellular processes as ameboid locomotion, cytokinesis (the separation of daughter cells during cell division), cytoplasmic streaming, and the extension and retraction of various cellular protuberances.

4. ANTIBODIES

All organisms are continually subject to attack by other organisms, including disease-causing microorganisms and viruses. In higher animals, these **pathogens** may breach the physical barrier presented by the skin and mucous membranes (a first line of defense) only to be identified as foreign invaders and destroyed by the **immune system.**

A. Overview of the Immune System

Two types of immunity have been distinguished:

1. **Cellular immunity,** which guards against virally infected cells, fungi, parasites, and foreign tissue, is mediated by **T lymphocytes** or **T cells,** so called because their development occurs in the thymus.

2. **Humoral immunity** (*humor* is an archaic term for fluid), which is most effective against bacterial infections and the extracellular phases of viral infections, is mediated by an enormously diverse collection of related proteins known as **antibodies** or **immunoglobulins.** Antibodies are produced by **B lymphocytes** or **B cells,** which in mammals mature in the bone marrow.

The following sections focus on antibody structure and function.

*The immune response is triggered by the presence of a foreign macromolecule, often a protein or carbohydrate, known as an **antigen.*** B cells display immunoglobulins on their surfaces. If a B cell encounters an antigen that binds to its particular immunoglobulin, it engulfs the antigen–antibody complex, degrades it, and displays the antigen fragments on the cell surface. T cells then stimulate the B cell to proliferate. Most of the B cell progeny are circulating cells that secrete large amounts of the antigen-specific antibody. These antibodies can bind to additional antigen molecules, thereby marking them for destruction by other components of the immune system. Although most B cells live only a few days unless stimulated by their corresponding antigen, a few long-lived **memory B cells** can recognize their corresponding antigen several weeks or even many years later and can mount a more rapid and massive immune response (called a secondary response) than B cells that have not yet encountered their corresponding antigen (Fig. 7-32).

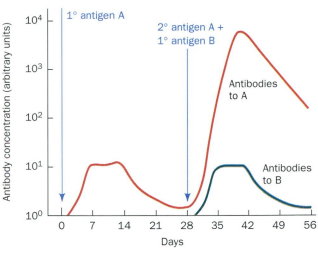

Figure 7-32. **Primary and secondary immune responses.** Antibodies to antigen A appear in the blood following primary immunization on day 0 and secondary immunization on day 28. Antigen B is included in the secondary immunization to demonstrate the specificity of immunological memory for antigen A. The secondary response to antigen A is both faster and greater than the primary response.

B. Antibody Structure

The immunoglobulins form a related but enormously diverse group of proteins. All immunoglobulins contain at least four subunits: two identical ~23-kD **light chains (L)** and two identical 53- to 75-kD **heavy chains (H).** These subunits associate by disulfide bonds and by noncovalent interactions to form a roughly Y-shaped symmetric molecule with the formula $(LH)_2$ (Fig. 7-33).

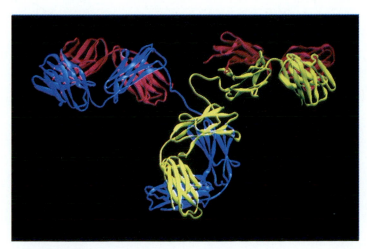

Figure 7-33. **X-Ray structure of a mouse antibody.** The heavy chains are yellow and blue, and the light chains are both red. The antigen-binding sites are located at the ends of the arms of the Y-shaped molecule. [Courtesy of Alexander McPherson, University of California at Riverside.]

● See the Interactive Exercises.

Table 7-2.   Classes of Human Immunoglobulins

Class	Heavy Chain	Light Chain	Subunit Structure	Molecular Mass (kD)
IgA	α	κ or λ	$(\alpha_2\kappa_2)_n J^a$ or $(\alpha_2\lambda_2)_n J^a$	360–720
IgD	δ	κ or λ	$\delta_2\kappa_2$ or $\delta_2\lambda_2$	160
IgE	ε	κ or λ	$\varepsilon_2\kappa_2$ or $\varepsilon_2\lambda_2$	190
IgG[b]	γ	κ or λ	$\gamma_2\kappa_2$ or $\gamma_2\lambda_2$	150
IgM	μ	κ or λ	$(\mu_2\kappa_2)_5 J$ or $(\mu_2\lambda_2)_5 J$	950

[a]$n = 1, 2,$ or 3.
[b]IgG has four subclasses, IgG1, IgG2, IgG3, and IgG4, which differ in their γ chains.

The five classes of immunoglobulin **(Ig)** differ in the type of heavy chain they contain and, in some cases, in their subunit structure (Table 7-2). For example, **IgM** consists of five Y-shaped molecules arranged around a central **J subunit; IgA** occurs as monomers, dimers, and trimers. The various immunoglobulin classes also have different physiological functions. IgM is most effective against microorganisms and is the first immunoglobulin to be secreted in response to an antigen. **IgG,** the most common immunoglobulin, is equally distributed between the blood and the extravascular fluid. IgA occurs predominantly in the intestinal tract and defends against invading pathogens by adhering to their antigenic sites so as to block their attachment to epithelial (outer) surfaces. **IgE,** which is normally present in the blood in minute concentrations, protects against parasites and has been implicated in allergic reactions. **IgD,** which is also present in small amounts, has no known function. Our discussion of antibody structure will focus on IgG.

IgG can be cleaved through limited proteolysis with the enzyme **papain** into three ~50-kD fragments: two identical **Fab fragments** and one **Fc fragment.** The Fab fragments are the "arms" of the Y-shaped antibody and contain an entire L chain and the N-terminal half of an H chain (Fig. 7-34).

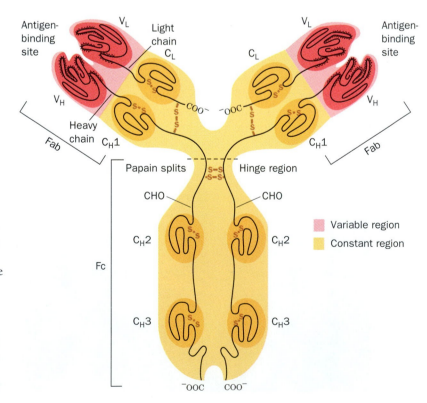

Figure 7-34.   *Key to Structure.* Diagram of human immunoglobulin G (IgG). Each light chain contains a variable (V_L) and a constant (C_L) region, and each heavy chain contains one variable (V_H) and three constant ($C_H 1$, $C_H 2$, and $C_H 3$) regions. Each of the variable and constant domains contains a disulfide bond, and the four polypeptide chains are linked by disulfide bonds. The proteolytic enzyme papain cleaves IgG at the hinge region to yield two Fab fragments and one Fc fragment. CHO represents carbohydrate chains. [Figure copyrighted © by Irving Geis.]

These fragments contain IgG's antigen-binding sites (the "ab" in Fab stands for *a*ntigen *b*inding). The Fc portion ("c" because it *c*rystallizes easily) derives from the "stem" of the antibody and consists of the C-terminal halves of two H chains. The arms of the Y are connected to the stem by a flexible hinge region. The hinge angles may vary, so an antibody molecule may not be perfectly symmetrical (see Fig. 7-33).

Although all IgG molecules have the same overall structure, IgGs that recognize different antigens have different amino acid sequences. The light chains of different antibodies differ mostly in their N-terminal halves. These polypeptides are therefore said to have a **variable region, V_L** (residues 1–108), and a **constant region, C_L** (residues 109–214). Comparisons of H chains, which have 446 residues, reveal that H chains also have a variable region, V_H, and a constant region, C_H. As indicated in Fig. 7-34, the C_H region consists of three ~110-residue segments, **C_H1, C_H2,** and **C_H3,** which are homologous to each other and to C_L. In fact, all the constant and variable regions resemble each other in sequence and in disulfide-bonding pattern. These similarities suggest that the six different homology units of an IgG evolved through the duplication of a primordial gene encoding an ~110-residue protein.

C. Antigen–Antibody Binding

The immunoglobulin homology units all have the same characteristic **immunoglobulin fold:** a sandwich composed of three- and four-stranded antiparallel β sheets that are linked by a disulfide bond (Fig. 7-35). Nevertheless, the basic immunoglobulin structure must accommodate an enormous variety of antigens. The ability to recognize antigens resides in three loops in the variable domain. Most of the amino acid variation among antibodies is concentrated in these three short segments, called **hypervariable** sequences. As hypothesized by Elvin Kabat, the hypervariable sequences line an immunoglobulin's antigen-binding site, so that their amino acids determine its binding specificity.

Scientists have determined the X-ray structures of Fab fragments from numerous **monoclonal antibodies** (see Box 7-4) and monospecific antibodies isolated from patients with **multiple myeloma** (a disease in which a cancerous *B* cell proliferates and produces massive amounts of a single immunoglobulin; immunoglobulins purified from ordinary blood are heterogeneous and hence cannot be used for detailed structural studies). As expected from the positions of the hypervariable sequences, the antigen-binding site is located at the tip of each Fab fragment in a crevice between its V_L and V_H domains.

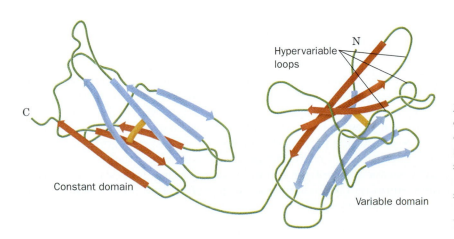

***Figure 7-35. Immunoglobulin folds in a light chain.** Both the constant and variable domains consist of a sandwich of a four-stranded antiparallel β sheet (blue) and a three-stranded antiparallel β sheet (orange) that are linked by a disulfide bond (yellow). The positions of the three hypervariable sequences in the variable domain are indicated. [After Schiffer, M., Girling, R.L., Ely, K.R., and Edmundson, A.B., Biochemistry **12**, 4628 (1973).]*

Box 7-4
BIOCHEMISTRY IN FOCUS

Monoclonal Antibodies

Introducing a foreign molecule into an animal induces the synthesis of large amounts of antigen-specific but heterogeneous antibodies. One might expect that a single lymphocyte from such an animal could be cloned (allowed to reproduce) to yield a harvest of homogeneous immunoglobulin molecules. Unfortunately, lymphocytes do not grow continuously in culture. In the late 1970s, however, César Milstein and Georges Köhler developed a technique for immortalizing such cells so that they can grow continuously and secrete virtually unlimited quantities of a specific antibody. Typically, lymphocytes from a mouse that has been immunized with a particular antigen are harvested and fused with mouse myeloma cells (a type of blood system cancer), which can multiply indefinitely (see figure). The cells are then incubated in a selective medium that inhibits the synthesis of purines, which are essential for myeloma growth (the myeloma cells lack an enzyme involved in an alternate pathway for the production of purine nucleotides). The only cells that can grow in the selective medium are fused cells, known as **hybridoma cells**, that combine the missing enzyme (it is supplied by the lymphocyte) with the immortal attributes of the myeloma cells. Clones derived from single fused cells are then screened for the presence of antibodies to the original antigen. Antibody-producing cells can be grown in large quantities in tissue culture or as semisolid tumors in mouse hosts.

Monoclonal antibodies are used to purify macromolecules (Section 5-2), to identify infectious diseases, and to test for the presence of drugs and other substances in body tissues. Because of their purity and specificity and, to some extent, their biocompatibility, monoclonal antibodies also hold considerable promise as therapeutic agents against cancer and other diseases.

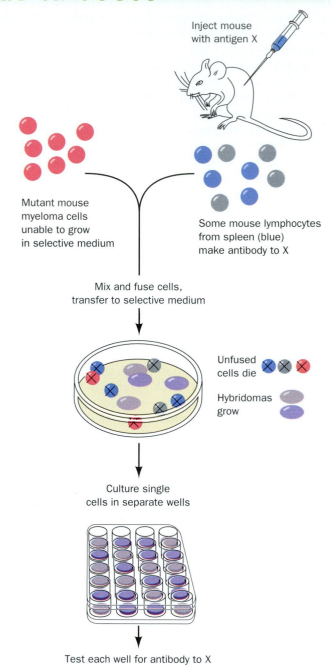

Inject mouse with antigen X

Mutant mouse myeloma cells unable to grow in selective medium

Some mouse lymphocytes from spleen (blue) make antibody to X

Mix and fuse cells, transfer to selective medium

Unfused cells die

Hybridomas grow

Culture single cells in separate wells

Test each well for antibody to X

The association between antibodies and their antigens involves van der Waals, hydrophobic, hydrogen bonding, and ionic interactions. Their dissociation constants range from 10^{-4} to 10^{-10} M, comparable (or even greater) in strength to the associations between enzymes and the small molecules with which they react. The specificity and strength of an antigen–antibody complex are a function of the exquisite structural complementarity between the antigen and the antibody (e.g., Fig. 7-36). These are also the features that make antibodies such useful laboratory reagents (see Fig. 5-3, for example).

Most immunoglobulins are divalent molecules; that is, they can bind two identical antigens simultaneously (IgM and IgA are multivalent). A foreign substance or organism usually has multiple antigenic regions, and a typical immune response generates a mixture of antibodies with different specificities. Divalent binding allows antibodies to cross-link antigens to form an extended lattice (Fig. 7-37), which hastens the removal of the antigen and triggers *B* cell proliferation.

D. Generating Antibody Diversity

A novel antigen does not direct a *B* cell to begin manufacturing a new immunoglobulin to which it can bind. Rather, *an antigen merely stimulates the proliferation of a pre-existing B cell whose antibodies happen to recognize the antigen*. The immune system has the potential to produce many billions of different antibodies, enough to react with almost any antigen the organism might encounter. However, the number of immunoglobulin genes is far too small to account for the observed level of antibody diversity. The diversity in antibody sequences arises instead from genetic changes during *B* lymphocyte development (Section 27-3C).

An individual synthesizes only a small fraction of its potential immunoglobulin repertoire over its lifetime. What is perhaps more remarkable is that the power of the immune system is unleashed only against foreign substances and not against any of the tens of thousands of endogenous (self) molecules of various sorts. Virtually all macromolecules are potentially antigenic, as can be demonstrated by transplanting tissues from one individual to another, even within a species. This incompatibility presents obvious challenges for therapies ranging from routine blood transfusions to multiple organ transplants.

The mechanism whereby an individual's immune system distinguishes self from non-self is but poorly understood. It begins to operate around the time of birth and must be ongoing, since new lymphocytes arise throughout an individual's lifetime. Occasionally, the immune system loses tolerance to some of its self-antigens, resulting in an **autoimmune disease,** which may produce symptoms ranging from mild to lethal (see Box 7-5).

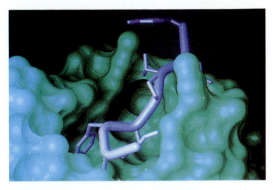

Figure 7-36. Interaction between an antigen and an antibody. This X-ray structure shows a portion of the solvent-accessible surface of a monoclonal antibody Fab fragment (*green*) with a stick model of a bound nine-residue fragment of its peptide antigen (*blue*). [Courtesy of Ian Wilson, The Scripps Research Institute, La Jolla, California.]

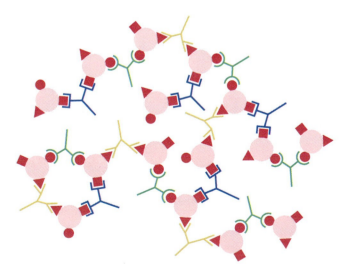

Figure 7-37. Antigen cross-linking by antibodies. A mixture of divalent antibodies that recognize the several different antigenic regions of an intruding particle such as a toxin molecule or a bacterium can form an extensive lattice of antigen and antibody molecules.

Box 7-5
BIOCHEMISTRY IN HEALTH AND DISEASE

Autoimmune Diseases

All the body's organ systems are theoretically susceptible to attack by an immune system that has lost its self-tolerance, but some tissues are attacked more often than others. Some of the most common autoimmune diseases are listed below. The symptoms of a particular disease reflect the type of tissue with which the autoantibodies react. In general, autoimmune diseases are chronic, often with periods of remission, and their clinical severity may differ among individuals.

The loss of tolerance to one's own antigens may result from an innate malfunctioning of the mechanism by which the immune system distinguishes self from non-self, possibly precipitated by an event, such as trauma or infection, in which tissues that are normally sequestered from the immune system are exposed to lymphocytes. For example, breaching the blood–brain barrier may allow lymphocytes access to the brain or spinal cord, and injury may allow access to the spaces at joints, which are not normally served by blood vessels. There is also evidence that some autoimmune diseases may be caused by antibodies to viral or bacterial antigens; such antibodies cross-react with endogenous substances because of chance antigenic similarities. Some diseases, such as systemic lupus erythematosus, represent a more generalized breakdown of the immune system, so that antibodies to many endogenous substances (e.g., DNA and phospholipids) may be generated.

The pathological effects of autoimmune diseases reflect the complexity of the immune system, which is designed to combat a wide variety of foreign substances in different ways. The agents that mediate tissue destruction as part of the normal battle against infection, for example, can cause unregulated damage in a runaway immune response such as in an autoimmune disease. Hormones also affect the course of an immune response, which may explain why multiple sclerosis and systemic lupus erythematosus are markedly more prevalent in women than men. The complexities of cellular and humoral immunity that make it difficult to discern the cause of autoimmune diseases also make these diseases difficult to treat. Research in this area has raised the intriguing possibility that autoimmunity may contribute to or complicate other common diseases such as atherosclerosis ("hardening of the arteries").

Disease	Target Tissue	Major Symptoms
Addison's disease	Adrenal cortex	Low blood glucose, muscle weakness, Na^+ loss, K^+ retention, increased susceptibility to stress
Crohn's disease	Intestinal lining	Intestinal inflammation, chronic diarrhea
Graves' disease	Thyroid gland	Oversecretion of thyroid hormone resulting in increased appetite accompanied by weight loss
Insulin-dependent diabetes mellitus	Pancreatic β cells	Loss of ability to make insulin
Multiple sclerosis	Myelin sheath of nerve fibers in brain and spinal cord	Progressive loss of motor control
Myasthenia gravis	Acetylcholine receptors at nerve–muscle synapses	Progressive muscle weakness
Psoriasis	Epidermis	Hyperproliferation of the skin
Rheumatoid arthritis	Connective tissue	Inflammation and degeneration of the joints
Systemic lupus erythematosus	DNA, phospholipids, other tissue components	Rash, joint and muscle pain, anemia, kidney damage, mental dysfunction

SUMMARY

1. Myoglobin, a monomeric heme-containing muscle protein, reversibly binds a single O_2 molecule.

2. Hemoglobin, a tetramer with pseudo-D_2 symmetry, has distinctly different conformations in its oxy and deoxy states.

3. Oxygen binds to hemoglobin in a sigmoidal fashion, indicating cooperative binding.

4. O_2 binding to a heme group induces a conformational change in the entire hemoglobin molecule that includes movements at the subunit interfaces and the disruption of ion pairs. The result is a shift from the T to the R state.

5. CO_2 promotes O_2 dissociation from hemoglobin through the Bohr effect. BPG decreases hemoglobin's O_2 affinity by binding to deoxyhemoglobin.

6. Hemoglobin variants have revealed structure–function relationships. Hemoglobin S produces the symptoms of sickle-cell anemia by forming rigid fibers in the deoxy form.

7. The symmetry and sequential models of allosterism explain how binding of a ligand at one site affects binding of another ligand at a different site.

8. Myofibrils consist of repeating arrays of thick filaments (made of myosin) and thin filaments (made of actin, tropomyosin, and troponin).

9. The heads of myosin molecules form bridges to actin in thin filaments such that the detachment and reattachment of the myosin heads causes the thick and thin filaments to slide past each other during muscle contraction. ATP hydrolysis provides the contractile force.

10. The immune system responds to foreign macromolecules through the production of antibodies (immunoglobulins).

11. The Y-shaped IgG molecule consists of two heavy and two light chains. The two antigen-binding sites are formed by the hypervariable sequences in the variable domains at the ends of a heavy and a light chain.

REFERENCES

Myoglobin and Hemoglobin

Ackers, G.K., Doyle, M.L., Myers, D., and Daugherty, M.A., Molecular code for cooperativity in hemoglobin, *Science* **255**, 54–63 (1992).

Dickerson, R.E. and Geis, I., *Hemoglobin,* Benjamin/Cummings (1983). [A beautifully written and illustrated treatise on the structure, function, and evolution of hemoglobin.]

Nagel, R.L. and Roth, E.F., Jr., Malaria and red cell genetic defects, *Blood* **74**, 1213–1221 (1989).

Perutz, M.F., Wilkinson, A.J., Paoli, M., and Dodson, G.G., The stereochemical mechanism of the cooperative effects in hemoglobin revisited, *Annu. Rev. Biophys. Biomol. Struct.* **27**, 1–34 (1998).

Perutz, M.F., Myoglobin and haemoglobin: Role of distal residues in reactions with haem ligands, *Trends Biochem. Sci.* **14**, 42–44 (1989).

Riggs, A., Hemoglobins, *Curr. Opin. Struct. Biol.* **1**, 915–921 (1991). [Includes a discussion of invertebrate hemoglobins.]

Urich, K., *Comparative Animal Biochemistry,* Chapter 7, Springer-Verlag (1994). [Discusses non-hemoglobin oxygen-transport proteins.]

Weatherall, D.J., Clegg, J.B., Higgs, D.R., and Wood, W.G., The hemoglobinopathies, *in* Scriver, C.R., Beaudet, A.L., Sly, W.S., Valle, D., Stanbury, J.B., Wyngaarden, J.B., and Frederickson, D.S. (Eds.), *The Metabolic and Molecular Bases of Inherited Disease* (7th ed.), pp. 3417–3484, McGraw–Hill (1995).

Myosin and Actin

Alberts, B., Bray, D., Lewis, J., Raff, M., Roberts, K., and Watson, J.D., *Molecular Biology of the Cell* (3rd ed.), Chapter 16, Garland Publishing (1994).

Block, S.M., Fifty ways to love your lever: Myosin motors, *Cell* **87**, 151–157 (1996).

Milligan, R.A., Protein–protein interactions in the rigor actomyosin complex, *Proc. Natl. Acad. Sci.* **93**, 21–26 (1996).

Milligan, R.A., Whittaker, M., and Safer, D., Molecular structure of F-actin and the location of surface binding sites, *Nature* **348**, 217–221 (1990).

Rayment, I. and Holden, H.M., Myosin subfragment-1: structure and function of a molecular motor, *Curr. Opin. Struct. Biol.* **3**, 944–952 (1993).

Spudich, J.A., How molecular motors work, *Nature* **372**, 515–518 (1994).

Antibodies

Davies, D.R. and Chacko, S., Antibody structure, *Acc. Chem. Res.* **26**, 421–427 (1993).

Davies, D.R. and Cohen, G.H., Interactions of protein antigens with antibodies, *Proc. Natl. Acad. Sci.* **93**, 7–12 (1996).

Harris, L.J., Larson, S.B., Hasel, K.W., Day, J., Greenwood, A., and McPherson, A., The three-dimensional structure of an intact monoclonal antibody for canine lymphoma, *Nature* **360**, 369–372 (1992). [The first and, as yet, only high-resolution X-ray structure of an intact IgG.]

Lodish, H., Baltimore, D., Berk, A., Zipursky, S.L., Matsudaria, P., and Darnell, J., *Molecular Cell Biology* (3rd ed.), Chapter 27, Scientific American Books (1995). [This chapter covers the immune system and antibody function.]

Male, D., Cooke, A., Owen, M., Trowsdale, J., and Champion, B., *Advanced Immunology* (3rd ed.), Mosby (1996). [Chapters 1 and 2 provide an overview of the immune system and antibody structure.]

Steinman, L., Autoimmune disease, *Sci. Am.* **269**(3), 107–114 (1993).

KEY TERMS

heme
oxygenation
Y_{O_2}
pO_2
hyperbolic curve
saturation
p_{50}
ligand
sigmoidal curve
cooperative binding
Hill equation
Hill constant
positive cooperativity
negative cooperativity

Perutz mechanism
T state
R state
Bohr effect
erythrocyte
variant
lyse
anemia
cyanosis
polycythemia
heterozygote
homozygote
hemoglobin S
sickle-cell anemia

allosteric interaction
striated muscle
myofibril
sarcomere
thick filament
thin filament
sliding filament model
ATPase
microfilament
immunofluorescence
  microscopy
treadmilling
pathogen
immune system

cellular immunity
humoral immunity
lymphocyte
immunoglobulin
memory *B* cell
Fab fragment
Fc fragment
variable region
immunoglobulin fold
hypervariability
monoclonal antibody
autoimmune disease

STUDY EXERCISES

1. Describe the structural and functional differences between myoglobin and hemoglobin.

2. Explain how hemoglobin effectively delivers O_2 to myoglobin in muscles.

3. Explain the structural basis for cooperative O_2 binding to hemoglobin.

4. What is the physiological relevance of the Bohr effect?

5. How does BPG affect O_2 binding to hemoglobin?

6. Differentiate the symmetry and sequential models of allosterism.

7. Draw a diagram of the components of a myofibril.

8. Explain the molecular basis of the sliding filament model of muscle contraction.

9. Identify the parts of an immunoglobulin molecule.

PROBLEMS

1. Estimate K from the following data describing ligand binding to a protein.

[Ligand] (mM)	Y
0.25	0.30
0.5	0.45
0.8	0.56
1.4	0.66
2.2	0.80
3.0	0.83
4.5	0.86
6.0	0.93

2. Which set of binding data is likely to represent cooperative ligand binding to an oligomeric protein?

(a) [Ligand] (mM)	Y	(b) [Ligand] (mM)	Y
0.1	0.3	0.2	0.1
0.2	0.5	0.3	0.3
0.4	0.7	0.4	0.6
0.7	0.9	0.6	0.8

3. Is the p_{50} higher or lower than normal in (a) hemoglobin Yakima and (b) hemoglobin Kansas? Explain.

4. Hemoglobin S homozygotes who are severely anemic often have elevated levels of BPG in their erythrocytes. Is this a beneficial effect?

5. The crocodile, which can remain under water without breathing for up to 1 h, drowns its air-breathing prey and then dines at its leisure. An adaptation that aids the crocodile in doing so is that it can utilize virtually 100% of the O_2 in its blood whereas humans, for example, can extract only ~65% of the O_2 in their blood. Crocodile Hb does not bind BPG. However, crocodile deoxyHb preferentially binds HCO_3^-. How does this help the crocodile obtain its dinner?

6. A myosin head can undergo five ATP hydrolysis cycles per second, each of which moves an actin monomer by ~60 Å. How is it possible for an entire sarcomere to shorten by 1000 Å in this same period?

7. **Rigor mortis,** the stiffening of muscles after death, is caused by depletion of cellular ATP. Describe the molecular basis of rigor.

8. Give the approximate molecular masses of an immunoglobulin G molecule analyzed by (a) gel filtration chromatography, (b) SDS-PAGE, and (c) SDS-PAGE in the presence of 2-mercaptoethanol.

9. Explain why the variation in V_L and V_H domains of immunoglobulins is largely confined to the hypervariable loops.

10. Why do antibodies raised against a native protein sometimes fail to bind to the corresponding denatured protein?

11. Antibodies raised to a macromolecular antigen usually produce an antigen–antibody precipitate when mixed with that antigen. Explain why no precipitate forms when (a) Fab fragments from these antibodies are mixed with the antigen; (b) antibodies raised against a small antigen are mixed with that small antigen; and (c) the antibody is in great excess over the antigen and vice versa.

Sugars are relatively simple molecules that can be linked together in various ways to form larger molecules, for example, starch. This storage form of carbohydrate is the primary source of energy in many foods, including bread, rice, and pasta.
[Charles D. Winters/Photo Researchers.]

CARBOHYDRATES

195

Carbohydrates or **saccharides** (Greek: *sakcharon,* sugar) are the most abundant biological molecules. They are chemically simpler than nucleotides or amino acids, containing just three elements—carbon, hydrogen, and oxygen—combined according to the formula $(C \cdot H_2O)_n$, where $n \geq 3$. The basic carbohydrate units are called **monosaccharides.** There are numerous different types of monosaccharides, which, as we discuss below, differ in their number of carbon atoms and in the arrangement of the H and O atoms attached to these carbons. Furthermore, monosaccharides can be strung together in almost limitless ways to form **polysaccharides.**

Until the 1960s, carbohydrates were thought to have only passive roles as energy sources (e.g., glucose and starch) and as structural materials (e.g., cellulose). Carbohydrates, as we shall see, do not catalyze complex chemical reactions as do proteins, nor do carbohydrates replicate themselves as do nucleic acids. And because polysaccharides are not built according to a genetic "blueprint," as are nucleic acids and proteins, they tend to be much more heterogeneous—both in size and in composition—than other biological molecules.

However, it has become clear that the innate structural variation in carbohydrates is fundamental to their biological activity. The apparently haphazard arrangements of carbohydrates on proteins and on the surfaces of cells are the key to many recognition events between proteins and between cells. An understanding of carbohydrate structure, from the simplest monosaccharides to the most complex branched polysaccharides, is essential for appreciating the varied functions of carbohydrates in biological systems.

1. MONOSACCHARIDES

Monosaccharides, or simple sugars, are synthesized from smaller precursors that are ultimately derived from CO_2 and H_2O by photosynthesis.

A. Classification of Monosaccharides

Monosaccharides are aldehyde or ketone derivatives of straight-chain polyhydroxy alcohols containing at least three carbon atoms. They are classified according to the chemical nature of their carbonyl group and the number of their C atoms. If the carbonyl group is an aldehyde, the sugar is an **aldose.** If the carbonyl group is a ketone, the sugar is a **ketose.** The smallest monosaccharides, those with three carbon atoms, are **trioses.** Those with four, five, six, seven, etc. C atoms are, respectively, **tetroses, pentoses, hexoses, heptoses,** etc.

The aldohexose **D-glucose** has the formula $(C \cdot H_2O)_6$.

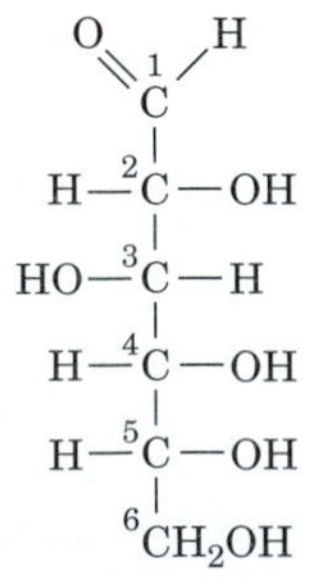

$$
\begin{array}{c}
\overset{1}{C}\!\!\diagup^{H}_{\diagdown O} \\
| \\
H-\overset{2}{C}-OH \\
| \\
HO-\overset{3}{C}-H \\
| \\
H-\overset{4}{C}-OH \\
| \\
H-\overset{5}{C}-OH \\
| \\
\overset{6}{C}H_2OH
\end{array}
$$

D-Glucose

All but two of its six C atoms, C1 and C6, are chiral centers, so D-glucose is one of $2^4 = 16$ possible stereoisomers. The stereochemistry and nomenclature of the D-aldoses are presented in Fig. 8-1. The assignment of D or L is made according to the Fischer convention (Section 4-2): *D sugars have the same absolute configuration at the asymmetric center farthest from their carbonyl group as does D-glyceraldehyde* (i.e., the —OH at C5 of D-glucose is on the right in a Fischer projection). The L sugars are the mirror images of their D counterparts.

Sugars that differ only by the configuration around one C atom are known as **epimers** of one another. Thus, D-glucose and **D-mannose** are epimers with respect to C2.

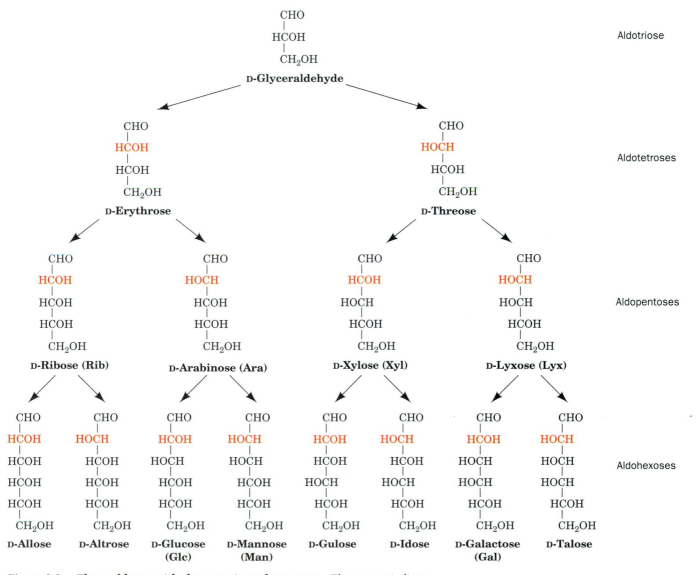

Figure 8-1. The D-aldoses with three to six carbon atoms. The arrows indicate stereochemical relationships (not biosynthetic pathways). The configuration around C2 (*red*) distinguishes the members of each pair of monosaccharides. The L counterparts of these 15 sugars are their mirror images.

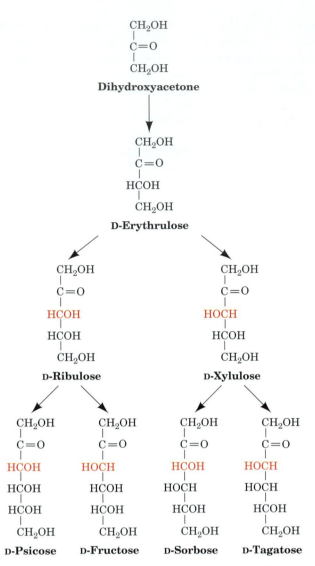

Figure 8-2.   The D-ketoses with three to six carbon atoms. The configuration around C3 (*red*) distinguishes the members of each pair.

The most common ketoses are those with their ketone function at C2 (Fig. 8-2). The position of their carbonyl group gives ketoses one less asymmetric center than their isomeric aldoses, so a ketohexose has only $2^3 = 8$ possible stereoisomers (4 D sugars and 4 L sugars). Note that some ketoses are named by inserting -*ul*- before the suffix -*ose* in the name of the corresponding aldose; thus, **D-xylulose** is the ketose corresponding to the aldose **D-xylose.**

Because L sugars are biologically much less abundant than D sugars, the D prefix is often omitted. The most important monosaccharides are the aldoses **glyceraldehyde, ribose, glucose, mannose,** and **galactose,** and the ketoses **dihydroxyacetone, ribulose,** and **fructose.** We shall encounter these substances in our studies of metabolism.

B. *Configuration and Conformation*

Alcohols react with the carbonyl groups of aldehydes and ketones to form **hemiacetals** and **hemiketals,** respectively.

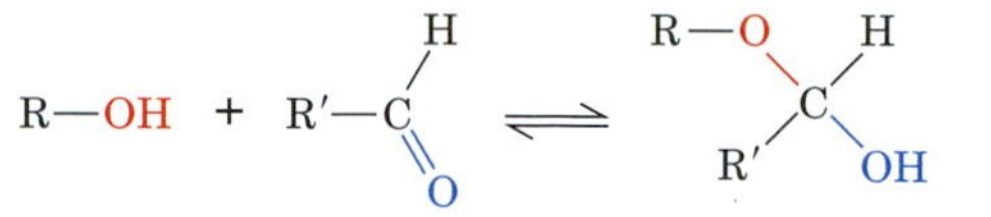

R—OH + R′—C(H)(=O) ⇌ R—O—C(H)(R′)(OH)

Alcohol **Aldehyde** **Hemiacetal**

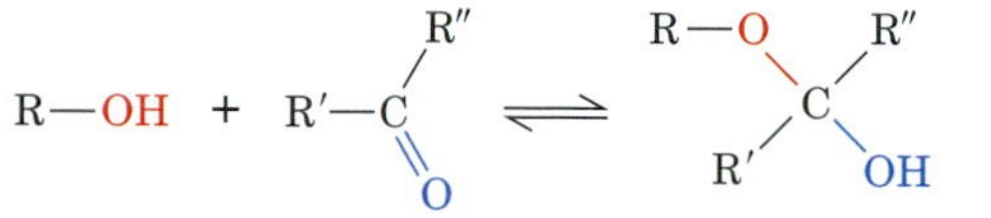

R—OH + R′—C(R″)(=O) ⇌ R—O—C(R″)(R′)(OH)

Alcohol **Ketone** **Hemiketal**

The hydroxyl and either the aldehyde or the ketone functions of mono-saccharides can likewise react intramolecularly to form cyclic hemiacetals and hemiketals (Fig. 8-3). The configurations of the substituents of each carbon atom in these sugar rings are conveniently represented by their **Haworth projections,** in which the darker ring bonds project in front of the plane of the paper and the lighter ring bonds project behind it.

A sugar with a six-membered ring is known as a **pyranose** in analogy with **pyran,** the simplest compound containing such a ring. Similarly, sug-

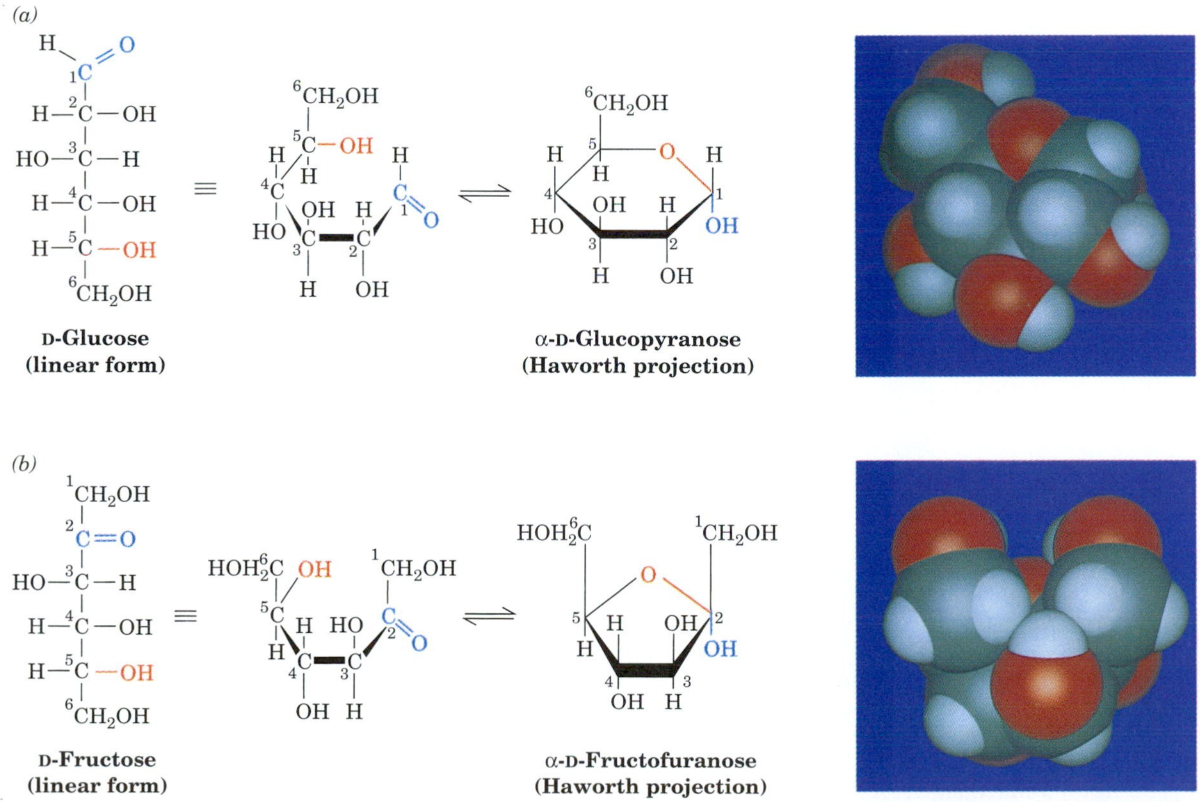

(a)

**D-Glucose
(linear form)**

**α-D-Glucopyranose
(Haworth projection)**

(b)

**D-Fructose
(linear form)**

**α-D-Fructofuranose
(Haworth projection)**

Figure 8-3. *Key to Structure.* Cyclization of glucose and fructose. (*a*) The linear form of D-glucose yields the cyclic hemiacetal α-D-glucopyranose. (*b*) The linear form of D-fructose yields the hemiketal α-D-fructofuranose. The cyclic sugars are shown as both Haworth projections and space-filling models. [Courtesy of Robert Stodola, Fox Chase Cancer Center.]

ars with five-membered rings are designated **furanoses** in analogy with **furan.**

Pyran   Furan

The cyclic forms of glucose and fructose with six- and five-membered rings are therefore known as **glucopyranose** and **fructofuranose,** respectively.

Cyclic Sugars Have Two Anomeric Forms

When a monosaccharide cyclizes, the carbonyl carbon, called the **anomeric carbon,** becomes a chiral center with two possible configurations. The pair of stereoisomers that differ in configuration at the anomeric carbon are called **anomers.** In the α anomer, the OH substituent of the anomeric carbon is on the opposite side of the sugar ring from the CH_2OH group at the chiral center that designates the D or L configuration (C5 in hexoses). The other anomer is known as the β form (Fig. 8-4).

The two anomers of D-glucose have slightly different physical and chemical properties, including different optical rotations (Section 4-2). *The anomers freely interconvert in aqueous solution,* so at equilibrium, D-glucose is a mixture of the β anomer (63.6%) and the α anomer (36.4%). The linear form is normally present in only minute amounts.

Sugars Are Conformationally Variable

A given hexose or pentose can assume pyranose or furanose forms. In principle, hexoses and larger sugars can form rings of seven or more atoms, but such rings are rarely observed because of the greater stabilities of the five- and six-membered rings. The internal strain of three- and four-membered rings makes them less stable than the linear forms.

The use of Haworth formulas may lead to the erroneous impression that furanose and pyranose rings are planar. This cannot be the case, however, because all the atomic orbitals in the ring atoms are tetrahedrally (sp^3) hybridized. The pyranose ring, like the cyclohexane ring, can assume a chair conformation, in which the substituents of each atom are arranged tetrahedrally. Of the two possible chair conformations, the one that predominates is the one in which the bulkiest ring substituents occupy **equa-**

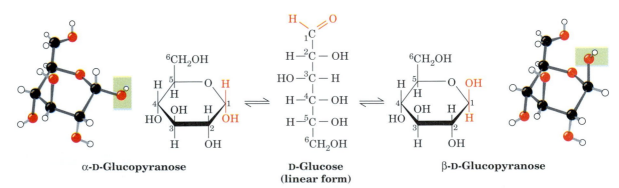

α-D-Glucopyranose   D-Glucose (linear form)   β-D-Glucopyranose

Figure 8-4.   **α and β anomers.** The monosaccharides α-D-glucopyranose and β-D-glucopyranose, drawn as Haworth projections and ball-and-stick models, interconvert through the linear form. They differ only by their configuration about the anomeric carbon, C1. ● See **Kinemage Exercise 7-1.**

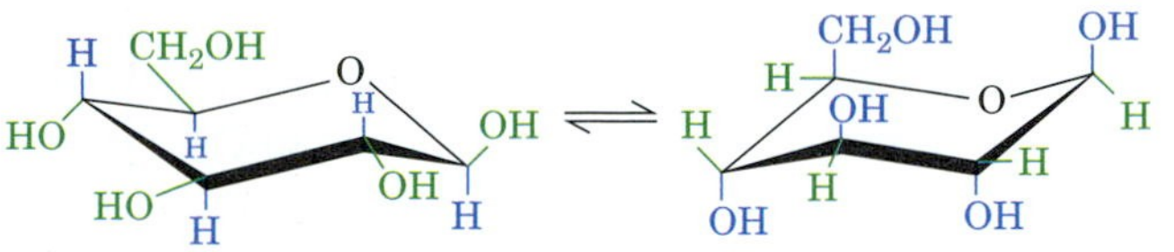

Figure 8-5. The two chair conformations of β-D-glucopyranose. In the conformation on the left, which predominates, the relatively bulky OH and CH_2OH
substituents all occupy equatorial positions, where they extend alternately above and
below the ring. In the conformation on the right (drawn in ball-and-stick form
in Fig. 8-4, *right*), the bulky groups occupy the more crowded axial (vertical)
positions. ● See Kinemage Exercise 7-1.

torial positions rather than at the more crowded **axial** positions (Fig. 8-5).
Only β-D-glucose can simultaneously have all five of its non-H substituents
in equatorial positions. Perhaps this is why glucose is the most abundant
monosaccharide in nature.

Furanose rings can also adopt different conformations, whose stabilities
depend on the arrangements of bulky substituents. Note that a monosaccharide can readily shift its *conformation,* because no bonds are broken in
the process. The shift in *configuration* between the α and β anomeric forms
or between the pyranose and furanose forms, which requires breaking and
re-forming bonds, occurs slowly in aqueous solution. Other changes in configuration, such as **epimerization,** do not occur under physiological conditions without the appropriate enzyme.

C. Sugar Derivatives

Because the cyclic and linear forms of aldoses and ketoses do interconvert,
these sugars undergo reactions typical of aldehydes and ketones.

1. Mild chemical or enzymatic oxidation of an aldose converts its aldehyde group to a carboxylic acid group, thereby yielding an **aldonic
   acid** such as **gluconic acid** (*at right*). Aldonic acids are named by appending the suffix *-onic acid* to the root name of the parent aldose.

2. The specific oxidation of the primary alcohol group of aldoses yields
   uronic acids, which are named by appending *-uronic acid* to the root
   name of the parent aldose, for example, D-**glucuronic acid** (*at right*).
   Uronic acids can assume the pyranose, furanose, and linear forms.

3. Aldoses and ketoses can be reduced under mild conditions, for
   example, by treatment with $NaBH_4$, to yield acyclic polyhydroxy alcohols known as **alditols,** which are named by appending the suffix
   -itol to the root name of the parent aldose. **Ribitol** is a component of
   flavin coenzymes (Fig. 3-3), and **glycerol** and the cyclic polyhydroxy
   alcohol ***myo*-inositol** are important lipid components (Section 9-1).
   Xylitol is a sweetener that is used in "sugarless" gum and candies.

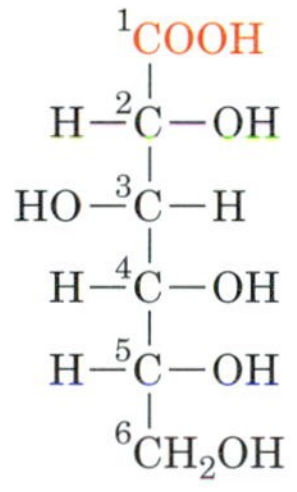

D-Gluconic acid

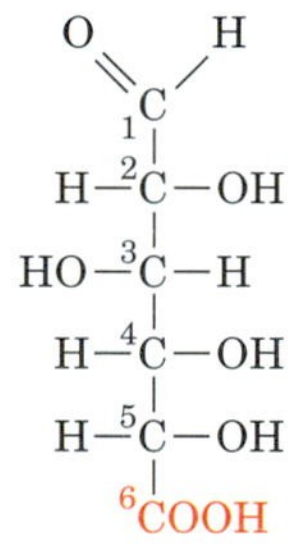

D-Glucuronic acid

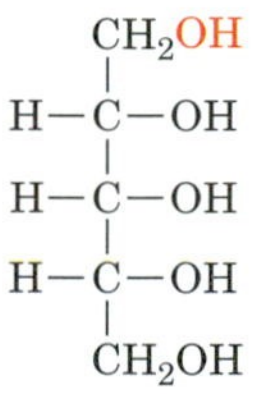

Ribitol 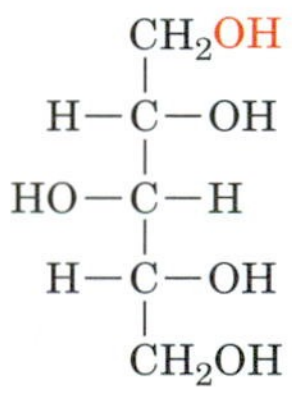 Xylitol 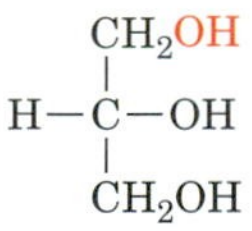 Glycerol 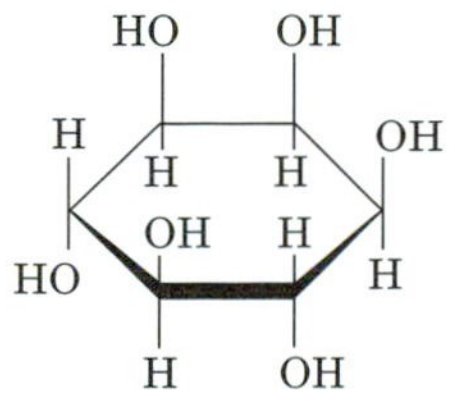 *myo*-Inositol

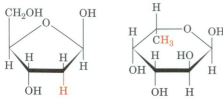

β-D-2-Deoxyribose   **α-L-Fucose**

4. Monosaccharide units in which an OH group is replaced by H are known as **deoxy sugars.** The biologically most important of these is **β-D-2-deoxyribose** (*at left*), the sugar component of DNA's sugar–phosphate backbone (Fig. 3-6). **L-Fucose** is one of the few L sugar components of polysaccharides.

5. In **amino sugars,** one or more OH groups have been replaced by an amino group, which is often acetylated. **D-Glucosamine** and **D-galactosamine** are the most common.

α-D-Glucosamine
(2-amino-2-deoxy-
α-D-glucopyranose)

α-D-Galactosamine
(2-amino-2-deoxy-
α-D-galactopyranose)

N-Acetylneuraminic acid, which is derived from **N-acetylmannosamine** and **pyruvic acid** (Fig. 8-6), is an important constituent of **glycoproteins** and **glycolipids** (proteins and lipids with covalently attached carbohydrate). *N*-Acetylneuraminic acid and its derivatives are often referred to as **sialic acids.**

6. The anomeric group of a sugar can condense with an alcohol to form **α-** and **β-glycosides** (Greek: *glykys,* sweet; Fig. 8-7), which are cyclic **acetals** or **ketals.**

A cyclic acetal   **A cyclic ketal**

The bond connecting the anomeric carbon to the alcohol oxygen is termed a **glycosidic bond.** *N*-Glycosidic bonds, which form between the anomeric carbon and an amine, are the bonds that link D-ribose to purines and pyrimidines in nucleic acids (Fig. 3-6). Like peptide

N-Acetylneuraminic acid
(linear form)

N-Acetylneuraminic acid
(pyranose form)

***Figure 8-6.* N-Acetylneuraminic acid.** In the cyclic form of this nine-carbon monosaccharide, the pyranose ring incorporates the pyruvic acid residue (*blue*) and part of the mannose moiety.

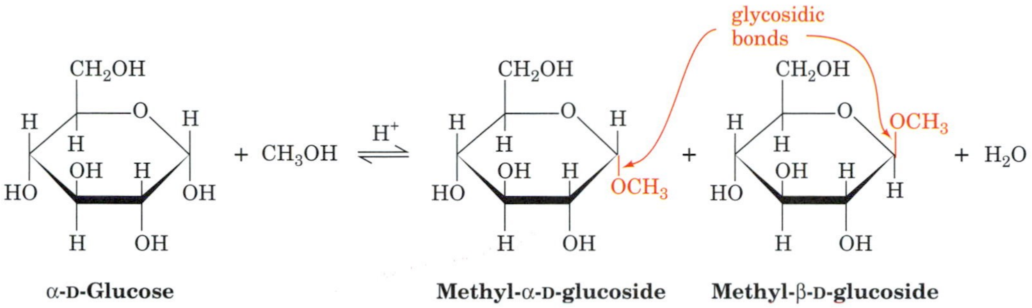

Figure 8-7. Formation of glycosides. The acid-catalyzed condensation of α-D-glucose with methanol yields an anomeric pair of **methyl-D-glucosides.**

bonds, glycosidic bonds hydrolyze extremely slowly under physiological conditions in the absence of appropriate hydrolytic enzymes.

2. POLYSACCHARIDES

Polysaccharides, which are also known as **glycans,** *consist of monosaccharides linked together by glycosidic bonds.* They are classified as **homopolysaccharides** or **heteropolysaccharides** if they consist of one type or more than one type of monosaccharide. Although the monosaccharide sequences of heteropolysaccharides can, in principle, be even more varied than those of proteins, many are composed of only a few types of monosaccharides that alternate in a repetitive sequence.

Polysaccharides, in contrast to proteins and nucleic acids, form branched as well as linear polymers. This is because glycosidic linkages can be made to any of the hydroxyl groups of a monosaccharide. Fortunately for structural biochemists, most polysaccharides are linear and those that branch do so in only a few well-defined ways.

A complete description of an **oligosaccharide** or polysaccharide includes the identities, anomeric forms, and linkages of all its component monosaccharide units. Some of this information can be gathered through the use of specific **exoglycosidases** and **endoglycosidases,** enzymes that hydrolyze monosaccharide units in much the same way that exopeptidases and endopeptidases cleave amino acid residues from polypeptides (Section 5-3). NMR measurements are also invaluable in determining both sequences and conformations of polysaccharides.

A. Disaccharides

The simplest polysaccharides are the **disaccharides. Lactose** (*at right*), for example, occurs naturally only in milk, where its concentration ranges from 0 to 7% depending on the species (see Box 8-1). The systematic name for lactose, *O*-β-D-galactopyranosyl-(1→4)-D-glucopyranose, specifies its monosaccharides, their ring types, and how they are linked together. The symbol (1→4) indicates that the glycosidic bond links C1 of galactose to C4 of glucose. Note that lactose has a free anomeric carbon on its glucose residue. Sugars bearing anomeric carbons that have not formed glycosides are termed **reducing sugars** because they readily reduce mild oxidizing agents. Polysaccharides that are reducing sugars generally have only one residue that lacks a glycosidic linkage, the so-called **reducing end,** because their component monosaccharide units are almost always glycosidically linked.

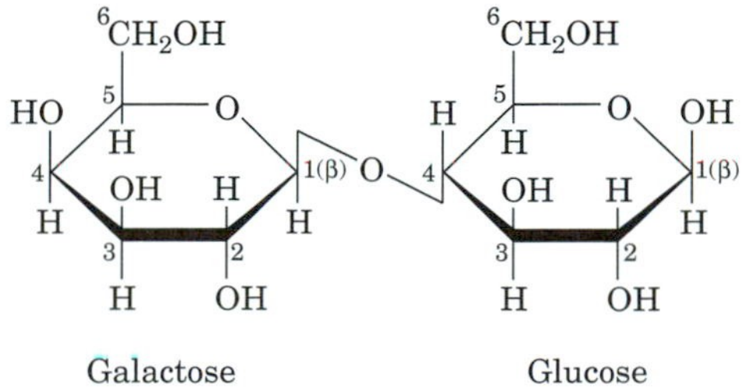

Galactose   Glucose

Lactose

Box 8-1

BIOCHEMISTRY IN HEALTH AND DISEASE

Lactose Intolerance

In infants, lactose (also known as milk sugar) is hydrolyzed by the intestinal enzyme **β-D-galactosidase** (or **lactase**) to its component monosaccharides for absorption into the bloodstream. The galactose is enzymatically converted (epimerized) to glucose, which is the primary metabolic fuel of many tissues.

Since mammals are unlikely to encounter lactose after they have been weaned, most adult mammals have low levels of β-galactosidase. Consequently, much of the lactose they might ingest moves through their digestive tract to the colon, where bacterial fermentation generates large quantities of

CO_2, H_2, and irritating organic acids. These products cause the embarrassing and often painful digestive upset known as **lactose intolerance**.

Lactose intolerance, which was once considered a metabolic disturbance, is actually the norm in adult humans, particularly those of African and Asian descent. Interestingly, however, β-galactosidase levels decrease only mildly with age in descendants of populations that have historically relied on dairy products for nutrition throughout life. Modern food technology has come to the aid of milk lovers who develop lactose intolerance: Milk in which the lactose has been hydrolyzed enzymatically is widely available.

Sucrose

Glucose   Fructose

The most abundant disaccharide is **sucrose** (*at left*), the major form in which carbohydrates are transported in plants. Sucrose is familiar to us as common table sugar (● **see Kinemage Exercise 7-2**). The systematic name for sucrose, *O*-α-D-glucopyranosyl-(1→2)-β-D-fructofuranoside, indicates that the anomeric carbon of each sugar (C1 in glucose and C2 in fructose) participates in the glycosidic bond.

Other common disaccharides occur as the hydrolysis products of larger polysaccharides. Only a few tri- and higher oligosaccharides occur in nature, all of them in plants.

B. Structural Polysaccharides: Cellulose and Chitin

Plants have rigid cell walls that can withstand osmotic pressure differences between the extracellular and intracellular spaces of up to 20 atm. In large plants, such as trees, the cell walls also have a load-bearing function. **Cellulose,** the primary structural component of plant cell walls (Fig. 8-8), accounts for over half of the carbon in the biosphere: Approximately 10^{15} kg of cellulose is estimated to be synthesized and degraded annually.

Cellulose is a linear polymer of up to 15,000 D-glucose residues linked by β(1→4) glycosidic bonds.

Glucose   Glucose

Cellulose

***Figure 8-8.* Electron micrograph of a cell wall**. The cellulose fibers in this sample of cell wall from the alga *Chaetomorpha* are arranged in layers. [Biophoto Associates/Photo Researchers.]

X-Ray studies of cellulose fibers have led Anatole Sarko to propose the structure diagrammed in Fig. 8-9. This highly cohesive, hydrogen-bonded structure gives cellulose fibers exceptional strength and makes them water insoluble despite their hydrophilicity. In plant cell walls, the cellulose fibers are embedded in and cross-linked by a matrix containing other polysaccharides and **lignin,** a plasticlike phenolic polymer. The resulting composite material can withstand large stresses because the matrix evenly distributes the stresses among the cellulose reinforcing elements.

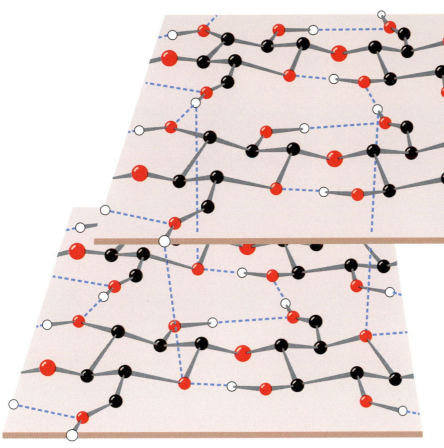

Figure 8-9. **Model of cellulose.** Cellulose fibers consist of ~40 parallel, extended glycan chains. Each of the β(1→4)-linked glucose units in a chain is rotated 180° with respect to its neighboring residues and is held in this position by intrachain hydrogen bonds *(dashed lines)*. The glycan chains line up laterally to form sheets, and these sheets stack vertically so that they are staggered by half the length of a glucose unit. The entire assembly is stabilized by intermolecular hydrogen bonds. Hydrogen atoms not participating in hydrogen bonds have been omitted for clarity.

Although vertebrates themselves do not possess an enzyme capable of hydrolyzing the β(1→4) linkages of cellulose, the digestive tracts of herbivores contain symbiotic microorganisms that secrete a series of enzymes, collectively known as **cellulases,** that do so. The same is true of termites. Nevertheless, the degradation of cellulose is a slow process because its tightly packed and hydrogen-bonded glycan chains are not easily accessible to cellulase and do not separate readily even after many of their glycosidic bonds have been hydrolyzed. Thus, cows must chew their cud and the decay of dead trees by fungi and other organisms generally takes many years.

Chitin is the principal structural component of the exoskeletons of invertebrates such as crustaceans, insects, and spiders and is also present in the cell walls of most fungi and many algae. It is therefore almost as abundant as cellulose. Chitin is a homopolymer of β(1→4)-linked *N*-acetyl-D-glucosamine residues.

Chitin

It differs chemically from cellulose only in that each C2 OH group is replaced by an acetamide function. X-Ray analysis indicates that chitin and cellulose have similar structures.

C. Storage Polysaccharides: Starch and Glycogen

Starch is a mixture of glycans that plants synthesize as their principal food reserve. It is deposited in the chloroplasts of plant cells as insoluble granules composed of **α-amylose** and **amylopectin.** α-Amylose is a linear polymer of several thousand glucose residues linked by α(1→4) bonds.

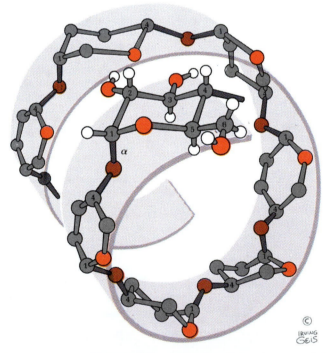

α-Amylose

Note that although α-amylose is an isomer of cellulose, it has very different structural properties. While cellulose's β-glycosidic linkages cause it to assume a tightly packed, fully extended conformation (Fig. 8-9), α-amylose's α-glycosidic bonds cause it to adopt an irregularly aggregating helically coiled conformation (Fig. 8-10).

Figure 8-10.   α-Amylose. This regularly repeating polymer forms a left-handed helix. Note the great differences in structure and properties that result from changing α-amylose's α(1→4) linkages to the β(1→4) linkages of cellulose (Fig. 8-9). [Figure copyrighted © by Irving Geis.]

Amylopectin consists mainly of $\alpha(1\rightarrow4)$-linked glucose residues but is a branched molecule with $\alpha(1\rightarrow6)$ branch points every 24 to 30 glucose residues on average.

Amylopectin

Amylopectin molecules contain up to 10^6 glucose residues, making them some of the largest molecules in nature. The storage of glucose as starch greatly reduces the large intracellular osmotic pressure that would result from its storage in monomeric form, because osmotic pressure is proportional to the number of solute molecules in a given volume (Section 2-1D).

The digestion of starch, the main carbohydrate source in the human diet, begins in the mouth. Saliva contains an **amylase,** which randomly hydrolyzes the $\alpha(1\rightarrow4)$ glycosidic bonds of starch. Starch digestion continues in the small intestine under the influence of pancreatic amylase, which degrades starch to a mixture of $\alpha(1\rightarrow4)$-linked glucose disaccharides (called **maltose**) and trisaccharides **(maltotriose),** and oligosaccharides known as **dextrins** that contain the $\alpha(1\rightarrow6)$ branches. These oligosaccharides are hydrolyzed to their component monosaccharides by an **α-glucosidase,** which removes one glucose residue at a time, and a **debranching enzyme,** which hydrolyzes $\alpha(1\rightarrow6)$ as well as $\alpha(1\rightarrow4)$ bonds. The resulting monosaccharides are absorbed by the intestine and transported to the bloodstream.

Glycogen, the storage polysaccharide of animals, is present in all cells but is most prevalent in skeletal muscle and in liver, where it occurs as cytoplasmic granules (Fig. 8-11). The primary structure of glycogen resembles that of amylopectin, but glycogen is more highly branched, with branch points occurring every 8 to 12 glucose residues. In the cell, glycogen is degraded for metabolic use by **glycogen phosphorylase,** which cleaves glycogen's $\alpha(1\rightarrow4)$ bonds sequentially inward from its nonreducing ends. *Glycogen's highly branched structure, which has many nonreducing ends, permits the rapid mobilization of glucose in times of metabolic need.* The $\alpha(1\rightarrow6)$ branches of glycogen are cleaved by **glycogen debranching enzyme** (glycogen breakdown is discussed further in Section 15-1).

D. Glycosaminoglycans

The extracellular spaces, particularly those of connective tissues such as cartilage, tendon, skin, and blood vessel walls, contain collagen (Section 6-1C) and other proteins embedded in a gel-like matrix that is composed largely of **glycosaminoglycans.** These unbranched polysaccharides consist of alternating uronic acid and hexosamine residues. Solutions of glycosaminoglycans have a slimy, mucuslike consistency that results from their high viscosity and elasticity.

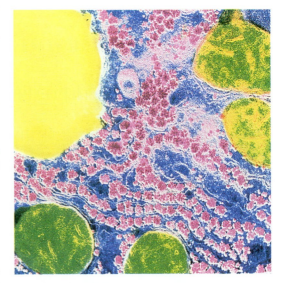

Figure 8-11. **Photomicrograph showing the glycogen granules (*pink*) of a liver cell.** The greenish objects are mitochondria, and the yellow object is a fat globule. The glycogen content of liver may reach 10% of its net weight. [CNRI/ Science Photo Library/Photo Researchers.]

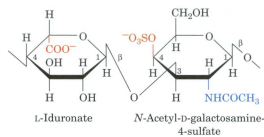

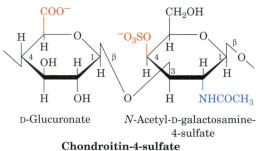

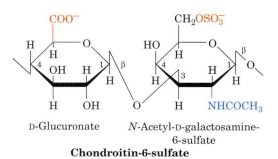

Figure 8-12.   Repeating disaccharide units of some glycosaminoglycans. The anionic groups are shown in red and the *N*-acetylamido groups are shown in blue. ● See Kinemage Exercise 7-3.

Hyaluronic acid is an important glycosaminoglycan component of connective tissue, synovial fluid (the fluid that lubricates joints), and the vitreous humor of the eye. Hyaluronic acid molecules are composed of 250 to 25,000 β(1→4)-linked disaccharide units that consist of D-glucuronic acid and **N-acetyl-D-glucosamine (GlcNAc)** linked by a β(1→3) bond (Fig. 8-12). The disaccharide units of hyaluronic acid are extended, forming a rigid molecule whose numerous repelling anionic groups bind cations and water molecules. In solution, hyaluronate occupies a volume ~1000 times that in its dry state.

Hyaluronate solutions have a viscosity that is shear dependent (an object under shear stress has equal and opposite forces applied across its opposite faces). At low shear rates, hyaluronate molecules form tangled masses that greatly impede flow; that is, the solution is quite viscous. As the shear stress increases, the stiff hyaluronate molecules tend to line up with the flow and thus offer less resistance to it. This viscoelastic behavior makes hyaluronate solutions excellent biological shock absorbers and lubricants.

The other common glycosaminoglycans shown in Fig. 8-12 consist of 50 to 1000 sulfated disaccharide units. **Chondroitin-4-sulfate** and **chondroitin-6-sulfate** differ only in the sulfation of their **N-acetylgalactosamine (GalNAc)** residues. **Dermatan sulfate** is derived from chondroitin by enzymatic epimerization of C5 of glucuronate residues to form **iduronate** residues. **Keratan sulfate** (not to be confused with the fibrous protein keratin; Section 6-1C) is the most heterogeneous of the major glycosaminoglycans in that its sulfate content is variable and it contains small amounts of fucose, mannose, GlcNAc, and sialic acid. **Heparin** is also variably sulfated, with an average of 2.5 sulfate residues per disaccharide unit, which makes it the most highly charged polymer in mammalian tissues.

In contrast to the other glycosaminoglycans, heparin is not a constituent of connective tissue but occurs almost exclusively in the intracellular granules of the mast cells that occur in arterial walls. It inhibits the clotting of blood, and its release, through injury, is thought to prevent runaway clot formation. Heparin is therefore in wide clinical use to inhibit blood clotting, for example, in postsurgical patients.

3. GLYCOPROTEINS

Many proteins are actually glycoproteins, with carbohydrate contents varying from <1% to >90% by weight. Glycoproteins occur in all forms of life and have functions that span the entire spectrum of protein activities, including those of enzymes, transport proteins, receptors, hormones, and structural proteins. The polypeptide chains of glycoproteins, like those of all proteins, are synthesized under genetic control. Their carbohydrate chains, in contrast, are enzymatically generated and covalently linked to the polypeptide without the rigid guidance of nucleic acid templates. For this reason, glycoproteins tend to have variable carbohydrate composition, a phenomenon known as **microheterogeneity.**

A. *Proteoglycans*

Proteins and glycosaminoglycans in the extracellular matrix aggregate covalently and noncovalently to form a diverse group of macromolecules known as **proteoglycans.** Electron micrographs (Fig. 8-13*a*) and other evidence indicate that proteoglycans have a bottlebrush-like molecular architecture, with "bristles" noncovalently attached to a filamentous hyaluronic acid "backbone." The bristles consist of a **core protein** to which glycosaminoglycans, most often keratan sulfate and chondroitin sulfate, are covalently linked (Fig. 8-13*b*). Smaller oligosaccharides are usually linked to the core protein near its site of attachment to hyaluronate. These oligosaccharides are glycosidically linked to the protein via the amide N of specific Asn residues (and are therefore known as ***N*-linked oligosaccharides;** Section 8-3C). The keratan sulfate and chondroitin sulfate chains are glycosidically linked to the core protein via oligosaccharides that are covalently bonded to side chain O atoms of specific Ser or Thr residues (i.e., ***O*-linked oligosaccharides).**

Altogether, a central strand of hyaluronic acid, which varies in length from 4000 to 40,000 Å, can have up to 100 associated core proteins, each of which binds ~50 keratan sulfate chains of up to 250 disaccharide units and ~100 chondroitin sulfate chains of up to 1000 disaccharide units each. This accounts for the enormous molecular masses of many proteoglycans, which range up to tens of millions of daltons.

The extended brushlike structure of proteoglycans, together with the polyanionic character of their keratan sulfate and chondroitin sulfate components, cause these complexes to be highly hydrated. Cartilage, which consists of a meshwork of collagen fibrils that is filled in by proteoglycans, is

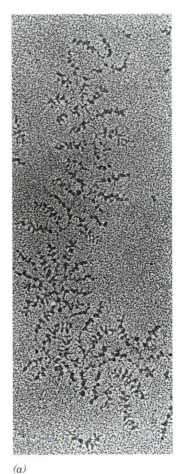

(a)

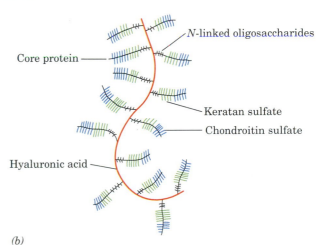

(b)

Figure 8-13.   A proteoglycan. (*a*) Electron micrograph showing a central strand of hyaluronic acid, which supports numerous projections. [From Caplan, A.I., *Sci. Am.* **251**(4), 87 (1984). Copyright © Scientific American, Inc. Used by permission.] (*b*) Bottlebrush model of the proteoglycan shown in *a*. Numerous core proteins are noncovalently linked to the central hyaluronic acid strand. Each core protein has three saccharide-binding regions.

characterized by its high resilience: The application of pressure on cartilage squeezes water away from the charged regions of its proteoglycans until charge–charge repulsions prevent further compression. When the pressure is released, the water returns. Indeed, the cartilage in the joints, which lacks blood vessels, is nourished by this flow of liquid brought about by body movements. This explains why long periods of inactivity cause cartilage to become thin and fragile.

B. Bacterial Cell Walls

Bacteria are surrounded by rigid cell walls (Fig. 1-6) that give them their characteristic shapes (Fig. 1-7) and permit them to live in **hypotonic** (less than intracellular salt concentration) environments that would otherwise cause them to swell osmotically until their plasma (cell) membranes lysed (burst). Bacterial cell walls are of considerable medical significance because they are responsible for bacterial **virulence** (disease-evoking power). In fact, the symptoms of many bacterial diseases can be elicited in animals merely by the injection of bacterial cell walls. Furthermore, the characteristic antigens of bacteria are components of their cell walls, so injecting bacterial cell wall preparations into an animal often invokes its immunity against these bacteria.

Bacteria are classified as **gram-positive** or **gram-negative** according to whether or not they take up Gram stain (a procedure developed in 1884 by Christian Gram in which heat-fixed cells are successively treated with the dye crystal violet and iodine and then destained by ethanol or acetone). Gram-positive bacteria (Fig. 8-14a) have a thick cell wall (~250 Å) surrounding their plasma membrane, whereas gram-negative bacteria (Fig. 8-14b) have a thin cell wall (~30 Å) covered by a complex outer membrane. This outer membrane functions, in part, to exclude substances toxic to the bacterium, including Gram stain. This accounts for the observation that gram-negative bacteria are more resistant to antibiotics than gram-positive bacteria.

The cell walls of bacteria consist of covalently linked polysaccharide and polypeptide chains, which form a baglike macromolecule that completely encases the cell. This framework, whose structure was elucidated in large part by Jack Strominger, is known as a **peptidoglycan.** Its polysaccharide component consists of linear chains of alternating $\beta(1\rightarrow4)$-linked GlcNAc and **N-acetylmuramic acid** (Latin: *murus,* wall). The lactic

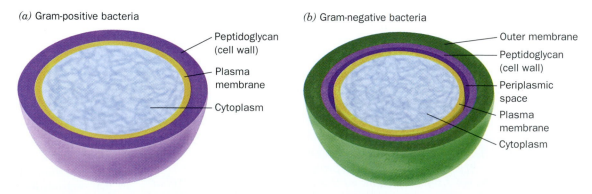

Figure 8-14. **Bacterial cell walls.** This diagram compares the cell envelopes of (a) gram-positive bacteria and (b) gram-negative bacteria.

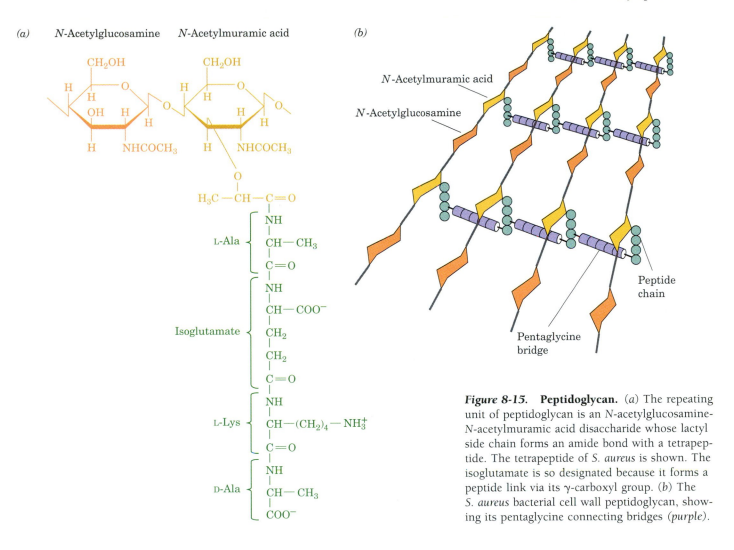

(a) N-Acetylglucosamine   N-Acetylmuramic acid

L-Ala

Isoglutamate

L-Lys

D-Ala

(b)

N-Acetylmuramic acid

N-Acetylglucosamine

Peptide chain

Pentaglycine bridge

Figure 8-15. Peptidoglycan. (*a*) The repeating unit of peptidoglycan is an N-acetylglucosamine-N-acetylmuramic acid disaccharide whose lactyl side chain forms an amide bond with a tetrapeptide. The tetrapeptide of *S. aureus* is shown. The isoglutamate is so designated because it forms a peptide link via its γ-carboxyl group. (*b*) The *S. aureus* bacterial cell wall peptidoglycan, showing its pentaglycine connecting bridges (*purple*).

acid group of N-acetylmuramic acid forms an amide bond with a D-amino acid–containing tetrapeptide to form the peptidoglycan repeating unit (Fig. 8-15). Neighboring parallel peptidoglycan chains are covalently cross-linked through their tetrapeptide side chains.

In the bacterium *Staphylococcus aureus,* whose tetrapeptide has the sequence L-Ala-D-isoglutamyl-L-Lys-D-Ala, the cross-link consists of a pentaglycine chain that extends from the terminal carboxyl group of one tetrapeptide to the ε-amino group of the Lys in a neighboring tetrapeptide. The bacterial cell wall consists of several concentric layers of peptidoglycan that are probably cross-linked in the third dimension. The cell walls of gram-positive bacteria also bear additional polysaccharide derivatives that help shield the cell.

The D-amino acids of peptidoglycans render them resistant to proteases, which are mostly specific for L-amino acids. However, **lysozyme,** an enzyme that is present in tears, mucus, and other body secretions, as well as in egg whites, catalyzes the hydrolysis of the β(1→4) glycosidic linkage between N-acetylmuramic acid and N-acetylglucosamine (the structure and mechanism of lysozyme are examined in detail in Section 11-4). The cell wall is also compromised by antibiotics such as penicillin, which inhibits bacterial cell wall biosynthesis (see Box 8-2).

Penicillin

In 1928, Alexander Fleming noticed that the chance contamination of a bacterial culture plate with the mold *Penicillium notatum* resulted in the lysis of the bacteria in the vicinity of the mold. This was caused by the presence of **penicillin,** an antibiotic secreted by the mold.

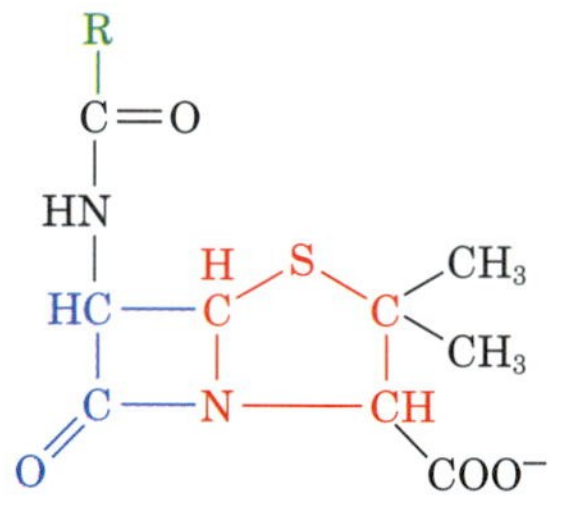

Penicillin

Penicillin contains a thiazolidine ring (*red*) fused to a β-lac-

tam ring (*blue*). A variable R group is bonded to the β-lactam ring via a peptide link.

Penicillin specifically binds to and inactivates enzymes that cross-link the peptidoglycan strands of bacterial cell walls. Since cell wall expansion in growing cells requires that their rigid cell walls be opened up for the insertion of new cell wall material, exposure of growing bacteria to penicillin results in cell lysis. However, since no human enzyme binds penicillin specifically, it is not toxic to humans and is therefore therapeutically useful.

Most bacteria that are resistant to penicillin secrete the enzyme **penicillinase,** which inactivates penicillin by cleaving the amide bond of its β-lactam ring. However, the observation that penicillinase activity varies with the nature of penicillin's R group led to the development of semisynthetic penicillins such as **ampicillin,** whose R group is an aminobenzyl group $[-CH(NH_2)C_6H_5]$ rather than a benzyl group. Such antibiotics are often clinically effective against penicillin-resistant strains of bacteria.

C. Glycosylated Proteins

Almost all the secreted and membrane-associated proteins of eukaryotic cells are **glycosylated.** Oligosaccharides are covalently attached to proteins by either *N*-glycosidic or *O*-glycosidic bonds.

N-Linked Oligosaccharides

In N-linked oligosaccharides, GlcNAc is invariably β-linked to the amide nitrogen of an Asn residue in the sequence Asn-X-Ser or Asn-X-Thr, where X is any amino acid except possibly Pro or Asp.

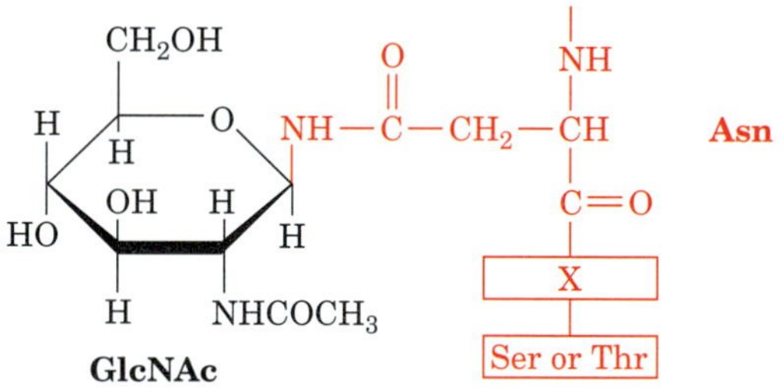

GlcNAc

N-Glycosylation occurs **cotranslationally,** that is, while the polypeptide is being synthesized. Proteins containing *N*-linked oligosaccharides typically are glycosylated and then **processed** as elucidated, in large part, by Stuart Kornfeld (Fig. 8-16):

1. An oligosaccharide containing 9 mannose residues, 3 glucose residues, and 2 GlcNAc residues is attached to the Asn of a growing polypeptide chain that is being synthesized by a ribosome associated with the endoplasmic reticulum (Section 10-2D).

2. Some of the sugars are removed during processing, which begins in the lumen (internal space) of the endoplasmic reticulum and contin-

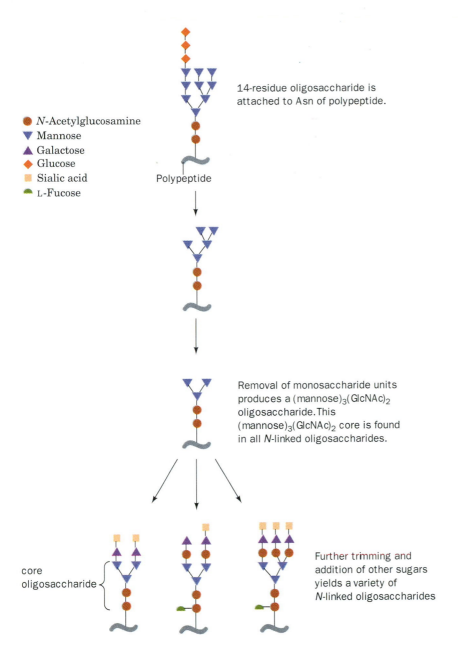

14-residue oligosaccharide is attached to Asn of polypeptide.

- ● *N*-Acetylglucosamine
- ▼ Mannose
- ▲ Galactose
- ◆ Glucose
- ■ Sialic acid
- ● L-Fucose

Polypeptide

Removal of monosaccharide units produces a (mannose)$_3$(GlcNAc)$_2$ oligosaccharide. This (mannose)$_3$(GlcNAc)$_2$ core is found in all *N*-linked oligosaccharides.

core oligosaccharide

Further trimming and addition of other sugars yields a variety of *N*-linked oligosaccharides

Figure 8-16. Synthesis of N-linked oligosaccharides. The addition of a (mannose)$_9$(glucose)$_3$(GlcNAc)$_2$ oligosaccharide is followed by removal of monosaccharides as catalyzed by glycosidases, and the addition of other monosaccharides as catalyzed by glycosyl-transferases. The core pentasaccharide occurs in all N-linked oligosaccharides. [Adapted from Kornfeld, R. and Kornfeld, S., *Annu. Rev. Biochem.* **54**, 640 (1985).] ● **See Kinemage Exercise 7-4.**

ues in the Golgi apparatus (Fig. 1-8). Enzymatic trimming is accomplished by glucosidases and mannosidases.

3. Additional monosaccharide residues, including GlcNAc, galactose, fucose, and sialic acid, are added by the action of specific **glycosyltransferases** in the Golgi apparatus.

The exact steps of *N*-linked oligosaccharide processing vary with the identity of the glycoprotein and the battery of endoglycosidases in the cell, but all *N*-linked oligosaccharides have a common core oligosaccharide with the following structure:

$$\begin{array}{c} \text{Man } \alpha \ (1\longrightarrow 6) \\ \\ \text{Man } \alpha \ (1\longrightarrow 3) \end{array} \Big\rangle \ \text{Man } \beta \ (1\longrightarrow 4) \ \text{GlcNAc } \beta \ (1\longrightarrow 4) \ \text{GlcNAc}—$$

In some glycoproteins, processing is brief, leaving "high-mannose" oligosaccharides; in other glycoproteins, extensive processing generates large oligosaccharides containing several kinds of sugar residues. *There is enormous diversity among the oligosaccharides of N-linked glycoproteins.* Indeed, even glycoproteins with a given polypeptide chain exhibit considerable microheterogeneity, presumably as a consequence of incomplete glycosylation and lack of absolute specificity on the part of glycosidases and glycosyltransferases.

O-Linked Oligosaccharides

The most common *O*-glycosidic attachment involves the disaccharide core *β-galactosyl-(1→3)-α-N-acetylgalactosamine linked to the OH group of either Ser or Thr.*

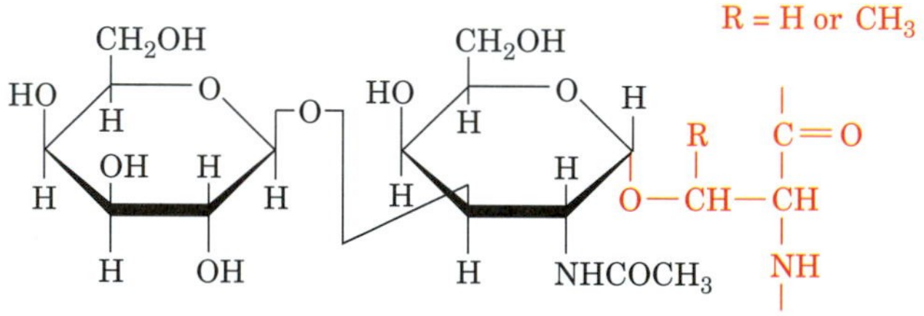

β-**Galactosyl-(1 ⟶ 3)-α-*N*-acetylgalactosaminyl-Ser/Thr**

Less commonly, galactose, mannose, and xylose form *O*-glycosides with Ser or Thr. Galactose also forms *O*-glycosidic bonds to the 5-hydroxylysyl residues of collagen (Section 6-1C). *O*-Linked oligosaccharides vary in size from a single galactose residue in collagen to the chains of up to 1000 disaccharide units in proteoglycans.

O-Linked oligosaccharides are synthesized in the Golgi apparatus by the serial addition of monosaccharide units to a completed polypeptide chain. Synthesis starts with the transfer of GalNAc to a Ser or Thr residue on the polypeptide. *N*-Linked oligosaccharides are transferred to an Asn in a specific amino acid sequence, but *O*-glycosylated Ser and Thr residues are not members of any common sequence. Instead, the locations of glycosylation sites are specified only by the secondary or tertiary structure of the polypeptide. *O*-Glycosylation continues with stepwise addition of sugars by the corresponding glycosyltransferases. The energetics and enzymology of oligosaccharide synthesis are discussed further in Section 15-5.

D. Functions of Oligosaccharides

A single protein may contain several *N*- and *O*-linked oligosaccharide chains, although different molecules of the same glycoprotein may differ in the sequences, locations, and numbers of covalently attached carbohydrates (the variant species of a glycoprotein are known as its **glycoforms**). This heterogeneity makes it difficult to assign discrete biological functions to oligosaccharide chains. In fact, certain glycoproteins synthesized by cells that lack particular oligosaccharide-processing enzymes appear to function normally despite abnormal or absent glycosylation. In other cases, however, glycosylation may affect a protein's structure, stability, or activity.

Structural Effects of Oligosaccharides

Oligosaccharides are usually attached to proteins at sequences that form surface loops or turns. Since sugars are hydrophilic, the oligosaccharides tend to project away from the protein surface. Nevertheless, some oligosaccharides may play structural roles by limiting the conformational freedom

Figure 8-17. **Model of oligosaccharide dynamics.** The allowed conformations of a (GlcNAc)$_2$(mannose)$_{5-9}$ oligosaccharide *(yellow)* attached to the bovine pancreatic enzyme **ribonuclease B** *(purple)* are shown in superimposed "snapshots." [Courtesy of Raymond Dwek, Oxford University, U.K.]

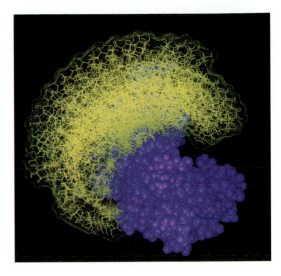

of their attached polypeptide chains (recall that *N*-linked oligosaccharides are added during protein synthesis, before a protein has assumed its final folded shape). *O*-Linked oligosaccharides, which are usually clustered in heavily glycosylated segments of a protein, may help stiffen and extend the polypeptide chain.

Because carbohydrate chains are often conformationally mobile, oligosaccharides attached to proteins can occupy time-averaged volumes of considerable size (Fig. 8-17). In this way, an oligosaccharide can shield a protein's surface, possibly modifying its activity or protecting it from proteolysis.

Oligosaccharides Mediate Recognition Events

The many possible ways that carbohydrates can be linked together to form branched structures gives them the potential to carry more biological information than either nucleic acids or proteins of similar size. For example, two different nucleotides can make only two distinct dinucleotides, but two different hexoses can combine in 36 different ways (although not all possibilities are necessarily realized in nature).

The first evidence that unique combinations of carbohydrates might be involved in intercellular communication came with the discovery that all cells are coated with sugars in the form of **glycoconjugates** such as glycoproteins and glycolipids. The oligosaccharides of glycoconjugates form a fuzzy layer up to 1400 Å thick in some cells (Fig. 8-18).

Additional evidence that cell-surface carbohydrates have recognition functions comes from **lectins** (proteins that bind carbohydrates), which are ubiquitous in nature and frequently appear on the surfaces of cells. Lectins are exquisitely specific: They can recognize individual monosaccharides in particular linkages to other sugars in an oligosaccharide (this property also makes lectins useful laboratory tools for isolating glycoproteins and oligosaccharides). Protein–carbohydrate interactions are typically characterized by extensive hydrogen bonding (often including bridging water molecules) and the van der Waals packing of hydrophobic sugar faces against aromatic side chains (Fig. 8-19).

The carbohydrates on cell surfaces are some of the best known immunochemical markers. For example, the **ABO blood group antigens** are oligosaccharide components of glycolipids on the surfaces of an individual's cells (not just red blood cells). Individuals with type A cells have A antigens on their cell surfaces and carry anti-B antibodies in their blood; those with type B cells, which bear B antigens, carry anti-A antibodies; those with type AB cells, which have both A and B antigens, carry neither

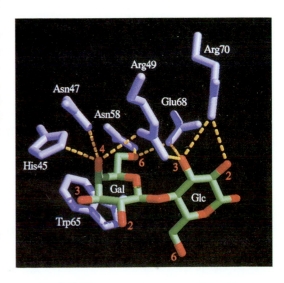

Figure 8-18. **Electron micrograph of the erythrocyte surface.** Its thick (up to 1400 Å) carbohydrate coat, which is called the **glycocalyx**, consists of closely packed oligosaccharides attached to cell-surface proteins and lipids. [Courtesy of Harrison Latta, UCLA.]

Figure 8-19. **Carbohydrate binding by a lectin.** Human galectin-2 binds β-galactosides, such as lactose, primarily through their galactose residue. The galactose and glucose residues are shown in green (with red O atoms), and the lectin amino acid side chains are shown in blue. Hydrogen bonds between the side chains and the sugar residues are shown as dashed yellow lines. [Courtesy of Hakon Leffler, University of California at San Francisco.]

Table 8-1. **Structures of the A, B, and H Antigenic Determinants in Erythrocytes**

Type	Antigen[a]
H	Galβ(1→4)GlcNAc··· ↑1,2 L-Fucα
A	GalNAcα(1→3)Galβ(1→4)GlcNAc··· ↑1,2 L-Fucα
B	Galα(1→3)Galβ(1→4)GlcNAc··· ↑1,2 L-Fucα

[a]Gal, Galactose; GalNAc, N-acetylgalactosamine; GlcNAc, N-acetylglucosamine, L-Fuc, L-fucose.

anti-A nor anti-B antibodies; and type O individuals, whose cells bear neither antigen, carry both anti-A and anti-B antibodies. Consequently, the transfusion of type A blood into a type B individual, for example, results in an anti-A antibody–A antigen reaction, which agglutinates (clumps together) the transfused erythrocytes, resulting in an often fatal blockage of blood vessels.

Table 8-1 lists the oligosaccharides found in the **A, B,** and **H antigens** (type O individuals have the H antigen). These occur at the nonreducing ends of the oligosaccharide components of glycolipids. The H antigen is the precursor oligosaccharide of A and B antigens. Type A individuals have a 303-residue glycosyltransferase that specifically adds a GalNAc residue to the terminal position of the H antigen. In type B individuals, this enzyme, which differs by four amino acid residues from that of type A individuals, instead adds a galactose residue. In type O individuals, the enzyme is inactive because its synthesis terminates after its 115th residue.

Cell-surface carbohydrates also mediate cell–cell binding events (see Box 8-3). Other intra- and extracellular processes that depend on protein–carbohydrate interactions include the following:

Box 8-3

BIOCHEMISTRY IN FOCUS

Selectins and Cell–Cell Interactions

Leukocytes are white blood cells that circulate in the bloodstream before entering tissues that have been damaged by infection or mechanical injury. In order to leave the circulation, leukocytes must attach to the surface of endothelial cells (the cells that line blood vessels) before migrating between the cells to reach the site of damage.

The initial attachment between leukocytes and endothelial cells involves proteins known as **selectins** and their glycoprotein ligands. Leukocytes constitutively (continually) express selectins on their surface; endothelial cells transiently display their own selectins in response to tissue damage.

All the selectins recognize and bind cell surface glycoproteins with the prototypical oligosaccharide structure:

Reciprocal selectin–oligosaccharide interactions between leukocytes and endothelial cells allow the endothelial cells to "capture" leukocytes. Additional interactions (between different types of proteins) prevent the leukocytes from detaching and being washed away. Firmly attached leukocytes then move to the underlying tissues.

The regulated expression of selectins helps target leukocytes to sites where they are needed. Although the carbohydrate ligands are not well characterized, subtle differences in oligosaccharide structure or conformation presumably further modulate the interaction between the two cell types. Drugs that interfere with selectin–carbohydrate binding may someday help treat chronic inflammatory diseases caused by over-reactive leukocytes.

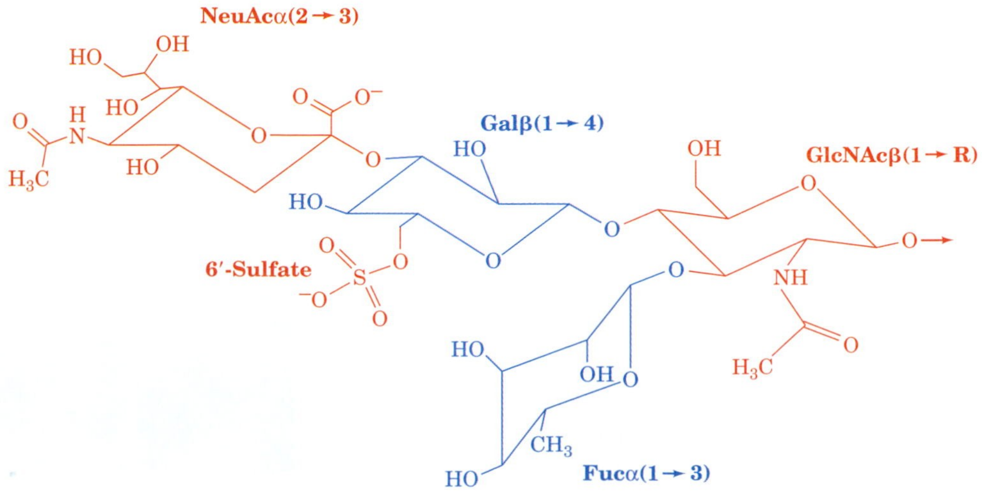

1. Newly synthesized proteins that are destined for **lysosomes** (organelles containing a variety of hydrolytic enzymes and which function as cellular recycling centers) contain *N*-linked oligosaccharides with **mannose-6-phosphate** residues. A mannose-6-phosphate receptor in the Golgi apparatus selects these proteins for transport to lysosomes.

2. Circulating glycoproteins that have lost their terminal sialic acid (or sialic acid and galactose) residues are selectively cleared from the blood by liver cells that recognize and bind the newly exposed galactose (or GlcNAc) residues. Thus, the range of glycoforms of a given glycoprotein probably ensures that it has a range of lifetimes in the blood.

3. The significant differences in the distributions of cell-surface carbohydrates in cancerous and noncancerous cells may contribute to the uncontrolled growth of cancer cells. Normal cells stop growing when they touch each other, a phenomenon known as **contact inhibition,** but cancer cells are under no such control and therefore form **malignant tumors.**

4. Mammalian spermatozoa recognize a glycoprotein component of the ovum's outer layer. Sperm proteins that recognize GlcNAc or galactose residues appear to mediate the binding and activation events that are necessary for fertilization.

5. Many viruses, bacteria, and eukaryotic parasites invade their target tissues by first binding to cell-surface carbohydrates (Fig. 8-20).

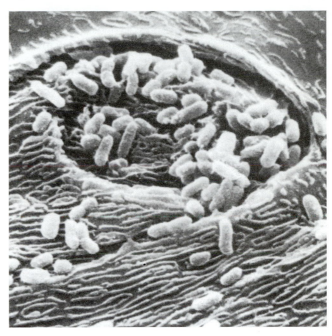

Figure 8-20. **Scanning electron micrograph of bacteria adhering to human cheek cells.** The white cylindrical objects are *E. coli* cells that are adhering to mannose residues on the plasma membrane of the cheek cells. This is the first step of a bacterial infection. [Courtesy of Frederic Silverblatt and Craig Kuehn, Veterans Administration Hospital, Sepulveda, California.]

SUMMARY

1. Monosaccharides, the simplest carbohydrates, are classified as aldoses or ketoses.

2. The cyclic hemiacetal and hemiketal forms of monosaccharides have either the α or β configuration at their anomeric carbon but are conformationally variable.

3. Monosaccharide derivatives include aldonic acids, uronic acids, alditols, deoxy sugars, amino sugars, and α- and β-glycosides.

4. Polysaccharides consist of monosaccharides linked by glycosidic bonds.

5. Cellulose and chitin are polysaccharides whose β(1→4) linkages cause them to adopt rigid and extended structures.

6. The storage polysaccharides starch and glycogen consist of α-glycosidically linked glucose residues.

7. Glycosaminoglycans are unbranched polysaccharides containing uronic acid and amino sugars that are often sulfated.

8. Proteoglycans are enormous molecules consisting of hyaluronic acid with attached core proteins that bear numerous glycosaminoglycans and oligosaccharides.

9. Bacterial cell walls are made of peptidoglycan, a network of polysaccharide and polypeptide chains.

10. Glycosylated proteins may contain *N*-linked oligosaccharides (attached to Asn) or *O*-linked oligosaccharides (attached to Ser or Thr) or both. Different molecules of a glycoprotein may contain different sequences and locations of oligosaccharides.

11. Oligosaccharides appear to play important roles in cell-surface recognition phenomena.

REFERENCES

Abeijon, C. and Hirschberg, C.B., Topography of glycosylation reactions in the endoplasmic reticulum, *Trends Biochem. Sci.* **17,** 32–36 (1992).

Allen, H.J. and Kisailus, E.C. (Eds.), *Glycoconjugates. Composition, Structure, and Function,* Marcel Dekker (1992). [Includes chapters on nomenclature, structural analysis, lectins, and glycoproteins.]

Brady, J.W., Theoretical studies of oligosaccharide structure and conformational dynamics, *Curr. Opin. Struct. Biol.* **1,** 711–715 (1991).

El Khadem, H.S., *Carbohydrate Chemistry. Monosaccharides and Their Oligomers,* Academic Press (1988).

Ghuysen, J.-M. and Hakenbeck, R. (Eds.), *Bacterial Cell Wall,* Elsevier (1994). [Chapters 1 and 2 provide a good introduction to peptidoglycan structure and function.]

Hart, G.W., Glycosylation, *Curr. Opin. Cell Biol.* **4,** 1017–1023 (1992).

Kjellén, L. and Lindahl, U., Proteoglycans: Structure and interactions, *Annu. Rev. Biochem.* **60,** 443–475 (1991).

Paulson, J.C., Glycoproteins: what are the sugar chains for? *Trends Biochem. Sci.* **14,** 272–276 (1989).

Sharon, N. and Lis, H., Carbohydrates in cell recognition, *Sci. Am.* **268**(1), 82–89 (1993).

Solomons, T.W.G., *Organic Chemistry* (6th ed.), Chapter 22, Wiley (1996). [A general discussion of carbohydrate nomenclature and chemistry. Other comprehensive organic chemistry textbooks have similar material.]

Weis, W.I., and Drickamer, K., Structural basis of lectin–carbohydrate recognition, *Annu. Rev. Biochem.* **65,** 441–473 (1996).

Woods, R.J., Three-dimensional structures of oligosaccharides, *Curr. Opin. Struct. Biol.* **5,** 591–598 (1995).

KEY TERMS

monosaccharide	α anomer	glycosidic bond	O-linked oligosaccharide
disaccharide	β anomer	glycan	gram-positive
oligosaccharide	aldonic acid	homopolysaccharide	gram-negative
polysaccharide	uronic acid	heteropolysaccharide	peptidoglycan
aldose	alditol	exoglycosidase	glycosylation
ketose	deoxy sugar	endoglycosidase	oligosaccharide processing
epimer	amino sugar	reducing sugar	glycoforms
hemiacetal	glycoprotein	glycosaminoglycan	glycoconjugate
hemiketal	glycolipid	microheterogeneity	lectin
pyranose	α-glycoside	proteoglycan	
furanose	β-glycoside	N-linked oligosaccharide	

STUDY EXERCISES

1. Show how aldoses and ketoses can form five- and six-membered rings.

2. Draw Fischer and Haworth projections for glucose.

3. Compare and contrast the structures and functions of cellulose, chitin, starch, and glycogen.

4. How do the physical properties of glycosaminoglycans and proteoglycans relate to their biological roles?

5. Explain the differences between N- and O-linked oligosaccharides in glycoproteins.

PROBLEMS

1. How many stereoisomers are possible for (a) a ketopentose, (b) a ketohexose, and (c) a ketoheptose?

2. Which of the following pairs of sugars are epimers of each other?
   (a) D-sorbose and D-psicose
   (b) D-sorbose and D-fructose
   (c) D-fructose and L-fructose
   (d) D-arabinose and D-ribose
   (e) D-ribose and D-ribulose

3. Draw the furanose and pyranose forms of D-ribose.

4. Are (a) D-glucitol, (b) D-galactitol, and (c) D-glycerol optically active?

5. Draw a Fischer projection of L-fucose. L-Fucose is the deoxy form of which L-hexose?

6. Deduce the structure of the disaccharide trehalose from the following information: Complete hydrolysis yields only D-glucose; it is hydrolyzed by α-glucosidase but not β-glucosidase; and it does not reduce Cu^{2+} to Cu^+.

7. How many different disaccharides of D-glucopyranose are possible?

8. How many reducing ends are in a molecule of glycogen that contains 10,000 residues with a branch every 10 residues?

9. Is amylose or amylopectin more likely to be a long-term storage polysaccharide in plants?

10. Calculate the net charge of a chondroitin-4-sulfate molecule containing 100 disaccharide units.

Because of their hydrophobicity, which prevents them from mixing freely with the aqueous phase, lipids have structures and functions that differ from those of other classes of biological molecules. The forces responsible for the separation of oil and water are also important determinants of cell structure. [Jeremy Burgess/Photo Researchers.]

LIPIDS

Lipids (Greek: *lipos,* fat) are the fourth major group of molecules found in all cells. Unlike nucleic acids, proteins, and polysaccharides, lipids are not polymeric. However, they do aggregate, and it is in this state that they perform their most obvious function as the structural matrix of biological membranes.

Lipids exhibit greater structural variety than the other classes of biological molecules. To a certain extent, lipids constitute a catchall category of substances that are similar only in that they are largely hydrophobic and only sparingly soluble in water. Because of the hydrophobicity of lipids, lipid analysis requires more effort than studies of more soluble (and therefore easier to work with) molecules. Many of the most interesting functions of lipids have come to light only since the mid-1980s or so. In general, lipids perform three biological functions (although certain lipids apparently serve more than one purpose in some cells):

1. Lipid molecules, in the form of a lipid bilayer, are essential components, together with proteins, of biological membranes.
2. Lipids containing hydrocarbon chains serve as energy stores.
3. Many intra- and intercellular signaling events involve lipid molecules.

In this brief chapter, we examine the structures and physical properties of the most common types of lipids and the properties of the lipid bilayer. The following chapter takes up the subject of biological membranes and some additional lipid functions.

1. LIPID CLASSIFICATION

Lipids are substances of biological origin that are soluble in organic solvents such as chloroform and methanol. Hence, they are easily separated from other biological materials by extraction into organic solvents. Fats, oils, certain vitamins and hormones, and most nonprotein membrane components are lipids. In this section, we discuss the structures and physical properties of the major classes of lipids.

A. Fatty Acids

Fatty acids are carboxylic acids with long-chain hydrocarbon side groups (Fig. 9-1). They usually occur in esterified form as major components of the various lipids described in this chapter. The more common biological fatty acids are listed in Table 9-1. In higher plants and animals, the predominant fatty acid residues are those of the C_{16} and C_{18} species: **palmitic, oleic, linoleic,** and **stearic acids.** Fatty acids with <14 or >20 carbon atoms are uncommon. Most fatty acids have an even number of carbon atoms because they are biosynthesized by the concatenation of C_2 units (Section 19-4).

Over half of the fatty acid residues of plant and animal lipids are **unsaturated** (contain double bonds) and are often **polyunsaturated** (contain two or more double bonds). Bacterial fatty acids are rarely polyunsaturated but are commonly branched, hydroxylated, or contain cyclopropane rings.

Table 9-1 indicates that the first double bond of an unsaturated fatty acid commonly occurs between its C9 and C10 atoms counting from the carboxyl C atom. This bond is called a Δ^9- or 9-double bond. In polyunsaturated fatty acids, the double bonds tend to occur at every third carbon atom (e.g., $-CH=CH-CH_2-CH=CH-$) and so are not conjugated (as in $-CH=CH-CH=CH-$).

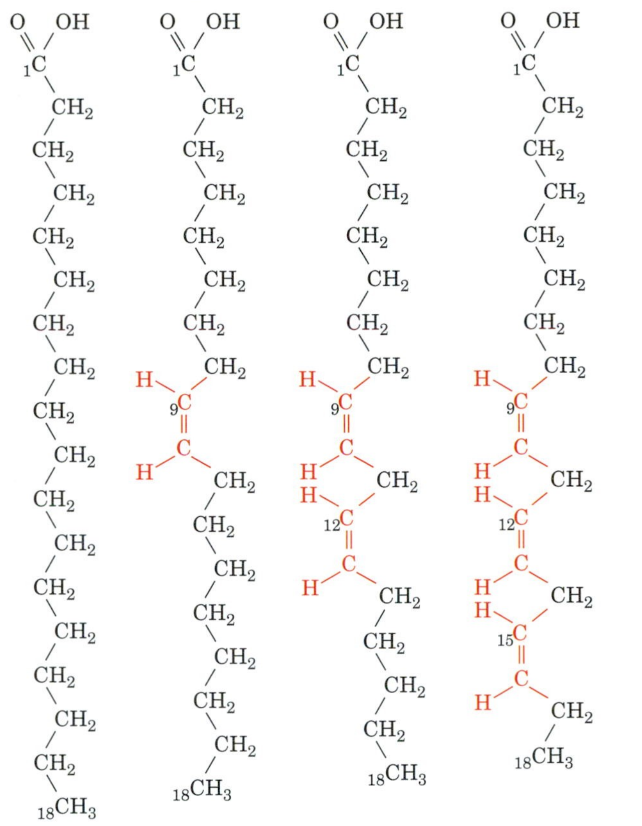

Stearic acid **Oleic acid** **Linoleic acid** **α-Linolenic acid**

Figure 9-1. **The structural formulas of some C_{18} fatty acids.** The double bonds all have the cis configuration.

Table 9-1. **The Common Biological Fatty Acids**

Symbol[a]	*Common Name*	*Systematic Name*	*Structure*	*mp (°C)*
Saturated fatty acids				
12:0	Lauric acid	Dodecanoic acid	$CH_3(CH_2)_{10}COOH$	44.2
14:0	Myristic acid	Tetradecanoic acid	$CH_3(CH_2)_{12}COOH$	52
16:0	Palmitic acid	Hexadecanoic acid	$CH_3(CH_2)_{14}COOH$	63.1
18:0	Stearic acid	Octadecanoic acid	$CH_3(CH_2)_{16}COOH$	69.1
20:0	Arachidic acid	Eicosanoic acid	$CH_3(CH_2)_{18}COOH$	75.4
22:0	Behenic acid	Docosanoic acid	$CH_3(CH_2)_{20}COOH$	81
24:0	Lignoceric acid	Tetracosanoic acid	$CH_3(CH_2)_{22}COOH$	84.2
Unsaturated fatty acids (all double bonds are cis)				
16:1	Palmitoleic acid	9-Hexadecenoic acid	$CH_3(CH_2)_5CH{=}CH(CH_2)_7COOH$	−0.5
18:1	Oleic acid	9-Octadecenoic acid	$CH_3(CH_2)_7CH{=}CH(CH_2)_7COOH$	13.2
18:2	Linoleic acid	9,12-Octadecadienoic acid	$CH_3(CH_2)_4(CH{=}CHCH_2)_2(CH_2)_6COOH$	−9
18:3	α-Linolenic acid	9,12,15-Octadecatrienoic acid	$CH_3CH_2(CH{=}CHCH_2)_3(CH_2)_6COOH$	−17
18:3	γ-Linolenic acid	6,9,12-Octadecatrienoic acid	$CH_3(CH_2)_4(CH{=}CHCH_2)_3(CH_2)_3COOH$	
20:4	Arachidonic acid	5,8,11,14-Eicosatetraenoic acid	$CH_3(CH_2)_4(CH{=}CHCH_2)_4(CH_2)_2COOH$	−49.5
20:5	EPA	5,8,11,14,17-Eicosapentaenoic acid	$CH_3CH_2(CH{=}CHCH_2)_5(CH_2)_2COOH$	−54
24:1	Nervonic acid	15-Tetracosenoic acid	$CH_3(CH_2)_7CH{=}CH(CH_2)_{13}COOH$	39

[a]Number of carbon atoms: Number of double bonds.

Source: Dawson, R.M.C., Elliott, D.C., Elliott, W.H., and Jones, K.M., *Data for Biochemical Research* (3rd ed.), Chapter 8, Clarendon Press (1986).

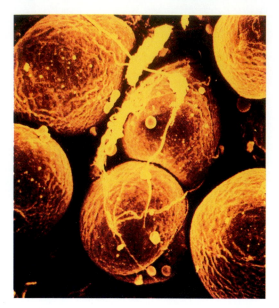

Figure 9-2. **Scanning electron micrograph of adipocytes.** Each adipocyte contains a fat globule that occupies nearly the entire cell. [Fred E. Hossler/Visuals Unlimited.]

Saturated fatty acids (which are fully reduced or "saturated" with hydrogen) are highly flexible molecules that can assume a wide range of conformations because there is relatively free rotation around each of their C—C bonds. Nevertheless, their lowest energy conformation is the fully extended conformation, which has the least amount of steric interference between neighboring methylene groups. The melting points (mp) of saturated fatty acids, like those of most substances, increase with their molecular mass (Table 9-1).

Fatty acid double bonds almost always have the cis configuration (Fig. 9-1). This puts a rigid 30° bend in the hydrocarbon chain. Consequently, unsaturated fatty acids pack together less efficiently than saturated fatty acids. The reduced van der Waals interactions of unsaturated fatty acids cause their melting points to decrease with the degree of unsaturation. The fluidity of lipids containing fatty acid residues likewise increases with the degree of unsaturation of the fatty acids. This phenomenon, as we shall see, has important consequences for biological membranes.

B. Triacylglycerols

The fats and oils that occur in plants and animals consist largely of mixtures of **triacylglycerols** (also called **triglycerides**). These nonpolar, water-insoluble substances are fatty acid triesters of **glycerol:**

Glycerol　　　**Triacylglycerol**

Triacylglycerols function as energy reservoirs in animals and are therefore their most abundant class of lipids even though they are not components of cellular membranes.

Triacylglycerols differ according to the identity and placement of their three fatty acid residues. Most triacylglycerols contain two or three different types of fatty acid residues and are named according to their placement on the glycerol moiety, for example, **1-palmitoleoyl-2-linoleoyl-3-stearoyl-glycerol** (*at left*). **Fats** and **oils** (which differ only in that fats are solid and oils are liquid at room temperature) are complex mixtures of triacylglycerols whose fatty acid compositions vary with the organism that produced them. Plant oils are usually richer in unsaturated fatty acid residues than animal fats, as the lower melting points of oils imply.

Triacylglycerols Function as Energy Reserves

Fats are a highly efficient form in which to store metabolic energy. This is because fats are less oxidized than carbohydrates or proteins and hence yield significantly more energy per unit mass on complete oxidation. Furthermore, fats, which are nonpolar, are stored in anhydrous form, whereas glycogen (Section 8-2C), for example, binds about twice its weight of water under physiological conditions. *Fats therefore provide about six times the metabolic energy of an equal weight of hydrated glycogen.*

In animals, **adipocytes** (fat cells; Fig. 9-2) are specialized for the synthesis and storage of triacylglycerols. Whereas other types of cells have only a few small droplets of fat dispersed in their cytosol, adipocytes may be al-

1-Palmitoleoyl-2-linoleoyl-3-stearoyl-glycerol

most entirely filled with fat globules. **Adipose tissue** is most abundant in a subcutaneous layer and in the abdominal cavity. The fat content of normal humans (21% for men, 26% for women) allows them to survive starvation for 2 or 3 months. In contrast, the body's glycogen supply, which functions as a short-term energy store, can provide for the body's energy needs for less than a day. The subcutaneous fat layer also provides thermal insulation, which is particularly important for warm-blooded aquatic animals, such as whales, seals, geese, and penguins, which are routinely exposed to low temperatures.

C. Glycerophospholipids

Glycerophospholipids (or **phosphoglycerides**) are the major lipid components of biological membranes. They consist of **glycerol-3-phosphate** whose C1 and C2 positions are esterified with fatty acids. In addition, the phosphoryl group is linked to another group, X (Fig. 9-3). *Glycerophospholipids are therefore amphiphilic molecules with nonpolar aliphatic (hydrocarbon) "tails" and polar phosphoryl-X "heads."*

The simplest glycerophospholipids, in which X = H, are **phosphatidic acids;** they are present in only small amounts in biological membranes. In the glycerophospholipids that commonly occur in biological membranes, the head groups are derived from polar alcohols (Table 9-2). Saturated C_{16} or C_{18} fatty acids usually occur at the C1 position of the glycerophospholipids, and the C2 position is often occupied by an unsaturated C_{16} to C_{20}

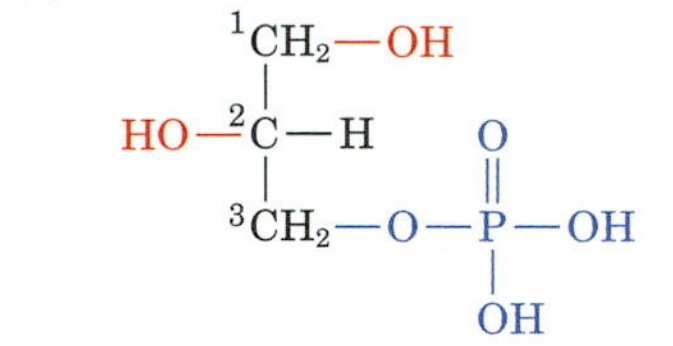

Glycerol-3-phosphate

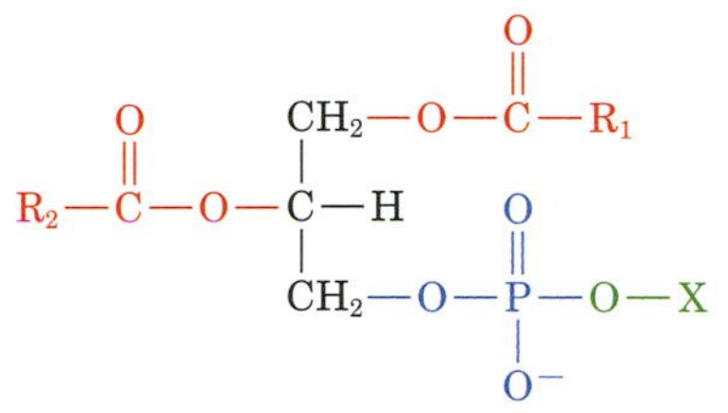

Glycerophospholipid

Figure 9-3. Structure of glycerophospholipids. (*a*) The backbone, L-glycerol-3-phosphate. (*b*) The general formula of the glycerophospholipids. R_1 and R_2 are the long-chain hydrocarbon tails of fatty acids, and X is derived from a polar alcohol (see Table 9-2).

Table 9-2. **The Common Classes of Glycerophospholipids**

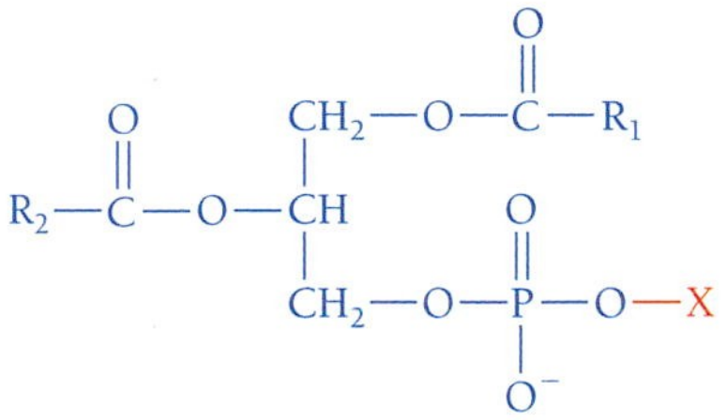

Name of X—OH	Formula of —X	Name of Phospholipid
Water	—H	Phosphatidic acid
Ethanolamine	—CH₂CH₂NH₃⁺	Phosphatidylethanolamine
Choline	—CH₂CH₂N(CH₃)₃⁺	Phosphatidylcholine (lecithin)
Serine	—CH₂CH(NH₃⁺)COO⁻	Phosphatidylserine
myo-Inositol	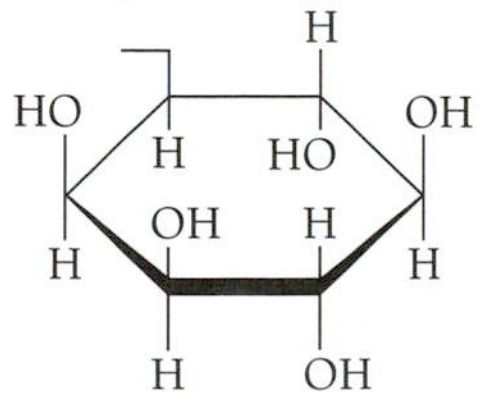 (structure)	Phosphatidylinositol
Glycerol	—CH₂CH(OH)CH₂OH	Phosphatidylglycerol
Phosphatidylglycerol	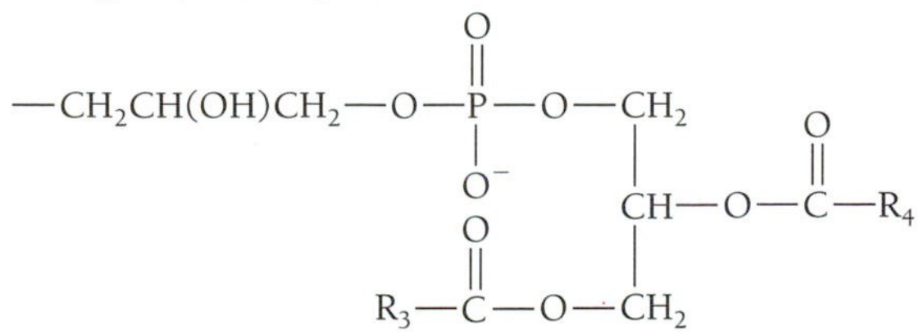 (structure)	Diphosphatidylglycerol (cardiolipin)

Figure 9-4.   The glycerophospholipid 1-stearoyl-2-oleoyl-3-phosphatidylcholine.
(*a*) Molecular formula in Fischer projection.
(*b*) Space-filling model with H white, C gray, O red, and P green. Note how the unsaturated oleoyl chain is bent compared to the saturated stearoyl chain. [Courtesy of Richard Pastor, FDA, Bethesda, Maryland.]

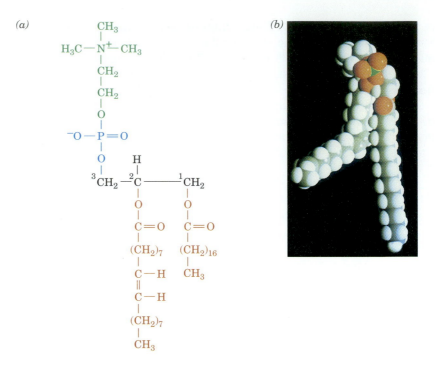

(*a*)

(*b*)

1-Stearoyl-2-oleoyl-3-phosphatidylcholine

fatty acid. Individual glycerophospholipids are named according to the identities of these fatty acid residues (e.g., Fig. 9-4). A glycerophospholipid containing two palmitoyl chains is an important component of **lung surfactant** (see Box 9-1).

Phospholipases Hydrolyze Glycerophospholipids

The chemical structures—including fatty acyl chains and head groups—of glycerophospholipids can be determined from the products of the hydrolytic reactions catalyzed by enzymes known as **phospholipases.** For example, **phospholipase A_2** hydrolytically excises the fatty acid residue at C2, leaving a **lysophospholipid** (Fig. 9-5). Lysophospholipids, as their name implies, are powerful detergents that disrupt cell membranes, thereby lysing cells. Bee and snake venoms are rich sources of phospholipase A_2. Other

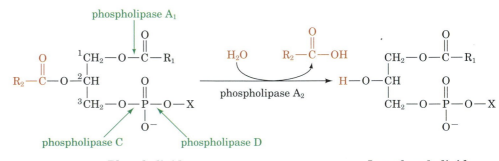

Phospholipid   **Lysophospholipid**

Figure 9-5.   Action of phospholipases. Phospholipase A_2 hydrolytically excises the C2 fatty acid residue from a triacylglycerol to yield the corresponding lysophospholipid. The bonds hydrolyzed by other types of phospholipases, which are named according to their specificities, are also indicated.

Box 9-1

BIOCHEMISTRY IN HEALTH AND DISEASE

Lung Surfactant

Dipalmitoyl phosphatidylcholine (DPPC) is the major lipid of lung surfactant, the protein–lipid mixture that is essential for normal pulmonary function. The surfaces of the cells that form the alveoli (small air spaces of the lung) are coated with surfactant, which decreases the alveolar surface tension. Lung surfactant contains 80 to 90% phospholipid by weight, and 70 to 80% of the phospholipid is phosphatidylcholine, mostly the dipalmitoyl species.

Because the palmitoyl chains of DPPC are fully saturated, they tend to extend straight out without bending. This allows close packing of DPPC molecules, which are oriented in a single layer with their nonpolar tails toward the air and their polar heads toward the alveolar cells. When air is expired from the lungs, the volume and surface area of the alveoli decrease. Complete collapse of the alveolar space is prevented by the surfactant, because the closely packed DPPC molecules resist compression. Reopening a collapsed air space requires a much greater force than expanding an already open air space.

Lung surfactant is continuously synthesized, secreted, and recycled by alveolar cells. Because surfactant production is low until just before birth, premature infants are at risk of developing **respiratory distress syndrome**, which is characterized by difficulty in breathing due to alveolar collapse. The syndrome can be treated by introducing exogenous surfactant into the lungs. A related condition in adults (**adult respiratory distress syndrome**) is characterized by insufficient surfactant, usually secondary to other lung injury. This condition, too, can be treated with exogenous surfactant.

---

types of phospholipases act at different sites in glycerophospholipids, as shown in Fig. 9-5.

Enzymes that act on lipids have fascinated biochemists because the enzymes must gain access to portions of the lipids that are buried in a nonaqueous environment. Phospholipases A_2, which constitute some of the best understood lipid-specific enzymes, are relatively small proteins (~14 kD, ~125 amino acid residues). The X-ray structure of phospholipase A_2 from cobra venom suggests that the enzyme binds a glycerophospholipid molecule such that its polar head group fits into the enzyme's active site, whereas the hydrophobic tails, which extend beyond the active site, interact with several aromatic side chains (Fig. 9-6).

Lipases specific for triacylglycerols and membrane lipids catalyze their degradation *in vivo*. Frequently, the hydrolysis products are not destined for further degradation but instead serve as intra- and extracellular signal molecules. For example, **lysophosphatidic acid (1-acyl-glycerol-3-phosphate),**

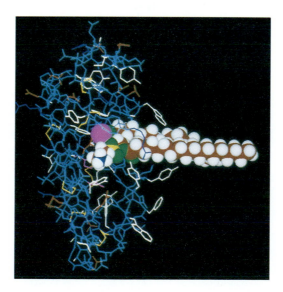

Figure 9-6. **Model of phospholipase A_2 and a glycerophospholipid.** The X-ray structure of the enzyme from cobra venom is shown with a space-filling model of dimyristoyl phosphatidylethanolamine in its active site as located by NMR methods. A Ca^{2+} ion in the active site is shown in magenta. [Courtesy of Edward A. Dennis, University of California at San Diego.] ● See the Interactive Exercises.

which is not actually lytic since it has a small head group (an unsubstituted phosphate group), is produced by hydrolysis of membrane lipids in blood platelets and injured cells and stimulates cell growth as part of the wound-repair process. **1,2-Diacylglycerol,** derived from membrane lipids by the action of **phospholipase C,** is an intracellular signal molecule that activates a **protein kinase** (Section 21-3D; kinases catalyze ATP-dependent phosphoryl-transfer reactions).

Plasmalogens Contain an Ether Linkage

Plasmalogens are glycerophospholipids in which the C1 substituent of the glycerol moiety is linked via an α,β-unsaturated ether linkage in the cis configuration rather than through an ester linkage.

A plasmalogen

Ethanolamine, choline, and serine (Table 9-2) form the most common plasmalogen head groups. The functions of most plasmalogens are not well understood.

D. *Sphingolipids*

Sphingolipids are also major membrane components. Most sphingolipids are derivatives of the C_{18} amino alcohol **sphingosine,** whose double bond has the trans configuration. The *N*-acyl fatty acid derivatives of sphingosine are known as **ceramides.**

Sphingosine   **A ceramide**

Ceramides are the parent compounds of the more abundant sphingolipids:

1. **Sphingomyelins,** the most common sphingolipids, are ceramides bearing either a phosphocholine (Fig. 9-7) or a phosphoethanolamine head group, so they can also be classified as **sphingophospholipids.** *Although sphingomyelins differ chemically from phosphatidylcholine and phosphatidylethanolamine, their conformations and charge distributions are quite similar* (compare Figs. 9-4 and 9-7). The membra-

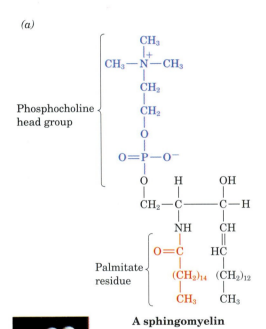

Phosphocholine head group

Palmitate residue

A sphingomyelin

(b)

Figure 9-7. A sphingomyelin. (*a*) Molecular formula. (*b*) Space-filling model with H white, C gray, N blue, and O red. [Courtesy of Richard Pastor, FDA, Bethesda, Maryland.]

nous myelin sheath that surrounds and electrically insulates many nerve cell axons is particularly rich in sphingomyelins (Fig. 9-8).

2. **Cerebrosides** are ceramides with head groups that consist of a single sugar residue. These lipids are therefore **glycosphingolipids. Galactocerebrosides** and **glucocerebrosides** are the most prevalent. Cerebrosides, in contrast to phospholipids, lack phosphate groups and hence are nonionic.

3. **Gangliosides** are the most complex glycosphingolipids. They are ceramides with attached oligosaccharides that include at least one sialic acid residue. The structures of **gangliosides G_{M1}, G_{M2},** and **G_{M3},** three of the over 60 that are known, are shown in Fig. 9-9. Gangliosides are primarily components of cell-surface membranes and constitute a significant fraction (6%) of brain lipids.

Gangliosides have considerable physiological and medical significance. Their complex carbohydrate head groups, which extend beyond the surfaces of cell membranes, act as specific receptors for certain pituitary glycoprotein hormones that regulate a number of important physiological functions. Gangliosides are also receptors for certain bacterial protein toxins such as **cholera toxin.** There is considerable evidence that gangliosides are specific determinants of cell–cell recognition, so they probably have an important role in the growth and differentiation of tissues as well as in carcinogenesis. Disorders of ganglioside breakdown are responsible for several hereditary **sphingolipid storage diseases,** such as **Tay-Sachs disease,** which are characterized by an invariably fatal neurological deterioration in early childhood.

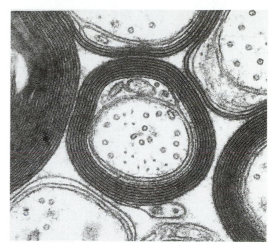

Figure 9-8. **Electron micrograph of myelinated nerve fibers.** This cross-sectional view shows the spirally wrapped membranes around each nerve axon. The myelin sheath may be 10–15 layers thick. Its high lipid content makes it an electrical insulator. [Courtesy of Cedric S. Raine, Albert Einstein College of Medicine.]

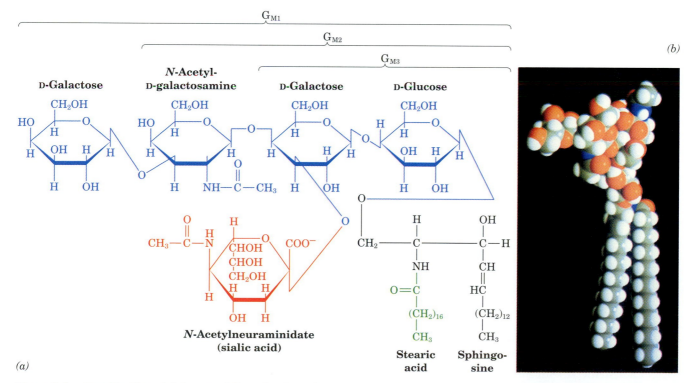

Figure 9-9. **Gangliosides.** (*a*) Structural formula of gangliosides G_{M1}, G_{M2}, and G_{M3}. Gangliosides G_{M2} and G_{M3} differ from G_{M1} only by the sequential absences of the terminal D-galactose and N-acetyl-D-galactosamine residues. Other gangliosides have different oligosaccharide head groups. (*b*) Space-filling model of G_{M1} with H white, C gray, N blue, and O red. [Courtesy of Richard Venable, FDA, Bethesda, Maryland.]

Sphingolipids, like glycerophospholipids, are a source of smaller lipids that have discrete signaling activity. Sphingomyelin itself, as well as the ceramide portions of more complex sphingolipids, appear to specifically modulate the activities of protein kinases and **protein phosphatases** (enzymes that remove phosphoryl groups from proteins) that are involved in regulating cell growth and differentiation.

E. Steroids

Steroids, which are mostly of eukaryotic origin, are derivatives of **cyclopentanoperhydrophenanthrene,**

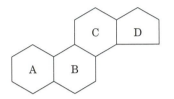

Cyclopentanoperhydrophenanthrene

a compound that consists of four fused, nonplanar rings (labeled A–D). The much maligned **cholesterol,** which is the most abundant steroid in animals, is further classified as a **sterol** because of its C3 OH group (Fig. 9-10). Cholesterol is a major component of animal plasma membranes. Its polar OH group gives it a weak amphiphilic character, whereas its fused ring system provides it with greater rigidity than other membrane lipids. Cholesterol can also be esterified to long-chain fatty acids to form **cholesteryl esters,** for example,

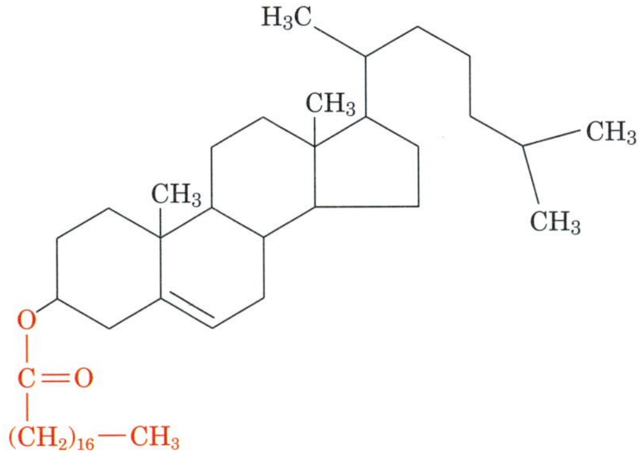

Cholesteryl stearate

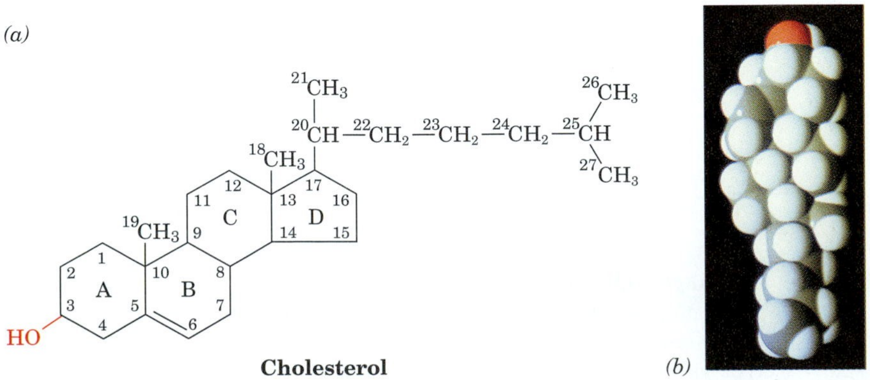

(a)

Cholesterol

Figure 9-10.   Cholesterol. (*a*) Structural formula with the standard numbering system. (*b*) Space-filling model with H white, C gray, and O red. [Courtesy of Richard Pastor, FDA, Bethesda, Maryland.]

(b)

Plants contain little cholesterol but synthesize other sterols. Yeast and fungi also synthesize sterols, which differ from cholesterol in their aliphatic side chains and number of double bonds. Prokaryotes contain little, if any, sterol.

In mammals, cholesterol is the metabolic precursor of **steroid hormones,** substances that regulate a great variety of physiological functions. The structures of some steroid hormones are shown in Fig. 9-11. Steroid hormones are classified according to the physiological responses they evoke:

1. The **glucocorticoids,** such as **cortisol** (a C_{21} compound), affect carbohydrate, protein, and lipid metabolism and influence a wide variety of other vital functions, including inflammatory reactions and the capacity to cope with stress.

2. **Aldosterone** and other **mineralocorticoids** regulate the excretion of salt and water by the kidneys.

3. The **androgens** and **estrogens** affect sexual development and function. **Testosterone,** a C_{19} compound, is the prototypic androgen (male sex hormone).

Glucocorticoids and mineralocorticoids are synthesized by the cortex (outer layer) of the adrenal gland. Both androgens and estrogens (female sex hormones) are synthesized by testes and ovaries (although androgens predominate in testes and estrogens predominate in ovaries) and, to a lesser extent, by the adrenal cortex. Because steroid hormones are water-insoluble, they bind to proteins for transport through the blood to their target tissues.

Impaired adrenocortical function, either through disease or trauma, results in Addison's disease (Box 7-5), which is characterized by hyperglycemia (elevated amounts of glucose in the blood), muscle weakness, Na^+ loss, K^+ retention, impaired cardiac function, and greatly increased sus-

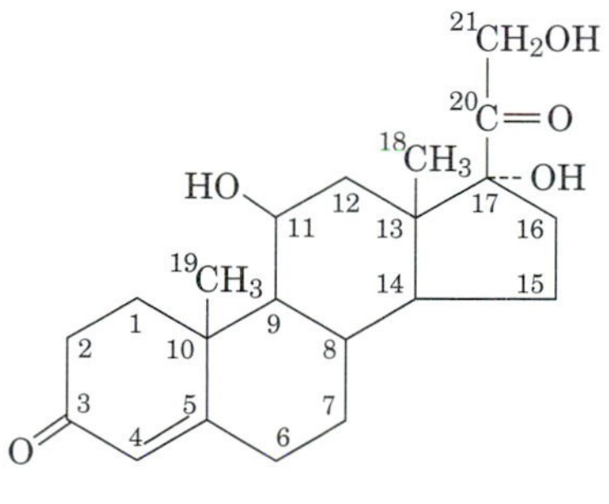

Cortisol (hydrocortisone)
(a glucocorticoid)

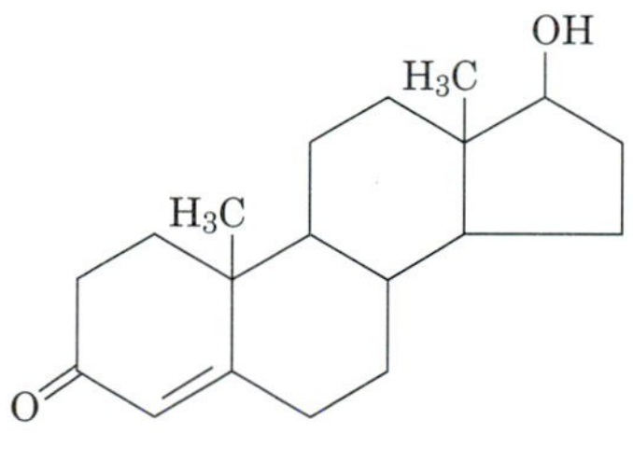

Testosterone
(an androgen)

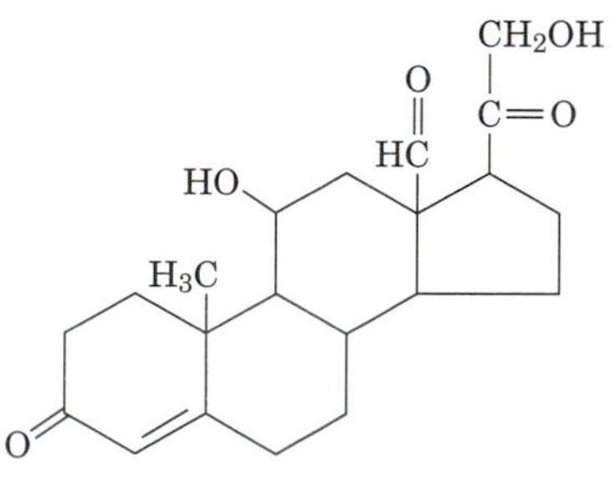

Aldosterone
(a mineralocorticoid)

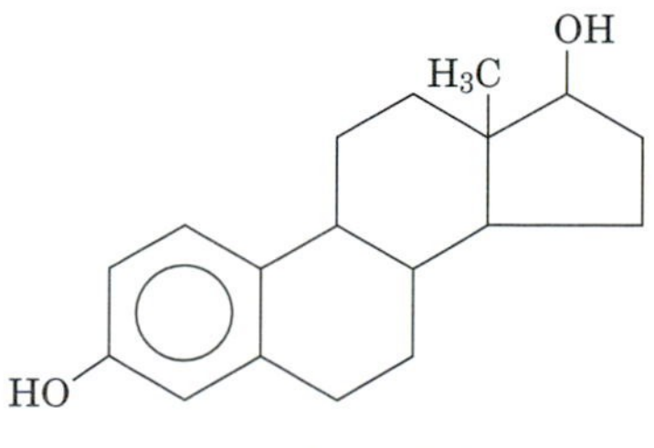

β-Estradiol
(an estrogen)

Figure 9-11. **Some representative steroid hormones.**

ceptibility to stress. The victim, unless treated by the administration of glucocorticoids and mineralocorticoids, slowly languishes and dies without any particular pain or distress.

Vitamin D Regulates Ca²⁺ Metabolism

The various forms of **vitamin D,** which are really hormones, are sterol derivatives in which the steroid B ring is disrupted between C9 and C10.

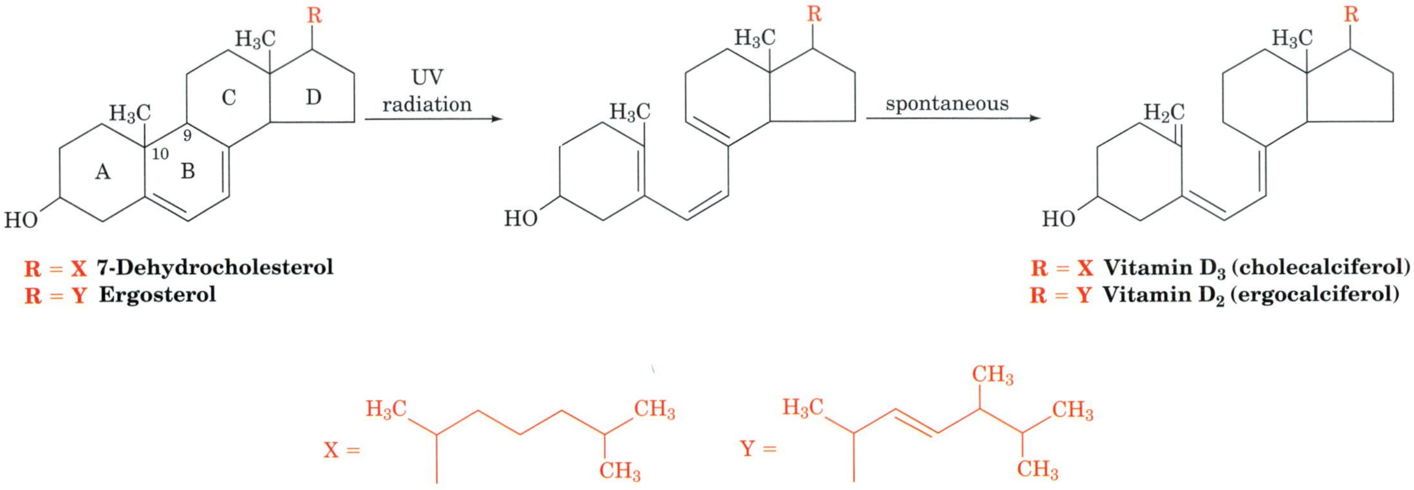

R = X **7-Dehydrocholesterol**
R = Y **Ergosterol**

R = X **Vitamin D₃ (cholecalciferol)**
R = Y **Vitamin D₂ (ergocalciferol)**

Vitamin D₂ (ergocalciferol) is nonenzymatically formed in the skin of animals through the photolytic action of UV light on the plant sterol **ergosterol,** a common milk additive, whereas the closely related **vitamin D₃ (cholecalciferol)** is similarly derived from **7-dehydrocholesterol** (hence the saying that sunlight provides vitamin D).

Vitamins D_2 and D_3 are inactive; the active forms are produced through their enzymatic hydroxylation (addition of an OH group) carried out by the liver (at C25) and by the kidney (at C1) to yield **1α,25-dihydroxycholecalciferol.**

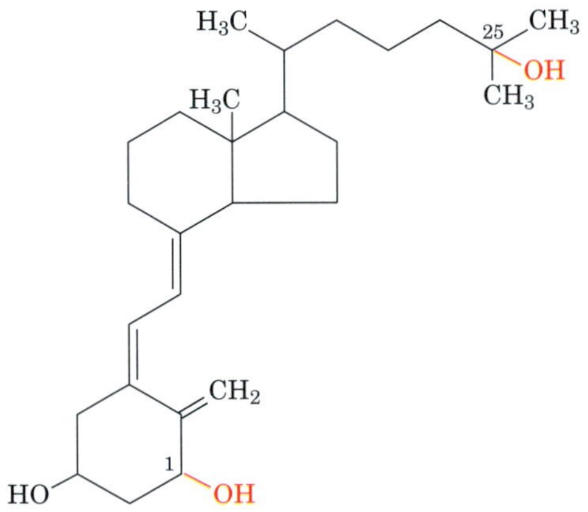

1α,25-Dihydroxycholecalciferol

Active vitamin D increases serum [Ca²⁺] by promoting the intestinal absorption of dietary Ca^{2+}. This increases the deposition of Ca^{2+} in bones and teeth. Vitamin D deficiency produces **rickets** in children, a disease characterized by stunted growth and deformed bones caused by insufficient bone mineralization. Although rickets was first described in 1645, it was not until the early twentieth century that animal fats, particularly fish liver oils, were shown to prevent this deficiency disease. Rickets can also be prevented by exposing children to sunlight or just ultraviolet light in the wavelength range 230 to 313 nm, regardless of their diets.

Since vitamin D is water insoluble, it can accumulate in fatty tissues. Excessive intake of vitamin D over long periods results in **vitamin D intoxication.** The consequent high serum $[Ca^{2+}]$ results in aberrant calcification of soft tissues and in the development of kidney stones, which can cause kidney failure. The observation that the level of skin pigmentation in indigenous human populations tends to increase with their proximity to the equator is explained by the hypothesis that skin pigmentation functions to prevent vitamin D intoxication by filtering out excessive solar radiation.

F. Other Lipids

In addition to the well-characterized lipids that are found in large amounts in cellular membranes, many organisms synthesize other compounds that are not membrane components but are classified as lipids on the basis of their physical properties. For example, lipids occur in the waxy coatings of plants, where they protect cells from desiccation by creating a water-impermeable barrier. Some of the lipids synthesized by plants are essential components of the human diet (see Box 9-2).

Box 9-2

BIOCHEMISTRY IN FOCUS

The Lipid Vitamins A, E, and K

Humans cannot synthesize the compounds known as vitamins A, E, and K and must therefore obtain them from food. Other than being lipids, these vitamins—whose structures are shown below—have little in common.

Vitamin A, or **retinol,** is derived mainly from plant pigments such as β-carotene. Retinol is oxidized to **retinal** (an aldehyde), which functions as the eye's photoreceptor at low light intensities. Light causes the retinal to isomerize, triggering an impulse through the optic nerve. A severe deficiency of vitamin A can lead to blindness.

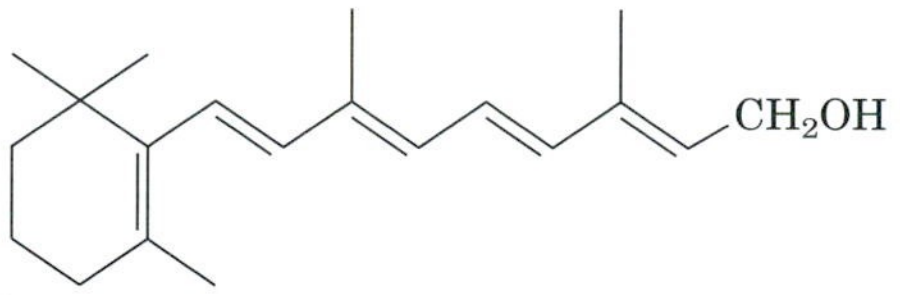

**Retinol
(vitamin A)**

Vitamin E is actually a group of compounds whose most abundant member is **α-tocopherol.** This highly hydrophobic molecule is incorporated into cell membranes, where it acts as an antioxidant to prevent oxidative damage to membrane proteins and lipids. A deficiency of vitamin E elicits a variety of nonspecific symptoms, which makes the deficiency difficult to detect. The popularity of vitamin E supplements rests on the idea that vitamin E may minimize the effects of aging (thought to result in part from oxidative damage to cells).

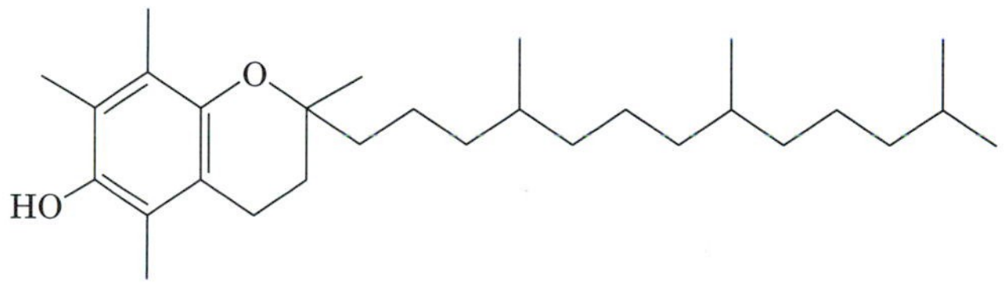

**α-Tocopherol
(vitamin E)**

Vitamin K is a lipid synthesized by plants (as **phylloquinone**) and bacteria (as **menaquinone**). About half of the daily requirement for humans is supplied by intestinal bacteria. Vitamin K participates in the carboxylation of Glu residues in some of the proteins involved in blood clotting. A vitamin K deficiency prevents this carboxylation, and the resulting inactive clotting proteins lead to excessive bleeding. Compounds that interfere with vitamin K function are active ingredients in some rodent poisons.

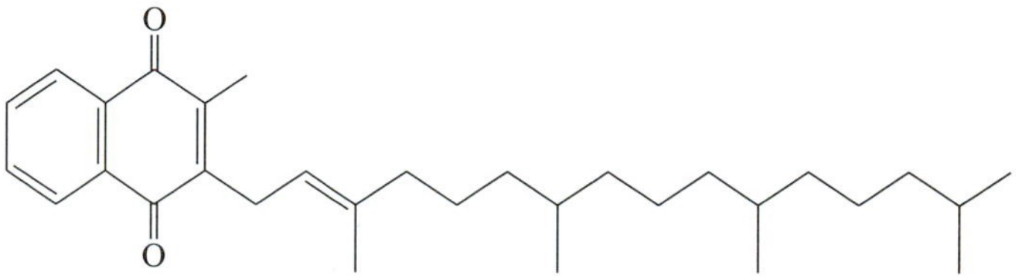

**Phylloquinone
(vitamin K)**

Eicosanoids Are Derived from Arachidonic Acid

Other less common lipids are derived from relatively abundant membrane lipids. **Prostaglandins** (e.g., Fig. 9-12) were discovered in the 1930s by Ulf von Euler, who thought they were produced by the prostate gland. Prostaglandins and related compounds—**prostacyclins, thromboxanes,** and **leukotrienes**—are known collectively as **eicosanoids** because they are all C_{20} compounds (Greek: *eikosi,* twenty). *Eicosanoids act at very low concentrations and are involved in the production of pain and fever, and in the regulation of blood pressure, blood coagulation, and reproduction.* Unlike hormones, eicosanoids are not transported by the bloodstream to their sites of action but tend to act locally, close to the cells that produce them. In fact, most eicosanoids decompose within seconds or minutes, which limits their effects on nearby tissues. Not surprisingly, the enzymes that synthesize the various eicosanoids are the targets of intensive pharmacological research.

In humans, the most important eicosanoid precursor is **arachidonic acid,** a polyunsaturated fatty acid with four double bonds. Arachidonate is stored in cell membranes as the C2 ester of **phosphatidylinositol** and other phospholipids. The fatty acid residue is released by the action of phospholipase A_2.

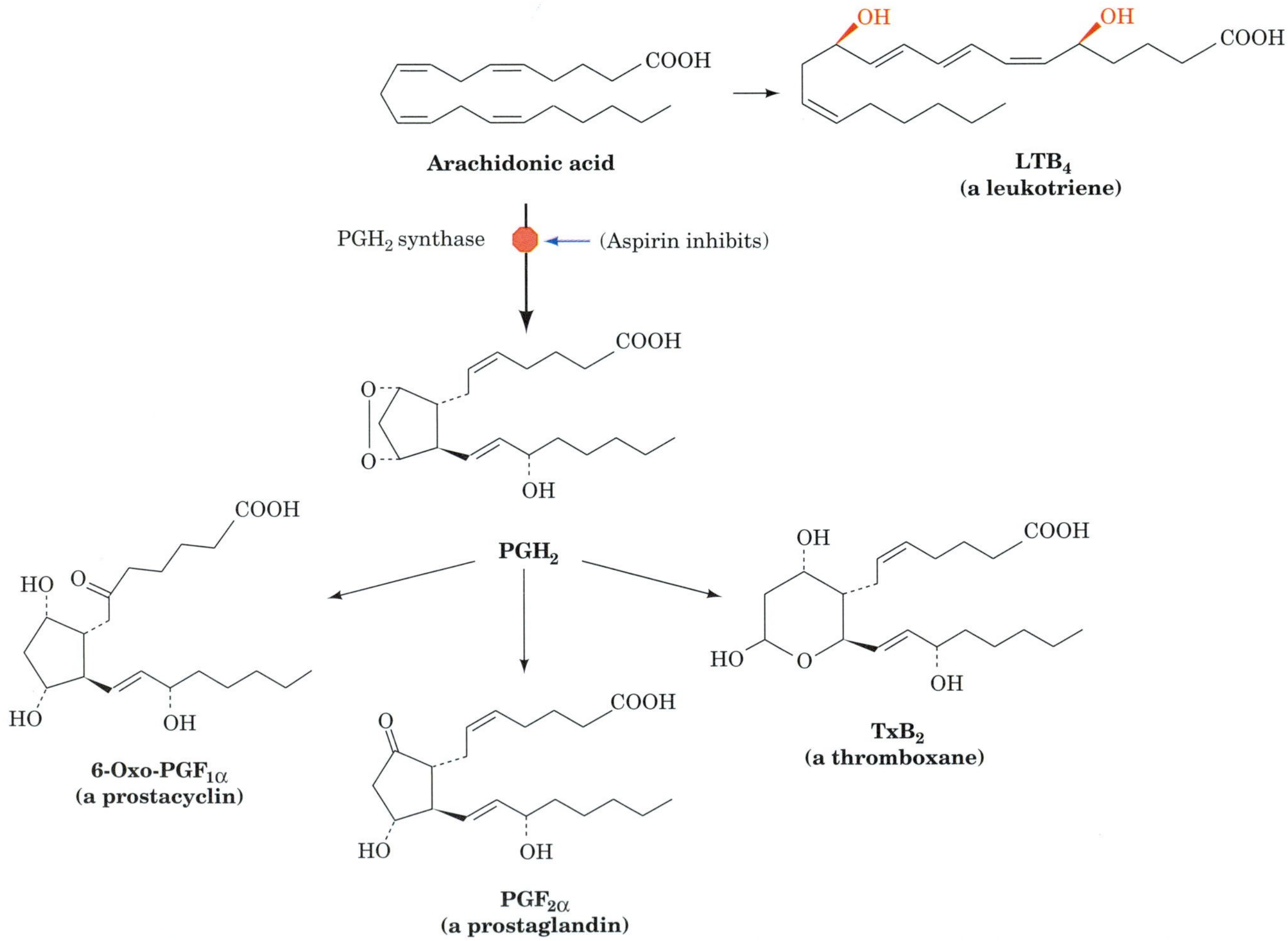

Figure 9-12.   Eicosanoids. Arachidonate is the precursor of prostaglandins (PG), prostacyclins, and thromboxanes (Tx). Arachidonate also leads to leukotrienes. Although only a single example of each type of eicosanoid is shown, each has numerous physiologically significant derivatives, which are designated by letters and subscripts (e.g., **PGH₂** for **prostaglandin H₂**).

The specific products of arachidonate metabolism are tissue-dependent. For example, platelets produce thromboxanes almost exclusively, but endothelial cells (which line the walls of blood vessels) predominantly synthesize prostacyclins. Interestingly, thromboxanes stimulate vasoconstriction and platelet aggregation, while prostacyclins elicit the opposite effects. Thus, the two substances act in opposition to maintain a balance in the cardiovascular system.

The use of **aspirin** as an analgesic (pain-relieving), antipyretic (fever-reducing), and anti-inflammatory agent has been widespread since the nineteenth century. Yet it was not until 1971 that John Vane discovered its mechanism of action: Aspirin inhibits the synthesis of prostaglandins, prostacyclins, and thromboxanes from arachidonate (Fig. 9-12) by acetylating a specific Ser residue of an enzyme known as **PGH$_2$ synthase,** which prevents arachidonate from reaching this enzyme's active site. **Nonsteroidal anti-inflammatory drugs (NSAIDs)** such as **ibuprofen** and **acetaminophen** noncovalently bind to this enzyme so as to similarly block its active site.

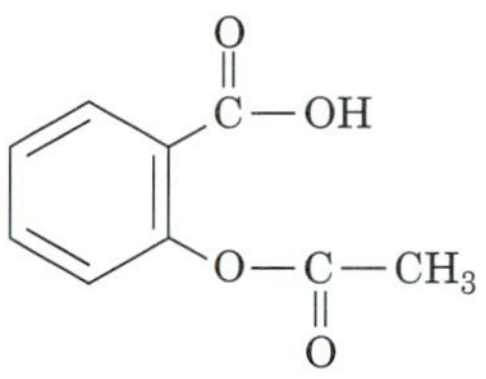

Aspirin
(acetylsalicylic acid)

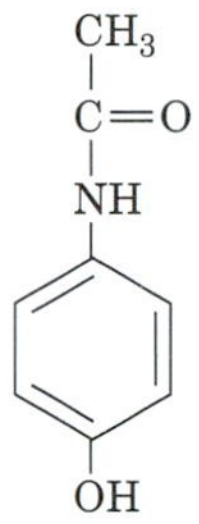

Acetaminophen

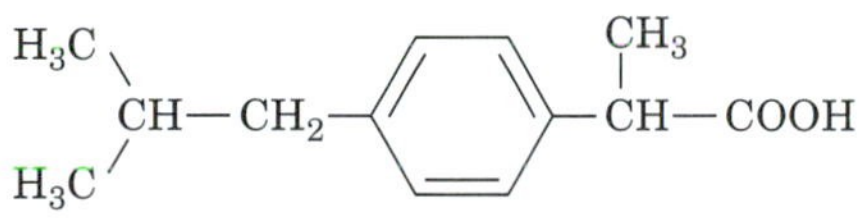

Ibuprofen

Leukotrienes are synthesized from arachidonate by an aspirin-insensitive pathway (Fig. 9-12) in many cell types. Various inflammatory and hypersensitivity disorders (such as asthma) are associated with elevated levels of leukotrienes.

2. LIPID BILAYERS

In living systems, lipids are seldom found as free molecules but instead associate with other molecules, usually other lipids. In this section, we discuss how lipids aggregate to form micelles and bilayers. We are concerned with the physical properties of lipid bilayers because these aggregates form the structural basis for biological membranes.

A. Why Bilayers Form

In aqueous solutions, amphiphilic molecules such as soaps and detergents form micelles (globular aggregates whose hydrocarbon groups are out of contact with water; Section 2-1C). This molecular arrangement eliminates unfavorable contacts between water and the hydrophobic tails of the amphiphiles and yet permits the solvation of the polar head groups.

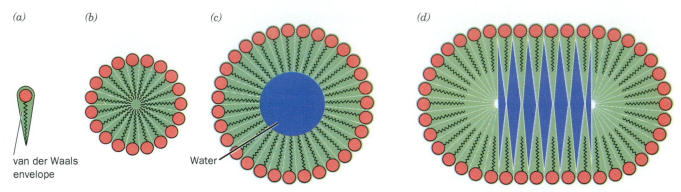

Figure 9-13. **Aggregates of single-tailed lipids.** The tapered van der Waals envelope of these lipids (*a*) permits them to pack efficiently to form a spheroidal micelle (*b*). The diameter of these micelles depends on the length of the tails. Spheroidal micelles composed of many more lipid molecules than the optimal number (*c*) would have an unfavorable water-filled center (*blue*). Such micelles could flatten out to collapse the hollow center, but as these ellipsoidal micelles become elongated, they also develop water-filled spaces (*d*).

The approximate size and shape of a micelle can be predicted from geometrical considerations. Single-tailed amphiphiles, such as soap anions, form spheroidal or ellipsoidal micelles because of their tapered shapes (their hydrated head groups are wider than their tails; Fig. 9-13*a,b*). The number of molecules in such a micelle depends on the amphiphile, but for many substances it is on the order of several hundred. Too few lipid molecules would expose the hydrophobic core of the micelle to water, whereas too many would give the micelle an energetically unfavorable hollow center (Fig. 9-13*c*). Of course, a large micelle could flatten out to eliminate this hollow center, but the resulting decrease of curvature at the flattened surfaces would also generate empty spaces (Fig. 9-13*d*).

The two hydrocarbon tails of glycerophospholipids and sphingolipids give these amphiphiles a somewhat rectangular cross section (Fig. 9-14*a*). The steric requirements of packing such molecules together yields large disklike micelles (Fig. 9-14*b*) that are really extended bimolecular leaflets. These **lipid bilayers** are ~60 Å thick, as measured by electron microscopy and X-ray diffraction techniques, the value expected for more or less fully extended hydrocarbon tails.

A suspension of phospholipids (glycerophospholipids or sphingomyelins) can form **liposomes**—closed, self-sealing solvent-filled vesicles that are bounded by only a single bilayer (Fig. 9-15). They usually have diameters of several hundred angstroms and, in a given preparation, are rather uniform in size. Once formed, liposomes are quite stable and can be purified by dialysis, gel filtration chromatography, or centrifugation. Lipo-

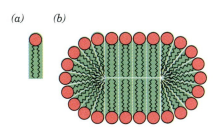

Figure 9-14. **Bilayer formation by phospholipids.** The cylindrical van der Waals envelope of these lipids (*a*) causes them to form extended disklike micelles (*b*) that are better described as lipid bilayers.

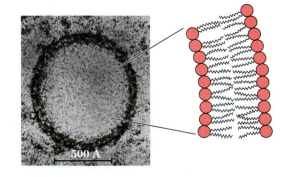

Figure 9-15. **Electron micrograph of a liposome.** Its wall, as the accompanying diagram indicates, consists of a lipid bilayer. [Courtesy of Walther Stoeckenius, University of California at San Francisco.]

(a) Transverse diffusion (flip-flop)

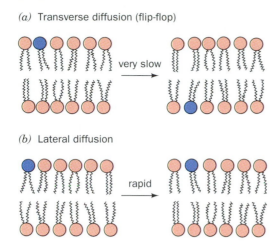

(b) Lateral diffusion

Figure 9-16. Phospholipid diffusion in a lipid bilayer. (*a*) Transverse diffusion (a flip-flop) is defined as the transfer of a phospholipid molecule from one bilayer leaflet to the other. (*b*) Lateral diffusion is defined as the pairwise exchange of neighboring phospholipid molecules in the same bilayer leaflet.

somes whose internal environment differs from the surrounding solution can therefore be readily prepared. Liposomes serve as models of biological membranes and also hold promise as vehicles for drug delivery since they are absorbed by many cells through fusion with the plasma membrane.

B. Lipid Mobility

The transfer of a lipid molecule across a bilayer (Fig. 9-16a), a process termed **transverse diffusion** *or a* **flip-flop,** *is an extremely rare event.* This is because a flip-flop requires the hydrated, polar head group of the lipid to pass through the anhydrous hydrocarbon core of the bilayer. The flip-flop rates of phospholipids have half-times of several days or more. In contrast to their low flip-flop rates, *lipids are highly mobile in the plane of the bilayer* (**lateral diffusion;** Fig. 9-16*b*). It has been estimated that lipids in a membrane can diffuse the 1-μm length of a bacterial cell in ~1 s. Because of the mobilities of the lipids, the lipid bilayer can be considered to be a two-dimensional fluid.

The interior of the lipid bilayer is in constant motion due to rotations around the C—C bonds of the lipid tails. Various physical measurements suggest that the interior of the bilayer has the viscosity of light machine oil. This feature of the bilayer core is evident in molecular dynamics simulations, in which the time-dependent positions of atoms are predicted from calculations of the forces acting on them (Fig. 9-17). The viscosity of the bilayer increases dramatically closer to the lipid head groups, whose rotation is limited and whose lateral mobility is more constrained by interactions between other polar or charged head groups. The outer surfaces of the lipid bilayer are, of course, flanked by several layers of ordered water molecules.

Note that the hydrophobic tails of the lipids shown in Fig. 9-17 are not stiffly regimented as Fig. 9-16 might suggest. That lipid tails bend and interdigitate is not surprising. A typical biological membrane includes many different lipid molecules, some of whose tails are of different lengths or are

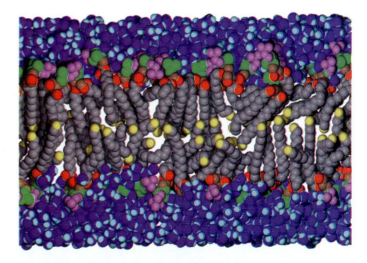

Figure 9-17. Model (snapshot) of a lipid bilayer at a particular instant in time. The conformations of dipalmitoyl phosphatidylcholine molecules in a bilayer surrounded by water were modeled by computer. Atom colors are chain C gray (except terminal methyl C yellow and glycerol C brown), ester O red, phosphate P and O green, choline C and N pale violet, water O dark blue, and water H light blue. Lipid hydrogens have been omitted. [Courtesy of Richard Pastor and Richard Venable, FDA, Bethesda, Maryland.]

kinked due to the presence of double bonds. Under physiological conditions, highly mobile chains fill any gaps that might form between lipids in the bilayer interior.

The Fluidity of a Lipid Bilayer Is Temperature-Dependent

As a lipid bilayer cools below a characteristic **transition temperature,** *it undergoes a sort of phase change in which it becomes a gel-like solid; that is, it loses its fluidity* (Fig. 9-18). Above the transition temperature, the highly mobile lipids are in a state known as a **liquid crystal** because they are ordered in some directions but not in others. The bilayer is thicker in the gel state than in the liquid crystal state due to the stiffening of the hydrocarbon tails at lower temperatures.

The transition temperature of a bilayer increases with the chain length and the degree of saturation of its component fatty acid residues for the same reasons that the melting points of fatty acids increase with these quantities. The transition temperatures of most biological membranes are in the range 10 to 40°C. Bacteria and cold-blooded animals such as fish modify (through lipid synthesis and degradation) the fatty acid compositions of their membrane lipids with ambient temperature so as to maintain a constant level of fluidity. Thus, the fluidity of biological membranes is one of their important physiological attributes.

Cholesterol Modulates Membrane Fluidity

Cholesterol, which by itself does not form a bilayer, decreases membrane fluidity because its rigid steroid ring system interferes with the motions of the fatty acid side chains in other membrane lipids. It also broadens the temperature range of the phase transition. This is because cholesterol inhibits the crystallization of fatty acid side chains by fitting in between them. Thus, cholesterol functions as a kind of membrane plasticizer.

Biological membranes contain both lipids and proteins that mutually interact. Membrane proteins have structural features that permit them to extend into or completely through the hydrophobic core of the bilayer. The physical properties of membrane lipids, in turn, may be constrained by the presence of proteins among them. As we shall see in the following chapter, the complexities of such relationships provide keys to understanding the various functions of membranes.

(a) Above transition temperature   *(b)* Below transition temperature

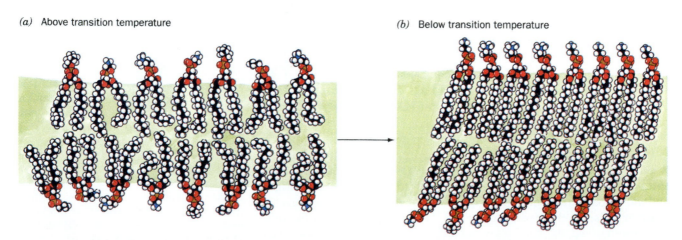

Figure 9-18. **Phase transition in a lipid bilayer.** (*a*) Above the transition temperature, both the lipid molecules as a whole and their nonpolar tails are highly mobile in the plane of the bilayer. (*b*) Below the transition temperature, the lipid molecules form a much more orderly array to yield a gel-like solid. [After Robertson, R.N., *The Lively Membranes,* pp. 69–70, Cambridge University Press (1983).]

SUMMARY

1. Lipids are a diverse group of molecules that are soluble in organic solvents and, in contrast to other major types of biomolecules, do not form polymers.

2. Fatty acids are carboxylic acids whose chain lengths and degrees of unsaturation vary.

3. Adipocytes and other cells contain stores of triacylglycerols, which consist of three fatty acids esterified to glycerol.

4. Glycerophospholipids are amphipathic molecules that contain two fatty acid chains and a polar head group.

5. The sphingolipids include sphingomyelins, cerebrosides, and gangliosides.

6. Cholesterol, steroid hormones, and vitamin D are all based on a four-ring structure.

7. The arachidonic acid derivatives prostaglandins, prostacyclins, thromboxanes, and leukotrienes are signaling molecules that have diverse physiological roles.

8. Glycerophospholipids and sphingolipids form bilayers in which their nonpolar tails associate with each other and their polar head groups are exposed to the aqueous solvent.

9. Although the transverse diffusion of a lipid across a bilayer is extremely slow, lipids rapidly diffuse in the plane of the bilayer. Bilayer fluidity varies with temperature and with the chain lengths and degree of saturation of its component fatty acid residues.

REFERENCES

Brown, M.S. and Goldstein, J.L., Koch's postulates for cholesterol, *Cell* **71,** 187–188 (1992). [Describes approaches in characterizing the involvement of cholesterol in atherosclerosis.]

Cullis, R.R. and Hope, M.J., Physical properties and functional roles of lipids in membranes, *in* Vance, D.E. and Vance, J. (Eds.), *Biochemistry of Lipids, Lipoproteins and Membranes,* Elsevier (1991).

Gelb, M.H., Jain, M.K., Hanel, A.M., and Berg, O.G., Interfacial enzymology of glycerolipid hydrolases: lessons from secreted phospholipases A₂, *Annu. Rev. Biochem.* **64,** 653–688 (1995). [This detailed review includes a description of the challenges of characterizing enzymes that act on lipids at the lipid–water interface.]

Gurr, M.I. and Harwood, J.L., *Lipid Biochemistry: An Introduction* (4th ed.), Chapman & Hall (1991).

Needleman, P., Turk, J., Jakschik, B.A., Morrison, A.R., and Lefkowith, J.B., Arachidonic acid metabolism, *Annu. Rev. Biochem.* **55,** 69–102 (1986).

Pastor, R.W., Molecular dynamics and Monte Carlo simulations of lipid bilayers, *Curr. Opin. Struct. Biol.* **4,** 486–492 (1994). [Discusses the structures and fluidity of lipid bilayers.]

Weissman, G., Aspirin, *Sci. Am.* **264**(1), 84–90 (1991).

KEY TERMS

lipid
fatty acid
saturation
triacylglycerol
fats
oils
adipocyte
glycerophospholipid
phosphatidic acid
phospholipase
lysophospholipid
plasmalogen
sphingolipid
ceramide
sphingomyelin
cerebroside
ganglioside
steroid
sterol
glucocorticoid
mineralocorticoid
androgen
estrogen
prostaglandin
eicosanoid
lipid bilayer
liposome
transverse diffusion
lateral diffusion
transition temperature

STUDY EXERCISES

1. How do lipids differ from the three other major classes of biological molecules?

2. How does unsaturation affect the physical properties of fatty acids or the membrane lipids to which they are esterified?

3. Compare the structures and physical properties of triacylglycerols, glycerophospholipids, and sphingolipids.

4. Summarize the functions of steroids and eicosanoids.

5. Explain why lateral diffusion is faster than transverse diffusion in membrane lipids.

PROBLEMS

1. Does *trans*-oleic acid have a higher or lower melting point than *cis*-oleic acid? Explain.

2. How many different types of triacylglycerols could incorporate the fatty acids shown in Fig. 9-1?

3. Which triacylglycerol yields more energy on oxidation: one containing three residues of linolenic acid or three residues of stearic acid?

4. What products are obtained when 1-palmitoyl-2-oleoyl-3-phosphatidylserine is hydrolyzed by (a) phospholipase A_1; (b) phospholipase A_2; (c) phospholipase C; (d) phospholipase D?

5. Does the phosphatidylglycerol "head group" of cardiolipin (Table 9-2) project out of a lipid bilayer like other glycerophospholipid head groups?

6. Most hormones, such as peptide hormones, exert their effects by binding to cell-surface receptors. However, steroid hormones do so by binding to cytosolic receptors. How is this possible?

7. Animals cannot synthesize linoleic acid (a precursor of arachidonic acid) and therefore must obtain this **essential fatty acid** from their diet. Explain why cultured animal cells can survive in the absence of linoleic acid.

8. Why can't triacylglycerols be significant components of lipid bilayers?

9. Why would a bilayer containing only gangliosides be unstable?

10. When bacteria growing at 20°C are warmed to 30°C, are they more likely to synthesize membrane lipids with (a) saturated or unsaturated fatty acids, and (b) short-chain or long-chain fatty acids? Explain.

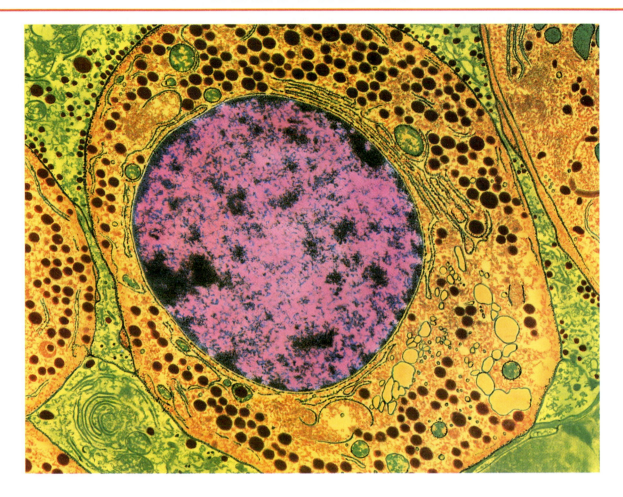

A cell's membranes delineate discrete metabolic compartments and, through the action of proteins embedded in the lipid bilayer, mediate the flow of molecules and information between the compartments. This electron micrograph of a pituitary gland cell shows a variety of membrane-enclosed organelles including the nucleus (pink), which is surrounded by the highly folded endoplasmic reticulum (thin green curves), mitochondria (round green objects), and numerous vesicles (brown) containing hormones to be secreted. [Quest/Science Photo Library/Photo Researchers.]

BIOLOGICAL MEMBRANES

The lipid bilayer presents a formidable barrier to the passage of polar molecules. Its ability to partition two aqueous compartments is of fundamental importance for the plasma membranes that surround all cells and for the membranes that define the organelles of eukaryotic cells. These membranes organize biological processes by compartmentalizing them. The impermeability of lipid bilayers to polar or charged molecules allows the solute concentrations on each side to differ dramatically. Thus, a bilayer-bounded compartment can maintain the complement of large and small molecules required to carry out a particular metabolic process. Yet the nature of biological processes, which are characterized by steady-state fluxes of energy and materials (Section 1-4E) and by the ability to respond to changing external conditions, demands that certain molecules and ions pass through the lipid bilayer.

Biological membranes contain proteins as well as lipids. Membrane proteins catalyze numerous chemical reactions, mediate the flow of nutrients and wastes, and participate in relaying information about the extracellular environment to various intracellular components. In this chapter, we examine the unique features of membrane proteins that allow them to interact with the hydrophobic interior of the lipid bilayer. We also discuss the dynamic organization of proteins and lipids within membranes and the structures and functions of lipoproteins. Finally, we introduce mechanisms for transporting various substances across membranes. Many critical cellular activities, including oxidative phosphorylation (Chapter 17) and photosynthesis (Chapter 18), occur on and in membranes and depend on the features of membranes and membrane proteins that are introduced in this chapter.

1. MEMBRANE PROTEINS

All biological membranes are composed of proteins associated with a lipid bilayer matrix, although the exact lipid and protein components vary with the identity of the membrane. Protein-to-lipid ratios vary considerably with membrane function. For example, the lipid-rich myelinated membranes that surround and insulate certain nerve axons (Fig. 9-8) have a protein-to-lipid ratio of 0.23, whereas the protein-rich inner membrane of mitochondria, which mediates numerous chemical reactions, has a protein-to-lipid ratio of 3.2. Eukaryotic membranes are typically ~50% protein.

Membrane proteins, like other proteins, come in a tremendous variety of sizes and shapes, but they can be classified roughly by their mode of interaction with the membrane as integral or peripheral membrane proteins or as lipid-linked proteins.

A. Integral Membrane Proteins

Integral or *intrinsic proteins* are tightly bound to membranes by hydrophobic interactions (Fig. 10-1) and can be separated from them only by treatment with agents that disrupt membranes. These include organic solvents, detergents (e.g., sodium dodecyl sulfate; Section 5-2D), and chaotropic agents (substances that disrupt water structure; Section 6-4B). Some integral proteins bind lipids so tenaciously that they can be freed from them only under denaturing conditions.

Integral proteins tend to aggregate and precipitate in aqueous solution unless they are solubilized by detergents or water-miscible organic solvents such as butanol or glycerol. Solubilized integral proteins can be purified by many of the protein fractionation methods discussed in Section 5-2.

Integral membrane protein

Figure 10-1.   Model of an integral membrane protein. The protein is solvated by membrane lipids through hydrophobic interactions between the protein and the lipids' nonpolar tails. The polar head groups may also associate with the protein through hydrogen bonding and salt bridges. [After Robertson, R.N., *The Lively Membranes,* p. 56, Cambridge University Press (1983).]

Integral Proteins Are Asymmetrically Oriented Amphiphiles

Integral proteins are amphiphiles; the protein segments immersed in a membrane's nonpolar interior have predominantly hydrophobic surface residues, whereas those portions that extend into the aqueous environment are by and large sheathed with polar residues. This was demonstrated through **surface labeling,** a technique employing agents that react with proteins but cannot penetrate membranes. For example, the extracellular domain of an integral protein binds antibodies elicited against it, but its cytoplasmic domain will do so only if the membrane has been ruptured. Membrane-impermeable protein-specific reagents that are fluorescent or radioactively labeled can be similarly employed. Alternatively, proteases, which digest only the solvent-exposed portions of an integral protein, may be used to identify the membrane-immersed portions of the protein. These techniques revealed, for example, that the erythrocyte membrane protein **glycophorin A** has three domains (Fig. 10-2): (1) a 72-residue externally located N-terminal domain that bears 16 carbohydrate chains; (2) a 19-residue sequence, consisting almost entirely of hydrophobic residues, that spans the erythrocyte cell membrane; and (3) a 40-residue cytoplasmic C-terminal domain that has a high proportion of charged and polar residues. Thus, glycophorin A is a **transmembrane protein;** that is, it completely spans the membrane.

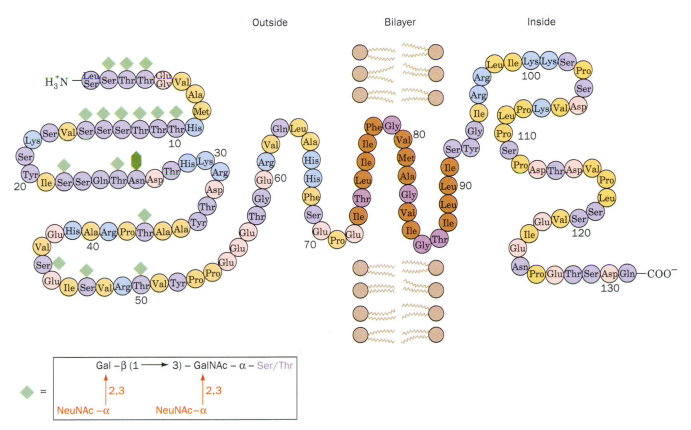

Figure 10-2. Human erythrocyte glycophorin A. The protein bears 15 *O*-linked oligosaccharides (*green diamonds*) and one that is *N*-linked (*dark green hexagon*) on its extracellular domain. The predominant sequence of the *O*-linked oligosaccharides is also shown (NeuNAc = *N*-acetylneuraminic acid). The protein's transmembrane portion (*brown and purple*) consists of 19 sequential predominantly hydrophobic residues. Its C-terminal portion, which is located on the membrane's cytoplasmic face, is rich in anionic (*pink*) and cationic (*blue*) residues. There are two common genetic variants of glycophorin A: Glycophorin A^M has Ser and Gly and positions 1 and 5, whereas glycophorin A^N has Leu and Glu at these positions. [After Marchesi, V.T., *Semin. Hematol.* **16,** 8 (1979).]

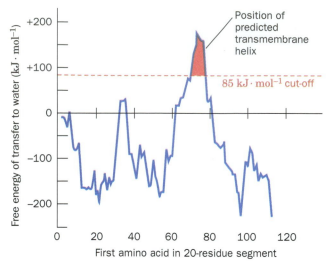

Figure 10-3.   Identification of glycophorin A's transmembrane domain. The calculated free energy change in transferring 20-residue-long α-helical segments from the interior of a membrane to water is plotted against the position of the segment's first residue. Peaks higher than +85 kJ · mol⁻¹ indicate a transmembrane helix. [After Engleman, D.M., Steitz, T.A., and Goldman, A., *Annu. Rev. Biophys. Biophys. Chem.* **15**, 343 (1986).]

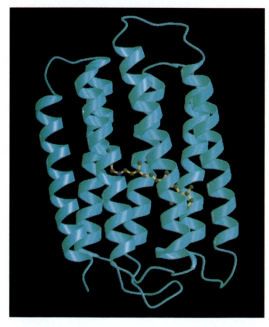

Figure 10-4.   The structure of bacterio-rhodopsin. The protein is shown in ribbon form (*cyan*) as viewed from within the membrane plane and with its covalently bound retinal shown in ball-and-stick form. [Courtesy of N. Grigorieff and Richard Henderson, MRC Laboratory of Molecular Biology, Cambridge, England.] ● **See Kinemage Exercise 8-1.**

Studies of a variety of biological membranes have established that *biological membranes are asymmetric in that a particular membrane protein is invariably located on only one particular face of a membrane, or in the case of a transmembrane protein, oriented in only one direction with respect to the membrane.* However, no protein is known to be completely buried in a membrane; that is, all membrane-associated proteins are at least partially exposed to the aqueous environment.

Glycophorin A's transmembrane domain, as is common in many integral membrane proteins, almost certainly forms an α helix, thereby satisfying the hydrogen bonding requirements of its polypeptide backbone. The existence of glycophorin A's single transmembrane helix is predicted by computing the free energy change in transferring α-helical polypeptide segments from the nonpolar interior of a membrane to water (Fig. 10-3) or, more simply, by identifying hydrophobic polypeptide segments using a hydropathy index such as that in Table 6-2. Similar analyses of other integral membrane proteins have identified their membrane-immersed helices.

Bacteriorhodopsin Contains Seven Transmembrane Segments

Many membrane proteins, such as **bacteriorhodopsin** (Fig. 10-4), contain multiple transmembrane α helices. Bacteriorhodopsin, a 247-residue protein isolated from the halophilic (salt-loving) bacterium *Halobacterium halobium* (it grows best in 4.3 M NaCl), is a light-driven proton pump: It generates a proton concentration gradient that powers ATP synthesis by a mechanism discussed in Section 17-3. Bacteriorhodopsin's light-absorbing group, **retinal,** which is covalently linked to Lys 216 of the protein, is also the light-sensitive element in vision.

Retinal residue

The cell membrane of the bacterium contains ~0.5-μm-wide patches, known as purple membrane, that consist of 75% bacteriorhodopsin and 25% lipid. The protein molecules are arranged in an ordered two-dimensional array (two-dimensional crystal) whose interstices are filled by lipid. Richard Henderson and Nigel Unwin took advantage of this unusual membrane structure to determine the structure of bacteriorhodopsin through **electron crystallography,** a technique resembling X-ray crystallography in which the electron beam of an electron microscope is used to elicit diffraction from a two-dimensional crystal.

Bacteriorhodopsin consists largely of a bundle of seven ~25-residue α-helical rods that span the lipid bilayer in directions almost perpendicular to the bilayer plane (Fig. 10-4). As expected, the amino acid side chains that contact the lipid tails are highly hydrophobic. Individual membrane-spanning helices are connected in head-to-tail fashion by hydrophilic loops of varying size. This arrangement places the protein's charged residues near the surfaces of the membrane in contact with the aqueous environment.

The Photosynthetic Reaction Center Contains Eleven Transmembrane Helices

The primary photochemical process of photosynthesis in purple photosynthetic bacteria is mediated by the so-called **photosynthetic reaction center** (Section 18-2B). This transmembrane protein consists of at least three nonidentical ~300-residue subunits that collectively bind four **chlorophyll** molecules, four other chromophores, and a nonheme Fe atom. The 1187-residue photosynthetic reaction center of *Rhodopseudomonas viridis,* whose X-ray structure was determined in 1984 by Hartmut Michel, Johann Deisenhofer, and Robert Huber, was the first transmembrane protein to be described in atomic detail (Fig. 10-5). The protein's membrane-spanning portion consists of 11 α helices that form a 45-Å-long cylinder with the expected hydrophobic surface.

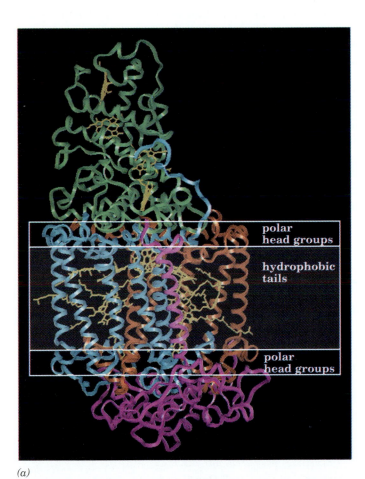

(a)

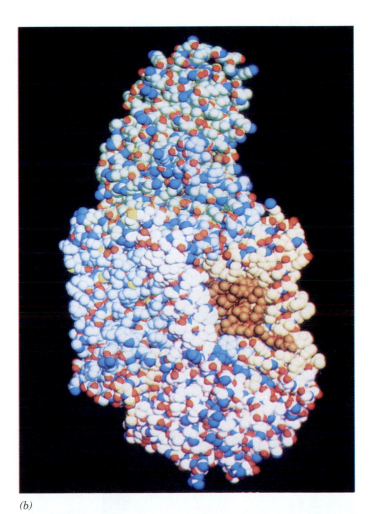

(b)

Figure 10-5. X-Ray structure of the photosynthetic reaction center of *Rhodopseudomonas viridis.* (*a*) A ribbon diagram in which only the C_α backbone and the light-absorbing groups (*yellow*) are shown. The H, M, and L subunits (*pink, blue,* and *orange,* respectively) collectively have 11 transmembrane helices. A *c*-type cytochrome (*green*) is bound to the external face of the complex. The position that the transmembrane protein is thought to occupy in the lipid bilayer is indicated schematically. [Based on an X-ray structure by Johann Deisenhofer, Robert Huber, and Hartmut Michel, Max-Planck-Institut für Biochemie, Germany.]

(*b*) A space-filling model in which nitrogen atoms are blue, oxygens are red, sulfurs are yellow, and the carbon atoms of the H, M, L, and cytochrome subunits are tinted pink, blue, orange, and green, respectively. Exposed portions of the chromophores are brown. Note how few polar groups (nitrogens and oxygens) are externally exposed in the portion of the protein that is immersed in the nonpolar region of the lipid bilayer. [From Deisenhofer, J. and Michel, H., *Les Prix Nobel* (1989).] ● See Kinemage Exercise 8-2.

(a)

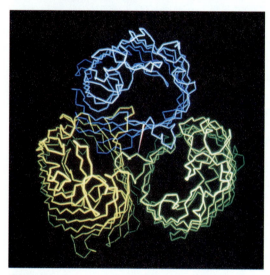

(b)

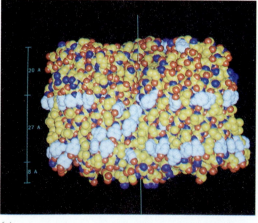

(c)

Hydrophobic interactions, as we saw in Section 6-4A, are the dominant forces stabilizing the three-dimensional structures of water-soluble globular proteins. However, since the transmembrane regions of integral membrane proteins are immersed in nonpolar environments, what stabilizes their structures? Analysis of the photosynthetic reaction center indicates that its interior residues have hydrophobicities comparable to those of water-soluble proteins. However, the membrane-exposed residues of the photosynthetic reaction center, on average, are even more hydrophobic than its interior residues. Thus, *the difference between integral proteins and water-soluble proteins is only skin-deep: Their interiors are similar, but their surface polarities are consistent with the polarities of their environments.*

Porins Contain Transmembrane β Barrels

A protein segment immersed in the nonpolar interior of a membrane must fold so that it satisfies the hydrogen bonding potential of its polypeptide backbone. An α helix can do this; so can an antiparallel β sheet that rolls up to form a barrel (a β barrel; Section 6-2B). Such β structures occur in **porins,** which are channel-forming proteins in the outer membranes of gram-negative bacteria (Section 8-3B). The outer membrane protects the bacteria from hostile environments while the porins permit the entry of small polar solutes such as nutrients. Porins also occur in eukaryotes in the outer membranes of mitochondria and chloroplasts (consistent with the descent of these organelles from free-living bacteria; Section 1-3B).

Bacterial porins are monomers or trimers of identical 30- to 50-kD subunits. X-Ray structural studies show that each porin subunit consists mainly of a 16- or 18-stranded antiparallel β barrel that forms a solvent-accessible central channel with a length of ~55 Å and a minimum diameter of ~7 Å (Fig. 10-6). As expected, the side chains at the protein's membrane-exposed surface are nonpolar, thereby forming a ~27-Å-high hydrophobic band encircling the trimer (Fig. 10-6c). In contrast, the side chains at the solvent-exposed surface of the protein, including those lining the walls of the aqueous channel, are polar. Possible mechanisms for the solute selectivity of porins are discussed in Section 10-4B.

B. Lipid-Linked Proteins

Some membrane-associated proteins contain covalently attached lipids that anchor the protein to the membrane. The lipid group, like any modifying group, may also mediate protein–protein interactions or modify the structure and activity of the protein to which it is attached. **Lipid-linked proteins** come in three varieties: prenylated proteins, fatty acylated proteins, and glycosylphosphatidylinositol-linked proteins. A single protein may contain more than one covalently linked lipid group.

Figure 10-6. **X-Ray structure of the *E. coli* OmpF porin.** (*a*) A ribbon diagram of the 16-stranded monomer. (*b*) The C_α backbone of the trimer viewed ~30° from its threefold axis of symmetry, showing the pore through each subunit. (*c*) A space-filling model of the trimer viewed perpendicular to its three-fold axis (*vertical green line*). N atoms are blue, O atoms are red, and C atoms are yellow, except those in the side chains of aromatic residues, which are white. The aromatic groups appear to delimit an ~27-Å-high hydrophobic band (*scale at left*) that is immersed in the nonpolar portion of the bacterial outer membrane (with the cell's exterior at the tops of *a* and *c*).Compare this hydrophobic band with that in Fig. 10-5*b*. [Part *a* based on an X-ray structure by and parts *b* and *c* courtesy of Tilman Schirmer and Johan Jansonius, University of Basel, Switzerland.] ● **See Kinemage Exercise 8-3.**

Prenylated proteins have covalently attached lipids that are built from **isoprene units** (**isoprene** is a C_5 compound). The most common **isoprenoid groups** are the C_{15} **farnesyl** and C_{20} **geranylgeranyl** residues:

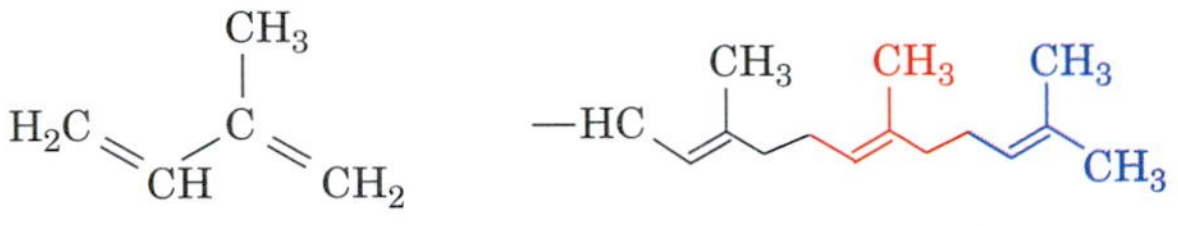

Isoprene

Farnesyl residue

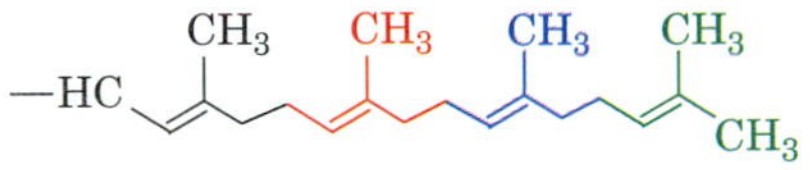

Geranylgeranyl residue

The most common prenylation site in proteins is the C-terminal tetrapeptide C-X-X-Y, where C is Cys and X is often an aliphatic amino acid residue. Residue Y influences the type of prenylation: Proteins are farnesylated when Y is Ala, Met, or Ser and geranylgeranylated when Y is Leu. In both cases, the prenyl group is enzymatically linked to the Cys sulfur atom via a thioether linkage. The X-X-Y tripeptide is then proteolytically excised, and the newly exposed terminal carboxyl group is esterified with a methyl group, producing a C-terminus with the structure

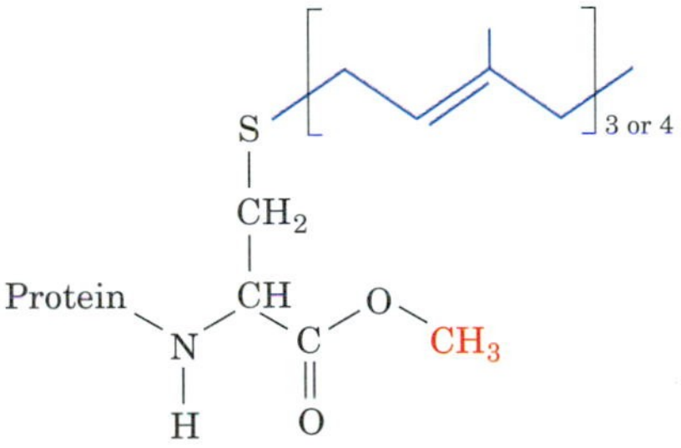

Two kinds of fatty acids, myristic acid and palmitic acid, are linked to membrane proteins. Myristic acid, a biologically rare saturated C_{14} fatty acid, is appended to a protein via an amide linkage to the α-amino group of an N-terminal Gly residue. **Myristoylation** is stable: The fatty acyl group remains attached to the protein throughout its lifetime. Myristoylated proteins are located in a number of subcellular compartments, including the cytosol, endoplasmic reticulum, plasma membrane, and the nucleus.

In **palmitoylation**, the saturated C_{16} fatty acid palmitic acid is joined in thioester linkage to a specific Cys residue. Palmitoylated proteins occur almost exclusively on the cytoplasmic face of the plasma membrane, where many participate in transmembrane signaling. The palmitoyl group can be removed by the action of **palmitoyl thioesterases,** suggesting that reversible palmitoylation may regulate the association of the protein with the membrane and thereby modulate the signaling processes.

Glycosylphosphatidylinositol-linked proteins (GPI-linked proteins) occur in all eukaryotes but are particularly abundant in some parasitic protozoa, which contain relatively few membrane proteins anchored by transmembrane polypeptide segments. Like glycoproteins and glycolipids, GPI-linked proteins are located only on the exterior surface of the plasma membrane.

The core structure of the GPI group consists of phosphatidylinositol (Table 9-2) glycosidically linked to a linear tetrasaccharide composed of

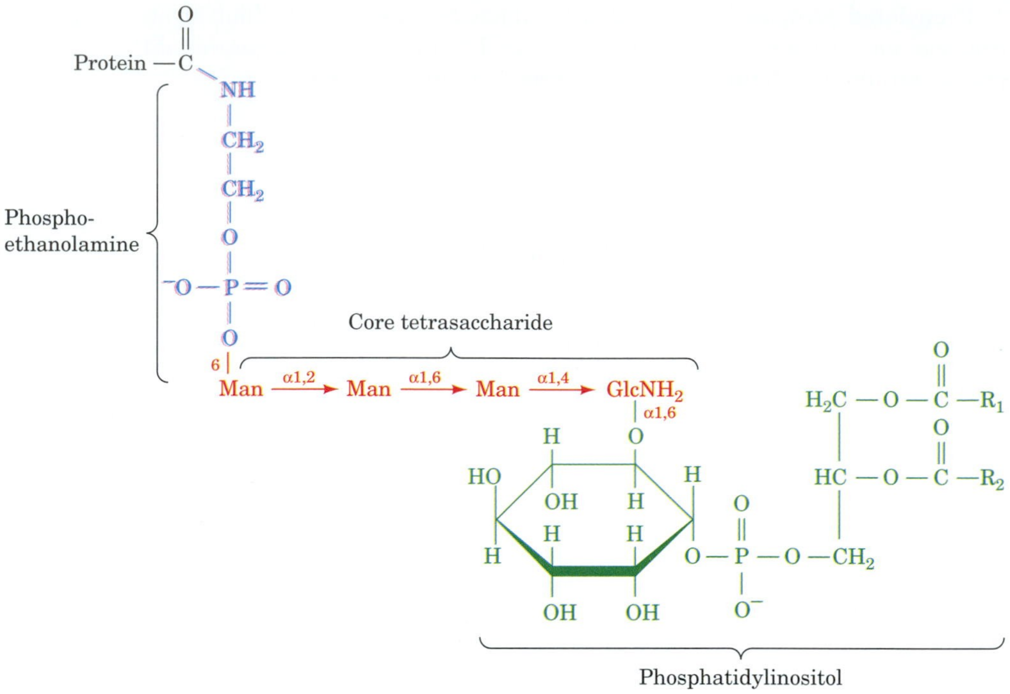

Figure 10-7.   The core structure of the GPI anchors of proteins. R_1 and R_2 represent fatty acid residues whose identities vary with the protein. The tetrasaccharide may have a variety of attached sugar residues whose identities also vary.

three mannose residues and one glucosaminyl residue (Fig. 10-7). The mannose at the nonreducing end of this assembly forms a phosphodiester bond with a phosphoethanolamine residue that is amide-linked to the protein's C-terminal carboxyl group. The core tetrasaccharide is generally substituted with a variety of sugar residues that vary with the identity of the protein. There is likewise considerable diversity in the fatty acid residues of the phosphatidylinositol group.

C. Peripheral Membrane Proteins

Peripheral or **extrinsic proteins,** unlike integral membrane proteins or lipid-linked proteins, can be dissociated from membranes by relatively mild procedures that leave the membrane intact, such as exposure to high ionic strength salt solutions or pH changes. Peripheral proteins do not bind lipid and, once purified, behave like water-soluble proteins. They associate with membranes by binding at their surfaces, most likely to integral proteins, through electrostatic and hydrogen bonding interactions. Cytochrome *c* (Sections 5-4A and 6-2C) is a peripheral membrane protein that is associated with the outer surface of the inner mitochondrial membrane.

2. MEMBRANE STRUCTURE AND ASSEMBLY

Membranes were once thought to consist of a phospholipid bilayer sandwiched between two layers of unfolded polypeptide. This sandwich model, which is improbable on thermodynamic grounds, was further discredited by the discovery of transmembrane proteins through **freeze-fracture electron microscopy.** In this technique, a membrane that is frozen in liquid nitrogen ($-196°C$) is shattered in a way that often splits the membrane between the

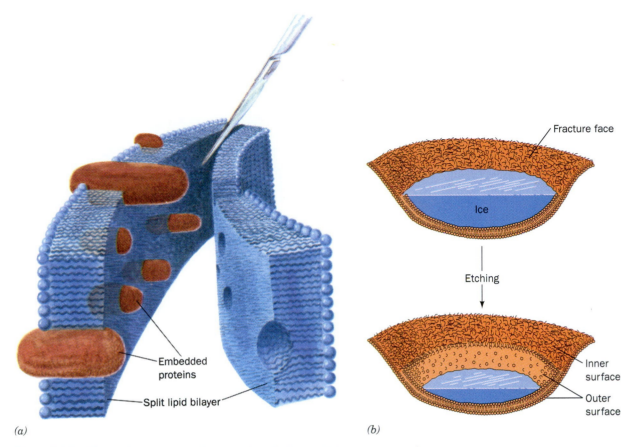

(a)

(b)

Figure 10-8. **Freeze-fracture and freeze-etch techniques.** (*a*) A membrane that has been split by freeze fracture exposes the interior of the lipid bilayer and its embedded proteins. (*b*) In the freeze-etch procedure, the ice that encases a freeze-fractured membrane (*top*) is partially sublimed away to expose the outer membrane surface (*bottom*) for electron microscopy.

leaflets of its bilayer (Fig. 10-8*a*). In a further elaboration of this technique, known as **freeze-etching,** the external surface of the membrane adjacent to the cleaved area is exposed by subliming (etching) away, at −100°C, some of the ice in which it is encased (Fig. 10-8*b*). Electron micrographs of these freeze-fractured and freeze-etched membranes (Fig. 10-9) reveal that the

Figure 10-9. **Electron micrograph of a human erythrocyte plasma membrane.** This sample was freeze-fractured and then freeze-etched to reveal the outer (polar) side of the membrane. The exposed interior face of the membrane is studded with numerous 50- to 85-Å-diameter globular particles that are integral membrane proteins. The outer surface appears smoother than the inner (nonpolar) surface because proteins do not project very far beyond the outer membrane surface. [Courtesy of Vincent Marchesi, Yale University.]

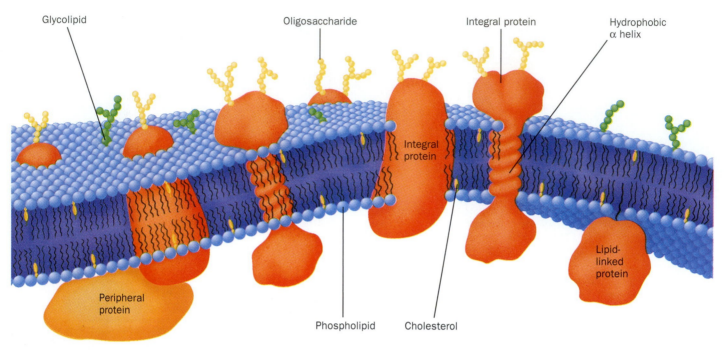

Figure 10-10.   Diagram of a plasma membrane. Integral proteins *(orange)* are embedded in a bilayer composed of phospholipids *(blue head groups attached to wiggly tails)* and cholesterol *(yellow).* The carbohydrate components *(green and yellow beads)* of glycoproteins and glycolipids occur on only the external face of the membrane.

interior face of the membrane is studded with globular particles that are integral membrane proteins, whereas the outer membrane surfaces have a relatively smooth appearance because integral membrane proteins do not protrude very far beyond them.

● See Guided Exploration 9:

Membrane Structure and the Fluid Mosaic Model.

A. The Fluid Mosaic Model

The demonstrated fluidity of artificial lipid bilayers (Section 9-2B) suggests that biological membranes have similar properties. This idea was proposed in 1972 by S. Jonathan Singer and Garth Nicolson in their unifying theory of membrane structure known as the **fluid mosaic model.** In this model, integral proteins are visualized as "icebergs" floating in a two-dimensional lipid "sea" (Fig. 10-10). A key element of the model is that integral proteins can diffuse laterally in the lipid matrix unless their movements are restricted by association with other cell components. This model of membrane fluidity explained the earlier experimental results of Michael Edidin, who fused cultured cells and observed the intermingling of their differently labeled cell-surface proteins (Fig. 10-11).

The rates of diffusion of proteins in membranes can be determined from **fluorescence photobleaching recovery** measurements. A **fluorophore** (fluorescent group) is specifically attached to a membrane component in an immobilized cell or an artificial membrane system. An intense laser pulse focused on a very small area ($\sim$3 μm^2) destroys (bleaches) the fluorophore there (Fig. 10-12). The rate at which the bleached area recovers its fluorescence, as monitored by fluorescence microscopy, indicates the rate at which unbleached and bleached fluorophore-labeled molecules laterally diffuse into and out of the bleached area.

Fluorescence photobleaching recovery measurements demonstrate that membrane proteins vary in their lateral diffusion rates. Some 30 to 90% of

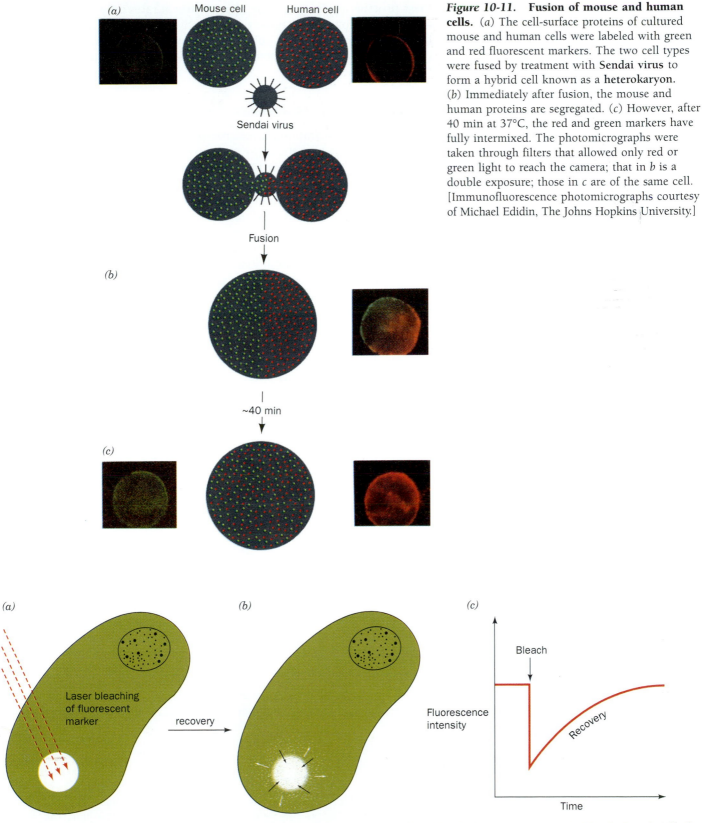

Figure 10-11. Fusion of mouse and human cells. (*a*) The cell-surface proteins of cultured mouse and human cells were labeled with green and red fluorescent markers. The two cell types were fused by treatment with **Sendai virus** to form a hybrid cell known as a **heterokaryon**. (*b*) Immediately after fusion, the mouse and human proteins are segregated. (*c*) However, after 40 min at 37°C, the red and green markers have fully intermixed. The photomicrographs were taken through filters that allowed only red or green light to reach the camera; that in *b* is a double exposure; those in *c* are of the same cell. [Immunofluorescence photomicrographs courtesy of Michael Edidin, The Johns Hopkins University.]

Figure 10-12. The fluorescence photobleaching recovery technique. (*a*) An intense laser light pulse bleaches the fluorescent markers (*green*) from a small region of an immobilized cell that has a fluorophore-labeled membrane component. (*b*) The fluorescence of the bleached area recovers as the bleached molecules laterally diffuse out of it and intact fluorescent molecules diffuse into it. (*c*) The fluorescence recovery rate depends on the diffusion rate of the labeled molecule.

these proteins are freely mobile; they diffuse at rates only an order of magnitude or so slower than those of the much smaller lipids, so they can diffuse the 20-μm length of a eukaryotic cell within an hour. Other proteins diffuse more slowly, and some are essentially immobile due to submembrane attachments.

B. The Erythrocyte Membrane

The erythrocyte membrane's relative simplicity, availability, and ease of isolation have made it the most extensively studied and best understood of biological membranes. It is therefore a model for the more complex membranes of other cell types. A mature mammalian erythrocyte is devoid of organelles and carries out few metabolic processes; it is essentially a membranous bag of hemoglobin. Erythrocyte membranes can be obtained by osmotic lysis, which causes the cell contents to leak out. The resulting membranous particles are known as erythrocyte **ghosts** because, on return to physiological conditions, they reseal to form colorless particles that retain their original shape but are devoid of cytoplasm.

Erythrocyte Membranes Contain a Variety of Proteins

The proteins of the erythrocyte membrane can be separated by SDS-PAGE (Section 5-2D) after first solubilizing the membrane in a 1% SDS solution. The resulting electrophoretogram of a human erythrocyte membrane (Fig. 10-13) exhibits a number of protein bands when treated with **Coomassie brilliant blue** (which stains proteins) or with **periodic acid–Schiff's reagent** (**PAS;** which stains carbohydrates). The polypeptides corresponding to bands 1, 2, 4.1, 4.2, 5, and 6 are readily extracted from the membrane by changes in ionic strength or pH and hence are peripheral proteins. These proteins are located on the inner side of the membrane,

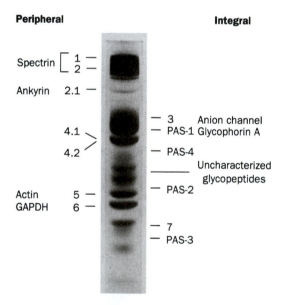

Figure 10-13.   Electrophoretogram of human erythrocyte membrane proteins. Following SDS-PAGE, proteins were stained by Coomassie brilliant blue. The minor bands are not labeled for the sake of simplicity. The positions of the four glycoproteins revealed by PAS staining are indicted. GAPDH is the glycolytic enzyme glyceraldehyde-3-phosphate dehydrogenase. [Courtesy of Vincent Marchesi, Yale University.]

consistent with the observation that they are not altered when intact erythrocytes or sealed ghosts are incubated with proteolytic enzymes or membrane-impermeable protein-labeling reagents. These proteins are altered, however, if "leaky" ghosts are so treated.

In contrast, bands 3 and 7 and all four PAS bands correspond to integral proteins; they can be released from the membrane only by extraction with detergents or organic solvents. Band 3 protein, which accounts for >30% of the membrane protein ($\sim$1 million molecules per cell), is an anion channel that facilitates the transmembrane movement of HCO_3^- (recall that the CO_2 transported in blood is converted in the erythrocyte to HCO_3^-; Section 7-2C).

The Erythrocyte's Membrane Skeleton Is Responsible for Its Shape

A normal erythrocyte's biconcave disklike shape (Fig. 7-16a) ensures the rapid diffusion of O_2 to its hemoglobin molecules by placing them no further than 1 μm from the cell surface. However, the rim and the dimple regions of an erythrocyte do not occupy fixed positions on the cell membrane. This can be demonstrated by anchoring an erythrocyte to a microscope slide by a small portion of its surface and inducing the cell to move laterally with a gentle flow of buffer. A point originally on the rim of the erythrocyte will move across the dimple to the rim on the opposite side of the cell. Evidently, the membrane rolls across the cell while maintaining its shape, much like the tread of a tractor. This remarkable mechanical property of the erythrocyte membrane results from the presence of a submembranous network of proteins that function as a membrane "skeleton."

The fluidity and flexibility imparted to an erythrocyte by its membrane skeleton has important physiological consequences. A slurry of solid particles of a size and concentration equal to that of red cells in blood has the flow characteristics approximating those of sand. Consequently, in order for blood to flow at all, much less for its erythrocytes to squeeze through capillary blood vessels smaller in diameter than they are, erythrocyte membranes, with their membrane skeletons, must be fluidlike and easily deformable.

The protein **spectrin,** so called because it was discovered in erythrocyte ghosts, accounts for $\sim$75% of the erythrocyte membrane skeleton. It is composed of two similar polypeptide chains, band 1 (α subunit; 280 kD) and band 2 (β subunit; 246 kD), which each consist of repeating 106-residue segments that are predicted to fold into triple-stranded α-helical coiled coils (Fig. 10-14). These large polypeptides are loosely intertwined to form a flexible wormlike $\alpha\beta$ dimer that is $\sim$1000 Å long. Two such heterodimers further associate in a head-to-head manner to form an $(\alpha\beta)_2$ tetramer. There are $\sim$100,000 spectrin tetramers per cell, and they are cross-linked at both ends by attachments to band 4.1 and 5 proteins to form a dense

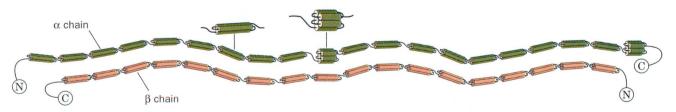

Figure 10-14.   Structure of an $\alpha\beta$ dimer of spectrin. Both of these antiparallel polypeptides contain multiple 106-residue repeats, which are thought to form triple-helical bundles that are flexibly connected by nonhelical segments. Two of these heterodimers join, head to head, to form an $(\alpha\beta)_2$ heterotetramer. [After Speicher, D.W. and Marchesi, V., *Nature* **311,** 177 (1984).]

and irregular protein meshwork that underlies the erythrocyte plasma membrane (Fig. 10-15). Band 5, a globular protein that forms filamentous oligomers, has been identified as actin, a common cytoskeletal component in other cells, including muscle (Section 7-3B). Spectrin also associates with band 2.1, an 1880-residue monomer known as **ankyrin,** which, in turn, binds to band 3, the anion channel protein. This attachment anchors the membrane skeleton to the membrane. Immunochemical studies have revealed spectrinlike, ankyrinlike, and band 4.1-like proteins in a variety of tissues.

The erythrocyte cytoskeleton is deformable, which provides these cells with the flexibility they need to squeeze through capillaries. Individuals with **hereditary spherocytosis** have spheroidal erythrocytes that are relatively fragile and inflexible. These individuals suffer from hemolytic anemia because the spleen, a labyrinthine organ with narrow passages that normally filters out aged erythrocytes (which lose flexibility toward the end of their ~120-day lifetimes), prematurely removes spherocytotic erythrocytes. The hemolytic anemia may be alleviated by surgically removing the spleen. However, the primary defect in spherocytotic cells is reduced

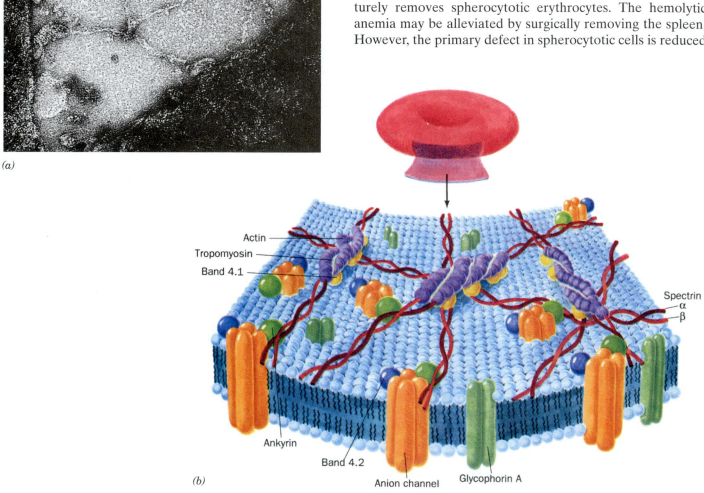

(a)

(b)

Actin
Tropomyosin
Band 4.1
Spectrin
α
β
Ankyrin
Band 4.2
Anion channel
Glycophorin A

Figure 10-15.   The human erythrocyte membrane skeleton.
(*a*) An electron micrograph of an erythrocyte membrane skeleton that has been stretched to an area 9 to 10 times greater than that of the native membrane. Stretching makes it possible to obtain clear images of the membrane skeleton, which in its native state is so densely packed and irregularly flexed that it is difficult to pick out individual molecules and to ascertain how they are interconnected. Note the predominantly hexagonal network composed of spectrin tetramers cross-linked by junctions containing actin and band 4.1 protein. [Courtesy of Daniel Branton, Harvard University.] (*b*) A model of the erythrocyte membrane skeleton. The so-called junctional complex, which is magnified in this drawing, contains actin, tropomyosin (Section 7-3B), and band 4.1 protein as well as other proteins. [After Goodman, S.R., Krebs, K.E., Whitfield, C.F., Riederer, B.M., and Zagen, I.S., *CRC Crit. Rev. Biochem.* **23**, 196 (1988).]

synthesis of spectrin, the production of an abnormal spectrin that binds band 4.1 protein with reduced affinity, or the absence of band 4.1 protein.

The Cytosol Contributes to Membrane Heterogeneity

Biological membranes, just as the fluid mosaic model postulates, are mosaics consisting of heterogeneous patches of lipids and proteins (e.g., the purple membrane, Section 10-1A). The uneven distribution of membrane components may result, in part, from the influence of the cytoskeleton. It is quite likely that some integral membrane proteins are firmly attached to elements of the cytoskeleton or are trapped within the spaces defined by those "fences." Other membrane proteins may be able to squeeze through gaps or "gates" between cytoskeletal components, whereas still other proteins can diffuse freely without interacting with the cytoskeleton at all (Fig. 10-16). Support for this **gates and fences model** comes from the finding that partial destruction of the cytoskeleton results in freer protein diffusion.

Proteins confined by the fences of cytoskeletal elements may form distinct membrane domains. Even the aggregation of a few proteins can result in a microdomain whose properties differ from those of surrounding areas. Specific interactions between proteins and the head groups or tails of nearby lipids can lead to heterogeneity in membrane lipid composition. Many membrane proteins are believed to be surrounded by a ring of specific lipids, called **annular lipids.** Some membrane domains appear to be formed primarily by the aggregation of particular species of lipids, producing, for example, regions that are rich in glycosphingolipids, sphingomyelin, and cholesterol but lacking glycerophospholipids.

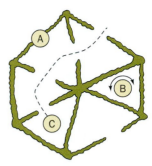

Figure 10-16. Model rationalizing the various mobilities of membrane proteins. Protein A, which interacts tightly with the underlying cytoskeleton, is immobile. Protein B is free to rotate within the confines of the cytoskeletal "fences." Protein C diffuses by traveling through "gates" in the cytoskeleton. The diffusion of some membrane proteins is not affected by the cytoskeleton. [After Edidin, M., *Trends Cell Biol.* **2**, 378 (1992).]

C. Lipid Asymmetry

Membrane components have specific orientations so that the two leaflets of the bilayer are not equivalent in composition or in function. For example, *membrane glycoproteins and glycolipids are invariably oriented with their carbohydrate moieties facing the cell's exterior.* The distribution of certain lipids between the inner and outer leaflets of a membrane can be established through the use of phospholipases (Section 9-1C). Phospholipases cannot pass through membranes, so phospholipids on only the external surfaces of intact cells are susceptible to hydrolysis by these enzymes. Such studies reveal that lipids in biological membranes are asymmetrically distributed (e.g., Fig. 10-17). How does this asymmetry arise?

In eukaryotes, the enzymes of membrane lipid biosynthesis are mostly integral membrane proteins of the **endoplasmic reticulum (ER,** the inter-

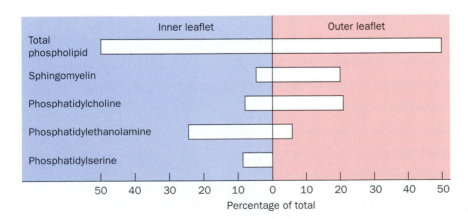

Figure 10-17. Asymmetric distribution of membrane phospholipids in the human erythrocyte membrane. The phospholipid content is expressed as mol %. [After Rothman, J.E. and Lenard, J., *Science* **194**, 1744 (1977).]

connected membranous vesicles that occupy much of the cytosol; Fig. 1-8). In prokaryotes, lipids are synthesized in the plasma membrane. Hence, membrane lipids are fabricated on site. Eugene Kennedy and James Rothman demonstrated this to be the case in bacteria through the use of selective labeling. They gave growing bacteria a 1-minute pulse of $^{32}PO_4^{3-}$ in order to radioactively label the phosphoryl groups of only the newly synthesized phospholipids. **Trinitrobenzenesulfonic acid (TNBS),** a membrane-impermeable reagent that combines with phosphatidylethanolamine (**PE;** Fig. 10-18), was then immediately added to the cell suspension. Analysis of the resulting doubly labeled membranes showed that none of the TNBS-labeled PE was radioactively labeled. This observation indicates that *newly made PE is synthesized on the cytoplasmic face of the membrane* (Fig. 10-19, *lower left*).

However, if an interval of only 3 minutes was allowed to elapse between the $^{32}PO_4^{3-}$ pulse and the TNBS addition, about half of the ^{32}P-labeled PE was also TNBS labeled (Fig. 10-19, *lower right*). This observation indicates that the flip-flop rate of PE in the bacterial membrane is ~100,000-fold greater than it is in bilayers consisting of only phospholipids (where the flip-flop rates have half-times of many days).

How do phospholipids synthesized on one side of the membrane reach its other side so quickly? Phospholipid flip-flops appear to be facilitated in two ways:

1. Membranes contain proteins known as **flipases** that catalyze the flip-flops of specific phospholipids. These proteins tend to equilibrate the distribution of their corresponding phospholipids across a bilayer; that is, the net transport of a phospholipid is from the side of the bilayer with the higher concentration of the phospholipid to the opposite side. Such a process, as we shall see in Section 10-4B, is a form of **facilitated diffusion.**

2. Membranes contain proteins known as **phospholipid translocases** that transport specific phospholipids across a bilayer in a process that is driven by ATP hydrolysis. These proteins can transport certain phos-

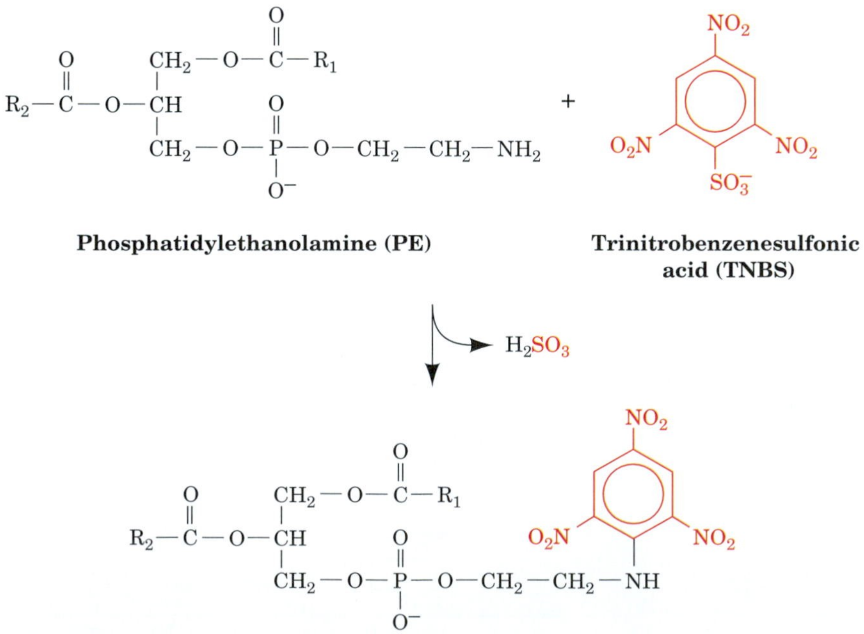

Figure 10-18. The reaction of TNBS with phosphatidylethanolamine.

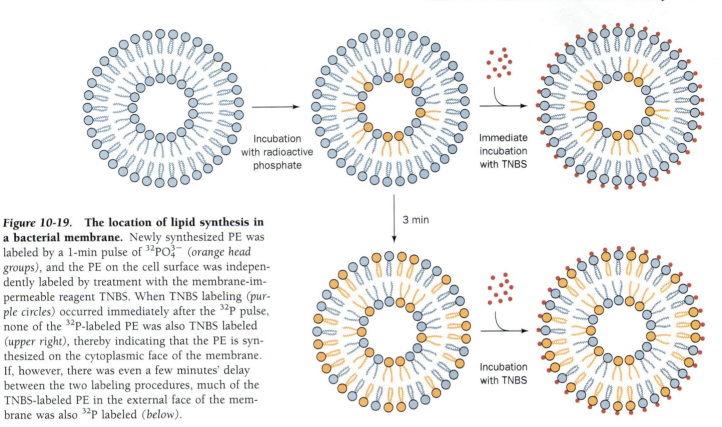

Figure 10-19. The location of lipid synthesis in a bacterial membrane. Newly synthesized PE was labeled by a 1-min pulse of $^{32}PO_4^{3-}$ *(orange head groups)*, and the PE on the cell surface was independently labeled by treatment with the membrane-impermeable reagent TNBS. When TNBS labeling *(purple circles)* occurred immediately after the ^{32}P pulse, none of the ^{32}P-labeled PE was also TNBS labeled *(upper right)*, thereby indicating that the PE is synthesized on the cytoplasmic face of the membrane. If, however, there was even a few minutes' delay between the two labeling procedures, much of the TNBS-labeled PE in the external face of the membrane was also ^{32}P labeled *(below)*.

Labels within figure: Incubation with radioactive phosphate · Immediate incubation with TNBS · 3 min · Incubation with TNBS

pholipids from the side of a bilayer that has the lower concentration of the phospholipid to the opposite side, thereby establishing a non-equilibrium distribution of the phospholipid. Such a process, as we shall see in Section 10-4C, is a form of **active transport.**

The observed distribution of phospholipids across membranes (e.g., Fig. 10-17) therefore appears to arise from the membrane orientations of the enzymes that synthesize phospholipids combined with the countervailing tendencies of ATP-dependent phospholipid translocases to generate asymmetric phospholipid distributions and those of flipases to equilibrate these distributions. The importance of these lipid transport systems is demonstrated by the observation that the presence of phosphatidylserine on the exteriors of many cells induces blood clotting (i.e., it is an indication of tissue damage) and, in erythrocytes, marks the cell for removal from the circulation.

In all cells, *new membranes are generated by the expansion of existing membranes.* In eukaryotic cells, lipids synthesized on the cytoplasmic face of the ER are transported to other parts of the cell by vesicles that bud off from the ER and fuse with other cellular membranes. These vesicles also carry membrane proteins.

D. The Secretory Pathway

Membrane proteins, as are all proteins, are ribosomally synthesized under the direction of messenger RNA templates (translation is discussed in Chapter 26). The polypeptide grows from its N-terminus to its C-terminus by the stepwise addition of amino acid residues. Ribosomes may be free in the cytosol or bound to the ER to form the **rough endoplasmic reticulum**

(**RER,** so called because of the knobby appearance its bound ribosomes give it; Fig. 1-8). *Free ribosomes synthesize mostly soluble and mitochondrial proteins, whereas membrane-bound ribosomes manufacture transmembrane proteins and proteins destined for secretion, operation within the ER, and incorporation into **lysosomes*** (Fig. 1-8; membranous vesicles containing a battery of hydrolytic enzymes that degrade and recycle cell components). These latter proteins initially appear in the ER.

The Signal Hypothesis

The **signal hypothesis,** which was formulated by Günter Blobel, César Milstein, and David Sabatini, partially explains how large, relatively polar polypeptides pass through the RER membrane as they are synthesized (Fig. 10-20):

1. *All secreted ER and lysosomal proteins, as well as many transmembrane proteins, are synthesized with N-terminal 13-to 36-residue **signal peptides.*** These signal peptides consist of a 7- to 13-residue hydrophobic core flanked by several relatively hydrophilic residues (Fig. 10-21). Signal peptides otherwise have little sequence homology.

2. The signal peptide first protrudes beyond the ribosomal surface after ~80 residues have been linked together. At this point, the **signal recognition particle (SRP),** a 325-kD complex of six different polypeptides and a 300-nucleotide RNA molecule, binds to the ribosome (the RNA may facilitate the binding of the SRP to the ribosome by binding to ribosomal RNA). This arrests further polypeptide growth.

3. The SRP–ribosome complex diffuses to the RER surface, where it is bound by a transmembrane protein, the **SRP receptor (docking**

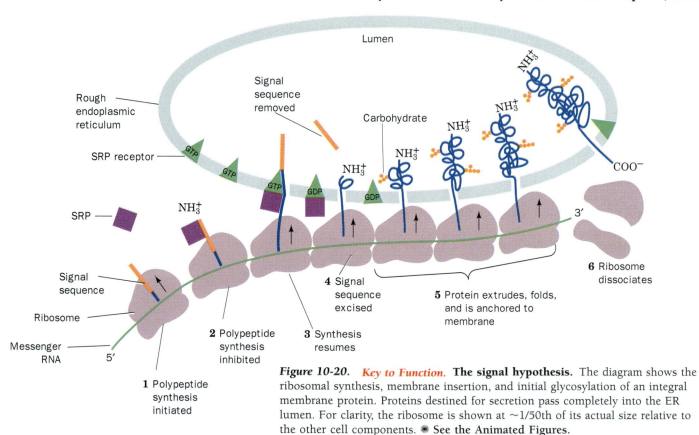

Figure 10-20. ***Key to Function.*** **The signal hypothesis.** The diagram shows the ribosomal synthesis, membrane insertion, and initial glycosylation of an integral membrane protein. Proteins destined for secretion pass completely into the ER lumen. For clarity, the ribosome is shown at ~1/50th of its actual size relative to the other cell components. ✴ See the Animated Figures.

Signal
peptidase
cleavage
site

Bovine growth hormone	M M A A G P R T S L L L A F A L L C L P W T Q V V G	A F P
Bovine proalbumin	M K W V T F I S L L L L F S S A Y S	R G V
Human proinsulin	M A L W M R L L P L L A L L A L W G P D P A A A	F V N
Human interferon-γ	M K Y T S Y I L A F Q L C I Y L G S L G	C Y C
Human α-fibrinogen	M F S M R I V C L V L S V V G T A W T	A D S
Human IgG heavy chain	M E F G L S W L F L V A I L K G V Q C	E V Q
Rat amylase	M K F V L L L S L I G F C W A	Q Y D
Murine α-fetoprotein	M K W I T P A S L I L L L L H F A A S K	A L H
Chicken lysozyme	M R S L L I L V L C F L P L A A L G	K V F
Zea mays rein protein 22.1	M A T K I L A L L A L L A L L V S A T N A	F I I

Figure 10-21.   The N-terminal sequences of some eukaryotic secretory preproteins. The hydrophobic cores *(brown)* of most signal peptides are preceded by basic residues *(blue)*. The one-letter code for amino acid residues is given in Table 4-1. [After Watson, M.E.E., *Nucleic Acids Res.* **12**, 5147–5156 (1984).]

protein). Binding stimulates the ribosome to resume polypeptide synthesis and facilitates the passage of the growing polypeptide's N-terminus into the lumen of the RER. The polypeptide passes through the membrane via a transmembrane protein channel. The signal peptide, SRP, and SRP receptor then dissociate. In the process, **guanosine triphosphate (GTP)** is hydrolyzed to **guanosine diphosphate (GDP)** plus inorganic phosphate (P_i). Most ribosomal processes, as we shall see in Section 26-4, are driven by GTP hydrolysis.

4. Shortly after the signal peptide enters the RER lumen, it is specifically cleaved from the growing polypeptide by a membrane-bound **signal peptidase.** The polypeptide with its attached signal peptide is therefore called a **preprotein.**

5. The nascent (growing) polypeptide starts to fold to its native conformation, a process that is facilitated by its interaction with the chaperone protein Hsp70 (Section 6-4C). Enzymes in the lumen then initiate **post-translational modification** of the polypeptide, including the attachment of core oligosaccharides to form glycoproteins (Section 8-3C) and the formation of disulfide bonds as catalyzed by protein disulfide isomerase (Section 6-4C).

6. When polypeptide synthesis is completed, the ribosome dissociates from the RER. Secretory, ER, and lysosomal proteins pass completely through the RER membrane into the lumen. Transmembrane proteins, in contrast, contain a hydrophobic ~20-residue "membrane anchor" or "stop-transfer" sequence that arrests the passage of the growing polypeptide chain through the membrane. *Transmembrane proteins therefore remain embedded in the ER membrane with their C-termini on its cytoplasmic side.*

The signal hypothesis also applies to bacteria. Proteins that traverse the bacterial plasma membrane have N-terminal signal peptides similar to those of eukaryotes and are synthesized by membrane-bound ribosomes. In gram-negative bacteria, the use of recombinant DNA techniques (Section 3-5) to add a signal peptide to a normally cytoplasmic protein causes the hybrid protein to be transported to the **periplasmic compartment** (the space between the plasma membrane and the cell wall; Fig. 8-14*b*). However, the signal hypothesis does not account for the orientation of all integral membrane proteins, for example, bacteriorhodopsin (Fig. 10-4), which has multiple membrane-spanning segments.

In eukaryotes, cytoplasmically synthesized proteins destined for the mitochondrial **matrix** (its inner compartment) must traverse two membranes (some chloroplast proteins must cross three chloroplast membranes). The translocation of some mitochondrial matrix proteins requires molecular chaperones such as Hsp70 (Section 6-4C) and an import receptor in the outer mitochondrial membrane. However, other mitochondrial proteins do not require these mediators and reach their final destination by other mechanisms.

Proteins Are Targeted to Specific Destinations

Sometime after their synthesis on the RER, partially processed transmembrane, secretory, and lysosomal proteins appear in the **Golgi apparatus** (Fig. 1-8), an organelle consisting of a stack of flattened and functionally distinct membranous sacs. Proteins embedded in the membrane or free in the lumen move from one Golgi compartment to another for further post-translational processing, mostly glycosylation (Fig. 10-22). In the final Golgi compartment, the processed proteins are sorted and sent to their final cellular destinations.

The vehicles in which proteins are transported from the Golgi apparatus to their final destinations are known as **coated vesicles.** This is because these membranous vesicles are enclosed on their outer (cytosolic) face by a polyhedral framework of the protein **clathrin** (Fig. 10-23), which is thought to act as a flexible scaffolding in promoting the budding off of the vesicle from its membrane of origin. However, vesicles that ferry proteins between the ER and the Golgi apparatus are not coated with clathrin but are instead surrounded with **COPI** or **COPII** (COP stands for *co*at *p*rotein). COPI- and COPII-coated vesicles have a fuzzy appearance under the electron microscope (Fig. 10-24) rather than the polyhedral shell of clathrin-coated vesicles (Fig. 10-23).

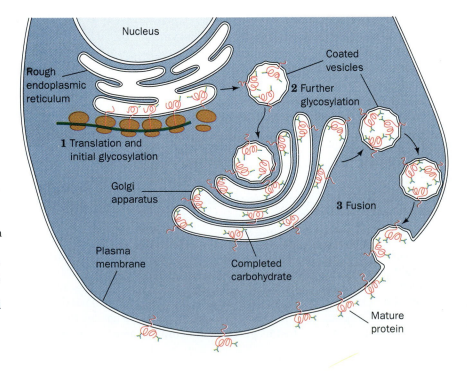

Figure 10-22. *Key to Function.* **The posttranslational processing of integral membrane proteins.** (1) As the protein is being synthesized by a ribosome, glycosylation is initiated in the lumen of the ER. (2) After ribosomal synthesis is completed, coated vesicles containing the protein bud off from the ER and move to the Golgi apparatus, where protein processing is completed. (3) Later, coated vesicles containing the mature protein bud off from the Golgi apparatus and fuse to the membrane for which the protein is targeted, here shown as the plasma membrane.

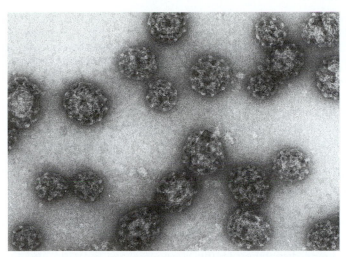

(a)

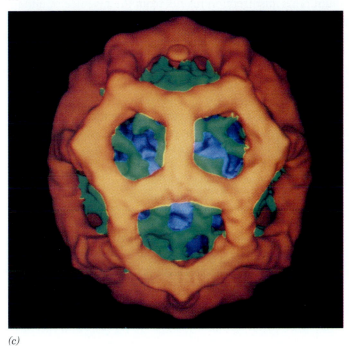

(c)

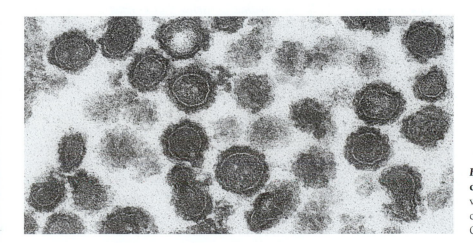

(b)

Figure 10-23. **Clathrin-coated vesicles.** A polyhedral framework of clathrin and its associated proteins form a cage around a membranous sac. (*a*) An electron micrograph of coated vesicles. [Courtesy of Barbara Pearse, Medical Research Council, Cambridge, U.K.] (*b*) An electron micrograph of **triskelions**, the three-legged protein complexes that assemble to form the clathrin cage. The variable orientations of the triskelion legs indicate their flexibility. [Courtesy of Daniel Branton, Harvard University.] (*c*) A three-dimensional map, generated from electron micrographs, of a clathrin coat. The polyhedral clathrin coat is shown in orange, the clathrin terminal domains are green, and an inner shell of accessory proteins is blue. Each vertex of the polyhedron is the center of a triskelion, and its edges, which are ~150 Å long, are formed by the overlapping legs of adjoining triskelions. Such frameworks, which consist of 12 pentagons and a variable number of hexagons, are the most parsimonious way of enclosing spheroidal objects in polyhedral cages. [Courtesy of Barbara Pearse, Medical Research Council, Cambridge, U.K.]

Figure 10-24. **Electron micrograph of COP-coated vesicles.** Note how the surface of the vesicles lacks the polyhedral quality of clathrin-coated vesicles (Fig. 10-23*a*). [Courtesy of Lelio Orci, University of Geneva, Switzerland.]

In all cases, *the budding of a vesicle and its fusion with another membrane preserves the orientation of its transmembrane proteins. Thus, the lumens of the ER and Golgi apparatus are topologically equivalent to the outside of the cell. This explains why the carbohydrate moieties of integral membrane glycoproteins occur only on the external surfaces of plasma membranes.*

How are proteins in the ER selected for transport to the Golgi apparatus and from there to their respective membranous destinations? A clue as to the nature of this process is provided by the human hereditary defect known as **I-cell disease,** which, in homozygotes, is characterized by progressive psychomotor retardation, skeletal deformities, and early death. The lysosomes in the connective tissue of I-cell disease victims contain large inclusions (after which the disease is named) of glycosaminoglycans and glycolipids as a result of the absence of several lysosomal hydrolases. These enzymes are synthesized on the ER with their correct amino acid sequences but, rather than being dispatched to the lysosomes, are secreted into the extracellular medium. This misdirection results from the absence of a mannose-6-phosphate recognition marker on the carbohydrate moieties of these hydrolases because of a deficiency in an enzyme required for mannose phosphorylation. The mannose-6-phosphate residues are bound by a receptor in the coated vesicles that transport lysosomal hydrolases from the Golgi apparatus to the lysosomes.

Most soluble ER-resident proteins in mammals have the C-terminal sequence Lys-Asp-Glu-Leu (KDEL, using the one-letter amino acid symbols; Table 4-1). Altering this sequence causes the protein to be secreted. KDEL proteins, like secretory and lysosomal proteins, readily leave the ER via membranous vesicles, but KDEL proteins are promptly retrieved from some later compartment and returned to the ER.

3. LIPOPROTEINS AND RECEPTOR-MEDIATED ENDOCYTOSIS

Lipids, such as phospholipids, triacylglycerols, and cholesterol, are only sparingly soluble in aqueous solution. Hence, *they are transported by the circulation in complex with proteins and are taken up by cells in a receptor-mediated process.*

A. Lipoprotein Structure

Lipoproteins are globular micellelike particles that consist of a nonpolar core of triacylglycerols and cholesteryl esters surrounded by an amphiphilic coating of protein, phospholipid, and cholesterol. There are five classes of lipoproteins (Table 10-1):

1. **Chylomicrons,** which transport exogenous (externally supplied; in this case, dietary) triacylglycerols and cholesterol from the intestines to the tissues.

2–4. **Very low density lipoproteins (VLDL), intermediate density lipoproteins (IDL),** and **low density lipoproteins (LDL),** a group of related particles that transport endogenous (internally produced) triacylglycerols and cholesterol from the liver to the tissues.

5. **High density lipoproteins (HDL),** which transport endogenous cholesterol from the tissues to the liver.

Table 10-1. **Characteristics of the Major Classes of Lipoproteins in Human Plasma**

	Chylomicrons	VLDL	IDL	LDL	HDL
Density (g·cm^{-3})	<0.95	<1.006	1.006–1.019	1.019–1.063	1.063–1.210
Particle diameter (Å)	750–12,000	300–800	250–350	180–250	50–120
Particle mass (kD)	400,000	10,000–80,000	5000–10,000	2300	175–360
% Protein[a]	1.5–2.5	5–10	15–20	20–25	40–55
% Phospholipids[a]	7–9	15–20	22	15–20	20–35
% Free cholesterol[a]	1–3	5–10	8	7–10	3–4
% Triacylglyceorols[b]	84–89	50–65	22	7–10	3–5
% Cholesteryl esters[b]	3–5	10–15	30	35–40	12
Major apolipoproteins	A-I, A-II, B-48, C-I, C-II, C-III, E	B-100, C-I, C-II, C-III, E	B-100, C-III, E	B-100	A-I, A-II, C-I, C-II, C-III, D, E

[a]Surface components.
[b]Core lipids.

The physiological roles of the various lipoproteins and their metabolic processing are described in detail in Section 19-1.

Each lipoprotein contains just enough protein, phospholipid, and cholesterol to form an ~20-Å-thick monolayer of these substances on the particle surface (Fig. 10-25). Lipoprotein densities increase with decreasing particle diameter because the density of their outer coating is greater than that of their inner core. Thus, the HDL, which are the most dense of the lipoproteins, are also the smallest.

Apolipoproteins Have Amphipathic Helices That Coat Lipoprotein Surfaces

The protein components of lipoproteins are known as **apolipoproteins** or just **apoproteins.** At least nine apolipoproteins are distributed in different amounts in the human lipoproteins (Table 10-1). Most of the apolipoproteins are water soluble and associate rather weakly with lipoproteins. These apolipoproteins also have a high helix content, which increases when they are incorporated into lipoproteins. Apparently, the helices are stabilized by a lipid environment, because helices fully satisfy the backbone's hydrogen bonding potential in the lipoprotein's water-free interior.

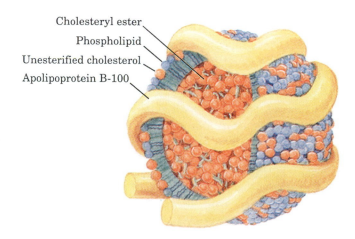

Cholesteryl ester
Phospholipid
Unesterified cholesterol
Apolipoprotein B-100

Figure 10-25. **Diagram of LDL, the major cholesterol carrier of the bloodstream.** This spheroidal particle consists of some 1500 cholesteryl ester molecules surrounded by an amphiphilic coat of ~800 phospholipid molecules, ~500 cholesterol molecules, and a single 550-kD molecule of apolipoprotein B-100.

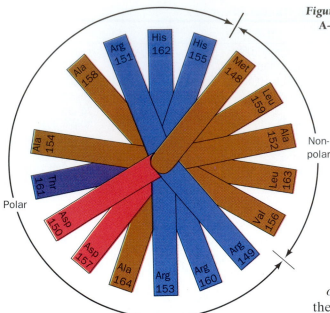

Polar

Non-polar

Figure 10-26. **A helical wheel projection of a portion of apolipoprotein A-I.** The postulated amphipathic α helix constitutes residues 148 to 164. In a helical wheel representation, the side chain positions are projected down the helix axis onto a plane. Note the segregation of nonpolar and polar residues to different sides of the helix, as well as the segregation of basic residues to the outer edges of the polar surface and the acidic residues to the center. Other apolipoprotein helices have similar polarity distributions. [After Kaiser, E.T., *in* Oxender, D.L. and Fox, C.F. (Eds.), *Protein Engineering,* p. 194, Liss (1987).]

Apolipoprotein A-I, which occurs in chylomicrons and HDL, is a 243-residue, 29-kD polypeptide. It consists largely of six tandem 22-residue segments of similar sequence that each have a high propensity for forming an α helix followed by a β turn (Section 6-1D). *These putative α helices, as well as similar helices that occur in most other apolipoproteins, have their hydrophobic and hydrophilic residues on opposite sides of the helical cylinder* (Fig. 10-26). Furthermore, the polar helix face has a dipolar character because its negatively charged residues project from the center of this face, whereas its positively charged residues are located at its edges. This suggests that *lipoprotein α helices float on phospholipid surfaces, much like logs on water.* The phospholipids are arrayed with their charged groups bound to oppositely charged residues on the polar face of the helix and with the first few methylene groups of their fatty acid residues in hydrophobic association with the nonpolar face of the helix.

Apolipoprotein B-100 (apoB-100), a 4536-residue monomer (and thus one of the largest monomeric proteins known), has a hydrophobicity approaching that of integral membrane proteins and contains relatively few amphipathic helices. Hence, in contrast to the other less hydrophobic plasma apolipoproteins, apoB-100 is not water soluble. Each LDL particle contains only one molecule of apoB-100, which immunoelectron microscopy indicates has an extended form that covers at least half of the particle surface (Fig. 10-25).

B. Receptor-Mediated Endocytosis of LDL

Cholesterol, as we have seen, is an essential component of animal cell membranes. Cells can obtain cholesterol either exogenously or, if this source is insufficient, through its internal synthesis (Section 19-7A). Michael Brown and Joseph Goldstein have demonstrated that *cells obtain exogenous cholesterol mainly through the **receptor-mediated endocytosis** (engulfment) of LDL in a process that occurs as follows* (Fig. 10-27): The LDL particles are sequestered by **LDL receptors,** cell-surface transmembrane glycoproteins that specifically bind apoB-100. LDL receptors cluster into **coated pits,** which gather the cell-surface receptors that are destined for endocytosis while excluding other cell-surface proteins. The coated pits, which have a clathrin backing, invaginate from the plasma membrane to form coated vesicles (Fig. 10-28). Next, the vesicles, minus their clathrin coats, fuse with **endosomes,** whose internal pH is ~5.0. Under these conditions, LDL dissociates from its receptor. The receptors are recycled back to the cell surface, while the endosome with enclosed LDL fuses with a lysosome. In the lysosome, LDL's apoB-100 is rapidly degraded to its component amino acids, and the cholesteryl esters are hydrolyzed to yield cholesterol and fatty acids.

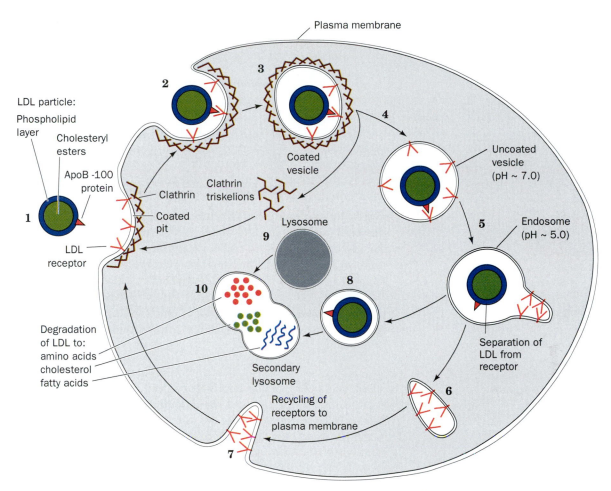

Figure 10-27. *Key to Function.* **Receptor-mediated endocytosis of LDL.** The LDL specifically binds to LDL receptors on coated pits (**1**). These bud into the cell (**2**) to form coated vesicles (**3**) whose clathrin coats depolymerize as triskelions, resulting in the formation of smooth-surfaced vesicles (**4**). These vesicles then fuse with vesicles called endosomes (**5**), which have an internal pH of ∼5.0. The acidity induces the LDL to dissociate from its receptor. The LDL accumulates in the vesicular portion of the endosome, whereas the LDL receptors concentrate in the membrane of an attached tubular structure, which then separates from the endosome (**6**) and subsequently recycles the LDL receptors to the plasma membrane (**7**). The vesicular portion of the endosome (**8**) fuses with a lysosome (**9**), yielding **a secondary lysosome** (**10**) wherein the apoB-100 component of the LDL and the cholesteryl esters are hydrolyzed.

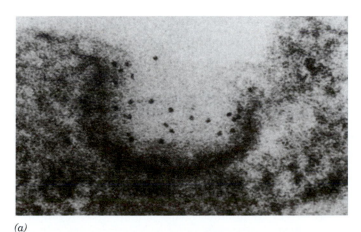

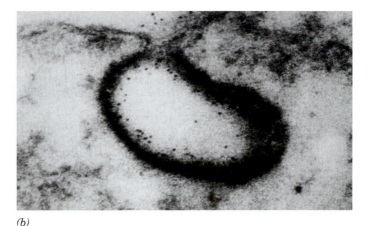

(a)

(b)

Figure 10-28. **Electron micrographs showing the endocytosis of LDL by cultured human fibroblasts.** The LDL was conjugated to ferritin (an iron-carrying protein) so that it appears as dark dots. (*a*) LDL binds to a coated pit on the cell surface.

(*b*) The coated pit invaginates and begins to pinch off from the cell membrane to form a coated vesicle enclosing the bound LDL. [From Anderson, R.G.W., Brown, M.S., and Goldstein, J.L., *Cell* **10**, 356 (1977). Copyright © by Cell Press.]

Box 10-1

BIOCHEMISTRY IN HEALTH AND DISEASE

Cholesterol and Atherosclerosis

The most common cause of human death in Western industrialized countries is **myocardial infarction (heart attack)**, in which the blockage of blood flow causes the death of heart tissue. Blockage of the coronary arteries is commonly the result of **atherosclerosis,** a progressive disease that begins as intracellular lipid deposits in the smooth muscle cells of the inner arterial wall. These lesions eventually become fibrous, calcified plaques that narrow and even block the arteries. The initial arterial thickenings contain almost pure cholesteryl ester. Accordingly, the development of atherosclerosis is strongly correlated with the level of plasma cholesterol.

Cholesterol delivered to the cells via LDL is incorporated into the cell membranes. Excess cholesterol is converted to cholesteryl esters and transported via HDL to the liver, where it is disposed of in the form of **bile acids** that are secreted into the small intestine (where these detergentlike molecules facilitate the digestion and absorption of lipids). The amount of circulating cholesterol that can contribute to atherosclerotic lesions depends on how cholesterol is partitioned between LDL and HDL. For example, women typically have HDL levels higher than men and also less heart disease. Cholesterol levels also vary with diet, stress, and the level of endogenous cholesterol synthesis.

The cells of homozygotes with the inherited disease **familial hypercholesterolemia (FH)** are unable to take up LDL because they genetically lack the LDL receptor protein. These individuals therefore have plasma cholesterol levels that are much greater than the average level of ~175 mg/100 mL. The excess cholesterol is deposited in the skin and tendons in the form of yellow nodules known as **xanthomas** and, more importantly, in the arteries. Consequently, these individuals develop symptoms of coronary artery disease in early childhood. Heterozygotes with FH (~1 person in 500) are less severely affected: They develop the symptoms of coronary artery disease after the age of 30.

Nongenetic risk factors for heart disease include cigarette smoking. Cigarette smoke oxidizes LDL, which promotes its uptake by macrophages (a type of white blood cell) in the arterial walls. Thus, smokers are more likely to develop atherosclerotic plaques than nonsmokers.

The clinical management of cholesterol levels is not simple because of the many factors that affect cholesterol synthesis, transport, and deposition in arteries. In some cases, reducing dietary cholesterol decreases plasma cholesterol levels. Substances that selectively absorb bile acids (the cholesterol degradation products that normally recycle between the liver and the intestine) may also help reduce cholesterol levels. By far the most effective treatment for atherosclerosis, although hardly a cure, is the administration of drugs that inhibit cholesterol biosynthesis (Section 19-7A).

Receptor-mediated endocytosis is a general mechanism whereby cells take up large molecules, each through a corresponding specific receptor. Many cell-surface receptors recycle between the plasma membrane and the endosomal compartment as does the LDL receptor, even in the absence of ligand. The LDL receptor cycles in and out of the cell about every 10 minutes. Defects in the LDL receptor system lead to abnormally high levels of circulating cholesterol with the attendant increased risk of heart disease (see Box 10-1).

4. TRANSPORT ACROSS MEMBRANES

Metabolic reactions occur within cells or organelles that are separated from their surroundings by membranes. The nonpolar cores of biological membranes make them highly impermeable to most ionic and polar substances, which can traverse membranes only through the action of specific **transport proteins.** Such proteins mediate all transmembrane movements of small inorganic ions as well as metabolites such as amino acids, sugars, and nucleotides. More complicated processes (e.g., endocytosis) are required to move larger substances such as proteins and macromolecular aggregates across membranes. In this section, we discuss the thermodynamics and chemical mechanisms of some membrane transport systems.

A. Thermodynamics of Transport

The diffusion of a substance between two sides of a membrane

$$A(out) \rightleftharpoons A(in)$$

thermodynamically resembles a chemical equilibration. We saw in Section 1-4D that the free energy of a solute, A, varies with its concentration:

$$\overline{G}_A - \overline{G}_A^{\circ\prime} = RT \ln [A] \qquad [10\text{-}1]$$

where $\overline{G}_A$ is the **chemical potential** (partial molar free energy) of A (the bar indicates quantity per mole) and $\overline{G}_A^{\circ\prime}$ is the chemical potential of its standard state. Thus, a difference in the concentrations of the substance on two sides of a membrane generates a **chemical potential difference:**

$$\Delta\overline{G}_A = \overline{G}_A(in) - \overline{G}_A(out) = RT \ln \left(\frac{[A]_{in}}{[A]_{out}}\right) \qquad [10\text{-}2]$$

Consequently, if the concentration of A outside the membrane is greater than that inside, $\Delta\overline{G}_A$ for the transfer of A from outside to inside will be negative and the spontaneous net flow of A will be inward. If, however, [A] is greater inside than outside, $\Delta\overline{G}_A$ is positive and an inward net flow of A can occur only if an exergonic process, such as ATP hydrolysis, is coupled to it to make the overall free energy change negative.

The transmembrane movement of ions also results in charge differences across the membrane, thereby generating an electrical potential difference, $\Delta\Psi = \Psi(in) - \Psi(out)$, where $\Delta\Psi$ is termed the **membrane potential.** Consequently, if A is ionic, Eq. 10-2 must be amended to include the electrical work required to transfer a mole of A across the membrane from outside to inside:

$$\Delta\overline{G}_A = RT \ln \left(\frac{[A]_{in}}{[A]_{out}}\right) + Z_A \mathscr{F} \Delta\Psi \qquad [10\text{-}3]$$

where Z_A is the ionic charge of A; $\mathscr{F}$, the Faraday constant, is the charge of a mole of electrons ($96,485 \ C \cdot mol^{-1}$; C is the symbol for coulomb); and $\overline{G}_A$ is now termed the **electrochemical potential** of A. The membrane potentials of living cells are commonly as high as $-100 \ mV$ (inside negative; note that $1 \ V = 1 \ J \cdot C^{-1}$). Hence, the last term in Eq. 10-3 is often significant for ionic substances, particularly in mitochondria and chloroplasts, as we shall see in Chapters 17 and 18.

Transport May Be Mediated or Nonmediated

There are two types of transport processes: **nonmediated transport** and **mediated transport.** Nonmediated transport occurs through simple diffusion. In contrast, mediated transport occurs through the action of specific carrier proteins. The driving force for the nonmediated flow of a substance through a medium is its chemical potential gradient. Thus, *the substance diffuses in the direction that eliminates its concentration gradient, at a rate proportional to the magnitude of this gradient. The rate of diffusion of a substance also depends on its solubility in the membrane's nonpolar core.* Thus, nonpolar molecules such as steroids and O_2 readily diffuse through biological membranes by nonmediated transport, according to their concentration gradients across the membranes. Surprisingly, *water diffuses easily across lipid bilayers.* This is due not just to the high concentration of water molecules but also to the presence of proteins called **aquaporins** that provide a path for water molecules to diffuse across the membrane.

Mediated transport is classified into two categories depending on the thermodynamics of the system:

1. **Passive-mediated transport,** or facilitated diffusion, in which a specific molecule flows from high concentration to low concentration.

2. **Active transport,** in which a specific molecule is transported from low concentration to high concentration, that is, against its concentration gradient. Such an endergonic process must be coupled to a sufficiently exergonic process to make it favorable (i.e., $\Delta G < 0$).

B. *Mechanisms of Mediated Transport*

Substances that are too large or too polar to diffuse across lipid bilayers on their own may be conveyed across membranes via proteins or other molecules that are variously called **carriers, permeases, channels,** and **transporters.**

Ionophores Facilitate Ion Diffusion

Some bacterial compounds are **ionophores,** molecules that increase the permeability of membranes to ions. These molecules often exert an antibiotic effect by discharging the ion concentration gradients that healthy cells actively maintain. **Carrier ionophores** bind ions and diffuse through the membrane to release them on the other side (Fig. 10-29a). **Channel-forming ionophores** are small proteins that form solvent-filled transmembrane channels or pores though which their selected ions can diffuse (Fig. 10-29b). Note that since ionophores passively permit ions to diffuse across a membrane in either direction, they can only dissipate—not generate—an ion concentration gradient.

Valinomycin, a well-characterized carrier ionophore, contains D- and L- amino acid residues linked in a circular arrangement. A K^+ ion is octahe-

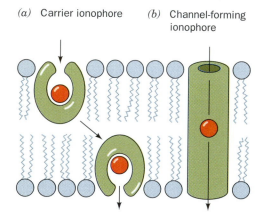

(a) Carrier ionophore   *(b)* Channel-forming ionophore

Figure 10-29.   Ionophore action. (*a*) Carrier ionophores transport ions by diffusing through the lipid bilayer. (*b*) Channel-forming ionophores span the membrane with a channel through which ions can diffuse.

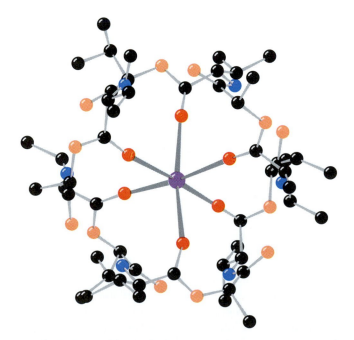

Figure 10-30.   X-Ray structure of valinomycin in complex with a K^+ ion. Six oxygen atoms (*dark red*) octahedrally coordinate the K^+ ion (*purple*). [After Neupert-Laves, K. and Dobler, M., *Helv. Chim. Acta* **58**, 439 (1975).]

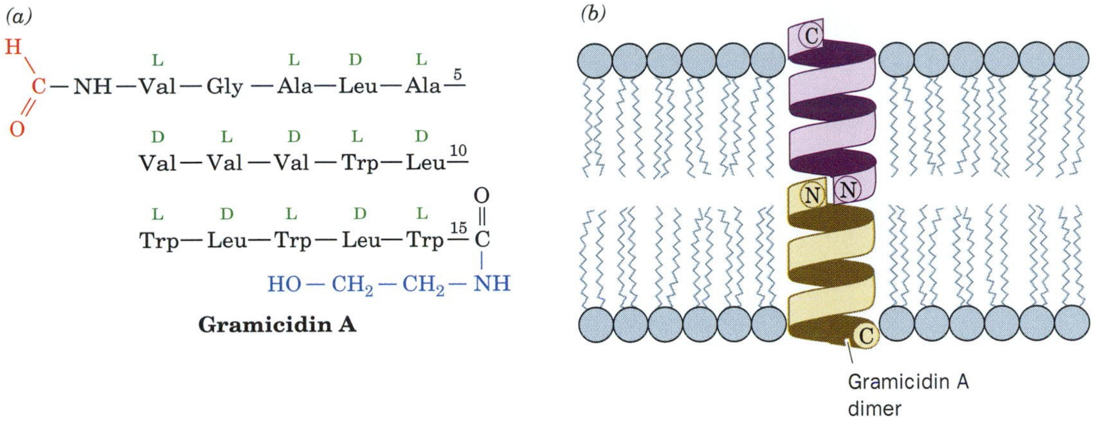

(a)

$$\begin{array}{c}
\underset{\parallel}{\overset{H}{\underset{O}{C}}}-NH-\underset{L}{Val}-Gly-\underset{L}{Ala}-\underset{D}{Leu}-\underset{L}{Ala}\underline{}^5
\end{array}$$

$$\underset{D}{Val}-\underset{L}{Val}-\underset{D}{Val}-\underset{L}{Trp}-\underset{D}{Leu}\underline{}^{10}$$

$$\underset{L}{Trp}-\underset{D}{Leu}-\underset{L}{Trp}-\underset{D}{Leu}-\underset{L}{Trp}\underline{}^{15}\overset{O}{\underset{\parallel}{C}}$$

$$HO-CH_2-CH_2-NH$$

Gramicidin A

(b)

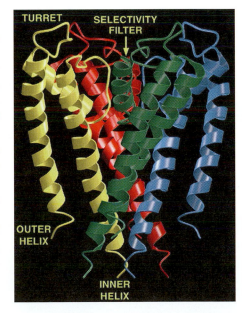

Gramicidin A
dimer

Figure 10-31.   Gramicidin A. (*a*) This 15-residue peptide of alternating D- and L-amino acids is chemically blocked at both its N- and C-termini. (*b*) Two membrane-embedded molecules of gramicidin A dimerize by hydrogen bonding between their N-formyl ends (N) to form a transmembrane channel.

drally coordinated by six carbonyl groups in the center of the ionophore (Fig. 10-30). Methyl and isopropyl side chains project outward to provide the complex with a nonpolar exterior that makes it soluble in the hydrophobic core of a lipid bilayer. The K^+ ion (with a diameter of 2.67 Å) fits snugly into valinomycin's coordination site, but the site is too large for Na^+ (diameter 1.90 Å) to coordinate with all six carbonyl oxygens. Valinomycin therefore has 10,000-fold greater binding affinity for K^+ than for Na^+. Even small amounts of the ionophore greatly increase the permeability of a membrane toward K^+, since a single molecule of valinomycin can transport up to 10^4 K^+ ions per second across the membrane.

The channel-forming ionophore **gramicidin A** is a 15-residue linear polypeptide consisting of alternating L and D residues, all of which are hydrophobic (Fig. 10-31*a*). Nuclear magnetic resonance and X-ray crystallographic evidence indicate that gramicidin A dimerizes in a head-to-head fashion to form a transmembrane channel (Fig. 10-31*b*). The alternating L and D residues of gramicidin A allow it to form a 4-Å-diameter helix with a nonpolar exterior and a polar central channel that facilitates the passage of Na^+ and K^+ ions.

Some small molecules cross membranes via porins. The subunits of the *E. coli* OmpF protein, for example (see Fig. 10-6), each form a transmembrane β barrel that surrounds an aqueous channel. The size of the channel and the residues that form its walls determine what types of molecules can pass through. With a prominent Lys residue in its channel, OmpF is specific for small cationic molecules.

The K^+ Channel Is Highly Selective

Virtually all cells contain ion-specific channels, exemplified by the K^+ channel from *Streptomyces lividans*. This integral membrane protein is highly specific for K^+ ions and mediates extremely fast diffusion of up to 10^8 ions per second. The channel's three-dimensional structure explains both of these functional features.

The K^+ channel is a homotetramer, with each subunit consisting mainly of two long α helices (Fig. 10-32). The inner helix forms part of the wall of a transmembrane pore oriented along the molecule's fourfold rotation axis, and the outer helix faces the membrane interior. Overall, the protein re-

Figure 10-32.   Structure of the K^+ channel from *S. lividans*. The four subunits are shown in different colors. Each subunit has an inner helix that forms part of the central pore, an outer helix that contacts the membrane interior, and a turret that projects out into the extracellular space. The selectivity filter at the wide end of the protein allows passage of K^+ ions but not Na^+ ions. [Courtesy of Roderick MacKinnon, Rockefeller University.]

Figure 10-33.   **Views of the pore of the K⁺ channel.** (*a*) Cross-sectional view of the protein at the level of the selectivity filter, looking down the axis of the pore. The pore (in the exact center) is lined by backbone carbonyl groups that co-ordinate dehydrated K^+ ions. The side chains of Tyr 78 in each subunit make hydrogen bonds and van der Waals contacts with surrounding residues, so that the selectivity filter cannot change confor-mation to accommodate Na^+ ions. (*b*) Contour surface of the pore within a stick model of the protein. The selectivity filter, the narrowest part of the pore, is at the top. A wider central cavity, lined with hydrophilic groups, helps decrease the energy barrier for transporting an ion through the long hydrophobic pore. [Courtesy of Roderick MacKinnon, Rockefeller University.]

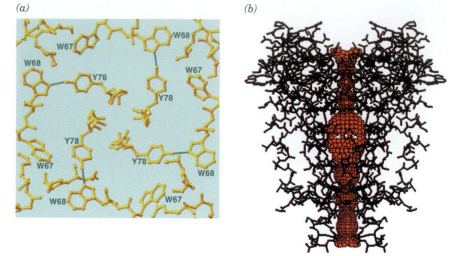

(*a*)   (*b*)

sembles a teepee with its wide end facing the extracellular space. Both the extracellular and intracellular mouths of the pore are flanked by acidic residues that attract K^+ and help prevent the entry of anions. The pore it-self is lined mostly with hydrophobic side chains.

A puzzling feature of the K^+ channel is that it is 10,000 times more permeant to K^+ ions than to Na^+ ions, even though Na^+ is smaller and should easily pass through the 6-Å-diameter pore. The answer to this paradox is that the wide end of the protein includes a structure called the selectivity filter (Fig. 10-33*a*). In this region, the protein folds such that backbone carbonyl groups project into the pore and the side chains point back into the protein interior. The carbonyl oxygen atoms are arranged with a geometry suitable for coordinating K^+ ions as they move through the pore. In effect, the carbonyl groups take the place of the water molecules that normally surround ions in solution. These groups cannot move inward to coordinate the smaller Na^+ ions because the pro-tein structure is rigidly held by hydrogen bonds and van der Waals in-teractions involving the side chains of the selectivity filter. This prevents Na^+ ions from passing through the selectivity filter. Interestingly, the selectivity filter appears to accommodate two K^+ ions. It is thought that as the second ion enters the pore, it exerts a repulsive force that keeps the first ion moving and prevents it from becoming stuck in a carbonyl coordination cage.

The rapid rate of ion movement through the K^+ channel also depends on a second structural feature—a hydrophilic cavity in the pore about mid-way across the bilayer, corresponding to the highest point of the energy barrier in the ion's transmembrane journey. This cavity is about 10 Å in di-ameter, large enough for a K^+ ion to be surrounded by a layer of water molecules (Fig. 10-33*b*). This aqueous oasis helps decrease the energetic barrier to transporting an ion through a long hydrophobic pore and there-fore increases the rate at which ions diffuse across the membrane.

Transport Proteins Alternate between Conformations

Not all proteins that mediate transmembrane traffic offer a discrete bilayer-spanning pore, as does the K^+ channel. Instead, these proteins un-

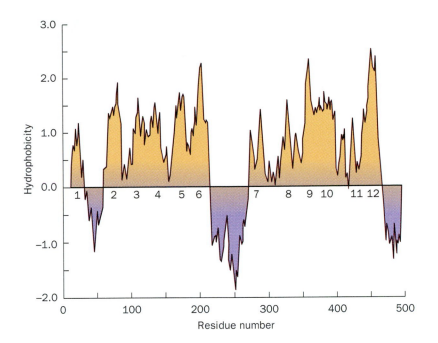

Figure 10-34. **Hydropathy plot of the human erythrocyte glucose transporter.** The plot was generated by averaging the hydropathies (Table 6-2) of the residues within a sliding window of 21 residues (enough to form a transmembrane α helix) and plotting these averages with respect to the position of the window's middle residue. The numbers refer to the 12 predicted membrane-spanning regions. [After Mueckler, M., Caruso, C., Baldwin, S.A., Panico, M., Blench, I., Morris, H.R., Allard, W.J., Lienhard, G.E., and Lodish, H.F., *Science* **229**, 944 (1985).]

dergo conformational changes to move substances from one side of the membrane to the other. The erythrocyte glucose transporter is such a protein. Sequence analysis reveals that this 55-kD glycoprotein has 12 membrane-spanning helices (Fig. 10-34). These helices probably form a hydrophobic cylinder surrounding a hydrophilic cavity that allows glucose to move from the extracellular fluid into the cytosol.

Biochemical evidence indicates that the transporter has glucose binding sites on both sides of the membrane. John Barnett showed that adding a propyl group to glucose C1 prevents glucose binding to the outer surface of the membrane, whereas adding a propyl group to C6 prevents binding to the inner surface. He therefore proposed that this transmembrane protein has two alternate conformations: one with the glucose site facing the external cell surface, requiring O1 contact and leaving O6 free, and the other with the glucose site facing the internal cell surface, requiring O6 contact and leaving O1 free. Transport apparently occurs as follows (Fig. 10-35):

1. Glucose binds to the protein on one face of the membrane.
2. A conformational change closes the first binding site and exposes the binding site on the other side of the membrane.
3. Glucose dissociates from the protein.
4. The transport cycle is completed by the reversion of the glucose transporter to its initial conformation in the absence of bound glucose.

This transport cycle can occur in either direction, according to the relative concentrations of intracellular and extracellular glucose. The glucose transporter provides a means of equilibrating the glucose concentration across the erythrocyte membrane without any accompanying leakage of small molecules or ions (as might occur through an always-open channel such as a porin).

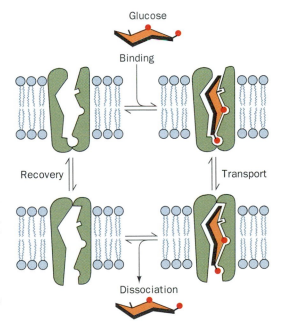

Figure 10-35. *Key to Function.* **Model for glucose transport.** The transport protein alternates between two mutually exclusive conformations. Glucose binds on one side of the membrane and is released on the other side after the protein conformation changes. The glucose molecule *(red)* is not drawn to scale. [After Baldwin, S.A. and Lienhard, G.E., *Trends Biochem. Sci.* **6**, 210 (1981).] ✴ See the Animated Figures.

Box 10-2

BIOCHEMISTRY IN FOCUS

Differentiating Mediated and Nonmediated Transport

Glucose and many other compounds can enter cells by a nonmediated pathway; that is, they slowly diffuse into cells at a rate proportional to their membrane solubility and their concentrations on either side of the membrane. This is a linear process: The **flux** (rate of transport per unit area) of a substance across the membrane increases with the magnitude of its concentration gradient (the difference between its internal and external concentrations). If the same substance, say glucose, moves across a membrane by means of a transport protein, its flux is no longer linear. This is one of four characteristics that distinguish mediated from nonmediated transport:

1. **Speed and specificity.** The solubilities of the chemically similar sugars D-glucose and D-mannitol in a synthetic lipid bilayer are similar. However, the rate at which glucose moves through the erythrocyte membrane is four orders of magnitude faster than that of D-mannitol. The erythrocyte membrane must therefore contain a system that transports glucose and that can distinguish D-glucose from D-mannitol.

2. **Saturation.** The rate of glucose transport into an erythrocyte does not increase infinitely as the external glucose concentration increases: The rate gradually approaches a maximum. Such an observation is evidence that a specific number of sites on the membrane are involved in the transport of glucose. At high [glucose], the transporters become saturated, much like myoglobin becomes saturated with O_2 at high pO_2 (Fig. 7-4). As expected, the following plot of glucose flux versus [glucose] is hyperbolic.

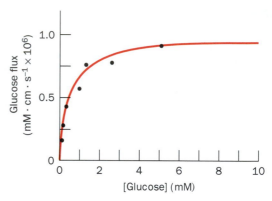

[Graph based on data from Stein, W.D., *Movement of Molecules across Membranes*, p. 134, Academic Press (1967).]

The nonmediated glucose flux increases linearly with [glucose] but would not visibly depart from the baseline on the scale of the above graph.

3. **Competition.** The above curve is shifted to the right in the presence of a substance that competes with glucose for binding to the transporter; for example, 6-O-benzyl-D-galactose has this effect. Competition is not a feature of nonmediated transport, since no transport protein is involved.

4. **Inactivation.** Reagents that chemically modify proteins and hence may affect their functions may eliminate the rapid, saturable flux of glucose into the erythrocyte. The susceptibility of the erythrocyte glucose transport system to protein-modifying reagents is additional proof that it is a protein.

All known transport proteins appear to be asymmetrically situated transmembrane proteins that alternate between two conformational states in which the ligand-binding sites are exposed, in turn, to opposite sides of the membrane. Such a mechanism is analogous to the T → R allosteric transition of proteins such as hemoglobin (Section 7-2). In fact, many of the features of ligand-binding proteins such as myoglobin and hemoglobin also apply to transport proteins (see Box 10-2).

Some transport proteins move more than one substance at a time. Hence, mediated transport can be categorized according to the stoichiometry of the transport process (Fig. 10-36):

1. A **uniport** involves the movement of a single molecule at a time. The erythrocyte glucose transporter is a uniport system.

2. A **symport** simultaneously transports two different molecules in the same direction.

3. An **antiport** simultaneously transports two different molecules in opposite directions.

C. ATP-Driven Active Transport

Since the glucose concentration in blood plasma is generally higher than that in cells, the erythrocyte glucose transporter normally transports glucose into the erythrocyte, where it is metabolized. Many substances, however, are available on one side of a membrane in lower concentrations than are required on the other side of the membrane. Such substances must be actively and selectively transported across the membrane against their concentration gradients.

Active transport is an endergonic process that, in most cases, is coupled to the hydrolysis of ATP. The elucidation of the mechanism by which the chemical energy released from ATP is used to drive a mechanical process has been a challenging biochemical problem. In this section, we examine membrane-bound ATPases that translocate cations; these proteins carry out **primary active transport. In secondary active transport,** the free energy of the electrochemical gradient generated by an ion-pumping ATPase is used to transport a neutral molecule against its concentration gradient.

(Na^+-K^+)–ATPase

One of the most thoroughly studied active transport systems is the **(Na^+-K^+)–ATPase** in the plasma membranes of higher eukaryotes, which was first characterized by Jens Skou. This transmembrane protein consists of two types of subunits: a 110-kD nonglycosylated α subunit that contains the enzyme's catalytic activity and ion-binding sites, and a 55-kD glycoprotein β subunit of unknown function. Sequence analysis suggests that the α subunit has eight transmembrane α-helical segments and two large cytoplasmic domains. The β subunit has a single transmembrane helix and a large extracellular domain. The protein may function as an $(\alpha\beta)_2$ tetramer *in vivo* (Fig. 10-37).

The (Na^+-K^+)–ATPase is often called the **(Na^+-K^+) pump** because it pumps Na^+ out of and K^+ into the cell with the concomitant hydrolysis of intracellular ATP. The overall stoichiometry of the reaction is

$$3Na^+(in) + 2K^+(out) + ATP + H_2O \rightleftharpoons$$
$$3Na^+(out) + 2K^+(in) + ADP + P_i$$

The (Na^+-K^+)–ATPase is an antiport that generates a charge separation across the membrane, since three positive charges exit the cell for every two that enter. This extrusion of Na^+ enables animal cells to control their water content osmotically; *without functioning (Na^+-K^+)-ATPases to maintain a low internal $[Na^+]$, water would osmotically rush in to such an extent that animal cells, which lack cell walls, would swell and burst.* The electrochemical gradient generated by the (Na^+-K^+)–ATPase is also responsible for the electrical excitability of nerve cells. In fact, all cells expend a large fraction of the ATP they produce (up to 70% in nerve cells) to maintain their required cytosolic Na^+ and K^+ concentrations.

The key to the (Na^+-K^+)–ATPase is the phosphorylation of a specific Asp residue of the transport protein. ATP phosphorylates the transporter only in the presence of Na^+, whereas the aspartyl phosphate residue (*at right*) is subject to hydrolysis only in the presence of K^+. This suggests that the (Na^+-K^+)–ATPase has two conformational states (called E_1 and E_2) with different structures, different catalytic activities, and different ligand

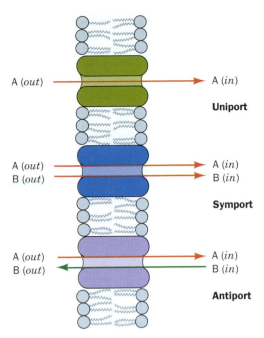

Figure 10-36. Uniport, symport, and antiport translocation systems.

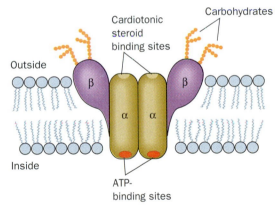

Figure 10-37. (Na^+-K^+)–ATPase. This diagram shows the transporter's putative dimeric structure and its orientation in the plasma membrane. Cardiotonic steroids (Box 10-3) bind to the external surface of the transporter, thereby inhibiting transport.

Aspartyl phosphate residue

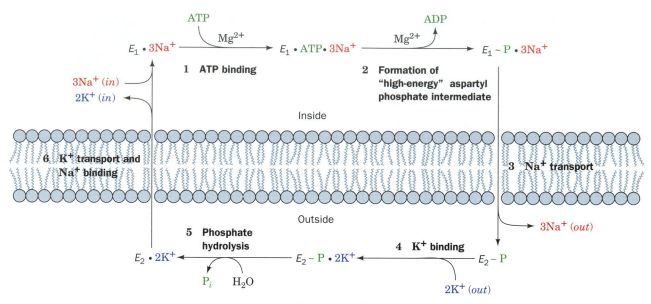

Figure 10-38.   *Key to Function.* Scheme for the active transport of Na$^+$ and K$^+$ by the (Na$^+$–K$^+$)–ATPase.

specificities. The protein is thought to operate in the following manner (Fig. 10-38):

1. The transporter in the E_1 state binds three Na$^+$ ions inside the cell and then binds ATP to yield an $E_1 \cdot$ATP$\cdot$3Na$^+$ complex.

2. ATP hydrolysis produces ADP and a "high-energy" aspartyl phosphate intermediate $E_1{\sim}$P$\cdot$3Na$^+$ (here "$\sim$" indicates a "high-energy" bond).

3. This "high-energy" intermediate relaxes to its "low-energy" conformation, E_2—P$\cdot$3Na$^+$, and releases its bound Na$^+$ outside the cell.

4. E_2—P binds two K$^+$ ions from outside the cell to form an E_2—P$\cdot$2K$^+$ complex.

5. The phosphate group is hydrolyzed, yielding $E_2 \cdot$2K$^+$.

6. $E_2 \cdot$2K$^+$ changes conformation, releases its two K$^+$ ions inside the cell, and replaces them with three Na$^+$ ions, thereby completing the transport cycle.

Although each of the above reaction steps is individually reversible, the cycle, as diagrammed in Fig. 10-38, circulates only in the clockwise direction under normal physiological conditions. This is because ATP hydrolysis and ion transport are coupled vectorial (unidirectional) processes. The vectorial nature of the reaction cycle results from the alternation of some of the steps of the exergonic ATP hydrolysis reaction (Steps 1 + 2 and Step 5) with some of the steps of the endergonic ion transport process (Steps 3 + 4 and Step 6). Thus, *neither reaction can go to completion unless the other one also does.* Study of the (Na$^+$–K$^+$)–ATPase has been greatly facilitated by the use of glycosides that inhibit the transporter (see Box 10-3).

Ca^{2+}–ATPase

Transient increases in cytosolic [Ca^{2+}] trigger numerous cellular responses including muscle contraction (Section 7-3C), release of neurotransmitters, and glycogen breakdown (Section 15-3). Moreover, Ca^{2+} is an important activator of oxidative metabolism (Section 16-4).

Box 10-3
BIOCHEMISTRY IN FOCUS

The Action of Cardiac Glycosides

The **cardiac glycosides** are natural products that increase the intensity of heart muscle contraction. Indeed, **digitalis**, an extract of purple foxglove leaves, which contains a mixture of cardiac glycosides including **digitoxin** (**digitalin**; see figure), has been used to treat congestive heart failure for centuries. The cardiac glycoside **ouabain** (pronounced wabane), a product of the East African Ouabio tree, has been long used as an arrow poison.

These two steroids, which are still among the most commonly prescribed cardiac drugs, inhibit the (Na^+-K^+)–ATPase by binding strongly to an externally exposed portion of the pro-

tein (Fig. 10-37) so as to block Step 5 in Fig. 10-38. The resultant increase in intracellular $[Na^+]$ stimulates the cardiac (Na^+-Ca^{2+}) antiport system, which pumps Na^+ out of and Ca^{2+} into the cell. The increased cytosolic $[Ca^{2+}]$ boosts the $[Ca^{2+}]$ in the sarcoplasmic reticulum. Thus, the release of Ca^{2+} to trigger muscle contraction (Section 7-3C) produces a larger than normal increase in cytosolic $[Ca^{2+}]$, thereby intensifying the force of cardiac muscle contraction. Ouabain, which was once thought to be produced only by plants, has recently been discovered to be an animal hormone that is secreted by the adrenal cortex and functions to regulate cellular $[Na^+]$ and overall body salt and water balance.

Digitoxin (digitalin)

Ouabain

The $[Ca^{2+}]$ in the cytosol ($\sim$0.1 μM) is four orders of magnitude less than it is in the extracellular spaces [$\sim$1500 μM; intracellular Ca^{2+} might otherwise combine with phosphate to form $Ca_3(PO_4)_2$, which has a maximum solubility of only 65 μM]. This large concentration gradient is maintained by the active transport of Ca^{2+} across the plasma membrane and the endoplasmic reticulum (the sarcoplasmic reticulum in muscle) by a **Ca^{2+}–ATPase (Ca^{2+} pump)** that actively pumps Ca^{2+} out of the cytosol at the expense of ATP hydrolysis (Ca^{2+} is also pumped into the mitochondrial matrix by a different system; Section 17-1B). The mechanism of the Ca^{2+}–ATPase (Fig. 10-39) resembles that of the (Na^+-K^+)–ATPase (Fig. 10-38).

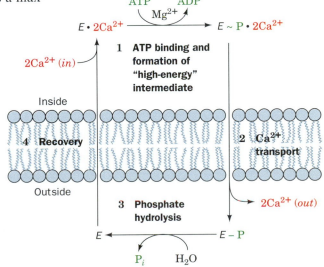

Figure 10-39. Scheme for the active transport of Ca^{2+} by the Ca^{2+}–ATPase. Here (*in*) refers to the cytosol and (*out*) refers to the outside of the cell for plasma membrane Ca^{2+}–ATPase or the lumen of the endoplasmic reticulum (or sarcoplasmic reticulum) for the Ca^{2+}–ATPase of that membrane.

Box 10-4

BIOCHEMISTRY IN HEALTH AND DISEASE

HCl and Peptic Ulcers

The extreme acidic conditions of the stomach are a necessary part of the digestive process: The low pH kills ingested microbes, denatures proteins, and activates the protease **pepsin**, which performs optimally at pH ~2.0. Pepsin has broad specificity for peptide bonds and is especially efficient at hydrolyzing the polypeptide chains of collagen.

The stomach itself is protected from its contents by a thick layer of mucus, which is a viscous mixture of water and heavily *O*-glycosylated proteins called **mucins**. Peptic ulcers may develop if stomach acid reaches the underlying gastric mucosa.

The (H^+-K^+)–ATPase of the gastric mucosa is activated by **histamine** stimulation of a cell-surface receptor. Compounds that block this process by competing with histamine for binding to the receptor can reduce HCl production. For example, **cimetidine** (trade name Tagamet) and its analogs, which resemble histamine,

bind to the histamine receptor but do not activate the (H^+-K^+)–ATPase. Such drugs are widely prescribed to alleviate the painful and otherwise often fatal symptoms of peptic ulcers.

Many ulcers are ultimately caused by infection with the recently discovered bacterium *Helicobacter pylori* (*below*), which thrives in the nutrient-rich gastric mucus. Because the bacterium is thus somewhat sequestered from the host's antimicrobial weaponry, it tends to induce a state of chronic inflammation of the stomach tissue, which then becomes susceptible to additional acid-induced damage. In such cases, antibiotics that eliminate the infection are therefore a better treatment for peptic ulcers than drugs such as cimetidine that merely relieve the symptoms (although these drugs are also widely used to eliminate the symptoms of heartburn, which is caused by the reflux of stomach contents into the acid-sensitive esophagus).

Cimetidine

Histamine

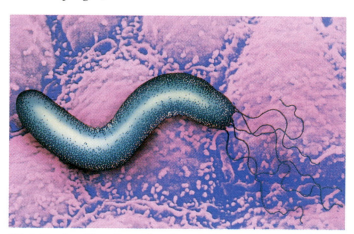

[Chris Bjomberg/Photo Researchers.]

(H^+-K^+)–ATPase of Gastric Mucosa

Cells in the wall of the mammalian stomach (the **gastric mucosa**) secrete HCl at a concentration of 0.15 M (pH 0.8). Since the cytosolic pH of these cells is 7.4, this represents a pH difference of over 6.0 units, the largest known in eukaryotic cells. The secreted protons are derived from the intracellular hydration of CO_2 by carbonic anhydrase:

$$CO_2 + H_2O \rightleftharpoons HCO_3^- + H^+$$

The secretion of H^+ involves an (H^+-K^+)–**ATPase,** an antiport with structure and properties similar to that of (Na^+-K^+)–ATPase. In this case, however, the K^+, which enters the cell as H^+ is pumped out, is subsequently externalized by its cotransport with Cl^-. HCl is therefore the overall transported product. Excess production of this acid can damage the gastric mucosa (see Box 10-4).

D. Ion Gradient–Driven Active Transport

Systems such as the (Na^+-K^+)–ATPase generate electrochemical gradients across membranes. The free energy stored in an electrochemical gra-

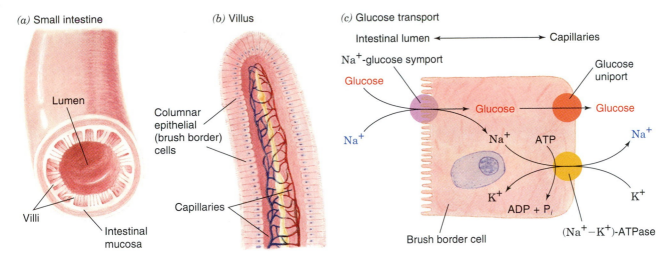

(a) Small intestine

Lumen

Villi

Intestinal
mucosa

(b) Villus

Columnar
epithelial
(brush border)
cells

Capillaries

(c) Glucose transport

Intestinal lumen ◄————————► Capillaries

Na^+-glucose symport

Glucose

Na^+

Glucose

Na^+

ATP

K^+

ADP + P_i

Glucose
uniport

Glucose

Na^+

K^+

$(Na^+–K^+)$-ATPase

Brush border cell

Figure 10-40.   Glucose transport in the intestinal epithelium. The brushlike villi lining the small intestine greatly increase its surface area (*a*), thereby facilitating the absorption of nutrients. The brush border cells from which the villi are formed (*b*) concentrate glucose from the intestinal lumen in symport with Na^+ (*c*), a process that is driven by the $(Na^+–K^+)$–ATPase, which is located on the capillary side of the cell and functions to maintain a low internal $[Na^+]$. The glucose is exported to the bloodstream via a separate passive-mediated uniport system like that in the erythrocyte.

dient (Eq. 10-3) can be harnessed to power various endergonic physiological processes. For example, cells of the intestinal epithelium take up dietary glucose by Na^+-dependent symport (Fig. 10-40). The immediate energy source for this "uphill" transport process is the Na^+ gradient. This process is an example of secondary active transport because *the Na^+ gradient in these cells is maintained by the $(Na^+–K^+)$–ATPase.* The Na^+–glucose transport system concentrates glucose inside the cell. Glucose is then transported into the capillaries through a passive-mediated glucose uniport (this transporter resembles the one in erythrocytes; Fig. 10-35). Thus, since glucose enhances Na^+ resorption, which in turn enhances water resorption, glucose is often fed to individuals suffering from salt and water losses due to diarrhea.

Lactose Permease Requires a Proton Gradient

Gram-negative bacteria such as *E. coli* contain several active transport systems for concentrating sugars. One extensively studied system, **lactose permease** (also known as **galactoside permease**), *utilizes the proton gradient across the bacterial cell membrane to cotransport H^+ and lactose* (Fig. 10-41). The proton gradient is metabolically generated through oxidative metabolism in a manner similar to that in mitochondria (Section 17-2). The electrochemical potential gradient created by these latter systems is used mainly to drive the synthesis of ATP.

Lactose permease, like the $(Na^+–K^+)$–ATPase, is a transmembrane protein with two major conformational states:

1. E-1, which has a low-affinity lactose-binding site facing the interior of the cell.
2. E-2, which has a high-affinity lactose-binding site facing the exterior of the cell.

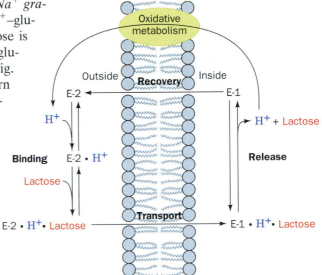

Oxidative
metabolism

Outside   Inside

Recovery

E-2 ◄———— E-1

H^+ H^+ + Lactose

Binding   E-2 · H^+ **Release**

Lactose

E-2 · H^+· Lactose   **Transport**   E-1 · H^+· Lactose

Figure 10-41.   Scheme for the cotransport of H^+ and lactose by lactose permease in *E. coli.* H^+ binds first to E-2 outside the cell, followed by lactose. They are released from E-1 inside the cell. E-2 must bind both lactose and H^+ in order to change conformation to E-1, thereby cotransporting these substances into the cell. E-1 changes conformation to E-2 when neither lactose nor H^+ is bound, thus completing the transport cycle.

E-1 and E-2 can interconvert only when their H^+- and lactose-binding sites are either both filled or both empty. This prevents dissipation of the H^+ gradient without cotransport of lactose into the cell. It also prevents transport of lactose out of the cell since this would require cotransport of H^+ against its concentration gradient.

SUMMARY

1. The proteins of biological membranes include integral (intrinsic) proteins that contain one or more transmembrane α helices or a β barrel. In all cases, the membrane-exposed surface of the protein is hydrophobic.

2. Other membrane-associated proteins may be anchored to the membrane via isoprenoid, fatty acid, or glycosylphosphatidylinositol groups. Peripheral (extrinsic) proteins are loosely associated with the membrane surface.

3. The fluid mosaic model of membrane structure accounts for the lateral diffusion of membrane proteins and lipids.

4. The arrangement of membrane proteins may depend on their interactions with an underlying protein skeleton, as in the erythrocyte.

5. Lipids synthesized on the cytoplasmic face of the ER membrane are distributed between the leaflets by the action of flipases and phospholipid translocases.

6. The synthesis of transmembrane, secretory, and lysosomal proteins begins on the RER. According to the signal hypothesis, a signal peptide directs the nascent polypeptide through the ER membrane. Coated vesicles transport membrane-embedded and lumenal proteins from the ER to the Golgi apparatus for processing, and from there to other membranes.

7. Lipoproteins, complexes of nonpolar lipids surrounded by a coat of amphiphilic lipids and apolipoproteins, transport lipids in vivo.

8. Cholesterol transported by LDL is taken into cells through the receptor-mediated endocytosis of LDL.

9. The mediated and nonmediated transport of a substance across a membrane is driven by its chemical potential difference.

10. Passive-mediated transporters include ionophores, porins, and certain transport proteins. These transport proteins alternate between two conformational states that expose the ligand-binding site to opposite sides of the membrane.

11. Active transport, in most cases, is driven by ATP hydrolysis. In the $(Na^+–K^+)$–ATPase and other ion transporters, ATP hydrolysis and ion transport are coupled and vectorial.

12. In secondary active transport, an ion gradient maintained by an ATPase drives the transport of another substance.

REFERENCES

Membrane Structure and Membrane Proteins

Edidin, M., Patches, posts and fences: proteins and plasma membrane domains, *Trends Cell Biol.* **2**, 376–380 (1992).

Luna, E.J. and Hitt, A.I., Cytoskeleton–plasma membrane interactions, *Science* **258**, 955–964 (1992).

Petty, H.R., *Molecular Biology of Membranes. Structure and Function,* Plenum Press (1993). [Summarizes the structures of membrane proteins and lipids and describes the biological functions of membranes.]

Udenfriend, S. and Kodukula, K., How glycosyl-phosphatidyl-inositol-anchored membrane proteins are made, *Annu. Rev. Biochem.* **64**, 563–591 (1995).

Zhang, F.L. and Casey, P.J., Protein prenylation: molecular mechanism and functional consequences, *Annu. Rev. Biochem.* **65**, 241–269 (1996).

Membrane Assembly and Targeting

Devaux, P.E., Protein involvement in transmembrane lipid asymmetry, *Annu. Rev. Biophys. Biomol. Struct.* **21**, 417–439 (1992).

Schatz, G. and Dobberstein, B., Common principles of protein translocation across membranes, *Science* **271**, 1519–1526 (1996).

Südhof, T.C., The synaptic vesicle cycle: a cascade of protein–protein interactions, *Nature* **375**, 645–653 (1995). [Describes membrane fusion, exocytosis, and endocytosis using as an example the cycling of neurotransmitters.]

Lipoproteins

Brown, M.S. and Goldstein, J.L., Koch's postulates for cholesterol, *Cell* **71**, 187–188 (1992). [Describes approaches to identifying cholesterol as the cause of atherosclerosis.]

Rosseneu, M. (Ed.), *Structure and Function of Apolipoproteins,* CRC Press (1992).

Schmid, S.L., The mechanism of receptor mediated endocytosis: More questions than answers, *BioEssays* **14**, 589–596 (1992).

Transport Proteins

Carafoli, E., The Ca^{2+} pump of the plasma membrane, *J. Biol. Chem.* **267**, 2115–2118 (1992).

Lingrel, J.B. and Kuntzweiler, T., Na$^+$,K$^+$-ATPase, *J. Biol. Chem.* **269,** 19659–19662 (1994).

Nikaido, H., Porins and specific diffusion channels in bacterial outer membranes, *J. Biol. Chem.* **269,** 3905–3908 (1994).

Silverman, M., Structure and function of hexose transporters, *Annu. Rev. Biochem.* **60,** 757–794 (1991). [Discusses both passive facilitative glucose transporters and the Na$^+$–glucose symport.]

KEY TERMS

integral (intrinsic) protein
transmembrane protein
prenylation
myristoylation
palmitoylation
GPI-linked protein
peripheral (extrinsic) protein
freeze-fracture electron
  microscopy

fluid mosaic model
fluorescence photobleaching
  recovery
cytoskeleton
gates and fences model
annular lipids
flipase
phospholipid translocase
signal hypothesis

signal peptide
coated vesicle
lipoprotein
transport protein
chemical potential difference
electrical potential difference
$\Delta\Psi$
nonmediated transport
mediated transport

passive-mediated transport
ionophore
uniport
symport
antiport
primary active transport
secondary active transport

STUDY EXERCISES

1. Explain the differences between integral and peripheral membrane proteins.

2. Describe the covalent modifications of lipid-linked proteins.

3. Describe the fluid mosaic model.

4. How can the cytoskeleton influence membrane protein distribution?

5. Describe the membrane translocation of lysosomal proteins.

6. Trace the route followed by a cell-surface glycoprotein, starting from its synthesis on a ribosome.

7. Describe how LDL transports cholesterol and is taken up by cells.

8. Explain the differences between mediated and nonmediated transport across membranes.

9. What are the similarities and differences among ionophores, porins, and passive-mediated transport proteins?

10. Distinguish passive-mediated transport, active transport, and secondary active transport.

PROBLEMS

1. (a) How many turns of an α helix are required to span a lipid bilayer ($\sim$30 Å across)? (b) What is the minimum number of residues required? (c) Why do most transmembrane helices contain more than the minimum number of residues?

2. The distance between the C_α atoms in a β sheet is $\sim$3.5 Å. Can a single 9-residue segment with a β conformation serve as the transmembrane portion of an integral membrane protein?

3. Are the following lipid samples likely to correspond to the inner or outer leaflet of a eukaryotic plasma membrane? (a) 20% Phosphatidylcholine, 15% phosphatidylserine, 65% other lipids. (b) 35% Phosphatidylcholine, 15% gangliosides, 5% cholesterol, 45% other lipids.

4. Describe the labeling pattern of glycophorin A when a membrane-impermeable protein-labeling reagent is added to (a) a preparation of solubilized erythrocyte proteins; (b) intact erythrocyte ghosts; and (c) erythrocyte ghosts that are initially leaky and then immediately sealed and transferred to a solution that does not contain the labeling reagent.

5. Predict the effect of a mutation in signal peptidase that narrows its specificity so that it cleaves only between two Leu residues.

6. Indicate whether the following compounds are likely to cross a membrane by nonmediated or mediated transport: (a) ethanol, (b) glycine, (c) cholesterol, (d) ATP.

7. (a) Calculate the chemical potential difference when intracellular [Na$^+$] = 10 mM and extracellular [Na$^+$] = 150 mM at 37°C. (b) What would the electrochemical potential be if the membrane potential were −60 mV (inside negative)?

8. (a) What happens to K$^+$ transport by valinomycin when the membrane is cooled below its transition temperature? (b) The N-terminus of gramicidin A is formylated (Fig. 10-32a). Could gramicidin A form a transmembrane channel if its N-terminus were not blocked in this fashion?

9. How long would it take 100 molecules of valinomycin to transport enough K$^+$ to change the concentration inside an erythrocyte of volume 100 μm^3 by 10 mM? (Assume that the valinomycin does not also transport any K$^+$ out of the cell, which it really does, and that the valinomycin molecules inside the cell are always saturated with K$^+$.)

10. If the ATP supply in the cell shown in Fig. 10-40c suddenly vanished, would the intracellular glucose concentration increase, decrease, or remain the same?

11. Endothelial cells and pericytes in the retina of the eye have different mechanisms for glucose uptake. The figure at the right shows the rate of glucose uptake for each type of cell in the presence of increasing amounts of sodium. What do these results reveal about the glucose transporter in each cell type? [Problem provided by Kathleen Cornely, Providence College.]

12. The transport of lactate into rat hepatocytes depends on the extracellular pH. Transport is rapid at pH 5 and slow at pH 8. What role might hydrogen ions play in lactate transport in these cells? There are several possibilities. [Problem provided by Kathleen Cornely, Providence College.]

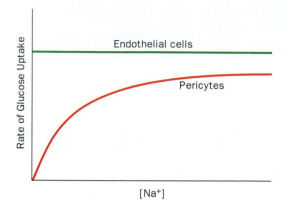

III

ENZYMES

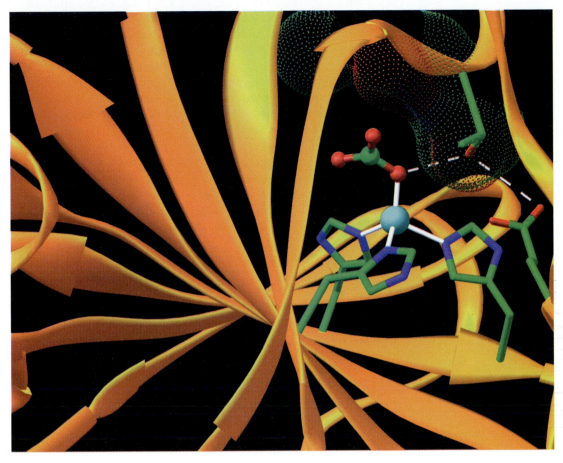

The active site of carbonic anhydrase includes three His residues to which a Zn^{2+} ion (cyan sphere) is liganded. The HCO_3^- substrate forms the Zn^{2+} ion's fourth ligand and interacts with the enzyme via van der Waals contacts and a hydrogen bonded network that includes several side chains. How are the functional groups of this and other enzyme active sites arranged so that they promote the conversion of a specific substrate to a specific product? What forces are harnessed to drive such reactions? [Based on an X-ray structure by K.K. Kannan, Bhabha Atomic Research Centre, Bombay, India.]

ENZYMATIC CATALYSIS

Living systems are shaped by an enormous variety of biochemical reactions, nearly all of which are mediated by a series of remarkable biological catalysts known as enzymes. **Enzymology,** the study of enzymes, has its roots in the early days of biochemistry; both disciplines evolved together from nineteenth century investigations of fermentation and digestion. Initially, the inability to reproduce most biochemical reactions in the laboratory led Louis Pasteur and others to assume that living systems were endowed with a "vital force" that permitted them to evade the laws of nature governing inanimate matter. Some investigators, however, notably Justus von Liebig, argued that biological processes were caused by the action of chemical substances that were then known as "ferments." Indeed, the name "enzyme" (Greek: *en,* in + *zyme,* yeast) was coined in 1878 in an effort to emphasize that there is something *in* yeast, as opposed to the yeast itself, that catalyzes the reactions of fermentation. Eventually, Eduard Buchner showed that a cell-free yeast extract could in fact carry out the synthesis of ethanol from glucose (**alcoholic fermentation;** Section 14-3B):

$$C_6H_{12}O_6 \longrightarrow 2\ CH_3CH_2OH + 2\ CO_2$$

This chemical transformation actually proceeds in 11 enzyme-catalyzed steps.

The chemical composition of enzymes was not firmly established until 1926, when James Sumner crystallized jack bean **urease,** which catalyzes the hydrolysis of urea to NH_3 and CO_2, and demonstrated that these crystals consist of protein. Enzymological experience since then has amply demonstrated that most enzymes are proteins (some species of RNA also have catalytic properties; Section 23-2E).

This chapter is concerned with one of the central questions of biochemistry: How do enzymes work? We shall see that enzymes increase the rates of chemical reactions by lowering the free energy barrier that separates the reactants and products. Enzymes accomplish this feat through various mechanisms that depend on the arrangement of functional groups in the enzyme's **active site,** the region of the enzyme where catalysis occurs. In this chapter, we describe these mechanisms, along with examples that illustrate how enzymes combine several mechanisms to catalyze biological reactions. The following chapter includes a discussion of enzyme kinetics, the study of the rates at which such reactions occur.

1. GENERAL PROPERTIES OF ENZYMES

Biochemical research since Pasteur's era has shown that, although enzymes are subject to the same laws of nature that govern the behavior of other substances, enzymes differ from ordinary chemical catalysts in several important respects:

1. **Higher reaction rates.** The rates of enzymatically catalyzed reactions are typically 10^6 to 10^{12} times greater than those of the corresponding uncatalyzed reactions (Table 11-1) and are at least several orders of magnitude greater than those of the corresponding chemically catalyzed reactions.

2. **Milder reaction conditions.** Enzymatically catalyzed reactions occur under relatively mild conditions: temperatures below 100°C, atmospheric pressure, and nearly neutral pH. In contrast, efficient chemical catalysis often requires elevated temperatures and pressures as well as extremes of pH.

Table 11-1. Catalytic Power of Some Enzymes

Enzyme	Nonenzymatic Reaction Rate (s^{-1})	Enzymatic Reaction Rate (s^{-1})	Rate Enhancement
Carbonic anhydrase	1.3×10^{-1}	1×10^6	7.7×10^6
Chorismate mutase	2.6×10^{-5}	50	1.9×10^6
Triose phosphate isomerase	4.3×10^{-6}	4300	1.0×10^9
Carboxypeptidase A	3.0×10^{-9}	578	1.9×10^{11}
AMP nucleosidase	1.0×10^{-11}	60	6.0×10^{12}
Staphylococcal nuclease	1.7×10^{-13}	95	5.6×10^{14}

Source: Radzicka, A. and Wolfenden, R., *Science* **267**, 91 (1995).

3. **Greater reaction specificity.** Enzymes have a vastly greater degree of specificity with respect to the identities of both their **substrates** (reactants) and their products than do chemical catalysts; that is, enzymatic reactions rarely have side products.

4. **Capacity for regulation.** The catalytic activities of many enzymes vary in response to the concentrations of substances other than their substrates. The mechanisms of these regulatory processes include allosteric control, covalent modification of enzymes, and variation of the amounts of enzymes synthesized.

A. Enzyme Nomenclature

Before delving further into the specific properties of enzymes, a word on nomenclature is in order. Enzymes are commonly named by appending the suffix *-ase* to the name of the enzyme's substrate or to a phrase describing the enzyme's catalytic action. Thus, urease catalyzes the hydrolysis of urea, and **alcohol dehydrogenase** catalyzes the oxidation of alcohols to their corresponding aldehydes. Since there were at first no systematic rules for naming enzymes, this practice occasionally resulted in two different names being used for the same enzyme or, conversely, in the same name being used for two different enzymes. Moreover, many enzymes, such as **catalase** (which mediates the dismutation of H_2O_2 to H_2O and O_2), were given names that provide no clue to their function. In an effort to eliminate this confusion and to provide rules for rationally naming the rapidly growing number of newly discovered enzymes, a scheme for the systematic functional classification and nomenclature of enzymes was adopted by the International Union of Biochemistry and Molecular Biology (IUBMB).

Enzymes are classified and named according to the nature of the chemical reactions they catalyze. There are six major classes of enzymatic reactions (Table 11-2), as well as subclasses and sub-subclasses. Each enzyme is assigned two names and a four-part classification number. Its **alternative name** is convenient for everyday use and is often an enzyme's previously

Table 11-2. Enzyme Classification According to Reaction Type

Classification	Type of Reaction Catalyzed
1. Oxidoreductases	Oxidation–reduction reactions
2. Transferases	Transfer of functional groups
3. Hydrolases	Hydrolysis reactions
4. Lyases	Group elimination to form double bonds
5. Isomerases	Isomerization
6. Ligases	Bond formation coupled with ATP hydrolysis

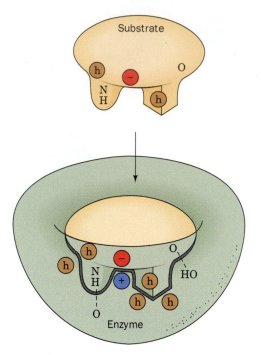

Figure 11-1.   An enzyme–substrate complex.
The geometric and the physical complementarity
between the enzyme and substrate depend on
noncovalent forces. Hydrophobic groups are
represented by an *h* in a brown circle, and dashed
lines represent hydrogen bonds.

used trivial name. Its **systematic name** is used when ambiguity must be minimized; it is the name of its substrate(s) followed by a word ending in *-ase* specifying the type of reaction the enzyme catalyzes according to its major group classification. For example, the enzyme whose recommended name is carboxypeptidase A (Section 5-3A) has the systematic name peptidyl-L-amino acid hydrolase and the **classification number** EC 3.4.17.1 ("EC" stands for Enzyme Commission, and the numbers represent the class, subclass, sub-subclass, and its arbitrarily assigned serial number in its sub-subclass). For our purposes, the recommended name of an enzyme is adequate. Systematic names and EC classification numbers can be obtained via the Internet (http://www.expasy.ch/sprot/enzyme.html).

B. Substrate Specificity

The noncovalent forces through which substrates and other molecules bind to enzymes are similar in character to the forces that dictate the conformations of the proteins themselves (Section 6-4A). Both involve van der Waals, electrostatic, hydrogen bonding, and hydrophobic interactions. In general, a substrate-binding site consists of an indentation or cleft on the surface of an enzyme molecule that is complementary in shape to the substrate **(geometric complementarity)**. Moreover, the amino acid residues that form the binding site are arranged to interact specifically with the substrate in an attractive manner **(electronic complementarity;** Fig. 11-1). Molecules that differ in shape or functional group distribution from the substrate cannot productively bind to the enzyme. X-Ray studies indicate that the substrate-binding sites of most enzymes are largely preformed but undergo some conformational change on substrate binding (a phenomenon called **induced fit).** The complementarity between enzymes and their substrates is the basis of the "lock-and-key" model of enzyme function first proposed by Emil Fischer in 1894. As we shall see, such specific binding is necessary but not sufficient for efficient catalysis.

Enzymes are Stereospecific

Enzymes are highly specific both in binding chiral substrates and in catalyzing their reactions. This **stereospecificity** arises because enzymes, by virtue of their inherent chirality (proteins consist of only L-amino acids), form asymmetric active sites. For example, yeast alcohol dehydrogenase **(YADH)** catalyzes the interconversion of ethanol and acetaldehyde:

$$CH_3CH_2OH \ + \ NAD^+ \ \underset{\overrightarrow{}}{\overset{YADH}{\rightleftharpoons}} \ CH_3\overset{\overset{\textstyle O}{\|}}{C}H \ + \ NADH \ + \ H^+$$

Ethanol   **Acetaldehyde**

In this reaction, the nicotinamide ring of NAD^+ (Fig. 3-4) is reduced, yielding NADH. Ethanol is a **prochiral** molecule; that is, it can become chiral through the substitution of one of its methylene hydrogens (these atoms are labeled "*pro-R*" and "*pro-S*" according to their configurations as described by the *RS* system; Box 4-1).

$$H_{pro\text{-}S} - \overset{\overset{\textstyle OH}{|}}{\underset{\underset{\textstyle CH_3}{|}}{C}} - H_{pro\text{-}R}$$

The methylene hydrogens of ethanol are chemically equivalent but occupy different positions relative to the CH_3 and OH substituents of the methylene carbon and can therefore be distinguished by the enzyme active site.

The stereospecificity of YADH was strikingly demonstrated in the following series of experiments by Frank Westheimer and Birgit Vennesland:

1. If the YADH reaction is carried out with deuterated ethanol, the NAD^+ is deuterated to form NADD:

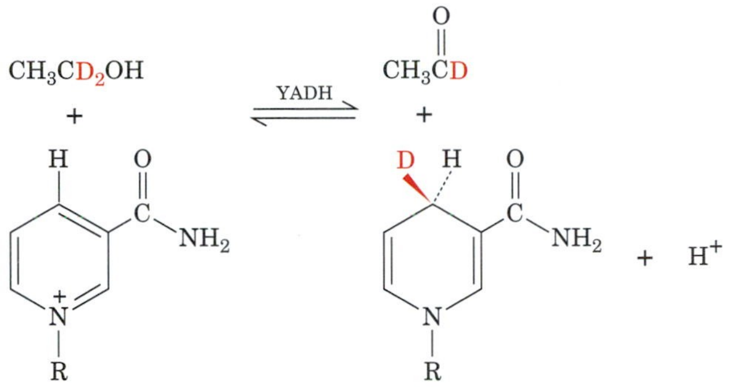

Note that the nicotinamide ring of NAD^+ is also prochiral.

2. On isolating this NADD and using it in the reverse reaction to reduce normal acetaldehyde, the deuterium is quantitatively transferred from the NADD to the acetaldehyde to form the product ethanol:

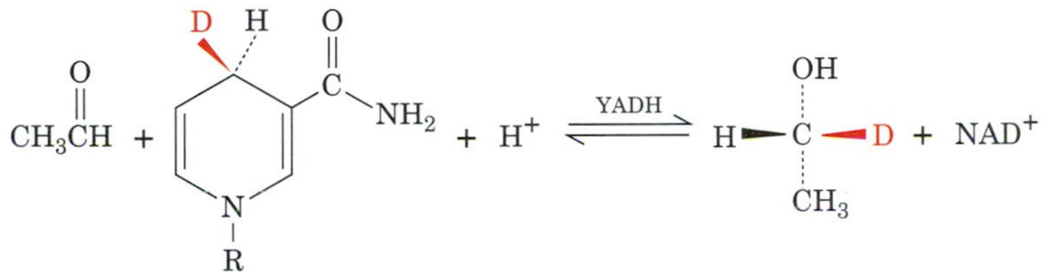

3. If the enantiomer of the foregoing CH_3CHDOH is made as follows:

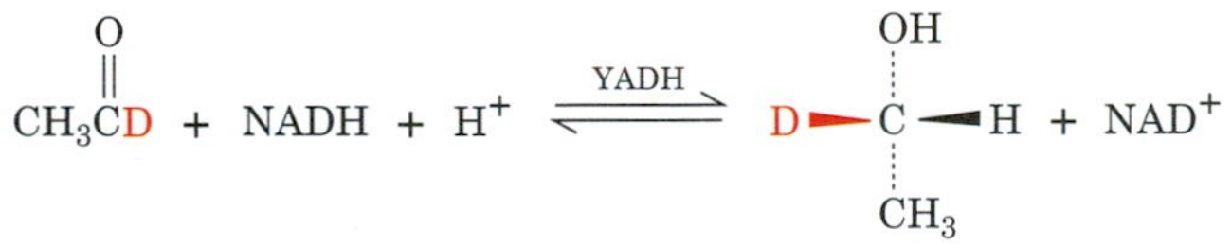

none of the deuterium is transferred from the product ethanol to NAD^+ in the reverse reaction.

These observations indicate that YADH distinguishes between the *pro-R* and *pro-S* hydrogens of ethanol and transfers only the *pro-R* hydrogen. The stereospecificity of YADH is by no means unusual. As we consider biochemical reactions, we shall find that *nearly all enzymes that participate in chiral reactions are absolutely stereospecific.*

Enzymes Vary in Geometric Specificity

The stereospecificity of enzymes is not particularly surprising in light of the complementarity of an enzyme's binding site for its substrate. A substance of the wrong chirality will not fit into an enzymatic binding site for much the same reason that you cannot fit your right hand into your left glove. In addition to their stereospecificity, however, most enzymes are quite selective about the identities of the chemical groups on their substrates. Indeed, such **geometric specificity** is a more stringent requirement than is stereospecificity.

Enzymes vary considerably in their degree of geometric specificity. A few enzymes are absolutely specific for only one compound. Most enzymes, however, catalyze the reactions of a small range of related compounds. For example, YADH catalyzes the oxidation of several small primary and secondary alcohols to their corresponding aldehydes or ketones but none so efficiently as ethanol. $NADP^+$, which differs from NAD^+ only by the addition of a phosphoryl group at the 2' position of its adenosine ribose group (Fig. 3-4), does not bind to YADH. On the other hand, many enzymes bind $NADP^+$ but not NAD^+.

Some enzymes, particularly digestive enzymes, are so permissive in their ranges of acceptable substrates that their geometric specificities are more accurately described as preferences. Some enzymes are not even very specific in the type of reaction they catalyze. For example, chymotrypsin, in addition to its ability to mediate peptide bond hydrolysis, also catalyzes ester bond hydrolysis.

$$RC\overset{O}{\underset{\|}{-}}NHR' \;+\; H_2O \;\xrightarrow{\text{chymotrypsin}}\; RC\overset{O}{\underset{\|}{-}}O^- \;+\; H_3\overset{+}{N}R'$$

Peptide

$$RC\overset{O}{\underset{\|}{-}}OR' \;+\; H_2O \;\xrightarrow{\text{chymotrypsin}}\; RC\overset{O}{\underset{\|}{-}}O^- \;+\; HOR'$$

Ester H^+

This property makes it convenient to measure chymotrypsin activity using small synthetic esters as substrates. Such permissiveness is much more the exception than the rule. Indeed, most intracellular enzymes function *in vivo* to catalyze a particular reaction on a particular substrate.

C. Cofactors and Coenzymes

The functional groups of proteins, as we shall see, can facilely participate in acid–base reactions, form certain types of transient covalent bonds, and take part in charge–charge interactions. They are, however, less suitable for catalyzing oxidation–reduction reactions and many types of group-transfer processes. Although enzymes catalyze such reactions, they can do so only in association with small molecule **cofactors,** which essentially act as the enzymes' "chemical teeth."

Cofactors may be metal ions, such as Cu^{2+}, Fe^{3+}, or Zn^{2+}. The essential nature of these cofactors explains why organisms require trace amounts of certain elements in their diets. It also explains, in part, the toxic effects of certain heavy metals. For example, Cd^{2+} and Hg^{2+} can replace Zn^{2+} (all are in the same group of the periodic table) in the active sites of certain enzymes, including RNA polymerase, and thereby render these enzymes inactive.

Cofactors may also be organic molecules known as **coenzymes,** such as the NAD^+ in YADH. Some cofactors (e.g., NAD^+) are only transiently associated with a given enzyme molecule, so that they function as **cosubstrates.** Other cofactors, known as **prosthetic groups,** are permanently associated with their protein, often by covalent bonds. For example, the heme prosthetic group of cytochrome *c* is tightly bound to the protein through extensive hydrophobic and hydrogen bonding interactions together with covalent bonds between the heme and specific protein side chains (Box 17-1).

A catalytically active enzyme–cofactor complex is called a **holoenzyme.** The enzymatically inactive protein resulting from the removal of a holoenzyme's cofactor is referred to as an **apoenzyme;** that is,

$$\text{apoenzyme } (inactive) + \text{cofactor} \rightleftharpoons \text{holoenzyme } (active)$$

Coenzymes Must Be Regenerated

Coenzymes are chemically changed by the enzymatic reactions in which they participate. In the YADH reaction, for example, NAD^+ is reduced to NADH. *In order to complete the catalytic cycle, the coenzyme must return to its original state* (in the YADH example, NADH must be oxidized back to NAD^+). For prosthetic groups, regeneration occurs in a separate phase of the enzymatic reaction sequence. For transiently bound coenzymes (cosubstrates), such as NAD^+, however, the regeneration reaction may be catalyzed by a different enzyme.

Many Vitamins Are Coenzyme Precursors

Table 11-3 lists the most common coenzymes, along with the types of reactions in which they participate (we shall describe the structures of these substances and their reaction mechanisms in the appropriate sections of the text). Many organisms are unable to synthesize certain portions of essential coenzymes. These substances must be present in the organism's diet and are therefore called **vitamins.** Table 11-3 also lists the vitamin precursors of the common coenzymes.

Many coenzymes were discovered as growth factors for microorganisms or as substances that cure nutritional deficiency diseases in humans and/or animals. For example, the NAD^+ component **nicotinamide,** or its carboxylic acid analog **nicotinic acid** (**niacin;** Fig. 11-2), relieves the ultimately fatal dietary deficiency disease in humans known as **pellagra.** The symptoms of pellagra include diarrhea, dermatitis, and dementia.

The vitamins in the human diet that are coenzyme precursors are all water-soluble vitamins. In contrast, the lipid-soluble vitamins, such as vitamins A and D, are not components of coenzymes, although they are also required in trace amounts in the diets of many higher animals. The distant ancestors of humans probably had the ability to synthesize the various vitamins, as do many modern plants and microorganisms. Yet since vitamins are normally available in the diets of higher animals, which all eat other organisms, or are synthesized by the bacteria that normally inhabit their digestive systems, it is believed that the superfluous cellular machinery to synthesize them was lost through evolution.

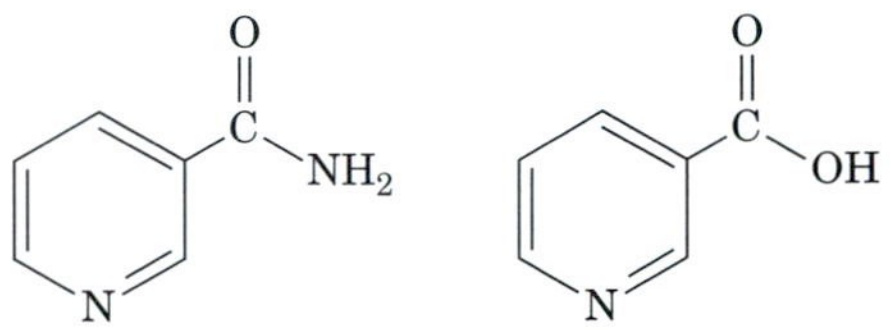

Nicotinamide   **Nicotinic acid**
(niacinamide)   **(niacin)**

Figure 11-2. **The structures of nicotinamide and nicotinic acid.** These vitamins form the redox-active components of the nicotinamide coenzymes NAD^+ and $NADP^+$ (compare with Fig. 3-4).

Table 11-3. **Characteristics of Common Coenzymes**

Coenzyme	Reaction Mediated	Vitamin Source	Human Deficiency Disease
Biocytin	Carboxylation	Biotin	*a*
Coenzyme A	Acyl transfer	Pantothenate	*a*
Cobalamin coenzymes	Alkylation	Cobalamin (B$_{12}$)	Pernicious anemia
Flavin coenzymes	Oxidation–reduction	Riboflavin (B$_2$)	*a*
Lipoic acid	Acyl transfer	—	*a*
Nicotinamide coenzymes	Oxidation–reduction	Nicotinamide (niacin)	Pellagra
Pyridoxal phosphate	Amino group transfer	Pyridoxine (B$_6$)	*a*
Tetrahydrofolate	One-carbon group transfer	Folic acid	Megaloblastic anemia
Thiamine pyrophosphate	Aldehyde transfer	Thiamine (B$_1$)	Beriberi

[a]No specific name; deficiency in humans is rare or unobserved.

2. ACTIVATION ENERGY AND THE REACTION COORDINATE

Much of our understanding of how enzymes catalyze chemical reactions comes from **transition state theory,** which was developed in the 1930s, principally by Henry Eyring. Consider a bimolecular reaction involving three atoms, such as the reaction of a hydrogen atom with diatomic hydrogen (H_2) to yield a new H_2 molecule and a different hydrogen atom:

$$H_A\text{—}H_B + H_C \longrightarrow H_A + H_B\text{—}H_C$$

In this reaction, H_C must approach the diatomic molecule H_A—H_B so that, at some point in the reaction, there exists a high-energy (unstable) complex represented as $H_A\cdots H_B\cdots H_C$. In this complex, the H_A—H_B covalent bond is in the process of breaking while the H_B—H_C bond is in the process of forming. The point of highest free energy is called the **transition state** of the system.

Reactants generally approach one another along the path of minimum free energy, their so-called **reaction coordinate.** A plot of free energy versus the reaction coordinate is called a **transition state diagram** or **reaction coordinate diagram** (Fig. 11-3). The reactants and products are states of minimum free energy, and the transition state corresponds to the highest point of the diagram. For the $H + H_2$ reaction, the reactants and products have the same free energy (Fig. 11-3a). If the atoms in the reacting system are of different types, such as in the reaction

$$A + B \longrightarrow X^\ddagger \longrightarrow P + Q$$

where A and B are the reactants, P and Q are the products, and $X^\ddagger$ represents the transition state, the transition state diagram is no longer symmetrical because there is a free energy difference between the reactants and products (Fig. 11-3b). In either case, $\Delta G^\ddagger$, the free energy of the transition state less that of the reactants, is known as the **free energy of activation.**

Passage through the transition state requires only 10^{-13} to 10^{-14} s, so the concentration of the transition state in a reacting system is small. Hence,

(a)

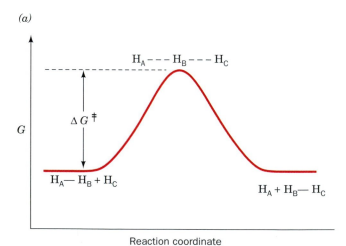

(b)

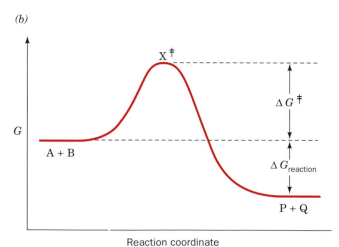

Figure 11-3.   Transition state diagram. (*a*) The $H + H_2$ reaction. The reactants and products correspond to low free energy structures. The point of highest free energy is the transition state, in which the reactants are partially converted to products. $\Delta G^\ddagger$ is the free energy of activation, the difference in free energy between the reactants and the transition state, $X^\ddagger$. (*b*) Transition state diagram for the reaction $A + B \rightarrow P + Q$. This is a spontaneous reaction; that is, $\Delta G_{reaction} < 0$ (the free energy of P + Q is less than the free energy of A + B).

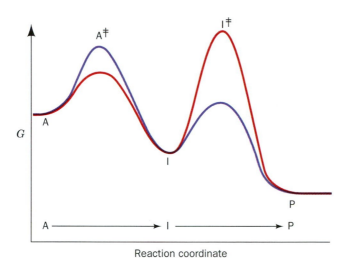

Figure 11-4. **Transition state diagram for a two-step reaction.** The blue curve represents a reaction (A → I → P) whose first step is rate determining, and the red curve represents a reaction whose second step is rate determining.

the decomposition of the transition state to products (or back to reactants) is postulated to be the rate-determining process of the overall reaction. Thermodynamic arguments lead to the conclusion that the reaction rate is proportional to $e^{-\Delta G^{\ddagger}/RT}$, where R is the gas constant and T is the absolute temperature. Thus, *the greater the value of $\Delta G^{\ddagger}$, the slower the reaction rate.* This is because the larger the $\Delta G^{\ddagger}$, the smaller the number of reactant molecules that have sufficient thermal energy to achieve the transition state free energy.

Chemical reactions commonly consist of several steps. For a two-step reaction such as

$$A \longrightarrow I \longrightarrow P$$

where I is an intermediate of the reaction, there are two transition states and two activation energy barriers. The shape of the transition state diagram for such a reaction reflects the relative rates of the two steps (Fig. 11-4). If the activation energy of the first step is greater than that of the second step, then the first step is slower than the second step, and conversely, if the activation energy of the second step is greater. In a multistep reaction, the step with the highest transition state free energy acts as a "bottleneck" and is therefore said to be the **rate-determining step** of the reaction.

Catalysts Reduce $\Delta G^{\ddagger}$

Catalysts act by lowering the transition state free energy for the reaction being catalyzed (Fig. 11-5). The difference between the values of $\Delta G^{\ddagger}$ for the uncatalyzed and catalyzed reactions, $\Delta\Delta G^{\ddagger}_{cat}$, indicates the efficiency of the catalyst. The **rate enhancement** (ratio of the rates of the catalyzed and uncatalyzed reactions) is given by $e^{\Delta\Delta G^{\ddagger}_{cat}/RT}$. Hence, a 10-fold rate enhancement requires a $\Delta\Delta G^{\ddagger}_{cat}$ of only 5.71 kJ·mol^{-1}, which is less than half the free energy of a typical hydrogen bond. Similarly, a millionfold rate acceleration occurs when $\Delta\Delta G^{\ddagger}_{cat} \approx 34$ kJ·mol^{-1}, a small fraction of the free energy of most covalent bonds. Thus, from a theoretical standpoint, tremendous catalytic efficiency seems within reach of the reactive groups that occur in the active sites of enzymes. How these groups actually function at the atomic level is the subject of much of this and other chapters.

Note that a catalyst lowers the free energy barrier by the same amount for both the forward and reverse reactions (Fig. 11-5). Consequently, a catalyst equally accelerates the forward and reverse reactions. Keep in

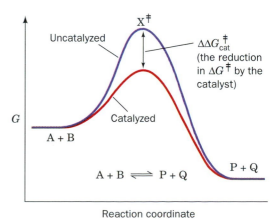

Figure 11-5. **Effect of a catalyst on the transition state diagram of a reaction.** Here $\Delta\Delta G^{\ddagger}_{cat} = \Delta G^{\ddagger}(uncat) - \Delta G^{\ddagger}(cat)$.

mind also that while a catalyst can accelerate the conversion of reactants to products (or products back to reactants), the likelihood of the net reaction occurring in one direction or the other depends only on the free energy difference between the reactants and the products. If $\Delta G_{\text{reaction}} < 0$, the reaction proceeds spontaneously from reactants toward products; if $\Delta G_{\text{reaction}} > 0$, the reverse reaction proceeds spontaneously. *An enzyme cannot alter $\Delta G_{\text{reaction}}$; it can only decrease $\Delta G^{\ddagger}$ to allow the reaction to approach equilibrium (where the rates of the forward and reverse reactions are equal) more quickly than it would in the absence of a catalyst.* The actual velocity with which reactants are converted to products is the subject of kinetics (Section 12-1).

3. CATALYTIC MECHANISMS

Enzymes achieve their enormous rate accelerations via the same catalytic mechanisms used by chemical catalysts. Enzymes have simply been better designed through evolution. Enzymes, like other catalysts, reduce the free energy of the transition state ($\Delta G^{\ddagger}$); that is, *they stabilize the transition state of the catalyzed reaction.* What makes enzymes such effective catalysts is their specificity of substrate binding combined with their arrangement of catalytic groups. As we shall see, however, the distinction between substrate-binding groups and catalytic groups is somewhat arbitrary.

Much can be learned about enzymatic reaction mechanisms by examining the corresponding nonenzymatic reactions of model compounds. Both types of reactions can be described using the **curved arrow convention** to trace the electron pair rearrangements that occur in going from reactants to products. The movement of an electron pair (which may be either a lone pair or a pair forming a covalent bond) is symbolized by a curved arrow emanating from the electron pair and pointing to the electron-deficient center attracting the electron pair. For example, imine **(Schiff base)** formation, a biochemically important reaction between an amine and an aldehyde or ketone, is represented as follows:

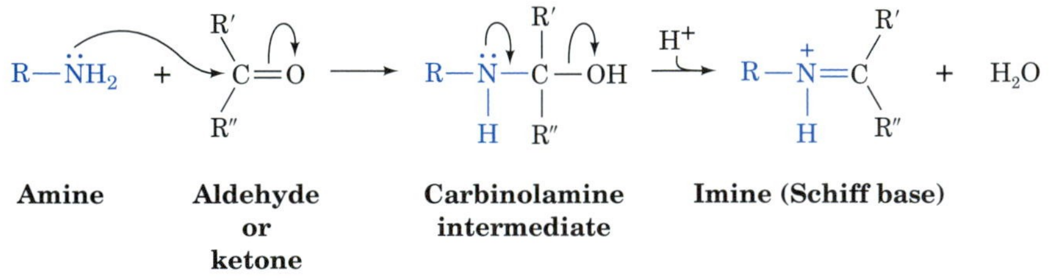

| Amine | Aldehyde or ketone | Carbinolamine intermediate | Imine (Schiff base) |

In the first reaction step, the amine's unshared electron pair adds to the electron-deficient carbonyl carbon while one electron pair from its C=O double bond transfers to the oxygen atom. In the second step, the unshared electron pair on the nitrogen atom adds to the electron-deficient carbon atom with the elimination of water. *At all times, the rules of chemical reason apply to the system:* For example, there are never five bonds to a carbon atom or two bonds to a hydrogen atom.

The types of catalytic mechanisms that enzymes employ have been classified as

1. Acid–base catalysis
2. Covalent catalysis
3. Metal ion catalysis
4. Electrostatic catalysis

5. Proximity and orientation effects

6. Preferential binding of the transition state complex

In this section, we consider each of these types of mechanisms in turn.

A. Acid–Base Catalysis

General acid catalysis is a process in which partial proton transfer from an acid lowers the free energy of a reaction's transition state. For example, an uncatalyzed keto–enol tautomerization reaction occurs quite slowly as a result of the high free energy of its carbanionlike transition state (Fig. 11-6*a*; the transition state is drawn in square brackets to indicate its instability). Proton donation to the oxygen atom (Fig. 11-6*b*), however, reduces the carbanion character of the transition state, thereby accelerating the reaction.

A reaction may also be stimulated by **general base catalysis** *if its rate is increased by partial proton abstraction by a base* (e.g., Fig. 11-6*c*). Some reactions may be simultaneously subject to both processes; these are **concerted acid–base catalyzed reactions.**

Many types of biochemical reactions are susceptible to acid and/or base catalysis. The side chains of the amino acid residues Asp, Glu, His, Cys, Tyr, and Lys have pK's in or near the physiological pH range (Table 4-1), which permits them to act as acid and/or base catalysts. Indeed, *the ability of enzymes to arrange several catalytic groups around their substrates makes concerted acid–base catalysis a common enzymatic mechanism.* The catalytic activity of these enzymes is sensitive to pH, since the pH influences the state of protonation of side chains at the active site (see Box 11-1).

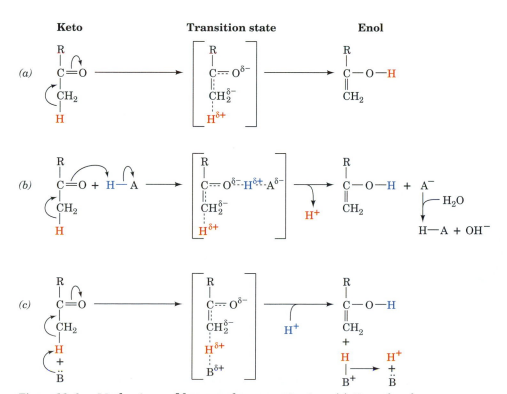

Figure 11-6. **Mechanisms of keto–enol tautomerization.** (*a*) Uncatalyzed. (*b*) General acid catalyzed. (*c*) General base catalyzed. The acid is represented as H—A and the base as :B.

Box 11-1

BIOCHEMISTRY IN FOCUS

Effects of pH

Most proteins are active within only a narrow pH range, typically 5 to 9. This is a result of the effects of pH on a combination of factors: (1) the binding of substrate to enzyme, (2) the ionization states of the amino acid residues involved in the catalytic activity of the enzyme, (3) the ionization of the substrate, and (4) the variation of protein structure (usually significant only at extremes of pH).

The rates of many enzymatic reactions exhibit bell-shaped curves as a function of pH. For example, the pH-dependence of the rate of the reaction catalyzed by **fumarase** (Section 16-3G) produces the following curve:

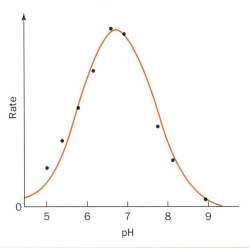

Such curves reflect the ionization of certain amino acid residues that must be in a specific ionization state for enzymatic activity. The observed pK's (the inflection points of the curve) often provide valuable clues to the identities of the amino acid residues essential for enzymatic activity. For example, an observed pK of ~4 suggests that an Asp or Glu residue is essential to the enzyme. Similarly, pK's of ~6 or ~10 suggest the participation of a His or a Lys residue, respectively. However, the pK of a given acid–base group may vary by as much as several pH units from its expected value, depending on its microenvironment (e.g., an Asp residue in a nonpolar environment or in close proximity to another Asp residue would attract protons more strongly than otherwise and hence have a higher pK). Furthermore, pH effects on an enzymatic rate may reflect denaturation of the enzyme rather than protonation or deprotonation of specific catalytic residues. The replacement of a particular residue by site-directed mutagenesis or comparisons of enzyme variants generated by evolution is a more reliable approach to identifying residues that are required for substrate binding or catalysis.

[Figure adapted from Tanford, C., *Physical Chemistry of Macromolecules*, p. 647, Wiley (1961).]

Figure 11-7.   The X-ray structure of bovine pancreatic RNase S. A nonhydrolyzable substrate analog, the dinucleotide phosphonate UpcA, is bound in the active site. RNase S is a catalytically active form of RNase A in which the peptide bond between residues 20 and 21 has been hydrolyzed. [Figure copyrighted © by Irving Geis.] ● See the Interactive Exercises.

The RNase A Reaction

Bovine pancreatic RNase A (RNase A) provides an example of enzymatically mediated acid–base catalysis. This digestive enzyme (Fig. 11-7) is secreted by the pancreas into the small intestine, where it hydrolyzes RNA to its component nucleotides. The isolation of 2′,3′-cyclic nucleotides from RNase A digests of RNA indicates that 2′,3′-cyclic nucleotides are intermediates in the RNase A reaction (Fig. 11-8). The pH-dependence of the rate of the RNase A reaction suggests the involvement of two ionizable residues with pK's of 5.4 and 6.4. This information, together with chemical derivatization and X-ray studies, indicates that RNase A has two essential His residues, His 12 and His 119, that act in a concerted manner as general acid and base catalysts. Evidently, the RNase A reaction is a two-step process (Fig. 11-8):

1. His 12, acting as a general base, abstracts a proton from an RNA 2′-OH group, thereby promoting its nucleophilic attack on the adjacent phosphorus atom. His 119, acting as a general acid, promotes bond scission by protonating the leaving group.

2. The 2′,3′ cyclic intermediate is hydrolyzed through what is essentially the reverse of the first step in which water replaces the leaving group. Thus, His 12 now acts as a general acid and His 119 as a general base to yield the hydrolyzed RNA and the enzyme in its original state.

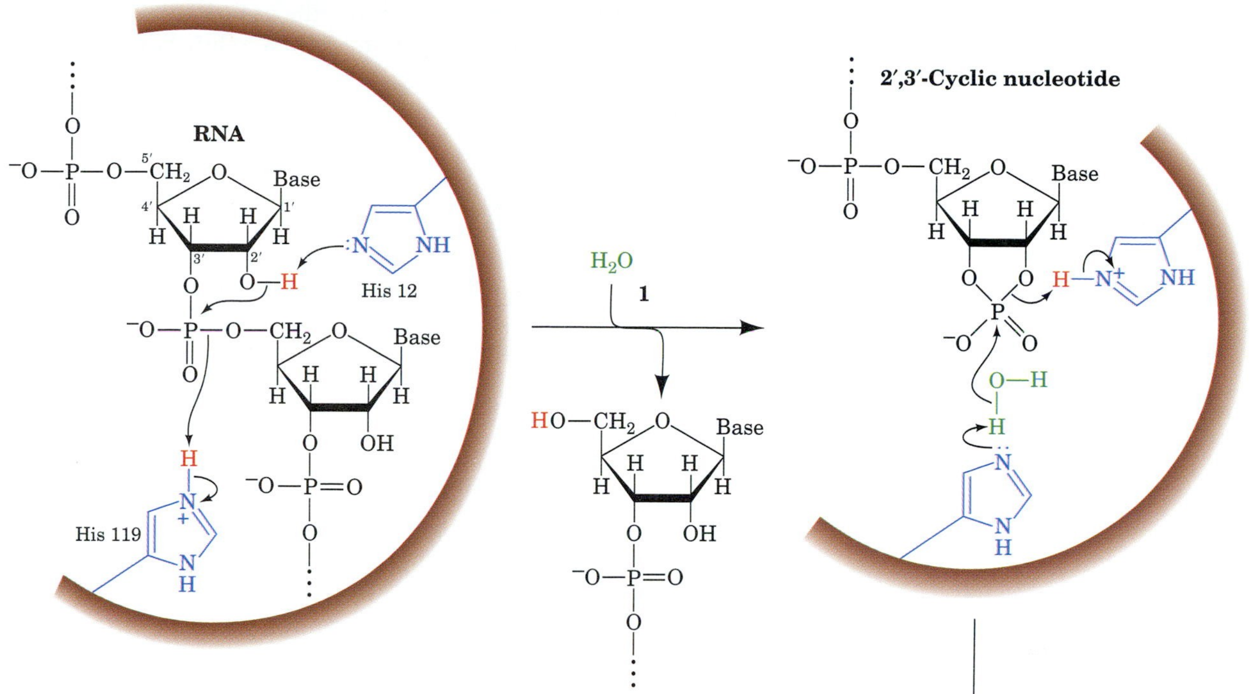

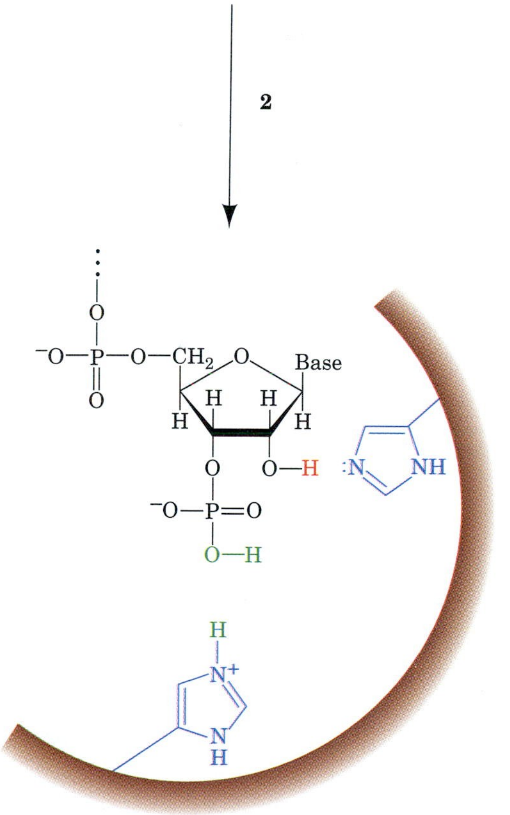

Figure 11-8. **The RNase A mechanism.** The bovine pancreatic RNase A–catalyzed hydrolysis of RNA is a two-step process with the intermediate formation of a 2′,3′-cyclic nucleotide.

B. Covalent Catalysis

Covalent catalysis accelerates reaction rates through the transient formation of a catalyst–substrate covalent bond. Usually, this covalent bond is formed by the reaction of a nucleophilic group on the catalyst with an electrophilic group on the substrate, and hence this form of catalysis is often also called **nucleophilic catalysis.** The decarboxylation of acetoacetate, as chemically catalyzed by primary amines, is an example of such a process (Fig. 11-9).

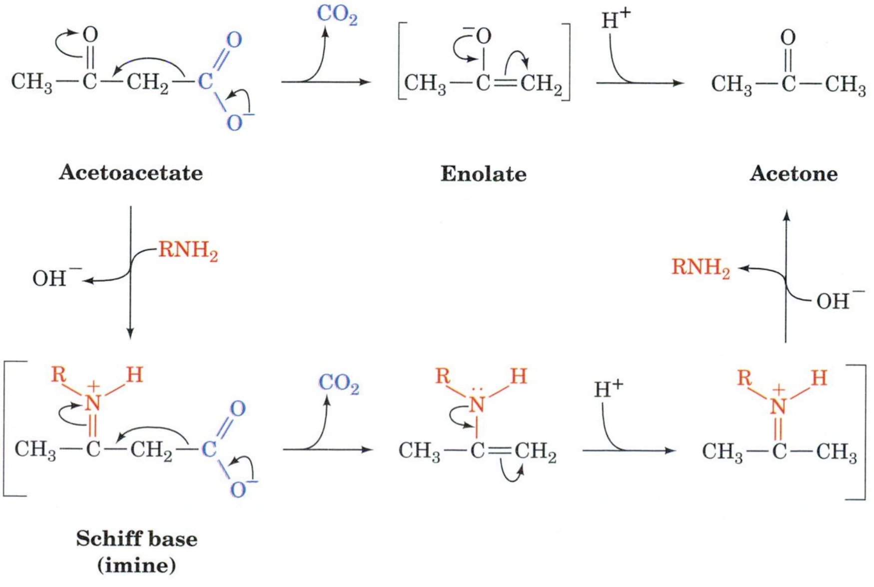

Figure 11-9. **The decarboxylation of acetoacetate.** The uncatalyzed reaction mechanism is at the top, and the mechanism as catalyzed by primary amines is at the bottom.

In the first stage of this reaction, the amine, a nucleophile, attacks the carbonyl group of acetoacetate to form a Schiff base (imine bond).

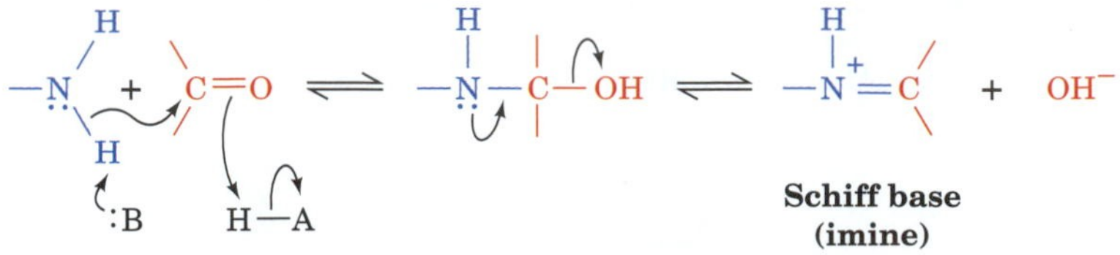

**Schiff base
(imine)**

The protonated nitrogen atom of the covalent intermediate then acts as an electron sink (Fig. 11-9, *bottom*) to reduce the high-energy enolate character of the transition state. The formation and decomposition of the Schiff base occurs quite rapidly so that it is not the rate-determining step of this reaction.

Covalent catalysis can be conceptually decomposed into three stages:

1. The nucleophilic reaction between the catalyst and the substrate to form a covalent bond.
2. The withdrawal of electrons from the reaction center by the now electrophilic catalyst.
3. The elimination of the catalyst, a reaction that is essentially the reverse of stage 1.

The nucleophilicity of a substance is closely related to its basicity. Indeed, the mechanism of nucleophilic catalysis resembles that of base catalysis except that, instead of abstracting a proton from the substrate, the catalyst nucleophilically attacks the substrate to form a covalent bond. Biologically important nucleophiles are negatively charged or contain unshared electron pairs that easily form covalent bonds with electron-deficient centers (Fig. 11-10*a*). Electrophiles, in contrast, include groups that are positively charged, contain an unfilled valence electron shell, or contain an electronegative atom (Fig. 11-10*b*).

An important aspect of covalent catalysis is that *the more stable the covalent bond formed, the less easily it can decompose in the final steps of a reaction.* A good covalent catalyst must therefore combine the seemingly contradictory properties of high nucleophilicity and the ability to form a

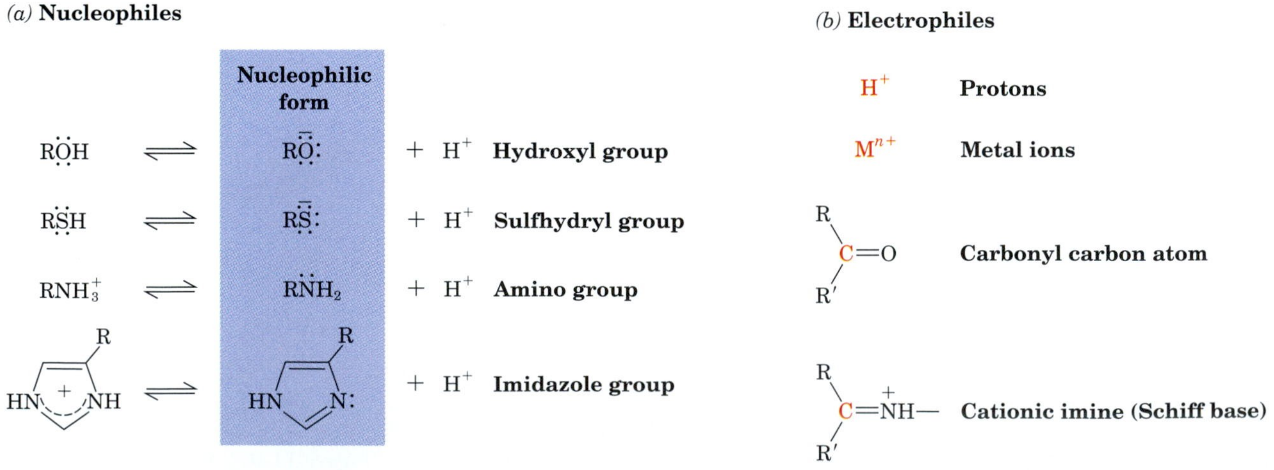

(a) **Nucleophiles**

(b) **Electrophiles**

Figure 11-10. **Biologically important nucleophilic and electrophilic groups.** (*a*) Nucleophilic groups such as hydroxyl, sulfhydryl, amino, and imidazole groups are nucleophiles in their basic forms. (*b*) Electrophilic groups contain an electron-deficient atom (*red*).

good leaving group, that is, to easily reverse the bond formation step. Groups with high polarizability (highly mobile electrons), such as imidazole and thiol functions, have these properties and hence make good covalent catalysts. Such functional groups in proteins include the ε-amino group of Lys, the imidazole group of His, the thiol group of Cys, the carboxyl group of Asp, and the hydroxyl group of Ser. In addition, several coenzymes, notably **thiamine pyrophosphate** (Section 14-3B) and **pyridoxal phosphate** (Section 20-2A), function in association with their apoenzymes as covalent catalysts. Enzymes commonly employ covalent catalytic mechanisms as is indicated by the large variety of covalently linked enzyme–substrate reaction intermediates that have been isolated.

C. Metal Ion Catalysis

Nearly one-third of all known enzymes require the presence of metal ions for catalytic activity. This group of enzymes includes the **metalloenzymes,** which contain tightly bound metal ion cofactors, most commonly transition metal ions such as Fe^{2+}, Fe^{3+}, Cu^{2+}, Zn^{2+}, Mn^{2+}, or Co^{2+}. **Metal-activated enzymes,** in contrast, loosely bind metal ions from solution, usually the alkali and alkaline earth metal ions Na^+, K^+, Mg^{2+}, or Ca^{2+}. In this group of enzymes, the ions often play a structural rather than a catalytic role.

Metal ions participate in the catalytic process in three major ways:

1. By binding to substrates to orient them properly for reaction.
2. By mediating oxidation–reduction reactions through reversible changes in the metal ion's oxidation state.
3. By electrostatically stabilizing or shielding negative charges.

In many metal ion–catalyzed reactions, the metal ion acts in much the same way as a proton to neutralize negative charge. Yet metal ions are often much more effective catalysts than protons because metal ions can be present in high concentrations at neutral pH and may have charges greater than +1.

A metal ion's charge also makes its bound water molecules more acidic than free H_2O and therefore a source of nucleophilic OH^- ions even below neutral pH. An excellent example of this phenomenon occurs in the catalytic mechanism of carbonic anhydrase (Box 2-2), a widely occurring enzyme that catalyzes the reaction

$$CO_2 + H_2O \rightleftharpoons HCO_3^- + H^+$$

Carbonic anhydrase contains an essential Zn^{2+} ion that is implicated in the enzyme's catalytic mechanism as follows:

1. The crystal structure of human carbonic anhydrase (Fig. 11-11) reveals that its Zn^{2+} lies at the bottom of a 15-Å-deep active site cleft, where it is tetrahedrally coordinated by three evolutionarily invariant His side chains and an H_2O molecule. This Zn^{2+}-polarized H_2O ionizes through base catalysis that is facilitated by a fourth His residue.
2. The resulting Zn^{2+}-bound OH^- nucleophilically attacks the nearby enzymatically bound CO_2, thereby converting it to HCO_3^- (*at right*).
3. The catalytic site is then regenerated by the binding and ionization of another H_2O at the Zn^{2+} (*at right*).

D. Electrostatic Catalysis

The binding of substrate generally excludes water from an enzyme's active site. The active site therefore has the polarity characteristics of an organic

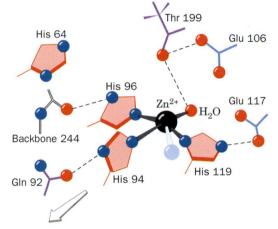

Figure 11-11. The active site of human carbonic anhydrase. The light blue ligand to the Zn^{2+} indicates a probable fifth Zn^{2+} coordination site. The arrow points toward the opening of the active site cavity. [After Sheridan, R.P. and Allen, L.C., *J. Am. Chem. Soc.* **103,** 1545 (1981).]
● See the Interactive Exercises.

Im = imidazole

solvent, where electrostatic interactions are much stronger than they are in aqueous solutions. Thus, the pK's of amino acid side chains in proteins may vary by several units from their normal values (Table 4-1) because of the proximity of charged groups.

Although experimental evidence and theoretical analyses on the subject are still sparse, *the charge distributions around the active sites of enzymes seem to be arranged so as to stabilize the transition states of the catalyzed reactions.* This mode of rate enhancement, which resembles metal ion catalysis, is termed **electrostatic catalysis.** Moreover, in several enzymes, charge distributions apparently guide polar substrates toward their binding sites to further enhance the reaction rate.

E. Catalysis through Proximity and Orientation Effects

Although enzymes employ catalytic mechanisms that resemble those of organic model reactions, they are far more catalytically efficient than these models. Such efficiency must arise from the specific physical conditions at enzyme catalytic sites that promote the corresponding chemical reactions. The most obvious effects are **proximity** and **orientation:** Reactants must come together with the proper spatial relationship for a reaction to occur. Consider the bimolecular reaction of imidazole with ***p*-nitrophenylacetate.**

$$CH_3-\overset{O}{\overset{\|}{C}}-O-\langle\text{aryl}\rangle-NO_2 \longrightarrow CH_3-\overset{O}{\overset{\|}{C}} + \quad {}^-O-\langle\text{aryl}\rangle-NO_2$$

***p*-Nitrophenylacetate**   ***p*-Nitrophenolate**

N-Acetylimidazolium

Imidazole

The progress of the reaction is conveniently monitored by the appearance of the intensely yellow ***p*-nitrophenolate** ion. The related intramolecular reaction

$$\longrightarrow \quad + \quad {}^-O-\langle\text{aryl}\rangle-NO_2$$

occurs about 24 times faster. Thus, when the imidazole catalyst is covalently attached to the reactant, it is 24 times more effective than when it is free in solution. This rate enhancement results from both proximity and orientation effects.

By simply binding their substrates, enzymes facilitate their catalyzed reactions in three ways:

1. Enzymes bring substrates into contact with their catalytic groups and, in reactions with more than one substrate, with each other. However, calculations based on simple model systems suggest that such proximity effects alone can enhance reaction rates by no more than a factor of ~5.

2. Enzymes bind their substrates in the proper orientations for reaction. Molecules are not equally reactive in all directions. Rather, *they react most readily if they have the proper relative orientation* (Fig. 11-12).

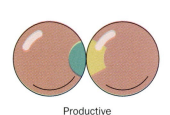

Productive

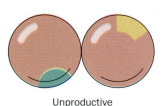

Unproductive

Figure 11-12. Orientation effects. Molecules are susceptible to chemical attack over only limited regions of their surfaces (represented by the colored areas). Without the proper relative orientation (*top*), reactions do not occur (*bottom*).

For example, in an S_N2 (bimolecular nucleophilic substitution) reaction, the incoming nucleophile optimally attacks its target along the direction opposite to that of the bond to the leaving group (backside attack). Reacting atoms whose approaches deviate by as little as 10° from this optimum direction are significantly less reactive. It is estimated that properly orienting substrates can increase reaction rates by a factor of up to ~100. Enzymes, as we shall see, align their substrates and catalytic groups so as to optimize reactivity.

3. Enzymes freeze out the relative translational and rotational motions of their substrates and catalytic groups. This is an important aspect of catalysis because, in the transition state, the reacting groups have little relative motion. Indeed, experiments with model compounds suggest that *this effect can promote rate enhancements of up to ~10^7!*

Bringing substrates and catalytic groups together in a reactive orientation orders them and therefore has a substantial entropic penalty. The free energy required to overcome this entropy loss is supplied by the binding energy of the substrate(s) to the enzyme.

F. Catalysis by Preferential Transition State Binding

The rate enhancements effected by enzymes are often greater than can be reasonably accounted for by the catalytic mechanisms discussed so far. However, we have not yet considered one of the most important mechanisms of enzymatic catalysis: *An enzyme may bind the transition state of the reaction it catalyzes with greater affinity than its substrates or products.* When taken together with the previously described catalytic mechanisms, preferential transition state binding explains the observed rates of enzyme-catalyzed reactions.

The original concept of transition state binding proposed that enzymes mechanically strained their substrates toward the transition state geometry through binding sites into which undistorted substrates did not properly fit. Such strain promotes many organic reactions. For example, the rate of the reaction

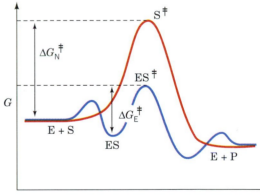

is 315 times faster when R is CH_3 rather than H because of the greater steric repulsion between the CH_3 groups and the reacting groups. The strained reactant more closely resembles the transition state of the reaction than does the corresponding unstrained reactant. Thus, as was first suggested by Linus Pauling and further amplified by Richard Wolfenden and Gustav Lienhard, *enzymes that preferentially bind the transition state structure increase its concentration and therefore proportionally increase the reaction rate.*

The more tightly an enzyme binds its reaction's transition state relative to the substrate, the greater is the rate of the catalyzed reaction relative to that of the uncatalyzed reaction; that is, catalysis results from the preferential binding and therefore the stabilization of the transition state relative to the substrate (Fig. 11-13). In other words, the free energy difference between an enzyme–substrate complex (ES) and an enzyme–transition state

Figure 11-13. Key to Function. Effect of preferential transition state binding. The reaction coordinate diagram for a hypothetical enzyme-catalyzed reaction involving a single substrate is blue, and the diagram for the corresponding uncatalyzed reaction is red. $\Delta G_N^{\ddagger}$ is the free energy of activation for the nonenzymatic reaction and $\Delta G_E^{\ddagger}$ is the free energy of activation for the enzyme-catalyzed reaction. The small dips in the reaction coordinate diagram for the enzyme-catalyzed reaction arise from the binding of substrate and product to the enzyme.
✳ See the Animated Figures.

complex ($ES^{\ddagger}$) is less than the free energy difference between S and $S^{\ddagger}$ in an uncatalyzed reaction.

As we saw in Section 11-2, the rate enhancement of a catalyzed reaction is given by $e^{\Delta\Delta G^{\ddagger}_{cat}/RT}$, where $\Delta\Delta G^{\ddagger}_{cat}$ is the difference in the values of $\Delta G^{\ddagger}$ for the uncatalyzed ($\Delta G^{\ddagger}_{N}$) and the catalyzed ($\Delta G^{\ddagger}_{E}$) reactions. Thus, a rate enhancement of 10^6, which requires that an enzyme bind its transition state complex with 10^6-fold higher affinity than its substrate, corresponds to a 34.2 kJ·mol^{-1} stabilization at 25°C, roughly the free energy of two hydrogen bonds. Consequently, the enzymatic binding of a transition state by two hydrogen bonds that cannot form when the substrate first binds to the enzyme should result in a rate enhancement of ~10^6 based on this effect alone.

It is commonly observed that an enzyme binds poor substrates, which have low reaction rates, as well as or even better than good ones, which have high reaction rates. Thus, a good substrate does not necessarily bind to its enzyme with high affinity, but it does so on activation to the transition state.

Transition State Analogs Are Enzyme Inhibitors

If an enzyme preferentially binds its transition state, then it can be expected that **transition state analogs,** *stable molecules that geometrically and electronically resemble the transition state, are potent inhibitors of the enzyme.* For example, the reaction catalyzed by **proline racemase** from *Clostridium sticklandii* is thought to occur via a planar transition state:

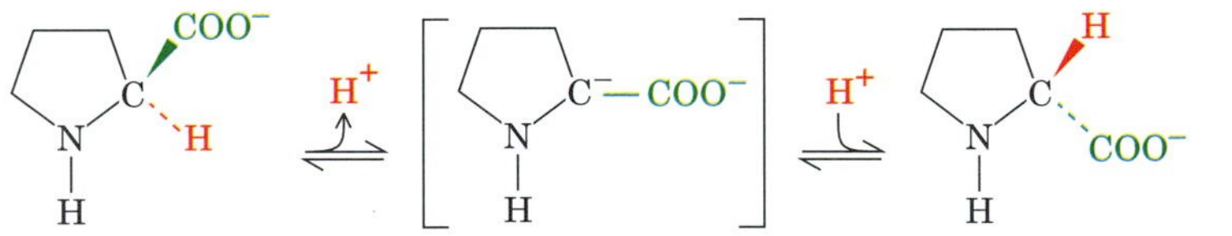

| L-Proline | Planar transition state | D-Proline |

Proline racemase is inhibited by the planar analogs of proline, **pyrrole-2-carboxylate** and **Δ-1-pyrroline-2-carboxylate,**

Pyrrole-2-carboxylate   Δ-1-Pyrroline-2-carboxylate

both of which bind to the enzyme with 160-fold greater affinity than does proline. These compounds are therefore thought to be analogs of the transition state in the proline racemase reaction.

Hundreds of transition state analogs for various enzymes have been reported. Some are naturally occurring antibiotics. Others were designed to investigate the mechanism of particular enzymes or to act as specific enzyme inhibitors for therapeutic or agricultural use. Indeed, *the theory that enzymes bind transition states with higher affinity than substrates has led to a rational basis for drug design based on the understanding of specific enzyme reaction mechanisms.* Transition state binding also explains why antibodies raised against putative transition state analogs are able to catalyze the corresponding reaction (see Box 11-2).

Box 11-2

BIOCHEMISTRY IN FOCUS

Catalytic Antibodies

Antibodies and enzymes share the ability to bind compounds with great specificity and high affinity. This property has been exploited in the development of antibodies with catalytic activity. A catalytic antibody, however, rather than tightly binding a substrate, instead tightly binds the structure that corresponds to the transition state of the reaction. The development of such catalytic antibodies, or **abzymes**, thus confirms the idea that stabilization of the transition state is the basis for much of an enzyme's catalytic power.

The first catalytic antibodies were raised against compounds designed to mimic the transition state of an esterolytic reaction:

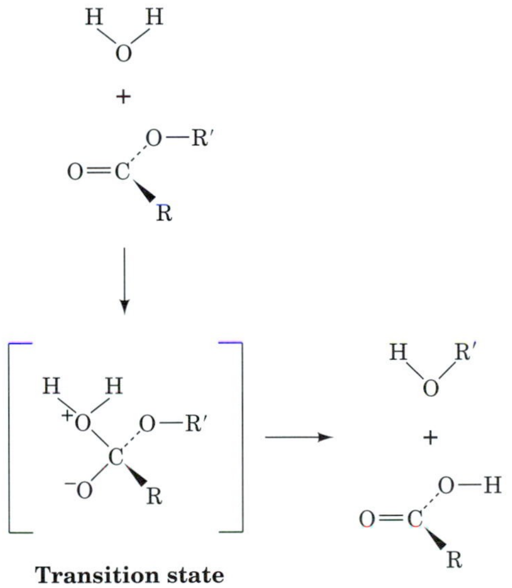

Transition state

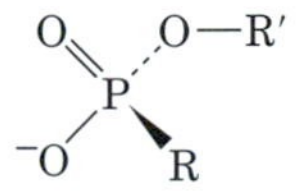

Stable analog of transition state

The tetrahedral anionic phosphonate compound, which resembles the transition state, clearly differs from the planar, neutral substrate or product. The phosphonate, coupled to a carrier protein (it is too small to be antigenic on its own), elicited antibodies with the binding characteristics of an enzyme. Some of the resulting antibodies increased the rate of hydrolysis of model substrates by a factor of 1000.

Many additional studies have shown that it is possible to generate antibodies that can catalyze a wide variety of reactions, in some cases as much as 10^8 times faster than the uncatalyzed reaction. This rate acceleration is typical of enzymes. Other features of antibody-catalyzed reactions are also

reminiscent of enzyme-catalyzed reactions. For example, the antigen-binding sites of antibodies, where catalysis occurs, contain ionizable amino acid side chains whose activities are therefore pH dependent. Antibodies exhibit the same kind of substrate specificity as do enzymes, and yield reaction products with defined stereochemistry. Furthermore, catalytic antibodies, like enzymes, can be inhibited by compounds that bind to the antibody but do not undergo a reaction. The similarities between catalytic antibodies and enzymes extends beyond their shared ability to bind transition states. Their catalytic mechanisms are often similar. For example, the X-ray structure of an antibody with the catalytic activity of a serine protease revealed that its "active site" contains a Ser and a His residue whose arrangement in space is closely superimposable on that of the Ser and His residues in the active sites of serine proteases (Fig. 11-24).

The catalytic repertoire of antibodies appears to be virtually limitless. In fact, careful design of an immunogenic compound can lead to antibodies capable of catalyzing reactions for which there is no known enzyme. One such reaction, which is important in organic synthesis, is the **Diels–Alder reaction**, the simplest example of which is the reaction of 1,3-butadiene with ethylene to yield cyclohexene.

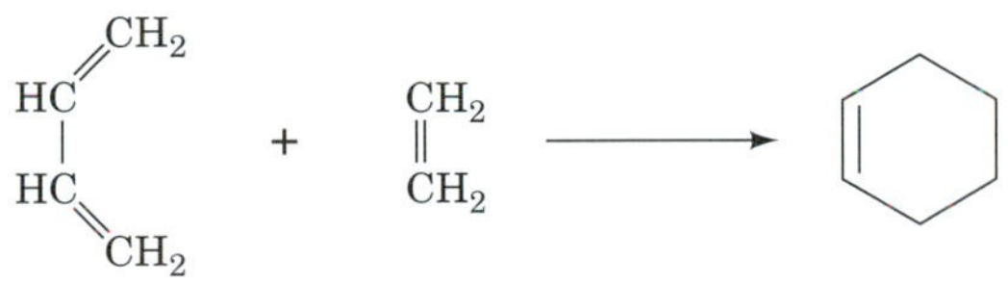

1,3-Butadiene   Ethylene   Cyclohexene

Antibodies, like enzymes, can catalyze an even wider range of reactions with the aid of cofactors. A cofactor-binding site can be constructed at an antigen-binding site by using a substrate–cofactor complex as the immunogen, by covalently linking the cofactor to the antibody, or by engineering a cofactor-binding site (e.g., three His residues that bind a metal ion) through recombinant DNA techniques.

The potential applications of catalytic antibodies include catalysis of difficult chemical transformations and the production of extremely specific peptidases that could be used to manipulate proteins in much the same way that restriction endonucleases are used to manipulate nucleic acids (Section 3-4A). One advantage of antibody catalysts over conventional enzymes is that novel catalysts can be generated relatively quickly (monoclonal antibodies can be obtained within a few weeks). Another is that because all antibody proteins are fundamentally similar, standard procedures can be used to synthesize them in bacterial expression systems and to purify them.

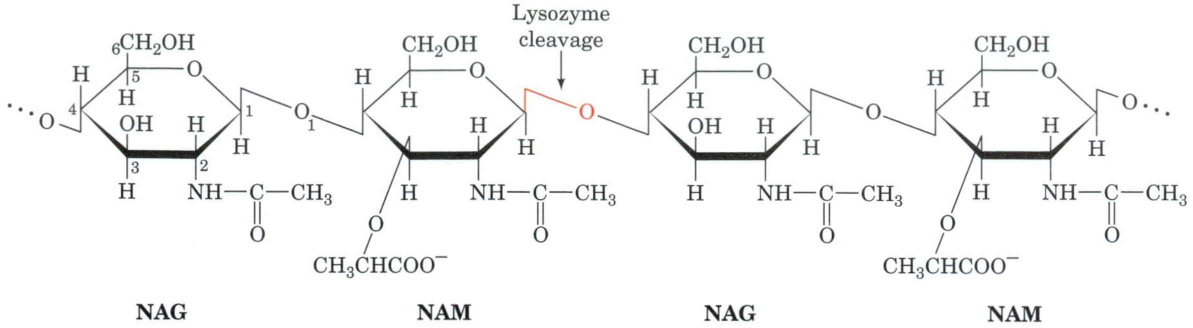

Figure 11-14. **The lysozyme cleavage site.** The enzyme cleaves after a β(1→4) linkage in the alternating NAG–NAM polysaccharide component of bacterial cell walls.

4. LYSOZYME

In the remainder of this chapter, we investigate the catalytic mechanisms of some well-characterized enzymes. In doing so, we shall see how enzymes apply the catalytic principles described in the preceding section.

Lysozyme is an enzyme that destroys bacterial cell walls. It does so by hydrolyzing the β(1→4) glycosidic linkages from *N*-acetylmuramic acid (**NAM** or **MurNAc**) to *N*-acetylglucosamine (**NAG** or **GlcNAc**) in cell wall peptidoglycans (Fig. 11-14 and Section 8-3B). It likewise hydrolyzes β(1→4)-linked poly(NAG) (chitin; Section 8-2B), a cell wall component of most fungi as well as the major component of the exoskeletons of insects and crustaceans. Lysozyme occurs widely in the cells and secretions of vertebrates, where it may function as a bactericidal agent or help dispose of bacteria after they have been killed by other means.

Hen egg white (HEW) lysozyme is the most widely studied species of lysozyme and is one of the mechanistically best understood enzymes. It is a rather small protein (14.6 kD) whose single polypeptide chain consists of 129 amino acid residues and is internally cross-linked by four disulfide bonds. Lysozyme catalyzes the hydrolysis of its substrate at a rate that is ~10^8-fold greater than that of the uncatalyzed reaction.

A. Enzyme Structure

The X-ray structure of HEW lysozyme, which was elucidated by David Phillips in 1965, shows that the protein molecule is roughly ellipsoidal in shape with dimensions $30 \times 30 \times 45$ Å (Fig. 11-15). *Its most striking feature is a prominent cleft, the substrate-binding site, that traverses one face of the molecule.* The polypeptide chain forms five helical segments as well as a three-stranded antiparallel β sheet that comprises one wall of the binding cleft (Fig. 11-15b). As expected, most of the nonpolar side chains are in the interior of the molecule, out of contact with the aqueous solvent.

Figure 11-15. **The X-ray structure of HEW lysozyme (opposite).** (*a*) The polypeptide chain is shown with a bound $(NAG)_6$ substrate (*green*). The positions of the backbone C_α atoms are indicated together with those of the side chains that line the substrate-binding site and form disulfide bonds. The substrate's sugar rings are designated A, at its nonreducing end (*right*), through F, at its reducing end (*left*). Lysozyme catalyzes the hydrolysis of the glycosidic bond between residues D and E. Rings A, B, and C are observed in the X-ray structure of the complex of $(NAG)_3$ with lysozyme; the positions of rings D, E, and F were inferred from model building studies. [Figure copyrighted © by Irving Geis.] (*b*) A ribbon diagram of lysozyme highlighting the protein's secondary structure and indicating the positions of its catalytically important side chains. (*c*) A computer-generated model showing the protein's molecular envelope (*purple*) and C_α backbone (*blue*). The side chains of the catalytic residues, Asp 52 (*upper*) and Glu 35 (*lower*), are colored yellow. Note the enzyme's prominent substrate-binding cleft. [Courtesy of Arthur Olson, The Scripps Research Institute, La Jolla, California.] Parts *a, b,* and *c* have approximately the same orientation. ● See the Interactive Exercises and Kinemage Exercise 9.

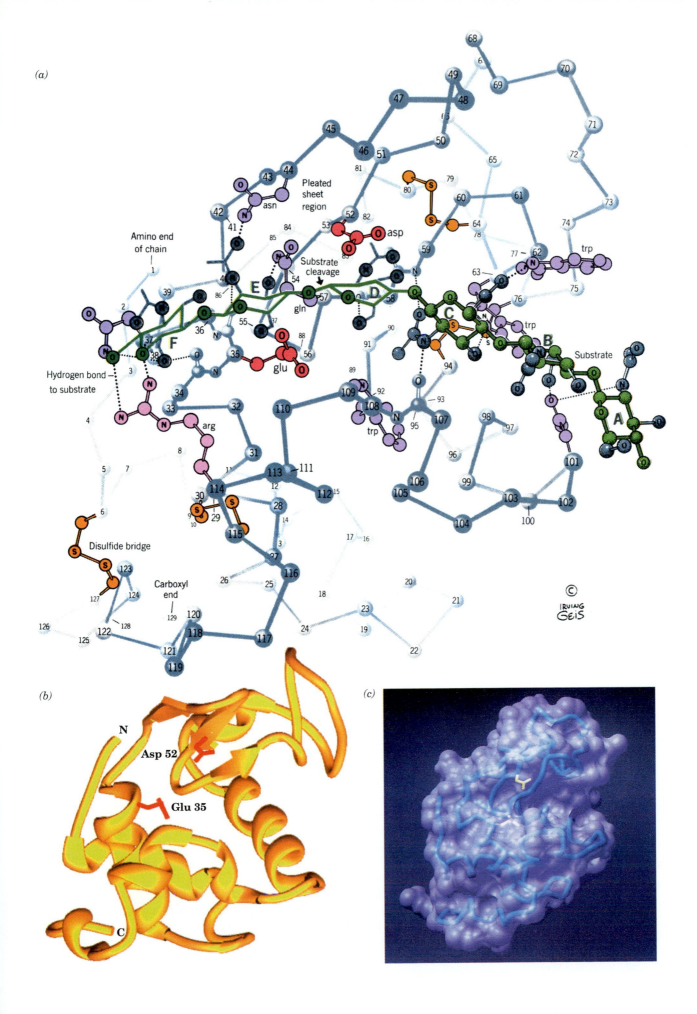

(a)

68
6
69
70
49
47
48
71
45
46
51
50
72
81
65
Pleated
sheet
region
73
43
44
79
asn
42
52
82
80
60
41
53
asp
64
59
78
61
Amino end
of chain
84
85
N
54
Substrate
cleavage
63
77
62 trp
1
86
E
O
57
58
N
76
75
39
36
D
C
S
N
trp
2
O
F
55
37
gln
91
B
Substrate
N
35
56
90
94
Hydrogen bond
to substrate
38
glu
89
109
92
A
34
O
95
108
93
3
33
32
110
trp
107
98
4
31
113
111
106
96
97
8
112
99
101
5
7
114
115
12
105
103
102
30
S
S
28
100
9
29
14
104
6
S
10
27
17
16
123
Disulfide bridge
124
116
20
26
25
18
21
126
122
128
125
121
118
119
117
120
23
24
19
22
127

Carboxyl
end
129

ⒸIRVING GEIS

(b)

N

Asp 52

Glu 35

C

(c)

Box 11-3

BIOCHEMISTRY IN CONTEXT

Observing Enzyme Action by X-Ray Crystallography

The atomic rearrangements that occur during catalysis are to some extent accessible to X-ray crystallographic analysis. Because protein crystals are mostly solvent, not only are the native structures of proteins preserved in the crystalline state, but often their catalytic functions are also intact. Substrates can diffuse through solvent channels in the crystal into the enzyme's active site. However, an enzyme binds its substrates only transiently before it catalyzes a reaction and releases the products. For this reason, the chemical transformations during an enzymatic reaction would be accessible only from "before and after" snapshots, that is, from the structure of the enzyme in the absence of substrate and, in some cases, the structure of the enzyme with its loosely bound products. Several approaches have been used to get around this limitation.

The X-ray structures of enzymes in complexes with slow-reacting substrates may be determined. In this approach, the substrate must remain stably bound to the enzyme for the day or more that is usually required to measure the crystal's X-ray diffraction intensities. Alternatively, an unreactive substrate analog can be used. In this case, the molecule binds much as a substrate would but is not subject to the catalytic reaction. This approach yields somewhat incomplete information, however, since some of the molecular interactions that allow catalysis to occur are missing or distorted. Nevertheless, such data, along with knowledge of nonenzymatic reaction mechanisms, often allows the enzymatic reaction mechanism to be deduced with a fair degree of certainty. Indeed, most of our present structural knowledge of enzyme–substrate interactions is based on this technique.

Enzymatic reactions, like all chemical reactions, are temperature sensitive. Therefore, cooling a crystallized enzyme can significantly slow the rate at which it reacts with a substrate that diffuses to it. For example, cooling a crystal to less than 40 K can slow reaction times from less than a microsecond to hours or days, long enough to measure a set of X-ray diffraction intensities.

Another approach, which has been successfully used to investigate the atomic events at the heme group of myoglobin, solves two problems inherent in analyzing rapid biochemical events. First, taking a "snapshot" of an enzyme in action requires a short exposure time, in analogy to conventional photography. Accordingly, very intense radiation must be used, in this case an X-ray beam generated by a synchrotron (a type of "atom smasher" in which electrons are accelerated around a circular track to near light speed, thereby emitting intense X-radiation). This radiation is many orders of magnitude more intense than that available from conventional X-ray generators. Second, all the molecules in the crystal must act simultaneously; otherwise, the data will be "blurry." In a study of CO dissociation from myoglobin (the CO binds to myoglobin in much the same way as does O_2 but much more tightly), the molecules were made to dissociate from the heme on cue by a flash of laser light with a duration of a few nanoseconds. Subsequent molecular motions, ultimately leading to the recombination of the CO with myoglobin, were monitored at intervals of microseconds to milliseconds. With refinements, this experimentally complicated technique may eventually prove useful for documenting the operations of enzymes, whose catalytic cycles are often complete within nanoseconds.

The elucidation of an enzyme's mechanism of action requires a knowledge of the structure of its enzyme–substrate complex. This is because, even if the active site residues have been identified through chemical and physical means, their three-dimensional arrangement relative to the substrate as well as to each other must be known in order to understand how the enzyme works. However, an enzyme binds its good substrates only transiently before it catalyzes a reaction and releases products. Consequently, *most of our knowledge of enzyme–substrate complexes derives from X-ray studies of enzymes in their complexes with inhibitors or poor substrates* that remain stably bound to the enzyme for the day or more required to measure a protein crystal's X-ray diffraction intensities (but see Box 11-3).

Lysozyme's Catalytic Site Was Identified through Model Building

X-Ray structural analysis revealed that the trisaccharide $(NAG)_3$, which is only slowly hydrolyzed by lysozyme, binds to the enzyme at a site corresponding to the site occupied by residues A, B, and C as drawn in Fig. 11-15a. Phillips used model building to investigate how a larger substrate

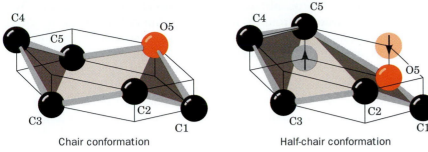

Figure 11-16. Chair and half-chair conformations. Hexose rings normally assume the chair conformation. It is postulated, however, that binding by lysozyme distorts the D ring into the half-chair conformation in which atoms C1, C2, C5, and O5 are coplanar. ✳ **See the Animated Figures.**

could bind to the enzyme. Lysozyme's active site cleft is long enough to accommodate an oligosaccharide of six residues (labeled A to F in Fig. 11-15a). However, the fourth residue (D) appeared unable to bind to the enzyme because its C6 and O6 atoms too closely contact protein side chains and residue C. This steric interference could be relieved by distorting the glucose ring from its normal chair conformation to that of a half-chair (Fig. 11-16). This distortion, which renders atoms C1, C2, C5, and O5 of residue D coplanar, moves the C6 group from its normal equatorial position to an axial position, where it makes no close contacts and can hydrogen bond to the backbone carbonyl group of Gln 57. Continuing the model building, Phillips found that residues E and F apparently bind to the enzyme without distortion and with a number of favorable hydrogen bonding and van der Waals contacts. Some of the hydrogen bonds through which the six substrate saccharide residues bind to the enzyme are diagrammed in Fig. 11-17.

We are almost in a position to identify lysozyme's catalytic site. In the enzyme's natural substrate, every second residue is an NAM. Model building, however, indicated that a lactyl side chain cannot be accommodated in the binding subsites of either residues C or E. Hence, the NAM residues must bind to the enzyme in subsites B, D, and F.

$$\cdots-NAG-NAM-NAG-NAM-NAG-NAM-\cdots\rightarrow\left(\begin{array}{c}\text{reducing}\\\text{end}\end{array}\right)$$
$$\quad\;A\quad\;\;B\quad\;\;\;C\quad\;\;\;D\quad\;\;\;E\quad\;\;\;F$$

The observation that lysozyme hydrolyzes β(1→4) linkages from NAM to NAG implies that bond cleavage occurs either between residues B and C or between residues D and E. Since (NAG)$_3$ is stably bound but not cleaved by the enzyme while spanning subsites B and C, the probable cleavage site is between residues D and E. In-

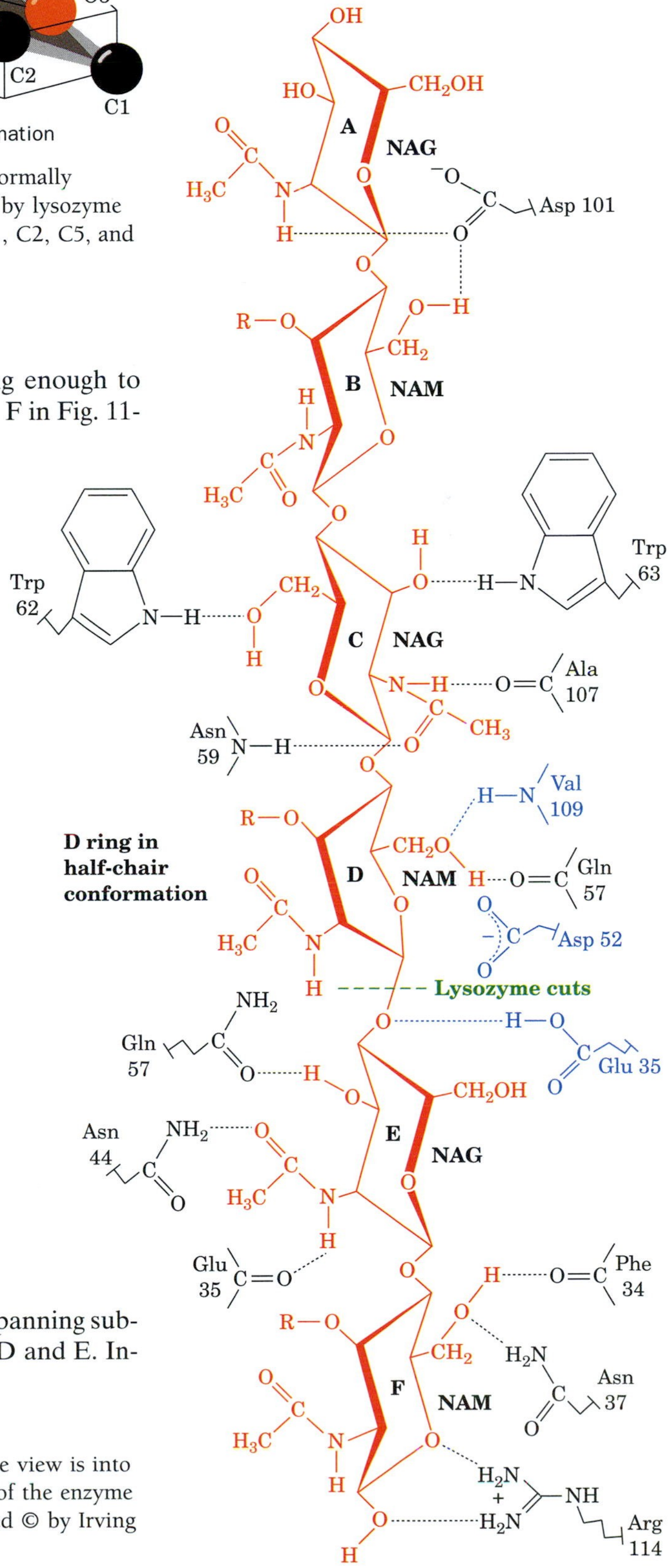

Figure 11-17. The interactions of lysozyme with its substrate. The view is into the binding cleft with the heavier edges of the rings facing the outside of the enzyme and the lighter ones against the bottom of the cleft. [Figure copyrighted © by Irving Geis.] 🔵 **See Kinemage Exercise 9.**

deed, lysozyme nearly quantitatively hydrolyzes $(NAG)_6$ between the second and third residues from its reducing terminus (the end with a free C1—OH), just as is expected if the enzyme has six saccharide-binding subsites and cleaves its bound substrate between subsites D and E.

The bond that lysozyme cleaves was identified by carrying out the lysozyme-catalyzed hydrolysis of $(NAG)_3$ in $H_2^{18}O$. The resulting product had ^{18}O bonded to the C1 atom of its newly liberated reducing terminus, thereby demonstrating that bond cleavage occurs between C1 and the bridge oxygen O1:

Thus, lysozyme catalyzes the hydrolysis of the C1—O1 bond of a bound substrate's D residue. Moreover, this reaction occurs with retention of configuration so that the D-ring product remains the β anomer.

B. Catalytic Mechanism

It remains to identify lysozyme's catalytic groups. The reaction catalyzed by lysozyme, the hydrolysis of a glycoside, is the conversion of an acetal to a hemiacetal. Nonenzymatic acetal hydrolysis is an acid-catalyzed reaction that involves the protonation of a reactant oxygen atom followed by cleavage of its C—O bond (Fig. 11-18). This results in the formation of a resonance-stabilized carbocation that is called an **oxonium ion.** To attain resonance stabilization, the oxonium ion's R and R′ groups must be coplanar with its C, O, and H atoms. The oxonium ion then adds water to yield the hemiacetal and regenerate the acid catalyst. In searching for catalytic groups on an enzyme that mediate acetal hydrolysis, we should therefore seek a potential acid catalyst and possibly a group that could further stabilize an oxonium ion intermediate.

Glu 35 and Asp 52 Are Lysozyme's Catalytic Residues

The only functional groups in the immediate vicinity of lysozyme's reactive center that have the required catalytic properties are the side chains of Glu 35 and Asp 52. These side chains, which are disposed to either side of the glycosidic linkage to be cleaved (Fig. 11-15), have markedly different environments. Asp 52 is surrounded by several conserved polar residues with which it forms a complex hydrogen bonding network. Asp 52 is therefore predicted to have a normal pK; that is, it should be unprotonated and hence negatively charged throughout the 3 to 8 pH range over which lysozyme is catalytically active. In contrast, the carboxyl group of Glu 35 is nestled in a predominantly nonpolar pocket where it is likely to remain protonated at unusually high pH's for carboxyl groups. The carboxyl O

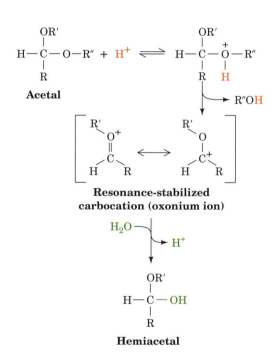

Figure 11-18.   The mechanism of the nonenzymatic acid-catalyzed hydrolysis of an acetal to a hemiacetal. The reaction involves the protonation of one of the acetal's oxygen atoms followed by cleavage of its C—O bond to form an alcohol (R″OH) and a resonance-stabilized carbocation (oxonium ion). The addition of water to the oxonium ion forms the hemiacetal and regenerates the H⁺ catalyst. Note that the oxonium ion's C, O, H, R, and R′ atoms all lie in the same plane.

atoms of both Asp 52 and Glu 35 are ~3 Å from the C1—O1 bond of residue D, which makes them prime candidates for electrostatic and acid catalysts, respectively. Other studies using protein-modifying reagents and site-directed mutagenesis (e.g., changing Asp 52 to Asn and Glu 35 to Gln) have verified that these residues are catalytically important.

The Phillips Mechanism

With much of the foregoing information, Phillips postulated the following enzymatic mechanism for lysozyme (Fig. 11-19):

1. Lysozyme attaches to a bacterial cell wall by binding to a hexasaccharide unit. In the process, *residue D is distorted toward the half-chair conformation* in response to the unfavorable contacts that its —C6H2OH group would otherwise make with the protein.

2. *Glu 35 transfers its proton to the O1 of the D ring, the only polar group in its vicinity (this is an example of general acid catalysis). The C1—O1 bond is thereby cleaved, generating a resonance-stabilized oxonium ion at C1.*

3. The ionized carboxyl group of Asp 52 *stabilizes the developing oxonium ion through charge–charge interactions (electrostatic catalysis).* This carboxylate group cannot form a covalent bond with the substrate because the 3.0-Å minimum distance between C1 and a carboxyl O atom of Asp 52 is much greater than the ~1.5-Å length of a C—O covalent bond (in fact, some lysozyme species lack this acidic residue altogether). The bond cleavage reaction is facilitated by the strain in the D ring that distorts it to the planar half-chair conformation. This is a result of the oxonium ion's required planarity; that is, *the initial binding conformation of the D ring resembles that of the reaction's transition state (catalysis by preferential binding of the transition state;* Fig. 11-20).

4. At this point, the enzyme releases the hydrolyzed E ring with its attached polysaccharide, yielding an enzyme-stabilized oxonium ion. This intermediate subsequently adds H_2O from solution in a reversal of the preceding steps to form product and to reprotonate Glu 35. The enzymatic cleft shields the oxonium ion so that H_2O can attack only from the side that permits retention of configuration. The enzyme then releases the D-ring product with its attached saccharide, thereby completing the catalytic cycle.

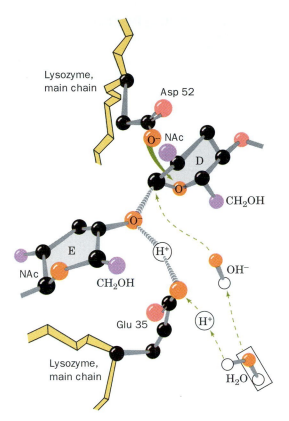

Figure 11-19. *Key to Function.* The Phillips mechanism for the lysozyme reaction. Cleavage of the glycosidic bond between the substrate D and E rings occurs through the protonation of the bridge oxygen atom by Glu 35. The resulting D-ring oxonium ion is stabilized by the proximity of the Asp 52 carboxylate group and the enzyme-induced distortion of the D ring. Once the E ring is released, H_2O from solution provides both an OH^- that combines with the oxonium ion and an H^+ that reprotonates Glu 35. NAc represents the N-acetylamino substituent at C2 of each glucose ring. ● See Kinemage Exercise 9. ✳ See the Animated Figures.

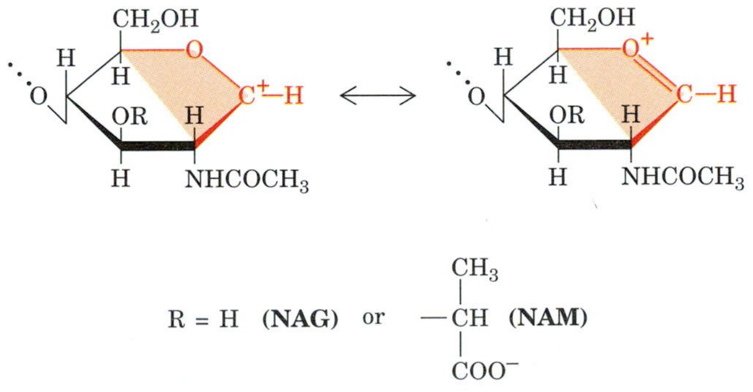

R = H (**NAG**) or —CH (**NAM**) with CH3 and COO⁻

Figure 11-20. Transition state for the lysozyme reaction. The oxonium ion form of the D ring is stabilized by resonance. This requires that atoms C1, C2, C5, and O5 be coplanar *(shading)*; that is, the hexose ring must assume the half-chair conformation.

***Table 11-4.* Binding Free Energies of HEW Lysozyme Subsites**

Site	Bound Saccharide	Binding Free Energy $(kJ \cdot mol^{-1})$
A	NAG	−7.5
B	NAM	−12.3
C	NAG	−23.8
D	**NAM**	**+12.1**
E	NAG	−7.1
F	NAM	−7.1

Source: Chipman, D.M. and Sharon, N., *Science* **165**, 459 (1969).

Role of Strain

Many of the mechanistic investigations of lysozyme have had the elusive goal of establishing the catalytic role of strain. Measurements of the binding equilibria of various oligosaccharides to lysozyme indicate that all saccharide residues except that binding to the D subsite contribute energetically toward the binding of substrate to lysozyme; only binding NAM in the D subsite requires a free energy input (Table 11-4). The Phillips mechanism explains this observation as being indicative of the energy penalty of straining the D ring from its preferred chair conformation toward the half-chair form.

Experiments with inhibitors also support the Phillips mechanism. As we have discussed in Section 11-3F, an enzyme that catalyzes a reaction by the preferential binding of its transition state has a greater binding affinity for an inhibitor that has the transition state geometry (transition state analog) than it does for its substrate. The δ-lactone analog of (NAG)$_4$ (Fig. 11-21) is a transition state analog of lysozyme since *this compound's lactone ring has the half-chair conformation that geometrically resembles the proposed oxonium ion transition state of the substrate's D ring*. X-Ray studies indicate, in accordance with prediction, that this inhibitor binds to lysozyme's A—B—C—D subsites such that the lactone ring occupies the D subsite in a half-chairlike conformation.

Despite the foregoing, *the role of substrate distortion in lysozyme catalysis has been questioned*. Theoretical studies by Michael Levitt and Arieh Warshel on substrate binding by lysozyme suggested that the protein is too flexible to mechanically distort the D ring of a bound substrate. Indeed, Nathan Sharon and David Chipman determined that the NAG lactone inhibitor (Fig. 11-21) binds to the D subsite with only 9.2 kJ·mol^{-1} greater affinity than does NAG. This quantity, as is explained in Section 11-3F, corresponds to no more than an ~40-fold rate enhancement of the lysozyme reaction as a result of strain (recall that the difference in binding energy between a transition state analog and a substrate is indicative of the enzyme's rate enhancement arising from the preferential binding of the transition state). Such an enhancement is hardly a major portion of lysozyme's ~10^8-fold rate enhancement (accounting for only ~20% of the reaction's $\Delta\Delta G^{\ddagger}_{cat}$). Moreover, an ***N*-acetylxylosamine (XylNAc)** residue,

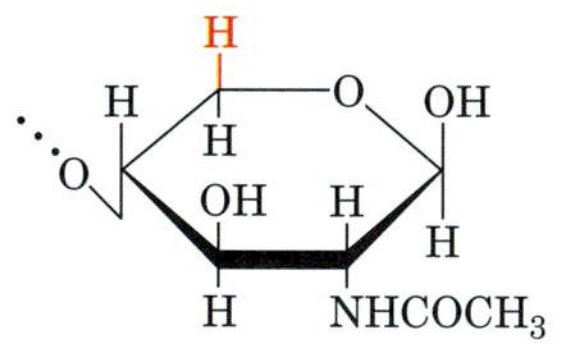

***N*-Acetylxylosamine residue**

which lacks the sterically hindered —C6H$_2$OH group of NAM and NAG, has only marginally greater binding affinity for the D subsite than does

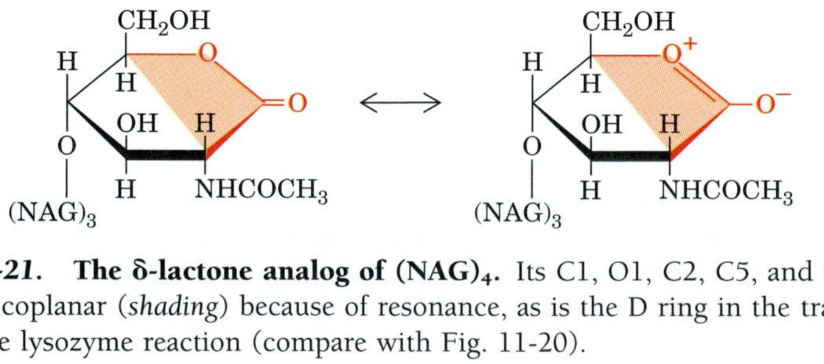

Figure 11-21.   The δ-lactone analog of (NAG)$_4$. Its C1, O1, C2, C5, and O5 atoms are coplanar (*shading*) because of resonance, as is the D ring in the transition state of the lysozyme reaction (compare with Fig. 11-20).

NAG. Yet, recall that the Phillips mechanism postulates that it is the unfavorable contacts made by this —C6H$_2$OH group that promote D-ring distortion.

The apparent inconsistencies among the foregoing experimental observations have been largely rationalized by Michael James' X-ray crystal structure determination of lysozyme in complex with NAM–NAG–NAM. This trisaccharide binds, as expected, to the B, C, and D subsites of lysozyme. *The NAM in the D subsite, in agreement with the Phillips mechanism, is distorted to the half-chair conformation with its —C6H$_2$OH group in a nearly axial position due to steric clashes that would otherwise occur with the acetamido group of the C-subsite NAG.* This strained conformation is stabilized by a strong hydrogen bond between the D ring O6 and the backbone NH of Val 109 (transition state stabilization). Indeed, the mutation of Val 109 to Pro, which lacks the NH group to make such a hydrogen bond, inactivates the enzyme.

The unexpectedly small free energy differences in binding NAG, NAG lactone, and XylNAc to the D subsite are explained by the fact that they are not the normal occupants of the D subsite and are able to bind there without strain. NAM, the normal occupant of this subsite, has a bulky lactyl side chain which prevents it from binding in an unstrained manner. Finally, a carboxyl O atom of Glu 35 forms a strong hydrogen bond with the O1 atom of the D-site NAM which, no doubt, facilitates the proton transfer step in the Phillips mechanism. Thus, the Phillips mechanism appears to be substantially correct.

5. SERINE PROTEASES

Our next example of enzymatic mechanisms is a diverse and widespread group of proteolytic enzymes known as the **serine proteases,** so named because they have a common catalytic mechanism involving a peculiarly reactive Ser residue. The serine proteases include digestive enzymes from prokaryotes and eukaryotes, as well as more specialized proteins that participate in development, blood coagulation (clotting), inflammation, and numerous other processes. In this section, we focus on some of the best studied serine proteases: chymotrypsin, trypsin, and elastase.

A. The Active Site

Chymotrypsin, trypsin, and elastase are digestive enzymes that are synthesized by the pancreas and secreted into the duodenum (the small intestine's upper loop). All these enzymes catalyze the hydrolysis of peptide (amide) bonds but with different specificities for the side chains flanking the scissile (to be cleaved) peptide bond. Chymotrypsin is specific for a bulky hydrophobic residue preceding the scissile bond, trypsin is specific for a positively charged residue, and elastase is specific for a small neutral residue (Table 5-5). Together, they form a potent digestive team.

Chymotrypsin's catalytically important groups have been identified by chemical labeling studies. A diagnostic test for the presence of the active site Ser of serine proteases is its reaction with **diisopropylphosphofluoridate (DIPF),** which irreversibly inactivates the enzyme (*at right*). Other Ser residues, including those on the same protein, do not react with DIPF. *DIPF reacts only with Ser 195 of chymotrypsin, thereby demonstrating that this residue is the enzyme's active site Ser.* This specificity makes DIPF and related compounds extremely toxic (see Box 11-4).

Diisopropylphospho-
fluoridate (DIPF)

DIP–Enzyme

Box 11-4

BIOCHEMISTRY IN HEALTH AND DISEASE

Nerve Poisons

The use of DIPF as an enzyme-inactivating agent came about through the discovery that organophosphorus compounds such as DIPF are potent nerve poisons. The neurotoxicity of DIPF arises from its ability to inactivate **acetylcholinesterase**, an enzyme that catalyzes the hydrolysis of **acetylcholine**.

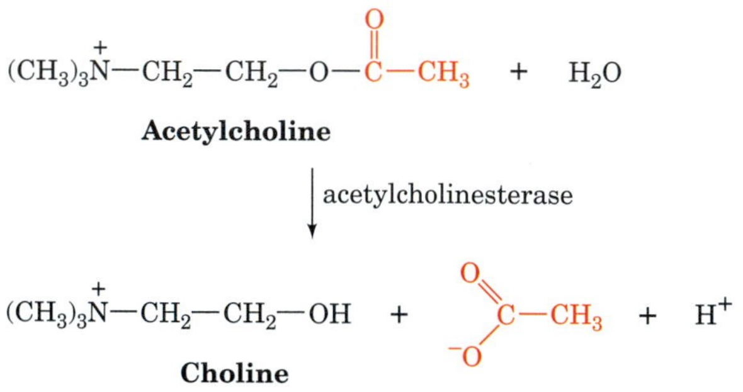

$$(CH_3)_3 \overset{+}{N}-CH_2-CH_2-O-\overset{\overset{O}{\|}}{C}-CH_3 \; + \; H_2O$$

Acetylcholine

acetylcholinesterase

$$(CH_3)_3 \overset{+}{N}-CH_2-CH_2-OH \; + \; \overset{\overset{O}{\|}}{\underset{-O}{C}}-CH_3 \; + \; H^+$$

Choline

The esterase activity of acetylcholinesterase, like that of chymotrypsin (Section 11-1B), requires a reactive Ser residue.

Acetylcholine is a **neurotransmitter**: It transmits nerve impulses across certain types of **synapses** (junctions between nerve cells). Acetylcholinesterase in the synapse normally degrades acetylcholine so that the nerve impulse has a duration of only a millisecond or so. The inactivation of acetylcholinesterase prevents hydrolysis of the neurotransmitter. As a result, the acetylcholine receptor, which is a Na^+–K^+ channel, remains open for longer than normal, thereby interfering with the regular sequence of nerve impulses. DIPF is so toxic to humans (death occurs through the inability to

breathe) that it has been used militarily as a nerve gas. Related compounds, such as **parathion** and **malathion**,

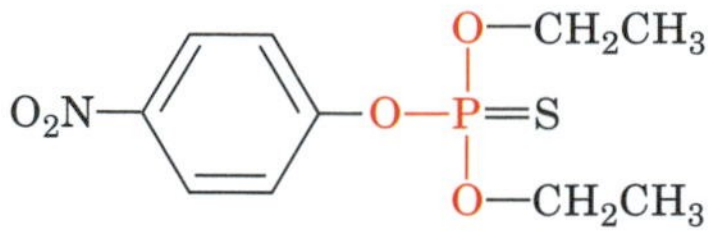

Parathion

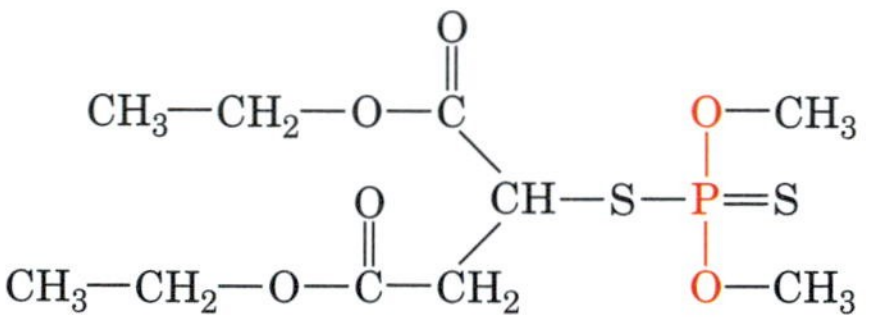

Malathion

are useful insecticides because they are far more toxic to insects than to mammals. Neurotoxins such as DIPF and **sarin** (which gained notoriety after its release by terrorists in a Tokyo subway in 1995)

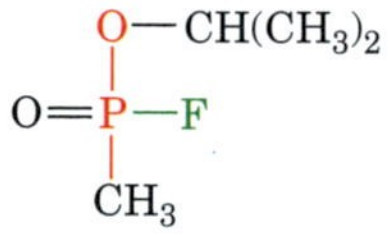

$$O=\overset{\overset{O-CH(CH_3)_2}{|}}{\underset{CH_3}{P}}-F$$

Sarin

are inactivated by the enzyme **paraoxonase**. This enzyme occurs as two isoforms (one has Arg at position 192 and the other has Gln) with different activities, and individuals express widely differing levels of the enzyme. These factors may account for the large observed differences in individuals' sensitivity to nerve poisons.

A second catalytically important residue, His 57, was discovered through **affinity labeling.** In this technique, a substrate analog bearing a reactive group specifically binds at the enzyme's active site, where it reacts to form a stable covalent bond with a nearby susceptible group (these reactive substrate analogs have been dubbed the "Trojan horses" of biochemistry). The affinity labeled group(s) can subsequently be identified.

Chymotrypsin specifically binds **tosyl-L-phenylalanine chloromethylketone (TPCK),**

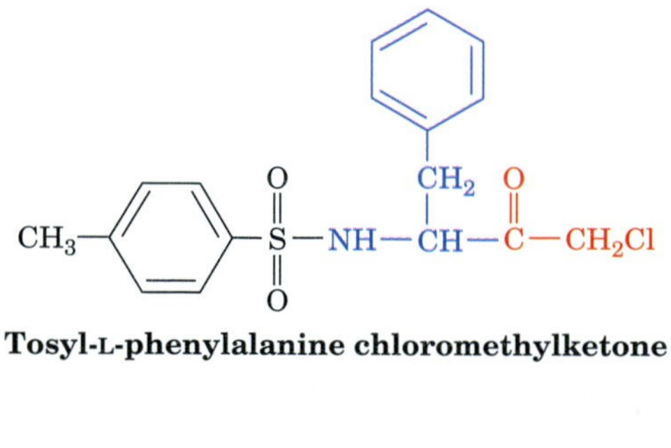

Tosyl-L-phenylalanine chloromethylketone

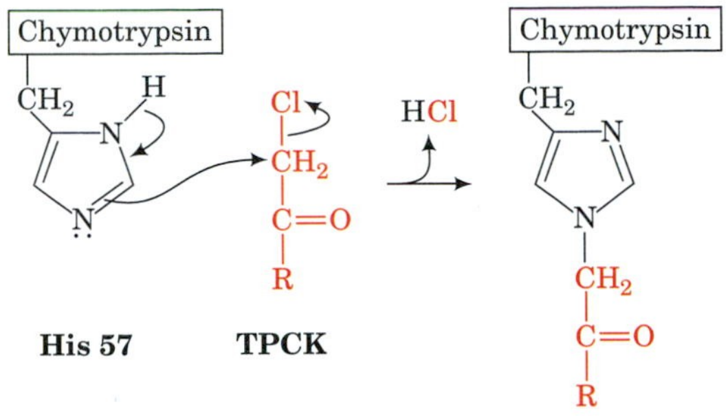

Figure 11-22. Reaction of TPCK with His 57 of chymotrypsin.

because of its resemblance to a Phe residue (one of chymotrypsin's pre-ferred substrate residues). Active site–bound TPCK's chloromethylketone group is a strong alkylating agent; it reacts only with His 57 (Fig. 11-22), thereby inactivating the enzyme. Trypsin, which prefers basic residues, is similarly inactivated by **tosyl-L-lysine chloromethylketone.**

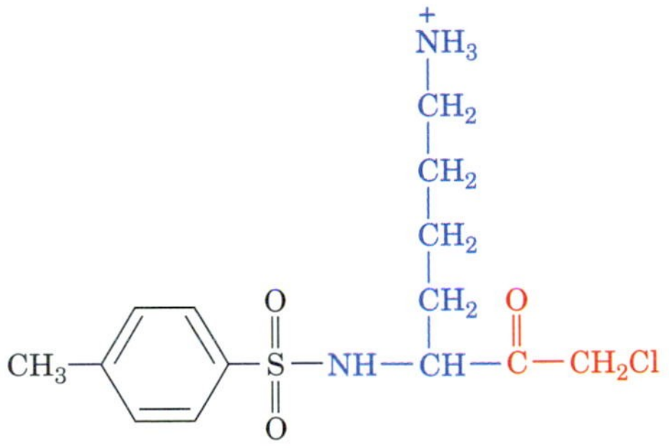

Tosyl-L-lysine chloromethylketone

B. X-Ray Structures

Chymotrypsin, trypsin, and elastase are strikingly similar: The primary structures of these ~240-residue enzymes are ~40% identical (for com-parison, the α and β chains of human hemoglobin have a 44% sequence identity). Furthermore, all these enzymes have a reactive Ser and a cat-alytically essential His. It therefore came as no surprise when their X-ray structures all proved to be closely related.

The structure of bovine chymotrypsin was elucidated in 1967 by David Blow. This was followed by the determination of the structures of bovine trypsin (Fig. 11-23) by Robert Stroud and Richard Dickerson, and porcine elastase by David Shotton and Herman Watson. Each of these proteins is folded into two domains, both of which have extensive regions of an-tiparallel β sheets in a barrel-like arrangement but contain little helix. For convenience in comparing the structures of these three enzymes, we shall assign them the same residue numbering system—that of bovine **chymotrypsinogen,** the 245-residue precursor of chymotrypsin (Section 11-5D).

In all three structures, the catalytically essential His 57 and Ser 195 residues are located in the enzyme's substrate-binding site (center of Fig. 11-23*a*). The X-ray structures also show that Asp 102, which is present in all serine proteases, is buried in a nearby solvent-inaccessible pocket. *These*

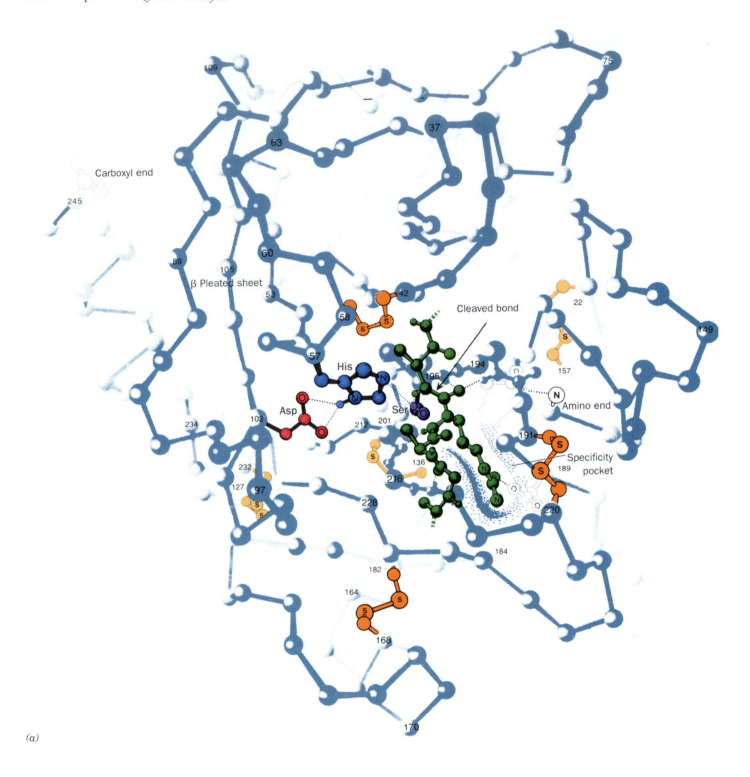

(a)

Figure 11-23.   The X-ray structure of bovine trypsin. (*a*) A drawing of the enzyme showing its disulfide bonds (*orange*) and the side chains of the catalytic triad, Ser 195 (*purple*), His 57 (*blue*), and Asp 102 (*red*). A polypeptide substrate (*green*) is shown with its Arg side chain occupying the enzyme's specificity pocket (*stippling*). The active sites of chymotrypsin and elastase contain almost identically arranged catalytic triads. [Figure copyrighted © by Irving Geis.] (*b*) (*opposite*) A ribbon diagram of trypsin highlighting its secondary structure and indicating the arrangement of its catalytic triad. (*c*) A computer-generated drawing showing the surface of trypsin (*blue*) superimposed on its polypeptide backbone (*purple*). The side chains of the catalytic triad are shown in green. [Courtesy of Arthur Olson, The Scripps Research Institute, La Jolla, California.] Parts *a*, *b*, and *c* have approximately the same orientation. ● **See Kinemage Exercise 10-1.**

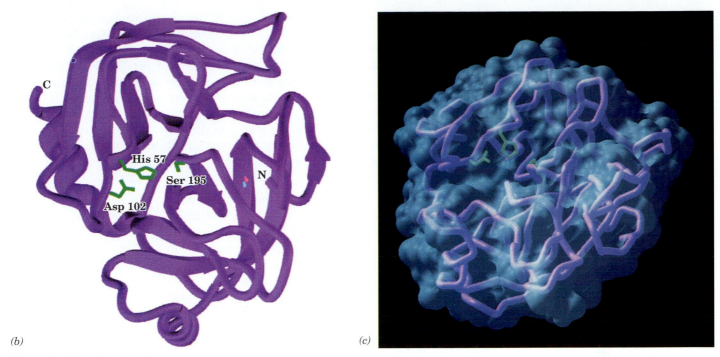

(b)

(c)

Figure 11-23. (*Continued*)

three invariant residues form a hydrogen bonded constellation referred to as the **catalytic triad** (Figs. 11-23 and 11-24).

Substrate Specificities Are Only Partially Rationalized

The X-ray structures of the above three enzymes suggest the basis for their differing substrate specificities:

1. In chymotrypsin, the bulky aromatic side chain of the preferred Phe, Trp, or Tyr residue that contributes the carbonyl group of the scissile peptide fits snugly into a slitlike hydrophobic pocket located near the catalytic groups.

2. In trypsin, the residue corresponding to chymotrypsin Ser 189, which lies at the back of the binding pocket, is the anionic residue Asp. The cationic side chains of trypsin's preferred residues, Arg and Lys, can therefore form ion pairs with this Asp residue. The rest of chymotrypsin's specificity pocket is preserved in trypsin so that it can accommodate the bulky side chains of Arg and Lys.

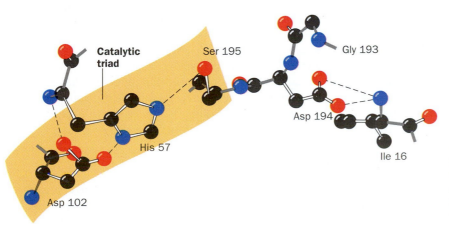

Figure 11-24. **The active site residues of chymotrypsin.** The view is in approximately the same direction as in Fig. 11-23. The catalytic triad consists of Ser 195, His 57, and Asp 102. [After Blow, D.M. and Steitz, T.A., *Annu. Rev. Biochem.* **39**, 86 (1970).]

3. Elastase is so named because it rapidly hydrolyzes the otherwise nearly indigestible Ala, Gly, and Val-rich protein **elastin** (a major connective tissue component). Elastase's binding pocket is largely occluded by the side chains of a Val and Thr residue that replace the Gly residues lining the specificity pockets in both chymotrypsin and trypsin. Consequently elastase, whose substrate-binding site is better described as merely a depression, specifically cleaves peptide bonds after small neutral residues, particularly Ala. In contrast, chymotrypsin and trypsin hydrolyze such peptide bonds extremely slowly because these small substrates cannot be sufficiently immobilized on the enzyme surface for efficient catalysis to occur.

Despite the foregoing, changing trypsin's Asp 189 to Ser by site-directed mutagenesis (Section 3-5D) does not switch its specificity to that of chymotrypsin but instead yields a poor, nonspecific protease. Replacing additional residues in trypsin's specificity pocket with those of chymotrypsin fails to significantly improve this catalytic activity. However, trypsin is converted to a reasonably active chymotrypsinlike enzyme when two surface loops that connect the walls of the specificity pocket (residues 185–188 and 221–225) are also replaced by those of chymotrypsin. These loops, which are conserved in each enzyme, are apparently necessary not for substrate binding per se but for properly positioning the scissile bond. These results highlight an important caveat for genetic engineers: Enzymes are so exquisitely tailored to their functions that they often respond to mutagenic tinkering in unexpected ways.

Evolutionary Relationships among Serine Proteases

We have seen that sequence and structural similarities among proteins reveal their evolutionary relationships (Sections 5-4 and 6-2C). *The great similarities among chymotrypsin, trypsin, and elastase indicate that these proteins arose through duplications of an ancestral serine protease gene followed by the divergent evolution of the resulting enzymes.* Indeed, the close structural resemblance of these pancreatic enzymes to certain bacterial proteases indicates that the primordial trypsin gene arose before the divergence of prokaryotes and eukaryotes.

There are two known serine proteases whose primary and tertiary structures bear no discernible relationship to each other or to chymotrypsin. Nevertheless, these proteins also contain catalytic triads at their active sites whose structures closely resemble that of chymotrypsin. These enzymes are **subtilisin,** an endopeptidase that was originally isolated from *Bacillus subtilis,* and wheat germ **serine carboxypeptidase II,** an exopeptidase. Since the orders of the corresponding active site residues in the amino acid sequences of the three types of serine proteases are quite different (Fig. 11-25), it seems highly improbable that they could have evolved from a common ancestor protein. These enzymes apparently constitute a remarkable example of **convergent evolution:** *Nature seems to have independently discovered the same catalytic mechanism at least three times.*

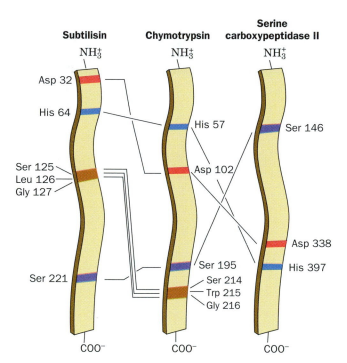

Figure 11-25.   A diagram indicating the relative positions of the active site residues of three unrelated serine proteases. The catalytic triads in subtilisin, chymotrypsin, and serine carboxypeptidase II each consist of Ser, His, and Asp residues. The peptide backbones of Ser 214, Trp 215, and Gly 216 in chymotrypsin, and their counterparts in subtilisin, participate in substrate-binding interactions. [After Robertus, J.D., Alden, R.A., Birktoft, J.J., Kraut, J., Powers, J.C., and Wilcox, P.E., *Biochemistry* **11,** 2449 (1972).]

🔵 See Kinemage Exercise 10-2.

🔵 See Guided Exploration 10:

The Catalytic Mechanism of Serine Proteases.

C. Catalytic Mechanism

A catalytic mechanism based on considerable chemical and structural data has been formulated and is given here in terms of chymotrypsin (Fig. 11-26), although it applies to all serine proteases and certain other hydrolytic enzymes:

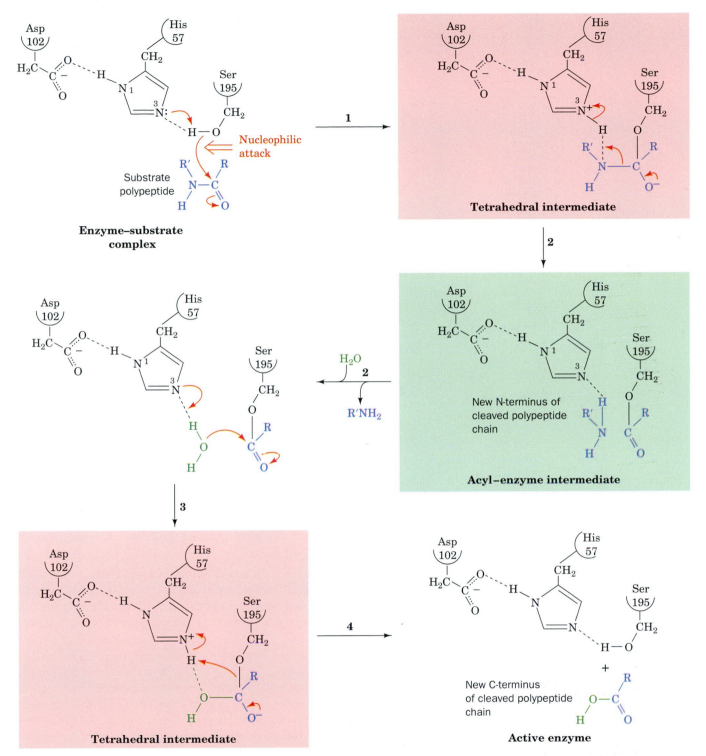

Figure 11-26.   *Key to Function*. The catalytic mechanism of the serine proteases. The reaction involves (**1**) the nucleophilic attack of the active site Ser on the carbonyl carbon atom of the scissile peptide bond to form the tetrahedral intermediate; (**2**) the decomposition of the tetrahedral intermediate to the acyl–enzyme intermediate through general acid catalysis by the active site Asp-polarized His, followed by loss of the amine product and its replacement by a water molecule; (**3**) the reversal of Step 2 to form a second tetrahedral intermediate, and (**4**) the reversal of Step 1 to yield the reaction's carboxyl product and the active enzyme.

1. After chymotrypsin has bound a substrate, Ser 195 nucleophilically attacks the scissile peptide's carbonyl group to form the **tetrahedral intermediate**, which resembles the reaction's transition state (covalent catalysis). X-Ray studies indicate that Ser 195 is ideally positioned to carry out this nucleophilic attack (i.e., catalysis also occurs by proximity and orientation effects). This nucleophilic attack involves transfer of a proton to the imidazole ring of His 57, thereby forming an imidazolium ion (general base catalysis). This process is aided by the polarizing effect of the unsolvated carboxylate ion of Asp 102, which is hydrogen bonded to His 57 (electrostatic catalysis). The tetrahedral intermediate has a well-defined, although transient, existence. We shall see that much of chymotrypsin's catalytic power derives from its preferential binding of this transition state (transition state binding catalysis).

2. The tetrahedral intermediate decomposes to the **acyl–enzyme intermediate** under the driving force of proton donation from N3 of His 57 (general acid catalysis). The amine leaving group ($R'NH_2$, the new N-terminal portion of the cleaved polypeptide chain) is released from the enzyme and replaced by water from the solvent. The acyl–enzyme intermediate is highly susceptible to hydrolytic cleavage. Despite its instability, elastase's acyl–enzyme intermediate has been visualized through X-ray crystallography at temperatures low enough to arrest catalysis.

3 & 4. The acyl–enzyme intermediate is deacylated by what is essentially the reversal of the previous steps followed by the release of the resulting carboxylate product (the new C-terminal portion of the cleaved polypeptide chain), thereby regenerating the active enzyme. In this process, water is the attacking nucleophile and Ser 195 is the leaving group.

Serine Proteases Preferentially Bind the Transition State

Detailed comparisons of the X-ray structures of several serine protease–inhibitor complexes have revealed a further structural basis for catalysis in these enzymes (Fig. 11-27):

1. The conformational distortion that occurs with the formation of the tetrahedral intermediate causes the now anionic carbonyl oxygen of the scissile peptide to move deeper into the active site so as to occupy a previously unoccupied position called the **oxyanion hole.**

2. There, it forms two hydrogen bonds with the enzyme that cannot form when the carbonyl group is in its normal trigonal conformation. The two enzymatic hydrogen bond donors were first noted by Joseph Kraut to occupy corresponding positions in chymotrypsin and subtilisin. He proposed the existence of the oxyanion hole on the basis of the premise that convergent evolution had made the active sites of these unrelated enzymes functionally identical.

3. The tetrahedral distortion, moreover, permits the formation of an otherwise unsatisfied hydrogen bond between the enzyme and the backbone NH group of the residue preceding the scissile peptide bond.

This preferential binding of the transition state (or the tetrahedral intermediate) over the enzyme–substrate complex or the acyl–enzyme intermediate is responsible for much of the catalytic efficiency of serine proteases. Thus, mutating any or all of the residues in chymotrypsin's catalytic triad yields en-

(a)

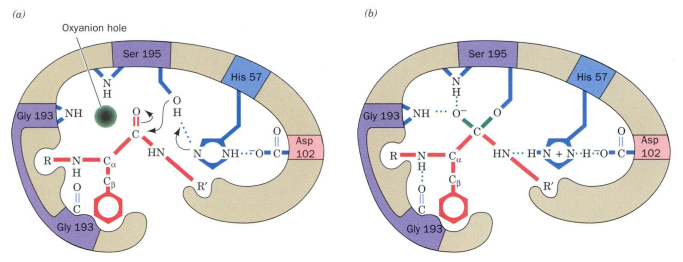

(b)

Figure 11-27. Transition state stabilization in the serine pro-teases. (*a*) When the substrate binds to the enzyme, the trigonal carbonyl carbon of the scissile peptide is conformationally constrained from binding in the oxyanion hole (*upper left*). (*b*) In the tetrahedral intermediate, the now charged carbonyl oxygen of the scissile peptide (the oxyanion) enters the oxyanion hole and hydrogen bonds to the backbone NH groups of Gly 193 and

Ser 195. The consequent conformational distortion permits the NH group of the residue preceding the scissile peptide bond to form an otherwise unsatisfied hydrogen bond to Gly 193. Serine proteases therefore preferentially bind the tetrahedral intermedi-ate. [After Robertus, J.D., Kraut, J., Alden, R.A., and Birktoft, J.J., *Biochemistry* **11**, 4302 (1972).] ● **See Kinemage Exercise 10-3.**

zymes that still enhance proteolysis by $\sim 5 \times 10^4$-fold over the uncatalyzed reaction (versus a rate enhancement of $\sim 10^{10}$ for the native enzyme). Sim-ilarly, the reason that DIPF is such an effective inhibitor of serine proteases is because its tetrahedral phosphate group makes this compound a transi-tion state analog.

The Tetrahedral Intermediate Resembles the Complex of Trypsin with Trypsin Inhibitor

Perhaps the most convincing structural evidence for the existence of the tetrahedral intermediate was provided by Robert Huber in an X-ray study of the complex between **bovine pancreatic trypsin inhibitor (BPTI)** and trypsin. The 58-residue BPTI binds to the active site region of trypsin to form a complex with a tightly packed interface and a network of hydrogen bonded cross-links. This interaction prevents any trypsin that is prematurely activated in the pancreas from digesting that organ (Section 11-5D). The complex's 10^{13} M^{-1} association constant, among the largest of any known protein–protein interaction, emphasizes BPTI's physiological importance.

The portion of BPTI in contact with the trypsin active site resembles bound substrate. A specific Lys side chain of BPTI occupies the trypsin specificity pocket (Fig. 11-28*a*), and the inhibitor's Lys-Ala peptide bond is

(a)

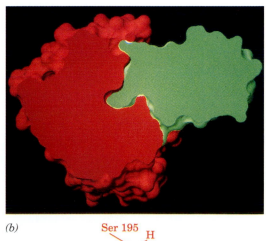

(b)

Figure 11-28. The trypsin–BPTI complex. (*a*) The X-ray structure is shown as a computer-generated cutaway drawing indicating how trypsin (*red*) binds BPTI (*green*). The green protrusion extending into the red cavity near the center of the figure represents the inhibitor's Lys 15 side chain occupying trypsin's specificity pocket. Note the close complementary fit of the two proteins. [Courtesy of Michael Connolly, New York University.] (*b*) Trypsin Ser 195 is in closer-than-van der Waals contact with the carbonyl carbon of BPTI's scissile peptide, which is pyramidally distorted toward Ser 195. The normal proteolytic reaction is apparently arrested somewhere along the reaction coordinate preceding the tetrahedral intermediate. ● See Kinemage Exercise 10-1.

positioned as if it were the scissile peptide bond (Fig. 11-28*b*). What is most remarkable about the BPTI–trypsin complex is that its conformation is well along the reaction coordinate toward the tetrahedral intermediate: The side chain oxygen of trypsin Ser 195, the active Ser, is in closer-than-van der Waals contact (2.6 Å) with the pyramidally distorted carbonyl carbon of BPTI's "scissile" peptide. However, the proteolytic reaction cannot proceed past this point because of the rigidity of the complex and because it is so tightly sealed that the leaving group cannot leave and water cannot enter the reaction site.

Protease inhibitors are common in nature, where they have protective and regulatory functions. For example, certain plants release protease inhibitors in response to insect bites, thereby causing the offending insect to starve by inactivating its digestive enzymes. Protease inhibitors constitute ~10% of the blood plasma proteins. For instance, **α_1-proteinase inhibitor,** which is secreted by the liver, inhibits **leukocyte elastase** (leukocytes are a type of white blood cell; the action of leukocyte elastase is thought to be part of the inflammatory process). Pathological variants of α_1-proteinase inhibitor with reduced activity are associated with **pulmonary emphysema,** a degenerative disease of the lungs resulting from the hydrolysis of its elastic fibers. Smokers also suffer from reduced activity of their α_1-proteinase inhibitor because smoking oxidizes its active site Met residue.

D. Zymogens

Proteolytic enzymes are usually biosynthesized as somewhat larger inactive precursors known as **zymogens** (enzyme precursors, in general, are known as **proenzymes**). In the case of digestive enzymes, the reason for this is clear: If these enzymes were synthesized in their active forms, they would digest the tissues that synthesized them. Indeed, **acute pancreatitis,** a painful and sometimes fatal condition that can be precipitated by pancreatic trauma, is characterized by the premature activation of the digestive enzymes synthesized by this organ.

The activation of **trypsinogen,** the zymogen of trypsin, occurs when trypsinogen enters the duodenum from the pancreas. **Enteropeptidase,** a serine protease whose secretion from the duodenal mucosa is under hormonal control, excises the N-terminal hexapeptide from trypsinogen by specifically cleaving its Lys 15—Ile 16 peptide bond (Fig. 11-29). Since this activating cleavage occurs at a trypsin-sensitive site (recall that trypsin cleaves after Arg and Lys residues), the small amount of trypsin produced by enteropeptidase also catalyzes trypsinogen activation, generating even

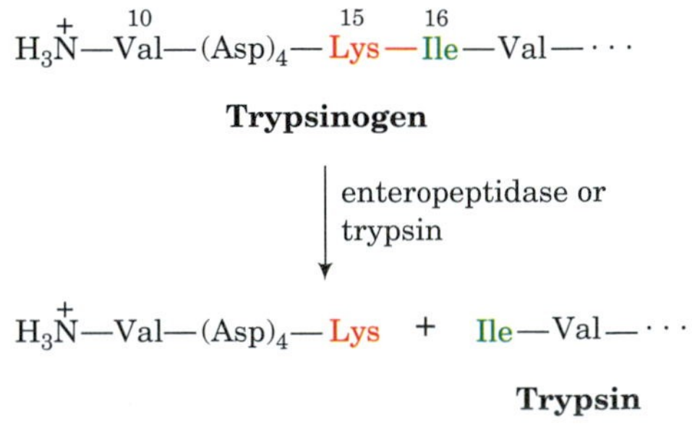

$$H_3\overset{+}{N}\text{—Val}\overset{10}{—}\text{(Asp)}_4\text{—}\overset{15}{\text{Lys}}\overset{16}{\text{—Ile—Val—}}\cdots$$

Trypsinogen

enteropeptidase or trypsin

$$H_3\overset{+}{N}\text{—Val—(Asp)}_4\text{—Lys}\quad+\quad\text{Ile—Val—}\cdots$$

Trypsin

Figure 11-29.   The activation of trypsinogen to trypsin. Proteolytic excision of the N-terminal hexapeptide is catalyzed by either enteropeptidase or trypsin. The chymotrypsinogen residue-numbering system is used here; that is, Val 10 is actually trypsinogen's N-terminus and Ile 16 is trypsin's N-terminus.

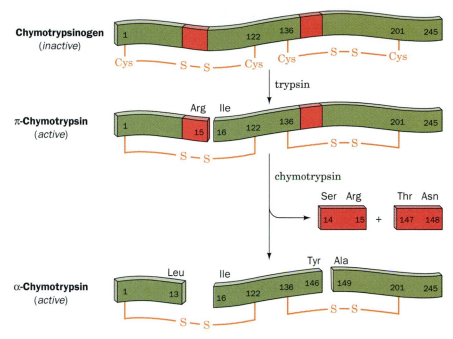

Figure 11-30.   The activation of chymotrypsinogen by proteolytic cleavage.
Both π- and α-chymotrypsin are enzymatically active. ● See Kinemage Exercise
10-4.

more trypsin, etc. Thus, trypsinogen activation is said to be **autocatalytic.**

Chymotrypsinogen is activated by trypsin-catalyzed cleavage of its
Arg 15—Ile 16 peptide bond, to form π-chymotrypsin (Fig. 11-30). π-
Chymotrypsin subsequently undergoes **autolysis** (self-digestion) to
specifically excise two dipeptides, Ser 14–Arg 15 and Thr 147–Asn 148,
thereby yielding the equally active enzyme α-chymotrypsin (usually referred to simply as chymotrypsin). The biological significance of this latter process, if any, is unknown.

Proelastase, the zymogen of elastase, is activated by a single tryptic
cleavage that excises a short N-terminal peptide. Trypsin also activates pancreatic **procarboxypeptidases A** and **B** and **prophospholipase A₂** (Section
9-1C). The autocatalytic nature of trypsinogen activation and the fact that
trypsin activates other hydrolytic enzymes makes it essential that trypsinogen not be activated in the pancreas. We have seen that the all but irreversible binding of trypsin inhibitors such as BPTI to trypsin is a defense
against trypsinogen's inappropriate activation. Two other protective mechanisms are

1. The trypsin-catalyzed activation of trypsinogen occurs quite slowly,
   presumably because the unusually large negative charge of its highly
   conserved N-terminal hexapeptide (Fig. 11-29) repels the Asp at the
   back of trypsin's specificity pocket.
2. Pancreatic zymogens are stored in intracellular vesicles known as
   zymogen granules, whose membranous walls are thought to be resistant to proteolytic degradation.

Sequential proenzyme activation makes it possible to quickly generate large
quantities of active enzymes in response to diverse physiological signals.
For example, the serine proteases that lead to blood clotting are synthesized as zymogens by the liver and circulate until they are activated by injury to a blood vessel (see Box 11-5).

Box 11-5

BIOCHEMISTRY IN HEALTH AND DISEASE

The Blood Coagulation Cascade

When a blood vessel is damaged, a clot forms as a result of the aggregation of platelets (small enucleated blood cells) and the formation of an insoluble **fibrin** network that traps additional blood cells. Fibrin is produced from the soluble circulating protein **fibrinogen** through the action of the serine protease **thrombin**. Thrombin is the last in a series of coagulation enzymes that are sequentially activated by proteolysis of their zymogen forms. The overall process is known as the **coagulation cascade**, although experimental evidence shows that the pathway is not strictly linear, as the waterfall analogy might suggest.

The various components of the coagulation cascade, which include enzymes as well as nonenzymatic protein cofactors, are assigned Roman numerals, largely for historical reasons that do not reflect their order of action *in vivo*. The suffix *a* denotes an active factor.

The catalytic domains of the coagulation proteases resemble trypsin in sequence and mechanism but are much more specific for their substrates. Additional domains mediate interactions with cofactors and help anchor the proteins to the platelet membrane, which serves as a stage for many of the coagulation reactions.

Coagulation is initiated when a protein released from tissues (**tissue factor**) forms a complex with circulating **factor**

Zymogens Have Distorted Active Sites

Since the zymogens of trypsin, chymotrypsin, and elastase have all their catalytic residues, why aren't they enzymatically active? Comparisons of the X-ray structures of trypsinogen with that of trypsin, and of chymotrypsinogen with that of chymotrypsin, show that on activation, the newly liberated N-terminal Ile 16 residue moves from the surface of the protein to an internal position, where its free cationic amino group forms an ion pair with the invariant anionic Asp 194, which is close to the catalytic triad (Fig. 11-24). Without this conformational change, the enzyme cannot properly bind its substrate or stabilize the tetrahedral intermediate because its specificity pocket and oxyanion hole are improperly formed. This provides further structural evidence favoring the role of transition state binding in the catalytic mechanism of serine proteases. Nevertheless, because their catalytic triads are structurally intact, the zymogens of serine proteases actually have low levels of enzymatic activity, an observation that was made only after the above structural comparisons suggested that this might be the case.

VII or VIIa (factor VIIa is generated from factor VII by trace amounts of other coagulation proteases, including factor VIIa itself). The tissue factor–VIIa complex proteolytically converts the zymogen **factor X** to factor Xa. Factor Xa then converts **prothrombin** to thrombin, which subsequently generates fibrin. The tissue factor–dependent steps of coagulation are known as the **extrinsic pathway** because the source of tissue factor is extravascular. The extrinsic pathway is quickly dampened through the action of a protein that inhibits factor VII once factor Xa has been generated.

Sustained thrombin activation requires the activity of the **intrinsic pathway** (so named because all its components are present in the circulation). The intrinsic pathway is stimulated by the tissue factor–VIIa complex, which converts **factor IX** to its active form, factor IXa. The ensuing thrombin activates a number of components of the intrinsic pathway, including **factor XI**, a protease that activates factor IX, to maintain coagulation in the absence of tissue factor or factor VIIa. Thrombin also activates **factors V and VIII**, which are cofactors rather than proteases. Factor Va promotes prothrombin activation by factor Xa, and factor VIIIa promotes factor X activation by factor IXa. Thus, thrombin promotes its own activation through a feedback mechanism that amplifies the preceding steps of the cascade. **Factor XIII**, an enzyme that chemically cross-links fibrin to form a strong fibrous network, is also activated by thrombin.

The intrinsic pathway of coagulation can be triggered by exposure to negatively charged surfaces such as glass. Con-

sequently, blood clots when it is collected in a clean glass test tube. In the absence of tissue factor, a fibrin clot may not appear for several minutes, but when tissue factor is present, a clot forms within a few seconds. This suggests that rapid blood clotting *in vivo* requires tissue factor as well as the proteins of the intrinsic pathway. Additional evidence for the importance of the extrinsic pathway is that individuals who are deficient in factor VII tend to bleed excessively. Abnormal bleeding also results from congenital defects in factor VIII (**hemophilia a**) or factor IX (**hemophilia b**). Interestingly, a factor XI deficiency causes only a mild bleeding disorder.

The sequential activation of zymogens in the coagulation cascade leads to a burst of thrombin activity, since trace amounts of factors VIIa, IXa, and Xa can activate much larger amounts of their respective substrates. Perhaps not surprisingly, thrombin eventually activates mechanisms that shut down clot formation, thereby limiting the duration of the clotting process and hence the extent of the clot. Such control of clotting is of extreme physiological importance since the formation of even one inappropriate blood clot within an individual's lifetime may have fatal consequences.

[Figure adapted from Davie, E.W., *Thromb. Haemost.* 74, 2 (1995).]

SUMMARY

1. Enzymes, almost all of which are proteins, are grouped into six mechanistic classes.

2. Enzymes accelerate reactions by factors of up to at least 10^{15}.

3. The substrate specificity of an enzyme depends on the geometric and electronic character of its active site.

4. Some enzymes catalyze reactions with the assistance of metal ion cofactors or organic coenzymes that function as reversibly bound cosubstrates or as permanently associated prosthetic groups. Many coenzymes are derived from vitamins.

5. Enzymes catalyze reactions by decreasing the activation free energy, $\Delta G^{\ddagger}$, which is the free energy required to reach the transition state, the point of highest free energy in the reaction.

6. Enzymes use the same catalytic mechanisms employed by chemical catalysts, including general acid and general base catalysis, covalent catalysis, and metal ion catalysis.

7. The arrangement of functional groups in an enzyme active site allows catalysis by electrostatic effects and proximity and orientation effects.

8. A particularly important mechanism of enzyme-mediated catalysis is the preferential binding of the transition state of the catalyzed reaction.

9. The catalytic mechanism of lysozyme involves Glu 35, in its uncharged form, acting as an acid catalyst to cleave the polysaccharide substrate between its D and E rings. The resulting positively charged planar oxonium ion is electrostatically stabilized by Asp 52 in its carboxylate form. The reaction is facilitated by the distortion of residue D to the planar half-chair conformation, which resembles the reaction's transition state.

10. Serine proteases contain a Ser–His–Asp catalytic triad near a binding pocket that helps determine the enzymes' substrate specificity.

11. Catalysis in the serine proteases occurs through acid–base catalysis, covalent catalysis, proximity and orientation effects, electrostatic catalysis, and by preferential transition state binding in the oxyanion hole.

12. Synthesis of pancreatic proteases as inactive zymogens protects the pancreas from self-digestion. Zymogens are activated by specific proteolytic cleavages.

REFERENCES

Catalytic Mechanisms

Fersht, A., *Enzyme Structure and Mechanism* (2nd ed.), Freeman (1985).

Gerlt, J.A., Protein engineering to study enzyme catalytic mechanisms, *Curr. Opin. Struct. Biol.* **4,** 593–600 (1994). [Describes how information can be gained from mutagenesis and structural analysis of enzymes.]

Hackney, D.D., Binding energy and catalysis, *in* Sigman, D.S. and Boyer, P.D. (Eds.), *The Enzymes* (3rd ed.), Vol. 19, pp. 1–36, Academic Press (1990).

Kraut, J., How do enzymes work? *Science* **242,** 533–540 (1988). [A brief and very readable review of transition state theory and applications.]

Kyte, J., *Mechanism in Protein Chemistry,* Garland Publishing (1995).

Schramm, V.L., Horenstein, B.A., and Kline, P.C., Transition state analysis and inhibitor design for enzymatic reactions, *J. Biol. Chem.* **269,** 18259–18262 (1994).

Lysozyme

Johnson, L.N., Cheetham, J., McLaughlin, P.J., Acharya, K.R., Barford, D., and Phillips, D.C., Protein–oligosaccharide interactions: Lysozyme, phosphorylase, amylases, *Curr. Top. Microbiol. Immunol.* **139,** 81–134 (1988).

Kirby, A.J., Mechanism and stereoelectronic effects in the lysozyme reaction, *CRC Crit. Rev. Biochem.* **22,** 283–315 (1987).

Strynadka, N.C.J. and James, M.N.G., Lysozyme revisited: crystallographic evidence for distortion of an *N*-acetylmuramic acid residue bound in site D, *J. Mol. Biol.* **220,** 401–424 (1991).

Serine Proteases

Craik, C.S., Roczniak, S., Largman, C., and Rutter, W.J., The catalytic role of the active site aspartic acid in serine proteases, *Science* **237,** 909–913 (1987).

Davie, E.W., Biochemical and molecular aspects of the coagulation cascade, *Thromb. Haemost.* **74,** 1–6 (1995). [A brief review by one of the pioneers of the cascade hypothesis.]

Hedstrom, L., Szilagyi, L., and Rutter, W.J., Converting trypsin to chymotrypsin: the role of surface loops, *Science* **255,** 1249–1253 (1992).

Perona, J.J. and Craik, C.S., Evolutionary divergence of substrate specificity within the chymotrypsin-like serine protease fold, *J. Biol. Chem.* **272,** 29987–29990 (1997).

Phillips, M.A. and Fletterick, R.J., Proteases, *Curr. Opin. Struct. Biol.* **2,** 713–720 (1992). [A review of the mechanisms and physiological importance of different types of proteases.]

KEY TERMS

active site	prosthetic group	general base catalysis	affinity labeling
substrate	holoenzyme	covalent catalysis	catalytic triad
EC classification	apoenzyme	metalloenzyme	convergent evolution
induced fit	vitamin	metal-activated enzyme	tetrahedral intermediate
prochirality	transition state	electrostatic catalysis	acyl–enzyme intermediate
cofactor	$\Delta G^{\ddagger}$	transition state analog	oxyanion hole
coenzyme	rate-determining step	oxonium ion	zymogen
cosubstrate	general acid catalysis	serine protease	

STUDY EXERCISES

1. What properties distinguish enzymes from other catalysts?

2. What factors influence an enzyme's substrate specificity?

3. Why are cofactors required for some enzymatic reactions?

4. Sketch and label the various parts of transition state diagrams for a reaction with and without a catalyst.

5. What is the relationship between ΔG and $\Delta G^{\ddagger}$?

6. Explain how nucleophiles act as covalent catalysts.

7. Why is it unlikely that nonenzymatic catalysts operate by transition state stabilization?

8. Describe the catalytic mechanisms in the lysozyme reaction.

9. What is the function of the oxyanion hole in serine proteases?

10. Describe the mechanisms that prevent the inappropriate activation of zymogens in the pancreas.

PROBLEMS

1. Which type of enzyme (Table 11-2) catalyzes the following reactions?

(a)

$$
\begin{array}{ccc}
\text{COO}^- & & \text{COO}^- \\
| & & | \\
\text{H}-\text{C}-\text{CH}_3 & \longrightarrow & \text{H}_3\text{C}-\text{C}-\text{H} \\
| & & | \\
\text{NH}_3^+ & & \text{NH}_3^+
\end{array}
$$

(b)

$$
\begin{array}{ccc}
\text{COO}^- & & \text{H} \\
| & & | \\
\text{C}=\text{O} \ + \ \text{H}^+ & \longrightarrow & \text{C}=\text{O} \ + \ \text{O}=\text{C}=\text{O} \\
| & & | \\
\text{CH}_3 & & \text{CH}_3
\end{array}
$$

(c)

$$
\begin{array}{c}
\text{COO}^- \\
| \\
\text{C}=\text{O} \ + \ \text{NADH} \ + \ \text{H}^+ \ \longrightarrow \\
| \\
\text{CH}_3
\end{array}
$$

$$
\begin{array}{c}
\text{COO}^- \\
| \\
\text{HO}-\text{C}-\text{H} \ + \ \text{NAD}^+ \\
| \\
\text{CH}_3
\end{array}
$$

(d)

$$
\begin{array}{c}
\text{COO}^- \\
| \\
\text{H}-\text{C}-(\text{CH}_2)_2-\text{C} \overset{\displaystyle O}{\underset{\displaystyle O^-}{\diagdown}} \ + \ \text{ATP} \ + \ \text{NH}_4^+ \longrightarrow \\
| \\
\text{NH}_3^+
\end{array}
$$

$$
\begin{array}{c}
\text{COO}^- \\
| \\
\text{H}-\text{C}-(\text{CH}_2)_2-\text{C} \overset{\displaystyle O}{\underset{\displaystyle NH_2}{\diagdown}} \ + \ \text{ADP} \ + \ \text{P}_i \\
| \\
\text{NH}_3^+
\end{array}
$$

2. Draw a transition state diagram of (a) a nonenzymatic reaction and the corresponding enzyme-catalyzed reaction in which (b) S binds loosely to the enzyme and (c) S binds very tightly to the enzyme. Compare $\Delta G^{\ddagger}$ for each case. Why is tight binding of S not advantageous?

3. Studies at different pH's show that an enzyme has two catalytically important residues whose pK's are ~4 and ~10. Chemical modification experiments indicate that a Glu and a Lys residue are essential for activity. Match the residues to their pK's and explain whether they are likely to act as acid or base catalysts.

4. The covalent catalytic mechanism of an enzyme depends on a single active-site Cys whose pK is 8. A mutation in a nearby residue alters the microenvironment so that this pK increases to 10. Would the mutation cause the reaction rate to increase or decrease? Explain.

5. Explain why RNase A cannot catalyze the hydrolysis of DNA.

6. Suggest a transition state analog for proline racemase that differs from those discussed in the text. Justify your suggestion.

7. Wolfenden has stated that it is meaningless to distinguish between the "binding sites" and the "catalytic sites" of enzymes. Explain.

8. Explain why lysozyme cleaves the artificial substrate (NAG)$_4$ ~4000 times more slowly than it cleaves (NAG)$_6$.

9. Lysozyme residues Asp 101 and Arg 114 are required for efficient catalysis, although they are located at some distance from the active site Glu 35 and Asp 52. Substituting Ala for either Asp 101 or Arg 114 does not significantly alter the enzyme's tertiary structure, but it significantly reduces its catalytic activity. Explain. [Problem by Bruce Wightman, Muhlenberg College.]

10. Design a chloromethylketone inhibitor of elastase.

11. The comparison of the active site geometries of chymotrypsin and subtilisin under the assumption that their similarities have catalytic significance has led to greater mechanistic understanding of both these enzymes. Discuss the validity of this strategy.

12. Predict the effect of mutating Asp 102 of trypsin to Asn (a) on substrate binding and (b) on catalysis.

13. Tofu (bean curd), a high-protein soybean product, is prepared in such a way as to remove the trypsin inhibitor present in soybeans. Explain the reason(s) for this treatment.

14. Explain why chymotrypsin is not self-activating as is trypsin.

The speed and efficiency of an enzyme can be quantified in order to help discern its mechanism and physiological role.
[Reza Estakhvian/Tony Stone Images/New York, Inc.]

ENZYME KINETICS, INHIBITION, AND REGULATION

Early enzymologists, often working with crude preparations of yeast or liver cells, could do little more than observe the conversion of substrates to products catalyzed by as yet unpurified enzymes. Measuring the rates of such reactions therefore came to be a powerful tool for characterizing enzyme activity. The application of simple mathematical models to enzyme activity under varying laboratory conditions, and in the presence of competing substrates or enzyme inhibitors, made it possible to deduce the probable physiological functions and regulatory mechanisms of various enzymes.

The study of enzymatic reaction rates, or **enzyme kinetics,** is no less important now than it was early in the twentieth century. In many cases, the rate of a reaction and how the rate changes in response to different conditions reveal the path followed by the reactants and are therefore indicative of the reaction mechanism. Kinetic data, combined with detailed information about an enzyme's structure and its catalytic mechanisms, provide some of the most powerful clues to the enzyme's biological function and may suggest ways to modify it for therapeutic purposes.

We begin our consideration of enzyme kinetics by reviewing chemical kinetics. Following that, we derive the basic equations of enzyme kinetics and describe the effects of inhibitors on enzymes. We also consider an example of enzyme regulation that highlights several aspects of enzyme function.

1. REACTION KINETICS

Kinetic measurements of enzymatically catalyzed reactions are among the most powerful techniques for elucidating the catalytic mechanisms of enzymes. Enzyme kinetics is a branch of chemical kinetics, so we begin this section by reviewing the principles of chemical kinetics.

A. Chemical Kinetics

A reaction of overall stoichiometry

$$A \longrightarrow P$$

where A represents reactants and P represents products, may actually occur through a sequence of **elementary reactions** (simple molecular processes) such as

$$A \longrightarrow I_1 \longrightarrow I_2 \longrightarrow P$$

Here, I_1 and I_2 symbolize **intermediates** in the reaction. *Characterization of each elementary reaction constitutes the mechanistic description of the overall reaction process.*

Reaction Order

At constant temperature, *the rate of an elementary reaction is proportional to the frequency with which the reacting molecules come together.* The proportionality constant is known as a **rate constant** and is symbolized k. For the elementary reaction $A \rightarrow P$, the instantaneous rate of appearance of product or disappearance of reactant, which is called the **velocity** (v) of the reaction, is

$$v = \frac{d[P]}{dt} = -\frac{d[A]}{dt} = k[A] \qquad [12\text{-}1]$$

In other words, the reaction velocity at any time point is proportional to the concentration of the reactant A. This is an example of a **first-order** reaction. The rate constant therefore has units of reciprocal seconds (s^{-1}). *The **reaction order** of an elementary reaction corresponds to the **molecularity** of the reaction, which is the number of molecules that must simultaneously collide to generate a product.* Thus, a first-order elementary reaction is a **unimolecular** reaction.

Consider the elementary reaction $2A \rightarrow P$. This **bimolecular** reaction is a **second-order** reaction, and its instantaneous velocity is described by

$$v = -\frac{d[A]}{dt} = k[A]^2 \qquad [12\text{-}2]$$

In this case, the reaction velocity is proportional to the square of the concentration of A, and the second-order rate constant k has units of $M^{-1} \cdot s^{-1}$.

The bimolecular reaction $A + B \rightarrow P$ is also a second-order reaction with an instantaneous velocity described by

$$v = -\frac{d[A]}{dt} = -\frac{d[B]}{dt} = k[A][B] \qquad [12\text{-}3]$$

Here, the reaction is said to be first order in [A] and first order in [B]. Unimolecular and bimolecular reactions are common. **Termolecular** reactions are unusual because the simultaneous collision of three molecules is a rare event. Fourth- and higher-order reactions are unknown.

Rate Equations

*A **rate equation** describes the progress of a reaction as a function of time* and can be derived from the equations that describe the instantaneous reaction velocity. Thus, a first-order rate equation is obtained by rearranging Eq. 12-1

$$\frac{d[A]}{[A]} = d\ln[A] = -k\,dt \qquad [12\text{-}4]$$

and integrating it from $[A]_o$, the initial concentration of A, to [A], the concentration of A at time t:

$$\int_{[A]_o}^{[A]} d\ln[A] = -k\int_0^t dt \qquad [12\text{-}5]$$

This results in

$$\boxed{\ln[A] = \ln[A]_o - kt} \qquad [12\text{-}6]$$

or, taking the antilog of both sides,

$$[A] = [A]_o\, e^{-kt} \qquad [12\text{-}7]$$

Equation 12-6 is a linear equation of the form $y = mx + b$ and can be plotted as in Fig. 12-1. Therefore, if a reaction is first order, a plot of $\ln[A]$ versus t will yield a straight line whose slope is $-k$ (the negative of the first-order rate constant) and whose intercept on the $\ln[A]$ axis is $\ln[A]_o$.

One of the hallmarks of a first-order reaction is that *the time for half of the reactant initially present to decompose, its **half-time** or **half-life**, $t_{1/2}$, is a constant and hence independent of the initial concentration of the reactant.* This is easily demonstrated by substituting the relationship $[A] = [A]_o/2$ when $t = t_{1/2}$ into Eq. 12-6 and rearranging:

$$\ln\!\left(\frac{[A]_o/2}{[A]_o}\right) = -kt_{1/2} \qquad [12\text{-}8]$$

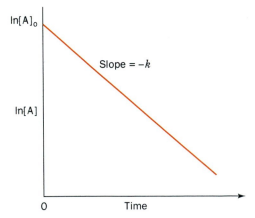

Figure 12-1.   A plot of a first-order rate equation. The slope of the line obtained when $\ln[A]$ is plotted against time gives the rate constant k.

Thus

$$t_{1/2} = \frac{\ln 2}{k} = \frac{0.693}{k} \qquad [12\text{-}9]$$

Substances that are inherently unstable, such as radioactive nuclei, decompose through first-order reactions (see Box 12-1).

Box 12-1

BIOCHEMISTRY IN FOCUS

Isotopic Labeling

In the laboratory, it is often useful to label large or small molecules so that they can be easily detected after chromatographic or electrophoretic separations or in various binding assays. One of the most common labeling techniques is to attach a radioactive isotope to a molecule or to synthesize the molecule so that it contains the radioactive isotope in place of the normally occurring isotope. Molecules labeled in this way can be detected in solution or in solid form by measuring the radioactivity emitted by the label. This method is more sensitive than spectroscopic measurements, and it is often easier to carry out than more laborious assays based on chemical or biological activities. Metabolites labeled with NMR-active isotopes such as ^{13}C can also be detected in living tissues by NMR techniques.

Some of the most common radioactive isotopes (**radionuclides**) used in biochemistry are listed below, along with their half-lives and the type of radioactivity emitted by the spontaneously disintegrating atomic nuclei.

Radionuclide	Half-life	Type of Radiation[a]
^{3}H	12 years	β
^{14}C	5715 years	β
^{24}Na	15 hours	β
^{32}P	14 days	β
^{35}S	87 days	β
^{40}K	1.25×10^{9} years	β
^{45}Ca	163 days	β
^{125}I	60 days	γ
^{131}I	8 days	β, γ

[a] β particles are emitted electrons, and γ rays are emitted photons.

Nucleic acids can be easily labeled by attaching a terminal nucleotide that contains ^{32}P in place of the normal nonradioactive ^{31}P. Proteins can be labeled by chemically or enzymatically linking ^{125}I to a Tyr residue. Assays for cell growth and division often measure the uptake of ^{3}H-labeled **thymidine** (thymidine is incorporated exclusively into DNA). Protein synthesis is similarly monitored by the appearance of ^{35}S-labeled Met in proteins. Of course, the choice of a particular isotopic label also depends on the time-course of the experiment and the method for detecting radioactivity.

A **Geiger counter**, which electronically detects the ionization of a gas caused by the passage of radiation, is not sensitive enough to detect low-energy emitters such as ^{3}H and ^{14}C. This limitation is circumvented through **liquid scintillation counting**. In this technique, a β-emitting sample is dissolved in a solvent that contains a fluorescent molecule. The β particles cause this fluor to emit light that can then be optically detected. Radioactive substances that emit γ rays are detected by **scintillation counters** when the γ rays dislodge electrons from a crystal of NaI in the counter. These electrons induce fluorescence that is measured. In **autoradiography**, a radioactive substance immobilized on paper or in an agarose or polyacrylamide gel is detected by laying X-ray film over the sample followed by incubation and development of the film (dark areas on the developed film correspond to areas exposed to radioactivity). Thin sections of tissue can also be prepared for **microradiography** by covering them with a layer of photographic emulsion and examining the developed emulsion under a microscope. Instruments that electronically measure radioactivity in solid samples without the use of film offer the advantage of digitized results and multiple exposure times (in contrast, film can be developed only once).

The use of radioactive isotopes as molecular labels is not without drawbacks. First and foremost is the danger of working with potentially mutagenic materials (irradiation can cause DNA damage). In addition, radioactive laboratory materials (samples as well as glassware) must be disposed of properly or the resulting contamination could cause errors in subsequent measurements of radioactivity as well as a health hazard. Scintillation fluid presents a particular problem for disposal because of the large volumes required (it also consists largely of organic solvents). The preceding table reveals that while disposal of short-lived radionuclides (such as ^{32}P, ^{35}S, and ^{125}I) can be accomplished mainly by storing the material until the radioactivity has decayed to insignificant levels, the safe disposal of long-lived species (such as ^{3}H and ^{14}C) is a problem that is unlikely to vanish any time soon. This is one reason why molecular labeling techniques that rely on chemical tags or fluorescent compounds are becoming increasingly popular.

In a second-order reaction with one type of reactant, $2\,A \rightarrow P$, the variation of [A] with time is quite different from that in a first-order reaction. Rearranging Eq. 12-2 and integrating it over the same limits used for the first-order reaction yields

$$\int_{[A]_o}^{[A]} -\frac{d[A]}{[A]^2} = k\int_0^t dt \qquad [12\text{-}10]$$

so that

$$\boxed{\frac{1}{[A]} = \frac{1}{[A]_o} + kt} \qquad [12\text{-}11]$$

Equation 12-11 is a linear equation in terms of the variables $1/[A]$ and t. *The half-time for a second-order reaction is expressed* $t_{1/2} = 1/k[A]_o$ *and therefore, in contrast to a first-order reaction, depends on the initial reactant concentration.* Equations 12-6 and 12-11 may be used to distinguish a first-order from a second-order reaction by plotting $\ln[A]$ versus t and $1/[A]$ versus t and observing which, if any, of these plots is a straight line.

To experimentally determine the rate constant for the second-order reaction $A + B \rightarrow P$, it is often convenient to increase the concentration of one reactant relative to the other, for example, $[B] \gg [A]$. Under these conditions, [B] does not change significantly over the course of the reaction. The reaction rate therefore depends only on [A], the concentration of the reactant that is present in limited amounts. Hence, the reaction appears to be first order with respect to A and is therefore said to be a **pseudo-first-order** reaction. The reaction is first order with respect to B when $[A] \gg [B]$.

> ◉ See Guided Exploration 11:
>
> Michaelis–Menten Kinetics, Lineweaver–Burk Plots, and Enzyme Inhibition.

B. Enzyme Kinetics

Enzymes catalyze a tremendous variety of reactions using different combinations of six basic catalytic mechanisms (Section 11-3). Some enzymes act on only a single substrate molecule; others act on two or more different substrate molecules whose order of binding may or may not be obligatory. Some enzymes form covalently bound intermediate complexes with their substrates; others do not. *Yet all enzymes can be analyzed such that their reaction rates as well as their overall efficiency can be quantified.*

The study of enzyme kinetics began in 1902 when Adrian Brown investigated the rates of hydrolysis of sucrose by the yeast enzyme **β-fructofuranosidase:**

$$\text{Sucrose} + H_2O \longrightarrow \text{glucose} + \text{fructose}$$

Brown found that when the sucrose concentration is much higher than that of the enzyme, the reaction rate becomes independent of the sucrose concentration; that is, the rate is **zero order** with respect to sucrose. He therefore proposed that the overall reaction is composed of two elementary reactions in which the substrate forms a complex with the enzyme that subsequently decomposes to products and enzyme:

$$E + S \underset{k_{-1}}{\overset{k_1}{\rightleftharpoons}} ES \overset{k_2}{\longrightarrow} P + E \qquad [12\text{-}12]$$

Here E, S, ES, and P symbolize the enzyme, substrate, **enzyme–substrate complex,** and products, respectively. According to this model, when the substrate concentration becomes high enough to entirely convert the enzyme to the ES form, the second step of the reaction becomes rate limiting and the overall reaction rate becomes insensitive to further increases in substrate concentration.

Each of the elementary reactions that make up the above enzymatic reaction is characterized by a rate constant: k_1 and k_{-1} are the forward and reverse rate constants for formation of the ES complex (the first reaction), and k_2 is the rate constant for the decomposition of ES to P (the second reaction). Here we assume, for the sake of mathematical simplicity, that the second reaction is irreversible.

The Michaelis–Menten Equation

In a complex kinetic scheme, the rate of formation of product can be expressed as the product of the rate constant of the reaction yielding product and the concentration of its immediately preceding intermediate. The general expression for the **velocity** (rate) of Reaction 12-12 is therefore

$$v = \frac{d[P]}{dt} = k_2[ES] \qquad [12\text{-}13]$$

The overall rate of production of ES is the difference between the rates of the elementary reactions leading to its appearance and those resulting in its disappearance:

$$\frac{d[ES]}{dt} = k_1[E][S] - k_{-1}[ES] - k_2[ES] \qquad [12\text{-}14]$$

This equation cannot be explicitly integrated, however, without simplifying assumptions. Two possibilities are

1. **Assumption of equilibrium.** In 1913, Leonor Michaelis and Maud Menten, building on the work of Victor Henri, assumed that $k_{-1} \gg k_2$, so that the first step of the reaction reaches equilibrium.

$$K_S = \frac{k_{-1}}{k_1} = \frac{[E][S]}{[ES]} \qquad [12\text{-}15]$$

Here K_S is the dissociation constant of the first step in the enzymatic reaction. With this assumption, Eq. 12-14 can be integrated. Although this assumption is often not correct, in recognition of the importance of this pioneering work, the enzyme–substrate complex, ES, is known as the **Michaelis complex.**

2. **Assumption of steady state.** Figure 12-2 illustrates the progress curves of the various participants in Reaction 12-12 under the physiologically common condition that substrate is in great excess over enzyme ($[S] \gg [E]$). With the exception of the initial stage of the reaction, which is usually over within milliseconds of mixing E and S, [ES] remains approximately constant until the substrate is nearly exhausted. Hence, the rate of synthesis of ES must equal its rate of consumption over most of the course of the reaction. In other words, ES maintains a **steady state** and [ES] can be treated as having a constant value:

$$\frac{d[ES]}{dt} = 0 \qquad [12\text{-}16]$$

This so-called **steady state assumption,** a more general condition than that of equilibrium, was first proposed in 1925 by G.E. Briggs and John B.S. Haldane.

In order to be useful, kinetic expressions for overall reactions must be formulated in terms of experimentally measurable quantities. The quanti-

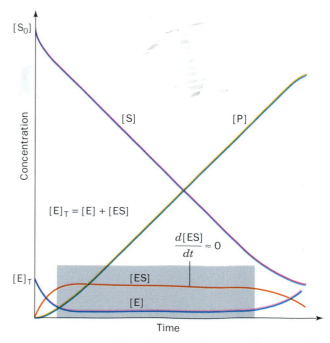

Figure 12-2. The progress curves for a simple enzyme-catalyzed reaction. With the exception of the initial phase of the reaction (before the shaded block), the slopes of the progress curves for [E] and [ES] are essentially zero as long as $[S] \gg [E]$ (within the shaded block). [After Segel, I.H., *Enzyme Kinetics*, p. 27, Wiley (1975).]
✳ See the Animated Figures.

ties [ES] and [E] are not, in general, directly measurable, but the total enzyme concentration

$$[E]_T = [E] + [ES] \qquad [12\text{-}17]$$

is usually readily determined. The rate equation for the overall enzymatic reaction can then be derived. First, Eq. 12-14 is combined with the steady state assumption (Eq. 12-16) to give

$$k_1[E][S] = k_{-1}[ES] + k_2[ES] \qquad [12\text{-}18]$$

Letting $[E] = [E]_T - [ES]$ and rearranging yields

$$\frac{([E]_T - [ES])[S]}{[ES]} = \frac{k_{-1} + k_2}{k_1} \qquad [12\text{-}19]$$

The **Michaelis constant, K_M,** is defined as

$$K_M = \frac{k_{-1} + k_2}{k_1} \qquad [12\text{-}20]$$

so Eq. 12-19 can then be rearranged to give

$$K_M[ES] = ([E]_T - [ES])[S] \qquad [12\text{-}21]$$

Solving for [ES],

$$[ES] = \frac{[E]_T[S]}{K_M + [S]} \qquad [12\text{-}22]$$

The expression for the **initial velocity** (v_o) of the reaction, the velocity (Eq. 12-13) at $t = 0$, thereby becomes

$$v_o = \left(\frac{d[P]}{dt}\right)_{t=0} = k_2[ES] = \frac{k_2[E]_T[S]}{K_M + [S]} \qquad [12\text{-}23]$$

Both $[E]_T$ and $[S]$ are experimentally measurable quantities. The use of the initial velocity (operationally taken as the velocity measured before more than ~10% of the substrate has been converted to product)—rather than just the velocity—minimizes such complicating factors as the effects of reversible reactions, inhibition of the enzyme by its product(s), and progressive inactivation of the enzyme. (This is also why the rate of the reverse reaction in Eq. 12-12 can be assumed to be zero.)

The **maximal velocity** of a reaction, V_{max}, occurs at high substrate concentrations when the enzyme is **saturated,** that is, when it is entirely in the ES form:

$$V_{max} = k_2[E]_T \qquad [12\text{-}24]$$

Therefore, combining Eqs. 12-23 and 12-24, we obtain

$$\boxed{v_o = \frac{V_{max}[S]}{K_M + [S]}} \qquad [12\text{-}25]$$

This expression, the **Michaelis–Menten equation,** *is the basic equation of enzyme kinetics.* It describes a rectangular hyperbola such as that plotted in Fig. 12-3. The saturation function for oxygen binding to myoglobin (Eq. 7-3) has the same algebraic form.

Significance of the Michaelis Constant

The Michaelis constant, K_M, has a simple operational definition. At the substrate concentration at which $[S] = K_M$, Eq. 12-25 yields $v_o = V_{max}/2$ so that K_M *is the substrate concentration at which the reaction velocity is half-*

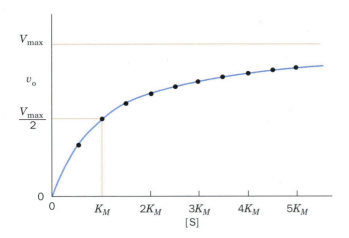

maximal. Therefore, if an enzyme has a small value of K_M, it achieves maximal catalytic efficiency at low substrate concentrations.

The K_M is unique for each enzyme–substrate pair: Different substrates that react with a given enzyme do so with different K_M values. Likewise, different enzymes that act on a single substrate have different K_M values. The magnitude of K_M varies widely with the identity of the enzyme and the nature of the substrate (Table 12-1). It is also a function of temperature and pH. The Michaelis constant (Eq. 12-20) can be expressed as

$$K_M = \frac{k_{-1}}{k_1} + \frac{k_2}{k_1} = K_S + \frac{k_2}{k_1} \qquad [12\text{-}26]$$

Since K_S is the dissociation constant of the Michaelis complex, as K_S decreases, the enzyme's affinity for substrate increases. K_M is therefore also a measure of the affinity of the enzyme for its substrate, provided k_2/k_1 is small compared to K_S, that is, $k_2 < k_{-1}$ so that the ES → P reaction proceeds more slowly than ES reverts to E + S.

k_{cat}/K_M Is a Measure of Catalytic Efficiency

We can define the **catalytic constant,** k_{cat}, of an enzyme as

$$k_{cat} = \frac{V_{max}}{[E]_T} \qquad [12\text{-}27]$$

This quantity is also known as the **turnover number** of an enzyme because it is the number of reaction processes (turnovers) that each active site catalyzes per unit time. The turnover numbers for a selection of enzymes are given in Table 12-1. Note that these quantities vary by over eight orders of

Table 12-1. **The Values of K_M, k_{cat}, and k_{cat}/K_M for Some Enzymes and Substrates**

Enzyme	Substrate	K_M (M)	k_{cat} (s^{-1})	k_{cat}/K_M (M$^{-1} \cdot$ s^{-1})
Acetylcholinesterase	Acetylcholine	9.5×10^{-5}	1.4×10^4	1.5×10^8
Carbonic anhydrase	CO_2	1.2×10^{-2}	1.0×10^6	8.3×10^7
	HCO_3^-	2.6×10^{-2}	4.0×10^5	1.5×10^7
Catalase	H_2O_2	2.5×10^{-2}	1.0×10^7	4.0×10^8
Chymotrypsin	N-Acetylglycine ethyl ester	4.4×10^{-1}	5.1×10^{-2}	1.2×10^{-1}
	N-Acetylvaline ethyl ester	8.8×10^{-2}	1.7×10^{-1}	1.9
	N-Acetyltryosine ethyl ester	6.6×10^{-4}	1.9×10^2	2.9×10^5
Fumarase	Fumarate	5.0×10^{-6}	8.0×10^2	1.6×10^8
	Malate	2.5×10^{-5}	9.0×10^2	3.6×10^7
Urease	Urea	2.5×10^{-2}	1.0×10^4	4.0×10^5

magnitude. Equation 12-24 indicates that for a simple system, such as the Michaelis–Menten model reaction (Eq. 12-12), $k_{cat} = k_2$. For enzymes with more complicated mechanisms (e.g., multiple substrates or multiple reaction intermediates), k_{cat} may be a function of several rate constants.

When $[S] \ll K_M$, very little ES is formed. Consequently, $[E] \approx [E]_T$, so Eq. 12-23 reduces to a second-order rate equation:

$$v_o \approx \left(\frac{k_2}{K_M}\right)[E]_T[S] \approx \left(\frac{k_{cat}}{K_M}\right)[E][S] \qquad [12\text{-}28]$$

Box 12-2
BIOCHEMISTRY IN FOCUS

Kinetics and Transition State Theory

How is the rate of a reaction related to its activation energy (Section 11-2)? Consider a bimolecular reaction that proceeds along the following pathway:

$$A + B \underset{}{\overset{K^\ddagger}{\rightleftharpoons}} X^\ddagger \xrightarrow{k'} P + Q$$

where $X^\ddagger$ represents the transition state. The rate of the reaction can be expressed as

$$\frac{d[P]}{dt} = k[A][B] = k'[X^\ddagger] \qquad [12\text{-A}]$$

where k is the ordinary rate constant of the elementary reaction and k' is the rate constant for the decomposition of $X^\ddagger$ to products.

Although $X^\ddagger$ is unstable, it is assumed to be in rapid equilibrium with the reactants; that is,

$$K^\ddagger = \frac{[X^\ddagger]}{[A][B]} \qquad [12\text{-B}]$$

where $K^\ddagger$ is an equilibrium constant. This central assumption of transition state theory permits the powerful formalism of thermodynamics to be applied to the theory of reaction rates.

Since $K^\ddagger$ is an equilibrium constant, it can be expressed as

$$-RT \ln K^\ddagger = \Delta G^\ddagger \qquad [12\text{-C}]$$

where T is the absolute temperature and R (8.3145 $J \cdot K^{-1} \cdot mol^{-1}$) is the gas constant (this relationship between equilibrium constants and free energy is derived in Section 1-4D). Combining the three preceding equations yields

$$\frac{d[P]}{dt} = k' \, e^{-\Delta G^\ddagger/RT} [A][B] \qquad [12\text{-D}]$$

This equation indicates that the rate of a reaction not only depends on the concentrations of its reactants, but also decreases exponentially with $\Delta G^\ddagger$. Thus, *the larger the difference between the free energy of the transition state and that of the reactants, that is, the less stable the transition state, the slower the reaction proceeds.*

We must now evaluate k', the rate at which $X^\ddagger$ decomposes. The transition state structure is held together by a bond that is assumed to be so weak that it flies apart during its first vibrational excursion. Therefore, k' is expressed

$$k' = \kappa\nu \qquad [12\text{-E}]$$

where ν is the vibrational frequency of the bond that breaks as $X^\ddagger$ decomposes to products, and κ, the **transmission coefficient**, is the probability that the breakdown of $X^\ddagger$ will be in the direction of product formation rather than back to reactants. For most spontaneous reactions, κ is assumed to be 1.0 (although this number, which must be between 0 and 1, can rarely be calculated with confidence).

Planck's law states that

$$\nu = \varepsilon/h \qquad [12\text{-F}]$$

where, in this case, ε is the average energy of the vibration that leads to the decomposition of $X^\ddagger$, and h ($=6.6261 \times 10^{-34} \, J \cdot s$) is **Planck's constant**. Statistical mechanics tells us that at a temperature T, the classical energy of an oscillator is

$$\varepsilon = k_B T \qquad [12\text{-G}]$$

where k_B ($=1.3807 \times 10^{-23} \, J \cdot K^{-1}$) is the **Boltzmann constant** and $k_B T$ is essentially the available thermal energy. Combining Eqs. 12-E through 12-G gives

$$k' = \frac{k_B T}{h} \qquad [12\text{-H}]$$

Thus, combining Eqs. 12-A, 12-D, and 12-H yields the expression for the rate constant of the elementary reaction:

$$k = \frac{k_B T}{h} e^{-\Delta G^\ddagger/RT} \qquad [12\text{-I}]$$

This equation indicates that as the temperature rises, so that there is increased thermal energy available to drive the reacting complex over the activation barrier, the reaction speeds up.

Here, k_{cat}/K_M is the apparent second-order rate constant of the enzymatic reaction; the rate of the reaction varies directly with how often enzyme and substrate encounter one another in solution. *The quantity k_{cat}/K_M is therefore a measure of an enzyme's catalytic efficiency.*

There is an upper limit to the value of k_{cat}/K_M: It can be no greater than k_1; that is, the decomposition of ES to E + P can occur no more frequently than E and S come together to form ES. The most efficient enzymes have k_{cat}/K_M values near the **diffusion-controlled limit** of 10^8 to 10^9 $M^{-1} \cdot s^{-1}$. These enzymes catalyze a reaction almost every time they encounter a substrate molecule and hence have achieved a state of virtual catalytic perfection. The relationship between the catalytic rate and the thermodynamics of the transition state can now be appreciated (see Box 12-2).

C. Analysis of Kinetic Data

There are several methods for determining the values of the parameters of the Michaelis–Menten equation (i.e., V_{max} and K_M). At very high values of [S], the initial velocity, v_o, asymptotically approaches V_{max}. In practice, however, it is very difficult to assess V_{max} accurately from direct plots of v_o versus [S] such as Fig. 12-3, because, even at substrate concentrations as high as [S] = 10 K_M, Eq. 12-25 indicates that v_o is only 91% of V_{max}, so that the value of V_{max} will almost certainly be underestimated.

A better method for determining the values of V_{max} and K_M, which was formulated by Hans Lineweaver and Dean Burk, uses the reciprocal of the Michaelis–Menten equation (Eq. 12-25):

$$\frac{1}{v_o} = \left(\frac{K_M}{V_{max}}\right)\frac{1}{[S]} + \frac{1}{V_{max}} \qquad [12\text{-}29]$$

This is a linear equation in $1/v_o$ and $1/[S]$. If these quantities are plotted to obtain the so-called **Lineweaver–Burk** or **double-reciprocal plot,** the slope of the line is K_M/V_{max}, the $1/v_o$ intercept is $1/V_{max}$, and the extrapolated $1/[S]$ intercept is $-1/K_M$ (Fig. 12-4).

As can be seen in Fig. 12-3, the best estimates of kinetic parameters are obtained by collecting data over a range of [S] from ~0.5 K_M to ~5 K_M. Thus, a disadvantage of the Lineweaver–Burk plots is that most experimental measurements of [S] are crowded onto the left side of the graph

SAMPLE CALCULATION

Determine K_M and V_{max} for an enzyme from the following data:

[S] (mM)	v_o ($\mu M \cdot s^{-1}$)
1	2.5
2	4.0
5	6.3
10	7.6
20	9.0

First, convert the data to reciprocal form ($1/[S]$ in units of mM^{-1}, and $1/v_o$ in units of $\mu M^{-1} \cdot s$). Next, make a plot of $1/v_o$ versus $1/[S]$. The x- and y-intercepts can be estimated by eye or calculated by linear regression. According to Eq. 12-29 and Fig. 12-4, the y-intercept, which has a value of ~0.1 $\mu M^{-1} \cdot s$, is equivalent to $1/V_{max}$, so V_{max} (the reciprocal of the y-intercept) is 10 $\mu M \cdot s^{-1}$. The x-intercept, -0.33 mM^{-1}, is equivalent to $-1/K_M$, so K_M (the negative reciprocal of the x-intercept) is equal to 3.0 mM.

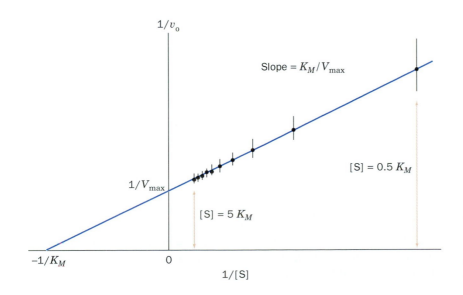

Figure 12-4. Key to Function. **A double-reciprocal (Lineweaver–Burk) plot.** The error bars represent ±0.05 V_{max}. The indicated points are the same as those in Fig. 12-3. Note the large effect of small errors at small [S] (large 1/[S]) and the crowding together of points at large [S].
✳ **See the Animated Figures.**

(Fig. 12-4). Moreover, for small values of [S], small errors in v_o lead to large errors in $1/v_o$ and hence to large errors in K_M and V_{max}.

Several other types of plots, each with its advantages and disadvantages, can also be used to determine K_M and V_{max} from kinetic data. However, kinetic data are now commonly analyzed by computer using mathematically sophisticated statistical treatments. Nevertheless, Lineweaver–Burk plots are valuable for the visual presentation of kinetic data.

Steady State Kinetics Cannot Unambiguously Establish a Reaction Mechanism

Although steady state kinetics provides valuable information about the rates of buildup and breakdown of ES, it provides little insight as to the nature of ES. Thus, an enzymatic reaction may, in reality, pass through several more or less stable intermediate states such as

$$\text{E + S} \rightleftharpoons \text{ES} \rightleftharpoons \text{EX} \rightleftharpoons \text{EP} \rightleftharpoons \text{E + P}$$

or take a more complex path such as

$$\text{E + S} \rightleftharpoons \text{ES} \begin{array}{c} \nearrow \text{EX} \searrow \\ \searrow \text{EY} \nearrow \end{array} \text{EP} \longrightarrow \text{E + P}$$

Unfortunately, steady state kinetic measurements are incapable of revealing the number of intermediates in an enzyme-catalyzed reaction. Thus, such measurements of a multistep reaction can be likened to a "black box" containing a system of water pipes with one inlet and one drain.

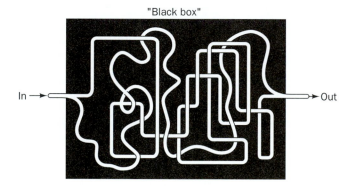

At steady state, that is, after the pipes have filled with water, the relationship between input pressure and output flow can be measured. However, such measurements yield no information concerning the detailed construction of the plumbing connecting the inlet to the drain. This would require additional information, such as opening the box and tracing the pipes. Likewise, steady state kinetic measurements can provide a phenomenological description of enzymatic behavior, but the nature of the intermediates remains indeterminate. The existence of intermediates must be verified independently, for example, by identifying them through the use of spectroscopic techniques.

The foregoing highlights a central principle of enzymology: *The steady state kinetic analysis of a reaction cannot unambiguously establish its mechanism.* This is because no matter how simple, elegant, or rational a postulated mechanism, there are an infinite number of alternative mechanisms that can also account for the kinetic data. Usually, it is the simpler and more elegant mechanism that turns out to be correct, but this is not always the case. However, *if kinetic data are not compatible with a given mechanism, then that mechanism must be rejected.* Therefore, although kinetics cannot be used to establish a mechanism unambiguously without confirming data, such as the physical demonstration of an intermediate's existence, the steady state kinetic analysis of a reaction is of great value because it can be used to eliminate proposed mechanisms.

D. Bisubstrate Reactions

We have heretofore been concerned with simple, single-substrate reactions that obey the Michaelis–Menten model (Reaction 12-12). Yet, enzymatic reactions requiring multiple substrates and yielding multiple products are far more common. Indeed, those involving two substrates and yielding two products

$$A + B \underset{E}{\rightleftharpoons} P + Q$$

account for ~60% of known biochemical reactions. Almost all of these so-called **bisubstrate reactions** are either **transferase** reactions in which the enzyme catalyzes the transfer of a specific functional group, X, from one of the substrates to the other:

$$P-X + B \underset{E}{\rightleftharpoons} P + B-X$$

or oxidation–reduction reactions in which reducing equivalents are transferred between the two substrates. For example, the hydrolysis of a peptide bond by trypsin (Section 11-5) is the transfer of the peptide carbonyl group from the peptide nitrogen atom to water (Fig. 12-5a), whereas in the alcohol dehydrogenase reaction (Section 11-1B), a hydride ion is formally transferred from ethanol to NAD^+ (Fig. 12-5b). Although bisubstrate reactions could, in principle, occur through a vast variety of mechanisms, only a few types are commonly observed.

Sequential Reactions

Reactions in which all substrates must combine with the enzyme before a reaction can occur and products be released are known as ***Sequential reactions.*** In such reactions, the group being transferred, X, is directly passed from A (= P—X) to B, yielding P and Q (= B—X). Hence, such reactions are also called **single-displacement reactions.**

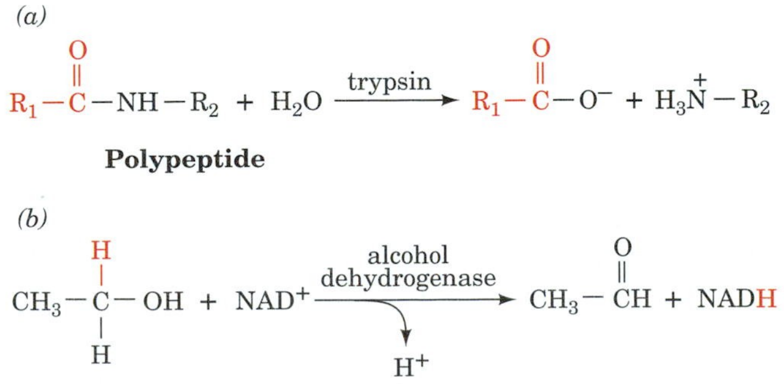

(a)

$$R_1-\overset{\overset{\displaystyle O}{\|}}{C}-NH-R_2 + H_2O \xrightarrow{\text{trypsin}} R_1-\overset{\overset{\displaystyle O}{\|}}{C}-O^- + H_3\overset{+}{N}-R_2$$

Polypeptide

(b)

$$CH_3-\overset{\overset{\displaystyle H}{|}}{\underset{\underset{\displaystyle H}{|}}{C}}-OH + NAD^+ \xrightarrow[\text{dehydrogenase}]{\text{alcohol}} CH_3-\overset{\overset{\displaystyle O}{\|}}{C}H + NADH$$

$$H^+$$

Figure 12-5.   Some bisubstrate reactions. (*a*) In the peptide hydrolysis reaction catalyzed by trypsin, the peptide carbonyl group, with its pendent polypeptide chain, R_1, is transferred from the peptide nitrogen atom to a water molecule. (*b*) In the alcohol dehydrogenase reaction, a hydride ion is formally transferred from ethanol to NAD^+.

Sequential reactions can be subclassified into those with a compulsory order of substrate addition to the enzyme, which are said to have an **Ordered mechanism,** and those with no preference for the order of substrate addition, which are described as having a **Random mechanism.** In the Ordered mechanism, the binding of the first substrate is apparently required for the enzyme to form the binding site for the second substrate, whereas in the Random mechanism, both binding sites are present on the free enzyme.

In a notation developed by W.W. Cleland, substrates are designated by the letters A and B in the order that they add to the enzyme, products are designated by P and Q in the order that they leave the enzyme, the enzyme is represented by a horizontal line, and successive additions of substrates and releases of products are denoted by vertical arrows. An Ordered bisubstrate reaction is thereby diagrammed:

where A and B are said to be the **leading** and **following** substrates, respectively. Many NAD^+- and $NADP^+$-requiring dehydrogenases follow an Ordered bisubstrate mechanism in which the coenzyme is the leading substrate.

A Random bisubstrate reaction is diagrammed:

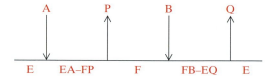

Some dehydrogenases and kinases operate through Random bisubstrate mechanisms.

Ping Pong Reactions

Group-transfer reactions in which one or more products are released before all substrates have been added are known as **Ping Pong reactions.** The Ping Pong bisubstrate reaction is represented by

Here, a functional group X of the first substrate A ($= P$—X) is displaced from the substrate by the enzyme E to yield the first product P and a stable enzyme form F ($= E$—X) in which X is tightly (often covalently) bound to the enzyme (Ping). In the second stage of the reaction, X is displaced from the enzyme by the second substrate B to yield the second product Q ($= B$—X), thereby regenerating the original form of the enzyme, E (Pong). Such reactions are therefore also known as **double-displacement reactions.** *Note that in Ping Pong reactions, the substrates A and B do not encounter one another on the surface of the enzyme.* Many enzymes, including trypsin (in which F is the acyl–enzyme intermediate; Section 11-5), transaminases, and some flavoenzymes, react with Ping Pong mechanisms.

Bisubstrate Mechanisms Can Be Distinguished by Kinetic Measurements

The rate equations that describe the foregoing bisubstrate mechanisms are considerably more complicated than the equation for a single-substrate reaction. In fact, the equations for bisubstrate mechanisms (which are beyond the scope of this text) contain as many as four kinetic constants versus two (V_{max} and K_M) for the Michaelis–Menten equation. Nevertheless, steady state kinetic measurements can be used to distinguish among the various bisubstrate mechanisms.

2. ENZYME INHIBITION

Many substances alter the activity of an enzyme by reversibly combining with it in a way that influences the binding of substrate and/or its turnover number. *Substances that reduce an enzyme's activity in this way are known as **inhibitors**.* A large part of the modern pharmaceutical arsenal consists of enzyme inhibitors. For example, AIDS is treated almost exclusively with drugs that inhibit the activities of certain viral enzymes (see Box 12-3).

Inhibitors act through a variety of mechanisms. Some enzyme inhibitors are substances that structurally resemble their enzyme's substrates but either do not react or react very slowly. These substances are commonly used to probe the chemical and conformational nature of an enzyme's active site in an effort to elucidate the enzyme's catalytic mechanism. Other inhibitors affect catalytic activity without interfering with substrate binding. Many do both. In this section, we discuss several of the simplest mechanisms for reversible inhibition and their effects on the kinetic behavior of enzymes that follow the Michaelis–Menten model.

A. Competitive Inhibition

A substance that competes directly with a normal substrate for an enzyme's substrate-binding site is known as a **competitive inhibitor.** Such an inhibitor usually resembles the substrate so that it specifically binds to the active site but differs from the substrate so that it cannot react as the substrate does. For example, **succinate dehydrogenase,** a citric acid cycle enzyme that converts **succinate** to **fumarate** (Section 16-3F), is competitively inhibited by **malonate,** which structurally resembles succinate but cannot be dehydrogenated.

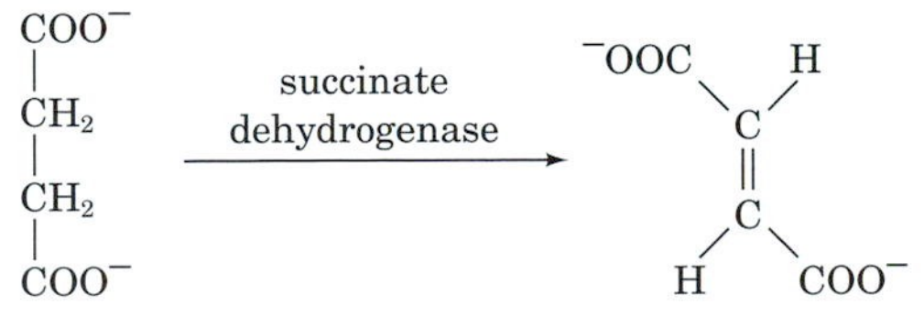

Succinate   **Fumarate**

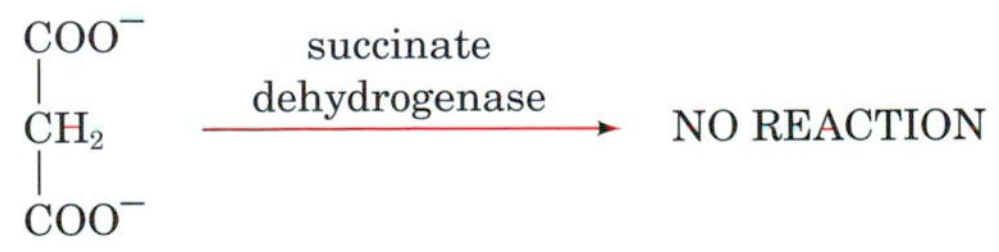

Malonate

Box 12-3
BIOCHEMISTRY IN HEALTH AND DISEASE

HIV Enzyme Inhibitors

The **human immunodeficiency virus (HIV)** causes **acquired immunodeficiency syndrome (AIDS)** by infecting and destroying the host's immune system. In the first steps of infection, HIV attaches to a target cell and injects its genetic material (which is RNA rather than DNA) into the host cell. The viral RNA is transcribed into DNA by a viral enzyme called **reverse transcriptase** (Box 24-2). After this DNA is integrated into the host's genome, the cell can produce more viral RNA and proteins for packaging it into new viral particles. Most of the viral proteins are synthesized as part of larger polypeptide precursors known as **polyproteins**. Consequently, proteolytic processing by the virally encoded **HIV protease** to release these viral proteins is necessary for viral reproduction. In the absence of an effective vaccine for HIV, efforts to prevent and treat AIDS have led to the development of compounds that inhibit HIV reverse transcriptase and HIV protease.

Several inhibitors of reverse transcriptase have been developed. The archetype is **AZT (3′-azido-3′-deoxythymidine; Zidovudine)**, which is taken up by cells, phosphorylated, and incorporated into the DNA chains synthesized from the HIV template by reverse transcriptase. Because AZT lacks a 3′-OH group, it acts as a chain terminator, as do the dideoxynucleotides used for DNA sequencing (Section 3-4C). Most cellular DNA polymerases have a low affinity for phosphorylated AZT, but reverse transcriptase has a high affinity for this drug, which makes AZT effective against viral replication.

Other nucleoside analogs that are used to treat HIV infection are **2′,3′-dideoxycytidine (ddC, Zalcitabine)** and **2′,3′-dideoxyinosine (ddI, Didanosine;** inosine nucleotides are metabolically converted to adenosine and guanosine nucleotides), which also act as chain terminators. Nonnucleoside compounds such as **nevirapine** do not bind to the reverse transcriptase active site but instead bind to a hydrophobic pocket elsewhere on the enzyme.

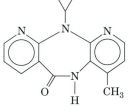

**3′-Azido-3′-deoxythymidine
(AZT; Zidovudine)**

**2′,3′-Dideoxycytidine
(ddC, Zalcitabine)**

**2′,3′-Dideoxyinosine
(ddI, Didanosine)**

Nevirapine

The effectiveness of malonate as a competitive inhibitor of succinate dehydrogenase strongly suggests that the enzyme's substrate-binding site is designed to bind both of the substrate's carboxylate groups, presumably through the influence of two appropriately placed positively charged residues.

The general model for competitive inhibition is given by the following reaction scheme:

$$\text{E} + \text{S} \underset{k_{-1}}{\overset{k_1}{\rightleftharpoons}} \text{ES} \overset{k_2}{\longrightarrow} \text{P} + \text{E}$$
$$+$$
$$\text{I}$$
$$K_\text{I} \Updownarrow$$
$$\text{EI} + \text{S} \longrightarrow \text{NO REACTION}$$

Here it is assumed that I, the inhibitor, binds reversibly to the enzyme and is in rapid equilibrium with it so that

$$K_\text{I} = \frac{[\text{E}][\text{I}]}{[\text{EI}]} \qquad [12\text{-}30]$$

and EI, the enzyme–inhibitor complex, is catalytically inactive. *A competitive inhibitor therefore reduces the concentration of free enzyme available for substrate binding.*

HIV protease is a homodimer of 99-residue subunits. Mechanistically, it is a so-called **aspartic protease**, a family of proteases that includes **pepsin** (the gastric protease that operates at low pH). Comparisons among the active sites of the aspartic proteases were instrumental in designing HIV protease inhibitors. HIV protease cleaves a number of specific peptide bonds, including Phe-Pro and Tyr-Pro peptide bonds in its physiological substrates, the HIV proteins (*see right*). Inhibitors based on these sequences should therefore selectively inhibit the viral protease. The **peptidomimetic** (peptide-imitating) drugs **ritonavir** and **saquinavir** contain phenyl and other bulky groups that bind in the HIV protease active site. Of perhaps even greater importance, these drugs have the geometry of the catalyzed reaction's tetrahedral transition state (*red*). The enzyme's peptide substrate is shown for comparison.

The efficacy of anti-HIV agents, like that of many drugs, is limited by several factors:

1. **Side effects.** Despite their preference for viral enzymes, anti-HIV drugs also interfere with normal cellular processes. For example, the inhibition of DNA synthesis by reverse transcriptase inhibitors in rapidly dividing cells, such as the bone marrow cells that give rise to erythrocytes, can lead to severe anemia. Other side effects include nausea, kidney stones, and rashes. Side effects are particularly troublesome in HIV infection, because drugs must be taken several times daily for many years, if not for a lifetime.

2. **Bioavailability.** Many drug candidates are poorly absorbed orally, and many of those that are absorbed by the body are quickly excreted or chemically modified, usually by the liver, to inactive forms (e.g., peptides containing an HIV target sequence are poorly absorbed and rapidly proteolyzed; hence the need for peptidomimetics). Drugs may have longer half-lives if they bind to blood proteins such as **serum albumin**, but they must also enter their target cells to be effective. For example, cells take up AZT, but its phosphorylated form does not cross the cell membrane. The central nervous system is particularly difficult to target because of the almost impenetrable matrix that lies between nervous tissue and the blood vessels that supply it (the so-called **blood–brain barrier**).

3. **Resistance.** Acquired resistance to antiviral agents is a significant problem in HIV infection because the error-prone reverse transcriptase allows HIV to mutate

—Phe—Pro—
HIV protease substrate

Saquinavir $K_I = 0.40$ nM

Ritonavir $K_I = 0.015$ nM

rapidly. Over 20 mutations in HIV are known to be associated with drug resistance.

In the case of HIV infection, it seems unlikely that a single drug will prove to be a "magic bullet," in part because HIV infects many cell types, but mostly because of its ability to rapidly evolve resistance against any one drug. The recent outstanding success of anti-HIV therapy rests on combination therapy, in which several different drugs are administered simultaneously. This successfully keeps AIDS at bay by reducing levels of HIV, in some cases to undetectable levels, thereby reducing the probability that HIV will evolve a drug-resistant variant. The advantages of using an inhibitor "cocktail" containing inhibitors of reverse transcriptase and HIV protease include (1) decreasing the likelihood that a viral strain will spontaneously develop resistance to each compound in the mix and (2) decreasing the doses and hence the side effects of the individual compounds.

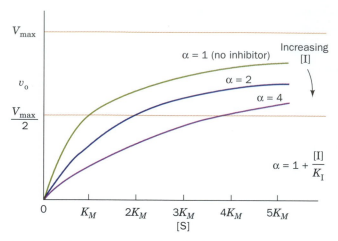

Figure 12-6.   A plot of v_o versus [S] for a Michaelis–Menten reaction in the presence of different concentrations of a competitive inhibitor.

The Michaelis–Menten equation for a competitively inhibited reaction is derived as before (Section 12-1B), but with an additional term to account for the fraction of $[E]_T$ that binds to I to form EI ($[E]_T = [E] + [ES] + [EI]$). The resulting equation,

$$v_o = \frac{V_{max}[S]}{\alpha K_M + [S]}$$ [12-31]

is the Michaelis–Menten equation that has been modified by a factor, α, which is defined as

$$\alpha = 1 + \frac{[I]}{K_I}$$ [12-32]

Note that α, a function of the inhibitor's concentration and its affinity for the enzyme, cannot be less than 1.

Figure 12-6 shows the hyperbolic plot of Eq. 12-31 for various values of α. Note that as [S] approaches infinity, v_o approaches V_{max} for any value of α. Thus, an infinitely high concentration of substrate can overcome the effects of the inhibitor (i.e., the inhibitor does not affect the turnover number of the enzyme). The presence of I makes [S] appear more dilute than it really is (makes K_M appear larger than it really is), a consequence of the binding of I and S to E being mutually exclusive.

Competitive inhibition is the principle behind the use of ethanol to treat methanol poisoning. Methanol itself is only mildly toxic. However, the liver enzyme alcohol dehydrogenase converts methanol to the highly toxic formaldehyde, only small amounts of which cause blindness and death. Ethanol competes with methanol for binding to the active site of liver alcohol dehydrogenase, thereby slowing the production of formaldehyde from methanol (the ethanol is converted to the readily metabolized acetaldehyde). Thus, through the administration of ethanol, a large portion of the methanol will be harmlessly excreted from the body in the urine before it can be converted to formaldehyde.

K_I Can Be Measured

Recasting Eq. 12-31 in the double-reciprocal form yields

$$\frac{1}{v_o} = \left(\frac{\alpha K_M}{V_{max}}\right)\frac{1}{[S]} + \frac{1}{V_{max}}$$ [12-33]

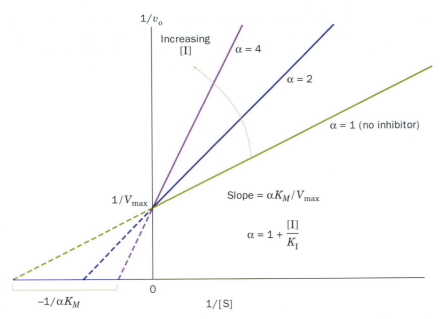

Figure 12-7.   A Lineweaver–Burk plot of the competitively inhibited Michaelis–Menten enzyme described by Fig. 12-6. Note that all lines intersect on the $1/v_o$ axis at $1/V_{max}$. The varying slopes indicate the effect of the inhibitor on K_M.
✳ **See the Animated Figures.**

A plot of this equation is linear and has a slope of $\alpha K_M/V_{max}$, a $1/[S]$ intercept of $-1/\alpha K_M$, and a $1/v_o$ intercept of $1/V_{max}$ (Fig. 12-7). The double-reciprocal plots for a competitive inhibitor at various concentrations of I intersect at $1/V_{max}$ on the $1/v_o$ axis, a property that is diagnostic of competitive inhibition.

By determining the values of α at different inhibitor concentrations for an enzyme of known K_M, the value of K_I can be found from Eq. 12-32. Comparing the K_I values of competitive inhibitors with different structures can provide information about the binding properties of an enzyme's active site and hence its catalytic mechanism. For example, to ascertain the importance of the various segments of an ATP molecule

for binding to the active site of an ATP-requiring enzyme, one might determine the K_I, say, for ADP, AMP, ribose, triphosphate, etc. Since many of these ATP components are unreactive, inhibition studies are the most convenient method of monitoring their binding to the enzyme.

Competitive inhibition studies are also used to determine the affinity of transition state analogs for an enzyme's active site (Section 11-3F). For example, the recently developed HIV protease inhibitors (Box 12-3) have been designed to mimic this enzyme's transition state and thus bind to the enzyme with high affinity. Inhibitor studies are the mainstays of such drug development.

B. Uncompetitive Inhibition

In **uncompetitive inhibition,** the inhibitor binds directly to the enzyme–substrate complex but not to the free enzyme:

$$\text{E} + \text{S} \underset{k_{-1}}{\overset{k_1}{\rightleftharpoons}} \text{ES} \overset{k_2}{\longrightarrow} \text{P} + \text{E}$$
$$+$$
$$\text{I}$$
$$K_I' \updownarrow$$
$$\text{ESI} \longrightarrow \text{NO REACTION}$$

In this case, the inhibitor binding step has the dissociation constant

$$K_I' = \frac{[\text{ES}][\text{I}]}{[\text{ESI}]} \tag{12-34}$$

The uncompetitive inhibitor, which need not resemble the substrate, presumably *distorts the active site, thereby rendering the enzyme catalytically inactive.*

The Michaelis–Menten equation for uncompetitive inhibition and the equation for its double-reciprocal plot are given in Table 12-2. The double-reciprocal plot consists of a family of parallel lines (Fig. 12-8) with slope K_M/V_{max}, $1/v_o$ intercepts of α'/V_{max}, and $1/[\text{S}]$ intercepts of $-\alpha'/K_M$. Note that in uncompetitive inhibition, both K_M and V_{max} are decreased, but the ratio V_{max}/K_M remains unchanged. In contrast to competitive inhibition, the effects of uncompetitive inhibition on V_{max} are not reversed by increasing the substrate concentration.

Uncompetitive inhibition requires that the inhibitor affect the catalytic function of the enzyme but not its substrate binding. This is difficult to envision for single-substrate enzymes. In actuality, uncompetitive inhibition is significant only in multisubstrate enzymes.

Table 12-2.　**Effects of Inhibitors on Michaelis–Menten Reactions**[a]

Type of Inhibition	Michaelis–Menten Equation	Lineweaver–Burk Equation	Effect of Inhibitor
None	$v_o = \dfrac{V_{max}[\text{S}]}{K_M + [\text{S}]}$	$\dfrac{1}{v_o} = \dfrac{K_M}{V_{max}} \dfrac{1}{[\text{S}]} + \dfrac{1}{V_{max}}$	None
Competitive	$v_o = \dfrac{V_{max}[\text{S}]}{\alpha K_M + [\text{S}]}$	$\dfrac{1}{v_o} = \dfrac{\alpha K_M}{V_{max}} \dfrac{1}{[\text{S}]} + \dfrac{1}{V_{max}}$	Increases K_M
Uncompetitive	$v_o = \dfrac{V_{max}[\text{S}]}{K_M + \alpha'[\text{S}]}$	$\dfrac{1}{v_o} = \dfrac{K_M}{V_{max}} \dfrac{1}{[\text{S}]} + \dfrac{\alpha'}{V_{max}}$	Decreases K_M and V_{max}
Mixed (noncompetitive)	$v_o = \dfrac{V_{max}[\text{S}]}{\alpha K_M + \alpha'[\text{S}]}$	$\dfrac{1}{v_o} = \dfrac{\alpha K_M}{V_{max}} \dfrac{1}{[\text{S}]} + \dfrac{\alpha'}{V_{max}}$	Decreases V_{max}; may increase or decrease K_M

[a]$\alpha = 1 + \dfrac{[\text{I}]}{K_I}$　and　$\alpha' = 1 + \dfrac{[\text{I}]}{K_I'}$

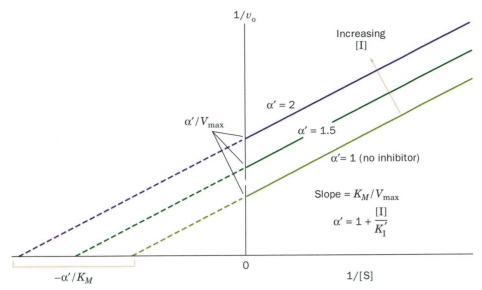

Figure 12-8. A Lineweaver–Burk plot of a Michaelis–Menten enzyme in the presence of an uncompetitive inhibitor. Note that all lines have identical slopes of K_M/V_{max}. ✳ See the Animated Figures.

C. Mixed Inhibition

If both the enzyme and the enzyme–substrate complex bind inhibitor, the following model results:

$$E + S \underset{k_{-1}}{\overset{k_1}{\rightleftharpoons}} ES \overset{k_2}{\longrightarrow} P + E$$

$$+ \qquad +$$
$$I \qquad I$$

$$K_I \updownarrow \qquad K_I' \updownarrow$$

$$EI \qquad ESI \longrightarrow NO\ REACTION$$

This phenomenon is known as **mixed inhibition** (alternatively, **noncompetitive inhibition**). Presumably, *a mixed inhibitor binds to enzyme sites that participate in both substrate binding and catalysis*. The two dissociation constants for inhibitor binding

$$K_I = \frac{[E][I]}{[EI]} \quad \text{and} \quad K_I' = \frac{[ES][I]}{[ESI]} \qquad [12\text{-}35]$$

are not necessarily equivalent.

The Michaelis–Menten equation and the corresponding double-reciprocal equation for mixed inhibition are given in Table 12-2. As in uncompetitive inhibition, the values of K_M and V_{max} are modulated by the presence of inhibitor. The name *mixed inhibition* arises from the fact that the denominator of the Michaelis–Menten equation has the factor α multiplying K_M as in competitive inhibition and the factor α' multiplying [S] as in uncompetitive inhibition.

Double-reciprocal plots for mixed inhibition consist of lines that have the slope $\alpha K_M/V_{max}$ with a $1/v_o$ intercept of α'/V_{max} and a $1/[S]$ intercept

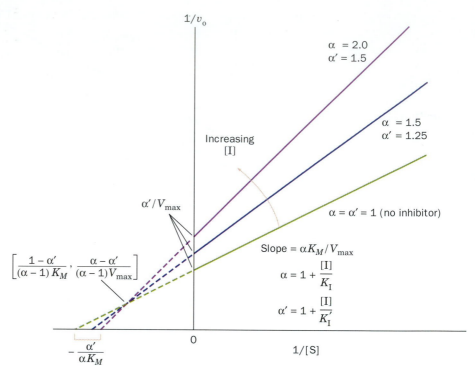

Figure 12-9. **A Lineweaver–Burk plot of a Michaelis–Menten enzyme in the presence of a mixed inhibitor.** Note that the lines all intersect to the left of the $1/v_o$ axis. The coordinates of this intersection point are given in brackets. When $K_I = K'_I$, $\alpha = \alpha'$ and the lines intersect on the $1/[S]$ axis at $-1/K_M$.
✴ See the Animated Figures.

of $-\alpha'/\alpha K_M$ (Fig. 12-9). The lines for different values of [I] intersect to the left of the $1/v_o$ axis. For the special case that $K_I = K'_I$ ($\alpha = \alpha'$), this intersection is, in addition, on the $1/[S]$ axis (a situation that, in an ambiguity of nomenclature, is sometimes described as noncompetitive inhibition). Mixed inhibition, like uncompetitive inhibition, is an important feature in the kinetics of multisubstrate enzymes.

Irreversible Inactivation Resembles Noncompetitive Inhibition

If an inhibitor binds irreversibly to an enzyme, the inhibitor is classified as an **inactivator.** Inactivators truly reduce the effective level of $[E]_T$ (and therefore V_{max}) at all values of [S] without changing K_M. The double-reciprocal plots for irreversible inactivation therefore resemble those for noncompetitive inhibition (the lines intersect on the $1/[S]$ axis). Reagents that chemically modify specific amino acid residues can act as inactivators. For example, the compounds used to identify the catalytic Ser and His residues of serine proteases (Section 11-5A) are inactivators of these enzymes.

3. REGULATION OF ENZYME ACTIVITY

An organism must be able to regulate the catalytic activities of its component enzymes so that it can coordinate its numerous metabolic processes, respond to changes in its environment, and grow and differentiate, all in an orderly manner. There are two ways that this may occur:

1. **Control of enzyme availability.** The amount of a given enzyme in a cell depends on both its rate of synthesis and its rate of degradation. Each of these rates is directly controlled by the cell and is subject to dramatic changes over time spans of minutes (in bacteria) to hours (in higher organisms).

2. **Control of enzyme activity.** An enzyme's catalytic activity can be directly regulated through structural alterations that influence the enzyme's substrate-binding affinity. Just as hemoglobin's oxygen affinity is allosterically regulated by the binding of ligands such as O_2, CO_2, H^+, and BPG (Section 7-2C), an enzyme's substrate-binding affinity may likewise vary with the binding of small molecules, called **allosteric effectors.** *Allosteric mechanisms can cause large changes in enzymatic activity.* The activities of some enzymes are similarly regulated by covalent modification, usually phosphorylation and dephosphorylation of specific Ser, Thr, or Tyr residues.

In this section, we consider the allosteric control of enzymatic activity by considering one example—**aspartate transcarbamoylase (ATCase)** from *E. coli*. We shall examine other examples of allosteric control as well as covalent modification in later chapters.

The Feedback Inhibition of ATCase Regulates Pyrimidine Synthesis

Aspartate transcarbamoylase catalyzes the formation of **N-carbamoyl aspartate** from **carbamoyl phosphate** and aspartate:

Carbamoyl phosphate   **Aspartate**

aspartate transcarbamoylase

$+ \quad H_2PO_4^-$

N-Carbamoyl aspartate

This reaction is the first step unique to the biosynthesis of pyrimidines (Section 22-2A). The allosteric behavior of *E. coli* ATCase has been investigated by John Gerhart and Howard Schachman, who demonstrated that both of its substrates bind cooperatively to the enzyme. Moreover, ATCase is allosterically inhibited by **cytidine triphosphate (CTP),** a pyrimidine nucleotide, and is allosterically activated by adenosine triphosphate (ATP), a purine nucleotide.

The v_o versus [S] curve for ATCase (Fig. 12-10) is sigmoidal, rather than hyperbolic as it is in enzymes that follow the Michaelis–Menten model. This is consistent with cooperative substrate binding (recall that hemoglobin's O_2-binding curve is also sigmoidal; Fig. 7-7). ATCase's allosteric effectors shift the entire curve to the right or the left: At a given substrate concentration, CTP decreases the enzyme's catalytic rate, whereas ATP increases it.

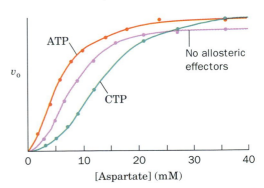

Figure 12-10. **Plot of v_o versus [Aspartate] for the ATCase reaction.** Reaction velocity was measured in the absence of allosteric effectors, in the presence of 0.4 mM CTP (an inhibitor), and in the presence of 2.0 mM ATP (an activator). [After Kantrowitz, E.R., Pastra-Landis, S.C., and Lipscomb, W.N., *Trends Biochem. Sci.* 5, 125 (1980).] ✳ See the Animated Figures.

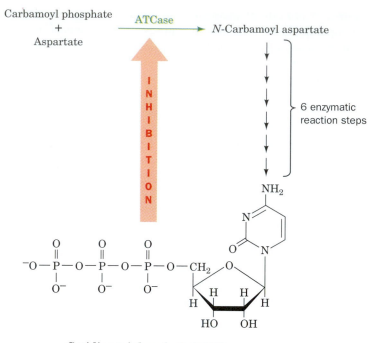

Cytidine triphosphate (CTP)

Figure 12-11.   **A schematic representation of the pyrimidine biosynthesis pathway.** CTP, the end product of the pathway, inhibits ATCase, which catalyzes the pathway's first step.

CTP, which is a product of the pyrimidine biosynthetic pathway, is an example of a **feedback inhibitor,** since *it inhibits an earlier step in its own biosynthesis* (Fig. 12-11). Thus, when CTP levels are high, CTP binds to ATCase, thereby reducing the rate of CTP synthesis. Conversely, when cellular [CTP] decreases, CTP dissociates from ATCase and CTP synthesis accelerates.

The metabolic significance of the ATP activation of ATCase is that it tends to coordinate the rates of synthesis of purine and pyrimidine nucleotides, which are required in roughly equal amounts in nucleic acid biosynthesis. For instance, if the ATP concentration is much greater than that of CTP, ATCase is activated to synthesize pyrimidine nucleotides until the concentrations of ATP and CTP become balanced. Conversely, if the CTP concentration is greater than that of ATP, CTP inhibition of ATCase permits purine nucleotide biosynthesis to balance the ATP and CTP concentrations.

Allosteric Changes Alter ATCase's Substrate-Binding Sites

E. coli ATCase (300 kD) has the subunit composition c_6r_6, where c and r represent its catalytic and regulatory subunits. The X-ray structure of ATCase (Fig. 12-12), determined by William Lipscomb, reveals that the catalytic subunits are arranged as two sets of trimers (c_3) in complex with three sets of regulatory dimers (r_2). Each regulatory dimer joins two catalytic subunits in different c_3 trimers.

The isolated catalytic trimers are catalytically active, have a maximum catalytic rate greater than that of intact ATCase, exhibit a noncooperative (hyperbolic) substrate saturation curve, and are unaffected by the presence of ATP or CTP. The isolated regulatory dimers bind these allosteric effec-

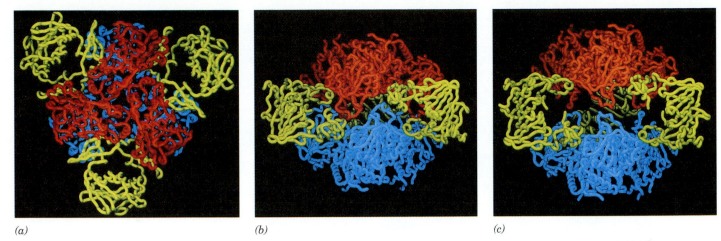

(a)   (b)   (c)

Figure 12-12. The X-ray structure of ATCase. The polypeptide backbones of the T-state enzyme are viewed (*a*) along the protein's molecular three-fold axis of symmetry and (*b*) along a molecular two-fold axis of symmetry perpendicular to the view in *a*. The regulatory dimers (*yellow*) join the upper catalytic trimer (*red*) to the lower catalytic trimer (*blue*). (*c*) The R-state enzyme viewed as in *b*. Note how the rotation of the regulatory dimers in the R → T transition causes the catalytic trimers to move apart along the three-fold axis. [Courtesy of Michael Pique, The Scripps Research Institute, La Jolla, California.] ● See **Kinemage Exercise 8-1.**

tors but are devoid of enzymatic activity. Evidently, *the regulatory subunits allosterically reduce the activity of the catalytic subunits in the intact enzyme.*

As allosteric theory predicts (Section 7-2E), the activator ATP preferentially binds to ATCase's active (R or high substrate affinity) state, whereas the inhibitor CTP preferentially binds to the enzyme's inactive (T or low substrate affinity) state. Similarly, the unreactive bisubstrate analog **N-(phosphonacetyl)-L-aspartate (PALA)**

$$
\begin{array}{cc}
\underset{\displaystyle \underset{\displaystyle \overset{|}{NH}}{\overset{O}{\parallel}}{C}-CH_2-PO_3^{2-} & \underset{\displaystyle \underset{\displaystyle \overset{|}{NH_3^+}}{}}{H_2N-\overset{O}{\overset{\parallel}{C}}-O-PO_3^{2-}} \\
^-OOC-CH_2-CH-COO^- & ^-OOC-CH_2-CH-COO^-
\end{array}
$$

N-(Phosphonacetyl)-   **Carbamoyl phosphate**
L-aspartate (PALA)   **+**
                                        Aspartate

binds tightly to R-state but not to T-state ATCase.

X-Ray structures have been determined for the T-state ATCase–CTP complex and the R-state ATCase–PALA complex (as a rule, unreactive substrate analogs, such as PALA, form complexes with an enzyme that are more amenable to structural analysis than complexes of the enzyme with rapidly reacting substrates; Box 11-3). Structural studies reveal that in the T → R transition, the enzyme's catalytic trimers separate along the molecular three-fold axis by ~11 Å and reorient around this axis relative to each other by 12° (Fig. 12-12*b*, *c*). In addition, the regulatory dimers rotate clockwise by 15° around their two-fold axes and separate by ~4 Å along the three-fold axis. Such large quaternary shifts are reminiscent of those in hemoglobin (Section 7-2C).

Each catalytic subunit of ATCase consists of a carbamoyl phosphate–binding domain and an aspartate-binding domain. The binding of PALA to the enzyme, which presumably mimics the binding of both substrates, induces a conformational change that swings the two domains together such

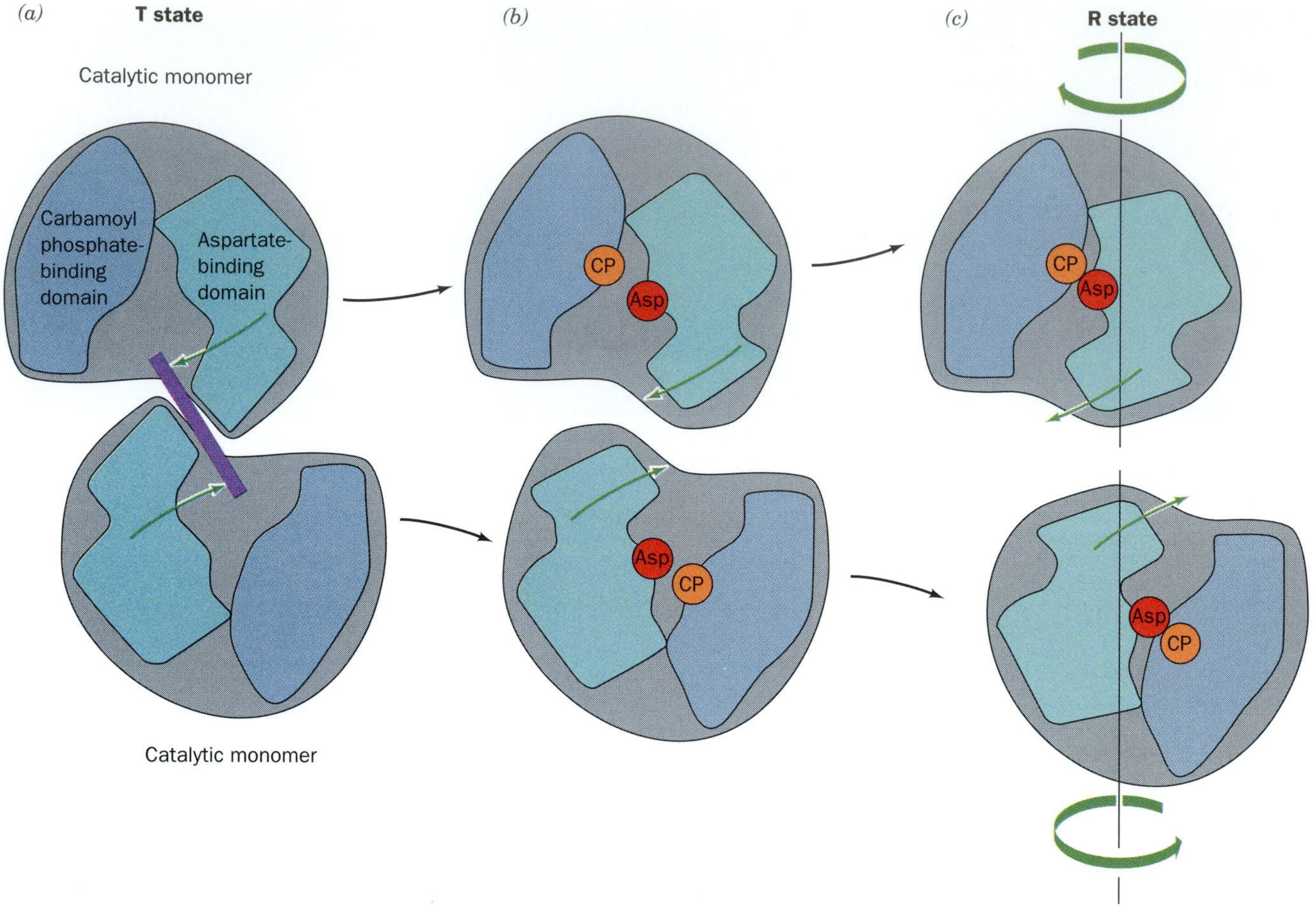

Figure 12-13.   A schematic diagram of tertiary and quaternary conformational changes in ATCase. Two of the six vertically interacting catalytic ATCase subunits are shown. (*a*) In the absence of bound substrate, the protein is held in the T state because the motions that bring together the two domains of each subunit (*green arrows*) are prevented by steric interference (*purple bar*) between the contacting aspartic acid–binding domains.

(*b*) The binding of carbamoyl phosphate (CP) followed by aspartic acid (Asp) to their respective binding sites causes the subunits to move apart and rotate with respect to each other so as to permit the T → R transition. (*c*) In the R state, the two domains of each subunit come together to promote the reaction of their bound substrates to form products. [Figure copyrighted © by Irving Geis.] ○ See Kinemage Exercises 8-1 and 8-2.

that their two bound substrates can react to form product (Fig. 12-13). The conformational changes—movements of up to 8 Å for some residues—in a single catalytic subunit triggers ATCase's T → R quaternary shift. ATCase's tertiary and quaternary shifts are tightly coupled (i.e., ATCase closely follows the symmetry model of allosterism; Section 7-2E). *The binding of substrate to one catalytic subunit therefore increases the substrate-binding affinity and catalytic activity of the other five catalytic subunits* and hence accounts for the enzyme's cooperative substrate binding.

The Structural Basis of Allosterism in ATCase

The structural basis for the effects of CTP and ATP on ATCase activity is gradually being unveiled. Both the inhibitor CTP and the activator ATP bind to the same site on the outer edge of the regulatory subunit, about 60 Å away from the nearest catalytic site. CTP binds preferentially to the T state, increasing its stability, whereas ATP binds preferentially to the R state, increasing its stability.

The binding of CTP and ATP to their less favored enzyme states also has structural consequences. When CTP binds to R-state ATCase, it induces a contraction in the regulatory dimer that causes the catalytic trimers to come together by 0.5 Å (become more T-like, that is, less active). This, in turn, reorients key residues in the enzyme's active sites, thereby decreasing the enzyme's catalytic activity. ATP has essentially opposite effects when binding to the T-state enzyme: It causes the catalytic trimers to move apart by 0.4 Å (become more R-like, that is, more active), thereby reorienting key residues in the enzyme's active sites so as to increase the enzyme's catalytic activity.

Allosteric Transitions in Other Enzymes Resemble Those of Hemoglobin and ATCase

Allosteric enzymes are widely distributed in nature and tend to occupy key regulatory positions in metabolic pathways. Such enzymes are symmetrical proteins containing at least two subunits. In all known cases, quaternary structural changes communicate binding and catalytic effects among all active sites in the enzyme. The quaternary shifts are primarily rotations of subunits relative to one another. Secondary structures are largely preserved in T → R transitions, which is probably important for mechanically transmitting allosteric effects over distances of tens of angstroms.

SUMMARY

1. Elementary chemical reactions may be first order, second order, or rarely, third order. In each case, a rate equation describes the progress of the reaction as a function of time.

2. The Michaelis–Menten equation describes the relationship between initial reaction velocity and substrate concentration under steady state conditions.

3. K_M is the substrate concentration at which the reaction velocity is half-maximal. The value of k_{cat}/K_M indicates an enzyme's catalytic efficiency.

4. Kinetic data can be plotted in double-reciprocal form to determine K_M and V_{max}.

5. Bisubstrate reactions are classified as Sequential (single displacement) or Ping Pong (double displacement). A Sequential reaction may proceed by an Ordered or Random mechanism.

6. Reversible inhibitors reduce an enzyme's activity by binding to the substrate-binding site (competitive inhibition), to the enzyme–substrate complex (uncompetitive inhibition), or to both the enzyme and the enzyme–substrate complex (mixed inhibition).

7. Enzyme activity may be regulated by allosteric effectors.

8. The activity of ATCase is increased by ATP and decreased by CTP, which alter the conformation of the catalytic sites by stabilizing the R and the T states of the enzyme, respectively.

REFERENCES

Cornish-Bowden, A., *Fundamentals of Enzyme Kinetics,* Butterworths (1979).

Evans, P.R., Structural aspects of allostery, *Curr. Opin. Struct. Biol.* **1,** 773–779 (1991).

Fersht, A., *Enzyme Structure and Mechanism* (2nd ed.), Chapters 3–7, Freeman (1985).

Kantrowitz, E.R. and Lipscomb, W.N., *Escherichia coli* aspartate transcarbamoylase: the molecular basis for a concerted allosteric transition, *Trends Biochem. Sci.* **15,** 53–59 (1990).

Lipsky, J.J., Antiretroviral drugs for AIDS, *Lancet* **348,** 800–803 (1996).

Perutz, M., *Mechanisms of Cooperativity and Allosteric Regulation in Proteins,* Cambridge University Press (1990).

Schulz, A.R., *Enzyme Kinetics,* Cambridge University Press (1994).

Segel, I.H., *Enzyme Kinetics,* Wiley (1975).

Tinoco, I., Jr., Sauer, K., and Wang, J.C., *Physical Chemistry. Principles and Applications for Biological Sciences* (3rd ed.), Chapters 7 and 8, Prentice–Hall (1995).

Wood, W.B., Wilson, J.H., Benbow, R.M., and Hood, L.E., *Biochemistry. A Problems Approach* (2nd ed.), Chapter 8, Benjamin/Cummings (1981). [Contains instructive problems on enzyme kinetics with answers worked out in detail.]

KEY TERMS

elementary reaction	k_{-1}	k_{cat}/K_M	inhibitor
k	k_2	diffusion-controlled limit	inactivator
v	steady state assumption	Lineweaver–Burk (double-	competitive inhibition
reaction order	K_M	reciprocal) plot	K_I
first-order reaction	v_o	Sequential reaction	uncompetitive inhibition
second-order reaction	V_{max}	Ordered mechanism	mixed inhibition
pseudo-first-order reaction	enzyme saturation	Random mechanism	allosteric effector
molecularity	Michaelis–Menten equation	Ping Pong reaction	feedback inhibition
rate equation	Michaelis complex	single-displacement	
$t_{1/2}$	k_{cat}	reaction	
ES complex	turnover number	double-displacement reaction	

STUDY EXERCISES

1. Write the rate equations for a first-order and a second-order reaction.

2. What are the differences between instantaneous velocity, initial velocity, and maximal velocity for an enzymatic reaction?

3. Derive the Michaelis–Menten equation.

4. What do the values of K_M and k_{cat}/K_M reveal about an enzyme?

5. Write the Lineweaver–Burk (double-reciprocal) equation and describe the features of a Lineweaver–Burk plot.

6. Use Cleland notation to describe Ordered and Random Sequential reactions and a Ping Pong reaction.

7. Describe the effects of different types of inhibitors on K_M and V_{max}.

8. What distinguishes an inhibitor from an inactivator?

9. List some mechanisms for regulating enzyme activity.

10. Explain the structural basis for cooperative substrate binding and allosteric regulation in ATCase.

PROBLEMS

1. If there are 10 μmol of the radioactive isotope ^{32}P (half-life 14 days) at $t = 0$, how much ^{32}P will remain at (a) 7 days, (b) 14 days, (c) 21 days, and (d) 70 days?

2. For each reaction below, determine whether the reaction is first order or second order and calculate the rate constant.

Time (s)	Reaction A reactant (mM)	Reaction B reactant (mM)
0	6.2	5.4
1	3.1	4.6
2	2.1	3.9
3	1.6	3.2
4	1.3	2.7
5	1.1	2.3

3. At what concentration of S (expressed as a multiple of K_M) will $v_o = 0.95\ V_{max}$?

4. Identify the enzymes in Table 12-1 whose catalytic efficiency is near the diffusion-controlled limit.

5. Explain why the following data sets from a Lineweaver–Burk plot are not individually ideal for determining K_M for an enzyme-catalyzed reaction.

Set A	1/[S] (mM^{-1})	1/v_o (μM^{-1}·s)
	0.5	2.4
	1.0	2.6
	1.5	2.9
	2.0	3.1

Set B	1/[S] (mM^{-1})	1/v_o (μM^{-1}·s)
	8	5.9
	10	6.8
	12	7.8
	14	8.7

6. Calculate K_M and V_{max} from the following data:

[S] (μM)	v_o (mM·s^{-1})
0.1	0.34
0.2	0.53
0.4	0.74
0.8	0.91
1.6	1.04

7. In a bisubstrate reaction, a small amount of the first product P is isotopically labeled (P*) and added to the enzyme and the first substrate A. No B or Q is present. Will A (= P—X) become isotopically labeled (A*) if the reaction follows (a) a Ping Pong mechanism or (b) a Sequential mechanism?

8. Determine the type of inhibition of an enzymatic reaction from the following data collected in the presence and absence of the inhibitor.

[S] (mM)	v_o (mM·min^{-1})	v_o with I present (mM·min^{-1})
1	1.3	0.8
2	2.0	1.2
4	2.8	1.7
8	3.6	2.2
12	4.0	2.4

9. Estimate K_I for a competitive inhibitor when [I] = 5 mM gives a value of K_M that is three times the K_M for the uninhibited reaction.

10. Enzyme X and enzyme Y catalyze the same reaction and exhibit the v_o versus [S] curves shown below. Which enzyme is more efficient at low [S]? Which is more efficient at high [S]?

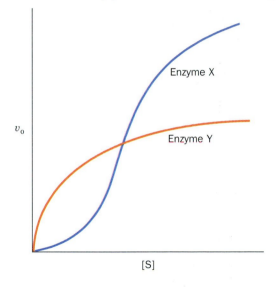

11. Sphingosine-1-phosphate (SPP) has recently been discovered to be important for cell survival. The synthesis of SPP from sphingosine and ATP is catalyzed by the enzyme sphingosine kinase. An understanding of the kinetics of the sphingosine kinase reaction may be important in the development of drugs to treat cancer. The velocity of the sphingosine kinase reaction was measured in the presence and absence of *threo*-sphingosine, a stereoisomer of sphingosine. The results are shown below.

[Sphingosine] (µM)	v_o (mg·min^{-1}) (no inhibitor)	v_o (mg·min^{-1}) (with threo-sphingosine)
2.5	32.3	8.5
3.5	40	11.5
5	50.8	14.6
10	72	25.4
20	87.7	43.9
50	115.4	70.8

Construct a Lineweaver–Burk plot to answer the following questions:

(a) What are the K_M and V_{max} in the presence and absence of the inhibitor?

(b) What kind of an inhibitor is *threo*-sphingosine? Explain.

[Problem provided by Kathleen Cornely, Providence College.]

IV

METABOLISM

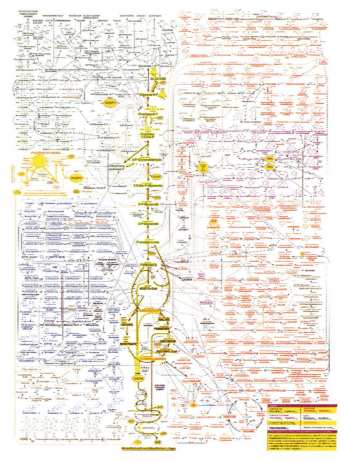

The processes by which biological molecules are broken down and resynthesized form a complex, yet highly regulated, network of interdependent enzymatic reactions that are collectively known as life. [Diagram designed by Donald E. Nicholson, Department of Biochemistry and Molecular Biology, The University of Leeds, England–and Sigma.]

INTRODUCTION TO METABOLISM

353

Understanding the chemical compositions and three-dimensional structures of biological molecules is not sufficient to understand how they are assembled into organisms or how they function to sustain life. We must therefore examine the reactions in which biological molecules are built and broken down. We must also consider how free energy is consumed in building cellular materials and carrying out cellular work and how free energy is generated from organic or other sources. **Metabolism,** the overall process through which living systems acquire and use free energy to carry out their various functions, is traditionally divided into two parts:

1. **Catabolism,** or degradation, in which nutrients and cell constituents are broken down to salvage their components and/or to generate energy.
2. **Anabolism,** or biosynthesis, in which biomolecules are synthesized from simpler components.

In general, catabolic reactions carry out the exergonic oxidation of nutrient molecules. The free energy thereby released is used to drive such endergonic processes as anabolic reactions, the performance of mechanical work, and the active transport of molecules against concentration gradients. Exergonic and endergonic processes are often coupled through the intermediate synthesis of a "high-energy" compound such as ATP. This simple principle underlies many of the chemical reactions presented in the following chapters. In this chapter, we introduce the general features of metabolic reactions and the roles of ATP and other compounds as energy carriers. We also examine some approaches to studying the dynamic metabolic transformations in organisms.

1. OVERVIEW OF METABOLISM

A bewildering array of chemical reactions occur in any living cell. Yet the principles that govern metabolism are the same in all organisms, a result of their common evolutionary origin and the constraints of the laws of thermodynamics. In fact, many of the specific reactions of metabolism are common to all organisms, with variations due primarily to differences in the source of the free energy that supports them.

A. Trophic Strategies

The nutritional requirements of an organism reflect its source of metabolic free energy. For example, some prokaryotes are **autotrophs** (Greek: *autos,* self + *trophos,* feeder), which can synthesize all their cellular constituents from simple molecules such as H_2O, CO_2, NH_3, and H_2S. There are two possible free energy sources for this process. **Chemolithotrophs** (Greek: *lithos,* stone) obtain their free energy through the oxidation of inorganic compounds such as NH_3, H_2S, or even Fe^{2+}:

$$2\ NH_3 + 4\ O_2 \longrightarrow 2\ HNO_3 + 2\ H_2O$$
$$H_2S + 2\ O_2 \longrightarrow H_2SO_4$$
$$4\ FeCO_3 + O_2 + 6\ H_2O \longrightarrow 4\ Fe(OH)_3 + 4\ CO_2$$

Photoautotrophs do so via photosynthesis, a process in which light energy powers the transfer of electrons from inorganic donors to CO_2 to produce carbohydrates, $(CH_2O)_n$, which are later oxidized to release free energy. **Heterotrophs** (Greek: *hetero,* other) obtain free energy through the oxidation of organic compounds (carbohydrates, lipids, and proteins) and hence ultimately depend on autotrophs for these substances.

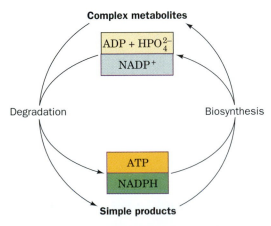

Figure 13-1.   Roles of ATP and NADP$^+$ in metabolism. ATP and NADPH generated through the degradation of complex metabolites are sources of free energy for biosynthetic and other reactions.

Organisms can be further classified by the identity of the oxidizing agent for nutrient breakdown. **Obligate aerobes** (which include animals) must use O_2, whereas **anaerobes** employ oxidizing agents such as sulfate or nitrate. **Facultative anaerobes,** such as *E. coli,* can grow in either the presence or the absence of O_2. **Obligate anaerobes,** in contrast, are poisoned by the presence of O_2. Their metabolisms are thought to resemble those of the earliest life forms, which arose over 3.5 billion years ago when the earth's atmosphere lacked O_2. Most of our discussion of metabolism will focus on aerobic processes.

B. Metabolic Pathways

Metabolic pathways are series of connected enzymatic reactions that produce specific products. Their reactants, intermediates, and products are referred to as **metabolites.** There are over 2000 known metabolic reactions, each catalyzed by a distinct enzyme. The types of enzymes and metabolites in a given cell vary with the identity of the organism, the cell type, its nutritional status, and its developmental stage. Many metabolic pathways are branched and interconnected, so delineating a pathway from a network of thousands of reactions is somewhat arbitrary and is driven by tradition as much as by chemical logic (see Box 13-1).

In general, catabolic and anabolic pathways are related as follows (Fig. 13-1; *opposite*): In catabolic pathways, complex metabolites are exergonically broken down into simpler products, in many cases, a two-carbon acetyl unit linked to **coenzyme A** to form **acetyl-coenzyme A (acetyl-CoA;** Section 13-2D). The free energy released in this degradative process is conserved by the synthesis of ATP from $ADP + P_i$ or by the reduction of the coenzyme $NADP^+$ (Fig. 3-4) to NADPH. ATP and NADPH are the major free energy sources for anabolic reactions. We shall take a closer look at the thermodynamic properties of acetyl-CoA, ATP, and NADPH later in this chapter.

A striking characteristic of degradative metabolism is that *the pathways for the catabolism of a large number of diverse substances (carbohydrates, lipids, and proteins) converge on a few common intermediates.* These intermediates are then further metabolized in a central oxidative pathway. Figure 13-2 outlines the breakdown of various foodstuffs to their monomeric units and then to acetyl-CoA. This is followed by the oxidation of the acetyl carbons to CO_2 by the **citric acid cycle** (Chapter 16). The reduced coenzymes **NADH** and **FADH$_2$** (Fig. 3-3) that this process yields then pass their electrons to O_2 to produce H_2O in the process of **oxidative phosphorylation** (Chapter 17).

Biosynthetic pathways carry out the opposite process. *Relatively few metabolites serve as starting materials for a host of varied products.* In the next several chapters, we discuss many catabolic and anabolic pathways in detail.

Metabolic Pathways Occur in Specific Cellular Locations

The compartmentation of the eukaryotic cytoplasm allows different metabolic pathways to operate in different locations. For example, oxidative phosphorylation occurs in the mitochondria, while **glycolysis** (a carbohydrate degradation pathway) and fatty acid biosynthesis occur in the cy-

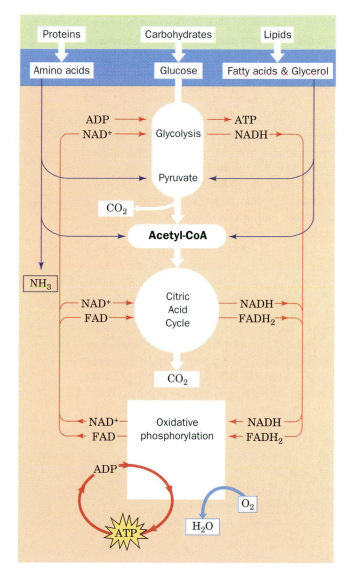

Figure 13-2.   *Key to Metabolism.* Overview of catabolism. Complex metabolites such as carbohydrates, proteins, and lipids are degraded first to their monomeric units, chiefly glucose, amino acids, fatty acids, and glycerol, and then to the common intermediate, acetyl-CoA. The acetyl group is oxidized to CO_2 via the citric acid cycle with the concomitant reduction of NAD^+ and FAD. Reoxidation of NADH and FADH$_2$ by O_2 during oxidative phosphorylation yields H_2O and ATP.

Box 13-1

B I O C H E M I S T R Y   I N   F O C U S

Organizing Metabolic Reactions

The task of cataloging all the enzymatic reactions that occur in a given organism is formidable and, in nearly all cases, far from complete. However, the metabolic reactions that constitute the major catabolic and anabolic pathways can be organized in a diagram. An example of such a metabolic map, color-coded by type of metabolite, is shown at right. Each point in the diagram represents a metabolite, and the lines between points correspond to the enzymatic reactions by which the metabolites are interconverted. The highly branched and interconnected nature of metabolic pathways is evident, as is the existence of cyclic pathways. Not all the reactions depicted here occur in all organisms, and there are innumerable enzymatic reactions that are not common enough to merit inclusion in a "universal" map.

Internet-accessible databases link metabolic enzymes (identified by EC number; Section 11-1A) to the corresponding gene sequences in different organisms and to the three-dimensional structures of the proteins, if known. Such linked databases are essential for navigating the ever-growing mass of information relating to proteins and their metabolic functions. Sequence and structural homology among enzymes that catalyze related reactions also makes it easier to deduce the probable metabolic functions of proteins encoded by genes that are discovered through genome sequencing (Section 3-4C). Presumably, diagrams such as the one shown here will facilitate the compilation of newly discovered enzymatic reactions into coherent metabolic pathways.

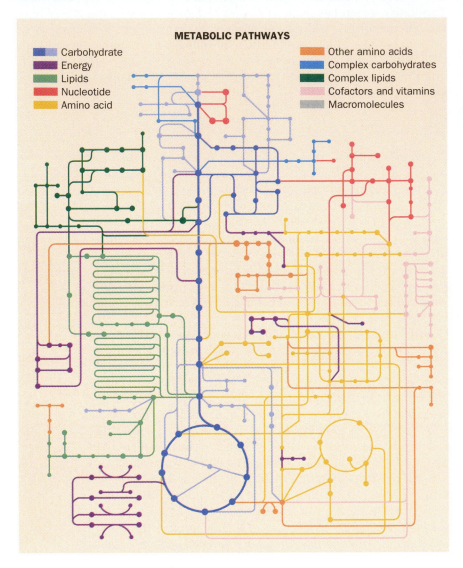

METABOLIC PATHWAYS

Carbohydrate
Energy
Lipids
Nucleotide
Amino acid

Other amino acids
Complex carbohydrates
Complex lipids
Cofactors and vitamins
Macromolecules

[Figure adapted from the Kyoto Encyclopedia of Genes and Genomes (www.genome.ad.jp/kegg/).]

tosol. Table 13-1 lists the major metabolic features of eukaryotic organelles. Metabolic processes in prokaryotes, which lack organelles, may be localized to particular areas of the cytosol.

The synthesis of metabolites in specific membrane-bounded compartments in eukaryotic cells requires mechanisms to transport these substances between compartments. Accordingly, transport proteins (Section 10-4) are essential components of many metabolic processes. For example, a transport protein is required to move ATP, which is generated in the mitochondria, to the cytosol, where most of it is consumed.

In multicellular organisms, compartmentation is carried a step further to the level of tissues and organs. The mammalian liver, for example, is

Table 13-1.   **Metabolic Functions of Eukaryotic Organelles**

Organelle	Function
Mitochondrion	Citric acid cycle, oxidative phosphorylation, fatty acid oxidation, amino acid breakdown
Cytosol	Glycolysis, pentose phosphate pathway, fatty acid biosynthesis, many reactions of gluconeogenesis
Lysosomes	Enzymatic digestion of cell components and ingested matter
Nucleus	DNA replication and transcription, RNA processing
Golgi apparatus	Posttranslational processing of membrane and secretory proteins; formation of plasma membrane and secretory vesicles
Rough endoplasmic reticulum	Synthesis of membrane-bound and secretory proteins
Smooth endoplasmic reticulum	Lipid and steroid biosynthesis
Peroxisomes (glyoxysomes in plants)	Oxidative reactions catalyzed by amino acid oxidases and catalase; glyoxylate cycle reactions in plants

largely responsible for the synthesis of glucose from noncarbohydrate precursors (**gluconeogenesis;** Section 15-4) so as to maintain a relatively constant level of glucose in the circulation, whereas adipose tissue is specialized for the storage of triacylglycerols. The interdependence of the metabolic functions of the various organs is the subject of Chapter 21.

An intriguing manifestation of specialization in tissues and subcellular compartments is the existence of **isozymes,** enzymes that catalyze the same reaction but are encoded by different genes and have different kinetic or regulatory properties. For example, vertebrates possess two homologs of the enzyme **lactate dehydrogenase (LDH):** the M type, which predominates in tissues subject to anaerobic conditions such as skeletal muscle and liver, and the H type, which predominates in aerobic tissues such as heart muscle. Lactate dehydrogenase catalyzes the interconversion of **pyruvate,** a product of glycolysis, and **lactate** (Section 14-3A). The M-type isozyme appears to mainly function in the reduction by NADH of pyruvate to lactate, whereas the H-type enzyme appears to be better adapted to catalyze the reverse reaction. The existence of isozymes allows for the testing of various illnesses. For example, heart attacks cause the death of heart muscle cells, which consequently rupture and release H-type LDH into the blood. A blood test indicating the presence of H-type LDH is therefore diagnostic of a heart attack.

C. Thermodynamic Considerations

Recall from Section 1-4D that the free energy change (ΔG) of a biochemical process, such as the reaction

$$A + B \rightleftharpoons C + D$$

is related to the standard free energy change ($\Delta G^{\circ\prime}$) and the concentrations of the reactants (Eq. 1-15):

$$\Delta G = \Delta G^{\circ\prime} + RT \ln \left(\frac{[C][D]}{[A][B]} \right) \qquad [13\text{-}1]$$

However, when the reactants are present at values close to their equilibrium values, $[C][D]/[A][B] \approx K_{eq}$, and $\Delta G \approx 0$. This is the case for many metabolic reactions, which are said to be **near-equilibrium reactions.** Because their ΔG values are close to zero, they can be relatively easily reversed by changing the ratio of products to reactants. When the reactants are in excess of their equilibrium concentrations, the net reaction proceeds in the forward direction until the excess reactants have been converted to products and equilibrium is attained. Conversely, when products are in excess, the net reaction proceeds in the reverse direction so as to convert products to reactants until the equilibrium concentration ratio is again achieved. *Enzymes that catalyze near-equilibrium reactions tend to act quickly to restore equilibrium concentrations, and the net rates of such reactions are effectively regulated by the relative concentrations of substrates and products.*

Other metabolic reactions function far from equilibrium; that is, they are irreversible. This is because an enzyme catalyzing such a reaction has insufficient activity to allow it to come to equilibrium. Reactants therefore accumulate in large excess of their equilibrium amounts, making $\Delta G \ll 0$. Changes in substrate concentrations therefore have relatively little effect on the rate of an irreversible reaction; the enzyme is essentially saturated. Only changes in the activity of the enzyme, through allosteric interactions, for example, can significantly alter this rate. The enzyme is therefore analogous to a dam on a river: *It controls the flow of substrate through the reaction by varying its activity, much as a dam controls the flow of a river by varying the opening of its floodgates.*

Understanding the **flux** (rate of flow) of metabolites through a metabolic pathway requires knowledge of which reactions are functioning near equilibrium and which are far from it. Most enzymes in a metabolic pathway operate near equilibrium and therefore have net rates that vary with their substrate concentrations. However, certain enzymes that operate far from equilibrium are strategically located in metabolic pathways. This has several important implications:

1. **Metabolic pathways are irreversible.** A highly exergonic reaction (one with $\Delta G \ll 0$) is irreversible; that is, it goes to completion. If such a reaction is part of a multistep pathway, it confers directionality on the pathway; that is, it makes the entire pathway irreversible.

2. **Every metabolic pathway has a first committed step.** Although most reactions in a metabolic pathway function close to equilibrium, there is generally an irreversible (exergonic) reaction early in the pathway that "commits" its product to continue down the pathway (likewise, water that has gone over a dam cannot return).

3. **Catabolic and anabolic pathways differ.** If a metabolite is converted to another metabolite by an exergonic process, free energy must be supplied to convert the second metabolite back to the first. This energetically "uphill" process requires a different pathway for at least some of the reaction steps.

The existence of independent interconversion routes, as we shall see, is an important property of metabolic pathways because it allows independent control of the two processes. If metabolite 2 is required by the cell, it is necessary to "turn off" the pathway from 2 to 1 while "turning on" the pathway from 1 to 2. Such independent control would be impossible without different pathways.

D. Control of Metabolic Flux

Living organisms are thermodynamically open systems that tend to maintain a steady state rather than reaching equilibrium (Section 1-4E). This is strikingly demonstrated by the observation that, over a 40-year time span, a normal human adult consumes literally tons of nutrients and imbibes over 20,000 L of water but does so without significant weight change. *The flux of intermediates through a metabolic pathway in a steady state is more or less constant; that is, the rates of synthesis and breakdown of each pathway intermediate maintain it at a constant concentration.* A steady state far from equilibrium is thermodynamically efficient, because only a nonequilibrium process ($\Delta G \neq 0$) can perform useful work. Indeed, living systems that have reached equilibrium are dead.

Since a metabolic pathway is a series of enzyme-catalyzed reactions, it is easiest to describe the flux of metabolites through the pathway by considering its reaction steps individually. The flux of metabolites, *J*, through each reaction step is the rate of the forward reaction, v_f, less that of the reverse reaction, v_r:

$$J = v_f - v_r \qquad [13\text{-}2]$$

At equilibrium, by definition, there is no net flux ($J = 0$), although v_f and v_r may be quite large. In reactions that are far from equilibrium, $v_f \gg v_r$, so the flux is essentially equal to the rate of the forward reaction ($J \approx v_f$). *The flux throughout a steady state pathway is constant and is set by the pathway's rate-determining step (or steps).* Clearly, flux through the rate-determining step must vary in response to the organism's metabolic requirements so as to reach a new steady state, and this change in flux must be communicated throughout the pathway.

By definition, the rate-determining step (which is often the first committed step) of a pathway is its slowest step. The product of the rate-determining step is therefore removed by succeeding steps in the pathway before it can equilibrate with reactant. Thus, *the rate-determining step functions far from equilibrium and has a large negative free energy change.* In an analogous manner, a dam creates a difference in water levels between its upstream and downstream sides, and a large negative free energy change results from the hydrostatic pressure difference. The dam can release water to generate electricity, varying the water flow according to the need for electrical power.

The relative insensitivity of the rate of a nonequilibrium reaction to variations in the concentrations of its substrates permits the establishment of a steady-state flux of metabolites through the pathway. Altering the rate of that reaction therefore alters the flux of material through the entire pathway, often by an order of magnitude or more. Several mechanisms may control flux through the rate-determining step:

1. **Allosteric control.** Many enzymes are allosterically regulated (Section 12-3) by effectors that are often substrates, products, or coenzymes of the pathway but not necessarily of the enzyme in question. For example, in negative feedback regulation, the product of a pathway inhibits an earlier step in the pathway:

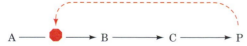

Thus, as we have seen, CTP, a product of pyrimidine biosynthesis, inhibits ATCase, which catalyzes the rate-determining step in this pathway (Fig. 12-11).

2. **Covalent modification (enzymatic interconversion).** Many enzymes that control pathway fluxes have specific sites that may be enzymatically phosphorylated and dephosphorylated or covalently modified in some other way.

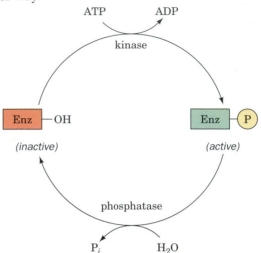

Such enzymatic modification processes, which are themselves subject to control, greatly alter the activities of the modified enzymes. This flux control mechanism is discussed in Section 14-4B.

3. **Substrate cycles.** If v_f and v_r represent the rates of two opposing nonequilibrium reactions that are catalyzed by different enzymes, v_f and v_r may be independently varied.

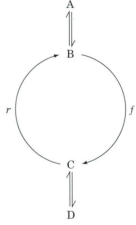

For example, flux $(v_f - v_r)$ can be increased not just by accelerating the forward reaction but by slowing the reverse reaction. The flux through such a **substrate cycle,** as we shall see in Section 14-4, is more sensitive to the concentrations of allosteric effectors than is the flux through a single unopposed nonequilibrium reaction.

4. **Genetic control.** Enzyme concentrations, and hence enzyme activities, may be altered by protein synthesis in response to metabolic needs. Genetic control of enzyme concentrations is a major concern of Part V of this text.

Mechanisms 1 to 3 can respond rapidly (within seconds or minutes) to external stimuli and are therefore classified as "short-term" control mechanisms. Mechanism 4 responds more slowly to changing conditions (within hours or days in higher organisms) and is therefore regarded as a "long-term" control mechanism.

Reactions that function near equilibrium respond rapidly to changes in substrate concentration. Thus, a series of near-equilibrium reactions downstream from the rate-determining step all have the same flux. Control of most metabolic pathways, however, is shared by several nonequilibrium steps, so the flux of material through a pathway may depend on multiple effectors whose relative importance reflects the overall metabolic needs of the organism at a given time.

2. "HIGH-ENERGY" COMPOUNDS

The complete oxidation of a metabolic fuel such as glucose

$$C_6H_{12}O_6 + 6\ O_2 \longrightarrow 6\ CO_2 + 6\ H_2O$$

releases considerable energy ($\Delta G^{\circ\prime} = -2850\ kJ \cdot mol^{-1}$). The complete oxidation of palmitate, a typical fatty acid,

$$C_{16}H_{32}O_2 + 23\ O_2 \longrightarrow 16\ CO_2 + 16\ H_2O$$

is even more exergonic ($\Delta G^{\circ\prime} = -9781\ kJ \cdot mol^{-1}$). Oxidative metabolism proceeds in a stepwise fashion, so the released free energy can be recovered in a manageable form at each exergonic step of the overall process. *These "packets" of energy are conserved by the synthesis of a few types of* **"high-energy"** *intermediates whose subsequent exergonic breakdown drives endergonic processes.* These intermediates therefore form a sort of free energy "currency" through which free energy–producing reactions such as glucose oxidation or fatty acid oxidation "pay for" the free energy–consuming processes in biological systems.

A. ATP and Phosphoryl Group Transfer

The "high-energy" intermediate adenosine triphosphate (ATP; Fig. 13-3), which occurs in all known life forms, is the primary cellular energy currency. ATP consists of an **adenosine** moiety (adenine + ribose) to which three phosphoryl ($-PO_3^{2-}$) groups are sequentially linked via a **phosphoester** bond followed by two **phosphoanhydride** bonds.

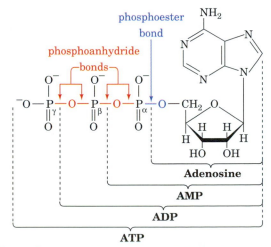

Figure 13-3.   *Key to Structure.* The structure of ATP indicating its relationship to ADP, AMP, and adenosine. The phosphoryl groups, starting from AMP, are referred to as the α-, β-, and γ-phosphates. Note the differences between phosphoester and phosphoanhydride bonds.

Table 13-2.   **Standard Free Energies of Phosphate Hydrolysis of Some Compounds of Biological Interest**

Compound	$\Delta G^{\circ\prime}$ (kJ·mol^{-1})
Phosphoenolpyruvate	-61.9
1,3-Bisphosphoglycerate	-49.4
Acetyl phosphate	-43.1
Phosphocreatine	-43.1
PP$_i$	-33.5
ATP ($\rightarrow$ AMP + PP$_i$)	-32.2
ATP ($\rightarrow$ ADP + P$_i$)	-30.5
Glucose-1-phosphate	-20.9
Fructose-6-phosphate	-13.8
Glucose-6-phosphate	-13.8
Glycerol-3-phosphate	-9.2

Source: Jencks, W.P., *in* Fasman, G.D. (Ed.), *Handbook of Biochemistry and Molecular Biology* (3rd ed.), Physical and Chemical Data, Vol. I, pp. 296–304, CRC Press (1976).

The biological importance of ATP rests in the large amount of free energy change that accompanies cleavage of its phosphoanhydride bonds. This occurs when either a phosphoryl group is transferred to another compound, leaving ADP, or a nucleotidyl (AMP) group is transferred, leaving **pyrophosphate** ($P_2O_7^{4-}$; **PP$_i$**). When the acceptor is water, the process is known as hydrolysis:

$$ATP + H_2O \rightleftharpoons ADP + P_i$$
$$ATP + H_2O \rightleftharpoons AMP + PP_i$$

Most biological group-transfer reactions involve acceptors other than water. However, knowing the free energy of hydrolysis of various phosphoryl compounds allows us to calculate the energy of transfer of phosphoryl groups to other acceptors by determining the difference in free energy of hydrolysis of the phosphoryl donor and acceptor.

The $\Delta G^{\circ\prime}$ values for hydrolysis of several phosphorylated compounds of biochemical importance are tabulated in Table 13-2. The negatives of these values are often referred to as **phosphoryl group-transfer potentials;** they are a measure of the tendency of phosphorylated compounds to transfer their phosphoryl groups to water. Note that ATP has an intermediate phosphoryl group-transfer potential. Under standard conditions, the compounds above ATP in Table 13-2 can spontaneously transfer a phosphoryl group to ADP to form ATP, which can, in turn, spontaneously transfer a phosphoryl group to the appropriate groups to form the compounds listed below it.

Rationalizing the "Energy" in "High-Energy" Compounds

Bonds whose hydrolysis proceeds with large negative values of $\Delta G^{\circ\prime}$ (customarily more than -25 kJ·mol^{-1}) are often referred to as **"high-energy" bonds** or **"energy-rich" bonds** and are frequently symbolized by the squiggle ($\sim$). Thus, ATP can be represented as AR—P$\sim$P$\sim$P, where A, R, and P symbolize adenyl, ribosyl, and phosphoryl groups, respectively. Yet the phosphoester bond joining the adenosyl group of ATP to its α-phosphoryl group appears to be not greatly different in electronic character from the so-called "high-energy" bonds bridging its α- and β- and its β- and γ-phosphoryl groups. In fact, none of these bonds has any unusual properties, so the term "high-energy" bond is somewhat of a misnomer (in any case, it should not be confused with the term "bond energy," which is defined as the energy required to break, not hydrolyze, a covalent bond). Why, then, are the phosphoryl group-transfer reactions of ATP so exergonic? Several factors appear to be responsible for the "high-energy" character of phosphoanhydride bonds such as those in ATP (Fig. 13-4):

1. The resonance stabilization of a phosphoanhydride bond is less than that of its hydrolysis products. This is because a phosphoanhydride's two strongly electron-withdrawing groups must compete for the π electrons of its bridging oxygen atom, whereas this competition is absent in the hydrolysis products. In other words, the electronic requirements of the phosphoryl groups are less satisfied in a phosphoanhydride than in its hydrolysis products.

2. Of perhaps greater importance is the destabilizing effect of the electrostatic repulsions between the charged groups of a phosphoanhydride compared to those of its hydrolysis products. In the physiological pH range, ATP has three to four negative charges whose mutual electrostatic repulsions are partially relieved by ATP hydrolysis.

3. Another destabilizing influence, which is difficult to assess, is the smaller solvation energy of a phosphoanhydride compared to that of

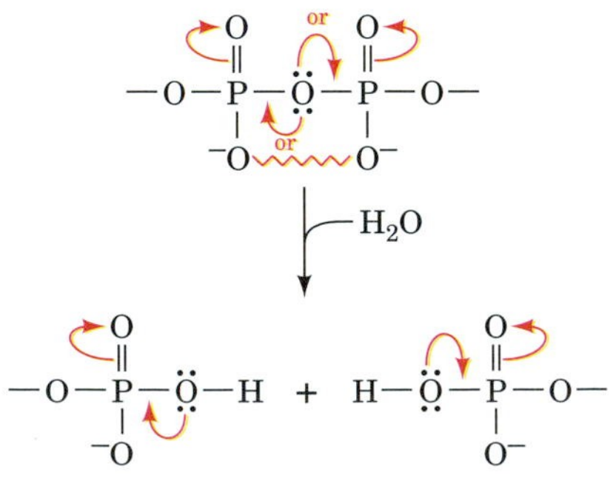

Figure 13-4.   Resonance and electrostatic stabilization in a phosphoanhydride and its hydrolytic products. The competing resonances (*curved arrows* from the central O) and charge–charge repulsions (*zigzag line*) between phosphoryl groups decrease the stability of a phosphoanhydride relative to its hydrolysis products.

ATP and ΔG

The standard conditions reflected in $\Delta G^{\circ\prime}$ values never occur in living organisms. Furthermore, other compounds that are present at high concentrations and that can potentially interact with the substrates and products of a metabolic reaction may dramatically affect ΔG values. For example, Mg^{2+} ions in cells partially neutralize the negative charges on the phosphate groups in ATP and its hydrolysis products, thereby diminishing the electrostatic repulsions that make ATP hydrolysis so exergonic. Similarly, changes in pH alter the ionic character of phosphorylated compounds and therefore alter their free energies.

In a given cell, the concentrations of many ions, coenzymes, and metabolites vary with both location and time, of-

ten by several orders of magnitude. Intracellular ATP concentrations are maintained within a relative narrow range, usually 2–10 mM, but the concentrations of ADP and P_i are more variable. Consider a typical cell with [ATP] = 3.0 mM, [ADP] = 0.8 mM, and [P_i] = 4.0 mM. Using Eq. 13-1, the actual free energy of ATP hydrolysis at 37°C is calculated below. This value is even greater than the standard free energy of ATP hydrolysis. However, because of the difficulty in accurately measuring the concentrations of particular chemical species in a cell or organelle, the ΔG's for most *in vivo* reactions are little more than estimates. For the sake of consistency, we shall, for the most part, use $\Delta G^{\circ\prime}$ values in this textbook.

$$\Delta G = \Delta G^{\circ\prime} + RT \ln\left(\frac{[ADP][P_i]}{[ATP]}\right)$$

$$= -30.5 \text{ kJ}\cdot\text{mol}^{-1} + (8.3145 \text{ J}\cdot\text{K}^{-1}\cdot\text{mol}^{-1})(310 \text{ K}) \ln\left(\frac{(0.8 \times 10^{-3} \text{ M})(4.0 \times 10^{-3} \text{ M})}{(3.0 \times 10^{-3} \text{ M})}\right)$$

$$= -30.5 \text{ kJ}\cdot\text{mol}^{-1} - 17.6 \text{ kJ}\cdot\text{mol}^{-1} = -48.1 \text{ kJ}\cdot\text{mol}^{-1}$$

its hydrolysis products. Some estimates suggest that this factor provides the dominant thermodynamic driving force for the hydrolysis of phosphoanhydrides.

Of course, the free energy change for any reaction, including phosphoryl group transfer from a "high-energy" compound, depends in part on the concentrations of the reactants and products (Eq. 13-1). Furthermore, because ATP and its hydrolysis products are ions, ΔG also depends on pH and ionic strength (see Box 13-2).

B. Coupled Reactions

The exergonic reactions of "high-energy" compounds can be coupled to endergonic processes to drive them to completion. The thermodynamic explanation for the coupling of an exergonic and an endergonic process is based on the additivity of free energy. Consider the following two-step reaction pathway:

$$\begin{align}(1)\quad & A + B \rightleftharpoons C + D \quad \Delta G_1\\(2)\quad & D + E \rightleftharpoons F + G \quad \Delta G_2\end{align}$$

If $\Delta G_1 \geq 0$, Reaction 1 will not occur spontaneously. However, if ΔG_2 is sufficiently exergonic so that $\Delta G_1 + \Delta G_2 < 0$, then although the equilibrium concentration of D in Reaction 1 will be relatively small, it will be larger than that in Reaction 2. As Reaction 2 converts D to products, Reaction 1 will operate in the forward direction to replenish the equilibrium concentration of D. The highly exergonic Reaction 2 therefore "drives" or "pulls" the endergonic Reaction 1, and the two reactions are said to be coupled through their common intermediate, D. That these coupled reactions proceed spontaneously can also be seen by summing Reactions 1 and 2 to yield the overall reaction

$$(1 + 2) \qquad A + B + E \rightleftharpoons C + F + G \qquad \Delta G_3$$

where $\Delta G_3 = \Delta G_1 + \Delta G_2 < 0$. *As long as the overall pathway is exergonic, it will operate in the forward direction.*

To illustrate this concept, let us consider two examples of phosphoryl group-transfer reactions. The initial step in the metabolism of glucose is its conversion to **glucose-6-phosphate** (Section 14-2A). Yet the direct reaction of glucose and P_i is thermodynamically unfavorable ($\Delta G^{\circ\prime} = +13.8$ kJ·mol^{-1}; Fig. 13-5a). In cells, however, this reaction is coupled to the exergonic cleavage of ATP (for ATP hydrolysis, $\Delta G^{\circ\prime} = -30.5$ kJ·mol^{-1}), so the overall reaction is thermodynamically favorable ($\Delta G^{\circ\prime} = +13.8 - 30.5 = -16.7$ kJ·mol^{-1}). ATP can be similarly regenerated ($\Delta G^{\circ\prime} = +30.5$ kJ·mol^{-1}) by coupling its synthesis from ADP and P_i to the even more exergonic cleavage of **phosphoenolpyruvate** ($\Delta G^{\circ\prime} = -61.9$ kJ·mol^{-1}; Fig. 13-5b and Section 14-2J).

Phosphoanhydride Hydrolysis Drives Some Biochemical Processes

The free energy of the phosphoanhydride bonds of "high-energy" compounds such as ATP can be used to drive reactions toward equilibrium even when the phosphoryl groups are not transferred to another organic compound. For example, ATP hydrolysis (i.e., phosphoryl group transfer directly to H_2O) provides the free energy for the operation of molecular chaperones (Section 6-4C), muscle contraction (Section 7-3), and transmembrane active transport (Section 10-4C). In these processes, proteins

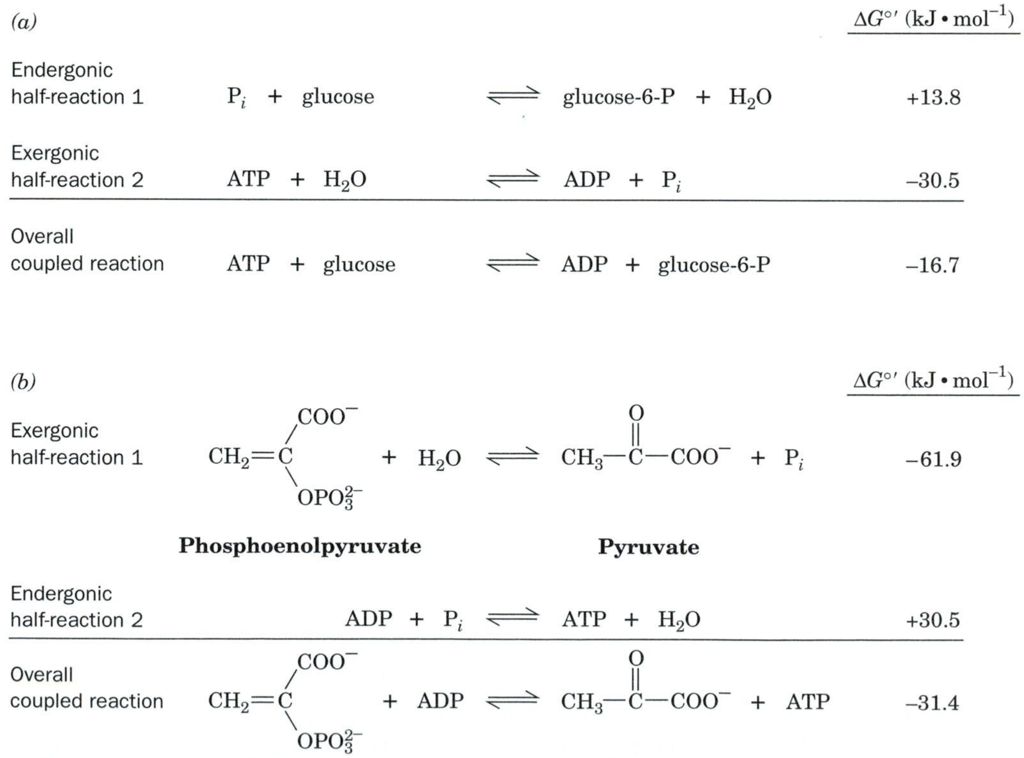

(a) $\Delta G^{\circ\prime}$ (kJ • mol^{-1})

Endergonic half-reaction 1	P_i + glucose $\rightleftharpoons$ glucose-6-P + H_2O		+13.8
Exergonic half-reaction 2	ATP + H_2O $\rightleftharpoons$ ADP + P_i		−30.5
Overall coupled reaction	ATP + glucose $\rightleftharpoons$ ADP + glucose-6-P		−16.7

(b) $\Delta G^{\circ\prime}$ (kJ • mol^{-1})

Exergonic half-reaction 1

$$\underset{\text{Phosphoenolpyruvate}}{CH_2{=}C{<}^{COO^-}_{OPO_3^{2-}}} + H_2O \rightleftharpoons \underset{\text{Pyruvate}}{CH_3{-}\overset{O}{\overset{\|}{C}}{-}COO^-} + P_i \qquad -61.9$$

Endergonic half-reaction 2

$$ADP + P_i \rightleftharpoons ATP + H_2O \qquad +30.5$$

Overall coupled reaction

$$CH_2{=}C{<}^{COO^-}_{OPO_3^{2-}} + ADP \rightleftharpoons CH_3{-}\overset{O}{\overset{\|}{C}}{-}COO^- + ATP \qquad -31.4$$

Figure 13-5. Some coupled reactions involving ATP. (*a*) The phosphorylation of glucose to form glucose-6-phosphate and ADP. (*b*) The phosphorylation of ADP by phosphoenolpyruvate to form ATP and pyruvate. Each reaction has been conceptually decomposed into a direct phosphorylation step (half-reaction 1) and a step in which ATP is hydrolyzed (half-reaction 2). Both half-reactions proceed in the direction that makes the overall reaction exergonic ($\Delta G < 0$).

undergo conformational changes in response to binding ATP. *The exergonic hydrolysis of ATP and release of ADP and P$_i$ renders these changes irreversible and thereby drives the processes forward.* GTP hydrolysis functions similarly to drive some of the reactions of signal transduction (Section 21-3B) and protein synthesis (Section 26-4).

In the absence of an appropriate enzyme, phosphoanhydride bonds are stable; that is, they hydrolyze quite slowly, despite the large amount of free energy released by these reactions. This is because these hydrolysis reactions have unusually high free energies of activation ($\Delta G^{\ddagger}$; Section 11-2). Consequently, *ATP hydrolysis is thermodynamically favored but kinetically disfavored.* For example, consider the reaction of glucose with ATP that yields glucose-6-phosphate (Fig. 13-5a). $\Delta G^{\ddagger}$ for the nonenzymatic transfer of a phosphoryl group from ATP to glucose is greater than that for ATP hydrolysis, so the hydrolysis reaction predominates (although neither reaction occurs at a biologically significant rate). However, in the presence of the enzyme **hexokinase** (Section 14-2A), glucose-6-phosphate is formed far more rapidly than ATP is hydrolyzed. This is because the catalytic influence of the enzyme reduces the activation energy for phosphoryl group transfer from ATP to glucose to less than the activation energy for ATP hydrolysis. This example underscores the point that even a thermodynamically favored reaction ($\Delta G < 0$) may not occur in a living system in the absence of a specific enzyme that catalyzes the reaction (i.e., lowers $\Delta G^{\ddagger}$ to increase the rate of product formation; Box 12-2).

Inorganic Pyrophosphatase Catalyzes Additional Phosphoanhydride Bond Cleavage

Although many reactions involving ATP yield ADP and P$_i$ (**orthophosphate cleavage**), others yield AMP and PP$_i$ (**pyrophosphate cleavage**). In these latter cases, the PP$_i$ is rapidly hydrolyzed to 2 P$_i$ by **inorganic pyrophosphatase** ($\Delta G^{\circ\prime} = -33.5$ kJ·mol^{-1}) so that *the pyrophosphate cleavage of ATP ultimately consumes two "high-energy" phosphoanhydride bonds.* The attachment of amino acids to tRNA molecules for protein synthesis is an example of this phenomenon (Fig. 13-6 and Section 26-2B). The two steps of the main reaction are readily reversible because the free energies of hydrolysis of the bonds formed are comparable to that of ATP hydrolysis. The overall reaction is driven to completion by the irreversible hydrolysis of PP$_i$. Nucleic acid biosynthesis from nucleoside triphosphates also releases PP$_i$ (Sections 24-1 and 25-1). The free energy changes of these reactions are around 0, so the subsequent hydrolysis of PP$_i$ is also essential for the synthesis of nucleic acids.

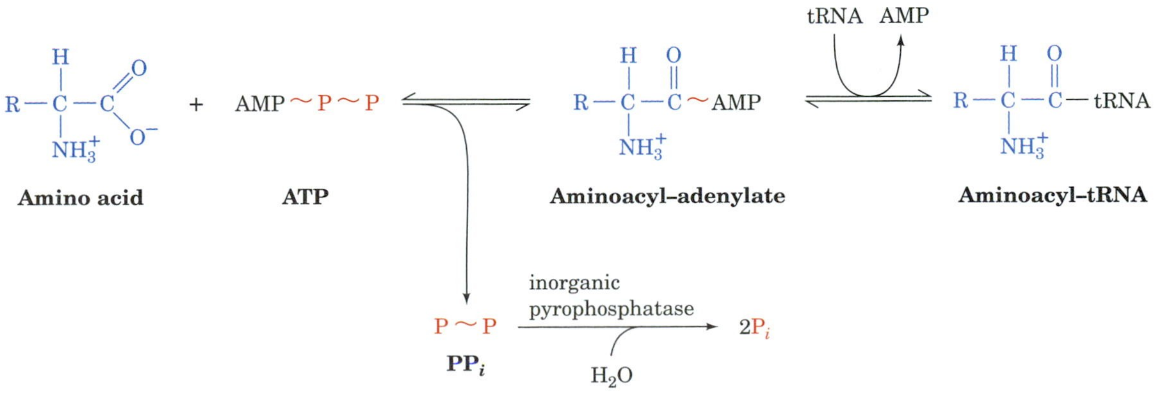

SAMPLE CALCULATION 1

Calculate the equilibrium constant for the hydrolysis of glucose-1-phosphate at 37°C.

$\Delta G^{\circ\prime}$ for the reaction

Glucose-1-phosphate + H$_2$O → glucose + P$_i$

is -20.9 kJ·mol^{-1} (Table 13-2). At equilibrium, $\Delta G = 0$ and Eq. 13-1 becomes

$\Delta G^{\circ\prime} = -RT \ln K$.

Therefore,

$K = e^{-\Delta G^{\circ\prime}/RT}$

$K = e^{-(-20,900\ \text{J}\cdot\text{mol}^{-1})/(8.3145\ \text{J}\cdot\text{K}^{-1}\cdot\text{mol}^{-1})(310\ \text{K})}$

$K = 3.3 \times 10^3$

Figure 13-6. Pyrophosphate cleavage in the synthesis of an aminoacyl–tRNA. In the first reaction step, the amino acid is adenylylated by ATP. In the second step, a tRNA molecule displaces the AMP moiety to form an aminoacyl–tRNA. The highly exergonic hydrolysis of pyrophosphate ($\Delta G^{\circ\prime} = -33.5$ kJ·mol^{-1}) drives the reaction forward.

C. Other Phosphorylated Compounds

"High-energy" compounds other than ATP are essential for energy metabolism, in part because they help maintain a relatively constant level of cellular ATP. *ATP is continually being hydrolyzed and regenerated.* Indeed, experimental evidence indicates that the metabolic half-life of an ATP molecule varies from seconds to minutes depending on the cell type and its metabolic activity. For instance, brain cells have only a few seconds' supply of ATP (which partly accounts for the rapid deterioration of brain tissue by oxygen deprivation). An average person at rest consumes and regenerates ATP at a rate of ~3 mol (1.5 kg) per hour and as much as an order of magnitude faster during strenuous activity.

Just as ATP drives endergonic reactions through the exergonic process of phosphoryl group transfer and phosphoanhydride hydrolysis, *ATP itself can be regenerated by coupling its formation to a more highly exergonic metabolic process.* As Table 13-2 indicates, in the thermodynamic hierarchy of phosphoryl-transfer agents, ATP occupies the middle rank. ATP can therefore be formed from ADP by direct transfer of a phosphoryl group from a "high-energy" compound (e.g., phosphoenolpyruvate; Fig. 13-5*b* and Section 14-2J). Such a reaction is referred to as a **substrate-level phosphorylation.** Other mechanisms generate ATP indirectly, using the energy supplied by transmembrane proton concentration gradients. In oxidative metabolism, this process is called **oxidative phosphorylation** (Section 17-3), whereas in photosynthesis, it is termed **photophosphorylation** (Section 18-2D).

The flow of energy from "high-energy" phosphate compounds to ATP and from ATP to "low-energy" phosphate compounds is diagrammed in Fig. 13-7. These reactions are catalyzed by enzymes known as **kinases,** which transfer phosphoryl groups from ATP to other compounds or from phosphorylated compounds to ADP. We shall revisit these processes in our discussion of carbohydrate metabolism in Chapters 14 and 15.

The compounds whose phosphoryl group-transfer potentials are greater than that of ATP have additional stabilizing effects. For example, the hydrolysis of **acyl phosphates** (mixed phosphoric–carboxylic anhydrides), such as **acetyl phosphate** and **1,3-bisphosphoglycerate,**

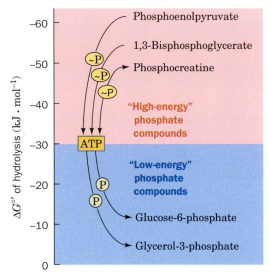

Figure 13-7. **Position of ATP relative to "high-energy" and "low-energy" phosphate compounds.** Phosphoryl groups flow from the "high-energy" donors, via the ATP–ADP system, to "low-energy" acceptors.

$$
\underset{\textbf{Acetyl phosphate}}{CH_3 - \overset{\overset{\displaystyle O}{\|}}{C} \sim OPO_3^{2-}} \qquad \underset{\textbf{1,3-Bisphosphoglycerate}}{^{-2}O_3POCH_2 - \overset{\overset{\displaystyle OH}{|}}{CH} - \overset{\overset{\displaystyle O}{\|}}{C} \sim OPO_3^{2-}}
$$

is driven by the same competing resonance and differential solvation effects that influence the hydrolysis of phosphoanhydrides (Fig. 13-4). Apparently, these effects are more pronounced for acyl phosphates than for phosphoanhydrides, as the rankings in Table 13-2 indicate.

In contrast, compounds such as glucose-6-phosphate and **glycerol-3-phosphate,**

$$
\underset{\textbf{\alpha-\textsc{d}-Glucose-6-phosphate}}{}\qquad \underset{\textbf{\textsc{l}-Glycerol-3-phosphate}}{HO - \overset{\overset{\displaystyle CH_2OH}{|}}{\underset{\underset{\displaystyle CH_2OPO_3^{2-}}{|}}{C}} - H}
$$

which are below ATP in Table 13-2, have no significantly different resonance stabilization or charge separation compared to their hydrolysis products. Their free energies of hydrolysis are therefore much less than those of the preceding "high-energy" compounds.

The high phosphoryl group-transfer potentials of **phosphoguanidines,** such as **phosphocreatine** and **phosphoarginine,** largely result from the competing resonances in the **guanidino** group, which are even more pronounced than they are in the phosphate group of phosphoanhydrides.

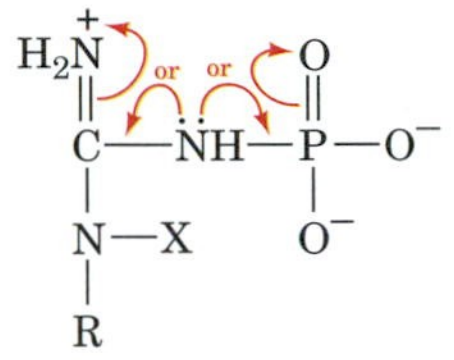

$$R = CH_2-CO_2^- \; ; \; X = CH_3 \qquad \textbf{Phosphocreatine}$$

$$R = CH_2-CH_2-CH_2-\overset{\overset{\displaystyle NH_3^+}{|}}{CH}-CO_2^- \; ; \; X = H \qquad \textbf{Phosphoarginine}$$

Consequently, phosphocreatine can transfer its phosphoryl group to ADP to form ATP.

Phosphocreatine Provides a "High-Energy" Reservoir for ATP Formation

Muscle and nerve cells, which have a high ATP turnover, rely on phosphoguanidines to regenerate ATP rapidly. In vertebrates, phosphocreatine is synthesized by the reversible phosphorylation of creatine by ATP as catalyzed by **creatine kinase:**

$$\text{ATP} + \text{creatine} \rightleftharpoons \text{phosphocreatine} + \text{ADP} \quad \Delta G^{\circ\prime} = +12.6 \text{ kJ} \cdot \text{mol}^{-1}$$

Note that this reaction is endergonic under standard conditions; however, *the intracellular concentrations of its reactants and products are such that it operates close to equilibrium* ($\Delta G \approx 0$). Accordingly, when the cell is in a resting state, so that [ATP] is relatively high, the reaction proceeds with net synthesis of phosphocreatine, whereas at times of high metabolic activity, when [ATP] is low, the equilibrium shifts so as to yield net synthesis of ATP from phosphocreatine and ADP. *Phosphocreatine thereby acts as an ATP "buffer" in cells that contain creatine kinase.* A resting vertebrate skeletal muscle normally has sufficient phosphocreatine to supply its free energy needs for several minutes (but for only a few seconds at maximum exertion). In the muscles of some invertebrates, such as lobsters, phosphoarginine performs the same function. These phosphoguanidines are collectively named **phosphagens.**

Nucleoside Triphosphates Are Freely Interconverted

Many biosynthetic processes, such as the synthesis of proteins and nucleic acids, require nucleoside triphosphates other than ATP. For example, RNA synthesis requires the ribonucleotides CTP, GTP, and UTP, along with ATP, and DNA synthesis requires dCTP, dGTP, dTTP, and dATP (Section 3-1). All these nucleoside triphosphates **(NTPs)** are synthesized from ATP and the corresponding nucleoside diphosphate **(NDP)** in a reaction catalyzed by the nonspecific enzyme **nucleoside diphosphate kinase:**

$$\text{ATP} + \text{NDP} \rightleftharpoons \text{ADP} + \text{NTP}$$

The $\Delta G^{\circ\prime}$ values for these reactions are nearly 0, as might be expected from the structural similarities among the NTPs. These reactions are driven by the depletion of the NTPs through their exergonic utilization in subsequent reactions.

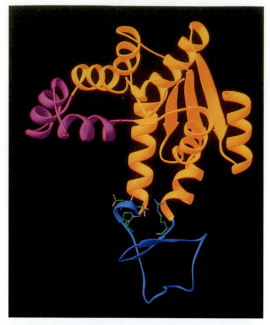

(a)

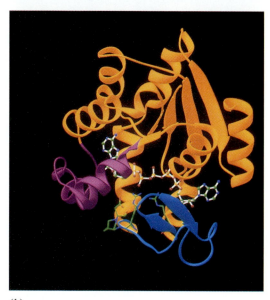

(b)

Figure 13-8. Conformational changes in adenylate kinase on binding substrate. (a) The unliganded enzyme. (b) The enzyme with the bound bisubstrate analog Ap$_5$A. The Ap$_5$A is shown in ball-and-stick form (C green, N blue, O red, and P yellow). Several of the protein's side chains that have been implicated in substrate binding are shown in stick form. The protein's magenta and cyan domains undergo extensive conformational changes on ligand binding, whereas the remainder of the protein (*gold*), whose orientation is the same in *a* and *b,* largely maintains its conformation. [Based on an X-ray structure by Georg Schulz, Institut für Organische Chemie und Biochemie, Freiburg, Germany.]
🔵 See the Interactive Exercises.

Other kinases reversibly convert nucleoside monophosphates to their diphosphate forms at the expense of ATP. One of these phosphoryl group-transfer reactions is catalyzed by **adenylate kinase:**

$$\text{AMP} + \text{ATP} \rightleftharpoons 2\ \text{ADP}$$

This enzyme is present in all tissues, where it functions to maintain equilibrium concentrations of the three nucleotides. When AMP accumulates, it is converted to ADP, which can be used to synthesize ATP through substrate-level phosphorylation, oxidative phosphorylation, or photophosphorylation. The reverse reaction helps restore cellular ATP as rapid consumption of ATP increases the level of ADP.

The X-ray structure of adenylate kinase, determined by Georg Schulz, reveals that, in the reaction catalyzed by this enzyme, two ~30-residue domains of the enzyme close over the substrates (Fig. 13-8), thereby tightly binding them and preventing water from entering the active site (which would lead to hydrolysis rather than phosphoryl group transfer). The movement of one of these domains depends on the presence of four invariant charged residues. Interactions between these groups and the bound substrates apparently trigger the rearrangements around the substrate-binding site (Fig. 13-8*b*).

Once the adenylate kinase reaction is complete, the tightly bound products must be rapidly released to maintain the enzyme's catalytic efficiency. Yet since the reaction is energetically neutral (the net number of phosphoanhydride bonds is unchanged), another source of free energy is required for rapid product release. The comparison of the X-ray structures of unliganded adenylate kinase and adenylate kinase in complex with the bisubstrate model compound **Ap$_5$A** (AMP and ATP connected by a fifth phosphate) show how the enzyme avoids the kinetic trap of tight-binding substrates and products: On binding substrate, a portion of the protein remote from the active site increases its chain mobility and thereby consumes some of the free energy of substrate binding. The region "resolidifies" when the binding site is opened and the products are released. This mechanism is thought to act as an "energetic counterweight" to help adenylate kinase maintain a high reaction rate.

D. Thioesters

The ubiquity of phosphorylated compounds in metabolism is consistent with their early evolutionary appearance. Yet phosphate is (and was) scarce in the abiotic world, which suggests that other kinds of molecules might have served as energy-rich compounds even before metabolic pathways became specialized for phosphorylated compounds. One candidate for a primitive "high-energy" compound is the **thioester,** which offers as its main recommendation its occurrence in the central metabolic pathways of all known organisms. Notably, the thioester bond is involved in substrate-level phosphorylation, an ATP-generating process that is independent of—and presumably arose before—oxidative phosphorylation.

The thioester bond appears in modern metabolic pathways as a reaction intermediate (involving a Cys residue in an enzyme active site) and in the form of acetyl-CoA (Fig. 13-9), the common product of carbohydrate, fatty acid, and amino acid catabolism. **Coenzyme A (CoASH or CoA)** consists of a β-mercaptoethylamine group bonded through an amide linkage to the vitamin **pantothenic acid,** which, in turn, is attached to a 3-phosphoadenosine moiety via a pyrophosphate bridge. The acetyl group of acetyl-CoA is bonded as a thioester to the sulfhydryl portion of the β-mercaptoethylamine group. *CoA thereby functions as a carrier of acetyl and other acyl groups*

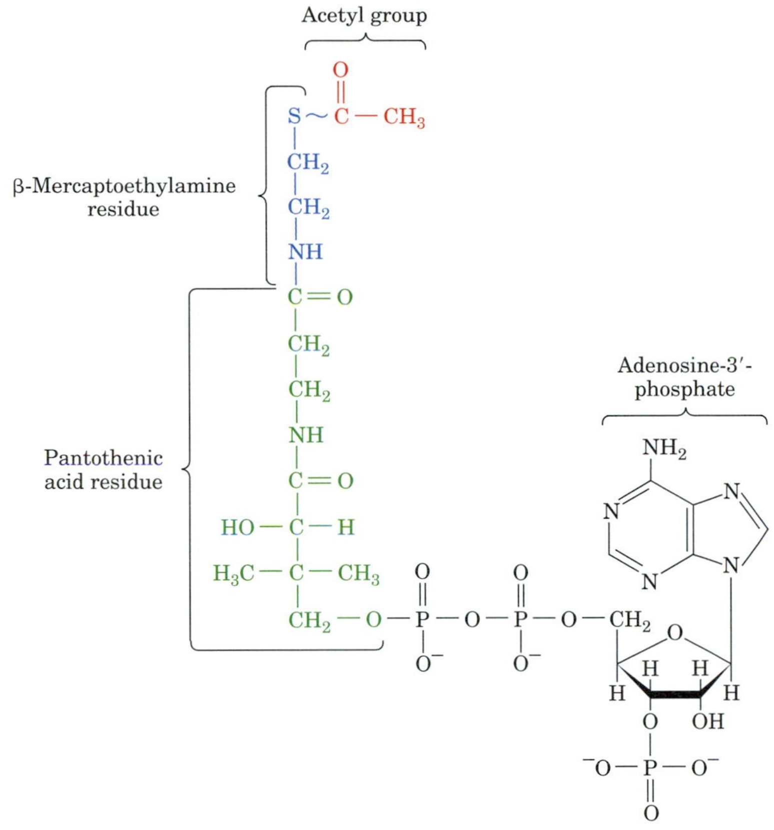

Acetyl group

$$
\begin{array}{c}
O \\
\parallel \\
S \sim C - CH_3 \\
\mid \\
CH_2 \\
\mid \\
CH_2 \\
\mid \\
NH \\
\mid \\
C = O \\
\mid \\
CH_2 \\
\mid \\
CH_2 \\
\mid \\
NH \\
\mid \\
C = O \\
\mid \\
HO - C - H \\
\mid \\
H_3C - C - CH_3 \\
\mid \\
CH_2
\end{array}
$$

β-Mercaptoethylamine residue

Pantothenic acid residue

Adenosine-3'-phosphate

Acetyl-coenzyme A (acetyl-CoA)

Figure 13-9. **The chemical structure of acetyl-CoA.** The thioester bond is drawn with a $\sim$ to indicate that it is a "high-energy" bond. In CoA, the acetyl group is replaced by hydrogen.

(the A of CoA stands for "Acetylation"). Thioesters also take the form of acyl chains bonded to a phosphopantetheine residue that is linked to a Ser OH group in a protein rather than to AMP, as in CoA.

Acetyl-CoA is a "high-energy" compound. The $\Delta G°'$ for the hydrolysis of its thioester bond is -31.5 kJ $\cdot$ mol^{-1}, which makes this reaction slightly (1 kJ $\cdot$ mol^{-1}) more exergonic than ATP hydrolysis. The hydrolysis of thioesters is more exergonic than that of ordinary esters because the thioester is less stabilized by resonance. This destabilization is a result of the large atomic radius of S, which reduces the electronic overlap between C and S compared to that between C and O.

The formation of a thioester bond in a metabolic intermediate conserves a portion of the free energy of oxidation of a metabolic fuel. That free energy can then be used to drive an exergonic process. In the citric acid cycle, for example, cleavage of a thioester **(succinyl-CoA)** releases sufficient free energy to synthesize GTP from GDP and P$_i$ (Section 16-3E).

3. OXIDATION–REDUCTION REACTIONS

As metabolic fuels are oxidized to CO_2, electrons are transferred to molecular carriers that, in aerobic organisms, ultimately transfer the electrons to molecular oxygen. The process of electron transport results in a transmembrane proton concentration gradient that drives ATP synthesis (oxidative phosphorylation; Section 17-3). Even obligate anaerobes, which do not carry out oxidative phosphorylation, rely on the oxidation of substrates to drive ATP synthesis. In fact, oxidation–reduction reactions (also known as **redox reactions**) supply living things with most of their free energy. In

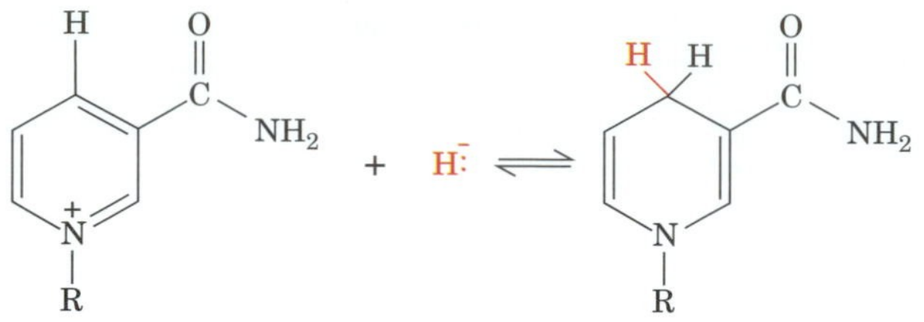

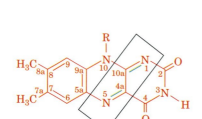

Figure 13-10.   Reduction of NAD$^+$ to NADH. R represents the ribose–pyrophosphoryl–adenosine portion of the coenzyme. Only the nicotinamide ring is affected by reduction, which is formally represented here as occurring by hydride transfer.

this section, we examine the thermodynamic basis for the conservation of free energy during substrate oxidation.

A. NAD$^+$ and FAD

Two of the most widely occurring electron carriers are the nucleotide coenzymes nicotinamide adenine dinucleotide (NAD$^+$) and flavin adenine dinucleotide (FAD). The nicotinamide portion of NAD$^+$ (and its phosphorylated counterpart NADP$^+$; Fig. 3-4) is the site of reversible reduction, which formally occurs as the transfer of a hydride ion (H$^-$; a proton with two electrons) as is shown in Fig. 13-10. The terminal electron acceptor in aerobic organisms, O$_2$, can accept only unpaired electrons; that is, electrons must be transferred to O$_2$ one at a time. Electrons that are removed from metabolites as pairs (e.g., with the two-electron reduction of NAD$^+$) must be transferred to other carriers that can undergo one-electron redox reactions. FAD (Fig. 3-3) is one such coenzyme; it can undergo both one-electron and two-electron transfers.

The conjugated ring system of FAD can accept one or two electrons to produce the stable radical (semiquinone) FADH · or the fully reduced (hydroquinone) FADH$_2$ (Fig. 13-11). The change in the electronic state of the ring system on reduction is reflected in a color change from brilliant yellow (in FAD) to pale yellow (in FADH$_2$). The metabolic functions of NAD$^+$ and FAD demand that they undergo reversible reduction so that they can accept electrons, pass them on to other electron carriers, and thereby be regenerated to participate in additional cycles of oxidation and reduction.

Humans cannot synthesize the flavin moiety of FAD but, rather, must obtain it from their diets, for example, in the form of riboflavin (vitamin B$_2$; Fig. 3-3). Nevertheless, riboflavin deficiency is quite rare in humans, in part because of the tight binding of flavin prosthetic groups to their apoenzymes. The symptoms of riboflavin deficiency, which are associated with general malnutrition or bizarre diets, include an inflamed tongue, lesions in the corners of the mouth, and dermatitis.

Flavin adenine dinucleotide (FAD)
(oxidized or quinone form)

H·

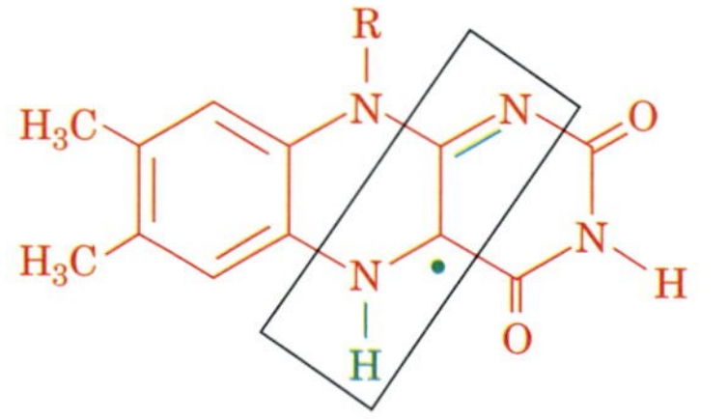

FADH · (radical or semiquinone form)

H·

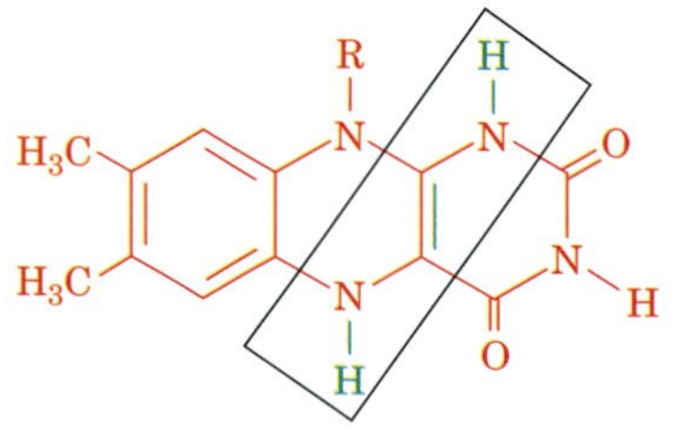

FADH$_2$ (reduced or hydroquinone form)

Figure 13-11.   Reduction of FAD to FADH$_2$. R represents the ribitol–pyrophosphoryl–adenosine portion of the coenzyme. The conjugated ring system of FAD undergoes either two sequential one-electron reductions or a two-electron transfer that bypasses the **semiquinone** state.

B. The Nernst Equation

Oxidation–reduction reactions resemble other types of group-transfer reactions except that the "groups" transferred are electrons, which are passed from an **electron donor** (**reductant** or **reducing agent**) to an **electron acceptor** (**oxidant** or **oxidizing agent**). For example, in the reaction

$$Fe^{3+} + Cu^+ \rightleftharpoons Fe^{2+} + Cu^{2+}$$

Cu^+, the reductant, is oxidized to Cu^{2+} while Fe^{3+}, the oxidant, is reduced to Fe^{2+}.

Redox reactions can be divided into two **half-reactions** or **redox couples,** such as

$$Fe^{3+} + e^- \rightleftharpoons Fe^{2+} \text{ (reduction)}$$
$$Cu^+ \rightleftharpoons Cu^{2+} + e^- \text{ (oxidation)}$$

whose sum is the whole reaction above. These particular half-reactions occur during the oxidation of cytochrome c oxidase in the mitochondrion (Section 17-2F). Note that for electrons to be transferred, both half-reactions must occur simultaneously. In fact, the electrons are the two half-reactions' common intermediate.

Electrochemical Cells

A half-reaction consists of an electron donor and its conjugate electron acceptor; in the oxidative half-reaction shown above, Cu^+ is the electron donor and Cu^{2+} is its conjugate electron acceptor. Together these constitute a **conjugate redox pair** analogous to a conjugate acid–base pair (HA and A^-; Section 2-2B). An important difference between redox pairs and acid–base pairs, however, is that *the two half-reactions of a redox reaction, each consisting of a conjugate redox pair, can be physically separated to form an electrochemical cell* (Fig. 13-12). In such a device, each half-reaction takes place in its separate **half-cell,** and electrons are passed between half-cells as an electric current in the wire connecting their two electrodes. A salt bridge is necessary to complete the electrical circuit by providing a conduit for ions to migrate and thereby maintain electrical neutrality.

The free energy of an oxidation–reduction reaction is particularly easy to determine by simply measuring the voltage difference between its two half-cells. Consider the general reaction

$$A_{ox}^{n+} + B_{red} \rightleftharpoons A_{red} + B_{ox}^{n+}$$

in which n electrons per mole of reactants are transferred from reductant (B_{red}) to oxidant (A_{ox}^{n+}). The free energy of this reaction is expressed as

$$\Delta G = \Delta G^\circ + RT \ln \left(\frac{[A_{red}][B_{ox}^{n+}]}{[A_{ox}^{n+}][B_{red}]} \right) \qquad [13\text{-}3]$$

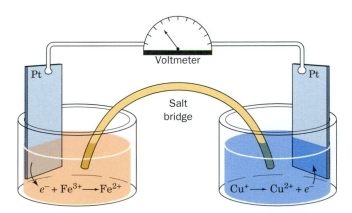

Figure 13-12.   An electrochemical cell. The half-cell undergoing oxidation (here $Cu^+ \rightarrow Cu^{2+} + e^-$) passes the liberated electrons through the wire to the half-cell undergoing reduction (here $e^- + Fe^{3+} \rightarrow Fe^{2+}$). Electroneutrality in the two half-cells is maintained by the transfer of ions through the electrolyte-containing salt bridge.

Under reversible conditions,

$$\Delta G = -w' = -w_{el} \qquad [13\text{-}4]$$

where w' is non-pressure–volume work. In this case, w' is equivalent to w_{el}, the electrical work required to transfer the n moles of electrons through the electrical potential difference $\Delta\mathscr{E}$ [where the units of $\mathscr{E}$ are volts (V), the number of joules (J) of work required to transfer 1 coulomb (C) of charge]. This, according to the laws of electrostatics, is

$$w_{el} = n\mathscr{F}\Delta\mathscr{E} \qquad [13\text{-}5]$$

where $\mathscr{F}$, the **faraday,** is the electrical charge of 1 mol of electrons (1 $\mathscr{F}$ = 96,485 C·mol^{-1} = 96,485 J·V^{-1}·mol^{-1}), and n is the number of moles of electrons transferred per mole of reactant converted. Thus, substituting Eq. 13-5 into Eq. 13-4,

$$\Delta G = -n\mathscr{F}\Delta\mathscr{E} \qquad [13\text{-}6]$$

Combining Eqs. 13-3 and 13-6, and making the analogous substitution for $\Delta G°$, yields the **Nernst equation:**

$$\boxed{\Delta\mathscr{E} = \Delta\mathscr{E}° - \frac{RT}{n\mathscr{F}}\ln\left(\frac{[A_{red}][B_{ox}^{n+}]}{[A_{ox}^{n+}][B_{red}]}\right)} \qquad [13\text{-}7]$$

which was originally formulated in 1881 by Walther Nernst. Here $\mathscr{E}$ is the **reduction potential.** $\Delta\mathscr{E}$, the **electromotive force (emf),** can be described as the "electron pressure" that the electrochemical cell exerts. The quantity $\mathscr{E}°$, the reduction potential when all components are in their standard states, is called the **standard reduction potential.** If these standard states refer to biochemical standard states (Section 1-4D), then $\mathscr{E}°$ is replaced by $\mathscr{E}°'$. Note that a positive $\Delta\mathscr{E}$ in Eq. 13-6 results in a negative ΔG; in other words, *a positive $\Delta\mathscr{E}$ indicates a spontaneous reaction, one that can do work.*

C. Measurements of Reduction Potential

Equation 13-6 shows that the free energy change of a redox reaction can be determined by directly measuring its reduction potential with a voltmeter (Fig. 13-12). Such measurements make it possible to determine the order of spontaneous electron transfers among a set of electron carriers such as those of the electron transport pathway that mediates oxidative phosphorylation in cells.

Any redox reaction can be divided into its component half-reactions:

$$A_{ox}^{n+} + n\,e^- \rightleftharpoons A_{red}$$
$$B_{ox}^{n+} + n\,e^- \rightleftharpoons B_{red}$$

where, by convention, both half-reactions are written as reductions. These half-reactions can be assigned reduction potentials, $\mathscr{E}_A$ and $\mathscr{E}_B$, in accordance with the Nernst equation:

$$\mathscr{E}_A = \mathscr{E}_A° - \frac{RT}{n\mathscr{F}}\ln\left(\frac{[A_{red}]}{[A_{ox}^{n+}]}\right) \qquad [13\text{-}8]$$

$$\mathscr{E}_B = \mathscr{E}_B° - \frac{RT}{n\mathscr{F}}\ln\left(\frac{[B_{red}]}{[B_{ox}^{n+}]}\right) \qquad [13\text{-}9]$$

For the overall redox reaction involving the two half-reactions,

$$\Delta\mathscr{E}° = \mathscr{E}°_{(e^-\text{acceptor})} - \mathscr{E}°_{(e^-\text{donor})} \qquad [13\text{-}10]$$

Thus, when the reaction proceeds with A as the electron acceptor and B as the electron donor, $\Delta\mathscr{E}° = \mathscr{E}_A° - \mathscr{E}_B°$, and $\Delta\mathscr{E} = \mathscr{E}_A - \mathscr{E}_B$.

Reduction potentials, like free energies, must be defined with respect to some arbitrary standard, in this case, the hydrogen half-reaction

$$2\,H^+ + 2\,e^- \rightleftharpoons H_2(g)$$

in which H^+ is in equilibrium with $H_2(g)$ that is in contact with a Pt electrode. This half-cell is arbitrarily assigned a standard reduction potential $\mathscr{E}°$ of 0 V (1 V = 1 J·C^{-1}) at pH 0, 25°C, and 1 atm. Under the biochemical convention, where the standard state is pH 7.0, the hydrogen half-reaction has a standard reduction potential $\mathscr{E}°'$ of -0.421 V.

When $\Delta\mathscr{E}$ is positive, ΔG is negative (Eq. 13-6), indicating a spontaneous process. In combining two half-reactions under standard conditions, the direction of spontaneity therefore involves the reduction of the redox couple with the more positive standard reduction potential. In other words, *the more positive the standard reduction potential, the higher the affinity of the redox couple's oxidized form for electrons, that is, the greater the tendency for the redox couple's oxidized form to accept electrons and thus become reduced.*

Biochemical Half-Reactions Are Physiologically Significant

The biochemical standard reduction potentials ($\Delta\mathscr{E}°'$) of some biochemically important half-reactions are listed in Table 13-3. The oxidized form of a redox couple with a large positive standard reduction potential has a

---

SAMPLE CALCULATION 2

Calculate $\Delta G°'$ for the oxidation of NADH by FAD.

Combining the relevant half-reactions gives

$$NADH + FAD + H^+ \longrightarrow NAD^+ + FADH_2$$

According to Eq. 13-10,

$$\Delta\mathscr{E}°' = \mathscr{E}°'_{(e^-\,\text{acceptor})} - \mathscr{E}°'_{(e^-\,\text{donor})}$$

$$= \mathscr{E}°'_{(FADH_2/FAD)} - \mathscr{E}°'_{(NADH/NAD^+)}$$

$$= (-0.219\,\text{V}) - (-0.315\,\text{V}) = 0.096\,\text{V}$$

$$= 0.096\,\text{J}\cdot\text{C}^{-1}$$

Because $\Delta G = -n\mathscr{F}\Delta\mathscr{E}$ (Eq. 13-6),

$\Delta G°' = -n\mathscr{F}\Delta\mathscr{E}°'$ and thus

$$\Delta G°' = -\,(2\,\text{mol }e^-/\text{mol reactant}) \times$$
$$(96{,}485\,\text{C}\cdot\text{mol}^{-1}\,e^-)(0.096\,\text{J}\cdot\text{C}^{-1})$$

$$= -18{,}500\,\text{J}\cdot\text{mol}^{-1}\,\text{reactant}$$

$$= -18.5\,\text{kJ}\cdot\text{mol}^{-1}\,\text{reactant}$$

Table 13-3. Standard Reduction Potentials of Some Biochemically Important Half-Reactions

Half-Reaction	$\mathscr{E}°'$ (V)
$\frac{1}{2}O_2 + 2\,H^+ + 2\,e^- \rightleftharpoons H_2O$	0.815
$SO_4^{2-} + 2\,H^+ + 2\,e^- \rightleftharpoons SO_3^{2-} + H_2O$	0.48
$NO_3^- + 2\,H^+ + 2\,e^- \rightleftharpoons NO_2^- + H_2O$	0.42
Cytochrome a_3 (Fe^{3+}) $+ e^- \rightleftharpoons$ cytochrome a_3 (Fe^{2+})	0.385
$O_2(g) + 2\,H^+ + 2\,e^- \rightleftharpoons H_2O_2$	0.295
Cytochrome a (Fe^{3+}) $+ e^- \rightleftharpoons$ cytochrome a (Fe^{2+})	0.29
Cytochrome c (Fe^{3+}) $+ e^- \rightleftharpoons$ cytochrome c (Fe^{2+})	0.235
Cytochrome c_1 (Fe^{3+}) $+ e^- \rightleftharpoons$ cytochrome c_1 (Fe^{2+})	0.22
Cytochrome b (Fe^{3+}) $+ e^- \rightleftharpoons$ cytochrome b (Fe^{2+}) (*mitochondrial*)	0.077
Ubiquinone $+ 2\,H^+ + 2\,e^- \rightleftharpoons$ ubiquinol	0.045
Fumarate$^- + 2\,H^+ + 2\,e^- \rightleftharpoons$ succinate$^-$	0.031
FAD $+ 2\,H^+ + 2\,e^- \rightleftharpoons$ FADH$_2$ (*in flavoproteins*)	~0.
Oxaloacetate$^- + 2\,H^+ + 2\,e^- \rightleftharpoons$ malate$^-$	-0.166
Pyruvate$^- + 2\,H^+ + 2\,e^- \rightleftharpoons$ lactate$^-$	-0.185
Acetaldehyde $+ 2\,H^+ + 2\,e^- \rightleftharpoons$ ethanol	-0.197
FAD $+ 2\,H^+ + 2\,e^- \rightleftharpoons$ FADH$_2$ (*free coenzyme*)	-0.219
$S + 2\,H^+ + 2\,e^- \rightleftharpoons H_2S$	-0.23
Lipoic acid $+ 2\,H^+ + 2\,e^- \rightleftharpoons$ dihydrolipoic acid	-0.29
$NAD^+ + H^+ + 2\,e^- \rightleftharpoons$ NADH	-0.315
$NADP^+ + H^+ + 2\,e^- \rightleftharpoons$ NADPH	-0.320
Cystine $+ 2\,H^+ + 2\,e^- \rightleftharpoons$ 2 cysteine	-0.340
Acetoacetate$^- + 2\,H^+ + 2\,e^- \rightleftharpoons$ β-hydroxybutyrate$^-$	-0.346
$H^+ + e^- \rightleftharpoons \frac{1}{2}H_2$	-0.421
Acetate$^- + 3\,H^+ + 2\,e^- \rightleftharpoons$ acetaldehyde $+ H_2O$	-0.581

Source: Mostly from Loach, P.A., *In* Fasman, G.D. (Ed.), *Handbook of Biochemistry and Molecular Biology* (3rd ed.), *Physical and Chemical Data*, Vol. I, pp. 123–130, CRC Press (1976).

high affinity for electrons and is a strong electron acceptor (oxidizing agent), whereas its conjugate reductant is a weak electron donor (reducing agent). For example, O_2 is the strongest oxidizing agent in Table 13-3, whereas H_2O, which tightly holds its electrons, is the weakest reducing agent listed. The converse is true of half-reactions with large negative standard reduction potentials.

Since electrons spontaneously flow from low to high reduction potentials, they are transferred, under standard conditions, from the reduced products in any half-reaction in Table 13-3 to the oxidized reactants of any half-reaction above it (although this may not occur at a measurable rate in the absence of a suitable enzyme). Note that Fe^{3+} ions of the various cytochromes listed in Table 13-3 have significantly different reduction potentials. This indicates that *the protein components of redox enzymes play active roles in electron-transfer reactions by modulating the reduction potentials of their bound redox-active centers.*

Electron-transfer reactions are of great biological importance. For example, in the mitochondrial electron-transport chain (Section 17-2), electrons are passed from NADH along a series of electron acceptors of increasing reduction potential (including FAD and others listed in Table 13-3) to O_2. ATP is generated from ADP and P_i by coupling its synthesis to this free energy cascade. *NADH thereby functions as an energy-rich electron-transfer coenzyme.* In fact, the oxidation of one NADH to NAD^+ supplies sufficient free energy to generate three ATPs. NAD^+ is an electron acceptor in many exergonic metabolite oxidations. In serving as the electron donor in ATP synthesis, it fulfills its cyclic role as a free energy conduit in a manner analogous to ATP (Fig. 13-7).

4. EXPERIMENTAL APPROACHES TO THE STUDY OF METABOLISM

A metabolic pathway can be understood at several levels:

1. In terms of the sequence of reactions by which a specific nutrient is converted to end products, and the energetics of these conversions.
2. In terms of the mechanisms by which each intermediate is converted to its successor. Such an analysis requires the isolation and characterization of the specific enzymes that catalyze each reaction.
3. In terms of the control mechanisms that regulate the flow of metabolites through the pathway. These include the interorgan relationships that adjust metabolic activity to the needs of the entire organism.

Elucidating a metabolic pathway on all these levels is a complex process, often requiring contributions from a variety of disciplines.

The outlines of the major metabolic pathways have been known for decades, although in many cases, the enzymology behind various steps of the pathways remains obscure. Likewise, the mechanisms that regulate pathway activity under different physiological conditions are not entirely understood. These areas are of great interest because of their potential to yield information that could be useful in improving human health and curing metabolic diseases. In addition, the unexplored metabolisms of unusual organisms, including recently discovered "extremophiles," hold the promise of novel biological materials and enzymatic processes that can be exploited for the environmentally sensitive production of industrial materials, foods, and therapeutic drugs.

Early metabolic studies used whole organisms, often yeast, but also mammals. For example, Frederick Banting and Charles Best established the role of the pancreas in diabetes in 1921; they surgically removed that organ from dogs and observed that the animals then developed the disease. Techniques for studying metabolic processes have since become more refined, progressing from whole-organ preparations and thin tissue slices to cultured cells and organelles isolated by ultracentrifugation (Section 5-2E). The most recent approaches include identifying active genes and cataloguing their protein products (see Box 13-3).

A. Tracing Metabolic Fates

A metabolic pathway in which one compound is converted to another can be followed by tracing a specifically labeled metabolite. Franz Knoop formulated this technique in 1904 to study fatty acid oxidation. He fed dogs fatty acids chemically labeled with phenyl groups and isolated the phenyl-substituted end products from the dogs' urine. From the differences in these products, depending on whether the phenyl-substituted starting material contained odd or even numbers of carbon atoms, Knoop deduced that fatty acids are degraded in two-carbon units (Section 19-2).

Chemical labeling has the disadvantage that the chemical properties of labeled metabolites differ from those of normal metabolites. This problem is eliminated by labeling molecules with isotopes. *The fate of an isotopically labeled atom in a metabolite can therefore be elucidated by following its progress through the metabolic pathway of interest.* The advent of isotopic labeling and tracing techniques in the 1940s therefore revolutionized the study of metabolism.

One of the early advances in metabolic understanding resulting from the use of isotopic tracers was the demonstration, by David Shemin and David Rittenberg in 1945, that the nitrogen atoms of heme (Fig. 7-2) are derived from glycine rather than from ammonia, glutamic acid, proline, or leucine (Section 20-6A). They showed this by feeding rats these ^{15}N-labeled nutrients, isolating the heme in their blood, and analyzing it for ^{15}N content. Only when the rats were fed [^{15}N]glycine did the heme contain ^{15}N. This technique was also used with the radioactive isotope ^{14}C to demonstrate that all of cholesterol's carbon atoms are derived from acetyl-CoA (Section 19-7A). Radioactive isotopes (Box 12-1) have become virtually indispensable for establishing the metabolic origins of complex metabolites.

Another method for tracing the fates of labeled metabolites is nuclear magnetic resonance (NMR), which detects specific isotopes, including ^1H, ^{13}C, ^{15}N, and ^{31}P, by their characteristic nuclear spins. Since the NMR spectrum of a particular nucleus varies with its immediate environment, it is possible to identify the peaks corresponding to specific atoms even in relatively complex mixtures. The development of magnets large enough to accommodate animals and humans, and to localize spectra to specific organs, has made it possible to study metabolic pathways noninvasively by NMR techniques. For example, ^{31}P NMR can be used to study energy metabolism in muscle by monitoring the levels of phosphorylated compounds such as ATP, ADP, and phosphocreatine.

Isotopically labeling specific atoms of metabolites with ^{13}C (which is only 1.10% naturally abundant) permits the metabolic progress of the labeled

Box 13-3

BIOCHEMISTRY IN FOCUS

Transcriptomics and Proteomics

The overall metabolic capabilities of an organism are encoded in its genes. In theory, it should be possible to reconstruct a cell's metabolic activities from its DNA sequences. At present, this can be done only in a general sense. For example, the sequenced genome of *Vibrio cholerae,* the bacterium that causes cholera, reveals a large repertoire of genes encoding transport proteins and enzymes for catabolizing a wide range of food molecules. This is consistent with the complicated lifestyle of *V. cholerae,* which can live on its own or in association with zooplankton as well as in the human gastrointestinal tract.

A precise accounting of an organism's entire metabolism will be possible only when the functions of all its genes can be assigned by identifying their products or by noting their homology to known genes from other organisms. Most sequenced genomes still contain a significant number of genes (up to 40% of the total) whose functions are not yet known. Even a partial list of an organism's genes is useful, but compiling a catalog of genes is not the same as understanding how the genes work. For example, some genes are expressed continuously at high levels; others are expressed rarely, perhaps only once during the organism's life-cycle.

Creating an accurate picture of gene expression is the goal of **transcriptomics**, the study of messenger RNA (mRNA) production. Identifying and quantifying all the mRNA transcripts from a single cell type reveals which genes are active. This can be done by assembling many short strands of DNA of different known sequences on a solid support and then allowing them to hybridize with fluorescent-labeled mRNA molecules from a cell preparation. The strength of fluorescence indicates how much mRNA binds to a particular complementary DNA sequence.

The DNA of a microarray may represent an entire genome or just a few selected genes. Because the detection and quantification of DNA–RNA hybrids is easily automated, it is possible to examine multiple preparations from cells at different developmental stages or from cells grown under different conditions. Differences in the expression of particular genes can then be linked to certain developmental processes or growth patterns. For example, cancer researchers use DNA chips to profile the patterns of gene expression in tumor cells because different types of tumors synthesize different proteins. This information can be considered in choosing how best to treat the cancer.

Unfortunately, the correlation between the amount of a particular mRNA and the amount of its protein product is not perfect. Some mRNAs are rapidly degraded whereas others are translated many times, yielding large quantities of the corresponding protein. Furthermore, many proteins are post-translationally modified, sometimes in different ways. This means that the number of proteins in a cell exceeds the number of mRNAs.

A more reliable way than transcriptomics to assess gene expression is to examine a cell's **proteome**, the complete set of proteins that the cell synthesizes. This **proteomics** approach requires that the proteins first be separated, usually by **two-dimensional gel electrophoresis**. In this technique, the proteins in a mixture are first separated by charge; then the gel is turned 90° and the proteins are separated by mass.

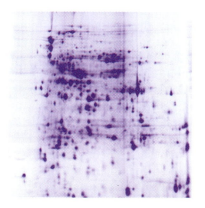

[Photo courtesy of Amersham Pharmacia Biotech.]

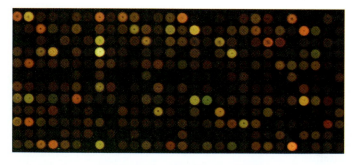

[Photo courtesy of Trey Ideker.]

Tens to hundreds of thousands of different DNA sequences can fit into a square centimeter, so the collection of DNA is called a **DNA microarray** or **DNA chip**.

Proteins in the resulting "spots" can then be identified by breaking them apart and measuring the masses of their peptide fragments. As with DNA microarrays, comparisons of proteins expressed under different conditions may reveal which enzymes and regulatory proteins are required for certain metabolic activities.

Protein-cataloging efforts are limited mostly by the technical problems of detecting minute quantities of thousands of different proteins using techniques that have not yet been fully automated, as they have been for DNA microarrays. In addition, nucleic acids can be amplified by the polymerase chain reaction (PCR); (Section 3-5C), but there is no comparable procedure for amplifying trace amounts of protein. Most likely, a profile of a cell's metabolic activities will come to light through a combination of transcriptomics, proteomics, and more traditional approaches.

atoms to be followed by ^{13}C NMR. Figure 13-13 shows the *in vivo* ^{13}C NMR spectra of a rat liver before and after an injection of D-[1-^{13}C]glucose. The ^{13}C can be seen entering the liver and then being incorporated into glycogen (the storage form of glucose; Section 15-2).

Isotopic Tracers Can Establish Precursor–Product Relationships

Defining the exact sequence of chemical transformations by which one metabolite is converted to another is sometimes difficult, and contradictory schemes still occasionally appear in the research literature. Consider the biosynthesis of the plasmalogens (Section 9-1C) and **alkylacylglycerophospholipids** (Section 19-6A). Alkylacylglycerophospholipids are ethers, whereas the closely related plasmalogens are vinyl ethers. Their similar

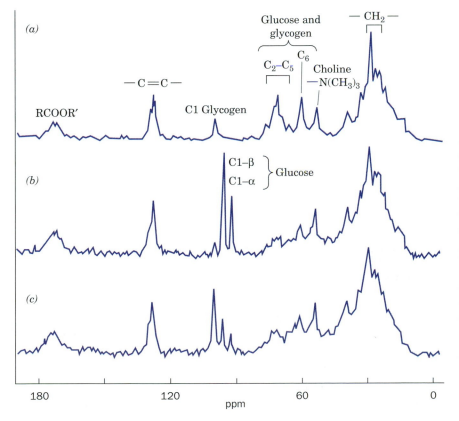

Figure 13-13. The conversion of [1-^{13}C]glucose to glycogen as observed by localized *in vivo* ^{13}C NMR. (*a*) The natural abundance ^{13}C NMR spectrum of the liver of a live rat. Note the resonance corresponding to C1 of glycogen. (*b*) The ^{13}C NMR spectrum of the liver of the same rat ~5 min after it was intravenously injected with 100 mg of [1-^{13}C]glucose (90% enriched). The resonances of the C1 atom of both the α and β anomers of glucose are clearly distinguishable from each other and from the resonance of the C1 atom of glycogen. (*c*) The ^{13}C NMR spectrum of the liver of the same rat ~30 min after the [1-^{13}C]glucose injection. The C1 resonances of both the α- and β-glucose anomers are much reduced while the C1 resonance of glycogen has increased. [After Reo, N.V., Siegfried, B.A., and Acherman, J.J.H., *J. Biol. Chem.* **259,** 13665 (1984).]

structures brings up the interesting question of their biosynthetic relationship: Which is the precursor and which is the product? Two possible modes of synthesis can be envisioned:

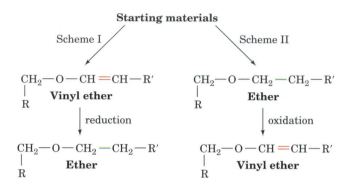

In Scheme I, the starting material is converted to the vinyl ether (plasmalogen), which is then reduced to yield the ether (alkylacylglycerophospholipid). Accordingly, the vinyl ether would be the precursor and the ether the product. In Scheme II, the ether is formed first and then oxidized to yield the vinyl ether. The ether would then be the precursor and the vinyl ether the product.

Precursor–product relationships can be most easily sorted out through the use of **radioactive tracers**. A small amount of the labeled starting material is administered in a pulse to an organism and is then converted to labeled products:

$$\text{Starting material*} \longrightarrow \text{A*} \longrightarrow \text{B*} \longrightarrow \text{end products*}$$

(here the * represents the radioactive label). The specific radioactivity of the products is followed over time.

A metabolic pathway normally operates in a steady state; that is, the throughput of metabolites in each of its reactions is equal. Moreover, the rates of most metabolic reactions are first order for a given substrate. Making these assumptions, the rate of change of B's radioactivity, [B*], is equal to the rate of passage of label from A* to B* less the rate of passage from B* to the pathway's next product:

$$\frac{d[\text{B*}]}{dt} = k[\text{A*}] - k[\text{B*}] = k([\text{A*}] - [\text{B*}]) \qquad [13\text{-}11]$$

where k is the pseudo-first-order rate constant for both the conversion of A to B and the conversion of B to its product, and t is time. Inspection of this equation indicates the criteria that must be met to establish that A is the precursor of B (Fig. 13-14):

1. While the radioactivity of the product is rising ($d[\text{B*}]/dt > 0$), it is less than that of its precursor ($[\text{A*}] > [\text{B*}]$).
2. When the radioactivity of the product is at its peak ($d[\text{B*}]/dt = 0$), it is equal to that of its precursor ($[\text{A*}] = [\text{B*}]$). This also implies that the radioactivity of a product peaks after that of its precursor.
3. After the radioactivity of the product has peaked ($d[\text{B*}]/dt < 0$), it remains greater than that of its precursor ($[\text{A*}] < [\text{B*}]$).

Such a demonstration of the precursor–product relationship between alkylacylglycerophospholipid and plasmalogen, using ^{14}C-labeled starting materials, indicated that the ether is the precursor and the vinyl ether is the product (Scheme II).

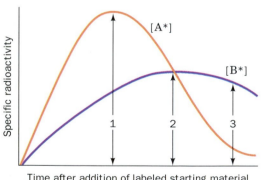

Figure 13-14.   The flow of a pulse of radioactivity from precursor to product. At point 1, the radioactivity of the product (B*, *purple*) is increasing and is less than that of its precursor (A*, *orange*); at point 2, product radioactivity is maximal and is equal to that of its precursor; and at point 3, product radioactivity is decreasing and is greater than that of its precursor.

B. Perturbing the System

Many of the techniques used to elucidate the intermediates and enzymes of metabolic pathways involve perturbing the system in some way and observing how this affects the activity of the pathway. One way to perturb a pathway is to add certain substances, called **metabolic inhibitors,** that block the pathway at specific points, thereby causing the preceding intermediates to build up. This approach was used in elucidating the conversion of glucose to ethanol in yeast by glycolysis (Section 14-2B). Similarly, the addition of substances that block electron transfer at different sites was used to deduce the sequence of electron carriers in the mitochondrial electron-transport chain (Section 17-2B).

Genetic Defects Also Cause Metabolic Intermediates to Accumulate

Archibald Garrod's realization, in the early 1900s, that human genetic diseases are the consequence of deficiencies in specific enzymes also contributed to the elucidation of metabolic pathways. For example, on the ingestion of either phenylalanine or tyrosine, individuals with the largely harmless inherited condition known as **alcaptonuria,** but not normal subjects, excrete **homogentisic acid** in their urine (Box 20-1). This is because the liver of alcaptonurics lacks an enzyme that catalyzes the breakdown of homogentisic acid (Fig. 13-15).

Genetic Manipulation Alters Metabolic Processes

Early studies of metabolism led to the astounding discovery that *the basic metabolic pathways in most organisms are essentially identical.* This metabolic uniformity has greatly facilitated the study of metabolic reactions. Thus, although a mutation that inactivates or deletes an enzyme in a pathway of interest may be unknown in higher organisms, it can be readily generated in a rapidly reproducing microorganism through the use of **mutagens** (chemical agents that induce genetic changes; Section 24-4A), X-rays, or, more recently, through genetic engineering techniques (Section 3-5). The desired mutants, which cannot synthesize the pathway's end product, can be identified by their requirement for that product in their culture medium.

Higher organisms that have been engineered to lack particular genes (i.e., gene "knockouts"; Section 3-5D) are useful, particularly in cases where the absence of a single gene product results in a metabolic defect that is not lethal. Genetic engineering techniques have advanced to the point where it is possible to selectively "knock out" a gene only in a particular tissue. This approach is necessary in cases where a gene product is required for development and therefore cannot be entirely deleted. In the opposite approach, techniques for constructing transgenic animals make it possible to express genes in tissues in which they were not originally present.

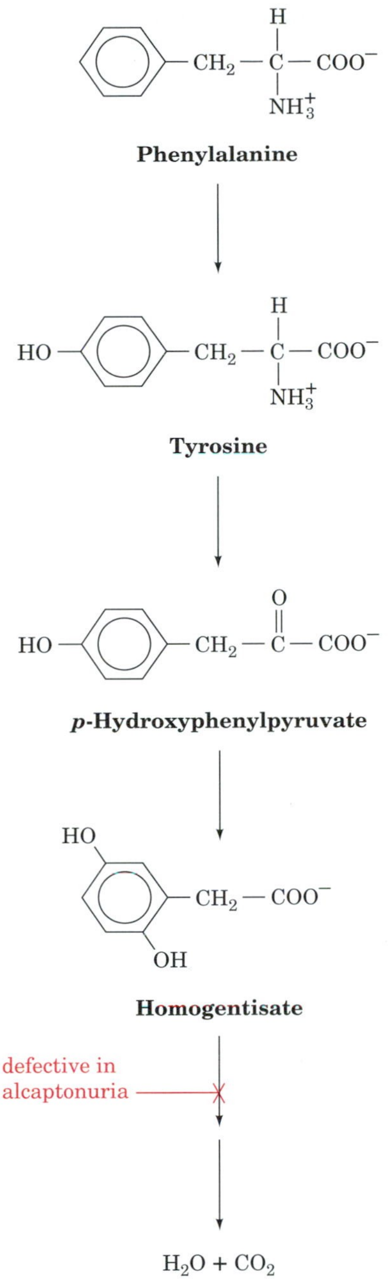

Figure 13-15. **Pathway for phenylalanine degradation.** Alcaptonurics lack the enzyme that breaks down homogentisate; therefore, this intermediate accumulates and is excreted in the urine.

S U M M A R Y

1. The free energy released from catabolic oxidation reactions is used to drive endergonic anabolic reactions.

2. Heterotrophic organisms obtain their free energy from compounds synthesized by chemolithotrophic or photoautotrophic organisms.

3. Sequences of reactions form metabolic pathways that operate in different cellular locations.

4. Near-equilibrium reactions are freely reversible, whereas reactions that function far from equilibrium serve as regulatory points and render metabolic pathways irreversible.

5. Flux through a metabolic pathway is controlled by regulating the activities of the enzymes that catalyze its rate-determining steps.

6. The free energy of the "high-energy" compound ATP is made available through cleavage of one or both of its phosphoanhydride bonds.

7. An exergonic reaction such as ATP or PP_i hydrolysis can be coupled to an endergonic reaction to make it more favorable.

8. Substrate-level phosphorylation is the synthesis of ATP from ADP by phosphoryl group transfer from another compound.

9. The common product of carbohydrate, lipid, and protein catabolism, acetyl-CoA, is a "high-energy" thioester.

10. The coenzymes NAD^+ and FAD are reversibly reduced during the oxidation of metabolites.

11. The Nernst equation relates the electromotive force of a redox reaction to the standard reduction potentials and concentrations of the electron donors and acceptors.

12. Electrons flow spontaneously from the compound with the more negative reduction potential to the compound with the more positive reduction potential.

13. Studies of metabolic pathways endeavor to determine the order of metabolic transformations, their enzymatic mechanisms, their regulation, and their relationships to metabolic processes in other tissues.

14. The steps of metabolic pathways can be determined through the use of isotopic tracers and by examining the effects of metabolic inhibitors, natural mutations, and genetically engineered changes.

REFERENCES

Goodridge, A.G., The new metabolism: molecular genetics in the analysis of metabolic regulation, *FASEB J.* **4,** 3099–3110 (1990).

Harold, F.M., *The Vital Force: A Study of Bioenergetics,* Chapters 1 and 2, Freeman (1986).

Jeffrey, F.M.H., Rajagopal, A., Malloy, C.R., and Sherry, A.D., ^{13}C-NMR: a simple yet comprehensive method for analysis of intermediary metabolism, *Trends Biochem. Sci.* **16,** 5–10 (1991).

Scriver, C.R., Beaudet, A.L., Sly, W.S., Valle, D., Stanbury, J.B., Wyngaarden, J.B., and Frederickson, D.S. (Eds.), *The Metabolic and Molecular Bases of Inherited Disease* (7th ed.), McGraw–Hill (1995). [Most chapters in this encyclopedic work include a review of a normal metabolic process that is disrupted by disease.]

Westheimer, F.H., Why nature chose phosphates, *Science* **235,** 1173–1178 (1987).

KEY TERMS

metabolism	anaerobic	substrate-level phosphoryla-tion	half-reaction
catabolism	metabolite	oxidative phosphorylation	conjugate redox pair
anabolism	isozyme	photophosphorylation	electrochemical cell
autotroph	near-equilibrium reaction	kinase	$\mathscr{F}$
chemolithotroph	substrate cycle	phosphagen	Nernst equation
photoautotroph	"high-energy" intermediate	reducing agent	$\mathscr{E}°'$
heterotroph	orthophosphate cleavage	oxidizing agent	radioactive tracer
aerobic	pyrophosphate cleavage		

STUDY EXERCISES

1. Describe the differences between autotrophs and heterotrophs.

2. Explain the metabolic significance of reactions that function near equilibrium and reactions that function far from equilibrium.

3. Why is ATP a "high-energy" compound?

4. Describe the ways an exergonic process can drive an endergonic process.

5. What is the metabolic role of reduced coenzymes?

6. Explain the terms of the Nernst equation.

7. How is $\Delta\mathscr{E}$ related to ΔG?

8. Explain how isotopically labeled compounds may reveal precursor–product relationships.

PROBLEMS

1. Assuming 100% efficiency of energy conservation, how many moles of ATP can be synthesized under standard conditions by the complete oxidation of (a) 1 mol of glucose and (b) 1 mol of palmitate?

2. Does the magnitude of the free energy change for ATP hydrolysis increase or decrease as the pH increases from 5 to 6?

3. The reaction for "activation" of a fatty acid ($RCOO^-$),

$$ATP + CoA + RCOO^- \rightleftharpoons RCO{-}CoA + AMP + PP_i$$

has $\Delta G^{\circ\prime} = +4.6$ kJ·mol^{-1}. What is the thermodynamic driving force for this reaction?

4. Predict whether the creatine kinase reaction will proceed in the direction of ATP synthesis or phosphocreatine synthesis at 25°C when [ATP] = 4 mM, [ADP] = 0.15 mM, [phosphocreatine] = 2.5 mM, and [creatine] = 1 mM.

5. If intracellular [ATP] = 5 mM, [ADP] = 0.5 mM, and [P_i] = 1.0 mM, calculate the concentration of AMP at pH 7 and 25°C under the condition that the adenylate kinase reaction is at equilibrium.

6. List the following substances in order of their decreasing oxidizing power: (a) acetoacetate, (b) cytochrome b (Fe^{3+}), (c) NAD^+, (d) SO_4^{2-}, and (e) pyruvate.

7. Write a balanced equation for the oxidation of ubiquinol by cytochrome c. Calculate $\Delta G^{\circ\prime}$ and $\Delta \mathscr{E}^{\circ\prime}$ for the reaction.

8. Under standard conditions, will the following reactions proceed spontaneously as written?

   (a) Fumarate + NADH + H$^+$ $\rightleftharpoons$ succinate + NAD$^+$

   (b) Cyto a (Fe^{2+}) + cyto b (Fe^{3+}) $\rightleftharpoons$ cyto a (Fe^{3+}) + cyto b (Fe^{2+})

9. Under standard conditions, is the oxidation of free FADH$_2$ by ubiquinone sufficiently exergonic to drive the synthesis of ATP?

10. A hypothetical three-step metabolic pathway consists of the intermediates W, X, Y, and Z and the enzymes A, B, and C. Deduce the order of the enzymatic steps in the pathway from the following information:

    1. Compound Q, a metabolic inhibitor of enzyme B, causes Z to build up.

    2. A mutant in enzyme C requires Y for growth.

    3. An inhibitor of enzyme A causes W, Y, and Z to accumulate.

    4. Compound P, a metabolic inhibitor of enzyme C, causes W and Z to build up.

CHAPTER 14

The combustion of a metabolic fuel such as glucose is accomplished through a multistep process so that energy can be conserved and used to perform cellular work. In trained runners, the delivery of glucose to the muscles is particularly efficient. [Copyright © Gamma Liaison.]

GLUCOSE CATABOLISM

The fermentation (anaerobic breakdown) of glucose to ethanol and CO_2 by yeast has been exploited for many centuries in baking and winemaking. However, scientific investigation of the chemistry of this catabolic pathway began only in the mid-nineteenth century, with the experiments of Louis Pasteur and others. Nearly a century would pass before the complete pathway was elucidated. During that interval, several important features of the pathway came to light:

1. In 1905, Arthur Harden and William Young discovered that phosphate is required for glucose fermentation.
2. Certain reagents, such as iodoacetic acid and fluoride ion, inhibit the formation of pathway products, thereby causing pathway intermediates to accumulate. Different substances cause the buildup of different intermediates and thereby reveal the sequence of molecular interconversions.
3. Studies of how different organisms break down glucose indicated that, with few exceptions, all of them do so the same way.

The efforts of many investigators came to fruition in 1940, when the complete pathway of glucose breakdown was described. This pathway, which is named **glycolysis** (Greek: *glykus,* sweet + *lysis,* loosening), is alternately known as the **Embden–Meyerhof–Parnas pathway** to commemorate the work of Gustav Embden, Otto Meyerhof, and Jacob Parnas in its elucidation.

Glycolysis, which is probably the most completely understood biochemical pathway, is a sequence of 10 enzymatic reactions in which one molecule of glucose is converted to two molecules of the three-carbon sugar pyruvate with the concomitant generation of 2 ATP. It plays a key role in energy metabolism by providing a significant portion of the free energy used by most organisms and by preparing glucose and other compounds for further oxidative degradation. Thus, it is fitting that we begin our discussion of specific metabolic pathways by considering glycolysis. We shall examine the sequence of reactions by which glucose is degraded, along with some of the relevant enzyme mechanisms. We will then examine the features that influence glycolytic flux and the ultimate fate of its products. Finally, we will discuss the catabolism of other hexoses and the **pentose phosphate pathway,** an alternative pathway for glucose catabolism that functions to provide biosynthetic precursors.

1. OVERVIEW OF GLYCOLYSIS

> ● See Guided Exploration 12:
> Glycolysis Overview.

Before beginning our detailed discussion of glycolysis, let us first take a moment to survey the overall pathway as it fits in with animal metabolism as a whole. Glucose usually arises in the blood as result of the breakdown of polysaccharides (e.g., liver glycogen or dietary starch and glycogen) or from its synthesis from noncarbohydrate precursors (**gluconeogenesis;** Section 15-4). Glucose enters most cells by a specific carrier that transports it from the exterior of the cell into the cytosol (Section 10-4B). The enzymes of glycolysis are located in the cytosol, where they are only loosely associated, if at all, with each other or with other cell structures.

Glycolysis converts glucose to two C_3 units (pyruvate). The free energy released in this process is harvested to synthesize ATP from ADP and P_i. Thus, glycolysis is a pathway of chemically coupled phosphorylation reactions (Section 13-2B). The 10 reactions of glycolysis are diagrammed in Fig.

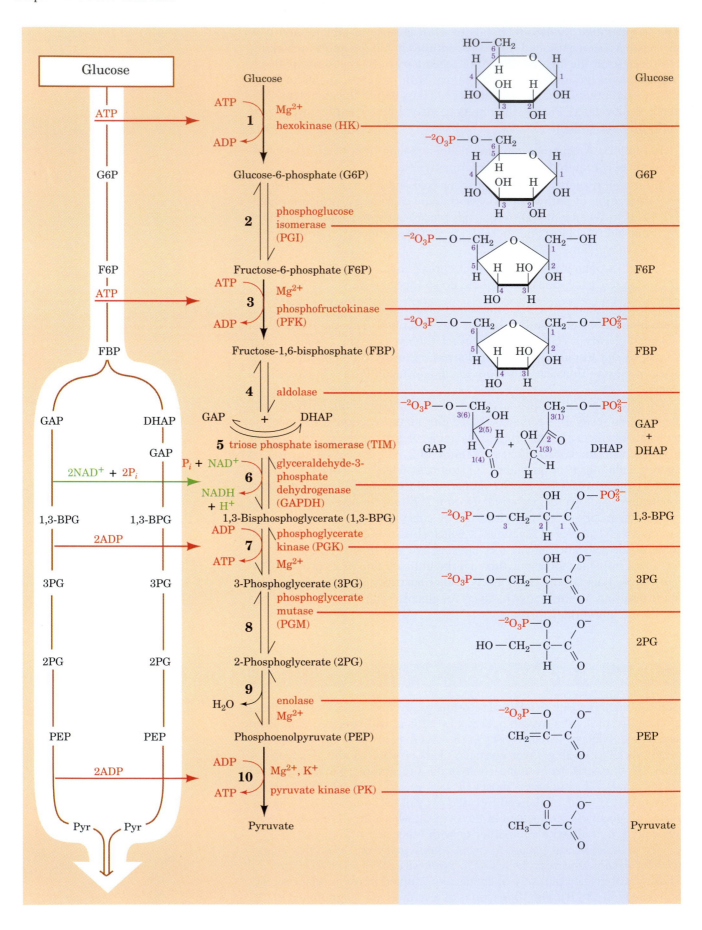

Figure 14-1 (*opposite*). *Key to Metabolism.* **Glycolysis.** In its first stage (Reactions 1–5), one molecule of glucose is converted to two glyceraldehyde-3-phosphate molecules in a series of reactions that consumes 2 ATP. In the second stage of glycolysis (Reactions 6–10), the two glyceraldehyde-3-phosphate molecules are converted to two pyruvate molecules, generating 4 ATP and 2 NADH. ✷ **See the Animated Figures.**

14-1. Note that ATP is used early in the pathway to synthesize phosphorylated compounds (Reactions 1 and 3) but is later resynthesized twice over (Reactions 7 and 10). Glycolysis can therefore be divided into two stages:

Stage I Energy investment (Reactions 1–5). In this preparatory stage, the hexose glucose is phosphorylated and cleaved to yield two molecules of the triose **glyceraldehyde-3-phosphate.** This process consumes 2 ATP.

Stage II Energy recovery (Reactions 6–10). The two molecules of glyceraldehyde-3-phosphate are converted to pyruvate, with concomitant generation of 4 ATP. Glycolysis therefore has a net "profit" of 2 ATP per glucose: Stage I consumes 2 ATP; Stage II produces 4 ATP.

The phosphoryl groups that are initially transferred from ATP to the hexose do not immediately result in "high-energy" compounds. However, subsequent enzymatic transformations convert these "low-energy" products to compounds with high phosphoryl group-transfer potentials, which are capable of phosphorylating ADP to form ATP. The overall reaction is

Glucose + 2 NAD^+ + 2 ADP + 2 P_i $\longrightarrow$
$$2 \text{ pyruvate} + 2 \text{ NADH} + 2 \text{ ATP} + 2 \text{ H}_2\text{O} + 4 \text{ H}^+$$

Hence, the NADH formed in the process must be continually reoxidized to keep the pathway supplied with its primary oxidizing agent, NAD^+. In Section 14-3, we shall examine how organisms do so under aerobic or anaerobic conditions.

2. THE REACTIONS OF GLYCOLYSIS

In this section, we examine the reactions of glycolysis more closely, describing the properties of the individual enzymes and their mechanisms. As we study the individual glycolytic enzymes, we shall encounter many of the catalytic mechanisms described in Section 11-3.

A. Hexokinase: First ATP Utilization

Reaction 1 of glycolysis is the transfer of a phosphoryl group from ATP to glucose to form **glucose-6-phosphate (G6P)** in a reaction catalyzed by **hexokinase** (*at right*).

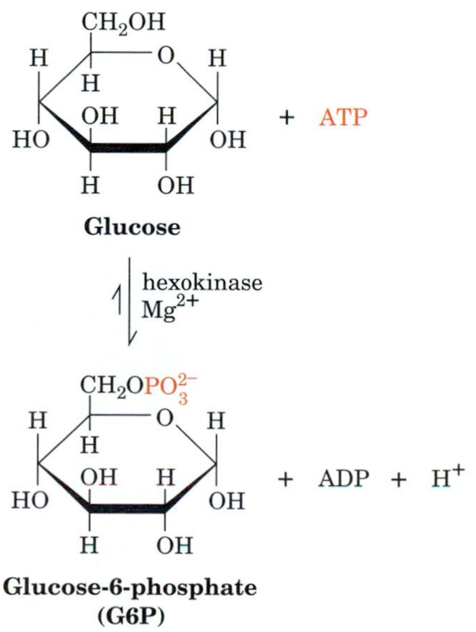

A kinase is an enzyme that transfers phosphoryl groups between ATP and a metabolite (Section 13-2C). The metabolite that serves as the phosphoryl group acceptor is indicated in the prefix of the kinase name. Hexokinase is a ubiquitous, relatively nonspecific enzyme that catalyzes the phosphorylation of hexoses such as D-glucose, D-mannose, and D-fructose. Liver cells also contain **glucokinase,** which catalyzes the same reaction but which is primarily involved in maintaining blood glucose levels (Section 21-1D).

Glucose

$$\begin{array}{c} \text{hexokinase} \\ \text{Mg}^{2+} \end{array}$$

Glucose-6-phosphate (G6P)

The second substrate for hexokinase, as with other kinases, is an Mg^{2+}–ATP complex. In fact, uncomplexed ATP is a potent competitive inhibitor of hexokinase. Although we do not always explicitly mention the participation of Mg^{2+}, it is essential for kinase activity. The Mg^{2+} shields the negative charges of the ATP's α- and β- or β- and γ-phosphate oxygen atoms, making the γ-phosphorus atom more accessible for nucleophilic attack by the C6-OH group of glucose.

ATP   **Glucose**

Comparison of the X-ray structures of yeast hexokinase and the glucose–hexokinase complex indicates that *glucose induces a large conformational change in hexokinase* (Fig. 14-2). The two lobes that form its active site cleft swing together by up to 8 Å so as to engulf the glucose in a manner that suggests the closing of jaws. *This movement places the ATP in close proximity to the* —C6H$_2$OH *group of glucose and excludes water from the active site (catalysis by proximity effects;* Section 11-3E). If the catalytic and reacting groups were in the proper position for reaction while the enzyme was in the open position (Fig. 14-2a), ATP hydrolysis (i.e., phosphoryl group transfer to water, which is thermodynamically favored) would almost certainly be the dominant reaction.

Clearly, the substrate-induced conformational change in hexokinase is responsible for the enzyme's specificity. In addition, the active site polarity is reduced by exclusion of water, thereby expediting the nucleophilic reac-

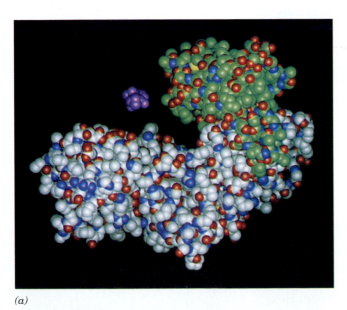

(a)

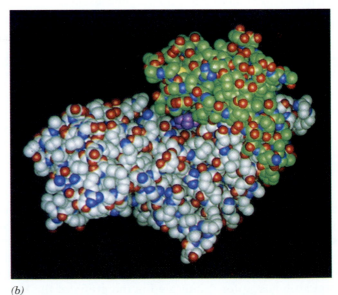

(b)

Figure 14-2.   Substrate-induced conformational changes in yeast hexokinase. (*a*) Space-filling model of a hexokinase subunit showing the prominent bilobal appearance of the free enzyme (the C atoms in the small lobe are shaded green, and those in the large lobe are light gray; the N and O atoms are blue and red). (*b*) Model of the hexokinase complex with glucose (*magenta*). The lobes have swung together to engulf the substrate. [Based on X-ray structures by Thomas Steitz, Yale University.]

● See the Interactive Exercises.

tion process. Other kinases have the same deeply clefted structure as hexokinase and undergo conformational changes on binding their substrates (e.g., adenylate kinase; Fig. 13-8).

B. Phosphoglucose Isomerase

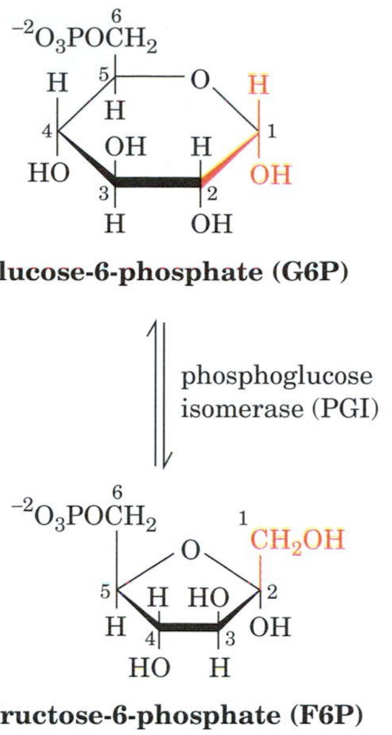

Reaction 2 of glycolysis is the conversion of G6P to **fructose-6-phosphate (F6P)** by **phosphoglucose isomerase (PGI;** *at right*). This is the isomerization of an aldose to a ketose.

Since G6P and F6P both exist predominantly in their cyclic forms, the reaction requires ring opening followed by isomerization and subsequent ring closure (the interconversions of cyclic and linear forms of hexoses are shown in Fig. 8-3).

A proposed reaction mechanism for the PGI reaction involves general acid–base catalysis by the enzyme (Fig. 14-3):

Step 1 The substrate binds.

Step 2 An enzymatic acid, probably the ε-amino group of a conserved Lys residue, catalyzes ring opening.

Step 3 A base, thought to be the carboxylate group of a conserved Glu

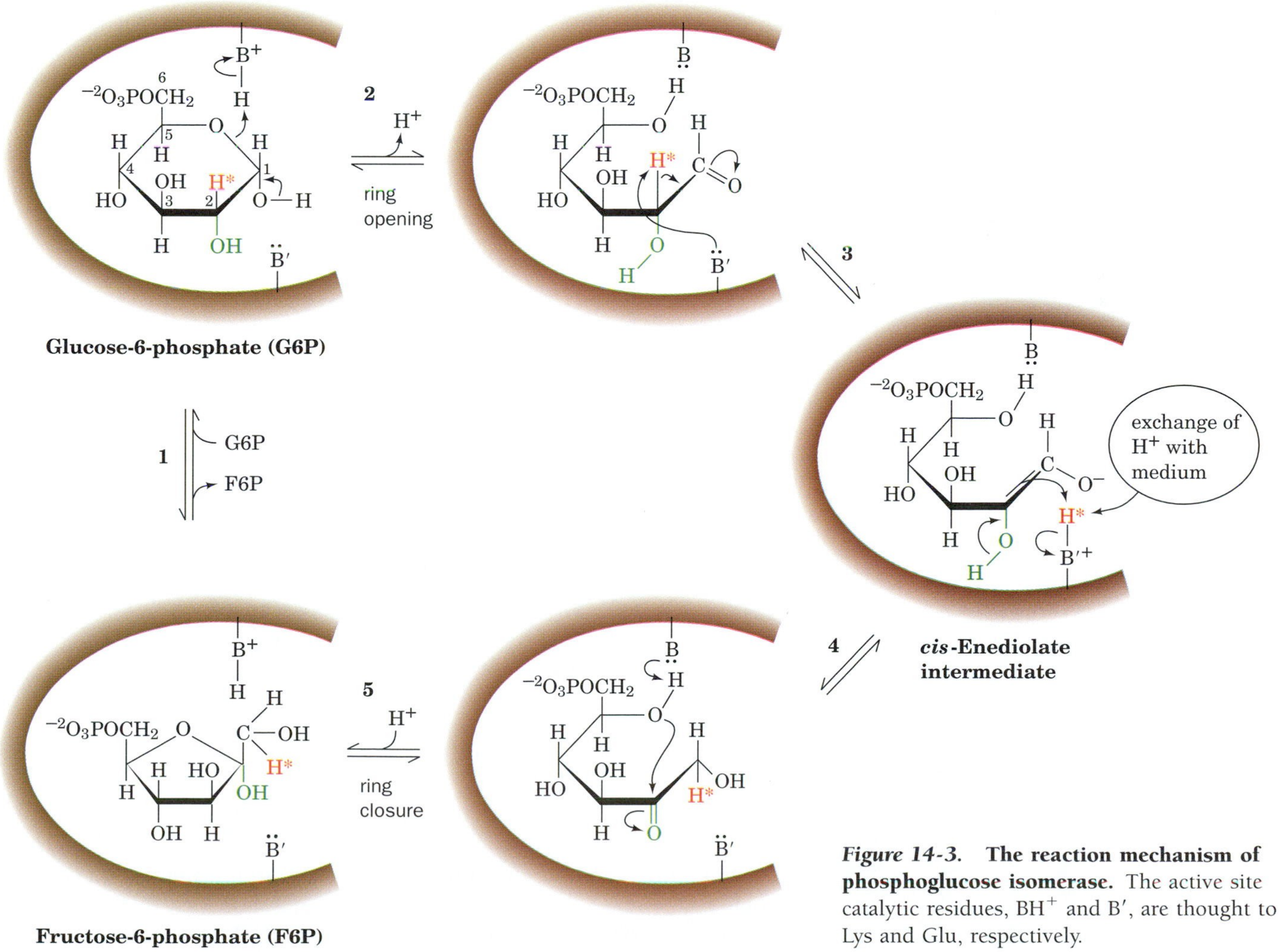

Glucose-6-phosphate (G6P)

Fructose-6-phosphate (F6P)

Figure 14-3. **The reaction mechanism of phosphoglucose isomerase.** The active site catalytic residues, BH⁺ and B′, are thought to be Lys and Glu, respectively.

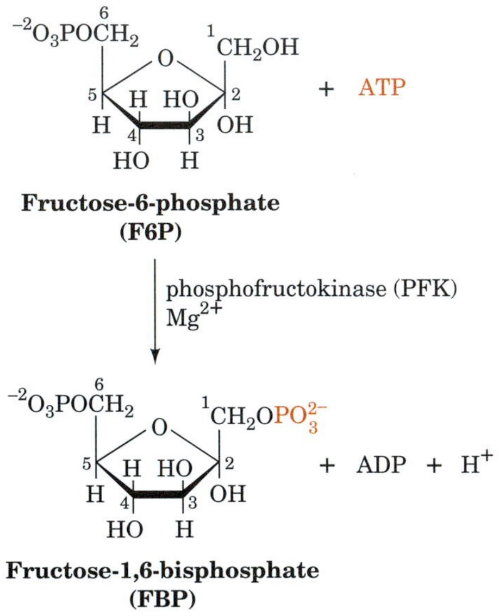

residue, abstracts the acidic proton from C2 to form a *cis*-enediolate intermediate (this proton is acidic because it is α to a carbonyl group).

Step 4 The proton is replaced on C1 in an overall proton transfer. Protons abstracted by bases rapidly exchange with solvent protons. Nevertheless, Irwin Rose confirmed this step by demonstrating that 2-[³H]G6P is occasionally converted to 1-[³H]F6P by intramolecular proton transfer before the ³H has had a chance to exchange with the medium.

Step 5 The ring closes to form the product, which is subsequently released to yield free enzyme, thereby completing the catalytic cycle.

C. Phosphofructokinase: Second ATP Utilization

In Reaction 3 of glycolysis, **phosphofructokinase (PFK)** phosphorylates F6P to yield **fructose-1,6-bisphosphate (FBP or F1,6P;** *at left*). (The product is a *bis*phosphate rather than a *di*phosphate because its two phosphate groups are not attached directly to each other.)

The PFK reaction is similar to the hexokinase reaction. The enzyme catalyzes the nucleophilic attack by the C1-OH group of F6P on the electrophilic γ-phosphorus atom of the Mg^{2+}–ATP complex.

Phosphofructokinase plays a central role in control of glycolysis because it catalyzes one of the pathway's rate-determining reactions. In many organisms, the activity of PFK is enhanced allosterically by several substances, including AMP, and inhibited allosterically by several other substances, including ATP and citrate. The regulatory properties of PFK are examined in Section 14-4A.

D. Aldolase

Aldolase catalyzes Reaction 4 of glycolysis, the cleavage of FBP to form the two trioses **glyceraldehyde-3-phosphate (GAP)** and **dihydroxyacetone phosphate (DHAP).**

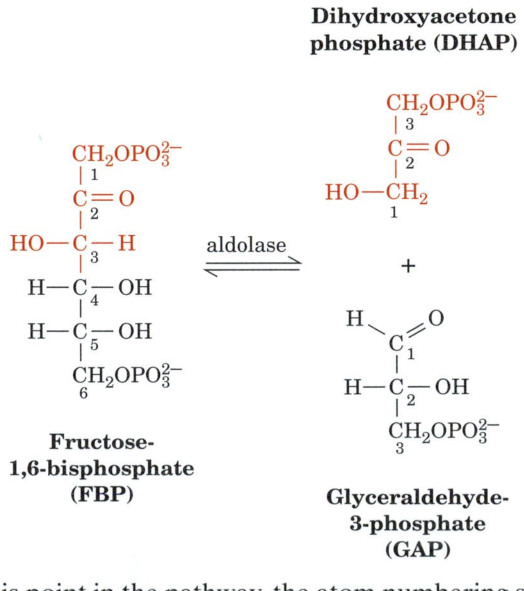

Note that at this point in the pathway, the atom numbering system changes. Atoms 1, 2, and 3 of glucose become atoms 3, 2, and 1 of DHAP, thus reversing order. Atoms 4, 5, and 6 become atoms 1, 2, and 3 of GAP.

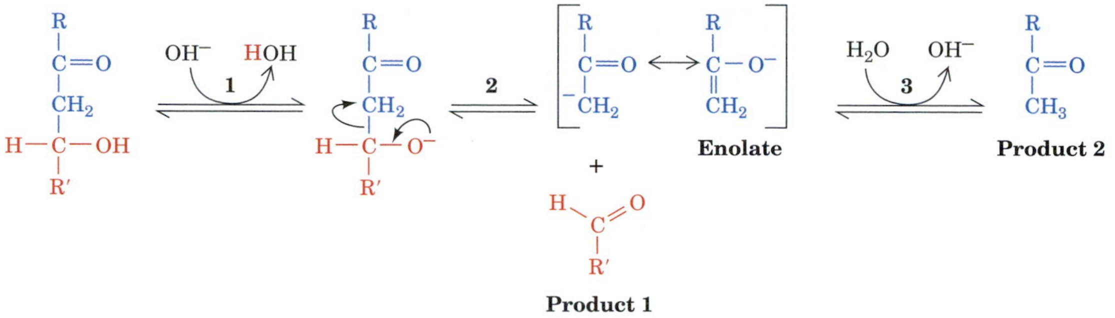

Figure 14-4. The mechanism of base-catalyzed aldol cleavage. Aldol condensation occurs by the reverse mechanism.

Reaction 4 is an **aldol cleavage (retro aldol condensation)** whose nonenzymatic base-catalyzed mechanism is shown in Fig. 14-4. The **enolate** intermediate is stabilized by resonance, as a result of the electron-withdrawing character of the carbonyl oxygen atom. Note that aldol cleavage between C3 and C4 of FBP requires a carbonyl at C2 and a hydroxyl at C4. Hence, the "logic" of Reaction 2 in the glycolytic pathway, the isomerization of G6P to F6P, is clear. Aldol cleavage of G6P would yield products of unequal carbon chain length, while *aldol cleavage of FBP results in two interconvertible C₃ compounds that can therefore enter a common degradative pathway.*

There Are Two Mechanistic Classes of Aldolases

Aldol cleavage is catalyzed by stabilizing its enolate intermediate through increased electron delocalization. There are two types of aldolases that are classified according to their mechanism of enolate stabilization. In Class I aldolases, which occur in animals and plants, the reaction occurs as follows (Fig. 14-5):

Step 1 Substrate binding.

Step 2 Reaction of the FBP carbonyl group with the ε-amino group of the active site Lys to form an iminium cation, that is, a protonated Schiff base.

Step 3 C3—C4 bond cleavage resulting in enamine formation and the release of GAP. The iminium ion is a better electron-withdrawing group than the oxygen atom of the precursor carbonyl group. Thus, catalysis occurs because the enamine intermediate (Fig. 14-5, Step 3) is more stable than the corresponding enolate intermediate of the base-catalyzed aldol cleavage reaction (Fig. 14-4, Step 2).

Step 4 Protonation of the enamine to an iminium cation.

Step 5 Hydrolysis of this iminium cation to release DHAP, with regeneration of the free enzyme.

Class II aldolases, which occur in fungi, algae, and some bacteria, do not form a Schiff base with the substrate. Rather, a divalent cation, usually Zn^{2+} or Fe^{2+}, polarizes the carbonyl oxygen of the substrate to stabilize the enolate intermediate of the reaction:

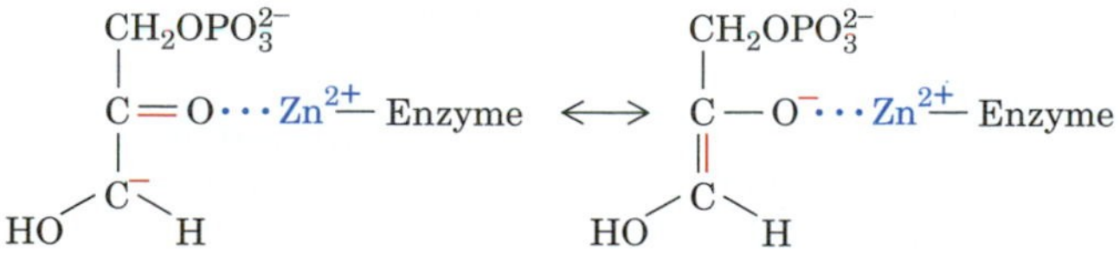

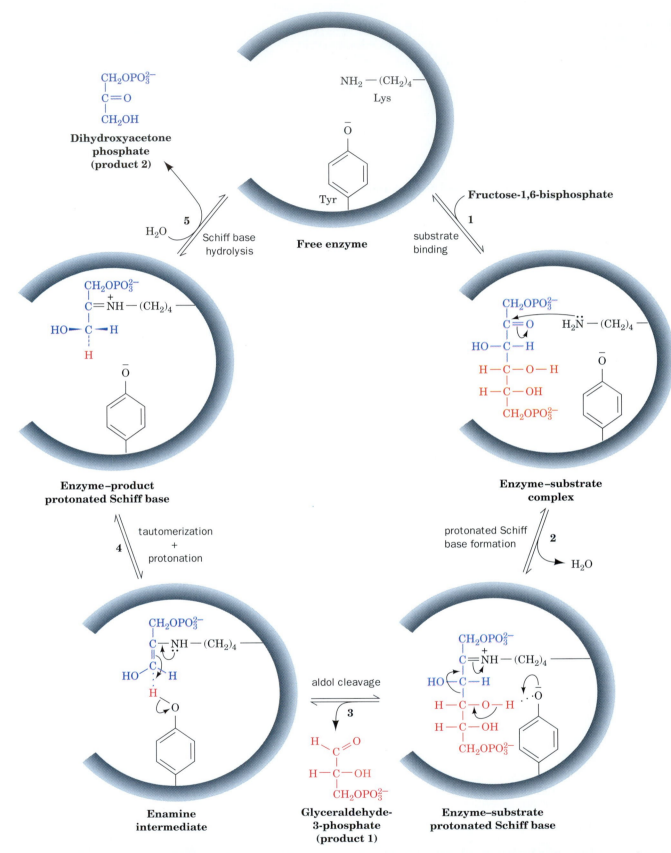

Figure 14-5.   The enzymatic mechanism of Class I aldolase.
The reaction involves (**1**) substrate binding; (**2**) Schiff base formation between the enzyme's active site Lys residue and the
open-chain form of FBP; (**3**) aldol cleavage to form an enamine
intermediate of the enzyme and DHAP, with release of GAP; (**4**)
tautomerization and protonation to the iminium form of the
Schiff base; and (**5**) hydrolysis of the Schiff base with release of
DHAP. ✳ See the Animated Figures.

E. Triose Phosphate Isomerase

Only one of the products of the aldol cleavage reaction, GAP, continues along the glycolytic pathway (Fig. 14-1). However, DHAP and GAP are ketose–aldose isomers (like F6P and G6P). They are interconverted by an isomerization reaction with an enediol (or enediolate) intermediate. **Triose phosphate isomerase (TIM** or **TPI)** catalyzes this process in Reaction 5 of glycolysis, the final reaction of Stage I.

Glyceraldehyde-3-phosphate (an aldose)

Dihydroxyacetone phosphate (a ketose)

Enediol intermediate

Support for this reaction scheme comes from the use of the transition state analogs **phosphoglycohydroxamate** and **2-phosphoglycolate,** stable compounds whose geometry resembles that of the proposed enediol or enediolate intermediate:

Phosphoglyco-hydroxamate **Proposed enediolate intermediate** **2-Phosphoglycolate**

Enzymes catalyze reactions by binding the transition state complex more tightly than the substrate (Section 11-3F), and, in fact, phosphoglycohydroxamate and 2-phosphoglycolate bind 155- and 100-fold more tightly to TIM than does either GAP or DHAP.

Glu 165 and His 95 Act as General Acids and Bases

Mechanistic considerations suggest that the conversion of GAP to the enediol intermediate is catalyzed by a general base, which abstracts a proton from C2 of GAP, and by a general acid, which protonates its carbonyl oxygen atom. X-Ray studies reveal that the Glu 165 side chain is ideally situated to abstract the C2 proton from GAP (Fig. 14-6). In fact, the mutagenic replacement of Glu 165 by Asp, which X-ray studies show withdraws the carboxylate group only ~1 Å farther away from the substrate than its position in the wild-type enzyme, reduces TIM's catalytic activity 1000-fold. X-Ray studies similarly indicate that His 95 is hydrogen bonded to and hence is properly positioned to protonate GAP's carbonyl oxygen. The positively charged side chain of Lys 12 is thought to electrostatically stabilize the negatively charged transition state in the reaction. In the con-

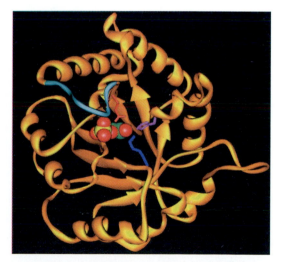

Figure 14-6. **A ribbon diagram of yeast TIM in complex with its transition state analog 2-phosphoglycolate.** A single subunit of this homodimeric enzyme is viewed roughly along the axis of its α/β barrel. The enzyme's flexible loop is cyan, and the side chains of the catalytic Lys, His, and Glu residues are blue, magenta, and red, respectively. The 2-phosphoglycolate is represented by a space-filling model colored according to atom type (C, green; O, red; P, yellow). [Based on an X-ray structure determined by Gregory Petsko, Brandeis University.] ● See the Interactive Exercises and Kinemage Exercises 12-1 and 12-2.

version of the enediol intermediate to DHAP, Glu 165 acts as a general acid to protonate C1 and His 95 acts as a general base to abstract the proton from the OH group, thereby restoring these catalytic groups to their initial protonation states.

A Flexible Loop Closes over the Active Site

The comparison of the X-ray structure of TIM with that of the enzyme–phosphoglycohydroxamate complex reveals that when substrate binds to TIM, a conserved 10-residue loop closes over the active site like a hinged lid, in a movement that involves main chain shifts of >7 Å (Fig. 14-6). A four-residue segment of this loop makes a hydrogen bond with the phosphate group of the substrate. Mutagenic excision of these four residues does not significantly distort the protein, so substrate binding is not greatly impaired. However, the catalytic power of the mutant enzyme is reduced 10^5-fold, and it only weakly binds phosphoglycohydroxamate. Evidently, loop closure preferentially stabilizes the enzymatic reaction's enediol-like transition state.

Loop closure in the TIM reaction also supplies a striking example of the so-called **stereoelectronic control** that enzymes can exert on a reaction. In solution, the enediol intermediate readily breaks down with the elimination of the phosphate at C3 to form the toxic compound **methylglyoxal:**

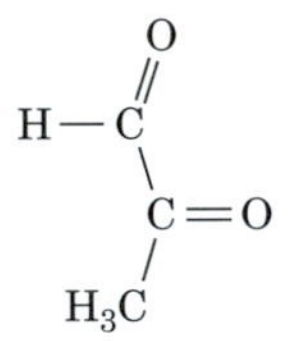

$$\begin{array}{c} O \\ \parallel \\ H-C \\ \diagdown \\ C=O \\ \diagup \\ H_3C \end{array}$$

Methylglyoxal

On the enzyme's surface, however, this reaction is prevented because the phosphate group is held by the flexible loop in a position that disfavors phosphate elimination. In the mutant enzyme lacking the flexible loop, the enediol is able to escape: ~85% of the enediol intermediate is released into solution where it rapidly decomposes to methylglyoxal and P_i. Thus, the flexible loop closure assures that substrate is efficiently transformed to product.

α/β Barrel Enzymes May Have Evolved by Divergent Evolution

TIM was the first protein known to contain an α/β barrel (also known as a TIM barrel), a cylinder of eight parallel β strands surrounded by eight parallel α helices (Figs. 6-27c and 6-28d). This striking structural motif has since been found in numerous different proteins, essentially all of which are enzymes (including the glycolytic enzymes aldolase, enolase, and pyruvate kinase). Intriguingly, the active sites of all known α/β barrel enzymes are located in the mouth of the barrel at the end that contains the C-terminal ends of the β strands, although there is no obvious structural rationale for this. Despite the fact that few of these proteins exhibit significant sequence similarity, it has been postulated that all of them have evolved from a common ancestor (divergent evolution). However, it has also been argued that the α/β barrel is a particularly stable arrangement that nature has independently discovered on several occasions (convergent evolution).

Triose Phosphate Isomerase Is a Perfect Enzyme

Jeremy Knowles has demonstrated that TIM has achieved **catalytic perfection.** This means that the rate of the bimolecular reaction between enzyme and substrate is diffusion controlled, so product formation occurs as rapidly as enzyme and substrate can collide in solution. Any increase in TIM's catalytic efficiency therefore would not increase its reaction rate.

GAP and DHAP are interconverted so efficiently that the concentrations of these two metabolites are maintained at their equilibrium values: $K = \text{[GAP]/[DHAP]} = 4.73 \times 10^{-2}$. At equilibrium, [DHAP] $\gg$ [GAP]. However, under the steady state conditions in a cell, GAP is consumed in the succeeding reactions of the glycolytic pathway. *As GAP is siphoned off in this manner, more DHAP is converted to GAP to maintain the equilibrium ratio.* In effect, DHAP follows GAP into the second stage of glycolysis, so a single pathway accounts for the metabolism of both products of the aldolase reaction.

Taking Stock of Glycolysis So Far

At this point in the glycolytic pathway, one molecule of glucose has been transformed into two molecules of GAP. This completes the first stage of glycolysis (Fig. 14-7). Note that 2 ATP have been consumed in generating the phosphorylated intermediates. This energy investment has not yet paid off, but with a little chemical artistry, the "low-energy" GAP can be converted to "high-energy" compounds whose free energies of hydrolysis can be coupled to ATP synthesis in the second stage of glycolysis.

F. Glyceraldehyde-3-Phosphate Dehydrogenase: First "High-Energy" Intermediate Formation

Reaction 6 of glycolysis is the oxidation and phosphorylation of GAP by NAD^+ and P_i as catalyzed by **glyceraldehyde-3-phosphate dehydrogenase (GAPDH;** Fig. 6-30). This is the first instance of the chemical artistry alluded to above. *In this reaction, aldehyde oxidation, an exergonic reaction, drives the synthesis of the "high-energy" acyl phosphate **1,3-bisphosphoglycerate (1,3-BPG).*** Recall that acyl phosphates are compounds with high phosphoryl group-transfer potential (Section 13-2C).

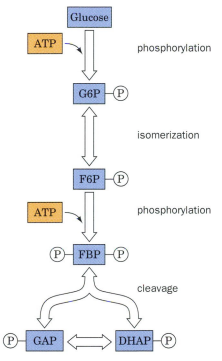

Figure 14-7. **Schematic view of the first stage of glycolysis.** In this series of reactions, a hexose is phosphorylated, isomerized, phosphorylated again, and then cleaved to two interconvertible triose phosphates. Two ATP are consumed in the process.

O=C—H
|
H—C—OH   + NAD$^+$ + P$_i$
|
CH$_2$OPO$_3^{2-}$

Glyceraldehyde-3-phosphate (GAP)

glyceraldehyde-3-phosphate dehydrogenase (GAPDH)

O=C—OPO$_3^{2-}$
|
H—C—OH   + NADH + H$^+$
|
CH$_2$OPO$_3^{2-}$

1,3-Bisphosphoglycerate (1,3-BPG)

Mechanistic Studies

Several key enzymological experiments have contributed to the elucidation of the GAPDH reaction mechanism:

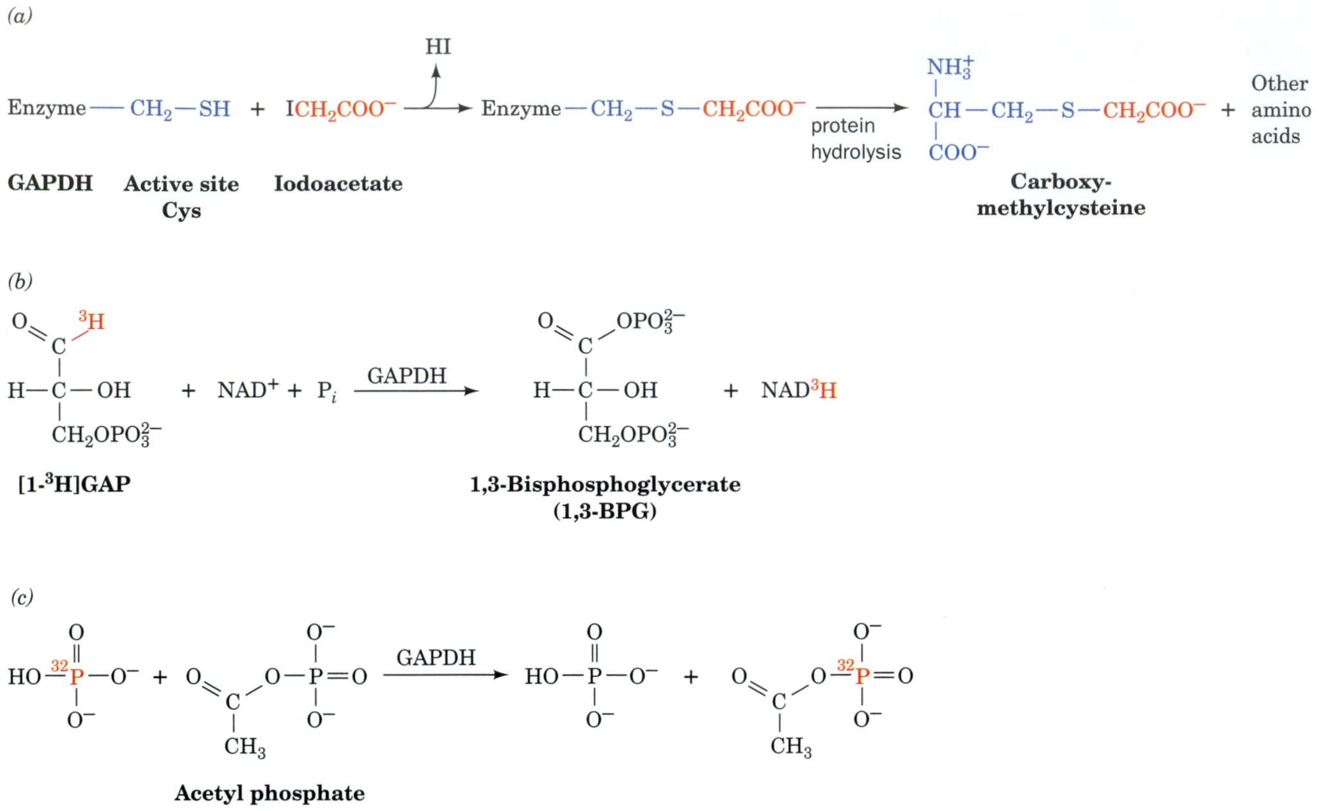

(a)

GAPDH   **Active site**   **Iodoacetate**
            Cys

                                                                        **Carboxy-
methylcysteine**

(b)

[1-³H]GAP   **1,3-Bisphosphoglycerate
(1,3-BPG)**

(c)

Acetyl phosphate

Figure 14-8. **Reactions that were used to elucidate the enzymatic mechanism of GAPDH.** (*a*) The reaction of iodoacetate with an active site Cys residue. (*b*) Quantitative tritium transfer from substrate to NAD⁺. (*c*) The enzyme-catalyzed exchange of ^{32}P from phosphate to acetyl phosphate.

1. GAPDH is inactivated by alkylation with stoichiometric amounts of iodoacetate. The presence of **carboxymethylcysteine** in the hydrolysate of the resulting alkylated enzyme (Fig. 14-8*a*) suggests that GAPDH has an active site Cys sulfhydryl group.

2. GAPDH quantitatively transfers ^{3}H from C1 of GAP to NAD⁺ (Fig. 14-8*b*), thereby establishing that this reaction occurs via direct hydride transfer.

3. GAPDH catalyzes exchange of ^{32}P between P$_i$ and the product analog **acetyl phosphate** (Fig. 14-8*c*). Such isotope exchange reactions are indicative of an acyl–enzyme intermediate; that is, the acetyl group forms a covalent complex with the enzyme, similar to the acyl–enzyme intermediate in the serine protease reaction mechanism (Section 11-5C).

David Trentham has proposed a mechanism for GAPDH based on this information and the results of kinetic studies (Fig. 14-9):

Step 1 GAP binds to the enzyme.

Step 2 The essential sulfhydryl group, acting as a nucleophile, attacks the aldehyde to form a **thiohemiacetal.**

Step 3 The thiohemiacetal undergoes oxidation to an **acyl thioester** by direct hydride transfer to NAD⁺. This intermediate, which has been isolated, has a large free energy of hydrolysis. Thus, *the energy of aldehyde oxidation has not been dissipated but has been conserved*

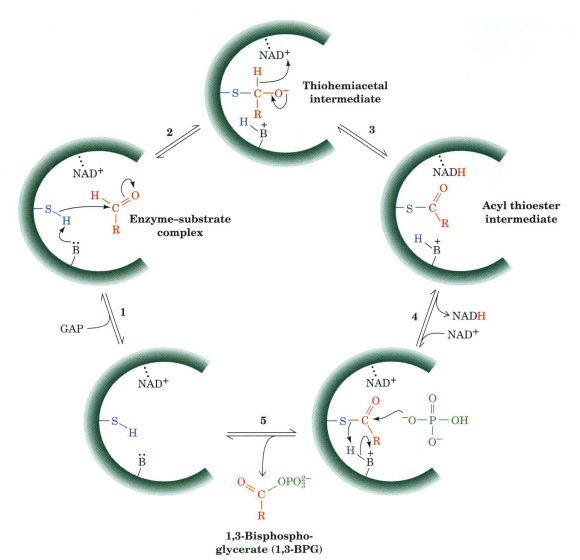

Figure 14-9.   The enzymatic mechanism of GAPDH. (1) GAP binds to the enzyme; (2) the active site sulfhydryl group forms a thiohemiacetal with the substrate; (3) NAD$^+$ oxidizes the thiohemiacetal to a thioester; (4) the newly formed NADH is replaced by NAD$^+$; and (5) P$_i$ attacks the thioester, forming the acyl phosphate product, 1,3-BPG, and regenerating the active enzyme. ✳ **See the Animated Figures.**

through the synthesis of the thioester and the reduction of NAD$^+$ to NADH.

Step 4 Another molecule of NAD$^+$ replaces NADH.

Step 5 The thioester intermediate undergoes nucleophilic attack by P$_i$ to regenerate free enzyme and form the "high-energy" mixed anhydride 1,3-BPG.

G. Phosphoglycerate Kinase: First ATP Generation

Reaction 7 of the glycolytic pathway yields ATP together with **3-phosphoglycerate (3PG)** in a reaction catalyzed by **phosphoglycerate kinase (PGK;** *at right*). (Note that this enzyme is called a "kinase" because the reverse reaction is phosphoryl group transfer from ATP to 3PG.)

1,3-Bisphosphoglycerate (1,3-BPG)

Mg^{2+} ⇅ phosphoglycerate kinase (PGK)

3-Phosphoglycerate (3PG)

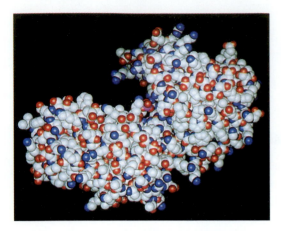

Figure 14-10. A space-filling model of yeast phosphoglycerate kinase. The substrate-binding site is at the bottom of a deep cleft between the two lobes of the protein. This site is marked by the P atom (*magenta*) of 3PG. Compare this structure with that of hexokinase (Fig. 14-2a). [Based on an X-ray structure by Herman Watson, University of Bristol, U.K.]

PGK (Fig. 14-10) is conspicuously bilobal in appearance. The Mg^{2+}–ADP binding site is located on one domain, ~10 Å from the 1,3-BPG binding site, which is on the other domain. Physical measurements suggest that, on substrate binding, the two domains of PGK swing together to permit the substrates to react in a water-free environment, as occurs in hexokinase (Section 14-2A). Indeed, the appearance of PGK is remarkably similar to that of hexokinase (Fig. 14-2), even though the structures of these proteins are otherwise unrelated.

Coupling between the GAPDH and PGK Reactions

As described in Section 13-2B, a slightly unfavorable reaction can be coupled to a highly favorable reaction so that both reactions proceed in the forward direction. In the case of the sixth and seventh reactions of glycolysis, *1,3-BPG is the common intermediate whose consumption in the PGK reaction "pulls" the GAPDH reaction forward.* The energetics of the overall reaction pair are

$$GAP + P_i + NAD^+ \longrightarrow 1,3\text{-BPG} + NADH \quad \Delta G^{\circ\prime} = +6.7 \text{ kJ} \cdot \text{mol}^{-1}$$
$$1,3\text{-BPG} + ADP \longrightarrow 3PG + ATP \quad \Delta G^{\circ\prime} = -18.8 \text{ kJ} \cdot \text{mol}^{-1}$$

$$GAP + P_i + NAD^+ + ADP \longrightarrow 3PG + NADH + ATP$$
$$\Delta G^{\circ\prime} = -12.1 \text{ kJ} \cdot \text{mol}^{-1}$$

Although the GAPDH reaction is endergonic, the strongly exergonic nature of the transfer of a phosphoryl group from 1,3-BPG to ADP makes the overall synthesis of NADH and ATP from GAP, P_i, NAD^+, and ADP favorable. *This production of ATP, which does not involve O_2, is an example of substrate-level phosphorylation.* The subsequent oxidation of the NADH produced in this reaction by O_2 generates additional ATP by oxidative phosphorylation, as we shall see in Section 17-3.

H. Phosphoglycerate Mutase

In Reaction 8 of glycolysis, 3PG is converted to **2-phosphoglycerate (2PG)** by **phosphoglycerate mutase (PGM):**

3-Phosphoglycerate (3PG) ⇌ [phosphoglycerate mutase (PGM)] ⇌ 2-Phosphoglycerate (2PG)

A **mutase** catalyzes the transfer of a functional group from one position to another on a molecule. This more or less energetically neutral reaction is necessary preparation for the next reaction in glycolysis, which generates a "high-energy" phosphoryl compound.

Reaction Mechanism of Phosphoglycerate Mutase

At first sight, the reaction catalyzed by phosphoglycerate mutase appears to be a simple intramolecular phosphoryl group transfer. This is not the case, however. The active enzyme has a phosphoryl group at its active site, attached to His 8 (*at left*). The phosphoryl group is transferred to the substrate to form a bisphospho intermediate. This intermediate then rephosphorylates the enzyme to form the product and regenerate the

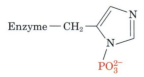

Phospho-His residue

active phosphoenzyme. The enzyme's X-ray structure shows the proximity of His 8 to the substrate (Fig. 14-11).

Catalysis by phosphoglycerate mutase occurs as follows (Fig. 14-12):

Step 1   3PG binds to the phosphoenzyme in which His 8 is phosphorylated.

Step 2   The enzyme's phosphoryl group is transferred to the substrate, resulting in an intermediate 2,3-bisphosphoglycerate–enzyme complex.

Steps 3 & 4   The complex decomposes to form the product 2PG and regenerate the phosphoenzyme.

The phosphoryl group of 3PG therefore ends up on C2 of the next 3PG to undergo reaction.

Occasionally, 2,3-bisphosphoglycerate (2,3-BPG) formed in Step 2 of the reaction dissociates from the dephosphoenzyme, leaving it in an inactive form. Trace amounts of 2,3-BPG must therefore always be available to regenerate the active phosphoenzyme by the reverse reaction. 2,3-BPG also specifically binds to deoxyhemoglobin, thereby decreasing its oxygen affinity (Section 7-2C). Consequently, erythrocytes require much more 2,3-BPG (5 mM) than the trace amounts that are used to prime phosphoglycerate mutase (see Box 14-1).

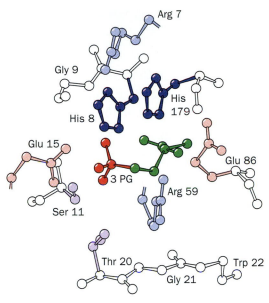

Figure 14-11. **The active site region of yeast phosphoglycerate mutase (dephospho form).** The substrate, 3PG, binds to an ionic pocket. His 8 is phosphorylated in the active enzyme. [After Winn, S.I., Watson, H.I., Harkins, R.N., and Fothergill, L.A., *Phil. Trans. R. Soc. London Ser. B* **293,** 126 (1981).]

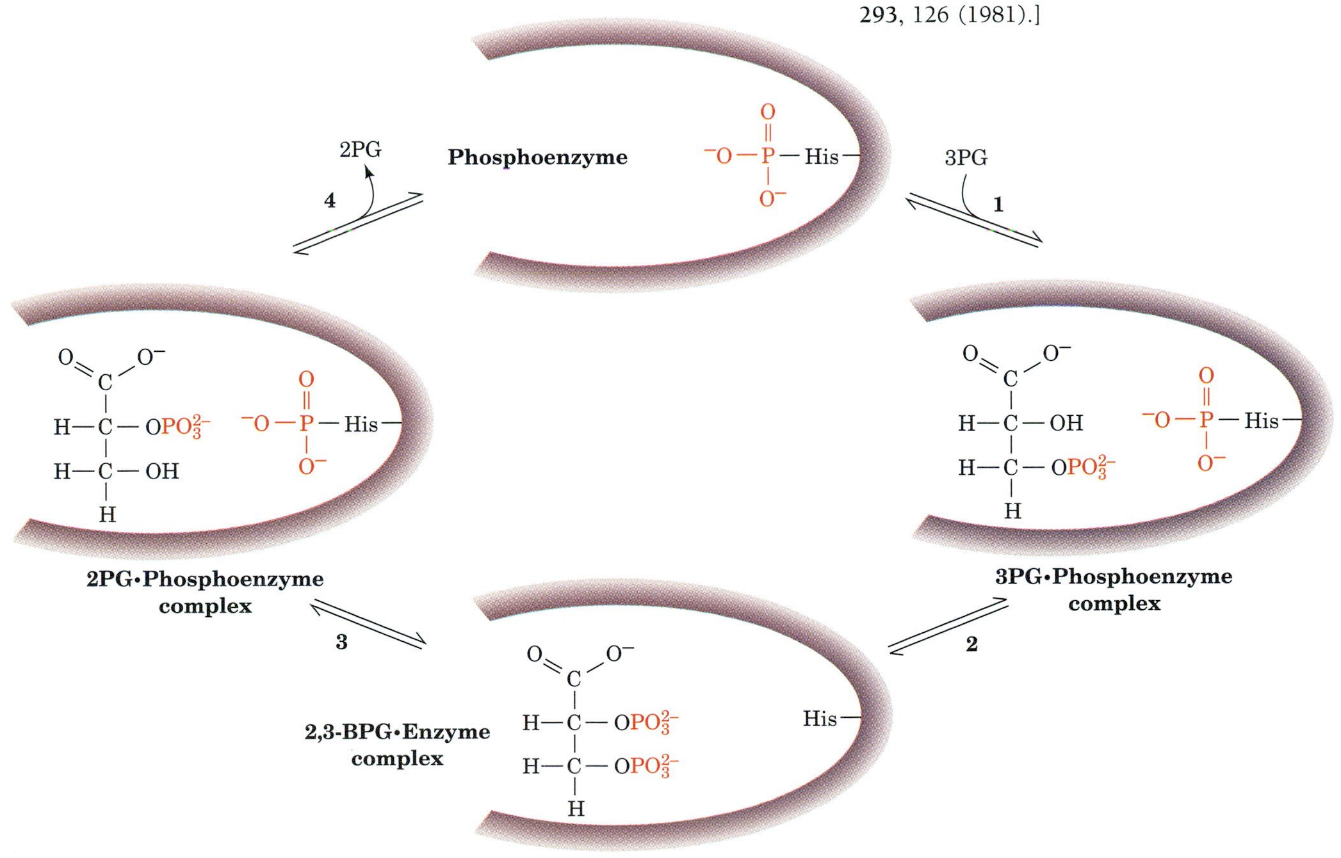

Figure 14-12. **A proposed reaction mechanism for phosphoglycerate mutase.** The active form of the enzyme contains a phospho-His residue at the active site. (**1**) Formation of an enzyme–substrate complex; (**2**) transfer of the enzyme-bound phosphoryl group to the substrate; (**3**) rephosphorylation of the enzyme by the other phosphoryl group of the substrate; and (**4**) release of product, regenerating the active phosphoenzyme.

Box 14-1

BIOCHEMISTRY IN FOCUS

Synthesis of 2,3-Bisphosphoglycerate in Erythrocytes and Its Effect on the Oxygen Carrying Capacity of the Blood

The specific binding of 2,3-bisphosphoglycerate (2,3-BPG) to deoxyhemoglobin decreases the oxygen affinity of hemoglobin (Section 7-2C). Erythrocytes synthesize and degrade 2,3-BPG by a detour from the glycolytic pathway.

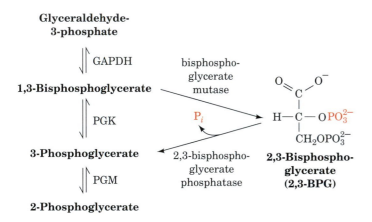

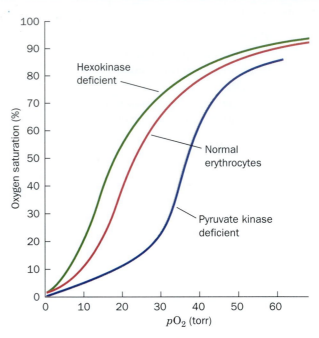

Bisphosphoglycerate mutase catalyzes the transfer of a phosphoryl group from C1 to C2 of 1,3-BPG. The resulting 2,3-BPG is hydrolyzed to 3PG by **2,3-bisphosphoglycerate phosphatase**. The 3PG then continues through the glycolytic pathway.

The level of available 2,3-BPG regulates hemoglobin's oxygen affinity. Consequently, inherited defects of glycolysis in erythrocytes alter the ability of the blood to carry oxygen as is indicated by the oxygen-saturation curve of its hemoglobin.

For example, in hexokinase-deficient erythrocytes, the concentrations of all the glycolytic intermediates are low (since hexokinase catalyzes the first step of glycolysis), thereby resulting in a diminished 2,3-BPG concentration and an increased hemoglobin oxygen affinity (*green curve*). Conversely, a deficiency in pyruvate kinase (which catalyzes the final reaction of glycolysis; Fig. 14-1) decreases hemoglobin's oxygen affinity (*purple curve*) through an increase in 2,3-BPG concentration resulting from this blockade. Thus, although erythrocytes, which lack nuclei and other organelles, have only a minimal metabolism, this metabolism is physiologically significant.

[Oxygen-saturation curves after Delivoria-Papadopoulos, M., Oski, F.A., and Gottlieb, A.J., *Science* **165**, 601 (1969).]

I. Enolase: Second "High-Energy" Intermediate Formation

In Reaction 9 of glycolysis, 2PG is dehydrated to **phosphoenolpyruvate (PEP)** in a reaction catalyzed by **enolase**.

$$\text{2-Phosphoglycerate (2PG)} \xrightleftharpoons{\text{enolase}} \text{Phosphoenolpyruvate (PEP)} + H_2O$$

2-Phosphoglycerate (2PG)   **Phosphoenolpyruvate (PEP)**

The enzyme forms a complex with a divalent cation such as Mg^{2+} before the substrate binds. Fluoride ion inhibits glycolysis by blocking enolase activity (F$^-$ was one of the metabolic inhibitors used in elucidating the glycolytic pathway). In the presence of P$_i$, F$^-$ blocks substrate binding to enolase by forming a bound complex with Mg^{2+} at the enzyme's active site. Enolase's substrate, 2PG, therefore builds up, and, through the action of PGM, 3PG also builds up.

J. Pyruvate Kinase: Second ATP Generation

In Reaction 10 of glycolysis, its final reaction, **pyruvate kinase (PK)** couples the free energy of PEP cleavage to the synthesis of ATP during the formation of pyruvate.

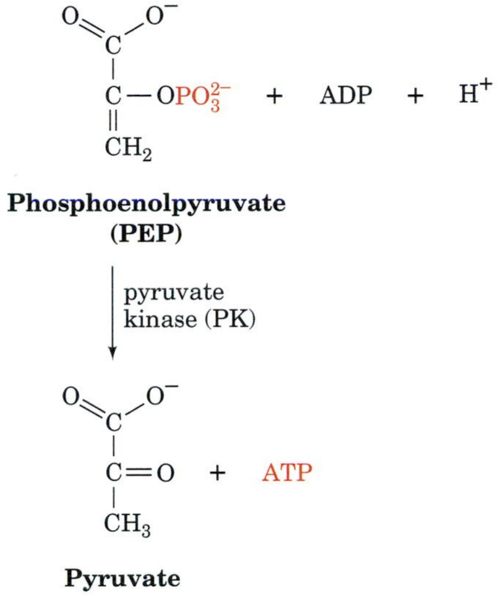

**Phosphoenolpyruvate
(PEP)**

pyruvate
kinase (PK)

Pyruvate

Catalytic Mechanism of Pyruvate Kinase

The PK reaction, which requires both monovalent (K$^+$) and divalent (Mg^{2+}) cations, occurs as follows (Fig. 14-13):

Step 1 A β-phosphoryl oxygen of ADP nucleophilically attacks the PEP phosphorus atom, thereby displacing **enolpyruvate** and forming ATP.

Step 2 Enolpyruvate tautomerizes to pyruvate.

The PK reaction is highly exergonic, supplying more than enough free energy to drive ATP synthesis (another example of substrate-level phosphorylation). At this point, the "logic" of the enolase reaction becomes clear. The standard free energy of hydrolysis of 2PG is only $-16 \text{ kJ} \cdot \text{mol}^{-1}$, which

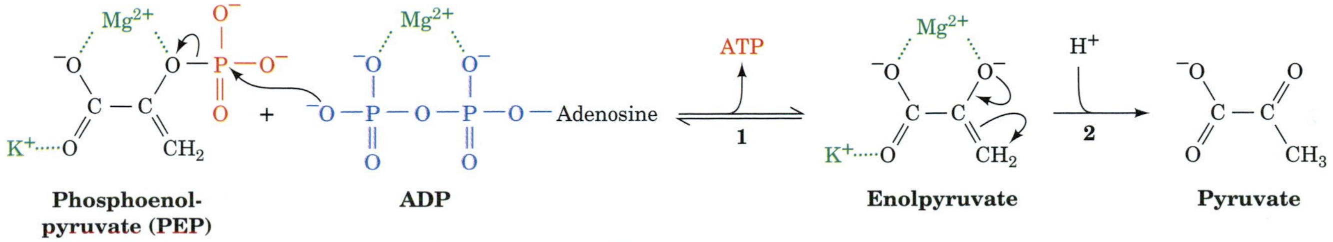

**Phosphoenol-
pyruvate (PEP)**   **ADP**   **Enolpyruvate**   **Pyruvate**

Figure 14-13.   The mechanism of the reaction catalyzed by pyruvate kinase. (1) Nucleophilic attack of an ADP β-phosphoryl oxygen atom on the phosphorus atom of PEP to form ATP and enolpyruvate; and (2) tautomerization of enolpyruvate to pyruvate.

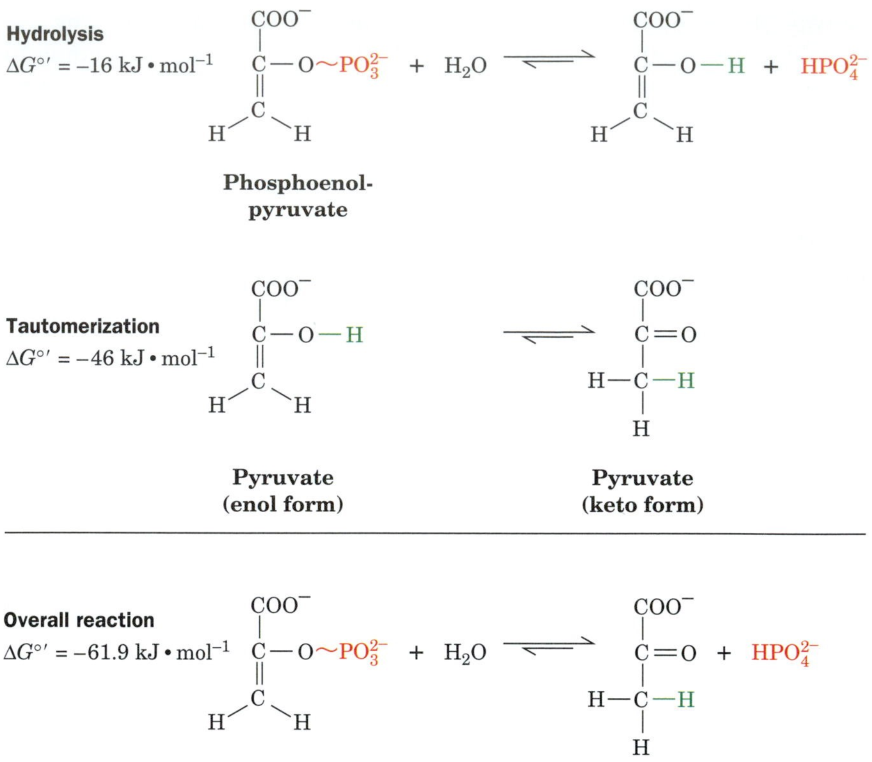

Figure 14-14. **The hydrolysis of PEP.** The reaction is broken down into two steps, hydrolysis and tautomerization. The overall $\Delta G^{\circ\prime}$ value is much greater than the $\Delta G^{\circ\prime}$ for ATP synthesis from ADP and P_i.

Hydrolysis
$\Delta G^{\circ\prime} = -16 \text{ kJ} \cdot \text{mol}^{-1}$

Phosphoenol-pyruvate

Tautomerization
$\Delta G^{\circ\prime} = -46 \text{ kJ} \cdot \text{mol}^{-1}$

Pyruvate (enol form)   **Pyruvate (keto form)**

Overall reaction
$\Delta G^{\circ\prime} = -61.9 \text{ kJ} \cdot \text{mol}^{-1}$

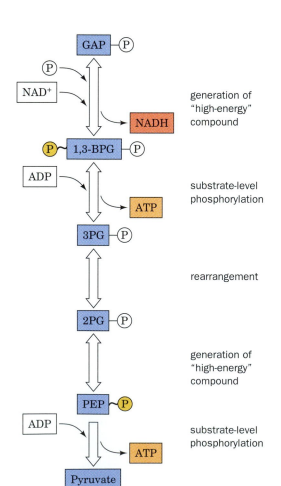

generation of "high-energy" compound

substrate-level phosphorylation

rearrangement

generation of "high-energy" compound

substrate-level phosphorylation

is insufficient to drive ATP synthesis from ADP ($\Delta G^{\circ\prime} = 30.5 \text{ kJ} \cdot \text{mol}^{-1}$). However, the dehydration of 2PG results in the formation of a "high-energy" compound capable of such synthesis. *The high phosphoryl group-transfer potential of PEP reflects the large release of free energy on converting the product enolpyruvate to its keto tautomer.* Consider the hydrolysis of PEP as a two-step reaction (Fig. 14-14). The tautomerization step supplies considerably more free energy than the phosphoryl group-transfer step.

Assessing Stage II of Glycolysis

The energy investment of the first stage of glycolysis (2 ATP consumed) is doubly repaid in the second stage of glycolysis because two phosphorylated C_3 units are transformed to two pyruvates with the coupled synthesis of 4 ATP. This process is shown schematically in Fig. 14-15.

The overall reaction of glycolysis, as we have seen, is

Glucose + 2 NAD^+ + 2 ADP + 2 $P_i \longrightarrow$
        2 pyruvate + 2 NADH + 2 ATP + 2 H_2O + 4 H^+

Let us consider each of the three products of glycolysis:

1. **ATP.** The initial investment of 2 ATP per glucose in Stage I and the subsequent generation of 4 ATP by substrate-level phosphorylation (two for each GAP that proceeds through Stage II) gives a net yield

Figure 14-15. **Schematic view of the second stage of glycolysis.** In this series of reactions, GAP undergoes phosphorylation and oxidation, followed by molecular rearrangements so that both phosphoryl groups have sufficient free energy to be transferred to ADP to produce ATP. Two molecules of GAP are converted to pyruvate for every molecule of glucose that enters Stage I of glycolysis.

of 2 ATP per glucose. In some tissues and organisms for which glucose is the primary metabolic fuel, ATP produced by glycolysis satisfies most of the cell's energy needs.

2. **NADH.** During its catabolism by the glycolytic pathway, glucose is oxidized to the extent that two NAD^+ are reduced to two NADH. As described in Section 13-3C, reduced coenzymes such as NADH represent a source of free energy than can be recovered by subsequent oxidation. Under aerobic conditions, electrons pass from reduced coenzymes through a series of electron carriers to the final oxidizing agent, O_2, in a process known as **electron transport** (Section 17-2). The free energy of electron transport drives the synthesis of ATP from ADP (oxidative phosphorylation; Section 17-3). In aerobic organisms, this sequence of events also serves to regenerate oxidized NAD^+ that can participate in further rounds of catalysis mediated by GAPDH. Under anaerobic conditions, NADH must be reoxidized by other means in order to keep the glycolytic pathway supplied with NAD^+ (Section 14-3).

3. **Pyruvate.** The two pyruvate molecules produced through the partial oxidation of each glucose are still relatively reduced molecules. Under aerobic conditions, complete oxidation of the pyruvate carbon atoms to CO_2 is mediated by the citric acid cycle (Chapter 16). The energy released in that process drives the synthesis of much more ATP than is generated by the limited oxidation of glucose by the glycolytic pathway alone. In anaerobic metabolism, pyruvate is metabolized to a lesser extent to regenerate NAD^+, as we shall see in the following section.

3. FERMENTATION: THE ANAEROBIC FATE OF PYRUVATE

The three common metabolic fates of pyruvate produced by glycolysis are outlined in Fig. 14-16:

1. *Under aerobic conditions, the pyruvate is completely oxidized via the citric acid cycle to CO_2 and H_2O.*

2. *Under anaerobic conditions, pyruvate must be converted to a reduced*

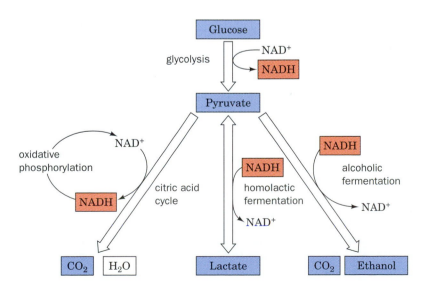

Figure 14-16. Metabolic fate of pyruvate. Under aerobic conditions (*left*), the pyruvate carbons are oxidized to CO_2 by the citric acid cycle and the electrons are eventually transferred to yield H_2O in oxidative phosphorylation. Under anaerobic conditions in muscle, pyruvate is reversibly converted to lactate (*middle*), whereas in yeast, it is converted to CO_2 and ethanol (*right*).

end product in order to reoxidize the NADH produced by the GAPDH reaction. This occurs in two ways:

(**a**) In yeast, pyruvate is decarboxylated to yield CO_2 and **acetaldehyde,** which is then reduced by NADH to yield NAD^+ and ethanol. This process is known as **alcoholic fermentation** (a fermentation is an anaerobic biological reaction process).

(**b**) Under anaerobic conditions in muscle, pyruvate is reduced to **lactate** to regenerate NAD^+ in a process known as **homolactic fermentation.**

Thus, in aerobic glycolysis, NADH acts as a "high-energy" compound, whereas in anaerobic glycolysis, its free energy of oxidation is dissipated as heat.

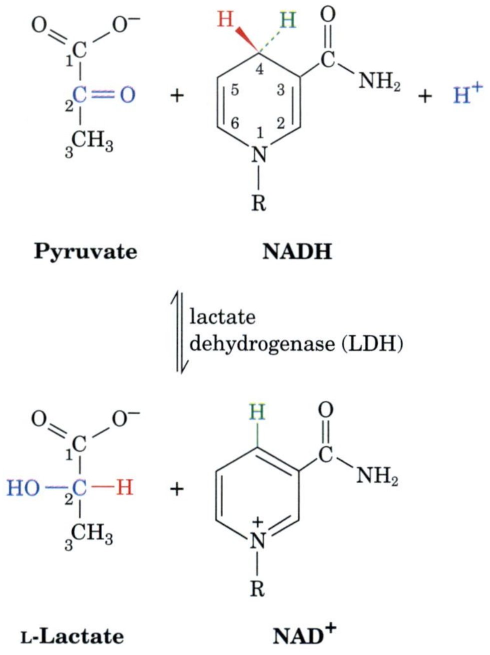

A. *Homolactic Fermentation*

In muscle, particularly during vigorous activity when the demand for ATP is high and oxygen is in short supply, **lactate dehydrogenase (LDH)** catalyzes the oxidation of NADH by pyruvate to yield NAD^+ and lactate *(at left)*. This reaction is often classified as Reaction 11 of glycolysis. The lactate dehydrogenase reaction is freely reversible, so *pyruvate and lactate concentrations are readily equilibrated.*

In the proposed mechanism for pyruvate reduction by LDH, a hydride ion is stereospecifically transferred from C4 of NADH to C2 of pyruvate with concomitant transfer of a proton from the imidazolium moiety of His 195.

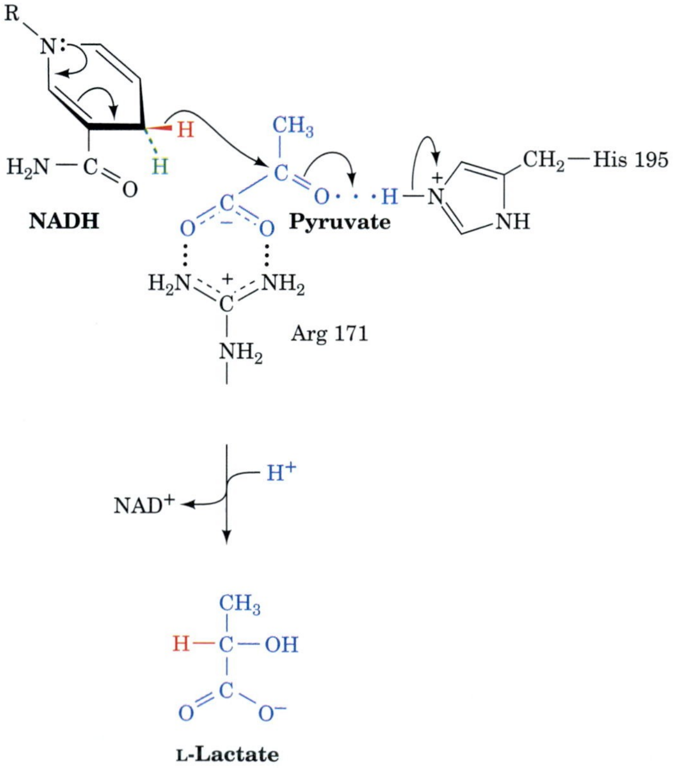

Both His 195 and Arg 171 interact electrostatically with the substrate carboxylate group to orient pyruvate (or lactate, in the reverse reaction) in the enzyme active site.

The overall process of anaerobic glycolysis in muscle can be represented as

$$\text{Glucose} + 2\ \text{ADP} + 2\ P_i \rightarrow 2\ \text{lactate} + 2\ \text{ATP} + 2\ H_2O + 2\ H^+$$

Lactate represents a sort of dead end for anaerobic glucose metabolism. The lactate can either be exported from the cell or converted back to pyruvate. Much of the lactate produced in skeletal muscle cells is carried by the blood to the liver, where it is used to synthesize glucose (Section 21-2A).

Contrary to widely held belief, it is not lactate buildup in the muscle per se that causes muscle fatigue and soreness but the accumulation of glycolytically generated acid (muscles can maintain their workload in the presence of high lactate concentrations if the pH is kept constant).

B. Alcoholic Fermentation

Under anaerobic conditions in yeast, NAD^+ for glycolysis is regenerated in a process that has been valued for thousands of years: the conversion of pyruvate to ethanol and CO_2. Ethanol is, of course, the active ingredient of wine and spirits; CO_2 so produced leavens bread.

Yeast (Fig. 14-17) produces ethanol and CO_2 via two consecutive reactions (Fig. 14-18):

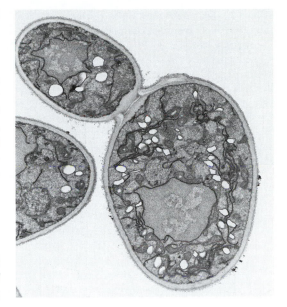

Figure 14-17. An electron micrograph of yeast cells. [Biophoto Associates Photo Researchers.]

1. The decarboxylation of pyruvate to form acetaldehyde and CO_2 as catalyzed by **pyruvate decarboxylase** (an enzyme not present in animals).

2. The reduction of acetaldehyde to ethanol by NADH as catalyzed by alcohol dehydrogenase (Section 11-1B), thereby regenerating NAD^+ for use in the GAPDH reaction of glycolysis.

TPP Is an Essential Cofactor of Pyruvate Decarboxylase

Pyruvate decarboxylase contains the coenzyme **thiamine pyrophosphate (TPP; also called thiamin diphosphate, ThDP).**

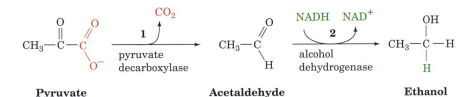

Figure 14-18. The two reactions of alcoholic fermentation. (1) Decarboxylation of pyruvate to form acetaldehyde; and (2) reduction of acetaldehyde to ethanol by NADH.

***Figure 14-19.   TPP binding to pyruvate
decarboxylase from *Saccharomyces uvarum*
(brewer's yeast).*** The TPP and the side chain of
Glu 51 are shown in skeletal form with C green,
N blue, O red, S yellow, and P orange. The TPP
binds in a cavity situated between the dimer's two
subunits (*cyan and magenta*) where it hydrogen
bonds to Glu 51. [Based on an X-ray structure by
William Furey and Martin Sax, Veterans
Administration Medical Center and University of
Pittsburgh.] ● **See the Interactive Exercises.**

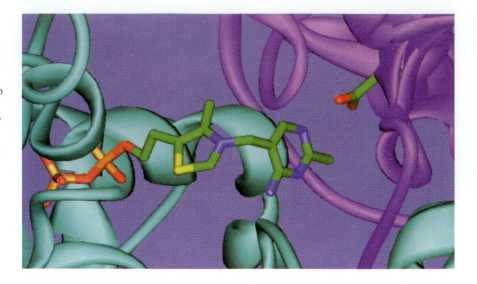

TPP binds tightly but noncovalently to pyruvate decarboxylase (Fig. 14-19).

The enzyme uses TPP because uncatalyzed decarboxylation of an α-keto acid such as pyruvate requires the buildup of negative charge on the carbonyl carbon atom in the transition state, an unstable situation:

$$ \text{transition state unstable} $$

This transition state can be stabilized by delocalizing the developing negative charge into a suitable "electron sink." The amino acid residues of proteins function poorly in this capacity but TPP does so easily.

TPP's catalytically active functional group is the ***thiazolium ring.*** The C2-H atom of this group is relatively acidic because of the adjacent positively charged quaternary nitrogen atom, which electrostatically stabilizes the carbanion formed when the proton dissociates. This dipolar carbanion (or **ylid**) is the active form of the coenzyme. Pyruvate decarboxylase operates as follows (Fig. 14-20):

Step 1 Nucleophilic attack by the ylid form of TPP on the carbonyl carbon of pyruvate.

Step 2 Departure of CO_2 to generate a resonance-stabilized carbanion adduct in which the thiazolium ring of the coenzyme acts as an electron sink.

Step 3 Protonation of the carbanion.

Step 4 Elimination of the TPP ylid to form acetaldehyde and regenerate the active enzyme.

This mechanism has been corroborated by the isolation of the **hydroxyethylthiamine pyrophosphate** intermediate.

Beriberi is a Thiamine Deficiency Disease

The ability of TPP's thiazolium ring to add to carbonyl groups and act as an electron sink makes it the coenzyme most utilized in α-keto acid decarboxylation reactions. Consequently, thiamine **(vitamin B₁),** which is neither synthesized nor stored in significant amounts by the tissues of most

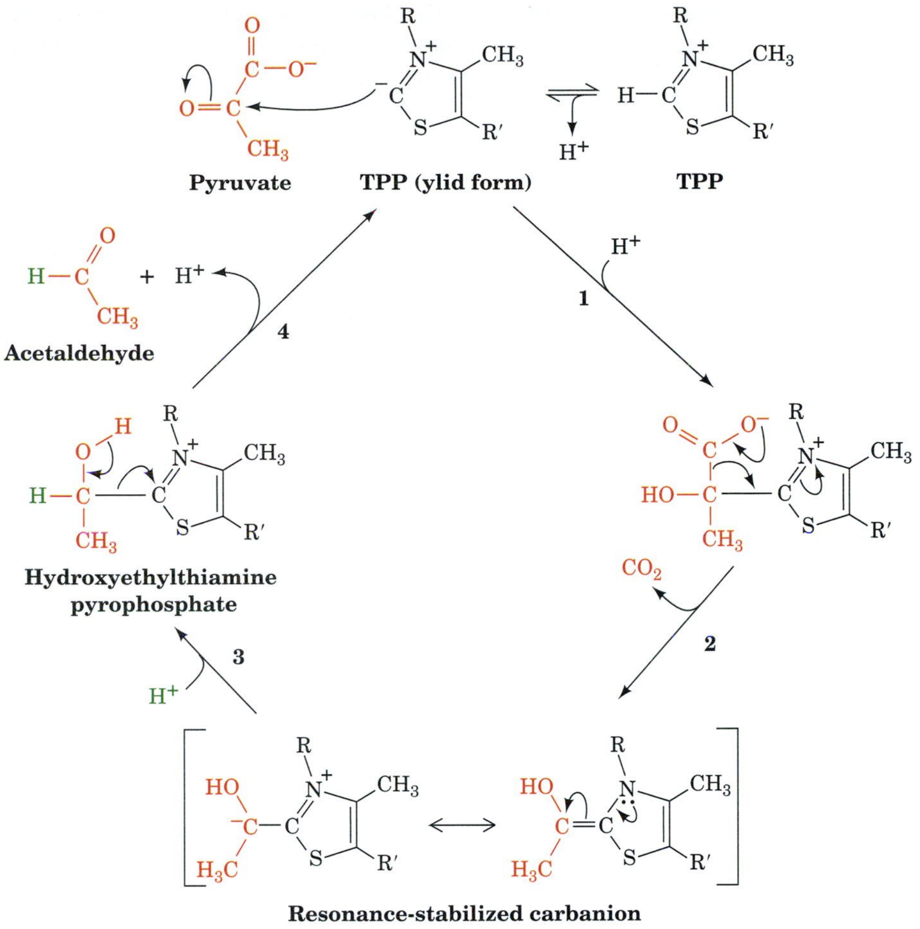

Figure 14-20.   The reaction mechanism of pyruvate decarboxylase. (1) Nucleophilic attack by the ylid form of TPP on the carbonyl carbon of pyruvate; (2) departure of CO_2 to generate a resonance-stabilized carbanion; (3) protonation of the carbanion; and (4) elimination of the TPP ylid and release of product.

vertebrates, is required in their diets. Thiamine deficiency in humans results in an ultimately fatal condition known as **beriberi** that is characterized by neurological disturbances causing pain, paralysis and atrophy (wasting) of the limbs, and/or cardiac failure resulting in edema (the accumulation of fluid in tissues and body cavities). Beriberi was particularly prevalent in the rice-consuming areas of Asia because of the custom of polishing this staple grain to remove its coarse but thiamine-containing outer layers. Beriberi frequently develops in chronic alcoholics as a consequence of their penchant for drinking but not eating.

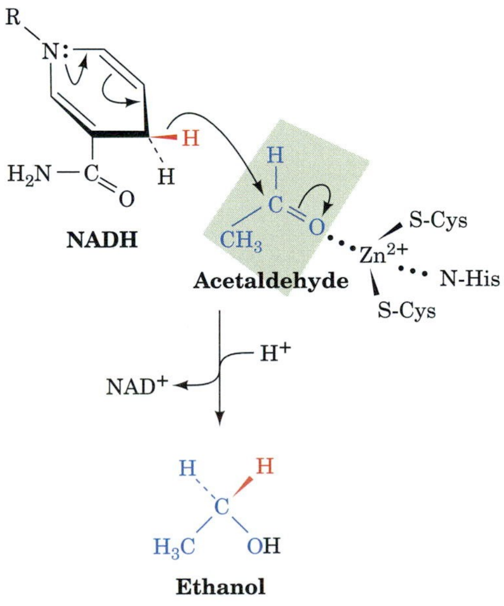

Reduction of Acetaldehyde and Regeneration of NAD$^+$

Yeast alcohol dehydrogenase (YADH), the enzyme that converts acetaldehyde to ethanol, is a tetramer, each subunit of which binds one Zn^{2+} ion. The Zn^{2+} polarizes the carbonyl group of acetaldehyde to stabilize the developing negative charge in the transition state of the reaction *(at right)*. This facilitates the stereospecific transfer of a hydrogen from NADH to acetaldehyde (Section 11-1B).

Mammalian liver alcohol dehydrogenase **(LADH)** metabolizes the alcohols anaerobically produced by the intestinal flora as well as those from external sources (the direction of the alcohol dehydrogenase reaction varies with the relative concentrations of ethanol and acetaldehyde). Mammalian LADH is a dimer with significant amino acid sequence similarity to YADH. It has two bound Zn^{2+} ions per monomer, only one of which functions like those in the yeast enzyme.

C. Energetics of Fermentation

Thermodynamics permits us to dissect the process of fermentation into its component parts and to account for the free energy changes that occur. This enables us to calculate the efficiency with which the free energy of glucose catabolism is used in the synthesis of ATP. For homolactic fermentation,

$$\text{Glucose} \longrightarrow 2 \text{ lactate} + 2 \text{ H}^+ \qquad \Delta G^{\circ\prime} = -196 \text{ kJ} \cdot \text{mol}^{-1}$$

For alcoholic fermentation,

$$\text{Glucose} \longrightarrow 2 \text{ CO}_2 + 2 \text{ ethanol} \qquad \Delta G^{\circ\prime} = -235 \text{ kJ} \cdot \text{mol}^{-1}$$

Each of these processes is coupled to the net formation of 2 ATP, which requires $\Delta G^{\circ\prime} = +61 \text{ kJ} \cdot \text{mol}^{-1}$ of glucose consumed. Dividing $\Delta G^{\circ\prime}$ of ATP formation by that of lactate formation indicates that homolactic fermentation is 31% "efficient"; that is, 31% of the free energy released by this process under standard biochemical conditions is sequestered in the form of ATP. The rest is dissipated as heat, thereby making the process irreversible. Likewise, alcoholic fermentation is 26% efficient under biochemical standard state conditions. *Under physiological conditions, where the concentrations of reactants and products differ from those of the standard state, these reactions have thermodynamic efficiencies of >50%.*

Anaerobic fermentation uses glucose in a profligate manner compared to oxidative phosphorylation: Fermentation results in the production of 2 ATP per glucose, whereas oxidative phosphorylation yields up to 38 ATP per glucose (Section 17-3C). This accounts for Pasteur's observation that yeast consume far more sugar when growing anaerobically than when growing aerobically (the **Pasteur effect**). However, *the rate of ATP production by anaerobic glycolysis can be up to 100 times faster than that of oxidative phosphorylation. Consequently, when tissues such as muscle are rapidly consuming ATP, they regenerate it almost entirely by anaerobic glycolysis.* (Homolactic fermentation does not really "waste" glucose since the lactate is aerobically reconverted to glucose by the liver; Section 21-2A.) Certain muscles are specialized for the rapid production of ATP by glycolysis (see Box 14-2).

4. CONTROL OF GLYCOLYSIS

Under steady state conditions, glycolysis operates continuously, although the glycolytic flux must vary to meet the needs of the organism. Elucidation of the flux control mechanisms of a given pathway, such as glycolysis, commonly involves three steps:

1. Identification of the rate-determining step(s) of the pathway by measuring the *in vivo* ΔG for each reaction. Enzymes that operate far from equilibrium are potential control points (Section 13-1D).

2. *In vitro* identification of allosteric modifiers of the enzymes catalyzing the rate-determining reactions. The mechanisms by which these compounds act are determined from their effects on the enzymes' kinetics.

Box 14-2
BIOCHEMISTRY IN FOCUS

Glycolytic ATP Production in Muscle

Skeletal muscle consists of both **slow-twitch** (Type I) and **fast-twitch** (Type II) **fibers.** Fast-twitch fibers, so called because they predominate in muscles capable of short bursts of rapid activity, are nearly devoid of mitochondria (where oxidative phosphorylation occurs). Consequently, they must obtain nearly all of their ATP through anaerobic glycolysis, for which they have a particularly large capacity. Muscles designed to contract slowly and steadily, in contrast, are enriched in slow-twitch fibers that are rich in mitochondria and obtain most of their ATP through oxidative phosphorylation.

Fast- and slow-twitch fibers were originally known as white and red fibers, respectively, because otherwise pale-colored muscle tissue, when enriched with mitochondria,

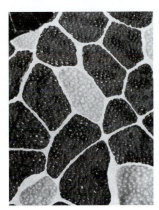

Slow-twitch muscle fiber

Fast-twitch muscle fiber

takes on the red color characteristic of their heme-containing cytochromes. However, fiber color is an imperfect indictor of muscle physiology.

In a familiar example, the flight muscles of migratory birds such as ducks and geese, which need a continuous energy supply, are rich in slow-twitch fibers. Therefore, these birds have dark breast meat. In contrast, the flight muscles of less ambitious fliers, such as chickens and turkeys, which are used only for short bursts (often to escape danger), consist mainly of fast-twitch fibers that form white meat. In humans, the muscles of sprinters are relatively rich in fast-twitch fibers, whereas distance runners have a greater proportion of slow-twitch fibers (although their muscles have the same color). [Photo courtesy of J.D. MacDougall, McMaster University, Canada.]

---

3. Measurement of the *in vivo* levels of the proposed regulators under various conditions to establish whether these concentration changes are consistent with the proposed control mechanism.

Let us examine the thermodynamics of glycolysis in muscle tissue with an eye toward understanding its control mechanisms (keep in mind that different tissues control glycolysis in different ways). Table 14-1 lists the standard free energy changes ($\Delta G^{\circ\prime}$) and the actual physiological free energy change (ΔG) associated with each reaction in the pathway. It is important to realize that the free energy changes associated with the reactions under standard conditions may differ dramatically from the actual values *in vivo*.

Table 14-1. $\Delta G^{\circ\prime}$ and ΔG for the Reactions of Glycolysis in Heart Muscle[a]

Reaction	Enzyme	$\Delta G^{\circ\prime}$ $(kJ \cdot mol^{-1})$	ΔG $(kJ \cdot mol^{-1})$
1	Hexokinase	−20.9	−27.2
2	PGI	+2.2	−1.4
3	PFK	−17.2	−25.9
4	Aldolase	+22.8	−5.9
5	TIM	+7.9	+4.4
6 + 7	GAPDH + PGK	−16.7	−1.1
8	PGM	+4.7	−0.6
9	Enolase	−3.2	−2.4
10	PK	−23.0	−13.9

[a]Calculated from data in Newsholme, E.A. and Start, C., *Regulation in Metabolism*, p. 97, Wiley (1973).

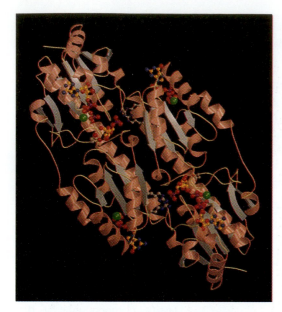

Figure 14-21. The X-ray structure of PFK from E. coli. Two subunits of the tetrameric enzyme are shown in ribbon form with helices pink, β strands gray, and the remaining chain segments white. Each subunit binds its substrates F6P (*near the center of each subunit*) and Mg^{2+}–ATP (*lower right and upper left; the green balls represent* Mg^{2+}), along with the activator Mg^{2+}–ADP (*top right and lower left, in the rear*). [Courtesy of Philip Evans, Cambridge University.] ● See Kinemage Exercise 13-1.

Only three reactions, those catalyzed by hexokinase, phosphofructokinase, and pyruvate kinase, function with large negative free energy changes in heart muscle under physiological conditions. These nonequilibrium reactions of glycolysis are candidates for flux-control points. The other glycolytic reactions function near equilibrium: Their forward and reverse rates are much faster than the actual flux through the pathway. Consequently, these equilibrium reactions are very sensitive to changes in the concentration of pathway intermediates and readily accommodate changes in flux generated at the rate-determining step(s) of the pathway.

A. Phosphofructokinase: The Major Flux-Controlling Enzyme of Glycolysis in Muscle

In vitro studies of hexokinase, phosphofructokinase, and pyruvate kinase indicate that each is controlled by a variety of compounds. Yet when the G6P source for glycolysis is glycogen, rather than glucose, as is often the case in skeletal muscle, the hexokinase reaction is not required (Section 15-1). Pyruvate kinase catalyzes the last reaction of glycolysis and is therefore unlikely to be the primary point for regulating flux through the entire pathway. Evidently, PFK, an elaborately regulated enzyme functioning far from equilibrium, is the major control point for glycolysis in muscle under most conditions.

PFK (Fig. 14-21) is a tetrameric enzyme with two conformational states, R and T, that are in equilibrium. ATP is both a substrate and an allosteric inhibitor of phosphofructokinase. Other compounds, including ADP, AMP, and **fructose-2,6-bisphosphate (F2,6P),** reverse the inhibitory effects of ATP and are therefore considered activators. Each PFK subunit has two binding sites for ATP: a substrate site and an inhibitor site. The substrate site binds ATP equally well in either conformation, but the inhibitor site binds ATP almost exclusively in the T state. The other substrate of PFK, F6P, preferentially binds to the R state. Consequently, at high concentrations, ATP acts as an allosteric inhibitor of PFK by binding to the T state, thereby shifting the T ⇌ R equilibrium in favor of the T state and thus decreasing PFK's affinity for F6P (this is similar to the action of 2,3-BPG in decreasing the affinity of hemoglobin for O_2; Section 7-2C). In graphical terms, at high concentrations of ATP, the hyperbolic (noncooperative) curve of PFK activity versus [F6P] is converted to the sigmoidal (cooperative) curve characteristic of allosteric enzymes (Fig. 14-22). For example, when [F6P] = 0.5 mM (the dashed line in Fig. 14-22), the enzyme is nearly maximally active, but in the presence of 1 mM ATP, the activity drops to 15% of its original level, a nearly 7-fold decrease. (Actually, the most potent allosteric effector of PFK is F2,6P, which we will discuss in Section 15-4C.)

Structural Basis for Allosterism in Phosphofructokinase

The X-ray structures of PFK from several organisms have been determined in both the R and the T states by Philip Evans. The R state of PFK is stabilized by the binding of its substrate F6P. In the R state of *Bacillus stearothermophilus* PFK, the side chain of Arg 162 forms an ion pair with the phosphoryl group of an F6P bound in an active site of another subunit (Fig. 14-23). However, Arg 162 is located at the end of a helical turn that unwinds on transition to the T state. The positively charged side chain of Arg 162 thereby swings away and is replaced by the negatively charged side chain of Glu 161. As a consequence, the doubly negative phosphoryl group of F6P has a greatly diminished affinity for the T-state enzyme. The unwinding of this helical turn, which is obligatory for the R → T transition, is prevented by the binding of the activator ADP to its effector site on the

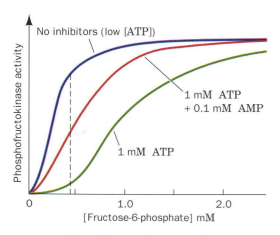

Figure 14-22. PFK activity versus F6P concentration. The various conditions are as follows: purple, no inhibitors or activators; green, 1 mM ATP; and red, 1 mM ATP + 0.1 mM AMP. [After data from Mansour, T.E. and Ahlfors, C.E., *J. Biol. Chem.* **243,** 2523–2533 (1968).] ✳ See the Animated Figures.

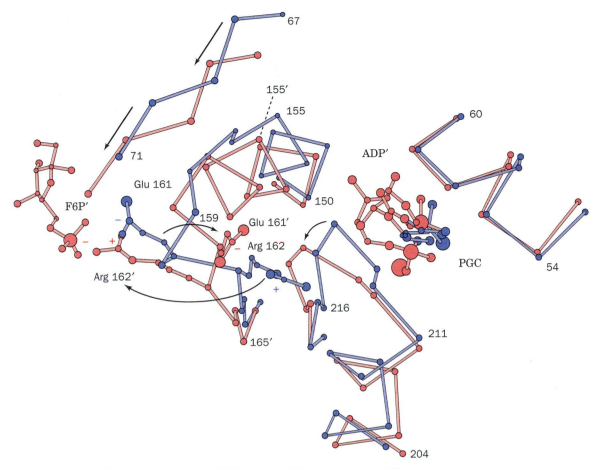

Figure 14-23.   Allosteric changes in PFK from *Bacillus stearothermophilus*.
Segments of the T state (*blue*) are superimposed on segments of the R state (*red*)
that undergo a large conformational rearrangement on the T → R allosteric transition (indicated by the arrows). Residues of the R-state structure are marked by a
prime. Note that in the R state, Arg 162′ forms an attractive ionic interaction with
F6P′, whereas in the T state, F6P′ is repelled by Glu 161. Also shown are bound ligands: the nonphysiological inhibitor 2-phosphoglycolate (PGC; a PEP analog) for
the T state, and the cooperative substrate F6P and the activator ADP for the R state.
[After Schirmer, T. and Evans, P.R., *Nature* **343**, 142 (1990).] ● **See Kinemage
Exercise 13-2.**

enzyme. Presumably, ATP can bind to this site only when the helical turn
is in its unwound conformation (the T state).

AMP Overcomes the ATP Inhibition of PFK

Direct allosteric regulation of PFK by ATP may superficially appear to
be the means by which glycolytic flux is controlled. After all, when [ATP]
is high as a result of low metabolic demand, PFK is inhibited and flux
through glycolysis is low; conversely when [ATP] is low, flux through the
pathway is high and ATP is synthesized to replenish the pool. Consideration of the physiological variation in ATP concentration, however, indicates
that the situation must be more complex. The metabolic flux through glycolysis may vary by 100-fold or more, depending on the metabolic demand
for ATP. However, *measurements of [ATP] in vivo at various levels of metabolic activity indicate that [ATP] varies <10% between rest and vigorous
exertion.* Yet there is no known allosteric mechanism that can account for
a 100-fold change in flux of a nonequilibrium reaction with only a 10%

change in effector concentration. Thus, some other mechanism(s) must be responsible for controlling glycolytic flux.

The inhibition of PFK by ATP is relieved by AMP as well as ADP. This results from AMP's preferential binding to the R state of PFK. If a PFK solution containing 1 mM ATP and 0.5 mM F6P is brought to 0.1 mM in AMP, the activity of PFK rises from 15 to 50% of its maximal activity, a 3-fold increase (Fig. 14-22).

The [ATP] decreases by only 10% in going from a resting state to one of vigorous activity because it is buffered by the action of two enzymes: creatine kinase and adenylate kinase (Section 13-2C). Adenylate kinase catalyzes the reaction

$$2\ \text{ADP} \rightleftharpoons \text{ATP} + \text{AMP} \qquad K = \frac{[\text{ATP}][\text{AMP}]}{[\text{ADP}]^2} = 0.44$$

which rapidly equilibrates the ADP resulting from ATP hydrolysis in muscle contraction with ATP and AMP.

In muscle, [ATP] is ~50 times greater than [AMP] and ~10 times greater than [ADP]. Consequently, *a change in [ATP] from, for example, 1 to 0.9 mM, a 10% decrease, can result in a 100% increase in [ADP] (from 0.1 to 0.2 mM) as a result of the adenylate kinase reaction, and a >400% increase in [AMP] (from 0.02 to ~0.1 mM)*. Therefore, a metabolic signal consisting of a decrease in [ATP] too small to relieve PFK inhibition is amplified significantly by the adenylate kinase reaction, which increases [AMP] by an amount that produces a much larger increase in PFK activity.

B. Substrate Cycling

Even a finely tuned allosteric mechanism like that of PFK can account for only a fraction of the 100-fold alterations in glycolytic flux. Recall from Section 13-1D that only a near-equilibrium reaction can undergo large changes in flux because, in a near-equilibrium reaction, $v_f - v_r \approx 0$ (where v_f and v_r are the forward and reverse reaction rates) and hence a small change in v_f will result in a large fractional change in $v_f - v_r$. However, this is not the case for the PFK reaction because, for such nonequilibrium reactions, v_r is negligible.

Nevertheless, *such equilibriumlike conditions may be imposed on a nonequilibrium reaction if a second enzyme catalyzes the regeneration of its substrate from its product in a thermodynamically favorable manner*. Then v_r is no longer negligible compared to v_f. This situation requires that the forward process (e.g., formation of FBP from F6P) and the reverse process (e.g., breakdown of FBP to F6P) be accomplished by different reactions since the laws of thermodynamics would otherwise be violated (i.e., for a single reaction, the forward and reverse reactions cannot simultaneously be favorable).

Under physiological conditions, the reaction catalyzed by PFK:

$$\text{F6P} + \text{ATP} \longrightarrow \text{FBP} + \text{ADP}$$

is highly exergonic ($\Delta G = -25.9\ \text{kJ} \cdot \text{mol}^{-1}$). Consequently, the back reaction has a negligible rate compared to the forward reaction. **Fructose-1,6-bisphosphatase (FBPase),** however, which is present in many mammalian tissues (and which is an essential enzyme in gluconeogenesis; Section 15-4B), catalyzes the exergonic hydrolysis of FBP ($\Delta G = -8.6\ \text{kJ} \cdot \text{mol}^{-1}$):

$$\text{FBP} + \text{H}_2\text{O} \longrightarrow \text{F6P} + \text{P}_i$$

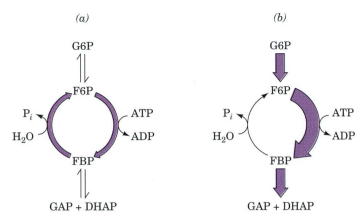

Figure 14-24. Substrate cycling in the regulation of PFK. (*a*) In resting muscle, both enzymes in the F6P/FBP substrate cycle are active, and glycolytic flux is low. (*b*) In active muscle, PFK activity increases while FBPase activity decreases. This dramatically increases the flux through PFK and therefore results in high glycolytic flux.

Note that the combined reactions catalyzed by PFK and FBPase result in net ATP hydrolysis:

$$ATP + H_2O \Longrightarrow ADP + P_i$$

Such a set of opposing reactions is known as a **substrate cycle** because it cycles a substrate to an intermediate and back again. When this set of reactions was discovered, it was referred to as a **futile cycle** since its net result seemed to be the useless consumption of ATP. In fact, the PFK activators AMP and F2,6P allosterically inhibit FBPase, which suggested that only one of the enzymes operates in a cell under a given set of conditions. However, it is now clear that both enzymes often function simultaneously at significant rates.

Eric Newsholme has proposed that substrate cycles are not at all "futile" but, rather, have a regulatory function. *The combined effects of allosteric effectors on the opposing reactions of a substrate cycle can produce a much greater fractional effect on pathway flux* ($v_f - v_r$) *than is possible through allosteric regulation of a single enzyme.* Substrate cycling does not increase the maximum flux through a pathway. On the contrary, it functions to decrease the minimum flux. In a sense, the substrate is put into a "holding pattern." In the PFK/FBPase example (Fig. 14-24), the cycling of substrate appears to be the energetic "price" that a muscle must pay to be able to change rapidly from a resting state (where $v_f - v_r$ is small), in which substrate cycling is maximal, to one of sustained high activity (where $v_f - v_r$ is large). The rate of substrate cycling itself may be under hormonal or neuronal control so as to increase the sensitivity of the metabolic system under conditions when high activity (fight or flight) is anticipated.

Pyruvate kinase, another enzyme whose rate helps control glycolytic flux, is allosterically regulated by some of the same factors that regulate PFK activity. For example, ATP reduces the affinity of pyruvate kinase for its substrate. In addition, pyruvate kinase is subject to **feed-forward activation** by FBP. This mechanism ensures that once metabolites pass the PFK step of glycolysis, they will continue through the pathway.

Substrate Cycling, Thermogenesis, and Obesity

Many animals, including adult humans, are thought to generate much of their body heat, particularly when it is cold, through substrate cycling in muscle and liver, a process known as **nonshivering thermogenesis** (the muscle contractions of shivering or any other movement also produce heat). Substrate cycling is stimulated by thyroid hormones (which stimulate metabolism in most tissues) as is indicated, for example, by the observation that rats lacking a functional thyroid gland do not survive at 5°C. Chronically obese individuals tend to have lower than normal metabolic rates, which is probably due, in part, to a reduced rate of nonshivering thermogenesis. Such individuals therefore tend to be cold sensitive. Indeed, whereas normal individuals increase their rate of thyroid hormone activation on exposure to cold, genetically obese animals and obese humans fail to do so.

5.   METABOLISM OF HEXOSES OTHER THAN GLUCOSE

Together with glucose, the hexoses fructose, galactose, and mannose are prominent metabolic fuels. After digestion, these monosaccharides enter the bloodstream, which carries them to various tissues. Fructose, galactose, and mannose are converted to glycolytic intermediates that are then metabolized by the glycolytic pathway (Fig. 14-25).

A. *Fructose*

Fructose is a major fuel source in diets that contain large amounts of fruit or sucrose (a disaccharide of fructose and glucose; Section 8-2A). There are two pathways for the metabolism of fructose; one occurs in muscle and the other occurs in liver. This dichotomy results from the different enzymes present in these tissues.

Fructose metabolism in muscle differs little from that of glucose. Hexokinase (Section 14-2A), which converts glucose to G6P, also phosphorylates fructose, yielding F6P (Fig. 14-26, *left*). The entry of fructose into glycolysis therefore involves only one reaction step.

Liver contains a hexokinase known as **glucokinase,** which has a low affinity for hexoses, including fructose (Section 21-1D). Fructose metabolism in liver must therefore differ from that in muscle. In fact, liver converts fructose to glycolytic intermediates through a pathway that involves seven enzymes (Fig. 14-26, *right*):

1. **Fructokinase** catalyzes the phosphorylation of fructose by ATP at C1 to form **fructose-1-phosphate.** Neither hexokinase nor PFK can phosphorylate fructose-1-phosphate at C6 to form the glycolytic intermediate FBP.

2. Class I aldolase (Section 14-2D) has several isozymic forms. Muscle contains Type A aldolase, which is specific for FBP. Liver, however, contains Type B aldolase, for which fructose-1-phosphate is also a substrate (Type B aldolase is sometimes called **fructose-1-phosphate aldolase**). In liver, fructose-1-phosphate therefore undergoes an aldol cleavage:

   Fructose-1-phosphate $\rightleftharpoons$

   dihydroxyacetone phosphate + glyceraldehyde

3. Direct phosphorylation of **glyceraldehyde** by ATP through the action of **glyceraldehyde kinase** forms the glycolytic intermediate GAP.

4–7. Alternatively, glyceraldehyde is converted to the glycolytic intermediate DHAP by its NADH-dependent reduction to glycerol as

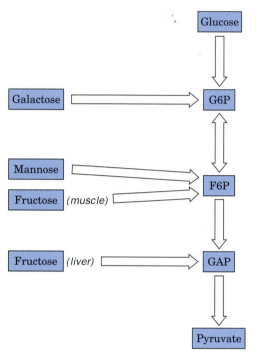

Figure 14-25.   Entry of other hexoses into glycolysis. Fructose (in muscle) and mannose are converted to F6P; liver fructose is converted to GAP; and galactose is converted to G6P.

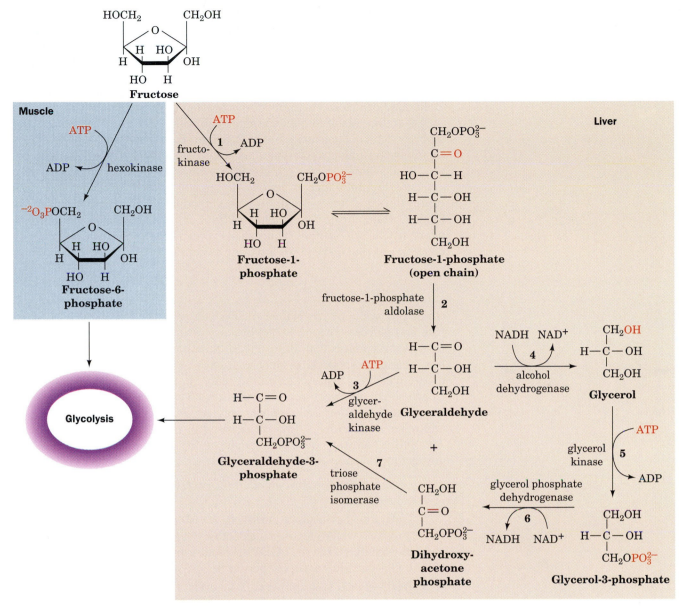

Figure 14-26. **The metabolism of fructose.** In muscle (*left*), the conversion of fructose to the glycolytic intermediate F6P involves only one enzyme, hexokinase. In liver (*right*), seven enzymes participate in the conversion of fructose to glycolytic intermediates: (**1**) fructokinase, (**2**) fructose-1-phosphate aldolase, (**3**) glyceraldehyde kinase, (**4**) alcohol dehydrogenase, (**5**) glycerol kinase, (**6**) glycerol phosphate dehydrogenase, and (**7**) triose phosphate isomerase.

catalyzed by alcohol dehydrogenase (Reaction 4), phosphorylation to **glycerol-3-phosphate** through the action of **glycerol kinase** (Reaction 5), and NAD^+-dependent reoxidation to DHAP catalyzed by **glycerol phosphate dehydrogenase** (Reaction 6). The DHAP is then converted to GAP by triose phosphate isomerase (Reaction 7).

The two pathways leading from glyceraldehyde to GAP have the same net cost: Both consume ATP, and although NADH is oxidized in Reaction 4, it is reduced again in Reaction 6. The longer pathway, however, produces glycerol-3-phosphate, which (along with DHAP) can become the glycerol backbone of glycerophospholipids and triacylglycerols (Section 19-6A).

Excessive Fructose Depletes Liver P*i*

At one time, fructose was thought to have advantages over glucose for intravenous feeding. The liver, however, encounters metabolic problems when the blood concentration of this sugar is too high (higher than can be attained by simply eating fructose-containing foods). When the fructose concentration is high, fructose-1-phosphate may be produced faster than Type B aldolase can cleave it. Intravenous feeding of large amounts of fructose may therefore result in high enough fructose-1-phosphate accumulation to severely deplete the liver's store of P_i. Under these conditions, [ATP] drops, thereby activating glycolysis and lactate production. The lactate concentration in the blood under such conditions can reach life-threatening levels.

Fructose intolerance, a genetic disease in which ingestion of fructose causes the same fructose-1-phosphate accumulation as with its intravenous feeding, results from a deficiency of Type B aldolase. This condition appears to be self-limiting: Individuals with fructose intolerance rapidly develop a strong distaste for anything sweet.

B. Galactose

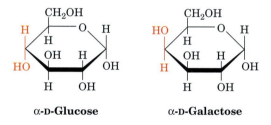

α-D-Glucose   α-D-Galactose

Galactose is obtained from the hydrolysis of lactose (a disaccharide of galactose and glucose; Section 8-2A) in dairy products. Galactose and glucose *(at left)* are epimers that differ only in their configuration at C4. Although hexokinase phosphorylates glucose, fructose, and mannose, it does not recognize galactose. An epimerization reaction must therefore occur before galactose enters glycolysis. This reaction takes place after the conversion of galactose to its **uridine diphosphate** derivative (the role of UDP–sugars and other nucleotidyl–sugars is discussed in more detail in Section 15-5). The entire pathway converting galactose to a glycolytic intermediate requires four reactions (Fig. 14-27):

1. Galactose is phosphorylated at C1 by ATP in a reaction catalyzed by **galactokinase.**

2. **Galactose-1-phosphate uridylyl transferase** transfers the uridylyl group of UDP–glucose to **galactose-1-phosphate** to yield **glucose-1-phosphate (G1P)** and **UDP–galactose** by the reversible cleavage of UDP–glucose's pyrophosphoryl bond.

3. **UDP–galactose-4-epimerase** converts UDP–galactose back to UDP–glucose. This enzyme has an associated NAD⁺, which suggests that the reaction involves the sequential oxidation and reduction of the hexose C4 atom.

CH₂OH   CH₂OH

UDP-Galactose   UDP-Glucose

NAD⁺   NAD⁺

NADH   CH₂OH   NADH

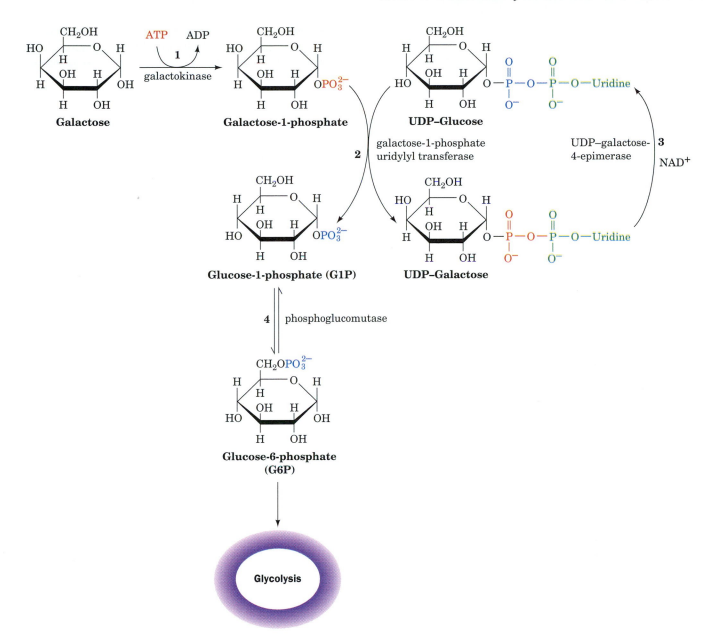

Figure 14-27. The metabolism of galactose. Four enzymes participate in the conversion of galactose to the glycolytic intermediate G6P: (**1**) galactokinase, (**2**) galactose-1-phosphate uridylyl transferase, (**3**) UDP–galactose-4-epimerase, and (**4**) phosphoglucomutase.

4. G1P is converted to the glycolytic intermediate G6P by the action of **phosphoglucomutase.**

Galactosemia

Galactosemia is a genetic disease characterized by the inability to convert galactose to glucose. Its symptoms include failure to thrive, mental retardation, and, in some instances, death from liver damage. Most cases of galactosemia involve a deficiency in the enzyme catalyzing Reaction 2 of the interconversion, galactose-1-phosphate uridylyl transferase. Formation of UDP–galactose from galactose-1-phosphate is thus prevented, leading to a buildup of toxic metabolic by-products. For example, the increased galactose concentration in the blood results in a higher galactose concentration in the lens of the eye, where this sugar is reduced to **galactitol.**

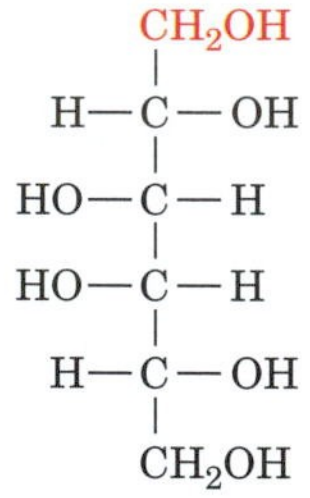

$$
\begin{array}{c}
CH_2OH \\
| \\
H-C-OH \\
| \\
HO-C-H \\
| \\
HO-C-H \\
| \\
H-C-OH \\
| \\
CH_2OH
\end{array}
$$

D-Galactitol

The presence of this sugar alcohol in the lens eventually causes cataract formation (clouding of the lens).

Galactosemia is treated by a galactose-free diet. Except for the mental retardation, this reverses all symptoms of the disease. The galactosyl units that are essential for the synthesis of glycoproteins (Section 8-3) and glycolipids (Section 9-1D) can be synthesized from glucose by a reversal of the epimerase reaction. These syntheses therefore do not require dietary galactose.

C. Mannose

Mannose, a product of digestion of polysaccharides and glycoproteins, is the C2 epimer of glucose:

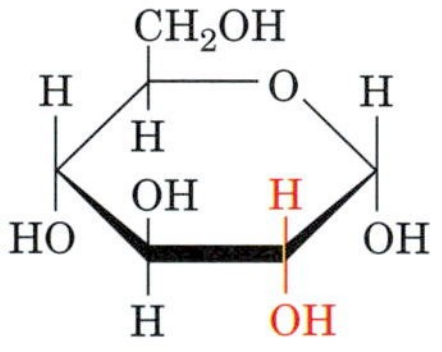

α-D-Glucose 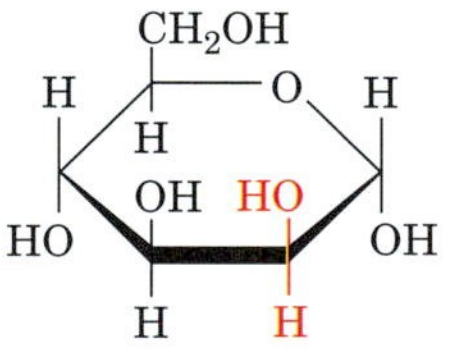   α-D-Mannose

Mannose enters the glycolytic pathway after its conversion to F6P via a two-reaction pathway (Fig. 14-28):

1. Hexokinase recognizes mannose and converts it to **mannose-6-phosphate.**
2. **Phosphomannose isomerase** then converts this aldose to the glycolytic intermediate F6P in a reaction whose mechanism resembles that of phosphoglucose isomerase (Section 14-2B).

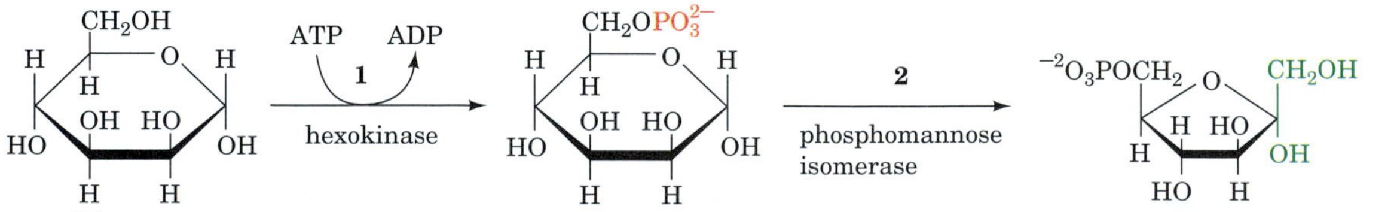

Figure 14-28. **The metabolism of mannose.** Two enzymes are required to convert mannose to the glycolytic intermediate F6P: (**1**) hexokinase and (**2**) phosphomannose isomerase.

6. THE PENTOSE PHOSPHATE PATHWAY

ATP is the cell's "energy currency"; its exergonic cleavage is coupled to many otherwise endergonic cell functions. *Cells also have a second currency, reducing power.* Many endergonic reactions, notably the reductive biosynthesis of fatty acids (Section 19-4) and cholesterol (Section 19-7A), require NADPH in addition to ATP. Despite their close chemical resemblance, *NADPH and NADH are not metabolically interchangeable.* Whereas NADH uses the free energy of metabolite oxidation to synthesize ATP (oxidative phosphorylation), NADPH uses the free energy of metabolite oxidation for reductive biosynthesis. This differentiation is possible because the dehydrogenases involved in oxidative and reductive metabolism are highly specific for their respective coenzymes. Indeed, cells normally maintain their $[NAD^+]/[NADH]$ ratio near 1000, which favors metabolite oxidation, while keeping their $[NADP^+]/[NADPH]$ ratio near 0.01, which favors reductive biosynthesis.

NADPH is generated by the oxidation of glucose-6-phosphate via an alternative pathway to glycolysis, the pentose phosphate pathway (also called the **hexose monophosphate shunt;** Fig. 14-29). Tissues most heavily involved in lipid biosynthesis (liver, mammary gland, adipose tissue, and adrenal cortex) are rich in pentose phosphate pathway enzymes. Indeed, some 30% of the glucose oxidation in liver occurs via the pentose phosphate pathway rather than glycolysis.

The overall reaction of the pentose phosphate pathway is

$$3 \text{ G6P} + 6 \text{ NADP}^+ + 3 \text{ H}_2\text{O} \rightleftharpoons$$
$$6 \text{ NADPH} + 6 \text{ H}^+ + 3 \text{ CO}_2 + 2 \text{ F6P} + \text{GAP}$$

However, the pathway can be considered to have three stages:

Stage 1 Oxidative reactions (Fig. 14-29, Reactions 1–3), which yield NADPH and **ribulose-5-phosphate (Ru5P).**

$$3 \text{ G6P} + 6 \text{ NADP}^+ + 3 \text{ H}_2\text{O} \longrightarrow$$
$$6 \text{ NADPH} + 6 \text{ H}^+ + 3 \text{ CO}_2 + 3 \text{ Ru5P}$$

Stage 2 Isomerization and epimerization reactions (Fig. 14-29, Reactions 4 and 5), which transform Ru5P either to **ribose-5-phosphate (R5P)** or **xylulose-5-phosphate (Xu5P).**

$$3 \text{ Ru5P} \rightleftharpoons \text{R5P} + 2 \text{ Xu5P}$$

Stage 3 A series of C—C bond cleavage and formation reactions (Fig. 14-29, Reactions 6–8) that convert two molecules of Xu5P and one molecule of R5P to two molecules of F6P and one molecule of GAP.

The reactions of Stages 2 and 3 are freely reversible, so the products of the pathway vary with the needs of the cell. In this section, we discuss the three stages of the pentose phosphate pathway and how this pathway is controlled.

A. Stage 1: Oxidative Reactions of NADPH Production

G6P is considered the starting point of the pentose phosphate pathway. This metabolite may arise through the action of hexokinase on glucose (Reaction 1 of glycolysis; Section 14-2A) or from glycogen breakdown (which produces G6P directly; Section 15-1). Only the first three reactions of the pentose phosphate pathway are involved in NADPH production (Fig. 14-29):

1. **Glucose-6-phosphate dehydrogenase (G6PD)** catalyzes net transfer of a hydride ion to NADP$^+$ from C1 of G6P to form **6-phospho-glucono-δ-lactone** *(opposite)*.

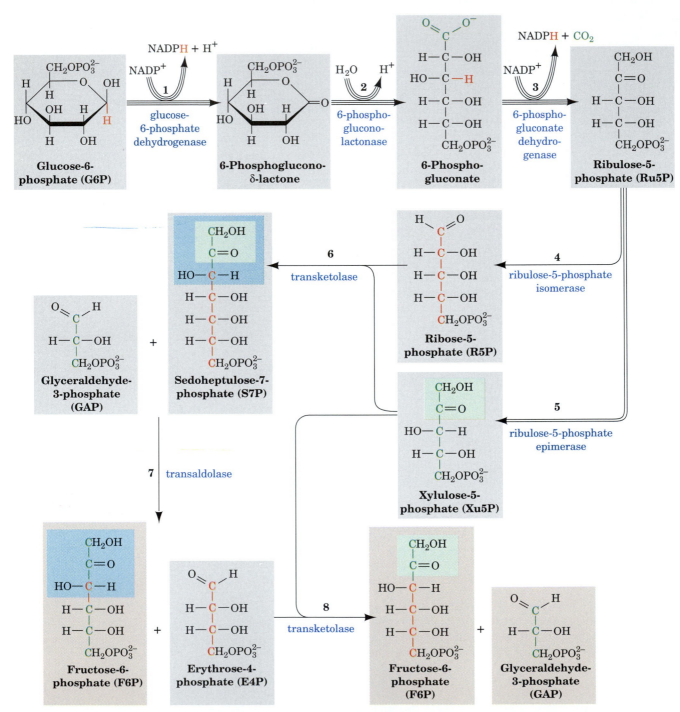

Figure 14-29. *Key to Metabolism.* **The pentose phosphate pathway.** The number of lines in an arrow represents the number of molecules reacting in one turn of the pathway so as to convert 3 G6P to 3 CO$_2$, 2 F6P, and 1 GAP. For the sake of clarity, sugars from Reaction 3 onward are shown in their linear forms. The car-bon skeleton of R5P and the atoms derived from it are drawn in red, and those from Xu5P are drawn in green. The C$_2$ units trans-ferred by transketolase are shaded in green, and the C$_3$ units transferred by transaldolase are shaded in blue.

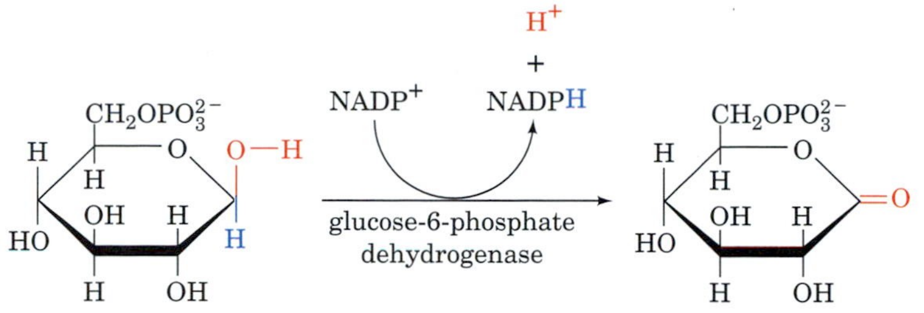

G6P → **6-Phosphoglucono-δ-lactone**

G6P, a cyclic hemiacetal with C1 in the aldehyde oxidation state, is thereby oxidized to a cyclic ester (lactone). The enzyme is specific for $NADP^+$ and is strongly inhibited by NADPH.

2. **6-Phosphogluconolactonase** increases the rate of hydrolysis of 6-phosphoglucono-δ-lactone to **6-phosphogluconate** (the nonenzymatic reaction occurs at a significant rate).

3. **6-Phosphogluconate dehydrogenase** catalyzes the oxidative decarboxylation of 6-phosphogluconate, a β-hydroxy acid, to Ru5P and CO_2 (Fig. 14-30). This reaction is thought to proceed via the formation of a β-keto acid intermediate. The keto group presumably facilitates decarboxylation by acting as an electron sink.

Formation of Ru5P completes the oxidative portion of the pentose phosphate pathway. *It generates two molecules of NADPH for each molecule of G6P that enters the pathway.*

B. Stage 2: Isomerization and Epimerization of Ribulose-5-Phosphate

Ru5P is converted to R5P by **ribulose-5-phosphate isomerase** (Fig. 14-29, Reaction 4) or to Xu5P by **ribulose-5-phosphate epimerase** (Fig. 14-29, Reaction 5). These isomerization and epimerization reactions, like the reaction catalyzed by triose phosphate isomerase (Section 14-2E), are thought to occur via enediolate intermediates.

The relative amounts of R5P and Xu5P produced from Ru5P depend on the needs of the cell. For example, R5P is an essential precursor in the biosynthesis of nucleotides (Chapter 22). Accordingly, R5P production is relatively high (in fact, the entire pentose phosphate pathway activity may

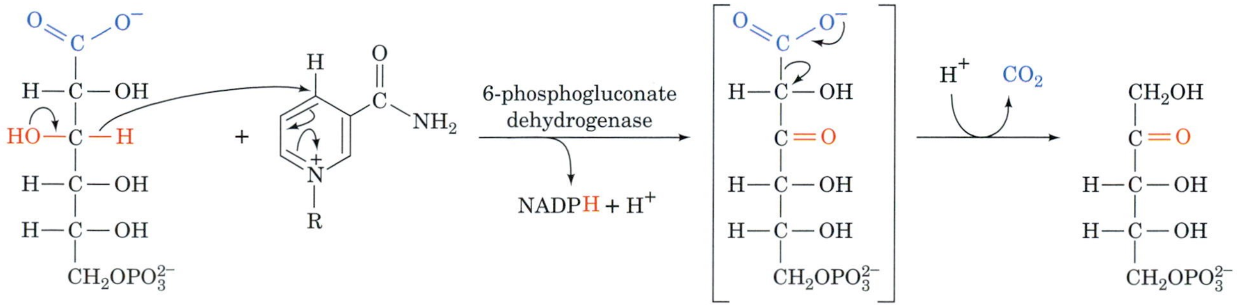

6-Phosphogluconate **NADP⁺** **β-Keto acid intermediate** **Ru5P**

Figure 14-30. The 6-phosphogluconate dehydrogenase reaction. Oxidation of the OH group forms an easily decarboxylated β-keto acid (although the proposed intermediate has not been isolated).

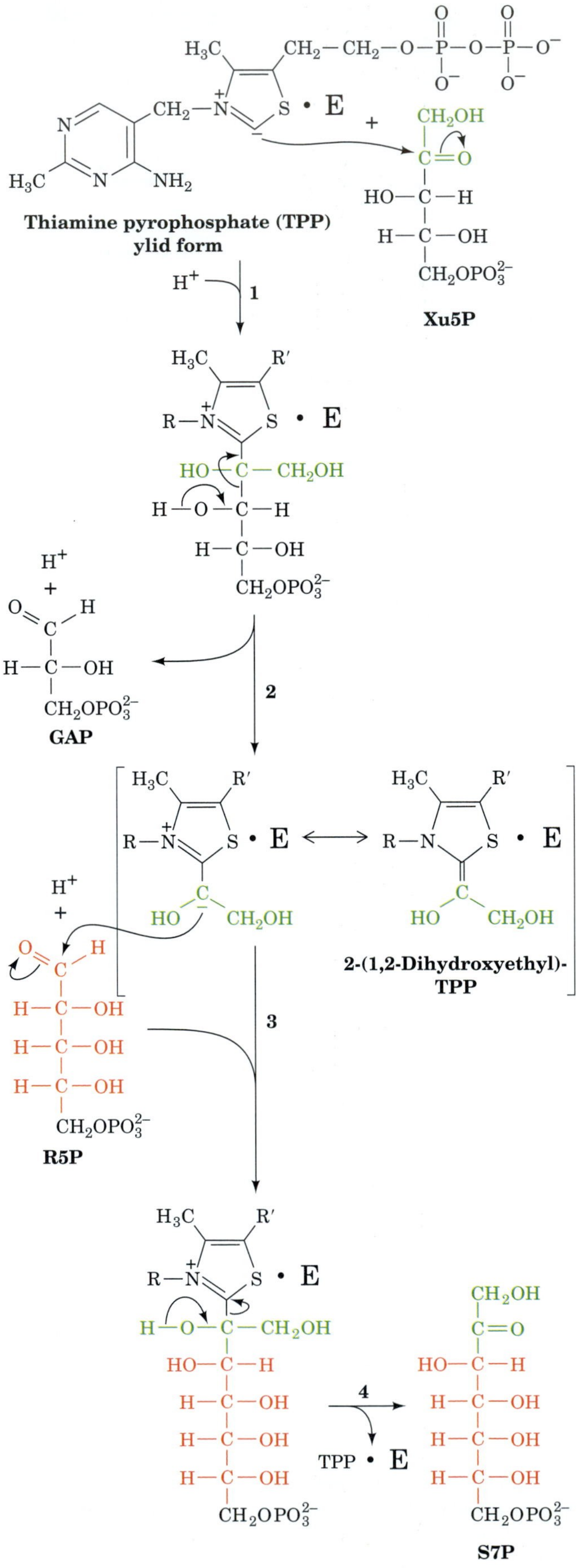

Figure 14-31.   Mechanism of transketolase.
Transketolase (represented by E) uses the coenzyme TPP to stabilize the carbanion formed on cleavage of the C2—C3 bond of Xu5P. The reaction occurs as follows: (**1**) The TPP ylid attacks the carbonyl group the Xu5P; (**2**) C2—C3 bond cleavage yields GAP and enzyme-bound 2-(1,2-dihydroxyethyl)-TPP, a resonance-stabilized carbanion; (**3**) the C2 carbanion attacks the aldehyde carbon of R5P to form an S7P–TPP adduct; (**4**) TPP is eliminated, yielding S7P and the regenerated TPP–enzyme.

be elevated) in rapidly dividing cells, in which the rate of DNA synthesis is increased. If the pathway is being used solely for NADPH production, Xu5P and R5P are produced in a 2:1 ratio for conversion to glycolytic intermediates in the third stage of the pentose phosphate pathway as is discussed below.

C. Stage 3: Carbon–Carbon Bond Cleavage and Formation Reactions

How is a five-carbon sugar transformed to a six-carbon sugar such as F6P? The rearrangements of carbon atoms in the third stage of the pentose phosphate pathway are easier to follow by considering the stoichiometry of the pathway. Every three G6P molecules that enter the pathway yield three Ru5P molecules in Stage 1. These three pentoses are then converted to one R5P and two Xu5P (Fig. 14-29, Reactions 4 and 5). The conversion of these three C_5 sugars to two C_6 sugars and one C_3 sugar involves a remarkable "juggling act" catalyzed by two enzymes, **transaldolase** and **transketolase.** These enzymes have mechanisms that involve the generation of stabilized carbanions and their addition to the electrophilic centers of aldehydes.

Transketolase Catalyzes the Transfer of C₂ Units

Transketolase, which has a thiamine pyrophosphate cofactor (TPP; Section 14-3B), catalyzes the transfer of a C_2 unit from Xu5P to R5P, yielding GAP and **sedoheptulose-7-phosphate (S7P;** Fig. 14-29, Reaction 6). The reaction intermediate is a covalent adduct between Xu5P and TPP (Fig. 14-31; *opposite*). The X-ray structure of the dimeric enzyme shows that the TPP binds in a deep cleft between the subunits so that residues from both subunits participate in its binding, just as in pyruvate decarboxylase (another TPP-requiring enzyme; Fig. 14-19). In fact, the structures are so similar that they likely diverged from a common ancestor.

Transaldolase Catalyzes the Transfer of C₃ Units

Transaldolase catalyzes the transfer of a C_3 unit from S7P to GAP yielding **erythrose-4-phosphate (E4P)** and F6P (Fig. 14-29, Reaction 7). The reaction occurs by aldol cleavage (Section 14-2D), which begins with the formation of a Schiff base between an ε-amino group of an essential Lys residue and the carbonyl group of S7P (Fig. 14-32).

A Second Transketolase Reaction Yields Glyceraldehyde-3-Phosphate and a Second Fructose-6-Phosphate Molecule

In a second transketolase reaction, a C_2 unit is transferred from a second molecule of Xu5P to E4P to form GAP and another molecule of F6P (Fig. 14-29, Reaction 8). The third stage of the pentose phosphate pathway thus transforms two molecules of Xu5P and one of R5P to two molecules

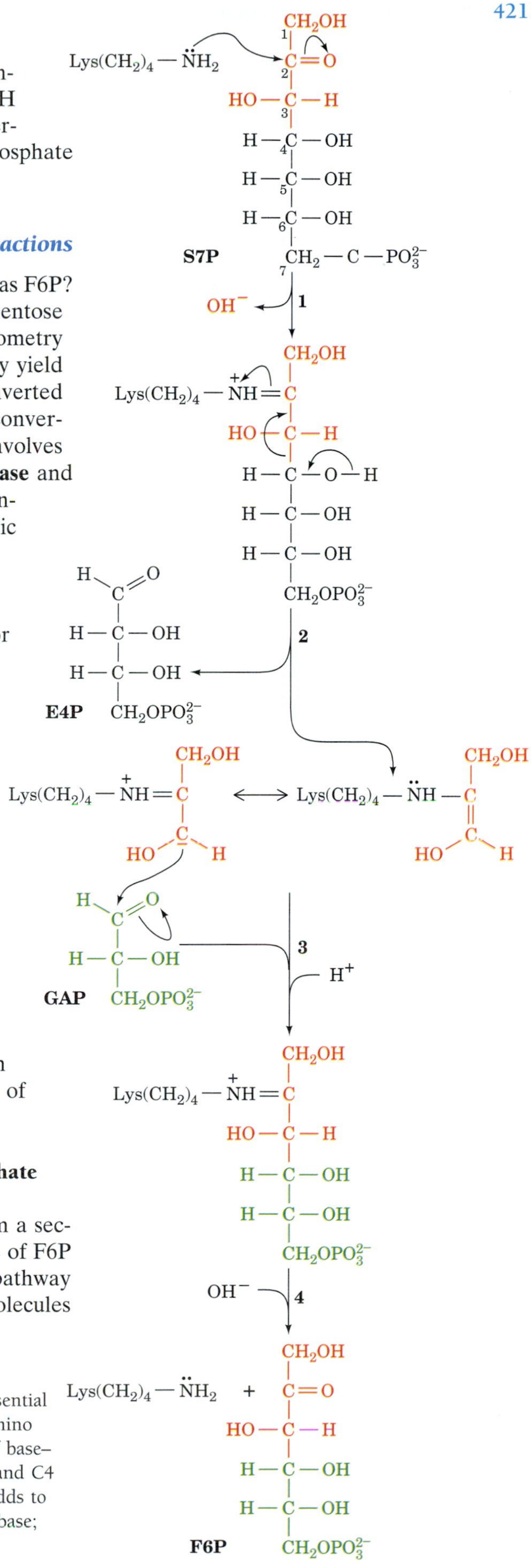

Figure 14-32.   Mechanism of transaldolase. Transaldolase contains an essential Lys residue that facilitates an aldol cleavage reaction as follows: (**1**) The ε-amino group of Lys forms a Schiff base with the carbonyl group of S7P; (**2**) a Schiff base–stabilized C3 carbanion is formed in an aldol cleavage reaction between C3 and C4 that eliminates E4P; (**3**) the enzyme-bound resonance-stabilized carbanion adds to the carbonyl C atom of GAP, forming F6P linked to the enzyme via a Schiff base; (**4**) the Schiff base hydrolyzes, regenerating active enzyme and releasing F6P.

$$(6) \quad C_5 + C_5 \rightleftharpoons C_7 + C_3$$

$$(7) \quad C_7 + C_3 \rightleftharpoons C_6 + C_4$$

$$(8) \quad C_5 + C_4 \rightleftharpoons C_6 + C_3$$

$$(\text{Sum}) \quad 3\,C_5 \rightleftharpoons 2\,C_6 + C_3$$

Figure 14-33. Summary of carbon skeleton rearrangements in the pentose phosphate pathway. A series of carbon–carbon bond formations and cleavages convert three C_5 sugars to two C_6 and one C_3 sugar. The number to the left of each reaction is keyed to the corresponding reaction in Fig. 14-29.

of F6P and one molecule of GAP. These carbon skeleton transformations (Fig. 14-29, Reactions 6–8) are summarized in Fig. 14-33.

D. Control of the Pentose Phosphate Pathway

The principal products of the pentose phosphate pathway are R5P and NADPH. The transaldolase and transketolase reactions convert excess R5P to glycolytic intermediates when the metabolic need for NADPH exceeds that of R5P in nucleotide biosynthesis. The resulting GAP and F6P can be consumed through glycolysis and oxidative phosphorylation or recycled by gluconeogenesis (Section 15-4) to form G6P.

When the need for R5P outstrips the need for NADPH, F6P and GAP can be diverted from the glycolytic pathway for use in the synthesis of R5P by reversal of the transaldolase and transketolase reactions. The relationship between glycolysis and the pentose phosphate pathway is diagrammed in Fig. 14-34.

Flux through the pentose phosphate pathway and thus the rate of NADPH production is controlled by the rate of the glucose-6-phosphate dehydrogenase reaction (Fig. 14-29, Reaction 1). The activity of this enzyme, which catalyzes the pathway's first committed step ($\Delta G = -17.6$ kJ·mol^{-1} in liver), is regulated by the NADP$^+$ concentration (i.e., regulation by substrate availability). When the cell consumes NADPH, the NADP$^+$ concentration rises, increasing the rate of the G6PD reaction and thereby stimulating NADPH regeneration. In some tissues, the amount of enzyme synthesized also appears to be under hormonal control. A deficiency in G6PD is the most common clinically significant enzyme defect of the pentose phosphate pathway (see Box 14-3).

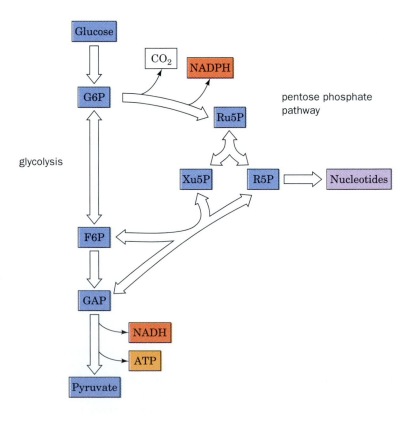

Figure 14-34. Relationship between glycolysis and the pentose phosphate pathway. The pentose phosphate pathway, which begins with G6P produced in Step 2 of glycolysis, generates NADPH for use in reductive reactions and R5P for nucleotide synthesis. Excess R5P is converted to glycolytic intermediates by a sequence of reactions that can operate in reverse to generate additional R5P, if needed.

BIOCHEMISTRY IN HEALTH AND DISEASE

Glucose-6-Phosphate Dehydrogenase Deficiency

NADPH is required for several reductive processes in addition to biosynthesis. For example, erythrocytes require a plentiful supply of reduced **glutathione (GSH)**, a Cys-containing tripeptide.

$$H_3\overset{+}{N}-CH-CH_2-CH_2-\overset{\overset{O}{\|}}{C}-NH-CH-\overset{\overset{O}{\|}}{C}-NH-CH_2-COO^-$$

with COO^- below the first CH, and CH_2 / SH below the middle CH.

Glutathione (GSH)
(γ-L-glutamyl-L-cysteinylglycine)

A major function of GSH in the erythrocyte is to reductively eliminate H_2O_2 and organic hydroperoxides, which are reactive oxygen metabolites that can irreversibly damage hemoglobin and cleave the C—C bonds in the phospholipid tails of cell membranes. The unchecked buildup of peroxides results in premature cell lysis. Peroxides are eliminated by reaction with glutathione, catalyzed by **glutathione peroxidase.**

$$2\,GSH + R-O-O-H \xrightarrow{\text{glutathione peroxidase}} GSSG + ROH + H_2O$$

Organic hydroperoxide

GSSG represents oxidized glutathione (two GSH molecules linked through a disulfide bond between their sulfhydryl groups). Reduced GSH is subsequently regenerated by the reduction of GSSG by NADPH as catalyzed by **glutathione reductase.**

$$GSSG + NADPH + H^+ \xrightarrow{\text{glutathione reductase}} 2\,GSH + NADP^+$$

A steady supply of NADPH is therefore vital for erythrocyte integrity.

The erythrocytes in individuals who are deficient in glucose-6-phosphate dehydrogenase (G6PD) are particularly sensitive to oxidative damage, although clinical symptoms may be absent. This enzyme deficiency, which is common in African, Asian, and Mediterranean populations, came to light through investigations of the hemolytic anemia that is induced in these individuals when they ingest drugs such as the antimalarial compound **primaquine**

$$NH-\overset{\overset{CH_3}{|}}{CH}-CH_2-CH_2-CH_2-NH_2$$

Primaquine

or eat **fava beans** (broad beans, *Vicia faba*), a staple Middle Eastern vegetable. Primaquine stimulates peroxide formation, thereby increasing the demand for NADPH to a level that the mutant cells cannot meet. Certain toxic glycosides present in small amounts in fava beans have the same effect, producing a condition known as **favism.**

The major reason for low enzymatic activity in affected cells appears to be an accelerated rate of breakdown of the mutant enzyme. This explains why patients with relatively mild forms of G6PD deficiency react to primaquine with hemolytic anemia but recover within a week despite continued primaquine treatment. Mature erythrocytes lack a nucleus and protein synthesizing machinery and therefore cannot synthesize new enzyme molecules to replace degraded ones (they likewise cannot synthesize new membrane components, which is why they are so sensitive to membrane damage in the first place). The initial primaquine treatments result in the lysis of old red blood cells whose defective G6PD has been largely degraded. Lysis products stimulate the release of young cells that contain more enzyme and are therefore better able to cope with primaquine stress.

It is estimated that ~400 million people are deficient in G6PD, which makes this condition the most common human enzyme deficiency. The high prevalence of defective G6PD in malarial areas of the world suggests that such mutations confer resistance to the malarial parasite, *Plasmodium falciparum.* Indeed, erythrocytes with G6PD deficiency appear to be less suitable hosts for plasmodia than normal cells. Thus, like the sickle-cell trait (Box 7-3), *a defective G6PD confers a selective advantage on individuals living where malaria is endemic.*

Curiously, however, only females who are heterozygous for this sex-linked trait are resistant to malaria. Plasmodia eventually adapt to living in G6PD-deficient erythrocytes (e.g., in males with a single X chromosome and hence a single copy of the defective gene, and in homozygous females, whose cells each contain two X chromosomes). Apparently, the parasite cannot adapt to conditions in heterozygous females, in whom roughly half the erythrocytes are G6PD deficient and the rest are normal.

The importance of NADPH in cells other than erythrocytes has been demonstrated through the development of mice in which the G6PD gene has been knocked out. All the cells in these animals are extremely sensitive to oxidative stress, even though they contain other mechanisms for eliminating reactive oxygen species.

SUMMARY

1. Glycolysis is a sequence of 10 enzyme-catalyzed reactions by which one molecule of glucose is converted to two molecules of pyruvate, with the net production of two ATP and the reduction of two NAD^+ to two NADH.

2. In the first stage of glycolysis, glucose is phosphorylated by hexokinase, isomerized by phosphoglucose isomerase (PGI), phosphorylated by phosphofructokinase (PFK), and cleaved by aldolase to yield the trioses glyceraldehyde-3-phosphate (GAP) and dihydroxyacetone phosphate (DHAP), which are interconverted by triose phosphate isomerase (TIM). These reactions consume 2 ATP per glucose.

3. In the second stage of glycolysis, GAP is oxidatively phosphorylated by glyceraldehyde-3-phosphate dehydrogenase (GAPDH), dephosphorylated by phosphoglycerate kinase (PGK) to produce ATP, isomerized by phosphoglycerate mutase (PGM), dehydrated by enolase, and dephosphorylated by pyruvate kinase to produce a second ATP and pyruvate. This stage produces 4 ATP per glucose for a net yield of 2 ATP per glucose.

4. Under anaerobic conditions, pyruvate is reduced to regenerate NAD^+ for glycolysis. In homolactic fermentation, pyruvate is reversibly reduced to lactate.

5. In alcoholic fermentation, pyruvate is decarboxylated by a thiamine pyrophosphate (TPP)-dependent mechanism, and the resulting acetaldehyde is reduced to ethanol.

6. The glycolytic reactions catalyzed by hexokinase, phosphofructokinase, and pyruvate kinase are metabolically irreversible.

7. Phosphofructokinase is the primary flux control point for glycolysis. ATP inhibition of this allosteric enzyme is relieved by AMP and ADP, whose concentrations change more dramatically than those of ATP.

8. The opposing reactions of the fructose-6-phosphate (F6P)/ fructose-1,6-bisphosphate (FBP) substrate cycle allow large changes in glycolytic flux.

9. Fructose, galactose, and mannose are enzymatically converted to glycolytic intermediates for catabolism.

10. In the pentose phosphate pathway, glucose-6-phosphate (G6P) is oxidized and decarboxylated to produce two NADPH, CO_2, and ribulose-5-phosphate (Ru5P).

11. Depending on the cell's needs, ribulose-5-phosphate may be isomerized to ribose-5-phosphate (R5P) for nucleotide synthesis or converted, via ribose-5-phosphate and xylulose-5-phosphate (Xu5P), to fructose-6-phosphate and glyceraldehyde-3-phosphate, which can re-enter the glycolytic pathway.

REFERENCES

Beutler, E., G6PDH Deficiency, *Blood* **84,** 3613–3636 (1994).

Fell, D.A., Metabolic control analysis: a survey of its theoretical and experimental development, *Biochem. J.* **286,** 313–330 (1992).

Gefflaut, T., Blonski, C., Perie, J., and Wilson, M., Class I aldolases: substrate specificity, mechanism, inhibitors and structural aspects, *Prog. Biophys. Molec. Biol.* **63,** 301–340 (1995).

Goldsmith, E.J. and Cobb, M.H., Protein kinases, *Curr. Opin. Struct. Biol.* **4,** 833–840 (1994).

Knowles, J.R., Enzyme catalysis: not different, just better, *Nature* **350,** 121–124 (1991). [A lucid discussion of the triose phosphate isomerase mechanism.]

Luzzato, L. and Mehta, A., Glucose-6-phosphate dehydrogenase deficiency, *in* Scriver, C.R., Beaudet, A.L., Sly, W.S., Valle, D., Stanbury, J.B., Wyngaarden, J.B., and Frederickson, D.S. (Eds.), *The Metabolic and Molecular Bases of Inherited Disease* (7th ed.), pp. 3367–3398, McGraw–Hill (1995).

Mattevi, A., Valentini, G., Rizzi, M., Speranza, M.L., Bolognesi, M., and Coda, A., Crystal structure of *Escherichia coli* pyruvate kinase type I: molecular basis of allosteric transition, *Structure* **3,** 729–741 (1995).

Muirhead, H. and Watson, H., Glycolytic enzymes; from hexose to pyruvate, *Curr. Opin. Struct. Biol.* **2,** 870–876 (1992). [A brief summary of the structures of glycolytic enzymes.]

Schirmer, T. and Evans, P.R., Structural basis of the allosteric behaviour of phosphofructokinase, *Nature* **343,** 140–145 (1990).

Wood, T., *The Pentose Phosphate Pathway*, Academic Press (1985).

KEY TERMS

glycolysis	enediol intermediate	alcoholic fermentation	Pasteur effect
pentose phosphate pathway	catalytic perfection	homolactic fermentation	futile cycle
aldol cleavage	mutase	TPP	

STUDY EXERCISES

1. Write the reactions of glycolysis, showing the structural formulas of the intermediates and the names of the enzymes that catalyze the reactions.

2. Describe the three possible fates of pyruvate.

3. Describe the mechanisms that regulate phosphofructokinase activity.

4. What is the metabolic advantage of a substrate cycle?

5. Describe how fructose, galactose, and mannose enter the gly-colytic pathway.

6. Outline the reactions of the pentose phosphate pathway.

7. How does flux through the pentose phosphate pathway change in response to the need for NADPH or ribose-5-phosphate?

PROBLEMS

1. The aldolase reaction can proceed in reverse as an enzymatic aldol condensation. If the enzyme were not stereospecific, how many different products would be obtained?

2. Arsenate (AsO_4^{3-}), a structural analog of phosphate, can act as a substrate for any reaction in which phosphate is a substrate. Arsenate esters, unlike phosphate esters, are kinetically as well as thermodynamically unstable and hydrolyze almost instantaneously. Write a balanced overall equation for the conversion of glucose to pyruvate in the presence of ATP, ADP, NAD^+, and either (a) phosphate or (b) arsenate. (c) Why is arsenate a poison?

3. Draw the enediolate intermediates of the ribulose-5-phosphate isomerase reaction (Ru5P → R5P) and the ribulose-5-phosphate epimerase reaction (Ru5P → Xu5P).

4. (a) Why is it possible for the ΔG values in Table 14-1 to differ from the $\Delta G^{\circ\prime}$ values? (b) If a reaction has a $\Delta G^{\circ\prime}$ value of at least -30.5 kJ·mol^{-1}, sufficient to drive the synthesis of ATP ($\Delta G^{\circ\prime} = 30.5$ kJ·mol^{-1}), can it still drive the synthesis of ATP *in vivo* when its ΔG is only -10 kJ·mol^{-1}? Explain.

5. $\Delta G^{\circ\prime}$ for the aldolase reaction is 22.8 kJ·mol^{-1}. In the cell at 37°C, [DHAP]/[GAP] = 5.5. Calculate the equilibrium ratio of [FBP]/[GAP] when [GAP] = 10^{-4} M.

6. The half-reactions involved in the lactate dehydrogenase reaction and their standard reduction potentials are

Pyruvate + 2 H$^+$ + 2 e^- $\longrightarrow$ lactate $\mathscr{E}^{\circ\prime} = -0.185$ V

NAD$^+$ + 2 H$^+$ + 2 e^- $\longrightarrow$ NADH + H$^+$ $\mathscr{E}^{\circ\prime} = -0.315$ V

Calculate ΔG at pH 7.0 for the reaction under the following conditions:

(a) [lactate]/[pyruvate] = 1 and [NAD$^+$]/[NADH] = 1
(b) [lactate]/[pyruvate] = 160 and [NAD$^+$]/[NADH] = 160
(c) [lactate]/[pyruvate] = 1000 and [NAD$^+$]/[NADH] = 1000

7. Since the PFK reaction is the primary regulatory point of glycolysis, describe the metabolic importance of also regulating flux through the pyruvate kinase reaction.

8. Compare the ATP yield of three glucose molecules that enter glycolysis and are converted to pyruvate with that of three glucose molecules that proceed through the pentose phosphate pathway such that their carbon skeletons (as two F6P and one GAP) re-enter glycolysis and are metabolized to pyruvate.

9. If G6P is labeled at its C2 position, where will the label appear in the products of the pentose phosphate pathway?

10. Explain why some tissues continue to respire (produce CO_2) in the presence of high concentrations of fluoride ion, which inhibits glycolysis.

11. The catalytic behavior of liver and brain phosphofructokinase-1 (PFK-1) was observed in the presence of AMP, phosphate, and fructose-2,6-bisphosphate. The following table lists the concentrations of each effector required to achieve 50% of the maximal velocity. Compare the response of the two isozymes to the three effectors and discuss the possible implications of their different responses.

PFK-1 isozyme	Phosphate	AMP	F2,6P
Liver	200 μM	10 μM	0.05 μM
Brain	350 μM	75 μM	4.5 μM

[Problem provided by Kathleen Cornely, Providence College.]

CHAPTER 15

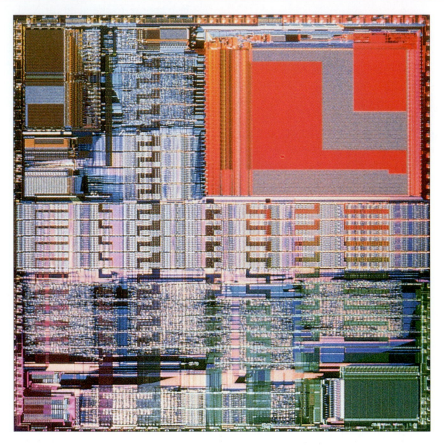

Cells are capable of carrying out a multitude of metabolic processes and, like computer chips, do so in a highly organized way in a small space. How are metabolic reactions such as glycogen metabolism and gluconeogenesis regulated to ensure net synthesis or degradation of particular molecules according to cellular needs?
[Courtesy of Thomas A. Way, I.B.M. Corporation.]

GLYCOGEN METABOLISM AND GLUCONEOGENESIS

Glycogen (in animals, fungi, and bacteria) and starch (in plants) function to stockpile glucose for later metabolic use. In animals, a constant supply of glucose is essential for tissues such as the brain and red blood cells, which depend almost entirely on glucose as an energy source (other tissues can also oxidize fatty acids for energy; Section 19-2). The mobilization of glucose from glycogen stores, primarily in the liver, provides a constant supply of glucose (~5 mM in blood) to all tissues. When glucose is plentiful, such as immediately after a meal, glycogen synthesis accelerates. Yet the liver's capacity to store glycogen is only sufficient to supply the brain with glucose for about half a day. Under fasting conditions, most of the body's glucose needs are met by **gluconeogenesis** (literally, new glucose synthesis) from noncarbohydrate precursors such as amino acids. Not surprisingly, the regulation of glucose synthesis, storage, mobilization, and catabolism by glycolysis (Section 14-2) or the pentose phosphate pathway (Section 14-6) is elaborate and sensitive to the immediate and long-term energy needs of the organism.

The importance of glycogen for glucose storage is plainly illustrated by the effects of deficiencies of the enzymes that release stored glucose. **McArdle's disease,** for example, is an inherited condition whose major symptom is painful muscle cramps on exertion. The muscles in afflicted individuals lack the enzyme required for glycogen breakdown to yield glucose. Although glycogen is synthesized normally, it cannot supply fuel for glycolysis to keep up with the demand for ATP.

Figure 15-1 summarizes the metabolic uses of glucose. Glucose-6-phosphate (G6P), a key branch point, is derived from free glucose through the action of hexokinase (Section 14-2A) or is the product of glycogen breakdown or gluconeogenesis. G6P has several possible fates: It can be used to synthesize glycogen; it can be catabolized via glycolysis to yield ATP and carbon atoms (as acetyl-CoA) that are further oxidized by the citric acid cycle; and it can be shunted through the pentose phosphate pathway to generate NADPH and/or ribose-5-phosphate. In the liver, G6P can be converted to glucose for export to other tissues via the bloodstream.

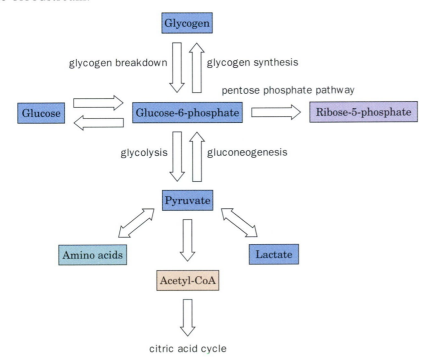

Figure 15-1.   Overview of glucose metabolism. Glucose-6-phosphate (G6P) is produced by the phosphorylation of free glucose, by glycogen degradation, and by gluconeogenesis. It is also a precursor for glycogen synthesis and the pentose phosphate pathway. The liver can hydrolyze G6P to glucose. Glucose is metabolized by glycolysis to pyruvate, which can be further broken down to acetyl-CoA for oxidation by the citric acid cycle. Lactate and amino acids, which are reversibly converted to pyruvate, are precursors for gluconeogenesis. ✳ **See the Animated Figures.**

(a)

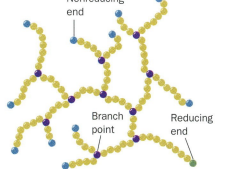

α(1 ⟶ 6) linkage

Reducing
end

Nonreducing
ends

Branch
point

α(1 ⟶ 4)
linkage

(b)

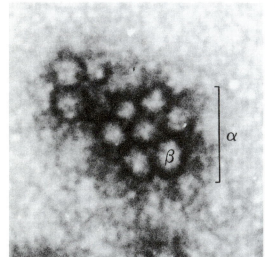

Nonreducing
end

Branch
point

Reducing
end

(c)

Figure 15-2. The structure of glycogen.
(a) Molecular formula. In the actual molecule, there are ~12 residues per chain. (b) Schematic diagram of glycogen's branched structure. Note that the molecule has many nonreducing ends but only one reducing end. (c) Electron micrograph of a glycogen granule from rat skeletal muscle. Each granule (labeled α) consists of several spherical glycogen molecules (β) and associated proteins. [From Calder, P.C., *Int. J. Biochem.* **23**, 1339 (1991). Copyright Elsevier Science. Used with permission.]

The opposing processes of glycogen synthesis and degradation, and of glycolysis and gluconeogenesis, are reciprocally regulated; that is, one is largely turned on while the other is largely turned off. In this chapter, we examine the enzymatic steps of glycogen metabolism and gluconeogenesis, paying particular attention to the regulatory mechanisms that ensure efficient operation of opposing metabolic pathways.

1. GLYCOGEN BREAKDOWN

Glycogen is a polymer of α(1→4)-linked D-glucose with α(1→6)-linked branches every 8–14 residues (Fig. 15-2a,b and Section 8-2C). Glycogen occurs as intracellular granules of 100- to 400-Å-diameter spheroidal molecules that each contain up to 120,000 glucose units (Fig. 15-2c). The granules are especially prominent in the cells that make the greatest use of glycogen: muscle (up to 1–2% glycogen by weight) and liver cells (up to 10% glycogen by weight; Fig. 8-11). Glycogen granules also contain the enzymes that catalyze glycogen synthesis and degradation as well as many of the proteins that regulate these processes.

Glucose units are mobilized by their sequential removal from the nonreducing ends of glycogen (ends lacking a C1 —OH group). Whereas glycogen has only one reducing end, there is a nonreducing end on every branch. *Glycogen's highly branched structure therefore permits rapid glucose mobilization through the simultaneous release of the glucose units at the end of every branch.*

Glycogen breakdown, or **glycogenolysis,** requires three enzymes:

1. **Glycogen phosphorylase** (or simply **phosphorylase**) catalyzes glycogen **phosphorolysis** (bond cleavage by the substitution of a phosphate group) to yield **glucose-1-phosphate (G1P).**

$$\begin{array}{ccc} \text{Glycogen} + P_i & \rightleftharpoons & \text{glycogen} + \text{G1P} \\ (n \text{ residues}) & & (n-1 \text{ residues}) \end{array}$$

This enzyme releases a glucose unit only if it is at least five units away from a branch point.

2. **Glycogen debranching enzyme** removes glycogen's branches, thereby making additional glucose residues accessible to glycogen phosphorylase.

3. Phosphoglucomutase converts G1P to G6P, which can have several metabolic fates (Fig. 15-1).

A. Glycogen Phosphorylase

Glycogen phosphorylase is a dimer of identical 842-residue (97-kD) subunits that catalyzes the rate-controlling step in glycogen breakdown. It is regulated both by allosteric interactions and by **covalent modification** (phosphorylation and dephosphorylation). The phosphorylated form of the enzyme, **phosphorylase *a*,** has a phosphoryl group esterified to Ser 14. The dephospho form is called **phosphorylase *b*.** Phosphorylase's allosteric inhibitors (ATP, G6P, and glucose) and its allosteric activator (AMP) interact differently with the phospho- and dephosphoenzymes, resulting in an extremely sensitive regulation process (Section 15-3).

The high-resolution X-ray structures of phosphorylase *a* and phosphorylase *b*, respectively determined by Robert Fletterick and Louise Johnson, are similar. Both structures have a large N-terminal domain (484 residues; the largest known domain) and a smaller C-terminal domain (Fig. 15-3). The N-terminal domain includes the phosphorylation site (Ser 14), the allosteric effector site, a glycogen-binding site (called the glycogen storage site), and all the intersubunit contacts in the dimer. The catalytic site is located at the center of the subunit.

An ~30-Å-long crevice on the surface of the phosphorylase monomer connects the glycogen storage site to the active site. *Since this crevice can accommodate four or five sugar residues in a chain but is too narrow to admit branched oligosaccharides, it provides a clear physical rationale for the inability of phosphorylase to cleave glycosyl residues closer than five units from a branch point.* Presumably, the glycogen storage site increases the catalytic efficiency of phosphorylase by permitting it to phosphorylyze many glucose residues on the same glycogen particle without having to dissociate and reassociate completely between catalytic cycles.

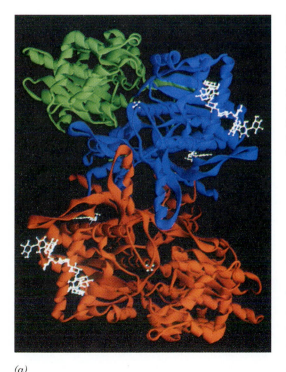

(a)

Figure 15-3.   The X-ray structure of rabbit muscle glycogen phosphorylase.
(*a*) Ribbon diagram of the phosphorylase *a* dimer viewed along its molecular twofold axis of symmetry. The bottom subunit is colored orange, and the top subunit's N-terminal and C-terminal domains are colored blue and green, respectively. The various bound ligands are white: The phosphate group at the center of each subunit marks the enzyme's catalytic site; maltoheptose (a glucose heptamer) is bound at each glycogen storage site; and the AMPs at the "back" of the protein identify the allosteric effector sites. [Courtesy of Stephen Sprang, University of Texas Southwestern Medical Center.] (*b*) An interpretive drawing of the structure in *a* showing the enzyme's various ligand-binding sites. ● **See Kinemage Exercise 14-1.**

(b)

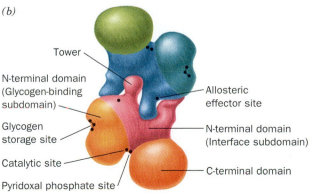

Tower

N-terminal domain (Glycogen-binding subdomain)

Glycogen storage site

Catalytic site

Pyridoxal phosphate site

Allosteric effector site

N-terminal domain (Interface subdomain)

C-terminal domain

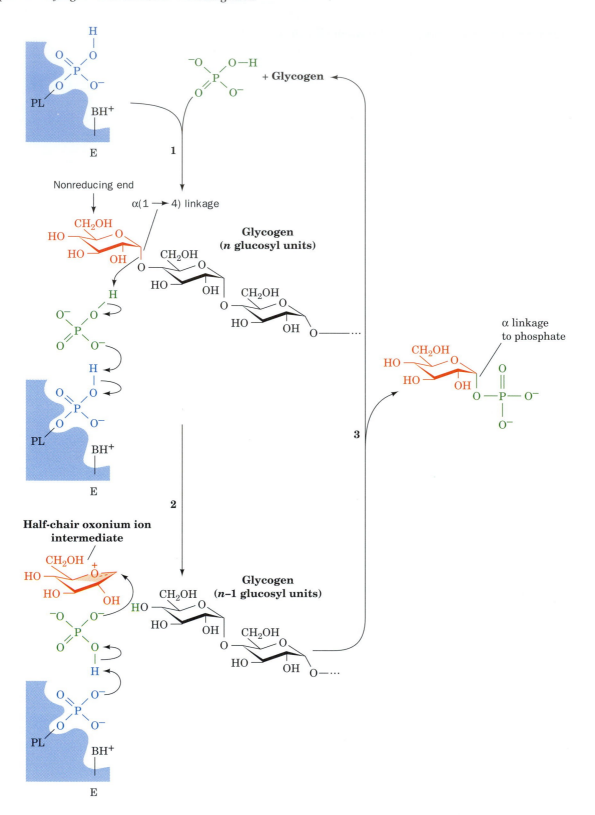

Figure 15-4. The reaction mechanism of glycogen phosphorylase. PL is an enzyme-bound pyridoxal group; BH^+ is a positively charged amino acid side chain, probably that of Lys 568, necessary for maintaining PLP electrical neutrality. (**1**) Formation of an $E \cdot P_i \cdot$ glycogen ternary complex. (**2**) Shielded oxonium ion intermediate formation from the α-linked terminal glucosyl residue involving acid catalysis by P_i as facilitated by proton transfer from PLP. The oxonium ion has the half-chair conformation. (**3**) Reaction of P_i with the oxonium ion with overall retention of configuration about C1 to form α-D-glucose-1-phosphate. The glycogen, which has one less residue than before, cycles back to Step 1.

Phosphorylase binds the cofactor **pyridoxal-5′-phosphate (PLP;** *at right*), which it requires for activity. This prosthetic group, a **vitamin B₆** deriva-tive, is covalently linked to the enzyme via a Schiff base (imine) formed between its aldehyde group and the ε-amino group of Lys 680. PLP also occurs in a variety of enzymes involved in amino acid metabolism, where PLP's conjugated ring system functions catalytically to delocalize electrons (Sections 20-2A and 20-4A). In phosphorylase, however, only the phosphate group participates in catalysis, where it acts as a general acid–base catalyst. Phosphorolysis of glycogen proceeds by a Random mechanism (Section 12-1D) involving an enzyme · P_i · glycogen ternary complex. An oxonium ion intermediate forms during C1—O1 bond cleavage, as in the reaction cat-alyzed by lysozyme (Section 11-4B). The reaction mechanism is dia-grammed in Fig. 15-4 *(opposite)*, which shows the participation of PLP's phosphate group as a general acid–base catalyst.

Pyridoxal-5′-phosphate (PLP)

Conformational Changes in Glycogen Phosphorylase

The structural differences between the active (R) and inactive (T) con-formations of phosphorylase (Fig. 15-5) are fairly well understood in terms of the symmetry model of allosterism (Section 7-2E). The T-state enzyme has a buried active site and hence a low affinity for its substrates, whereas the R-state enzyme has an accessible catalytic site and a high-affinity phos-phate-binding site.

AMP promotes phosphorylase's T (*inactive*) → R (*active*) conformational shift by binding to the R state of the enzyme at its allosteric effector site. In

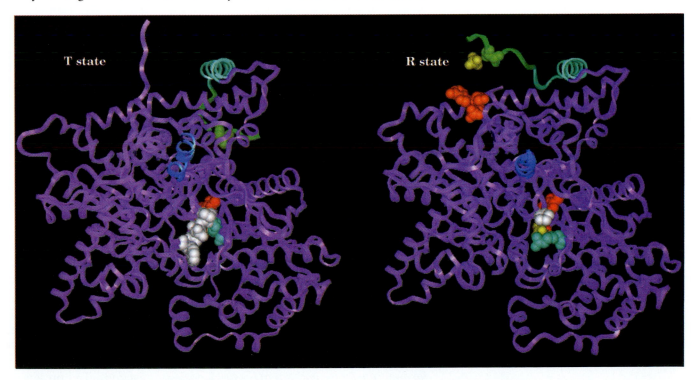

Figure 15-5. Conformational changes in glycogen phosphory-lase. One subunit of the dimeric phosphorylase *b* is shown (*left*) in the T state in the absence of allosteric effectors and (*right*) in the R state with bound AMP. The view is of the lower (*orange*) subunit in Fig. 15-3 as seen from the top of the page. The tower helix is blue, the N-terminal helix is cyan, and the N-terminal residues that change conformation on AMP binding are green. Of the groups that are shown in space-filling representation, Ser 14, the phosphorylation site, is light green; AMP is orange; the active site PLP is red; the Arg 569 side chain, which reorients in the T → R transition so as to interact with the substrate phosphate, is cyan; loop residues 282 to 284, which in the R state are mostly disordered and hence not seen, are white; and the phosphates, both at the active site and at the R state Ser 14 phosphorylation site (not present in phosphorylase *b* but shown for position), are yellow. ● See Kinemage Exercises 14-2 and 14-3.

doing so, AMP's adenine, ribose, and phosphate groups bind to separate segments of the polypeptide chain to link the active site, the subunit interface, and the N-terminal region, the latter having undergone a large conformational shift (involving a 36 Å movement of Ser 14) from its position in the T-state enzyme. AMP binding also causes glycogen phosphorylase's tower helices (Figs. 15-3 and 15-5) to tilt and pull apart so as to pack more favorably. These tertiary movements trigger a concerted T → R transition, which largely consists of an ~10° relative rotation of the two subunits.

The movement of the tower helices displaces and disorders a loop that covers the T-state active site so as to prevent substrate access. Tower movement also causes the Arg 569 side chain, which is located in the active site near the PLP and the P_i-binding site, to rotate in a way that increases the enzyme's binding affinity for its anionic P_i substrate (Fig. 15-5).

ATP also binds to the allosteric effector site, but in the T state, so that it inhibits rather than promotes the T → R conformational shift. This is because the β- and γ-phosphate groups of ATP prevent the proper alignment of its ribose and α-phosphate groups that result in the conformational changes elicited by AMP.

Phosphorylation and dephosphorylation can alter enzymatic activity in a manner reminiscent of allosteric regulation. The phosphate group has a double negative charge (a property not shared by naturally occurring amino acid residues) and its covalent attachment to a protein can induce dramatic conformational changes. In phosphorylase, the phosphorylation of Ser 14 causes tertiary and quaternary changes as the N-terminal segment moves to allow the phospho-Ser to ion pair with two cationic Arg residues. *The presence of the Ser 14–phosphoryl group causes similar conformational changes as does the binding of AMP, thereby shifting the enzyme's T ⇌ R equilibrium in favor of the R state.* This accounts for the observation that phosphorylase *b* requires AMP for activity and that the *a* form is active without AMP. We shall return to the regulation of phosphorylase activity when we discuss the mechanisms that balance glycogen synthesis against glycogen degradation (Section 15-3).

B. Glycogen Debranching Enzyme

Phosphorolysis proceeds along a glycogen branch until it approaches to within four or five residues of an α(1→6) branch point, leaving a "limit branch." Glycogen debranching enzyme acts as an **α(1→4) transglycosylase** (glycosyltransferase) by transferring an α(1→4)-linked trisaccharide unit from a limit branch of glycogen to the nonreducing end of another branch (Fig. 15-6). This reaction forms a new α(1→4) linkage with three more units available for phosphorylase-catalyzed phosphorolysis. The α(1→6) bond linking the remaining glycosyl residue in the branch to the main chain is hydrolyzed (not phosphorylyzed) by the same debranching enzyme to yield glucose and debranched glycogen. About 10% of the residues in glycogen (those at the branch points) are therefore converted to glucose rather than G1P. *Debranching enzyme has separate active sites for the transferase and the α(1→6)-glucosidase reactions.* The presence of two independent catalytic activities on the same enzyme no doubt improves the efficiency of the debranching process.

The maximal rate of the glycogen phosphorylase reaction is much greater than that of the glycogen debranching reaction. Consequently, the outermost branches of glycogen, which constitute nearly half of its residues, are degraded in muscle in a few seconds under conditions of high metabolic demand. Glycogen degradation beyond this point requires debranching and hence occurs more slowly. This, in part, accounts for the fact that a muscle can sustain its maximum exertion for only a few seconds.

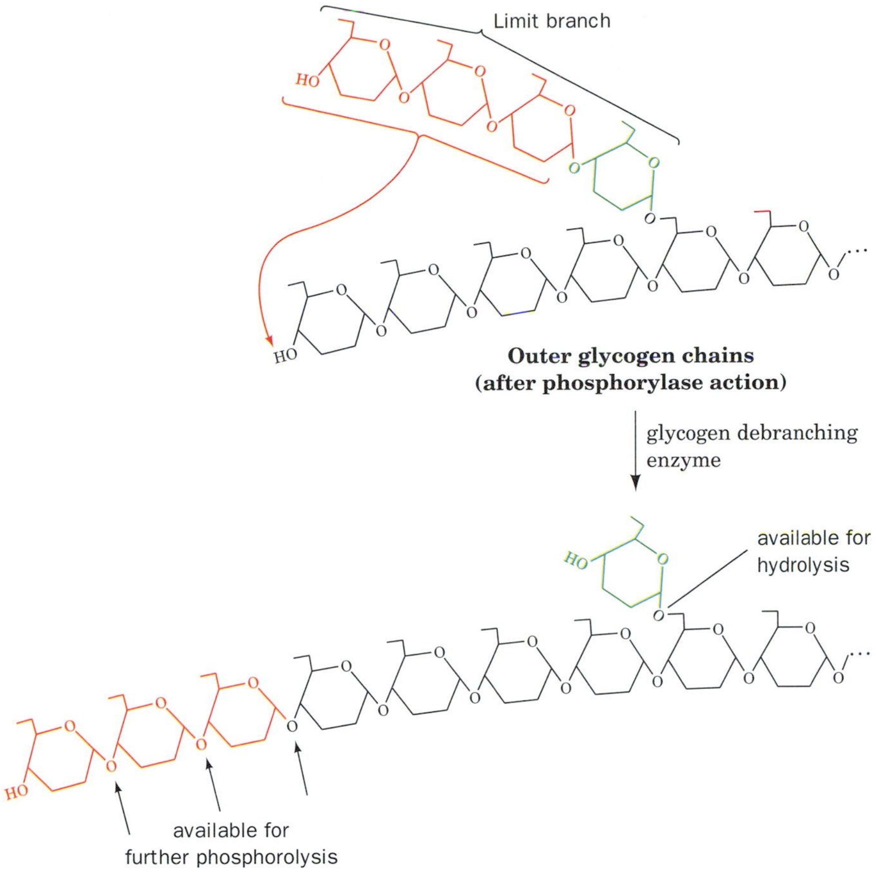

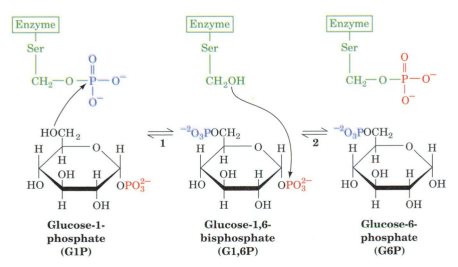

Figure 15-6. The reactions catalyzed by debranching enzyme. The enzyme transfers the terminal three α(1→4)-linked glucose residues from a "limit branch" of glycogen to the nonreducing end of another branch. The α(1→6) bond of the residue remaining at the branch point is hydrolyzed by further action of debranching enzyme to yield free glucose. The newly elongated branch is subject to degradation by glycogen phosphorylase.

C. Phosphoglucomutase

Phosphorylase converts the glucosyl units of glycogen to G1P, which, in turn, is converted by phosphoglucomutase to G6P. The phosphoglucomutase reaction is similar to that catalyzed by phosphoglycerate mutase (Section 14-2H). A phosphoryl group is transferred from the active phosphoenzyme to G1P, forming **glucose-1,6-bisphosphate (G1,6P),** which then rephosphorylates the enzyme to yield G6P (Fig. 15-7; this near-equilibrium reaction also functions in reverse). An important difference between this enzyme and phosphoglycerate mutase is that the phosphoryl group in phos-

Figure 15-7. The mechanism of phosphoglucomutase. (1) The OH group at C6 of G1P attacks the phosphoenzyme to form a dephosphoenzyme–G1,6P intermediate. (2) The Ser OH group on the dephosphoenzyme attacks the phosphoryl group at C1 to regenerate the phosphoenzyme with the formation of G6P.

phoglucomutase is covalently bound to a Ser hydroxyl group rather than to a His imidazole nitrogen.

Glucose-6-Phosphatase Generates Glucose in the Liver

The G6P produced by glycogen breakdown can continue along the glycolytic pathway or the pentose phosphate pathway (note that the glucose is already phosphorylated, so that the ATP-consuming hexokinase-catalyzed phosphorylation of glucose is bypassed). In the liver, G6P is also made available for use by other tissues. Because G6P cannot pass through the cell membrane, it is first hydrolyzed by **glucose-6-phosphatase:**

$$G6P + H_2O \longrightarrow glucose + P_i$$

The resulting glucose leaves the cell and is carried by the blood to other tissues. Muscle and other tissues lack glucose-6-phosphatase and therefore retain their G6P.

2. GLYCOGEN SYNTHESIS

The $\Delta G^{\circ\prime}$ for the glycogen phosphorylase reaction is $+3.1$ kJ·mol^{-1}, but under physiological conditions, glycogen breakdown is exergonic ($\Delta G = -5$ to -8 kJ·mol^{-1}). The synthesis of glycogen from G1P under physiological conditions is therefore thermodynamically unfavorable without free energy input. Consequently, *glycogen synthesis and breakdown must occur by separate pathways.* This recurrent metabolic strategy—that biosynthetic and degradative pathways of metabolism are different—is particularly important when both pathways must operate under similar physiological conditions. This situation is thermodynamically impossible if one pathway is just the reverse of the other.

It was not thermodynamics, however, that led to recognition of the separation of synthetic and degradative pathways for glycogen, but McArdle's disease. Individuals with the disease lack muscle glycogen phosphorylase activity and therefore cannot break down glycogen. Yet their muscles contain moderately high quantities of normal glycogen. Clearly, glycogen synthesis does not require glycogen phosphorylase. In this section, we describe the three enzymes that participate in glycogen synthesis: **UDP–glucose pyrophosphorylase, glycogen synthase,** and **glycogen branching enzyme.** The opposing reactions of glycogen synthesis and degradation are diagrammed in Fig. 15-8.

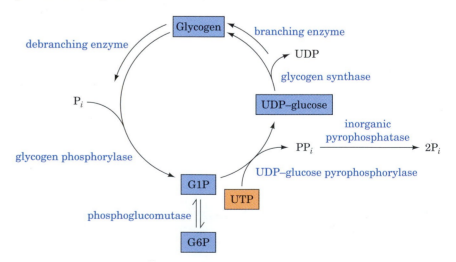

Figure 15-8. **Opposing pathways of glycogen synthesis and degradation.** The exergonic process of glycogen breakdown is reversed by a process that uses UTP to generate a UDP–glucose intermediate.

A. UDP–Glucose Pyrophosphorylase

Since the direct conversion of G1P to glycogen and P_i is thermodynamically unfavorable (positive ΔG) under physiological conditions, glycogen biosynthesis requires an exergonic step. This is accomplished, as Luis Leloir discovered in 1957, by combining G1P with uridine triphosphate (UTP) in a reaction catalyzed by UDP–glucose pyrophosphorylase (Fig. 15-9). The product of this reaction, **uridine diphosphate glucose (UDP–glucose or UDPG),** is an "activated" compound that can donate a glucosyl unit to the growing glycogen chain. The formation of UDPG itself has $\Delta G^{\circ\prime} \approx 0$ (it is a phosphoanhydride exchange reaction), but the subsequent exergonic hydrolysis of PP_i by the omnipresent enzyme inorganic pyrophosphatase makes the overall reaction exergonic:

$$
\begin{array}{lr}
 & \Delta G^{\circ\prime}\ (\text{kJ} \cdot \text{mol}^{-1}) \\
\text{G1P} + \text{UTP} \rightleftharpoons \text{UDPG} + PP_i & \sim 0 \\
\text{H}_2\text{O} + PP_i \longrightarrow 2\ P_i & -33.5 \\
\hline
\text{Overall} \quad \text{G1P} + \text{UTP} + \text{H}_2\text{O} \longrightarrow \text{UDPG} + 2\ P_i & -33.5
\end{array}
$$

This is an example of the common biosynthetic strategy of cleaving a nucleoside triphosphate to form PP_i. The free energy of PP_i hydrolysis can

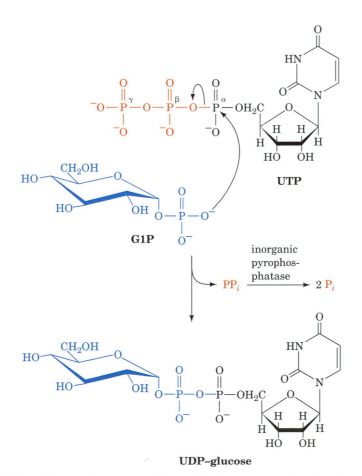

Figure 15-9. The reaction catalyzed by UDP–glucose pyrophosphorylase. In this phosphoanhydride exchange reaction, the phosphoryl oxygen of G1P attacks the α-phosphorus atom of UTP to form UDP–glucose and PP_i. The PP_i is rapidly hydrolyzed by inorganic pyrophosphatase.

then be used to drive an otherwise unfavorable reaction to completion (Section 13-2B).

B. Glycogen Synthase

In the next step of glycogen synthesis, the glycogen synthase reaction, the glucosyl unit of UDPG is transferred to the C4-OH group on one of glycogen's nonreducing ends to form an $\alpha(1\rightarrow4)$ glycosidic bond. The $\Delta G^{\circ\prime}$ for the glycogen synthase reaction

$$\begin{array}{ccc} \text{UDPG} + \text{glycogen} & \longrightarrow & \text{UDP} + \text{glycogen} \\ (n \text{ residues}) & & (n + 1 \text{ residues}) \end{array}$$

is -13.4 kJ $\cdot$ mol^{-1}, making the overall reaction spontaneous under the same conditions that glycogen breakdown by glycogen phosphorylase is also spontaneous. However, glycogen synthesis does have an energetic price. Combining the first two reactions of glycogen synthesis gives

$$\begin{array}{ccc} \text{Glycogen} + \text{G1P} + \text{UTP} & \longrightarrow & \text{glycogen} + \text{UDP} + 2\,\text{P}_i \\ (n \text{ residues}) & & (n + 1 \text{ residues}) \end{array}$$

Thus, *one molecule of UTP is cleaved to UDP for each glucose residue incorporated into glycogen.* The UTP is replenished through a phosphoryl-transfer reaction mediated by nucleoside diphosphate kinase (Section 13-2C):

$$\text{UDP} + \text{ATP} \rightleftharpoons \text{UTP} + \text{ADP}$$

so that UTP consumption is energetically equivalent to ATP consumption.

The transfer of a glucosyl unit from UDPG to a growing glycogen chain involves the formation of a glycosyl oxonium ion by the elimination of UDP, a good leaving group (Fig. 15-10). The enzyme is inhibited by **1,5-gluconolactone,**

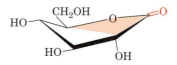

1,5-Gluconolactone

an analog that mimics the oxonium ion's half-chair geometry. The same analog inhibits both glycogen phosphorylase (Section 15-1A) and lysozyme (Section 11-4B), which have similar mechanisms.

Human muscle glycogen synthase is a homotetramer of 737-residue subunits (the liver isozyme has 703-residue subunits). Like glycogen phosphorylase, it has two enzymatically interconvertible forms; in this case, however, the phosphorylated *b* form is less active, and the original (dephosphorylated) *a* form is more active. (Note: For enzymes subject to covalent modification, "*a*" refers to the more active form and "*b*" refers to the less active form.)

Glycogen synthase is under allosteric control; it is strongly inhibited by physiological concentrations of ATP, ADP, and P$_i$. In fact, the phosphorylated enzyme is almost totally inactive *in vivo*. The dephosphorylated enzyme, however, can be activated by G6P, so the cell's glycogen synthase activity varies with [G6P] and the fraction of the enzyme in its dephosphorylated form. The mechanistic details of the interconversion of phosphorylated and dephosphorylated forms of glycogen synthase are complex and are not as well understood as those of glycogen phosphory-

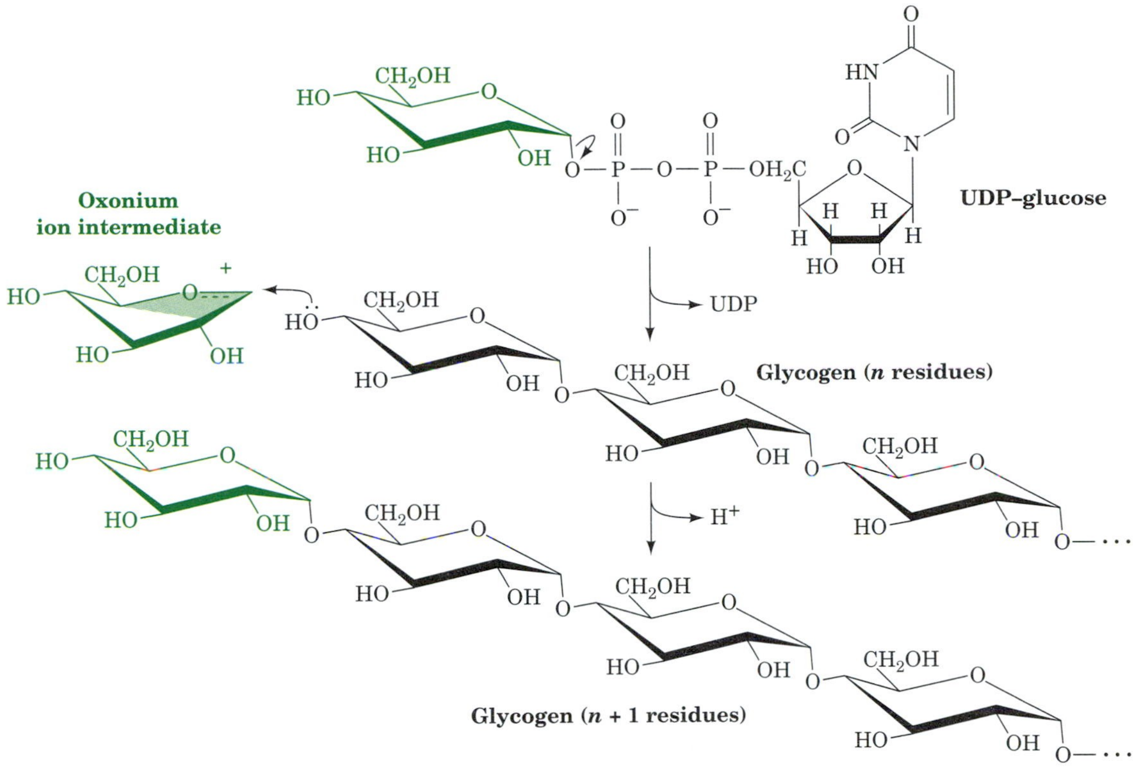

Figure 15-10. The reaction catalyzed by glycogen synthase. This reaction involves a glucosyl oxonium ion intermediate.

lase (for one thing, glycogen synthase has multiple phosphorylation sites). We shall discuss the regulation of glycogen synthase further in Section 15-3B.

Glycogenin Primes Glycogen Synthesis

Glycogen synthase cannot simply link together two glucose residues; it can only extend an already existing $\alpha(1\rightarrow4)$-linked glucan chain. How, then, is glycogen synthesis initiated? In the first step of glycogen synthesis, a **tyrosine glucosyltransferase** attaches a glucose residue to the Tyr 194 OH group of a 37-kD protein named **glycogenin.** Glycogenin then autocatalytically extends the glucan chain by up to seven additional residues donated by UDPG, forming a glycogen "primer." Only at this point does glycogen synthase commence glycogen synthesis by extending the primer. Analysis of glycogen granules suggests that each glycogen molecule is associated with only one molecule each of glycogenin and glycogen synthase.

C. Glycogen Branching Enzyme

Glycogen synthase generates only $\alpha(1\rightarrow4)$ linkages to yield α-amylose. Branching to form glycogen is accomplished by a separate enzyme, **amylo-(1,4$\rightarrow$1,6)-transglycosylase (branching enzyme),** which is distinct from glycogen debranching enzyme (Section 15-1B). A branch is created by transferring a 7-residue segment from the end of a chain to the C6-OH group of a glucose residue on the same or another glycogen chain (Fig.

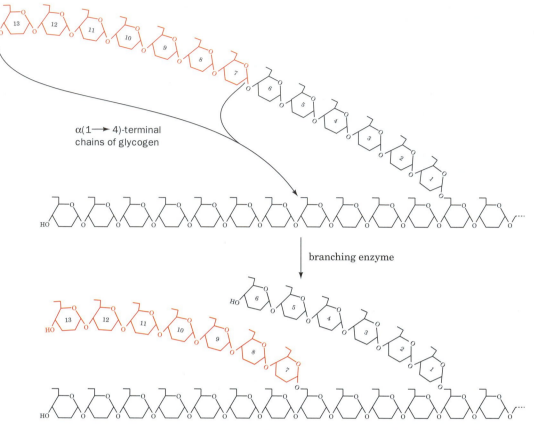

α(1→4)-terminal
chains of glycogen

branching enzyme

Figure 15-11. **The branching of glycogen.** Branches are formed by transferring a seven-residue terminal segment from an α(1→4)-linked glucan chain to the C6-OH group of a glucose residue on the same or another chain.

15-11). Each transferred segment must come from a chain of at least 11 residues, and the new branch point must be at least 4 residues away from other branch points. The branching pattern of glycogen has been optimized by evolution for the efficient storage and mobilization of glucose (see Box 15-1).

3. CONTROL OF GLYCOGEN METABOLISM

If glycogen synthesis and breakdown proceed simultaneously, all that is achieved is the wasteful hydrolysis of UTP. Glycogen metabolism must therefore be controlled according to cellular needs. *The regulation of glycogen metabolism involves allosteric control as well as hormonal control by covalent modification of the pathway's regulatory enzymes.*

A. Direct Allosteric Control of Glycogen Phosphorylase and Glycogen Synthase

As we saw in Sections 13-1D and 14-4B, the net flux, J, of reactants through a step in a metabolic pathway is the difference between the forward and reverse reaction velocities, v_f and v_r. However, the flux varies dramatically with substrate concentration as the reaction approaches equilibrium ($v_f \approx v_r$). The flux through a near-equilibrium reaction is therefore all but uncontrollable.

Box 15-1

BIOCHEMISTRY IN CONTEXT

Optimizing Glycogen Structure

The function of glycogen in animal cells is to store the metabolic fuel glucose and to release it rapidly when needed. Glycogen must be a polymer, because glucose itself could not be stored without a drastic increase in intracellular osmotic pressure (Section 2-1D). It has been estimated that the total concentration of glucose residues stored as glycogen in a liver cell is ~0.4 M, whereas the concentration of glycogen is only ~10 nM. This huge difference mitigates osmotic stress.

To fulfill its biological function, the glycogen polymer must store the largest amount of glucose in the smallest possible volume while maximizing both the amount of glucose available for release by glycogen phosphorylase and the number of nonreducing ends (to maximize the rate at which glucose residues can be mobilized). All these criteria must be met by optimizing just two variables: the degree of branching and chain length.

In a glycogen molecule, shown schematically here,

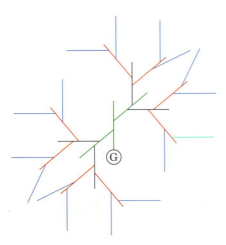

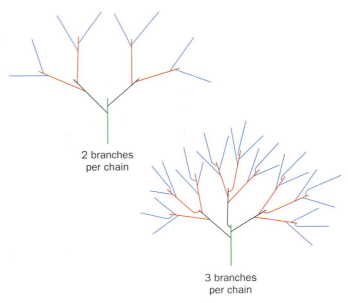

2 branches
per chain

3 branches
per chain

the glycogen chains, beginning with the innermost chain attached to glycogenin (G), have two branches (the outermost chains are unbranched). The entire molecule is roughly spherical and is organized in tiers. There are an estimated 12 tiers in mature glycogen (only 4 are shown above).

With two branches per chain, the number of chains in a given tier is twice the number of the preceding tier, and the outermost tier contains about half of the total glucose residues (regardless of the number of tiers). When the degree of branching increases, for example, to three branches per chain, the proportion of residues in the outermost tier increases, but so does the density of glucose residues. This severely limits the maximum size of the glycogen particle and the number of glucose residues it can accommodate. Thus, glycogen has around two branches per chain.

Mathematical analysis of the other variable, chain length, yields an optimal value of 13, which is in good agreement with the actual length of glycogen chains in cells (8–14 residues). Consider the two simplified glycogen molecules shown below, which contain the same number of glucose residues (the same total length of line segments) and the same branching pattern.

The molecule with the shorter chains packs more glucose in a given volume and has more points for phosphorylase attack, but only about half the amount of glucose can be released before debranching must occur (debranching is much slower than phosphorolysis). In the less dense molecule, the longer chains increase the number of residues that can be continuously phosphorylyzed; however, there are fewer points of attack. Thirteen residues is apparently a compromise for mobilizing the largest amount of glucose in the shortest time.

Amylopectin (Section 8-2C), which is chemically similar to glycogen, is a much larger molecule and has longer chains. Amylose lacks branches altogether. Evidently, starch, unlike glycogen, is not designed for rapid mobilization of metabolic fuel.

[Figures adapted from Meléndez-Hevia, E., Waddell, T.G., and Shelton, E.D., *Biochem. J.* **295**, 477–483 (1993).]

Precise flux control of a pathway is possible when an enzyme functioning far from equilibrium is opposed by a separately controlled enzyme. Then, v_f and v_r vary independently and v_r can be larger or smaller than v_f, allowing control of both rate and direction. Exactly this situation occurs in glycogen metabolism through the opposition of the glycogen phosphorylase and glycogen synthase reactions.

Both glycogen phosphorylase and glycogen synthase are under allosteric control by effectors that include ATP, G6P, and AMP. Muscle glycogen phosphorylase is activated by AMP and inhibited by ATP and G6P. Glycogen synthase, on the other hand, is activated by G6P. This suggests that when there is high demand for ATP (low [ATP], low [G6P], and high [AMP]), glycogen phosphorylase is stimulated and glycogen synthase is inhibited, which favors glycogen breakdown. Conversely, when [ATP] and [G6P] are high, glycogen synthesis is favored.

In vivo, this allosteric scheme is superimposed on an additional control system based on covalent modification. For example, phosphorylase *a* is active even without AMP stimulation (Section 15-1A), and glycogen synthase is essentially inactive (Section 15-2B) unless it is dephosphorylated and G6P is present. *Thus, covalent modification (phosphorylation and dephosphorylation) of glycogen phosphorylase and glycogen synthase provides a more sophisticated control system that modulates the responsiveness of these enzymes to their allosteric effectors.*

● See Guided Exploration 13:

Control of Glycogen Metabolism.

B. Covalent Modification of Glycogen Phosphorylase and Glycogen Synthase

The interconversion of the *a* and *b* forms of glycogen synthase and glycogen phosphorylase is accomplished through enzyme-catalyzed phosphorylation and dephosphorylation, a process that is under hormonal control (Section 15-3C). Enzymatically interconvertible enzyme systems can therefore respond to a greater number of effectors than simple allosteric systems. Furthermore, if the enzymes that covalently modify a target enzyme are themselves under allosteric control, it is possible for a small change in the concentration of an allosteric effector of a modifying enzyme (resulting from hormonal stimulation, for example) to cause a large change in the activity of the modified target enzyme. A set of kinases and phosphatases linked in cascade fashion therefore has enormous potential for signal amplification and regulatory flexibility in response to different metabolic signals. Note that the correlation between phosphorylation and enzyme activity varies with the enzyme. For example, glycogen phosphorylase is activated by phosphorylation ($b \rightarrow a$), whereas glycogen synthase is inactivated by phosphorylation ($a \rightarrow b$). Conversely, dephosphorylation inactivates glycogen phosphorylase and activates glycogen synthase.

Glycogen Phosphorylase Is Activated by Phosphorylation

The cascade that governs enzymatic interconversion of glycogen phosphorylase involves three enzymes (Fig. 15-12):

1. **Phosphorylase kinase,** which specifically phosphorylates Ser 14 of glycogen phosphorylase *b*.
2. **cAMP-dependent protein kinase (cAPK),** which phosphorylates and thereby activates phosphorylase kinase.
3. **Phosphoprotein phosphatase-1,** which dephosphorylates and thereby deactivates both glycogen phosphorylase *a* and phosphorylase kinase.

Phosphorylase *b* is sensitive to allosteric effectors, but phosphorylase *a* is much less so (Fig. 15-13). In the resting cell, the concentrations of ATP

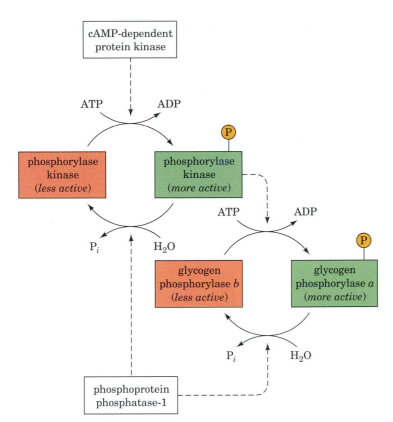

Figure 15-12.   The glycogen phosphorylase interconvertible enzyme system. Conversion of phosphorylase *b* (the less active form) to phosphorylase *a* (the more active form) is accomplished through phosphorylation catalyzed by phosphorylase kinase, which is itself subject to activation through phosphorylation by cAMP-dependent protein kinase (cAPK). Both glycogen phosphorylase *a* and phosphorylase kinase are dephosphorylated by phosphoprotein phosphatase-1.

and G6P are high enough to inhibit phosphorylase *b*. The level of phosphorylase activity is therefore largely determined by the fraction of enzyme present as phosphorylase *a*. The steady state fraction of the phosphorylated enzyme depends on the relative activities of phosphorylase kinase, cAPK, and phosphoprotein phosphatase-1. Let us examine the factors that regu-

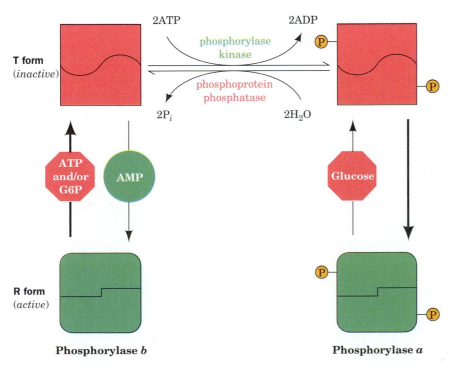

Figure 15-13.   The control of glycogen phosphorylase activity. The enzyme may assume the enzymatically inactive T conformation (*above*) or the catalytically active R form (*below*). The conformation of phosphorylase *b* is allosterically controlled by effectors such as AMP, ATP, and G6P and is mostly in the T state under physiological conditions. In contrast, the modified form of the enzyme, phosphorylase *a*, is largely unresponsive to these effectors and is mostly in the R state unless there is a high level of glucose. Under usual physiological conditions, the enzymatic activity of glycogen phosphorylase is essentially determined by its rates of modification and demodification. Note that only the T-state enzyme is subject to phosphorylation and dephosphorylation, so effector binding influences the rates of these modification/demodification events.

late the activities of these enzymes before we return to the regulation of glycogen synthase activity.

cAMP-Dependent Protein Kinase

The primary intracellular signal for glycogen phosphorylase activation by phosphorylase kinase is **adenosine-3′,5′-cyclic monophosphate (cyclic AMP or cAMP).** The cAMP concentration in a cell is a function of the ratio of its rate of synthesis from ATP by **adenylate cyclase** and its rate of breakdown to AMP by a specific **phosphodiesterase.**

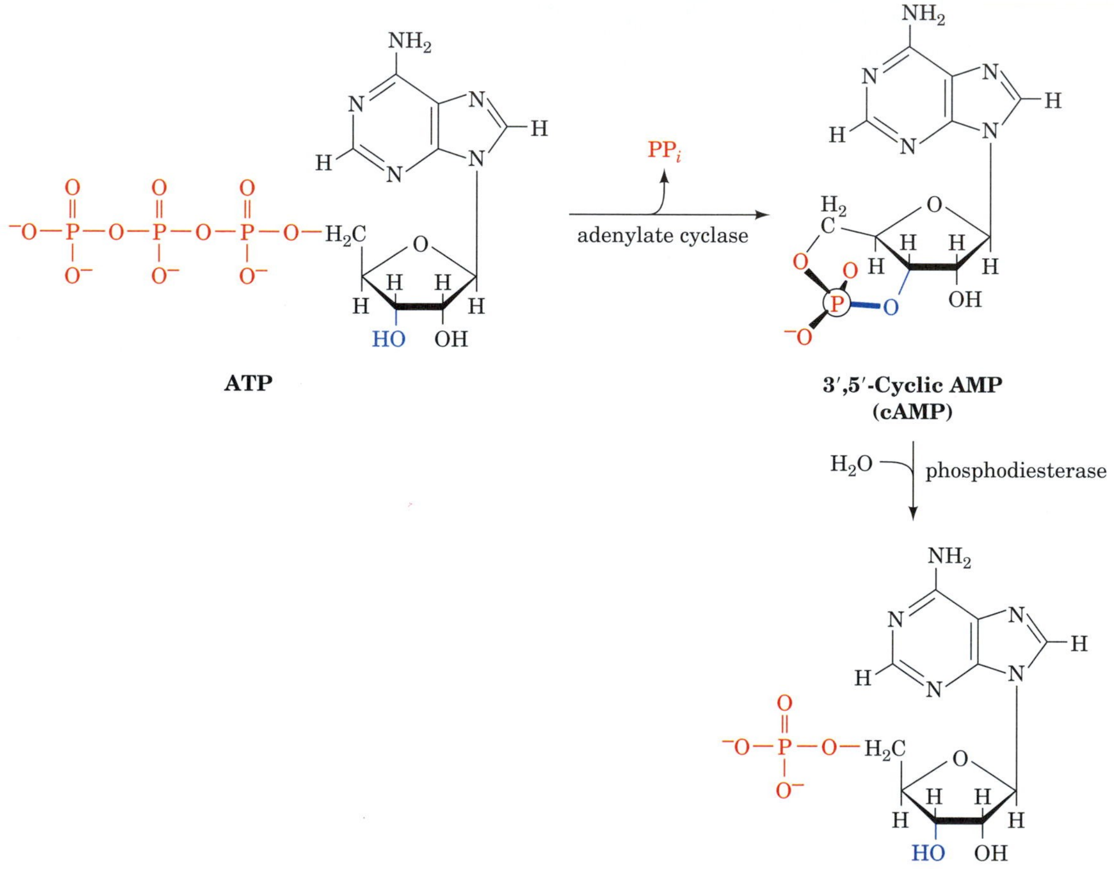

ATP

3′,5′-Cyclic AMP (cAMP)

AMP

Adenylate cyclase, a transmembrane protein, is stimulated by the binding of certain hormones (Section 15-3C) to their cell-surface receptors, to catalyze the synthesis of cAMP inside the cell.

Cyclic AMP is absolutely required for the activity of cAPK, an enzyme that phosphorylates specific Ser or Thr residues of numerous cellular proteins, including phosphorylase kinase and glycogen synthase. These proteins all contain a consensus kinase-recognition sequence, Arg-Arg-X-Ser/Thr-Y, where Ser/Thr is the phosphorylation site, X is any small residue, and Y is a large hydrophobic residue. In the absence of cAMP, cAPK is an inactive tetramer consisting of two regulatory and two catalytic subunits, R_2C_2. The cAMP binds to the regulatory subunit to cause the dissociation of active catalytic monomers:

$$R_2C_2 \; + \; 4 \, cAMP \rightleftharpoons 2 \, C \; + \; R_2(cAMP)_4$$
$$\text{\textit{(inactive)}} \qquad\qquad\qquad \text{\textit{(active)}}$$

The intracellular concentration of cAMP therefore determines the fraction of cAPK in its active form and thus the rate at which it phosphorylates its substrates.

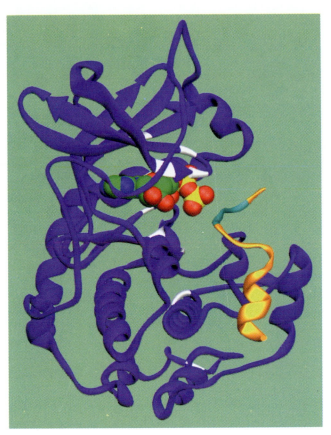

Figure 15-14. **Structure of the C subunit of mouse cAPK.** This model shows the protein in complex with ATP and a 20-residue segment of a naturally occurring inhibitor. The protein's main chain is purple with the 11 residues that are highly conserved in all known protein kinases colored white. The polypeptide inhibitor is gold, and its pseudo-target sequence, Arg-Arg-Asn-Ala-Ile, in which Ala has replaced the phosphorylatable Ser/Thr, is cyan. The ATP is shown in space-filling representation colored according to atom type (C, green; N, blue; O, red; and P, yellow). Note that the inhibitor's pseudo-target sequence is close to ATP's γ-phosphate group, the group that the enzyme transfers. [Based on an X-ray structure by Susan Taylor and Janusz Sowadski, University of California at San Diego.] ⊙ See the Interactive Exercises and Kinemage Exercise 15.

The X-ray structure of the 350-residue C subunit of mouse cAPK in complex with ATP and a 20-residue inhibitor peptide, which was determined by Susan Taylor and Janusz Sowadski, is shown in Fig. 15-14. The C subunit, like other kinases (Figs. 14-2 and 14-10), is bilobal. In cAPK, the deep cleft between the lobes is occupied by ATP and a segment of the inhibitor peptide that resembles the 5-residue consensus sequence for phosphorylation except that the phosphorylated Ser/Thr is replaced by Ala. This cleft must therefore contain the kinase catalytic site.

A large family of protein kinases play key roles in the signaling pathways by which many hormones, growth factors, neurotransmitters, and toxins affect the functions of their target cells (Section 21-3). Indeed, ~10% of the proteins in mammalian cells are phosphorylated.

Phosphorylase Kinase

Phosphorylase kinase is a 1300-kD protein with four nonidentical subunits, labeled α, β, γ, and δ. The γ subunit contains the catalytic site, and the other three subunits have regulatory functions. *Phosphorylase kinase is maximally activated by Ca²⁺ and by the phosphorylation of its α and β subunits.*

The γ subunit of phosphorylase kinase contains a 386-residue kinase domain, which is 36% identical in sequence to cAPK and has a similar structure (Fig. 15-15). The γ subunit is not subject to phosphorylation, as are some other protein kinases, because the Ser, Thr, or Tyr residue that is phosphorylated to activate these other kinases is replaced by a Glu residue in the γ subunit. The negative charge of the Glu is thought to mimic the presence of a phosphate group and interact with a conserved Arg residue near the active site. However, full catalytic activity of the γ subunit is prevented by its C-terminal segment (not present in Fig. 15-15), which binds

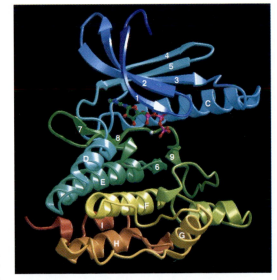

Figure 15-15. **Structure of the kinase domain of phosphorylase kinase.** The protein is color ramped in rainbow order with its N-terminus blue and its C-terminus orange. Bound ATP is shown with its atoms colored by type (C, green; N, blue; O, red; and P, magenta). One of two catalytically important manganese ions is shown as a silver sphere. [Courtesy of Louise Johnson, Oxford University, Oxford, U.K.]

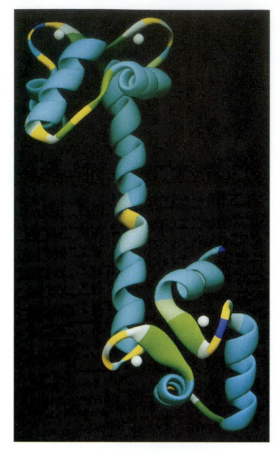

Figure 15-16.   The X-ray structure of rat testis calmodulin. A seven-turn α helix separates two remarkably similar globular domains. The residues are color-coded according to their backbone conformation angles (φ and ψ; Fig. 6-6): cyan, α helical; green, β sheet; yellow, between helix and sheet; and magenta, left-handed helix. The Gly residues are white and the N-terminus is blue. The two Ca^{2+} ions bound to each domain are represented by white spheres. NMR studies of CaM show that in solution the middle portion of the central helix is actually nonhelical. [Courtesy of Mike Carson, University of Alabama at Birmingham. X-Ray structure determined by Charles Bugg, University of Alabama at Birmingham.] ● See Kinemage Exercise 16-1.

in the kinase's active site. This polypeptide segment thereby acts as a "pseudosubstrate" to block the kinase active site, much as the inhibitory peptide blocks the active site of cAPK (Fig. 15-14).

Ca^{2+} ion, in concentrations as low as 10^{-7} M, activates phosphorylase kinase through its binding to the δ subunit, which is also known as **calmodulin (CaM).** Ca^{2+} binding, as is discussed below, induces a conformational change in CaM that unmasks a hydrophobic binding site for the pseudosubstrate. Thus, when binding Ca^{2+}, CaM extracts the pseudosubstrate from the γ subunit's catalytic site, thereby activating the kinase. Phosphorylation of the α and β subunits of phosphorylase kinase causes the enzyme to become activated at much lower Ca^{2+} concentrations, although the mechanism by which the α and β subunits modulate the activity of the γ subunit is not understood.

The conversion of glycogen phosphorylase *b* to glycogen phosphorylase *a* through the action of phosphorylase kinase increases the rate of glycogen breakdown. The physiological significance of the Ca^{2+} trigger for this activation is that muscle contraction is also triggered by a transient increase in the level of cytosolic Ca^{2+} (Section 7-3C). The rate of glycogen breakdown is thereby linked to the rate of muscle contraction. This is critical because glycogen provides fuel for glycolysis to generate the ATP required for muscle contraction. Since Ca^{2+} release occurs in response to nerve impulses, whereas the phosphorylation of phosphorylase kinase ultimately occurs in response to the presence of certain hormones, these two signals act synergistically in muscle cells to stimulate glycogenolysis.

Calmodulin

CaM is a ubiquitous, often free-floating eukaryotic Ca^{2+}-binding protein that participates in numerous cellular regulatory processes. The X-ray structure of this highly conserved 148-residue protein has a curious dumbbell-like shape in which two structurally similar globular domains are connected by a seven-turn α helix (Fig. 15-16).

CaM's two globular domains each contain two high-affinity Ca^{2+}-binding sites. The Ca^{2+} ion in each of these sites is octahedrally coordinated by oxygen atoms from the backbone and side chains as well as from a protein-associated water molecule. Each of these Ca^{2+}-binding sites is formed by nearly superimposable helix–loop–helix motifs known as **EF hands** (Fig. 15-17) that also form the Ca^{2+}-binding sites in numerous other Ca^{2+}-sensing proteins of known structure.

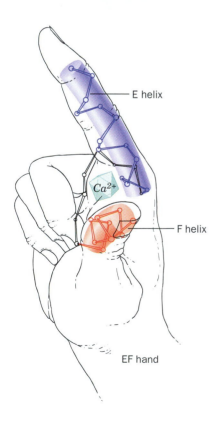

Figure 15-17.   The EF hand. The Ca^{2+}-binding sites in many proteins that sense the level of Ca^{2+} are formed by helix–loop–helix motifs named EF hands. [After Kretsinger, R.H., *Annu. Rev. Biochem.* **45,** 241 (1976).] ● See Kinemage Exercise 16-1.

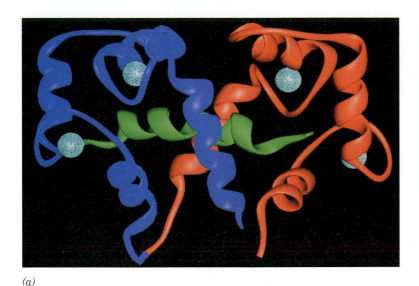

(a)

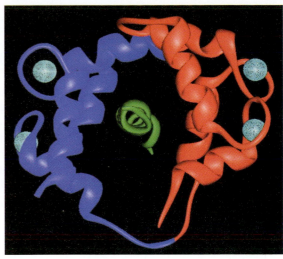

(b)

Figure 15-18. **The NMR structure of calmodulin in complex with a 26-residue target polypeptide.** The N-terminal domain of CaM (from the fruit fly *Drosophila melanogaster*) is blue, its C-terminal domain is red, the target polypeptide is green, and the Ca^{2+} ions are represented by cyan spheres. (*a*) A view of the complex in which the N-terminus of the target polypeptide is on the right. (*b*) The perpendicular view as seen from the right side of the structure shown in *a*. In both views, the pseudo–two-fold axis relating the N- and C-terminal domains of CaM is approximately vertical. Note how the segment that joins the two domains is unwound and bent (bottom loop in *b*) so that CaM forms a globular protein that largely encloses the helical target polypeptide within a hydrophobic tunnel in a manner resembling two hands holding a rope. [Based on an NMR structure by Marius Clore, Angela Gronenborn, and Ad Bax, National Institutes of Health.] ● **See Kinemage Exercise 16-2.**

The binding of Ca^{2+} to either domain of CaM induces a conformational change in that domain, which exposes an otherwise buried Met-rich hydrophobic patch. This patch, in turn, binds with high affinity to the CaM-binding domain of the phosphorylase kinase γ subunit (which contains the kinase pseudosubstrate) and to the CaM-binding domains of many other Ca^{2+}-regulated proteins. These CaM-binding domains have little mutual sequence similarity but are all basic amphiphilic α helices.

Despite uncomplexed CaM's extended appearance (Fig. 15-16), a variety of studies indicate that both of its globular domains bind to a single target helix. This was confirmed by both the NMR and X-ray structures of Ca^{2+}–CaM in complex with target polypeptides (Fig. 15-18). Thus, CaM's central α helix serves as a flexible tether rather than a rigid spacer, a property that probably extends the range of sequences to which CaM can bind. Indeed, both of CaM's globular domains are required for CaM to activate its targets: Domains that have been separated by tryptic cleavage bind to their target peptides but do not cause enzyme activation.

Phosphoprotein Phosphatase-1

The level of activity of many phosphorylatable enzymes is maintained by the opposition of phosphorylation, as catalyzed by a corresponding kinase, and hydrolytic dephosphorylation, as catalyzed by a phosphatase. Phosphoprotein phosphatase-1 removes the phosphoryl groups from glycogen phosphorylase *a* and the α and β subunits of phosphorylase kinase (Fig. 15-12), as well as those of other proteins involved in glycogen metabolism (see below).

Phosphoprotein phosphatase-1 is controlled differently in muscle and in liver. In muscle, phosphoprotein phosphatase-1 is active only when it is bound to glycogen through its glycogen-binding **G subunit.** The activity of protein phosphatase-1 and its affinity for the G subunit are regulated by phosphorylation of the G subunit at two separate sites (Fig. 15-19). Phosphorylation of site 1 by an **insulin-stimulated protein kinase** (a homolog of cAPK and the γ subunit of phosphorylase kinase) activates phosphoprotein phosphatase-1, whereas phosphorylation of site 2 by cAPK (which can also phosphorylate site 1) causes the enzyme to be released into the cytoplasm, where it cannot dephosphorylate the glycogen-bound enzymes of glycogen metabolism.

In the cytosol, phosphoprotein phosphatase-1 is also inhibited by its binding to the protein **phosphoprotein phosphatase inhibitor 1 (inhibitor-1).** This latter protein provides yet another example of control by covalent modification: It too is activated by cAPK and deactivated by phosphoprotein phosphatase-1 (Fig. 15-20, *lower left*). *The concentration of cAMP therefore controls the fraction of an enzyme in its phosphorylated form, not only by increasing the rate at which it is phosphorylated, but also by decreasing the rate at which it is dephosphorylated.* In the case of glycogen phosphorylase, an increase in [cAMP] not only increases this enzyme's rate of activation, but also decreases its rate of deactivation.

In liver, the activity of phosphoprotein phosphatase-1 is controlled by its binding to phosphorylase *a.* Both the R and T forms of phosphorylase *a* strongly bind phosphoprotein phosphatase-1, but only in the T state is the Ser 14 phosphoryl group accessible for hydrolysis (in the R state, the Ser 14 phosphoryl group is buried at the dimer interface; Fig. 15-5). Consequently, when phosphorylase *a* is in its active R form, it effectively sequesters phosphoprotein phosphatase-1. However, under conditions in which phosphorylase *a* converts to the T state (see below), phosphoprotein phosphatase-1 hydrolyzes the now exposed Ser 14 phosphoryl group, thereby converting phosphorylase *a* to phosphorylase *b,* which has only a low affinity for phosphoprotein phosphatase-1. One effect of phosphorylase *a* dephosphorylation, therefore, is to relieve the inhibition of phosphoprotein phosphatase-1. Since liver cells contain 10 times more glycogen phosphorylase than phosphoprotein phosphatase-1, the phosphatase is not released until more than ~90% of the glycogen phosphorylase is in the *b*

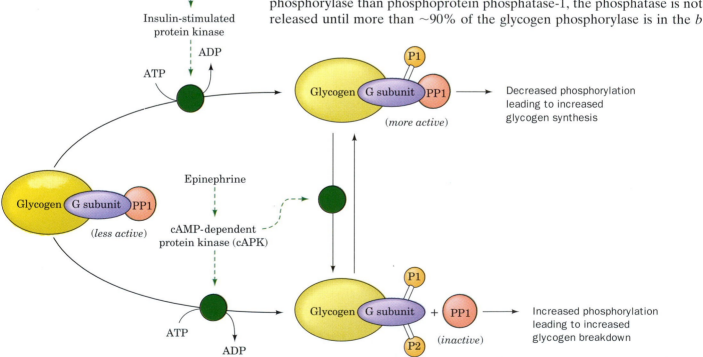

Figure 15-19.   Regulation of phosphoprotein phosphatase-1 in muscle. The antagonistic effects of insulin and epinephrine on glycogen metabolism in muscle occur through their effects on the glycogen-bound G subunit of phosphoprotein phosphatase-1 (PP1). Green circles and dashed arrows indicate activation.

form. Only then can phosphoprotein phosphatase-1 dephosphorylate its other target proteins, including glycogen synthase.

Glucose is an allosteric inhibitor of phosphorylase *a* (Fig. 15-13). Consequently, when the concentration of glucose is high, phosphorylase *a* converts to its T form, thereby leading to its dephosphorylation and the dephosphorylation of glycogen synthase. Glucose is therefore thought to be important in the control of glycogen metabolism in the liver.

Glycogen Synthase

Phosphorylase kinase, which activates glycogen phosphorylase, also phosphorylates and thereby inactivates glycogen synthase. Six other protein kinases, including cAPK, are known to at least partially deactivate human muscle glycogen synthase by phosphorylating one or more of the nine Ser residues on its subunits (Fig. 15-20). The reason for this elaborate regulation of glycogen synthase is unclear.

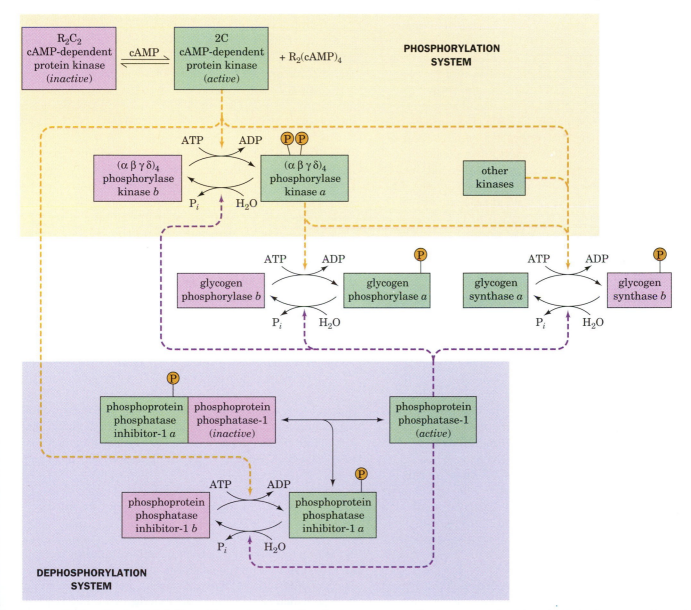

Figure 15-20. *Key to Metabolism.* **The major phosphorylation and dephosphorylation systems that regulate glycogen metabolism in muscle.** Activated enzymes are shaded green, and deactivated enzymes are shaded pink. Dashed arrows indicate facilitation of a phosphorylation or dephosphorylation reaction. ✳ See the Animated Figures.

The balance between net synthesis and degradation of glycogen and the rates of these processes depend on the relative activities of glycogen synthase and glycogen phosphorylase. To a large extent, the rates of phosphorylation and dephosphorylation of these enzymes control glycogen synthesis and breakdown. The two processes are linked by cAPK and phosphorylase kinase, which, through phosphorylation, activate phosphorylase as they inactivate glycogen synthase (Fig. 15-20). They are also linked by phosphoprotein phosphatase-1, which in liver is inhibited by phosphor-

---

Box 15-2
BIOCHEMISTRY IN HEALTH AND DISEASE

Glycogen Storage Diseases

Glycogen storage diseases are inherited disorders that affect glycogen metabolism, producing glycogen that is abnormal in either quantity or quality. Studies of the genetic defects that underlie these diseases have helped elucidate the complexities of glycogen metabolism (e.g., McArdle's disease). Conversely, the biochemical characterization of the pathways affected by a genetic disease often leads to useful strategies for its treatment. The table on p. 449 lists the enzyme deficiencies associated with each type of glycogen storage disease.

Glycogen storage diseases that mainly affect the liver generally produce **hepatomegaly** (enlarged liver) and **hypoglycemia** (low blood sugar), whereas glycogen storage diseases that affect the muscles cause muscle cramps and weakness. Both types of disease may also cause cardiovascular and renal disturbances.

Type I: Glucose-6-Phosphatase Deficiency (von Gierke's Disease)

Glucose-6-phosphatase catalyzes the final step leading to the release of glucose into the bloodstream by the liver. Deficiency of this enzyme results in an increase of intracellular [G6P], which leads to a large accumulation of glycogen in the liver and kidney (recall that G6P activates glycogen synthase) and an inability to increase blood glucose concentration in response to glucagon or epinephrine. The symptoms of Type I glycogen storage disease include severe hepatomegaly and hypoglycemia and a general failure to thrive. Treatment of the disease has included drug-induced inhibition of glucose uptake by the liver (to increase blood [glucose]), continuous intragastric feeding overnight (again to increase blood [glucose]), surgical transposition of the portal vein, which ordinarily feeds the liver directly from the intestines (to allow this glucose-rich blood to reach peripheral tissues before it reaches the liver), and liver transplantation.

Type II: α-1,4-Glucosidase Deficiency (Pompe's Disease)

α-1,4-Glucosidase deficiency is the most devastating of the glycogen storage diseases. It results in a large accumulation of glycogen of normal structure in the lysosomes of all cells and causes death by cardiorespiratory failure, usually before the age of 1 year. α-1,4-Glucosidase is not involved in the main pathways of glycogen metabolism. It occurs in lysosomes, where it hydrolyzes maltose (a glucose disaccharide) and other linear oligosaccharides, as well as the outer branches of glycogen, thereby yielding free glucose. Normally, this alternative pathway of glycogen metabolism is not quantitatively important, and its physiological significance is not known.

Type III: Amylo-1,6-Glucosidase (Debranching Enzyme) Deficiency (Cori's Disease)

In Cori's disease, glycogen of abnormal structure containing very short outer chains accumulates in both liver and muscle since, in the absence of debranching enzyme, the glycogen cannot be further degraded. The resulting hypoglycemia is not as severe as in von Gierke's disease (Type I) and can be treated with frequent feedings and a high-protein diet (to offset the loss of amino acids used for gluconeogenesis). For unknown reasons, the symptoms of Cori's disease often disappear at puberty.

Type IV: Amylo-(1,4→1,6)-Transglycosylase (Branching Enzyme) Deficiency (Andersen's Disease)

Andersen's disease is one of the most severe glycogen storage diseases; victims rarely survive past the age of 4 years because of liver dysfunction. Liver glycogen is present in normal concentrations, but it contains long unbranched chains that greatly reduce its solubility. The liver dysfunction may be caused by a "foreign body" immune reaction to the abnormal glycogen.

Type V: Muscle Phosphorylase Deficiency (McArdle's Disease)

The symptoms of McArdle's disease, painful muscle cramps on exertion, typically do not appear until early adulthood and can be prevented by avoiding strenuous exercise. This condition affects glycogen metabolism in muscle but not in liver, which contains normal amounts of a different phosphorylase isozyme.

ylase *a* and therefore unable to activate (dephosphorylate) glycogen synthase unless it first inactivates (also by dephosphorylation) phosphorylase *a*. Of course, control by allosteric effectors is superimposed on control by covalent modification so that, for example, the availability of the substrate G6P (which activates glycogen synthase) also influences the rate at which glucose residues are incorporated into glycogen. Inherited deficiencies of enzymes can disrupt the fine control of glycogen metabolism, leading to various diseases (see Box 15-2).

Type VI: Liver Phosphorylase Deficiency (Hers' Disease)

Patients with a deficiency of liver phosphorylase have symptoms similar to those with mild forms of Type I glycogen storage disease. The hypoglycemia in this case results from the inability of liver glycogen phosphorylase to respond to the need for circulating glucose.

Type VII: Muscle Phosphofructokinase Deficiency (Tarui's Disease)

The result of a deficiency of the glycolytic enzyme PFK in muscle is an abnormal buildup of the glycolytic metabolites G6P and F6P. High concentrations of G6P increase the activities of glycogen synthase and UDP–glucose pyrophosphorylase (G6P is in equilibrium with G1P, a substrate for UDP–glucose pyrophosphorylase) so that glycogen accumulates in muscle. Other symptoms are similar to those of muscle phosphorylase deficiency, since PFK deficiency prevents glycolysis from keeping up with the ATP demand in contracting muscle.

Type VIII: X-Linked Phosphorylase Kinase Deficiency

Some individuals with symptoms of Type VI glycogen storage disease have normal phosphorylase enzymes but a defective phosphorylase kinase, which results in their inability to convert phosphorylase *b* to phosphorylase *a*. The α subunit of phosphorylase kinase is encoded by a gene on the X chromosome, so Type VIII disease is X-linked rather than autosomal recessive, as are the other glycogen storage diseases.

Type IX: Phosphorylase Kinase Deficiency

Phosphorylase kinase deficiency, an autosomal recessive disease, results from a mutation in one of the genes that encode the β, γ, and δ subunits of phosphorylase kinase. Because different tissues contain different phosphorylase kinase isozymes, the symptoms and severity of the disease vary according to the affected organs. Techniques for identifying genetic lesions are therefore more reliable than clinical symptoms for diagnosing a particular glycogen storage disease.

Type 0: Liver Glycogen Synthase Deficiency

Liver glycogen synthase deficiency is the only disease of glycogen metabolism in which there is a deficiency rather than an overabundance of glycogen. The activity of liver glycogen synthase is extremely low in individuals with Type 0 disease, who exhibit hyperglycemia after meals and hypoglycemia at other times. Some individuals, however, are asymptomatic, which suggests that there may be multiple forms of this autosomal recessive disorder.

Hereditary Glycogen Storage Diseases

Type	Enzyme Deficiency	Tissue	Common Name	Glycogen Structure
I	Glucose-6-phosphatase	Liver	von Gierke's disease	Normal
II	α-1,4-Glucosidase	All lysosomes	Pompe's disease	Normal
III	Amylo-1,6-glucosidase (debranching enzyme)	All organs	Cori's disease	Outer chains missing or very short
IV	Amylo-(1,4→1,6)-transglycosylase (branching enzyme)	Liver, probably all organs	Andersen's disease	Very long unbranched chains
V	Glycogen phosphorylase	Muscle	McArdle's disease	Normal
VI	Glycogen phosphorylase	Liver	Hers' disease	Normal
VII	Phosphofructokinase	Muscle	Tarui's disease	Normal
VIII	Phosphorylase kinase	Liver	X-linked phosphorylase kinase deficiency	Normal
IX	Phosphorylase kinase	All organs		Normal
0	Glycogen synthase	Liver		Normal, deficient in quantity

C. Hormonal Effects on Glycogen Metabolism

Glycogen metabolism in the liver is ultimately controlled by the polypeptide hormone **glucagon,**

$$\overset{+}{H_3N}—His—Ser—Glu—Gly—Thr—Phe—Thr—Ser—Asp—Tyr—10$$

$$Ser—Lys—Tyr—Leu—Asp—Ser—Arg—Arg—Ala—Gln—20$$

$$Asp—Phe—Val—Gln—Trp—Leu—Met—Asn—Thr—COO^- 29$$

Glucagon

which, like insulin, is synthesized by the pancreas in response to the concentration of glucose in the blood. In muscles and various tissues, control is exerted by insulin (Fig. 5-1) and by the adrenal hormones **epinephrine (adrenalin)** and **norepinephrine (noradrenalin).**

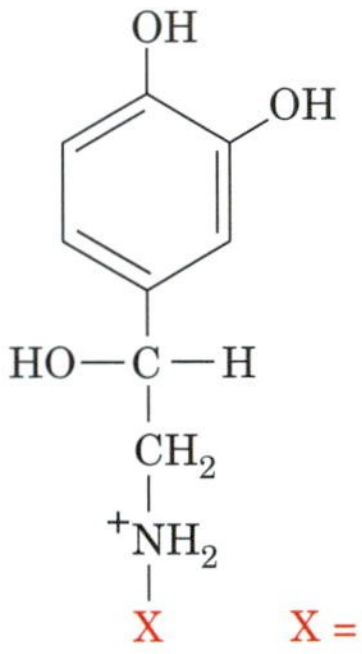

$$X = CH_3 \quad \textbf{Epinephrine}$$
$$X = H \quad \textbf{Norepinephrine}$$

These hormones affect metabolism in their target tissues by ultimately stimulating covalent modification (phosphorylation) of regulatory enzymes. They do so by binding to transmembrane **receptors** on the surface of cells. Different cell types have different complements of receptors and therefore respond to different sets of hormones. The responses involve the release inside the cell of molecules collectively known as **second messengers,** that is, intracellular mediators of the externally received hormonal message. Different receptors release different second messengers. cAMP, identified by Earl Sutherland in the 1950s, was the first second messenger discovered. Ca^{2+}, as released from intracellular reservoirs into the cytosol, is also a common second messenger. Receptors and second messengers are discussed in greater depth in Section 21-3.

When hormonal stimulation increases the intracellular cAMP concentration, cAPK activity increases, increasing the rates of phosphorylation of many proteins and decreasing their dephosphorylation rates as well. Because of the cascade nature of the regulatory system diagrammed in Fig. 15-20, *a small change in [cAMP] results in a large change in the fraction of phosphorylated enzymes.* When a large fraction of the glycogen metabolism enzymes are phosphorylated, the metabolic flux is in the direction of glycogen breakdown, since glycogen phosphorylase is active and glycogen synthase is inactive. When [cAMP] decreases, phosphorylation rates decrease, dephosphorylation rates increase, and the fraction of enzymes in their dephospho forms increases. The resulting activation of glycogen synthase and inhibition of glycogen phosphorylase cause the flux to shift to net glycogen synthesis.

Glucagon binding to its receptor on liver cells, which generates intracellular cAMP, results in glucose mobilization from stored glycogen (Fig. 15-21). Glucagon is released from the pancreas when the concentration of

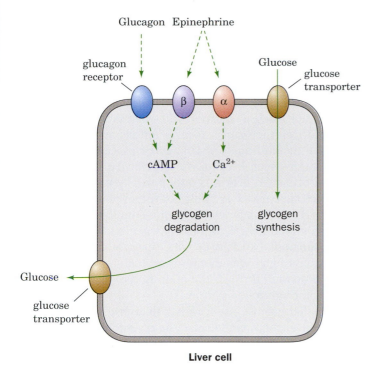

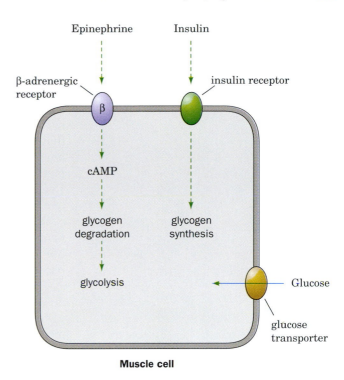

Figure 15-21.   Hormonal control of glycogen metabolism.
Epinephrine binding to β-adrenergic receptors on liver and muscle cells increases intracellular [cAMP], which promotes glycogen
degradation to G6P for glycolysis (in muscle) or to glucose for
export (in liver). The liver responds similarly to glucagon.
Epinephrine binding to α-adrenergic receptors on liver cells leads
to increased cytosolic [Ca^{2+}], which also promotes glycogen
degradation. When circulating glucose is plentiful, insulin
stimulates glycogen synthesis in muscle cells. The liver responds
directly to increased glucose by increasing glycogen synthesis.
✳ See the Animated Figures.

circulating glucose decreases to less than ~5 mM, such as during exercise
or several hours after a meal has been digested. Glucagon is therefore critical for the liver's function in supplying glucose to tissues that depend primarily on glycolysis for their energy needs. Muscle cells do not respond to
glucagon because they lack the appropriate receptor.

Epinephrine and norepinephrine, which are often called the "fight or
flight" hormones, are released into the bloodstream by the adrenal glands
in response to stress. There are two types of receptors for these hormones:
the **β-adrenergic receptor,** which is linked to the adenylate cyclase system,
and the **α-adrenergic receptor,** whose second messenger causes intracellular [Ca^{2+}] to increase (Section 21-3B). Muscle cells, which have the β-
adrenergic receptor (Fig. 15-21), respond to epinephrine by breaking down
glycogen for glycolysis, thereby generating ATP and helping the muscles
cope with the stress that triggered the epinephrine release.

Liver cells respond to epinephrine directly and indirectly. Epinephrine
promotes the release of glucagon from the pancreas, and glucagon binding
to its receptor on liver cells stimulates glycogen breakdown as described
above. Epinephrine also binds directly to both α- and β-adrenergic receptors on the surfaces of liver cells (Fig. 15-21). Binding to the β-adrenergic
receptor results in increased intracellular cAMP, which leads to glycogen
breakdown. Epinephrine binding to the α-adrenergic receptor stimulates
an increase in intracellular [Ca^{2+}], which reinforces the cells' response to
cAMP (recall that phosphorylase kinase, which activates glycogen phosphorylase and inactivates glycogen synthase, is fully active only when phosphorylated and in the presence of increased [Ca^{2+}]). In addition, glycogen

synthase is inactivated through phosphorylation catalyzed by several Ca^{2+}-dependent protein kinases.

Insulin and Epinephrine Are Antagonists

Insulin is released from the pancreas in response to high levels of circulating glucose (e.g., immediately after a meal). Hormonal stimulation by insulin increases the rate of glucose transport into the many types of cells that have insulin receptors on their surfaces (e.g., muscle and fat cells, but not liver and brain cells). In addition, [cAMP] decreases, causing glycogen metabolism to shift from glycogen breakdown to glycogen synthesis (Fig. 15-21). The mechanism of insulin action is only partially understood, but one of its target enzymes appears to be phosphoprotein phosphatase-1.

As outlined in Fig. 15-19, insulin activates insulin-stimulated protein kinase in muscle to phosphorylate site 1 on the glycogen-binding G subunit of phosphoprotein phosphatase-1 so as to activate this protein and thus dephosphorylate the enzymes of glycogen metabolism. The storage of glucose as glycogen is thereby promoted through the inhibition of glycogen breakdown and the stimulation of glycogen synthesis.

In liver, it is thought that glucose itself, rather than insulin, may be the messenger to which the glycogen metabolism system responds. *Glucose inhibits phosphorylase a by binding to the enzyme's inactive T state and thereby shifting the $T \rightleftharpoons R$ equilibrium toward the T state* (Fig. 15-13). This conformational shift exposes the Ser 14 phosphoryl group to dephosphorylation. An increase in glucose concentration therefore promotes inactivation of glycogen phosphorylase *a* through its conversion to phosphorylase *b*. Thus, when glucose is plentiful, the liver can store the excess as glycogen.

4. GLUCONEOGENESIS

When dietary sources of glucose are not available and when the liver has exhausted its supply of glycogen, glucose is synthesized from noncarbohydrate precursors by gluconeogenesis. In fact, gluconeogenesis provides a substantial fraction of the glucose produced in fasting humans, even within a few hours of eating. Gluconeogenesis occurs in liver and, to a lesser extent, in kidney.

The noncarbohydrate precursors that can be converted to glucose include the glycolysis products lactate and pyruvate, citric acid cycle intermediates, and the carbon skeletons of most amino acids. First, however, all these substances must be converted to the four-carbon compound **oxaloacetate,**

$$\overset{O}{\underset{{}^-O}{\diagdown}}C-CH_2-\overset{\overset{O}{\|}}{C}-\overset{\diagup O}{\underset{O^-}{C}}$$

Oxaloacetate

which itself is a citric acid cycle intermediate (Section 16-1). The only amino acids that cannot be converted to oxaloacetate in animals are leucine and lysine because their breakdown yields only acetyl-CoA (Section 20-4E) and because *there is no pathway in animals for the net conversion of acetyl-CoA to oxaloacetate.* Likewise, fatty acids cannot serve as glucose precursors in animals because most fatty acids are degraded completely to acetyl-CoA (Section 19-2).

For convenience, we consider gluconeogenesis to be the pathway by which pyruvate is converted to glucose. Most of the reactions of gluco-

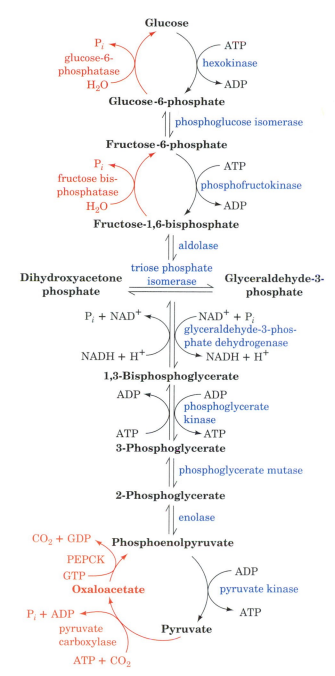

Figure 15-22. ***Key to Metabolism.*** **Comparison of the pathways of gluconeogenesis and glycolysis.** The red arrows represent the steps that are catalyzed by different enzymes in gluconeogenesis. The other seven reaction steps of gluconeogenesis are catalyzed by glycolytic enzymes that function near equilibrium.

✳ See the Animated Figures.

neogenesis are glycolytic reactions that proceed in reverse (Fig. 15-22). However, the glycolytic enzymes hexokinase, phosphofructokinase, and pyruvate kinase catalyze reactions with large negative free energy changes. These reactions must therefore be replaced in gluconeogenesis by reactions that make glucose synthesis thermodynamically favorable.

A. *Pyruvate to Phosphoenolpyruvate*

We begin our examination of the reactions unique to gluconeogenesis with the conversion of pyruvate to phosphoenolpyruvate (PEP). Because this step is the reverse of the highly exergonic reaction catalyzed by pyruvate kinase (Section 14-2J), it requires free energy input. This is accomplished by first converting the pyruvate to oxaloacetate. Oxaloacetate is a "high-energy" intermediate because its exergonic decarboxylation provides the

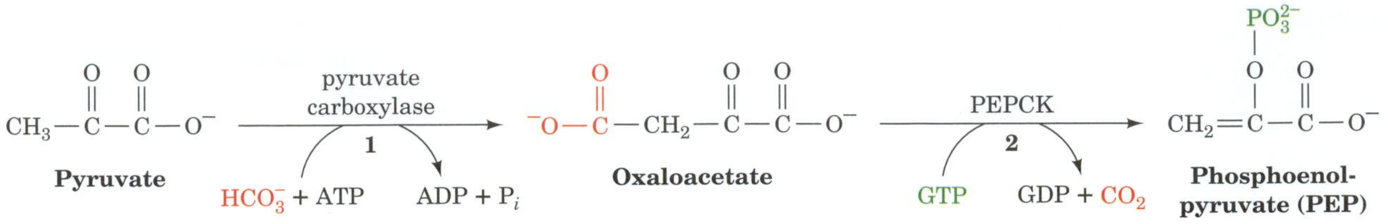

Figure 15-23.   The conversion of pyruvate to phosphoenolpyruvate (PEP). This process requires (1) pyruvate carboxylase to convert pyruvate to oxaloacetate and (2) PEP carboxykinase (PEPCK) to convert oxaloacetate to PEP.

free energy necessary for PEP synthesis. The process requires two enzymes (Fig. 15-23):

1. **Pyruvate carboxylase** catalyzes the ATP-driven formation of oxaloacetate from pyruvate and HCO_3^-.

2. **PEP carboxykinase (PEPCK)** converts oxaloacetate to PEP in a reaction that uses GTP as a phosphoryl-group donor.

Pyruvate Carboxylase Has a Biotin Prosthetic Group

Pyruvate carboxylase is a tetrameric protein of identical ~120-kD subunits, each of which has a **biotin** prosthetic group. Biotin (Fig. 15-24a) functions as a CO_2 carrier by forming a carboxyl substituent at its **ureido group** (Fig. 15-24b). Biotin is covalently bound to an enzyme Lys residue to form a **biocytin** (alternatively, **biotinyllysine**) residue (Fig. 15-24b). The biotin ring system is therefore at the end of a 14-Å-long flexible arm. Biotin, which was first identified in 1935 as a growth factor in yeast, is an essential human nutrient. Its nutritional deficiency is rare, however, because biotin occurs in many foods and is synthesized by intestinal bacteria.

The pyruvate carboxylase reaction occurs in two phases (Fig. 15-25):

Phase I   The cleavage of ATP to ADP acts to dehydrate bicarbonate via the formation of a "high-energy" carboxyphosphate intermediate. The reaction of the resulting CO_2 with biotin is exergonic. The biotin-bound carboxyl group is therefore "activated" relative to bicarbonate and can be transferred to another molecule without further free energy input.

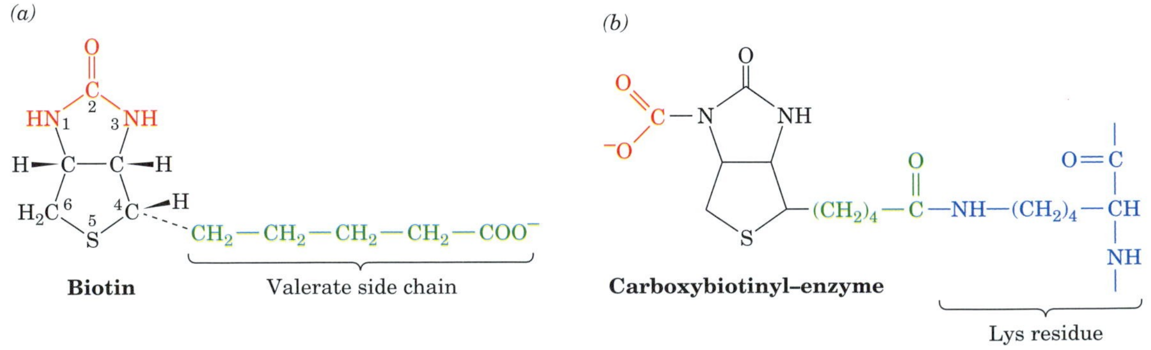

Figure 15-24.   Biotin and carboxybiotinyl–enzyme. (a) Biotin consists of an imidazoline ring that is cis-fused to a tetrahydrothiophene ring bearing a valerate side chain. Positions 1, 2, and 3 constitute a ureido group. (b) Biotin is covalently attached to carboxylases by an amide linkage between its valeryl carboxyl group and an ε-amino group of an enzyme Lys side chain. The carboxybiotinyl–enzyme forms when N1 of the biotin ureido group is carboxylated.

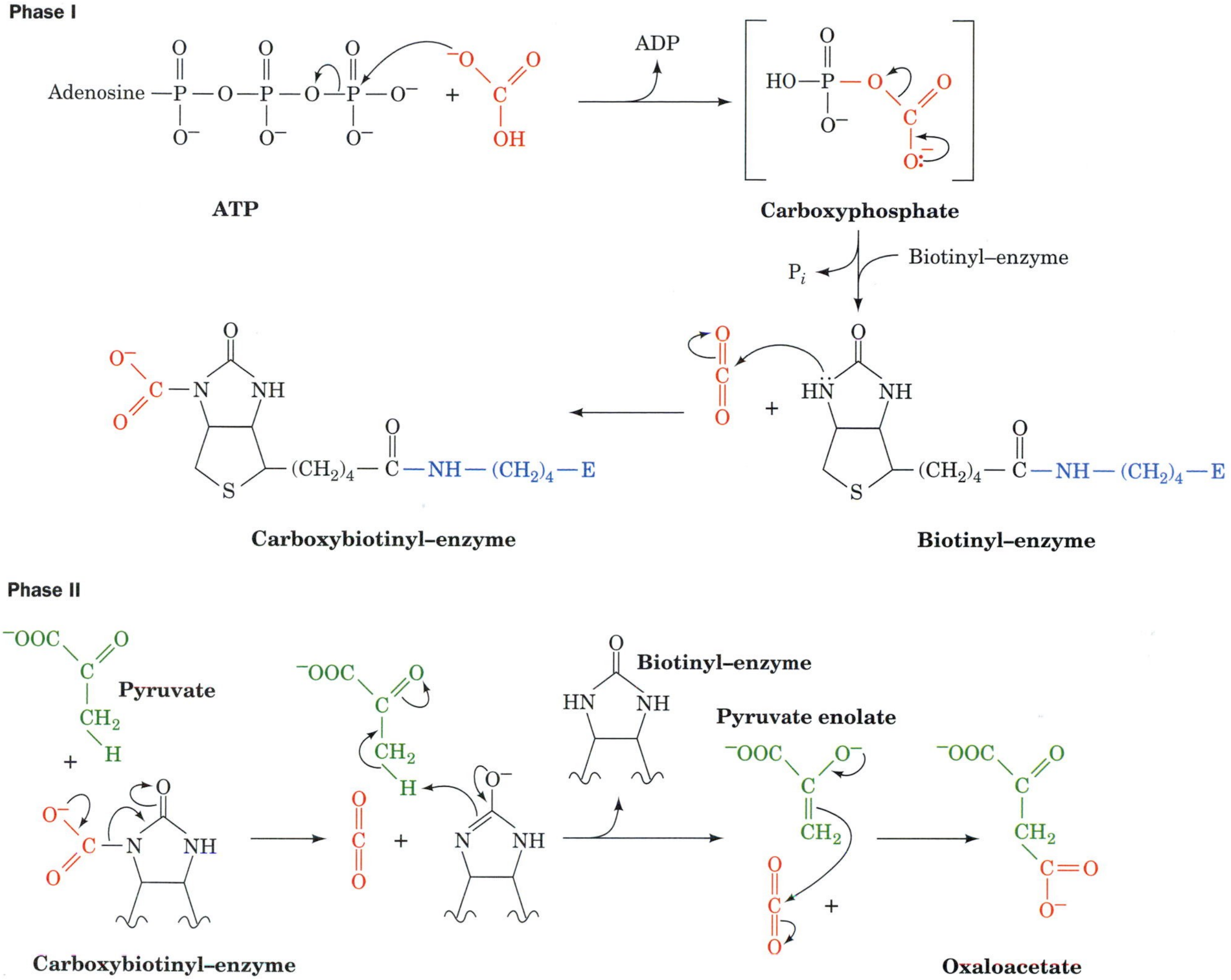

Phase I

Phase II

Figure 15-25. The two-phase reaction mechanism of pyruvate carboxylase.
Phase I is a three-step reaction in which carboxyphosphate is formed from bicarbonate and ATP, followed by the generation of CO_2, which then carboxylates biotin. Phase II is a three-step reaction in which CO_2 is produced at the active site via the elimination of the biotinyl–enzyme, which accepts a proton from pyruvate to generate pyruvate enolate. This enolate, in turn, nucleophilically attacks the CO_2, yielding oxaloacetate. [After Knowles, J.R., *Annu. Rev. Biochem.* **58**, 217 (1989).]

Phase II The activated carboxyl group is transferred from carboxybiotin to pyruvate in a three-step reaction to form oxaloacetate.

These two reaction phases occur on different subsites of the same enzyme; the 14-Å-long flexible arm of biocytin transfers the biotin ring between the two sites.

Oxaloacetate is both a precursor for gluconeogenesis and an intermediate of the citric acid cycle (Section 16-3). When the citric acid cycle substrate acetyl-CoA accumulates, it allosterically activates pyruvate carboxylase, thereby increasing the amount of oxaloacetate that can participate in the citric acid cycle. When citric acid cycle activity is low, oxaloacetate instead enters the gluconeogenic pathway.

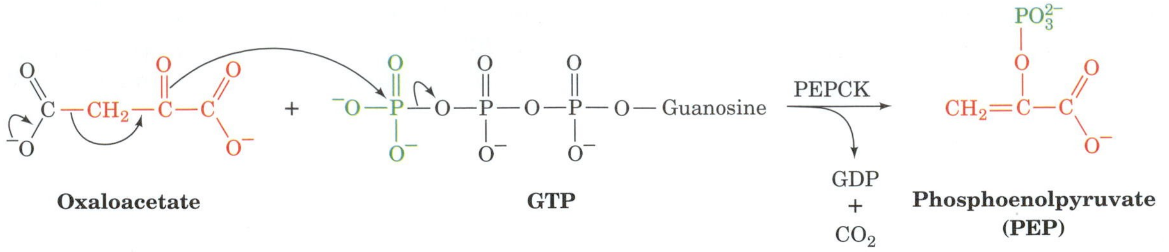

Figure 15-26. The PEPCK mechanism. Decarboxylation of oxaloacetate (a β-keto acid) forms a resonance-stabilized enolate anion whose oxygen atom attacks the γ-phosphoryl group of GTP, forming PEP and GDP.

PEP Carboxykinase

PEPCK, a monomeric 74-kD enzyme, catalyzes the GTP-driven decarboxylation of oxaloacetate to form PEP and GDP (Fig. 15-26). Note that the CO_2 that carboxylates pyruvate to yield oxaloacetate is eliminated in the formation of PEP. *Oxaloacetate can therefore be considered as "activated" pyruvate, with CO_2 and biotin facilitating the activation at the expense of ATP.*

Gluconeogenesis Requires Metabolite Transport between Mitochondria and Cytosol

The generation of oxaloacetate from pyruvate or citric acid cycle intermediates occurs only in the mitochondrion, whereas the enzymes that convert PEP to glucose are cytosolic. The cellular location of PEPCK varies: In some species, it is mitochondrial; in some, it is cytosolic; and in some (including humans) it is equally distributed between the two compartments. In order for gluconeogenesis to occur, either oxaloacetate must leave the mitochondrion for conversion to PEP, or the PEP formed there must enter the cytosol.

PEP is transported across the mitochondrial membrane by specific membrane transport proteins. There is, however, no such transport system for oxaloacetate. *In species with cytosolic PEPCK, oxaloacetate must first be converted either to aspartate* (Fig. 15-27, Route 1) *or to **malate*** (Fig. 15-27, Route 2), for which mitochondrial transport systems exist. The difference between these two routes involves the transport of NADH **reducing equivalents** (in the transport of reducing equivalents, the electrons—but not the electron carrier—cross the membrane). The **malate dehydrogenase** route (Route 2) results in the transport of reducing equivalents from the mitochondrion to the cytosol, since it uses mitochondrial NADH and produces cytosolic NADH. The **aspartate aminotransferase** route (Route 1) does not involve NADH. Cytosolic NADH is required for gluconeogenesis, so, under most conditions, the route through malate is a necessity. However, when the gluconeogenic precursor is lactate, its oxidation to pyruvate generates cytosolic NADH, and either transport system can then be used. All the reactions shown in Fig. 15-27 are freely reversible, so that, under appropriate conditions, the malate–aspartate shuttle system also operates to transport NADH reducing equivalents into the mitochondrion for electron transport and oxidative phosphorylation (Section 17-1B).

B. Hydrolytic Reactions

The route from PEP to fructose-1,6-bisphosphate (FBP) is catalyzed by the enzymes of glycolysis operating in reverse. However, *the glycolytic*

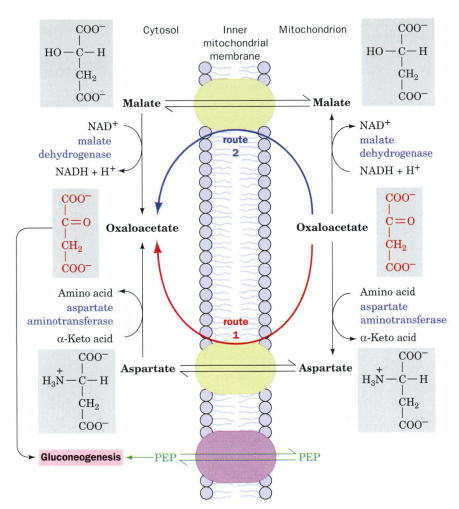

Figure 15-27.   The transport of PEP and oxaloacetate from the mitochondrion to the cytosol. PEP is directly transported between these compartments. Oxaloacetate, however, must first be converted to either aspartate through the action of aspartate aminotransferase (Route 1) or to malate by malate dehydrogenase (Route 2). Route 2 involves the mitochondrial oxidation of NADH followed by the cytosolic reduction of NAD^+ and therefore also transfers NADH reducing equivalents from the mitochondrion to the cytosol. ✳ See the Animated Figures.

reactions catalyzed by phosphofructokinase (PFK) and hexokinase are endergonic in the gluconeogenesis direction and hence must be bypassed by different gluconeogenic enzymes. FBP is hydrolyzed by fructose-1,6-bisphosphatase (FBPase). The resulting fructose-6-phosphate (F6P) is isomerized to G6P, which is then hydrolyzed by glucose-6-phosphatase, the same enzyme that converts glycogen-derived G6P to glucose (Section 15-1C) and which is present only in liver and kidney. Note that these two hydrolytic reactions release P_i rather than reversing the ATP → ADP reactions that occur at this point in the glycolytic pathway.

   The net energetic cost of converting two pyruvate molecules to one glucose molecule by gluconeogenesis is six ATP equivalents: two each at the steps catalyzed by pyruvate carboxylase, PEPCK, and phosphoglycerate kinase. Since the energetic profit of converting one glucose molecule to two pyruvate molecules via glycolysis is two ATP, (Section 14-1), the energetic cost of the futile cycle in which glucose is converted to pyruvate and then resynthesized is four ATP equivalents. Such free energy losses are the ther-

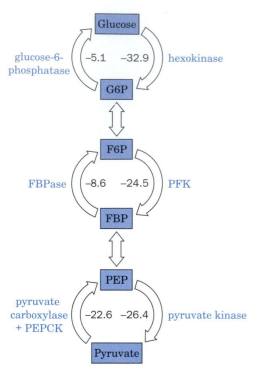

Figure 15-28. Substrate cycles in glucose metabolism. The interconversions of glucose and G6P, F6P and FBP, and PEP and pyruvate are catalyzed by different enzymes in the forward and reverse directions so that all reactions are exergonic (the ΔG values for the reactions in liver are given in kJ·mol^{-1}). [ΔG's obtained from Newsholme, E.A. and Leech, A.R., *Biochemistry for the Medical Sciences*, p. 448, Wiley (1983).]

modynamic price that must be paid to maintain the independent regulation of two opposing pathways.

Although glucose is considered the end point of the gluconeogenic pathway, it is possible for pathway intermediates to be directed elsewhere, for example, through the transketolase and transaldolase reactions of the pentose phosphate pathway (Section 14-6C) to produce ribose-5-phosphate. The G6P produced by gluconeogenesis may not be hydrolyzed to glucose but may instead be converted to G1P for incorporation into glycogen.

C. Regulation of Gluconeogenesis

The opposing pathways of gluconeogenesis and glycolysis, like glycogen synthesis and degradation, do not proceed simultaneously *in vivo*. Instead, these pathways are reciprocally regulated to meet the needs of the organism. There are three substrate cycles and therefore three potential points for regulating glycolytic versus gluconeogenic flux (Fig. 15-28).

Fructose-2,6-Bisphosphate Activates Phosphofructokinase and Inhibits Fructose-1,6-Bisphosphatase

The net flux through the substrate cycle created by the opposing actions of PFK and FBPase (described in Section 14-4B) is determined by the concentration of fructose-2,6-bisphosphate (F2,6P).

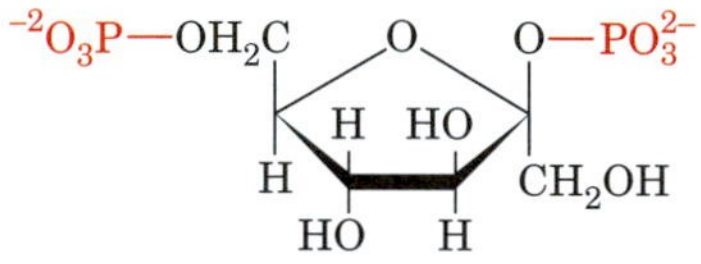

β-D-Fructose-2,6-bisphosphate (F2,6P)

F2,6P, which is not a glycolytic intermediate, is an extremely potent allosteric activator of PFK and an inhibitor of FBPase.

The concentration of F2,6P in the cell depends on the balance between its rates of synthesis and degradation by **phosphofructokinase-2 (PFK-2)** and **fructose bisphosphatase-2 (FBPase-2)**, respectively (Fig. 15-29). These enzyme activities are located on different domains of the same ~100-kD homodimeric protein. The bifunctional enzyme is regulated by a variety of allosteric effectors and by phosphorylation and dephosphorylation as cat-

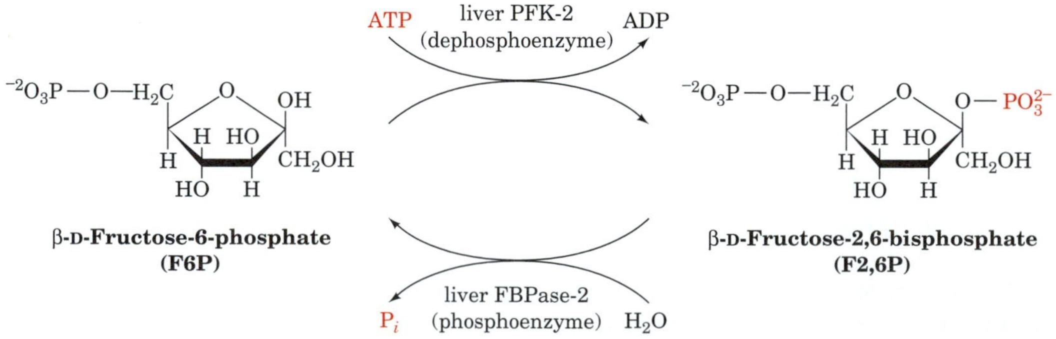

β-D-Fructose-6-phosphate (F6P)

β-D-Fructose-2,6-bisphosphate (F2,6P)

Figure 15-29. The formation and degradation of β-D-fructose-2,6-bisphosphate (F2,6P). The enzymatic activities of phosphofructokinase-2 (PFK-2) and fructose bisphosphatase-2 (FBPase-2) occur on different domains of the same protein molecule. The dephosphorylation of the liver enzyme activates PFK-2 while deactivating FBPase-2.

alyzed by cAPK and a phosphoprotein phosphatase. Thus, the balance between gluconeogenesis and glycolysis is under hormonal control.

For example, when [glucose] is low, glucagon stimulates the production of cAMP in liver cells. This activates cAPK to phosphorylate and thereby inactivate PFK-2, and to phosphorylate and thereby activate FBPase-2. The net result is a decrease in F2,6P concentration, which shifts the balance between the PFK and FBPase reactions in favor of FBP synthesis and hence increases gluconeogenic flux (Fig. 15-30). The concurrent increases in gluconeogenesis and glycogen breakdown allow the liver to release glucose into the circulation. Conversely, when the blood [glucose] is high, cAMP levels decrease, and the resulting increase in [F2,6P] promotes glycolysis.

In muscle, which is not a gluconeogenic tissue, the F2,6P control system functions quite differently from that in liver due to the presence of different PFK-2/FBPase-2 isozymes. For example, hormones that stimulate glycogen breakdown in heart muscle lead to phosphorylation of a site on the bifunctional enzyme that activates rather than inhibits PFK-2. The resulting increase in F2,6P stimulates glycolysis so that glycogen breakdown and glycolysis are coordinated. The skeletal muscle isozyme lacks a phosphorylation site altogether and is therefore not subject to cAMP-dependent control.

Other Allosteric Effectors Influence Gluconeogenic Flux

Acetyl-CoA activates pyruvate carboxylase (Section 15-4A), but there are no known allosteric effectors of PEPCK, which together with pyruvate carboxylase reverses the pyruvate kinase reaction. Pyruvate kinase, however, is allosterically inhibited in the liver by alanine, a major gluconeogenic precursor. Alanine is converted to pyruvate by the transfer of its amino group to an α-keto acid to yield a new amino acid and the α-keto acid pyruvate,

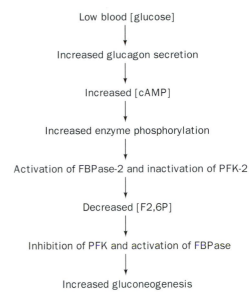

Alanine   **Pyruvate**

Figure 15-30. Sequence of metabolic events linking low blood [glucose] to gluconeogenesis in liver.

a process termed **transamination** (which is discussed in Section 20-2A). Liver pyruvate kinase is also inactivated by phosphorylation, further increasing gluconeogenic flux. Since phosphorylation also activates glycogen phosphorylase, the pathways of gluconeogenesis and glycogen breakdown both flow toward G6P, which is converted to glucose for export from the liver.

The activity of hexokinase (or glucokinase, the liver isozyme) is also controlled, as we shall see in Section 21-1D. The activity of glucose-6-phosphatase, in contrast, responds only to the levels of available substrate and is therefore not an important regulatory point.

Long-term regulation of glucose metabolism occurs not just through allosteric effectors, but also through changes in the amounts of enzymes synthesized. Pancreatic and adrenal hormones influence the rates of transcription and the stabilities of the mRNAs encoding many of the regulatory proteins of glucose metabolism. For example, low concentrations of insulin or high concentrations of intracellular cAMP promote transcription of the genes for PEPCK, FBPase, and glucose-6-phosphatase, and repress transcription of the genes for glucokinase, PFK, and the PFK-2/FBPase-2 bifunctional enzyme.

5. OTHER CARBOHYDRATE BIOSYNTHETIC PATHWAYS

The liver, by virtue of its mass and its metabolic machinery, is primarily responsible for maintaining a constant level of glucose in the circulation. Glucose produced by gluconeogenesis or from glycogen breakdown is released from the liver for use by other tissues as an energy source. Of course, glucose has other uses in the liver and elsewhere, for example, in the synthesis of lactose (see Box 15-3).

Nucleotide Sugars Power the Formation of Glycosidic Bonds

Glucose and other monosaccharides (principally mannose, N-acetylglucosamine, fucose, galactose, N-acetylneuraminic acid, and N-acetylgalactosamine) occur in glycoproteins and glycolipids. Formation of the glycosidic bonds that link sugars to each other and to other molecules requires free energy input under physiological conditions ($\Delta G^{\circ\prime} = 16$ kJ·mol^{-1}). This free energy, as we have seen in glycogen synthesis (Section 15-2A), is acquired through the cleavage of the phosphate ester bond of a nucleotide sugar, releasing PP$_i$, whose exergonic hydrolysis drives the reaction. The nucleoside diphosphate at a sugar's anomeric carbon atom is a good leaving group and thereby facilitates formation of a glycosidic bond to a second sugar in a reaction catalyzed by a glycosyltransferase (Fig. 15-31). In

---

Box 15-3
BIOCHEMISTRY IN FOCUS

Lactose Synthesis

Like sucrose in plants, lactose is a disaccharide that is synthesized for later use as a metabolic fuel, in this case, after digestion by very young mammals. Lactose, or milk sugar, is produced in the mammary gland by **lactose synthase**. In this reaction, the donor sugar is UDP–galactose, which is formed by the epimerization of UDP–glucose (Fig. 14-27). The acceptor sugar is glucose:

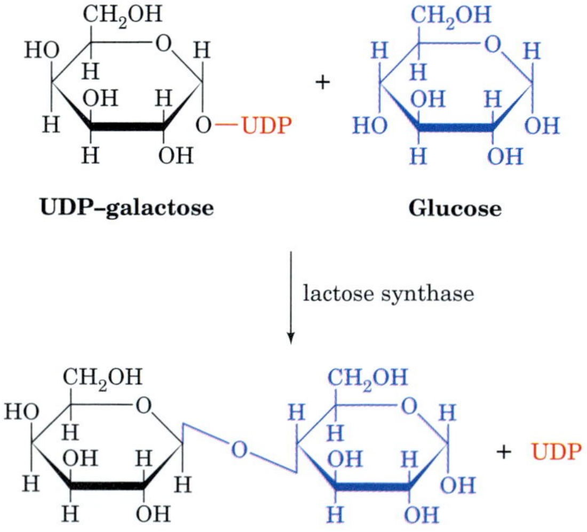

UDP–galactose   **Glucose**

lactose synthase

Lactose
[β-galactosyl-(1 ⟶ 4)-glucose]

Thus, both saccharide units of lactose are ultimately derived from glucose. Lactose synthase consists of two subunits:

1. **Galactosyltransferase**, the catalytic subunit, occurs in many tissues, where it catalyzes the reaction of UDP–galactose and N-acetylglucosamine to yield **N-acetyllactosamine**,

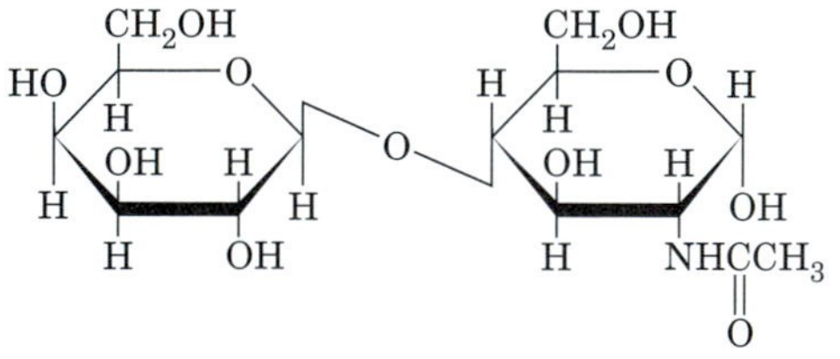

N-Acetyllactosamine

a constituent of many complex oligosaccharides.

2. **α-Lactalbumin**, a mammary gland protein with no catalytic activity, alters the specificity of galactosyltransferase so that it uses glucose as an acceptor, rather than N-acetylglucosamine, to form lactose instead of N-acetyllactosamine.

Synthesis of α-lactalbumin, whose sequence is ~37% identical to that of lysozyme (which also participates in reactions involving sugars), is triggered by hormonal changes at parturition (birth), thereby promoting lactose synthesis for milk production.

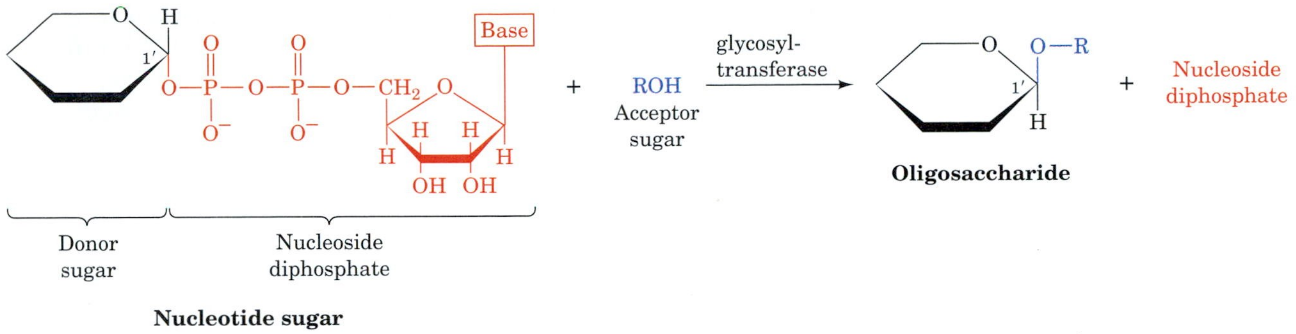

Figure 15-31.   Role of nucleotide sugars. These compounds are the glycosyl donors in oligosaccharide biosynthetic reactions as catalyzed by glycosyltransferases.

mammals, most glycosyl groups are donated by UDP–sugars, but fucose and mannose are carried by GDP, and sialic acid by CMP. In plants, starch is built from glucose units donated by **ADP–glucose,** and cellulose synthesis relies on ADP–glucose or **CDP–glucose.**

O-Linked Oligosaccharides Are Posttranslationally Formed

Nucleotide sugars are the donors in the synthesis of *O*-linked oligosaccharides and in the processing of the *N*-linked oligosaccharides of glycoproteins (Section 8-3C). *O*-linked oligosaccharides are synthesized in the Golgi apparatus by the serial addition of monosaccharide units to a completed polypeptide chain (Fig. 15-32). Synthesis begins with the transfer, as catalyzed by **GalNAc transferase,** of *N*-acetylgalactosamine (GalNAc) from UDP–GalNAc to a Ser or Thr residue on the polypeptide. The location of the glycosylation site is thought to be specified by the secondary or tertiary structure of the polypeptide. Glycosylation continues with the stepwise addition of sugars such as galactose, sialic acid, *N*-acetylglucosamine, and fucose. In each case, the sugar residue is transferred from its nucleotide diphosphate derivative by a corresponding glycosyltransferase.

N-Linked Oligosaccharides Are Constructed on Dolichol Carriers

The synthesis of *N*-linked oligosaccharides is more complicated than that of *O*-linked oligosaccharides. In the early stages of *N*-linked oligosaccharide synthesis, sugar residues are sequentially added to a lipid carrier, **dolichol pyrophosphate** (Fig. 15-33). Dolichol is a long-chain polyisoprenoid containing 17 to 21 isoprene units in animals and 14 to 25 units in fungi and plants. It anchors the growing oligosaccharide to the endoplasmic reticulum membrane, where the initial glycosylation reactions take place.

Although nucleotide sugars are the most common monosaccharide donors in glycosyltransferase reactions, several mannosyl and glucosyl

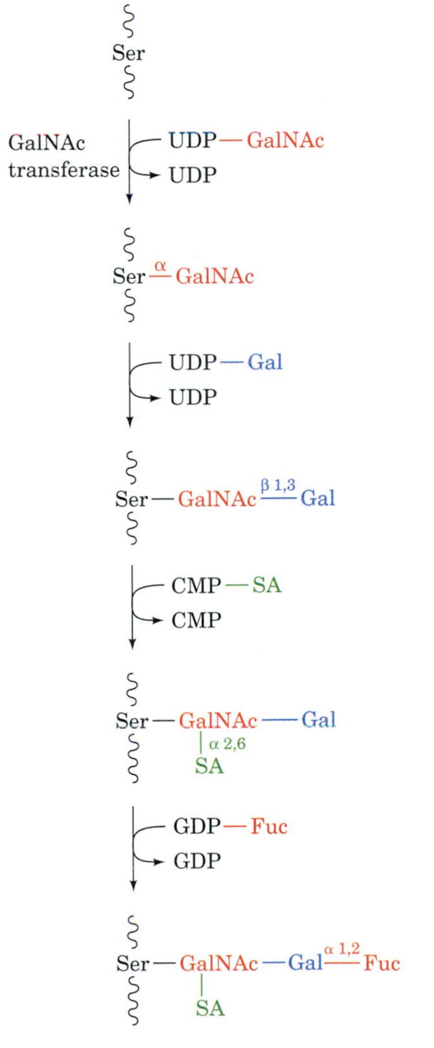

Figure 15-32.   Synthesis of an O-linked oligosaccharide chain. This pathway shows the proposed steps in the assembly of a carbohydrate moiety in canine submaxillary mucin. SA is sialic acid.

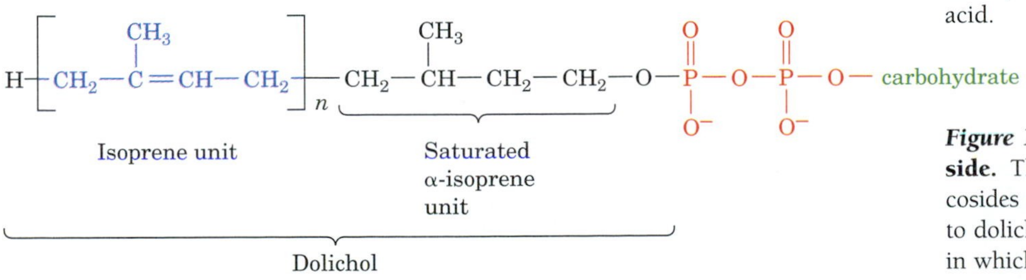

Figure 15-33.   Dolichol pyrophosphate glycoside. The carbohydrate precursors of *N*-linked glycosides are synthesized as oligosaccharides attached to dolichol, a long-chain polyisoprenol ($n = 14$–25) in which the α-isoprene unit is saturated.

residues are transferred to growing dolichol-PP-oligosaccharides from their corresponding dolichol-P derivatives. Dolichol phosphate "activates" a sugar residue for subsequent transfer, as does a nucleoside diphosphate.

The construction of an *N*-linked oligosaccharide begins, as is described in Section 8-3C, by the synthesis of an oligosaccharide with the composition (*N*-acetylglucosamine)$_2$(mannose)$_9$(glucose)$_3$. This occurs on a dolichol carrier in a 12-step process catalyzed by a series of specific glycosyltransferases (Fig. 15-34). Note that some of these reactions take place on the lumenal surface of the endoplasmic reticulum, whereas others occur on its cytoplasmic surface. Hence, on four occasions (Reactions 3, 5, 8, and 11 in Fig. 15-34), dolichol and its attached hydrophilic group are translocated,

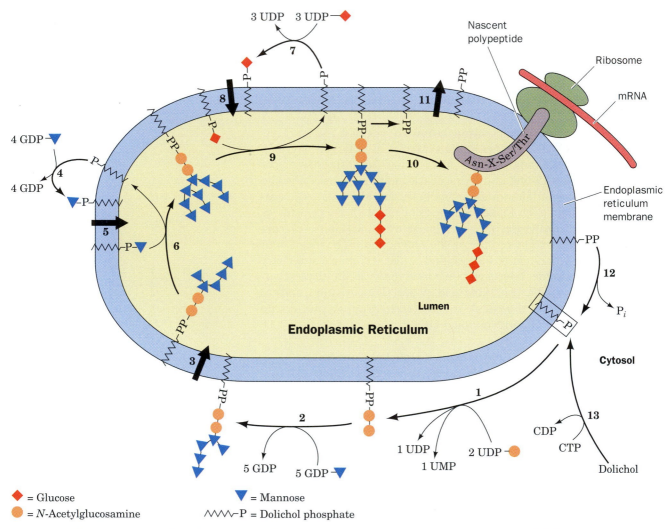

Figure 15-34. The pathway of dolichol-PP-oligosaccharide synthesis. (1) Addition of *N*-acetylglucosamine-1-P and a second *N*-acetylglucosamine to dolichol-P. (2) Addition of five mannosyl residues from GDP–mannose in reactions catalyzed by five different mannosyltransferases. (3) Membrane translocation of dolichol-PP-(*N*-acetylglucosamine)$_2$(mannose)$_5$ to the lumen of the endoplasmic reticulum (ER). (4) Cytosolic synthesis of dolichol-P-mannose from GDP–mannose and dolichol-P. (5) Membrane translocation of dolichol-P-mannose to the lumen of the ER. (6) Addition of four mannosyl residues from dolichol-P-mannose in reactions catalyzed by four different mannosyltransferases. (7) Cytosolic synthesis of dolichol-P-glucose from UDPG and dolichol-P. (8) Membrane translocation of dolichol-P-glucose to the lumen of the ER. (9) Addition of three glucosyl residues from dolichol-P-glucose. (10) Transfer of the oligosaccharide from dolichol-PP to the polypeptide chain at an Asn residue in the sequence Asn-X-Ser/Thr, releasing dolichol-PP. (11) Translocation of dolichol-PP to the cytoplasmic surface of the ER membrane. (12) Hydrolysis of dolichol-PP to dolichol-P. (13) Dolichol-P can also be formed by phosphorylation of dolichol by CTP. [Modified from Abeijon, C. and Hirschberg, C.B., *Trends Biochem. Sci.* **17**, 34 (1992).] ✴ **See the Animated Figures.**

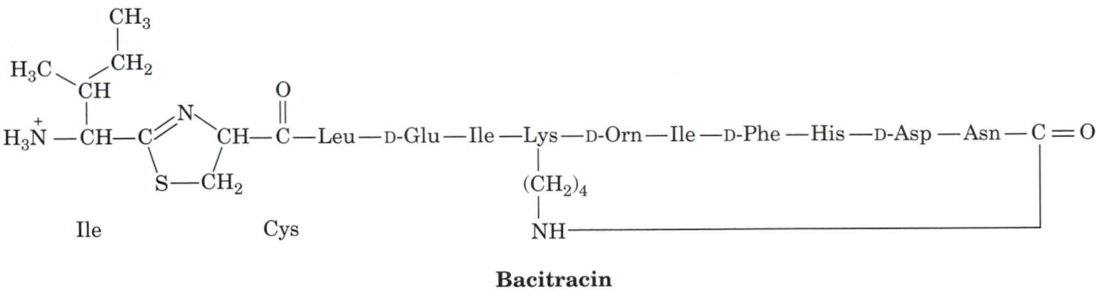

Bacitracin

Figure 15-35. **The chemical structure of bacitracin.** Note that this dodecapeptide has four D-amino acid residues and two unusual intrachain linkages. "Orn" represents the nonstandard amino acid ornithine (Fig. 20-8).

via unknown mechanisms, across the endoplasmic reticulum membrane. In the final steps of this process, the oligosaccharide is transferred to the Asn residue in a segment of sequence Asn-X-Ser/Thr (where X is any residue except Pro and possibly Asp) on a growing polypeptide chain. The resulting dolichol pyrophosphate is hydrolyzed to dolichol phosphate and P_i, a process similar to the pyrophosphatase cleavage of PP_i to 2 P_i. Further processing of the oligosaccharide takes place, as described in Section 8-3C, first in the endoplasmic reticulum and then in the Golgi apparatus (Fig. 8-16), where certain monosaccharide residues are trimmed away by specific glycosylases and others are added by specific nucleotide sugar–requiring glycosyltransferases.

Bacitracin Interferes with the Dephosphorylation of Dolichol Pyrophosphate

A number of compounds block the actions of specific glycosylation enzymes, including **bacitracin** (Fig. 15-35), a cyclic polypeptide that is a widely used antibiotic. Bacitracin forms a complex with dolichol pyrophosphate that inhibits its dephosphorylation (Fig. 15-34, Reaction 12), thereby preventing the synthesis of glycoproteins from dolichol-linked oligosaccharide precursors. Bacitracin is clinically useful because it inhibits bacterial cell wall synthesis (which also involves dolichol-linked oligosaccharides) but does not affect animal cells since it cannot cross cell membranes (bacterial cell wall synthesis is an extracellular process).

SUMMARY

1. Glycogen breakdown requires three enzymes. Glycogen phosphorylase converts the glucosyl units at the nonreducing ends of glycogen to glucose-1-phosphate (G1P). Debranching enzyme transfers an $\alpha(1\rightarrow4)$-linked trisaccharide to a nonreducing end and hydrolyzes the $\alpha(1\rightarrow6)$ linkage. Phosphoglucomutase converts G1P to glucose-6-phosphate (G6P). In liver, G6P is hydrolyzed by glucose-6-phosphatase to glucose for export to the tissues.

2. Glycogen synthesis requires a different pathway in which G1P is activated by reaction with UTP to form UDP–glucose. Glycogen synthase adds glucosyl units to the nonreducing ends of a growing glycogen molecule that has been primed by glycogenin. Branching enzyme removes an $\alpha(1\rightarrow4)$-linked seven-residue segment and reattaches it through an $\alpha(1\rightarrow6)$ linkage to form a branched chain.

3. Glycogen metabolism is controlled in part by allosteric effectors such as AMP, ATP, and G6P. Covalent modification of glycogen phosphorylase and glycogen synthase shifts their $T \rightleftharpoons R$ equilibria and therefore alters their sensitivity to allosteric effectors.

4. The ratio of phosphorylase *a* to phosphorylase *b* depends on the activity of phosphorylase kinase, which is regulated by the activity of cAMP-dependent protein kinase (cAPK), and on the activity of phosphoprotein phosphatase-1. Glycogen phosphorylase is activated by phosphorylation, whereas glycogen synthase is activated by dephosphorylation.

5. Hormones such as glucagon, epinephrine, and insulin control glycogen metabolism. Hormone signals that generate cAMP as a second messenger or that elevate intracellular Ca^{2+}, which binds to the calmodulin subunit of phosphorylase kinase, promote glycogen breakdown. Insulin stimulates glycogen synthesis in part by activating phosphoprotein phosphatase-1.

6. Compounds that can be converted to oxaloacetate can subsequently be converted to glucose. The conversion of pyruvate to glucose by gluconeogenesis requires enzymes that bypass the three exergonic steps of glycolysis: Pyruvate carboxylase and PEP carboxykinase (PEPCK) bypass pyruvate kinase, fructose-1,6-bisphosphatase (FBPase) bypasses phosphofructokinase, and glucose-6-phosphatase bypasses hexokinase.

7. Gluconeogenesis is regulated by changes in enzyme synthesis and by allosteric effectors, including fructose-2,6-bisphosphate (F2,6P), which inhibits FBPase and activates phosphofructokinase (PFK) and whose synthesis depends on the phosphorylation state of the bifunctional enzyme phosphofructokinase-2/fructose bisphosphatase-2 (PFK-2/FBPase-2).

8. Formation of glycosidic bonds requires nucleotide sugars.

REFERENCES

Barford, D. and Johnson, L.N., Electrostatic effects in the control of glycogen phosphorylase by phosphorylation, *Protein Sci.* **3,** 1726–1730 (1994).

Browner, M.F. and Fletterick, R.J., Phosphorylase: a biological transducer, *Trends Biochem. Sci.* **17,** 66–71 (1992).

Calder, P., Glycogen structure and biogenesis, *Int. J. Biochem.* **23,** 1335–1352 (1991).

Chen, Y.-T. and Burchell, A., Glycogen storage diseases, *in* Scriver, C.R., Beaudet, A.L., Sly, W.S., Valle, D., Stanbury, J.B., Wyngaarden, J.B., and Frederickson, D.S. (Eds.), *The Metabolic and Molecular Bases of Inherited Disease* (7th ed.), Chapter 24, McGraw–Hill, New York (1995). [Includes a review of glycogen metabolism.]

Johnson, L.N., Noble, M.E., and Owen, D.J., Active and inactive protein kinases: structural basis for regulation, *Cell* **85,** 149–158 (1996).

Meléndez-Hevia, E., Waddell, T.G., and Shelton, E.D., Optimization of molecular design in the evolution of metabolism: the glycogen molecule, *Biochem. J.* **295,** 477–483 (1993).

Pilkis, S.J. and Granner, D.K., Molecular physiology of the regulation of hepatic gluconeogenesis and glycolysis, *Annu. Rev. Physiol.* **54,** 885–909 (1992).

Pilkis, S.J., Claus, T.H., Kurland, I.J., and Lange, A.J., 6-Phosphofructo-2-kinase/fructose-2,6-bisphosphatase: a metabolic signaling enzyme, *Annu. Rev. Biochem.* **64,** 799–835 (1995).

Shulman, R.G., Bloch, G., and Rothman, D.L., *In vivo* regulation of muscle glycogen synthase and the control of glycogen synthesis, *Proc. Natl. Acad. Sci.* **92,** 8535–8542 (1995).

Zhang, M., Tanaka, T., and Ikura, M., Calcium-induced conformational transition revealed by the solution structure of apo calmodulin, *Nature Struct. Biol.* **2,** 758–767 (1995).

KEY TERMS

glycogenolysis	nucleotide sugar	CaM	reducing equivalent
phosphorolysis	interconvertible enzymes	second messenger	glycogen storage disease
debranching	cAMP	gluconeogenesis	

STUDY EXERCISES

1. List the metabolic sources and products of G6P.

2. How does the structure of glycogen relate to its metabolic function?

3. Describe the enzymatic degradation and synthesis of glycogen.

4. Why must opposing biosynthetic and degradative pathways differ in at least one enzyme?

5. Why does a phosphorylation/dephosphorylation system allow more sensitive regulation of a metabolic process than a simple allosteric system?

6. How does regulation of glycogen metabolism differ between liver and muscle?

7. Describe the reactions of gluconeogenesis.

8. Why is the malate–aspartate shuttle system important for gluconeogenesis?

9. Describe the role of fructose-2,6-bisphosphate in regulating gluconeogenesis.

PROBLEMS

1. Write the balanced equation for (a) the sequential conversion of glucose to pyruvate and of pyruvate to glucose and (b) the catabolism of six molecules of G6P by the pentose phosphate pathway followed by conversion of ribulose-5-phosphate back to G6P by gluconeogenesis.

2. Phosphoglucokinase catalyzes the phosphorylation of the C6 OH group of G1P. Why is this enzyme important for the normal function of phosphoglucomutase?

3. The free energy of hydrolysis of an $\alpha(1\rightarrow4)$ glycosidic bond is -15.5 kJ $\cdot$ mol^{-1}, whereas that of an $\alpha(1\rightarrow6)$ glycosidic bond is -7.1 kJ $\cdot$ mol^{-1}. Use these data to explain why glycogen debranching includes three reactions [breaking and re-forming $\alpha(1\rightarrow4)$ bonds and hydrolyzing $\alpha(1\rightarrow6)$ bonds], while glycogen branching requires only two reactions [breaking $\alpha(1\rightarrow4)$ bonds and forming $\alpha(1\rightarrow6)$ bonds].

4. Calculations based on the volume of a glucose residue and the branching pattern of cellular glycogen indicate that a glycogen molecule could have up to 28 branching tiers before becoming impossibly dense. What are the advantages of such a molecule and why is it not found *in vivo*?

5. One molecule of dietary glucose can be oxidized through glycolysis and the citric acid cycle to generate a maximum of 38 molecules of ATP. Calculate the fraction of this energy that is lost when the glucose is stored as glycogen before it is catabolized.

6. Many diabetics do not respond to insulin because of a deficiency of insulin receptors on their cells. How does this affect (a) the levels of circulating glucose immediately after a meal and (b) the rate of glycogen synthesis in muscle?

7. Caffeine inhibits cAMP phosphodiesterase. How does this affect metabolic responses to epinephrine?

8. Glucose-6-phosphatase is located inside the endoplasmic reticulum. Describe the probable symptoms of a defect in G6P transport across the endoplasmic reticulum membrane.

9. Individuals with McArdle's disease often experience a "second wind" resulting from cardiovascular adjustments that allow glucose mobilized from liver glycogen to fuel muscle contraction. Explain why the amount of ATP derived in the muscle from circulating glucose is less than the amount of ATP that would be obtained by mobilizing the same amount of glucose from muscle glycogen.

10. A sample of glycogen from a patient with liver disease is incubated with P$_i$, normal glycogen phosphorylase, and normal debranching enzyme. The ratio of G1P to glucose formed in this reaction mixture is 100. What is the patient's most probable enzymatic deficiency?

The synthesis and degradation of numerous biological materials depends on the flow of molecules and energy through the citric acid cycle, which has been likened to a metabolic water wheel. [© *Lyle Leduc/Gamma Liaison.*]

CITRIC ACID CYCLE

In the preceding two chapters, we examined the catabolism of glucose and its synthesis, storage, and mobilization. Although glucose is a source of energy for nearly all cells, it is not the only metabolic fuel, nor is glycolysis the only energy-yielding catabolic pathway. Cells that rely exclusively on glycolysis to meet their energy requirements actually waste most of the chemical potential energy of carbohydrates. When glucose is converted to lactate or ethanol, a relatively reduced product leaves the cell. If the end product of glycolysis is instead further oxidized, the cell can recover considerably more energy.

The oxidation of an organic compound requires an electron acceptor, such as NO_3^-, SO_4^{2-}, Fe^{3+}, or O_2, all of which are exploited as oxidants in different organisms. In aerobic organisms, the electrons produced by oxidative metabolism are ultimately transferred to O_2. Oxidation of metabolic fuels is carried out by the citric acid cycle, a sequence of reactions that arose sometime after levels of atmospheric oxygen became significant, about 3 billion years ago. As the reduced carbon atoms of metabolic fuels are oxidized to CO_2, electrons are transferred to electron carriers that are subsequently oxidized by O_2. In this chapter, we examine the oxidation reactions of the citric acid cycle itself. In the following chapter, we examine the fate of the electrons and see how their energy is used to drive the synthesis of ATP.

It is sometimes convenient to think of the citric acid cycle as an addendum to glycolysis. Pyruvate derived from glucose can be split into CO_2 and a two-carbon fragment that enters the cycle for oxidation as acetyl-CoA (Fig. 16-1). However, it is really misleading to think of the citric acid cycle as merely a continuation of carbohydrate catabolism. *The citric acid cycle is a central pathway for recovering energy from several metabolic fuels, including carbohydrates, fatty acids, and amino acids, that are broken down to acetyl-CoA for oxidation.* In fact, under some conditions, the principal function of the citric acid cycle is to recover energy from fatty acids. We shall also see that the citric acid cycle supplies the reactants for a variety of biosynthetic pathways.

We begin this chapter with an overview of the citric acid cycle. Next, we explore how acetyl-CoA, its starting compound, is formed from pyruvate. After discussing the reactions catalyzed by each of the enzymes of the cycle, we consider the regulation of these enzymes. Finally, we examine the links between citric acid cycle intermediates and other metabolic processes.

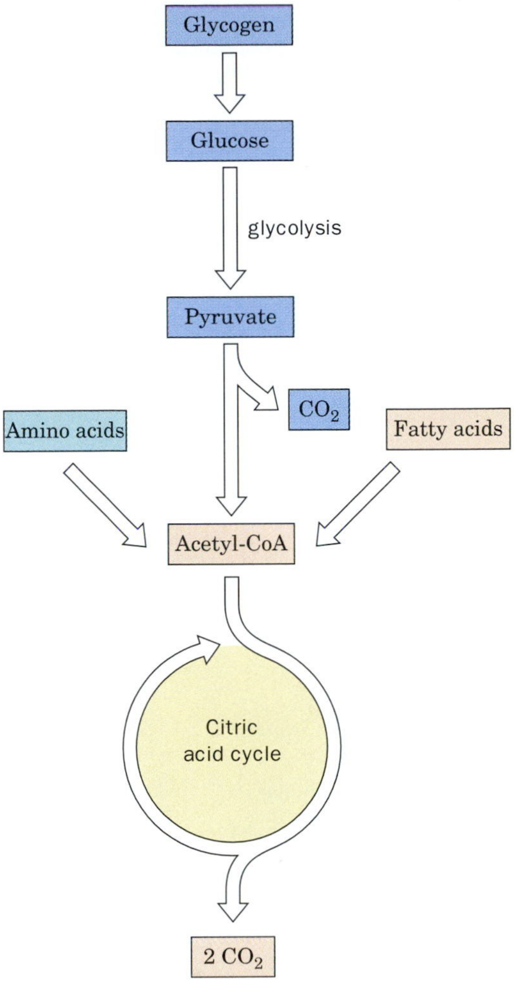

Figure 16-1.   Overview of oxidative fuel metabolism. Acetyl groups derived from carbohydrates, amino acids, and fatty acids enter the citric acid cycle, where they are oxidized to CO_2. ✳ See the Animated Figures.

1. OVERVIEW OF THE CITRIC ACID CYCLE

The citric acid cycle (Fig. 16-2) is an ingenious series of eight reactions that oxidizes the acetyl group of acetyl-CoA to two molecules of CO_2 in a manner that conserves the liberated free energy in the reduced compounds NADH and $FADH_2$. The cycle is named after the product of its first reaction, **citrate.** One complete round of the cycle yields two molecules of CO_2, three NADH, one $FADH_2$, and one "high-energy" compound (GTP or ATP).

The citric acid cycle first came to light in the 1930s. Until then, the mechanism of glucose oxidation and its relationship to cellular respiration (oxygen uptake) was a mystery. It was known that various dicarboxylates (**α-ketoglutarate, succinate,** and **malate**) and a tricarboxylate (citrate) are rapidly oxidized by muscle tissue during respiration. In 1935, Albert Szent-Györgyi found that cellular respiration is dramatically accelerated by small amounts of succinate, fumarate, malate, or oxaloacetate. In fact, the addi-

⦿ See Guided Exploration 14:

Citric Acid Cycle Overview.

468

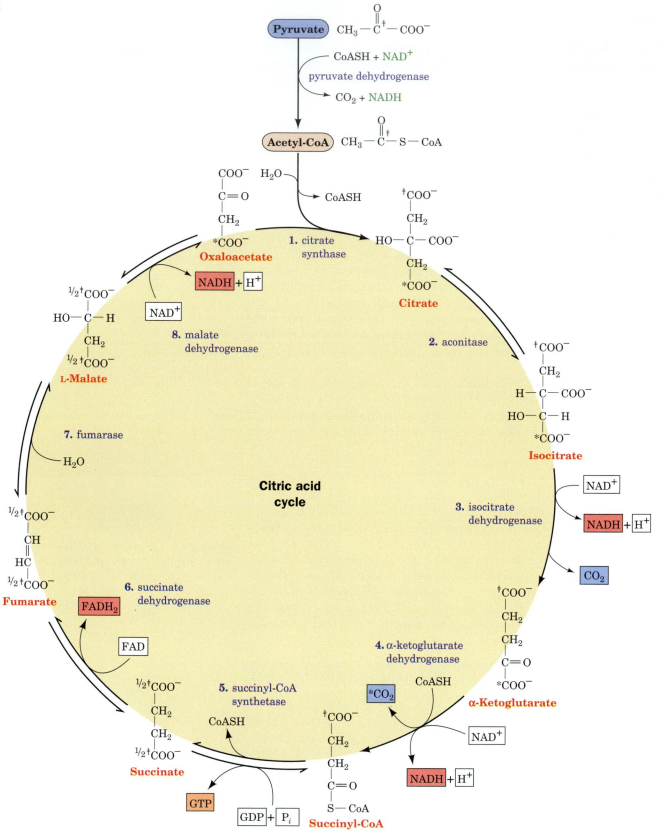

Figure 16-2. *Key to Metabolism.* **The reactions of the citric acid cycle.** The reactants and products of this catalytic cycle are boxed. The pyruvate → acetyl-CoA reaction (*top*) supplies the cycle's substrate via carbohydrate metabolism but is not considered to be part of the cycle. An isotopic label at C4 of oxaloacetate (*) becomes C1 of α-ketoglutarate and is released as CO_2 in Reaction 4. An isotopic label at C1 of acetyl-CoA (†) becomes C5 of α-ketoglutarate and is scrambled in Reaction 5 between C1 and C4 of succinate (½†). ✳ See the **Animated Figures.**

tion of these compounds stimulates O_2 uptake and CO_2 production far in excess of what would be expected for the direct oxidation of these compounds. *These findings are consistent with substrate oxidation by a catalytic system that repeatedly returns to its starting point.* However, the discovery of the circular nature of the pathway required additional studies to determine the sequence of interconversion of the eight intermediates. The circle was closed in principle in 1936 when Carl Martius and Franz Knoop showed that citrate (the first intermediate) can be formed nonenzymatically from oxaloacetate (the eighth intermediate) and pyruvate.

Hans Krebs used this chemical model as a point of departure for the biochemical experiments that led, in 1937, to his proposal of the citric acid cycle, a contribution that ranks as one of the most important achievements of metabolic chemistry. The idea of a catalytic cycle was not new to Krebs. In 1932, he and Kurt Henseleit had elucidated the outline of the urea cycle, which converts CO_2 and ammonia to urea for excretion (Section 20-3). Nevertheless, some major gaps remained in the complete elucidation of the citric acid cycle. For example, the mechanism of citrate formation from pyruvate and oxaloacetate was unknown. In fact, coenzyme A was not discovered until 1945, and only in 1951 was acetyl-CoA shown to be the intermediate that condenses with oxaloacetate to form citrate. Although the enzymes and intermediates of the citric acid cycle are now well established, many investigators continue to explore the molecular mechanisms of the enzymes and how the enzymes are regulated for optimal performance under varying metabolic conditions in different organisms.

Before we examine each of the reactions in detail, we should emphasize some general features of the citric acid cycle:

1. The circular pathway, which is also called the **Krebs cycle** or the **tricarboxylic acid (TCA) cycle,** oxidizes acetyl groups from many sources, not just pyruvate. Because it accounts for the major portion of carbohydrate, fatty acid, and amino acid oxidation, the citric acid cycle is often considered the "hub" of cellular metabolism.

2. The net reaction of the citric acid cycle is

$$3 \text{ NAD}^+ + \text{FAD} + \text{GDP} + \text{P}_i + \text{acetyl-CoA} + 2 \text{ H}_2\text{O} \longrightarrow$$
$$3 \text{ NADH} + \text{FADH}_2 + \text{GTP} + \text{CoA} + 2 \text{ CO}_2 + 3\text{H}^+$$

The oxaloacetate that is consumed in the first step of the citric acid cycle is regenerated in the last step of the cycle. Thus, *the citric acid cycle acts as a multistep catalyst that can oxidize an unlimited number of acetyl groups.*

3. In eukaryotes, all the enzymes of the citric acid cycle are located in the mitochondrion, so all substrates, including NAD^+ and GDP, must be generated in the mitochondria or be transported into mitochondria from the cytosol. Similarly, all the products of the citric acid cycle must be consumed in the mitochondria or transported into the cytosol.

4. The carbon atoms of the two molecules of CO_2 produced in one round of the cycle are not the two carbons of the acetyl group that began the round (Fig. 16-2). These acetyl carbon atoms are lost in subsequent rounds of the cycle. However, the net effect of each round of the cycle is the oxidation of one acetyl group to 2 CO_2.

5. Citric acid cycle intermediates are precursors for the biosynthesis of other compounds (e.g., oxaloacetate for gluconeogenesis; Section 15-4).

6. The oxidation of an acetyl group to 2 CO_2 requires the transfer of

four pairs of electrons. The reduction of 3 NAD^+ to 3 NADH accounts for three pairs of electrons; the reduction of FAD to $FADH_2$ accounts for the fourth pair. Much of the free energy of oxidation of the acetyl group is conserved in these reduced coenzymes. Energy is also recovered in GTP (or ATP). In Section 17-3C, we shall see that 11 ATP are formed when the four pairs of electrons are eventually transferred to O_2.

2. SYNTHESIS OF ACETYL-COENZYME A

Acetyl groups enter the citric acid cycle as part of the "high-energy" compound acetyl-CoA (recall that thioesters have high free energies of hydrolysis; Section 13-2D). Although acetyl-CoA can also be derived from fatty acids (Section 19-2) and some amino acids (Section 20-4), we shall focus here on the production of acetyl-CoA from pyruvate derived from carbohydrates.

As we saw in Section 14-3, the end product of glycolysis under anaerobic conditions is lactate or ethanol. However, under aerobic conditions, when the NADH generated by glycolysis is reoxidized in the mitochondria, the final product is pyruvate. A transport protein imports pyruvate along with H^+ (i.e., a pyruvate–H^+ symport) into the mitochondrion for further oxidation.

A. The Pyruvate Dehydrogenase Multienzyme Complex

Multienzyme complexes are groups of noncovalently associated enzymes that catalyze two or more sequential steps in a metabolic pathway. Virtually all organisms contain multienzyme complexes, which represent a step forward in the evolution of catalytic efficiency because they offer the following advantages:

1. Enzymatic reaction rates are limited by the frequency with which enzymes collide with their substrates (Section 11-3E). When a series of reactions occurs within a multienzyme complex, the distance that substrates must diffuse between active sites is minimized, thereby enhancing the reaction rate.

2. The channeling of metabolic intermediates between successive enzymes in a metabolic pathway reduces the opportunity for these intermediates to react with other molecules, thereby minimizing side reactions.

3. The reactions catalyzed by a multienzyme complex can be coordinately controlled.

Acetyl-CoA is formed from pyruvate through oxidative decarboxylation by a multienzyme complex named **pyruvate dehydrogenase.** This complex contains multiple copies of three enzymes: **pyruvate dehydrogenase (E_1), dihydrolipoyl transacetylase (E_2),** and **dihydrolipoyl dehydrogenase (E_3).**

The *E. coli* pyruvate dehydrogenase complex is an ~4600-kD particle with a diameter of about 300 Å (Fig. 16-3a). The core of the particle is made of 24 E_2 proteins arranged in a cube (Fig. 16-3b). Twenty-four E_1 proteins and 12 E_3 proteins are arranged around the E_2 core (Fig. 16-4).

Eukaryotic pyruvate dehydrogenase complexes are more complicated than the *E. coli* complex, although they catalyze the same reactions with homologous enzymes and by similar mechanisms. The eukaryotic complex is a dodecahedron (a regular polyhedron that has 12 pentagonal faces) made

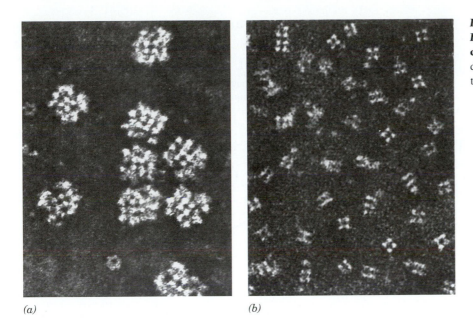

(a) (b)

Figure 16-3. **Electron micrographs of the *E. coli* pyruvate dehydrogenase multienzyme complex.** (*a*) The intact complex. (*b*) The dihydrolipoyl transacetylase (E₂) core complex. [Courtesy of Lester Reed, University of Texas at Austin.]

of 60 E_1 proteins and 12 E_3 proteins surrounding a core of 60 E_2 proteins. The mammalian pyruvate dehydrogenase complex also contains about six copies of the so-called **protein X** (a catalytically inactive E_2-like protein that may help bind E_3 to the complex) and one to three copies each of **pyruvate dehydrogenase kinase** and **pyruvate dehydrogenase phosphatase** (which regulate the activity of the complex by a phosphorylation/dephosphorylation mechanism; Section 16-4A).

B. The Reactions of the Pyruvate Dehydrogenase Complex

The pyruvate dehydrogenase complex catalyzes five sequential reactions with the overall stoichiometry

$$\text{Pyruvate} + \text{CoA} + \text{NAD}^+ \longrightarrow \text{acetyl-CoA} + \text{CO}_2 + \text{NADH}$$

Five different coenzymes are required: thiamine pyrophosphate (TPP; Section 14-3B), **lipoamide**, coenzyme A (Fig. 13-9), FAD (Fig. 3-3), and NAD^+ (Fig. 3-4). The coenzymes and their mechanistic functions are listed in Table

(a) (b) (c)

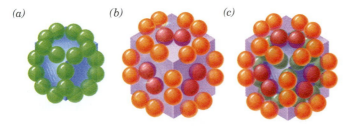

Figure 16-4. Structural organization of the *E. coli* pyruvate dehydrogenase multienzyme complex. (*a*) The dihydrolipoyl transacetylase (E₂) core. The 24 E₂ proteins (*green spheres*) associate as trimers at the corners of a cube. (*b*) The 24 pyruvate dehydrogenase (E₁) proteins (*orange spheres*) form dimers that associate with the E₂ core (*shaded cube*) along its 12 edges. The 12 dihydrolipoyl dehydrogenase (E₃) proteins (*purple spheres*) form dimers that attach to the six faces of the E₂ cube. (*c*) Parts *a* and *b* combined form the entire 60-subunit complex.

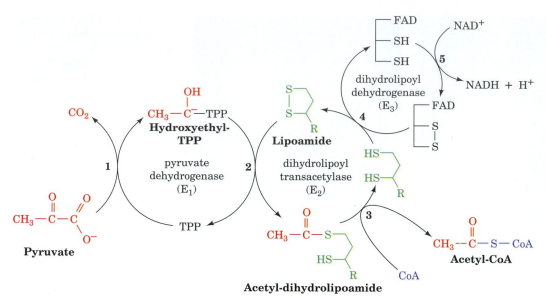

Figure 16-5. **The five reactions of the pyruvate dehydrogenase multienzyme complex.** E_1 (pyruvate dehydrogenase) contains TPP and catalyzes Reactions 1 and 2. E_2 (dihydrolipoyl transacetylase) contains lipoamide and catalyzes Reaction 3. E_3 (dihydrolipoyl dehydrogenase) contains FAD and a redox-active disulfide and catalyzes Reactions 4 and 5.

16-1. The sequence of reactions catalyzed by the pyruvate dehydrogenase complex is as follows (Fig. 16-5):

1. Pyruvate dehydrogenase (E_1), a TPP-requiring enzyme, decarboxylates pyruvate with the formation of a hydroxyethyl-TPP intermediate.

Table 16-1. The Coenzymes and Prosthetic Groups of Pyruvate Dehydrogenase

Cofactor	Location	Function
Thiamine pyrophosphate (TPP)	Bound to E_1	Decarboxylates pyruvate yielding a hydroxyethyl-TPP carbanion
Lipoic acid	Covalently linked to a Lys on E_2 (lipoamide)	Accepts the hydroxyethyl carbanion from TPP as an acetyl group
Coenzyme A (CoA)	Substrate for E_2	Accepts the acetyl group from lipoamide
Flavin adenine dinucleotide (FAD)	Bound to E_3	Reduced by lipoamide
Nicotinamide adenine dinucleotide (NAD^+)	Substrate for E_3	Reduced by $FADH_2$

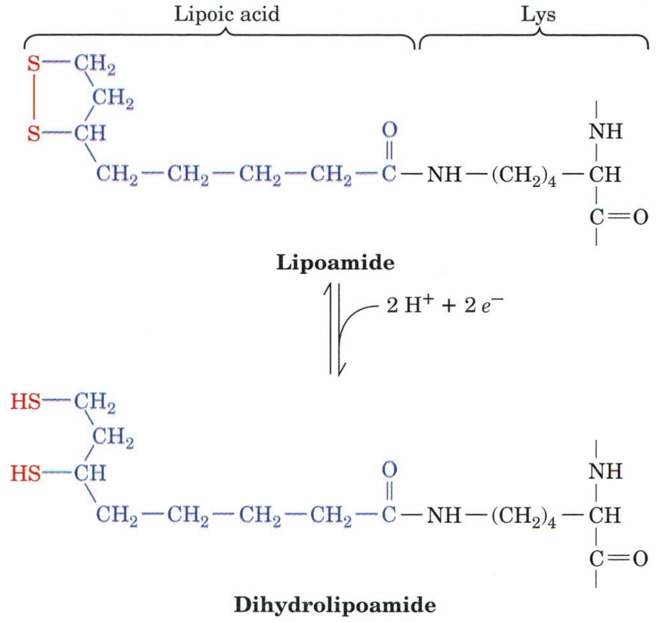

Figure 16-6.   Interconversion of lipoamide and dihydrolipoamide. Lipoamide consists of lipoic acid covalently joined to the ε-amino group of a Lys residue via an amide bond.

This reaction is identical to that catalyzed by yeast pyruvate decarboxylase (Section 14-3B).

2. The hydroxyethyl group is transferred to the next enzyme, dihydrolipoyl transacetylase (E_2), which contains a lipoamide group. Lipoamide consists of **lipoic acid** linked via an amide bond to the ε-amino group of a Lys residue (Fig. 16-6). The reactive center of lipoamide is a cyclic disulfide that can be reversibly reduced to yield **dihydrolipoamide.** The hydroxyethyl group derived from pyruvate attacks the lipoamide disulfide, and TPP is eliminated. The hydroxyethyl carbanion is thereby oxidized to an acetyl group as the lipoamide disulfide is reduced.

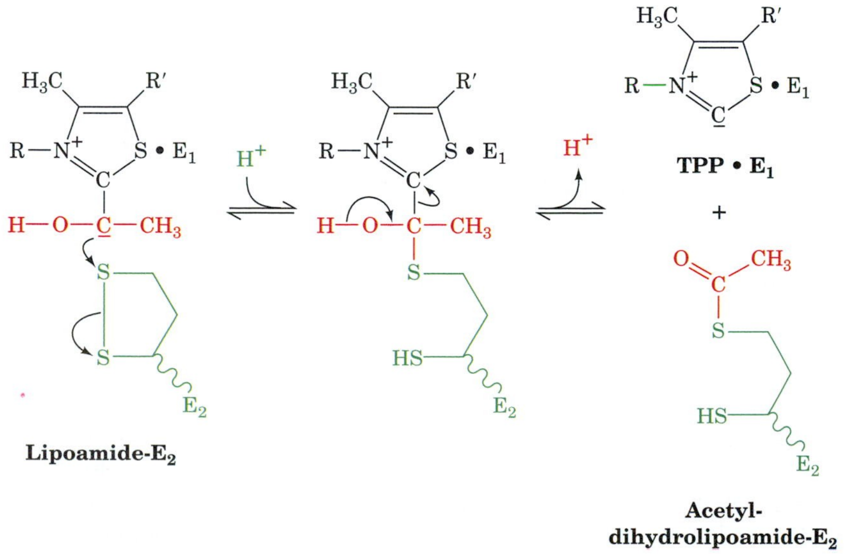

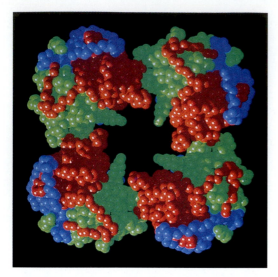

Figure 16-7. The X-ray structure of the A. vinelandii dihydrolipoyl transacetylase (E₂) catalytic domain. Each residue is represented by a sphere. The 24 proteins are arranged as 8 trimers at the corners of a cube (only the forward half of the complex is visible). The edge of the cube is ~125 Å long. Substrates and cofactors fit in the spaces between the E_2 proteins and in the interior of the cube. [Courtesy of Wim Hol and Andrea Mattevi, University of Washington.]

3. E_2 then catalyzes a transesterification reaction in which the acetyl group is transferred to CoA, yielding acetyl-CoA and dihydrolipoamide-E_2.

Acetyl-CoA

Acetyl-dihydrolipoamide-E₂

Dihydrolipoamide-E₂

4. Acetyl-CoA has now been formed, but the lipoamide group of E_2 must be regenerated. Dihydrolipoyl dehydrogenase (E_3) reoxidizes dihydrolipoamide to complete the catalytic cycle of E_2. Oxidized E_3 contains a reactive Cys—Cys disulfide group and a tightly bound FAD. The oxidation of dihydrolipoamide is a disulfide interchange reaction.

E_3 (oxidized)   **E_2**   **E_3 (reduced)**   **E_2**

5. Finally, reduced E_3 is reoxidized. The sulfhydryl groups are reoxidized by FAD, and the resulting FADH$_2$ is oxidized by NAD$^+$, producing NADH.

$$NAD^+ \quad NADH + H^+$$

E_3 (oxidized)

How are reaction intermediates channeled between E_2 (the core of the pyruvate dehydrogenase complex) and the E_1 and E_3 proteins on the outside? The key is the lipoamide group of E_2. The lipoic acid residue and the side chain of the Lys residue to which it is attached have a combined length of about 14 Å. This **lipoyllysyl arm** *(at left)* apparently acts as a long tether that swings the disulfide group from E_1 (where it picks up a hydroxyethyl group), to the E_2 active site (where the hydroxyethyl group is transferred to form acetyl-CoA), and from there to E_3 (where the reduced disulfide is reoxidized). The domains of E_2 that carry the lipoyllysyl arms are linked to the rest of the E_2 protein by a Pro- and Ala-rich flexible hinge. Because of the flexibility and reach of the lipoyllysyl arms, one E_1 protein can acetylate numerous E_2 proteins, and one E_3 protein can reoxidize several dihydrolipoamide groups.

The X-ray structure of the catalytic domain of E_2 from the bacterium *Azotobacter vinelandii*, in agreement with electron micrographs (Fig. 16-3*b*), shows that the 24 E_2 proteins are arranged in trimers at the corners of the pyruvate dehydrogenase core cube (Fig. 16-7). The core contains channels large enough for substrates to diffuse in and out. In fact, CoA must ap-

14 Å

**Lipoyllysyl arm
(fully extended)**

Box 16-1

BIOCHEMISTRY IN HEALTH AND DISEASE

Arsenic Poisoning

The toxicity of arsenic has been known since ancient times. As(III) compounds such as **arsenite** (AsO_3^{3-}) and **organic arsenicals** are toxic because they bind to sulfhydryl compounds (including lipoamide) that can form bidentate adducts.

Arsenite **Dihydrolipoamide**

Organic arsenical

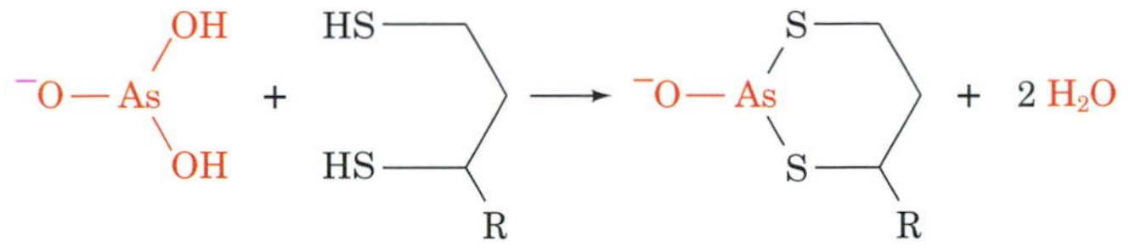

The inactivation of lipoamide-containing enzymes by arsenite, especially the pyruvate dehydrogenase and the α-ketoglutarate dehydrogenase complexes, brings respiration to a halt. However, organic arsenicals are more toxic to microorganisms than they are to humans, apparently because of differences in the sensitivities of their various enzymes to these compounds. This differential toxicity is the basis for the early twentieth century use of organic arsenicals in the treatment of **syphilis** (a bacterial disease) and **trypanosomiasis** (a parasitic disease). These compounds were actually the first antibiotics, although, not surprisingly, they produced severe side effects.

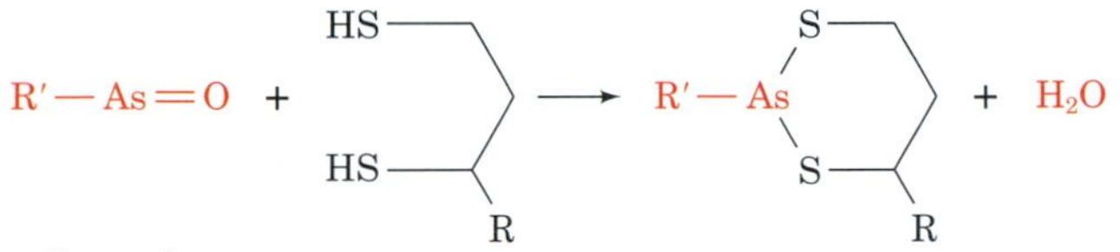

Arsenic is often suspected as a poison in untimely deaths. It was long thought that Napoleon Bonaparte died from arsenic poisoning while in exile on the island of St. Helena, a suspicion that is strongly supported by the recent finding that a lock of his hair contains high levels of arsenic. But was it murder or environmental pollution? Arsenic-containing dyes were used in wallpaper at the time, and it was eventually determined that in damp weather, fungi convert the arsenic to a volatile compound. Samples of the wallpaper from Napoleon's room in fact contain arsenic. Napoleon's arsenic poisoning may therefore have been unintentional.

Charles Darwin may also have been an unwitting victim of chronic arsenic poisoning. In the years following his epic voyage on the *Beagle,* Darwin was plagued by eczema, vertigo, headaches, gout, and nausea—all symptoms of arsenic poisoning. Fowler's solution, a widely used nineteenth century "tonic," contained 10 mg of arsenite per mL. Many individuals, quite possibly Darwin himself, took this "medication" for years.

---

proach its binding site from inside the cube. The lipoyllysyl arms probably protrude into the hollow interior of the E_2 core and swing around in order to "visit" the active sites of E_1, E_2, and E_3. The entire pyruvate dehydrogenase complex can be inactivated by the reaction of the lipoamide group with certain arsenic-containing compounds (see Box 16-1).

3. ENZYMES OF THE CITRIC ACID CYCLE

In this section, we discuss the eight enzymes of the citric acid cycle. The elucidation of the mechanisms for each of these enzymes is the result of an enormous amount of experimental work. Even so, there remain questions about the mechanistic details of the enzymes and their regulatory properties.

A. *Citrate Synthase*

Citrate synthase catalyzes the condensation of acetyl-CoA and oxaloacetate. This initial reaction of the citric acid cycle is the point at which carbon atoms (from carbohydrates, fatty acids, and amino acids) are "fed into the furnace" as acetyl-CoA. The citrate synthase reaction proceeds with an Ordered Sequential kinetic mechanism in which oxaloacetate binds before acetyl-CoA.

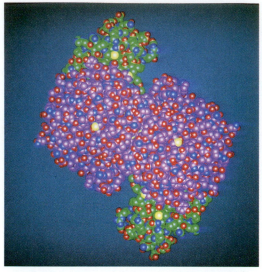

(a)

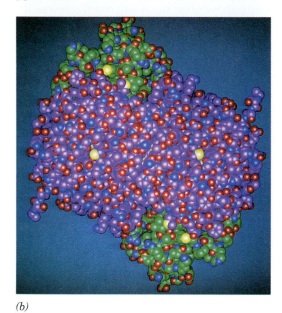

(b)

Figure 16-8.   Conformational changes in cit-rate synthase. (a) The open conformation. (b) The closed, substrate-binding conformation. The C atoms of the small domain in each subunit of the enzyme are green, and those of the large domain are magenta. N, O, and S atoms in both domains are blue, red, and yellow. The large conformational shift between the open and closed forms entails relative interatomic movements of up to 15 Å. [Courtesy of Anne Dallas, University of Pennsylvania; and Helen Berman, Fox Chase Cancer Center. Based on X-ray structures determined by James Remington and Robert Huber, Max-Planck-Institut für Biochemie, Germany.]
🔵 **See the Interactive Exercises.**

X-Ray studies show that the free enzyme (a dimer) is in an "open" form, with two domains that form a cleft containing the oxaloacetate binding site (Fig. 16-8a). When oxaloacetate binds, the smaller domain undergoes a remarkable 18° rotation, which closes the cleft (Fig. 16-8b). The existence of the "open" and "closed" forms explains the enzyme's Ordered Sequential kinetic behavior. *The conformational change generates the acetyl-CoA binding site and seals the oxaloacetate binding site so that solvent cannot reach the bound substrate.*

In the reaction mechanism proposed by James Remington, three ionizable side chains of citrate synthase participate in catalysis (Fig. 16-9):

1. The enol of acetyl-CoA is generated in the rate-limiting step of the reaction when Asp 375 (a base) removes a proton from the methyl group and His 274 (an acid) protonates the enolate oxygen.

2. **Citryl-CoA** is formed in a second concerted acid–base catalyzed step, in which the acetyl-CoA enol (a nucleophile) attacks oxaloacetate. His 274 abstracts the previously donated proton, and His 320 (an acid) donates a proton to oxaloacetate's carbonyl group. The citryl-CoA intermediate remains bound to the enzyme. Citrate synthase is one of the few enzymes that can directly form a carbon–carbon bond without the assistance of a metal ion cofactor.

3. Citryl-CoA is hydrolyzed to citrate and CoA. This hydrolysis provides the reaction's thermodynamic driving force ($\Delta G°' = -31.5$ kJ·mol^{-1}). We shall see later why this reaction requires such a large, seemingly wasteful, expenditure of free energy.

B. Aconitase

Aconitase catalyzes the reversible isomerization of citrate and **isocitrate,** with **cis-aconitate** as an intermediate.

$$
\begin{array}{ccccc}
\text{COO}^- & & \text{COO}^- & & \text{COO}^- \\
| & & | & & | \\
\text{CH}_2 & & \text{CH}_2 & & \text{CH}_2 \\
| & \xrightarrow{\text{H}_2\text{O}} & | & \xrightarrow{\text{H}_2\text{O}} & | \\
\text{HO—C—COO}^- & \rightleftharpoons & \text{C—COO}^- & \rightleftharpoons & \text{H—C—COO}^- \\
| & & \| & & | \\
\text{CH}_2 & & \text{CH} & & \text{HO—C—H} \\
| & & | & & | \\
\text{COO}^- & & \text{COO}^- & & \text{COO}^- \\
\textbf{Citrate} & & \textit{cis}\textbf{-Aconitate} & & \textbf{Isocitrate}
\end{array}
$$

The reaction begins with a dehydration step in which a proton and an OH group are removed. Since citrate has two carboxymethyl groups substituent to its central C atom, it is prochiral rather than chiral. Thus, although water might conceivably be eliminated from either of the two carboxymethyl arms, aconitase removes water only from citrate's lower (*pro-R*) arm (i.e., such that the product molecule has the *R* configuration; see Box 4-1).

Aconitase contains a **[4Fe–4S] iron–sulfur cluster** (an arrangement of four iron atoms and four sulfur atoms) that presumably coordinates the OH group of citrate to facilitate its elimination. Iron–sulfur clusters normally participate in redox processes (Section 17-2C); aconitase is an intriguing exception.

The second stage of the aconitase reaction is rehydration of the double bond of *cis*-aconitate to form isocitrate. Although addition of water across the double bond of *cis*-aconitate could potentially yield four stereoisomers, aconitase catalyzes the stereospecific addition of OH$^-$ and H$^+$ to produce only one isocitrate stereoisomer.

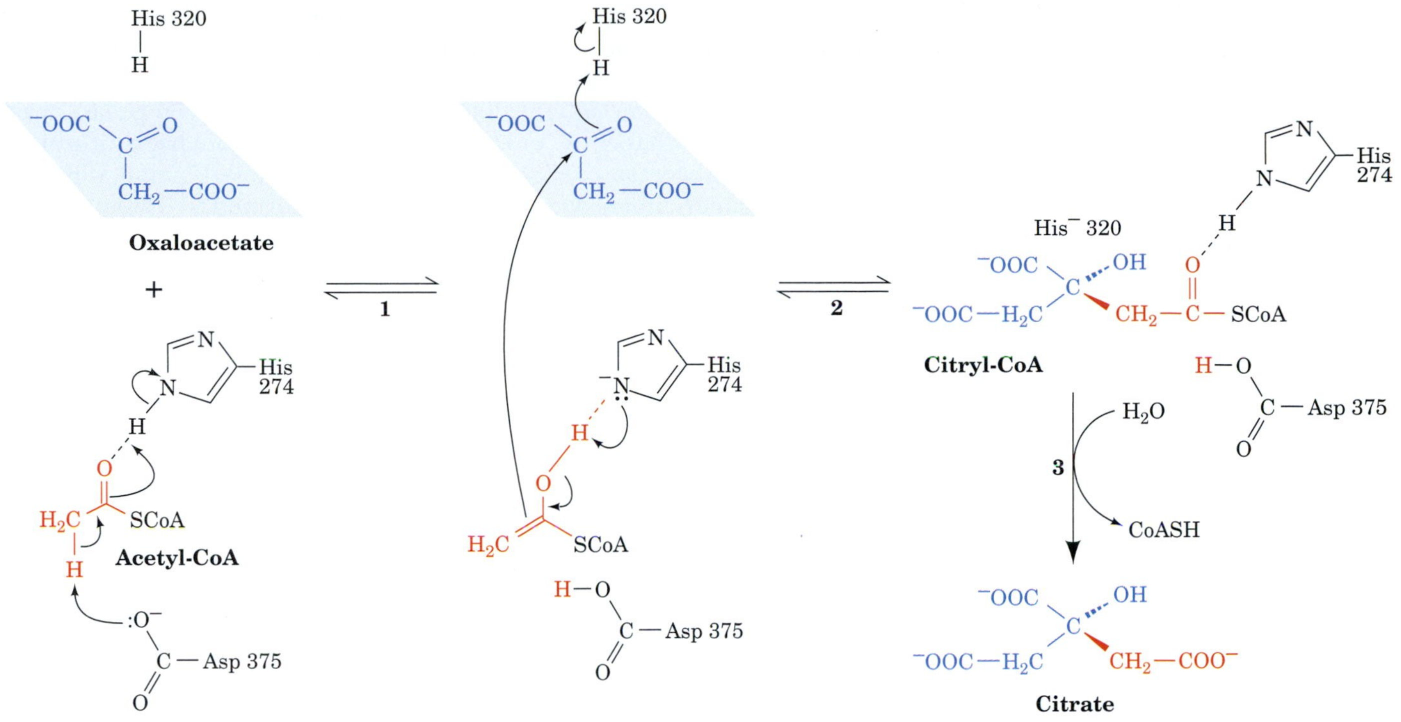

Figure 16-9.   The mechanism of the citrate synthase reaction. His 274 and His 320 in their neutral forms and Asp 375 have been implicated as general acid–base catalysts. The rate-limiting step is the formation of the acetyl-CoA enol, which is stabilized by a hydrogen bond to the resulting His 274 anion. The acetyl-

CoA enol then nucleophilically attacks oxaloacetate's carbonyl carbon. The resulting intermediate, citryl-CoA, is hydrolyzed to yield citrate and CoA. [Mostly after Remington, J.S., *Curr. Opin. Struct. Biol.* **2**, 732 (1992).]

C. NAD⁺-Dependent Isocitrate Dehydrogenase

Isocitrate dehydrogenase catalyzes the oxidative decarboxylation of isocitrate to α-ketoglutarate. This reaction produces the first CO_2 and NADH of the citric acid cycle. Note that this CO_2 began the citric acid cycle as a component of oxaloacetate, not of acetyl-CoA (Fig. 16-2). (Mammalian tissues also contain an isocitrate dehydrogenase isozyme that uses $NADP^+$ as a cofactor.)

NAD⁺-dependent isocitrate dehydrogenase, which also requires a Mn^{2+} or Mg^{2+} cofactor, catalyzes the oxidation of a secondary alcohol (isocitrate) to a ketone **(oxalosuccinate)** followed by the decarboxylation of the carboxyl group β to the ketone (Fig. 16-10). Mn^{2+} helps polarize the newly

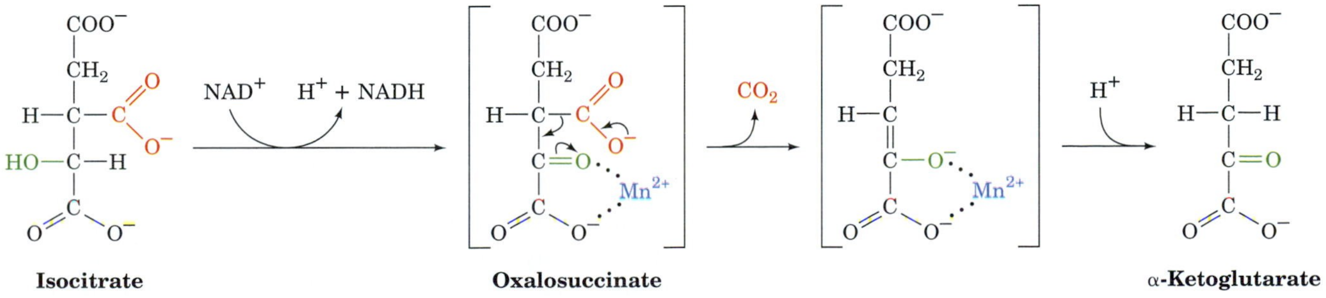

Figure 16-10.   The reaction mechanism of isocitrate dehydrogenase. Oxalosuccinate is shown in brackets because it does not dissociate from the enzyme.

formed carbonyl group. The isocitrate dehydrogenase reaction mechanism is similar to that of phosphogluconate dehydrogenase in the pentose phosphate pathway (Section 14-6A).

The oxalosuccinate intermediate of the isocitrate dehydrogenase reaction exists only transiently, and its existence had therefore been difficult to confirm. However, an enzymatic reaction can be slowed by mutating catalytically important residues—in this case, Tyr 160 and Lys 230—to create kinetic "bottlenecks" so that reaction intermediates accumulate. Accordingly, crystals of mutant isocitrate dehydrogenase were exposed to the substrate isocitrate and immediately visualized via X-ray crystallography using recently developed rapid X-ray intensity measurement techniques that require the highly intense X-rays generated by a synchrotron. These studies revealed the oxalosuccinate intermediate in the active site of the enzyme.

D. α-Ketoglutarate Dehydrogenase

α-Ketoglutarate dehydrogenase catalyzes the oxidative decarboxylation of an α-keto acid (α-ketoglutarate). This reaction produces the second CO_2 and NADH of the citric acid cycle.

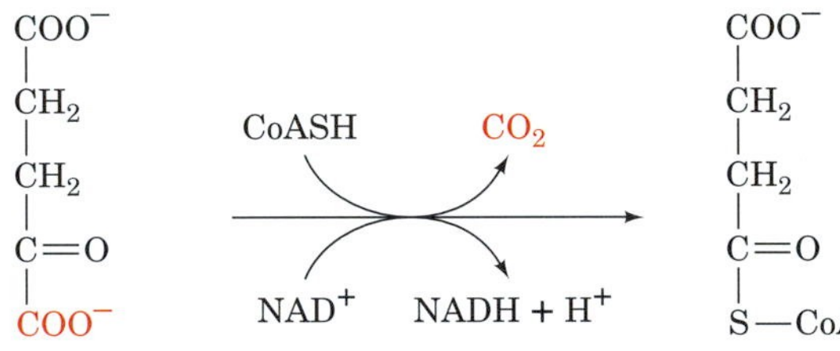

α-**Ketoglutarate**   **Succinyl-CoA**

Again, this CO_2 entered the citric acid cycle as a component of oxaloacetate rather than of acetyl-CoA (Fig. 16-2). Thus, although each round of the citric acid cycle oxidizes two C atoms to CO_2, the C atoms of the entering acetyl groups are not oxidized to CO_2 until subsequent rounds of the cycle.

The α-ketoglutarate dehydrogenase reaction chemically resembles the reaction catalyzed by the pyruvate dehydrogenase multienzyme complex. α-Ketoglutarate dehydrogenase is a multienzyme complex containing **α-ketoglutarate dehydrogenase (E_1), dihydrolipoyl transsuccinylase (E_2),** and **dihydrolipoyl dehydrogenase (E_3).** Indeed, this E_3 is identical to the E_3 of the pyruvate dehydrogenase complex (a third member of the **2-keto acid dehydrogenase** family of multienzyme complexes is **branched-chain α-keto acid dehydrogenase,** which participates in the degradation of isoleucine, leucine, and valine; Section 20-4D). The reactions catalyzed by the α-ketoglutarate dehydrogenase complex occur by mechanisms identical to those of the pyruvate dehydrogenase complex. Again, the product is a "high-energy" thioester, in this case, **succinyl-CoA.**

E. Succinyl-CoA Synthetase

Succinyl-CoA synthetase (also called **succinate thiokinase**) couples the cleavage of the "high-energy" succinyl-CoA to the synthesis of a "high-energy" nucleoside triphosphate (both names for the enzyme reflect the reverse reaction). GTP is usually synthesized from GDP + P_i by the mammalian enzyme; plant and bacterial enzymes usually use ADP + P_i to form ATP. These reactions are nevertheless energetically equivalent since ATP and GTP are rapidly interconverted through the action of nucleoside diphosphate kinase (Section 13-2C):

$$GTP + ADP \rightleftharpoons GDP + ATP \qquad \Delta G^{\circ\prime} = 0$$

How does succinyl-CoA synthetase couple the exergonic cleavage of succinyl-CoA ($\Delta G^{\circ\prime} = -32.6 \text{ kJ} \cdot \text{mol}^{-1}$) to the endergonic formation of a nucleoside triphosphate ($\Delta G^{\circ\prime} = 30.5 \text{ kJ} \cdot \text{mol}^{-1}$) from the corresponding nucleoside diphosphate and P_i? This question was answered by an experiment with isotopically labeled ADP. In the absence of succinyl-CoA, the spinach enzyme catalyzes the transfer of the γ-phosphoryl group from ATP to [^{14}C]ADP, producing [^{14}C]ATP. Such an isotope-exchange reaction suggests the participation of a phosphoryl-enzyme intermediate that mediates the reaction sequence

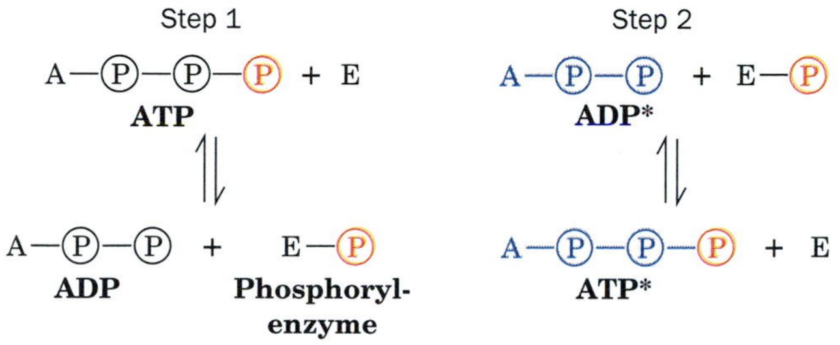

This information led to the isolation of a kinetically active phosphoryl-enzyme in which the phosphoryl group is covalently linked to the N3 position of a His residue. A three-step mechanism for succinyl-CoA synthetase is shown in Fig. 16-11.

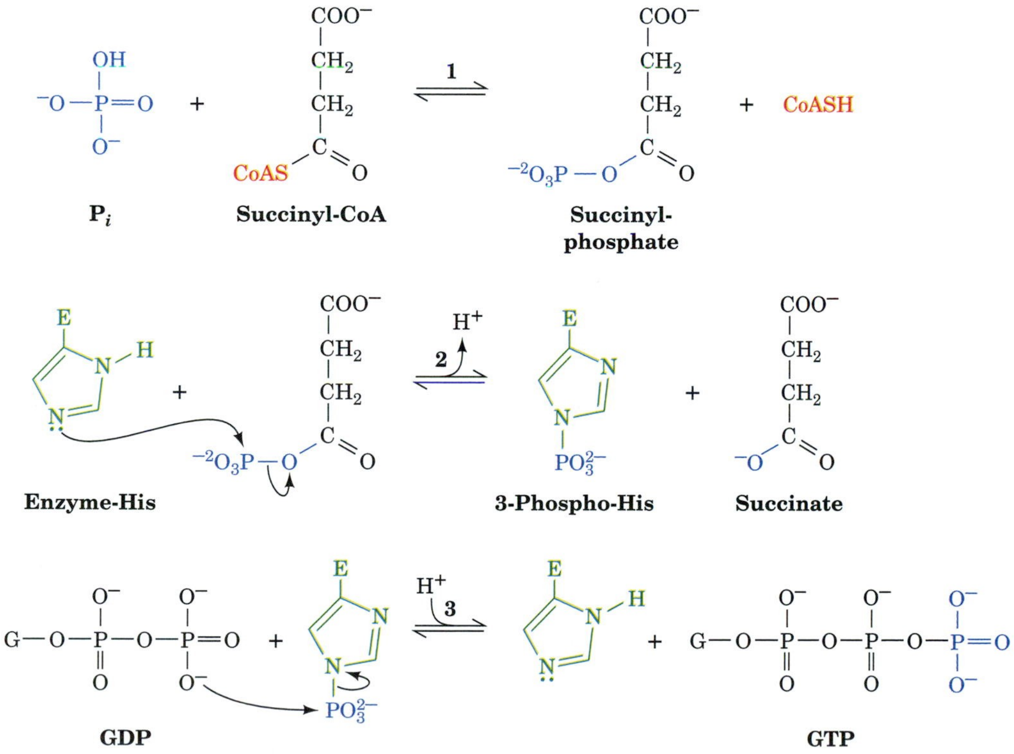

Figure 16-11.   The reaction catalyzed by succinyl-CoA synthetase. (1) Formation of succinyl-phosphate, a "high-energy" mixed anhydride. (2) Formation of phosphoryl-His, a "high-energy" intermediate. (3) Transfer of the phosphoryl group to GDP, forming GTP.

1. Succinyl-CoA reacts with P_i to form **succinyl-phosphate** and CoA.

2. The phosphoryl group is then transferred from succinyl-phosphate to a His residue of the enzyme, releasing succinate.

3. The phosphoryl group on the enzyme is transferred to GDP, forming GTP.

Note that in each of these steps, *the energy of succinyl-CoA is conserved through the formation of "high-energy" compounds: first, succinyl-phosphate, then a 3-phospho-His residue, and finally GTP.* The process is reminiscent of passing a hot potato.

By this point in the citric acid cycle, one acetyl equivalent has been completely oxidized to two CO_2. Two NADH and one GTP (equivalent to one ATP) have also been generated. In order to complete the cycle, succinate must be converted back to oxaloacetate. This is accomplished by the cycle's remaining three reactions. The reactions of the citric acid cycle and their stereospecificity have been confirmed through the use of radioactive tracer experiments (see Box 16-2).

F. Succinate Dehydrogenase

Succinate dehydrogenase catalyzes the stereospecific dehydrogenation of succinate to fumarate.

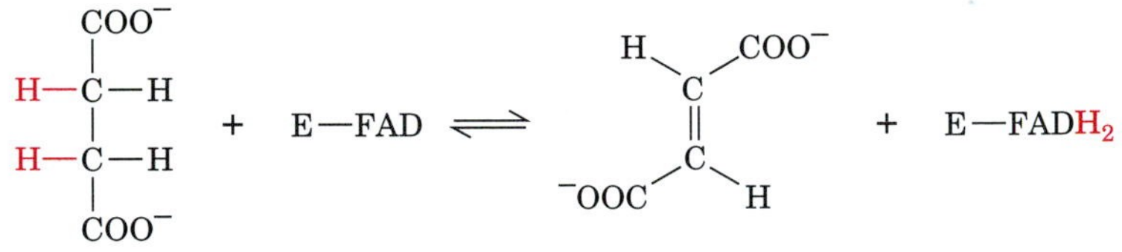

This enzyme is strongly inhibited by malonate *(at left)*, a structural analog of succinate and a classic example of a competitive inhibitor. When Krebs was formulating his theory of the citric acid cycle, the inhibition of cellular respiration by malonate provided one of the clues that succinate plays a catalytic role in oxidizing substrates and is not just another substrate.

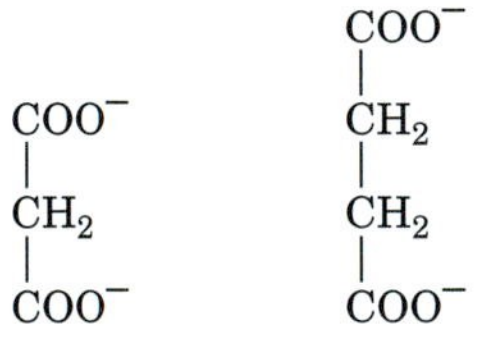

Succinate dehydrogenase contains an FAD prosthetic group that is covalently linked to the enzyme via a His residue (Fig. 16-12; in most other FAD-containing enzymes, the FAD is held tightly but noncovalently). In

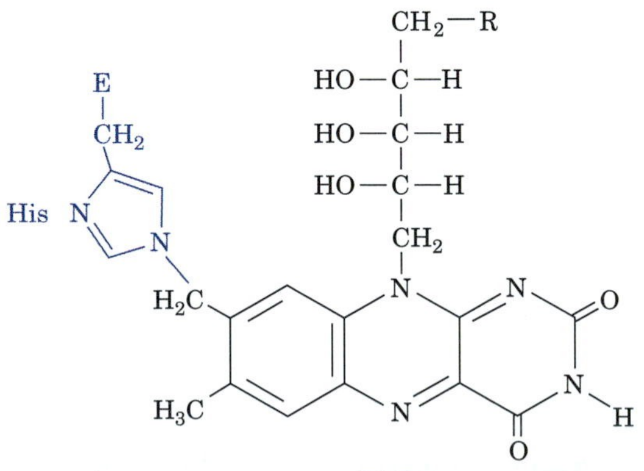

Malonate   Succinate

Figure 16-12.   The covalent attachment of FAD to a His residue of succinate dehydrogenase. R represents the ADP moiety.

Box 16-2

BIOCHEMISTRY IN FOCUS

The Stereospecificity of Citric Acid Cycle Reactions

Experiments with radioactive metabolites became possible in the late 1930s and early 1940s, when ^{11}C (half-life 20 min) and ^{14}C (half-life 5715 years) became available. At that time, the concepts of prochirality and the stereospecificity of enzymes were poorly understood. For example, it was widely believed that the two chemically equivalent halves of citrate (which has a plane of symmetry) are indistinguishable. It was therefore assumed that radioactivity originally located at C4 in oxaloacetate would be scrambled in citrate so that its C1 and C6 atoms would be equally labeled, resulting in α-ketoglutarate labeled at both C1 and C5:

Oxaloacetate   Citrate   α-Ketoglutarate

Thus, the finding that only C1 of α-ketoglutarate was radioactive threw the identity of the condensation product of oxaloacetate and acetyl-CoA into doubt. How could it be the "symmetrical" citrate molecule in light of such conclusive labeling experiments? This problem of which tricarboxylic acid was the cycle's original condensation product resulted in a name change from the "citric acid cycle" (proposed by Krebs) to the "tricarboxylic acid (TCA) cycle," a name that still persists.

It is now realized that citrate is a prochiral molecule and that aconitase can distinguish between its two carboxymethyl groups. In 1948, Alexander Ogston pointed out that citrate

can interact asymmetrically with the surface of aconitase by making a three-point attachment.

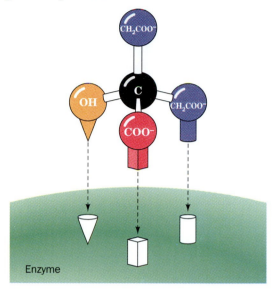

Because there is only one way the substrate can bind, only one of its two —CH_2COO^- groups will react to form isocitrate.

The labeled atoms of oxaloacetate are eventually scrambled, but not until after the reaction catalyzed by succinyl-CoA synthetase. The product, succinate, is two-fold rotationally symmetric rather than prochiral. Consequently, a label at C1 of succinyl-CoA will appear at both C1 and C4 of fumarate:

Succinate   Fumarate

general, FAD functions biochemically to oxidize alkanes (such as succinate) to alkenes (such as fumarate), whereas NAD^+ participates in the more exergonic oxidation of alcohols to aldehydes or ketones. The dehydrogenation of succinate produces $FADH_2$, which must be reoxidized before succinate dehydrogenase can undertake another catalytic cycle. *The reoxidation of this $FADH_2$ occurs in the mitochondrial electron-transport chain,* which we shall examine in more detail in Section 17-2. Succinate dehydrogenase, which is the only membrane-bound enzyme of the citric acid cycle (the others are components of the mitochondrial matrix), feeds electrons directly into the electron-transport machinery of the mitochondrial membrane. In the process, $FADH_2$ is reoxidized to FAD.

G. Fumarase

Fumarase (fumarate hydratase) catalyzes the hydration of the double bond of fumarate to form malate. The hydration reaction proceeds via a carbanion transition state. OH^- addition occurs before H^+ addition.

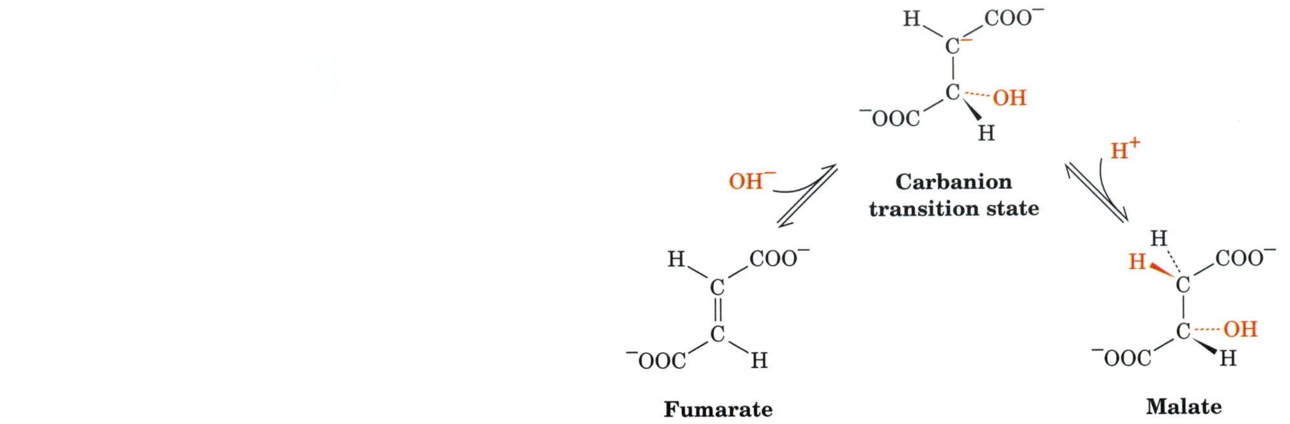

Carbanion transition state

Fumarate   **Malate**

H. Malate Dehydrogenase

Malate dehydrogenase catalyzes the final reaction of the citric acid cycle, the regeneration of oxaloacetate. The hydroxyl group of malate is oxidized in an NAD^+-dependent reaction.

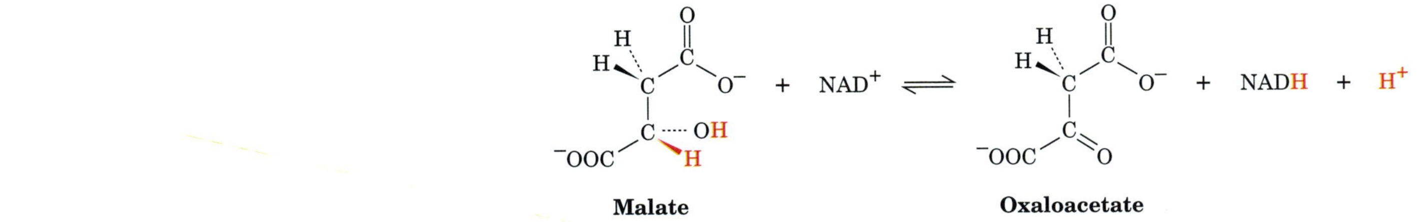

Malate   **Oxaloacetate**

Transfer of the hydride ion to NAD^+ occurs by the same mechanism used for hydride ion transfer in lactate dehydrogenase and alcohol dehydrogenase (Section 14-3). X-Ray crystallographic comparisons of the NAD^+-binding domains of these three enzymes indicate that they are remarkably similar and are consistent with the proposal that all NAD^+-binding domains evolved from a common ancestor.

The $\Delta G^{\circ\prime}$ value for the malate dehydrogenase reaction is $+29.7 \text{ kJ} \cdot \text{mol}^{-1}$; therefore, the concentration of oxaloacetate at equilibrium (and under cellular conditions) is very low relative to malate. Recall, however, that the reaction catalyzed by citrate synthase, the first reaction of the citric acid cycle, is highly exergonic ($\Delta G^{\circ\prime} = -31.5 \text{ kJ} \cdot \text{mol}^{-1}$) because of the cleavage of the thioester bond of citryl-CoA. We can now understand the necessity for such a seemingly wasteful process. It allows citrate formation to be exergonic even at the low oxaloacetate concentrations present in cells and thus helps keep the citric acid cycle rolling.

4. REGULATION OF THE CITRIC ACID CYCLE

The capacity of the citric acid cycle to generate energy for cellular needs is closely regulated. The availability of substrates, the need for citric acid cycle intermediates as biosynthetic precursors, and the demand for ATP all influence the operation of the cycle. There is some evidence that the enzymes of the citric acid cycle are physically associated, which might contribute to their coordinated regulation (see Box 16-3). Before we examine the various mechanisms for regulating the citric acid cycle, let us briefly consider the energy-generating capacity of the cycle.

The oxidation of one acetyl group to two molecules of CO_2 is a four-electron pair process (but keep in mind that it is not the carbon atoms of the incoming acetyl group that are oxidized). For every acetyl-CoA that enters the cycle, three molecules of NAD^+ are reduced to NADH, which accounts for three of the electron pairs, and one molecule of FAD is reduced to $FADH_2$, which accounts for the fourth electron pair. In addition, one GTP (or ATP) is produced.

The electrons carried by NADH and $FADH_2$ are funneled into the electron-transport chain, which culminates with the reduction of O_2 to H_2O. The energy of electron transport is conserved in the synthesis of ATP by oxidative phosphorylation (Section 17-3). For every NADH that passes its electrons on, approximately 3 ATP are produced from ADP + P_i. For every $FADH_2$, approximately 2 ATP are produced. Thus, one turn of the citric acid cycle ultimately generates approximately 12 ATP.

When glucose is converted to two molecules of pyruvate by glycolysis, two molecules of ATP are generated and two molecules of NAD^+ are reduced (Section 14-1). These NADH molecules yield approximately six molecules of ATP on passing their electrons to the electron-transport chain. When the two pyruvate molecules are converted to two acetyl-CoA by the pyruvate dehydrogenase complex, the two molecules of NADH produced in that process also eventually give rise to ~6 ATP. Two turns of the citric acid cycle (one for each acetyl group) generate ~24 ATP. Thus, one mol-

---

Box 16-3

BIOCHEMISTRY IN CONTEXT

The Metabolon Hypothesis

The evidence for channeling of intermediates between active sites in certain multienzyme complexes, such as the pyruvate dehydrogenase complex, has prompted some investigators to question whether channeling on a grander scale might occur between the active sites of enzymes in a metabolic pathway such as the citric acid cycle. According to the **metabolon hypothesis**, enzyme molecules are noncovalently associated in an assembly called a **metabolon** that is localized to a particular area in the cytoplasm or in an organelle. Presumably, the proximity of enzymes catalyzing sequential reactions increases the catalytic efficiency of the pathway and helps coordinate the overall regulation of the pathway.

Support for the existence of metabolons should center on three criteria:

1. The enzymes must associate specifically. Proteins are somewhat "sticky," so the weak interactions that link enzymes in a metabolon must be distinguished from nonspecific protein–protein interactions. In fact, some isolated enzymes do bind to each other but not to enzymes from unrelated pathways.

2. The association of enzymes must be visualized *in vivo*. Developments in electron microscopy of fixed cells and immunofluorescence microscopy of living cells may ultimately provide more concrete evidence that the enzymes participating in a metabolic pathway are not randomly distributed in a cell or organelle. Citric acid cycle enzymes, for example, apparently bind to the inner surface of the inner mitochondrial membrane. Furthermore, partially solubilized mitochondrial enzymes catalyze the reactions of the citric acid cycle several times faster than completely solubilized enzymes.

3. The metabolon must provide some metabolic advantage, perhaps related to rapid and sensitive regulation or more efficient catalysis. Channeling between active sites that are not located in the same enzyme has not yet been proved. However, computer simulations of two members of the putative citric acid cycle metabolon, malate dehydrogenase and citrate synthase, suggest that electronic forces can rapidly guide oxaloacetate from one active site to the other when the enzymes are in close proximity.

ecule of glucose can potentially yield ~38 molecules of ATP under aerobic conditions, when the citric acid cycle is operating. In contrast, only two molecules of ATP are produced per glucose molecule under anaerobic conditions.

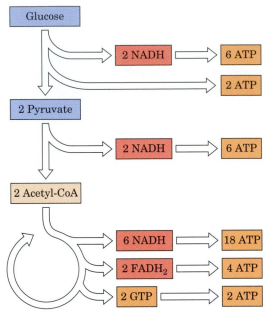

A. Regulation of Pyruvate Dehydrogenase

Given the large amount of ATP that can potentially be generated from carbohydrate catabolism via the citric acid cycle, it is not surprising that the entry of acetyl units derived from carbohydrate sources is regulated. The decarboxylation of pyruvate by the pyruvate dehydrogenase complex is irreversible, and since there are no other pathways in mammals for the synthesis of acetyl-CoA from pyruvate, it is crucial that the reaction be precisely controlled. Two regulatory systems are used:

1. **Product inhibition by NADH and acetyl-CoA.** These compounds compete with NAD^+ and CoA for binding sites on their respective enzymes. They also drive the reversible transacetylase (E_2) and dihydrolipoyl dehydrogenase (E_3) reactions backward (Fig. 16-5). High $[NADH]/[NAD^+]$ and $[acetyl\text{-}CoA]/[CoA]$ ratios therefore maintain E_2 in the acetylated form, incapable of accepting the hydroxyethyl group from the TPP on E_1. This, in turn, ties up the TPP on the E_1 subunit in its hydroxyethyl form, decreasing the rate of pyruvate decarboxylation.

2. **Covalent modification by phosphorylation/dephosphorylation of E_1.** In eukaryotes, the products of the pyruvate dehydrogenase reaction, NADH and acetyl-CoA, also activate the pyruvate dehydrogenase kinase associated with the enzyme complex. The resulting phosphorylation of a specific dehydrogenase Ser residue inactivates the pyruvate dehydrogenase complex (Fig. 16-13). Insulin, the hormone that signals fuel abundance, reverses the inactivation by activating pyruvate dehydrogenase phosphatase, which removes the phosphate groups from pyruvate dehydrogenase. Recall that insulin also activates glycogen synthesis by activating phosphoprotein phosphatase (Section 15-3B). Thus, in response to increases in blood [glucose], insulin promotes the synthesis of acetyl-CoA as well as glycogen.

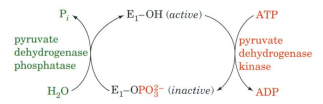

Figure 16-13.   Covalent modification of eukaryotic pyruvate dehydrogenase. E_1 is inactivated by the specific phosphorylation of one of its Ser residues in a reaction catalyzed by pyruvate dehydrogenase kinase. This phosphoryl group is hydrolyzed through the action of pyruvate dehydrogenase phosphatase, thereby reactivating E_1.

Other regulators of the pyruvate dehydrogenase system include pyruvate and ADP, which inhibit pyruvate dehydrogenase kinase, and Ca^{2+}, which inhibits pyruvate dehydrogenase kinase and activates pyruvate dehydrogenase phosphatase. In contrast to the glycogen metabolism control system (Section 15-3B), pyruvate dehydrogenase activity is unaffected by cAMP.

B. The Rate-Controlling Enzymes of the Citric Acid Cycle

To understand how a metabolic pathway is controlled, we must identify the enzymes that catalyze its rate-determining steps, the *in vitro* effectors of the enzymes, and the *in vivo* concentrations of these substances. *A proposed mechanism of flux control must operate within the physiological concentration range of the effector.*

Identifying the rate-determining steps of the citric acid cycle is more difficult than it is for glycolysis because most of the cycle's metabolites are present in both mitochondria and cytosol and we do not know their distribution between these two compartments (recall that identifying a pathway's rate-determining steps requires determining the ΔG of each of its reactions from the concentrations of its substrates and products). However, we shall assume that the compartments are in equilibrium and use the total cell concentrations of these substances to estimate their mitochondrial concentrations. Table 16-2 gives the standard free energy changes for the eight citric acid cycle enzyme and estimates of the physiological free energy changes for the reactions in heart muscle or liver tissue. We can see that *three of*

Table 16-2.   Standard Free Energy Changes ($\Delta G^{\circ\prime}$) and Physiological Free Energy Changes (ΔG) of Citric Acid Cycle Reactions

Reaction	Enzyme	$\Delta G^{\circ\prime}$ $(kJ \cdot mol^{-1})$	ΔG $(kJ \cdot mol^{-1})$
1	Citrate synthase	−31.5	Negative
2	Aconitase	~5	~0
3	Isocitrate dehydrogenase	−21	Negative
4	α-Ketoglutarate dehydrogenase multienzyme complex	−33	Negative
5	Succinyl-CoA synthetase	−2.1	~0
6	Succinate dehydrogenase	+6	~0
7	Fumarase	−3.4	~0
8	Malate dehydrogenase	+29.7	~0

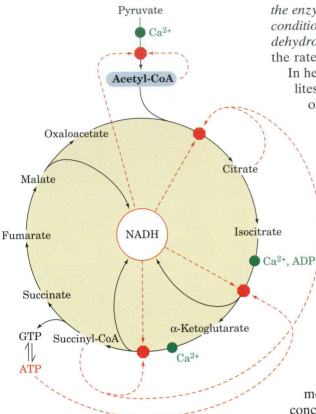

Figure 16-14. Regulation of the citric acid cycle. This diagram of the citric acid cycle, which includes the pyruvate dehydrogenase reaction, indicates points of inhibition (*red octagons*) and the pathway intermediates that function as inhibitors (*dashed red arrows*). ADP and Ca²⁺ (*green dots*) are activators. ✳ See the Animated Figures.

the enzymes are likely to function far from equilibrium under physiological conditions (negative ΔG): citrate synthase, NAD⁺-dependent isocitrate dehydrogenase, and α-ketoglutarate dehydrogenase. These are therefore the rate-determining enzymes of the cycle.

In heart muscle, where the citric acid cycle is active, the flux of metabolites through the citric acid cycle is proportional to the rate of cellular oxygen consumption. *Because oxygen consumption, NADH reoxidation, and ATP production are tightly coupled (Section 17-3), the citric acid cycle must be regulated by feedback mechanisms that coordinate NADH production with energy expenditure.* Unlike the rate-limiting enzymes of glycolysis and glycogen metabolism, which regulate flux by elaborate systems of allosteric control, substrate cycles, and covalent modification, the regulatory enzymes of the citric acid cycle seem to control flux primarily by three simple mechanisms: (1) substrate availability, (2) product inhibition, and (3) competitive feedback inhibition by intermediates further along the cycle. Some of the major regulatory mechanisms are diagrammed in Fig. 16-14. There is no single flux-control point in the citric acid cycle; rather, flux control is distributed among several enzymes.

Perhaps the most crucial regulators of the citric acid cycle are its substrates, acetyl-CoA and oxaloacetate, and its product, NADH. Both acetyl-CoA and oxaloacetate are present in mitochondria at concentrations that do not saturate citrate synthase. The metabolic flux through the enzyme therefore varies with substrate concentration and is controlled by substrate availability. We have already seen that the production of acetyl-CoA from pyruvate is regulated by the activity of pyruvate dehydrogenase. The concentration of oxaloacetate, which is in equilibrium with malate, fluctuates with the [NADH]/[NAD⁺] ratio according to the equilibrium expression

$$K = \frac{[\text{oxaloacetate}][\text{NADH}]}{[\text{malate}][\text{NAD}^+]}$$

If, for example, the muscle workload and respiration rate increase, mitochondrial [NADH] decreases. The consequent increase in [oxaloacetate] stimulates the citrate synthase reaction, which controls the rate of citrate formation.

Aconitase functions close to equilibrium, so the rate of citrate consumption depends on the activity of NAD⁺-dependent isocitrate dehydrogenase, which is strongly inhibited *in vitro* by its product NADH. Citrate synthase is also inhibited by NADH but is less sensitive than isocitrate dehydrogenase to changes in [NADH].

Other instances of product inhibition in the citric acid cycle are the inhibition of citrate synthase by citrate (citrate competes with oxaloacetate) and the inhibition of α-ketoglutarate dehydrogenase by NADH and succinyl-CoA. Succinyl-CoA also competes with acetyl-CoA in the citrate synthase reaction (competitive feedback inhibition). This interlocking system helps keep the citric acid cycle coordinately regulated.

Additional Regulatory Mechanisms

In vitro studies of citric acid cycle enzymes have identified a few allosteric activators and inhibitors. ADP is an allosteric activator of isocitrate dehydrogenase, whereas ATP inhibits this enzyme. Ca²⁺, in addition to its many other cellular functions, regulates the citric acid cycle at several points. It activates pyruvate dehydrogenase phosphatase (Fig. 16-13), which in turn activates the pyruvate dehydrogenase complex to produce acetyl-CoA. Ca^{2+} also activates both isocitrate dehydrogenase and α-ketoglutarate dehydrogenase (Fig. 16-14). Thus Ca^{2+}, the signal that stimulates muscle contraction, also stimulates the production of the ATP to fuel it.

In *E. coli,* isocitrate dehydrogenase is inactivated by the phosphorylation of a Ser residue. However, in contrast to most other enzymes that are regulated by phosphorylation/dephosphorylation, *E. coli* isocitrate dehydrogenase is not phosphorylated at an allosteric site but at the active site. The structure of the phosphorylated enzyme is similar to that of the unmodified enzyme. Presumably, electrostatic repulsion prevents the phosphorylated enzyme from binding its anionic substrate isocitrate.

5. REACTIONS RELATED TO THE CITRIC ACID CYCLE

At first glance, a metabolic pathway appears to be either catabolic, with the release and conservation of free energy, or anabolic, with a requirement for free energy. The citric acid cycle is catabolic, of course, because it involves degradation and is a major free-energy conservation system in most organisms. Cycle intermediates are required in only catalytic amounts to maintain the degradative function of the cycle. However, several biosynthetic pathways use citric acid cycle intermediates as starting materials for anabolic reactions. The citric acid cycle is therefore **amphibolic** (both anabolic and catabolic). In this section, we examine some of the reactions that feed intermediates into the citric acid cycle or draw them off; we also examine the **glyoxylate pathway,** a variation of the citric acid cycle that occurs only in plants and converts acetyl-CoA to oxaloacetate. Some of the reactions that use and replenish citric acid cycle intermediates are summarized in Fig. 16-15.

A. Pathways That Use Citric Acid Cycle Intermediates

Citric acid cycle intermediates are precursors in the biosynthesis of carbohydrates, fatty acids, and amino acids:

1. We have already seen that glucose can be synthesized from oxaloacetate (Section 15-4). Because gluconeogenesis takes place in the cytosol, oxaloacetate must be converted to malate or aspartate for transport out of the mitochondrion (Fig. 15-27).

2. Fatty acid synthesis is a cytosolic process that requires acetyl-CoA. Acetyl-CoA is generated in the mitochondria and is not transported across the mitochondrial membrane. *Cytosolic acetyl-CoA is therefore generated by the breakdown of citrate, which can cross the membrane, in a reaction catalyzed by **ATP-citrate lyase** (Section 19-4A).* This reaction uses the free energy of ATP to "undo" the citrate synthase reaction:

$$ATP + citrate + CoA \longrightarrow ADP + P_i + oxaloacetate + acetyl\text{-}CoA$$

3. *Amino acid biosynthesis uses α-ketoglutarate and oxaloacetate as starting materials.* For example, α-ketoglutarate is converted to glutamate

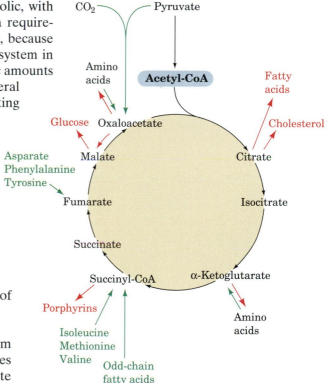

Figure 16-15. Amphibolic functions of the citric acid cycle. The diagram indicates the positions at which intermediates are drawn off for use in anabolic pathways (*red arrows*) and the points where anaplerotic reactions replenish cycle intermediates (*green arrows*). Reactions involving amino acid transamination and deamination are reversible, so their direction varies with metabolic demand. ✹ **See the Animated Figures.**

by reductive amination catalyzed by a **glutamate dehydrogenase** that requires either NADH or NADPH.

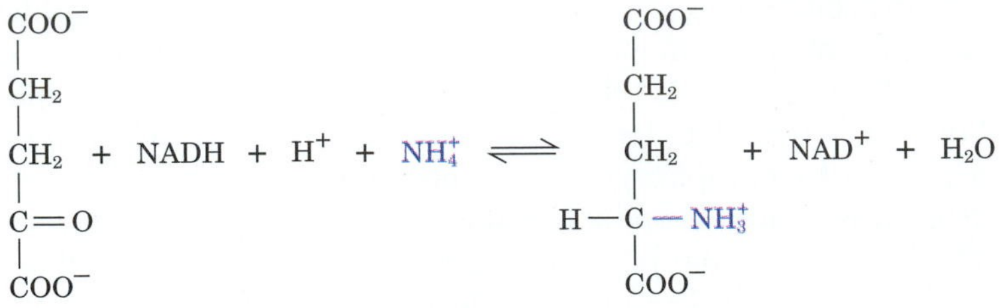

$$
\begin{array}{l}
\text{COO}^- \\
| \\
\text{CH}_2 \\
| \\
\text{CH}_2 \\
| \\
\text{C}=\text{O} \\
| \\
\text{COO}^-
\end{array}
+ \text{NADH} + \text{H}^+ + \text{NH}_4^+ \rightleftharpoons
\begin{array}{l}
\text{COO}^- \\
| \\
\text{CH}_2 \\
| \\
\text{CH}_2 \\
| \\
\text{H}-\text{C}-\text{NH}_3^+ \\
| \\
\text{COO}^-
\end{array}
+ \text{NAD}^+ + \text{H}_2\text{O}
$$

α-Ketoglutarate   **Glutamate**

Oxaloacetate undergoes transamination with alanine to produce aspartate and pyruvate (Section 20-2A).

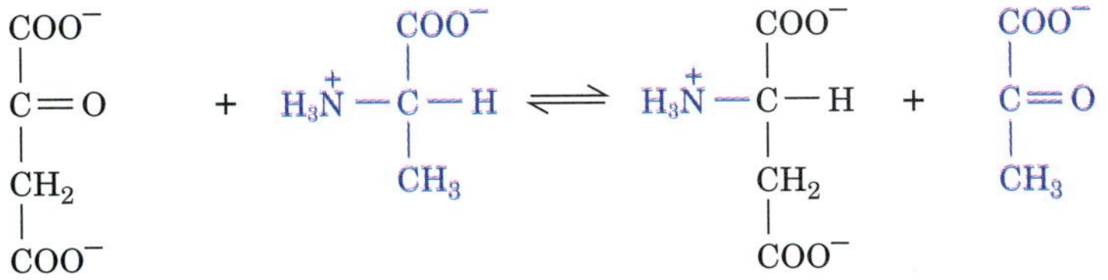

$$
\begin{array}{l}
\text{COO}^- \\
| \\
\text{C}=\text{O} \\
| \\
\text{CH}_2 \\
| \\
\text{COO}^-
\end{array}
+
\begin{array}{l}
\text{COO}^- \\
| \\
\text{H}_3\overset{+}{\text{N}}-\text{C}-\text{H} \\
| \\
\text{CH}_3
\end{array}
\rightleftharpoons
\begin{array}{l}
\text{COO}^- \\
| \\
\text{H}_3\overset{+}{\text{N}}-\text{C}-\text{H} \\
| \\
\text{CH}_2 \\
| \\
\text{COO}^-
\end{array}
+
\begin{array}{l}
\text{COO}^- \\
| \\
\text{C}=\text{O} \\
| \\
\text{CH}_3
\end{array}
$$

Oxaloacetate   **Alanine**   **Aspartate**   **Pyruvate**

B. Reactions That Replenish Citric Acid Cycle Intermediates

In aerobic organisms, the citric acid cycle is the major source of free energy, and hence the catabolic function of the citric acid cycle cannot be interrupted: Cycle intermediates that have been siphoned off must be replenished. The replenishing reactions are called **anaplerotic reactions** (filling up, Greek: *ana,* up + *plerotikos,* to fill). The most important of these reactions is catalyzed by pyruvate carboxylase, which produces oxaloacetate from pyruvate:

$$\text{Pyruvate} + \text{CO}_2 + \text{ATP} + \text{H}_2\text{O} \longrightarrow \text{oxaloacetate} + \text{ADP} + \text{P}_i$$

(This is also one of the first steps of gluconeogenesis; Section 15-4A). Pyruvate carboxylase "senses" the need for more citric acid cycle intermediates through its activator, acetyl-CoA. *Any decrease in the rate of the cycle caused by insufficient oxaloacetate or other intermediates allows the concentration of acetyl-CoA to rise.* This activates pyruvate carboxylase, which replenishes oxaloacetate. The reactions of the citric acid cycle convert the oxaloacetate to citrate, α-ketoglutarate, succinyl-CoA, and so on, until all the intermediates are restored to appropriate levels.

Other metabolites that feed into the citric acid cycle are succinyl-CoA, a product of the degradation of odd-chain fatty acids (Section 19-2E) and certain amino acids (Section 20-4), and α-ketoglutarate and oxaloacetate, which are formed by the reversible transamination of certain amino acids, as indicated above.

C. The Glyoxylate Pathway

Plants, but not animals, possess enzymes that mediate the net conversion of acetyl-CoA to oxaloacetate, which can be used for gluconeogenesis. These enzymes constitute the glyoxylate pathway (Fig. 16-16), which operates in two cellular compartments: the mitochondrion and the **glyoxysome,** a membrane-bounded plant organelle that is a specialized peroxisome. Most of

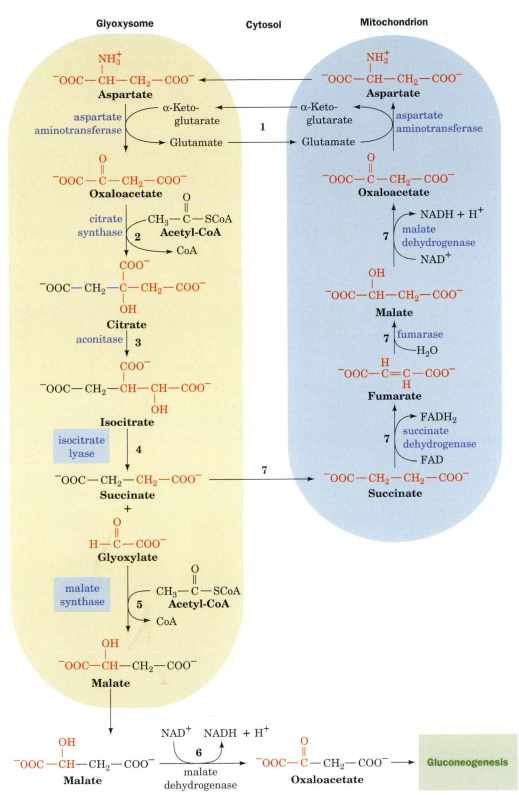

Figure 16-16. *Key to Metabolism.* The glyoxylate pathway.
Both mitochondrial and glyoxysomal enzymes are required. Isocitrate lyase and malate synthase, enzymes unique to plant glyoxysomes, are boxed. The pathway results in the net conversion
of two acetyl-CoA to oxaloacetate. (1) Mitochondrial oxaloacetate
is converted to aspartate, transported to the glyoxysome, and reconverted to oxaloacetate. (2) Oxaloacetate is condensed with
acetyl-CoA to form citrate. (3) Aconitase catalyzes the conversion

of citrate to isocitrate. (4) Isocitrate lyase catalyzes the cleavage of
isocitrate to succinate and glyoxylate. (5) Malate synthase catalyzes the condensation of glyoxylate with acetyl-CoA to form
malate. (6) After transport to the cytosol, malate is oxidized to
oxaloacetate, which can then be used in gluconeogenesis.
(7) Succinate is transported to the mitochondrion, where it is reconverted to oxaloacetate via the citric acid cycle.

the enzymes of the glyoxylate pathway are the same as those of the citric acid cycle.

1. Mitochondrial oxaloacetate is converted to aspartate and transported to the glyoxysome, where it is reconverted to oxaloacetate.
2. The oxaloacetate condenses with acetyl-CoA to form citrate.
3. Citrate is converted to isocitrate, as in the citric acid cycle.
4. Glyoxysomal **isocitrate lyase** cleaves isocitrate to succinate and **glyoxylate.** Succinate is transported to the mitochondrion where it enters the citric acid cycle for conversion back to oxaloacetate, completing the cycle. *The glyoxylate pathway therefore results in the net conversion of acetyl-CoA to glyoxylate instead of to two molecules of CO_2, as in the citric acid cycle.*
5. **Malate synthase,** a glyoxysomal enzyme, condenses glyoxylate with a second molecule of acetyl-CoA to form malate.
6. Malate exits the glyoxysome, and cytosolic malate dehydrogenase catalyzes the oxidation of malate to oxaloacetate by NAD^+.

The overall reaction of the glyoxylate cycle is therefore the formation of oxaloacetate from two molecules of acetyl-CoA:

$$2 \text{ Acetyl-CoA} + 2 \text{ NAD}^+ + \text{FAD} + 3 \text{ H}_2\text{O} \longrightarrow$$
$$\text{oxaloacetate} + 2 \text{ CoA} + 2 \text{ NADH} + \text{FADH}_2 + 4 \text{ H}^+$$

Isocitrate lyase and malate synthase occur only in plants. These enzymes enable germinating seeds to convert their stored triacylglycerols to acetyl-CoA and then to glucose. Organisms that lack the glyoxylate pathway cannot undertake the net synthesis of glucose from acetyl-CoA.

SUMMARY

1. The eight enzymes of the citric acid cycle function in a multistep catalytic cycle to oxidize an acetyl group to two CO_2 molecules with the concomitant generation of three NADH, one $FADH_2$, and one GTP. The free energy released when the reduced coenzymes ultimately reduce O_2 is used to generate ATP.

2. Acetyl groups enter the citric acid cycle as acetyl-CoA. The pyruvate dehydrogenase multienzyme complex, which contains three types of enzymes and five types of coenzymes, generates acetyl-CoA from the glycolytic product pyruvate. The lipoyllysyl arm of E_2 acts as a tether that swings reactive groups between enzymes in the complex.

3. Citrate synthase catalyzes the condensation of acetyl-CoA and oxaloacetate in a highly exergonic reaction.

4. Aconitase catalyzes the isomerization of citrate to isocitrate, and isocitrate dehydrogenase catalyzes the oxidative decarboxylation of isocitrate to α-ketoglutarate to produce the citric acid cycle's first CO_2 and NADH.

5. α-Ketoglutarate dehydrogenase catalyzes the oxidative decarboxylation of α-ketoglutarate to produce succinyl-CoA and the citric acid cycle's second CO_2 and NADH.

6. Succinyl-CoA synthetase couples the cleavage of succinyl-CoA to the synthesis of GTP (or in some organisms, ATP) via a phosphoryl-protein intermediate.

7. The citric acid cycle's remaining three reactions, catalyzed by succinate dehydrogenase, fumarase, and malate dehydrogenase, regenerate oxaloacetate to continue the citric acid cycle.

8. Entry of acetyl-CoA into the citric acid cycle is regulated at the pyruvate dehydrogenase step by product inhibition (by NADH and acetyl-CoA) and by covalent modification.

9. The citric acid cycle itself is regulated at the steps catalyzed by citrate synthase, NAD^+-dependent isocitrate dehydrogenase, and α-ketoglutarate dehydrogenase. Regulation is accomplished mainly by substrate availability, product inhibition, and feedback inhibition.

10. Some citric acid cycle intermediates are substrates for gluconeogenesis, fatty acid synthesis, and amino acid synthesis. Anaplerotic reactions such as the pyruvate carboxylase reaction replenish citric acid cycle intermediates.

11. The glyoxylate pathway, which operates only in plants, requires the glyoxysomal enzymes isocitrate lyase and malate synthase. This variation of the citric acid cycle permits net synthesis of glucose from acetyl-CoA.

REFERENCES

Barry, J.M., Enzymes and symmetrical molecules, *Trends Biochem. Sci.* **22**, 228–230 (1997).

Bolduc, J.M., Dyer, D.H., Scott, W.G., Singer, P., Sweet, R.M., Koshland, D.E., Jr., and Stoddard, B.L., Mutagenesis and Laue structures of enzyme intermediates: isocitrate dehydrogenase, *Science* **268**, 1312–1318 (1995).

Mattevi, A., de Kok, A., and Perham, R.N., The pyruvate dehydrogenase multienzyme complex, *Curr. Opin. Struct. Biol.* **2**, 877–887 (1992).

Remington, J.S., Mechanisms of citrate synthase and related enzymes (triose phosphate isomerase and mandelate racemase), *Curr. Opin. Struct. Biol.* **2**, 730–735 (1992).

Srere, P.A., Wanderings (wonderings) in metabolism, *Biol. Chem. Hoppe Seyler* **374**, 833–842 (1993). [A brief review of the metabolon hypothesis and some supporting evidence.]

KEY TERMS

multienzyme complex	catalytic cycle	anaplerotic reaction	glyoxylate pathway
lipoyllysyl arm	amphibolic pathway		

STUDY EXERCISES

1. Describe the five reactions of the pyruvate dehydrogenase multienzyme complex.

2. Draw the structures of the eight intermediates of the citric acid cycle and name the enzymes that catalyze their interconversion.

3. Write the net equations for oxidation of pyruvate, acetyl-CoA, and glucose to CO_2.

4. Which steps of the citric acid cycle regulate flux through the cycle?

5. Describe the role of Ca^{2+}, acetyl-CoA, and NADH in regulating pyruvate dehydrogenase and the citric acid cycle.

6. Explain how a catalytic cycle can supply precursors for other metabolic pathways without depleting its own intermediates.

7. Describe the reactions of the glyoxylate pathway.

PROBLEMS

1. (a) Explain why obligate anaerobes contain some citric acid cycle enzymes. (b) Why don't these organisms have a complete citric acid cycle?

2. The first organisms on earth may have been chemoautotrophs in which the citric acid cycle operated in reverse to "fix" atmospheric CO_2 in organic compounds. Complete a catalytic cycle that begins with the overall reaction succinate + 2 CO_2 → citrate.

3. The CO_2 produced in one round of the citric acid cycle does not originate in the acetyl carbons that entered that round. (a) If acetyl-CoA is labeled with ^{14}C at the carbonyl carbon, how many rounds of the cycle are required before $^{14}CO_2$ is released? (b) How many rounds are required if acetyl-CoA is labeled at its methyl group?

4. The branched-chain α-keto acid dehydrogenase complex, which participates in amino acid catabolism, contains the same three types of enzymes as are in the pyruvate dehydrogenase and the α-ketoglutarate dehydrogenase complexes. Draw the reaction product when valine is deaminated as in the glutamate ⇌ α-ketoglutarate reaction (Section 16-5A) and then is acted on by the branched-chain α-keto acid dehydrogenase.

5. Refer to Table 13-3 to explain why FAD rather than NAD^+ is used in the succinate dehydrogenase reaction.

6. Malonate is a competitive inhibitor of succinate in the succinate dehydrogenase reaction. Explain why increasing the oxaloacetate concentration can overcome malonate inhibition.

7. Anaplerotic reactions permit the citric acid cycle to supply intermediates to biosynthetic pathways while maintaining the proper levels of cycle intermediates. Write the equation for the net synthesis of citrate from pyruvate.

8. Given the following information, calculate the physiological ΔG of the isocitrate dehydrogenase reaction at 25°C and pH 7.0: $[NAD^+]/[NADH] = 8$, [α-ketoglutarate] = 0.1 mM; and [isocitrate] = 0.02 mM. Assume standard conditions for CO_2 ($\Delta G^{\circ\prime}$ is given in Table 16-2). Is this reaction a likely site for metabolic control?

9. In metabolic studies with isotopic tracers, labeled atoms that originate in asymmetric molecules do not always become evenly "scrambled" after passing though a symmetric intermediate. Is this observation contrary to the metabolon hypothesis? Explain.

10. Although animals cannot synthesize glucose from acetyl-CoA, if a rat is fed ^{14}C-labeled acetate, some of the label appears in glycogen extracted from its muscles. Explain.

Mitochondria oxidize metabolic fuels to generate energy in a form cells can use, analogously to the way that power plants use fuels to generate the electrical energy that industries and households require. How does the cellular machinery do so, and how does it adjust its energy output to meet the cell's demands? [© David Jeffrey/The Image Bank.]

ELECTRON TRANSPORT AND OXIDATIVE PHOSPHORYLATION

Aerobic organisms consume oxygen and generate carbon dioxide in the process of oxidizing metabolic fuels. The complete oxidation of glucose ($C_6H_{12}O_6$), for example, by molecular oxygen

$$C_6H_{12}O_6 + 6\,O_2 \longrightarrow 6\,CO_2 + 6\,H_2O$$

can be broken down into two half-reactions that the metabolic machinery carries out. In the first, glucose carbon atoms are oxidized:

$$C_6H_{12}O_6 + 6\,H_2O \longrightarrow 6\,CO_2 + 24\,H^+ + 24\,e^-$$

and in the second, molecular oxygen is reduced:

$$6\,O_2 + 24\,H^+ + 24\,e^- \longrightarrow 12\,H_2O$$

We have already seen that the first half-reaction is mediated by the enzymatic reactions of glycolysis and the citric acid cycle (the breakdown of fatty acids—the other major type of metabolic fuel—also requires the citric acid cycle). In this chapter, we describe the pathway by which the electrons from reduced fuel molecules are transferred to molecular oxygen. We also examine how the energy of fuel oxidation is conserved and used to synthesize ATP.

As we have seen, the 12 electron pairs released during glucose oxidation are not transferred directly to O_2. Rather, they are transferred to the coenzymes NAD^+ and FAD to form 10 NADH and 2 $FADH_2$ (Fig. 17-1) in the reactions catalyzed by the glycolytic enzyme glyceraldehyde-3-phosphate dehydrogenase (Section 14-2F), pyruvate dehydrogenase (Section 16-2B), and the citric acid cycle enzymes isocitrate dehydrogenase, α-ketoglutarate dehydrogenase, succinate dehydrogenase, and malate dehydrogenase (Section 16-3). *The electrons then pass into the* **mitochondrial electron-transport chain,** *a system of linked electron carriers.* The following events occur during the electron-transport process:

1. By transferring their electrons to other substances, the NADH and $FADH_2$ are reoxidized to NAD^+ and FAD so that they can participate in additional substrate oxidation reactions.

2. The transferred electrons participate in the sequential oxidation–reduction of over 10 **redox centers** (groups that undergo oxidation–reduction reactions) in four enzyme complexes before reducing O_2 to H_2O.

3. During electron transfer, protons are expelled from the mitochondrion, producing a proton gradient across the mitochondrial membrane. *The free energy stored in this electrochemical gradient drives the synthesis of ATP from ADP and P_i through* **oxidative phosphorylation.**

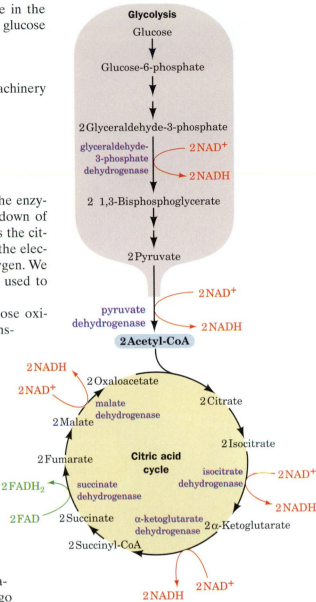

Figure 17-1. **The sites of electron transfer that form NADH and $FADH_2$ in glycolysis and the citric acid cycle.**

1. THE MITOCHONDRION

The mitochondrion (Greek: *mitos,* thread + *chondros,* granule) is the site of eukaryotic oxidative metabolism. Mitochondria contain pyruvate dehydrogenase, the citric acid cycle enzymes, the enzymes catalyzing fatty acid oxidation (Section 19-2), and the enzymes and redox proteins involved in electron transport and oxidative phosphorylation. It is therefore with good reason that the mitochondrion is often described as the cell's "power plant."

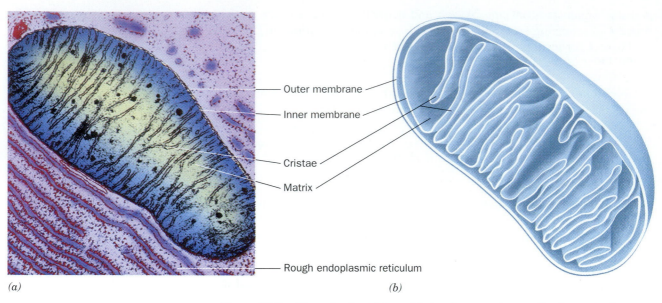

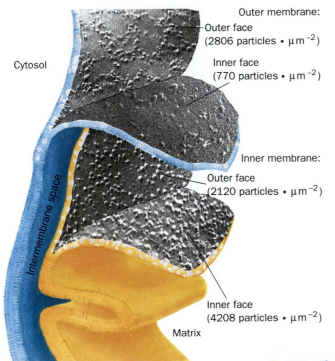

Figure 17-2.   The mitochondrion. (*a*) An electron micrograph of an animal mitochondrion. [K.R. Porter/Photo Researchers, Inc.] (*b*) Cutaway diagram of a mitochondrion.

Figure 17-3 labels:
Outer membrane:
Outer face (2806 particles · μm^{-2})
Inner face (770 particles · μm^{-2})
Cytosol
Inner membrane:
Outer face (2120 particles · μm^{-2})
Intermembrane space
Inner face (4208 particles · μm^{-2})
Matrix

Figure 17-3.   Freeze-fracture and freeze-etch electron micrographs of the inner and outer mitochondrial membranes. The inner membrane contains about twice the density of embedded particles as does the outer membrane. [Courtesy of L. Packer, University of California at Berkeley.]

A. Mitochondrial Anatomy

Mitochondria vary in size and shape, depending on their source and metabolic state, but they are often ellipsoidal with dimensions of around 0.5×1.0 μm—about the size of a bacterium. A eukaryotic cell typically contains ~2000 mitochondria, which occupy roughly one-fifth of its total cell volume. A mitochondrion is bounded by a smooth outer membrane and contains an extensively invaginated inner membrane (Fig. 17-2). The number of invaginations, called **cristae** (Latin: crests), reflects the respiratory activity of the cell. The proteins mediating electron transport and oxidative phosphorylation are bound in the inner mitochondrial membrane, so the respiration rate varies with membrane surface area.

The inner membrane divides the mitochondrion into two compartments, the **intermembrane space** and the internal **matrix.** The matrix is a gel-like solution that contains extremely high concentrations of the soluble enzymes of oxidative metabolism as well as substrates, nucleotide cofactors, and inorganic ions. The matrix also contains the mitochondrial genetic machinery—DNA, RNA, and ribosomes—that generates several (but by no means all) mitochondrial proteins.

B. Mitochondrial Transport Systems

Like bacterial outer membranes, the outer mitochondrial membrane contains porins, proteins that permit the free diffusion of molecules of up to 10 kD (Section 10-4B). *The intermembrane space is therefore equivalent to the cytosol in its concentrations of metabolites and ions.* The inner membrane, which is ~75% protein by mass, is considerably richer in proteins than is the outer membrane (Fig. 17-3). It is freely permeable only to O_2,

CO_2, and H_2O and contains, in addition to respiratory chain proteins, numerous transport proteins that control the passage of metabolites such as ATP, ADP, pyruvate, Ca^{2+}, and phosphate. *The controlled impermeability of the inner mitochondrial membrane to most ions and metabolites permits the generation of ion gradients across this barrier and results in the compartmentalization of metabolic functions between cytosol and mitochondria.*

Cytosolic Reducing Equivalents Are "Transported" into Mitochondria

The NADH produced in the cytosol by glycolysis must gain access to the mitochondrial electron-transport chain for aerobic oxidation. However, the inner mitochondrial membrane lacks an NADH transport protein. *Only the electrons from cytosolic NADH are transported into the mitochondrion by one of several ingenious "shuttle" systems.* We have already discussed the **malate–aspartate shuttle** (Fig. 15-27), in which cytosolic oxaloacetate is reduced to malate for transport into the mitochondrion. When malate is reoxidized in the matrix, it gives up the reducing equivalents that originated in the cytosol.

In the **glycerophosphate shuttle** (Fig. 17-4) of insect flight muscle (the tissue with the largest known sustained power output—about the same power-to-weight ratio as a small automobile engine), **3-phosphoglycerol dehydrogenase** catalyzes the oxidation of cytosolic NADH by dihydroxyacetone phosphate to yield NAD^+, which reenters glycolysis. The electrons of the resulting **3-phosphoglycerol** are transferred to **flavoprotein dehydrogenase** to form $FADH_2$. This enzyme, which is situated on the inner mitochondrial membrane's outer surface, supplies electrons directly to the electron-transport chain.

The ADP–ATP Translocator

Most of the ATP generated in the mitochondrial matrix through oxidative phosphorylation is used in the cytosol. The inner mitochondrial membrane contains an **ADP–ATP translocator** (also called the **adenine nucleotide translocase**) that transports ATP out of the matrix in exchange for ADP produced in the cytosol by ATP hydrolysis.

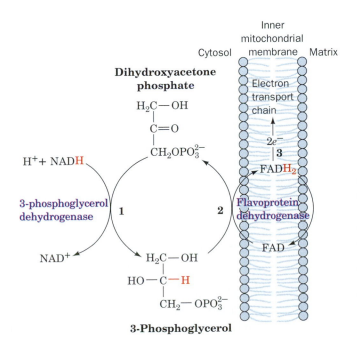

Figure 17-4. The glycerophosphate shuttle. The electrons of cytosolic NADH are transported to the mitochondrial electron-transport chain in three steps (shown in red as hydride transfers): (1) Cytosolic oxidation of NADH by dihydroxyacetone phosphate catalyzed by 3-phosphoglycerol dehydrogenase. (2) Oxidation of 3-phosphoglycerol by flavoprotein dehydrogenase with reduction of FAD to $FADH_2$. (3) Reoxidation of $FADH_2$ with passage of electrons into the electron-transport chain.

Figure 17-5. Conformational mechanism of the ADP–ATP translocator. An adenine nucleotide–binding site located in the intersubunit contact area of the translocator dimer is alternately exposed to the two sides of the membrane.

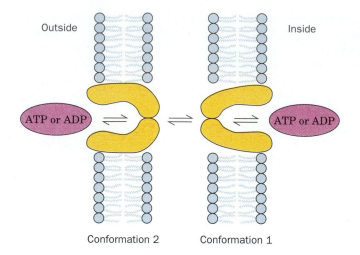

The ADP–ATP translocator, a dimer of identical 30-kD subunits, has one binding site for which ADP and ATP compete. It has two major conformations: one with its ATP–ADP binding site facing the inside of the mitochondrion, and the other with this site facing outward (Fig. 17-5). The translocator must bind ligand to change from one conformation to the other at a physiologically reasonable rate. In this respect, it differs from the glucose transporter (Fig. 10-35), which can change its conformation in the absence of ligand. Note that the export of ATP (net charge -4) and the import of ADP (net charge -3) results in the export of one negative charge per transport cycle. This **electrogenic** antiport is driven by the membrane potential difference, $\Delta\Psi$, across the inner mitochondrial membrane (positive outside), which is a consequence of the transmembrane proton gradient.

The P_i that is also required for synthesis of ATP in the matrix is imported from the cytosol via a P_i–H^+ symport system. The transmembrane proton gradient generated by the electron-transport machinery of the inner mitochondrial membrane thus not only provides the thermodynamic driving force for ATP synthesis (Section 17-3), it also drives the transport of the raw materials—ADP and P_i—for that process.

Ca^{2+} Transport

Separate systems in the inner mitochondrial membrane mediate the influx and efflux of Ca^{2+} (Fig. 17-6). The Ca^{2+} influx is driven by the membrane potential ($\Delta\Psi$, negative inside), which attracts positively charged ions. The rate of influx varies with the external $[Ca^{2+}]$ because the K_M for Ca^{2+} transport by this system is greater than the cytosolic Ca^{2+} concentration.

Ca^{2+} exits the matrix only in exchange for Na^+ (antiport). This exchange process normally operates at its maximum velocity. *Mitochondria (as well as endoplasmic reticulum and sarcoplasmic reticulum) therefore can act as a "buffer" for cytosolic Ca^{2+}:* If cytosolic $[Ca^{2+}]$ rises, the mitochondrial Ca^{2+} influx increases while Ca^{2+} efflux remains constant, resulting in a net influx of Ca^{2+}. The mitochondrial $[Ca^{2+}]$ therefore increases while the cytosolic $[Ca^{2+}]$ decreases to its original level (its set-point). Conversely, a decrease in cytosolic $[Ca^{2+}]$ reduces the influx, causing net efflux of Ca^{2+} from the mitochondrion and an increase of cytosolic $[Ca^{2+}]$ back to the set-point. When cytoplasmic $[Ca^{2+}]$ rises, for example, during increased muscle activity, the matrix $[Ca^{2+}]$ also rises, thereby activating the enzymes of

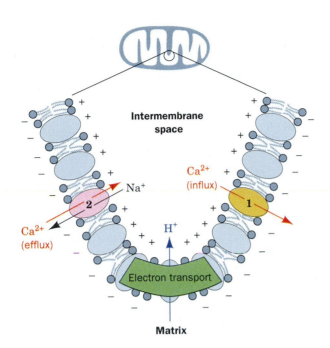

Figure 17-6.   **The two mitochondrial Ca^{2+} transport systems.** System 1 mediates Ca^{2+} influx to the matrix in response to the membrane potential (negative inside). System 2 mediates Ca^{2+} efflux in exchange for Na$^+$.

the citric acid cycle (Section 16-4B). This leads to an increase in the level of NADH, whose reoxidation by the mitochondrial electron-transport system generates the ATP needed for muscle contraction.

2. ELECTRON TRANSPORT

The electron carriers that ferry electrons from NADH and FADH$_2$ to O$_2$ are associated with the inner mitochondrial membrane. Some of these redox centers are mobile, and others are components of integral membrane protein complexes. The sequence of electron carriers roughly reflects their relative reduction potentials, so that the overall process of electron transport is exergonic. We begin this section by examining the thermodynamics of electron transport. We then consider the molecular characteristics of the various electron carriers.

A. Thermodynamics of Electron Transport

We can estimate the thermodynamic efficiency of electron transport by inspecting the standard reduction potentials of the redox centers. As we saw in our thermodynamic considerations of oxidation–reduction reactions (Section 13-3), an oxidized substrate's affinity for electrons increases with its standard reduction potential, $\mathscr{E}°'$ (Table 13-3 lists the standard reduction potentials of some biologically important half-reactions). The standard reduction potential difference, $\Delta\mathscr{E}°'$, for a redox reaction involving any two half-reactions is expressed

$$\Delta\mathscr{E}°' = \mathscr{E}°'_{(e^-\ acceptor)} - \mathscr{E}°'_{(e^-\ donor)}$$

NADH Oxidation Is Highly Exergonic

The half-reactions for oxidation of NADH by O$_2$ are

$$NAD^+ + H^+ + 2\,e^- \rightleftharpoons NADH \qquad \Delta\mathscr{E}°' = -0.315\ V$$

and

$$\tfrac{1}{2}\,O_2 + 2\,H^+ + 2\,e^- \rightleftharpoons H_2O \qquad \Delta\mathscr{E}°' = 0.815\ V$$

Since the O_2/H_2O half-reaction has the greater standard reduction potential and therefore the higher affinity for electrons, the NADH half-reaction is reversed so that NADH is the electron donor in this couple and O_2 the electron acceptor. The overall reaction is

$$\tfrac{1}{2} O_2 + NADH + H^+ \rightleftharpoons H_2O + NAD^+$$

so that

$$\Delta \mathscr{E}^{\circ\prime} = 0.815 \text{ V} - (-0.315 \text{ V}) = 1.130 \text{ V}$$

The standard free energy change for the reaction can then be calculated from Eq. 13-6:

$$\Delta G^{\circ\prime} = -n\mathscr{F}\Delta\mathscr{E}^{\circ\prime}$$

For NADH oxidation, $\Delta G^{\circ\prime} = -218 \text{ kJ} \cdot \text{mol}^{-1}$. In other words, the oxidation of 1 mol of NADH by O_2 (the transfer of 2 mol e^-) under standard biochemical conditions is associated with the release of 218 kJ of free energy.

Because the standard free energy required to synthesize 1 mol of ATP from ADP + P_i is 30.5 $\text{kJ} \cdot \text{mol}^{-1}$, the oxidation of NADH by O_2 is theoretically able to drive the formation of several moles of ATP. In mitochondria, the coupling of NADH oxidation to ATP synthesis is achieved by an electron-transport chain in which electrons pass through three protein complexes. *This allows the overall free energy change to be broken into three smaller parcels, each of which contributes to ATP synthesis by oxidative phosphorylation. Oxidation of one NADH results in the synthesis of approximately 3 ATP* (we shall see later why the relationship is not strictly stoichiometric). The thermodynamic efficiency of oxidative phosphorylation is therefore $3 \times 30.5 \text{ kJ} \cdot \text{mol}^{-1} \times 100/218 \text{ kJ} \cdot \text{mol}^{-1} = 42\%$ under standard biochemical conditions. However, under physiological conditions in active mitochondria (where the reactant and product concentrations as well as the pH deviate from standard conditions), this thermodynamic efficiency is thought to be ~70%. In comparison, the energy efficiency of a typical automobile engine is <30%.

B. The Sequence of Electron Transport

Oxidation of NADH and FADH$_2$ is carried out by the electron-transport chain, a set of protein complexes containing redox centers with progressively greater affinities for electrons (increasing standard reduction potentials). Electrons travel through this chain from lower to higher standard reduction potentials (Fig. 17-7). Electrons are carried from **Complexes I** and **II** to **Complex III** by **coenzyme Q** (**CoQ** or **ubiquinone;** so named because of its ubiquity in respiring organisms), and from Complex III to **Complex IV** by the peripheral membrane protein **cytochrome c.**

Complex I catalyzes oxidation of NADH by CoQ:

$$NADH + CoQ \ (oxidized) \longrightarrow NAD^+ + CoQ \ (reduced)$$
$$\Delta\mathscr{E}^{\circ\prime} = 0.360 \text{ V} \qquad \Delta G^{\circ\prime} = -69.5 \text{ kJ} \cdot \text{mol}^{-1}$$

Complex III catalyzes oxidation of CoQ (reduced) by cytochrome c:

$$CoQ \ (reduced) + 2 \ \text{cytochrome } c \ (oxidized) \longrightarrow$$
$$CoQ \ (oxidized) + 2 \ \text{cytochrome } c \ (reduced)$$
$$\Delta\mathscr{E}^{\circ\prime} = 0.190 \text{ V} \qquad \Delta G^{\circ\prime} = -36.7 \text{ kJ} \cdot \text{mol}^{-1}$$

Complex IV catalyzes oxidation of reduced cytochrome c by O$_2$, the terminal electron acceptor of the electron-transport process.

● See Guided Exploration 15:

Electron Transport and Oxidative Phosphorylation Overview.

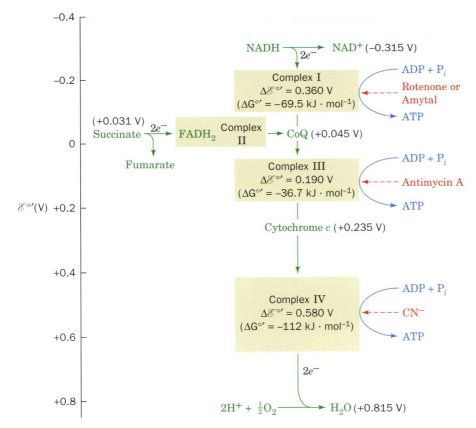

Figure 17-7. Overview of electron transport in the mitochondrion. The standard reduction potentials of its most mobile components (*green*) are indicated, as are the points where sufficient free energy is released to synthesize ATP (*blue*) and the sites of action of several respiratory inhibitors (*red*). Complexes I, III, and IV do not directly synthesize ATP but sequester the free energy necessary to do so by pumping protons outside the mitochondrion to form a proton gradient.

$$2 \text{ cytochrome } c \ (reduced) + \tfrac{1}{2} O_2 \longrightarrow 2 \text{ cytochrome } c \ (oxidized) + H_2O$$
$$\Delta\mathscr{E}^{\circ\prime} = 0.580 \text{ V} \qquad \Delta G^{\circ\prime} = -112 \text{ kJ} \cdot \text{mol}^{-1}$$

As an electron pair successively traverses Complexes I, III, and IV, sufficient free energy is released at each step to power the synthesis of an ATP molecule.

Complex II catalyzes the oxidation of $FADH_2$ by CoQ:

$$\text{FADH}_2 + \text{CoQ } (oxidized) \longrightarrow \text{FAD} + \text{CoQ } (reduced)$$
$$\Delta\mathscr{E}^{\circ\prime} = 0.085 \text{ V} \qquad \Delta G^{\circ\prime} = -16.4 \text{ kJ} \cdot \text{mol}^{-1}$$

This redox reaction does not release sufficient free energy to synthesize ATP; it functions only to inject the electrons from $FADH_2$ into the electron-transport chain.

Inhibitors Reveal the Workings of the Electron-Transport Chain

The sequence of events in electron transport was elucidated largely through the use of specific inhibitors and later corroborated by measurements of the standard reduction potentials of the redox components. The rate at which O_2 is consumed by a suspension of mitochondria is a sensitive measure of the activity of the electron-transport chain. Compounds that inhibit electron transport, as judged by their effect on O_2 consumption, include **rotenone** (a plant toxin used by Amazonian Indians to poison fish and which is also used as an insecticide), **amytal** (a barbiturate), **antimycin A** (an antibiotic), and **cyanide.**

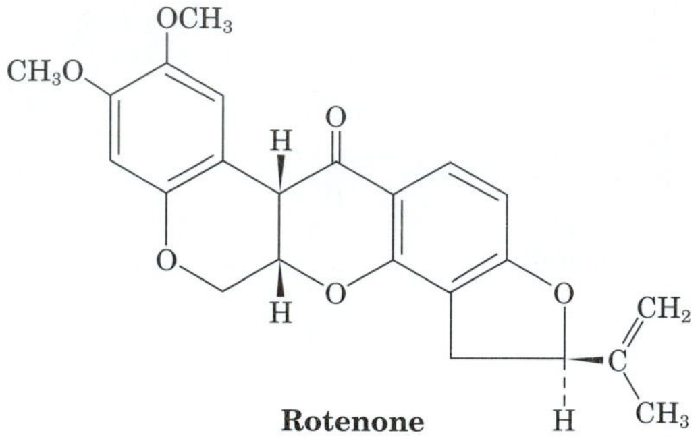

Rotenone

Amytal    **Cyanide**

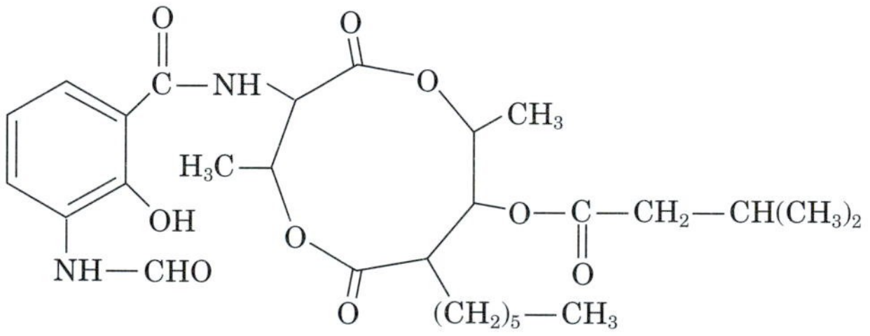

Antimycin A

Adding rotenone or amytal to a suspension of mitochondria blocks electron transport in Complex I; antimycin A blocks Complex III, and CN^- blocks electron transport in Complex IV (Fig. 17-7). Each of these inhibitors also halts O_2 consumption. Oxygen consumption resumes following addition of a substance whose electrons enter the electron-transport chain "downstream" of the block. For example, the addition of succinate to rotenone-blocked mitochondria restores electron transport and O_2 consumption. Experiments with inhibitors of electron transport thus reveal the points of entry of electrons from various substrates.

Each of the four respiratory complexes of the electron-transport chain consists of several protein components that are associated with a variety of redox-active prosthetic groups with successively increasing reduction potentials (Table 17-1). The complexes are all laterally mobile within the inner mitochondrial membrane; they do not appear to form any stable higher structures. Indeed, they are not even present in equimolar amounts. In the following sections, we examine their structures and the molecules that transfer electrons between them. Their relationships are summarized in Fig. 17-8.

C. Complex I (NADH–Coenzyme Q Oxidoreductase)

Complex I, which passes electrons from NADH to CoQ, may be the largest protein complex in the inner mitochondrial membrane, containing as many

Table 17-1. **Reduction Potentials of Electron-Transport Chain Components in Resting Mitochondria**

Component	$\mathcal{E}°'$ (V)
NADH	−0.315
Complex I (NADH–CoQ reductase; 850 kD, 43 subunits):	
FMN	?
(Fe–S)N-1a	−0.380
(Fe–S)N-1b	−0.250
(Fe–S)N-2	−0.030
(Fe–S)N-3,4	−0.245
(Fe–S)N-5,6	−0.270
Succinate	0.031
Complex II (succinate–CoQ reductase; 127 kD, 5 subunits):	
FAD	−0.040
(Fe–S)S-1	−0.030
(Fe–S)S-2	−0.245
(Fe–S)S-3	0.060
Cytochrome b_{560}	−0.080
Coenzyme Q	0.045
Complex III (CoQ–cytochrome c reductase; 248 kD, 11 subunits):	
Cytochrome b_H (b_{562})	0.030
Cytochrome b_L (b_{566})	−0.030
(Fe–S)	0.280
Cytochrome c_1	0.215
Cytochrome c	0.235
Complex IV (cytochrome c oxidase; ~200 kD, 6–13 subunits):	
Cytochrome a	0.210
Cu_A center	0.245
Cu_B	0.340
Cytochrome a_3	0.385
O_2	0.815

Source: Wilson, D.F., Erecińska, M., and Dutton, P.L., *Annu. Rev. Biophys. Bioeng.* **3**, 205 and 208 (1974); *and* Wilson, D.F., *In* Bittar, E.E. (Ed.), *Membrane Structure and Function,* Vol. 1, p. 160, Wiley (1980).

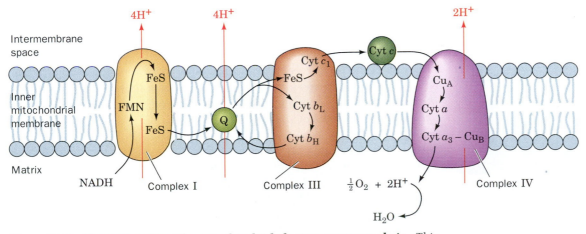

Figure 17-8. *Key to Function.* **The mitochondrial electron-transport chain.** This diagram indicates the pathways of electron transfer (*black*) and proton translocation (*red*). Electrons are transferred between Complexes I and III by the membrane-soluble coenzyme Q (Q) and between Complexes III and IV by the peripheral membrane protein cytochrome c. Complex II (not shown) transfers electrons from succinate to coenzyme Q. ✴ See the Animated Figures.

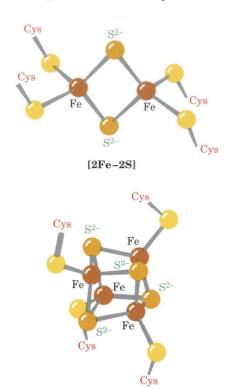

[2Fe–2S]

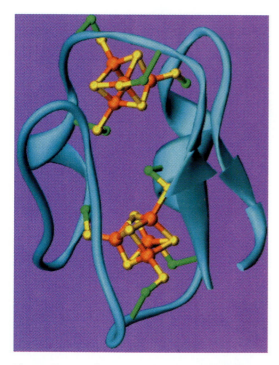

[4Fe–4S]

Figure 17-9.   The X-ray structure of ferredoxin from Peptococcus aerogenes. This monomeric 54-residue protein contains two [4Fe–4S] clusters. The C_β atoms of the four Cys residues liganding each cluster are green, the Fe atoms are orange, and the S atoms are yellow. [Based on an X-ray structure by Elinor Adman, Larry Sieker, and Lyle Jensen, University of Washington.]

● See the Interactive Exercises.

as 43 polypeptides with a total mass of 850 kD. It contains one molecule of **flavin mononucleotide** (**FMN,** a redox-active prosthetic group that differs from FAD only by the absence of the AMP group) and six to seven **iron–sulfur clusters** that participate in electron transport.

The Coenzymes of Complex I

Iron–sulfur clusters occur as the prosthetic groups of **iron–sulfur proteins** (also called **nonheme iron proteins**). The two most common types, designated **[2Fe–2S]** and **[4Fe–4S] clusters** *(at left)*, consist of equal numbers of iron and sulfide ions and are both coordinated to four protein Cys sulfhydryl groups. Note that the Fe atoms in both types of clusters are each coordinated by four S atoms, which are more or less tetrahedrally disposed around the Fe. The protein **ferredoxin** from bacteria contains two [4Fe–4S] clusters (Fig. 17-9).

Iron–sulfur clusters can undergo one-electron oxidation and reduction. *The oxidized and reduced states of all iron–sulfur clusters differ by one formal charge regardless of their number of Fe atoms.* This is because the Fe atoms in each cluster form a conjugated system and thus can have oxidation states between the +2 and +3 values possible for individual Fe atoms.

FMN and CoQ can each adopt three oxidation states (Fig. 17-10). They are capable of accepting and donating either one or two electrons because their semiquinone forms are stable (these semiquinones are stable **free radicals,** molecules with an unpaired electron). FMN is tightly bound to proteins; however, CoQ has a hydrophobic tail that makes it soluble in the inner mitochondrial membrane's lipid bilayer. In mammals, this tail consists of 10 C_5 isoprenoid units and hence the coenzyme is designated Q_{10}. In other organisms, CoQ may have only 6 (Q_6) or 8 (Q_8) isoprenoid units.

Electron Transfer in Complex I

The transit of electrons from NADH to CoQ presumably occurs by a stepwise mechanism, according to the reduction potential of the various redox centers in Complex I (Table 17-1). This process involves the transient reduction of each group as it binds electrons and its reoxidation when it passes the electrons to the next group. The exact sequence of electron movements is not known, in part because reducing equivalents appear to distribute among the iron–sulfur clusters faster than they are donated by NADH.

NADH can participate in only a two-electron transfer reaction. In contrast, the cytochromes of Complex III (see below), to which reduced CoQ passes its electrons, are capable of only one-electron reactions. *FMN and CoQ, which can transfer one or two electrons at a time, therefore provide an electron conduit between the two-electron donor NADH and the one-electron acceptors, the cytochromes.*

Proton Translocation

As electrons are transferred between the redox centers of Complex I, four protons are translocated from the matrix to the intermembrane space. The fate of protons donated by NADH is uncertain: They may be among those pumped across the membrane or used in the reduction of CoQ to its hydroquinone form, $CoQH_2$. The proton-pumping mechanism of Complex I is not well understood but almost certainly is driven by conformational changes induced by changes in the redox state of the protein (Fig. 17-11).

A model for proton pumping via conformational changes is provided by bacteriorhodopsin, an integral membrane protein of *Halobacter halobium* that contains seven membrane-spanning helical segments (Section 10-1A).

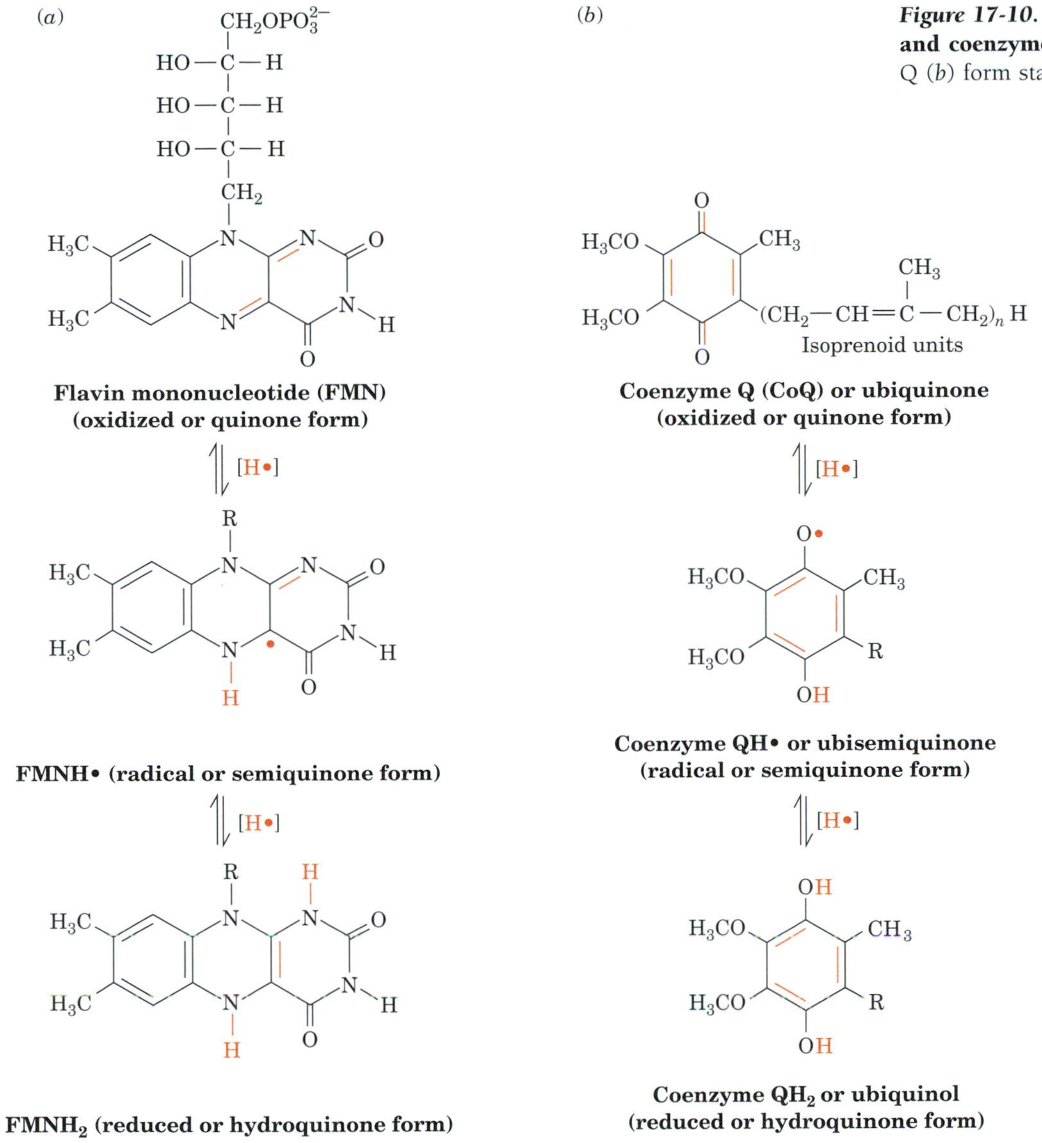

(a)

Flavin mononucleotide (FMN)
(oxidized or quinone form)

[H•]

FMNH• (radical or semiquinone form)

[H•]

FMNH$_2$ (reduced or hydroquinone form)

(b)

Isoprenoid units

Coenzyme Q (CoQ) or ubiquinone
(oxidized or quinone form)

[H•]

Coenzyme QH• or ubisemiquinone
(radical or semiquinone form)

[H•]

Coenzyme QH$_2$ or ubiquinol
(reduced or hydroquinone form)

Figure 17-10. The oxidation states of FMN and coenzyme Q. Both FMN (*a*) and coenzyme Q (*b*) form stable semiquinone free-radical states.

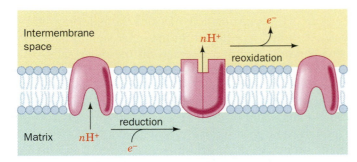

Figure 17-11. A model for electron transport–linked proton pumping. In the oxidized conformation, protons bind to amino acid side chains on the matrix side of the membrane. Reduction causes a conformational change that exposes protonated groups to the cytosolic side of the membrane and reduces their p*K*'s, causing the protons to dissociate. Reoxidation results in a conformational change that restores the protein to its original conformation. In Complex I, four protons are pumped across the inner mitochondrial membrane for each pair of electrons donated by NADH.

Box 17-1

BIOCHEMISTRY IN FOCUS

Cytochromes Are Electron-Transport Heme Proteins

Cytochromes, whose function was elucidated in 1925 by David Keilin, are redox-active proteins that occur in all organisms except a few types of obligate anaerobes. These proteins contain heme groups that alternate between their Fe(II) and Fe(III) oxidation states during electron transport.

The heme groups of the reduced Fe(II) cytochromes have prominent visible absorption spectra consisting of three peaks: the α, β, and γ (**Soret**) bands. The spectrum for cytochrome c is shown below.

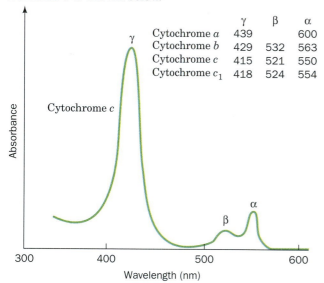

	γ	β	α
Cytochrome a	439		600
Cytochrome b	429	532	563
Cytochrome c	415	521	550
Cytochrome c_1	418	524	554

The wavelength of the α peak, which varies characteristically with the reduced cytochrome species (it is absent in oxidized cytochromes), is used to differentiate the various cytochromes (*top right of figure*).

Each group of cytochromes contains a differently substituted heme group coordinated with the redox-active iron atom (*right*). The b-type cytochromes contain **protoporphyrin IX**, which also occurs in myoglobin and hemoglobin (Section 7-1A). The heme group of c-type cytochromes differs from protoporphyrin IX in that its vinyl groups have added Cys sulfhydryls across their double bonds to form thioether linkages to the protein. Heme a contains a long hydrophobic tail of isoprene units attached to the porphyrin, as well as a formyl group in place of a methyl substituent in hemes b and c. The axial ligands of the heme iron also vary with the cytochrome type. In cytochromes a and b, both ligands are His residues, whereas in cytochrome c, one is His and the other is the S atom of Met.

Within each group of cytochromes, different heme group environments may be characterized by slightly different α peak wavelengths. For this reason, it is convenient to identify cytochromes by the wavelength (in nm) at which its α band absorbance is maximal (e.g., cytochrome b_{560} in Complex II). Cytochromes are also identified nondescriptively with either numbers or letters.

Reduced heme groups are highly reactive entities; they can transfer electrons over distances of 10 to 20 Å at physiologically significant rates. Hence cytochromes, in a sense, have the opposite function of enzymes: Instead of persuading un-

Bacteriorhodopsin is a light-driven proton pump: It obtains the free energy required for pumping protons through the absorption of light by its retinal prosthetic group. The conformation of bacteriorhodopsin alternates between two states so that a proton from the inner side of the membrane is bound in one conformation and released on the outer side of the membrane when the protein changes its conformation. The conformational change in bacteriorhodopsin is induced by a conformational change in retinal that is triggered by light absorption. Studies of this relatively small (26 kD) protein lend validity to the theory that molecular events at one site (the light-absorbing group) can affect the overall protein structure. The proton-binding groups of the much larger Complex I and their relationships to the redox centers have yet to be identified.

D. Complex II (Succinate–Coenzyme Q Oxidoreductase)

Complex II, which contains the citric acid cycle enzyme succinate dehydrogenase (Section 16-3F) and three other small hydrophobic subunits, passes electrons from succinate to CoQ. Its redox groups include succinyl dehydrogenase's covalently bound FAD (Fig. 16-12) to which electrons are

reactive substrates to react, they must prevent their hemes from transferring electrons nonspecifically to other cellular components. This, no doubt, is why these hemes are almost entirely enveloped by protein. However, cytochromes must also provide a path for electron transfer to an appropriate partner. Since electron transfer occurs far more efficiently through bonds than through space, protein structure appears to be an important determinant of the rate of electron transfer between proteins.

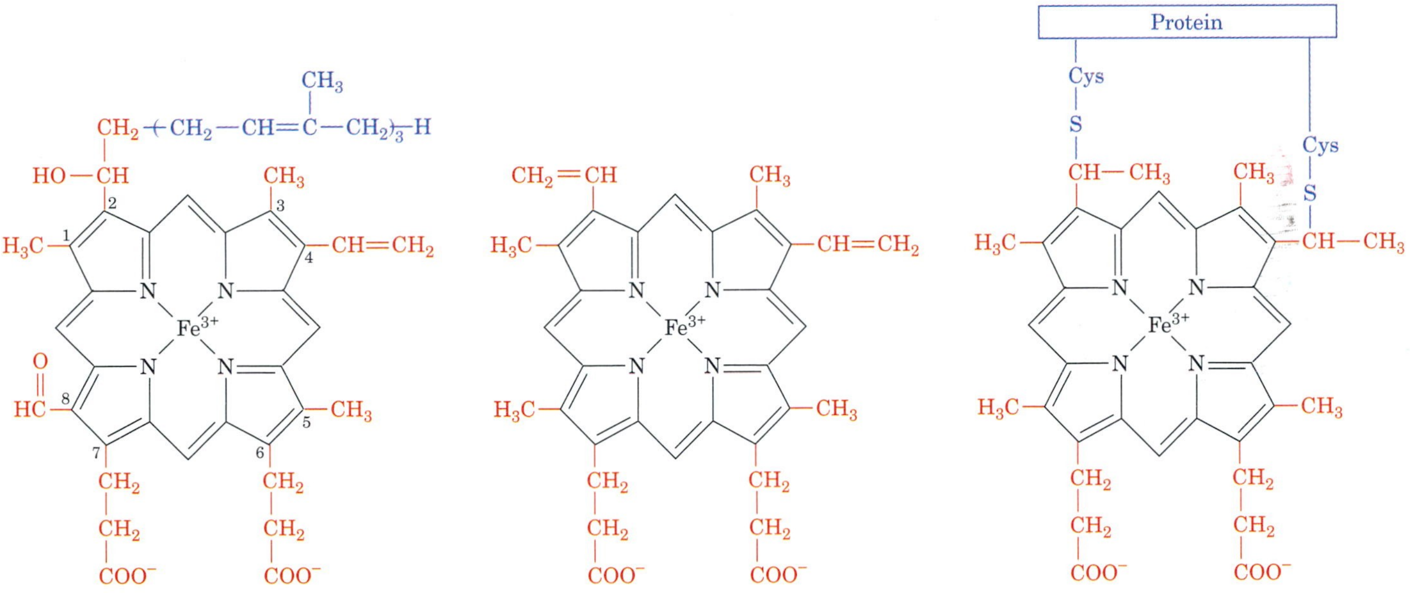

Heme *a*

Heme *b* (iron–protoporphyrin IX)

Heme *c*

initially passed, one [4Fe–4S] cluster, two [2Fe–2S] clusters, and one **cytochrome b_{560}** (cytochromes are discussed in Box 17-1).

The free energy for electron transfer from succinate to CoQ (Fig. 17-7) is insufficient to drive ATP synthesis. The complex is nevertheless important because it allows relatively high-potential electrons to enter the electron-transport chain by bypassing Complex I.

Note that Complexes I and II, despite their names, do not operate in series. But both accomplish the same result: the transfer of electrons to CoQ from reduced substrates (NADH or succinate). *CoQ, which diffuses in the lipid bilayer among the respiratory complexes, therefore serves as a sort of collection point for electrons.* As we shall see in Section 19-2C, the first step in fatty acid oxidation generates electrons that enter the electron-transport chain at the level of CoQ. CoQ also collects electrons from the $FADH_2$ produced by the glycerophosphate shuttle (Fig. 17-4).

E. Complex III (Coenzyme Q–Cytochrome c Oxidoreductase)

Complex III (also known as **cytochrome bc_1**) passes electrons from reduced CoQ to cytochrome *c*. It contains two **b-type cytochromes,** one **cytochrome**

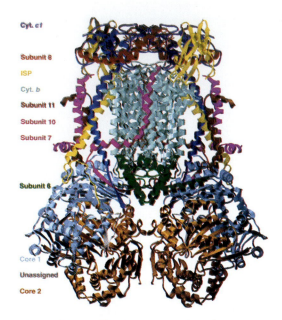

Cyt. c1
Subunit 8
ISP
Cyt. b
Subunit 11
Subunit 10
Subunit 7
Subunit 6
Core 1
Unassigned
Core 2

Figure 17-12.   The X-ray structure of the cytochrome bc_1 (Complex III). The view of this dimeric protein, from bovine heart, is from within the inner mitochondrial membrane with the intermembrane space at the top. The polypeptide chains from each subunit are drawn as colored ribbons, and the hemes are drawn in stick form. The top section of the complex extends into the intermembrane space, the middle region, which is composed largely of parallel α helices, spans the inner mitochondrial membrane, and the lower region extends into the matrix. [Courtesy of Johann Deisenhofer, University of Texas Southwestern Medical Center.] ● **See the Interactive Exercises.**

● See Guided Exploration 16:

The Q Cycle.

c_1, and one [2Fe–2S] cluster (known as the **Rieske center**). Complex III from bovine heart mitochondria is a dimer with 11 subunits in each 248-kD monomer. Its X-ray structure (Fig. 17-12), determined by Johann Deisenhofer, reveals a pear-shaped dimer whose widest part extends 75 Å into the mitochondrial matrix. The 42-Å-thick transmembrane portion consists of 13 transmembrane helices per monomer, most of which are tilted with respect to the plane of the membrane. Eight of these helices belong to the **cytochrome b** subunit, which binds both b-type cytochrome hemes, b_{562} (or b_H, for high potential, which lies near the intermembrane space) and b_{566} (or b_L, for low potential, which lies near the matrix). The **iron–sulfur protein (ISP)**, which binds the Rieske center, is anchored in the transmembrane region via its N-terminal two helices and extends into the intermembrane space, where the Rieske center is located. The **cytochrome c_1** subunit is similarly anchored to the transmembrane region via its C-terminal helix and extends a globular and relatively mobile c-type heme-containing domain 38 Å into the intermembrane space. The portion of the complex that extends into the mitochondrial matrix consists mainly of two structurally similar subunits named Core 1 and Core 2.

Electron Transport in Complex III: The Q Cycle

Complex III functions to permit one molecule of $CoQH_2$, a two-electron carrier, to reduce two molecules of cytochrome c, a one-electron carrier. This occurs by an unexpected bifurcation of the flow of electrons from $CoQH_2$ to cytochrome c_1 and to cytochrome b (in which the latter flow is also cyclic). It is this so-called **Q cycle** that permits Complex III to pump protons from the matrix to the intermembrane space.

The essence of the Q cycle is that *$CoQH_2$ undergoes a two-cycle reoxidation in which the semiquinone, $CoQ^{\cdot-}$, is a stable intermediate.* This involves two independent binding sites for coenzyme Q: Q_o, which binds $CoQH_2$ and is located between the Rieske [2Fe–2S] center and heme b_L in proximity to the intermembrane space; and Q_i, which binds both $CoQ^{\cdot-}$ and CoQ and is located near heme b_H in proximity to the matrix. In the first cycle (Fig. 17-13a), $CoQH_2$ from Complex I (**1** and **2**) binds to the Q_o site, where it transfers one of its electrons to the ISP (**3**), releasing its two protons into the intermembrane space and yielding $CoQ^{\cdot-}$. The ISP goes on to reduce cytochrome c_1, whereas the $CoQ^{\cdot-}$ transfers its remaining electron to cytochrome b_L (**4**), yielding fully oxidized CoQ. Cytochrome b_L then reduces cytochrome b_H (**6**). The CoQ from Step 4 is released from the Q_o site and rebinds to the Q_i site (**5**), where it picks up the electron from cytochrome b_H (**7**), reverting to the semiquinone form, $Q^{\cdot-}$. Thus, the reaction for this first cycle is

$$CoQH_2 + \text{cytochrome } c_1(Fe^{3+}) \longrightarrow$$
$$CoQ^{\cdot-} + \text{cytochrome } c_1(Fe^{2+}) + 2\,H^+ \text{ (cytosolic)}$$

In the second cycle (Fig. 17-13b), another $CoQH_2$ from Complex I repeats Steps 1 through 6: One electron reduces the ISP and then cytochrome c_1, and the other electron sequentially reduces cytochrome b_L and then cytochrome b_H. This second electron then reduces the $CoQ^{\cdot-}$ at the Q_i site produced in the first cycle (**8**), yielding $CoQH_2$. The protons consumed in this last step originate in the mitochondrial matrix. The reaction for the second cycle is therefore

$$CoQH_2 + CoQ^{\cdot-} + \text{cytochrome } c_1(Fe^{3+}) + 2\,H^+ \text{ (matrix)} \longrightarrow$$
$$CoQ + CoQH_2 + \text{cytochrome } c_1(Fe^{2+}) + 2\,H^+ \text{ (cytosolic)}$$

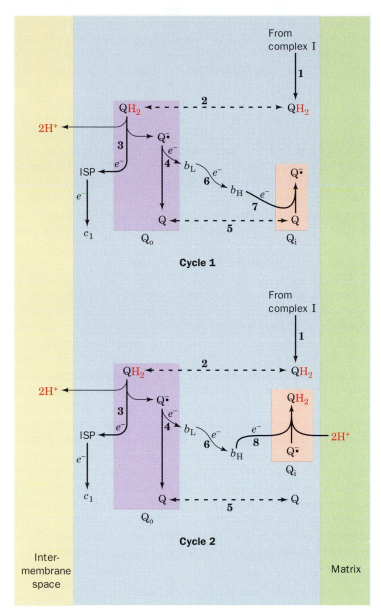

Figure 17-13. The Q cycle. The overall cycle is actually two cycles, the first requiring Reactions 1 through 7 and the second requiring Reactions 1 through 6 and 8. **(1)** Coenzyme QH_2 is supplied by Complex I on the matrix side of the membrane. **(2)** QH_2 diffuses to the cytosolic side of the membrane, where it binds in the Q_o site on the cytochrome b subunit of Complex III. **(3)** QH_2 reduces the Rieske iron–sulfur protein (ISP), forming CoQ^- semiquinone and releasing $2H^+$. The ISP goes on to reduce heme c_1. **(4)** CoQ^- reduces heme b_L to form coenzyme Q. **(5)** Q in Cycle 1 and Q^- in Cycle 2 diffuses to the matrix side, where it binds in the Q_i site on cytochrome b. **(6)** Heme b_L reduces heme b_H. **(7, Cycle 1 only)** Q is reduced to CoQ^- by heme b_H. **(8, Cycle 2 only)** CoQ^- bound in the Q_i site is reduced to $CoQH_2$ by heme b_H. The net reaction is the transfer of two electrons from $CoQH_2$ to cytochrome c_1 and the translocation of four protons from the matrix to the intermembrane space. [After Trumpower, B.L., *J. Biol. Chem.* **265**, 11410 (1990).]

For every two $CoQH_2$ that enter the Q cycle, one $CoQH_2$ is regenerated. The combination of both cycles, in which two electrons are transferred from $CoQH_2$ to cytochrome c_1, results in the overall reaction

$$CoQH_2 + 2 \text{ cytochrome } c_1(Fe^{3+}) + 2 H^+ \ (matrix) \longrightarrow$$
$$CoQ + 2 \text{ cytochrome } c_1(Fe^{2+}) + 4 H^+ \ (cytosolic)$$

The X-ray studies of Complex III provide direct evidence for the independent existence of the Q_o and Q_i sites. The antifungal agent **myxothiazol,**

Myxothiazol

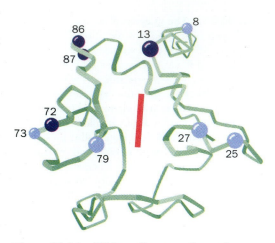

Figure 17-14. **Ribbon diagram of cytochrome *c* showing the Lys residues involved in intermolecular complex formation.** Dark and light blue balls, respectively, mark the position of Lys residues whose ε-amino groups are strongly and less strongly protected by cytochrome c_1 or cytochrome *c* oxidase against acetylation. Note that these Lys residues form a ring around the heme (*solid bar*) on one face of the protein. [After Mathews, F.S., *Prog. Biophys. Mol. Biol.* **45,** 45 (1986).] ● **See the Interactive Exercises and Kinemage Exercise 5.**

which is known to inhibit electron flow from $CoQH_2$ to the ISP and to heme b_L (Steps 3 and 4 of both cycles), binds in a pocket within cytochrome *b* midway between the iron positions of the Rieske [2Fe–2S] center and heme b_L. Thus, this binding pocket is likely to overlap the Q_o site. Similarly, antimycin A, which has been shown to block electron flow from heme b_H to CoQ or CoQ^- (Step 7 of Cycle 1 and Step 8 of Cycle 2), binds in a pocket near heme b_H, thereby identifying this pocket as site Q_i.

The circuitous route of electron transfer in Complex III is tied to the ability of coenzyme Q to diffuse within the hydrophobic core of the membrane in order to bind to both the Q_o and Q_i sites. *When $CoQH_2$ is oxidized, two reduced cytochrome c molecules and four protons appear on the outer side of the membrane.* Proton transport by the Q cycle thus differs from the proton-pumping mechanism of Complexes I and IV (see below): In the Q cycle, a redox center itself (CoQ) is the proton carrier.

Cytochrome *c*: A Soluble Electron Carrier

The electrons that flow to cytochrome c_1 are transferred to cytochrome *c*, which, unlike the other cytochromes of the respiratory electron-transport chain, is a peripheral membrane protein. It shuttles electrons between Complexes III and IV on the outer surface of the inner mitochondrial membrane. The structure and evolution of cytochrome *c* are discussed in Sections 5-4A and 6-2C. Several invariant Lys residues in cytochrome *c* lie in a ring around the exposed edge of its otherwise buried heme group (Fig. 17-14). These residues constitute a binding site that was identified by **differential labeling:** Treatment of cytochrome *c* with acetic anhydride (which acetylates Lys residues) in the presence and absence of cytochrome c_1 demonstrated that cytochrome c_1 completely shields these cytochrome *c* Lys residues. The reactivities of other cytochrome *c* Lys residues that are distant from the exposed heme edge are unaffected by the presence of cytochrome c_1. Nearly identical results were obtained when cytochrome c_1 was replaced by cytochrome *c* oxidase. Evidently, both these proteins have negatively charged sites that are complementary to the ring of positively charged Lys residues on cytochrome *c*. Such ion pairing interactions probably align redox groups for optimal electron transfer.

F. Complex IV (Cytochrome *c* Oxidase)

Cytochrome *c* oxidase catalyzes the one-electron oxidations of four consecutive reduced cytochrome *c* molecules and the concomitant four-electron reduction of one O_2 molecule:

$$4 \text{ Cytochrome } c(\text{Fe}^{2+}) + 4\,\text{H}^+ + \text{O}_2 \longrightarrow 4 \text{ cytochrome } c(\text{Fe}^{3+}) + 2\,\text{H}_2\text{O}$$

Mammalian Complex IV is a dimer whose ~200-kD component monomers are each composed of 13 subunits. The X-ray structure of Complex IV from bovine heart mitochondria, determined by Shinya Yoshikawa, reveals that ten of its subunits are transmembrane proteins that contain a total of 28 membrane-spanning α helices (Fig. 17-15). The core of Complex IV consists of its three largest and most hydrophobic subunits, I, II, and III, which are encoded by mitochondrial DNA (the remaining subunits are nuclearly encoded and must be transported into the mitochondrion). A concave area on the surface of the protein that faces the intermembrane space contains numerous acidic amino acids that can potentially interact with the ring of Lys residues in cytochrome *c*, the electron donor for Complex IV.

Complex IV contains four redox centers: **cytochrome *a*, cytochrome *a₃*,** a copper atom known as **Cu$_B$,** and a pair of copper atoms known as the

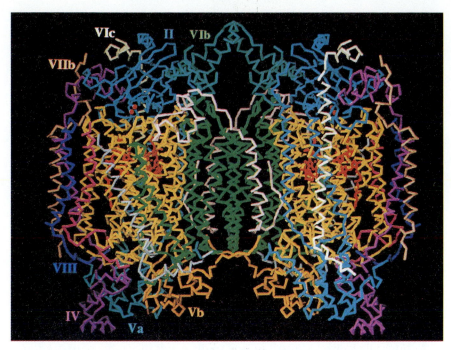

Figure 17-15.   The X-ray structure of the bovine heart cytochrome c oxidase dimer. Each of the 13 subunits is shown as a C_α trace, of a different color. The view is from within the membrane, with the intermembrane space at the top. [Courtesy of Shinya Yoshikawa, Osaka University, Japan.] ● See the Interactive Exercises.

Cu_A center. In addition, there is a Mg^{2+} ion and a Zn^{2+} ion (Fig. 17-16). The Cu_A center, which binds to Subunit II, lies 8 Å above the membrane surface. Its two copper ions are bridged by two sulfur atoms of Cys residues, giving it a geometry similar to that of a [2Fe–2S] cluster. The other redox groups—Cu_B and cytochromes a and a_3—all bind to Subunit I and lie ~13 Å below the membrane surface. The Mg^{2+} ion may participate in electron transfer or stabilize the arrangement of redox centers; the Zn^{2+}, which is part of a zinc finger motif (Section 6-4A) and is far from the redox centers, almost certainly plays a structural rather than a catalytic role.

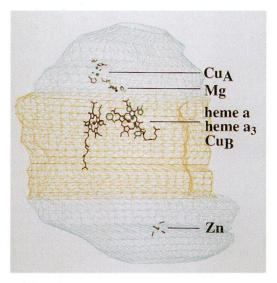

Figure 17-16.   Locations of redox centers in beef heart cytochrome c oxidase. The protein surface is shown as a cage with the membrane-spanning portion in yellow and the portions that protrude into the intermembrane space (*top*) and mitochondrial matrix (*bottom*) in cyan. Heme a (*left*) and heme a_3 (*right*) are red, the Cu ions are green, the Mg^{2+} ion is orange, a water oxygen to which the Mg^{2+} is liganded is blue, and the amino acid side chains that ligand metal ions are green and gray. The Zn^{2+} ion at the bottom of the structure does not participate in redox reactions. [Courtesy of Shinya Yoshikawa, Osaka University, Japan.]

Spectroscopic studies have shown that electron transfer in Complex IV is linear, proceeding from cytochrome *c* to the Cu_A center, then to heme *a*, and finally to heme a_3 and Cu_B. The Fe of heme a_3 lies only 4.5 Å from Cu_B; these redox groups, which may be bridged by a water molecule, really form a single binuclear complex. Electrons appear to travel between the redox centers of Complex IV via a hydrogen bond network involving amino acid side chains, the polypeptide backbone, and the propionate side chains of the heme groups.

Reduction of O_2 by Cytochrome *c* Oxidase

The reduction of O_2 to 2 H_2O by cytochrome *c* oxidase takes place at the cytochrome a_3–Cu_B binuclear complex. This reaction, which goes to completion in ~1 ms at room temperature, involves four consecutive one-electron transfers from the Cu_A and cytochrome *a* sites and occurs as follows (Fig. 17-17):

1 & 2. The binuclear $Fe(III)_{a3}$–$Cu(II)_B$ complex is reduced, by two one-electron transfers from cytochrome *c* via cytochrome *a* and the Cu_A center, to its $Fe(II)_{a3}$–$Cu(I)_B$ form.

3. O_2 binds to the reduced binuclear complex so as to bridge its Fe(II) and Cu(I) atoms.

4. Internal electron redistribution rapidly yields the stable peroxy adduct $Fe(III)—O^-—O^-$ Cu(II).

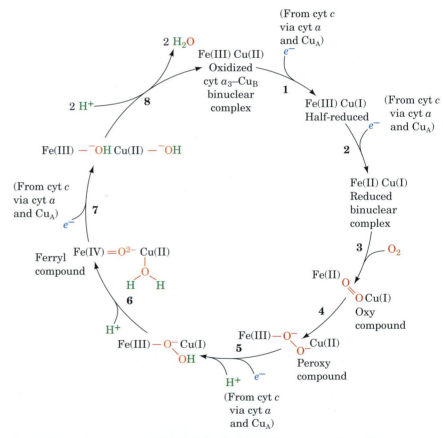

Figure 17-17.   The cytochrome *c* oxidase reaction. A total of four electrons ultimately donated by cytochrome *c*, along with four protons, are required to reduce O_2 to 2 H_2O at the cytochrome a_3–Cu_B binuclear complex. [Modified from Vartosis, C., Zhang, Y., Appelman, E.H., and Babcock, G.T., *Proc. Natl. Acad. Sci.* **90**, 240 (1993).]

5. A further one-electron transfer together with the acquisition of a proton converts the adduct to Fe(III)—O$^-$—OH Cu(I).

6. The acquisition of a second proton and an electronic rearrangement results in Fe(IV)=O^{2-} H$_2$O—Cu(II) [where Fe(IV) is said to have the **ferryl** oxidation state].

7. The fourth one-electron transfer together with a proton rearrangement then yields Fe(III)—OH$^-$ $^-$HO—Cu(II).

8. Finally, the acquisition of two more protons yields 2 H$_2$O together with the Fe(III)$_{a3}$–Cu(II)$_B$ complex, thereby completing the cycle.

Four protons are consumed in the reduction of O$_2$ by cytochrome c oxidase. These protons originate in the mitochondrial matrix and probably reach the heme a_3–Cu$_B$ complex via a hydrogen bond network that includes groups on the protein as well as water molecules that occupy cavities in the protein structure. The product of the reaction, water, exits the enzyme via a hydrophilic channel stretching from heme a_3, between subunits I and II, to the cytosolic surface of the enzyme. The route by which O$_2$ reaches the oxygen reduction site is not apparent in the X-ray structure of the fully oxidized enzyme. A conformational change that occurs on reduction of the Fe(III)$_{a3}$–Cu(II)$_B$ complex may open an O$_2$ channel, presumably lined with hydrophobic groups, possibly including membrane lipids that bind tightly to the protein.

In addition to the four protons used to reduce O$_2$, four protons from the matrix are translocated to the intermembrane space (thus, for every two electrons that traverse Complex IV, two protons are translocated; Fig. 17-8). The proposed pathways for proton transfer in Complex IV involve hydrogen bond networks of amino acid side chains and water molecules within the protein. Note that proton pumping in Complex IV more closely resembles that of Complex I than that of Complex III, in which CoQ ferries protons as it undergoes oxidation.

3. OXIDATIVE PHOSPHORYLATION

The endergonic synthesis of ATP from ADP and P$_i$ in mitochondria is catalyzed by an **ATP synthase** (also known as **Complex V**) that is driven by the electron-transport process. *The free energy released by electron transport through Complexes I–IV must be conserved in a form that ATP synthase can use.* Such energy conservation is referred to as **energy coupling.**

The physical characterization of energy coupling proved to be surprisingly elusive; many sensible and often ingenious ideas failed to withstand the test of experimental scrutiny. For example, one theory—now abandoned—was that electron transport yields a "high-energy" intermediate whose subsequent breakdown drives ATP synthesis. No such intermediate has ever been identified. In fact, ATP synthesis is coupled to electron transport through the formation of a transmembrane proton gradient during electron transport by Complexes I, III, and IV. In this section, we explore this coupling mechanism and the operation of ATP synthase.

A. The Chemiosmotic Theory

The **chemiosmotic theory,** proposed in 1961 by Peter Mitchell, spurred considerable controversy before becoming widely accepted. Mitchell's theory states that *the free energy of electron transport is conserved by pumping H$^+$ from the mitochondrial matrix to the intermembrane space to create an*

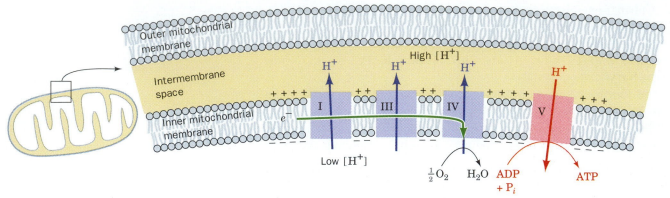

Figure 17-18. **The coupling of electron transport and ATP synthesis.** Electron transport (*green arrow*) generates a proton electrochemical gradient across the inner mitochondrial membrane. H^+ is pumped out of the mitochondrion during electron transport (*blue arrows*) and its exergonic return powers the syn-

thesis of ATP (*red arrows*). Note that the intermembrane space is topologically equivalent to the cytosol because the outer mitochondrial membrane is permeable to H^+. ✳ **See the Animated Figures.**

electrochemical H^+ gradient across the inner mitochondrial membrane. The electrochemical potential of this gradient is harnessed to synthesize ATP (Fig. 17-18). Several key observations are explained by the chemiosmotic theory:

1. Oxidative phosphorylation requires an intact inner mitochondrial membrane.

2. The inner mitochondrial membrane is impermeable to ions such as H^+, OH^-, K^+, and Cl^-, whose free diffusion would discharge an electrochemical gradient.

3. Electron transport results in the transport of H^+ out of intact mitochondria (the intermembrane space is equivalent to the cytosol), thereby creating a measurable electrochemical gradient across the inner mitochondrial membrane.

4. Compounds that increase the permeability of the inner mitochondrial membrane to protons, and thereby dissipate the electrochemical gradient, allow electron transport (from NADH and succinate oxidation) to continue but inhibit ATP synthesis; that is, they "uncouple" electron transport from oxidative phosphorylation. Conversely, increasing the acidity outside the inner mitochondrial membrane stimulates ATP synthesis.

An entirely analogous process occurs in bacteria, whose electron-transporting machinery is located in their plasma membranes (see Box 17-2).

Electron Transport Generates a Proton Gradient

Electron transport, as we have seen, causes Complexes I, III, and IV to transport protons across the inner mitochondrial membrane from the matrix, a region of low [H^+], to the intermembrane space (which is in contact with the cytosol), a region of high [H^+] (Fig. 17-8). The free energy sequestered by the resulting electrochemical gradient (also called the **protonmotive force, pmf**) powers ATP synthesis.

The free energy change of transporting a proton out of the mitochondrion has a chemical as well as an electrical component, since H^+ is an ion. ΔG is therefore expressed by Eq. 10-3, which in terms of pH is

$$\Delta G = 2.3\, RT\, [\text{pH } (in) - \text{pH } (out)] + Z\mathscr{F}\Delta\Psi \qquad [17\text{-}1]$$

Box 17-2

BIOCHEMISTRY IN CONTEXT

Bacterial Electron Transport and Oxidative Phosphorylation

It comes as no surprise that aerobic bacteria, whose ancestors gave rise to mitochondria, use similar machinery to oxidize reduced coenzymes and conserve their energy in ATP synthesis. In bacteria, the components of the respiratory electron-transport chain are located in the plasma membrane, and protons are pumped from the cytosol to the outside of the plasma membrane. Protons flow back into the cell via an ATP synthase, whose catalytic component is oriented toward the cytosol. This is exactly the arrangement expected if bacteria and mitochondria are evolutionarily related.

The oxidation of $CoQH_2$ is universal in aerobic organisms. In mitochondria, CoQ collects electrons donated by NADH (via Complex I), succinate (via Complex II), and fatty acids. In bacteria, CoQ is the collection point for electrons extracted by dehydrogenases specific for a wide variety of substrates. In bacteria, as in mitochondria, electrons flow from CoQ through cytochrome-based oxidoreductases before reaching O_2. In some species, two protein complexes (analogous to mitochondrial Complexes III and IV) carry out this process. In other species, including *E. coli*, a single type of enzyme, **quinol oxidase**, uses the electrons donated by CoQ to reduce O_2.

The advantage of a multicomplex electron-transport pathway is that it affords more opportunities for proton translocation across the bacterial membrane, so the ATP yield per electron is greater. However, the shorter electron-transport pathways may confer a selective advantage in the presence of toxins that inactivate the bacterial counterpart of mitochondrial Complex III. Multiple routes for electron transport probably also allow bacteria to adjust oxidative phosphorylation to the availability of different energy sources and to balance ATP synthesis against the regeneration of various reduced coenzymes. For example, in facultative anaerobic bacteria (which can grow in either the absence or presence of O_2), when energy needs are met through anaerobic fermentation, electron transport can be adjusted to regenerate NADH without synthesizing ATP by oxidative phosphorylation.

A variety of cytochrome-containing protein complexes occur in bacterial plasma membranes. Some of these proteins represent more streamlined versions of the mitochondrial complexes since they lack the additional subunits encoded by the nuclear genome of eukaryotes. However, this is not a universal feature of respiratory complexes, and many bacterial proteins (e.g., **cytochrome *d***) have no counterparts encoded by either the mitochondrial or nuclear genomes.

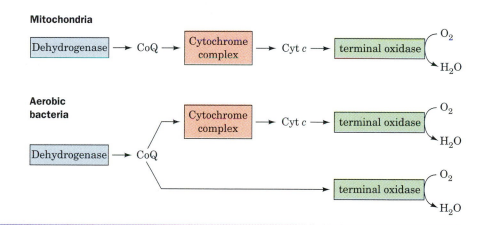

where Z is the charge on the proton (including sign), $\mathscr{F}$ is the Faraday constant, and $\Delta\Psi$ is the membrane potential. The sign convention for $\Delta\Psi$ is that when a proton is transported from negative to positive, $\Delta\Psi$ is positive. Since pH (*out*) is less than pH (*in*), *the export of protons from the mitochondrial matrix (against the proton gradient) is an endergonic process.*

The measured membrane potential across the inner membrane of a liver mitochondrion, for example, is 0.168 V (inside negative). The pH of its matrix is 0.75 units higher than that of its intermembrane space. ΔG for proton transport out of this mitochondrial matrix is therefore 21.5 kJ·mol^{-1}. Because formation of the proton gradient is an endergonic process, discharge of the gradient is exergonic. This free energy is harnessed by ATP synthase to drive the phosphorylation of ADP.

An ATP molecule's estimated physiological free energy of synthesis, around +40 to +50 kJ·mol^{-1}, is too large for ATP synthesis to be driven by the passage of a single proton back into the mitochondrial matrix; at least two protons are required. In fact, most experimental measurements (which are difficult to precisely quantitate) indicate that around three protons are required per ATP synthesized.

⦿ See Guided Exploration 17:

F_1F_0-ATP Synthase and the Binding Change Mechanism.

B. ATP Synthase

ATP synthase, also known as **proton-pumping ATP synthase** and **F_1F_0-ATPase,** is a multisubunit transmembrane protein with a total molecular mass of 450 kD. Efraim Racker discovered that mitochondrial ATP synthase is composed of two functional units, **F_0** and **F_1.** F_0 is a water-insoluble transmembrane proton channel containing at least eight different types of subunits. F_1 is a water-soluble peripheral membrane protein, composed of five types of subunits, that is easily and reversibly dissociated from F_0 by treatment with urea. Solubilized F_1 hydrolyzes ATP but cannot synthesize it (hence the name ATPase).

In electron micrographs, the matrix surface of the inner mitochondrial membrane is studded with molecules of ATP synthase (Fig. 17-19a). The F_1 component is connected to the membrane-embedded F_0 component by a protein stalk, giving the F_1 units a lollipoplike appearance (Fig. 17-19b,c).

The F_0 Component

The structure of the F_0 component is not known in detail. **Dicyclohexylcarbodiimide (DCCD)**

**Dicyclohexylcarbodiimide
(DCCD)**

is a lipid-soluble carboxyl reagent that inhibits proton transport through mammalian F_0 by reacting with a single Glu residue on one of the F_0 subunits. Reaction with DCCD usually implies that a carboxylic acid group is

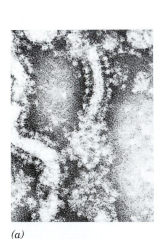

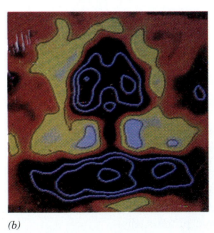

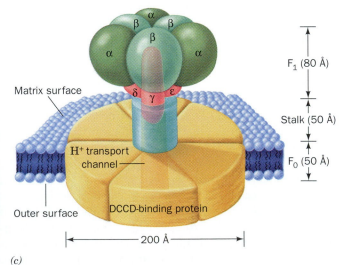

(a)   *(b)*   *(c)*

Figure 17-19.   Structure of ATP synthase. (a) An electron micrograph of cristae from a mitochondrion showing their F_1 "lollipops" projecting into the matrix. [From Parson, D.F., *Science* **140**, 985 (1963). Copyright © 1963 American Association for the Advancement of Science. Used by permission.] (b) A cryoelectron micrograph of *E. coli* F_1F_0-ATPase. The disklike F_0 component (viewed edge-on) is at the bottom, and the more spherical F_1 component is at the top. The *E. coli* ATP synthase resembles its eukaryotic counterpart but contains fewer subunits. [Courtesy of Edward Gogol and Roderick Capaldi, University of Oregon.] (c) An interpretive drawing showing the positions of the component subunits in the mitochondrial F_1F_0-ATPase.

located in a lipid environment; that is, it is buried in the membrane. Mammalian F_0 contains six copies of this **DCCD-binding protein,** which are thought to associate like staves of a barrel to form a polar H^+ transport channel that contains the buried Glu residues. The antibiotic **oligomycin B**

Oligomycin B

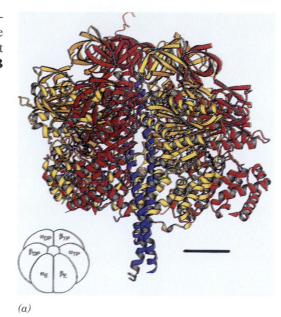

(a)

also binds to a subunit of F_0 to inhibit H^+ transport. Inhibition of proton flow from the intermembrane space back into the matrix (down its concentration gradient) blocks ATP synthesis.

The F_1 Component

The F_1 component of ATP synthase has the subunit composition $\alpha_3\beta_3\gamma\delta\varepsilon$. The X-ray structure of the F_1 subunit from bovine heart mitochondria has been determined by John Walker and Andrew Leslie. This 3440-residue (371-kD) protein consists of an 80-Å-high and 100-Å-wide spheroid that is mounted on a 30-Å-long stem (Fig. 17-20a). The α and β subunits, which are 20% identical in sequence and have nearly identical folds, are arranged alternately, like the segments of an orange, around the upper portion of a 90-Å-long α helix formed by the C-terminal segment of the γ subunit (Fig. 17-20b). The lower portion of the helix forms a bent left-handed antiparallel coiled coil with the N-terminal segment of the γ subunit. This coiled coil is almost certainly part of the stalk that links F_1 to F_0. The remainder of the γ subunit as well as the entire δ and ε subunits are not visible in this X-ray structure.

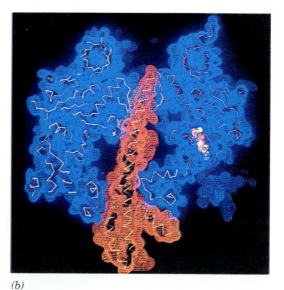

(b)

Figure 17-20.   The X-ray structure of F_1-ATPase from bovine heart mitochondria. (a) A ribbon diagram in which the α, β, and γ subunits are red, yellow, and blue, respectively. The inset drawing indicates the orientation of these subunits in this view. The bar is 20 Å long. (b) Cross section through the electron density map of the protein (the α and β subunits are blue, and the γ subunit is orange). The superimposed C_α backbones of these subunits are yellow, and a bound ATP analog is represented in space-filling form (C yellow, N blue, O red). (c) Pseudosymmetrical arrangement of the $\alpha_3\beta_3$ assembly as viewed from the top of Parts a and b. The surface is colored according to its electrical potential, with positive potentials blue and negative potentials red. Note the absence of charge on the inner surface of this sleeve. The portion of the γ subunit's C-terminal helix that contacts this sleeve is similarly devoid of charge. [From Abrahams, J.P., Leslie, A.G.W., Lutter, R., and Walker, J.E., *Nature* **370,** 623 and 627 (1994). Used with permission.] ● See the Interactive Exercises.

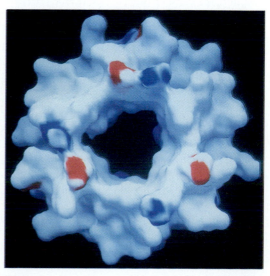

(c)

The cyclical arrangement and structural similarities of F_1's α and β subunits give it both pseudo-three-fold and pseudo-six-fold rotational symmetry (Fig. 17-20c). Nevertheless, the protein is asymmetric due to the presence of the γ subunit and also because each pair of α and β subunits adopts a different conformation (each with a different substrate affinity). The β subunits catalyze the ATP synthesis reaction although the α subunits also bind ATP.

The Binding Change Mechanism

The mechanism of ATP synthesis by proton-translocating ATP synthase can be conceptually broken down into three phases:

1. Translocation of protons carried out by F_0.
2. Catalysis of formation of the phosphoanhydride bond of ATP carried out by F_1.
3. Coupling of the dissipation of the proton gradient with ATP synthesis, which requires interaction of F_1 and F_0.

The available evidence supports a mechanism for ATP formation proposed by Paul Boyer. According to this **binding change mechanism,** F_1 has three interacting catalytic protomers ($\alpha\beta$ units), each in a different conformational state: one that binds substrates and products loosely (L state), one that binds them tightly (T state), and one that does not bind them at all (open or O state). *The free energy released on proton translocation is harnessed to interconvert these three states.* The phosphoanhydride bond of ATP is synthesized only in the T state, and ATP is released only in the O state. The reaction involves three steps (Fig. 17-21):

1. ADP and P_i bind to the loose (L) binding site (designated β_{DP} in Fig. 17-20).
2. A free energy–driven conformational change converts the L site to a tight (T) binding site (designated β_{TP}) that catalyzes the formation of ATP. This step also involves conformational changes of the other two protomers that convert the ATP-containing T site to an open (O) site (designated β_E) and convert the O site to an L site.
3. ATP is synthesized at the T site on one subunit while ATP dissociates from the O site on another subunit. The free energy supplied by the proton flow primarily facilitates the release of the newly synthesized ATP from the enzyme; that is, it drives the T → O transition,

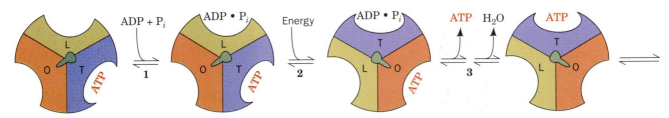

Figure 17-21.   *Key to Function.* The binding change mechanism for ATP synthase. F_1 has three chemically identical but conformationally distinct interacting $\alpha\beta$ protomers: O, the open conformation, has very low affinity for ligands and is catalytically inactive; L binds ligands loosely and is catalytically inactive; T binds ligands tightly and is catalytically active. ATP synthesis occurs in three steps: (1) ADP and P_i bind to site L. (2) An energy-dependent conformational change converts binding site L to T,

T to O, and O to L. (3) ATP is synthesized at site T and ATP is released from site O. The enzyme returns to its initial state after two more passes of this reaction sequence. The energy that drives the conformational change is apparently transmitted to the catalytic $\alpha_3\beta_3$ assembly via the rotation of the $\gamma\delta\epsilon$ assembly, here represented by the centrally located asymmetric object (*green*). [After Cross, R.L., *Annu. Rev. Biochem.* **50,** 687 (1980).]
✳ **See the Animated Figures.**

thereby disrupting the enzyme–ATP interactions that had previously promoted the spontaneous formation of ATP from ADP + P_i in the T site.

How is the free energy of proton transfer coupled to the synthesis of ATP? The cyclic nature of the binding change mechanism led Boyer to propose that *the binding changes are driven by the rotation of the catalytic assembly, $\alpha_3\beta_3$, with respect to other portions of the F_1F_0-ATPase.* This hypothesis is supported by the direct observation of this rotation made by attaching the top of F_1's $\alpha_3\beta_3\gamma$ assembly to a glass plate and observing, under a fluorescence microscope, a fluorescently labeled fiber of the muscle protein actin that had been linked to the γ subunit's stalk region. In the presence of ATP, the actin fibers were seen to rotate and always in the clockwise direction.

How can the γ subunit rotate with respect to the $\alpha_3\beta_3$ assembly, and how does this rotation promote the synthesis of ATP? The X-ray structure of F_1 indicates that the γ subunit's C-terminal segment is centered in a circular arrangement of α and β subunits, reminiscent of a cylindrical bearing rotating in a sleeve (Fig. 17-20*a,b*). Indeed, the contacting hydrophobic surfaces in this assembly are devoid of the hydrogen bonding and ionic interactions that would interfere with their free rotation (Fig. 17-20*c*) and thus appear to be "lubricated." In addition, the conformational differences between F_1's three catalytic sites are correlated with the position of the asymmetric base of the γ subunit. Apparently, the rotation of the γ subunit, responding to the influx of protons through the F_0 channel, acts much like an eccentric cam in an automobile engine to mechanically induce the conformational changes in the catalytic sites of the β subunits. Passage of around one proton through the F_0 channel induces one conformational change in F_1 (a 120° rotation), so that about three protons are required for the synthesis of one ATP from ADP + P_i.

C. The P/O Ratio

ATP synthesis is tightly coupled to the proton gradient; that is, ATP synthesis requires the discharge of the proton gradient, and the proton gradient cannot be discharged without the synthesis of ATP. The proton gradient is established through the activity of the electron-transporting complexes of the inner mitochondrial membrane. Therefore, it is possible to express the amount of ATP synthesized in terms of substrate molecules oxidized. Experiments with isolated mitochondria show that the oxidations of NADH, $FADH_2$, and **tetramethyl-*p*-phenylenediamine** (which donates an electron pair directly to Complex IV)

Tetramethyl-*p*-phenylenediamine

are associated with the synthesis of around 3, 2, and 1 ATP, respectively. This stoichiometry, called the **P/O ratio** (it relates the amount of ATP synthesized to the amount of oxygen reduced), is compatible with the chemiosmotic theory: The flow of two electrons through Complexes I, III, and IV results in the translocation of 10 protons into the intermembrane space (Fig. 17-8). Influx of these 10 protons through the F_1F_0-ATPase is sufficient to drive the synthesis of ~3 ATP. Electrons that enter the electron-transport chain as $FADH_2$ at Complex II bypass Complex I and therefore lead to the transmembrane movement of only 6 protons, enough to synthesize ~2 ATP. The transit of two electrons through Complex IV alone contributes 2 protons to the gradient, enough for ~1 ATP.

In actively respiring mitochondria, *P/O ratios are almost certainly not integral numbers.* This is also consistent with the chemiosmotic theory. Oxidation of a physiological substrate contributes to the transmembrane proton gradient at several points, but the gradient is tapped at only a single point, the F_1F_0-ATPase. Therefore, the number of protons translocated out of the mitochondrion by any component of the electron-transport chain need not be an integral multiple of the number of protons required to synthesize ATP from ADP + P_i. Moreover, the proton gradient is dissipated to some extent by the nonspecific leakage of protons back into the matrix and by the consumption of protons for other purposes, such as the transport of P_i into the matrix (Section 17-1B). Taking P_i transport into account gives a stoichiometry of four protons consumed per ATP synthesized from ADP + P_i.

Peter Hinkle's re-examination of P/O ratios suggests that the values given above are actually closer to 2.5, 1.5, and 1. If these values are correct, then the number of ATPs that are synthesized per molecule of glucose oxidized is 2.5 ATP/NADH × 10 NADH/glucose + 1.5 ATP/$FADH_2$ × 2 $FADH_2$/glucose + 2 ATP/glucose from the citric acid cycle + 2 ATP/glucose from glycolysis = 32 ATP/glucose rather than the conventional value of 38 implied by P/O ratios of 3, 2, and 1. For the sake of consistency, we shall use the value of 38 ATP/glucose throughout this textbook, but keep in mind that this value is disputed.

D. Uncoupling Oxidative Phosphorylation

Electron transport (the oxidation of NADH and $FADH_2$ by O_2) and oxidative phosphorylation (the proton gradient–driven synthesis of ATP) are normally tightly coupled. *This coupling depends on the impermeability of the inner mitochondrial membrane, which allows an electrochemical gradient to be established across this membrane by H^+ translocation during electron transport.* Virtually the only way for H^+ to re-enter the matrix is through the F_0 portion of ATP synthase. In the resting state, when oxidative phosphorylation is minimal, the electrochemical gradient across the inner mitochondrial membrane builds up to the extent that it prevents further proton pumping and therefore inhibits electron transport. When ATP synthesis increases, the electrochemical gradient dissipates, allowing electron transport to resume.

Over the years, compounds such as **2,4-dinitrophenol (DNP)** have been found to "uncouple" electron transport and ATP synthesis. DNP is a lipophilic weak acid that readily passes through membranes in its neutral, protonated state. In a pH gradient, it binds protons on the acidic side of the membrane, diffuses through the membrane, and releases the protons on the membrane's alkaline side, thereby acting as a proton-transporting ionophore (Section 10-4B) and dissipating the gradient (Fig. 17-22). The chemiosmotic theory provides a rationale for understanding the action of such **uncouplers.** *The presence in the inner mitochondrial membrane of an agent that increases its permeability to H^+ uncouples oxidative phosphorylation from electron transport by providing a route for the dissipation of the proton electrochemical gradient that does not require ATP synthesis.* Uncoupling therefore allows electron transport to proceed unchecked even when ATP synthesis is inhibited. Consequently, in the 1920s, DNP was used as a "diet pill," a practice that was effective in inducing weight loss but often caused fatal side effects. Under physiological conditions, *the dissipation*

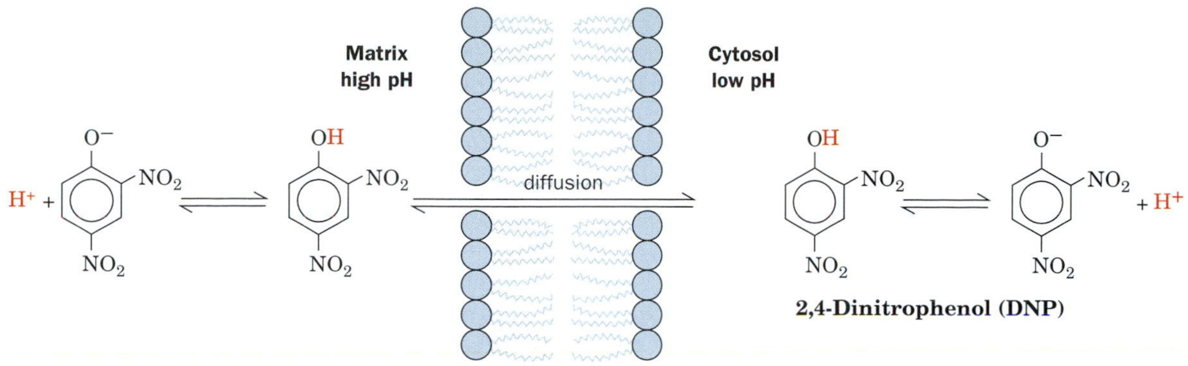

Figure 17-22. Action of 2,4-dinitrophenol. A proton-transporting ionophore such as DNP uncouples oxidative phosphorylation from electron transport by discharging the electrochemical proton gradient generated by electron transport.

of an electrochemical H^+ gradient, which is generated by electron transport and uncoupled from ATP synthesis, produces heat (see Box 17-3).

4. CONTROL OF ATP PRODUCTION

An adult woman requires some 1500 to 1800 kcal (6300–7500 kJ) of metabolic energy per day. This corresponds to the free energy of hydrolysis of over 200 mol of ATP to ADP and P_i. Yet the total amount of ATP present in the body at any one time is <0.1 mol; obviously, this sparse supply of ATP must be continually recycled. As we have seen, when carbohydrates serve as the energy supply and aerobic conditions prevail, this recycling involves glycogenolysis, glycolysis, the citric acid cycle, and oxidative phosphorylation.

Of course, the need for ATP is not constant. There is a 100-fold change in the rate of ATP consumption between sleep and vigorous activity. *The activities of the pathways that produce ATP are under strict coordinated control so that ATP is never produced more rapidly than necessary.* We have already discussed the control mechanisms of glycolysis, glycogenolysis, and the citric acid cycle (Sections 14-4, 15-3, and 16-4). In this section, we discuss the mechanisms that control the rate of oxidative phosphorylation.

A. Control of Oxidative Phosphorylation

In our discussions of metabolic pathways, we have seen that most of their reactions function close to equilibrium. The few irreversible reactions constitute potential control points of the pathways and usually are catalyzed by regulatory enzymes that are under allosteric control. In the case of oxidative phosphorylation, the pathway from NADH to cytochrome *c* functions near equilibrium:

$$\tfrac{1}{2} \text{NADH} + \text{cytochrome } c(\text{Fe}^{3+}) + \text{ADP} + P_i \rightleftharpoons$$
$$\tfrac{1}{2} \text{NAD}^+ + \text{cytochrome } c(\text{Fe}^{2+}) + \text{ATP} \qquad \Delta G^{\circ\prime} \approx 0$$

and hence

$$K_{eq} = \left(\frac{[\text{NAD}^+]}{[\text{NADH}]}\right)^{1/2} \frac{[c^{2+}]}{[c^{3+}]} \frac{[\text{ATP}]}{[\text{ADP}][P_i]}$$

This pathway is therefore readily reversed by the addition of its product, ATP. However, *the cytochrome c oxidase reaction (the terminal step of the*

Box 17-3

BIOCHEMISTRY IN FOCUS

Uncoupling in Brown Adipose Tissue Generates Heat

Heat generation is the physiological function of **brown adipose tissue (brown fat)**. This tissue is unlike typical (white) adipose tissue in that it contains numerous mitochondria whose cytochromes cause its brown color. Newborn mammals that lack fur, such as humans, as well as hibernating mammals, all contain brown fat in their neck and upper back that generates heat by **nonshivering thermogenesis** (the ATP hydrolysis that occurs during the muscle contraction of shivering—or any other movement—also produces heat).

The mechanism of heat generation in brown fat involves the regulated uncoupling of oxidative phosphorylation. Brown fat mitochondria contain a proton channel known as **uncoupling protein (UCP;** also called **thermogenin)**, a protein that is absent in the mitochondria of other tissues. In cold-adapted animals, UCP constitutes up to 15% of the protein in the inner mitochondrial membranes of brown fat. The flow of protons through UCP is inhibited by physiological concentrations of purine nucleotides (ADP, ATP, GDP, GTP), but this inhibition can be overcome by free fatty acids.

Thermogenesis in brown fat mitochondria is under hormonal control. Norepinephrine (**1**; noradrenalin) induces the production of the second messenger cAMP (**2**) and thereby activates cAMP-dependent protein kinase (**3**; Section 15-3C). The kinase then activates **hormone-sensitive triacylglycerol lipase** (**4**) by phosphorylating it. The activated lipase hydrolyzes triacylglycerols (**5**) to yield free fatty acids that counteract the inhibitory effect of the purine nucleotides on UCP (**6**). The resulting flow of protons through UCP dissipates the proton gradient across the inner mitochondrial membrane. This allows substrate oxidation to proceed (and generate heat) without the synthesis of ATP.

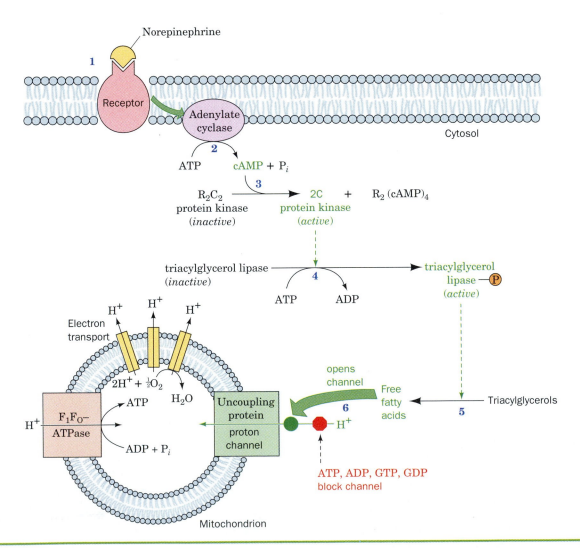

electron-transport chain) is irreversible and is therefore a potential control site. Cytochrome *c* oxidase, in contrast to most regulatory enzyme systems, appears to be controlled primarily by the availability of one of its substrates, reduced cytochrome *c* (c^{2+}). Since c^{2+} is in equilibrium with the rest of the coupled oxidative phosphorylation system, the concentration of c^{2+} ultimately depends on the intramitochondrial ratios of [NADH]/[NAD$^+$] and [ATP]/[ADP][P$_i$] (this latter quantity is known as the **ATP mass action ratio**). We can see by rearranging the foregoing equilibrium expression:

$$\frac{[c^{2+}]}{[c^{3+}]} = \left(\frac{[\text{NADH}]}{[\text{NAD}^+]}\right)^{1/2} \frac{[\text{ADP}][\text{P}_i]}{[\text{ATP}]} K_{\text{eq}}$$

The higher the [NADH]/[NAD$^+$] ratio and the lower the ATP mass action ratio, the higher the concentration of reduced cytochrome *c* and thus the higher the cytochrome *c* oxidase activity.

How is this system affected by changes in physical activity? In an individual at rest, the rate of ATP hydrolysis to ADP and P$_i$ is minimal and the ATP mass action ratio is high; the concentration of reduced cytochrome *c* is therefore low and the rate of oxidative phosphorylation is minimal. Increased activity results in hydrolysis of ATP to ADP and P$_i$, thereby decreasing the ATP mass action ratio and increasing the concentration of reduced cytochrome *c*. This results in an increase in the rate of electron transport and ADP phosphorylation.

The concentrations of ATP, ADP, and P$_i$ in the mitochondrial matrix depend on the activities of the transport proteins that import these substances from the cytosol. Thus, the ADP–ATP translocator and the P$_i$ transporter may play a part in regulating oxidative phosphorylation. There is also some evidence that Ca^{2+} stimulates the electron-transport complexes and possibly ATP synthase itself. This is consistent with the many other instances in which Ca^{2+} directly stimulates oxidative metabolic processes.

B. Coordinated Control of Oxidative Metabolism

The primary sources of the electrons that enter the mitochondrial electron-transport chain are glycolysis, fatty acid degradation, and the citric acid cycle. For example, 10 molecules of NAD$^+$ are converted to NADH per molecule of glucose oxidized (Fig. 17-1). Not surprisingly, the control of glycolysis and the citric acid cycle is coordinated with the demand for oxidative phosphorylation. An adequate supply of electrons to feed the electron-transport chain is provided by regulation of the control points of glycolysis and the citric acid cycle (phosphofructokinase, pyruvate dehydrogenase, citrate synthase, isocitrate dehydrogenase, and α-ketoglutarate dehydrogenase) by adenine nucleotides or NADH or both, as well as by certain metabolites (Fig. 17-23).

Figure 17-23. **The coordinated control of glycolysis and the citric acid cycle.** The diagram shows the effects of ATP, ADP, AMP, P$_i$, Ca^{2+}, and the [NADH]/[NAD$^+$] ratio (the vertical arrows indicate increases in this ratio). Here a green dot signifies activation and a red octagon represents inhibition. [After Newsholme, E.A. and Leech, A.R., *Biochemistry for the Medical Sciences*, pp. 316, 320, Wiley (1983).] ✳ **See the Animated Figures.**

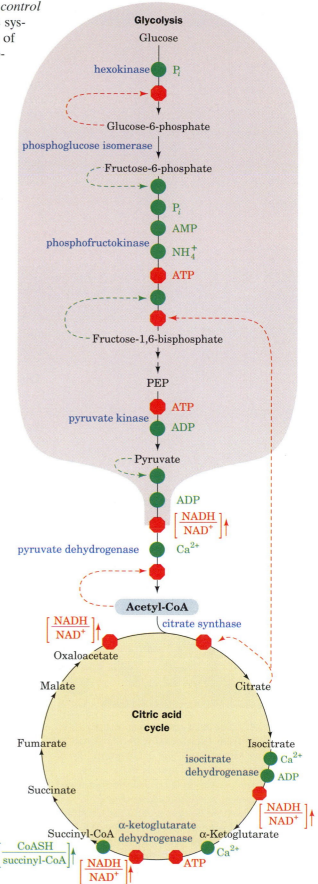

One particularly interesting regulatory effect is the inhibition of phosphofructokinase (PFK) by citrate. When demand for ATP decreases, [ATP] increases and [ADP] decreases. Because isocitrate dehydrogenase is activated by ADP and α-ketoglutarate dehydrogenase is inhibited by ATP, the citric acid cycle slows down. This causes the citrate concentration to build up. Citrate leaves the mitochondrion via a specific transport system and, *once in the cytosol, acts to restrain further carbohydrate breakdown by inhibiting PFK.*

5. PHYSIOLOGICAL IMPLICATIONS OF AEROBIC METABOLISM

Not all organisms carry out oxidative phosphorylation. However, those that do are able to extract considerably more energy from a given amount of a metabolic fuel. This principle is illustrated by the Pasteur effect (Section 14-3C): When anaerobically growing yeast are exposed to oxygen, their glucose consumption drops precipitously. An analogous effect is observed in mammalian muscle; the concentration of lactic acid (the anaerobic product of muscle glycolysis; Section 14-3A) drops dramatically when cells switch to aerobic metabolism. These effects are easily understood by examining the stoichiometries of anaerobic and aerobic breakdown of glucose (Section 16-4):

Anaerobic glycolysis:

$$C_6H_{12}O_6 + 2\ ADP + 2\ P_i \longrightarrow 2\ lactate + 2\ H^+ + 2\ H_2O + 2\ ATP$$

Aerobic metabolism of glucose:

$$C_6H_{12}O_6 + 38\ ADP + 38\ P_i + 6\ O_2 \longrightarrow 6\ CO_2 + 44\ H_2O + 38\ ATP$$

Thus, *aerobic metabolism is up to 19 times more efficient than anaerobic glycolysis in producing ATP.*

Aerobic metabolism has its drawbacks, however. Many organisms and tissues depend exclusively on aerobic metabolism and suffer irreversible damage during oxygen deprivation (see Box 17-4). Oxidative metabolism is also accompanied by the production of low levels of reactive oxygen metabolites that, over time, may damage cellular components. Evidently, the organisms that have existed during the last 2 billion years (the period in which the earth's atmosphere has contained significant amounts of O_2) exhibit physiological and biochemical adaptations that permit them to take advantage of the oxidizing power of O_2 while minimizing the potential dangers of oxygen itself.

A. Cytochrome P450

The presence of molecular oxygen has permitted the evolution of an extensive family of oxidative detoxification enzymes known collectively as **cytochromes P450.** These enzymes hydroxylate hydrophobic molecules to more soluble products. Several hundred types of cytochrome P450 enzymes are known, at least from their DNA sequences. This enormous family of enzymes catalyzes hydroxylation reactions using NADPH and O_2:

$$RH + O_2 + NADPH + H^+ \longrightarrow ROH + H_2O + NADP^+$$

where RH represents any of a wide variety of substrates.

The X-ray structure of cytochrome P450 from the bacterium *Pseudomonas putida* is shown in Fig. 17-24. All the known cytochromes P450 contain 400 to 530 amino acids and a heme group at the active site.

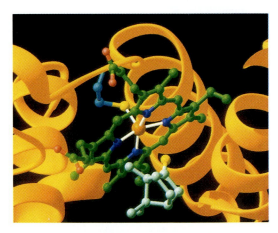

Figure 17-24.   The X-ray structure of cytochrome P450 from *Pseudomonas putida* showing its active site region. The heme group, the Cys side chain that axially ligands its Fe atom, and the enzyme's hydrophobic substrate **thiocamphor** are shown in ball-and-stick form with N blue, O red, S yellow, Fe orange, and the C atoms of the heme, its liganding Cys side chain, and the thiocamphor green, cyan, and light green, respectively. The bonds liganding the Fe are white. [Based on an X-ray structure by Thomas Poulos, University of California at Irvine.]

Box 17-4

BIOCHEMISTRY IN HEALTH AND DISEASE

Oxygen Deprivation in Heart Attack and Stroke

As outlined in Boxes 2-1 and 7-1, organisms larger than 1 mm thick require circulatory systems to deliver nutrients to cells and dispose of cellular wastes. In addition, the circulatory fluid in most larger organisms contains proteins specialized for oxygen transport (e.g., hemoglobin). The sophistication of oxygen-delivery systems and their elaborate regulation are consistent with their essential nature and their long period of evolution.

What happens during oxygen deprivation? Consider two common causes of human death, **myocardial infarction** (heart attack) and **stroke**, which result from interruption of the blood (O_2) supply to a portion of the heart or the brain, respectively. In the absence of O_2, a cell, which must then rely only on glycolysis for ATP production, rapidly depletes its stores of phosphocreatine (a source of rapid ATP production; Section 13-2C) and glycogen. As the rate of ATP pro-

duction falls below the level required by membrane ion pumps for maintaining proper intracellular ion concentrations, osmotic balance is disrupted so that the cell and its membrane-enveloped organelles begin to swell. The resulting overstretched membranes become permeable, thereby leaking their enclosed contents. For this reason, a useful diagnostic criterion for myocardial infarction is the presence in the blood of heart-specific enzymes, such as the H-type isozyme of lactate dehydrogenase (Section 13-1B), which leak out of necrotic (dead) heart tissue. Moreover, the decreased intracellular pH that accompanies anaerobic glycolysis (because of lactic acid production) permits the released lysosomal enzymes (which are active only at acidic pH's) to degrade the cell contents. Thus, O_2 deprivation leads not only to cessation of cellular activity but to irreversible cell damage and cell death. Rapidly respiring tissues, such as the heart and brain, are particularly susceptible to damage by oxygen deprivation.

The heme Fe atom is axially liganded to both a conserved Cys residue and a water molecule. When substrate binds to the enzyme, it displaces this water molecule, allowing O_2 to also bind. When CO is bound to the Fe, the enzyme absorbs light of wavelength 450 nm, giving the enzyme its name.

Many of the substrates of the different cytochromes P450 are lipophilic compounds whose accumulation could be toxic to the cell. This group of molecules includes polycyclic aromatic hydrocarbons (some of which are carcinogens), polychlorinated biphenyls (PCBs), phenobarbital, and steroids. *Hydroxylation catalyzed by cytochrome P450 converts these molecules to more soluble substances for excretion.*

One of the most interesting aspects of cytochrome P450 is that its substrates often induce the enzyme's synthesis, usually by increasing the rate of gene transcription. This self-induction provides an efficient detoxification mechanism. Some lines of research suggest that the cytochrome P450 enzymes have evolved as defense mechanisms in interspecies chemical warfare. Alternatively, the principal function of these enzymes may be more mundane, for example, to convert fatty acids or steroids to more biologically active forms. Of course, this same activity may also convert relatively harmless substances into more toxic compounds. For example, cytochrome P450 can oxidize the widely used analgesic **acetaminophen:**

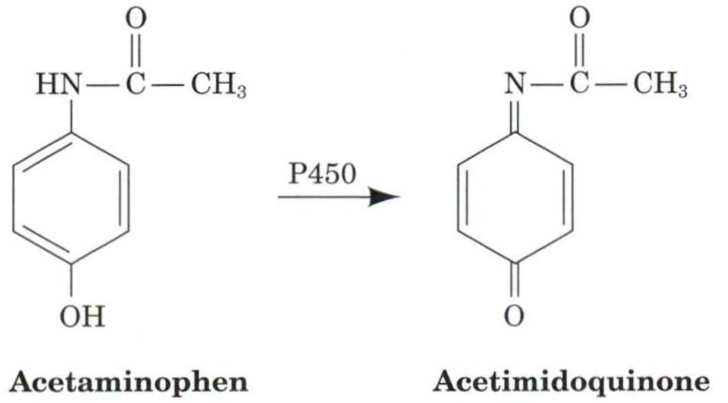

Acetaminophen   **Acetimidoquinone**

The product of this reaction, a strong electrophile that reacts with the thiol groups of glutathione (Box 14-3) and cellular proteins, is responsible for the liver toxicity of high doses of acetaminophen (the above reaction is not significant at therapeutic doses of the drug).

B. Reactive Oxygen Species

Although the four-electron reduction of O_2 by cytochrome *c* oxidase is nearly always orchestrated with great rapidity and precision, *O_2 is occasionally only partially reduced, yielding oxygen species that readily react with a variety of cellular components.* The best known reactive oxygen species is the **superoxide radical:**

$$O_2 + e^- \longrightarrow O_2^- \cdot$$

Superoxide radical is a precursor of other reactive species. Protonation of $O_2^- \cdot$ yields $HO_2 \cdot$, a much stronger oxidant than $O_2^- \cdot$. The most potent oxygen species in biological systems is probably the hydroxyl radical, which forms from the relatively harmless hydrogen peroxide (H_2O_2):

$$H_2O_2 + Fe^{2+} \longrightarrow \cdot OH + OH^- + Fe^{3+}$$

The hydroxyl radical also forms through the reaction of superoxide with H_2O_2:

$$O_2^- \cdot + H_2O_2 \longrightarrow O_2 + H_2O + \cdot OH$$

Although most free radicals are extremely short-lived (the half-life of $O_2^- \cdot$ is 1×10^{-6} s, and that of $\cdot OH$ is 1×10^{-9} s), they readily extract electrons from other molecules, converting them to free radicals and thereby initiating a chain reaction.

The random nature of free-radical attacks makes it difficult to characterize their reaction products, but all classes of biological molecules are susceptible to oxidative damage caused by free radicals. The oxidation of polyunsaturated lipids in cells may disrupt the structures of biological membranes, and oxidative damage to DNA may result in point mutations. Enzyme function may also be compromised through radical reactions with amino acid side chains. Because the mitochondrion is the site of the bulk of the cell's oxidative metabolism, its lipids, DNA, and proteins probably bear the brunt of free radical–related damage.

Several degenerative diseases, including **Parkinson's, Alzheimer's,** and **Huntington's diseases,** are associated with oxidative damage to mitochondria. Such observations have led to the free-radical theory of aging, which holds that *free-radical reactions arising during the course of normal oxidative metabolism are at least partially responsible for the aging process.* In fact, individuals with congenital defects in their mitochondrial DNA suffer from a variety of symptoms typical of old age, including neuromotor difficulties, deafness, and dementia. Their genetic defects may make their mitochondria all the more susceptible to the reactive oxygen species generated by the electron-transport machinery.

C. Antioxidant Mechanisms

Antioxidants destroy oxidative free radicals such as $O_2^- \cdot$ and $\cdot OH$. In 1969, Irwin Fridovich discovered that the enzyme **superoxide dismutase (SOD),** which is present in nearly all cells, catalyzes the conversion of $O_2^- \cdot$ to H_2O_2.

$$2\ O_2^- \cdot + 2\ H^+ \longrightarrow H_2O_2 + O_2$$

Mitochondrial and bacterial SOD are both Mn-containing tetramers; eukaryotic cytosolic SOD is a dimer containing copper and zinc ions. The rate

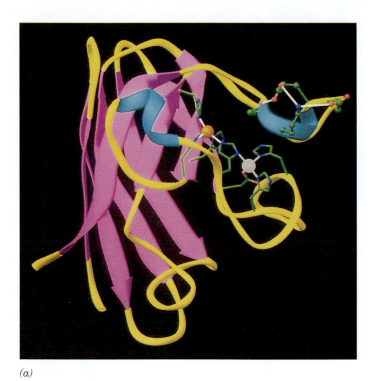

(a)

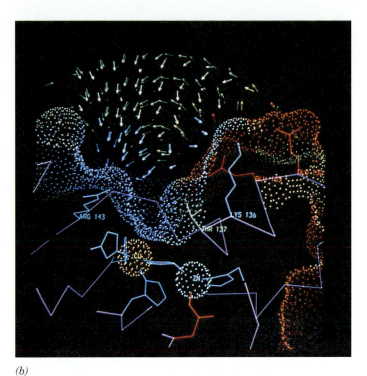

(b)

Figure 17-25. Electrostatic effects in human Cu,Zn-superoxide dismutase. (a) The X-ray structure of a subunit of the dimeric enzyme. The polypeptide backbone is colored according to its secondary structure, the Cu and Zn ions are represented by orange and silver balls, the side chains that ligand these metal ions are drawn in stick form, and the side chains forming the hydrogen-bonded network at the entrance to the active site pocket are shown in ball-and-stick form. Metal–ligand bonds and hydrogen bonds are represented by thin white lines. [Based on an X-ray structure by John Tainer, The Scripps Research Institute, La Jolla, California.] (b) Cross section of the active site channel of Cu,Zn-SOD. Its molecular surface is represented by a dot surface that is colored according to charge: red, most negative; yellow, negative; green, neutral; light blue, positive; dark blue, most positive. The electrostatic field vectors are represented by similarly colored arrows. The O_2^-·-binding site is located between the Cu ion and the side chain of Arg 143. [Courtesy of Elizabeth Getzoff, The Scripps Research Institute, La Jolla, California.]

of nonenzymatic superoxide breakdown is ~2×10^5 M^{-1}·s^{-1}, whereas the rate of the Cu,Zn-SOD–catalyzed reaction is ~2×10^9 M^{-1}·s^{-1}. This rate enhancement, which is close to the diffusion-controlled limit (Section 12-1B), is apparently accomplished by electrostatic guidance of the negatively charged superoxide substrate into the enzyme's active site (Fig. 17-25). The active-site Cu ion lies at the bottom of a deep pocket in each enzyme subunit. A hydrogen-bonded network of Glu 123, Glu 133, Lys 136, and Thr 137 at the entrance to the pocket facilitates the diffusion of O_2^-· to a site between the Cu ion and an Arg residue.

Some individuals with hereditary **amyotrophic lateral sclerosis (ALS; Lou Gehrig's disease)** have a mutant Cu,Zn-SOD. Most mutations in enzymes lead to a loss of function. However, the defect in ALS is inherited in a dominant manner, which is consistent with the gain of some toxic activity. Indeed, the mutant SOD acts as a peroxidase and is thought to oxidize lipids, leading to the motor neuron degeneration characteristic of the disease. Research into this debilitating condition has been aided by the development of genetically engineered mice that contain a copy of the mutant human SOD gene. Compounds that ameliorate the effects of the SOD mutation can thus be tested *in vivo* in these animal models.

SOD is considered a first-line defense against reactive oxygen species. The H_2O_2 produced in the reaction, which can potentially react to yield other reactive oxygen species, is degraded to water and oxygen by enzymes such as **catalase,** which catalyzes the reaction

$$2 H_2O_2 \longrightarrow 2 H_2O + O_2$$

and glutathione peroxidase (Box 14-3), which uses glutathione (GSH) as the reducing agent:

$$2 GSH + H_2O_2 \longrightarrow GSSG + 2 H_2O$$

This latter enzyme also catalyzes the breakdown of organic hydroperoxides. Some types of glutathione peroxidase require Se for activity; this is one reason why Se appears to have antioxidant activity.

SUMMARY

1. Electrons from the reduced coenzymes NADH and FADH$_2$ pass through a series of redox centers in the electron-transport chain before reducing O$_2$. During electron transfer, protons are translocated out of the mitochondrion to form an electrochemical gradient whose free energy drives ATP synthesis.

2. The mitochondrion contains soluble and membrane-bound enzymes for oxidative metabolism. Reducing equivalents are imported from the cytosol via a shuttle system. Specific transporters mediate the transmembrane movements of ADP, ATP, P$_i$, and Ca^{2+}.

3. Electrons flow from redox centers with more negative reduction potentials to those with more positive reduction potentials. Inhibitors have been used to reveal the sequence of electron carriers and the points of entry of electrons into the electron transport chain.

4. Electron transport is mediated by one-electron carriers (Fe–S clusters, cytochromes, and Cu ions) and two-electron carriers (CoQ, FMN, FAD).

5. Complex I transfers two electrons from NADH to CoQ and translocates four protons to the intermembrane space.

6. Complex II transfers electrons from succinate through FAD to CoQ.

7. Complex III transfers two electrons from CoQH$_2$ to two molecules of cytochrome c. The concomitant operation of the Q cycle translocates four protons to the intermembrane space.

8. Complex IV reduces O$_2$ to 2 H$_2$O using four electrons donated by cytochrome c and four protons from the matrix. Two protons are translocated to the intermembrane space for every two electrons that reduce oxygen.

9. As explained by the chemiosmotic theory, protons translocated into the intermembrane space during electron transport through Complexes I, III, and IV establish an electrochemical gradient across the inner mitochondrial membrane.

10. The influx of protons through the F$_0$ component of ATP synthase (F$_1$F$_0$-ATPase) drives its F$_1$ component to synthesize ATP from ADP + P$_i$ via the binding change mechanism, a process that is mechanically driven by the F$_0$-mediated rotation of F$_1$'s γ subunit with respect to its catalytic $\alpha_3\beta_3$ assembly.

11. The P/O ratio, the number of ATPs synthesized per oxygen reduced, need not be an integral number.

12. Agents that discharge the proton gradient can uncouple oxidative phosphorylation from electron transport.

13. Oxidative phosphorylation is controlled by the ratio [NADH]/[NAD$^+$] and by the ATP mass action ratio. Glycolysis and the citric acid cycle are coordinately regulated according to the need for oxidative phosphorylation.

14. The oxidizing power of O$_2$ permits organisms to extract more energy from metabolic fuels than is available anaerobically and to use O$_2$ in detoxification reactions catalyzed by cytochrome P450.

15. The incomplete reduction of O$_2$ during aerobic metabolism generates reactive oxygen species such as the superoxide and hydroxyl radicals. Oxidative damage resulting from free-radical reactions may contribute to the degeneration of aging and certain diseases. Superoxide dismutase helps protect against free radicals.

REFERENCES

Abrahams, J.P., Leslie, A.G.W., Lutter, R., and Walker, J.E., Structure at 2.8 Å resolution of F$_1$-ATPase from bovine heart mitochondria, *Nature* **370,** 621–628 (1994).

Beinert, H., Holm, R.H., and Münck, E., Iron-sulfur clusters: Nature's modular multipurpose structures, *Science* **277,** 653–659 (1997).

Boyer, P.D., The ATP synthase—A splendid molecular machine, *Annu. Rev. Biochem.* **66,** 717–749 (1997).

Brown, G.C., Control of respiration and ATP synthesis in mammalian mitochondria, *Biochem. J.* **284,** 1–13 (1992).

Cramer, W.A. and Knaff, D.B., *Energy Transduction in Biological Membranes,* Springer-Verlag (1991). [The first three chapters of this advanced textbook summarize the thermodynamics of biochemical processes in general and redox reactions in particular.]

Friedrich, T., Steinmüller, K., and Weiss, H., The proton-pumping respiratory complex I of bacteria and mitochondria and its homologue of chloroplasts, *FEBS Lett.* **367,** 107–111 (1995).

Hinkle, P.C., Kumar, M.A., Resetar, A., and Harris, D.L., Mechanistic stoichiometry of mitochondrial oxidative phosphorylation, *Biochemistry* **30,** 3576–3582 (1991). [Describes measurements of the P/O ratios indicating that their values are 2.5, 1.5, and 1.]

Junge, W., Lill, H., and Englebrecht, S., ATP synthase: An electrochemical transducer with rotatory mechanics, *Trends Biochem. Sci.* **22,** 420–423 (1997).

Lanyi, J.K., Bacteriorhodopsin as a model for proton pumps, *Nature* **375,** 461–463 (1995).

Michel, H., Behr, J., Harrenga, A., and Kannt, A., Cytochrome *c* oxidase: Structure and spectroscopy, *Annu. Rev. Biophys. Biomol. Struct.* **27,** 329–356 (1998).

Nicholls, D.G. and Ferguson, S.J., *Bioenergetics 2,* Academic Press (1992). [An authoritative monograph on the mechanism of oxidative phosphorylation and the techniques used to elucidate it.]

Trumpower, B.L. and Gennis, R.B., Energy transduction by cytochrome complexes in mitochondrial and bacterial respiration: The enzymology of coupling electron transfer reactions to transmembrane proton translocation, *Annu. Rev. Biochem.* **63,** 675–716 (1994).

Tsukihara, T., Aoyama, H., Yamashita, E., Tomizaki, T., Yamaguchi, H., Shinzawa-Itoh, K., Nakashima, R., Yaono, R., and Yoshikawa, S., The whole structure of the 13-subunit oxidized cytochrome *c* oxidase at 2.8 Å, *Science* **272,** 1136–1144 (1996); *and* Yoshikawa, S., Beef heart cytochrome *c* oxidase, *Curr. Opin. Struct. Biol.* **7,** 574–579 (1997).

Walker, J.E., The regulation of catalysis in ATP synthase, *Curr. Opin. Struct. Biol.* **4,** 912–918 (1994).

Xia, D., Yu, C.-A., Kim, H., Xia, J.-Z., Kachurin, A.M., Zhang, L., Yu, L., and Deisenhofer, J., Crystal structure of the cytochrome bc₁ complex from bovine heart mitochondria, *Science* **277,** 60–66 (1997).

Yu, B.P., Cellular defenses against damage from reactive oxygen species, *Physiol. Rev.* **74,** 139–162 (1994).

KEY TERMS

redox center	ADP–ATP translocator	free radical	protonmotive force
cristae	electrogenic transport	Q cycle	P/O ratio
mitochondrial matrix	iron–sulfur protein	differential labeling	uncoupler
intermembrane space	cytochrome	F_1F_0-ATPase	ATP mass action ratio
malate–aspartate shuttle	coenzyme Q	energy coupling	superoxide radical
glycerophosphate shuttle			

STUDY EXERCISES

1. Describe the route followed by electrons from glucose to O_2.

2. Describe how a pair of electrons from NADH is transferred to the one-electron carrier cytochrome *c*.

3. How does the mechanism of proton translocation in Complexes I and IV differ from that in Complex III?

4. Summarize the chemiosmotic theory.

5. Describe the binding change mechanism of F_1F_0-ATP synthase.

6. Explain why the P/O ratio for a given substrate is not necessarily an integer.

7. Explain how oxidative phosphorylation is linked to electron transport and how the two processes can be uncoupled.

8. What are the advantages and disadvantages of O_2-based metabolism?

PROBLEMS

1. Explain why a liver cell mitochondrion contains fewer cristae than a mitochondrion from a heart muscle cell.

2. How many ATPs are synthesized for every cytoplasmic NADH that participates in the glycerophosphate shuttle in insect flight muscle? How does this compare to the ATP yield when NADH reducing equivalents are transferred into the matrix via the malate–aspartate shuttle?

3. Calculate $\Delta G^{\circ\prime}$ for the oxidation of free $FADH_2$ by O_2. What is the maximum number of ATPs that can be synthesized, assuming standard conditions and 100% conservation of energy?

4. The O_2-consumption curve of a dilute, well-buffered suspension of mitochondria containing an excess of ADP and P_i takes the form

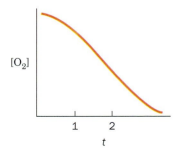

Sketch the curves obtained when (a) amytal is added at time $t = 1$; (b) amytal is added at $t = 1$ and succinate is added at $t = 2$; (c) CN^- is added at $t = 1$ and succinate is added at $t = 2$; (d) oligomycin is added at $t = 1$ and DNP at $t = 2$.

5. Why is it possible for electrons to flow from a redox center with a more positive $\mathscr{E}°'$ to one with a more negative $\mathscr{E}°'$ within an electron-transfer complex?

6. Bombarding a suspension of mitochondria with high frequency sound waves **(sonication)** produces **submitochondrial particles** derived from the inner mitochondrial membrane. These membranous vesicles seal inside out, so that the intermembrane space of the mitochondrion becomes the lumen of the submitochondrial particle. (a) Diagram the process of electron transfer and oxidative phosphorylation in these particles. (b) Assuming all the substrates for oxidative phosphorylation are present in excess, does ATP synthesis increase or decrease with an increase in the pH of the fluid in which the submitochondrial particles are suspended?

7. Explain why compounds such as DNP increase metabolic rates.

8. What is the advantage of hormones activating a lipase to stimulate nonshivering thermogenesis in brown fat rather than activating UCP directly (see Box 17-3)?

9. Describe the changes in [NADH]/[NAD$^+$] and [ATP]/[ADP] that occur during the switch from anaerobic to aerobic metabolism. How do these ratios influence the activity of glycolysis and the citric acid cycle?

10. Activated neutrophils and macrophages (types of white blood cells) fight invading bacteria by releasing superoxide. These cells contain an **NADPH oxidase** that catalyzes the reaction

$$2\ O_2 + NADPH \longrightarrow 2\ O_2^{-\cdot} + NADP^+ + H^+$$

Explain why flux through the glucose-6-phosphate dehydrogenase reaction increases in these cells.

Plants, such as these sunflowers, can survive literally on water, air, and light. They oxidize H_2O to O_2 and convert CO_2 from air into carbohydrates in a process that is driven by light.
[© Eric Horan/Gamma Liaison.]

PHOTOSYNTHESIS

The notion that plants obtain nourishment from such insubstantial things as light and air was not validated until the eighteenth century. Evidence that plants produce a vital substance—O_2—was not obtained until Joseph Priestly noted that the air in a jar in which a candle had burnt out could be "restored" by introducing a small plant into the jar. In the presence of sunlight, plants and cyanobacteria (formerly called blue-green algae) consume CO_2 and H_2O and produce O_2 and "fixed" carbon in the form of carbohydrate:

$$CO_2 + H_2O \xrightarrow{\text{light}} (CH_2O) + O_2$$

Photosynthesis, in which light energy drives the reduction of carbon, is essentially the reverse of oxidative carbohydrate metabolism. Photosynthetically produced carbohydrates therefore serve as an energy source for the organism that produces them as well as for nonphotosynthetic organisms that directly or indirectly consume photosynthetic organisms. It is estimated that photosynthesis annually fixes $\sim 10^{11}$ tons of carbon, which represents the storage of over 10^{18} kJ of energy. The process by which light energy is converted to chemical energy has its roots early in evolution, and its complexity is consistent with its long history. Our discussion focuses first on purple photosynthetic bacteria, because of the relative simplicity of their photosynthetic machinery, and then on plants, whose chloroplasts are the site of photosynthesis.

Early in the twentieth century, it was mistakenly thought that light absorbed by photosynthetic pigments directly reduced CO_2, which then combined with water to form carbohydrate. In fact, photosynthesis in plants is a two-stage process in which light energy is harnessed to oxidize H_2O:

$$2\ H_2O \xrightarrow{\text{light}} O_2 + 4\ [H\cdot]$$

and the resulting reducing agent, here symbolized [H·], subsequently reduces CO_2:

$$4\ [H\cdot] + CO_2 \longrightarrow (CH_2O) + H_2O$$

The two stages of photosynthesis are traditionally referred to as the **light reactions** and **dark reactions:**

1. In the light reactions, specialized pigment molecules capture light energy and are thereby oxidized. A series of electron-transfer reactions, which culminate with the reduction of $NADP^+$ to NADPH, generate a transmembrane proton gradient whose energy is tapped to synthesize ATP from $ADP + P_i$. The oxidized pigment molecules are reduced by H_2O, thereby generating O_2.

2. The dark reactions use NADPH and ATP to reduce CO_2 and incorporate it into three-carbon precursors of carbohydrates.

As we shall see, both processes occur in the light and are therefore better described as light-dependent and light-independent reactions. After describing the chloroplast and its contents, we shall consider the light reactions and dark reactions in turn.

1. CHLOROPLASTS

The site of photosynthesis in eukaryotes (algae and higher plants) is the **chloroplast.** Cells contain 1 to 1000 chloroplasts, which vary considerably in size and shape but are typically ~5-μm-long ellipsoids. These organelles presumably evolved from photosynthetic bacteria.

A. Chloroplast Anatomy

Like mitochondria, which they resemble in many ways, chloroplasts have a highly permeable outer membrane and a nearly impermeable inner membrane separated by a narrow intermembrane space (Fig. 18-1). The inner membrane encloses the **stroma,** a concentrated solution of enzymes, including those required for carbohydrate synthesis. The stroma also contains the DNA, RNA, and ribosomes involved in the synthesis of several chloroplast proteins.

The stroma, in turn, surrounds a third membranous compartment, the **thylakoid** (Greek: *thylakos,* a sac or pouch). The thylakoid is probably a single highly folded vesicle, although in most organisms it appears to consist of stacks of disklike sacs named **grana,** which are interconnected by unstacked **stromal lamellae.** A chloroplast usually contains 10 to 100 grana. Thylakoid membranes arise from invaginations in the inner membrane of developing chloroplasts and therefore resemble mitochondrial cristae. The thylakoid membrane contains protein complexes involved in harvesting light energy, transporting electrons, and synthesizing ATP. In photosynthetic bacteria, the machinery for the light reactions is located in the plasma membrane, which often forms invaginations or multilammelar structures that resemble grana.

B. Light-Absorbing Pigments

The principal photoreceptor in photosynthesis is **chlorophyll.** This cyclic tetrapyrrole, like the heme group of globins and cytochromes (Section 7-1A and Box 17-1), is derived biosynthetically from protoporphyrin IX. Chlorophyll molecules, however, differ from heme in several respects

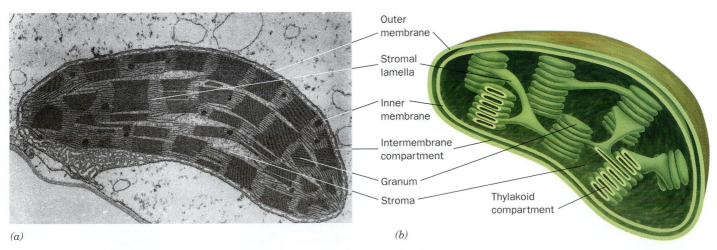

(a)   *(b)*

Figure 18-1.   Chloroplast from corn. (*a*) An electron micrograph. [Courtesy of T. Elliot Weier.] (*b*) Schematic diagram.

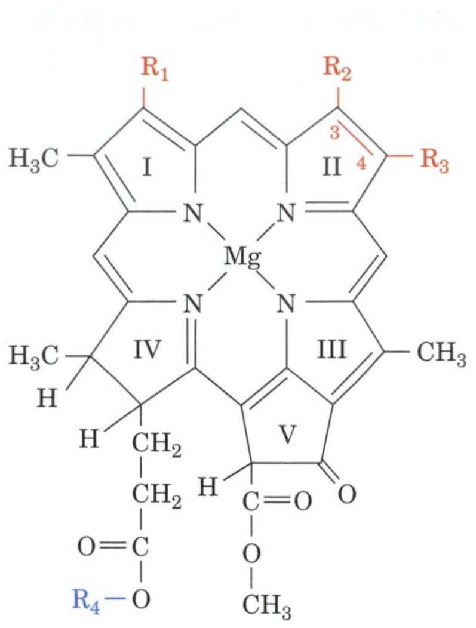

***Figure 18-2.* Chlorophyll structures.** The molecular formulas of chlorophylls *a* and *b* and bacteriochlorophylls *a* and *b* are compared to that of iron–protoporphyrin IX (heme). The isoprenoid phytyl and geranylgeranyl tails presumably increase the chlorophylls' solubility in nonpolar media.

Chlorophyll 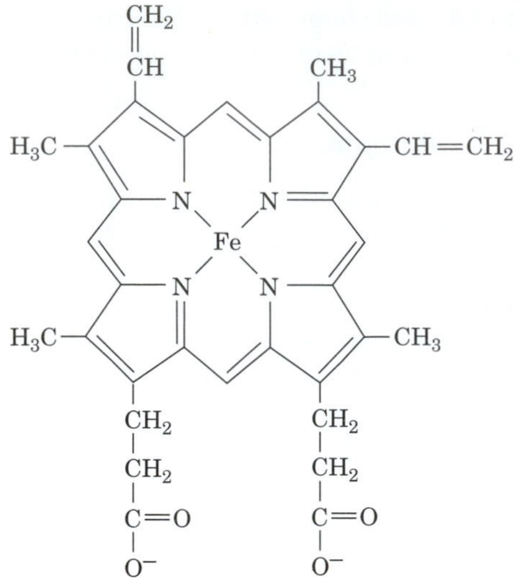   **Iron–protoporphyrin IX**

	R_1	R_2	R_3	R_4
Chlorophyll *a*	$-CH=CH_2$	$-CH_3$	$-CH_2-CH_3$	P
Chlorophyll *b*	$-CH=CH_2$	$-\overset{O}{\overset{\|}{C}}-H$	$-CH_2-CH_3$	P
Bacteriochlorophyll *a*	$-\overset{O}{\overset{\|}{C}}-CH_3$	$-CH_3{}^a$	$-CH_2-CH_3{}^a$	P or G
Bacteriochlorophyll *b*	$-\overset{O}{\overset{\|}{C}}-CH_3$	$-CH_3{}^a$	$=CH-CH_3{}^a$	P

a No double bond between positions C3 and C4.

P = $-CH_2$

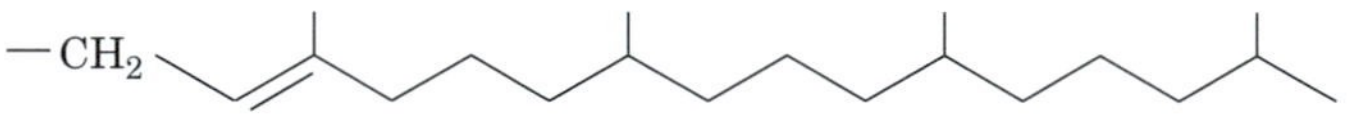

Phytyl side chain

G = $-CH_2$

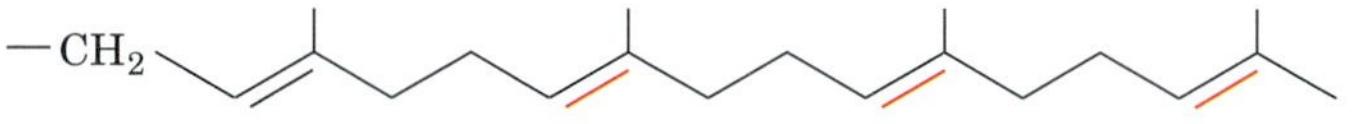

Geranylgeranyl side chain

(Fig. 18-2). In chlorophyll, the central metal ion is Mg^{2+} rather than Fe(II) or Fe(III), and a cyclopentanone ring, Ring V, is fused to pyrrole Ring III. The major chlorophyll forms in plants, **chlorophyll *a* (Chl *a*)** and **chlorophyll *b* (Chl *b*),** and the major forms in cyanobacteria, **bacteriochlorophyll *a* (BChl *a*)** and **bacteriochlorophyll *b* (BChl *b*),** also differ from heme and from each other in the degree of saturation of Rings II and IV and in the substituents of Rings I, II, and IV.

The highly conjugated chlorophyll molecules, along with other photosynthetic pigments, strongly absorb visible light (the most intense form of the solar radiation reaching the earth's surface; Fig. 18-3). The relatively small chemical differences among the various chlorophylls greatly affect their absorption spectra.

Light-Harvesting Complexes Contain Multiple Pigments

The primary reactions of photosynthesis, as is explained in Section 18-2B, take place at **photosynthetic reaction centers.** *Yet photosynthetic as-*

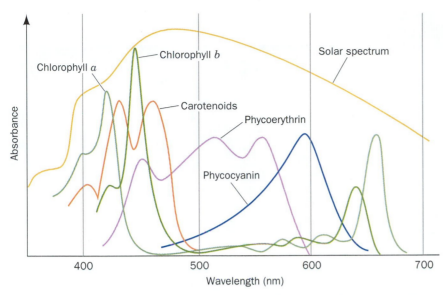

Figure 18-3. **The absorption spectra of various photosynthetic pigments.** The chlorophylls have two absorption bands, one in the red (long wavelength) and one in the blue (short wavelength). Phycoerythrin absorbs blue and green light, whereas phycocyanin absorbs yellow light. Together, these pigments absorb most of the visible light in the solar spectrum. [After a drawing by Govindjee, University of Illinois.]

semblies contain far more chlorophyll molecules than are contained in re-action centers. This is because most chlorophyll molecules do not participate directly in photochemical reactions but function to gather light; that is, *they act as light-harvesting antennas.* These **antenna chlorophylls** pass the energy of absorbed **photons** (units of light) from molecule to molecule until it reaches a photosynthetic reaction center (Fig. 18-4).

Transfer of energy from the antenna system to a reaction center occurs in $<10^{-10}$ s with an efficiency of >90%. This high efficiency depends on the chlorophyll molecules having appropriate spacings and relative orien-

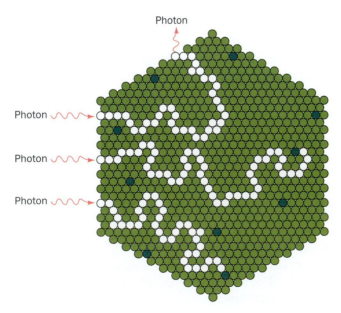

Figure 18-4. **Flow of energy through a photosynthetic antenna complex.** The energy of an absorbed photon randomly migrates among the molecules of the antenna complex (*light green circles*) until it reaches a reaction center chlorophyll (*dark green circles*) or, less frequently, is re-emitted (fluorescence).

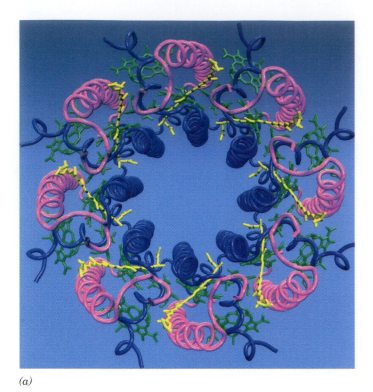

(a)

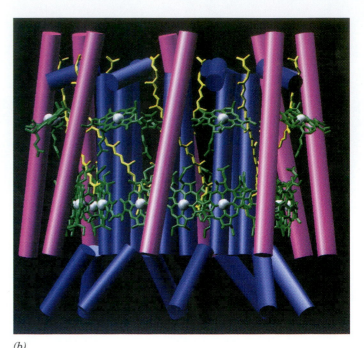

(b)

Figure 18-5.   The X-ray structure of the light-harvesting complex LH-2 from *Rps. molischianum*. (*a*) View perpendicular to the photosynthetic membrane showing that the α subunits (*blue;* 56 residues) and the β subunits (*magenta;* 45 residues), as represented by their C_α backbones, are arranged in two concentric eightfold symmetric rings. Thirty-two bacteriochlorophyll *a* (BChl *a; green*) and eight **lycopene** (a carotenoid; *yellow*) molecules are sandwiched between the protein rings. (*b*) View from the plane of the membrane, using the same colors as in Part *a*, in which the α helical portions of the proteins are represented by cylinders and the Mg^{2+} ions are represented by white spheres. Note that 8 of

the BChl *a* molecules are bound near the top of the complex with their ring systems nearly parallel to the plane of the membrane, whereas the remaining 16 BChl *a* molecules are bound near the bottom of the complex with their ring systems approximately perpendicular to the plane of the membrane. This arrangement, together with that of the lycopene molecules, presumably optimizes the light-absorbing and excitation-transmitting capability of this antenna system. [Courtesy of Juergen Koepke and Hartmut Michel, Max-Planck Institut für Biochemie, Germany.] ⬤ **See the Interactive Exercises.**

tations. Even in bright sunlight, a reaction center directly intercepts only ∼1 photon per second, a metabolically insignificant rate. Hence, a complex of antenna pigments, or **light-harvesting complex (LHC),** is essential.

LHCs consist of arrays of membrane-bound hydrophobic proteins that each contain numerous, often symmetrically arranged pigment molecules. For example, **LH-2** from the purple photosynthetic bacterium *Rhodospirillum molischianum* is an integral membrane protein that consists of eight α subunits and eight β subunits arranged in two eight-fold symmetric concentric rings between which are sandwiched 32 pigment molecules (Fig. 18-5). Other LHCs vary widely in their structure and complement of light-harvesting pigments. The number and arrangement of pigment molecules of each LHC have presumably been optimized for efficient energy transfer throughout the LHC.

Most LHCs contain other light-absorbing substances besides chlorophyll. These **accessory pigments** "fill in" the absorption spectra of the antenna complexes, covering the spectral regions where chlorophylls do not absorb strongly (Fig. 18-3). For example, **carotenoids,** which are linear polyenes such as **β-carotene,**

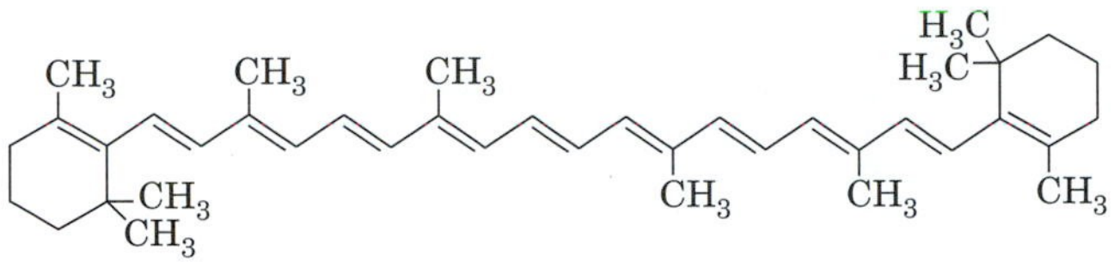

β-Carotene

are components of all green plants and many photosynthetic bacteria and are therefore the most common accessory pigments (they are largely responsible for the brilliant fall colors of deciduous trees as well as for the orange color of carrots, after which they were named).

Water-dwelling photosynthetic organisms, which undertake nearly half of the photosynthesis on earth, additionally contain other types of accessory pigments. This is because light outside the wavelengths 450 to 550 nm (blue and green light) is absorbed almost completely by passage through more than 10 m of water. In red algae and cyanobacteria, Chl *a* therefore is replaced as an antenna pigment by a set of linear tetrapyrroles, notably the red **phycoerythrin** and the blue **phycocyanin** (their spectra are shown in Fig. 18-3).

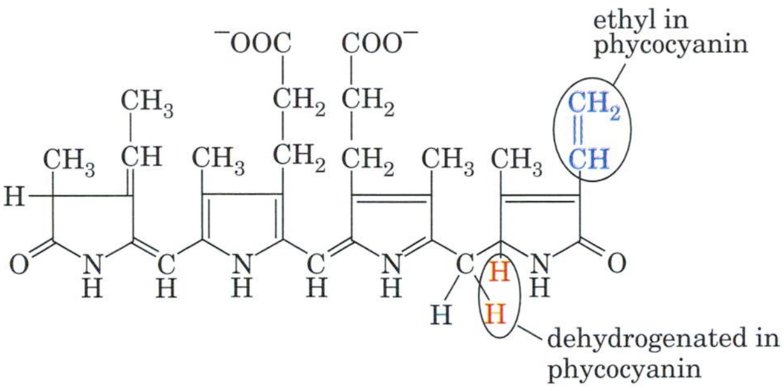

Phycoerythrin

2. THE LIGHT REACTIONS

Photosynthesis is a process in which electrons from excited chlorophyll molecules are passed through a series of acceptors that convert electronic energy to chemical energy. We can thus ask two questions: (1) What is the mechanism of energy transduction; and (2) How do photooxidized chlorophyll molecules regain their lost electrons?

A. The Interaction of Light and Matter

Electromagnetic radiation is propagated as discrete **quanta** (photons) whose energy E is given by **Planck's law:**

$$E = h\nu = \frac{hc}{\lambda}$$

where h is **Planck's constant** (6.626×10^{-34} J·s), c is the speed of light (2.998×10^8 m·s^{-1} in vacuum), ν is the frequency of the radiation, and λ is its wavelength.

When a molecule absorbs a photon, one of its electrons is promoted from its ground (lowest energy) state molecular orbital to one of higher energy. However, *a given molecule can absorb photons of only certain wavelengths because, as is required by the law of conservation of energy, the en-*

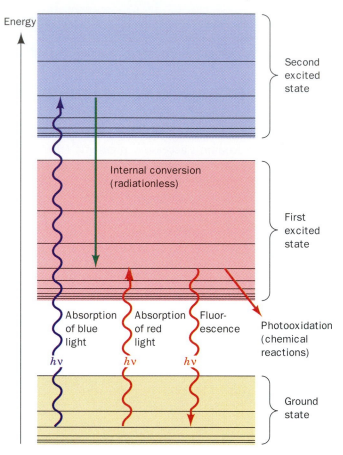

Figure 18-6.   An energy diagram indicating the electronic states of chlorophyll and their most important modes of interconversion. The wiggly arrows represent the absorption of photons or their fluorescent emission. Excitation energy may also be dissipated in radiationless processes such as internal conversion (heat production) and chemical reactions. ✳ See the Animated Figures.

ergy difference between the two states must exactly match the energy of the absorbed photon.

An electronically excited molecule can dissipate its excitation energy in several ways (Fig. 18-6):

1. **Internal conversion,** a common mode of decay in which electronic energy is converted to the kinetic energy of molecular motion, that is, to heat. Many molecules relax in this manner to their ground states. Chlorophyll molecules, however, usually relax only to their lowest excited states. Consequently, the photosynthetically applicable excitation energy of a chlorophyll molecule that has absorbed a photon in its short-wavelength band, which corresponds to its second excited state, is no different than if it had absorbed a photon in its less energetic long-wavelength band.

2. **Fluorescence,** in which an electronically excited molecule decays to its ground state by emitting a photon. A fluorescently emitted photon generally has a longer wavelength (lower energy) than that initially absorbed. Fluorescence accounts for the dissipation of only 3 to 6% of the light energy absorbed by living plants.

3. **Exciton transfer** (also known as **resonance energy transfer**), in which an excited molecule directly transfers its excitation energy to nearby

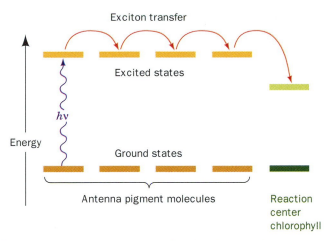

Figure 18-7.   Excitation energy trapping by the photosynthetic reaction center. Light energy that has been passed among pigment molecules by exciton transfer is trapped by the reaction center chlorophyll because its lowest excited state has a lower energy than those of the antenna pigment molecules.

unexcited molecules with similar electronic properties. This process occurs through interactions between the molecular orbitals of the participating molecules. *Light energy is funneled to photosynthetic reaction centers through exciton transfer among antenna pigments.* The energy (excitation) is trapped at the reaction center chlorophylls because they have slightly lower excited state energies than the antenna chlorophylls (Fig. 18-7).

4. **Photooxidation,** in which a light-excited donor molecule is oxidized by transferring an electron to an acceptor molecule, which is thereby reduced. This process occurs because the transferred electron is less tightly bound to the donor in its excited state than it is to the ground state. In photosynthesis, excited chlorophyll (Chl*) is such a donor. *The energy of the absorbed photon is thereby chemically transferred to the photosynthetic reaction system.* Photooxidized chlorophyll, Chl$^+$, a cationic free radical, eventually returns to its reduced state by oxidizing some other molecule.

B. Electron Transport in Photosynthetic Bacteria

In purple photosynthetic bacteria, a membrane-bound bacteriochlorophyll complex undergoes photooxidation when illuminated with red light. The excited electron is transferred along a series of carriers until it returns to the original bacteriochlorophyll complex. During the electron-transfer process, cytoplasmic protons are translocated across the plasma membrane. Dissipation of the resulting proton gradient drives ATP synthesis. The relatively simple photosynthetic reaction center of purple bacteria illustrates some general principles of the photochemical events that occur in the more complicated photosynthetic apparatus of plants and cyanobacteria (Section 18-2C).

The Photosynthetic Reaction Center Is a Transmembrane Protein

The photosynthetic reaction centers from several species of purple photosynthetic bacteria each contain three hydrophobic subunits known as H, L, and M. The L and M subunits collectively bind four molecules of bac-

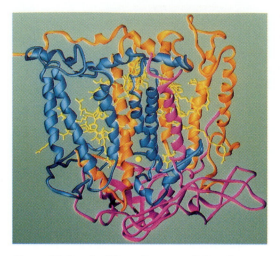

Figure 18-8. **A ribbon diagram of the photosynthetic reaction center from *Rb. sphaeroides*.** The H, M, and L subunits of this protein, as viewed from within the plane of the plasma membrane (cytoplasm below) are magenta, cyan, and orange, respectively. The prosthetic groups are yellow and are shown in skeletal form. The Fe(II) atom is represented by a sphere. The 11 largely vertical helices that form the central portion of the protein constitute its transmembrane region. Compare this structure with that of the photosynthetic reaction center from *Rps. viridis* (Fig. 10-5), whose H, M, and L subunits are 39, 50, and 59% identical to those of *Rb. sphaeroides*. Note that the *Rb. sphaeroides* protein lacks the *c*-type cytochrome (*green* in Fig. 10-5) on its periplasmic surface. [Based on an X-ray structure by Marianne Schiffer, Argonne National Laboratory.] ◉ **See the Interactive Exercises and Kinemage Exercise 8-2.**

teriochlorophyll, two molecules of **bacteriopheophytin** (**BPheo;** bacteriochlorophyll in which the Mg^{2+} ion is replaced by two protons), one Fe(II) ion, and two molecules of the redox coenzyme ubiquinone (Fig. 17-10*b*) or one molecule of ubiquinone and one of the related **menaquinone:**

Menaquinone

The photosynthetic reaction center from *Rhodopseudomonas* (*Rps.*) *viridis* was the first transmembrane protein to be described in atomic detail (Fig. 10-5). Its arrangement of 11 membrane-spanning helices closely resembles that of the reaction center from *Rhodobacter* (*Rb.*) *sphaeroides* (Fig. 18-8), although the *Rb. sphaeroides* protein lacks the *c*-type cytochrome of the *Rps. viridis* protein.

The disposition of prosthetic groups in the *Rps. viridis* protein is shown in Fig. 18-9. *The most striking aspect of the reaction center is that these groups are arranged with nearly perfect two-fold symmetry.* Two of the BChl molecules, the so-called **special pair,** are closely associated; they are nearly parallel and have an Mg–Mg distance of ~7 Å. The special pair is often named for its wavelength (in nm) of maximum absorbance [e.g., **P870;** photosynthetic bacteria tend to inhabit murky stagnant ponds where visible light (400–800 nm) does not penetrate; they require a near-infrared–absorbing species of chlorophyll]. Each member of the special pair contacts another BChl molecule that, in turn, associates with a BPheo molecule. The menaquinone is close to the L subunit BPheo (Fig. 18-9, *right*), whereas the ubiquinone associates with the M subunit BPheo *a* (Fig. 18-9, *left*). The Fe(II) is positioned between the menaquinone and the ubiquinone rings. Curiously, the two symmetry-related sets of prosthetic groups are not functionally equivalent; electrons are almost exclusively transferred through the L subunit (the right sides of Figs. 18-8 and 18-9). This effect is generally attributed to subtle structural differences between the L and M subunits.

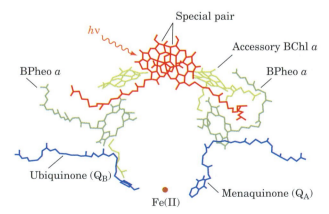

Figure 18-9. **Disposition of prosthetic groups in the photosynthetic reaction center of *Rps. viridis*.** Note that their rings, but not their aliphatic side chains, are arranged with close to two-fold symmetry. Photons are absorbed by the special pair of BChl *a* molecules (*red*). ◉ **See Kinemage Exercise 8-2.**

Photon Absorption Rapidly Photooxidizes the Special Pair

The photochemical events mediated by the photosynthetic reaction center occur as follows:

1. The primary photochemical event of bacterial photosynthesis is the absorption of a photon by the special pair (e.g., P870). The excited electron is delocalized over both its BChl molecules.

2. P870*, the excited state of P870, has but a fleeting existence. Within ~3 picoseconds (ps; 10^{-12} s), P870* transfers an electron to the BPheo on the right in Fig. 18-9 to yield P870$^+$ BPheo a^- (the intervening BChl group probably plays a role in conveying electrons, although it is not itself reduced; it is therefore known as the **accessory BChl**).

3. During the next 200 ps, the electron migrates to the menaquinone (or, in many species, the second ubiquinone), designated Q_A, to form the anionic semiquinone Q_A^-.

Rapid removal of the excited electron from the vicinity of P870$^+$ is an essential feature of the photosynthetic reaction center; this prevents return of the electron to P870$^+$, which would lead to the wasteful internal conversion of its excitation energy to heat. In fact, *electron transfer in the reaction center is so efficient that its overall* **quantum yield** *(ratio of molecules reacted to photons absorbed) is virtually 100%.* No man-made device has yet approached this level of efficiency.

Electrons Are Returned to the Photooxidized Special Pair via an Electron-Transport Chain

Q_A^-, which occupies a hydrophobic pocket in the photosynthetic reaction center, transfers its excited electron to the more solvent-exposed ubiquinone, Q_B, to form Q_B^- [the Fe(II) ion positioned between Q_A and Q_B does not directly participate in these redox reactions]. Q_A never becomes fully reduced; it shuttles between its oxidized and semiquinone forms.

When the reaction center is again excited, it transfers a second electron to Q_B^- to form the fully reduced Q_B^{2-}. This anionic quinol takes up two protons from the cytoplasmic side of the plasma membrane to form Q_BH_2. Thus, Q_B *is a molecular transducer that converts two light-driven one-electron excitations to a two-electron chemical reduction.*

The electrons taken up by Q_BH_2 are eventually returned to P870$^+$ via an electron-transport chain (Fig. 18-10). The details of this process are highly species dependent. The available redox carriers include a membrane-bound pool of ubiquinone molecules, a **cytochrome bc_1 complex,** and **cytochrome c_2.** The electron-transport pathway leads from Q_BH_2 through the ubiquinone pool, with which Q_BH_2 exchanges, to cytochrome bc_1, and then to cytochrome c_2. The reduced cytochrome c_2, which closely resembles mitochondrial cytochrome c, carries an electron back to P870$^+$. The reaction center is thereby reduced and prepared to absorb another photon.

Since electron transport in purple photosynthetic bacteria is a cyclic process (Fig. 18-10), *it results in no net oxidation–reduction.* However, when QH_2 transfers its electrons to cytochrome bc_1, its protons are translocated across the plasma membrane. Cytochrome bc_1 is a transmembrane protein complex containing a [2Fe–2S] iron–sulfur protein, a cytochrome c_1, and a cytochrome b that contains two hemes, b_H and b_L (H and L for high and low potential). Note that cytochrome bc_1 is strikingly similar to the proton-translocating Complex III of mitochondria (which is also called

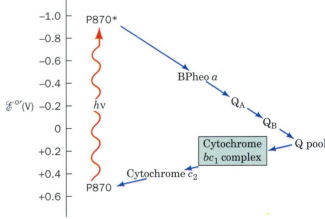

Figure 18-10. **The photosynthetic electron-transport system of purple photosynthetic bacteria.** Electrons liberated by the absorption of photons by P870 pass through BPheo a and Q_A before reaching Q_B, which exchanges with a pool of free ubiquinone. Electrons from QH_2 pass through cytochrome bc_1 to cytochrome c_2, which then reduces P870$^+$. Note that two photons are required for the two-electron reduction of Q to QH_2 and that cytochrome c_2 carries one electron at a time back to the photosynthetic reaction center. The overall process is essentially irreversible because electrons are transferred to progressively lower energy states (more positive standard reduction potentials).

cytochrome bc_1; Section 17-2E). In fact, electron transfer from QH_2 (a two-electron carrier) to the one-electron acceptor cytochrome c_2 occurs in a two-stage Q cycle, exactly as occurs in mitochondrial electron transport (Fig. 17-13). The net result is that for every two electrons transferred from QH_2 to cytochrome c_2, four protons enter the periplasmic space. Thus, photon absorption by the photosynthetic reaction center generates a trans-membrane H^+ gradient. Light-dependent synthesis of ATP, a process known as **photophosphorylation,** is driven by the dissipation of this gradient (Section 18-2D).

C. Two-Center Electron Transport

In plants and cyanobacteria, *photosynthesis is a noncyclic process that uses the reducing power generated by the light-driven oxidation of H_2O to produce NADPH.* This multistep process involves two reaction centers that each bear some resemblance to the bacterial photosynthetic reaction center. The reaction centers of green plants are **photosystem I (PSI),** which reduces $NADP^+$, and **photosystem II (PSII),** which oxidizes H_2O. Each photosystem is independently activated by light, but electrons flow from PSII to PSI. *PSI and PSII therefore operate in series to couple H_2O oxidation with $NADP^+$ reduction.* Evidence for the existence of two photosystems came from observations that in the presence of both red light (which activates only PSI) and yellow-green light (which also activates PSII), plants produce O_2 (i.e., oxidized H_2O) at a greater rate than the sum of the rates for each light acting alone. The herbicide **3-(3,4-dichlorophenyl)-1,1-dimethylurea (DCMU)** *(at left)* blocks electron flow from PSII to PSI so that even with adequate illumination (i.e., activation of both PSI and PSII), PSI is not supplied with electrons, PSII cannot be reoxidized, and photosynthetic oxygen production ceases.

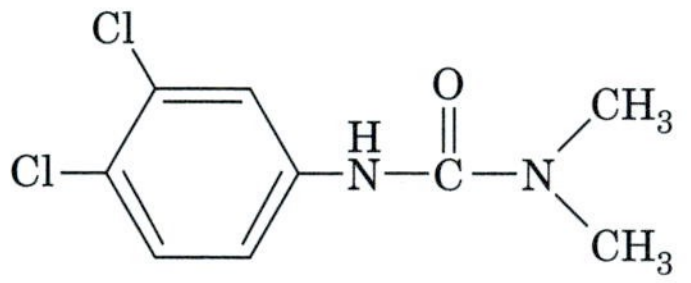

The pathway of electron transport in the chloroplast is more elaborate than in photosynthetic bacteria. *The components involved in electron transport from H_2O to $NADP^+$ are largely organized into three thylakoid membrane-bound particles (Fig. 18-11): PSII, a* **cytochrome b_6f complex,** *and* PSI. Electrons are transferred between these complexes via mobile electron carriers, much as occurs in the respiratory electron-transport chain. The ubiquinone analog **plastoquinone (Q),** via its reduction to **plastoquinol (QH$_2$),**

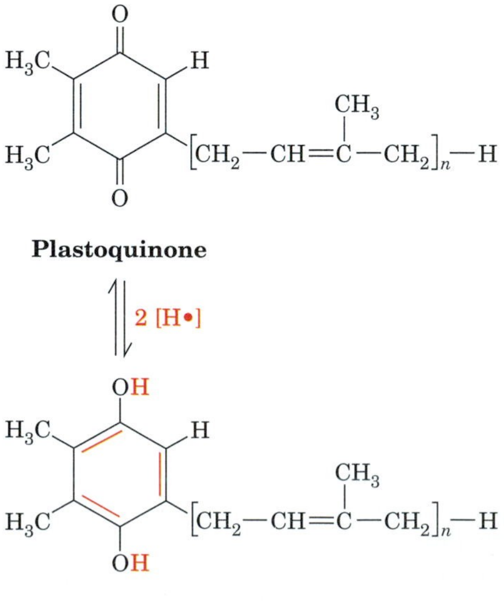

● See Guided Exploration 18:

Two-Center Photosynthesis (Z-scheme) Overview.

3-(3,4-Dichlorophenyl)-1,1-dimethylurea (DCMU)

Plastoquinone

2 [H•]

Plastoquinol

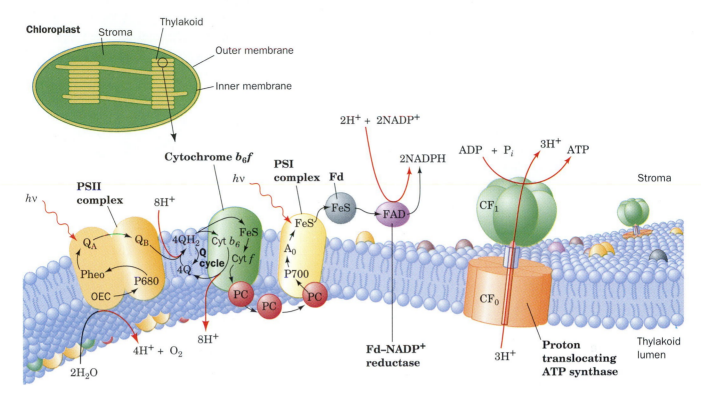

Figure 18-11. *Key to Function.* **A model of the thylakoid membrane.** The electron-transport system consists of three protein complexes: PSII, the cytochrome b_6f complex, and PSI, which are electrically "connected" by the diffusion of the electron carriers plastoquinone (Q) and plastocyanin (PC). Light-driven transport of electrons (*black arrows*) from H_2O to $NADP^+$ motivates the transport of protons (*red arrows*) into the thylakoid lumen. Additional protons are split off from water by the oxygen-evolving center (OEC), yielding O_2. The resulting proton gradient powers the synthesis of ATP by the CF_1CF_0 proton-translocating ATP synthase. The membrane also contains light-harvesting complexes (not shown) whose component pigments transfer their excitations to PSI and PSII. Fd represents ferredoxin. [After Ort, D.R. and Good, N.E., *Trends Biochem. Sci.* **13**, 469 (1988).]

links PSII to the cytochrome b_6f complex, which, in turn, interacts with PSI through the mobile protein **plastocyanin (PC).** Electrons eventually reach **ferredoxin–NADP$^+$ reductase,** where they are used to reduce $NADP^+$. The oxidation of water and the passage of electrons through a Q cycle generate a transmembrane proton gradient, with the greater $[H^+]$ in the thylakoid lumen and the lesser $[H^+]$ in the stroma. The free energy of this proton gradient is tapped by chloroplast ATP synthase (Section 18-2D).

The various prosthetic groups of the photosynthetic apparatus of plants can be arranged in a diagram known as the **Z-scheme** (Fig. 18-12). As in other electron-transport systems, electrons flow from low to high reduction potentials. The zig-zag nature of the Z-scheme reflects the two loci for photochemical events (one at PSI, one at PSII) that are required to drive electrons from H_2O to $NADP^+$. The evolutionary origins of the two-center photosynthetic process remain obscure (see Box 18-1).

O$_2$ Is Generated by a Five-Stage Water-Splitting Reaction

The **oxygen-evolving center (OEC)** of PSII is also known as the **water-splitting enzyme** because two water molecules are broken down to two O_2, four protons, and four electrons. Insight into this process was garnered by Pierre Joliet and Bessel Kok, who analyzed the production of O_2 by dark-

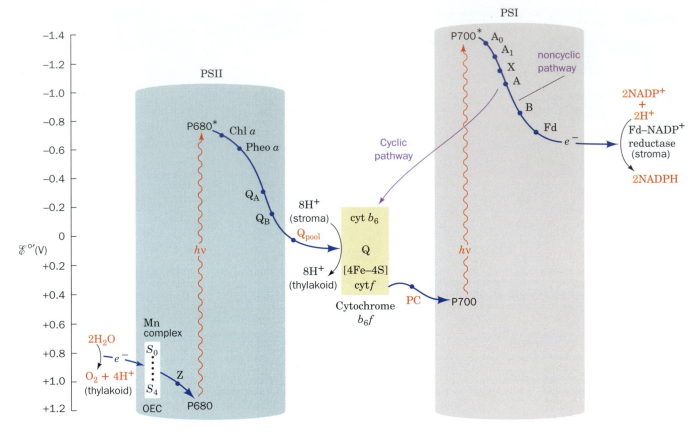

Figure 18-12.   The Z-scheme of photosynthesis. Electrons ejected from P680 in PSII by the absorption of photons are replaced with electrons abstracted from H_2O by an Mn complex (the oxygen-evolving center; OEC), thereby forming O_2 and four H^+. Each ejected electron passes through a chain of electron carriers to a pool of plastoquinone molecules (Q). The resulting plastoquinol, in turn, reduces the cytochrome b_6f particle (*yellow box*) with the concomitant translocation of protons into the thylakoid lumen. Cytochrome b_6f then transfers the electrons to plastocyanin (PC). The plastocyanin regenerates photooxidized P700 in PSI. The electron ejected from P700, through the intermediacy of a chain of electron carriers, reduces $NADP^+$ to NADPH in noncyclic electron transport. Alternatively, the electron may return to the cytochrome b_6f complex in a cyclic process that translocates additional protons into the thylakoid lumen. The reduction potentials increase downward so that electrons flow spontaneously in this direction.

adapted chloroplasts that were exposed to a series of short flashes of light. O_2 was evolved with a peculiar oscillatory pattern (Fig. 18-13). There is virtually no O_2 evolved by the first two flashes. The third flash results in the maximum O_2 yield. Thereafter, the amount of O_2 produced peaks with

Figure 18-13.   The O_2 yield per flash in dark-adapted spinach chloroplasts. Note that the yield peaks on the third flash and then on every fourth flash thereafter until the curve eventually damps out to its average value. [After Forbush, B., Kok, B., and McGloin, M.P., *Photochem. Photobiol.* **14**, 309 (1971).]

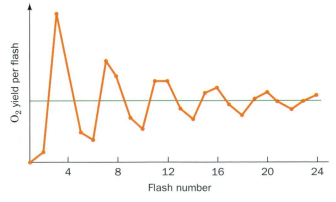

Box 18-1
BIOCHEMISTRY IN CONTEXT

Evolution of Photosynthetic Systems

Modern photosynthetic bacteria, which do not generate reducing equivalents by the light-driven oxidation of H_2O, are thought to resemble the original photosynthetic organisms. These presumably arose very early in the history of cellular life, when environmentally supplied sources of high-energy compounds were dwindling but reducing agents such as H_2S were still plentiful. In these photosynthetic bacteria, the photosynthetic reaction takes the form

$$CO_2 + 2\ H_2S \xrightarrow{light} (CH_2O) + 2\ S + H_2O$$

Apparently, the success of these life forms eventually caused them to exhaust the available reductive resources. The ancestors of modern cyanobacteria adapted to this situation by evolving a photosynthetic system with sufficient reduction potential to abstract electrons from H_2O. The gradual accumulation of the resulting toxic waste product, O_2, forced photosynthetic bacteria, most of which are obligate anaerobes, into the narrow ecological niches to which they are presently confined.

PSII appears to be a derivative of the photosynthetic reaction center of purple photosynthetic bacteria, with the added functionality of the oxygen-evolving center. There is evidence that at some point in evolutionary history, PSII may have been sufficient for eukaryotic photosynthesis. Certain mutant strains of the alga *Chlamydomonas reinhardtii* which lack PSI are able to channel water-derived electrons from PSII to NADP$^+$ by an unknown mechanism that probably requires the cytochrome $b_6 f$ complex. These organisms function and grow nearly normally (at 60–70% the rate of wild-type cells in bright light), even while short-circuiting the Z-scheme.

The origin of PSI is more problematic. It does not appear to be genetically related to PSII or to the photosynthetic reaction center of purple photosynthetic bacteria, but it is related to that of green sulfur bacteria, a second class of photosynthetic bacteria. PSI's participation in electron transfer from water to NADP$^+$ must confer some selective advantage—possibly finer regulatory control of the light reactions—despite the fact that the two-photosystem arrangement actually requires twice as many photons per electron-transfer event than does an arrangement based on a single reaction center.

---

every fourth flash until the oscillations damp out to a steady state. This periodicity indicates that each OEC must undergo four light-dependent reactions before releasing O_2.

The OEC is thought to cycle through five different states, S_0 through S_4 (Fig. 18-14). O_2 is released in the transition between S_4 and S_0. The observation that O_2 evolution peaks at the third rather than the fourth flash indicates that the OEC's resting state is predominantly S_1 rather than S_0. The oscillations gradually damp out because a small fraction of the reaction centers fail to be excited or become doubly excited by a given flash of light, so that the reaction centers eventually lose synchrony. The complete reaction sequence releases a total of four water-derived protons into the inner thylakoid space in a stepwise manner.

Since the OEC abstracts electrons from H_2O, its five states must have extraordinarily high reduction potentials (recall from Table 13-3 that the O_2/H_2O half-reaction has a standard reduction potential of 0.815 V). PSII must also stabilize the highly reactive intermediates for extended periods (as much as minutes) in close proximity to water. We are just beginning to understand how this occurs. The OEC contains four protein-bound Mn ions, along with two or three Ca^{2+} ions and four or five Cl^- ions. The Mn ions, whose arrangement is unknown, apparently cycle through Mn(III) and Mn(IV) oxidation states while abstracting protons and electrons from H_2O. The water-splitting reaction is driven by the excitation of the PSII reaction center.

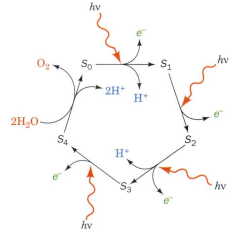

Figure 18-14. **The schematic mechanism of O_2 generation in chloroplasts.** Four electrons are stripped, one at a time, in light-driven reactions ($S_0 \rightarrow S_4$), from two bound H_2O molecules. In the recovery step ($S_4 \rightarrow S_0$), which is light independent, O_2 is released and two more H_2O molecules are bound. Three of these five steps release protons into the thylakoid lumen.

The PSII Reaction Center Transfers Electrons to Plastoquinone

The photon-absorbing center of PSII consists of Chl *a* (possibly a special pair), which is named **P680** after the wavelength of its absorption maximum. The P680$^+$ formed by light excitation is among the most powerful biological oxidants known. It abstracts electrons from H$_2$O via the four *S* states of the OEC. A tyrosine radical known as **Z** relays electrons from the Mn complex to the reaction center.

The chain of electron carriers on the reducing side of P680 (left side of Fig. 18-12) resembles that in the bacterial photosynthetic reaction center (Section 18-2B). Indeed, *the two sets of proteins have similar amino acid sequences, indicating that they arose from a common ancestor.* A single electron is transferred from P680* to a molecule of **pheophytin *a*** (**Pheo *a***; Chl *a* with its Mg^{2+} replaced by two protons), probably via a Chl *a* molecule, and then to a plastoquinone–Fe(II) complex designated Q$_A$. Ultimately, two electrons are transferred, one at a time, to a second plastoquinone molecule, Q$_B$, which then takes up two protons at the stromal surface of the thylakoid membrane. The resulting plastoquinol, Q$_B$H$_2$, exchanges with a membrane-bound pool of plastoquinone molecules. DCMU, as well as many other herbicides, competes with plastoquinone for the Q$_B$-binding site of PSII, thereby inhibiting photosynthesis.

Electron Transport through the Cytochrome *b₆f* Complex Generates a Proton Gradient

From the plastoquinone pool, electrons pass through the cytochrome *b₆f* complex. This integral membrane assembly, which closely resembles cytochrome *bc*$_1$, its bacterial counterpart (Section 18-2B), as well as Complex III of the mitochondrial electron-transport chain (Section 17-2E), contains one molecule of **cytochrome *f*** (*f* for *feuille*, French for leaf), one cytochrome *b₆* containing two hemes, one [2Fe–2S] protein, and one bound plastoquinol. Electron flow through the cytochrome *b₆f* complex presumably occurs through a Q cycle (Fig. 17-13). Accordingly, two protons are translocated across the thylakoid membrane for every electron transported. The four electrons derived from 2 H$_2$O by the OEC therefore lead to the translocation of 8 H$^+$ from the stroma to the thylakoid lumen. *Electron transport via the cytochrome b₆f complex generates much of the electrochemical proton gradient that drives the synthesis of ATP in chloroplasts.*

The 285-residue cytochrome *f* from turnip, the largest of the four polypeptides in the cytochrome *b₆f* complex, contains a single transmembrane segment near its C-terminus (residues 251–270); the protein's N-terminal 250 residues presumably extend into the thylakoid lumen. The X-ray structure of the 252-residue N-terminal segment of cytochrome *f* reveals an elongated two-domain protein (Fig. 18-15). Although cytochrome *f* contains a heme *c* group, its predominantly β-sheet structure differs markedly from the structures of other *c*-type cytochromes (e.g., Figs. 6-26, 6-31, and 10-5).

Electron transfer between cytochrome *f*, the terminal electron carrier of the cytochrome *b₆f* complex, and PSI is mediated by plastocyanin, a peripheral membrane protein located on the thylakoid luminal surface (Fig. 18-11). The Cu-containing redox center of this mobile 10.5-kD monomer cycles between its Cu(I) and Cu(II) oxidation states. The X-ray structure

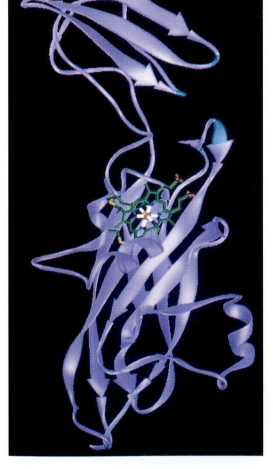

Figure 18-15.   A ribbon diagram of turnip cytochrome *f*. The heme group and the groups that covalently link it to the protein (Cys 21, Cys 24, His 25, and the N-terminal amino group) are shown in stick form with their C, N, O, and S atoms colored green, blue, red, and yellow; the heme's Fe atom is represented by an orange sphere. The five Lys and Arg residues that form a positively charged patch on the protein surface are cyan. [Based on an X-ray structure by Janet Smith, Purdue University.]

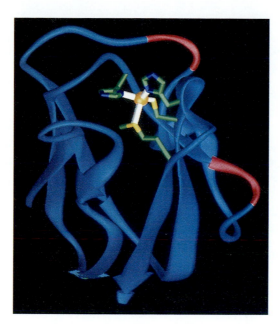

Figure 18-16. **A ribbon diagram of plasto-cyanin (PC) from poplar leaves.** This 99-residue monomeric protein, a member of the family of **blue copper proteins**, folds into a β sandwich. Its Cu atom (*orange sphere*), which alternates between its Cu(I) and Cu(II) oxidation states, is tetrahedrally liganded by the side chains of His 37, Cys 84, His 87, and Met 92, which are shown in stick form with their C, N, and S atoms green, blue, and yellow. Six conserved Asp and Glu residues that form a negatively charged patch on the protein's surface are red. [Based on an X-ray structure by Mitchell Guss and Hans Freeman, University of Sydney, Australia.]

of PC from poplar leaves shows that the Cu atom is coordinated with distorted tetrahedral geometry by a Cys, a Met, and two His residues (Fig. 18-16). Cu(II) complexes with four ligands normally adopt a square planar coordination geometry, whereas those of Cu(I) are generally tetrahedral. Evidently, the strain of Cu(II)'s protein-imposed tetrahedral coordination in PC promotes its reduction to Cu(I). This hypothesis accounts for PC's high standard reduction potential (0.370 V) compared to that of the normal Cu(II)/Cu(I) half-reaction (0.158 V) and illustrates how proteins can modulate the reduction potentials of their redox centers. In the case of plastocyanin, this facilitates electron transfer from the cytochrome $b_6 f$ complex to PSI.

The structures of cytochrome f and PC suggest how these proteins associate. Cytochrome f's Lys 187, a member of a conserved group of five positively charged residues on the protein's surface, can be cross-linked to Asp 44 on PC, which occupies a conserved negatively charged surface patch. Quite possibly, the two proteins associate through electrostatic interactions, much like cytochrome c is thought to interact with its redox partners in the mitochondrial electron transport chain (Section 17-2E).

PSI-Activated Electrons Follow Either of Two Pathways

PSI consists of two large, ~45% identical, ~83 kD protein subunits named **PsaA** and **PsaB**, nine small subunits (4–16 kD), numerous chlorophyll a molecules (~100 in cyanobacteria and ~200 in eukaryotes), 10 to 25 carotenoid molecules, three [4Fe–4S] clusters, and two molecules of **phylloquinone.**

Phylloquinone

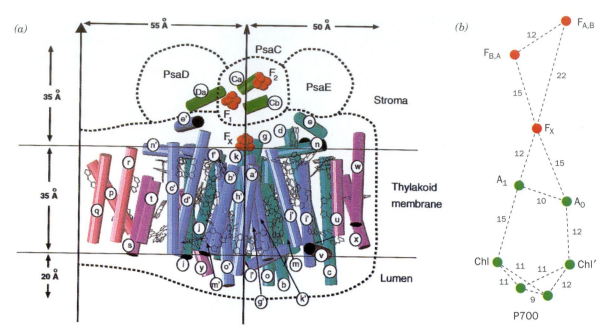

Figure 18-17. **The low-resolution structure of a subunit of PSI from the cyanobacterium *Synechococcus elongatus*.** (*a*) A view from within the thylakoid membrane with the stromal side above. The blue and turquoise transmembrane helices belong to subunits PsaA and PsaB. PsaC, PsaD, and PsaE are stromal subunits of PSI. The porphyrin rings of Chl *a* are shown as wire models, and the [Fe–S] clusters are shown as red spheres. The pseudo-twofold symmetry axis centered on F_X is indicated by the barbed arrow (*center*), whereas the 3-fold axis of the PSI trimer is marked by the solid arrow (*left*). [Courtesy of Norbert Krauss and Wolfram Saenger, Freie Universität Berlin, Germany.] (*b*) The arrangements of the centers of the electron carriers as viewed approximately in Part *a*. The distances between centers are given in Å.

The low-resolution (4-Å) X-ray structure of PSI from the thermophilic cyanobacterium *Synechococcus elongatus* has been determined by Wolfram Saenger (Fig. 18-17*a*). The protein is a symmetric trimer, each of whose monomeric units contains the above 11 proteins. Altogether, 31 transmembrane helices can be identified in the monomeric unit. Most of the helices are roughly perpendicular to the plane of the membrane.

The prosthetic groups at the core of PSI are arranged with approximate two-fold symmetry (Fig. 18-17*b*), owing to the homology of its PsaA and PsaB subunits. Two Chl *a* molecules, which are parallel, 9 Å apart, and close to the local two-fold axis, are assumed to be **P700,** PSI's photon-absorbing center. These Chl *a* molecules resemble the special pair in the bacterial photosynthetic reaction center. Interestingly, P700 is flanked by two symmetry-related Chl *a* molecules in a manner reminiscent of the arrangement of the accessory BChl molecules in the bacterial photosynthetic reaction center (Fig. 18-9). However, electron transport in PSI differs from that in purple photosynthetic bacteria.

Photooxidation of P700 yields $P700^+$, a weak oxidant that accepts one electron directly from plastocyanin. On the reducing side of P700, the electron passes through a chain of electron carriers of increasing reduction potential (right side of Fig. 18-12). The first of these carriers, designated A_0, appears to be Chl *a*, whereas the second carrier, A_1, is probably a phylloquinone. The electron finally proceeds through three [4Fe–4S] clusters designated F_X, F_A, and F_B. The numerous remaining chlorophylls and carotenoids in PSI appear to serve as a built-in light-harvesting complex that funnels light energy to P700.

Electrons ejected from PSI may follow either of two alternative pathways:

1. Most electrons follow a **noncyclic pathway** from PSI to an 11-kD, [2Fe–2S]-containing soluble **ferredoxin (Fd)** that is located in the stroma. Two reduced Fd molecules, in turn, deliver one electron each to the 314-residue monomeric enzyme ferredoxin–NADP$^+$ reductase (**FNR;** Fig. 18-18). The FAD group of FNR sequentially assumes the neutral semiquinone and fully reduced states before transferring the electrons to NADP$^+$ in a two-electron reduction. This reaction sequence yields the final product of the chloroplast light reactions, NADPH.

2. Some electrons return from PSI, via cytochrome b_6, to the plastoquinone pool, thereby traversing a **cyclic pathway** that translocates protons across the thylakoid membrane (Fig. 18-12). Note that the cyclic pathway is independent of the action of PSII and hence does not result in the evolution of O_2.

Cyclic electron flow presumably increases the level of ATP synthesis relative to that of NADPH and thus permits the cell to adjust the production of these two substances according to its needs. However, the mechanism that apportions electrons between the cyclic and noncyclic pathways is unknown. Fine-tuning of the light reactions also depends on the segregation of PSI and PSII in distinct portions of the thylakoid membrane (see Box 18-2).

D. Photophosphorylation

Chloroplasts generate ATP in much the same way as mitochondria, that is, by coupling the dissipation of a proton gradient to the enzymatic synthesis of ATP (Section 17-3). Photophosphorylation, like oxidative phosphorylation, requires an intact thylakoid membrane and can be uncoupled from light-driven electron transport by compounds such as 2,4-dinitrophenol (Fig. 17-22).

Electron micrographs of thylakoid membrane stromal surfaces and bacterial plasma membrane inner surfaces reveal lollipop-shaped structures (Fig. 18-19). These closely resemble the F_1 units of the proton-translocating ATP synthase in mitochondria (Fig. 17-19). In fact, the chloroplast ATP synthase, which is termed the **CF$_1$CF$_0$** complex (C for chloroplast), is remarkably similar to the mitochondrial F_1F_0 complex. For example,

1. Both the F_0 and the CF$_0$ units are hydrophobic transmembrane proteins that contain a proton-translocating channel.
2. Both the F_1 and the CF$_1$ units are hydrophilic peripheral membrane proteins of subunit composition $\alpha_3\beta_3\gamma\delta\epsilon$, of which β is a reversible ATPase.
3. Both ATP synthases are inhibited by oligomycin and by dicyclohexylcarbodiimide (DCCD).

Clearly, proton-translocating ATP synthase must have evolved very early in the history of cellular life. Note, however, that whereas chloroplast ATP synthase translocates protons out of the thylakoid space into the stroma (Fig. 18-11), mitochondrial ATP synthase conducts them from the intermembrane space into the matrix space (Fig. 17-18). This is because the stroma is topologically analogous to the mitochondrial matrix.

Photosynthesis with Noncyclic Electron Transport Produces Around 1.25 ATP per Absorbed Photon

At saturating light intensities, chloroplasts generate proton gradients of ~3.5 pH units across their thylakoid membranes as a result of two processes:

1. The evolution of a molecule of O_2 from two H_2O molecules releases four protons into the thylakoid lumen.

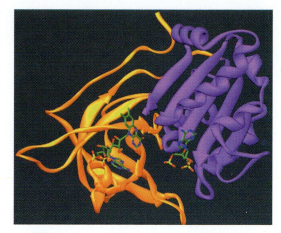

Figure 18-18.   A ribbon drawing of ferredoxin–NADP$^+$ reductase from spinach. Residues 1 to 161 (*gold*) fold into an antiparallel β barrel, which contains the FAD-binding site. Residues 162 to 314 (*magenta*), which provide the bulk of the NADP$^+$-binding site, form a dinucleotide-binding fold (Fig. 6-29). FAD and 2'-phospho-AMP (representing a portion of the NADP$^+$ structure) are shown in stick form with their C, N, O, and P atoms colored green, blue, red, and yellow, respectively. The cleft between the domains that faces the viewer appears likely to be the ferredoxin-binding site. [Based on an X-ray structure by Andrew Karplus, Cornell University, and Jon Herriott, University of Washington.] ● **See the Interactive Exercises.**

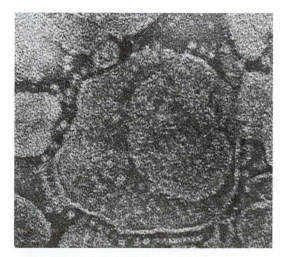

Figure 18-19.   Electron micrograph of thylakoids. The CF$_1$ "lollipops" of their ATP synthases project from their stromal surfaces. Compare this with Fig. 17-19*a*. [Courtesy of Peter Hinkle, Cornell University.]

Box 18-2

BIOCHEMISTRY IN FOCUS

Segregation of PSI and PSII

Freeze-fracture electron microscopy (Section 10-2) has revealed that the protein complexes of the thylakoid membrane have characteristic distributions (*see Figure below*).

1. PSI occurs mainly in the unstacked stromal lamellae, in contact with the stroma, where it has access to NADP$^+$.
2. PSII is located almost exclusively between the closely stacked grana, out of direct contact with the stroma.
3. Cytochrome b_6f is uniformly distributed throughout the membrane.

The high mobilities of plastoquinone and plastocyanin, the electron carriers that shuttle electrons between these particles, permit photosynthesis to proceed at a reasonable rate.

What function is served by the segregation of PSI and PSII? If these two photosystems were in close proximity, the higher excitation energy of PSII (P680 versus P700) would cause it to pass a large fraction of its absorbed photons to PSI via exciton transfer; that is, PSII would act as a light-harvesting antenna for PSI. The separation of the particles by around 100 Å eliminates this difficulty.

The physical separation of PSI and PSII also permits the chloroplast to respond to changes in illumination. The relative amounts of light absorbed by the two photosystems vary with how the light-harvesting complexes are distributed between the stacked and unstacked portions of the thylakoid membrane. Under high illumination (normally direct sunlight, which contains a high proportion of short-wavelength blue light), PSII absorbs more light than PSI. PSI is then unable to take up electrons as fast as PSII can supply them, so the plastoquinone is predominantly in its reduced state. The reduced plastoquinone activates a protein kinase to phosphorylate specific Thr residues of the LHCs, which, in response, migrate to the unstacked regions of the thylakoid membrane, where they associate with PSI. A greater fraction of the incident light is thereby funneled to PSI.

Under low illumination (normally shady light, which contains a high proportion of long-wavelength red light), PSI takes up electrons faster than PSII can provide them so that plastoquinone predominantly assumes its oxidized form. The LHCs are consequently dephosphorylated and migrate to the stacked portions of the thylakoid membrane, where they associate with PSII. The chloroplast therefore maintains the balance between its two photosystems by a light-activated feedback mechanism.

[Figure based on Anderson, J.M. and Anderson, B., *Trends Biochem. Sci.* **7**, 291 (1982).]

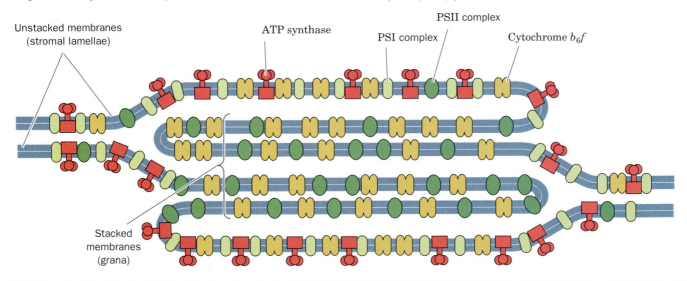

Unstacked membranes (stromal lamellae) **ATP synthase** **PSI complex** **PSII complex** **Cytochrome b_6f**

Stacked membranes (grana)

---

2. The transport of the liberated four electrons through the cytochrome b_6f complex occurs with the translocation of eight protons from the stroma to the thylakoid lumen.

Altogether, ~12 protons enter the lumen per molecule of O_2 produced by noncyclic electron transport.

The thylakoid membrane, in contrast to the inner mitochondrial membrane, is permeable to ions such as Mg^{2+} and Cl^-. Translocation of protons and electrons across the thylakoid membrane is consequently accompanied by the passage of these ions so as to maintain electrical neutrality (Mg^{2+} out and Cl^- in). This all but eliminates the membrane potential, $\Delta\Psi$. *The electrochemical gradient in chloroplasts is therefore almost entirely a result of the pH (concentration) gradient.*

Chloroplast ATP synthase, according to most estimates, produces one ATP for every three protons it transports from the thylakoid lumen to the stroma. Noncyclic electron transport in chloroplasts therefore results in the production of $\sim 12/3 = 4$ molecules of ATP per molecule of O_2 evolved (cyclic electron transport generates more ATP because more protons are translocated to the thylakoid lumen via the Q cycle mediated by cytochrome $b_6 f$).

Noncyclic electron transport, of course, also yields NADPH (2 NADPH for every 4 electrons liberated from 2 H_2O by the OEC). Each NADPH has the free energy to produce 3 ATP (Section 17-3C), for a total of 6 more ATP equivalents per O_2 produced. Consequently, a total of 10 ATP are generated per O_2 produced. A minimum of two photons is required for each electron traversing the system from H_2O to NADPH, that is, eight photons per O_2 produced. This is confirmed by experimental measurements which indicate that plants and algae require 8 to 10 photons of visible light to produce one molecule of O_2. Thus, the overall efficiency of the light reactions is 10 ATP/8–10 photons, or approximately 1.25 ATP per absorbed photon.

3. THE DARK REACTIONS

In the previous section we saw how plants harness light energy to generate ATP and NADPH. In this section we discuss how these products are used to synthesize carbohydrates and other substances from CO_2.

A. The Calvin Cycle

The metabolic pathway by which plants incorporate CO_2 into carbohydrates was elucidated between 1946 and 1953 by Melvin Calvin, James Bassham, and Andrew Benson. They did so by tracing the metabolic fate of the radioactive label from $^{14}CO_2$ in cultures of algal cells. Some of Calvin's earliest experiments indicated that algae exposed to $^{14}CO_2$ for a minute or more synthesize a complex mixture of labeled metabolites, including sugars and amino acids. Analysis of the algae within 5 s of their exposure to $^{14}CO_2$, however, showed that *the first stable radioactive compound formed is **3-phosphoglycerate (3PG),** which is initially labeled only in its carboxyl group.* This result immediately suggested that the 3PG was formed by the carboxylation of a C_2 compound. Yet no such precursor was found. The actual carboxylation reaction involves the pentose **ribulose-5-phosphate (Ru5P).**

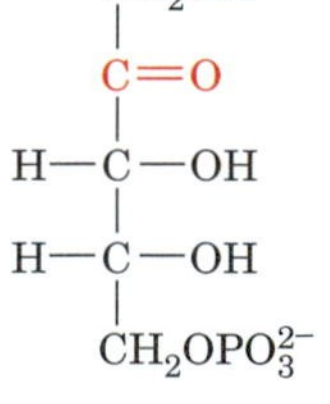

$$
\begin{array}{c}
CH_2OH \\
| \\
C{=}O \\
| \\
H{-}C{-}OH \\
| \\
H{-}C{-}OH \\
| \\
CH_2OPO_3^{2-}
\end{array}
$$

Ribulose-5-phosphate (Ru5P)

The resulting C_6 product splits into two C_3 compounds, both of which turn out to be 3PG. The overall pathway, diagrammed in Fig. 18-20, is known as the **Calvin cycle** or the **reductive pentose phosphate cycle.** It involves the carboxylation of a pentose, the formation of carbohydrate products, and the regeneration of Ru5P.

During the search for the carboxylation substrate, several other photosynthetic intermediates had been identified and their labeling patterns elucidated. For example, the hexose fructose-1,6-bisphosphate (FBP) is initially labeled only at its C3 and C4 positions but later becomes labeled to a lesser degree at its other atoms. A consideration of the flow of labeled carbon through the various tetrose, pentose, hexose, and heptose phosphates led, in what is a milestone of metabolic biochemistry, to the deduction of the Calvin cycle as is diagrammed in Fig. 18-20. The existence of many of its postulated reactions was eventually confirmed by *in vitro* studies using purified enzymes.

The Calvin Cycle Generates GAP from CO_2 via a Two-Stage Process

The Calvin cycle can be considered to have two stages:

Stage 1 The production phase (top line of Fig. 18-20), in which three molecules of Ru5P react with three molecules of CO_2 to yield six molecules of glyceraldehyde-3-phosphate (GAP) at the expense of nine ATP and six NADPH molecules. *The cyclic nature of the pathway makes this process equivalent to the synthesis of one GAP from three CO_2 molecules.* At this point, GAP can be bled off from the cycle for use in biosynthesis.

Stage 2 The recovery phase (bottom lines of Fig. 18-20), in which the carbon atoms of the remaining five GAPs are shuffled in a remarkable series of reactions, similar to those of the pentose phosphate pathway (Section 14-6), to re-form the three Ru5Ps with which the cycle began. This stage can be conceptually decomposed into four sets of reactions (with the numbers keyed to the corresponding reactions in Fig. 18-20):

$$\textbf{6. } C_3 + C_3 \longrightarrow C_6$$
$$\textbf{8. } C_3 + C_6 \longrightarrow C_4 + C_5$$
$$\textbf{9. } C_3 + C_4 \longrightarrow C_7$$
$$\textbf{11. } C_3 + C_7 \longrightarrow C_5 + C_5$$

The overall stoichiometry for this process is therefore

$$5\,C_3 \longrightarrow 3\,C_5$$

Note that this stage of the Calvin cycle occurs without further input of free energy (ATP) or reducing power (NADPH).

The first reaction of the Calvin cycle is the phosphorylation of Ru5P by **phosphoribulokinase** to form **ribulose-1,5-bisphosphate (RuBP).** Following the carboxylation of RuBP (Reaction 2; discussed below), the result-

Figure 18-20 *(Opposite).* **Key to Metabolism. The Calvin cycle.** The number of lines in an arrow indicates the number of molecules reacting in that step for a single turn of the cycle that converts three CO_2 molecules to one GAP molecule. For the sake of clarity, the sugars are all shown in their linear forms, although the hexoses and heptoses predominantly exist in their cyclic forms. The ^{14}C-labeling patterns generated in one turn of the cycle through the use of $^{14}CO_2$ are indicated in red. Note that two of the product Ru5Ps are labeled only at C3, whereas the third Ru5P is equally labeled at C1, C2, and C3. ✳ See the Animated Figures.

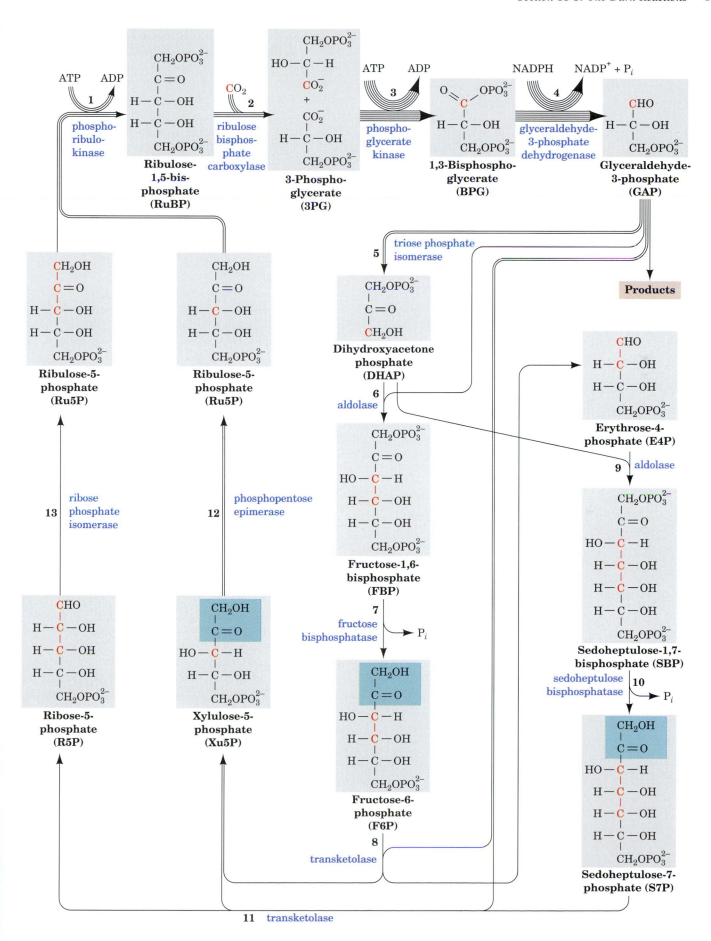

ing 3PG is converted first to 1,3-bisphosphoglycerate (BPG) and then to GAP. This latter sequence is the reverse of two consecutive glycolytic reactions (Sections 14-2G and 14-2F) except that the Calvin cycle reaction uses NADPH rather than NADH.

The second stage of the Calvin cycle begins with the reverse of a familiar glycolytic reaction, the isomerization of GAP to dihydroxyacetone phosphate (DHAP) by triose phosphate isomerase (Section 14-2E). Following this, DHAP is directed along two analogous paths: Reactions 6–8 or Reactions 9–11. Reactions 6 and 9 are aldolase-catalyzed aldol condensations in which DHAP is linked to an aldehyde. Reaction 6 is also the reverse of a glycolytic reaction (Section 14-2D). Reactions 7 and 10 are phosphate hydrolysis reactions that are catalyzed, respectively, by fructose bisphosphatase (FBPase; Section 14-4B) and **sedoheptulose bisphosphatase (SBPase).** The remaining Calvin cycle reactions are catalyzed by enzymes that also participate in the pentose phosphate pathway (Section 14-6; whose concurrent elucidation provided considerable supporting evidence for the Calvin cycle). In Reactions 8 and 11, both catalyzed by transketolase, a C_2 keto unit (shaded green in Fig. 18-20) is transferred from a ketose to GAP to form xylulose-5-phosphate (Xu5P), leaving the aldoses erythrose-4-phosphate (E4P) in Reaction 8 and ribose-5-phosphate (R5P) in Reaction 11. The E4P produced by Reaction 8 feeds into Reaction 9. The Xu5Ps produced by Reactions 8 and 11 are converted to Ru5P by **phosphopentose epimerase** in Reaction 12. The R5P from Reaction 11 is also converted to Ru5P by **ribose phosphate isomerase** in Reaction 13, thereby completing a turn of the Calvin cycle. Only 3 of the 11 Calvin cycle enzymes—phosphoribulokinase, the carboxylation enzyme **ribulose bisphosphate carboxylase,** and SBPase—have no equivalents in animal tissues.

RuBP Carboxylase Catalyzes CO_2 Fixation

The enzyme that catalyzes CO_2 fixation, ribulose bisphosphate carboxylase **(RuBP carboxylase),** is arguably the world's most important enzyme since nearly all life on earth ultimately depends on its action. This protein, presumably as a consequence of its low catalytic efficiency ($k_{cat} \approx 3\ s^{-1}$), accounts for up to 50% of leaf proteins and is therefore the most abundant protein in the biosphere. RuBP carboxylase from higher plants and most photosynthetic microorganisms consists of eight large (L) subunits (477 residues in tobacco leaves) encoded by chloroplast DNA, and eight small (S) subunits (123 residues) specified by a nuclear gene (the RuBP carboxylase from certain photosynthetic bacteria is an L_2 dimer whose L subunit has 28% sequence identity with and is structurally similar to that of the L_8S_8 enzyme). X-Ray studies by Carl-Ivar Brändén and by David Eisenberg demonstrated that the L_8S_8 enzyme has the symmetry of a square prism (Fig. 18-21a). The L subunit is made of a β sheet domain and an α/β barrel domain that contains the enzyme's catalytic site (Fig. 18-21b). The function of the S subunit is unknown; attempts to show that is has a regulatory role, in analogy with other enzymes, have been unsuccessful.

The accepted mechanism of RuBP carboxylase, which was largely formulated by Calvin, is indicated in Fig. 18-22. Abstraction of the C3 proton of RuBP, the reaction's rate-determining step, generates an enediolate that nucleophilically attacks CO_2. The resulting β-keto acid is rapidly attacked at its C3 position by H_2O to yield an adduct that splits, by a reaction similar to aldol cleavage, to yield the two product 3PG molecules. *The driving force for the overall reaction, which is highly exergonic ($\Delta G^{\circ\prime} = -35.1\ kJ \cdot mol^{-1}$), is provided by the cleavage of the β-keto acid intermediate to yield an additional resonance-stabilized carboxylate group.*

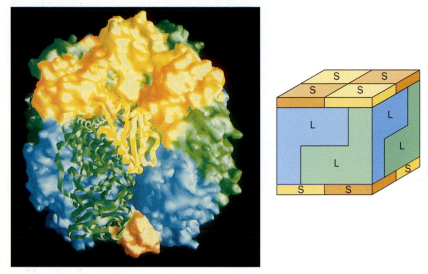

(a)

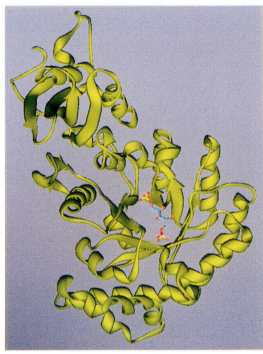

(b)

Figure 18-21.   The X-ray structure of RuBP carboxylase. (*a*) The quaternary structure of the L_8S_8 protein. One L and one S subunit are drawn as ribbons with the remainder represented by their solvent-accessible surfaces. The protein, which has D_4 symmetry (the symmetry of a square prism; Fig. 6-33*b*), is viewed with its four-fold axis tipped toward the viewer. As the accompanying diagram schematically indicates, the elongated L subunits (6 are clearly visible in the structural drawing) can be considered to associate as two interdigitated tetramers, with that extending from the top green and that extending from the bottom cyan. The members of the S_4 tetramers that cap the top and bottom of the complex are alternately colored yellow and orange (only one subunit of the lower S_4 tetramer is visible). [Based on an X-ray structure by Yasushi Kai, Osaka University, Japan.] (*b*) An L subunit. The transition state inhibitor 2-carboxyarabinitol-1,5-bisphosphate, represented in stick form, is positioned in the substrate-binding site in the mouth of the enzyme's α/β barrel. The subunit is oriented by a rotation about the vertical axis relative to that drawn in ribbon form in *a*. [Based on an X-ray structure by David Eisenberg, UCLA.]

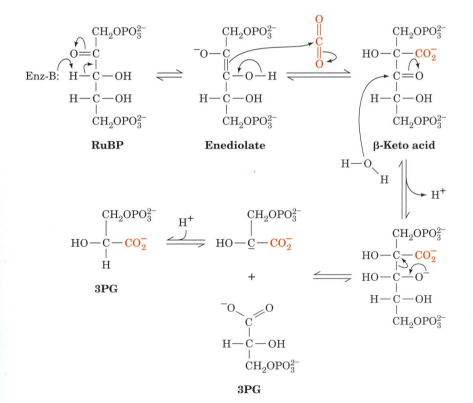

Figure 18-22.   The mechanism of the RuBP carboxylase reaction. The reaction proceeds via an enediolate intermediate that nucleophilically attacks CO_2 to form a β-keto acid. This intermediate reacts with water to yield two molecules of 3PG. ✻ See the Animated Figures.

RuBP carboxylase activity requires Mg^{2+}, which probably stabilizes developing negative charges during catalysis. The Mg^{2+} is, in part, bound to the enzyme by a catalytically important carbamate group ($-NH-COO^-$) that is generated by the reaction of a nonsubstrate CO_2 with the ε-amino group of Lys 201. This essential reaction is catalyzed *in vivo* by the enzyme **RuBP carboxylase activase** in an ATP-driven process.

GAP Is the Precursor of Glucose-1-Phosphate and Other Biosynthetic Products

The overall stoichiometry of the Calvin cycle is

$$3\ CO_2 + 9\ ATP + 6\ NADPH \longrightarrow GAP + 9\ ADP + 8\ P_i + 6\ NADP^+$$

GAP, the primary product of photosynthesis, is used in a variety of biosynthetic pathways, both inside and outside the chloroplast. For example, it can be converted to fructose-6-phosphate by the further action of Calvin cycle enzymes and then to glucose-1-phosphate (G1P) by phosphoglucose isomerase and phosphoglucomutase (Section 15-1C). *G1P is the precursor of the higher order carbohydrates characteristic of plants.* These most notably include sucrose (Section 8-2A), their major transport sugar for delivering carbohydrates to nonphotosynthesizing cells; starch (Section 8-2C), their chief storage polysaccharide; and cellulose (Section 8-2B), the primary structural component of their cell walls. In the synthesis of all these substances, G1P is activated by the formation of either ADP–, CDP–, GDP–, or UDP–glucose (Section 15-5), depending on the species and the pathway. The glucose unit is then transferred to the nonreducing end of a growing polysaccharide chain, as occurs in the synthesis of glycogen (Section 15-2B).

B. Control of the Calvin Cycle

During the day, plants satisfy their energy needs via the light and dark reactions of photosynthesis. At night, however, like other organisms, they must use their nutritional reserves to generate ATP and NADPH through glycolysis, oxidative phosphorylation, and the pentose phosphate pathway. Since the stroma contains the enzymes of glycolysis and the pentose phosphate pathway as well as those of the Calvin cycle, *plants must have a light-sensitive control mechanism to prevent the Calvin cycle from consuming this catabolically produced ATP and NADPH in a wasteful futile cycle.*

As we have seen, the control of flux in a metabolic pathway occurs at enzymatic steps that are far from equilibrium (large negative value of ΔG). Inspection of Table 18-1 indicates that the three best candidates for flux control in the Calvin cycle are the reactions catalyzed by RuBP carboxylase, FBPase, and SBPase (Reactions 2, 7, and 10 of Fig. 18-20). In fact, the catalytic efficiency of these three enzymes all vary *in vivo* with the level of illumination.

The activity of RuBP carboxylase responds to three light-dependent factors:

1. pH. On illumination, the pH of the stroma increases from ~7.0 to ~8.0 as protons are pumped from the stroma into the thylakoid lumen. RuBP carboxylase has a sharp pH optimum near pH 8.0.
2. $[Mg^{2+}]$. Recall that the light-induced influx of protons to the thylakoid lumen is accompanied by the efflux of Mg^{2+} to the stroma (Section 18-2D). This Mg^{2+} stimulates RuBP carboxylase.

Table 18-1. **Standard and Physiological Free Energy Changes for the Reactions of the Calvin Cycle**

Step[a]	Enzyme	$\Delta G°'$ $(kJ \cdot mol^{-1})$	ΔG $(kJ \cdot mol^{-1})$
1	Phosphoribulokinase	−21.8	−15.9
2	Ribulose bisphosphate carboxylase	−35.1	−41.0
3 + 4	Phosphoglycerate kinase + glyceraldehyde-3-phosphate dehydrogenase	+18.0	−6.7
5	Triose phosphate isomerase	−7.5	−0.8
6	Aldolase	−21.8	−1.7
7	Fructose bisphosphatase	−14.2	−27.2
8	Transketolase	+6.3	−3.8
9	Aldolase	−23.4	−0.8
10	Sedoheptulose bisphosphatase	−14.2	−29.7
11	Transketolase	+0.4	−5.9
12	Phosphopentose epimerase	+0.8	−0.4
13	Ribose phosphate isomerase	+2.1	−0.4

[a]Refer to Fig. 18-20.

Source: Bassham, J.A. and Buchanan, B.B., in Govindjee (Ed.), Photosynthesis, Vol. II, p. 155, Academic Press (1982).

3. The transition state analog **2-carboxyarabinitol-1-phosphate (CA1P).**

$$
\begin{array}{c}
CH_2OPO_3^{2-} \\
| \\
HO-C-CO_2^- \\
| \\
H-C-OH \\
| \\
H-C-OH \\
| \\
CH_2OH
\end{array}
$$

2-Carboxyarabinitol-1-phosphate (CA1P)

Many plants synthesize this compound, which inhibits RuBP carboxylase, only in the dark. RuBP carboxylase activase facilitates the release of the tight-biding CA1P from RuBP carboxylase as well as catalyzing its carbamoylation (Section 18-3A).

FBPase and SBPase are also activated by increased pH and $[Mg^{2+}]$, and by NADPH as well. The action of these factors is complemented by a second regulatory system that responds to the redox potential of the stroma. **Thioredoxin,** a 12-kD protein that occurs in many types of cells, contains a reversibly reducible cystine disulfide group. Reduced thioredoxin activates both FBPase and SBPase by a disulfide interchange reaction (Fig. 18-23). The redox level of thioredoxin is maintained by a second disulfide-containing enzyme, **ferredoxin–thioredoxin reductase,** which directly responds to the redox state of the soluble ferredoxin in the stroma. This in turn varies with the illumination level. The thioredoxin system also deactivates phosphofructokinase (PFK), the main flux-generating enzyme of glycolysis (Section 14-4A). Thus, in plants, *light stimulates the Calvin cycle while deactivating glycolysis, whereas darkness has the opposite effect* (that is, the so-called dark reactions do not occur in the dark).

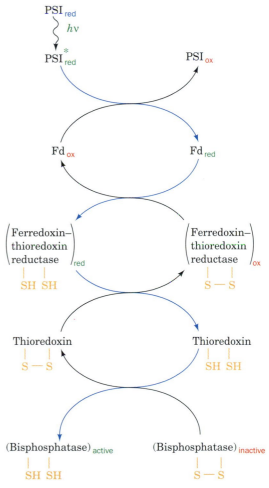

Figure 18-23. The light-activation mechanism of FBPase and SBPase. Photoactivated PSI reduces soluble ferredoxin (Fd), which reduces ferredoxin–thioredoxin reductase, which, in turn, reduces the disulfide linkage of thioredoxin. Reduced thioredoxin reacts with the inactive bisphosphatases by disulfide interchange, thereby activating these flux-controlling Calvin cycle enzymes.

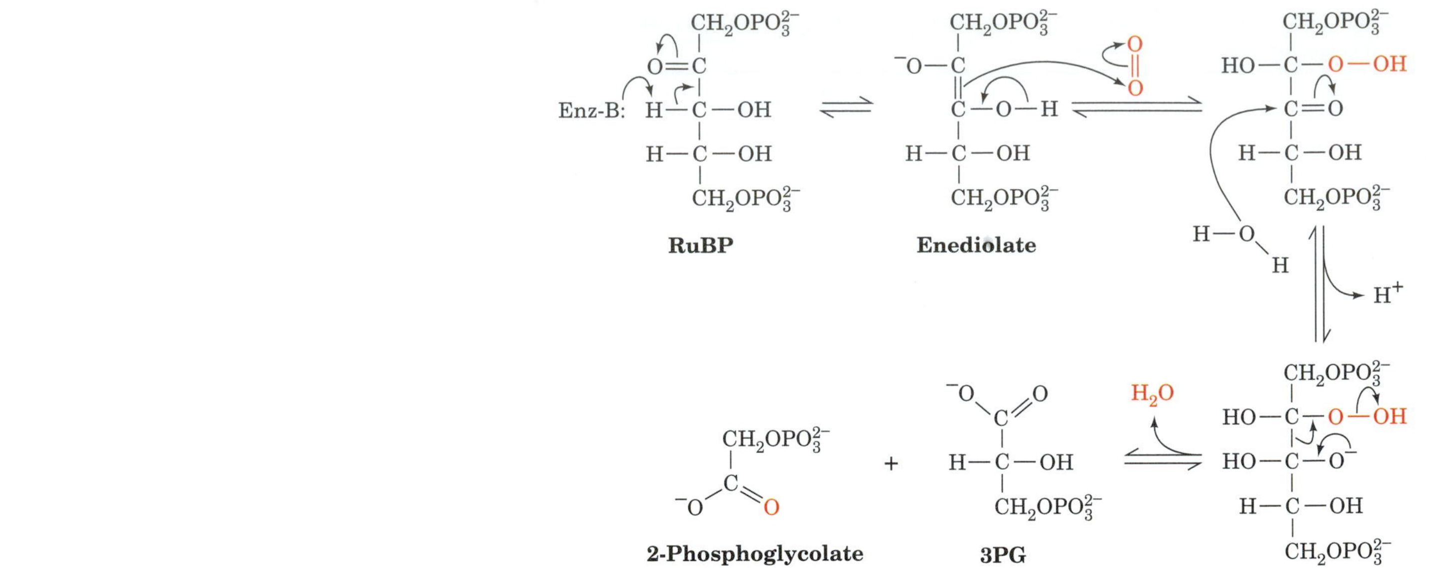

Figure 18-24.   **The probable mechanism of the oxygenase reaction catalyzed by RuBP carboxylase–oxygenase.** Note the similarity of this mechanism to that of the carboxylase reaction catalyzed by the same enzyme (Fig. 18-22).

C. Photorespiration

It has been known since the 1960s that *illuminated plants consume O_2 and evolve CO_2 in a pathway distinct from oxidative phosphorylation. In fact, at low CO_2 and high O_2 levels, this* **photorespiration** *process can outstrip photosynthetic CO_2 fixation.* The basis of photorespiration was unexpected: *O_2 competes with CO_2 as a substrate for RuBP carboxylase* (RuBP carboxylase is therefore also called **RuBP carboxylase–oxygenase** or **rubisco**). In the oxygenase reaction, O_2 reacts with the enzyme's other substrate, RuBP, to form 3PG and **2-phosphoglycolate** (Fig. 18-24). The 2-phosphoglycolate is hydrolyzed to **glycolate** by **glycolate phosphatase** and, as described below, is partially oxidized to yield CO_2 by a series of enzymatic reactions that occur in the **peroxisome** and the mitochondrion. Thus, photorespiration is a seemingly wasteful process that undoes some of the work of photosynthesis. In this section we discuss the biochemical basis of photorespiration and how certain plants manage to evade its deleterious effects.

Photorespiration Dissipates ATP and NADPH

The photorespiration pathway is outlined in Fig. 18-25. Glycolate is exported from the chloroplast to the peroxisome (also called the glyoxysome; Section 16-5C), where it is oxidized by **glycolate oxidase** to glyoxylate and H_2O_2. The H_2O_2, a potentially harmful oxidizing agent, is converted to H_2O and O_2 by the heme-containing enzyme catalase (Section 17-5C). The glyoxylate can be converted to glycine in a transamination reaction, as is discussed in Section 20-2A, and exported to the mitochondrion. There, two molecules of glycine are converted to one molecule of serine and one of CO_2. *This is the origin of the CO_2 generated by photorespiration.* The serine is transported back to the peroxisome, where a transamination reaction converts it to **hydroxypyruvate.** This substance is reduced to **glycerate** and phosphorylated in the cytosol to 3PG, which re-enters the chloroplast and is reconverted to RuBP in the Calvin cycle. *The net result of this com-*

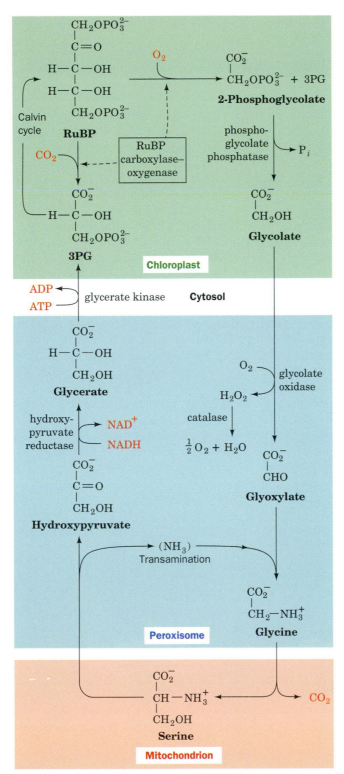

Figure 18-25. **Photorespiration.** This pathway metabolizes the phosphoglycolate produced by the RuBP carboxylase–catalyzed oxidation of RuBP. The reactions occur, as indicated, in the chloroplast, the peroxisome, the mitochondrion, and the cytosol. Note that two glycines are required to form serine + CO_2.

plex photorespiration cycle is that some of the ATP and NADPH generated by the light reactions is uselessly dissipated.

Although photorespiration has no known metabolic function, the RuBP carboxylases from the great variety of photosynthetic organisms so far tested all exhibit oxygenase activity. Yet, over the eons, the forces of evolution must have optimized the function of this important enzyme. Photorespiration may confer a selective advantage by protecting the photosynthetic apparatus from photooxidative damage when insufficient CO_2 is available to otherwise dissipate its absorbed light energy. This hypothesis is supported by the observation that when chloroplasts or leaf cells are brightly illuminated in the absence of both CO_2 and O_2, their photosynthetic capacity is rapidly and irreversibly lost.

C_4 Plants Concentrate CO_2

On a hot bright day, when photosynthesis has depleted the level of CO_2 at the chloroplast and raised that of O_2, the rate of photorespiration approaches the rate of photosynthesis. This phenomenon is a major limitation on the growth of many plants (and is therefore an important agricultural problem that is being attacked through genetic engineering studies—none of which has yet been successful). However, *certain species of plants, such as sugarcane, corn, and most important weeds, have a metabolic cycle that concentrates CO_2 in their photosynthetic cells, thereby almost totally preventing photorespiration.* The leaves of plants that have this so-called **C_4 cycle** have a characteristic anatomy. Their fine veins are concentrically surrounded by a single layer of so-called **bundle-sheath cells,** which in turn are surrounded by a layer of **mesophyll cells.**

The C_4 cycle (Fig. 18-26) was elucidated in the 1960s by Marshall Hatch and Rodger Slack. It begins when mesophyll cells, which lack RuBP carboxylase, take up atmospheric CO_2 by condensing it as HCO_3^- with phosphoenolpyruvate (PEP) to yield oxaloacetate. The oxaloacetate is reduced by NADPH to malate, which is exported to the bundle-sheath cells (the name C_4 refers to these four-carbon acids). There, the malate is oxidatively decarboxylated by $NADP^+$ to form CO_2, pyruvate, and NADPH. The CO_2, which has been concentrated by this process, enters the Calvin cycle. The

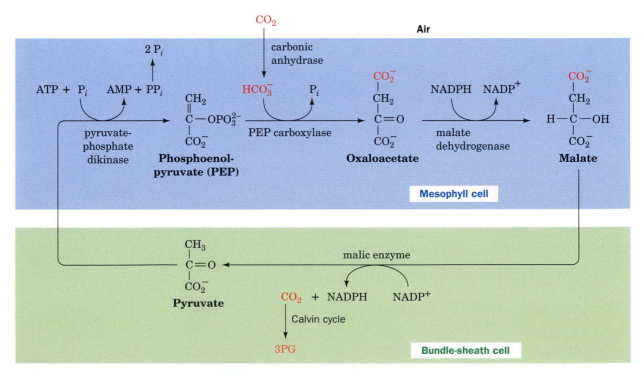

Figure 18-26. The C_4 pathway. CO_2 is concentrated in the mesophyll cells and transported to the bundle-sheath cells for entry into the Calvin cycle.

pyruvate is returned to the mesophyll cells, where it is phosphorylated to regenerate PEP. The enzyme that mediates this reaction, **pyruvate-phosphate dikinase,** has the unusual action of activating a phosphate group through the hydrolysis of ATP to AMP + PP$_i$. This PP$_i$ is further hydrolyzed to two P$_i$, which is tantamount to the consumption of a second ATP. *CO$_2$ is thereby concentrated in the bundle-sheath cells at the expense of 2 ATP per CO$_2$. Photosynthesis in C$_4$ plants therefore consumes a total of 5 ATP per CO$_2$ fixed versus the 3 ATP required by the Calvin cycle alone.*

C$_4$ plants occur largely in tropical regions because they grow faster under hot and sunny conditions than other, so-called **C$_3$ plants** (so named because they initially fix CO$_2$ in the form of three-carbon acids). In cooler climates, where photorespiration is less of a burden, C$_3$ plants have the advantage because they require less energy to fix CO$_2$.

CAM Plants Store CO$_2$ through a Variant of the C$_4$ Cycle

A variant of the C$_4$ cycle that separates CO$_2$ acquisition and the Calvin cycle in time rather than in space occurs in many desert-dwelling succulent plants. If these plants opened their stomata (pores in the leaves) by day to acquire CO$_2$, as most plants do, they would lose an unacceptable amount of water by evaporation. To minimize this loss, these succulents absorb CO$_2$ only at night and use the reactions of the C$_4$ pathway (Fig. 18-26) to store it as malate. This process is known as **Crassulacean acid metabolism (CAM),** so named because it was first discovered in plants of the family Crassulaceae. The large amount of PEP necessary to store a day's supply of CO$_2$ is obtained by the breakdown of starch via glycolysis. During the course of the day, the malate is broken down to CO$_2$, which enters the Calvin cycle, and pyruvate, which is used to resynthesize starch. CAM plants are thus able to carry out photosynthesis with minimal water loss.

SUMMARY

1. Photosynthesis is the process by which light energy drives the reduction of CO$_2$ to yield carbohydrates. In plants and cyanobacteria, photosynthesis oxidizes water to O$_2$.

2. In plants, the photosynthetic machinery consists of proteins embedded in the thylakoid membrane and dissolved in the stroma of chloroplasts.

3. Chlorophyll and other light-absorbing pigments are organized in light-harvesting complexes that funnel light energy to photosynthetic reaction centers.

4. The purple bacterial photosynthetic reaction center undergoes photooxidation when it absorbs a photon. The excited electron passes through a series of electron carriers and returns to the reaction center. During electron transport, protons are translocated across the plasma membrane.

5. In plants and cyanobacteria, photosystems I and II operate in series in an arrangement known as the Z-scheme. The oxidation of water, driven by the photooxidation of PSII, releases electrons that flow from PSII, through a cytochrome b_6f complex, through PSI, and, finally, to NADP$^+$.

6. Electron flow in PSI can be noncyclic, which results in NADP$^+$ reduction, or cyclic, which causes additional protons to be translocated into the thylakoid lumen.

7. Protons released by the oxidation of H$_2$O and proton translocation into the thylakoid lumen generate a transmembrane proton gradient that is tapped by chloroplast ATP synthase to drive the phosphorylation of ADP.

8. The dark reactions use the ATP and NADPH produced in the light reactions to power the synthesis of carbohydrates from CO$_2$. In the first phase of the Calvin cycle, CO$_2$ reacts with ribulose-1,5-bisphosphate (RuBP) to ultimately yield glyceraldehyde-3-phosphate (GAP). The remaining reactions of the cycle regenerate the RuBP acceptor of CO$_2$.

9. RuBP carboxylase, the key enzyme of the dark reactions, is regulated by pH, [Mg^{2+}], and the inhibitory compound 2-carboxyarabinitol-1-phosphate (CA1P). The two bisphosphatases of the Calvin cycle are controlled by the redox state of the chloroplast via disulfide interchange reactions mediated in part by thioredoxin.

10. Photorespiration, in which plants consume O$_2$ and evolve CO$_2$, uses the ATP and NADPH produced by the light reactions. C$_4$ plants minimize the oxygenase activity of RuBP carboxylase by concentrating CO$_2$ in their photosynthetic cells. CAM plants use a related mechanism to conserve water.

REFERENCES

Barber, J. and Anderson, B., Revealing the blueprint of photosynthesis, *Nature* **370**, 31–34 (1994).

Cramer, W.A., Martinez, S.E., Furbacher, P.N., Huang, D., and Smith, J.L., The cytochrome $b_6 f$ complex, *Curr. Opin. Struct. Biol.* **4**, 536–544 (1994).

El-Kabbani, O., Chang, C.-H., Tiede, D., Norris, J., and Schiffer, M., Comparison of reaction centers from *Rhodobacter sphaeroides* and *Rhodopseudomonas viridis:* Overall architecture and protein-pigment interactions, *Biochemistry* **30**, 5361–5369 (1991).

Fromme, P., Structure and function of photosystem I, *Curr. Opin. Struct. Biol.* **6**, 473–484 (1996).

Krauss, N., Schubert, W.-D., Klukas, O., Fromme, P., Witt, H.T., and Saenger, W., Photosystem I at 4 Å resolution represents the first structural model of a joint photosynthetic reaction centre and core antenna system, *Nature Struct. Biol.* **3**, 965–973 (1996).

Nugent, J.H.A., Oxygenic photosynthesis: electron transfer in photosystem I and photosystem II, *Eur. J. Biochem.* **237**, 519–531 (1996). [A useful review of photosynthesis in plants.]

Prince, R.C., Photosynthesis: the Z-scheme revised, *Trends Biochem. Sci.* **21**, 121–122 (1996). [A short summary of $NADP^+$ reduction by PSII in the absence of PSI.]

Schneider, G., Lindqvist, Y., and Brändén, C.-I., RUBISCO: Structure and mechanism, *Annu. Rev. Biophys. Biomol. Struct.* **21**, 119–143 (1992).

KEY TERMS

photosynthesis	Planck's law	quantum yield	C_3 plant
stroma	internal conversion	photophosphorylation	CAM
thylakoid	fluorescence	PSI	
grana	exciton transfer	PSII	
stromal lamellae	photooxidation	Z-scheme	
antenna chlorophyll	photosynthetic reaction	Calvin cycle	
LHC	center	photorespiration	
accessory pigments	special pair	C_4 plant	

STUDY EXERCISES

1. Explain the relationship between the light and dark reactions.

2. Why are light-harvesting complexes important?

3. How do molecules dissipate absorbed light energy?

4. Describe the pathway of electron transfer in the purple bacterial reaction center.

5. Describe the Z-scheme.

6. What are the implications of cyclic and noncyclic electron transfer in PSI?

7. Compare and contrast photophosphorylation and oxidative phosphorylation.

8. Summarize the two stages of the Calvin cycle.

9. How is photosynthesis regulated?

10. How do plants minimize photorespiration?

PROBLEMS

1. The "red tide" is a massive proliferation of certain algal species that causes seawater to become visibly red. Describe the spectral characteristics of the dominant photosynthetic pigments in these algae.

2. $H_2^{18}O$ is added to a suspension of chloroplasts capable of photosynthesis. Where does the label appear when the suspension is exposed to sunlight?

3. (a) Calculate the energy of one mole of photons of red light ($\lambda = 700$ nm). (b) How many moles of ATP could theoretically be synthesized using this energy?

4. (a) Calculate $\Delta \mathscr{E}^{\circ\prime}$ and $\Delta G^{\circ\prime}$ for the light reactions in plants, that is, the four-electron oxidation of H_2O by $NADP^+$. (b) Use the solution of Problem 18-3 to calculate how many moles of photons of red light ($\lambda = 700$ nm) are theoretically required to drive this process. (c) How many moles of photons of UV light ($\lambda = 220$ nm) would be required?

5. Describe the functional similarities between the bacterial photosynthetic reaction center and eukaryotic PSI.

6. Why is it possible for chloroplasts to absorb much more than 8–10 photons per O_2 molecule evolved?

7. Describe the effects of an increase in oxygen pressure on the dark reactions of photosynthesis.

8. Chloroplasts are illuminated until the levels of the Calvin cycle intermediates reach a steady state. The light is then turned off. How do the levels of RuBP and 3PG vary after this point?

9. Calculate the energy cost of the Calvin cycle combined with glycolysis and oxidative phosphorylation, that is, the ratio of the energy spent synthesizing starch from CO_2 and photosynthetically produced NADPH and ATP to the energy generated by the complete oxidation of starch. Assume that each NADPH is energetically equivalent to 3 ATP and that starch biosynthesis and breakdown are mechanistically identical to glycogen synthesis and breakdown.

10. The leaves of some species of desert plants taste sour in the early morning, but, as the day wears on, they become tasteless and then bitter. Explain.

Stored lipids can be mobilized for oxidation as metabolic fuels. [©Chris Cole/Allsport.]

LIPID METABOLISM

Most cells contain a wide variety of lipids, but many of these structurally distinct molecules are functionally similar. For example, most cells can tolerate variations in the lipid composition of their membranes, provided that membrane fluidity, which is largely a property of their component fatty acid chains, is maintained (Section 9-2B).

Even greater variation is exhibited in the cellular content of lipids stored as energy reserves. Triacylglycerol stores are built up and gradually depleted in response to changing physiological demands. The camel's hump is a well-known example of a fat depot that supplies energy as well as metabolic water for periods of days or weeks. Other organisms that undergo dramatic changes in body fat content are hibernating mammals and birds that migrate long distances without refueling. Lipid metabolism in these cases is notable for the sheer amount of material that flows through the few relatively simple biosynthetic and degradative pathways.

In this chapter, we acknowledge the central function of lipids in energy metabolism by first examining the absorption and transport of their component fatty acids and the oxidation of these fatty acids to produce energy. The second part of this chapter examines the synthesis of fatty acids and other lipids, including glycerophospholipids, sphingolipids, and cholesterol.

1. LIPID DIGESTION, ABSORPTION, AND TRANSPORT

Triacylglycerols (also called fats or triglycerides) constitute ~90% of the dietary lipid and are the major form of metabolic energy storage in humans. Triacylglycerols consist of glycerol triesters of fatty acids such as palmitic and oleic acids

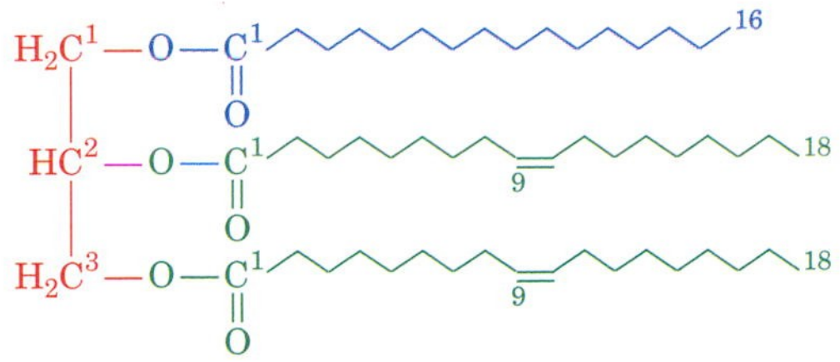

1-Palmitoyl-2,3-dioleoyl-glycerol

(the names and structural formulas of some biologically common fatty acids are listed in Table 9-1). The mechanisms for digesting, absorbing, and transporting triacylglycerols from the intestine to the tissues must accommodate their inherent hydrophobicity.

A. Digestion and Absorption

Since triacylglycerols are water insoluble, whereas digestive enzymes are water soluble, triacylglycerol digestion takes place at lipid–water interfaces. The rate of triacylglycerol digestion therefore depends on the surface area of the interface, which is greatly increased by the churning peristaltic movements of the intestine combined with the emulsifying action of **bile acids.** The bile acids (also called **bile salts**) are amphipathic detergentlike molecules that act to solubilize fat globules. Bile acids are cholesterol derivatives that are synthesized by the liver and secreted as glycine or **taurine**

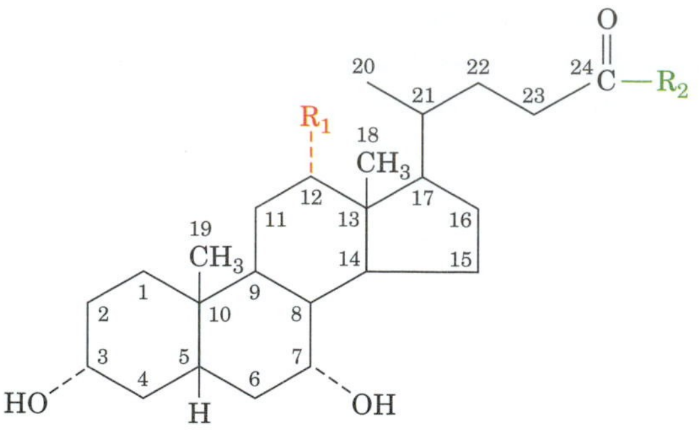

	R₁ = OH	R₁ = H

R_2 = OH	Cholic acid	Chenodeoxycholic acid
R_2 = NH—CH₂—COOH	Glycocholic acid	Glycochenodeoxycholic acid
R_2 = NH—CH₂—CH₂—SO₃H	Taurocholic acid	Taurochenodeoxycholic acid

Figure 19-1. **Structures of the major bile acids and their glycine and taurine conjugates.**

conjugates (Fig. 19-1) into the gallbladder for storage. From there, they are secreted into the small intestine, where lipid digestion and absorption mainly take place.

Lipases Act at the Lipid–Water Interface

Pancreatic **lipase (triacylglycerol lipase)** catalyzes the hydrolysis of triacylglycerols at their 1 and 3 positions to form sequentially **1,2-diacylglycerols** and **2-acylglycerols,** together with the Na⁺ and K⁺ salts of fatty acids (soaps). The enzymatic activity of pancreatic lipase greatly increases when it contacts the lipid–water interface, a phenomenon known as **interfacial activation.** Binding to the lipid–water interface requires pancreatic **colipase,** a 90-residue protein that forms a 1:1 complex with lipase. The X-ray structures, determined by Christian Cambillau, of pancreatic lipase–colipase complexes reveal the structural basis of the interfacial activation of lipase as well as how colipase helps lipase bind to the lipid–water interface (Fig. 19-2).

The active site of the 449-residue pancreatic lipase, which is contained in the enzyme's N-terminal domain (residues 1–336), contains a catalytic triad that closely resembles that in serine proteases (Section 11-5B; recall that ester hydrolysis is mechanistically similar to peptide hydrolysis). In the absence of lipid micelles, lipase's active site is covered by a 25-residue helical "lid." However, in the presence of the micelles, the lid undergoes a complex structural reorganization that exposes the active site. Simultaneously, a 10-residue loop, called the β5 loop, changes conformation in a way that forms the active enzyme's oxyanion hole and generates a hydrophobic surface near the entrance to the active site.

Colipase binds to the C-terminal domain of lipase (residues 337–449) such that the hydrophobic tips of its three loops extend from the complex. This creates a continuous hydrophobic plateau, extending >50 Å past the lipase active site, that presumably helps bind the complex to the lipid surface. Colipase also forms three hydrogen bonds to the opened lid, thereby stabilizing it in this conformation.

Other lipases, such as phospholipase A₂ (Fig. 9-6), also preferentially catalyze reactions at interfaces. Instead of changing its conformation, however, phospholipase A₂ contains a hydrophobic channel that provides the

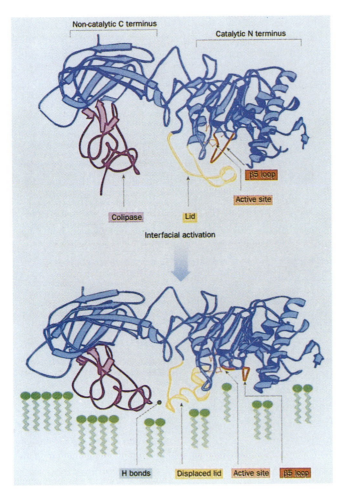

Figure 19-2. The mechanism of interfacial activation of triacylglycerol lipase. The enzyme is in complex with procolipase (the precursor of colipase; *purple*). On binding to a phospholipid micelle (*green*), the 25-residue lid (*yellow*) covering the enzyme's active site (*tan*) changes conformation so as to expose its hydrophobic residues, thereby uncovering the active site. This causes the 10-residue β5 loop (*brown*) to move aside in a way that forms the enzyme's oxyanion hole. The procolipase also changes its conformation so as to hydrogen bond to the "open" lid, thereby stabilizing it in this conformation and, together with lipase, forming an extended hydrophobic surface. [From *Nature* **362**, 793 (1993). Reproduced with permission.]

substrate with direct access from the phospholipid aggregate (micelle or membrane) surface to the bound enzyme's active site (Fig. 19-3). Hence, on leaving its micelle to bind to the enzyme, the substrate need not become solvated and then desolvated. In contrast, soluble and dispersed phospholipids must first surmount these significant kinetic barriers in order to bind to the enzyme.

Bile Acids and Fatty Acid–Binding Protein Facilitate the Intestinal Absorption of Lipids

The mixture of fatty acids and mono- and diacylglycerols produced by lipid digestion is absorbed by the cells lining the small intestine (the intestinal **mucosa**). *Bile acids not only aid lipid digestion; they are essential for the absorption of the digestion products.* The micelles formed by the bile acids take up the nonpolar lipid degradation products so as to permit their

(a)

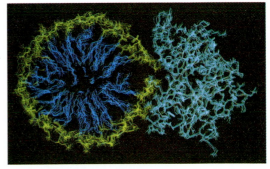

(b)

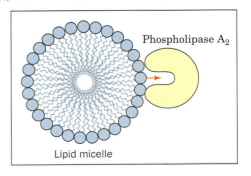

Figure 19-3. Substrate binding to phospholipase A₂. (*a*) Hypothetical model of phospholipase A₂ in complex with a micelle of lysophosphatidylethanolamine as shown in cross section. The protein is drawn in cyan, the phospholipid head groups are yellow, and their hydrocarbon tails are blue. The calculated atomic motions of the assembly are indicated through a series of superimposed images taken at 5-ps intervals. [Courtesy of Raymond Salemme, E.I. du Pont de Nemours & Company.] (*b*) Schematic diagram of a phospholipid contained in a micelle entering the hydrophobic phospholipid channel in phospholipase A₂ (*red arrow*).

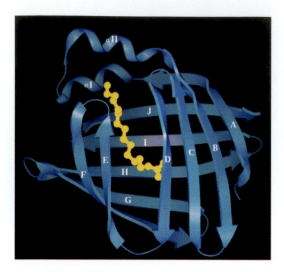

Figure 19-4.   The X-ray structure of rat intestinal fatty acid–binding protein, shown in ribbon form (*blue*), in complex with palmitate, shown in ball-and-stick form (*yellow*). [Courtesy of James Sacchettini, Texas A&M University.]

transport across the unstirred aqueous boundary layer at the intestinal wall. The importance of this process is demonstrated in individuals with obstructed bile ducts: They absorb little of their ingested lipids but, rather, eliminate them in hydrolyzed form in their feces. Bile acids are likewise required for the efficient intestinal absorption of the lipid-soluble vitamins A, D, E, and K.

Inside the intestinal cells, fatty acids form complexes with **intestinal fatty acid–binding protein (I-FABP),** a cytoplasmic protein that increases the effective solubility of these water-insoluble substances and also protects the cell from their detergentlike effects. The X-ray structure of rat I-FABP, determined by James Sacchettini, shows that this monomeric, 131-residue protein consists largely of 10 antiparallel β strands stacked in two approximately orthogonal β sheets (Fig. 19-4). A fatty acid molecule occupies a gap between two of the β strands, lying between the β sheets so that it is more or less parallel to the gapped β strands (a structure that has been dubbed a "β-clam"). The fatty acid's carboxyl group interacts with Arg 106, Gln 115, and two bound water molecules, whereas its tail is encased by the side chains of several hydrophobic, mostly aromatic residues.

B. Lipid Transport

The fatty acid products of lipid digestion that are absorbed by the intestinal mucosa are converted to triacylglycerols and packaged into lipoprotein particles called **chylomicrons.** These, in turn, are released into the intestinal lymph and transported through the lymphatic vessels before draining into the large body veins. The bloodstream then delivers chylomicrons to other tissues.

Other lipoproteins, whose structures are described in Section 10-3A, are synthesized by the liver. Although each class of lipoprotein has a different physiological purpose, they all function to maintain their otherwise insoluble lipid components in aqueous solution.

Chylomicrons Are Delipidated in the Capillaries of Peripheral Tissues

Chylomicrons, which contain exogenous (dietary) triacylglycerols and cholesterol, adhere to binding sites on the inner surface (endothelium) of the capillaries in skeletal muscle and adipose tissue. The chylomicron's component triacylglycerols are hydrolyzed through the action of the extracellular enzyme **lipoprotein lipase.** The tissues then take up the liberated monoacylglycerol and fatty acids. The chylomicrons shrink as their triacylglycerols are progressively hydrolyzed until they are reduced to cholesterol-enriched **chylomicron remnants.** The remnants dissociate from the capillary endothelium and reenter the circulation to be taken up by the liver. *Chylomicrons therefore deliver dietary triacylglycerols to muscle and adipose tissue, and dietary cholesterol to the liver* (Fig. 19-5, *left*).

VLDL Are Gradually Degraded

Very low density lipoproteins (VLDL), which are synthesized in the liver to transport endogenous triacylglycerols and cholesterol, are also degraded by lipoprotein lipase in the capillaries of adipose tissue and muscle (Fig. 19-5, *right*). The released fatty acids are taken up by the cells and oxidized for energy or used to resynthesize triacylglycerols. The glycerol backbone of triacylglycerols is transported to the liver or kidneys and converted to the glycolytic intermediate dihydroxyacetone phosphate. Oxidation of this three-carbon unit, however, yields only a fraction of the energy available from oxidizing the three fatty acyl chains of a triacylglycerol.

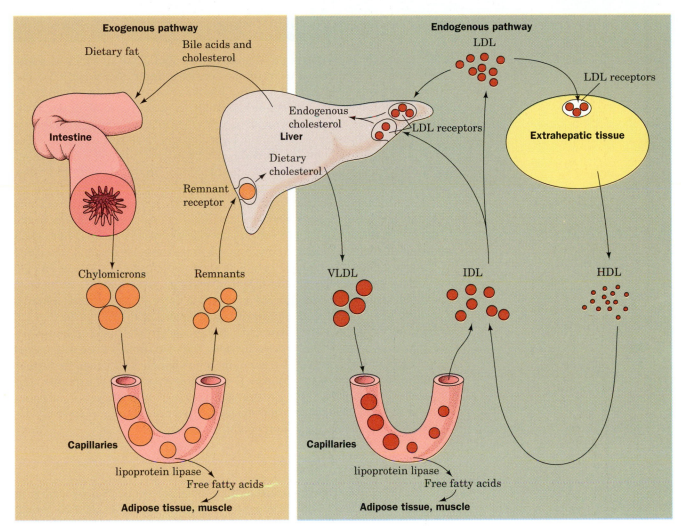

Figure 19-5.   A model for plasma triacylglycerol and cholesterol transport in humans.   [After Brown, M.S. and Goldstein, J.L., *in* Brunwald, E., Isselbacher, K.J., Petersdorf, R.G., Wilson, J.D., Martin, J.B., and Fauci, A.S. (Eds.), *Harrison's Principles of Internal Medicine* (11th ed.), p. 1652, McGraw–Hill (1987).] ✳ See the Animated Figures.

After giving up their triacylglycerols, the VLDL remnants, which have also lost some of their apolipoproteins (Section 10-3A), appear in the circulation first as **intermediate density lipoproteins (IDL)** and then as **low density lipoproteins (LDL).** About half of the VLDL, after degradation to IDL and LDL, are taken up by the liver via receptor-mediated endocytosis (Section 10-3B).

HDL Transports Cholesterol from the Tissues to the Liver

High density lipoproteins (HDL) have essentially the opposite function of LDL: *They remove cholesterol from the tissues.* HDL is assembled in the plasma from components largely obtained through the degradation of other lipoproteins. *Circulating HDL probably acquires its cholesterol by extracting it from cell surface membranes and converts it to cholesteryl esters.* HDL then transfers these cholesteryl esters to VLDL in a poorly understood process. There are indications that the liver directly takes up HDL through the agency of a specific HDL receptor.

2. FATTY ACID OXIDATION

The triacylglycerols stored in adipocytes are mobilized in times of metabolic need by the action of **hormone-sensitive lipase** (Section 19-5). The free fatty acids are released into the bloodstream, where they bind to albumin, a soluble 66-kD monomeric protein. In the absence of albumin, the maximum solubility of fatty acids is $\sim 10^{-6}$ M; the effective solubility of fatty acids in complex with albumin is as high as 2 mM. Nevertheless, those rare individuals with **analbuminemia** (severely depressed levels of albumin) suffer no apparent adverse symptoms; evidently, their fatty acids are transported in complex with other serum proteins.

The biochemical strategy of fatty acid oxidation was understood long before the oxidative enzymes were purified. In 1904, Franz Knoop, in the first use of chemical labels to trace metabolic pathways, fed dogs fatty acids labeled at their ω (last) carbon atom by a benzene ring and isolated the phenyl-containing metabolic products from their urine. Dogs fed labeled odd-chain fatty acids excreted **hippuric acid,** the glycine amide of **benzoic acid,** whereas those fed labeled even-chain fatty acids excreted **phenylaceturic acid,** the glycine amide of **phenylacetic acid** (Fig. 19-6). Knoop therefore deduced that fatty acids are progressively degraded by two-carbon units and that this process involves the oxidation of the carbon atom β to the carboxyl group. Otherwise, the phenylacetic acid would be further oxidized to benzoic acid. Knoop's **β oxidation** hypothesis was finally confirmed in the 1950s. The β-oxidation pathway is a series of enzyme-catalyzed reactions that operates in a repetitive fashion to progressively degrade fatty acids by removing two-carbon units.

A. Fatty Acid Activation

Before fatty acids can be oxidized, they must be "primed" for reaction in an ATP-dependent acylation reaction to form fatty acyl-CoA. The activation process is catalyzed by a family of at least three **acyl-CoA synthetases** (also called **thiokinases**) that differ in their chain-length specificities. These

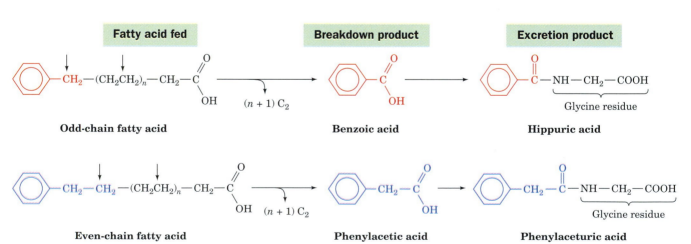

Figure 19-6.   Franz Knoop's classic experiment. The results indicated that fatty acids are metabolically oxidized at their β-carbon atom. ω-Phenyl-labeled fatty acids containing an odd number of carbon atoms are oxidized to the phenyl-labeled C_1 product, benzoic acid, whereas those with an even number of carbon atoms are oxidized to the phenyl-labeled C_2 product, phenylacetic acid. These products are excreted as their respective glycine amides, hippuric and phenylaceturic acids. The vertical arrows indicate the deduced sites of carbon oxidation. The intermediate C_2 products are oxidized to CO_2 and H_2O and were therefore not isolated.

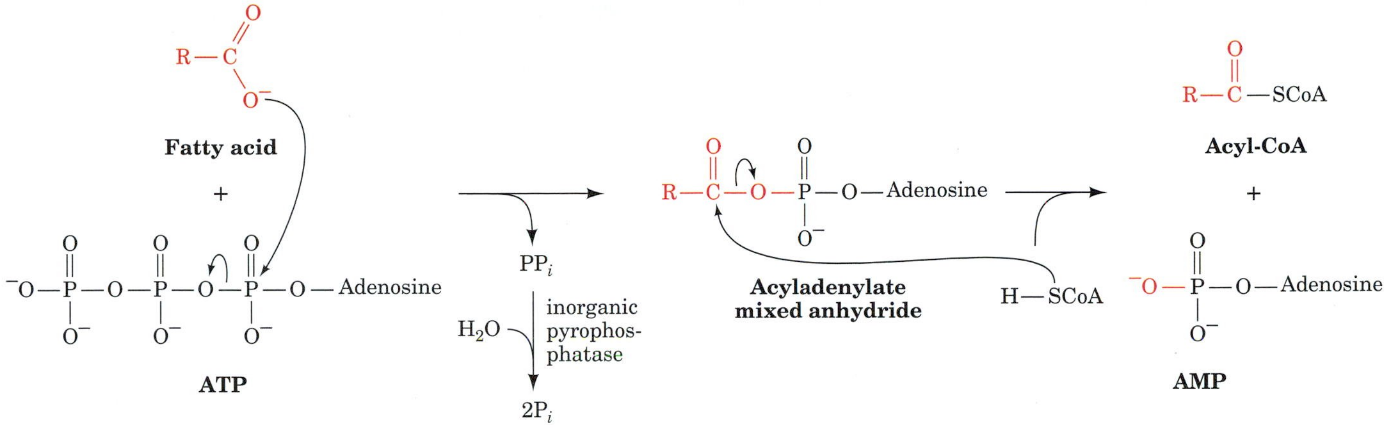

Figure 19-7. **The mechanism of fatty acid activation catalyzed by acyl-CoA synthetase.** Formation of acyl-CoA involves an intermediate acyladenylate mixed anhydride.

enzymes, which are associated with either the endoplasmic reticulum or the outer mitochondrial membrane, all catalyze the reaction

$$\text{Fatty acid} + \text{CoA} + \text{ATP} \rightleftharpoons \text{acyl-CoA} + \text{AMP} + \text{PP}_i$$

This reaction proceeds via an acyladenylate mixed anhydride intermediate that is attacked by the sulfhydryl group of CoA to form the thioester product (Fig. 19-7), thereby preserving the free energy of ATP hydrolysis in the "high-energy" thioester bond (Section 13-2D). The overall reaction is driven to completion by the highly exergonic hydrolysis of pyrophosphate catalyzed by inorganic pyrophosphatase.

B. Transport across the Mitochondrial Membrane

Although fatty acids are activated for oxidation in the cytosol, they are oxidized in the mitochondrion, as Eugene Kennedy and Albert Lehninger established in 1950. We must therefore consider how fatty acyl-CoA is transported across the inner mitochondrial membrane. A long-chain fatty acyl-CoA cannot directly cross the inner mitochondrial membrane. Instead, its acyl portion is first transferred to **carnitine,** a compound that occurs in both plant and animal tissues.

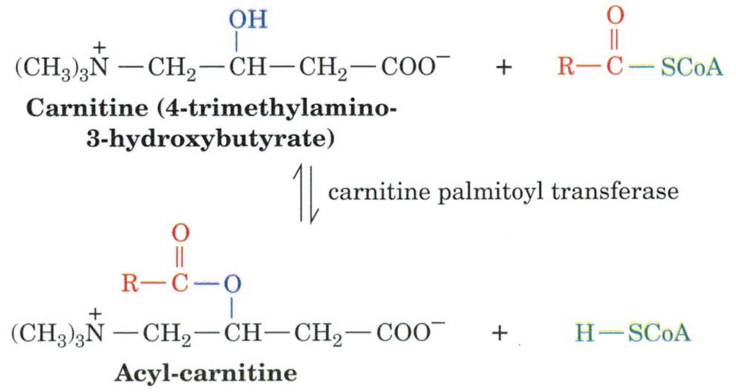

This transesterification reaction has an equilibrium constant close to 1, which indicates that the *O*-acyl bond of **acyl-carnitine** has a free energy of hydrolysis similar to that of acyl-CoA's thioester bond. **Carnitine palmitoyl transferases I** and **II,** which can transfer a variety of acyl groups (not just

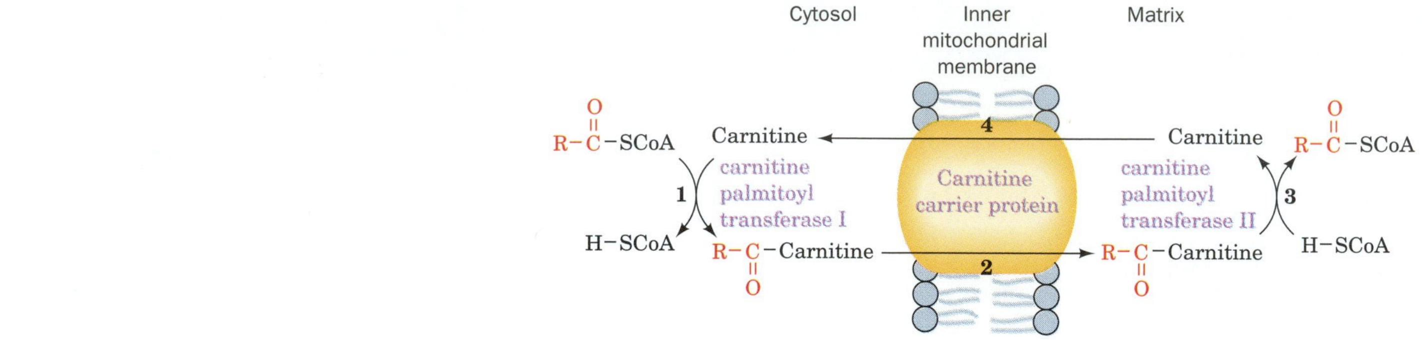

Cytosol   Inner   Matrix
mitochondrial
membrane

Figure 19-8. **The transport of fatty acids into the mitochondrion.** (1) The acyl group of a cytosolic acyl-CoA is transferred to carnitine, thereby releasing the CoA to its cytosolic pool. (2) The resulting acyl-carnitine is transported into the mitochondrial matrix by the carrier protein. (3) The acyl group is transferred to a CoA molecule from the mitochondrial pool. (4) The product carnitine is returned to the cytosol.

palmitoyl groups), are located, respectively, on the external and internal surfaces of the inner mitochondrial membrane. The translocation process itself is mediated by a specific carrier protein that transports acyl-carnitine into the mitochondrion while transporting free carnitine in the opposite direction. The acyl-CoA transport system is diagrammed in Fig. 19-8.

C. β Oxidation

The degradation of fatty acyl-CoA via β oxidation occurs in four reactions (Fig. 19-9):

1. Formation of a trans-α,β double bond through dehydrogenation by the flavoenzyme **acyl-CoA dehydrogenase.**
2. Hydration of the double bond by **enoyl-CoA hydratase** to form a **3-L-hydroxyacyl-CoA.**
3. NAD$^+$-dependent dehydrogenation of this β-hydroxyacyl-CoA by **3-L-hydroxyacyl-CoA dehydrogenase** to form the corresponding β-ketoacyl-CoA.
4. C$_\alpha$—C$_\beta$ cleavage in a thiolysis reaction with CoA as catalyzed by **β-ketoacyl-CoA thiolase** (also called just **thiolase**) to form acetyl-CoA and a new acyl-CoA containing two fewer C atoms than the original one.

Acyl-CoA Dehydrogenase Is Linked to the Electron-Transport Chain

Mitochondria contain three acyl-CoA dehydrogenases, with specificities for short (C$_4$–C$_6$), medium (C$_4$–C$_{12}$), and long chain (C$_8$–C$_{20}$) fatty acyl-CoAs. The reaction catalyzed by these enzymes is thought to involve removal of a proton at C$_\alpha$ and transfer of a hydride ion equivalent from C$_\beta$ to FAD (Fig. 19-9, Reaction 1). The X-ray structure of **medium-chain acyl-CoA dehydrogenase** in complex with **octanoyl-CoA,** determined by Jung-Ja Kim, clearly shows how the enzyme orients a basic group (Glu 376), the substrate C$_\alpha$—C$_\beta$ bond, and the FAD prosthetic group for reaction (Fig. 19-10).

The FADH$_2$ resulting from the oxidation of the fatty acyl-CoA substrate is reoxidized by the mitochondrial electron-transport chain through a series of electron-transfer reactions. **Electron-transfer flavoprotein (ETF)**

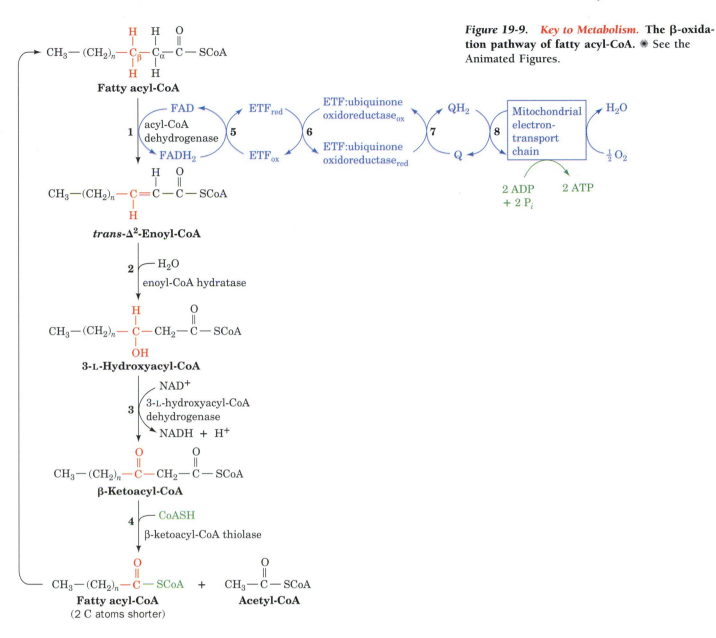

Figure 19-9. *Key to Metabolism.* **The β-oxidation pathway of fatty acyl-CoA.** ✳ See the Animated Figures.

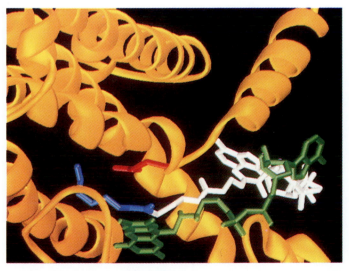

Figure 19-10. **A ribbon diagram of the active site region of medium-chain acyl-CoA dehydrogenase.** The enzyme, from pig liver mitochondria, is a tetramer of identical 385-residue subunits, each of which binds an FAD prosthetic group (*green*) and an octanoyl-CoA substrate (whose octanoyl and CoA moieties are blue and white) in largely extended conformations. The octanoyl-CoA binds such that its C_α—C_β bond is sandwiched between the carboxylate group of Glu 376 (*red*) and the flavin ring (*green*), consistent with the proposal that Glu 376 is the general base that abstracts the α proton in the α,β dehydrogenation reaction catalyzed by the enzyme. [Based on an X-ray structure by Jung-Ja Kim, Medical College of Wisconsin.] ◉ See the Interactive Exercises.

transfers an electron pair from FADH$_2$ to the flavo-iron–sulfur protein **ETF:ubiquinone oxidoreductase,** which in turn transfers an electron pair to the mitochondrial electron-transport chain by reducing coenzyme Q (CoQ; Fig. 19-9, Reactions 5–8). Reduction of O$_2$ to H$_2$O by the electron-transport chain beginning at the CoQ stage results in the synthesis of 2 ATP per electron pair transferred (Section 17-3C).

The Thiolase Reaction Occurs via Claisen Ester Cleavage

The fourth step of β oxidation is the thiolase reaction (Fig. 19-9, Reaction 4), which yields acetyl-CoA and a new acyl-CoA that is two carbon atoms shorter than the one that began the cycle (Fig. 19-11):

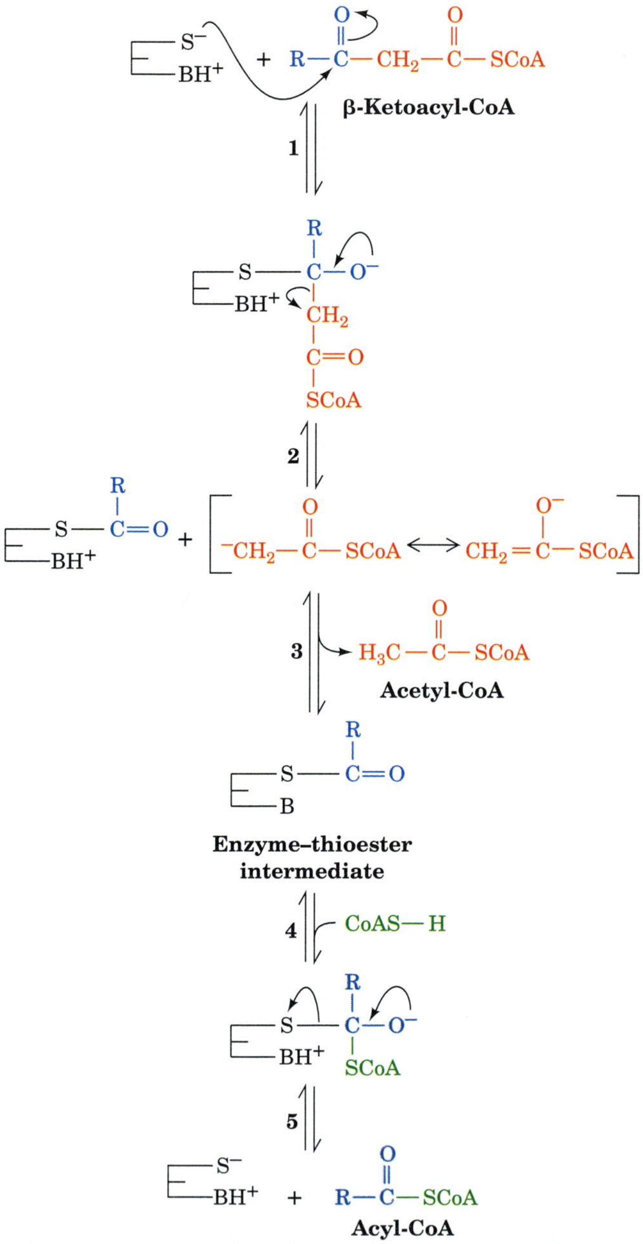

Figure 19-11.   The mechanism of action of β-ketoacyl-CoA thiolase. An active site Cys residue participates in the formation of an enzyme–thioester intermediate.

1. An active site thiol group adds to the β-keto group of the substrate acyl-CoA.

2. Carbon–carbon bond cleavage yields a thioester between the acyl-CoA substrate and the active site thiol group, together with an acetyl-CoA carbanion intermediate that is stabilized by electron withdrawal into this thioester's carbonyl group. This type of reaction is known as a Claisen ester cleavage (the reverse of a Claisen condensation). The citric acid cycle enzyme citrate synthase also catalyzes a reaction that involves a stabilized acetyl-CoA carbanion intermediate (Section 16-3A).

3. An enzyme acidic group protonates the acetyl-CoA carbanion, yielding acetyl-CoA.

4 & 5. Finally, CoA displaces the enzyme thiol group from the enzyme–thioester intermediate, yielding an acyl-CoA that is shortened by two C atoms.

Fatty Acid Oxidation Is Highly Exergonic

The function of fatty acid oxidation is, of course, to generate metabolic energy. Each round of β oxidation produces one NADH, one $FADH_2$, and one acetyl-CoA. Oxidation of acetyl-CoA via the citric acid cycle generates additional $FADH_2$ and 3 NADH, which are reoxidized through oxidative phosphorylation to form ATP. Complete oxidation of a fatty acid molecule is therefore a highly exergonic process that yields numerous ATPs. For example, oxidation of palmitoyl-CoA (which has a C_{16} fatty acyl group) involves seven rounds of β oxidation, yielding 7 $FADH_2$, 7 NADH, and 8 acetyl-CoA. Oxidation of the 8 acetyl-CoA, in turn, yields 8 GTP, 24 NADH, and 8 $FADH_2$. Since oxidative phosphorylation of the 31 NADH molecules yields 93 ATP and that of the 15 $FADH_2$ yields 30 ATP, subtracting the 2 ATP equivalents required for fatty acid acyl-CoA formation (Section 19-2A), *the oxidation of one palmitate molecule has a net yield of 129 ATP.*

D. Oxidation of Unsaturated Fatty Acids

Almost all unsaturated fatty acids of biological origin contain only cis double bonds, which most often begin between C9 and C10 (referred to as a Δ^9 or 9-double bond; Table 9-1). Additional double bonds, if any, occur at three-carbon intervals and are therefore never conjugated. Two examples of unsaturated fatty acids are oleic acid and linoleic acid.

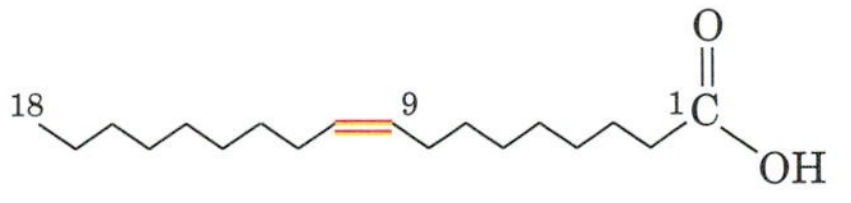

Oleic acid
(**9-*cis*-octadecenoic acid**)

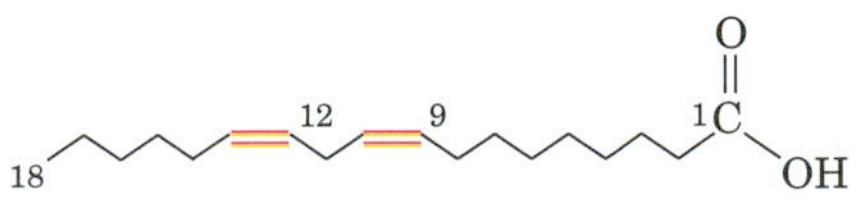

Linoleic acid
(**9,12-*cis*-octadecadienoic acid**)

Note that one of the double bonds in linoleic acid is at an odd-numbered carbon atom and the other is at an even-numbered carbon atom. The double bonds in fatty acids such as linoleic acid pose two problems for the β-

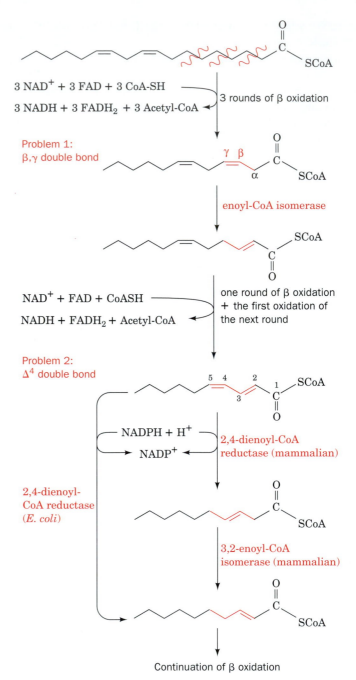

Figure 19-12.   The oxidation of unsaturated fatty acids. β oxidation of a fatty acid such as linoleic acid presents two problems. The first problem, a β,γ double bond, is solved by converting it to a trans-α,β double bond. The second problem, that a 2,4-dienoyl-CoA is a poor substrate for enoyl-CoA hydratase, is eliminated by the NADPH-dependent reduction of the 4-double bond to yield the β-oxidation substrate *trans*-2-enoyl-CoA. This step requires one enzyme in *E. coli* and two in mammals.

oxidation pathway that are solved through the actions of three additional enzymes (Fig. 19-12).

Problem 1: A β,γ Double Bond

The first enzymatic difficulty occurs after the third round of β oxidation: The resulting cis-β,γ double bond-containing enoyl-CoA is not a substrate for enoyl-CoA hydratase. **Enoyl-CoA isomerase,** however, converts the cis-Δ^3 double bond to the trans-Δ^2 form. The Δ^2 compound is the normal substrate of enoyl-CoA hydratase, so β oxidation can continue.

Problem 2: A Δ^4 Double Bond Inhibits Enoyl-CoA Hydratase

The next difficulty arises in the fifth round of β oxidation: The presence of a double bond at an even-numbered carbon atom results in the formation of 2,4-dienoyl-CoA, which is a poor substrate for enoyl-CoA hydratase. However, NADPH-dependent **2,4-dienoyl-CoA reductase** reduces the Δ^4 double bond. The *E. coli* reductase produces *trans*-2-enoyl-CoA, a normal substrate of β oxidation. The mammalian reductase, however, yields *trans*-3-enoyl-CoA, which, to proceed along the β-oxidation pathway, must first be isomerized to *trans*-2-enoyl-CoA by **3,2-enoyl-CoA isomerase.**

E. Oxidation of Odd-Chain Fatty Acids

Most fatty acids, for reasons explained in Section 19-4, have even numbers of carbon atoms and are therefore completely converted to acetyl-CoA. Some plants and marine organisms, however, synthesize fatty acids with an odd number of carbon atoms. *The final round of β oxidation of these fatty acids yields propionyl-CoA, which is converted to succinyl-CoA for entry into the citric acid cycle.* Propionate and propionyl-CoA are also produced by the oxidation of the amino acids isoleucine, valine, and methionine (Section 20-4D).

The conversion of propionyl-CoA to succinyl-CoA involves three enzymes (Fig. 19-13). The first reaction, catalyzed by **propionyl-CoA carboxylase,** requires a biotin prosthetic group and is driven by the hydroly-

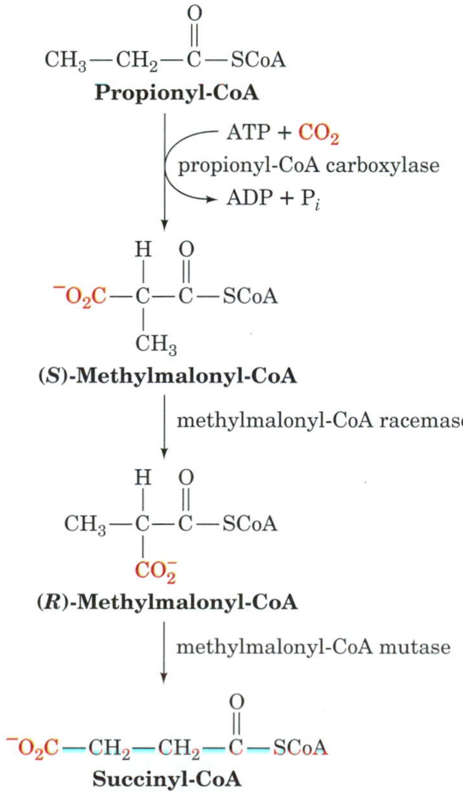

Figure 19-13. The conversion of propionyl-CoA to succinyl-CoA.

sis of ATP to ADP + P$_i$. This reaction resembles that of pyruvate carboxylase (Fig. 15-25).

The (S)-methylmalonyl-CoA product of the carboxylase reaction is converted to the R form by **methylmalonyl-CoA racemase.** (R)-Methylmalonyl-CoA is a substrate for **methylmalonyl-CoA mutase,** which catalyzes the third reaction in Fig. 19-13.

Methylmalonyl-CoA mutase catalyzes an unusual carbon skeleton rearrangement.

(R)-Methylmalonyl-CoA   **Succinyl-CoA**

The enzyme uses a **5′-deoxyadenosylcobalamin** prosthetic group (coenzyme B$_{12}$, a derivative of **cobalamin,** or **vitamin B$_{12}$;** see Box 19-1). Dorothy Hodgkin determined the structure of this complex molecule (Fig. 19-14) in 1956 through X-ray crystallographic analysis combined with chemical degradation studies, a landmark achievement.

5′-Deoxyadenosylcobalamin contains a hemelike **corrin** ring whose four pyrrole N atoms each ligand a six-coordinate Co ion. The fifth Co ligand is an N atom of a **5,6-dimethylbenzimidazole (DMB)** nucleotide that is covalently linked to the corrin D ring. The sixth ligand is a 5′-deoxyadenosyl group in which the deoxyribose C5′ atom forms a covalent C—Co bond, *one of only two carbon–metal bonds known in biology* (the other is a C—Ni bond in the bacterial enzyme **carbon monoxide dehydrogenase**). In some cobalamin-dependent enzymes, the sixth ligand instead is a CH$_3$ group that likewise forms a C—Co bond. There are only about a dozen known cobal-

---

Box 19-1

BIOCHEMISTRY IN HEALTH AND DISEASE

Vitamin B$_{12}$ Deficiency

The existence of vitamin B$_{12}$ came to light in 1926 when George Minot and William Murphy discovered that **pernicious anemia,** an often fatal disease of the elderly characterized by decreased numbers of red blood cells, low hemoglobin levels, and progressive neurological deterioration, can be treated by the daily consumption of large amounts of raw liver. Nevertheless, the antipernicious anemia factor—vitamin B$_{12}$—was not isolated until 1948.

Vitamin B$_{12}$ is synthesized by neither plants nor animals but only by a few species of bacteria. Herbivores obtain their vitamin B$_{12}$ from the bacteria that inhabit their gut (in fact, some animals, such as rabbits, must periodically eat some of their feces to obtain sufficient amounts of this essential substance). Humans, however, obtain almost all their vitamin

B$_{12}$ directly from their diet, particularly from meat. In the intestine, the glycoprotein **intrinsic factor,** which is secreted by the stomach, specifically binds vitamin B$_{12}$, and the protein–vitamin complex is absorbed via a receptor in the intestinal mucosa. The complex dissociates and the liberated vitamin B$_{12}$ is transported to the bloodstream. At least three different plasma proteins, called **transcobalamins,** bind the vitamin and facilitate its uptake by the tissues.

Pernicious anemia is not usually a dietary deficiency disease but, rather, results from insufficient secretion of intrinsic factor. The normal human requirement for cobalamin is very small, ~3 μg · day^{-1}, and the liver stores a 3- to 5-year supply of this vitamin. This accounts for the insidious onset of pernicious anemia and the fact that true dietary deficiency of vitamin B$_{12}$, even among strict vegetarians, is extremely rare.

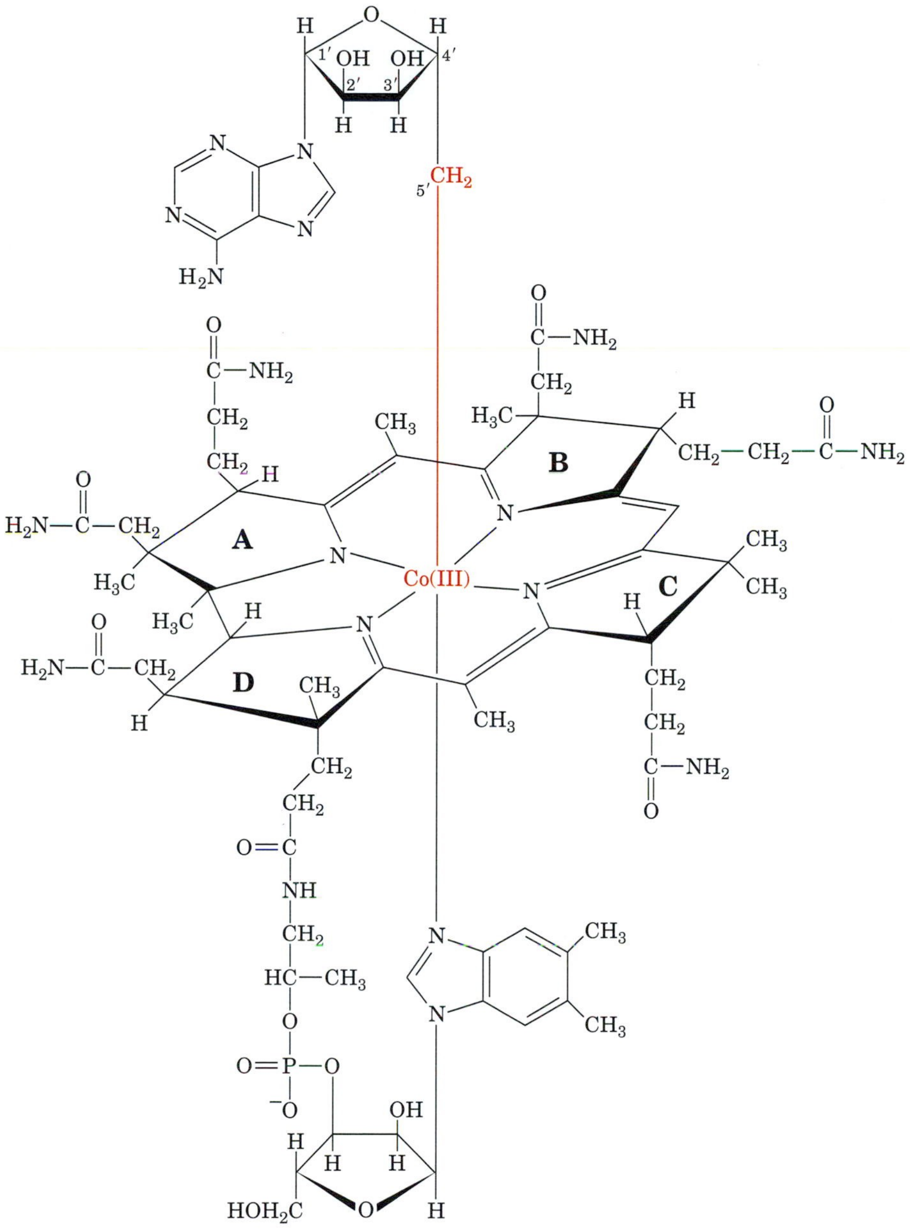

5'-Deoxyadenosylcobalamin (coenzyme B₁₂)

Figure 19-14. **The structure of 5'-deoxyadenosylcobalamin.**

amin-dependent enzymes, which catalyze molecular rearrangements or methyl-group transfer reactions.

Methylmalonyl-CoA Mutase Stabilizes and Protects Free Radical Intermediates

The proposed methylmalonyl-CoA mutase reaction mechanism (Fig. 19-15) begins with the **homolytic cleavage** of the cobalamin C—Co bond (the C and Co atoms each acquire one of the electrons that formed the cleaved electron pair bond). The Co ion therefore alternates between its Co(III) and Co(II) oxidation states and hence *functions as a reversible free radical generator*. The C—Co(III) bond is well suited to this function because it is inherently weak (dissociation energy 109 kJ·mol⁻¹) and is fur-

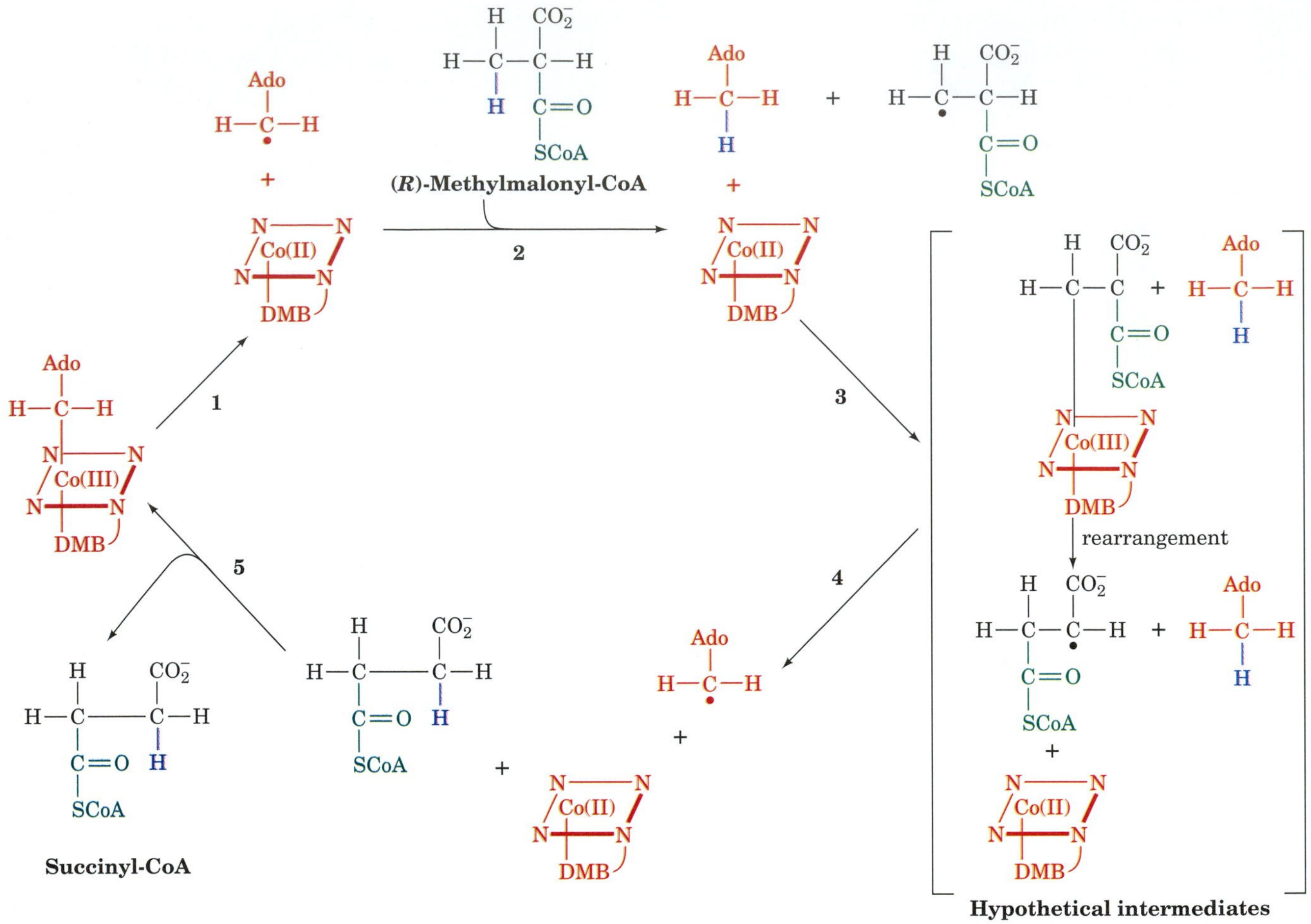

Figure 19-15.   The proposed mechanism of methylmalonyl-CoA mutase. (1) The homolytic cleavage of the C—Co(III) bond yields a 5′-deoxyadenosyl (Ado) radical and cobalamin in its Co(II) oxidation state. (2) The 5′-deoxyadenosyl radical abstracts a hydrogen atom from methylmalonyl-CoA, thereby generating a methylmalonyl-CoA radical. (3) A C—Co bond forms between the methylmalonyl-CoA radical and the coenzyme, followed by carbon skeleton rearrangement to form a succinyl-CoA radical. (4) The succinyl-CoA radical abstracts a hydrogen atom from 5′-deoxyadenosine to regenerate the 5′-deoxyadenosyl radical. (5) The release of succinyl-CoA reforms the coenzyme.

ther weakened through steric interactions with the enzyme (see below). Note that a homolytic cleavage reaction is unusual in biology; most other biological bond-cleavage reactions occur via **heterolytic cleavage** (in which the electron pair forming the cleaved bond is fully acquired by one of the separating atoms). The resulting 5′-deoxyadenosyl radical abstracts a hydrogen atom from the methylmalonyl-CoA substrate, which subsequently rearranges to yield succinyl-CoA.

The X-ray structure of methylmalonyl-CoA mutase from *Propionibacterium shermanii* in complex with the partial substrate **desulfo-CoA** (which lacks the substrate's S atom and its methylmalonyl group) was determined by Philip Evans. Coenzyme B_{12} is packed against the bottom of an α/β barrel. In free coenzyme B_{12}, the Co atom is axially liganded by the 5′ CH_2 group of its adenosyl residue and by an N atom of the DMB group (Fig. 19-14). However, in the enzyme, the DMB group has been replaced by a protein His side chain, and the adenosyl residue is not visible (Fig. 19-16). Spectroscopic measurements indicate that the Co atom in methylmalonyl-

CoA mutase is in the Co(II) state, thereby confirming that it has no sixth ligand (as occurs during its catalytic cycle; Fig. 19-15). Protein-induced strain makes the His N—Co bond extremely long (2.5 Å versus 1.9–2.0 Å in various other B_{12}-containing structures), stabilizing the Co(II) species relative to Co(III) and thereby favoring the formation of the adenosyl radical.

The desulfo-CoA binds to the enzyme with its pantetheine chain extended along a narrow tunnel down the center of the α/β barrel, which would put the methylmalonyl group of an intact substrate in close proximity to the top (unliganded) face of the cobalamin ring (Fig. 19-16). This tunnel provides the only direct access to the active site cavity, which protects the highly reactive free radical reaction intermediates from side reactions. The tunnel is lined by small hydrophilic residues (Ser and Thr), in contrast to the ~60 other known α/β barrel-containing enzymes in which the center of the barrel is occluded by large, often branched, hydrophobic side chains.

Succinyl-CoA Is Not Directly Consumed by the Citric Acid Cycle

Methylmalonyl-CoA mutase catalyzes the conversion of a metabolite to a citric acid cycle intermediate other than acetyl-CoA. However, such C_4 intermediates are actually catalysts, not substrates, of the citric acid cycle. *In order for succinyl-CoA to undergo net oxidation by the citric acid cycle, it must first be converted to pyruvate and thence to acetyl-CoA.* This is accomplished by converting succinyl-CoA to malate (Reactions 5–7 of the citric acid cycle; Fig. 16-2) followed by the oxidative decarboxylation of malate to pyruvate and CO_2 by **malic enzyme**

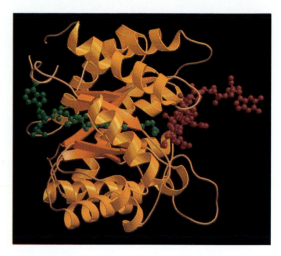

Figure 19-16. The X-ray structure of the N-terminal domain of the catalytically active α subunit of *P. shermanii* methylmalonyl-CoA mutase. The α/β barrel (*yellow*) is drawn in ribbon form and is shown in side view; its bound desulfo-CoA (*green*) and coenzyme B_{12} (*brown*) are shown in ball-and-stick form. [Courtesy of Philip Evans, MRC Laboratory of Molecular Biology, Cambridge, U.K.] ● See the Interactive Exercises.

$$\text{Malate} \xrightarrow[\text{NADP}^+ \quad \text{NADPH}]{\text{H}^+ +} \text{Pyruvate} + CO_2$$

Malate   **Pyruvate**

(this enzyme also functions in the C_4 cycle of photosynthesis; Fig. 18-26). Pyruvate is then completely oxidized via pyruvate dehydrogenase and the citric acid cycle.

F. Peroxisomal β Oxidation

The β oxidation of fatty acids occurs in the peroxisome as well as in the mitochondrion. *Peroxisomal β oxidation in animals shortens very long chain fatty acids (>22 C atoms), which are then fully degraded by the mitochondrial β-oxidation system.* In plants, fatty acid oxidation occurs exclusively in the peroxisomes and glyoxysomes (which are specialized peroxisomes).

Very long chain fatty acids diffuse into the peroxisome (no carnitine is required) and are activated by a long chain acyl-CoA synthetase. Peroxisomal β oxidation results in the same chemical changes to fatty acids as in the mitochondrial pathway but requires only three enzymes:

1. Acyl-CoA oxidase catalyzes the reaction

$$\text{Fatty acyl-CoA} + O_2 \longrightarrow \textit{trans-}\Delta^2\text{-enoyl-CoA} + H_2O_2$$

This enzyme uses an FAD cofactor, but the abstracted electrons are transferred directly to O_2 rather than passing through the electron-transport chain with its concomitant oxidative phosphorylation (Fig.

19-9, Reactions 5–8). Peroxisomal fatty acid oxidation therefore generates two fewer ATP per C_2 cycle than mitochondrial fatty acid oxidation. Catalase converts the H_2O_2 produced in the oxidase reaction to $H_2O + O_2$.

2. Peroxisomal enoyl-CoA hydratase and 3-L-hydroxyacyl-CoA dehydrogenase activities occur on a single polypeptide. The reactions catalyzed are identical to those of the mitochondrial system (Fig. 19-9, Reactions 2 and 3).

3. Peroxisomal thiolase catalyzes the final step of oxidation. This enzyme is almost inactive with acyl-CoAs of length C_8 or less, so peroxisomes incompletely oxidize fatty acids.

The peroxisome contains both a carnitine acetyltransferase and a transferase specific for longer chain acyl groups. Acyl-CoAs that have been chain-shortened by peroxisomal β oxidation are thereby converted to their carnitine esters. These substances, for the most part, passively diffuse out of the peroxisome to the mitochondrion, where they are oxidized further.

3. KETONE BODIES

The acetyl-CoA produced by oxidation of fatty acids can be further oxidized via the citric acid cycle. In liver mitochondria, however, a significant fraction of this acetyl-CoA has another fate. By a process known as **ketogenesis,** acetyl-CoA is converted to **acetoacetate** or D-β-hydroxybutyrate. These compounds together with acetone are somewhat inaccurately referred to as **ketone bodies:**

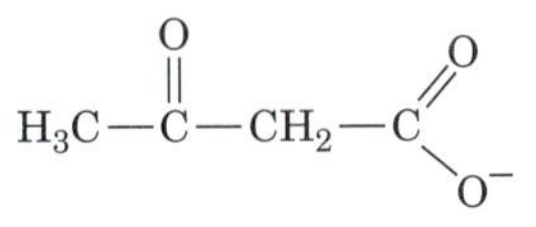

Acetoacetate 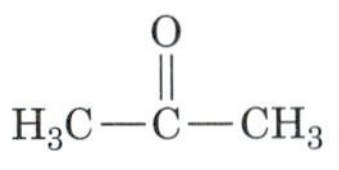   Acetone 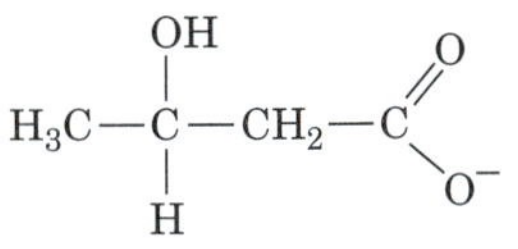   D-β-Hydroxybutyrate

Ketone bodies are important metabolic fuels for many peripheral tissues, particularly heart and skeletal muscle. The brain, under normal circumstances, uses only glucose as its energy source (fatty acids are unable to pass through the blood–brain barrier), but during starvation, the small, water-soluble ketone bodies become the brain's major fuel source (Section 21-4A).

Acetoacetate formation occurs in three reactions (Fig. 19-17):

1. Two molecules of acetyl-CoA are condensed to **acetoacetyl-CoA** by thiolase (also called **acetyl-CoA acetyltransferase**) working in the reverse direction from the way it does in the final step of β oxidation (Fig. 19-9, Reaction 4).

2. Condensation of the acetoacetyl-CoA with a third acetyl-CoA by **HMG-CoA synthase** forms **β-hydroxy-β-methylglutaryl-CoA (HMG-CoA).** The mechanism of this reaction resembles the reverse of the thiolase reaction (Fig. 19-11) in that an active site thiol group forms an acyl–thioester intermediate.

3. HMG-CoA is degraded to acetoacetate and acetyl-CoA in a mixed aldol–Claisen ester cleavage by **HMG-CoA lyase.** The mechanism of this reaction is analogous to the reverse of the citrate synthase reaction (Fig. 16-9). HMG-CoA is also a precursor in cholesterol biosynthesis (Section 19-7A).

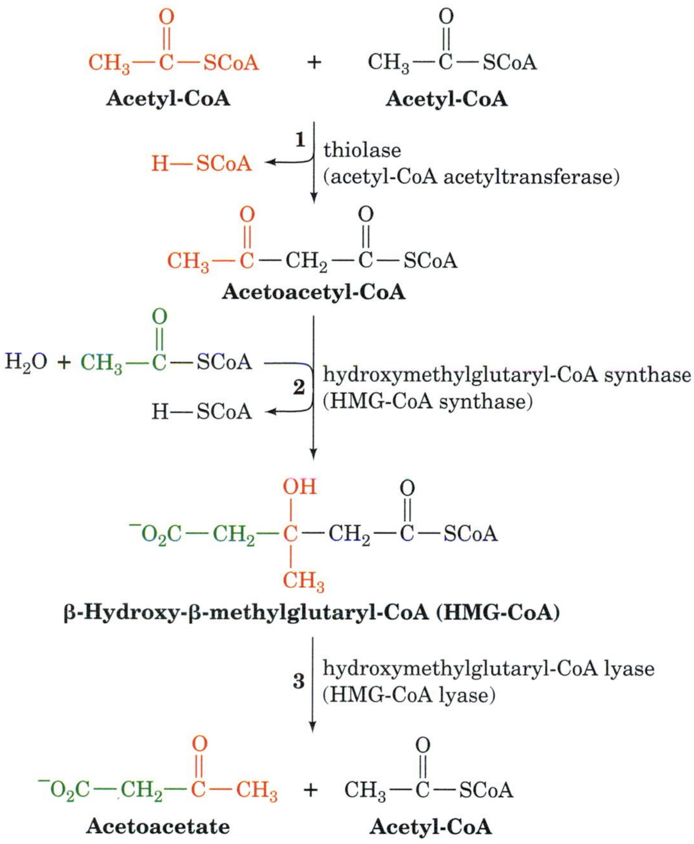

Figure 19-17. Ketogenesis. Acetoacetate is formed from acetyl-CoA in three steps. (1) Two molecules of acetyl-CoA condense to form acetoacetyl-CoA. (2) A Claisen ester condensation of the acetoacetyl-CoA with a third acetyl-CoA forms β-hydroxy-β-methylglutaryl-CoA (HMG-CoA). (3) HMG-CoA is degraded to acetoacetate and acetyl-CoA in a mixed aldol–Claisen ester cleavage.

Acetoacetate may be reduced to D-β-hydroxybutyrate by **β-hydroxybutyrate dehydrogenase:**

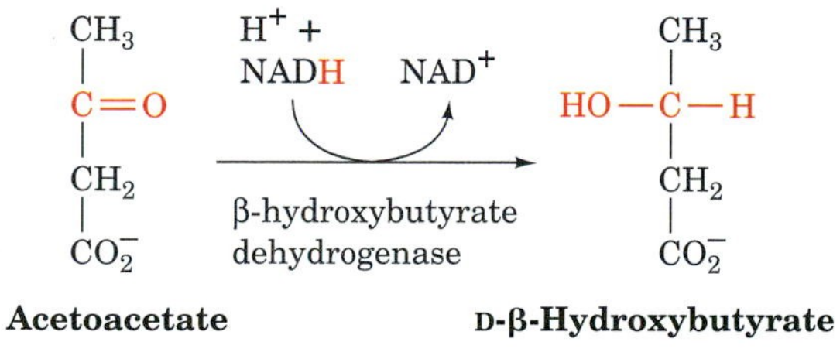

Acetoacetate, a β-keto acid, also undergoes relatively facile nonenzymatic decarboxylation to acetone and CO_2. Indeed, in individuals with **ketosis,** a pathological condition in which acetoacetate is produced faster than it can be metabolized (a symptom of diabetes; Section 21-4B), the breath has the characteristic sweet smell of acetone.

The liver releases acetoacetate and β-hydroxybutyrate, which are carried by the bloodstream to the peripheral tissues for use as alternative fuels. There, these products are converted to two acetyl-CoA as is diagrammed in Fig. 19-18. Succinyl-CoA, which acts as the CoA donor in this process, can also be converted to succinate with the coupled synthesis of GTP in the succinyl-CoA synthetase reaction of the citric acid cycle (Sec-

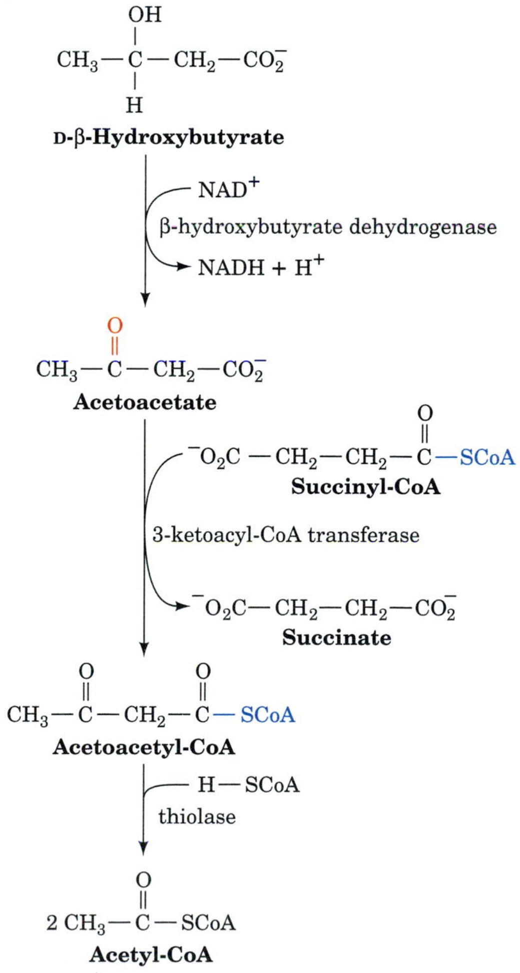

Figure 19-18. The metabolic conversion of ketone bodies to acetyl-CoA.

tion 16-3E). The "activation" of acetoacetate bypasses this step and therefore "costs" the free energy of GTP hydrolysis.

4. FATTY ACID BIOSYNTHESIS

Fatty acid biosynthesis occurs through condensation of C_2 units, the reverse of the β-oxidation process. Through isotopic labeling techniques, David Rittenberg and Konrad Bloch demonstrated, in 1945, that these condensation units are derived from acetic acid. Subsequent research showed that both acetyl-CoA and bicarbonate are required, and that a C_3 unit, **malonyl-CoA,** is an intermediate of fatty acid biosynthesis.

The pathway of fatty acid synthesis differs from that of fatty acid oxidation. This situation, as we saw in Section 15-3, is typical of opposing biosynthetic and degradative pathways because it permits them both to be thermodynamically favorable and independently regulated under similar physiological conditions. Figure 19-19 outlines fatty acid oxidation and synthesis with emphasis on the differences between these pathways, including the cellular locations of the pathways, the redox coenzymes, and the manner in which C_2 units are removed or added to the fatty acyl chain.

A. Transport of Mitochondrial Acetyl-CoA into the Cytosol

Acetyl-CoA, the starting material for fatty acid synthesis, is generated in the mitochondrion by the oxidative decarboxylation of pyruvate as catalyzed by pyruvate dehydrogenase (Section 16-2B) as well as by the oxidation of fatty acids. When the demand for ATP is low, so that the oxida-

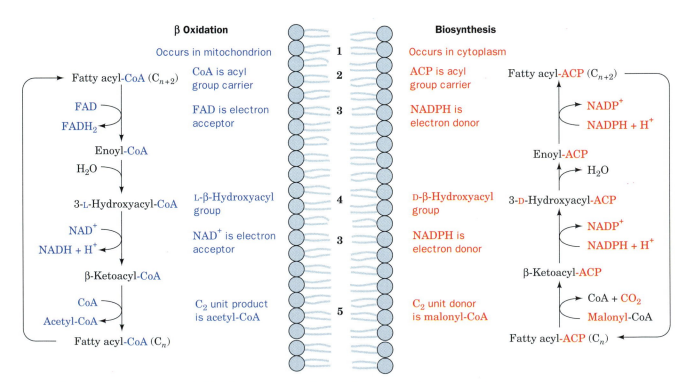

Figure 19-19.   A comparison of fatty acid β oxidation and fatty acid biosynthesis. Differences occur in (1) cellular location; (2) acyl group carrier; (3) electron acceptor/donor; (4) stereochemistry of the hydration/dehydration reaction; and (5) the form in which C_2 units are produced/donated.
✳ **See the Animated Figures.**

tion of acetyl-CoA via the citric acid cycle and oxidative phosphorylation is minimal, this mitochondrial acetyl-CoA may be stored for future use as fat. Fatty acid biosynthesis occurs in the cytosol, however, and the mitochondrial membrane is essentially impermeable to acetyl-CoA. *Acetyl-CoA enters the cytosol in the form of citrate via the **tricarboxylate transport system*** (Fig. 19-20). **ATP-citrate lyase** then catalyzes the reaction

$$\text{Citrate} + \text{CoA} + \text{ATP} \longrightarrow \text{acetyl-CoA} + \text{oxaloacetate} + \text{ADP} + \text{P}_i$$

which resembles the reverse of the citrate synthase reaction (Fig. 16-9) except that ATP hydrolysis is required to drive the synthesis of the thioester bond. Oxaloacetate is then reduced to malate by malate dehydrogenase. Malate is oxidatively decarboxylated to pyruvate by malic enzyme and returned in this form to the mitochondrion. This reaction involves the reoxidation of malate to oxaloacetate, a β-keto acid, which is then decarboxy-

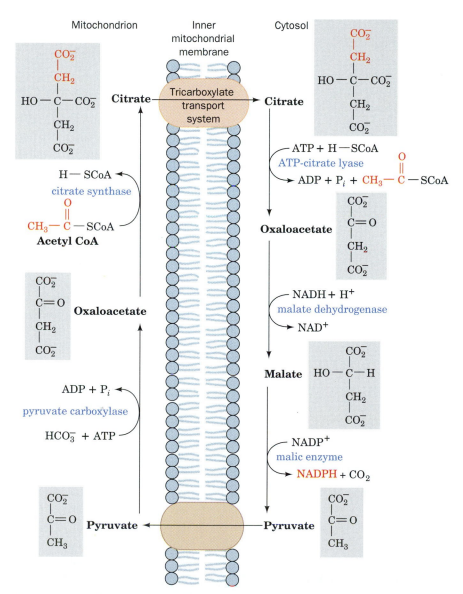

Figure 19-20. The tricarboxylate transport system. This sequence of reactions transfers acetyl-CoA from the mitochondrion to the cytosol.

lated, a reaction reminiscent of the isocitrate dehydrogenase reaction in the citric acid cycle (Section 16-3C). The NADPH produced is used in the reductive reactions of fatty acid biosynthesis.

B. Acetyl-CoA Carboxylase

Acetyl-CoA carboxylase catalyzes the first committed step of fatty acid biosynthesis and one of its rate-controlling steps. The mechanism of this biotin-dependent enzyme is similar to those of propionyl-CoA carboxylase (Section 19-2E) and pyruvate carboxylase (Fig. 15-25). The reaction occurs in two steps, a CO_2 activation and a carboxylation:

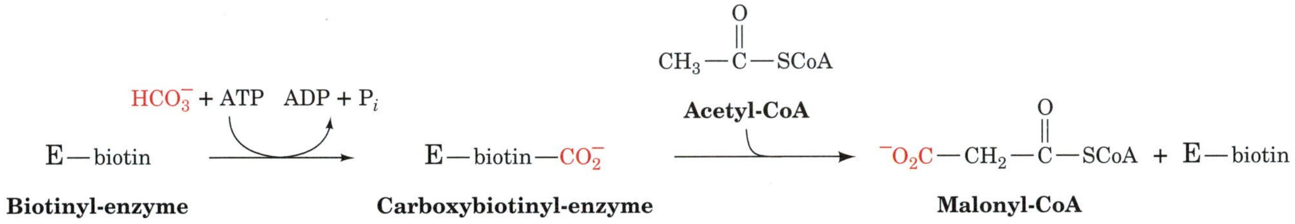

$$
\text{E—biotin} \xrightarrow[\text{HCO}_3^- + \text{ATP} \quad \text{ADP} + P_i]{} \text{E—biotin—CO}_2^- \xrightarrow[\text{Acetyl-CoA: CH}_3\text{—C(=O)—SCoA}]{} {}^-\text{O}_2\text{C—CH}_2\text{—C(=O)—SCoA} + \text{E—biotin}
$$

Biotinyl-enzyme   **Carboxybiotinyl-enzyme**   **Malonyl-CoA**

The result is a three-carbon (malonyl) group linked as a thioester to CoA.

Mammalian acetyl-CoA carboxylase, a 230-kD polypeptide, is subject to allosteric and hormonal control. For example, citrate stimulates acetyl-CoA carboxylase, possibly by reorienting its biotin prosthetic group so as to increase V_{max}. Long chain fatty acyl-CoAs are feedback inhibitors of the enzyme. Fine-tuning of enzyme activity is accomplished through covalent modification. Acetyl-CoA carboxylase is a substrate for several kinases. It has six phosphorylation sites, but phosphorylation of only one (Ser 79) is clearly associated with enzyme inactivation. Ser 79 is phosphorylated by **AMP-dependent protein kinase** in a cAMP-independent pathway. However, glucagon as well as epinephrine, which act through cAMP-dependent protein kinase (cAPK; Section 15-3C), promote the phosphorylation of Ser 79, possibly by inhibiting its dephosphorylation (recall that this occurs in glycogen metabolism when the cAPK-mediated phosphorylation of phosphoprotein phosphatase inhibitor-1 inhibits dephosphorylation). Insulin, on the other hand, stimulates dephosphorylation of acetyl-CoA carboxylase and thereby activates the enzyme.

The *E. coli* acetyl-CoA carboxylase, which is a multisubunit protein, is regulated by guanine nucleotides so that fatty acid synthesis is coordinated

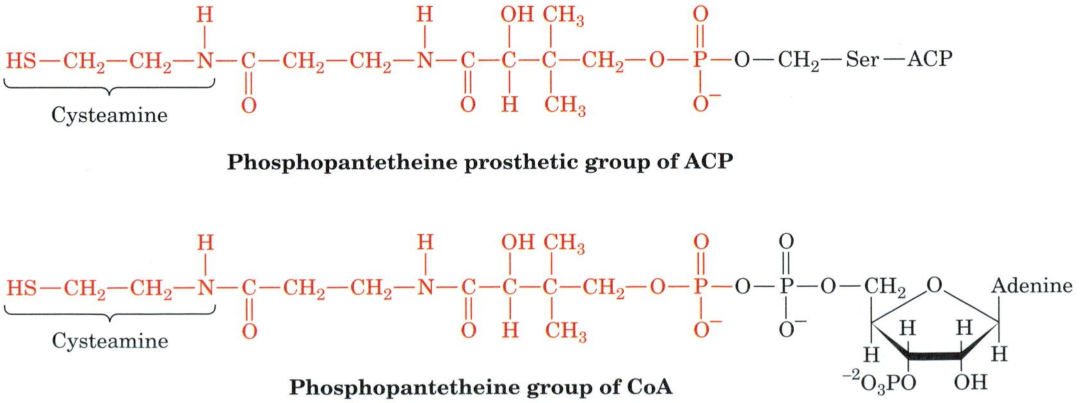

Phosphopantetheine prosthetic group of ACP

Phosphopantetheine group of CoA

Figure 19-21. **The phosphopantetheine group in acyl-carrier protein (ACP) and in CoA.**

with cell growth. In prokaryotes, fatty acids serve primarily as phospholipid precursors, since these organisms do not synthesize triacylglycerols for energy storage.

C. Fatty Acid Synthase

The synthesis of fatty acids, mainly palmitic acid, from acetyl-CoA and malonyl-CoA involves seven enzymatic reactions. These reactions were first studied in cell-free extracts of *E. coli*, in which they are catalyzed by independent enzymes. Individual enzymes with these activities also occur in chloroplasts (plant fatty acid synthesis does not occur in the cytosol). In yeast, **fatty acid synthase** is a cytosolic, 2500-kD multifunctional enzyme with the composition $\alpha_6\beta_6$, whereas in animals it is a 534-kD multifunctional enzyme consisting of two identical polypeptide chains. Presumably, such proteins evolved by the joining of previously independent enzymes.

Although fatty acid synthesis begins with the synthesis of a CoA ester, malonyl-CoA, the growing fatty acid is anchored to **acyl-carrier protein** (**ACP**; Fig. 19-21). ACP, like CoA, contains a phosphopantetheine group that forms a thioester with an acyl group. The phosphopantetheine phosphoryl group is esterified with a Ser OH group of ACP, whereas in CoA it is linked to AMP. In *E. coli*, ACP is a 10-kD polypeptide, whereas in animals it is part of the multifunctional fatty acid synthase.

The reactions catalyzed by mammalian fatty acid synthase are diagrammed in Fig. 19-22:

1 & 2. These are priming reactions in which the synthase is "loaded" with the condensation reaction precursors: An acetyl group originally linked as a thioester in acetyl-CoA is transferred first to ACP (**1**) and then to an enzyme Cys residue (**2a**); similarly, a malonyl group is transferred from malonyl-CoA to malonyl-ACP (**2b**).

3. In the condensation reaction, the malonyl-ACP is decarboxylated with the resulting carbanion attacking the acetyl-thioester to form a four-carbon β-ketoacyl-ACP. The decarboxylation reaction drives the condensation reaction.

4–6. Two reductions and a dehydration convert the β-keto group to an alkyl group. The coenzyme in both reductive steps is NADPH. In β oxidation, the analogs of Reactions 4 and 6, respectively, use NAD^+ and FAD (Fig. 19-9, Reactions 3 and 1). Moreover, Reaction 5 requires a D-β-hydroxyacyl substrate, whereas the analogous reaction in β oxidation forms the corresponding L isomer.

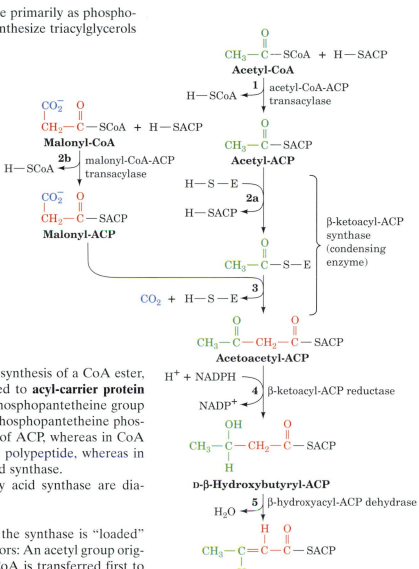

Figure 19-22. *Key to Metabolism.* **The reaction sequence for the biosynthesis of fatty acids.** In forming palmitate, the pathway is repeated for seven cycles of C_2 elongation followed by a final hydrolysis step. ✳ See the Animated Figures.

At this point, the acyl group, originally an acetyl group, has been elongated by a C_2 unit. This butyryl group is then transferred from ACP to the Cys-SH of the enzyme (a repeat of Reaction 2a) so that it can be extended by additional rounds of the fatty acid synthase reaction sequence. Note that the malonyl-CoA synthesized by the acetyl-CoA carboxylase reaction is decarboxylated in the condensation reaction. The formation of a C—C bond is an endergonic process requiring an activated precursor. Malonyl-CoA is a β-keto ester whose exergonic decarboxylation yields the acetyl-CoA carbanion required for C—C bond formation. The free energy required for the overall reaction is supplied by the ATP hydrolysis in the acetyl-CoA carboxylase reaction, which generates the malonyl-CoA. This carboxylation–decarboxylation sequence is similar to the activation of pyruvate to oxaloacetate for conversion to phosphoenolpyruvate in gluconeogenesis (Section 15-4A).

In each reaction cycle, the ACP is "reloaded" with a malonyl group, and the acyl chain grows by two carbon atoms. Seven such cycles are required to form palmitoyl-ACP. The thioester bond is then hydrolyzed by **palmitoyl thioesterase** (Fig. 19-22, Reaction 7), yielding palmitate, the normal product of the fatty acid synthase pathway, and regenerating the enzyme for a new round of synthesis.

The stoichiometry of palmitate synthesis is therefore

$$\text{Acetyl-CoA} + 7\text{ malonyl-CoA} + 14\text{ NADPH} + 14\text{ H}^+ \longrightarrow$$
$$\text{palmitate} + 7\text{ CO}_2 + 14\text{ NADP}^+ + 8\text{ CoA} + 6\text{ H}_2\text{O}$$

Since the 7 malonyl-CoA are derived from acetyl-CoA as follows:

$$7\text{ Acetyl-CoA} + 7\text{ CO}_2 + 7\text{ ATP} \longrightarrow$$
$$7\text{ malonyl-CoA} + 7\text{ ADP} + 7\text{ P}_i + 7\text{ H}^+$$

the overall stoichiometry for palmitate biosynthesis is

$$8\text{ Acetyl-CoA} + 14\text{ NADPH} + 7\text{ ATP} + 7\text{ H}^+ \longrightarrow$$
$$\text{palmitate} + 14\text{ NADP}^+ + 8\text{ CoA} + 6\text{ H}_2\text{O} + 7\text{ ADP} + 7\text{ P}_i$$

The Two Halves of Fatty Acid Synthase Operate in Concert

All the enzyme activities in fatty acid synthase except those catalyzing Reactions 2a and 4 of Fig. 19-22 remain functional when the native dimeric enzyme is dissociated into monomers. Electron microscopy and limited proteolysis of fatty acid synthase provide evidence that *contiguous stretches of its polypeptide chain fold to form a series of autonomous domains, each with a different catalytic activity*. Several other enzymes exhibit similar multifunctionality, but none has as many separate catalytic activities as does animal fatty acid synthase.

Since the condensation reaction requires the juxtaposition of the sulfhydryl groups of the ACP phosphopantetheine and an enzyme Cys residue, it has been proposed that these groups are on separate subunits that interact in a head-to-tail manner (Fig. 19-23). Thus, a Cys-linked acetyl group on one subunit is positioned for reaction near the ACP-linked malonyl group on the other subunit. The long, flexible phosphopantetheine chain of ACP is thought to transport the substrate between the various catalytic sites on each subunit. The flexible lipoyllysyl arms of the pyruvate dehydrogenase complex (Section 16-2B) perform a similar function in that enzyme. Because the fatty acid synthase dimer is symmetrical, two fatty acids can be synthesized simultaneously.

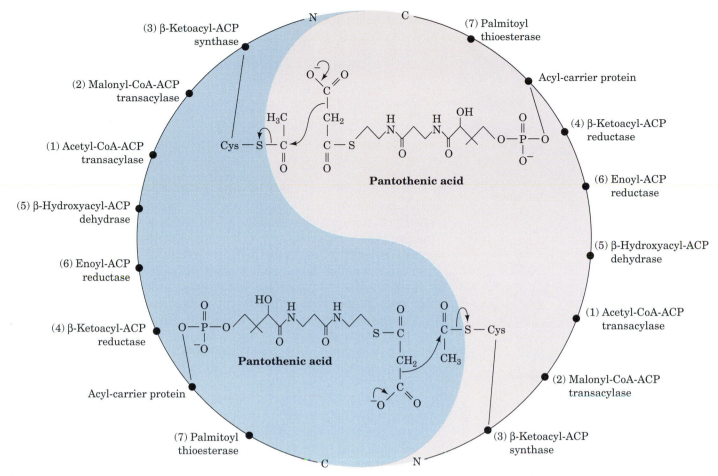

Figure 19-23. Schematic representation of animal fatty acid synthase. Two multifunctional subunits, in head-to-tail association, form the active dimer. The numbers refer to the reactions in Fig. 19-22. Acyl-carrier protein (ACP) and the essential Cys residue are shown in their acylated forms undergoing the condensation reaction catalyzed by β-ketoacyl-ACP synthase. [After Wakil, S.J., Stoops, J.K., and Joshi, V.C., *Annu. Rev. Biochem.* **52**, 556 (1983).]

D. Elongases and Desaturases

Palmitate, a saturated C_{16} fatty acid, is converted to longer chain saturated and unsaturated fatty acids through the actions of **elongases** and **desaturases.** Elongases are present in both the mitochondria and the endoplasmic reticulum, but the mechanisms of elongation at the two sites differ. Mitochondrial elongation (a process independent of the fatty acid synthase pathway in the cytosol) occurs by the successive addition and reduction of acetyl units in a reversal of fatty acid oxidation; the only chemical difference between these two pathways occurs in the final reduction step in which NADPH takes the place of $FADH_2$ as the terminal redox coenzyme (Fig. 19-24). Elongation in the endoplasmic reticulum involves the successive condensations of malonyl-CoA with acyl-CoA. These reactions are each followed by NADPH-dependent reductions similar to those catalyzed by fatty acid synthase, the only difference being that the fatty acid is elongated as its CoA derivative rather than as its ACP derivative.

Unsaturated fatty acids are produced by **terminal desaturases.** Mammalian systems contain four terminal desaturases of broad chain-length specificities designated Δ^9-, Δ^6-, Δ^5-, and Δ^4-**fatty acyl-CoA desaturases.**

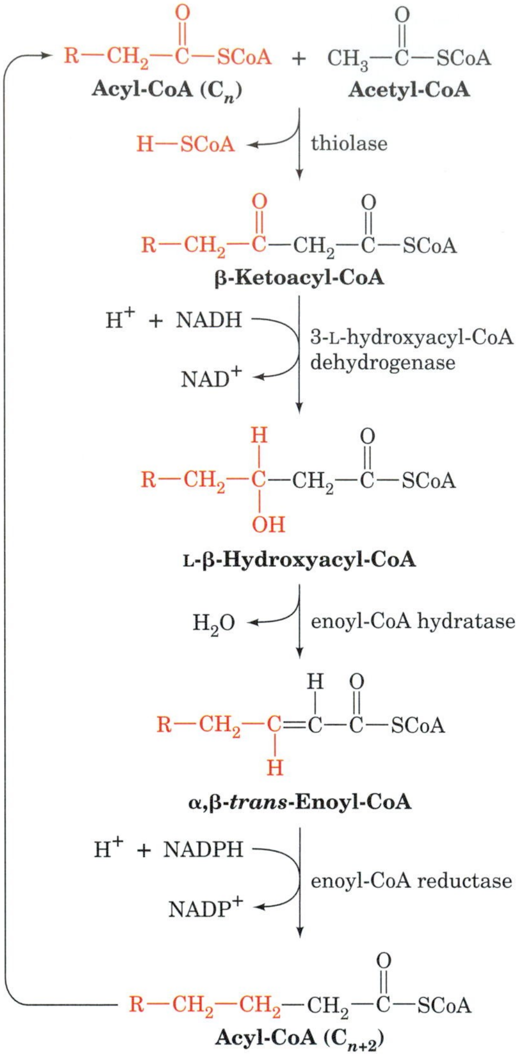

Figure 19-24. **Mitochondrial fatty acid elongation.** This process is the reverse of fatty acid oxidation (Fig. 19-9) except that the final reaction employs NADPH rather than FADH$_2$ as its redox coenzyme.

These nonheme iron–containing enzymes catalyze the general reaction

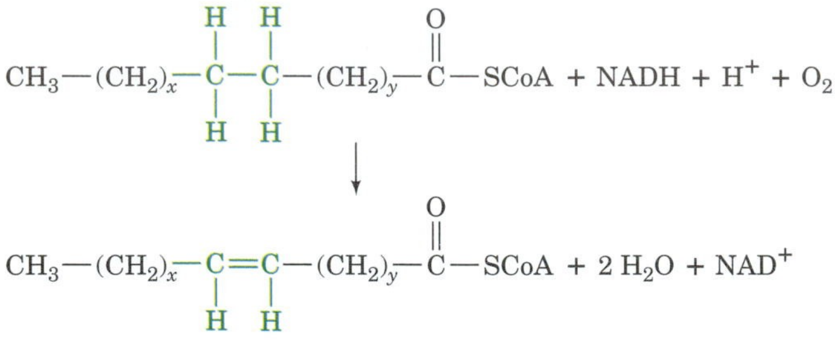

where $x \geq 5$ and where $(CH_2)_x$ can contain one or more double bonds. The $(CH_2)_y$ portion of the substrate is always saturated. Double bonds are inserted between existing double bonds in the $(CH_2)_x$ portion of the substrate and the CoA group such that the new double bond is three carbon atoms closer to the CoA group than the next double bond (not conjugated to an existing double bond) and, in animals, never at positions beyond C9.

A variety of unsaturated fatty acids can be synthesized by combinations of elongation and desaturation reactions. However, since palmitic acid is the shortest available fatty acid in animals, the above rules preclude the formation of the Δ^{12} double bond of linoleic acid ($\Delta^{9,12}$-octadecadienoic acid), a required precursor of **prostaglandins** (Section 9-1F). *Linoleic acid must consequently be obtained in the diet (ultimately from plants that have Δ^{12}- and Δ^{15}-desaturases) and is therefore an **essential fatty acid.*** Indeed, animals maintained on a fat-free diet develop an ultimately fatal condition that is initially characterized by poor growth, poor wound healing, and dermatitis. Linoleic acid is also an important constituent of epidermal sphingolipids that function as the skin's water permeability barrier: Animals deprived of linoleic acid must drink far more water than those with an adequate diet.

E. Synthesis of Triacylglycerols

Triacylglycerols are synthesized from fatty acyl-CoA esters and glycerol-3-phosphate or dihydroxyacetone phosphate (Fig. 19-25). The initial step in this process is catalyzed either by **glycerol-3-phosphate acyltransferase** in mitochondria and the endoplasmic reticulum, or by **dihydroxyacetone phosphate acyltransferase** in the endoplasmic reticulum or peroxisomes. In the latter case, the product acyl-dihydroxyacetone phosphate is reduced to the corresponding **lysophosphatidic acid** by an NADPH-dependent reductase. The lysophosphatidic acid is converted to a triacylglycerol by the successive actions of **1-acylglycerol-3-phosphate acyltransferase, phosphatidic acid phosphatase,** and **diacylglycerol acyltransferase.** The intermediate phosphatidic acid and 1,2-diacylglycerol can also be converted to phospholipids by the pathways described in Section 19-6A. The acyltransferases are not completely specific for particular fatty acyl-CoAs, either in chain length or degree of unsaturation, but in human adipose tissue triacylglycerols, palmitate tends to be concentrated at position 1 and oleate at position 2.

At this point, we can appreciate how triacylglycerols synthesized from fatty acids built from two-carbon acetyl units can be broken back down into acetyl units. In the liver, the resulting acetyl-CoA may be shunted to the formation of ketone bodies and later converted back to acetyl-CoA by another tissue. The acetyl-CoA can then either be used to build fatty acids that are stored as triacylglycerols or be oxidized by the citric acid cycle to release considerable ATP by oxidative phosphorylation. As we shall see,

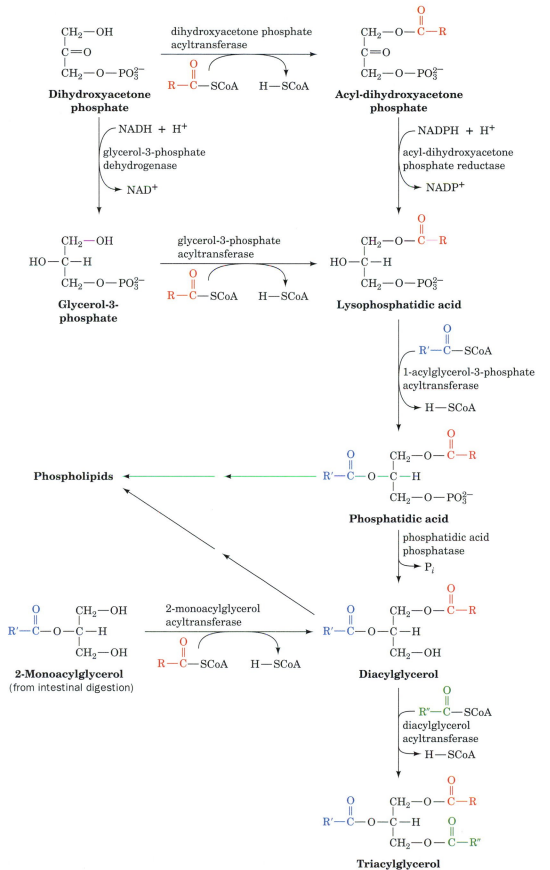

Figure 19-25. **The reactions of triacylglycerol biosynthesis.**

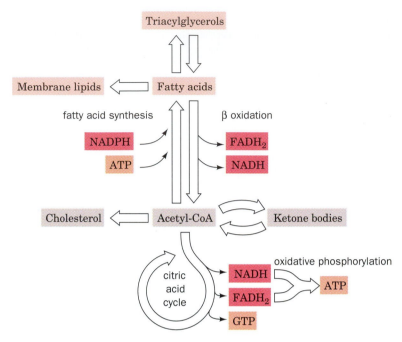

Figure 19-26.   A summary of lipid metabolism.

the flux of material in the direction of triacylglycerol synthesis or triacylglycerol degradation depends on the metabolic energy needs of the organism and the need for synthesis of other compounds, such as membrane lipids (Section 19-6) and cholesterol (Section 19-7A). These key features of lipid metabolism are summarized in Fig. 19-26.

5. REGULATION OF FATTY ACID METABOLISM

Discussions of metabolic control are usually concerned with the regulation of metabolite flow through a pathway in response to the differing energy needs and dietary states of an organism. In mammals, glycogen and triacylglycerols serve as primary fuels for energy-requiring processes and are synthesized in times of plenty for future use. In plants, starch and oils perform similar functions (see Box 19-2).

Synthesis and breakdown of glycogen and triacylglycerols are processes that concern the whole organism, with its organs and tissues forming an interdependent network connected by the bloodstream. The blood carries the metabolites responsible for energy production: triacylglycerols in the form of chylomicrons and VLDL, fatty acids as their albumin complexes, ketone bodies, amino acids, lactate, and glucose. The pancreatic α and β cells sense the organism's dietary and energetic state mainly through the glucose concentration in the blood. The α cells respond to low blood glucose concentrations (e.g., the fasting state) by secreting glucagon, whereas the β cells respond to high blood glucose concentrations (e.g., the fed state) by secreting insulin. We have already seen how these hormones influence glycogen metabolism (Section 15-3C). *They also regulate the rates of the opposing pathways of lipid metabolism and thereby control whether fatty acids will be oxidized or synthesized.* The major control mechanisms are summarized in Fig. 19-27.

Fatty acid oxidation is regulated largely by the concentration of fatty acids in the blood, which is, in turn, controlled by the hydrolysis rate of tri-

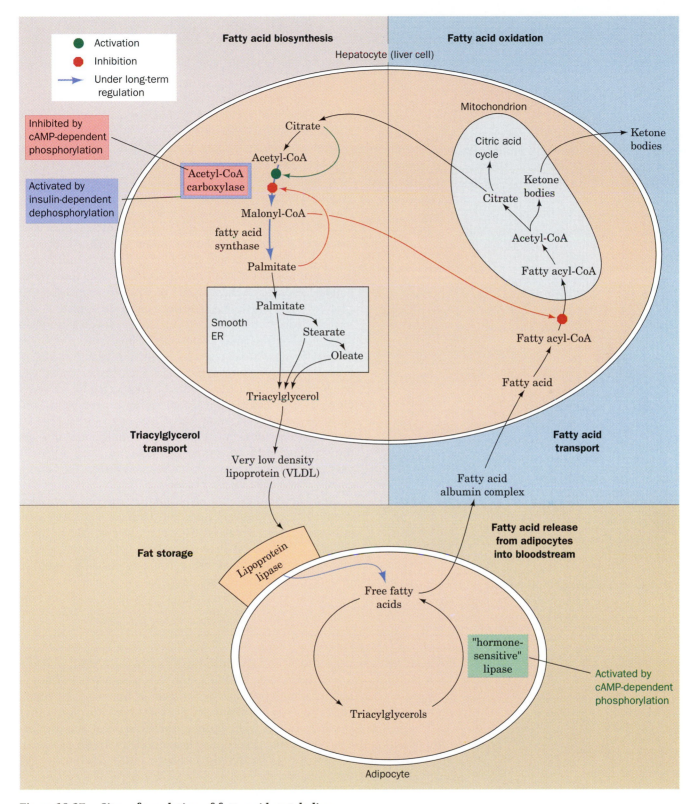

Figure 19-27.   Sites of regulation of fatty acid metabolism.

acylglycerols in adipose tissue by **hormone-sensitive triacylglycerol lipase.**
This enzyme is so named because it is susceptible to regulation by phosphorylation and dephosphorylation in response to hormonally controlled
cAMP levels. Glucagon, epinephrine, and norepinephrine increase adipose

Box 19-2

BIOCHEMISTRY IN CONTEXT

Plant Oils

Plants contain triacylglycerols that liquefy at lower temperatures than those in animal fats. Nevertheless, these oils, like fats, function as metabolic fuel reserves. In seeds, the proportion of oil to carbohydrate and protein varies widely; for example, it is 1–2% in wheat and ~60% in castor beans. Large amounts of lipid are stored not in a single droplet, as in animal fat cells (e.g., Fig. 9-2), but as ~1-μm-diameter **oil bodies** (the gray particles in the photo below; the clear areas are vacuoles, and the darker particles are various organelles).

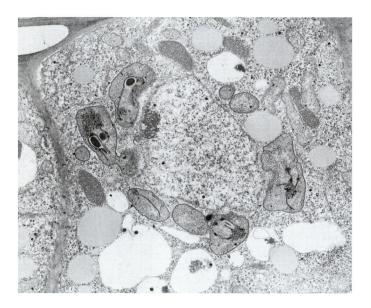

A single seed may contain over a thousand oil bodies, which are surrounded by a lipid monolayer containing proteins and hence are analogous to serum lipoproteins. During germination, lipases make the fatty acids available for β oxidation. The resulting acetyl-CoA can be further oxidized for energy via the citric acid cycle or converted to carbohydrate via the glyoxylate pathway (Section 16-5C).

Plant membranes contain relatively few types of fatty acids, but oil composition is far more variable in terms of acyl chain length, degree of unsaturation, and presence of functional groups such as hydroxy or epoxy groups. Because plant oils are widely used for cooking and industrial applications, the enzymes responsible for their synthesis have been the target of considerable study and manipulation. Approximately one-third of the world's harvest of vegetable oil is used in the manufacture of lubricants, plasticizers, and surfactants. Biological sources of these agents are often less expensive and certainly more renewable than petrochemical sources.

Knowledge of the enzymatic machinery involved in plant lipid metabolism makes it possible to manipulate the production of oils or to design novel oils. For example, a plant oil with improved nutritional value (i.e., higher content of polyunsaturated acyl chains) can be engineered by introducing a gene for yeast stearoyl-CoA desaturase into a plant. Similarly, the stability of an oil (its resistance to oxidation) can be increased by reducing its content of double bonds through the suppression of desaturases. In addition, such a genetic engineering approach may prove less costly than the current chemical hydrogenation methods.

Many developments in plant lipid engineering have used mammalian or yeast enzymes and familiar laboratory subjects such as tobacco plants. However, the cost-effective production of engineered plant oils will probably require greater knowledge of plant lipid metabolism and hosts that produce seeds with high oil content.

[Photo courtesy of Richard Trelease, Arizona State University.]

tissue cAMP concentrations. cAMP allosterically activates cAPK, which in turn phosphorylates certain enzymes. Phosphorylation activates hormone-sensitive lipase, thereby stimulating lipolysis in adipose tissue, raising blood fatty acid levels, and ultimately activating the β-oxidation pathway in other tissues such as liver and muscle. In liver, this process leads to the production of ketone bodies that are secreted into the bloodstream for use as an alternative fuel to glucose by peripheral tissues. cAPK also inactivates acetyl-CoA carboxylase (Section 19-4B), so *cAMP-dependent phosphorylation simultaneously stimulates fatty acid oxidation and inhibits fatty acid synthesis.*

Insulin has the opposite effect of glucagon and epinephrine: It stimulates the formation of glycogen and triacylglycerols. Insulin decreases cAMP levels, leading to the dephosphorylation and thus the inactivation of hormone-sensitive lipase. This reduces the amount of fatty acid available for oxidation. Insulin also activates acetyl-CoA carboxylase (Section

19-4B). *The glucagon:insulin ratio therefore determines the rate and direction of fatty acid metabolism.*

Another mechanism that inhibits fatty acid oxidation when fatty acid synthesis is stimulated is the inhibition of carnitine palmitoyl transferase I by malonyl-CoA. This inhibition keeps the newly synthesized fatty acids out of the mitochondria (Section 19-2B) and thus away from the β-oxidation system.

Factors such as substrate availability, allosteric interactions, and covalent modification (phosphorylation) regulate enzyme activity with response times of minutes or less. Such **short-term regulation** is complemented by **long-term regulation,** which requires hours or days and governs a pathway's regulatory enzyme by altering the amount of enzyme present. This is accomplished through changes in the rates of protein synthesis and/or breakdown.

The long-term regulation of lipid metabolism includes stimulation by insulin and inhibition by starvation of the synthesis of acetyl-CoA carboxylase and fatty acid synthase. The amount of adipose tissue lipoprotein lipase, the enzyme that initiates the entry of lipoprotein-packaged fatty acids into adipose tissue for storage (Section 19-1B), is also increased by insulin and decreased by starvation. Thus, an abundance of glucose, reflected in the level of insulin, promotes fatty acid synthesis and the storage of fatty acids by adipocytes, whereas starvation, when glucose is unavailable, decreases fatty acid synthesis and the uptake of fatty acids by adipocytes.

6. MEMBRANE LIPID SYNTHESIS

Most membrane lipids are dual-tailed amphipathic molecules composed of either 1,2-diacylglycerol or *N*-acylsphingosine (ceramide) linked to a polar head group that is either a carbohydrate or a phosphate ester (Fig. 19-28). In this section, we describe the biosynthesis of these complex lipids from their simpler components. These lipids are synthesized in membranes, mostly on the cytosolic face of the endoplasmic reticulum, and from there are transported in vesicles to their final cellular destinations (Section 10-2C).

A. *Glycerophospholipids*

Glycerophospholipids have significant asymmetry in their C1- and C2-linked fatty acyl groups: C1 substituents are mostly saturated fatty acids, whereas those at C2 are by and large unsaturated fatty acids.

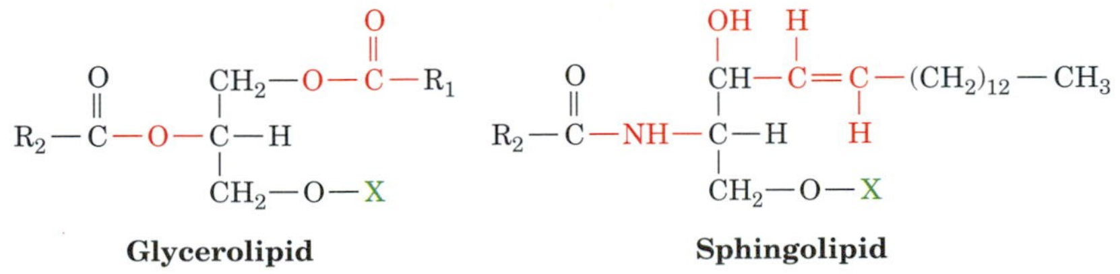

X = H	1,2-Diacylglycerol	*N*-Acylsphingosine (ceramide)
X = Carbohydrate	Glyceroglycolipid	Sphingoglycolipid (glycosphingolipid)
X = Phosphate ester	Glycerophospholipid	Sphingophospholipid

Figure 19-28. **The glycerolipids and sphingolipids.** The structures of the common head groups, X, are presented in Table 9-2. Plant membranes are particularly rich in glyceroglycolipids.

Biosynthesis of Diacylglycerophospholipids

The triacylglycerol precursors 1,2-diacylglycerol and phosphatidic acid are also the precursors of glycerophospholipids (Fig. 19-25). The polar head groups of glycerophospholipids are linked to C3 of the glycerol via a phosphodiester bond. In mammals, the head groups **ethanolamine** and **choline** are activated before being attached to the lipid (Fig. 19-29):

1. ATP phosphorylates the OH group of choline or ethanolamine.

2. The phosphoryl group of the resulting **phosphoethanolamine** or **phosphocholine** then attacks CTP, displacing PP_i, to form the corresponding CDP derivatives, which are activated phosphate esters of the polar head group.

3. The C3-OH group of 1,2-diacylglycerol attacks the phosphoryl group of the activated CDP–ethanolamine or CDP–choline, displacing CMP to yield the corresponding glycerophospholipid.

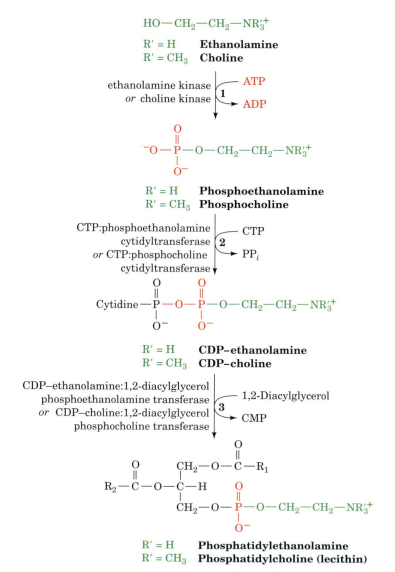

Figure 19-29.   The biosynthesis of phosphatidylethanolamine and phosphatidylcholine. In mammals, CDP–ethanolamine and CDP–choline are the precursors of the head groups.

In liver, phosphatidylethanolamine can also be converted to phosphatidyl-choline by the addition of methyl groups.

Phosphatidylserine is synthesized from phosphatidylethanolamine by a head group exchange reaction catalyzed by **phosphatidylethanolamine transferase** in which serine's OH group attacks the donor's phosphoryl group. The original head group is then eliminated, forming phosphatidylserine.

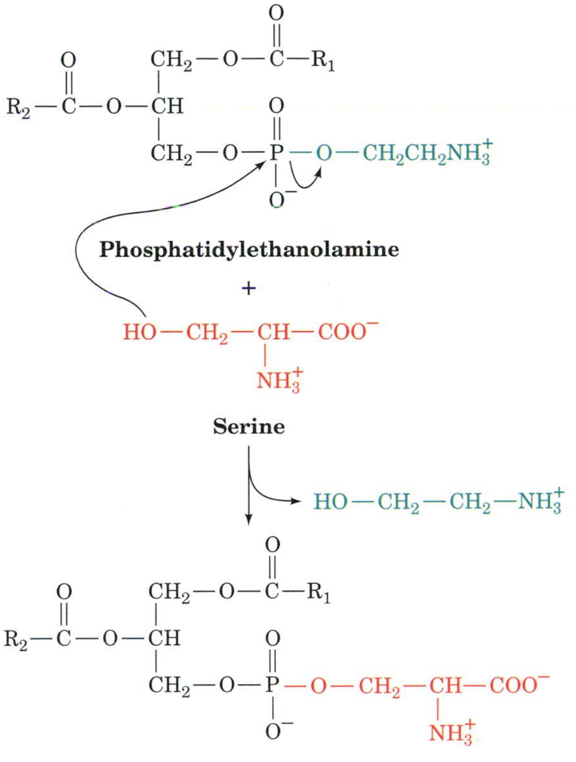

Phosphatidylethanolamine

+

Serine

Phosphatidylserine

In the synthesis of **phosphatidylinositol** and **phosphatidylglycerol,** the hydrophobic tail is activated rather than the polar head group. Phosphatidic acid, the precursor of 1,2-diacylglycerol (Fig. 19-25), attacks the α-phosphoryl group of CTP to form the activated **CDP–diacylglycerol** and PP$_i$ (Fig. 19-30). Phosphatidylinositol results from the attack of inositol on CDP–diacylglycerol. Phosphatidylglycerol is formed in two reactions: (1) attack of the C1 OH group of glycerol-3-phosphate on CDP–diacylglycerol, yielding **phosphatidylglycerol phosphate;** and (2) hydrolysis of the phosphoryl group to form phosphatidylglycerol.

Cardiolipin forms by the condensation of two molecules of phosphatidylglycerol with the elimination of one molecule of glycerol.

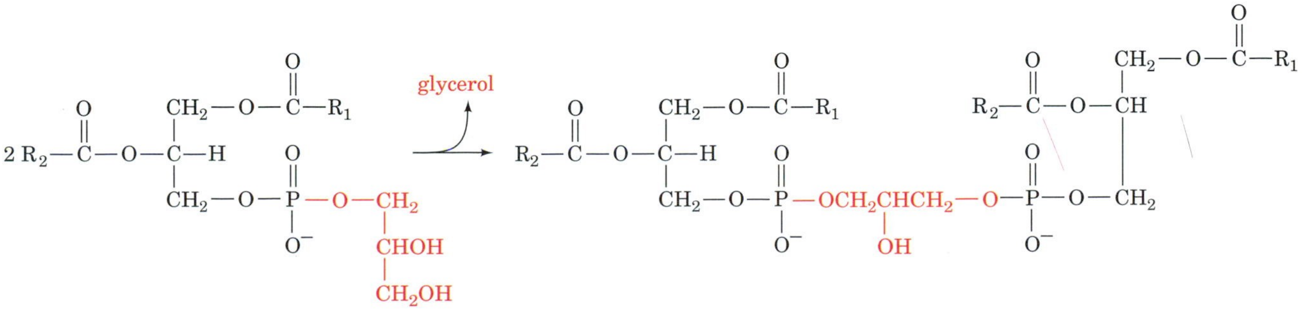

Phosphatidylglycerol   **Cardiolipin**

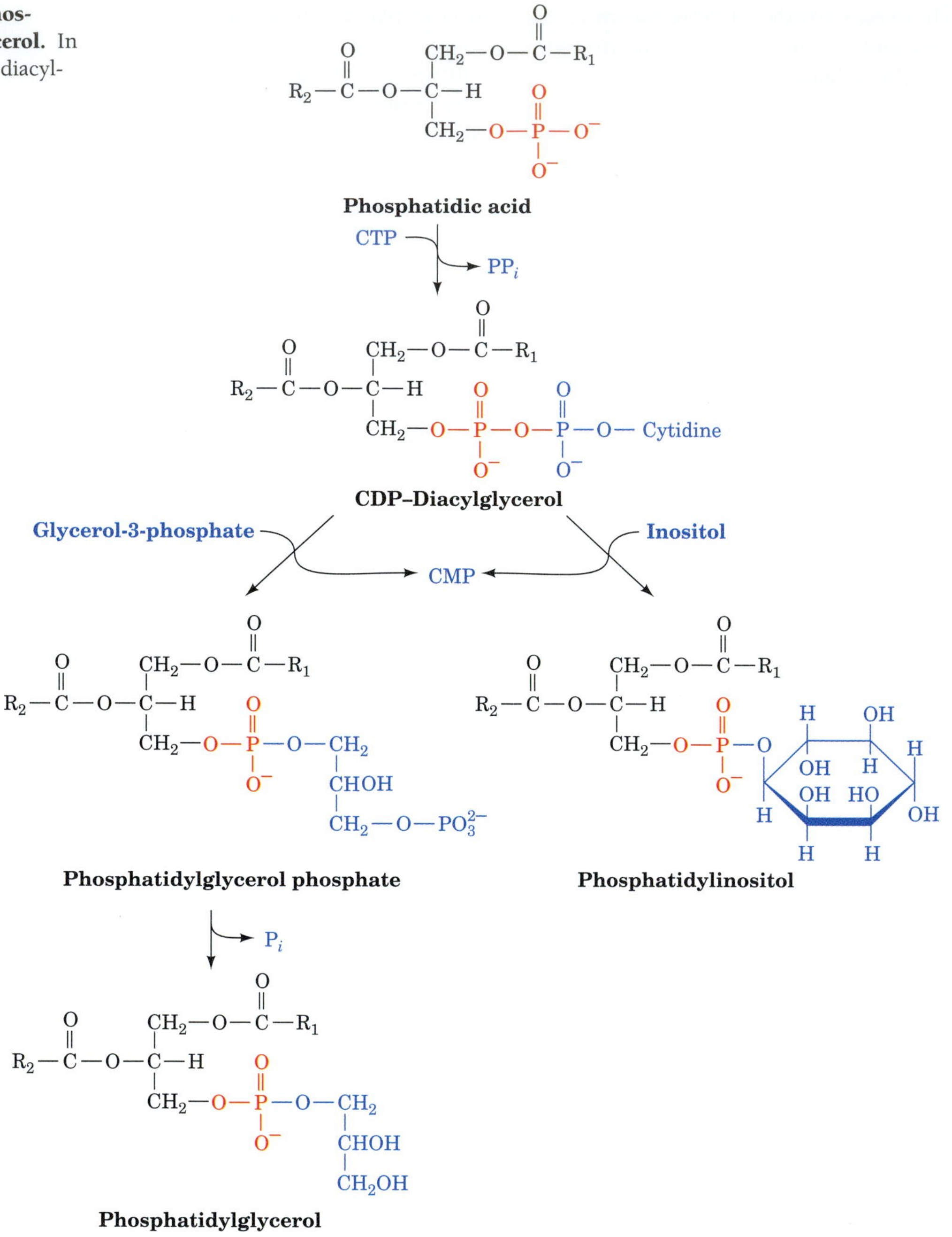

Figure 19-30.   The biosynthesis of phosphatidylinositol and phosphatidylglycerol. In mammals, this process involves a CDP–diacylglycerol intermediate.

Enzymes that synthesize phosphatidic acid have a general preference for saturated fatty acids at C1 and for unsaturated fatty acids at C2. Yet this general preference cannot account, for example, for the observations that ~80% of brain phosphatidylinositol has a stearoyl group (18:0) at C1 and an arachidonoyl group (20:4) at C2, and that ~40% of lung phosphatidylcholine has palmitoyl groups (16:0) at both positions (this latter substance is the major component of the surfactant that prevents the lung from collapsing when air is expelled; its deficiency in premature infants is responsible for respiratory distress syndrome; see Box 9-1). *William Lands showed that such side chain specificity results from "remodeling" reactions in which specific acyl groups of individual glycerophospholipids are exchanged by specific phospholipases and acyltransferases.*

Biosynthesis of Plasmalogens and Alkylacylglycerophospholipids

Eukaryotic membranes contain significant amounts of two other types of glycerophospholipids: **plasmalogens,** which contain a hydrocarbon chain linked to glycerol C1 via a vinyl ether linkage, and **alkylacylglycerophospholipids,** in which the alkyl substituent at glycerol C1 is attached via an ether linkage.

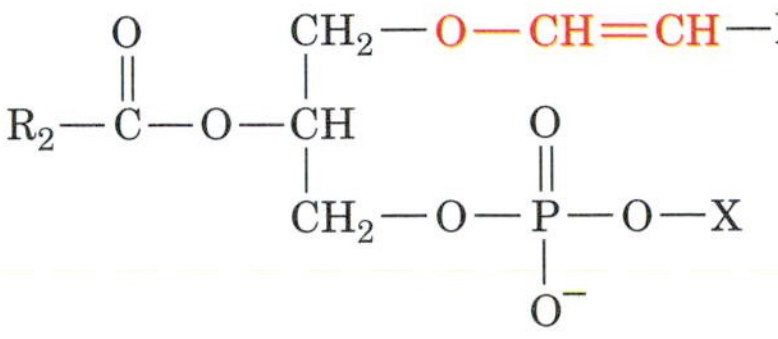

$$CoA-S-\overset{\displaystyle O}{\overset{\|}{C}}-CH_2-CH_2-(CH_2)_{12}-CH_3 \quad + \quad H_2N-\overset{\displaystyle CO_2^-}{\underset{\displaystyle CH_2OH}{\overset{|}{\underset{|}{C}}}}-H$$

Palmitoyl-CoA   **Serine**

A plasmalogen / An alkylacyl-glycerophospholipid structures shown.

A plasmalogen 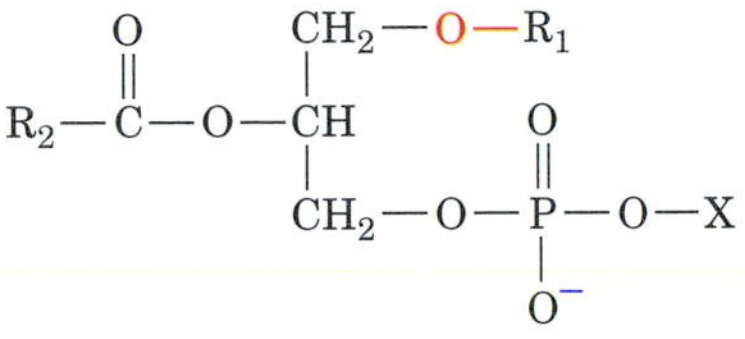   **An alkylacyl-glycerophospholipid**

About 20% of mammalian glycerophospholipids are plasmalogens, but the exact percentage varies both among species and among tissues within a given organism. For example, plasmalogens account for only 0.8% of the phospholipids in human liver but 23% of those in human nervous tissue. The alkylacylglycerophospholipids are much less abundant than the plasmalogens. Recall that the precursor–product relationship between the alkylacylglycerophospholipids and plasmalogens was established through studies using ^{14}C tracers (Section 13-4A). The alkylacylglycerophospholipid is synthesized first and this ether is then oxidized to a vinyl ether through the action of a desaturase.

1 3-ketosphinganine synthase
        $CO_2^- + CoASH$

3-Ketosphinganine
(3-ketodihydrosphingosine)

B. *Sphingolipids*

Most sphingolipids are **sphingoglycolipids;** that is, their polar head groups consist of carbohydrate units. **Cerebrosides** are ceramide monosaccharides, whereas **gangliosides** are sialic acid–containing ceramide oligosaccharides (Section 9-1D). These lipids are synthesized by attaching carbohydrate units to the C1-OH group of ceramide (*N*-acylsphingosine).

N-Acylsphingosine is synthesized in four reactions from the precursors palmitoyl-CoA and serine (Fig. 19-31):

2 NADPH + H^+
        3-ketosphinganine reductase
        $NADP^+$

Sphinganine
(dihydrosphingosine)

1. **3-Ketosphinganine synthase** catalyzes condensation of palmitoyl-CoA with serine, yielding **3-ketosphinganine.**

2. **3-Ketosphinganine reductase** catalyzes the NADPH-dependent reduction of 3-ketosphinganine's keto group to form **sphinganine (dihydrosphingosine).**

3. **Dihydroceramide** is formed by transfer of an acyl group from an acyl-CoA to sphinganine's 2-amino group, forming an amide bond.

4. **Dihydroceramide reductase** converts dihydroceramide to ceramide by an FAD-dependent oxidation reaction.

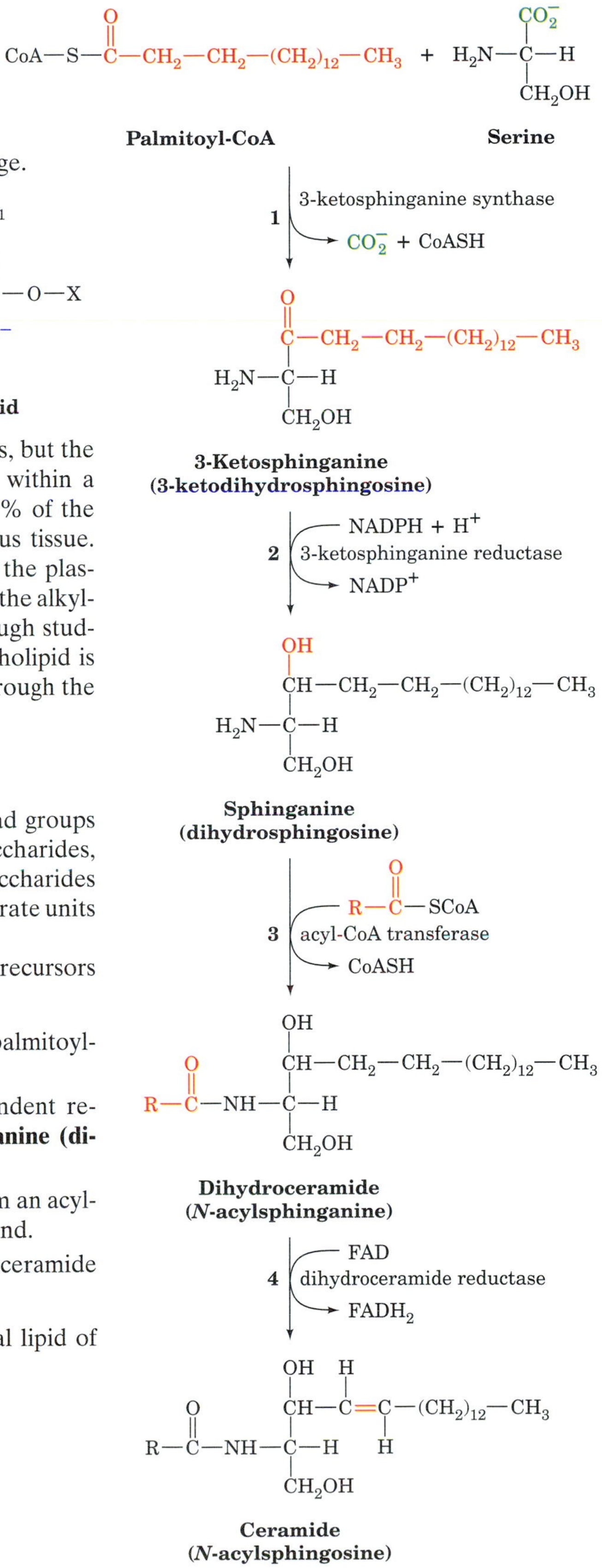

3 $R-\overset{O}{\overset{\|}{C}}-SCoA$
        acyl-CoA transferase
        CoASH

Dihydroceramide
(*N*-acylsphinganine)

The nonglycosylated lipid sphingomyelin, an important structural lipid of nerve cell membranes,

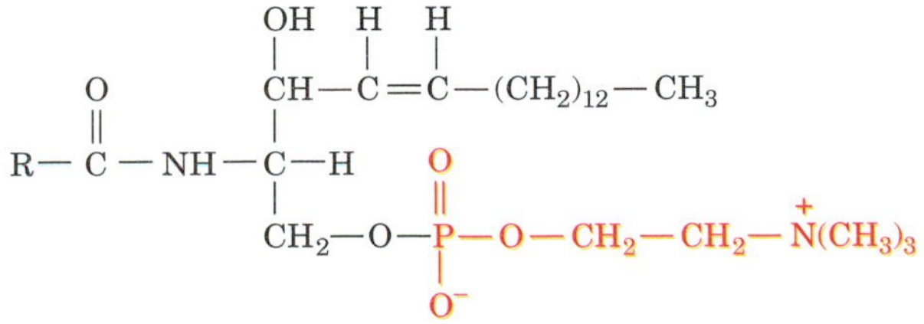

4 FAD
        dihydroceramide reductase
        $FADH_2$

Ceramide
(*N*-acylsphingosine)

Figure 19-31. **The biosynthesis of ceramide (*N*-acylsphingosine).**

Sphingomyelin

Figure 19-32. **The biosynthesis of globosides and G_M gangliosides.**

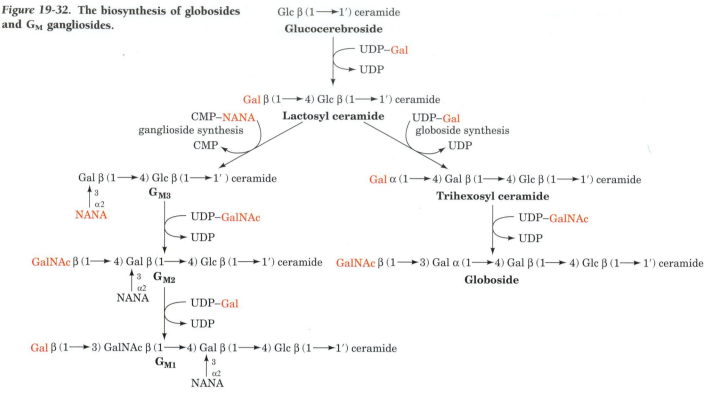

is the product of a reaction in which phosphatidylcholine donates its phosphocholine group to the C1-OH group of *N*-acylsphingosine.

Cerebrosides, which are most commonly 1-β-galactoceramide or 1-β-glucoceramide, are synthesized from ceramide by the addition of the glycosyl unit from the corresponding UDP–hexose to ceramide's C1-OH group. The more elaborate head groups of gangliosides are constructed through the action of a series of glycosyltransferases. The synthetic pathways begin with the transfer of a galactosyl unit from UDP–galactose to glucocerebroside to form a β(1→4) linkage (Fig. 19-32). Since this bond is the same as that linking glucose and galactose in lactose, this glycolipid is often referred to as **lactosyl ceramide.** Lactosyl ceramide is the precursor of both gangliosides and **globosides,** which are neutral ceramide oligosaccharides. To form a globoside, one galactosyl and one *N*-acetylgalactosaminyl (GalNAc) unit are sequentially added to lactosyl ceramide from UDP–Gal and UDP–GalNAc, respectively. The G_M gangliosides (Fig. 9-9) are formed by addition of ***N*-acetylneuraminic acid (NANA, sialic acid)**

N-Acetylneuraminic acid
(NANA, sialic acid)

from CMP–NANA to lactosyl ceramide, yielding G_{M3}. The sequential addition to G_{M3} of *N*-acetylgalactosamine and galactose units yields gangliosides G_{M2} and G_{M1}. Other gangliosides are synthesized through different sequences of glycosyl transfer reactions. Defects in the pathways for degrading these complex lipids are responsible for certain lipid storage diseases (see Box 19-3).

7. CHOLESTEROL METABOLISM

Cholesterol is a vital constituent of cell membranes and the precursor of steroid hormones and bile acids. It is clearly essential to life, yet its deposition in arteries is associated with cardiovascular disease and stroke, two leading causes of death in humans. *In a healthy organism, an intricate balance is maintained between the biosynthesis, utilization, and transport of cholesterol, keeping its harmful deposition to a minimum.* In this section, we study the pathways of cholesterol biosynthesis and transport and how they are controlled.

A. Cholesterol Biosynthesis

Isotopic labeling studies have shown that all the carbon atoms of cholesterol are derived from acetate.

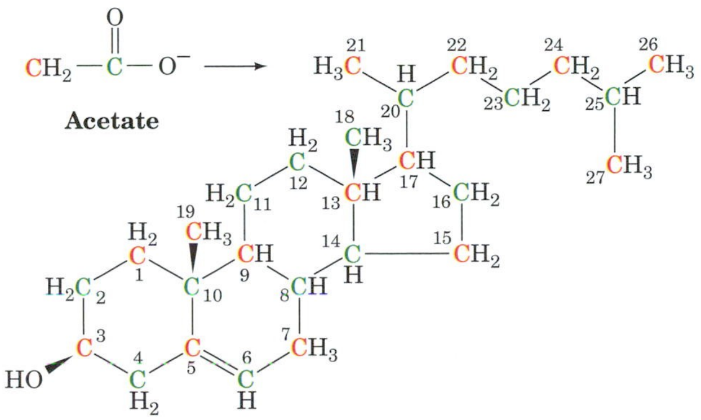

Cholesterol biosynthesis follows a lengthy pathway, first outlined by Konrad Bloch, in which acetate is converted to **isoprene units** that have the carbon skeleton of **isoprene.**

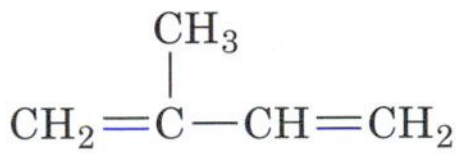

Isoprene
(2-methyl-1,3-butadiene) 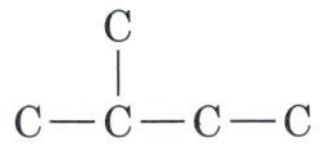   **An isoprene unit**

The isoprene units then condense to form a linear molecule with 30 carbons that cyclizes to form the four-ring structure of cholesterol.

HMG-CoA Is a Key Cholesterol Precursor

Acetyl-CoA is converted to isoprene units by a series of reactions that begins with formation of hydroxymethylglutaryl-CoA (HMG-CoA; this compound is also an intermediate in ketone body synthesis; Fig. 19-17). HMG-CoA synthesis requires thiolase and HMG-CoA synthase. In mitochondria, these two enzymes form HMG-CoA for ketone body synthesis. Cytosolic isozymes of these two proteins generate the HMG-CoA that is used in cholesterol biosynthesis. Four additional reactions convert HMG-CoA to the isoprenoid intermediate **isopentenyl pyrophosphate** (Fig. 19-33):

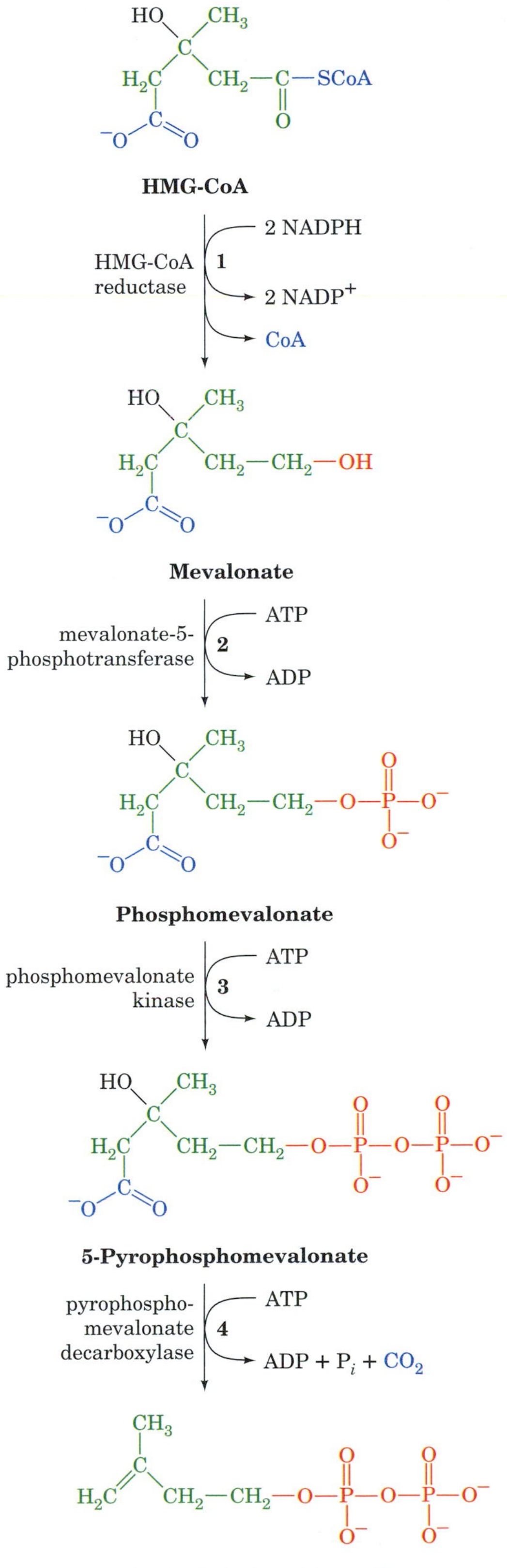

Figure 19-33. The formation of isopentenyl pyrophosphate from HMG-CoA.

Sphingolipid Degradation and Lipid Storage Diseases

Sphingoglycolipids are lysosomally degraded by a series of enzymatically mediated hydrolytic reactions (*below*).

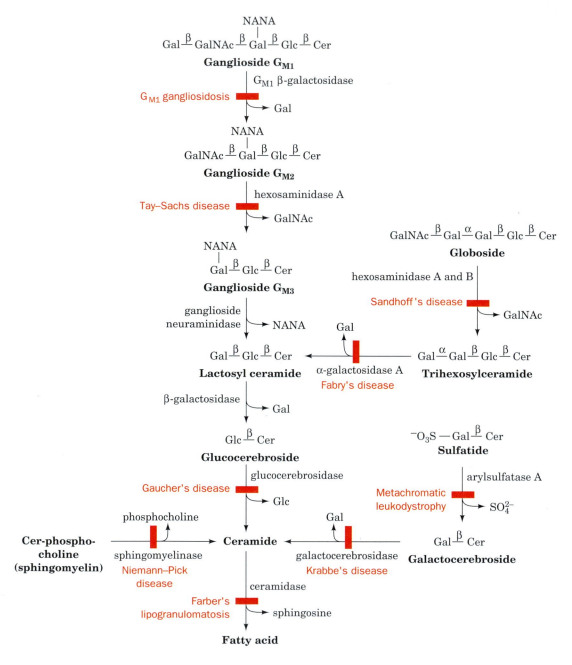

A hereditary defect in one of these enzymes (indicated by a red bar) results in a **sphingolipid storage disease**. The substrate of the missing enzyme therefore accumulates, often with disastrous consequences. In many cases, affected individuals suffer from mental retardation and die in infancy or early childhood. One of the most common lipid storage diseases is **Tay–Sachs disease**, an autosomal recessive deficiency in **hexosaminidase A**, which hydrolyzes *N*-acetylgalactosamine from ganglioside G_{M2}. The absence of hexosaminidase A activity results in the accumulation of G_{M2} as shell-like inclusions in neuronal cells.

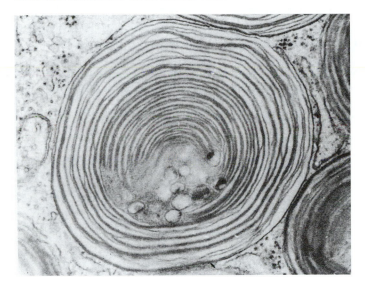

Although infants born with Tay–Sachs disease at first appear normal, by ~1 year of age, when sufficient G_{M2} has accumulated to interfere with neuronal function, they become progressively weaker, retarded, and blinded until they die, usually by the age of 3 years. It is possible, however, to screen potential carriers of this disease by a simple serum assay.

Experiments using a mouse model of Tay–Sachs disease suggest that G_{M2} accumulation can be reduced by inhibiting its synthesis. This is accomplished in mice by administering **N-butyldeoxynojirimycin,**

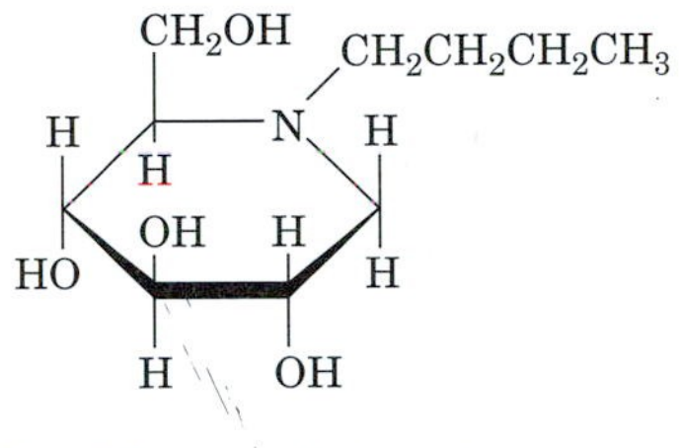

N-Butyldeoxynojirimycin

which inhibits the glycosyltransferase in the first step of glucoceramide synthesis. Similar "substrate deprivation" approaches may be effective in treating other lipid storage diseases, particularly when the defective but essential enzyme has some residual activity.

[Photo courtesy of John S. O'Brien, University of California at San Diego.]

1. The CoA thioester group of HMG-CoA is reduced to an alcohol in an NADPH-dependent four-electron reduction catalyzed by **HMG-CoA reductase,** yielding **mevalonate,** a C_6 compound.

2. The new OH group is phosphorylated by **mevalonate-5-phosphotransferase.**

3. The phosphate group is converted to a pyrophosphate by **phosphomevalonate kinase.**

4. The molecule undergoes an ATP-dependent decarboxylation–dehydration reaction catalyzed by **pyrophosphomevalonate decarboxylase.**

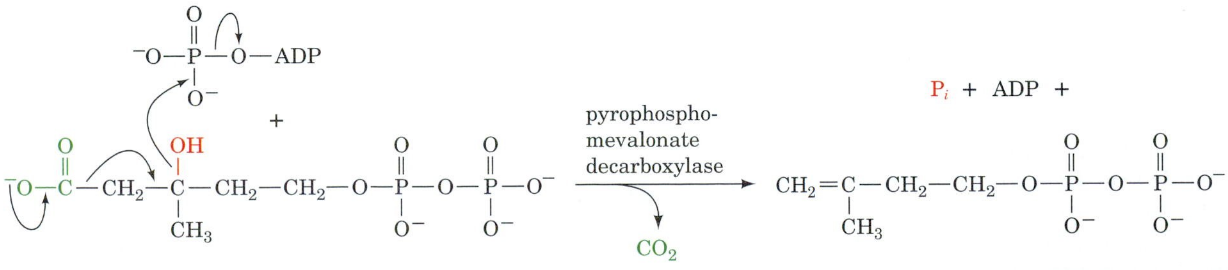

5-Pyrophosphomevalonate → (pyrophosphomevalonate decarboxylase) CO_2 → **Isopentenyl pyrophosphate**

P_i + ADP +

Squalene Is Formed by the Condensation of Six Isoprene Units

Isopentenyl pyrophosphate is converted to **dimethylallyl pyrophosphate** by **isopentenyl pyrophosphate isomerase.**

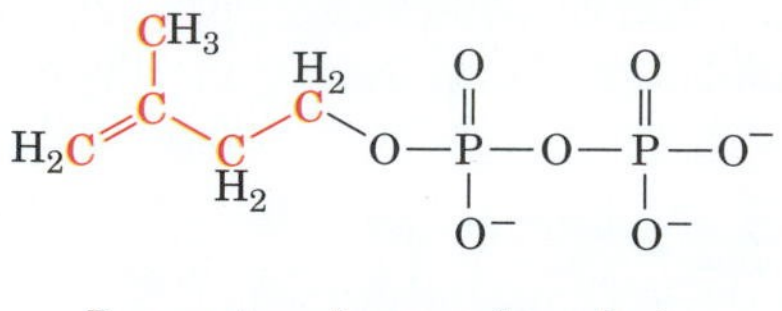

Isopentenyl pyrophosphate 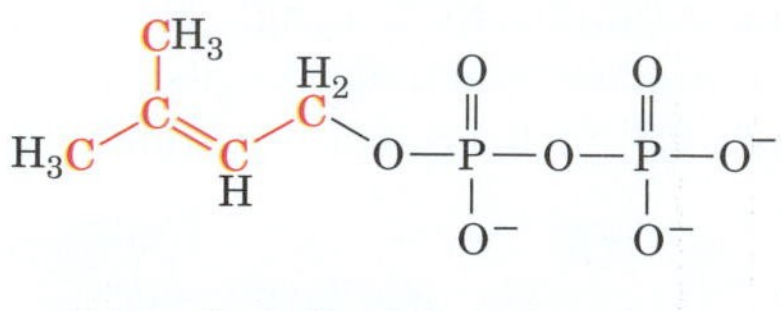   **Dimethylallyl pyrophosphate**

Four isopentenyl pyrophosphates and two dimethylallyl pyrophosphates condense to form the C_{30} cholesterol precursor **squalene** in three reactions catalyzed by two enzymes (Fig. 19-34):

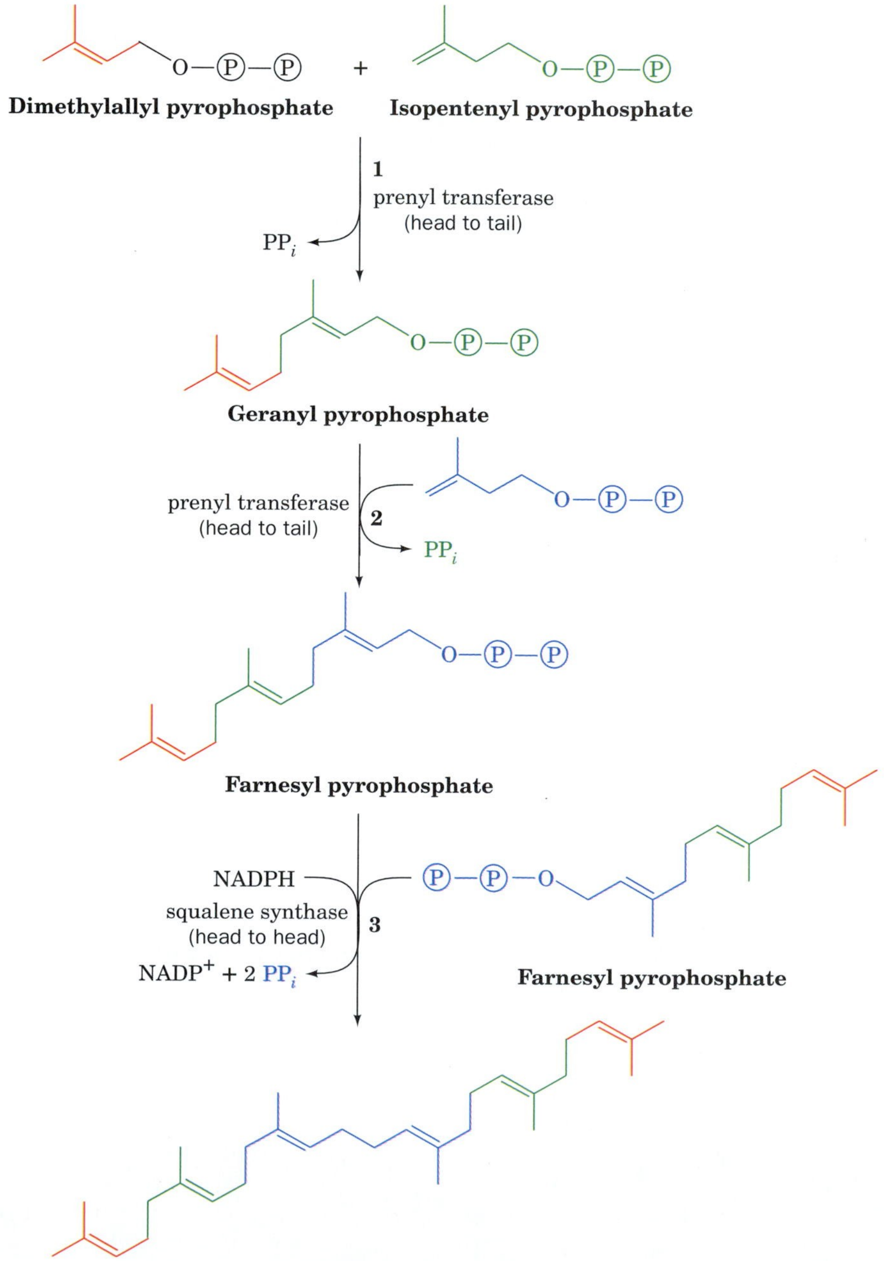

Figure 19-34.   The formation of squalene from isopentenyl pyrophosphate and dimethylallyl pyrophosphate. The pathway involves two head-to-tail condensations catalyzed by prenyl transferase and a head-to-head condensation catalyzed by squalene synthase.

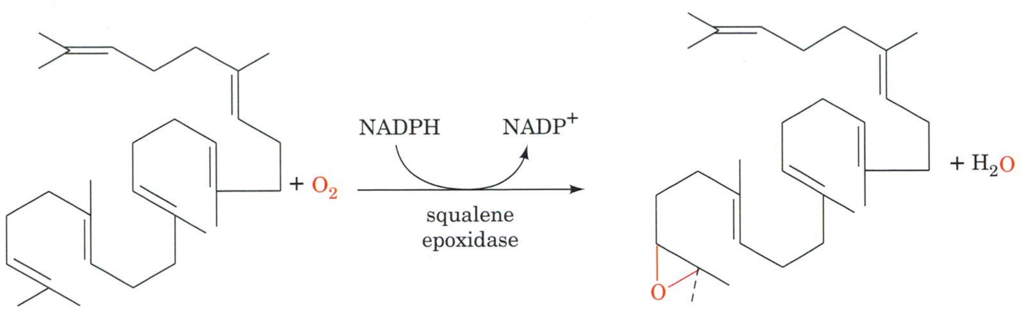

Squalene   **2,3-Oxidosqualene**

Figure 19-35. The squalene epoxidase reaction.

1. **Prenyl transferase** catalyzes the head-to-tail condensation of dimethylallyl pyrophosphate and isopentenyl pyrophosphate to yield the C_{10} compound **geranyl pyrophosphate.**

2. Prenyl transferase catalyzes a second head-to-tail condensation of geranyl pyrophosphate and isopentenyl pyrophosphate to yield the C_{15} compound **farnesyl pyrophosphate.** Prenyl transferase is one of the few enzymes that has been shown to proceed via an S_N1 reaction to form a carbocation intermediate with an ionization–condensation–elimination mechanism:

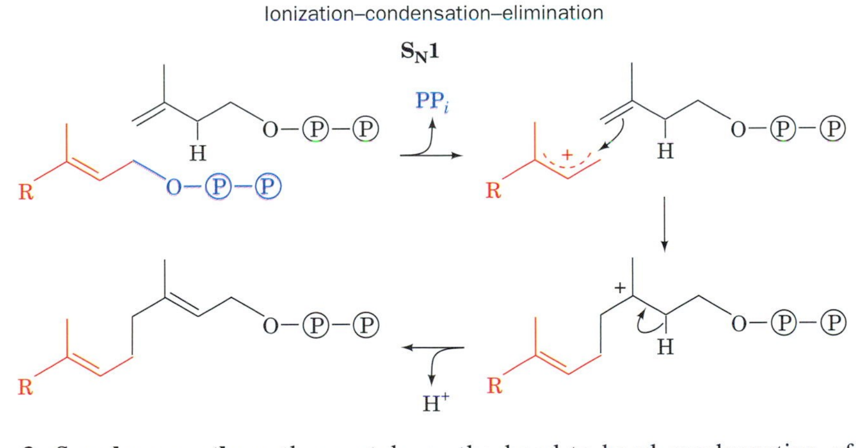

3. **Squalene synthase** then catalyzes the head-to-head condensation of two farnesyl pyrophosphate molecules to form squalene.

Farnesyl pyrophosphate is also the precursor of other isoprenoid compounds in mammals, including ubiquinone (Fig. 17-10) and the isoprenoid tails of some lipid-linked membrane proteins (Section 10-1B).

Squalene Cyclization Eventually Yields Cholesterol

Squalene, a linear hydrocarbon, cyclizes to form the tetracyclic steroid skeleton in two steps. **Squalene epoxidase** catalyzes oxidation of squalene to form **2,3-oxidosqualene** (Fig. 19-35). **Squalene oxidocyclase** converts this epoxide to the steroid **lanosterol.** The reaction is a chemically complex process involving cyclization of 2,3-oxidosqualene to a **protosterol** cation and rearrangement of this cation to lanosterol by a series of 1,2 hydride and methyl shifts (Fig. 19-36).

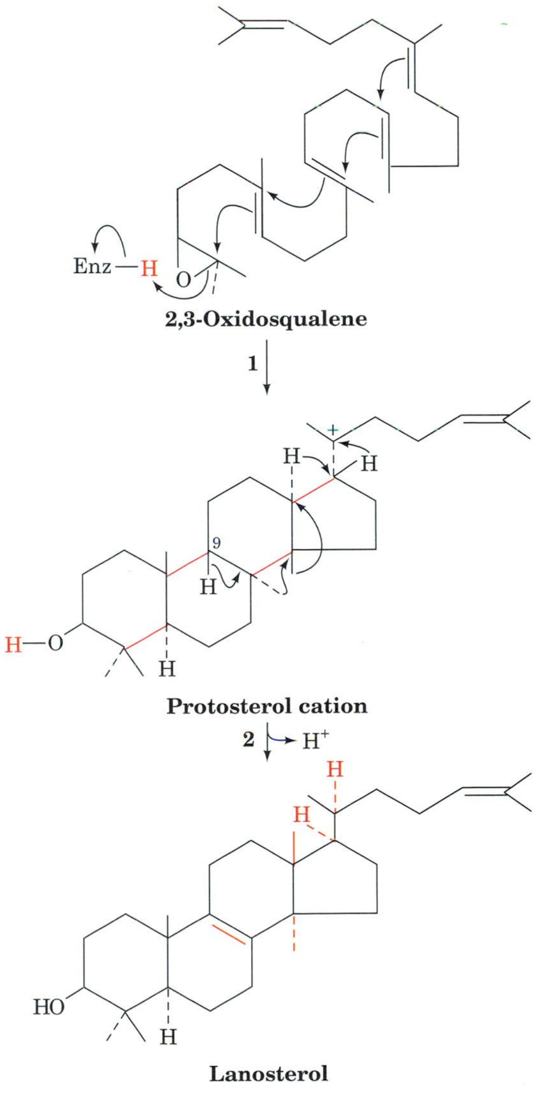

Figure 19-36. **The squalene oxidocyclase reaction.** (1) 2,3-Oxidosqualene is cyclized to the protosterol cation in a process that is initiated by the enzyme-mediated protonation of the squalene epoxide oxygen. The opening of the epoxide leaves an electron-deficient center whose migration drives the series of cyclizations that form the protosterol cation. (2) The elimination of a proton from C9 of the sterol to form a double bond initiates a series of methyl and hydride migrations that ultimately yields neutral lanosterol.

Conversion of lanosterol to cholesterol

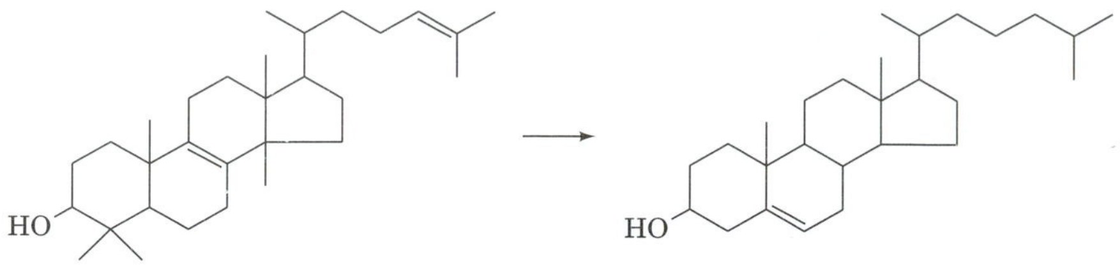

Lanosterol   **Cholesterol**

is a 19-step process that involves an oxidation and the loss of three methyl groups. The enzymes required for this process are embedded in the endoplasmic reticulum membrane.

B. Cholesterol Transport

Cholesterol synthesized by the liver is either converted to bile acids (Fig. 19-1) or esterified by **acyl-CoA:cholesterol acyltransferase (ACAT)** to form cholesteryl esters.

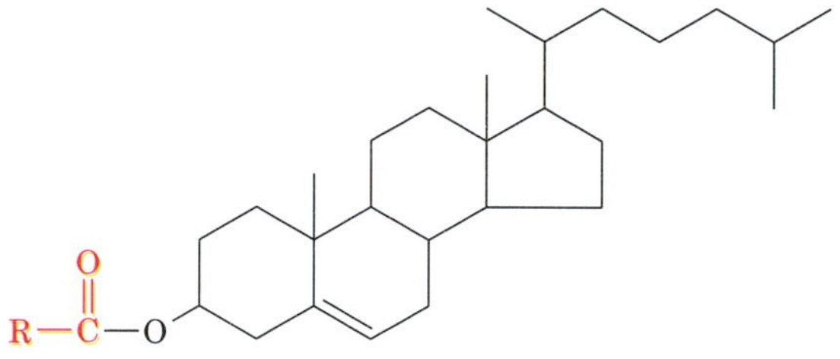

Cholesteryl ester

These highly hydrophobic compounds are transported throughout the body in lipoprotein complexes (Section 19-1B). They are first packaged in VLDL and enter the bloodstream. As the VLDL circulate, they gradually become IDL and then LDL as their component triacylglycerols and many of their apolipoproteins are removed. Peripheral tissues normally obtain most of their exogenous cholesterol from LDL by receptor-mediated endocytosis (Fig. 19-37; Section 10-3B). Inside the cell, cholesteryl esters are hydrolyzed by a lysosomal lipase to free cholesterol, which is either incorporated into cell membranes or reesterified by ACAT for storage as cholesteryl ester droplets.

Dietary cholesterol is transported from the intestine by chylomicrons. Thus, liver and peripheral tissues can either synthesize cholesterol or obtain it from circulating lipoproteins. Cholesterol circulates continuously between the liver and peripheral tissues: Whereas LDL transports cholesterol from the liver, cholesterol is transported back to the liver by HDL. Surplus cholesterol is disposed of by the liver as bile acids, the only significant mechanism the body has to dispose of this water-insoluble substance.

C. Control of Cholesterol Metabolism

Cholesterol biosynthesis and transport are regulated by controlling (1) the activity of HMG-CoA reductase, (2) the rate of LDL receptor synthesis, and (3) the rate of esterification of cholesterol by ACAT.

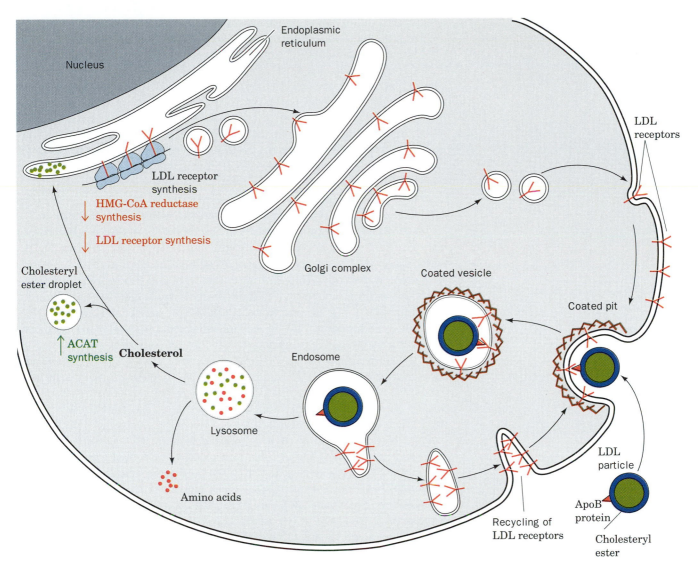

Figure 19-37.   LDL receptor-mediated endocytosis in mammalian cells. LDL receptor is synthesized on the endoplasmic reticulum, processed in the Golgi apparatus, and inserted into the plasma membrane as a component of coated pits. The apolipoprotein B (apoB) component of LDL binds specifically to receptors on the coated pits so that LDL is brought into the cell. Endosomes deliver LDL to lysosomes and recycle LDL receptor to the plasma membrane (Section 10-3B). Lysosomal degradation of LDL releases cholesterol, whose presence decreases the rate of synthesis of HMG-CoA reductase and LDL receptor (*down arrows*) while increasing that of acyl-CoA:cholesterol acyltransferase (ACAT; *up arrow*). [After Brown, M.S. and Goldstein, J.L., *Curr. Top. Cell. Reg.* **26**, 7 (1985).] ✳ **See the Animated Figures.**

HMG-CoA Reductase Is the Primary Control Site for Cholesterol Biosynthesis

HMG-CoA reductase is the rate-limiting enzyme in cholesterol biosynthesis and, as expected, is the pathway's main regulatory site. HMG-CoA reductase, an 887-residue membrane-bound enzyme of the endoplasmic reticulum, is subject to short-term regulation by reversible phosphorylation. However, *the primary regulatory mechanism is long-term feedback control of the amount of enzyme present in the cell.* When either LDL–cholesterol or mevalonate levels fall, the level of HMG-CoA reductase can rise as much as 200-fold, due to an increase in enzyme synthesis combined with a decrease in its degradation. These effects are readily reversed.

HMG-CoA reductase exists in interconvertible more active and less active forms, as do glycogen phosphorylase, glycogen synthase (Section 15-3B), and other enzymes. The unmodified form of HMG-CoA reductase is more active; the phosphorylated form is less active. HMG-CoA reductase is phosphorylated (inactivated) at its Ser 871 by the same covalently modifiable AMP-dependent protein kinase that acts on acetyl-CoA carboxylase (Section 19-4B). It appears that this control mechanism conserves energy when ATP levels fall and AMP levels rise, by generally inhibiting biosynthetic pathways.

LDL Receptor Activity Governs Cholesterol Removal from the Blood

The concentration of any species in the blood results from a balance of its rate of production and its rate of removal. LDL is produced from VLDL, which is secreted by the liver to carry cholesteryl esters and triacylglycerols to the rest of the body. Both limiting cholesterol and triacylglycerol biosynthesis and maintaining a low-fat and low-cholesterol diet decrease the rate of production of VLDL and hence LDL. Removal of LDL from the circulation is mediated by LDL receptors, which therefore play an important role in the maintenance of plasma LDL–cholesterol levels. At a given rate of VLDL production, *the serum concentration of LDL depends on the rate that the liver removes IDL and LDL from the circulation, which, in turn, depends on the number of functioning LDL receptors on the liver cell surface* (both IDL and LDL contain apolipoproteins that specifically bind to the LDL receptor; Fig. 19-38*a*).

High blood cholesterol **(hypercholesterolemia),** which results from the overproduction and/or underutilization of LDL, is known to be caused by two metabolic irregularities: (1) the genetic disease **familial hypercholesterolemia (FH;** Box 10-1) and (2) the consumption of a high-cholesterol diet. FH homozygotes, who lack functional LDL receptors, have plasma LDL–cholesterol levels three to five times higher than average (Fig. 19-38*b*). FH heterozygotes, which are far more common, have about half the normal number of functional LDL receptors and plasma LDL–cholesterol levels of about twice the average.

The ingestion of a high-cholesterol diet has a similar but milder effect (Fig. 19-38*c*). Excessive dietary cholesterol enters the liver cells in chylomicron remnants. High concentrations of intracellular cholesterol suppress the synthesis of LDL receptor protein. The resulting insufficiency of LDL receptors on the liver cell surface leads to high levels of circulating LDL.

Two strategies are used to counteract hypercholesterolemia besides following a low-cholesterol diet:

1. Ingestion of resins that bind bile acids, thereby preventing their intestinal absorption. Bile acids, which are derived from cholesterol, are normally efficiently reabsorbed and recycled by the liver and may circulate between the intestine and the liver several times each day. Elimination of resin-bound cholesterol in the feces forces the liver to convert more cholesterol to bile acids. The consequent decrease in the serum cholesterol concentration induces synthesis of LDL receptors (of course, not in FH homozygotes). Unfortunately, the decreased serum cholesterol level also induces the synthesis of HMG-CoA reductase, which increases the rate of cholesterol biosynthesis. Ingestion of bile acid–binding resins therefore provides only a 15 to 20% drop in serum cholesterol levels.

(a) **Normal**

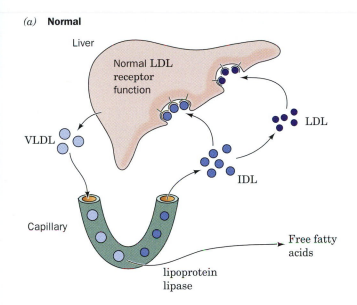

Figure 19-38. Uptake of plasma LDL and control of LDL production by liver LDL receptors. (*a*) In normal human subjects, VLDL is secreted by the liver and converted to IDL in the capillaries of the peripheral tissues. About half of the plasma IDL particles bind to the LDL receptor and are taken up by the liver. The remainder are converted to LDL at the peripheral tissues. (*b*) In individuals with the genetic defect known as familial hypercholesterolemia (FH), liver LDL receptors are diminished or absent. (*c*) In normal individuals who ingest a high-cholesterol diet, the liver is filled with cholesterol, which represses the rate of LDL receptor production. Receptor deficiency, whether of genetic or dietary origin, raises the plasma LDL level by increasing the rate of LDL production and decreasing the rate of LDL uptake. [After Goldstein, J.L. and Brown, M.S., *J. Lipid Res.* **25**, 1457 (1984).]

(b) **Familial hypercholesterolemia**

(c) **High cholesterol diet**

2. Treatment with competitive inhibitors of HMG-CoA reductase, notably the fungal products **compactin** and **lovastatin,** which resemble mevalonate.

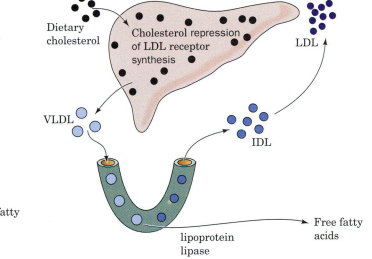

Mevalonate

R = H   X = H   **Compactin**
R = CH₃   X = H   **Lovastatin (Mevacor™)**
R = OH   X = H   **Pravastatin (Pravachol™)**
R = CH₃   X = CH₃   **Simvastatin (Zocor™)**

These inhibitors decrease the rate of cholesterol biosynthesis. The low cellular cholesterol supply is again met by induction of the LDL receptor and HMG-CoA reductase. Lovastatin-treated FH heterozygotes nevertheless routinely show a serum cholesterol decrease of 30%.

The combined use of lovastatin and cholesterol-binding resins results in a clinically dramatic 50 to 60% decrease in serum cholesterol levels. Other more recently developed HMG-CoA reductase inhibitors such as **pravastatin** and **simvastatin,** drugs collectively known as **statins,** yield even more efficacious results.

S U M M A R Y

1. Triacylglycerol digestion depends on the emulsifying activity of bile acids and the activation of lipases at the lipid–water interface. After they are absorbed, the lipid digestion products are packaged in lipoproteins for transport to the tissues via the bloodstream.

2. Fatty acids are released from triacylglycerols in their storage sites in adipose tissue by hormone-sensitive triacylglycerol lipase and carried in the blood as albumin complexes to their sites of oxidation.

3. Fatty acid oxidation begins with the activation of the acyl group by formation of a thioester with CoA. The acyl group is transferred to carnitine for transport into the mitochondria, where it is reesterified to CoA.

4. β oxidation occurs in four reactions: (1) formation of an α,β double bond, (2) hydration of the double bond, (3) dehydrogenation to form a β-ketoacyl-CoA, and (4) thiolysis by CoA to produce acetyl-CoA and an acyl-CoA shortened by two carbons. This process is repeated until fatty acids with even numbers of carbon atoms are converted to acetyl-CoA and the fatty acids with odd numbers of carbon atoms are converted to acetyl-CoA and one molecule of propionyl-CoA. The acetyl-CoA is oxidized by the citric acid cycle and oxidative phosphorylation to generate ATP.

5. The oxidation of unsaturated fatty acids requires an isomerase to convert Δ^3 double bonds to Δ^2 double bonds and a reductase to remove Δ^4 double bonds. The oxidation of odd-chain fatty acids yields propionyl-CoA, which is converted to succinyl-CoA through a cobalamin (B_{12})-dependent pathway. Very long chain fatty acids are partially oxidized by a three-enzyme system in peroxisomes.

6. The liver uses acetyl-CoA to synthesize acetoacetate and β-hydroxybutyrate, which are released into the bloodstream. Tissues that use these ketone bodies for fuel convert them back to acetyl-CoA.

7. In fatty acid synthesis, mitochondrial acetyl-CoA is shuttled to the cytosol via the tricarboxylate transport system and activated to malonyl-CoA by the action of acetyl-CoA carboxylase.

8. A series of seven enzymatic activities, which in mammals are contained in a multifunctional homodimeric enzyme, extend acyl-ACP chains by two carbons at a time. An enzyme-bound acyl group and malonyl-CoA condense to form a β-ketoacyl intermediate and CO_2. Two reductions and a dehydration yield an acyl-ACP in a series of reactions that resemble the reverse of β oxidation but are catalyzed by separate enzymes in the cytosol. Palmitate (C_{16}), the normal product of fatty acid biosynthesis, is synthesized in seven such reaction cycles and is then cleaved from ACP by a thioesterase.

9. Other fatty acids are synthesized from palmitate through the action of elongases and desaturases. Human triacylglycerols synthesized from fatty acyl-CoA and glyceraldehyde-3-phosphate or dihydroxyacetone phosphate tend to contain saturated fatty acids at C1 and unsaturated fatty acids at C2.

10. The opposing pathways of fatty acid degradation and synthesis are hormonally regulated. Glucagon and epinephrine activate hormone-sensitive lipase in adipose tissue, thereby increasing the supply of fatty acids for oxidation in other tissues, and inactivate acyl-CoA carboxylase. Insulin has the opposite effect. Hormones also regulate the levels of acetyl-CoA carboxylase and fatty acid synthase by controlling their rates of synthesis.

11. Mammalian phosphatidylethanolamine and phosphatidylcholine are synthesized from 1,2-diacylglycerol and CDP derivatives of the head groups. Phosphatidylinositol, phosphatidylglycerol, and cardiolipin synthesis begin with CDP–diacylglycerol.

12. Sphingoglycolipids are synthesized from ceramide (*N*-acylsphingosine, a derivative of palmitate and serine) by the addition of glycosyl units donated by nucleotide sugars.

13. Cholesterol is synthesized from acetyl units that pass through HMG-CoA and mevalonate intermediates on the way to being converted to a C_5 isoprene unit. Six isoprene units condense to form the C_{30} compound squalene, which cyclizes to yield lanosterol, the steroid precursor of cholesterol.

14. HMG-CoA reductase, the rate-limiting enzyme of cholesterol biosynthesis, is regulated through long-term control of its rate of synthesis, as well as through short-term control by phosphorylation. The level of cellular cholesterol also depends on its conversion to cholesteryl esters or other compounds and on the endocytosis of cholesterol-bearing LDL, which varies with the amount of LDL receptor on the cell surface.

REFERENCES

Derewenda, Z.S., Structure and function of lipases, *Adv. Prot. Chem.* **45,** 1–52 (1994).

Dowhan, W., Molecular basis for membrane phospholipid diversity: Why are there so many lipids? *Annu. Rev. Biochem.* **66,** 199–232 (1997).

Eaton, S., Bartlett, K., and Pourfarzam, M., Mammalian mitochondrial β-oxidation, *Biochem. J.* **320,** 345–357 (1996).

Goldstein, J.L. and Brown, M.S., Regulation of the mevalonate pathway, *Nature* **343,** 425–430 (1990).

Kent, C., Eukaryotic phospholipid biosynthesis, *Annu. Rev. Biochem.* **64,** 315–343 (1995).

Ludwig, M.L. and Matthews, R.G., Structure-based perspectives on B$_{12}$-dependent enzymes, *Annu. Rev. Biochem.* **66,** 266–313 (1997).

Mannaerts, G.P. and Van Veldhoven, P.P., Metabolic pathways in mammalian peroxisomes, *Biochimie* **75,** 147–158 (1993). [A brief review of peroxidase-specific metabolic pathways, including β oxidation.]

Platt, F.M., Neises, G.R., Reinkensmeier, G., Townsend, M.J., Perry, V.H., Proia, R.L., Winchester, B., Dwek, R.A., and Butters, T.D., Prevention of lysosomal storage in Tay–Sachs mice treated with *N*-butyldeoxynojirimycin, *Science* **276,** 426–431 (1997).

Thompson, G.A., *The Regulation of Membrane Lipid Metabolism* (2nd ed.), CRC Press (1992).

Töpfer, R., Martini, N., and Schell, J., Modification of plant lipid synthesis, *Science* **268,** 681–684 (1995).

Vance, D.E. and Vance, J. (Eds.), *Biochemistry of Lipids, Lipoproteins and Membranes,* Elsevier (1996).

van Echten, G. and Sandhoff, K., Ganglioside metabolism: enzymology, topology, and regulation, *J. Biol. Chem.* **268,** 5341–5344 (1993).

KEY TERMS

interfacial activation	HDL	ketosis	hormone-sensitive lipase
chylomicron	β oxidation	acyl-carrier protein	short-term regulation
VLDL	homolytic cleavage	elongase	long-term regulation
IDL	heterolytic cleavage	desaturase	isoprene unit
LDL	ketogenesis	essential fatty acid	

STUDY EXERCISES

1. How do bile salts aid in the digestion and absorption of lipids?

2. Describe the roles of the lipoproteins in humans.

3. Summarize the chemical transformations that occur during fatty acid activation and its degradation to acetyl-CoA.

4. What additional steps are required to oxidize unsaturated and odd-chain fatty acids?

5. How are ketone bodies synthesized and degraded?

6. Describe the shuttle systems for transporting fatty acids into the mitochondria and acetyl-CoA into the cytosol.

7. Summarize the similarities and differences between fatty acid oxidation and biosynthesis.

8. Describe the major mechanisms of regulating fatty acid metabolism in humans.

9. How are triacylglycerols, glycerophospholipids, and sphingolipids synthesized?

10. Summarize the chemical events in cholesterol synthesis.

11. Why do serum cholesterol levels depend on LDL receptor activity?

PROBLEMS

1. Explain why individuals with a hereditary deficiency of carnitine palmitoyl transferase II have muscle weakness. Why are these symptoms more severe during fasting?

2. The first three steps of β oxidation (Fig. 19-9) chemically resemble three successive steps of the citric acid cycle. Which steps are these?

3. Certain branched-chain fatty acids such as **pristanic acid**

$$CH_3 \!-\!\!\left(\!CH\!-\!CH_2\!-\!CH_2\!-\!CH_2\!\right)_{\!3}\!CH\!-\!C\overset{\displaystyle O}{\underset{\displaystyle O^-}{\big\backslash}}$$

with CH$_3$ branches

Pristanic acid

can undergo β oxidation. (a) How many cycles of β oxidation are required to completely degrade this fatty acid? (b) How many C_2 (acetyl-CoA), C_3 (propionyl-CoA), and C_4 (methylpropionyl-CoA) products are obtained?

4. Why are unsaturated fats preferable to saturated fats for an individual whose caloric intake must be limited?

5. Why is it important that liver cells lack 3-ketoacyl-CoA transferase (Fig. 19-18)?

6. The tricarboxylate transport system supplies cytosolic acetyl-CoA for palmitate synthesis. What percentage of the NADPH required for palmitate synthesis is thereby provided?

7. On what carbon atoms does the $^{14}CO_2$ used to synthesize malonyl-CoA from acetyl-CoA appear in palmitate?

8. Is the fatty acid shown below likely to be synthesized in animals? Explain.

$$CH_3 \left(CH_2 - CH = CH \right)_3 \left(CH_2 \right)_7 - C \overset{\displaystyle O}{\underset{\displaystyle O^-}{\diagup}}$$

9. Compare the energy cost, in ATP equivalents, of synthesizing stearate from mitochondrial acetyl-CoA to the energy recovered by degrading stearate (a) to acetyl-CoA and (b) to CO_2.

10. An animal is fed palmitate with a ^{14}C-labeled carboxyl group. (a) Under ketogenic conditions, where would the label appear in acetoacetate? (b) Under conditions of membrane lipid synthesis, where would the label appear in sphinganine?

All organisms need a source of nitrogen. Complex metabolic pathways convert a few nitrogen-containing compounds into many others, including amino acids for protein synthesis. These pea plants grow in symbiotic association with microorganisms that make nitrogen available to them; other crops require a nitrogen-containing fertilizer.
[Hans Reinhard/OKAPIA/Photo Researchers, Inc.]

AMINO ACID METABOLISM

The metabolism of amino acids comprises a wide array of synthetic and degradative reactions by which amino acids are assembled as precursors of polypeptides or other compounds and broken down to recover metabolic energy. The chemical transformations of amino acids are distinct from those of carbohydrates or lipids in that they involve the element nitrogen. We must therefore examine the origin of nitrogen in biological systems and its disposal.

The bulk of the cell's amino acids are incorporated into proteins, which are constantly being synthesized and degraded. Aside from this dynamic pool of polymerized amino acids, there is no true storage form of amino acids analogous to glycogen or triacylglycerols. Mammals synthesize certain amino acids and obtain the rest from their diet. *Excess dietary amino acids are not simply excreted but are converted to common metabolites that are precursors of glucose, fatty acids, and ketone bodies and are therefore metabolic fuels.*

In this chapter, we consider the pathways of amino acid metabolism, beginning with the degradation of proteins and the **deamination** (amino group removal) of their component amino acids. We then examine the incorporation of nitrogen into urea for excretion. Next, we examine the pathways by which the carbon skeletons of individual amino acids are broken down and synthesized. We conclude with a brief examination of some other biosynthetic pathways involving amino acids and **nitrogen fixation,** a process that converts atmospheric N_2 to a biologically useful form.

1. INTRACELLULAR PROTEIN DEGRADATION

The components of living cells are constantly turning over. Proteins have lifetimes that range from as short as a few minutes to weeks or more. In any case, *cells continuously synthesize proteins from and degrade them to amino acids.* This seemingly wasteful process has three functions: (1) to store nutrients in the form of proteins and to break them down in times of metabolic need, processes that are most significant in muscle tissue; (2) to eliminate abnormal proteins whose accumulation would be harmful to the cell; and (3) to permit the regulation of cellular metabolism by eliminating superfluous enzymes and regulatory proteins. *Controlling a protein's rate of degradation is therefore as important to the cellular and organismal economy as is controlling its rate of synthesis.*

The half-lives of different enzymes in a given tissue vary substantially, as is indicated for rat liver in Table 20-1. Remarkably, *the most rapidly degraded enzymes all occupy important metabolic control points, whereas the relatively stable enzymes have nearly constant catalytic activities under all physiological conditions.* The susceptibilities of enzymes to degradation have evidently evolved along with their catalytic and allosteric properties so that cells can respond efficiently to environmental changes and metabolic requirements. The rate of protein degradation in a cell also varies with its nutritional and hormonal state. For example, under conditions of nutritional deprivation, cells increase their rate of protein degradation so as to provide the necessary nutrients for indispensable metabolic processes.

A. Lysosomal Degradation

Lysosomes contain ~50 hydrolytic enzymes, including a variety of proteases. The lysosome maintains an internal pH of ~5, and its enzymes have acidic pH optima. This situation presumably protects the cell against acci-

Table 20-1.   Half-Lives of Some Rat Liver Enzymes

Enzyme	Half-Life (h)
Short-Lived Enzymes	
Ornithine decarboxylase	0.2
RNA polymerase I	1.3
Tyrosine aminotransferase	2.0
Serine dehydratase	4.0
PEP carboxylase	5.0
Long-Lived Enzymes	
Aldolase	118
GAPDH	130
Cytochrome *b*	130
LDH	130
Cytochrome *c*	150

Source: Dice, J.F. and Goldberg, A.L., *Arch. Biochem. Biophys.* **170,** 214 (1975).

dental lysosomal leakage since lysosomal enzymes are largely inactive at cytosolic pH's.

Lysosomes degrade substances that the cell takes up via endocytosis (Section 10-3B). They also recycle intracellular constituents that are enclosed within vacuoles that fuse with lysosomes. *In well-nourished cells, lysosomal protein degradation is nonselective.* In starving cells, however, this degradation would deplete essential enzymes and regulatory proteins. Lysosomes therefore also have a selective pathway, which is activated only after a prolonged fast, that imports and degrades cytosolic proteins containing the pentapeptide Lys-Phe-Glu-Arg-Gln (KFERQ) or a closely related sequence. Such **KFERQ** proteins are selectively lost from tissues that atrophy in response to fasting (e.g., liver and kidney) but not from tissues that do not do so (e.g., brain and testes). Many normal and pathological processes are associated with increased lysosomal activity, for example, the muscle wastage caused by disuse, denervation, or traumatic injury. The regression of the uterus after childbirth, in which this muscular organ reduces its mass from 2 kg to 50 g in nine days, is a striking example of this process. Many chronic inflammatory diseases such as **rheumatoid arthritis** involve the extracellular release of lysosomal enzymes, which break down surrounding tissues.

B. Ubiquitin

Protein breakdown in eukaryotic cells also occurs in an ATP-requiring process that is independent of lysosomes. This process involves **ubiquitin** (Fig. 20-1), a 76-residue monomeric protein named for its ubiquity and abundance. It is one of the most highly conserved eukaryotic proteins known (it is identical in such diverse organisms as humans, trout, and *Drosophila*), suggesting that it is uniquely suited for some essential cellular function.

Proteins are marked for degradation by covalently linking them to ubiquitin. This process occurs in three steps elucidated notably by Avram Hershko (Fig. 20-2).

1. In an ATP-requiring reaction, ubiquitin's terminal carboxyl group is conjugated, via a thioester bond, to **ubiquitin-activating enzyme (E1).**

2. The ubiquitin is then transferred to a specific sulfhydryl group on one of numerous homologous proteins named **ubiquitin-conjugating enzymes (E2's).**

3. **Ubiquitin-protein ligase (E3)** transfers the activated ubiquitin from E2 to a Lys ε-amino group of a previously bound protein, thereby forming an **isopeptide bond.** E3 therefore appears to have a key role in selecting the protein to be degraded. However, the large number of different E2's in a cell (>10 in yeast and >20 in the plant *Arabidopsis thaliana*) suggests that these proteins also function in target

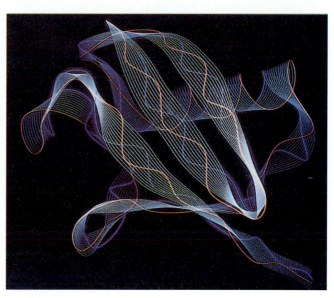

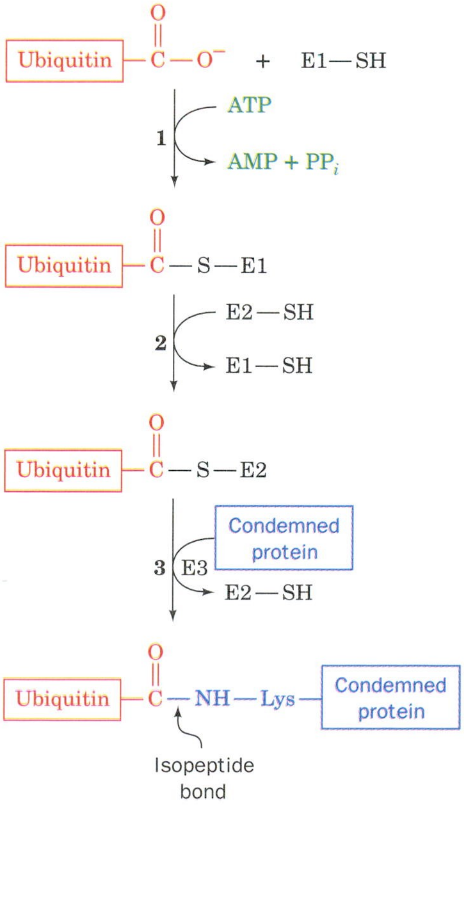

Figure 20-1.   **The X-ray structure of ubiquitin.** The white ribbon represents the polypeptide backbone and the red and blue curves, respectively, indicate the edges of the β strands toward which the carbonyl O's and the amide H's point. [Courtesy of Mike Carson, University of Alabama at Birmingham. X-Ray structure determined by Charles Bugg, University of Alabama at Birmingham.] ● See the Interactive Exercises.

Figure 20-2.   **The reactions involved in protein ubiquitination.** Ubiquitin's terminal carboxyl group is first joined, via a thioester linkage, to E1 in a reaction driven by ATP hydrolysis. The activated ubiquitin is subsequently transferred to a sulfhydryl group of E2 and then, in a reaction catalyzed by E3, to a Lys ε-amino group on a condemned protein, thereby marking the protein for proteolytic degradation.

protein selection. Indeed, some E2's transfer ubiquitin directly to target proteins.

Usually several ubiquitin molecules are linked to a condemned protein. In addition, 50 or more ubiquitin molecules may form a multiubiquitin chain linked to a target protein, in which Lys 48 of each ubiquitin forms an isopeptide bond with the C-terminal carboxyl group of the following ubiquitin. Indeed, multiubiquitination appears to be essential for the degradation of at least some proteins. Ubiquitinated proteins are dynamic entities, with ubiquitin molecules being rapidly attached and removed (the latter by **ubiquitin isopeptidases**).

The structural features by which at least native proteins are selected for destruction appear to be remarkably simple. The half-life of a cytoplasmic protein, as Alexander Varshavsky discovered, varies with the identity of its N-terminal residues via the so-called **N-end rule:** Proteins with N-terminal Asp, Arg, Leu, Lys, and Phe residues have half-lives of only 2–3 minutes, whereas those with N-terminal Ala, Gly, Met, Ser, and Val residues have half-lives of >10 hours in prokaryotes and >20 hours in eukaryotes. Evidently, E3 interacts with the N-terminal residue of a protein in selecting it for ubiquitination. Since the N-end rule holds for both prokaryotes and eukaryotes, this suggests that the system that selects proteins for degradation is conserved in prokaryotes and eukaryotes, even though prokaryotes lack ubiquitin. Nevertheless, it is clear that other, more complex signals are also important in the selection of proteins for degradation. For example, proteins with segments rich in Pro (P), Glu (E), Ser (S), and Thr (T) residues are rapidly degraded. Deleting these so-called **PEST** sequences prolongs the proteins' lifetimes, although the way in which they are recognized is unknown. Likewise, the criteria by which cells select defective proteins for degradation is unknown.

C. The Proteasome

Ubiquitinated proteins are proteolytically degraded in an ATP-dependent process mediated by a large (2000 kD, 26S) multiprotein complex named the **26S proteasome** (Fig. 20-3). The 26S proteasome consists of a hollow cylindrical core, known as the **20S proteasome,** which is covered at each end by a 19S "cap." The yeast 20S proteasome, which is closely similar to other eukaryotic 20S proteasomes, is composed of seven different types of α-like subunits and seven different types of β-like subunits. The X-ray structure of this enormous (6182-residue, 700-kD) protein complex, determined by Robert Huber, reveals that it consists of four stacked rings of subunits with its outer and inner rings, respectively, consisting of seven different α-type subunits and seven different β-type subunits (Fig. 20-4a). The various α-type subunits have folds that are similar to one another and likewise for the various β-type subunits. Consequently, this 28-subunit complex has exact two-fold rotational symmetry relating the two pairs of rings, but only quasi-seven-fold rotational symmetry relating the subunits within each ring (the 24,444 nonhydrogen atoms in the unique half of this complex make it the largest asymmetric unit whose structure is known in atomic detail).

Although the α-type subunits and the β-type subunits are also structurally similar, only the β-type subunits have proteolytic activity. The X-ray structure together with enzymological studies reveal that the β-type subunits catalyze peptide bond hydrolysis via a poorly understood mechanism that is unlike that of other known proteolytic enzymes and which involves the participation of the β-type subunits' N-terminal Thr residues. The active sites are located inside the barrel of the 20S proteasome, thereby preventing this omnivorous protein-dismantling machine from indiscrimi-

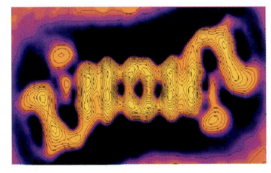

Figure 20-3.   An electron micrograph–based image of a 26S proteasome. The complex is around 450 × 190 Å and is seen with an estimated resolution of 25 Å. The central portion of this two-fold symmetric multiprotein complex, the 20S proteasome, consists of four stacked seven-membered rings of subunits that form a hollow barrel in which the proteolysis of ubiquitin-labeled proteins occurs. [Courtesy of Wolfgang Baumeister, Max-Planck-Institut für Biochemie, Germany.]

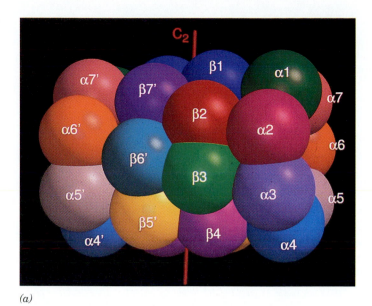

(a)

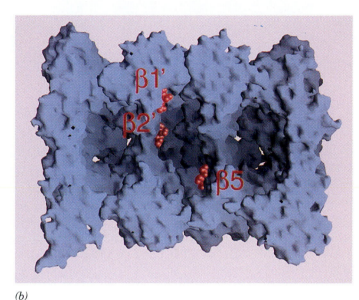

(b)

Figure 20-4.   The X-ray structure of the yeast 20S proteasome. (*a*) The arrangement of the 28 subunits shown as spheres. Four rings of seven subunits each are stacked to form a barrel (the ends of the barrel are at the right and left) with the α-type and β-type subunits forming the outer and inner rings, respec- tively. The complex's twofold (C_2) axis of symmetry is repre- sented by the vertical red line. (*b*) Surface view of the proteasome core cut along its cylindrical axis. Three bound protease inhibitor molecules are shown in red as space-filling models. [Courtesy of Robert Huber, Max-Planck-Institut für Biochemie, Germany.]

nately hydrolyzing the proteins in its vicinity. The seven different β-type subunits presumably confer a broad range of proteolytic specificities on the 20S proteasome. Nevertheless, the 20S proteasome cleaves its polypeptide substrates into ~8-residue fragments, which then diffuse out of the pro- teasome. Cytosolic peptidases then degrade these peptides to their com- ponent amino acids. The ubiquitin molecules attached to the target protein are not degraded, however, but are returned to the cell and reused.

The axial apertures in the α rings, through which unfolded polypeptides are thought to enter the 20S proteasome's hydrolytic chamber, are closed in the X-ray structure (Fig. 20-4*b*). This suggests that the 19S caps of the 26S proteasome, which have been shown to activate the 20S proteasome, control the access to it by inducing conformational changes in its α rings. Moreover, since the ATP-dependent degradation of ubiquitinated proteins is mediated by the 26S proteasome, whereas the 20S proteasome can only hydrolyze unfolded proteins in an ATP-independent fashion, it appears that the 19S caps also function to select and unfold an incoming ubiquitinated protein substrate. Although proteasomes are present in all cells, prokary- otes lack ubiquitin and contain only 20S proteasomes.

2. AMINO ACID DEAMINATION

Free amino acids originate from the degradation of cellular proteins and from the digestion of dietary proteins. The gastric protease **pepsin,** the pan- creatic enzymes trypsin, chymotrypsin, and elastase (discussed in Sections 5-3B and 11-5), and a host of other endo- and exopeptidases degrade polypeptides to oligopeptides and amino acids. These substances are ab- sorbed by the intestinal mucosa and transported via the bloodstream to be absorbed by other tissues.

The further degradation of amino acids takes place intracellularly and includes a step in which the α-amino group is removed. In many cases, the

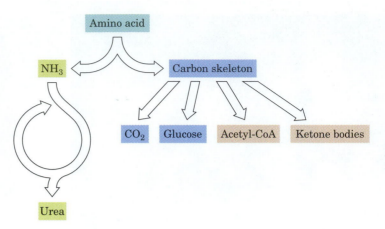

Figure 20-5.   Overview of amino acid catabolism. The amino group is removed and incorporated into urea for disposal. The remaining carbon skeleton (α-keto acid) can be broken down to CO_2 and H_2O or converted to glucose, acetyl-CoA, or ketone bodies.

amino group appears in glutamate and is then incorporated into urea for excretion (Section 20-3). The remaining carbon skeleton (α-keto acid) of the amino acid can be broken down to other compounds (Section 20-4). This metabolic theme is outlined in Fig. 20-5.

A.   Transamination

Most amino acids are deaminated by **transamination,** the transfer of their amino group to an α-keto acid to yield the α-keto acid of the original amino acid and a new amino acid. The predominant amino group acceptor is α-ketoglutarate, producing glutamate and the new α-keto acid:

$$\underset{\textbf{Amino acid}}{R-\overset{\overset{+}{N}H_3}{\underset{|}{C}}H-COO^-} \quad + \quad \underset{\textbf{α-Ketoglutarate}}{{}^-OOC-CH_2-CH_2-\overset{O}{\overset{||}{C}}-COO^-}$$

$$\Updownarrow$$

$$\underset{\textbf{α-Keto acid}}{R-\overset{O}{\overset{||}{C}}-COO^-} \quad + \quad \underset{\textbf{Glutamate}}{{}^-OOC-CH_2-CH_2-\overset{\overset{+}{N}H_3}{\underset{|}{C}}H-COO^-}$$

Glutamate's amino group, in turn, can be transferred to oxaloacetate in a second transamination reaction, yielding aspartate and re-forming α-ketoglutarate:

$$\underset{\textbf{Glutamate}}{{}^-OOC-CH_2-CH_2-\overset{\overset{+}{N}H_3}{\underset{|}{C}}H-COO^-} \quad + \quad \underset{\textbf{Oxaloacetate}}{{}^-OOC-CH_2-\overset{O}{\overset{||}{C}}-COO^-}$$

$$\Updownarrow$$

$$\underset{\textbf{α-Ketoglutarate}}{{}^-OOC-CH_2-CH_2-\overset{O}{\overset{||}{C}}-COO^-} \quad + \quad \underset{\textbf{Aspartate}}{{}^-OOC-CH_2-\overset{\overset{+}{N}H_3}{\underset{|}{C}}H-COO^-}$$

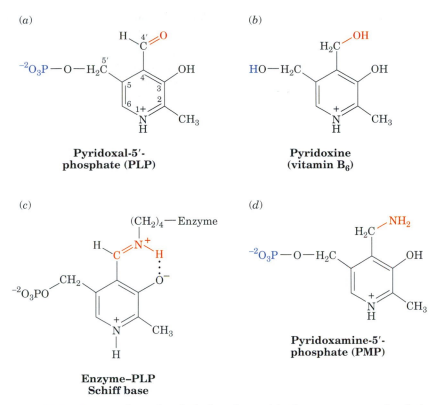

Figure 20-6. Forms of pyridoxal-5′-phosphate. (*a*) The coenzyme pyridoxal-5′-phosphate (PLP). (*b*) Pyridoxine (vitamin B₆). (*c*) The Schiff base that forms between PLP and an enzyme ε-amino group. (*d*) Pyridoxamine-5′-phosphate (PMP).

The enzymes that catalyze transamination, which are called **aminotransferases** or **transaminases,** require the coenzyme **pyridoxal-5′-phosphate** (**PLP;** Fig. 20-6*a*). PLP is a derivative of **pyridoxine** (**vitamin B₆;** Fig. 20-6*b*). The coenzyme is covalently attached to the enzyme via a Schiff base (imine) linkage formed by the condensation of its aldehyde group with the ε-amino group of an enzyme Lys residue (Fig. 20-6*c*). The Schiff base, which is conjugated to the pyridinium ring, is the center of the coenzyme's activity. When PLP accepts the amino group from an amino acid as described below, it becomes **pyridoxamine-5′-phosphate** (**PMP;** Fig. 20-6*d*).

Esmond Snell, Alexander Braunstein, and David Metzler demonstrated that the aminotransferase reaction occurs via a Ping Pong mechanism (Section 12-1D) whose two stages consist of three steps each (Fig. 20-7):

Stage I: Conversion of an Amino Acid to a Keto Acid

1. The amino acid's nucleophilic amino group attacks the enzyme–PLP Schiff base carbon atom in a **transimination** reaction to form an amino acid–Schiff base (an aldimine), with concomitant release of the enzyme's Lys amino group. This Lys is then free to act as a general base at the active site.

2. The amino acid–PLP Schiff base tautomerizes to an α-keto acid–PMP Schiff base (a ketimine) by the active-site Lys–catalyzed removal of the amino acid α-hydrogen and protonation of PLP atom C4′ via a resonance-stabilized carbanion intermediate. This resonance stabilization facilitates the cleavage of the C_α—H bond.

Steps 1 & 1′: Transimination:

Steps 2 & 2′: Tautomerization:

Steps 3 & 3′: Hydrolysis:

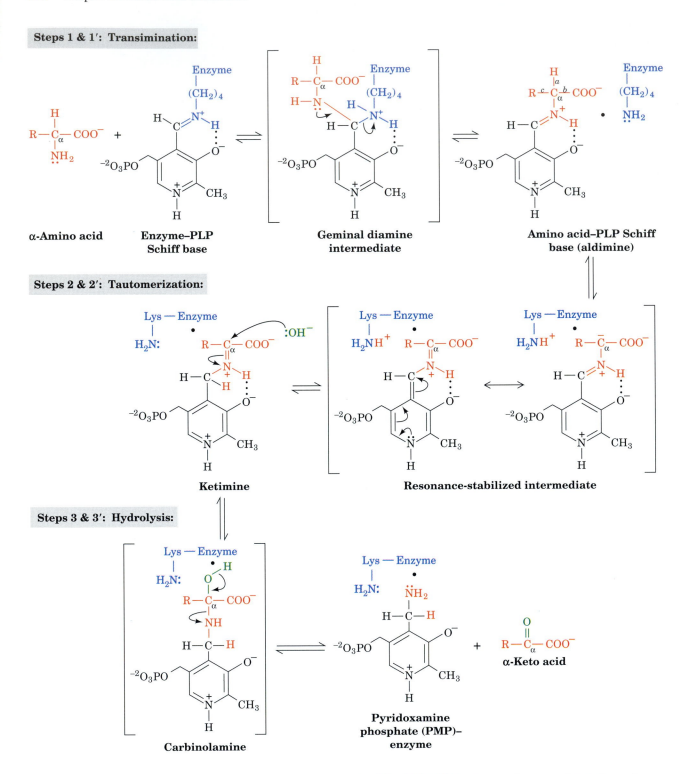

Figure 20-7. *Key to Function.* **The mechanism of PLP-dependent enzyme-catalyzed transamination.** The first stage of the reaction, in which the α-amino group of an amino acid is transferred to PLP yielding an α-keto acid and PMP, consists of three steps: (1) transimination; (2) tautomerization, in which the Lys released during the transimination reaction acts as a general acid–base catalyst; and (3) hydrolysis. The second stage of the reaction, in which the amino group of PMP is transferred to a different α-keto acid to yield a new α-amino acid and PLP, is essentially the reverse of the first stage: Steps 3′, 2′, and 1′ are, respectively, the reverse of Steps 3, 2, and 1. ✳ See the Animated Figures.

3. The α-keto acid–PMP Schiff base is hydrolyzed to PMP and an α-keto acid.

Stage II: Conversion of an α-Keto Acid to an Amino Acid

To complete the aminotransferase's catalytic cycle, the coenzyme must be converted from PMP back to the enzyme–PLP Schiff base. This involves the same three steps as above, but in reverse order:

3′. PMP reacts with an α-keto acid to form a Schiff base.

2′. The α-keto acid–PMP Schiff base tautomerizes to form an amino acid–PLP Schiff base.

1′. The ε-amino group of the active site Lys residue attacks the amino acid–PLP Schiff base in a transimination reaction to regenerate the active enzyme–PLP Schiff base and release the newly formed amino acid.

Note that removal of the substrate amino acid's amino group produces a resonance-stabilized C_α carbanion whose electrons are delocalized all the way to the coenzyme's protonated pyridinium nitrogen atom; that is, *PLP functions as an electron sink.* For transamination reactions, this electron-withdrawing capacity facilitates removal of the α proton (*a* bond cleavage, top right of Fig. 20-7) during tautomerization. PLP functions similarly in enzymatic reactions involving *b* and *c* bond cleavage.

Aminotransferases differ in their specificity for amino acid substrates in the first stage of the transamination reaction, thereby producing the correspondingly different α-keto acid products. Most aminotransferases, however, accept only α-ketoglutarate or (to a lesser extent) oxaloacetate as the α-keto acid substrate in the second stage of the reaction, thereby yielding glutamate or aspartate as their only amino acid product. *The amino groups from most amino acids are consequently funneled into the formation of glutamate or aspartate.*

B. Oxidative Deamination

Transamination, of course, does not result in any net deamination. Glutamate, however, is oxidatively deaminated by **glutamate dehydrogenase,** yielding ammonia and regenerating α-ketoglutarate for use in additional transamination reactions.

Glutamate dehydrogenase, a mitochondrial enzyme, is the only known enzyme that can accept either NAD^+ or $NADP^+$ as its redox coenzyme. Oxidation is thought to occur with transfer of a hydride ion from glutamate's C_α to $NAD(P)^+$, thereby forming α-iminoglutarate, which is hydrolyzed to α-ketoglutarate and ammonia:

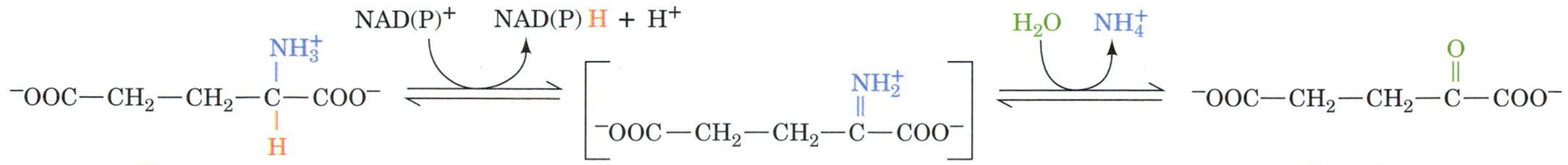

$\Delta G°′$ for this reaction is ~30 kJ·mol^{-1}, but glutamate dehydrogenase functions close to equilibrium ($\Delta G \approx 0$) *in vivo.* Flux is probably controlled by the concentrations of substrates and products. Because high concentrations of ammonia are toxic, the equilibrium position is physiologically important. The ammonia produced is converted to urea for disposal.

3. THE UREA CYCLE

Living organisms excrete the excess nitrogen arising from the metabolic breakdown of amino acids in one of three ways. Many aquatic animals simply excrete ammonia. Where water is less plentiful, however, processes have evolved that convert ammonia to less toxic waste products that require less water for excretion. One such product is urea, which is produced by most terrestrial vertebrates; another is **uric acid,** which is excreted by birds and terrestrial reptiles.

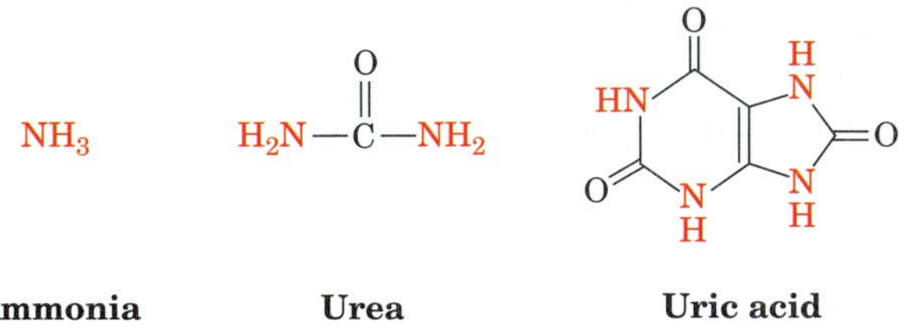

| Ammonia | Urea | Uric acid |

In this section, we focus our attention on urea formation. Uric acid biosynthesis is discussed in Section 22-4A.

Urea is synthesized in the liver by the enzymes of the **urea cycle.** It is then secreted into the bloodstream and sequestered by the kidneys for excretion in the urine. The urea cycle was outlined in 1932 by Hans Krebs and Kurt Henseleit (the first known metabolic cycle; Krebs did not elucidate the citric acid cycle until 1937). Its individual reactions were later described in detail by Sarah Ratner and Philip Cohen. The overall urea cycle reaction is

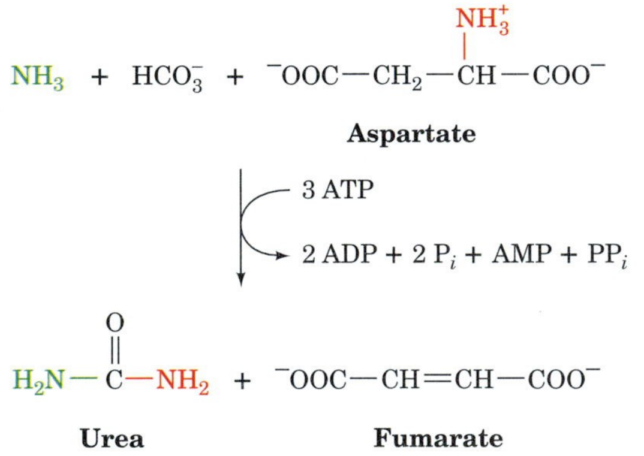

Thus, urea's two nitrogen atoms are contributed by ammonia and aspartate, whereas its carbon atom comes from HCO_3^-.

A. Reactions of the Urea Cycle

Five enzymatic reactions are involved in the urea cycle, two of which are mitochondrial and three cytosolic (Fig. 20-8):

Figure 20-8 (Opposite). **Key to Metabolism. The urea cycle.** Five enzymes participate in the urea cycle: (**1**) carbamoyl phosphate synthetase, (**2**) ornithine transcarbamoylase, (**3**) argininosuccinate synthetase, (**4**) argininosuccinase, and (**5**) arginase. Enzymes 1 and 2 are mitochondrial and enzymes 3–5 are cytosolic. Ornithine and citrulline must therefore be transported across the mitochondrial membrane by specific transport systems (*yellow circles*). The urea amino groups arise from the deamination of amino acids. One amino group (*green*) originates as ammonia that is generated by the glutamate dehydrogenase reaction (*top*). The other amino group (*red*) is transferred from an amino acid to oxaloacetate via transamination (*right*). The oxaloacetate is derived from the fumarate product of the argininosuccinase reaction via the same reactions that occur in the citric acid cycle but which take place in the cytosol (*bottom*). ✳ **See the Animated Figures.**

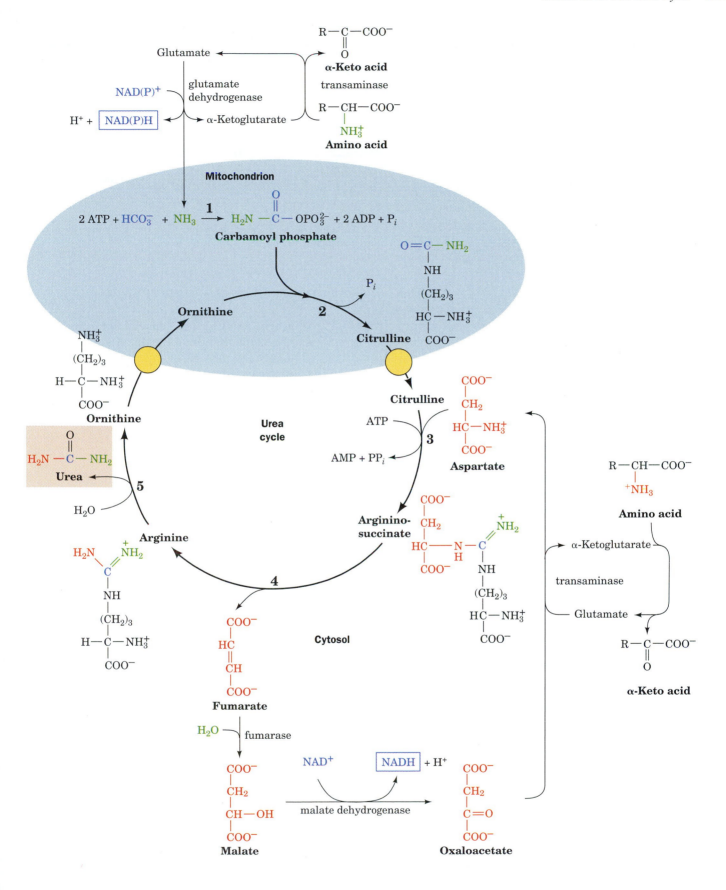

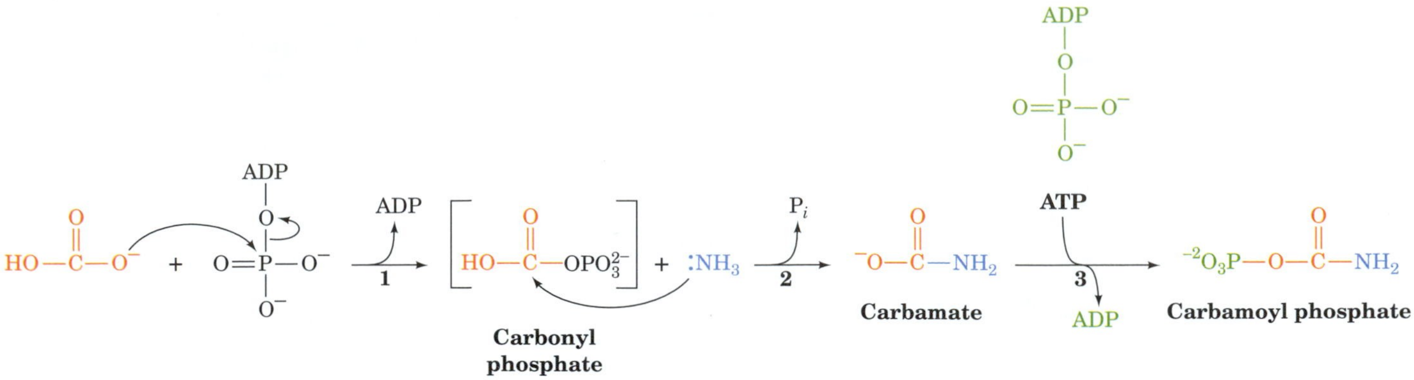

Figure 20-9. **The mechanism of action of CPS I.** (1) Phosphorylation activates HCO_3^- to form the postulated intermediate, carbonyl phosphate. (2) NH_3 attacks carbonyl phosphate to form carbamate. (3) ATP phosphorylates carbamate, yielding carbamoyl phosphate.

1. Carbamoyl Phosphate Synthetase: Acquisition of the First Urea Nitrogen Atom

Carbamoyl phosphate synthetase (CPS) is technically not a member of the urea cycle. It catalyzes the condensation and activation of NH_4^+ and HCO_3^- to form **carbamoyl phosphate,** the first of the cycle's two nitrogen-containing substrates, with the concomitant cleavage of 2 ATP. Eukaryotes have two forms of CPS: Mitochondrial **CPS I** uses ammonia as its nitrogen donor and participates in urea biosynthesis, whereas cytosolic **CPS II** uses glutamine as its nitrogen donor and is involved in pyrimidine biosynthesis (Section 22-2A). CPS I catalyzes an essentially irreversible reaction that is the rate-limiting step of the urea cycle (Fig. 20-9):

1. ATP activates HCO_3^- to form **carbonyl phosphate** and ADP.
2. Ammonia attacks carbonyl phosphate, displacing the phosphate to form **carbamate** and P_i.
3. A second ATP phosphorylates carbamate to form carbamoyl phosphate and ADP.

2. Ornithine Transcarbamoylase

Ornithine transcarbamoylase transfers the carbamoyl group of carbamoyl phosphate to **ornithine,** yielding **citrulline** (Fig. 20-8, Reaction 2). Note that both of these latter compounds are "nonstandard" α-amino acids that do not occur in proteins. The transcarbamoylase reaction occurs in the mitochondrion, so ornithine, which is produced in the cytosol, must enter the mitochondrion via a specific transport system. Likewise, since the remaining urea cycle reactions occur in the cytosol, citrulline must be exported from the mitochondrion.

3. Argininosuccinate Synthetase: Acquisition of the Second Urea Nitrogen Atom

Urea's second nitrogen atom is introduced by the condensation of citrulline's ureido group with an aspartate amino group by **argininosuccinate synthetase** (Fig. 20-10). ATP activates the ureido oxygen atom as a leaving group through formation of a citrullyl–AMP intermediate, and AMP is subsequently displaced by the aspartate amino group. The PP_i formed in this reaction is hydrolyzed to 2 P_i, so the reaction consumes two ATP equivalents.

4. Argininosuccinase

With the formation of argininosuccinate, all of the urea molecule components have been assembled. However, the amino group donated by as-

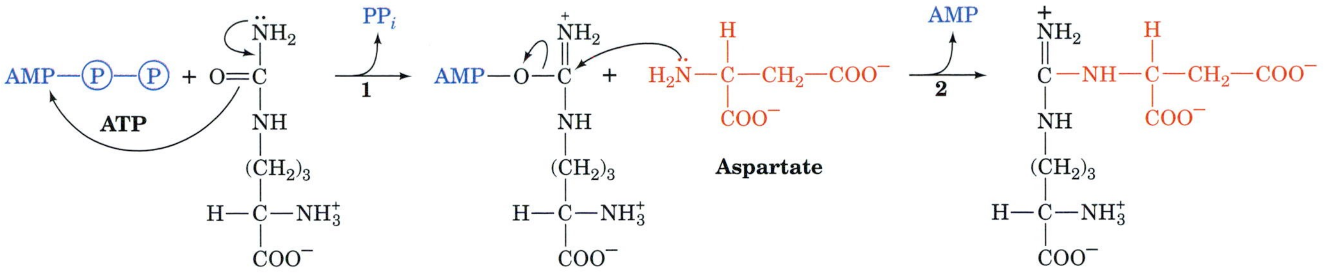

Figure 20-10. **The mechanism of action of argininosuccinate synthetase.**
(1) The formation of citrullyl–AMP activates the ureido oxygen of citrulline.
(2) The α-amino group of aspartate displaces AMP.

partate is still attached to the aspartate carbon skeleton. This situation is remedied by the **argininosuccinase**-catalyzed elimination of fumarate, leaving arginine (Fig. 20-8, Reaction 4). Arginine is urea's immediate precursor. The fumarate produced in the argininosuccinase reaction can be reconverted to aspartate for reuse in the argininosuccinate synthetase reaction (Fig. 20-8, *lower right*). This occurs via the fumarase and malate dehydrogenase reactions to form oxaloacetate followed by transamination to aspartate. The first two reactions are the same as those that occur in the citric acid cycle, although they take place in the cytosol rather than in the mitochondrion.

5. Arginase

The urea cycle's final reaction is the **arginase**-catalyzed hydrolysis of arginine to yield urea and regenerate ornithine (Fig. 20-8, Reaction 5). Ornithine is then returned to the mitochondrion for another round of the cycle.

The urea cycle thus converts two amino groups, one from ammonia and one from aspartate, and a carbon atom from HCO_3^- to the relatively nontoxic product, urea, at the cost of four "high-energy" phosphate bonds. The energy spent is more than recovered, however, during the formation of urea cycle substrates. The ammonia substrate for CPS I is a product of the glutamate dehydrogenase reaction, along with NAD(P)H (Fig. 20-8, *upper left*). The reconversion of fumarate to oxaloacetate (Fig. 20-8, *lower right*) also produces NADH. Mitochondrial reoxidation of these reduced coenzymes yields 6 ATP.

B. Regulation of the Urea Cycle

Carbamoyl phosphate synthetase I, which catalyzes the first committed step of the urea cycle, is allosterically activated by **N-acetylglutamate** *(at right)*. This metabolite is synthesized from glutamate and acetyl-CoA by **N-acetylglutamate synthase.** When amino acid breakdown rates increase, the concentration of glutamate increases as a result of transamination. The increased glutamate stimulates N-acetylglutamate synthesis. The resulting activation of carbamoyl phosphate synthetase increases the rate of urea production. Thus, the excess nitrogen produced by amino acid breakdown is efficiently excreted.

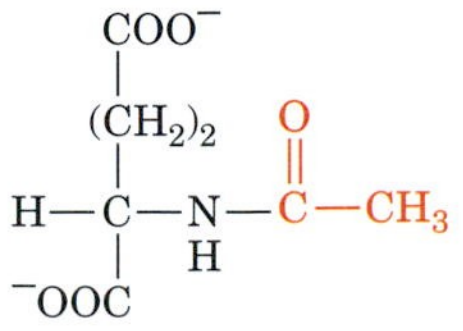

The remaining enzymes of the urea cycle are controlled by the concentrations of their substrates. In individuals with inherited deficiencies in urea cycle enzymes other than arginase, the corresponding substrate builds up, increasing the rate of the deficient reaction so that the rate of urea pro-

***N*-Acetylglutamate**

duction is normal (the total lack of a urea cycle enzyme, however, is lethal). The anomalous substrate buildup is not without cost, however. The substrate concentrations become elevated all the way back up the cycle to ammonia, resulting in **hyperammonemia** (elevated levels of ammonia in the blood). Although the root cause of ammonia toxicity is not completely understood, it is clear that the brain is particularly sensitive to high ammonia concentrations (symptoms of urea cycle enzyme deficiencies include mental retardation and lethargy).

4. BREAKDOWN OF AMINO ACIDS

Amino acids are degraded to compounds that can be metabolized to CO_2 and H_2O or used in gluconeogenesis. Indeed, oxidative breakdown of amino acids typically accounts for 10 to 15% of the metabolic energy generated by animals. In this section we consider how the carbon skeletons of the 20 "standard" amino acids are catabolized. We shall not describe in detail all of the many reactions involved. Rather, we shall consider how these pathways are organized and focus on a few reactions of chemical and/or medical interest.

"Standard" amino acids are degraded to one of seven metabolic intermediates: pyruvate, α-ketoglutarate, succinyl-CoA, fumarate, oxaloacetate, acetyl-CoA, or acetoacetate (Fig. 20-11). The amino acids can therefore be divided into two groups on the basis of their catabolic pathways:

1. Glucogenic amino acids, which are degraded to pyruvate, α-ketoglu-

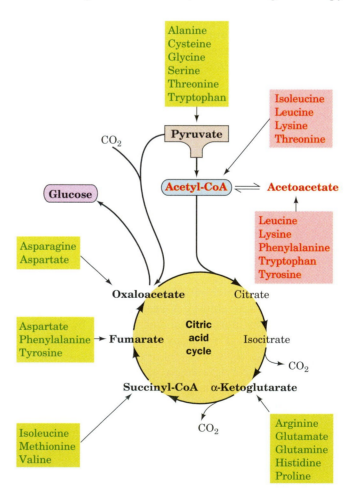

Figure 20-11. **Degradation of amino acids to one of seven common metabolic intermediates.** Glucogenic and ketogenic degradations are indicated in green and red, respectively.

tarate, succinyl-CoA, fumarate, or oxaloacetate and are therefore glucose precursors (Section 15-4).

2. **Ketogenic amino acids,** which are broken down to acetyl-CoA or acetoacetate and can thus be converted to fatty acids or ketone bodies (Section 19-3).

Some amino acids are precursors of both carbohydrates and ketone bodies. Since animals lack any metabolic pathways for the net conversion of acetyl-CoA or acetoacetate to gluconeogenic precursors, *no net synthesis of carbohydrates is possible from the purely ketogenic amino acids.*

In studying the specific pathways of amino acid breakdown, we shall organize the amino acids into groups that are degraded to each of the seven metabolites mentioned above.

A. Alanine, Cysteine, Glycine, Serine, and Threonine Are Degraded to Pyruvate

Five amino acids—alanine, cysteine, glycine, serine, and threonine—are broken down to yield pyruvate (Fig. 20-12). Alanine is straightforwardly transaminated to pyruvate. Serine is converted to pyruvate through dehy-

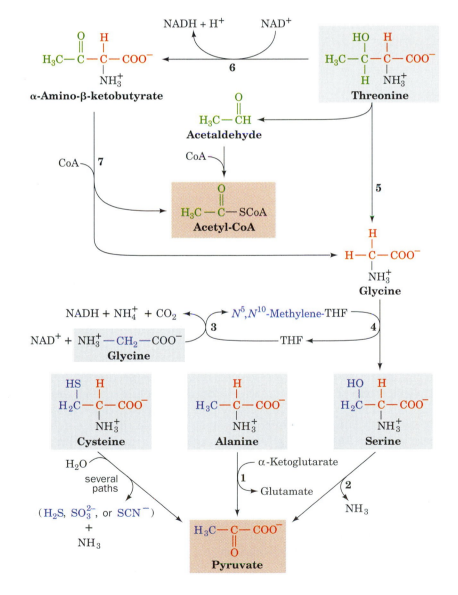

Figure 20-12. **The pathways converting alanine, cysteine, glycine, serine, and threonine to pyruvate.** The enzymes involved are (1) alanine aminotransferase, (2) serine dehydratase, (3) glycine cleavage system, (4 and 5) serine hydroxymethyltransferase, (6) threonine dehydrogenase, and (7) α-amino-β-ketobutyrate lyase.

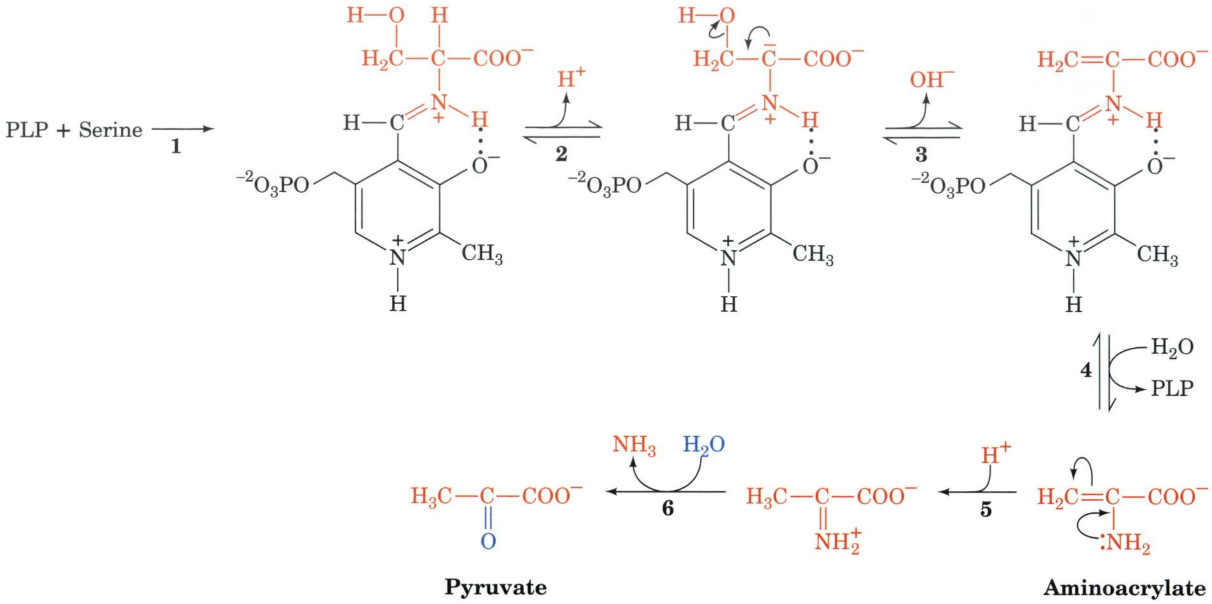

Figure 20-13.   The serine dehydratase reaction. This PLP-dependent enzyme catalyzes the elimination of water from serine in six steps: (1) formation of a serine–PLP Schiff base, (2) removal of the α-H atom of serine to form a resonance-stabilized carbanion, (3) β elimination of OH^-, (4) hydrolysis of the Schiff base to yield the PLP–enzyme and aminoacrylate, (5) nonenzymatic tautomerization to the imine, and (6) nonenzymatic hydrolysis to form pyruvate and ammonia.

dration by **serine dehydratase.** This PLP-dependent enzyme, like the aminotransferases (Section 20-2A), forms a PLP–amino acid Schiff base which facilitates the removal of the amino acid's α-hydrogen atom. In the serine dehydratase reaction, however, the C_α carbanion breaks down with the elimination of the amino acid's C_β OH, rather than with tautomerization (Fig. 20-7, Step 2), so that the substrate undergoes α,β elimination of H_2O rather than deamination (Fig. 20-13). The product of the dehydration, the enamine **aminoacrylate,** tautomerizes nonenzymatically to the corresponding imine, which spontaneously hydrolyzes to pyruvate and ammonia.

Cysteine can be converted to pyruvate via several routes in which the sulfhydryl group is released as H_2S, SO_3^{2-}, or SCN^-.

Glycine is converted to pyruvate by first being converted to serine by the enzyme **serine hydroxymethyltransferase,** another PLP-containing enzyme (Fig. 20-12, Reaction 4). This enzyme uses N^5,N^{10}-**methylene-tetrahydrofolate (N^5,N^{10}-methylene-THF)** as a one-carbon donor (the structure and chemistry of THF cofactors are described in Section 20-4D). The methylene group of the THF cofactor is obtained from a second glycine in Reaction 3 of Fig. 20-12, which is catalyzed by the **glycine cleavage system.** This enzyme is a multiprotein complex that resembles pyruvate dehydrogenase (Section 16-2). The glycine cleavage system mediates the major route of glycine degradation in mammalian tissues. An inherited deficiency of the glycine cleavage system causes the disease **nonketotic hyperglycinemia,** which is characterized by mental retardation and accumulation of large amounts of glycine in body fluids.

Threonine is both glucogenic and ketogenic since it generates both pyruvate and acetyl-CoA. Its major route of breakdown is through **threonine dehydrogenase** (Fig. 20-12, Reaction 6), producing **α-amino-β-keto-butyrate,** which is converted to acetyl-CoA and glycine by **α-amino-β-**

ketobutyrate lyase (Fig. 20-12, Reaction 7). The glycine can be converted, through serine, to pyruvate.

Serine Hydroxymethyltransferase Catalyzes PLP-Dependent C_α—C_β Bond Formation and Cleavage

Threonine can also be converted directly to glycine and acetaldehyde (which is subsequently oxidized to acetyl-CoA) via Reaction 5 of Fig. 20-12, which breaks threonine's C_α—C_β bond. This PLP-dependent reaction is catalyzed by serine hydroxymethyltransferase, the same enzyme that adds a hydroxymethyl group to glycine to produce serine (Fig. 20-12, Reaction 4). In the glycine → serine reaction, the amino acid's C_α—H bond is cleaved (as occurs in transamination; Fig. 20-7) and a C_α—C_β bond is formed. In contrast, the degradation of threonine to glycine by serine hydroxymethyltransferase acts in reverse, beginning with C_α—C_β bond cleavage.

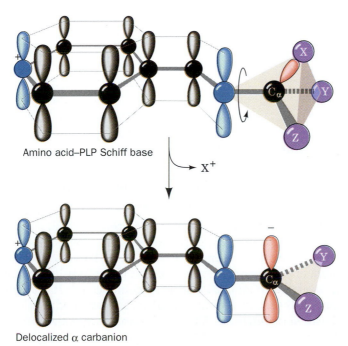

With the cleavage of any of the bonds to C_α, the PLP group delocalizes the electrons of the resulting carbanion. This feature of PLP action is the key to understanding how the same amino acid–PLP Schiff base can undergo cleavage of different bonds to C_α in different enzymes (bonds *a*, *b*, or *c* in the upper right of Fig. 20-7). The bond that is cleaved is the one that lies in the plane perpendicular to that of the π-orbital system of the PLP (Fig. 20-14). This arrangement allows the PLP π-orbital system to

Amino acid–PLP Schiff base

X^+

Delocalized α carbanion

Figure 20-14. **The π-orbital framework of a PLP–amino acid Schiff base.** The bond from X to C_α is in a plane perpendicular to the plane of the PLP π-orbital system (*top*) and is therefore labile. The broken bond's electron pair (*bottom*) is delocalized over the conjugated molecule.

overlap the bonding orbital containing the electron pair being delocalized. Any other geometry would result in the newly formed double bond being twisted out of planarity, a high-energy arrangement. Different bonds to C_α can be positioned for cleavage by rotation around the C_α—N bond. Evidently, *each enzyme binds its amino acid–PLP Schiff base adduct with the appropriate geometry for bond cleavage.*

B. Asparagine and Aspartate Are Degraded to Oxaloacetate

Transamination of aspartate leads directly to oxaloacetate:

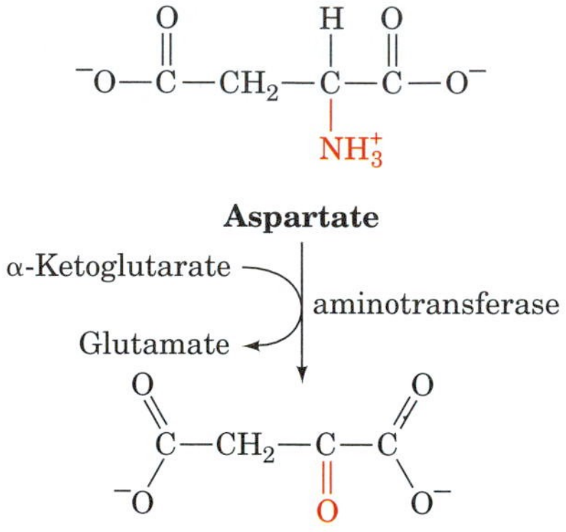

Aspartate

α-Ketoglutarate ⟍
                    ⟩ aminotransferase
Glutamate ⟋

Oxaloacetate

Asparagine is also converted to oxaloacetate in this manner after its hydrolysis to aspartate by **L-asparaginase:**

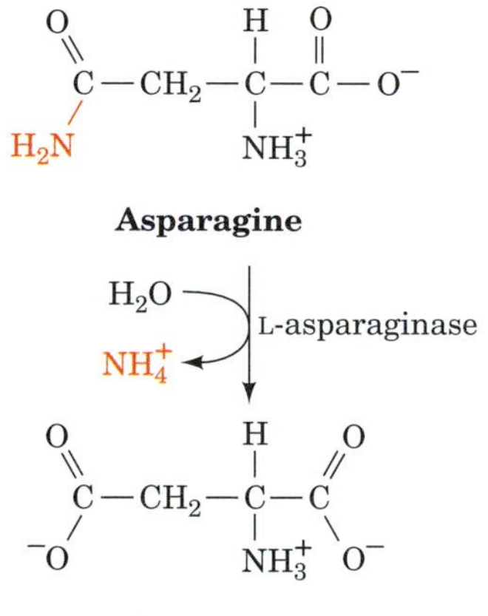

Asparagine

H_2O ⟍
         ⟩ L-asparaginase
NH_4^+ ⟋

Aspartate

Interestingly, L-asparaginase is an effective chemotherapeutic agent in the treatment of cancers that must obtain asparagine from the blood, particularly **acute lymphoblastic leukemia.**

C. Arginine, Glutamate, Glutamine, Histidine, and Proline Are Degraded to α-Ketoglutarate

Arginine, glutamine, histidine, and proline are all degraded by conversion to glutamate (Fig. 20-15), which in turn is oxidized to α-ketoglutarate by glutamate dehydrogenase (Section 20-2B). Conversion of glutamine to glutamate involves only one reaction: hydrolysis by **glutaminase.** Histidine's conversion to glutamate is more complicated: It is nonoxidatively deaminated, then it is hydrated, and its imidazole ring is cleaved to form **N-formiminoglutamate.** The formimino group is then transferred to tetrahy-

Figure 20-15. The degradation of arginine, glutamate, glutamine, histidine, and proline to α-ketoglutarate. The enzymes catalyzing the reactions are (1) glutamate dehydrogenase, (2) glutaminase, (3) arginase, (4) ornithine-δ-aminotransferase, (5) glutamate-5-semialdehyde dehydrogenase, (6) proline oxidase, (7) spontaneous, (8) histidine ammonia-lyase, (9) urocanate hydratase, (10) imidazalone propionase, and (11) glutamate formiminotransferase.

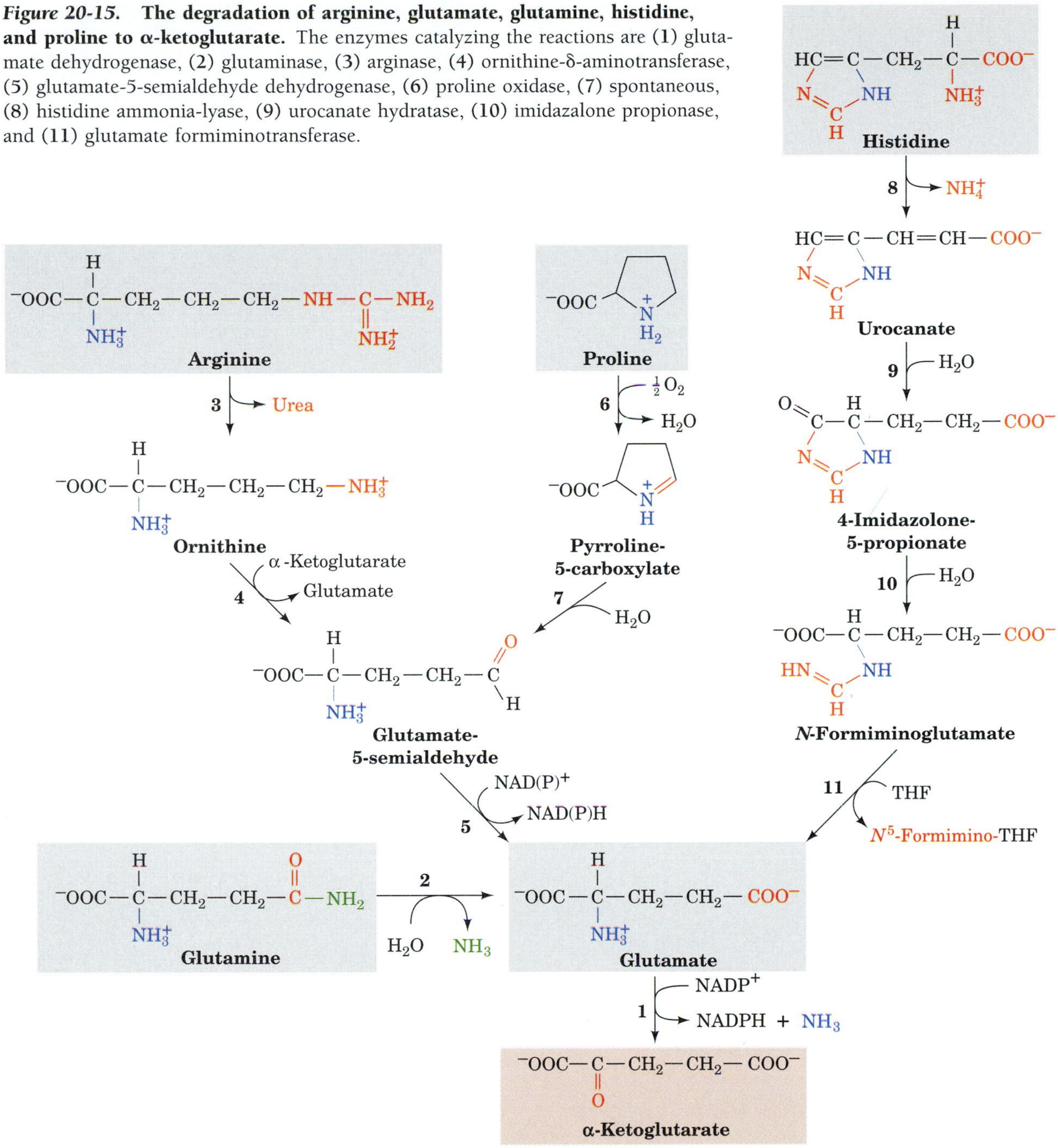

drofolate forming glutamate and N^5-**formimino-tetrahydrofolate** (Section 20-4D). Both arginine and proline are converted to glutamate through the intermediate formation of **glutamate-5-semialdehyde.**

D. *Isoleucine, Methionine, and Valine Are Degraded to Succinyl-CoA*

Isoleucine, methionine, and valine have complex degradative pathways that all yield propionyl-CoA, which is also a product of odd-chain fatty acid degradation. Propionyl-CoA is converted to succinyl-CoA by a series of reactions requiring biotin and coenzyme B_{12} (Section 19-2E).

Methionine Breakdown Involves Synthesis of *S*-Adenosylmethionine and Cysteine

Methionine degradation (Fig. 20-16) begins with its reaction with ATP to form **S-adenosylmethionine (SAM;** alternatively **AdoMet).** *This sulfonium ion's highly reactive methyl group makes it an important biological*

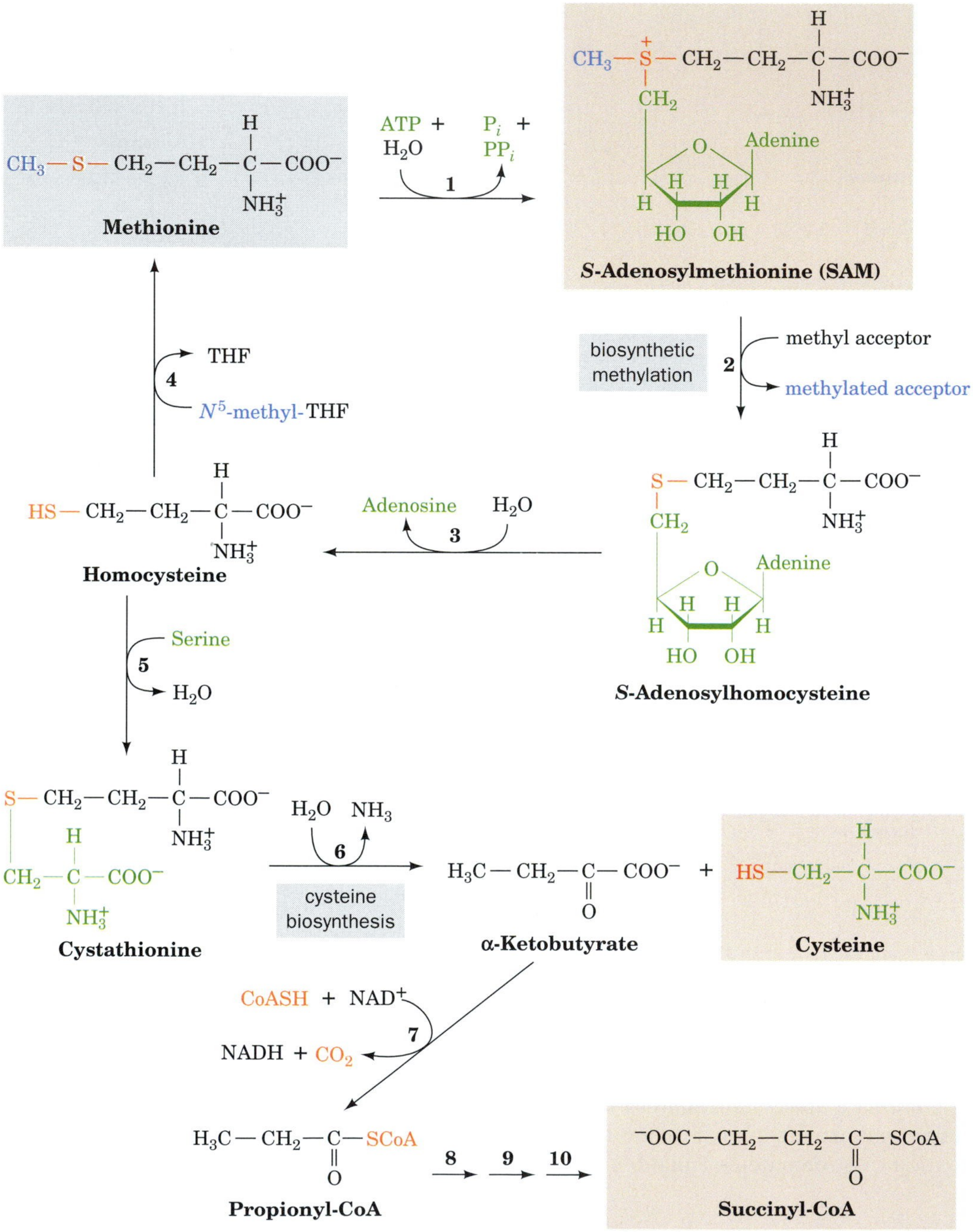

Figure 20-16.   Methionine degradation. This pathway yields cysteine and succinyl-CoA. The enzymes are (**1**) methionine adenosyltransferase in a reaction that yields the biological methylating agent *S*-adenosylmethionine (SAM), (**2**) methylase, (**3**) adenosylhomocysteinase, (**4**) methionine synthase (a coenzyme B_{12}–dependent enzyme), (**5**) cystathionine β-synthase (a PLP-dependent enzyme), (**6**) cystathionine γ-lyase, (**7**) α-keto acid dehydrogenase, (**8**) propionyl-CoA carboxylase, (**9**) methylmalonyl-CoA racemase, and (**10**) methylmalonyl-CoA mutase (a coenzyme B_{12}–dependent enzyme). Reactions 8–10 are discussed in Section 19-2E.

methylating agent. For instance, SAM is the methyl donor in the synthesis of phosphatidylcholine from phosphatidylethanolamine (Section 19-6A).

Donation of a methyl group from SAM leaves **S-adenosylhomocysteine,** which is then hydrolyzed to adenosine and **homocysteine.** The homocysteine can be methylated to re-form methionine via a reaction in which N^5-**methyltetrahydrofolate** (see below) is the methyl donor. Alternatively, the homocysteine can combine with serine to yield **cystathionine,** which subsequently forms cysteine (cysteine biosynthesis) and **α-ketobutyrate.** The α-ketobutyrate continues along the degradative pathway to propionyl-CoA and then succinyl-CoA.

Tetrahydrofolates Are One-Carbon Carriers

Many biosynthetic processes involve the addition of a C_1 unit to a metabolic precursor. In most carboxylation reactions (e.g., pyruvate carboxylase; Fig. 15-25), the enzyme uses a biotin cofactor. In some reactions, S-adenosylmethionine (Fig. 20-16) functions as a methylating agent. However, tetrahydrofolate (THF) is more versatile than either of these cofactors because it can transfer C_1 units in several oxidation states.

THF is a 6-methylpterin derivative linked in sequence to a **p-aminobenzoic acid** and a Glu residue.

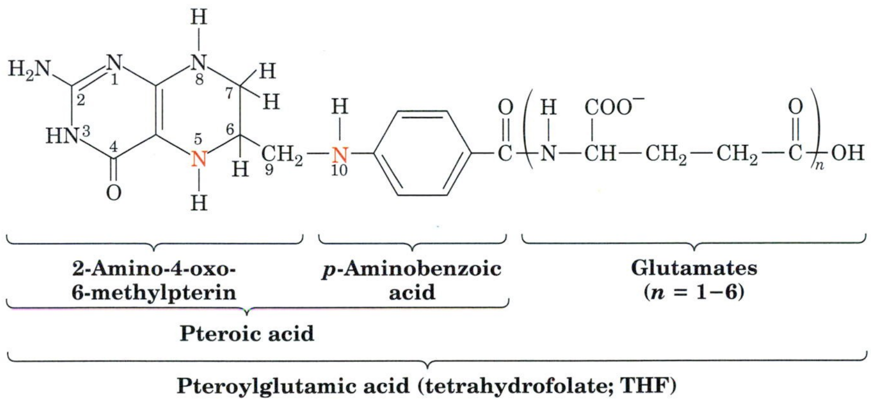

Pteroylglutamic acid (tetrahydrofolate; THF)

Up to five additional Glu residues are linked to the first glutamate via isopeptide bonds to form a polyglutamyl tail. THF is derived from the vitamin **folic acid** (Latin: *folium,* leaf), a doubly oxidized form of THF that must be enzymatically reduced before it becomes an active coenzyme (Fig. 20-17). Both reductions are catalyzed by **dihydrofolate reductase (DHFR).** Mammals cannot synthesize folic acid, so it must be provided in the diet or by intestinal microorganisms.

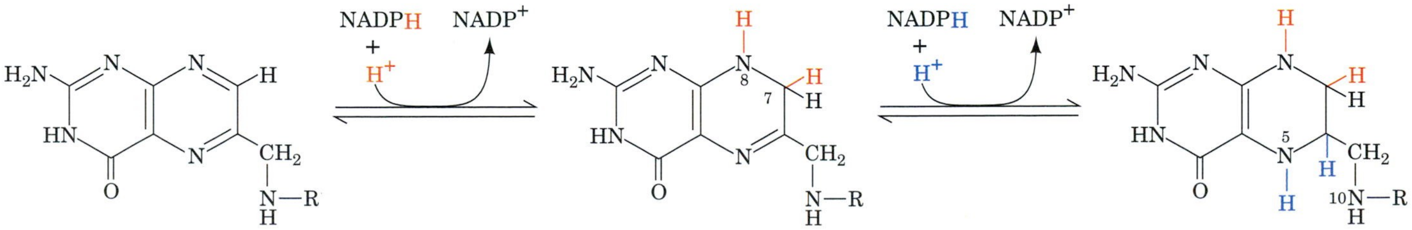

Figure 20-17.   The two-stage reduction of folate to THF. Both reactions are catalyzed by dihydrofolate reductase.

C_1 units are covalently attached to THF at positions N5, N10, or both N5 and N10. These C_1 units, which may be at the oxidation levels of formate, formaldehyde, or methanol (Table 20-2), are all interconvertible by enzymatic redox reactions (Fig. 20-18).

THF acquires C_1 units in the conversion of serine to glycine by serine hydroxymethyltransferase (the reverse of Reaction 4, Fig. 20-12), in the cleavage of glycine (Fig. 20-12, Reaction 3), and in histidine breakdown (Fig. 20-15, Reaction 11). The C_1 units carried by THF are used in the synthesis of methionine from homocysteine (Fig. 20-16) and in the synthesis of thymine nucleotides (Section 22-3B).

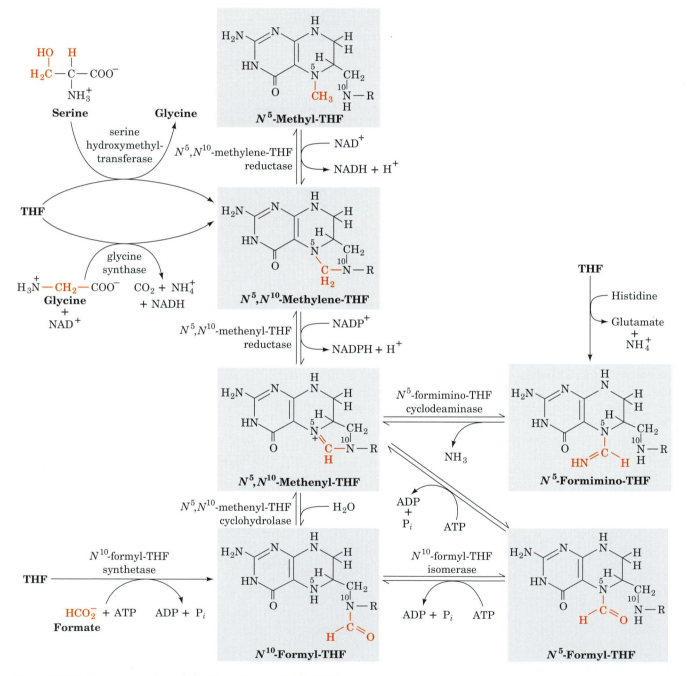

Figure 20-18. Interconversion of the C_1 units carried by THF.

Table 20-2. Oxidation Levels of C₁ Groups Carried by THF

Oxidation Level	Group Carried	THF Derivative(s)
Methanol	Methyl (—CH₃)	N^5-Methyl-THF
Formaldehyde	Methylene (—CH₂—)	N^5,N^{10}-Methylene-THF
Formate	Formyl (—CH=O)	N^5-Formyl-THF, N^{10}-formyl-THF
	Formimino (—CH=NH)	N^5-Formimino-THF
	Methenyl (—CH=)	N^5,N^{10}-Methenyl-THF

Sulfonamides (sulfa drugs) such as **sulfanilamide** are antibiotics that are structural analogs of the *p*-aminobenzoic acid constituent of THF.

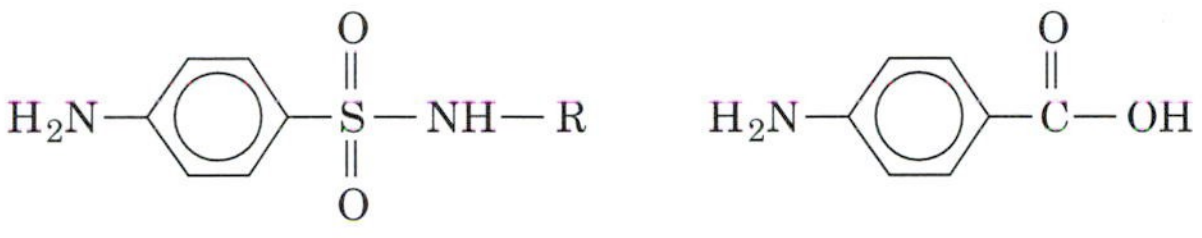

Sulfonamides
(R = H, sulfanilamide)

***p*-Aminobenzoic acid**

They competitively inhibit bacterial synthesis of THF at the *p*-aminobenzoic acid incorporation step, thereby blocking THF-requiring reactions. The inability of mammals to synthesize folic acid leaves them unaffected by sulfonamides, which accounts for the medical utility of these widely used antibacterial agents.

Branched-Chain Amino Acid Degradation Involves Acyl-CoA Oxidation

Degradation of the branched-chain amino acids isoleucine, leucine, and valine begins with three reactions that employ common enzymes (Fig. 20-19):

1. Transamination to the corresponding α-keto acid.
2. Oxidative decarboxylation to the corresponding acyl-CoA.
3. Dehydrogenation by FAD to form a double bond.

The remaining reactions of the isoleucine degradation pathway are analogous to those of fatty acid oxidation (Section 19-2C), thereby yielding acetyl-CoA and propionyl-CoA, which is subsequently converted via three reactions to succinyl-CoA. The degradation of valine, which contains one less carbon than isoleucine, yields CO_2 and propionyl-CoA, which is then converted to succinyl-CoA. The further degradation of leucine, which yields acetoacetate instead of propionyl-CoA, is considered in Section 20-4E.

Branched-chain α-keto acid dehydrogenase (BCKDH), which catalyzes Reaction 2 of Fig. 20-19, is a multienzyme complex that closely resembles the pyruvate dehydrogenase and α-ketoglutarate dehydrogenase complexes (Sections 16-2 and 16-3D). Indeed, all three multienzyme complexes share a common subunit, E_3 (dihydrolipoamide dehydrogenase), and employ the coenzymes TPP, lipoamide, and FAD in addition to their terminal oxidizing agent, NAD^+.

A genetic deficiency in BCKDH causes **maple syrup urine disease,** so named because the consequent buildup of branched-chain α-keto acids imparts the urine with the characteristic odor of maple syrup. Unless promptly treated by a diet low in branched-chain amino acids, maple syrup urine disease is rapidly fatal.

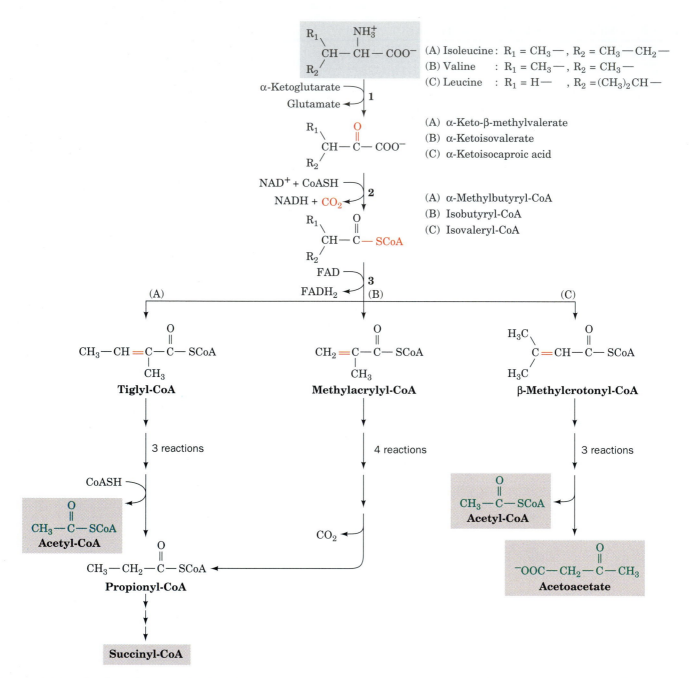

Figure 20-19.   The degradation of the branched-chain amino acids. Isoleucine (A), valine (B), and leucine (C) follow an initial common pathway utilizing three enzymes: (**1**) branched-chain amino acid aminotransferase, (**2**) branched-chain α-keto acid dehydrogenase (BCKDH), and (**3**) acyl-CoA dehydrogenase. Isoleucine degradation then continues (*left*) to yield acetyl-CoA and succinyl-CoA; valine degradation continues (*center*) to yield succinyl-CoA; and leucine degradation continues (*right*) to yield acetyl-CoA and acetoacetate.

E. Leucine and Lysine Are Degraded to Acetoacetate and/or Acetyl-CoA

Leucine degradation begins in the same manner as isoleucine and valine degradation (Fig. 20-19), but the dehydrogenated CoA adduct β-methylcrotonyl-CoA is converted to acetyl-CoA and acetoacetate, a ketone body.

The predominant pathway for lysine degradation in mammalian liver proceeds via formation of the α-ketoglutarate–lysine adduct **saccharopine** (Fig. 20-20). This pathway is worth examining in detail because we have

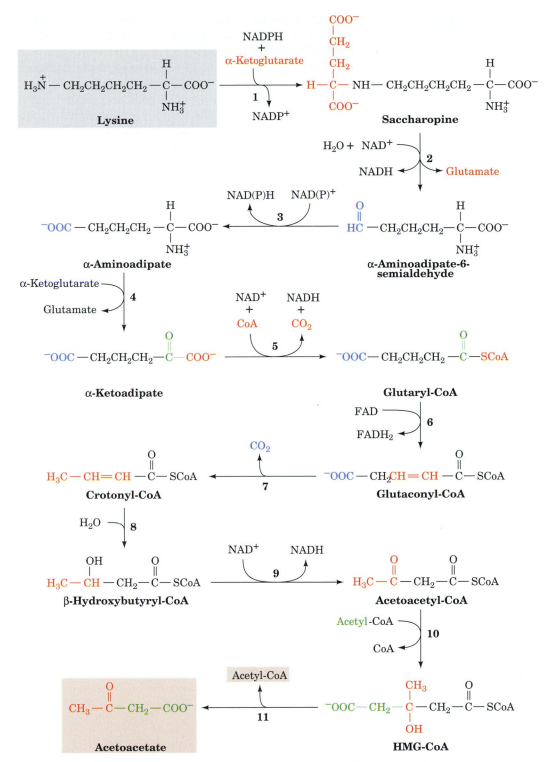

Figure 20-20. **The pathway of lysine degradation in mammalian liver.** The enzymes are (**1**) saccharopine dehydrogenase (NADP⁺, lysine forming), (**2**) saccharopine dehydrogenase (NAD⁺, glutamate forming), (**3**) aminoadipate-semialdehyde dehydrogenase, (**4**) aminoadipate aminotransferase (a PLP-dependent enzyme), (**5**) α-keto acid dehydrogenase, (**6**) glutaryl-CoA dehydrogenase, (**7**) decarboxylase, (**8**) enoyl-CoA hydratase, (**9**) β-hydroxyacyl-CoA dehydrogenase, (**10**) HMG-CoA synthase, and (**11**) HMG-CoA lyase. Reactions 10 and 11 are discussed in Section 19-3.

encountered 7 of its 11 reactions in other pathways. Reaction 4 is a PLP-dependent transamination. Reaction 5 is the oxidative decarboxylation of an α-keto acid by a multienzyme complex similar to pyruvate dehydrogenase (Section 16-2). Reactions 6, 8, and 9 are standard reactions of fatty acyl-CoA oxidation: dehydrogenation by FAD, hydration, and dehydrogenation by NAD^+. Reactions 10 and 11 are standard reactions in ketone body formation. Two moles of CO_2 are produced at Reactions 5 and 7 of the pathway.

The saccharopine pathway is thought to predominate in mammals because a genetic defect in the enzyme that catalyzes Reaction 1 in the sequence results in **hyperlysinemia** and **hyperlysinuria** (elevated levels of lysine in the blood and urine, respectively) along with mental and physical retardation. This is yet another example of how the study of rare inherited disorders has helped to trace metabolic pathways.

F. Tryptophan Is Degraded to Alanine and Acetoacetate

The complexity of the major tryptophan degradation pathway (outlined in Fig. 20-21) precludes a detailed discussion of all its reactions. However, one reaction is of particular interest: The fourth reaction is catalyzed by

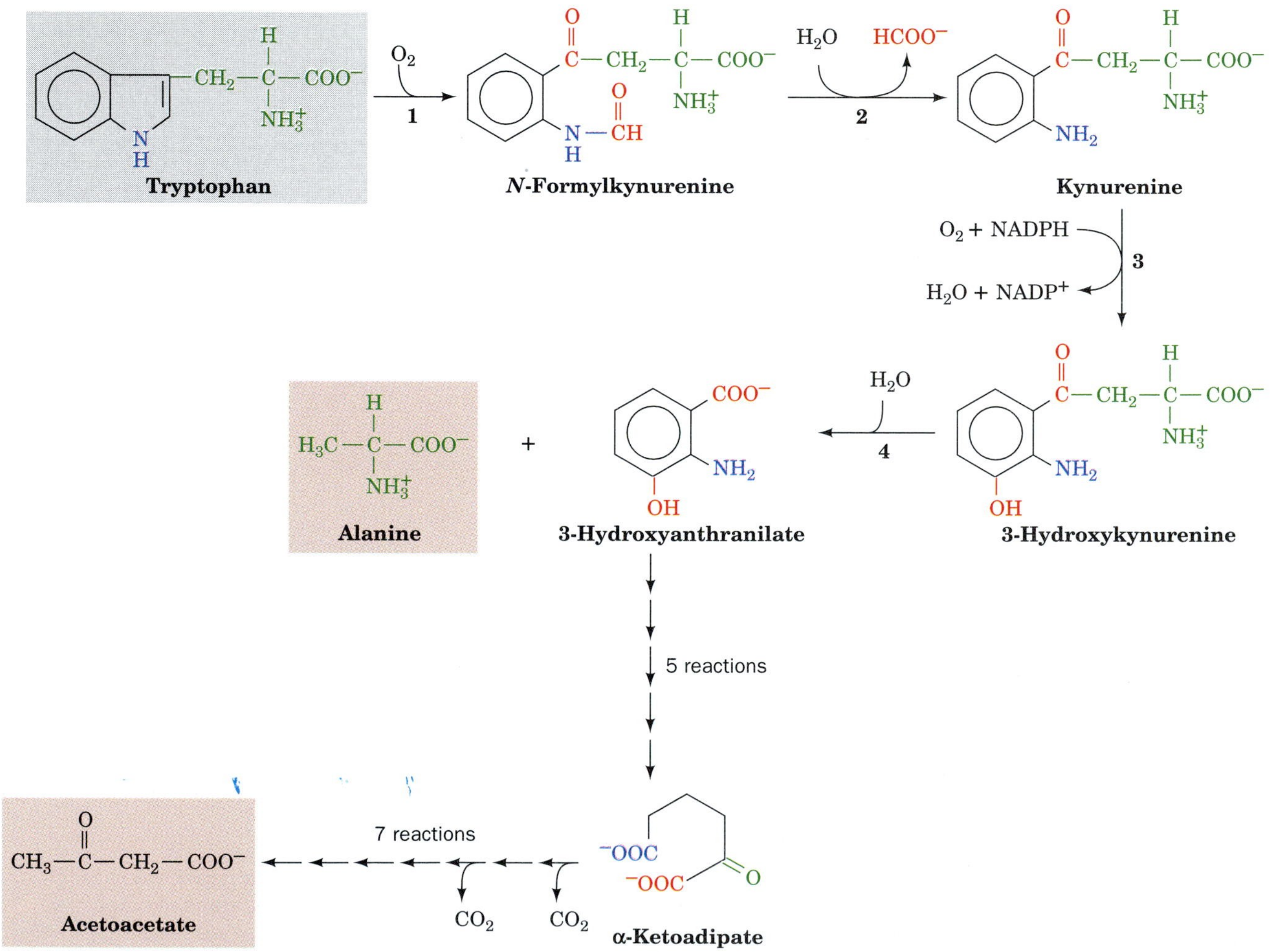

Figure 20-21. **The pathway of tryptophan degradation.** The enzymatic reactions shown are catalyzed by (**1**) tryptophan-2,3-dioxygenase, (**2**) formamidase, (**3**) kynurenine-3-monooxygenase, and (**4**) kynureninase (a PLP-dependent enzyme). Five reactions convert 3-hydroxyanthranilate to α-ketoadipate, which is converted to acetyl-CoA and acetoacetate in seven reactions as shown in Fig. 20-20, Reactions 5–11.

kynureninase, whose PLP group facilitates cleavage of the C_β—C_γ bond to release alanine. The kynureninase reaction follows the same initial steps as transamination (Fig. 20-7), but an enzyme nucleophilic group then attacks C_γ of the resonance-stabilized intermediate, resulting in C_β—C_γ bond cleavage. The remainder of the tryptophan skeleton is converted in five reactions to α-ketoadipate, which is also an intermediate in lysine degradation. α-Ketoadipate is broken down to acetyl-CoA and acetoacetate in seven reactions, as shown in Fig. 20-20.

G. Phenylalanine and Tyrosine Are Degraded to Fumarate and Acetoacetate

Since the first reaction in phenylalanine degradation is its hydroxylation to tyrosine, a single pathway (Fig. 20-22) is responsible for the breakdown of both of these amino acids. The final products of the six-reaction degradation are fumarate, a citric acid cycle intermediate, and acetoacetate, a ketone body. Defects in the enzymes that catalyze Reactions 1 and 4 cause disease (see Box 20-1).

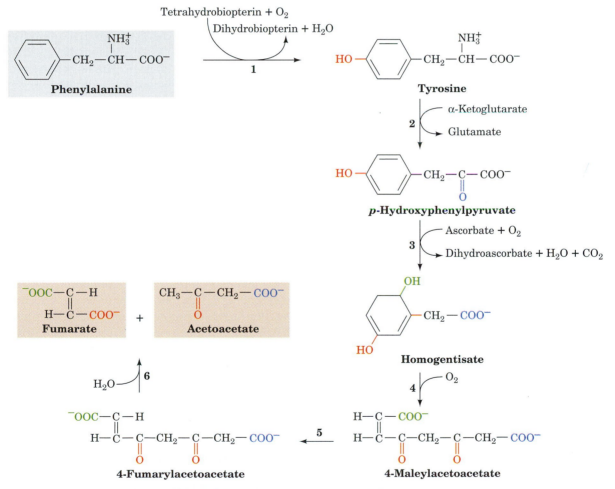

Figure 20-22.   The pathway of phenylalanine degradation. The enzymes involved are (**1**) phenylalanine hydroxylase, (**2**) aminotransferase, (**3**) *p*-hydroxyphenylpyruvate dioxygenase, (**4**) homogentisate dioxygenase, (**5**) maleylacetoacetate isomerase, and (**6**) fumarylacetoacetase.

Box 20-1

BIOCHEMISTRY IN HEALTH AND DISEASE

Phenylketonuria and Alcaptonuria Result from Defects in Phenylalanine Degradation

Archibald Garrod realized in the early 1900s that human genetic diseases result from specific enzyme deficiencies. We have repeatedly seen how this realization has contributed to the elucidation of metabolic pathways. The first such disease Garrod recognized was **alcaptonuria**, which results in the excretion of large quantities of **homogentisic acid** (Section 13-4B). This condition results from deficiency of **homogentisate dioxygenase** (Fig. 20-22, Reaction 4). Alcaptonurics suffer no ill effects other than arthritis later in life (although their urine darkens alarmingly because of the rapid air oxidation of the homogentisate they excrete).

Individuals suffering from **phenylketonuria (PKU)** are not so fortunate. Severe mental retardation occurs within a few months of birth if the disease is not detected and treated immediately. PKU is caused by the inability to hydroxylate phenylalanine (Fig. 20-22, Reaction 1) and therefore results in increased blood levels of phenylalanine (**hyperphenylalaninemia**). The excess phenylalanine is transaminated to **phenylpyruvate**

Phenylpyruvate

by an otherwise minor pathway. The "spillover" of phenylpyruvate (a phenylketone) into the urine was the first observation connected with the disease and gave the disease its name. All babies born in the United States are now screened for PKU immediately after birth by testing for elevated levels of phenylalanine in the blood.

Classic PKU results from a deficiency in phenylalanine hydroxylase. When this was established in 1947, PKU was the first inborn error of metabolism whose basic biochemical defect had been identified. Since all of the tyrosine breakdown enzymes are normal, treatment consists in providing the patient with a low-phenylalanine diet and monitoring the blood level of phenylalanine to ensure that it remains within normal limits for the first 5 to 10 years of life (the adverse effects of hyperphenylalaninemia seem to disappear after that age). **Aspartame (Nutrasweet)**, the main sweetening ingredient in diet soft drinks and many other dietetic food products, is Asp-Phe-methyl ester and is therefore a source of dietary phenylalanine. Consequently, a warning label for phenylketonurics appears on all these products.

Phenylalanine hydroxylase deficiency also accounts for another common symptom of PKU: Its victims have lighter hair and skin color than their siblings. This is because elevated phenylalanine levels inhibit tyrosine hydroxylation, the first reaction in the formation of the skin pigment **melanin** (Fig. 20-37).

Other types of hyperphenylalaninemia have been discovered since the introduction of infant screening techniques. These are caused by deficiencies in the enzymes catalyzing the formation or regeneration of 5,6,7,8-tetrahydrobiopterin, the phenylalanine hydroxylase cofactor (Fig. 20-24).

Pterins Are Redox Cofactors

The hydroxylation of phenylalanine by the Fe(III)-containing enzyme **phenylalanine hydroxylase** (Fig. 20-22, Reaction 1) requires the cofactor **biopterin**, a **pterin** derivative. Pterins are compounds that contain the **pteridine** ring (Fig. 20-23). Note the resemblance between the pteridine ring and the isoalloxazine ring of the flavin coenzymes (Fig. 13-11); the positions of the nitrogen atoms in pteridine are identical to those of the B and C rings of isoalloxazine. Folate derivatives also contain the pterin ring (Section 20-4D).

Pterins, like flavins, participate in biological oxidations. The active form of biopterin is the fully reduced form, **5,6,7,8-tetrahydrobiopterin.** It is produced from **7,8-dihydrobiopterin** and NADH, in what may be considered a priming reaction, by dihydrofolate reductase (Fig. 20-24), which simultaneously reduces dihydrofolate to tetrahydrofolate (Fig. 20-17). In the phenylalanine hydroxylase reaction, 5,6,7,8-tetrahydrobiopterin is oxidized to 7,8-dihydrobiopterin **(quinoid form).** This quinoid is subsequently reduced by the NADH-requiring enzyme **dihydropteridine reductase** to regenerate the active cofactor. Although dihydrofolate reductase and dihy-

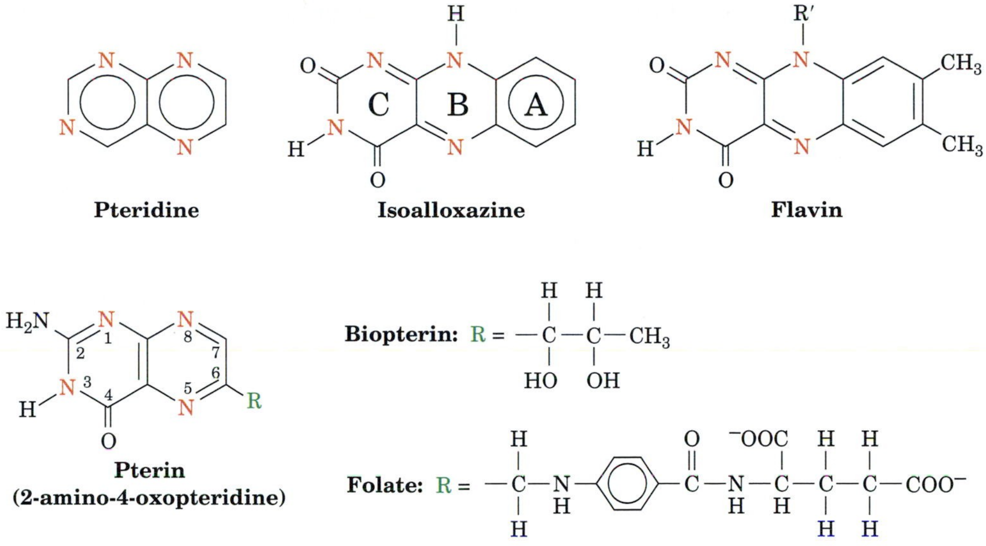

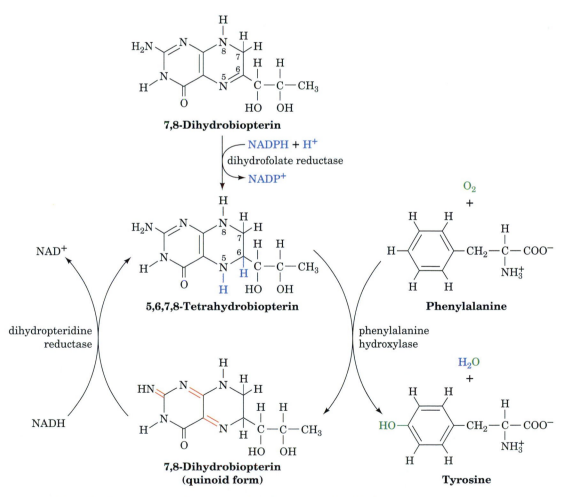

Pteridine

Isoalloxazine

Flavin

Pterin
(2-amino-4-oxopteridine)

Biopterin: R =

Folate: R =

Figure 20-23. **The pteridine ring nucleus of biopterin and folate.** Note the similar structures of pteridine and the isoalloxazine ring of flavin coenzymes.

7,8-Dihydrobiopterin

NADPH + H⁺
dihydrofolate reductase
NADP⁺

5,6,7,8-Tetrahydrobiopterin

Phenylalanine

O₂
+

NAD⁺

dihydropteridine
reductase

phenylalanine
hydroxylase

H₂O
+

NADH

7,8-Dihydrobiopterin
(quinoid form)

Tyrosine

Figure 20-24. **The formation, utilization, and regeneration of 5,6,7,8-tetrahydrobiopterin in the phenylalanine hydroxylase reaction.**

dropteridine reductase produce the same product, they use different tautomers of the substrate.

5. AMINO ACID BIOSYNTHESIS

Many amino acids are synthesized by pathways that are present only in plants and in microorganisms. Since mammals must obtain these amino acids in their diets, these substances are known as **essential amino acids.** The other amino acids, which can be synthesized by mammals from common intermediates, are termed **nonessential amino acids.** The essential and nonessential amino acids for humans are listed in Table 20-3. Arginine is classified as essential, even though it is synthesized by the urea cycle (Section 20-3A), because it is required in greater amounts than can be produced by this route during the normal growth and development of children (but not adults).

In this section we study the pathways involved in the formation of the nonessential amino acids. We also briefly consider such pathways for the essential amino acids as they occur in plants and microorganisms. Keep in mind that *there is considerable variation in these pathways among different species. In contrast, the basic pathways of carbohydrate and lipid metabolism are all but universal.*

A. Biosynthesis of the Nonessential Amino Acids

All the nonessential amino acids except tyrosine are synthesized by simple pathways leading from one of four common metabolic intermediates: pyruvate, oxaloacetate, α-ketoglutarate, and 3-phosphoglycerate. Tyrosine, which is really misclassified as being nonessential, is synthesized by the one-step hydroxylation of the essential amino acid phenylalanine (Fig. 20-22). Indeed, the dietary requirement for phenylalanine reflects the need for tyrosine as well. The presence of dietary tyrosine therefore decreases the need for phenylalanine.

Alanine, Asparagine, Aspartate, Glutamate, and Glutamine Are Synthesized from Pyruvate, Oxaloacetate, and α-Ketoglutarate

Pyruvate, oxaloacetate, and α-ketoglutarate are the α-keto acids (the so-called carbon skeletons) that correspond to alanine, aspartate, and glutamate, respectively. Indeed, the synthesis of each of these amino acids is a one-step transamination reaction (Fig. 20-25, Reactions 1–3). Asparagine and glutamine are, respectively, synthesized from aspartate and glutamate by ATP-dependent amidation. In the **glutamine synthetase** reaction (Fig. 20-25, Reaction 5), glutamate is first activated by reaction with ATP to form a **γ-glutamylphosphate** intermediate. NH_3 then displaces the phosphate group to produce glutamine. Curiously, aspartate amidation by **asparagine synthetase** to form asparagine follows a different route; it uses glutamine as its amino group donor and cleaves ATP to AMP + PP_i (Fig. 20-25, Reaction 4).

Glutamine Synthetase Is a Central Control Point in Nitrogen Metabolism

Glutamine is the amino group donor in the formation of many biosynthetic products as well as being a storage form of ammonia. The control of

Table 20-3.   **Essential and Nonessential Amino Acids in Humans**

Essential	Nonessential
Arginine[a]	Alanine
Histidine	Asparagine
Isoleucine	Aspartate
Leucine	Cysteine
Lysine	Glutamate
Methionine	Glutamine
Phenylalanine	Glycine
Threonine	Proline
Tryptophan	Serine
Valine	Tyrosine

[a]Although mammals synthesize arginine, they cleave most of it to form urea (Section 20-3A).

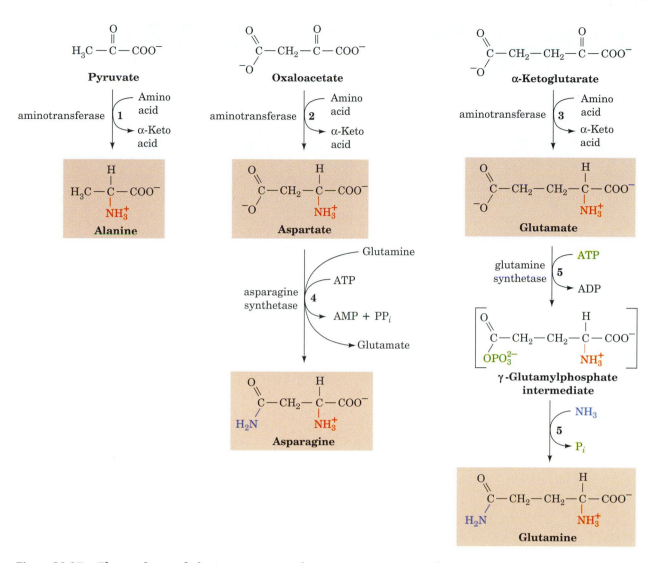

Figure 20-25.   The syntheses of alanine, aspartate, glutamate, asparagine, and glutamine. These reactions involve, respectively, transamination of (1) pyruvate, (2) oxaloacetate, and (3) α-ketoglutarate, and amidation of (4) aspartate and (5) glutamate.

glutamine synthetase is therefore vital for regulating nitrogen metabolism. Mammalian glutamine synthetases are activated by α-ketoglutarate, the product of glutamate's oxidative deamination (Section 20-2B). This control presumably prevents the accumulation of the ammonia produced by that reaction.

Bacterial glutamine synthetase, as Earl Stadtman showed, has a much more elaborate control system. This enzyme, which consists of 12 identical 469-residue subunits arranged at the corners of a hexagonal prism (Fig. 20-26), is regulated by several allosteric effectors as well as by covalent modification. Several aspects of its control system bear note. *Nine allosteric feedback inhibitors, each with its own binding site, control the activity of bacterial glutamine synthetase in a cumulative manner.* Six of these effectors—histidine, tryptophan, carbamoyl phosphate (as synthesized by carbamoyl phosphate synthetase II), glucosamine-6-phosphate, AMP, and CTP—are all end products of pathways leading from glutamine. The other three—alanine, serine, and glycine—reflect the cell's nitrogen level.

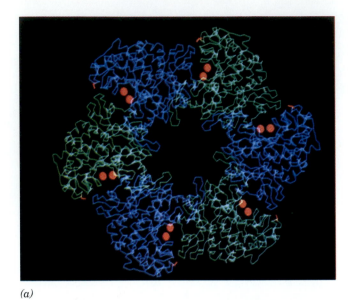

(a)

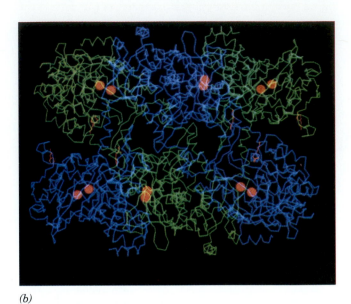

(b)

Figure 20-26.   The X-ray structure of glutamine synthetase from the bacterium *Salmonella typhimurium*. The enzyme consists of 12 identical subunits, here represented by their C_α backbones, arranged with D_6 symmetry (the symmetry of a hexagonal prism). (*a*) View down the six-fold axis of symmetry showing only the six subunits of the upper ring in alternating blue and green. The subunits of the lower ring are roughly directly below those of the upper ring. The protein, including its side chains

(not shown), has a diameter of 143 Å. Pairs of Mg^{2+} ions (*red spheres*) that are required for enzymatic activity are shown in each active site. Each adenylylation site, Tyr 397 (*red*), lies between two subunits. (*b*) Side view along one of the two-fold axes showing only the six nearest subunits. The molecule extends 103 Å along the six-fold axis, which is vertical in this view. [Courtesy of David Eisenberg, UCLA.]

E. coli glutamine synthetase is covalently modified by **adenylylation** (addition of an AMP group) of a specific Tyr residue (Fig. 20-27). The enzyme's susceptibility to cumulative feedback inhibition increases, and its activity therefore decreases, with its degree of adenylylation. The level of adenylylation is controlled by a complex metabolic cascade that is conceptually similar to that controlling glycogen phosphorylase (Section 15-3B). Both adenylylation and deadenylylation of glutamine synthetase are catalyzed by **adenylyltransferase** in complex with a tetrameric regulatory protein, **P_{II}.** This complex deadenylylates glutamine synthetase when P_{II} is **uridylylated** (also at a Tyr residue) and adenylylates glutamine synthetase when P_{II} lacks UMP residues. The level of P_{II} uridylylation, in turn, depends on the relative activities of two enzymatic activities located on the same protein: a **uridylyltransferase** that uridylylates P_{II} and a **uridylyl-removing enzyme** that hydrolytically excises the attached UMP groups of P_{II}. The uridylyltransferase is activated by α-ketoglutarate and ATP and inhibited by glutamine and P_i, whereas uridylyl-removing enzyme is insensitive to these metabolites. This intricate metabolic cascade therefore renders the activity of *E. coli* glutamine synthetase extremely responsive to the cell's nitrogen requirements.

Glutamate Is the Precursor of Proline, Ornithine, and Arginine

Conversion of glutamate to proline (Fig. 20-28, Reactions 1–4) involves the reduction of the γ-carboxyl group to an aldehyde followed by the formation of an internal Schiff base whose further reduction yields proline. Reduction of the glutamate γ-carboxyl group to an aldehyde is an endergonic process that is facilitated by first phosphorylating the carboxyl group in a reaction catalyzed by **γ-glutamyl kinase.** The unstable product, **glutamate-5-phosphate,** has not been isolated from reaction mixtures but is pre-

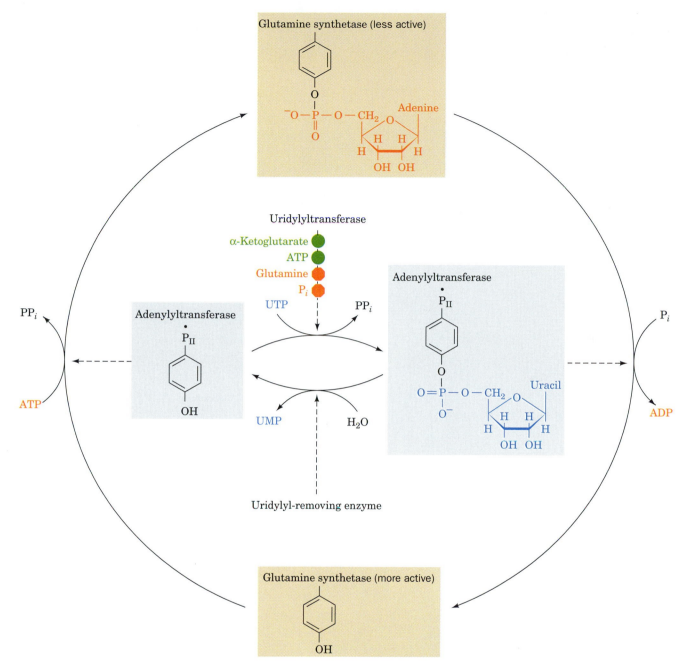

Figure 20-27. **The regulation of bacterial glutamine synthetase.** The adenylylation/deadenylylation of a specific Tyr residue is controlled by the level of uridylylation of a specific adenylyltransferase · P_{II} Tyr residue. This uridylylation level, in turn, is controlled by the relative activities of uridylyltransferase, which is sensitive to the levels of a variety of nitrogen metabolites, and uridylyl-removing enzyme, whose activity is independent of these metabolite levels.

sumed to be the substrate for the reduction that follows. The resulting **glutamate-5-semialdehyde** (which is also a product of arginine and proline degradation; Fig. 20-15) cyclizes spontaneously to form the internal Schiff base **Δ^1-pyrroline-5-carboxylate.** The final reduction to proline is catalyzed by **pyrroline-5-carboxylate reductase.** Whether the enzyme requires NADH or NADPH is unclear.

In humans, a three-step pathway leads from glutamate to ornithine via a branch from proline biosynthesis after Step 2 (Fig. 20-28). Glutamate-5-semialdehyde, which is in equilibrium with Δ^1-pyrroline-5-carboxylate, is

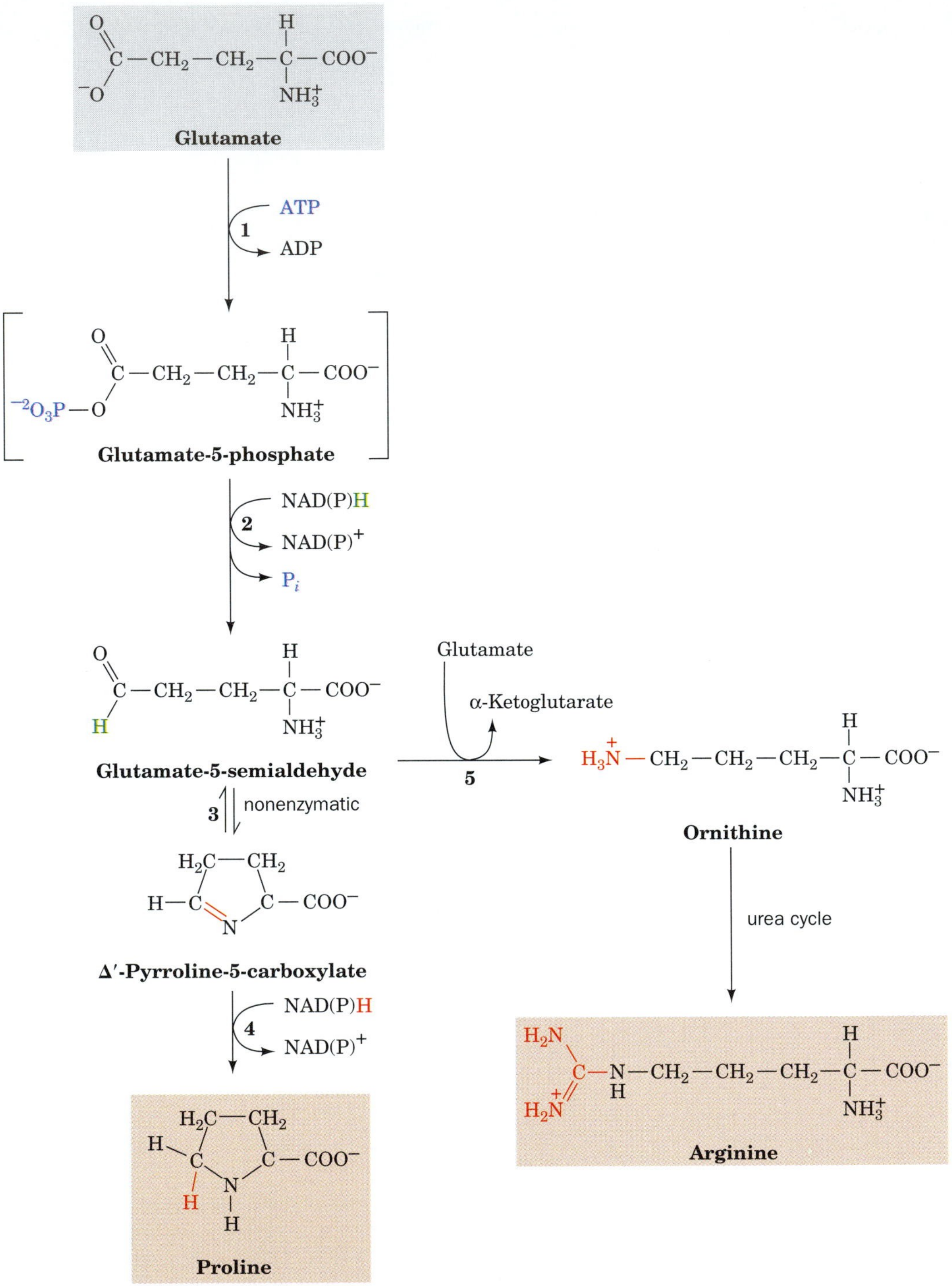

***Figure 20-28.*   The biosynthesis of the glutamate family of amino acids: argi-
nine, ornithine, and proline.** The catalysts for proline biosynthesis are (1) γ-glutamyl kinase, (2) dehydrogenase, (3) nonenzymatic, and (4) pyrroline-5-carboxylate
reductase. In mammals, ornithine is produced from glutamate-5-semialdehyde by the
action of ornithine-δ-aminotransferase (5). Ornithine is converted to arginine via the
urea cycle (Section 20-3A).

directly transaminated to yield ornithine in a reaction catalyzed by **ornithine-δ-aminotransferase** (Fig. 20-28, Reaction 5). Ornithine is converted
to arginine by the reactions of the urea cycle (Fig. 20-8).

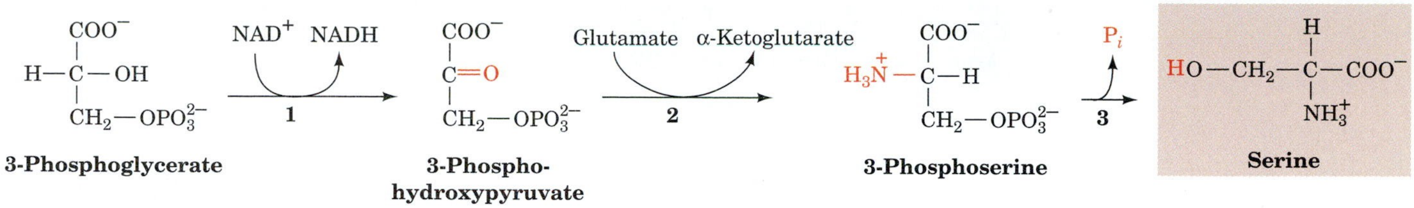

Figure 20-29. **The conversion of 3-phosphoglycerate to serine.** The pathway enzymes are (1) 3-phosphoglycerate dehydrogenase, (2) a PLP-dependent aminotransferase, and (3) phosphoserine phosphatase.

Serine, Cysteine, and Glycine Are Derived from 3-Phosphoglycerate

Serine is formed from the glycolytic intermediate 3-phosphoglycerate in a three-reaction pathway (Fig. 20-29):

1. Conversion of 3-phosphoglycerate's 2-OH group to a ketone, yielding **3-phosphohydroxypyruvate,** serine's phosphorylated keto acid analog.
2. Transamination of 3-phosphohydroxypyruvate to phosphoserine.
3. Hydrolysis of phosphoserine to serine.

Serine participates in glycine synthesis in two ways:

1. Direct conversion of serine to glycine by serine hydroxymethyltransferase in a reaction that also yields N^5,N^{10}-methylene-THF (Fig. 20-12, Reaction 4 in reverse).
2. Condensation of the N^5,N^{10}-methylene-THF with CO_2 and NH_4^+ by glycine synthase (Fig. 20-12, Reaction 3 in reverse).

In animals, cysteine is synthesized from serine and homocysteine, a breakdown product of methionine (Fig. 20-16, Reactions 5 and 6). Homocysteine combines with serine to yield cystathionine, which subsequently forms cysteine and α-ketobutyrate. Since cysteine's sulfhydryl group is derived from the essential amino acid methionine, cysteine can be considered to be an essential amino acid.

B. Biosynthesis of the Essential Amino Acids

Essential amino acids, like nonessential amino acids, are synthesized from familiar metabolic precursors. Their synthetic pathways are present only in microorganisms and in plants, however, and usually involve more steps than those of the nonessential amino acids. The enzymes that synthesize essential amino acids were apparently lost early in animal evolution, possibly because of the ready availability of these amino acids in the diet. We shall focus on only a few of the many reactions in the biosynthesis of essential amino acids.

The Aspartate Family: Lysine, Methionine, and Threonine

In bacteria, aspartate is the common precursor of lysine, methionine, and threonine (Fig. 20-30). The biosyntheses of these essential amino acids all begin with the **aspartokinase**-catalyzed phosphorylation of aspartate to yield **aspartyl-β-phosphate.** We have seen that the control of metabolic pathways commonly occurs at the first committed step of the pathway. One might therefore expect lysine, methionine, and threonine biosynthesis to be controlled as a group. Each of these pathways is, in fact, independently controlled. *E. coli* has three isozymes of aspartokinase that respond dif-

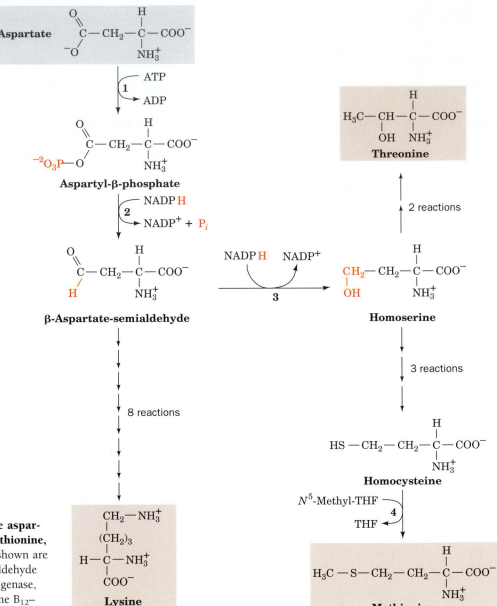

Figure 20-30.   **The biosynthesis of the aspartate family of amino acids: lysine, methionine, and threonine.** The pathway enzymes shown are (1) aspartokinase, (2) β-aspartate-semialdehyde dehydrogenase, (3) homoserine dehydrogenase, and (4) methionine synthase (a coenzyme B_{12}–dependent enzyme).

ferently to the three amino acids in terms both of feedback inhibition of enzyme activity and repression of enzyme synthesis. In addition, the pathway direction is controlled by feedback inhibition at the branch points by the amino acid products of the branches.

Methionine synthase (alternatively **homocysteine methyltransferase**) catalyzes the methylation of homocysteine to form methionine using N^5-methyl-THF as its methyl group donor (Reaction 4 in both Figs. 20-16 and 20-30). Methionine synthase is the only coenzyme B_{12}–associated enzyme in mammals besides methylmalonyl-CoA mutase (Section 19-2E). However, the coenzyme B_{12}'s Co ion in methionine synthase is axially liganded to a methyl group to form **methylcobalamin** rather than by a 5′-adenosyl group as in methylmalonyl-CoA mutase (Fig. 19-14). The X-ray structure of the 246-residue methylcobalamin-binding segment of the 1227-residue monomeric *E. coli* methionine synthase, determined by Martha Ludwig and Rowena Matthews, has a fold similar to that observed in the X-ray structure of methylmalonyl-CoA mutase (Fig. 19-16). In both enzyme structures, the coenzyme's 5,6-dimethylbenzimidazole (DMB) moiety (Fig. 19-14) is not liganded to the Co ion as in the free coenzyme but, rather, has swung aside to bind in a separate pocket and has been replaced by an enzyme His side chain.

High levels of homocysteine in the blood have recently been shown to be a significant risk factor in cardiovascular disease. Eating adequate quantities of folate, the vitamin precursor of THF (Section 20-4D), is therefore likely to be effective in alleviating this condition.

The Pyruvate Family: Leucine, Isoleucine, and Valine

Valine and isoleucine follow the same biosynthetic pathway utilizing pyruvate as a starting reactant, the only difference being in the first step of the series (Fig. 20-31). In this thiamine pyrophosphate–dependent reac-

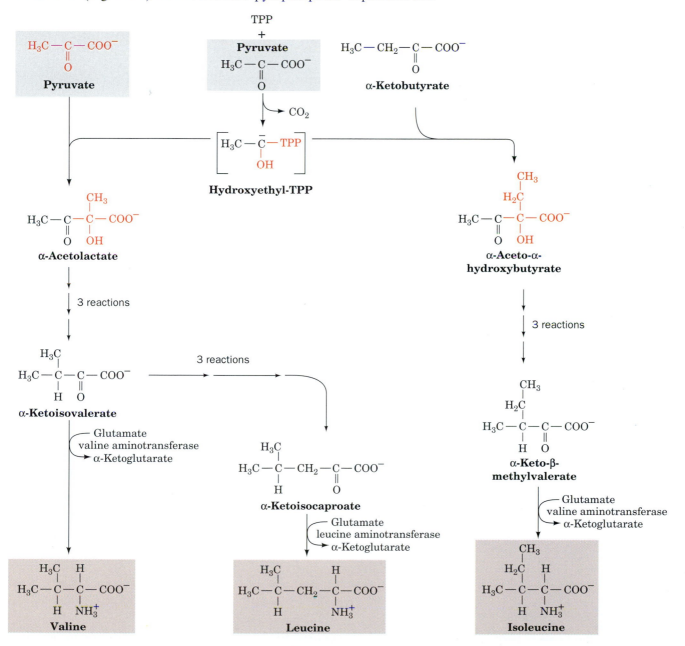

Figure 20-31. **The biosynthesis of the pyruvate family of amino acids: isoleucine, leucine, and valine.** The first enzyme, acetolactate synthase (a TPP enzyme), catalyzes two reactions, one leading to valine and leucine, and the other to isoleucine. Note also that **valine aminotransferase** catalyzes the formation of both valine and isoleucine from their respective α-keto acids.

tion, which resembles those catalyzed by pyruvate decarboxylase (Fig. 14-20) and transketolase (Fig. 14-31), pyruvate forms an adduct with TPP that is decarboxylated to hydroxyethyl-TPP. This resonance-stabilized carbanion adds either to the keto group of a second pyruvate to form **acetolactate** on the way to valine, or to the keto group of **α-ketobutyrate** to form **α-aceto-α-hydroxybutyrate** on the way to isoleucine. The leucine biosynthetic

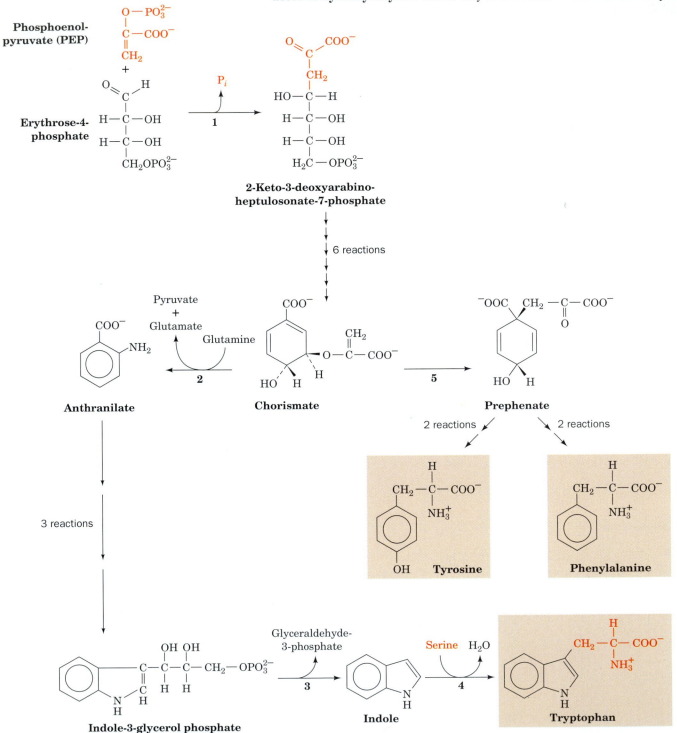

Figure 20-32. **The biosynthesis of phenylalanine, tryptophan, and tyrosine.** Some of the enzymes involved are (1) 2-keto-3-deoxy-D-arabinoheptulosonate-7-phosphate synthase, (2) anthra-nilate synthase, (3) tryptophan synthase, α subunit, (4) tryptophan synthase, β subunit (a PLP-dependent enzyme), and (5) chorismate mutase.

pathway branches off from the valine pathway. The final step in each of the three pathways, which begin with pyruvate rather than an amino acid, is the PLP-dependent transfer of an amino group from glutamate to form the amino acid.

The Aromatic Amino Acids: Phenylalanine, Tyrosine, and Tryptophan

The precursors of the aromatic amino acids are the glycolytic intermediate phosphoenolpyruvate (PEP) and erythrose-4-phosphate (an intermediate in the pentose phosphate pathway; Fig. 14-29). Their condensation forms **2-keto-3-deoxy-D-arabinoheptulosonate-7-phosphate** (Fig. 20-32). This C_7 compound cyclizes and is ultimately converted to **chorismate,** the branch point for tryptophan synthesis. Chorismate is converted to either **anthranilate** and then to tryptophan, or to **prephenate** and on to tyrosine or phenylalanine. Although mammals synthesize tyrosine by the hydroxylation of phenylalanine (Fig. 20-22), many microorganisms synthesize it directly from prephenate. The last step in the synthesis of tyrosine and phenylalanine is the addition of an amino group through transamination. In tryptophan synthesis, the amino group is part of the serine molecule that is added to **indole.**

Indole Is Channeled between Two Active Sites in Tryptophan Synthase

The final two reactions of tryptophan biosynthesis (Reactions 3 and 4 in Fig. 20-32) are both catalyzed by **tryptophan synthase:**

1. The α subunit (29 kD) of this $\alpha_2\beta_2$ bifunctional enzyme cleaves **indole-3-glycerol phosphate,** yielding indole and glyceraldehyde-3-phosphate.

2. The β subunit (43 kD) joins indole with serine in a PLP-dependent reaction to form tryptophan.

Either subunit alone is enzymatically active, but when the subunits are joined in the $\alpha_2\beta_2$ tetramer, the rates of both reactions and their substrate affinities increase by 1 to 2 orders of magnitude. Indole, the intermediate product, does not appear free in solution; the enzyme apparently sequesters it.

The X-ray structure of tryptophan synthase from *Salmonella typhimurium,* determined by Craig Hyde, Edith Miles, and David Davies, explains the latter observation. The protein forms a 150-Å-long, two-fold symmetric α-β-β-α complex in which the active sites of neighboring α and β subunits are separated by ~25 Å (Fig. 20-33). *These ac-*

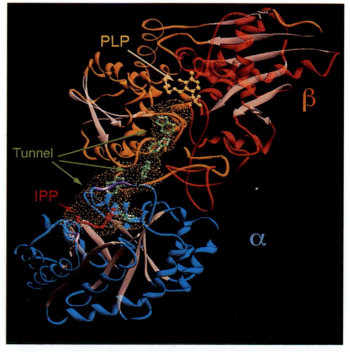

Figure 20-33. **A ribbon diagram of the bifunctional enzyme tryptophan synthase from *S. typhimurium.*** Only one αβ subunit of the αββα heterotetramer is shown. The α subunit is blue, the β subunit's N-terminal domain is orange, its C-terminal domain is red-orange, and all β sheets are tan. The active site of the α subunit is identified by its bound competitive inhibitor, **indolepropanol phosphate** (IPP; *red ball-and-stick model*), whereas that of the β subunit is marked by its PLP coenzyme (*yellow ball-and-stick model*). The solvent-accessible surface of the ~25-Å-long "tunnel" connecting the α and β active sites is outlined by yellow dots. Several indole molecules (*green ball-and-stick models*) have been modeled into the tunnel to show that it is wide enough for indole to pass from one active site to the other. [Courtesy of Craig Hyde, National Institutes of Health.] ● See the Interactive Exercises.

tive sites are joined by a solvent-filled tunnel that is wide enough to permit the passage of the intermediate substrate, indole. This suggests the following series of events: The indole-3-glycerol phosphate substrate binds to the α subunit through an opening into its active site, its "front door," and the glyceraldehyde-3-phosphate product leaves via the same route. Similarly, the β subunit active site has a "front door" opening to the solvent through which serine enters and tryptophan leaves. Both active sites also have "back doors" that are connected by the tunnel. *The indole intermediate presumably diffuses between the two active sites via the tunnel and hence does not escape to the solvent.*

This phenomenon, in which the intermediate of two reactions is directly transferred from one enzyme active site to another, is called **channeling.** Channeling increases the rate of a metabolic pathway by preventing the loss of its intermediate products as well as by protecting this intermediate from degradation. Channeling may be particularly important for indole since this nonpolar molecule otherwise can escape the bacterial cell by diffusing through its plasma and outer membranes.

In order for channeling to increase tryptophan synthase's catalytic efficiency, (1) its connected active sites must be coupled such that their catalyzed reactions occur in phase, and (2) after substrate has bound to the α subunit, its active site ("front door") must close off to ensure that the product indole passes through the tunnel ("back door") to the β subunit rather than escaping into solution. A variety of experimental evidence indicates that this series of events is facilitated through allosteric signals derived from covalent transformations at the β subunit's active site. These switch the enzyme between an open, low-activity conformation to which substrates bind, and a closed, high-activity conformation from which indole cannot escape.

Histidine Biosynthesis

Five of histidine's six C atoms are derived from **5-phosphoribosyl-α-pyrophosphate** (**PRPP;** Fig. 20-34), a phospho-sugar intermediate that is also involved in the biosynthesis of purine and pyrimidine nucleotides (Sections 22-1A and 22-2A). The histidine's sixth carbon originates from ATP. The ATP atoms that are not incorporated into histidine are eliminated as **5-aminoimidazole-4-carboxamide ribonucleotide** (Fig. 20-34, Reaction 2), which is also an intermediate in purine biosynthesis.

The unusual biosynthesis of histidine from a purine (N^1-5′-phosphoribosyl ATP, the product of Reaction 1 in Fig. 20-34) has been cited as evidence supporting the hypothesis that life was originally RNA based (Section 3-3C). His residues, as we have seen, are often components of enzyme active sites, where they act as nucleophiles and/or general acid–base catalysts. The discovery that RNA can have catalytic properties therefore suggests that the imidazole moiety of purines plays a similar role in these RNA enzymes. This further suggests that the histidine biosynthetic pathway is a "fossil" of the transition to more efficient protein-based life forms.

6. OTHER PRODUCTS OF AMINO ACID METABOLISM

Certain amino acids, in addition to their major function as protein building blocks, are essential precursors of a variety of important biomolecules, including nucleotides and nucleotide coenzymes, heme, and various hormones and neurotransmitters. In this section, we consider the pathways

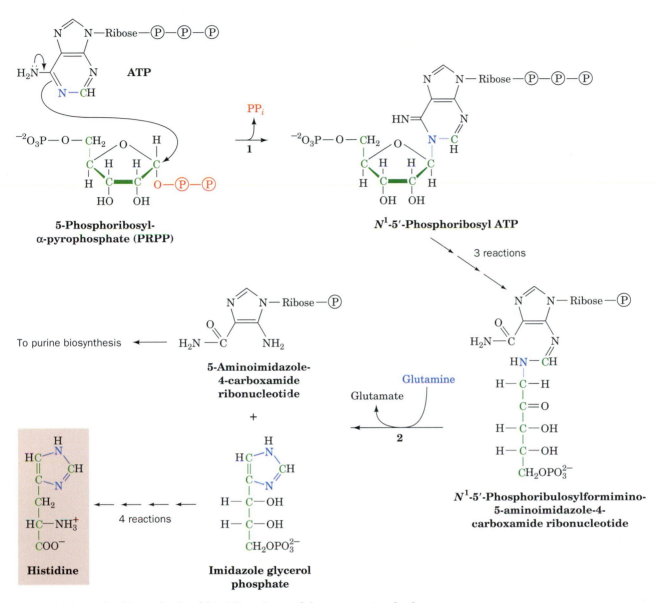

Figure 20-34. The biosynthesis of histidine. Some of the enzymes involved are (1) ATP phosphoribosyltransferase and (2) glutamine amidotransferase.

leading to some of these substances. The biosynthesis of nucleotides is considered in Chapter 22.

A. Heme Biosynthesis and Degradation

Heme, as we have seen, is an Fe-containing prosthetic group that is an essential component of many proteins, notably hemoglobin, myoglobin, and the cytochromes. The initial reactions of heme biosynthesis are common to the formation of other tetrapyrroles including chlorophyll in plants and bacteria (Fig. 18-2) and coenzyme B_{12} in bacteria (Fig. 19-14).

Elucidation of the heme biosynthetic pathway involved some interesting detective work. David Shemin and David Rittenberg, who were among the first to use isotopic tracers in the elucidation of metabolic pathways,

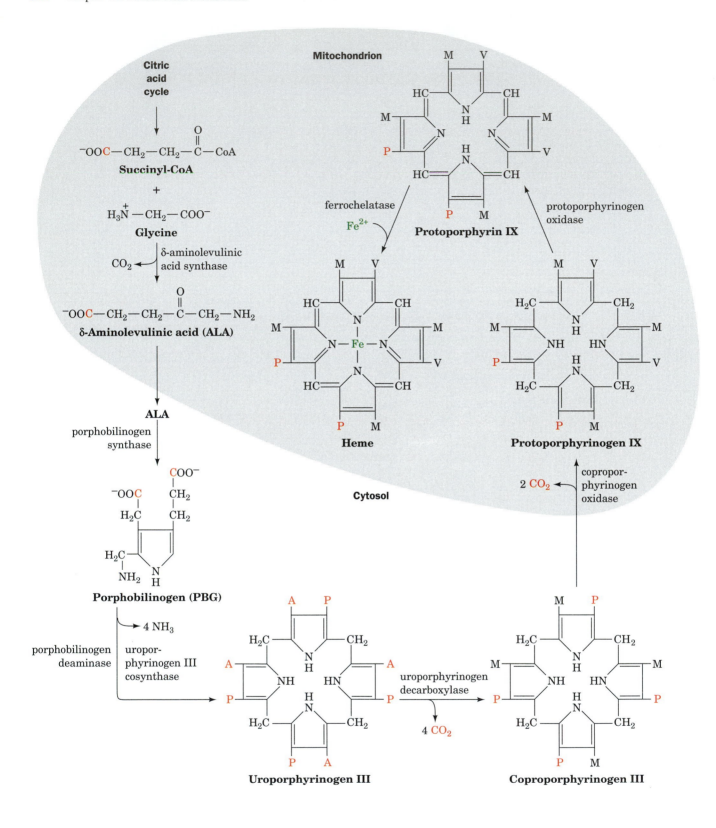

Figure 20-35. The pathway of heme biosynthesis. δ-Amino-levulinic acid (ALA) is synthesized in the mitochondrion from succinyl-CoA and glycine by ALA synthase. ALA is transported to the cytosol, where two molecules condense to form PBG, four molecules of which condense to form a porphyrin ring. The next three reactions involve oxidation of the pyrrole ring substituents, yielding protoporphyrinogen IX, which is transported back into the mitochondrion during its formation. After oxidation of the methylene groups, ferrochelatase catalyzes the insertion of Fe^{2+} to yield heme. A, P, M, and V, respectively, represent acetyl, propionyl, methyl, and vinyl ($-CH_2=CH_2$) groups. C atoms originating as the carboxyl group of acetate are red.

demonstrated, in 1945, that *all of heme's C and N atoms can be derived from acetate and glycine.* Heme biosynthesis takes place partly in the mitochondrion and partly in the cytosol (Fig. 20-35). Mitochondrial acetate is metabolized via the citric acid cycle to succinyl-CoA, which condenses with glycine in a reaction that produces CO_2 and **δ-aminolevulinic acid (ALA).** ALA is transported to the cytosol, where it combines with a second ALA to yield **porphobilinogen (PBG).** The reaction is catalyzed by the Zn-requiring enzyme **porphobilinogen synthase.**

Inhibition of PBG synthase by lead is one of the major manifestations of acute lead poisoning. Indeed, it has been suggested that the accumulation, in the blood, of ALA, which resembles the neurotransmitter **γ-aminobutyric acid** (Section 20-6B), is responsible for the psychosis that often accompanies lead poisoning.

The next phase of heme biosynthesis is the condensation of four PBG molecules to form **uroporphyrinogen III,** the porphyrin nucleus, in a series of reactions catalyzed by **porphobilinogen deaminase** (also called **uroporphyrinogen synthase**) and **uroporphyrinogen III cosynthase.** The initial product, **hydroxymethylbilane** *(at right),* is a linear tetrapyrrole that cyclizes. **Protoporphyrin IX,** to which Fe is added to form heme, is produced from uroporphyrinogen III in a series of reactions catalyzed by (1) **uroporphyrinogen decarboxylase,** which decarboxylates all four acetate side chains (A) to form methyl groups (M); (2) **coproporphyrinogen oxidase,** which oxidatively decarboxylates two of the propionate side chains (P) to vinyl groups (V); and (3) **protoporphyrinogen oxidase,** which oxidizes the methylene groups linking the pyrrole rings to methenyl groups. During the coproporphyrinogen oxidase reaction, the porphyrin is transported back into the mitochondrion. In the final reaction of heme biosynthesis, **ferrochelatase** inserts Fe(II) into protoporphyrin IX.

A = acetyl

P = propionyl

Hydroxymethylbilane

Regulation of Heme Biosynthesis

The two major sites of heme biosynthesis are erythroid cells, which synthesize ~85% of the body's heme groups, and the liver, which synthesizes most of the remainder. In liver, the level of heme synthesis must be adjusted according to metabolic conditions. For example, the synthesis of the heme-containing cytochrome P450 (Section 17-5A) fluctuates with the need for detoxification. In contrast, heme synthesis in erythroid cells is a one-time event; heme and protein synthesis ceases when the cell matures, so that the hemoglobin must last the erythrocyte's lifetime (~120 days).

In liver, the main control target in heme biosynthesis is ALA synthase, the enzyme catalyzing the pathway's first committed step. Heme, or its Fe(III) oxidation product **hemin,** controls this enzyme's activity through feedback inhibition, inhibition of the transport of ALA synthase from its site of synthesis in the cytosol to its reaction site in the mitochondrion (Fig. 20-35), and repression of ALA synthase synthesis.

In erythroid cells, heme exerts quite a different effect on its biosynthesis. Heme stimulates, rather than represses, protein synthesis in **reticulocytes** (immature erythrocytes). Although the vast majority of the protein synthesized by reticulocytes is globin, heme may also induce these cells to synthesize the enzymes of heme biosynthesis. Moreover, the rate-determining steps of heme biosynthesis in erythroid cells may be the ferrochelatase and porphobilinogen deaminase reactions rather than the ALA synthase reaction. This is consistent with the supposition that when erythroid heme biosynthesis is "switched on," all of its steps function at their maximal rates rather than any one step limiting the flow through the pathway. Heme-stimulated synthesis of globin also ensures that heme and globin are synthesized in the correct ratio for assembly into hemoglobin (Sec-

Box 20-2
BIOCHEMISTRY IN HEALTH AND DISEASE

The Porphyrias

Defects in heme biosynthesis in liver or erythroid cells result in the accumulation of porphyrin and/or its precursors and are therefore known as porphyrias. Two such defects are known to affect erythroid cells: uroporphyrinogen III cosynthase deficiency (**congenital erythropoietic porphyria**) and ferrochelatase deficiency (**erythropoietic protoporphyria**). The former results in accumulation of uroporphyrinogen derivatives. Excretion of these compounds colors the urine red; their deposition in the teeth turns them reddish brown; and their accumulation in the skin renders it extremely photosensitive, so that it ulcerates and forms disfiguring scars. Increased hair growth is also observed in afflicted individuals; fine hair may cover much of the face and extremities. These symptoms have prompted speculation that the werewolf legend has a biochemical basis.

The most common porphyria that primarily affects liver is porphobilinogen deaminase deficiency (**acute intermittent porphyria**). This disease is marked by intermittent attacks of abdominal pain and neurological dysfunction. Excessive amounts of ALA and PBG are excreted in the urine during and after such attacks. The urine may become red resulting from the excretion of excess porphyrins synthesized from PBG in nonhepatic cells, although the skin does not become unusually photosensitive. King George III, who ruled England during the American Revolution, and who has been widely portrayed as being mad, in fact had attacks characteristic of acute intermittent porphyria; he was reported to have urine the color of port wine and had several descendants who were diagnosed as having this disease. American history might have been quite different had George III not inherited this metabolic defect.

---

tion 27-3D). Genetic defects in heme biosynthesis cause conditions known as **porphyrias** (see Box 20-2).

Heme Degradation

At the end of their lifetime, red cells are removed from the circulation and their components degraded. Heme catabolism (Fig. 20-36) begins with oxidative cleavage, by heme oxygenase, of the porphyrin between rings A and B to form **biliverdin,** a green linear tetrapyrrole. Biliverdin's central methenyl bridge (between rings C and D) is then reduced to form the red-orange **bilirubin.** The changing colors of a healing bruise are a visible manifestation of heme degradation.

In the reaction forming biliverdin, the methenyl bridge carbon between rings A and B is released as CO, which is a tenacious heme ligand (with 200-fold greater affinity for hemoglobin than O_2; Section 7-1A). Consequently, ~1% of hemoglobin's binding sites are blocked by CO even in the absence of air pollution.

The highly lipophilic bilirubin is insoluble in aqueous solutions. Like other lipophilic metabolites, such as free fatty acids, it is transported in the blood in complex with serum albumin. Bilirubin derivatives are secreted in the bile and for the most part are further degraded by bacterial enzymes in the large intestine. Some of the resulting **urobilinogen** is reabsorbed and transported via the bloodstream to the kidney, where it is converted to the yellow **urobilin** and excreted, thus giving urine its characteristic color. Most urobilinogen, however, is microbially converted to the deeply red-brown **stercobilin,** the major pigment of feces.

When the blood contains excessive amounts of bilirubin, the deposition of this highly insoluble substance colors the skin and the whites of the eyes yellow. This condition, called **jaundice** (French: *jaune,* yellow), signals either an abnormally high rate of red cell destruction, liver dysfunction, or bile duct obstruction. Newborn infants, particularly when premature, often become jaundiced because they lack an enzyme that degrades bilirubin. Jaundiced infants are treated by bathing them with light from a fluorescent

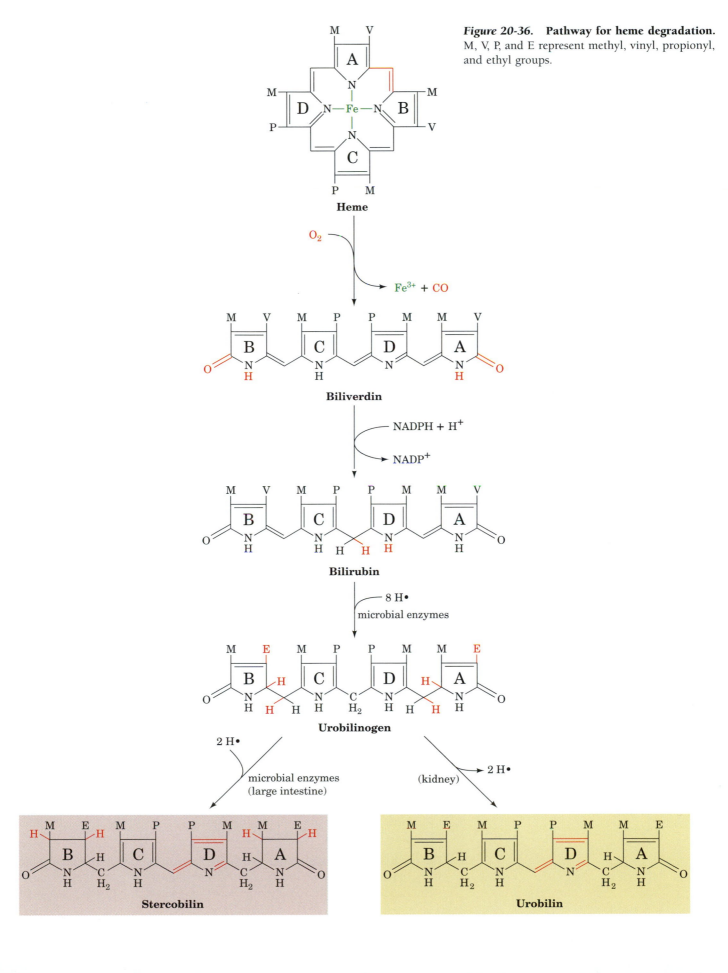

Figure 20-36. Pathway for heme degradation.
M, V, P, and E represent methyl, vinyl, propionyl, and ethyl groups.

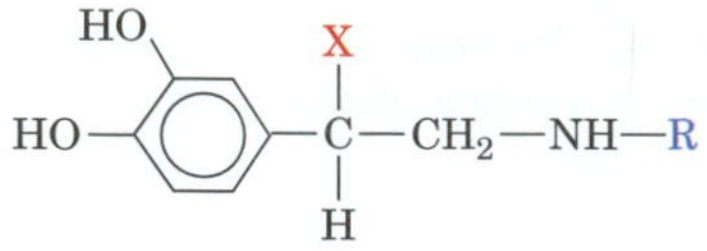

HO\
HO—◯—C(X)(H)—CH₂—NH—R

X = OH, R = CH₃ **Epinephrine (Adrenalin)**
X = OH, R = H   **Norepinephrine**
X = H,   R = H   **Dopamine**

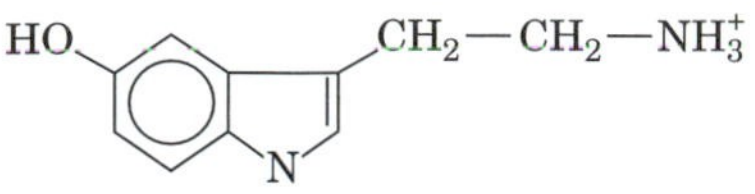

HO—[indole ring]—CH₂—CH₂—NH₃⁺

**Serotonin
(5-hydroxytryptamine)**

$^-OOC-CH_2-CH_2-CH_2-NH_3^+$

γ-Aminobutyric acid (GABA)

[imidazole ring]—CH₂—CH₂—NH₃⁺

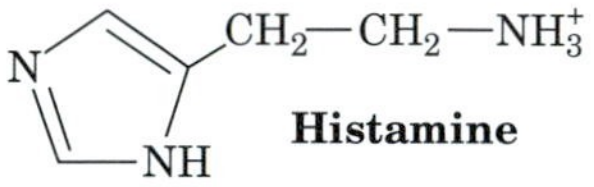

Histamine

lamp; this photochemically converts bilirubin to more soluble isomers that the infant can degrade and excrete.

B. Biosynthesis of Physiologically Active Amines

Epinephrine (adrenalin), norepinephrine, dopamine, serotonin (5-hydroxytryptamine), γ-aminobutyric acid (GABA), and **histamine** *(at left)* are hormones and/or neurotransmitters derived from amino acids. Epinephrine, as we have seen, activates muscle adenylate cyclase, thereby stimulating glycogen breakdown (Section 15-3C); deficiency in dopamine production in certain areas of the brain is associated with **Parkinson's disease,** a degenerative condition causing "shaking palsy"; serotonin causes smooth muscle contraction; GABA is one of the brain's major inhibitory neurotransmitters; and histamine is involved in allergic responses (as allergy sufferers who take antihistamines will realize), as well as in the control of acid secretion by the stomach (Box 10-4).

The biosynthesis of each of these physiologically active amines involves decarboxylation of the corresponding precursor amino acid. Amino acid decarboxylases are PLP-dependent enzymes that form a PLP–Schiff base with the substrate so as to stabilize the C_α carbanion formed on C_α—COO^- bond cleavage (Section 20-2A).

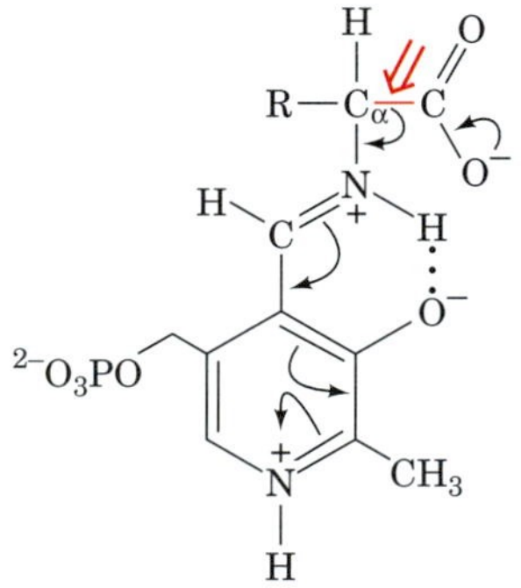

[reaction mechanism diagram]

Formation of histamine (from histidine) and GABA (from glutamate) are one-step processes; the synthesis of serotonin from tryptophan requires a hydroxylation step as well as decarboxylation. The various **catecholamines** —dopamine, norepinephrine, and epinephrine—are related to **catechol**

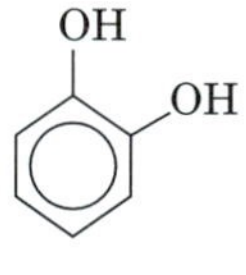

OH
◯—OH

Catechol

and are sequentially synthesized from tyrosine (Fig. 20-37):

1. Tyrosine is hydroxylated to **3,4-dihydroxyphenylalanine (L-DOPA)** in a reaction that requires 5,6,7,8-tetrahydrobiopterin (Fig. 20-24).
2. L-DOPA is decarboxylated to dopamine.
3. A second hydroxylation yields norepinephrine.
4. Methylation of norepinephrine's amino group by *S*-adenosylmethionine (Fig. 20-16) produces epinephrine.

The specific catecholamine that a cell produces depends on which enzymes of the pathway are present. In adrenal medulla, for example, epinephrine is the predominant product. In some areas of the brain, norepinephrine is

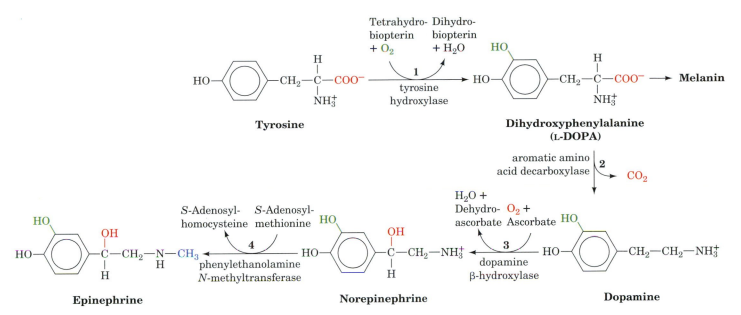

Figure 20-37.   The sequential synthesis of L-DOPA, dopamine, norepinephrine, and epinephrine from tyrosine. L-DOPA is also the precursor of the skin pigment melanin.

more common. In other areas, the pathway stops at dopamine. In melanocytes, L-DOPA is the precursor of red and black melanins, which are irregular cross-linked polymers that give hair and skin much of their color.

C. Nitric Oxide

Arginine is the precursor of a substance that was originally called **endothelium-derived relaxing factor (EDRF)** because it was synthesized by vascular endothelial cells and caused the underlying smooth muscle to relax. The signal for vasodilation was not a peptide, as expected, but the stable free radical nitric oxide, NO. The identification of NO as a vasodilator came in part from studies that identified NO as the decomposition product that mediates the vasodilating effects of compounds such as **nitroglycerin** *(at right)*. Nitroglycerin is often administered to individuals suffering from **angina pectoris** (a disease caused by insufficient blood flow to the heart muscle) to rapidly but temporarily relieve their chest pain.

The reaction that converts arginine to NO and citrulline is catalyzed by **nitric oxide synthase (NOS):**

$$\begin{array}{ccc}
CH_2 & CH & CH_2 \\
| & | & | \\
O & O & O \\
| & | & | \\
NO_2 & NO_2 & NO_2
\end{array}$$

Nitroglycerin

The reaction proceeds via an enzyme-bound hydroxyarginine intermediate and requires an array of redox coenzymes. NOS is a homodimeric protein of 125- to 160-kD subunits, and each subunit contains one FMN, one FAD, one tetrahydrobiopterin (Fig. 20-24), and one Fe(III)-heme. These cofactors facilitate the five-electron oxidation of arginine to produce NO.

Because NO is a gas, it rapidly diffuses across cell membranes, although its high reactivity (half-life ~5 s) prevents it from acting much further than ~1 mm from its site of synthesis. NO is produced by endothelial cells in response to a wide variety of agents and physiological conditions. Neuronal cells also synthesize NO (neuronal NOS is ~55% homologous to endothelial NOS). This endothelium-independent NO synthesis dilates cerebral and other arteries and is responsible for penile erection. The brain contains more NOS than any other tissue in the body, suggesting that NO is essential for the function of the central nervous system. A third type of NOS is found in leukocytes (white blood cells). These cells produce NO as part of their cytotoxic arsenal. NO combines with superoxide (Section 17-5B) to produce the highly reactive hydroxyl radical, which kills invading bacteria. The sustained release of NO has been implicated in **endotoxic shock** (an often fatal immune system overreaction to bacterial infection), in inflammation-related tissue damage, and in the damage to neurons in the vicinity of but not directly killed by a stroke (which often does greater harm than the stroke itself).

7. NITROGEN FIXATION

The most prominent chemical elements in living systems are O, H, C, N, and P. The elements O, H, and P occur widely in metabolically available forms (H_2O, O_2, and P_i). However, the major forms of C and N, CO_2 and N_2, are extremely stable (unreactive); for example, the $N\equiv N$ triple bond has a bond energy of 945 $kJ \cdot mol^{-1}$ (versus 351 $kJ \cdot mol^{-1}$ for a C—O single bond). CO_2, with minor exceptions, is metabolized (fixed) only by photosynthetic organisms (Chapter 18). *N_2 fixation is even less common; this element is converted to metabolically useful forms by only a few strains of bacteria, called **diazatrophs.***

Diazatrophs of the genus *Rhizobium* live symbiotically with root nodule cells of legumes (plants belonging to the pea family, including beans, clover, and alfalfa; Fig. 20-38), where they convert N_2 to NH_3.

$$N_2 + 8\,H^+ + 8\,e^- + 16\,ATP + 16\,H_2O \longrightarrow$$
$$2\,NH_3 + H_2 + 16\,ADP + 16\,P_i$$

The NH_3 thus formed can be incorporated either into glutamate by glutamate dehydrogenase (Section 20-2B) or into glutamine by glutamine synthetase (Section 20-5A). This nitrogen-fixing system produces more metabolically useful nitrogen than the legume needs; the excess is excreted into the soil, enriching it. It is therefore common agricultural practice to plant a field with alfalfa every few years to build up the supply of usable nitrogen in the soil for later use in growing other crops.

Nitrogenase Contains Several Novel Redox Centers

Nitrogenase, which catalyzes the reduction of N_2 to NH_3, consists of two proteins:

1. The **Fe-protein,** an ~64-kD dimer of identical subunits that contains one [4Fe–4S] cluster and two ATP-binding sites (Fig. 20-39).

Figure 20-38.   A photograph of the root nodules of a pea plant. [Vu/Cabisco/Visuals Unlimited.]

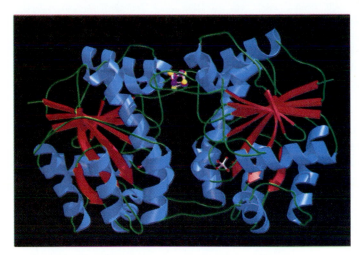

Figure 20-39. **The X-ray structure of the *Azotobacter vinelandii* nitrogenase Fe-protein dimer as viewed with its twofold molecular axis vertical in the plane of the paper.** The protein is drawn in ribbon form with its helices blue, its β sheets red, and its other segments green. The [4Fe–4S] cluster (*upper center*) and MoO_4^{2-} (*right of center*) are shown in ball-and-stick form (the MoO_4^{2-} ion presumably occupies the phosphate-binding site. [Courtesy of Douglas Rees, California Institute of Technology.] ● See the Interactive Exercises.

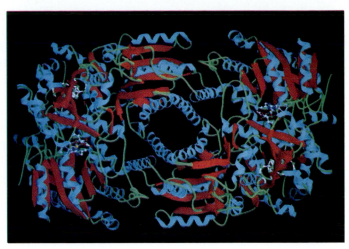

Figure 20-40. **The X-ray structure of the *A. vinelandii* nitrogenase MoFe-protein $\alpha_2\beta_2$ tetramer as viewed down its molecular twofold axis.** The protein is drawn in ribbon form and is colored as in Fig. 20-39. The P-cluster (*upper left and lower right*) and the FeMo-cofactor (*lower left and upper right*) are shown in ball-and-stick form. [Courtesy of Douglas Rees, California Institute of Technology.] ● See the Interactive Exercises.

2. The **MoFe-protein,** an ~220-kD protein of subunit structure $\alpha_2\beta_2$ that contains Fe and Mo (Fig. 20-40). Each αβ dimer contains two bound redox centers: (1) the **P-cluster** (Fig. 20-41*a*), which consists of a [4Fe–4S] cluster linked to a [4Fe–3S] cluster that are collectively liganded by the S atoms of 6 Cys residues, a Ser side chain O, and a backbone N atom; and (2) the **FeMo cofactor** (Fig. 20-41*b*), which consists of a [4Fe–3S] and a [Mo–3Fe–3S] cluster bridged by 3 sulfide ions. The Mo ion is coordinated by 3 sulfide ions, a His nitrogen, and 2 oxygen atoms of **homocitrate** (*at right*), an essential component of the FeMo cofactor. N_2 is thought to bind in the central cavity of the FeMo cofactor.

$$
\begin{array}{c}
COO^- \\
| \\
CH_2 \\
| \\
CH_2 \\
| \\
HO-C-COO^- \\
| \\
CH_2 \\
| \\
COO^-
\end{array}
$$

Homocitrate

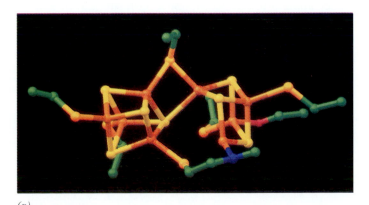

(*a*)

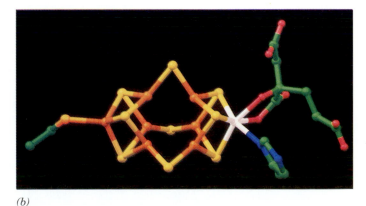

(*b*)

Figure 20-41. **The prosthetic groups of the *A. vinelandii* nitrogenase MoFe-protein.** (*a*) The P-cluster, in its oxidized state, and (*b*) the FeMo cofactor. The atoms are colored with C green, N blue, O red, sulfide ions yellow, Cys S atoms gold, Fe brown, and Mo white. [Based on an X-ray structure by Douglas Rees, California Institute of Technology.]

N_2 Reduction Is Energetically Costly

Nitrogen fixation by nitrogenase requires, in addition to N_2, a source of electrons and ATP. Electrons are generated either oxidatively or photosynthetically, depending on the organism. These electrons are transferred to ferredoxin, a [4Fe–4S]-containing electron carrier that transfers an electron to the Fe-protein of nitrogenase, beginning the nitrogen fixation process (Fig. 20-42). Two molecules of ATP bind to the reduced Fe-protein and are hydrolyzed as the electron is passed from the Fe-protein to the MoFe-protein. ATP hydrolysis is thought to cause a conformational change in the Fe-protein that alters its redox potential from -0.29 to -0.40 V, making the electron capable of N_2 reduction ($\mathscr{E}°'$ for the reaction $N_2 + 6\,H^+ + 6\,e^- \rightleftharpoons 2\,NH_3$ is -0.34 V).

The actual reduction of N_2 occurs on the MoFe-protein in three discrete steps, each involving an electron pair:

$$2\,H^+ + 2\,e^- \qquad 2\,H^+ + 2\,e^- \qquad 2\,H^+ + 2\,e^-$$

$$N\!\equiv\!N \longrightarrow H\!-\!N\!=\!N\!-\!H \longrightarrow \underset{H}{\overset{H}{N}}\!-\!\underset{H}{\overset{H}{N}} \longrightarrow 2\,NH_3$$

Diimine   **Hydrazine**

An electron transfer must occur six times per N_2 molecule fixed, so a total of 12 ATP are required to fix one N_2 molecule. However, nitrogenase also reduces H_2O to H_2, which in turn reacts with **diimine** to re-form N_2.

$$HN\!=\!NH + H_2 \longrightarrow N_2 + 2\,H_2$$

This futile cycle is favored when the ATP level is low and/or the reduction of the Fe-protein is sluggish. Even when ATP is plentiful, however, the cycle cannot be suppressed beyond about one H_2 molecule produced per N_2 reduced, and hence the cycle appears to be a requirement for the nitrogenase reaction. The minimum cost of N_2 reduction is therefore 8 electrons transferred and 16 ATP hydrolyzed (physiologically, 20–30 ATP). Hence, nitrogen fixation is an energetically expensive process.

Although atmospheric N_2 is the ultimate nitrogen source for all living things, most plants do not support the symbiotic growth of nitrogen-fixing bacteria. They must therefore depend on a source of "prefixed" nitrogen such as nitrate or ammonia. These nutrients come from lightning discharges (the source of ~10% of naturally fixed N_2), decaying organic matter in the soil, or from fertilizer applied to it. One major long-term goal of genetic engineering is to induce agriculturally useful nonleguminous plants to fix their own nitrogen. This would free farmers, particularly those in developing countries, from either purchasing fertilizers, periodically letting their

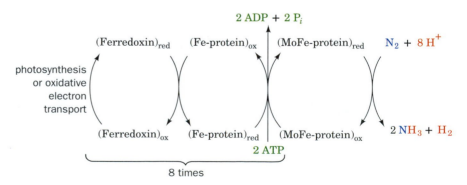

Figure 20-42.   The flow of electrons in the nitrogenase-catalyzed reduction of N_2.

fields lie fallow (giving legumes the opportunity to grow), or following the slash-and-burn techniques that are rapidly destroying the world's tropical forests.

SUMMARY

1. Intracellular proteins are degraded by lysosomal proteins or, after being ubiquitinated, by the action of proteasomes.

2. The degradation of an amino acid almost always begins with the removal of its amino group in a PLP-facilitated transamination reaction.

3. In the urea cycle, a nitrogen atom from ammonia (a product of the oxidative deamination of glutamate) and a nitrogen atom from aspartate combine with HCO_3^- to form urea for excretion. The rate-limiting step of this process is catalyzed by carbamoyl phosphate synthetase.

4. The 20 "standard" amino acids are degraded to compounds that give rise either to glucose or to ketone bodies or fatty acids: pyruvate, α-ketoglutarate, succinyl-CoA, fumarate, oxaloacetate, acetyl-CoA, or acetoacetate.

5. The nonessential amino acids are synthesized in simple pathways from pyruvate, oxaloacetate, α-ketoglutarate, and 3-phosphoglycerate. The pathways for the syntheses of essential amino acids are more complicated and vary among organisms.

6. Amino acids are the precursors of various biomolecules. Heme is synthesized from glycine and acetate. Various hormones and neurotransmitters are synthesized by the decarboxylation and hydroxylation of histidine, glutamate, tryptophan, and tyrosine. The five-electron oxidation of arginine yields the bioactive stable radical nitric oxide.

7. Nitrogen fixation in bacteria, which requires 8 electrons and at least 16 ATP, is catalyzed by nitrogenase, a multisubunit protein with redox centers containing Fe, S, and Mo.

REFERENCES

Groll, M., Ditzel, L., Löwe, J., Stock, D., Bochtler, M., Batunik, H.D., and Huber, R., Structure of 20S proteasome from yeast at 2.4 Å resolution, *Nature* **386,** 463–471 (1997).

Hayashi, H., Pyridoxal enzymes: mechanistic diversity and uniformity, *J. Biochem.* **118,** 463–473 (1995).

Hershko, A. and Ciechanover, A., The ubiquitin system, *Annu. Rev. Biochem.* **67,** 425–479 (1998).

Kim, J. and Rees, D.C., Nitrogenase and biological nitrogen fixation, *Biochemistry* **33,** 389–397 (1994).

Meijer, A.J., Lamers, W.H., and Chamuleau, R.A.F.M., Nitrogen metabolism and ornithine cycle function, *Physiol. Rev.* **70,** 701–748 (1990). [A review of the urea cycle and related topics.]

Peters, J.W., Stowell, M.H.B., Soltis, S.M., Finnegan, M.G., Johnson, M.K., and Rees, D.C., Redox-dependent structural changes in the nitrogenase P-cluster, *Biochemistry* **36,** 1181–1187 (1997).

Scriver, C.R., Beaudet, A.L., Sly, W.S., Valle, D., Stanbury, J.B., Wyngaarden, J.B., and Frederickson, D.S. (Eds.), *The Metabolic and Molecular Bases of Inherited Disease* (7th ed.), Chapters 27–38, McGraw–Hill (1995). [These chapters describe the normal and abnormal pathways of amino acid metabolism.]

Verhoef, P., Stampfer, M.J., and Rimm, E.B., Folate and coronary heart disease, *Curr. Opin. Lipidology* **9,** 17–22 (1998).

Warren, M.J., Jay, M., Hunt, D.M., Elder, G.H., and Röhl, S.C.G., The maddening business of King George III and porphyria, *Trends Biochem. Sci.* **21,** 229–234 (1996). [A fascinating account of retrospective medical detective work.]

Warren, M.J. and Scott, A.I., Tetrapyrrole assembly and modification into the ligands of biologically functional cofactors, *Trends Biochem. Sci.* **15,** 486–491 (1990).

KEY TERMS

ubiquitin	PLP	SAM	channeling
isopeptide bond	urea cycle	biopterin	porphyria
N-end rule	hyperammonemia	essential amino acid	jaundice
proteasome	glucogenic amino acid	nonessential amino acid	catecholamine
transamination	ketogenic amino acid	adenylylation	nitrogen fixation
deamination	THF	uridylylation	

STUDY EXERCISES

1. Describe the ubiquitin-dependent pathway of protein degradation.

2. How does PLP facilitate amino acid deamination?

3. Summarize the steps of the urea cycle.

4. Describe the two general metabolic fates of the carbon skeletons of amino acids.

5. What are the metabolic precursors of the nonessential amino acids?

6. Why is nitrogen fixation so energetically costly?

PROBLEMS

1. Explain why protein degradation by proteasomes requires ATP even though proteolysis is an exergonic process.

2. Explain why the symptoms of a partial deficiency in a urea cycle enzyme can be attenuated by a low-protein diet.

3. Explain why a high concentration of ammonia decreases the rate of the citric acid cycle.

4. In the degradation pathway for isoleucine (Fig. 20-19), draw the reactions that convert tiglyl-CoA to acetyl-CoA and propionyl-CoA.

5. Draw the amino acid–Schiff base that forms in the breakdown of 3-hydroxykynurenine to form 3-hydroxyanthranilate in the tryptophan degradation pathway (Fig. 20-21, Reaction 4) and indicate which bond is to be cleaved.

6. Which of the 20 "standard" amino acids are (a) purely gluco- genic, (b) purely ketogenic, and (c) both glucogenic and ke- togenic?

7. Alanine, cysteine, glycine, serine, and threonine are amino acids whose breakdown yields pyruvate. Which, if any, of the remaining 15 amino acids also do so?

8. What are the metabolic consequences of a defective uridyl- yl-removing enzyme in *E. coli?*

9. Many of the most widely used herbicides inhibit the synthe- sis of aromatic amino acids. Explain why these compounds are safe to use near animals.

10. One of the symptoms of **kwashiorkor,** the dietary protein de- ficiency disease in children, is the depigmentation of the skin and hair. Explain the biochemical basis of this symptom.

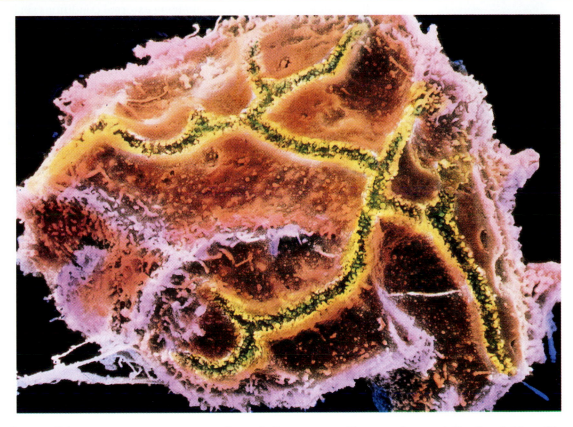

Hepatocytes (liver cells) carry out a great variety of metabolic processes. How are the specialized activities of hepatocytes and other types of cells coordinated for maximum metabolic efficiency and adaptability? How does faulty regulation result in disease?
[© P.M. Motta, T. Fujita, and M. Muto/Photo Researchers.]

MAMMALIAN FUEL METABOLISM: INTEGRATION AND REGULATION

1. ORGAN SPECIALIZATION
   A. The Brain
   B. Muscle
   C. Adipose Tissue
   D. Liver
2. INTERORGAN METABOLIC PATHWAYS
   A. The Cori Cycle
   B. The Glucose–Alanine Cycle
   C. Glucose Transporters

3. MECHANISMS OF HORMONE ACTION: SIGNAL TRANSDUCTION
   A. Hormonal Regulation of Fuel Metabolism
   B. The Adenylate Cyclase Signaling System
   C. Receptor Tyrosine Kinases
   D. The Phosphoinositide Pathway
4. DISTURBANCES IN FUEL METABOLISM
   A. Starvation
   B. Diabetes Mellitus
   C. Obesity

In even the simplest prokaryotic cell, metabolic processes must be coordinated so that opposing pathways do not operate simultaneously and so that the organism can respond to changing external conditions such as the availability of nutrients. In addition, the organism's metabolic activities must meet the demands set by genetically programmed growth and reproduction. The challenges of coordinating energy acquisition and expenditure are markedly more complex in multicellular organisms, in which individual cells must cooperate. In animals and plants, this task is simplified by the division of metabolic labor among tissues.

In animals, the interconnectedness of various tissues is ensured by neuronal circuits and by hormones. Such regulatory systems do not simply switch cells on and off but elicit an almost infinite array of responses. The exact response of a cell to a given regulatory signal depends on the cell's ability to recognize the signal and on the presence of synergistic or antagonistic signals.

Our examination of mammalian carbohydrate, lipid, and amino acid metabolism (Chapters 14–20) would be incomplete without a discussion of how such processes are coordinated at the molecular level and how their malfunctions produce disease. In this chapter, we summarize the specialized metabolism of different organs and the pathways that link them. We also examine the mechanisms by which extracellular hormones influence intracellular events. We conclude with a discussion of disruptions in mammalian fuel metabolism.

1. ORGAN SPECIALIZATION

Many of the metabolic pathways discussed so far have to do with the oxidation of metabolic fuels for the production of ATP. These pathways, which encompass the synthesis and breakdown of glucose, fatty acids, and amino acids, are summarized in Fig. 21-1.

1. **Glycolysis.** The metabolic degradation of glucose begins with its conversion to two molecules of pyruvate with the net generation of two molecules of ATP (Section 14-1).

2. **Gluconeogenesis.** Mammals can synthesize glucose from a variety of precursors, such as pyruvate, via a series of reactions that largely reverse the path of glycolysis (Section 15-4).

3. **Glycogen degradation and synthesis.** The opposing processes catalyzed by glycogen phosphorylase and glycogen synthase are reciprocally regulated by hormonally controlled phosphorylation and dephosphorylation (Section 15-3).

4. **Fatty acid synthesis and degradation.** Fatty acids are broken down through β oxidation to form acetyl-CoA (Section 19-2), which, through its conversion to malonyl-CoA, is also the substrate for fatty acid synthesis (Section 19-4).

5. **The citric acid cycle.** The citric acid cycle (Section 16-1) oxidizes acetyl-CoA to CO_2 and H_2O with the concomitant production of reduced coenzymes whose reoxidation drives ATP synthesis. Many glucogenic amino acids can be oxidized via the citric acid cycle following their breakdown to one of its intermediates (Section 20-4), which, in turn, are broken down to pyruvate and then to acetyl-CoA, the cycle's only substrate.

6. **Oxidative phosphorylation.** This mitochondrial pathway couples the oxidation of NADH and $FADH_2$ produced by glycolysis, β oxidation,

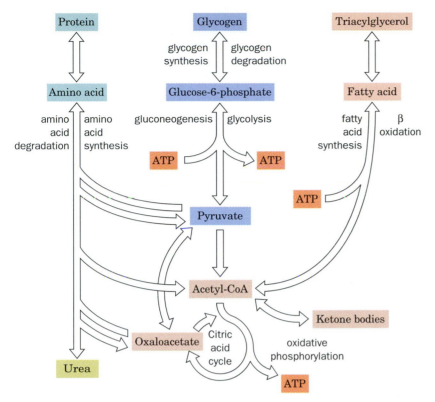

Figure 21-1. **The major pathways of fuel metabolism in mammals.** Proteins, glycogen, and triacylglycerols are built up from and broken down to smaller units: amino acids, glucose-6-phosphate, and fatty acids. Oxidation of these fuels yields metabolic energy in the form of ATP. Pyruvate (a product of glucose and amino acid degradation) and acetyl-CoA (a product of glucose, amino acid, and fatty acid degradation) occupy central positions in mammalian fuel metabolism. Compounds that give rise to pyruvate, such as oxaloacetate, can be used for gluconeogenesis; acetyl-CoA can give rise to ketone bodies but not glucose. Not all the pathways shown here occur in all cells or occur simultaneously in a given cell.

and the citric acid cycle to the phosphorylation of ADP (Section 17-3).

7. **Amino acid synthesis and degradation.** Excess amino acids are degraded to metabolic intermediates of glycolysis and the citric acid cycle (Section 20-4). The amino group is disposed of through urea synthesis (Section 20-3). Nonessential amino acids are synthesized via pathways that begin with common metabolites (Section 20-5A).

Two compounds lie at the crossroads of the major metabolic pathways: acetyl-CoA and pyruvate (Fig. 21-1). Acetyl-CoA is the common degradation product of glucose, fatty acids, and ketogenic amino acids. Its acetyl group can be oxidized to CO_2 and H_2O via the citric acid cycle and oxidative phosphorylation or used to synthesize ketone bodies or fatty acids. Pyruvate is the product of glycolysis and the breakdown of glucogenic amino acids. It can be oxidatively decarboxylated to yield acetyl-CoA, thereby committing its atoms either to oxidation or to the biosynthesis of fatty acids. Alternatively, pyruvate can be carboxylated via the pyruvate carboxylase reaction to form oxaloacetate, which can either replenish citric acid cycle intermediates or give rise to glucose or certain amino acids.

Only a few tissues, such as liver, can carry out all the reactions shown in Fig. 21-1, and in a given cell only a small portion of all possible metabolic

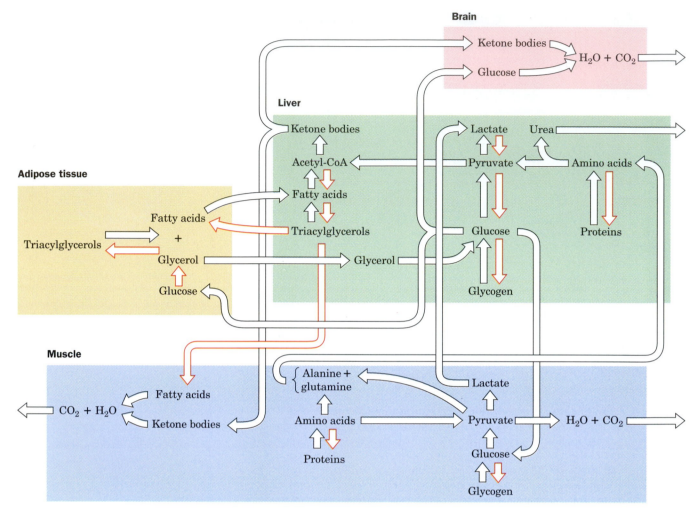

Figure 21-2.   *Key to Metabolism.* The metabolic interrelationships among brain, adipose tissue, muscle, and liver. The red arrows indicate pathways that predominate in the well-fed state.

reactions occur at a significant rate. The flux through any sequence of reactions depends on the presence of the appropriate enzyme catalysts and on the organism's need for the reaction products.

We shall consider the metabolism of four mammalian organs: brain, muscle, adipose tissue, and liver. Metabolites flow between these organs in well-defined pathways in which flux varies with the nutritional state of the animal (Fig. 21-2). For example, immediately following a meal, glucose, amino acids, and fatty acids are directly available from the intestine. Later, when these fuels have been exhausted, the liver supplies other tissues with glucose and ketone bodies, whereas adipose tissue provides them with fatty acids. All these organs are connected via the bloodstream.

A. The Brain

Brain tissue has a remarkably high respiration rate. Although the human brain constitutes only ~2% of the adult body mass, it is responsible for ~20% of its resting O_2 consumption. Most of the brain's energy production powers the plasma membrane $(Na^+ - K^+)$–ATPase (Section 10-4C), which maintains the membrane potential required for nerve impulse transmission.

Under usual conditions, glucose is the brain's primary fuel (although during an extended fast, the brain gradually switches to ketone bodies; Section 21-4A). Since brain cells store very little glycogen, *they require a steady supply of glucose from the blood*. A blood glucose concentration of less than half the normal value of ~5 mM results in brain dysfunction. Levels much below this result in coma, irreversible damage, and ultimately death.

B. Muscle

Muscle's major fuels are glucose (from glycogen), fatty acids, and ketone bodies. Rested, well-fed muscle synthesizes a glycogen store comprising 1 to 2% of its mass. Although triacylglycerols are a more efficient form of energy storage (Section 9-1B), the metabolic effort of synthesizing glycogen is cost-effective because glycogen can be mobilized more rapidly than fat and because glucose can be metabolized anaerobically, whereas fatty acids cannot.

In muscle, glycogen is converted to glucose-6-phosphate (G6P) for entry into glycolysis. Muscle cannot export glucose, however, because it lacks glucose-6-phosphatase. Furthermore, although muscle can synthesize glycogen from glucose, it does not participate in gluconeogenesis because it lacks the required enzymatic machinery. Consequently, *muscle carbohydrate metabolism serves only muscle*.

Muscle Contraction Is Anaerobic Under Conditions of High Exertion

Muscle contraction is driven by ATP hydrolysis (Section 7-3C) and therefore requires either an aerobic or an anaerobic ATP regeneration system. Respiration (the citric acid cycle and oxidative phosphorylation) is the body's major source of ATP resupply. Skeletal muscle at rest uses ~30% of the O_2 consumed by the human body. A muscle's respiration rate may increase in response to a heavy workload by as much as 25-fold. Yet, its rate of ATP hydrolysis can increase by a much greater amount. The ATP is initially regenerated by the reaction of phosphocreatine (Section 13-2C):

$$\text{Phosphocreatine} + \text{ADP} \rightleftharpoons \text{creatine} + \text{ATP}$$

(phosphocreatine is resynthesized in resting muscle by the reversal of this reaction). Under conditions of maximum exertion, however, such as during a sprint, a muscle has only about a 4-s supply of phosphocreatine. It must then shift to ATP production via glycolysis of G6P, a process whose maximum flux greatly exceeds those of the citric acid cycle and oxidative phosphorylation. Much of the G6P is therefore degraded anaerobically to lactate (Section 14-3A). As we shall see in Section 21-2A, export of lactate relieves much of the muscle's respiratory burden. Muscle fatigue, which occurs after ~20 s of maximal exertion, is not caused by exhaustion of the muscle's glycogen supply but by the drop in pH that results from the buildup of lactate (Section 14-3A). This phenomenon may be an adaptation that prevents muscle cells from committing suicide by fully depleting their ATP supply.

The Heart Is Largely Aerobic

The heart is a muscular organ that acts continuously rather than intermittently. Therefore, heart muscle relies entirely on aerobic metabolism and is richly endowed with mitochondria; they occupy up to 40% of its cytoplasmic space. The heart can metabolize fatty acids, ketone bodies, glucose, pyruvate, and lactate. Fatty acids are the resting heart's fuel of choice, but during heavy work, the heart greatly increases its consumption of glucose, which is derived mostly from its relatively limited glycogen store.

C. Adipose Tissue

The function of adipose tissue is to store and release fatty acids as needed for fuel. Adipose tissue is widely distributed throughout the body but occurs most prominently under the skin, in the abdominal cavity, and in skeletal muscle. The adipose tissue of a normal 70-kg man contains ~15 kg of fat. This amount represents some 590,000 kJ of energy (141,000 dieter's Calories), which is sufficient to maintain life for ~3 months.

Adipose tissue obtains most of its fatty acids for storage from circulating lipoproteins as described in Section 19-1B. Fatty acids are activated by the formation of the corresponding fatty acyl-CoA and then esterified with glycerol-3-phosphate to form the stored triacylglycerols. The glycerol-3-phosphate arises from the reduction of dihydroxyacetone phosphate, which must be glycolytically generated from glucose.

In times of metabolic need, adipocytes hydrolyze triacylglycerols to fatty acids and glycerol through the action of hormone-sensitive lipase (Section 19-5). If glycerol-3-phosphate is abundant, many of the fatty acids so formed are reesterified to triacylglycerols. If glycerol-3-phosphate is in short supply, the fatty acids are released into the bloodstream. Thus, *fatty acid mobilization depends in part on the rate of glucose uptake since glucose is the precursor of glycerol-3-phosphate.* Metabolic need is signaled directly by a decrease in [glucose] as well as by hormonal stimulation.

D. Liver

The liver is the body's central metabolic clearinghouse. It maintains the proper levels of circulating fuels for use by the brain, muscles, and other tissues. The liver is uniquely situated to carry out this task because all the nutrients absorbed by the intestines except fatty acids are released into the portal vein, which drains directly into the liver.

Glucokinase Converts Blood Glucose to Glucose-6-Phosphate

One of the liver's major functions is to act as a blood glucose "buffer." It does so by taking up and releasing glucose in response to hormones and to the concentration of glucose itself. After a carbohydrate-containing meal, when the blood glucose concentration reaches ~6 mM, the liver takes up glucose by converting it to G6P. This reaction is catalyzed by **glucokinase,** a liver isozyme of hexokinase. The hexokinases in most cells obey Michaelis–Menten kinetics, have a high glucose affinity ($K_M < 0.1$ mM), and are inhibited by their reaction product (G6P). Glucokinase, in contrast, has much lower glucose affinity (reaching half-maximal velocity at ~5 mM) and displays sigmoidal kinetics. Consequently, *glucokinase activity increases rapidly with blood [glucose] over the normal physiological range* (Fig. 21-3). Glucokinase, moreover, is not inhibited by physiological concentrations of G6P. Therefore, the higher the blood [glucose], the faster the liver converts glucose to G6P. At low blood [glucose], the liver does not compete with other tissues for the available glucose, whereas at high blood [glucose], when the glucose needs of these tissues are met, the liver can take up the excess glucose at a rate roughly proportional to the blood glucose concentration.

Glucokinase is a monomeric enzyme, so its sigmoidal kinetic behavior is somewhat puzzling (models of allosteric interactions do not explain cooperative behavior in a monomeric protein; Section 7-2E). Glucokinase is subject to metabolic control, however. Emile Van Schaftingen has isolated

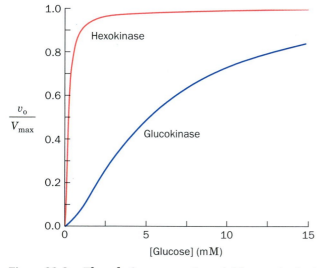

Figure 21-3. The relative enzymatic activities of hexokinase and glucokinase over the physiological blood glucose range. Glucokinase has much lower affinity for glucose ($K_M \approx 5$ mM) than does hexokinase ($K_M = 0.1$ mM) and exhibits sigmoidal rather than hyperbolic variation with [glucose]. [The glucokinase curve was generated using the Hill equation (Eq. 7-4) with $K = 10$ mM and $n = 1.5$ as obtained from Cardenas, M.L., Rabajille, E., and Niemeyer, H., *Eur. J. Biochem.* **145,** 163–171 (1984).]

a **glucokinase regulatory protein** from rat liver, which, in the presence of the glycolytic intermediate fructose-6-phosphate (F6P), is a competitive inhibitor of glucokinase. Since F6P and the glucokinase product G6P are equilibrated in liver cells by phosphoglucose isomerase, glucokinase is, in effect, inhibited by its product. Fructose-1-phosphate (F1P), an intermediate in liver fructose metabolism (Section 14-5A), overcomes this inhibition. Since fructose is normally available only from dietary sources, fructose may be the signal that triggers the uptake of dietary glucose by the liver.

Glucose-6-Phosphate Is at the Crossroads of Carbohydrate Metabolism

G6P has several alternative fates in the liver, depending on the glucose demand (Fig. 21-4):

1. G6P can be converted to glucose, by the action of glucose-6-phosphatase, for transport via the bloodstream to the peripheral organs. This occurs only when the blood [glucose] drops below ~5 mM. During exercise or fasting, low concentrations of blood glucose cause the pancreas to secrete glucagon. Glucagon receptors on the liver cell surface respond by activating adenylate cyclase. The resulting increase in intracellular [cAMP] triggers glycogen breakdown (Section 15-3).

2. G6P can be converted to glycogen (Section 15-2) when the body's demand for glucose is low.

3. G6P can be converted to acetyl-CoA via glycolysis and the action of pyruvate dehydrogenase. This glucose-derived acetyl-CoA, if it is not oxidized via the citric acid cycle and oxidative phosphorylation to generate ATP, can be used to synthesize fatty acids (Section 19-4), phospholipids (Section 19-6), and cholesterol (Section 19-7A).

4. G6P can be degraded via the pentose phosphate pathway (Section 14-6) to generate the NADPH required for the biosynthesis of fatty acids and other compounds.

The Liver Can Synthesize or Degrade Triacylglycerols

Fatty acids are also subject to alternative metabolic fates in the liver. When the demand for metabolic fuels is high, fatty acids are degraded to acetyl-CoA and then to ketone bodies for export to the peripheral tissues. The liver itself cannot use ketone bodies as fuel, because liver cells lack 3-ketoacyl-CoA transferase, an enzyme required to convert ketone bodies back to acetyl-CoA (Section 19-3). Fatty acids rather than glucose or ketone bodies are therefore the liver's major acetyl-CoA source under conditions of high metabolic demand. The liver generates its ATP from this acetyl-CoA through the citric acid cycle and oxidative phosphorylation.

When the demand for metabolic fuels is low, fatty acids are incorporated into triacylglycerols that are secreted into the bloodstream as VLDL for uptake by adipose tissue. Fatty acids can also be incorporated into phospholipids (Section 19-6). Under these conditions, fatty acids synthesized in the liver are not oxidized to acetyl-CoA because fatty acid synthesis (in the cytosol) is separated from fatty acid oxidation (in the mitochondria).

Amino Acids Are Metabolic Fuels

The liver degrades amino acids to a variety of metabolic intermediates that can be completely oxidized to CO_2 and H_2O or converted to glucose or ketone bodies (Section 20-4). Oxidation of amino acids provides a significant fraction of metabolic energy immediately after feeding, when the amino acids are present in relatively high concentrations in the blood. During a fast, when other fuels become scarce, glucose is produced from amino acids arising mostly from muscle protein degradation to alanine and glut-

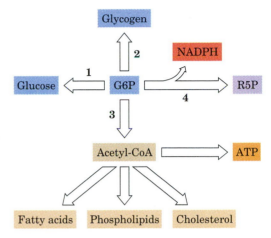

Figure 21-4.   Metabolic fate of glucose-6-phosphate in liver. G6P can be converted (1) to glucose for export or (2) to glycogen for storage. Acetyl-CoA derived from G6P degradation (3) is the starting material for lipid biosynthesis. It is also consumed in generating ATP by respiration. Degradation of G6P via the pentose phosphate pathway (4) yields NADPH.

amine. Thus, *proteins, in addition to their structural and functional roles, are important fuel reserves.*

2. INTERORGAN METABOLIC PATHWAYS

The ability of the liver to supply other tissues with glucose or ketone bodies, or the ability of adipocytes to make fatty acids available to other tissues, depends, of course, on the circulatory system, which transports metabolic fuels, intermediates, and waste products between tissues. In addition, several important metabolic pathways are composed of reactions occurring in multiple tissues. In this section, we describe two well-known interorgan pathways, along with the properties of the transporters that modulate the passage of glucose into and out of cells.

A. The Cori Cycle

The ATP that powers muscle contraction is generated through oxidative phosphorylation (in mitochondrion-rich slow-twitch muscle fibers; Box 14-2) or by rapid catabolism of glucose to lactate (in fast-twitch muscle fibers). Slow-twitch fibers also produce lactate when ATP demand exceeds oxidative flux. The lactate is transferred via the bloodstream to the liver, where it is reconverted to pyruvate by lactate dehydrogenase and then to glucose by gluconeogenesis. Thus, liver and muscle are linked by the bloodstream in a metabolic cycle known as the **Cori cycle** (Fig. 21-5) in honor of Carl and Gerty Cori, who first described it.

The ATP-consuming glycolysis/gluconeogenesis cycle would be a futile cycle if it occurred within a single cell. In this case, the two halves of the pathway occur in different organs. Liver ATP is used to resynthesize glucose from lactate produced in muscle. The resynthesized glucose returns to the muscle, where it may be stored as glycogen or catabolized immediately to generate ATP for muscle contraction.

The ATP consumed by the liver during the operation of the Cori cycle is regenerated by oxidative phosphorylation. After vigorous exertion, it may take at least 30 min for the oxygen consumption rate to decrease to its resting level. The elevated O_2 consumption pays off the **oxygen debt** created by the demand for ATP to drive gluconeogenesis.

B. The Glucose–Alanine Cycle

In a pathway similar to the Cori cycle, alanine rather than lactate travels from muscle to the liver. In muscle, certain aminotransferases use pyruvate

*Figure 21-5.   **The Cori cycle.** Lactate produced by muscle glycolysis is transported by the bloodstream to the liver, where it is converted to glucose by gluconeogenesis. The bloodstream carries the glucose back to the muscle, where it may be stored as glycogen. ✳ **See the Animated Figures.***

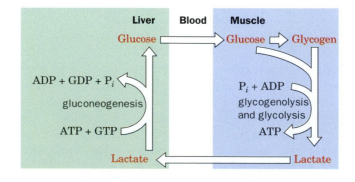

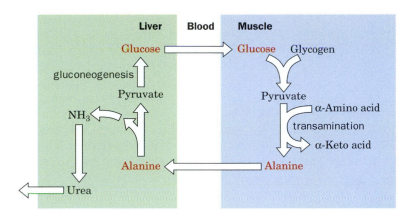

Figure 21-6. The glucose–alanine cycle.
Pyruvate produced by muscle glycolysis is the
amino-group acceptor for muscle aminotransferases. The resulting alanine is transported by the
bloodstream to the liver, where it is converted
back to pyruvate (its amino group is disposed of
via urea synthesis). The pyruvate is a substrate for
gluconeogenesis, and the bloodstream carries the
resulting glucose back to the muscles.
✳ See the Animated Figures.

as their α-keto acid substrate rather than α-ketoglutarate or oxaloacetate
(Section 20-2A):

$$\underset{\textbf{Amino acid}}{R-\overset{\overset{\textstyle +}{NH_3}}{\underset{|}{C}}H-COO^-} + \underset{\textbf{Pyruvate}}{H_3C-\overset{\overset{\textstyle O}{\|}}{C}-COO^-} \rightleftharpoons \underset{\textbf{α-Keto acid}}{R-\overset{\overset{\textstyle O}{\|}}{C}-COO^-} + \underset{\textbf{Alanine}}{H_3C-\overset{\overset{\textstyle +}{NH_3}}{\underset{|}{C}}H-COO^-}$$

The product amino acid, alanine, is released into the bloodstream and transported to the liver, where it undergoes transamination back to pyruvate.
This pyruvate is a substrate for gluconeogenesis, and the resulting glucose
can be returned to the muscles to be glycolytically degraded. This is the
glucose–alanine cycle (Fig. 21-6). The amino group carried by alanine ends
up in either ammonia or aspartate and can be used for urea biosynthesis
(which occurs only in the liver). Thus, *the glucose–alanine cycle is a mechanism for transporting nitrogen from muscle to liver.*

During fasting, the glucose formed in the liver by this route is also used
by other tissues, breaking the cycle. Because the pyruvate originates from
muscle protein degradation, muscle supplies glucose to other tissues even
though it does not carry out gluconeogenesis.

C. Glucose Transporters

The passage of metabolites between tissues requires mechanisms for transporting these substances into and out of cells. Amino acids, for example,
which are ionic, are transported by several carrier systems with relatively
broad substrate specificity. Glucose is likewise transported into cells via
protein carriers. We have already examined the passive glucose transporter
of erythrocytes (Section 10-4B), also known as **GLUT1.** This transporter
occurs in many tissues, including brain, muscle, and fat, but is expressed
only at low levels in liver. Five other related glucose transporters, **GLUT2,
GLUT3, GLUT4, GLUT5,** and **GLUT7,** are expressed in a tissue-specific
fashion.

GLUT2 occurs primarily in liver and in pancreatic β cells. Its high K_M
for glucose (~60 mM) allows glucose flux into and out of hepatocytes to
vary essentially linearly with glucose concentration. In other words, *GLUT2
is never saturated with glucose and therefore never limits the rate of glucose
transport.* Defects in GLUT2 produce symptoms resembling those of Type
I glycogen storage disease (Box 15-2).

GLUT3 is expressed in brain and nerve tissues, which exhibit a high glucose demand. It may operate in concert with GLUT1 to maximize glucose
uptake.

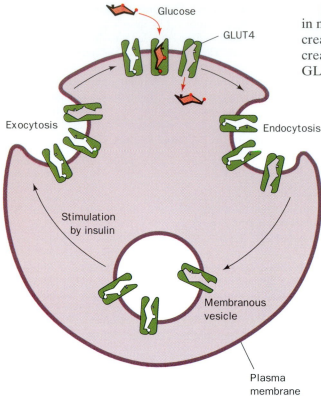

Figure 21-7.   GLUT4 activity. Glucose uptake in muscle and fat cells is regulated by the insulin-stimulated exocytosis of membranous vesicles containing GLUT4 (*left*). On insulin withdrawal, the process reverses itself through endocytosis (*right*). ✳ **See the Animated Figures.**

GLUT4 is an insulin-sensitive glucose transporter that occurs only in muscle and adipose tissue. Insulin stimulates GLUT4 activity by increasing its V_{max}. An increase in V_{max} can be effected through an increase in the intrinsic activity of a transporter, but in the case of GLUT4, the increase in V_{max} is accomplished through the appearance of additional transporter molecules in the plasma membrane (Fig. 21-7). In the absence of insulin, GLUT4 is localized in intracellular vesicles. Insulin promotes the translocation of GLUT4 to the plasma membrane. A relatively low K_M for glucose (2–5 mM) allows cells containing GLUT4 to rapidly take up glucose from the blood. On insulin withdrawal, the glucose transporters are gradually sequestered through endocytosis.

GLUT5 occurs primarily in the small intestine. Glucose transport from the intestinal lumen into mucosal cells is mediated predominantly by an unrelated Na^+-dependent glucose transporter (Fig. 10-40). GLUT5 may be responsible for the passage of glucose into the bloodstream. Alternatively, since GLUT5 has high affinity for fructose, its physiological importance may be related to the uptake of dietary fructose.

GLUT7, which is 68% identical in sequence to GLUT2, occurs in the endoplasmic reticulum of liver cells. In liver, G6P is dephosphorylated by the action of glucose-6-phosphatase, whose active site is in the lumen of the endoplasmic reticulum. Accordingly, the product of the reaction, glucose, must first cross the endoplasmic reticulum membrane before it can be exported from the cell via GLUT2.

3. MECHANISMS OF HORMONE ACTION: SIGNAL TRANSDUCTION

Living things coordinate their activities at every level of their organization through complex signaling systems involving chemical messengers known as **hormones.** In higher animals, **endocrine glands** synthesize and release hormones, which are carried by the bloodstream to their target cells (Fig. 21-8). We have already discussed steroid hormones (Section 9-1E), the peptide hormones insulin and glucagon (Section 15-3C), and

Figure 21-8.   Endocrine signaling. Hormones produced by endocrine cells reach their target cells via the bloodstream. Only cells that display the appropriate receptors can respond to the hormones.

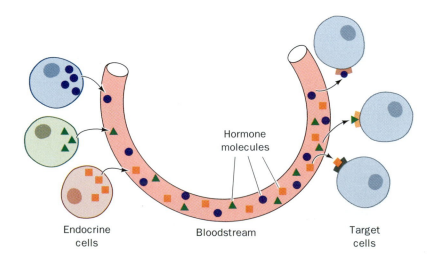

Endocrine cells   Bloodstream   Target cells

Hormone molecules

the catecholamines epinephrine and norepinephrine (Section 20-6B). With the exception of steroids, which can diffuse through cell membranes and interact directly with intracellular components (and which we shall discuss further in Section 27-3B), these extracellular signals must first bind to a cell-surface receptor. A **receptor** in this context is defined as a binding protein that is specific for its ligand and behaves in such a way that it elicits a discrete biochemical effect when its ligand is bound, a feature that differentiates a receptor from a simple binding or transport protein such as myoglobin (Section 7-1) or GLUT1 (Section 21-2C).

The binding of an extracellular ligand such as a hormone to its receptor causes a signal to be **transduced** (transmitted) to the cell interior. This signal-transduction event sets in motion a series of biochemical reactions that produce a biological response such as altered metabolism, cell differentiation, or cell growth and division. The exact nature of the response depends on many factors. Cells respond to a ligand only if they display the appropriate receptor. Specific intracellular responses are modulated by the number, type, and cellular locations of the elements of the signaling systems. Furthermore, a given cell typically contains receptors for many different ligands, so the response to one particular ligand may depend on the level of engagement of the other ligands with the cell's signal-transduction machinery.

The complexity of the signal-transduction pathways means that *cells can react to discrete signals or combinations of signals with variations in the magnitude and duration of the cellular response.* The signaling processes described below are multistep pathways offering multiple opportunities to amplify or dampen the response. Furthermore, these pathways can be activated as well as deactivated, usually as the result of the phosphorylation and dephosphorylation of cellular proteins in a way that alters their intrinsic activity or their interactions with other proteins.

A. Hormonal Regulation of Fuel Metabolism

The human endocrine system secretes a wide variety of hormones that enable the body to

1. Maintain **homeostasis** (a steady state; e.g., insulin and glucagon maintain the blood glucose level within rigid limits during feast or famine).

2. Respond to a wide variety of external stimuli (such as the preparation for "fight-or-flight" by epinephrine and norepinephrine).

3. Follow various cyclic and developmental programs (for instance, sex hormones regulate sexual differentiation, maturation, the menstrual cycle, and pregnancy).

The various endocrine glands are not just a collection of independent secretory organs but form a complex and highly interdependent control system. We shall limit our discussion to the major regulatory hormones of fuel metabolism.

Pancreatic Islet Hormones Regulate the Storage and Release of Glucose and Fatty Acids

The pancreas is a large glandular organ, the bulk of which is dedicated to producing digestive enzymes such as trypsin, RNase A, α-amylase, and phospholipase A_2. These proteins are secreted via the pancreatic duct into the small intestine. However, ~1 to 2% of pancreatic tissue consists of scattered clumps of cells known as **islets of Langerhans** (Fig. 21-9), which secrete polypeptide hormones into the bloodstream. These hormones, like

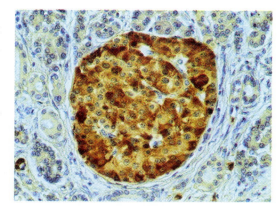

Figure 21-9. Pancreatic islet cells. The hormone-producing islet cells (*brown stain*) constitute only a small fraction of the pancreatic cells, most of which synthesize digestive enzymes. [© Parviz M. Pour/Photo Researchers, Inc.]

other proteins destined for secretion, are ribosomally synthesized as inactive precursors, processed in the rough endoplasmic reticulum and Golgi apparatus to form the mature hormones, and then packaged in secretory granules to await the signal for their release by exocytosis (Section 10-2D).

The **β cells** of the pancreatic islets secrete insulin (51 residues; Fig. 5-1) in response to high blood glucose levels. This stimulates muscle, liver, and adipose cells to store fuel for later use by synthesizing glycogen, protein, and fat. The **α cells** of the pancreatic islets secrete glucagon (29 residues; Section 15-3C) in response to low blood glucose. Glucagon has essentially the opposite physiological effects: It stimulates liver to release glucose through glycogenolysis and gluconeogenesis, and it stimulates adipose tissue to release fatty acids through lipolysis.

The Adrenal Medulla Secretes Catecholamines

The adrenal glands consist of two distinct types of tissue: the **medulla** (core), which is really an extension of the nervous system, and the more typically glandular **cortex** (outer layer), which synthesizes and secretes steroid hormones. The adrenal medulla synthesizes norepinephrine and its methyl derivative epinephrine from tyrosine as described in Section 20-6B. These catecholamines are stored in granules to await their exocytotic release under the control of the nervous system.

The biological effects of catecholamines are mediated by two classes of receptors, the **α- and β-adrenergic receptors.** These receptors, which occur in separate tissues in mammals, generally respond differently and often oppositely to catecholamines. For instance, β-adrenergic receptors, which activate adenylate cyclase (Section 21-3B), stimulate glycogenolysis and gluconeogenesis in liver, lipolysis in adipose tissue, smooth muscle relaxation in the bronchi and the blood vessels supplying the skeletal muscles, and increased heart action. In contrast, α-adrenergic receptors, whose intracellular effects are mediated either by the inhibition of adenylate cyclase or via the phosphoinositide cascade (Section 21-3D), stimulate smooth muscle contraction in blood vessels supplying peripheral organs such as skin and kidney, smooth muscle relaxation in the gastrointestinal tract, and blood platelet aggregation. *Most of these diverse effects are directed toward a common end: to mobilize energy resources and shunt them to where they are most needed to prepare the body for sudden action.*

◉ See Guided Exploration 19:

Mechanisms of Hormone Signaling Involving the Adenylate Cyclase System.

B. The Adenylate Cyclase Signaling System

Many hormones and other extracellular signals exert their effects through integral membrane glycoprotein receptors that contain seven transmembrane α helices. Bacteriorhodopsin (Fig. 10-4) is such a protein although it is not a hormone receptor. The β-adrenergic receptor (Fig. 21-10) is a member of this class of receptor. The adrenal hormones epinephrine and norepinephrine bind to the N-terminal extracellular domain of the receptor.

The strength of binding of any ligand L to its receptor R to form a complex RL (R + L ⇌ RL) can be described in terms of the dissociation constant of the RL complex

$$K_D = \frac{[\text{R}][\text{L}]}{[\text{RL}]} \qquad [21\text{-}1]$$

The dissociation constant for epinephrine binding to the β-adrenergic receptor is 5×10^{-6} M. Ligand binding to the β-adrenergic receptor extracellular domain probably results in a conformational change that is transmitted to the receptor's intracellular domain.

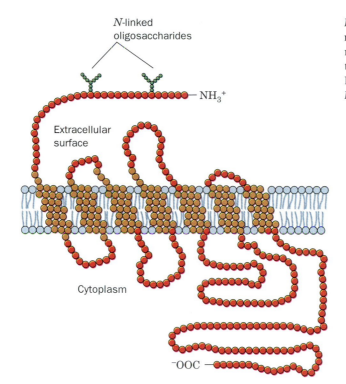

Figure 21-10.   The human β-adrenergic receptor. The seven segments of ~24 hydrophobic residues (*brown circles*) suggest that this glycoprotein has seven membrane-spanning helices. [After Dohlman, H.G., Caron, M.G., and Lefkowitz, R.J., *Biochemistry* **26**, 2660 (1987).]

Cell-surface receptors are believed to function much like allosteric proteins such as hemoglobin (Section 7-2E). *By alternating between two discrete conformations, one with ligand bound and one without, the receptor can transmit an extracellular signal to the cell interior.* This model of receptor action is similar to the operation of membrane transport proteins (e.g., Fig. 10-35); in fact, some membrane-bound receptors are ion channels that switch between the open and closed conformations in response to ligand binding.

Heterotrimeric G-Proteins Mediate Intracellular Responses

Many adrenergic receptors and polypeptide hormone receptors in eukaryotes activate intracellular **heterotrimeric G-proteins,** so named because they bind the guanine nucleotides GTP and GDP and consist of α, β, and γ subunits (45, 37, and 9 kD, respectively). The heterotrimeric G-protein family contains at least 20 members built from one each of the 15 α, 5 β, and 6 γ subunits that have yet been identified. However, all G-proteins are thought to have a structure similar to that shown in Fig. 21-11. The protein is tethered to the membrane through prenylation of the γ subunit C-terminus and through myristoylation and, in some cases, palmitoylation of the α subunit N-terminus (Section 10-1B). The N-terminus of the α subunit lies next to the β subunit on one side of the G-protein, leaving the bulk of the α subunit free to interact with other proteins.

When GDP is bound to the G_α subunit, the G-protein is in its inactive form. Following ligand binding to an appropriate receptor, the G-protein interacts with the intracellular portion of the receptor. This interaction induces G_α to release its bound GDP and take up GTP, which in turn causes the heterotrimeric G-protein to dissociate to its G_α and $G_{\beta\gamma}$ components, each of which can then activate other cellular components (see below).

The effect of G-protein activation is short-lived, however, because G_α is also a **GTPase,** hydrolyzing GTP to GDP + P_i at a rate of 2 to 3 min^{-1}.

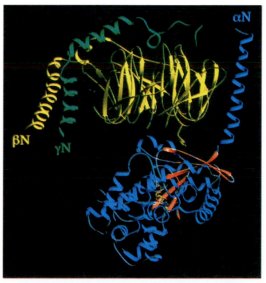

Figure 21-11.   Ribbon model of a G-protein. The α subunit is blue (with its so-called switch II region in red), the β subunit is yellow, the γ subunit is green, and the GDP molecule bound to the α subunit is shown as a stick model. The N-termini of the three subunits are marked. [Courtesy of Mark Wall and Stephen Sprang, University of Texas Southwestern Medical Center.]

● See the Interactive Exercises.

GTP hydrolysis causes the G-protein to reassemble as the inactive GDP-binding heterotrimer.

$$G_{\alpha\beta\gamma} \cdot GDP \xrightarrow{\quad GDP \quad GTP \quad} G_{\alpha} \cdot GTP + G_{\beta\gamma} \longrightarrow G_{\alpha\beta\gamma} \cdot GDP + P_i$$
$$\text{(inactive)} \qquad\qquad \text{(active)} \quad \text{(active)} \qquad \text{(inactive)}$$

The rate of GTP hydrolysis is apparently fast enough so that it does not create a runaway response (a single receptor molecule can activate several G-proteins, each of which can act on several other proteins). Continued occupancy of the receptor by its ligand, of course, results in repeated cycles of G-protein activation and hence a prolonged cellular response.

Cyclic AMP Is a Second Messenger

In the case of the β-adrenergic receptor, the $G_{\alpha} \cdot GTP$ complex activates the plasma membrane-bound enzyme adenylate cyclase, which converts ATP to cAMP (Section 15-3B). The cAMP concentration in a cell depends on its relative rates of synthesis (catalyzed by adenylate cyclase) and degradation (catalyzed by a specific phosphodiesterase). cAMP, a polar, freely diffusing cytoplasmic molecule, is called a second messenger since it mediates the hormonal (primary) message within the cell.

cAMP is absolutely required for the activity of cAMP-dependent protein kinase (cAPK), an enzyme that phosphorylates Ser or Thr residues of cellular proteins. The targets of cAPK include enzymes involved in glycogen metabolism (Section 15-3). Thus, when epinephrine binds to the β-adrenergic receptor of a muscle cell, for example, the sequential activation of a G-protein, adenylate cyclase, and cAPK leads to the activation of glycogen phosphorylase, thereby making glucose-6-phosphate available for glycolysis in a "fight-or-flight" response.

The Adenylate Cyclase System Is Regulated

Some hormone receptors inhibit rather than stimulate adenylate cyclase. The inhibitory effect is mediated by an "inhibitory" G-protein (G_i), which often has the same β and γ subunits as a "stimulatory" G-protein (G_s) but a different α subunit. G_i acts analogously to G_s in that, on binding to its corresponding ligand–receptor complex, its α subunit exchanges GDP for GTP. The $G_{i\alpha}$ subunit inhibits rather than activates adenylate cyclase through direct interactions and possibly also because the liberated $G_{\beta\gamma}$ unit binds to and sequesters $G_{s\alpha}$. The complexity of signaling systems centered on adenylate cyclase is only partially illustrated in Fig. 21-12.

The nature and magnitude of the cellular response ultimately reflects the presence and degree of activation or inhibition of all the preceding components of the signal-transduction pathway. The adenylate cyclase signaling pathways can also be limited or reversed, for example, through modulation of the GTPase activity of G-proteins and the activity of the phosphodiesterase that hydrolyzes cAMP to AMP. In addition, reactions catalyzed by cAPK are reversed by the action of protein phosphatases that dephosphorylate proteins containing phospho-Ser and phospho-Thr. Many drugs and toxins exert their effects by modifying components of the adenylate cyclase system (see Box 21-1).

Another hallmark of biological signaling systems is that they adapt to long-term stimuli by reducing their response to them, a process named **desensitization.** These signaling systems therefore respond to changes in stimulation levels rather than to their absolute values. In the case of the β-adrenergic receptor, continuous exposure to epinephrine leads to the

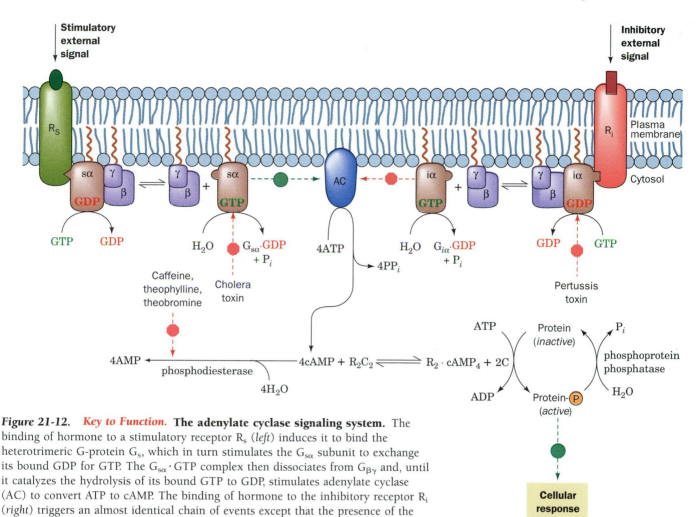

Figure 21-12. *Key to Function.* **The adenylate cyclase signaling system.** The binding of hormone to a stimulatory receptor R_s (*left*) induces it to bind the heterotrimeric G-protein G_s, which in turn stimulates the $G_{s\alpha}$ subunit to exchange its bound GDP for GTP. The $G_{s\alpha} \cdot$ GTP complex then dissociates from $G_{\beta\gamma}$ and, until it catalyzes the hydrolysis of its bound GTP to GDP, stimulates adenylate cyclase (AC) to convert ATP to cAMP. The binding of hormone to the inhibitory receptor R_i (*right*) triggers an almost identical chain of events except that the presence of the $G_{i\alpha} \cdot$ GTP complex inhibits adenylate cyclase. R_2C_2 represents cAMP-dependent protein kinase whose catalytic subunit C, when activated by the dissociation of the regulatory dimer as $R_2 \cdot$ cAMP$_4$, activates various cellular proteins by catalyzing their phosphorylation. The sites of action of certain drugs and toxins are indicated.

phosphorylation of one or more of the receptor's Ser residues. This phosphorylation, which is catalyzed by a specific kinase that acts on the hormone–receptor complex but not on the receptor alone, reduces the receptor's affinity for epinephrine. If the epinephrine level is reduced, the receptor is slowly dephosphorylated, thereby restoring the cell's epinephrine sensitivity.

C. Receptor Tyrosine Kinases

Many protein hormones known as **growth factors** stimulate the proliferation and differentiation of their target cells by binding to receptors whose C-terminal intracellular domains have **tyrosine kinase** activity. Such **receptor tyrosine kinases (RTKs)** typically contain only a single transmembrane segment and are monomers in the unliganded state. These structural features make it unlikely that ligand binding to an extracellular domain manifests itself as a conformational change in an intracellular domain (such a conformational shift seems more likely to occur in receptors with multiple transmembrane segments, such as the β-adrenergic receptor; Fig. 21-10).

> ● See Guided Exploration 20:
>
> **Mechanisms of Hormone Signaling Involving the Receptor Tyrosine Kinase System.**

Drugs and Toxins That Affect Cell Signaling

Complex processes such as the adenylate cyclase signaling system can be sabotaged by a variety of agents. For example, the methylated purine derivatives **caffeine** (an ingredient of coffee), **theophylline** (found in tea), and **theobromine** (found in chocolate)

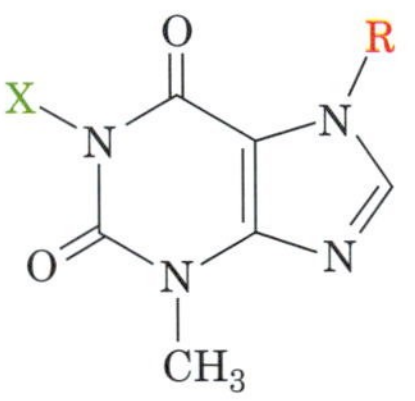

R = CH₃	X = CH₃	**Caffeine (1,3,7-trimethylxanthine)**
R = H	X = CH₃	**Theophylline (1,3-dimethylxanthine)**
R = CH₃	X = H	**Theobromine (1,7-dimethylxanthine)**

are stimulants, in part, because they inhibit cAMP phosphodiesterase, thereby prolonging the effects of the hormones responsible for generating cAMP.

Deadlier effects result from certain bacterial toxins that interfere with heterotrimeric G-protein function. The toxin released by *Vibrio cholerae* (the bacterium causing cholera) causes massive fluid loss of over a liter per hour from diarrhea. Victims die from dehydration unless their lost water and salts are replaced. **Cholera toxin**, an 87-kD protein of sub-

unit composition AB₅, binds to ganglioside G_{M1} (Fig. 9-9) on the surface of intestinal cells via its B subunits. This permits the toxin to enter the cell, probably via receptor-mediated endocytosis, where an ~195-residue proteolytic fragment of its A subunit is released. This fragment catalyzes the transfer of the ADP–ribose unit from NAD^+ to a specific Arg side chain of $G_{s\alpha}$ (*below*).

ADP-ribosylated $G_{s\alpha} \cdot GTP$ can activate adenylate cyclase but cannot hydrolyze its bound GTP (Fig. 21-12). As a consequence, the adenylate cyclase is locked in its active state and cellular cAMP levels increase ~100-fold. Intestinal cells, which normally respond to small increases in cAMP by secreting digestive fluid (an HCO_3^--rich salt solution), pour out enormous quantities of this fluid in response to the elevated cAMP concentrations.

Other bacterial toxins act similarly. Certain strains of *E. coli* cause a diarrheal disease similar to but less serious than cholera through their production of **heat-labile enterotoxin**, a protein that is closely similar to cholera toxin (their A and B subunits are >80% identical) and has the same mechanism of action. **Pertussis toxin** (secreted by *Bordetella pertussis*, the bacterium that causes **pertussis**, or whooping cough, which is still responsible for ~400,000 infant deaths per year worldwide) is an AB₅ protein homologous to cholera toxin that ADP-ribosylates a specific Cys residue of $G_{i\alpha}$. The modified $G_{i\alpha}$ cannot exchange its bound GDP for GTP and therefore cannot inhibit adenylate cyclase (Fig. 21-12).

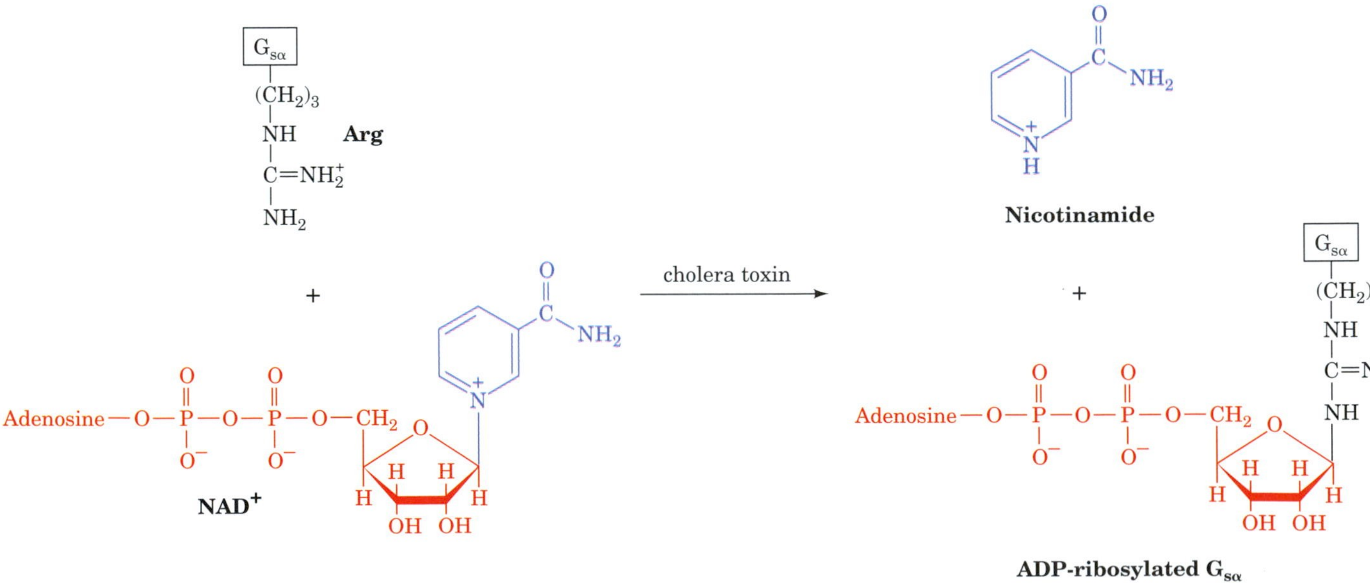

Indeed, the most common mechanism for activating RTKs appears to be ligand-induced dimerization of receptor proteins (i.e., ligand binding causes two monomeric receptors to form a dimer).

The **human growth hormone receptor** (which lacks tyrosine kinase activity) serves as a model for ligand-induced receptor dimerization. An X-ray crystal structure of the 191-residue **growth hormone** and the 238-residue extracellular domain of its binding protein shows that two identical molecules of the receptor bind to a single molecule of growth hormone (Fig. 21-13). Interestingly, the two receptor monomers in the complex bind to different portions of the asymmetric hormone molecule via the same ligand-binding sites so that the two receptor monomers in the complex are related by nearly exact two-fold symmetry.

Dimerization brings the receptor's intracellular domains together in a way that activates them. For RTKs, activation is accomplished by mutual cross-phosphorylation of Tyr residues in the two tyrosine kinase domains. This **autophosphorylation** enables the tyrosine kinase to phosphorylate Tyr residues in additional cytoplasmic proteins, which may be protein kinases themselves. Activated RTKs, as we shall see, also modify the activities of other proteins without phosphorylating them.

Many of the diverse cytoplasmic proteins that bind to autophosphorylated receptors contain one or two conserved ~100-residue modules known as **Src homology 2 (SH2) domains** (because they are similar to the sequence of a domain in the protein known as **Src**). SH2 domains bind phospho-Tyr residues with high affinity but do not bind the far more abundant phospho-Ser and phospho-Thr residues. This specificity has a simple explanation. X-Ray structural studies reveal that phospho-Tyr interacts with an Arg at the bottom of a deep pocket (Fig. 21-14). The side chains of Ser and Thr are too short to interact with this residue.

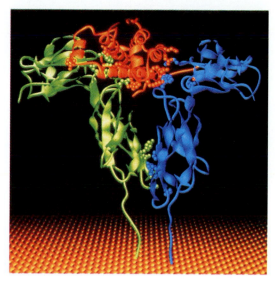

Figure 21-13.   The X-ray structure of the complex of human growth hormone and its receptor. Two identical molecules of the receptor's extracellular domain (*blue and green ribbon models*) bind a single molecule of growth hormone (*red*). Some side chains involved in intersubunit interactions are shown in space-filling form. The yellow pebbled surface represents the cell membrane. [Courtesy of Abraham de Vos and Anthony Kossiakoff, Genentech Inc., South San Francisco, California.] ● **See the Interactive Exercises.**

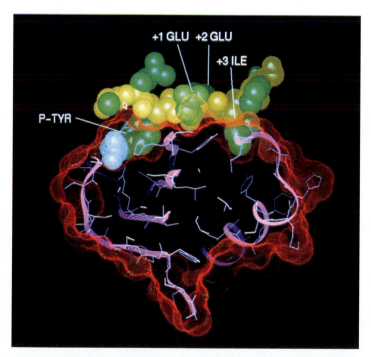

Figure 21-14.   The X-ray structure of the Src SH2 domain. An 11-residue polypeptide containing the protein's phospho-Tyr-Glu-Glu-Ile target tetrapeptide is bound to the SH2 domain. In this cutaway view, the protein surface is represented by red dots, the protein backbone (*purple*) is shown in ribbon form with its side chains in stick form, and the bound polypeptide's N-terminal 8-residue segment is shown in space-filling form with its backbone yellow, its side chains green, and its phosphate group white. [Courtesy of John Kuriyan, The Rockefeller University.]

Figure 21-15. **Receptor tyrosine kinase activation.** Growth factor binding results in dimerization of its cell-surface receptor. The intracellular tyrosine kinase domains phosphorylate each other (autophosphorylation), which causes them to bind to SH2-containing cellular proteins. SH2 binding to phospho-Tyr activates the SH2-containing protein. In some cases, specific Tyr residues of the SH2-containing protein are phosphorylated by the receptor tyrosine kinase. ✳ **See the Animated Figures.**

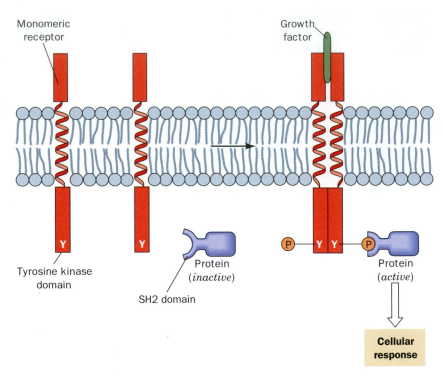

The interaction between a receptor tyrosine kinase and an SH2-containing protein positions the protein for phosphorylation or directly alters its activity (Fig. 21-15). Some growth factor receptors that do not have tyrosine kinase activity of their own associate with and thereby activate intracellular tyrosine kinases following ligand binding and receptor dimerization. Although these kinases are associated with different receptors, they phosphorylate overlapping sets of target proteins. This complex web of interactions explains why different extracellular signals often activate some of the same intracellular pathways.

Like other intracellular signals, cytoplasmic tyrosine kinases must be "turned off" after the system has delivered its message. The off switch is provided by **protein tyrosine phosphatases,** enzymes that dephosphorylate phospho-Tyr residues.

Ras Mediates a Multistep Signaling Pathway

Molecular genetic analysis of signaling in a variety of distantly related organisms has revealed a remarkably conserved pathway in which RTKs funnel the signal that they have bound ligand to a monomeric G-protein named **Ras,** which in turn relays the signal, via a "kinase cascade," to the transcriptional apparatus in the nucleus. This system regulates such essential functions as cell growth and differentiation.

The binding of a growth factor to its RTK leads to autophosphorylation of the RTK, which then interacts with an SH2-containing protein (Fig. 21-16, *left*). Many proteins that contain SH2 domains also have one or more unrelated 50- to 75-residue **SH3 domains.** SH3 domains, which bind Pro-rich sequences of 9 or 10 residues, are also present in some proteins that lack SH2 domains. Both types of domain mediate the interactions between kinases and regulatory proteins. In the signaling cascade shown in Fig. 21-16, a 217-residue mammalian protein known as **Grb2 (Sem-5** in the nematode *Caenorhabditis elegans*), which consists almost entirely of an SH2 domain flanked by two SH3 domains, forms a complex with the 1596-residue **Sos protein.** Sos contains a Pro-rich sequence that binds specifically to SH3 domains. The Grb2–Sos complex bridges the activated RTK and Ras in a

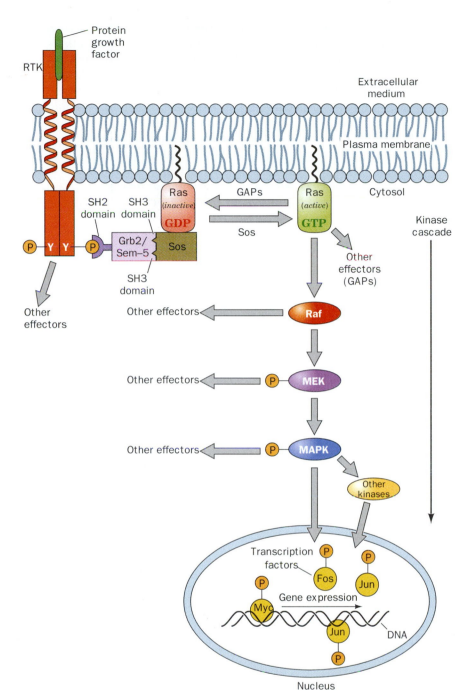

Figure 21-16. The Ras signaling cascade. RTK binding to its cognate growth factor induces the autophosphorylation of the RTK's cytosolic domain. Grb2/Sem-5 binds to the resulting phospho-Tyr–containing peptide segment via its SH2 domain and simultaneously binds to Pro-rich segments on Sos via its two SH3 domains. This activates Sos to exchange Ras's bound GDP for GTP, which activates Ras to bind to Raf. Then, in a so-called kinase cascade, Raf, a Ser/Thr kinase, phosphorylates MEK, which in turn phosphorylates MAPK, which then migrates to the nucleus, where it phosphorylates transcription factors such as Fos, Jun, and Myc, thereby modulating gene expression. [After Egan, S.E. and Weinberg, R.A., *Nature* **365**, 782 (1993).] ✳ See the Animated Figures.

way that induces Ras to exchange its bound GDP for GTP, thereby activating it. Only Ras·GTP is capable of further relaying the signal from the RTK. However, Ras, like other G-proteins, hydrolyzes GTP to GDP + P_i, thereby limiting the magnitude of the cell's response to the growth factor. Ras by itself hydrolyzes GTP with a half-life of 1 to 5 hours, too slowly for effective signal transduction. This led to the discovery of **GTPase activating protein (GAP),** a 120-kD protein that accelerates Ras's rate of GTP hydrolysis by up to five orders of magnitude.

The signaling pathway downstream of Ras consists of a linear cascade of protein kinases (Fig. 21-16, *right*). The Ser/Thr kinase **Raf,** which is activated by direct interaction with Ras·GTP, phosphorylates a protein alternatively known as **MEK** or **MAP kinase kinase,** thereby activating it as a kinase. Activated MEK phosphorylates a family of proteins variously termed **mitogen-activated protein kinases (MAPKs)** or **extracellular-signal-regulated kinases (ERKs).** MAPK must be phosphorylated at both its Thr and Tyr residues in the sequence Thr-Glu-Tyr for full activity. MEK (which stands for *M*AP kinase/*E*RK kinase-activating *k*inase) catalyzes both phosphorylations; it is therefore a Ser/Thr kinase as well as a Tyr kinase.

The activated MAP kinases migrate from the cytosol to the nucleus, where they phosphorylate a variety of proteins, including **Fos, Jun,** and **Myc.** These proteins are **transcription factors** (proteins that induce the transcription of their target genes; Section 25-2C): In their activated forms they stimulate various genes to produce the effects commissioned by the extracellular presence of the growth factor that initiated the signaling cascade. Variant proteins encoded by **oncogenes** interfere with such signaling pathways so as to induce uncontrolled cell growth (see Box 21-2).

Box 21-2

BIOCHEMISTRY IN HEALTH AND DISEASE

Oncogenes and Cancer

The growth and differentiation of cells in the body are normally strictly controlled. Thus, with few exceptions (e.g., blood-forming cells and hair follicles), cells in the adult body are largely quiescent. However, for a variety of reasons, a cell may be made to proliferate uncontrollably to form a tumor. **Malignant tumors (cancers)** grow in an invasive manner and are almost invariably life threatening. They are responsible for 20% of the mortalities in the United States.

Among the many causes of cancer are viruses that carry oncogenes (Greek: *onkos,* mass or tumor). For example, the **Rous sarcoma virus (RSV),** which induces the formation of **sarcomas** (cancers arising from connective tissues) in chickens, contains four genes. Three of these genes are essential for viral replication, whereas the fourth, **v-*src*** ("v" for viral, "*src*" for sarcoma), an oncogene, induces tumor formation. What is the origin of v-*src*, and how does it function? Hybridization studies by Michael Bishop and Harold Varmus in 1976 led to the remarkable discovery that uninfected chicken cells contain a gene, **c-*src*** ("c" for cellular), that is homologous to v-*src* and that is highly conserved in a wide variety of eukaryotes, suggesting that it is an essential cellular gene. Apparently, v-*src* was originally acquired from a cellular source by a non-tumor-forming ancestor of RSV. Both v-*src* and c-*src* encode a 60-kD tyrosine kinase. However, whereas the activity of c-*src* is strictly regulated, that of v-*src* is under no such control and hence its presence maintains the host cell in a proliferative state. Since cells are not killed by an RSV infection, this presumably enhances the viral replication rate.

Other oncogenes have been similarly linked to processes that regulate cell growth. For example, the **v-*erbB*** oncogene specifies a truncated version of the **epidermal growth factor (EGF)** receptor, which lacks the EGF-binding domain but retains its transmembrane segment and its tyrosine kinase domain. This kinase phosphorylates its target proteins in the absence of an extracellular signal, thereby driving uncontrolled cell proliferation.

The **v-*ras*** oncogene encodes a 21-kD protein, **p21$^{v\text{-}ras}$,** that resembles cellular Ras but hydrolyzes GTP much more slowly. The reduced braking effect of GTP hydrolysis on the rate of protein phosphorylation leads to increased activation of the kinases downstream of Ras (Fig. 21-16).

The transcription factors that respond to Ras-mediated signaling (e.g., Fos and Jun) are also encoded by **proto-oncogenes,** the normal cellular analogs of oncogenes. The viral genes v-*fos* and v-*jun* encode proteins that are nearly identical to their cellular counterparts and mimic their effects on host cells but in an uncontrolled fashion.

Oncogenes are not necessarily of viral origin. Indeed, few human cancers are caused by viruses. Rather, they are caused by proto-oncogenes that have mutated to form oncogenes. For example, a mutation in the **c-*ras*** gene, which converts Gly 12 of Ras to Val, reduces Ras's GTPase activity without affecting its ability to stimulate protein phosphorylation. This prolongs the time that Ras is in the "on" state, thereby inducing uncontrolled cell proliferation. In fact, oncogenic versions of c-*ras* are among the most commonly implicated oncogenes in human cancers.

To date, ~50 oncogenes have been identified. The subversive effects of oncogene products arise through their differences from the corresponding normal cellular proteins: They may have different rates of synthesis and/or degradation; they may have altered cellular functions; or they may resist control by cellular regulatory mechanisms. However, in order for a normal cell to undergo a **malignant transformation** (become a cancer cell), it must undergo several (an average of five) independent oncogenic events. This is a reflection of the complexity of cellular signaling networks (cells respond to a variety of hormones, growth factors, and transcription factors in partially overlapping ways) and explains why the incidence of cancer increases with age.

The Insulin Receptor Is a Dimer

The insulin receptor is a transmembrane glycoprotein with tyrosine kinase activity. It is present in nearly all vertebrate cells, with the highest receptor density in hepatocytes and adipocytes. Unlike other receptor tyrosine kinases, the insulin receptor is a dimer in its unliganded state. Its two protomeric units are synthesized as single polypeptides that are then proteolyzed to yield a mature receptor with the subunit composition $\alpha_2\beta_2$ (Fig. 21-17a). Insulin binds to the α subunits, which are entirely extracellular. The intracellular portion of the transmembrane β subunit contains its tyrosine kinase domain, whose X-ray structure (Fig. 21-17b) resembles those of Ser/Thr-specific protein kinases such as cAPK (Fig. 15-14). Comparison of the active sites of the insulin RTK and cAPK reveals why the insulin RTK phosphorylates Tyr but not Ser or Thr residues: Ser and Thr residues are simply too short to reach into the insulin RTK's phosphotransfer site.

When insulin binds to one or both α subunits, the intracellular domains of the receptor undergo a conformational change. The receptor autophosphorylates (Fig. 21-17b), and its tyrosine kinase activity thereby increases. The insulin receptor does not appear to associate directly with SH2-containing proteins. However, the activated receptor phosphorylates other proteins, including one known as **insulin receptor substrate-1 (IRS-1),** which then binds to certain SH2-containing proteins.

IRS-1 is required for many of insulin's biological effects, although the exact mechanisms are not all understood. The target proteins of IRS-1 include kinases that participate in multistep phosphorylation cascades that eventually alter the activities of the enzymes involved in glycogen and fatty acid metabolism.

D. The Phosphoinositide Pathway

Phosphatidylinositol-4,5-bisphosphate (PIP$_2$; Fig. 21-18), a phosphorylated glycerophospholipid that is a minor component of the plasma membrane's inner leaflet, is part of a second messenger system that transduces many hormone signals. Like the adenylate cyclase system, this **phosphoinositide pathway** includes a receptor with seven transmembrane segments, a heterotrimeric G-protein, and a specific protein kinase. Ligand binding to the receptor activates a G-protein, **G$_q$,** whose membrane-anchored α subunit in complex with GTP diffuses laterally along the plasma membrane to activate the membrane-bound enzyme **phospholipase C.** The activated en-

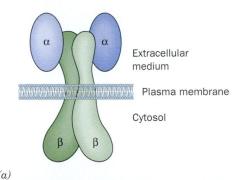

(a)

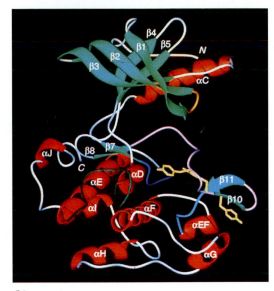

(b)

Figure 21-17.   The insulin receptor. (a) The receptor consists of two transmembrane β subunits and two extracellular α subunits. Insulin binds to a site on the α subunit, and the cytosolic portion of the β subunit contains tyrosine kinase activity. (b) The X-ray structure of the tyrosine kinase domain of the insulin receptor. The α helices are shown in red, the β sheets in green, and mechanistically implicated loops are shown in various colors. The three Tyr side chains that are autophosphorylated, most probably by the opposite β subunit, are yellow. [Courtesy of Wayne Hendrickson, Columbia University.]
  See the Interactive Exercises.

Figure 21-18.   Phosphatidylinositol-4,5-bisphosphate (PIP$_2$) and its hydrolysis products. PIP$_2$ is cleaved by phospholipase C to produce diacylglycerol (DG) and inositol-1,4,5-trisphosphate (IP$_3$), both of which are second messengers. (The *bis* and *tris* prefixes denote, respectively, two and three phosphoryl groups that are linked separately to the inositol; in di- and triphosphates, the phosphoryl groups are linked sequentially.)

Diacylglycerol (DG)

Phosphatidylinositol-4,5-bisphosphate (PIP$_2$)

phospholipase C

H_2O

Inositol-1,4,5-trisphosphate (IP$_3$)

zyme catalyzes the hydrolysis of PIP_2 at its glycero-phospho bond (Section 9-1C), yielding **inositol-1,4,5-trisphosphate (IP_3)** and **1,2-diacylglycerol (DG;** Fig. 21-18).

The charged IP_3 molecule is a water-soluble second messenger that diffuses through the cytoplasm to the endoplasmic reticulum. There, it binds to and induces the opening of a Ca^{2+} transport channel (an example of a receptor that is also an ion channel), thereby allowing the efflux of Ca^{2+} from the endoplasmic reticulum. This causes the cytosolic $[Ca^{2+}]$ to increase from ~0.1 μM to as much as 10 μM, which triggers such diverse cellular processes as glucose mobilization and muscle contraction through the intermediacy of calmodulin (Section 15-3B) and its homologs.

The nonpolar diacylglycerol product of phospholipase C action is a lipid-soluble second messenger. It remains embedded in the plasma membrane, where it activates **protein kinase C** to phosphorylate and thereby modulate the activities of several different cellular proteins. The phosphoinositide signaling system is diagrammed in Fig. 21-19.

Multiple protein kinase C enzymes are known; they differ in tissue expression, intracellular location, and their requirement for the diacylglycerol that activates them. Protein kinase C is a phosphorylated, cytosolic

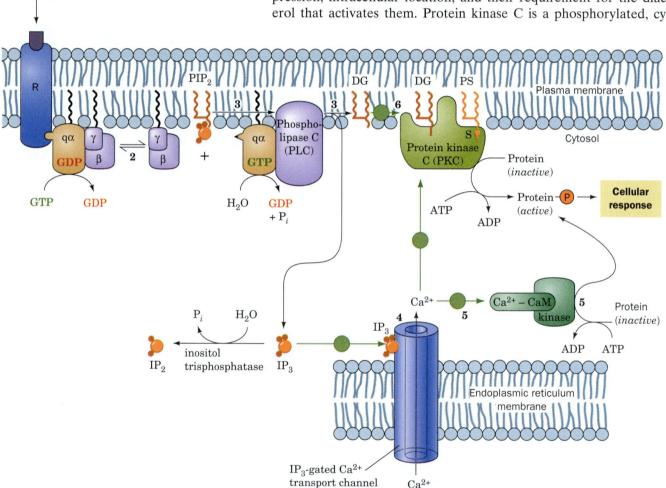

Figure 21-19.   *Key to Function.* The phosphoinositide signaling system. (1) Ligand binding to a cell-surface receptor R activates phospholipase C through the heterotrimeric G-protein G_q (2). Phospholipase C catalyzes the hydrolysis of PIP_2 to IP_3 and DG (3). The water-soluble IP_3 stimulates the release of Ca^{2+} sequestered in the endoplasmic reticulum (4), which in turn activates numerous cellular processes through the intermediacy of calmodulin (CaM; 5). The nonpolar DG remains associated with the membrane, where it activates protein kinase C to phosphorylate and thereby modulate the activities of a number of cellular proteins (6). Protein kinase C activation also requires the presence of the membrane lipid phosphatidylserine (PS) and Ca^{2+}.
✳ See the Animated Figures.

protein in its resting state. Diacylglycerol increases the membrane affinity of protein kinase C and also helps stabilize its active conformation. Full activation requires phosphatidylserine (which is present only in the cytoplasmic leaflet of the plasma membrane) and, in some cases, Ca^{2+} (presumably made available through the action of the IP_3 second messenger).

Choline-containing phospholipids hydrolyzed by phospholipase C yield diacylglycerols that differ not only from those released from PIP_2 but also in their effects on protein kinase C. Another lipid second messenger, sphingosine released from sphingolipids, inhibits protein kinase C. The catalytic activities of protein kinase C and cAPK are similar: Both kinases phosphorylate Ser and Thr residues.

The phosphoinositide signaling pathway in some cells yields a diacylglycerol that is predominantly 1-stearoyl-2-arachidonoyl-glycerol. This molecule is further degraded to yield arachidonate, the precursor of the bioactive eicosanoids (prostaglandins and thromboxanes; Section 9-1F). In other cells, IP_3 and diacylglycerol are rapidly recycled to re-form PIP_2 in the inner leaflet of the membrane. Some receptor tyrosine kinases activate a form of phospholipase C that contains two SH2 domains. This is one example of **cross talk,** the interaction of different signal-transduction pathways. Like other signaling systems, the phosphoinositide system is limited by the destruction of its second messengers, for example, through the action of **inositol trisphosphatase** (Fig. 21-19).

4. DISTURBANCES IN FUEL METABOLISM

The complexity of the mechanisms that regulate mammalian fuel metabolism permit the body to respond efficiently to changing energy demands and to accommodate changes in the availabilities of various fuels. Such complex systems can also malfunction, producing acute or chronic diseases of variable severity. Considerable effort has been directed at elucidating the molecular basis of conditions such as diabetes and obesity, both of which are essentially disorders of fuel metabolism. In this section, we examine the metabolic changes that occur in starvation, diabetes, and obesity.

A. Starvation

Because humans do not eat continuously, the disposition of dietary fuels and the mobilization of fuel stores shifts dramatically during the few hours between meals. Yet humans can survive fasts of up to a few months by adjusting their fuel metabolism. Such metabolic flexibility certainly evolved before modern humans became accustomed to thrice-daily meals.

Absorbed Fuels Are Allocated Immediately

When a meal is digested, nutrients are broken down to small, usually monomeric, units for absorption by the intestinal mucosa. From there, the products of digestion pass through the circulation to the rest of the body. Dietary proteins, for example, are broken down to amino acids for absorption. When the amino acids reach the liver via the portal vein, they may be used for protein synthesis or, if present in excess, oxidized to produce energy. There is no dedicated storage depot for amino acids; whatever the liver does not metabolize circulates to peripheral tissues to be catabolized or used for protein synthesis.

Dietary carbohydrates, like proteins, are degraded in the intestine, and the absorbed monomeric products (e.g., glucose derived from dietary

Table 21-1. **Fuel Reserves for a Normal 70-kg Man**

Fuel	Mass (kg)	Calories[a]
Tissues		
Fat (adipose triacyglycerols)	15	141,000
Protein (mainly muscle)	6	24,000
Glycogen (muscle)	0.150	600
Glycogen (liver)	0.075	300
Circulating fuels		
Glucose (extracellular fluid)	0.020	80
Free fatty acids (plasma)	0.0003	3
Triacylglycerols (plasma)	0.003	30
Total		*166,000*

[a]1 (dieter's) Calorie = 1 kcal = 4.184 kJ.
Source: Cahill, G.F., Jr., *New Engl. J. Med.* **282**, 669 (1970).

starch) are delivered to the liver via the portal vein. As much as one-third of the dietary glucose is immediately converted to glycogen in the liver; at least half of the remainder is converted to glycogen in muscle cells, and the rest is oxidized by these and other tissues for immediate energy needs. Both glucose uptake and glycogen synthesis are stimulated by insulin, whose concentration in the blood increases in response to high blood [glucose].

Dietary fatty acids are packaged as triacylglycerols in chylomicrons (Section 19-1), which circulate first in the lymph and then in the bloodstream and therefore are not delivered directly to the liver as are absorbed amino acids and carbohydrates. Instead, a significant portion of the dietary fatty acids are taken up by adipose tissue. Lipoprotein lipase first hydrolyzes the triacylglycerols, and the released fatty acids are absorbed and esterified in the adipocytes.

Blood Glucose Remains Constant

As tissues take up and metabolize glucose, blood [glucose] drops, thereby causing the pancreatic α cells release glucagon. This hormone stimulates glycogen breakdown and the release of glucose from the liver. It also promotes gluconeogenesis from amino acids and lactate. *The reciprocal effects of insulin and glucagon, which both respond to and regulate blood [glucose], ensure that the concentration of glucose available to extrahepatic tissues remains relatively constant.*

However, the body stores less than a day's supply of carbohydrate (Table 21-1). After an overnight fast, the combination of increased glucagon secretion and decreased insulin secretion promotes the mobilization of fatty acids from adipose tissue (Section 19-5). The diminished insulin also inhibits glucose uptake by muscle tissue. Muscles therefore switch from glucose to fatty acid metabolism for energy production. This adaptation spares glucose for use by tissues, such as the brain, that cannot utilize fatty acids.

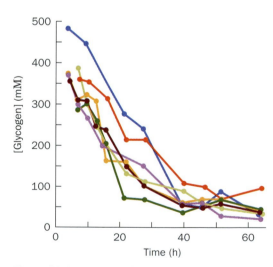

Figure 21-20. **Liver glycogen depletion during fasting.** The liver glycogen content in seven subjects was measured using ^{13}C NMR over the course of a 64-hour fast. [After Rothman, D.L., Magnusson, I., Katz, L.D., Shulman, R.G., and Shulman, G.I., *Science* **254**, 575 (1991).]

Gluconeogenesis Supplies Glucose during Starvation

After a lengthy fast, the liver's store of glycogen becomes depleted (Fig. 21-20). Under these conditions, the rate of gluconeogenesis increases. Gluconeogenesis supplies ~96% of the glucose produced by the liver after 40 hours of fasting. In animals, glucose cannot be synthesized from fatty acids. This is because neither pyruvate nor oxaloacetate, the precursors of glucose in gluconeogenesis, can be synthesized in a net manner from acetyl-

CoA. During starvation, glucose must therefore be synthesized from the glycerol product of triacylglycerol breakdown and, more importantly, from the amino acids derived from the proteolytic degradation of proteins, the major source of which is muscle. The breakdown of muscle cannot continue indefinitely, however, since loss of muscle mass would eventually prevent an animal from moving about in search of food. The organism must therefore make alternate metabolic arrangements.

After several days of starvation, the liver directs acetyl-CoA, which is derived from fatty acid β oxidation, to the synthesis of ketone bodies (Section 19-3). These fuels are then released into the blood. The brain gradually adapts to using ketone bodies as fuel through the synthesis of the appropriate enzymes: After a 3-day fast, only about one-third of the brain's energy requirements are satisfied by ketone bodies, but after 40 days of starvation, ~70% of its energy needs are so met. The rate of muscle breakdown during prolonged starvation consequently decreases to ~25% of its rate after a several-day fast. *The survival time of a starving individual therefore depends much more on the size of fat reserves than on muscle mass.* Indeed, highly obese individuals can survive a year or more without eating (and have done so in clinically supervised weight-reduction programs).

B. Diabetes Mellitus

In the disease **diabetes mellitus** (often called just **diabetes**), which is the third leading cause of death in the United States after heart disease and cancer, insulin either is not secreted in sufficient amounts or does not efficiently stimulate its target cells. As a consequence, blood glucose levels become so elevated that the glucose "spills over" into the urine, providing a convenient diagnostic test for the disease. Yet, despite these high blood glucose levels, cells "starve" since insulin-stimulated glucose entry into cells is impaired. Triacylglycerol hydrolysis, fatty acid oxidation, gluconeogenesis, and ketone body formation are accelerated and, in a condition known as **ketosis,** ketone body levels in the blood become abnormally high. Since ketone bodies are acids, their high concentration puts a strain on the buffering capacity of the blood and on the kidney, which controls blood pH by excreting the excess H^+ into the urine. This H^+ excretion is accompanied by Na^+, K^+, P_i, and H_2O excretion, causing severe dehydration (excessive thirst is a classic symptom of diabetes) and a decrease in blood volume—ultimately life-threatening situations.

There are two major forms of diabetes mellitus:

1. **Insulin-dependent** or **juvenile-onset diabetes mellitus,** which most often strikes suddenly in childhood.
2. **Non-insulin-dependent** or **maturity-onset diabetes mellitus,** which usually develops gradually after the age of 40.

Insulin-Dependent Diabetes Is Caused by a Deficiency of Pancreatic β Cells

In insulin-dependent (type I) diabetes mellitus, insulin is absent or nearly so because the pancreas lacks or has defective β cells. This condition usually results from an autoimmune response that selectively destroys pancreatic β cells. Individuals with insulin-dependent diabetes, as Frederick Banting and Charles Best first demonstrated in 1921, require daily insulin injections to survive and must follow carefully balanced diet and exercise regimens. Their life spans are, nevertheless, reduced by up to one-third as a result of degenerative complications such as kidney malfunction, nerve impairment, and cardiovascular disease that apparently arise from the imprecise metabolic control provided by periodic insulin injections. The **hyperglycemia** (high blood [glucose]) of diabetes mellitus also leads to blindness through retinal degeneration and the glucosylation of lens proteins, which causes cataracts.

The usually rapid onset of the symptoms of insulin-dependent diabetes had suggested that the autoimmune attack on the pancreatic β cells is of short duration. Typically, however, the disease develops over several years as the immune system slowly destroys the β cells. Only when >80% of these cells have been eliminated do the classic symptoms of diabetes suddenly emerge.

Non-Insulin-Dependent Diabetes May Be Caused by a Deficiency of Insulin Receptors

Non-insulin-dependent (type II) diabetes mellitus, which accounts for over 90% of the diagnosed cases of diabetes and affects 18% of the population over 65 years of age, usually occurs in obese individuals with a genetic predisposition for this condition. These individuals have normal or even greatly elevated insulin levels. Their symptoms may arise from a paucity of insulin receptors on normally insulin-responsive cells. These cells do not respond normally to insulin and are therefore said to be **insulin resistant.** As a result, blood glucose concentrations are much higher than normal, particularly after a meal (Fig. 21-21).

The hyperglycemia that accompanies insulin resistance induces the pancreatic β cells to increase their production of insulin. Yet the high basal level of insulin secretion diminishes the ability of the β cells to respond to further increases in blood glucose. Consequently, the hyperglycemia and its attendant complications tend to worsen over time.

A number of mutations in the insulin receptor have been associated with non-insulin-dependent diabetes. These defects produce alterations in insulin binding (α subunit mutations) or tyrosine kinase activity (β subunit mutations). However, genetic defects have been identified in only ~5% of non-insulin-dependent diabetics. It is therefore likely that many factors play a role in the development of this disease. For example, the increased insulin production resulting from overeating may eventually suppress the synthesis of insulin receptors. This hypothesis accounts for the observation that diet alone often decreases the severity of the disease.

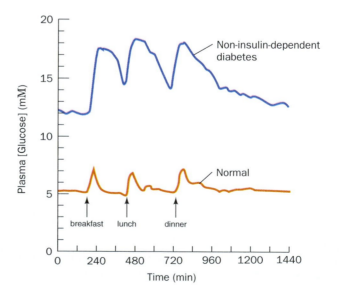

Figure 21-21. **Twenty-four-hour plasma glucose profiles in normal and non-insulin-dependent diabetic subjects.** The basal level of glucose and the peaks following meals are higher in the diabetic individuals. [After Bell, G.I., Pilkis, S.J., Weber, I.T., and Polonsky, K.S., *Annu. Rev. Physiol.* **58**, 178 (1996).]

Box 21-3

BIOCHEMISTRY IN FOCUS

How Do β Cells Respond to Blood Glucose Levels?

In normal humans, the pancreas responds to increases in the concentration of blood glucose by secreting insulin. Pancreatic β cells are most sensitive to glucose at concentrations of 5.5 to 6.0 mM (normal blood glucose concentrations range from 3.6 to 5.8 mM). There is no evidence for a cell-surface glucose "receptor" that might relay a signal to the secretory machinery in the β cell. In fact, glucose cannot act as a "hormone" because β cells contain GLUT2, the high-K_M glucose transporter that also occurs in liver. Accordingly, glucose readily enters β cells at a rate more or less proportional to its plasma concentration. The metabolism of the glucose that enters the β cell generates the signal for insulin secretion.

The rate-limiting step of β cell glucose metabolism is the reaction catalyzed by glucokinase (the same enzyme that occurs in hepatocytes). For this reason, glucokinase is considered the glucose "sensor." In β cells, G6P is not used to synthesize glycogen, and the activity of the pentose phosphate pathway is minor. Furthermore, lactate dehydrogenase activity is low. As a result, essentially all the G6P produced by the glucokinase reaction is degraded to pyruvate and then converted to acetyl-CoA for oxidation by the citric acid cycle. This one-way, linear catabolic pathway for glucose directly links the β cell's rate of oxidative phosphorylation to the amount of available glucose. By mechanisms that are not entirely understood, the overall level of respiratory activity regulates insulin synthesis and secretion.

The fact that even a partial decrease in glucokinase activity causes diabetes indicates that glucokinase acts as a gatekeeper for the insulin-secretion pathway. The involvement of mitochondrial processes (e.g., transport of cytosolic reducing equivalents into mitochondria or the operation of ATP synthase) in insulin production is implicated by evidence linking the age-dependent decline in mitochondrial oxidative capacity (Section 17-5B) to the age-dependent onset of non-insulin-dependent diabetes mellitus. It is possible that the chronic overproduction of insulin reflects the inability of aged cells to properly regulate insulin production. This would also help explain why most cases of non-insulin-dependent diabetes are not linked to mutations in the insulin receptor.

A rare form of non-insulin-dependent diabetes, called **maturity-onset diabetes of the young (MODY)**, appears by age 25 and is usually the result of a mutation in glucokinase. This form of diabetes is transmitted in a dominant fashion. Most diseases caused by enzyme defects are inherited in a recessive manner, because individuals can generally tolerate a 50% decrease in the activity of a given enzyme. A 50% decrease in glucokinase activity, however, has drastic consequences for fuel metabolism. This is strong evidence that glucokinase is an essential part of the apparatus that coordinates insulin release with glucose supply (see Box 21-3).

C. Obesity

The human body regulates glycogen and protein levels within relatively narrow limits, but fat reserves, which are much larger, can become enormous. The accumulation of fatty acids as triacylglycerols in adipose tissue is largely a result of excess fat intake compared to fat oxidation. Although the acetyl-CoA derived from carbohydrates can be used to synthesize fatty acids, this is an insignificant source of fatty acids in most Americans, whose diets contain abundant fat. Fat synthesis from carbohydrates occurs only when the carbohydrate intake is so high that glycogen stores, to which excess carbohydrate is normally directed, approach their maximum capacity.

A chronic imbalance between fat consumption and utilization increases the mass of adipose tissue through an increase in the number of adipocytes or their size (once formed, adipocytes are not lost, although their size may increase or decrease). The increase in adipose tissue mass increases the pool of fatty acids that can be mobilized to generate metabolic energy. Eventually, a steady state is achieved in which the mass of adipose tissue no longer increases and fat storage is balanced by fat mobilization. Studies in animals and humans show that the percentage of body fat roughly

Figure 21-22.   Normal (OB/OB, *left*) and obese (*ob/ob, right*) mice. [Courtesy of Richard D. Palmiter, University of Washington.]

corresponds to the dietary fat content. This phenomenon explains in part the high incidence of obesity in affluent societies, where fat-rich foods are plentiful. Considerable evidence suggests that behavior (e.g., eating habits and levels of physical activity) influences an individual's body composition. Yet some cases of obesity are clearly also the result of innate disturbances in the individual's capacity to metabolize fuels.

A Genetic Basis for Obesity

Most animals, including humans, tend to have stable weights; that is, if they are given free access to food, they eat just enough to maintain their so-called "set-point" weight. However, a strain of mice that are homozygous for defects in the *obese* gene (known as *ob/ob* mice) are over twice the weight of normal (OB/OB) mice (Fig. 21-22) and overeat when given access to unlimited quantities of food. The obese mice lack the protein **leptin** (Greek: *leptos,* thin), the product of the *obese* gene. Leptin is a 16-kD polypeptide that is normally produced by adipocytes (Fig. 21-23). When leptin is injected into *ob/ob* mice, they eat less and lose weight. Leptin has therefore been considered a "satiety" signal that affects the appetite control system of the brain.

The simple genetic basis for obesity in *ob/ob* mice does not appear to apply to most obese humans. Leptin levels in humans increase with the percentage of body fat, consistent with the synthesis of leptin by adipocytes. Thus, obesity in humans is apparently the result not of faulty leptin production but of "leptin resistance," perhaps due to a decrease in the level of a leptin receptor in the brain or a defect in the mechanism that transports leptin across the blood–brain barrier into the central nervous system.

A diminished response to leptin leads to high concentrations of **neuropeptide Y,**

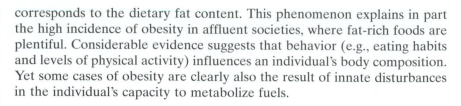

Neuropeptide Y

a 36-residue peptide released by the **hypothalamus,** a part of the brain that controls many physiological functions. Neuropeptide Y stimulates appetite and promotes insulin secretion, which in turn leads to fat accumulation. The resulting chronic hyperinsulinemia contributes to the development of insulin resistance and may therefore cause non-insulin-dependent diabetes (Section 21-4B). The discovery of leptin in 1994 revealed a link between appetite and adiposity. Although the leptin signal-transduction pathway is not yet understood, it represents a potential point for therapeutic intervention in cases where obesity cannot be otherwise controlled.

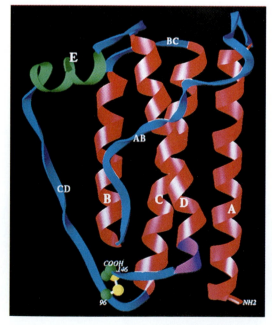

Figure 21-23.   The X-ray structure of human leptin-E100. This mutant form of leptin (Trp 100 → Glu) has comparable biological activity to the wild-type protein but crystallizes more readily. Leptin's residues pack into a bundle of four α helices (labeled A–D). A disulfide bond (atoms shown in green and yellow) links Cys 146 to Cys 96. [Courtesy of Faming Zhang, Eli Lilly & Co., Indianapolis, Indiana.]
● See the Interactive Exercises.

SUMMARY

1. The pathways for the synthesis and degradation of the major metabolic fuels (glucose, fatty acids, and amino acids) converge on acetyl-CoA and pyruvate. In mammals, flux through these pathways is tissue specific.

2. The brain uses glucose as its primary metabolic fuel. Muscle can oxidize a variety of fuels but depends on anaerobic glycolysis for maximum exertion. Adipose tissue stores excess fatty acids as triacylglycerols and mobilizes them as needed.

3. The liver maintains the concentrations of circulating fuels. The action of glucokinase allows liver to take up excess glucose, which can then be directed to several metabolic fates. The liver also converts fatty acids to ketone bodies and metabolizes amino acids derived from the diet or from protein breakdown.

4. The Cori cycle and the glucose–alanine cycle are multiorgan pathways through which the liver and muscle exchange metabolic intermediates.

5. The GLUT family of transporters mediate the passage of glucose into and out of cells. The high K_M of liver GLUT2 allows it to respond to a wide range of glucose concentrations. GLUT4 is the insulin-responsive glucose transporter in muscle and adipose tissue.

6. Hormones transmit regulatory signals to target tissues by binding to receptors that transduce the signal to responses in the interior of the cell. The major hormones involved in mammalian fuel metabolism are the pancreatic hormones insulin and glucagon and the adrenal medulla hormones epinephrine and norepinephrine.

7. The adenylate cyclase signaling system consists of a receptor, a heterotrimeric G-protein, adenylate cyclase, and cAMP-dependent protein kinase. G-proteins may inhibit or stimulate adenylate cyclase.

8. Receptor tyrosine kinases such as the insulin receptor phosphorylate specific tyrosine residues in response to ligand binding.

9. In the phosphoinositide pathway, hormone binding leads to the hydrolysis of phosphatidylinositol-4,5-bisphosphate (PIP_2) to yield inositol-1,4,5-trisphosphate (IP_3), which opens Ca^{2+} channels, and diacylglycerol (DG), which activates protein kinase C.

10. During starvation, when dietary fuels are unavailable, the liver releases glucose first by glycogen breakdown and then by gluconeogenesis from amino acid precursors. Eventually, ketone bodies supplied by fatty acid breakdown meet most of the body's energy needs.

11. Diabetes mellitus causes hyperglycemia and other physiological difficulties resulting from destruction of insulin-producing pancreatic β cells or from insulin resistance (loss of insulin receptors).

12. Obesity, the result of an imbalance between fat intake and fat oxidation, has genetic and behavioral causes.

REFERENCES

Fuel Metabolism

Bell, G.I., Pilkis, S.J., Weber, I.T., and Polonsky, K.S., Glucokinase mutations, insulin secretion, and diabetes mellitus, *Annu. Rev. Physiol.* **58**, 171–186 (1996).

Flatt, J.-P., Use and storage of carbohydrate and fat, *Am. J. Clin. Nutr.* **61**(suppl.), 952S–959S (1995). [A concise review of fuel metabolism and the development of obesity.]

Gould, G.W. and Holman, G.D., The glucose transporter family: structure, function and tissue-specific expression, *Biochem. J.* **295**, 329–341 (1993).

Hamann, A. and Matthaei, S., Regulation of energy balance by leptin, *Exp. Clin. Endocrinol. Diabetes* **104**, 293–300 (1996).

Lissner, L. and Heitmann, B.L., Dietary fat and obesity: evidence from epidemiology, *Eur. J. Clin. Nutr.* **49**, 79–90 (1995).

Matchinsky, F.M. and Collins, H.W., Essential biochemical design features of the fuel-sensing system in pancreatic β cells, *Chem. Biol.* **4**, 249–257 (1997).

Sacks, D.B. and McDonald, J.M., The pathogenesis of type II diabetes mellitus: a polygenic disease, *Am. J. Clin. Pathol.* **105**, 149–156 (1996).

Signal Transduction

Barford, D., Protein phosphatases, *Curr. Opin. Struct. Biol.* **5**, 728–734 (1995).

Hamm, H.E. and Gilchrist, A., Heterotrimeric G proteins, *Curr. Opin. Cell Biol.* **8**, 189–196 (1996).

Nishizuka, Y., Intracellular signaling by hydrolysis of phospholipids and activation of protein kinase C, *Science* **258**, 607–614 (1992).

Pawson, T., Protein modules and signalling networks, *Nature* **373**, 573–579 (1995). [Provides an overview of the proteins involved in signaling via protein tyrosine kinases.]

Sprang, S.R., G protein mechanisms: Insights from structural analysis, *Annu. Rev. Biochem.* **66**, 639–678 (1997).

Taussig, R. and Gilman, A.G., Mammalian membrane-bound adenylate cyclases, *J. Biol. Chem.* **270**, 1–4 (1995).

White, M.F. and Kahn, C.R., The insulin signaling system, *J. Biol. Chem.* **269**, 1–4 (1994).

KEY TERMS

Cori cycle
oxygen debt
glucose–alanine cycle
hormone
receptor
signal transduction

homeostasis
islets of Langerhans
adrenergic receptor
heterotrimeric G-protein
GTPase

desensitization
growth factor
tyrosine kinase
autophosphorylation
phosphoinositide pathway

cross talk
diabetes
ketosis
hyperglycemia
insulin resistance

STUDY EXERCISES

1. Summarize the major features of fuel metabolism in the brain, muscle, adipose tissue, and liver.

2. Explain why the high K_M of glucokinase is important for the role of the liver in buffering blood glucose.

3. Describe the conditions under which the Cori cycle and the glucose–alanine cycle operate.

4. Why do different tissues contain different members of the GLUT family of glucose transporters?

5. Describe the activity of each component of the signaling pathways based on adenylate cyclase, receptor tyrosine kinases, and phosphoinositides.

6. How do heterotrimeric G-proteins work?

7. Describe the metabolic changes that occur during starvation.

8. Distinguish insulin-dependent and non-insulin-dependent diabetes mellitus.

9. How is obesity related to non-insulin-dependent diabetes mellitus?

PROBLEMS

1. Predict the effect of an overdose of insulin on brain function in a normal person.

2. Why would oxidative metabolism, which generates ATP, cease when a cell's ATP supply is exhausted?

3. Calculate the percentage of receptors occupied by ligand when the total ligand concentration is 2 μM, the total receptor concentration is 0.3 μM, and $K = 1 \times 10^{-6}$ M.

4. How does the presence of the poorly hydrolyzable GTP analog GTPγS (in which an O atom on the terminal phosphate is replaced by an S atom) affect cAMP production by adenylate cyclase?

5. Phosphatidylethanolamine and PIP$_2$ containing identical fatty acyl residues can be hydrolyzed with the same efficiency by a certain phospholipase C. Will the hydrolysis products of the two lipids have the same effect on protein kinase C? Explain.

6. A growth factor that acts through a receptor tyrosine kinase stimulates cell division. Predict the effect of a viral protein that inhibits protein tyrosine phosphatase.

7. Explain why insulin is required for adipocytes to synthesize triacylglycerols from fatty acids.

8. Experienced runners know that it is poor practice to ingest large amounts of glucose immediately before running a marathon. What is the metabolic basis for this apparent paradox?

9. After several days of starvation, the capacity of the liver to metabolize acetyl-CoA via the citric acid cycle is greatly diminished. Explain.

10. If the circulatory system of an *ob/ob* mouse is surgically joined to that of a normal mouse, what will be the effect on the appetite and weight of the *ob/ob* mouse?

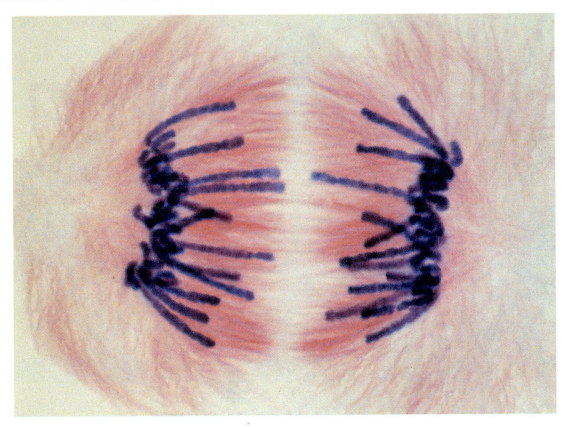

The synthesis of two complete sets of chromosomes requires large amounts of newly synthesized nucleotides. During cell division (mitosis), sets of identical chromosomes (purple) are pulled apart. [Courtesy of Andrew Bajer, University of Oregon.]

NUCLEOTIDE METABOLISM

Nucleotides are phosphate esters of a pentose (ribose or deoxyribose) in which a purine or pyrimidine base is linked to C1′ of the sugar (Section 3-1). Nucleoside triphosphates are the monomeric units that act as precursors of nucleic acids; nucleotides also perform a wide range of other biochemical functions. For example, we have seen how the cleavage of "high-energy" compounds such as ATP provides the free energy that makes various reactions thermodynamically favorable. We have also seen that nucleotides are components of some of the central cofactors of metabolism, including FAD, NAD^+, and coenzyme A. The importance of nucleotides in cellular metabolism is indicated by the observation that nearly all cells can synthesize them both *de novo* (anew) and from the degradation products of nucleic acids. However, unlike carbohydrates, amino acids, and fatty acids, nucleotides do not provide a significant source of metabolic energy.

In this chapter, we consider the nature of the nucleotide biosynthetic pathways. In doing so, we shall examine how they are regulated and the consequences of their blockade, both by genetic defects and through the administration of chemotherapeutic agents. We then discuss how nucleotides are degraded. In following the general chemical themes of nucleotide metabolism, we shall break our discussion into sections on purines, pyrimidines, and deoxynucleotides (including thymidylate). The nomenclature of the bases, nucleosides, and nucleotides is provided in Table 3-1.

1. SYNTHESIS OF PURINE RIBONUCLEOTIDES

In 1948, John Buchanan obtained the first clues to the *de novo* synthesis of purine nucleotides by feeding a variety of isotopically labeled compounds to pigeons and chemically determining the positions of the labeled atoms in their excreted **uric acid** (a purine).

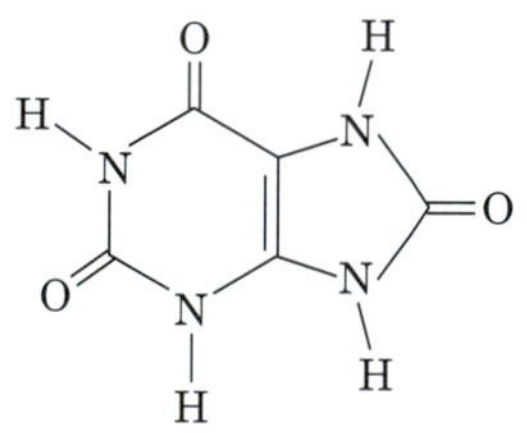

Uric acid

The results of his studies demonstrated that N1 of purines arises from the amino group of aspartate; C2 and C8 originate from formate; N3 and N9 are contributed by the amide group of glutamine; C4, C5, and N7 are derived from glycine (strongly suggesting that this molecule is wholly incorporated into the purine ring); and C6 comes from HCO_3^-.

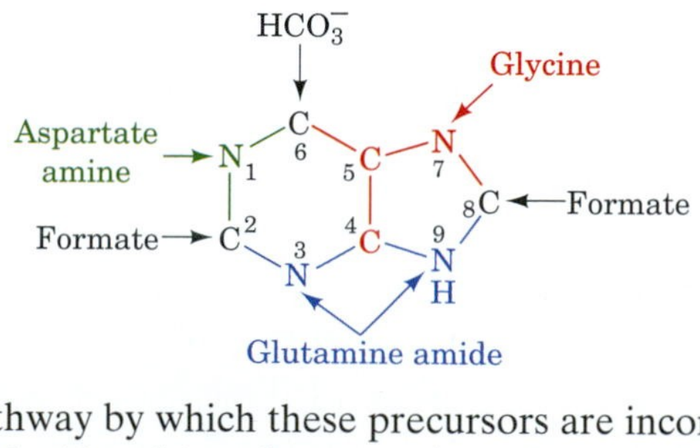

The actual pathway by which these precursors are incorporated into the purine ring was elucidated in subsequent investigations performed largely by Buchanan and G. Robert Greenberg. The initially synthesized purine derivative is **inosine monophosphate (IMP),**

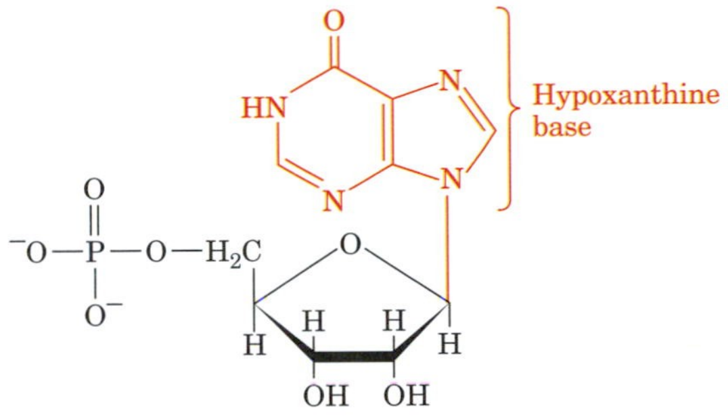

Inosine monophosphate (IMP)

the nucleotide of the base **hypoxanthine.** IMP is the precursor of both AMP and GMP. Thus, contrary to expectation, *purines are initially formed as ribonucleotides rather than as free bases.* Additional studies have demonstrated that such widely divergent organisms as *E. coli,* yeast, pigeons, and humans have virtually identical pathways for the biosynthesis of purine nucleotides, thereby further demonstrating the biochemical unity of life.

A. Synthesis of Inosine Monophosphate

IMP is synthesized in a pathway composed of 11 reactions (Fig. 22-1):

1. **Activation of ribose-5-phosphate**
   The starting material for purine biosynthesis is α-D-ribose-5-phosphate, a product of the pentose phosphate pathway (Section 14-6). In the first step of purine biosynthesis, **ribose phosphate pyrophosphokinase** activates the ribose by reacting it with ATP to form **5-phosphoribosyl-α-pyrophosphate (PRPP).** This compound is also a precursor in the biosynthesis of pyrimidine nucleotides (Section 22-2A) and the amino acids histidine and tryptophan (Section 20-5B). As is expected for an enzyme at such an important biosynthetic crossroads, the activity of ribose phosphate pyrophosphokinase is carefully regulated.

2. **Acquisition of purine atom N9**
   In the first reaction unique to purine biosynthesis, **amidophosphoribosyl transferase** catalyzes the displacement of PRPP's pyrophosphate group by glutamine's amide nitrogen. The reaction occurs with inversion of the α configuration at C1 of PRPP, thereby forming **β-5-phosphoribosylamine** and establishing the anomeric form of the future nucleotide. The reaction, which is driven to completion by the subsequent hydrolysis of the released PP_i, is the pathway's flux-controlling step.

3. **Acquisition of purine atoms C4, C5, and N7**
   Glycine's carboxyl group forms an amide with the amino group of phosphoribosylamine, yielding **glycinamide ribotide (GAR).** This reaction is reversible, despite its concomitant hydrolysis of ATP to ADP + P_i. It is the only step of the purine biosynthetic pathway in which more than one purine ring atom is acquired.

4. **Acquisition of purine atom C8**
   GAR's free α-amino group is formylated to yield **formylglycinamide ribotide (FGAR).** The formyl donor in this reaction is N^{10}-formyltetrahydrofolate (N^{10}-formyl-THF), a coenzyme that transfers C_1 units (THF cofactors are described in Section 20-4D). The X-ray

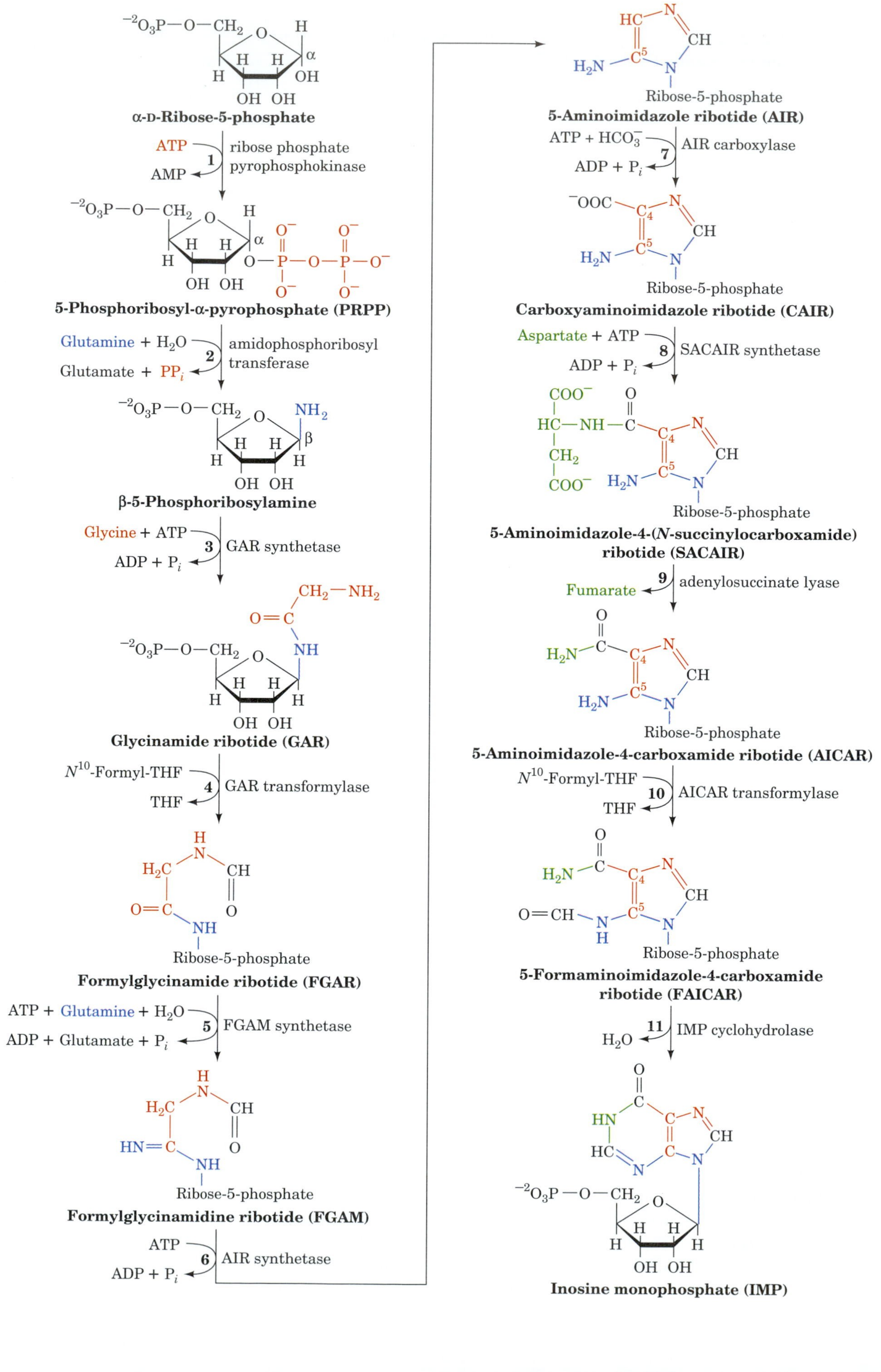

$^{-2}O_3P-O-CH_2$ ... H

α-D-Ribose-5-phosphate

ATP — | ribose phosphate
1 | pyrophosphokinase
AMP ◀

$^{-2}O_3P-O-CH_2$ ... H

5-Phosphoribosyl-α-pyrophosphate (PRPP)

Glutamine + H_2O — | amidophosphoribosyl
2 | transferase
Glutamate + PP_i ◀

$^{-2}O_3P-O-CH_2$ ... NH_2

β-5-Phosphoribosylamine

Glycine + ATP — | **3** GAR synthetase
ADP + P_i ◀

Glycinamide ribotide (GAR)

N^{10}-Formyl-THF — | **4** GAR transformylase
THF ◀

Formylglycinamide ribotide (FGAR)

ATP + Glutamine + H_2O — | **5** FGAM synthetase
ADP + Glutamate + P_i ◀

Formylglycinamidine ribotide (FGAM)

ATP — | **6** AIR synthetase
ADP + P_i ◀

5-Aminoimidazole ribotide (AIR)

ATP + HCO_3^- — | **7** AIR carboxylase
ADP + P_i ◀

Carboxyaminoimidazole ribotide (CAIR)

Aspartate + ATP — | **8** SACAIR synthetase
ADP + P_i ◀

5-Aminoimidazole-4-(N-succinylocarboxamide) ribotide (SACAIR)

Fumarate ◀ | **9** adenylosuccinate lyase

5-Aminoimidazole-4-carboxamide ribotide (AICAR)

N^{10}-Formyl-THF — | **10** AICAR transformylase
THF ◀

5-Formaminoimidazole-4-carboxamide ribotide (FAICAR)

H_2O ◀ | **11** IMP cyclohydrolase

Inosine monophosphate (IMP)

Figure 22-1 (*Opposite*).   *Key to Metabolism.* **The metabolic pathway for the** *de novo* **biosynthesis of IMP.** Here the purine residue is built up on a ribose ring in 11 enzyme-catalyzed reactions. ✳ See the Animated Figures.

structure of the enzyme catalyzing this reaction, **GAR transformylase,** in complex with GAR and the THF analog **5-deazatetrahydrofolate (5dTHF)** was determined by Robert Almassy (Fig. 22-2). Note the proximity of the GAR amine group to N10 of 5dTHF. This supports enzymatic studies suggesting that the GAR transformylase reaction proceeds via the nucleophilic attack of the GAR amine group on the formyl carbon of N^{10}-formyl-THF to yield a tetrahedral intermediate.

5. Acquisition of purine atom N3

The amide amino group of a second glutamine is transferred to the growing purine ring to form **formylglycinamidine ribotide (FGAM).** This reaction is driven by the coupled hydrolysis of ATP to ADP + P_i.

6. Formation of the purine imidazole ring

The purine imidazole ring is closed in an ATP-requiring intramolecular condensation that yields **5-aminoimidazole ribotide (AIR).** The aromatization of the imidazole ring is facilitated by the tautomeric shift of the reactant from its imine to its enamine form.

7. Acquisition of C6

Purine C6 is introduced as HCO_3^- in a reaction catalyzed by **AIR carboxylase** that yields **carboxyaminoimidazole ribotide (CAIR).** In *E. coli,* AIR carboxylase consists of two proteins called **PurE** and **PurK.** Although PurE alone can catalyze the carboxylation reaction, its K_M for HCO_3^- is ~110 mM, so the reaction would require an unphysiologically high HCO_3^- concentration (~100 mM) to proceed. PurK decreases the HCO_3^- concentration required for the PurE reaction by >1000-fold but at the expense of ATP hydrolysis.

8. Acquisition of N1

Purine atom N1 is contributed by aspartate in an amide-forming condensation reaction yielding **5-aminoimidazole-4-(*N*-succinylocarboxamide) ribotide (SACAIR).** This reaction, which is driven by the hydrolysis of ATP, chemically resembles Reaction 3.

9. Elimination of fumarate

SACAIR is cleaved with the release of fumarate, yielding **5-aminoimidazole-4-carboxamide ribotide (AICAR).** Reactions 8 and 9 chemically resemble the reactions in the urea cycle in which citrulline is aminated to form arginine (Section 20-3A). In both pathways, aspartate's amino group is transferred to an acceptor through an ATP-driven coupling reaction followed by the elimination of the aspartate carbon skeleton as fumarate.

10. Acquisition of C2

The final purine ring atom is acquired through formylation by N^{10}-formyl-THF, yielding **5-formaminoimidazole-4-carboxamide ribotide (FAICAR).** This reaction and Reaction 4 of purine biosynthesis are inhibited indirectly by **sulfonamides,** structural analogs of the *p*-aminobenzoic acid constituent of THF (Section 20-4D).

11. Cyclization to form IMP

The final reaction in the purine biosynthetic pathway, ring closure to form IMP, occurs through the elimination of water. In contrast to Reaction 6, the cyclization that forms the imidazole ring, this reaction does not require ATP hydrolysis.

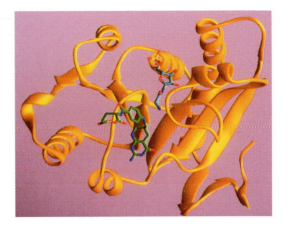

Figure 22-2. **Ribbon diagram of one subunit of *E. coli* GAR transformylase.** The enzyme is in complex with GAR (*upper right*) and 5dTHF (*lower left*). The C atoms of GAR and 5dTHF are cyan and green, whereas their N, O, and P atoms are blue, red, and yellow, respectively. [Based on an X-ray structure by Robert Almassy, Agouron Pharmaceuticals, LaJolla, California.]

B. Synthesis of Adenine and Guanine Ribonucleotides

IMP does not accumulate in the cell but is rapidly converted to AMP and GMP. AMP, which differs from IMP only in the replacement of its 6-keto group by an amino group, is synthesized in a two-reaction pathway (Fig. 22-3, *left*). In the first reaction, aspartate's amino group is linked to IMP in a reaction powered by the hydrolysis of GTP to GDP + P$_i$ to yield **adenylosuccinate.** In the second reaction, **adenylosuccinate lyase** eliminates fumarate from adenylosuccinate to form AMP. The same enzyme catalyzes Reaction 9 of the IMP pathway (Fig. 22-1).

GMP is also synthesized from IMP in a two-reaction pathway (Fig. 22-3, *right*). In the first reaction, IMP is dehydrogenated via the reduction of NAD$^+$ to form **xanthosine monophosphate (XMP;** the ribonucleotide of the base **xanthine**). XMP is then converted to GMP by the transfer of the glutamine amide nitrogen in a reaction driven by the hydrolysis of ATP to AMP + PP$_i$ (and subsequently to 2 P$_i$).

Nucleoside Diphosphates and Triphosphates Are Synthesized by the Phosphorylation of Nucleoside Monophosphates

In order to participate in nucleic acid synthesis, nucleoside monophosphates must first be converted to the corresponding nucleoside triphosphates.

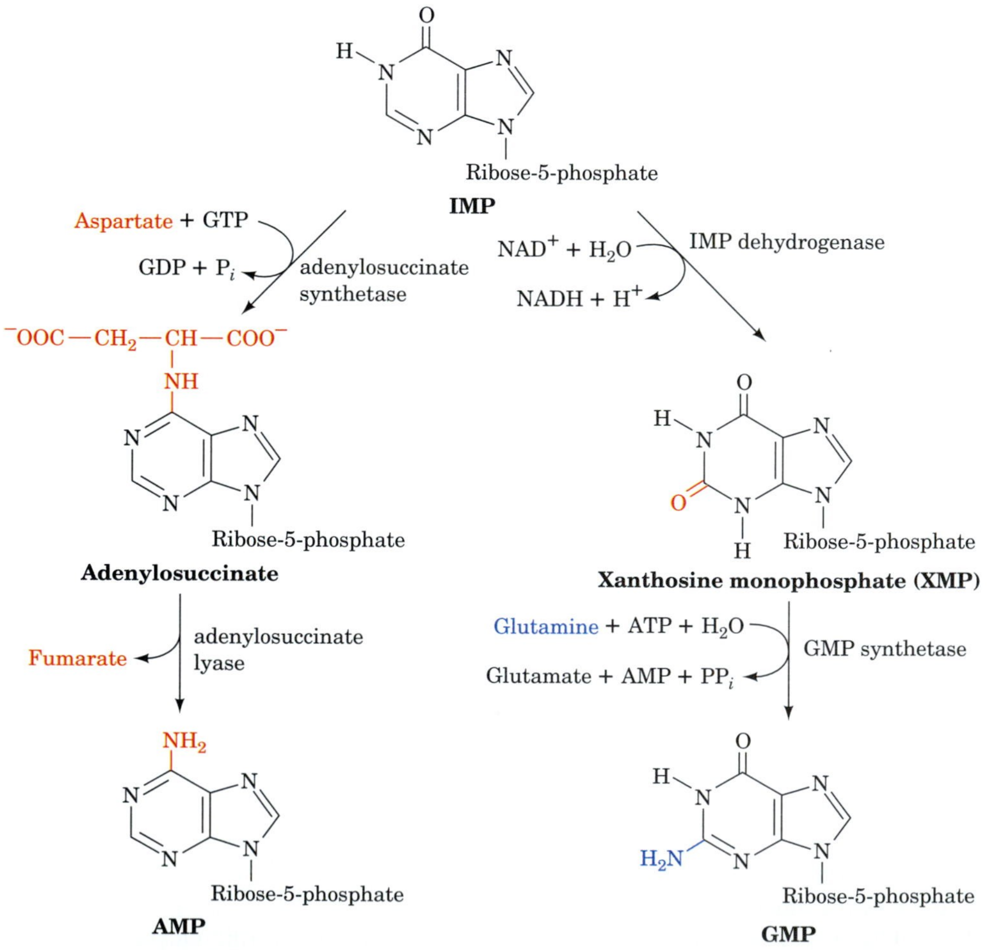

Figure 22-3. Conversion of IMP to AMP or GMP in separate two-reaction pathways.

First, nucleoside diphosphates are synthesized from the corresponding nucleoside monophosphates by base-specific **nucleoside monophosphate kinases.** For example, adenylate kinase (Section 13-2C) catalyzes the phosphorylation of AMP to ADP:

$$AMP + ATP \rightleftharpoons 2\ ADP$$

Similarly, GDP is produced by a guanine-specific enzyme:

$$GMP + ATP \rightleftharpoons GDP + ADP$$

These nucleoside monophosphate kinases do not discriminate between ribose and deoxyribose in the substrate.

Nucleoside diphosphates are converted to the corresponding triphosphates by **nucleoside diphosphate kinase;** for instance,

$$GDP + ATP \rightleftharpoons GTP + ADP$$

Although the reaction is written with ATP as the phosphoryl donor, this enzyme exhibits no preference for the bases of its substrates or for ribose over deoxyribose. Furthermore, the nucleoside diphosphate kinase reaction, as might be expected from the nearly identical structures of its substrates and products, normally operates close to equilibrium ($\Delta G \approx 0$). ADP is, of course, also converted to ATP by a variety of energy-releasing reactions such as those of glycolysis and oxidative phosphorylation. Indeed, it is these reactions that ultimately drive the foregoing kinase reactions.

C. Regulation of Purine Nucleotide Biosynthesis

The pathways synthesizing IMP, ATP, and GTP are individually regulated in most cells so as to control the total amounts of purine nucleotides available for nucleic acid synthesis, as well as the relative amounts of ATP and GTP. This control network is diagrammed in Fig. 22-4.

The IMP pathway is regulated at its first two reactions: those catalyzing the synthesis of PRPP and 5-phosphoribosylamine. Ribose phosphate pyrophosphokinase, the enzyme catalyzing Reaction 1 of the IMP pathway (Fig. 22-1), is inhibited by both ADP and GDP. Amidophosphoribosyl transferase, the enzyme catalyzing the first committed step of the IMP pathway (Reaction 2) is likewise subject to feedback inhibition. In this case, the enzyme binds ATP, ADP, and AMP at one inhibitory site and GTP, GDP, and GMP at another. *The rate of IMP production is therefore independently but synergistically controlled by the levels of adenine nucleotides and guanine nucleotides.* Moreover, amidophosphoribosyl transferase is allosterically stimulated by PRPP **(feedforward activation).**

A second level of regulation occurs immediately below the branch point leading from IMP to AMP and GMP. AMP and GMP are each competitive inhibitors of IMP in their own synthesis, which prevents excessive buildup of the pathway products. In addition, the rates of adenine and guanine nucleotide synthesis are coordinated. Recall that GTP powers the synthesis of AMP from IMP, whereas ATP powers the synthesis of GMP from IMP (Fig. 22-3). This reciprocity balances the production of AMP and GMP (which are required in roughly equal amounts in nucleic acid biosynthesis): *The rate of synthesis of GMP increases with [ATP], whereas that of AMP increases with [GTP].*

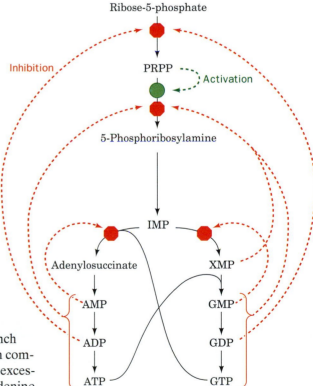

Figure 22-4. Control of the purine biosynthesis pathway. Red octagons and green dots indicate control points. Feedback inhibition is indicated by dashed red arrows, and feedforward activation is represented by dashed green arrows. ✳ See the Animated Figures.

D. Salvage of Purines

In most cells, the turnover of nucleic acids, particularly some types of RNA, releases adenine, guanine, and hypoxanthine (Section 22-4A). These free purines are reconverted to their corresponding nucleotides through **salvage pathways.** In contrast to the *de novo* purine nucleotide synthetic pathway, which is virtually identical in all cells, salvage pathways are diverse in character and distribution. In mammals, purines are mostly salvaged by two different enzymes. **Adenine phosphoribosyltransferase (APRT)** mediates AMP formation using PRPP:

$$\text{Adenine} + \text{PRPP} \rightleftharpoons \text{AMP} + \text{PP}_i$$

Hypoxanthine–guanine phosphoribosyltransferase (HGPRT) catalyzes the analogous reaction for both hypoxanthine and guanine:

$$\text{Hypoxanthine} + \text{PRPP} \rightleftharpoons \text{IMP} + \text{PP}_i$$
$$\text{Guanine} + \text{PRPP} \rightleftharpoons \text{GMP} + \text{PP}_i$$

Lesch–Nyhan Syndrome Results from HGPRT Deficiency

The symptoms of **Lesch–Nyhan syndrome,** which is caused by a severe HGPRT deficiency, indicate that purine salvage reactions have functions other than conservation of the energy required for *de novo* purine biosynthesis. This sex-linked congenital defect (it affects mostly males) results in excessive uric acid production (uric acid is a purine degradation product; Section 22-4A) and neurological abnormalities such as spasticity, mental retardation, and highly aggressive and destructive behavior, including a bizarre compulsion toward self-mutilation. For example, many children with Lesch–Nyhan syndrome have such an irresistible urge to bite their lips and fingers that they must be restrained. If the restraints are removed, communicative patients will plead that they be replaced, even as they attempt to injure themselves.

The excessive uric acid production in patients with Lesch–Nyhan syndrome is readily explained. The lack of HGPRT activity leads to an accumulation of the PRPP that would normally be used to salvage hypoxanthine and guanine. The excess PRPP activates amidophosphoribosyl transferase (which catalyzes Reaction 2 of the IMP biosynthetic pathway), thereby greatly accelerating the synthesis of purine nucleotides and thus the formation of their degradation product, uric acid. Yet the physiological basis of the associated neurological abnormalities remains obscure. That a defect in a single enzyme can cause such profound but well-defined behavioral changes nevertheless has important psychiatric implications.

2. SYNTHESIS OF PYRIMIDINE RIBONUCLEOTIDES

The biosynthesis of pyrimidines is simpler than that of purines. Isotopic labeling experiments have shown that atoms N1, C4, C5, and C6 of the pyrimidine ring are all derived from aspartic acid, C2 arises from HCO_3^-, and N3 is contributed by glutamine.

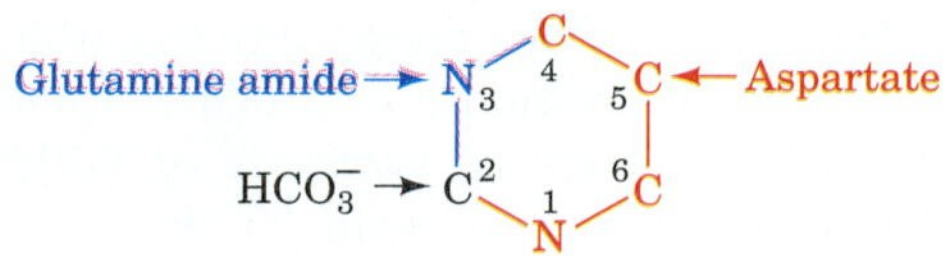

Glutamine amide → N3 C4 C5 ← Aspartate
HCO_3^- → C2 N1 C6

A. Synthesis of UMP

UMP, which is also the precursor of CMP, is synthesized in a six-reaction pathway (Fig. 22-5). In contrast to purine nucleotide synthesis, the pyrimidine ring is coupled to the ribose-5-phosphate moiety *after* the ring has been synthesized.

1. Synthesis of carbamoyl phosphate
   The first reaction of pyrimidine biosynthesis is the synthesis of **carbamoyl phosphate** from HCO_3^- and the amide nitrogen of glutamine by the cytosolic enzyme **carbamoyl phosphate synthetase II.** This reaction consumes two molecules of ATP: One provides a phosphate group and the other energizes the reaction. Carbamoyl phosphate is also synthesized by the urea cycle (Section 20-3A). In that reaction, catalyzed by the mitochondrial enzyme carbamoyl phosphate synthetase I, ammonia is the nitrogen source.

2. Synthesis of carbamoyl aspartate
   Condensation of carbamoyl phosphate with aspartate to form **car-**

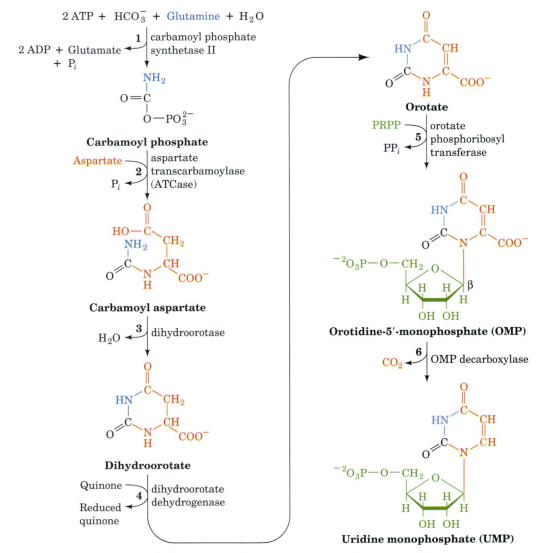

Figure 22-5. *Key to Metabolism.* The *de novo* synthesis of UMP. This metabolic pathway consists of six enzyme-catalyzed reactions. ✱ See the **Animated Figures.**

bamoyl aspartate is catalyzed by **aspartate transcarbamoylase (ATCase).** This reaction proceeds without ATP hydrolysis because carbamoyl phosphate is already "activated." The structure and regulation of *E. coli* ATCase are discussed in Section 12-3.

3. **Ring closure to form dihydroorotate**
   The third reaction of the pathway is an intramolecular condensation catalyzed by **dihydroorotase** to yield **dihydroorotate.**

4. **Oxidation of dihydroorotate**
   Dihydroorotate is irreversibly oxidized to **orotate** by **dihydroorotate dehydrogenase.** The eukaryotic enzyme, which contains FMN and nonheme Fe, is located on the outer surface of the inner mitochondrial membrane, where quinones supply its oxidizing power. The other five enzymes of pyrimidine nucleotide biosynthesis are cytosolic in animal cells.

5. **Acquisition of the ribose phosphate moiety**
   Orotate reacts with PRPP to yield **orotidine-5′-monophosphate (OMP)** in a reaction catalyzed by **orotate phosphoribosyl transferase.** This reaction, which is driven by the hydrolysis of the eliminated PP_i, fixes the anomeric form of pyrimidine nucleotides in the β configuration. Orotate phosphoribosyl transferase also salvages other pyrimidine bases, such as uracil and cytosine, by converting them to their corresponding nucleotides.

6. **Decarboxylation to form UMP**
   The final reaction of the pathway is the decarboxylation of OMP by **OMP decarboxylase** to form UMP. This is an unusual reaction in that it requires no cofactors.

In bacteria, the six enzymes of UMP biosynthesis occur as independent proteins. In animals, however, as Mary Ellen Jones demonstrated, the first three enzymatic activities of the pathway—carbamoyl phosphate synthetase II, ATCase, and dihydroorotase—occur on a single 210-kD polypeptide chain. Similarly, Reactions 5 and 6 of the animal pyrimidine pathway are catalyzed by a single polypeptide. In animal purine biosynthesis (Section 22-1A), single polypeptides catalyze Reactions 3, 4, and 6, Reactions 7 and 8, and Reactions 10 and 11. *The intermediate products of these multifunctional enzymes are not readily released to the medium but are channeled to the succeeding enzymatic activities of the pathway.* As in the pyruvate dehydrogenase complex (Section 16-2), fatty acid synthase (Section 19-4C), and tryptophan synthase (Section 20-5B), channeling in the nucleotide synthetic pathways increases the overall rate of these multistep processes and protects intermediates from degradation by other cellular enzymes.

B. Synthesis of UTP and CTP

The synthesis of UTP from UMP is analogous to the synthesis of purine nucleoside triphosphates (Section 22-1B). The process occurs by the sequential actions of a nucleoside monophosphate kinase and nucleoside diphosphate kinase:

$$UMP + ATP \rightleftharpoons UDP + ADP$$
$$UDP + ATP \rightleftharpoons UTP + ADP$$

CTP is formed by amination of UTP by **CTP synthetase** (Fig. 22-6). In animals, the amino group is donated by glutamine, whereas in bacteria it is supplied directly by ammonia.

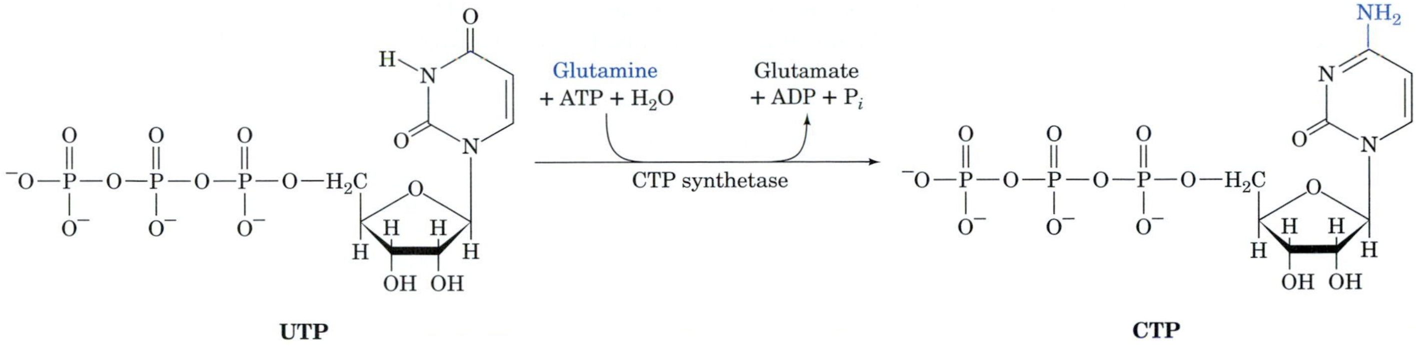

UTP   **CTP**

Figure 22-6. The synthesis of CTP from UTP.

C. Regulation of Pyrimidine Nucleotide Biosynthesis

In bacteria, the pyrimidine biosynthetic pathway is primarily regulated at Reaction 2, the ATCase reaction (Fig. 22-7a). In *E. coli,* control is exerted through the allosteric stimulation of ATCase by ATP and its inhibition by CTP (Section 12-3). In many bacteria, however, UTP is the major ATCase inhibitor.

In animals, ATCase is not a regulatory enzyme. Rather, pyrimidine biosynthesis is controlled by the activity of carbamoyl phosphate synthetase II, which is inhibited by UDP and UTP and activated by ATP and PRPP (Fig. 22-7b). A second level of control in the mammalian pathway occurs

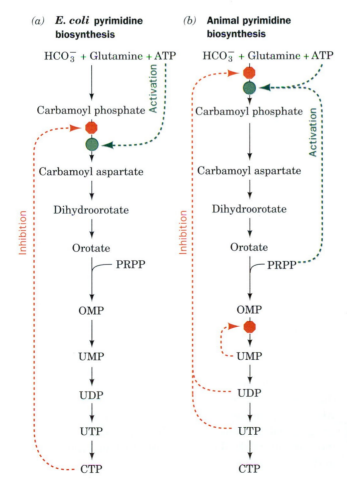

Figure 22-7. Regulation of pyrimidine biosynthesis. The control networks are shown for (*a*) *E. coli* and (*b*) animals. Red octagons and green dots indicate control points. Feedback inhibition is represented by dashed red arrows, and activation is indicated by dashed green arrows. ✳ **See the Animated Figures.**

at OMP decarboxylase, for which UMP and to a lesser extent CMP are competitive inhibitors. In all organisms, the rate of OMP production varies with the availability of its precursor, PRPP. Recall that the PRPP level depends on the activity of ribose phosphate pyrophosphokinase (Fig. 22-1, Reaction 1), which is inhibited by ADP and GDP (Section 22-1C).

Orotic Aciduria Results from an Inherited Enzyme Deficiency

   Orotic aciduria, an inherited human disease, is characterized by the urinary excretion of large amounts of orotic acid, retarded growth, and severe anemia. It results from a deficiency in the bifunctional enzyme catalyzing Reactions 5 and 6 of pyrimidine nucleotide biosynthesis. Consideration of the biochemistry of this situation led to its effective treatment: the administration of uridine and/or cytidine. The UMP formed through the phosphorylation of these nucleosides, besides replacing that normally synthesized, inhibits carbamoyl phosphate synthetase II so as to attenuate the rate of orotic acid synthesis. No other genetic deficiency in pyrimidine nucleotide biosynthesis is known in humans, presumably because such defects are lethal *in utero*.

3. FORMATION OF DEOXYRIBONUCLEOTIDES

DNA differs chemically from RNA in two major respects: (1) Its nucleotides contain 2'-deoxyribose residues rather than ribose residues, and (2) it contains the base thymine (5-methyluracil) rather than uracil. In this section, we consider the biosynthesis of these DNA components.

A. Production of Deoxyribose Residues

Deoxyribonucleotides are synthesized from their corresponding ribonucleotides by the reduction of their C2' position rather than by their de novo synthesis from deoxyribose-containing precursors.

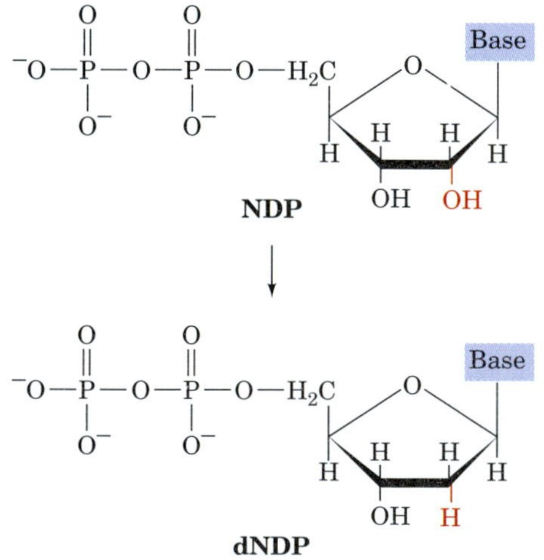

Enzymes that catalyze the formation of deoxyribonucleotides by the reduction of the corresponding ribonucleotides are named **ribonucleotide reductases.** There are as many as four classes of ribonucleotide reductases, which differ in their prosthetic groups, although they all replace the 2'-OH group of ribose with H via a free-radical mechanism. We shall discuss the mechanism of the Fe-containing enzyme that occurs in most eukaryotes and some prokaryotes.

Fe-containing ribonucleotide reductases reduce ribonucleoside diphosphates (NDPs) to the corresponding deoxyribonucleoside diphosphates (dNDPs). *E. coli* ribonucleotide reductase, as Peter Reichard demonstrated, is an $\alpha_2\beta_2$ tetramer that can be decomposed to two catalytically inactive homodimers, R1 (α_2) and R2 (β_2; Fig. 22-8*a*). Each α subunit contains a substrate-binding site that includes several redox-active thiol groups. The α subunit also contains two independent effector-binding sites that control the enzyme's catalytic activity as well as its substrate specificity.

The X-ray structure of R2 (Fig. 22-8*b*), which was determined by Hans Eklund, reveals that the β subunits are bundles of eight unusually long helices. Each β subunit contains a novel binuclear Fe(III) prosthetic group whose Fe(III) ions are liganded by a variety of groups including an O^{2-} ion (Fig. 22-8*c*). The Fe(III) complex interacts with Tyr 122 to form an unusual tyrosyl free radical that is 5 Å from the closest Fe atom and is buried 10 Å beneath the surface of the protein, where it is out of contact with solvent and any oxidizable side chain.

(a)

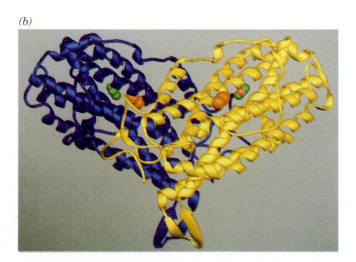

R1 (α_2) subunit

Allosteric sites
- Specificity site (ATP, dATP, dGTP, dTTP)
- Activity site (ATP, dATP)

Substrate-binding site (ATP, GDP, UDP, CDP)

SH SH   SH SH

Tyr   Tyr

R2 (β_2) subunit

(b)

(c)

Tyr 122

Figure 22-8. **E. coli ribonucleotide reductase.** (*a*) Schematic diagram of the quaternary structure. The enzyme consists of two identical pairs of subunits, R1 (α_2) and R2 (β_2). Each β subunit contains a binuclear Fe(III) complex that generates a phenoxy radical at Tyr 122. The α subunits each contain two different allosteric effector sites and five catalytically important Cys residues. The enzyme's two active sites occur at the interface between the α and β subunits. (*b*) A ribbon diagram of R2 viewed perpendicularly to its two-fold axis with the β subunits shown in blue and yellow. The Fe(III) ions are represented by orange spheres, and the radical-harboring Tyr 122 side chains are shown in space-filling representation with their C and O atoms green and red. (*c*) The binuclear Fe(III) complex of R2. Each Fe(III) ion is octahedrally coordinated by a His N atom and five O atoms, including those of the O^{2-} ion and the Glu carboxyl group that bridge the two Fe(III) ions. [Part *b* based on an X-ray structure by Hans Eklund, Swedish University of Agricultural Sciences.] ● **See the Interactive Exercises.**

Glu 238

Asp 84   H$_2$O O^{2-} H$_2$O   Glu 204

His 118   Fe 1   Fe 2   His 241

Glu 115

JoAnne Stubbe has proposed the following catalytic mechanism for *E. coli* ribonucleotide reductase (Fig. 22-9):

1. Ribonucleotide reductase's free radical (X·) abstracts an H atom from C3′ of the substrate in the reaction's rate-determining step.

2 & 3. Acid-catalyzed cleavage of the C2′—OH bond releases H_2O to yield a radical–cation intermediate. The 3′-OH group's unshared electron pair stabilizes the C2′ cation. This accounts for the radical's catalytic role.

4. The radical–cation intermediate is reduced by the enzyme's redox-active sulfhydryl pair to yield a 3′-deoxynucleotide radical and a protein disulfide group (this group must eventually be oxidized to regenerate the enzyme's activity).

5. The 3′ radical abstracts an H atom from the protein to yield the product deoxynucleoside diphosphate and restore the enzyme to its radical state.

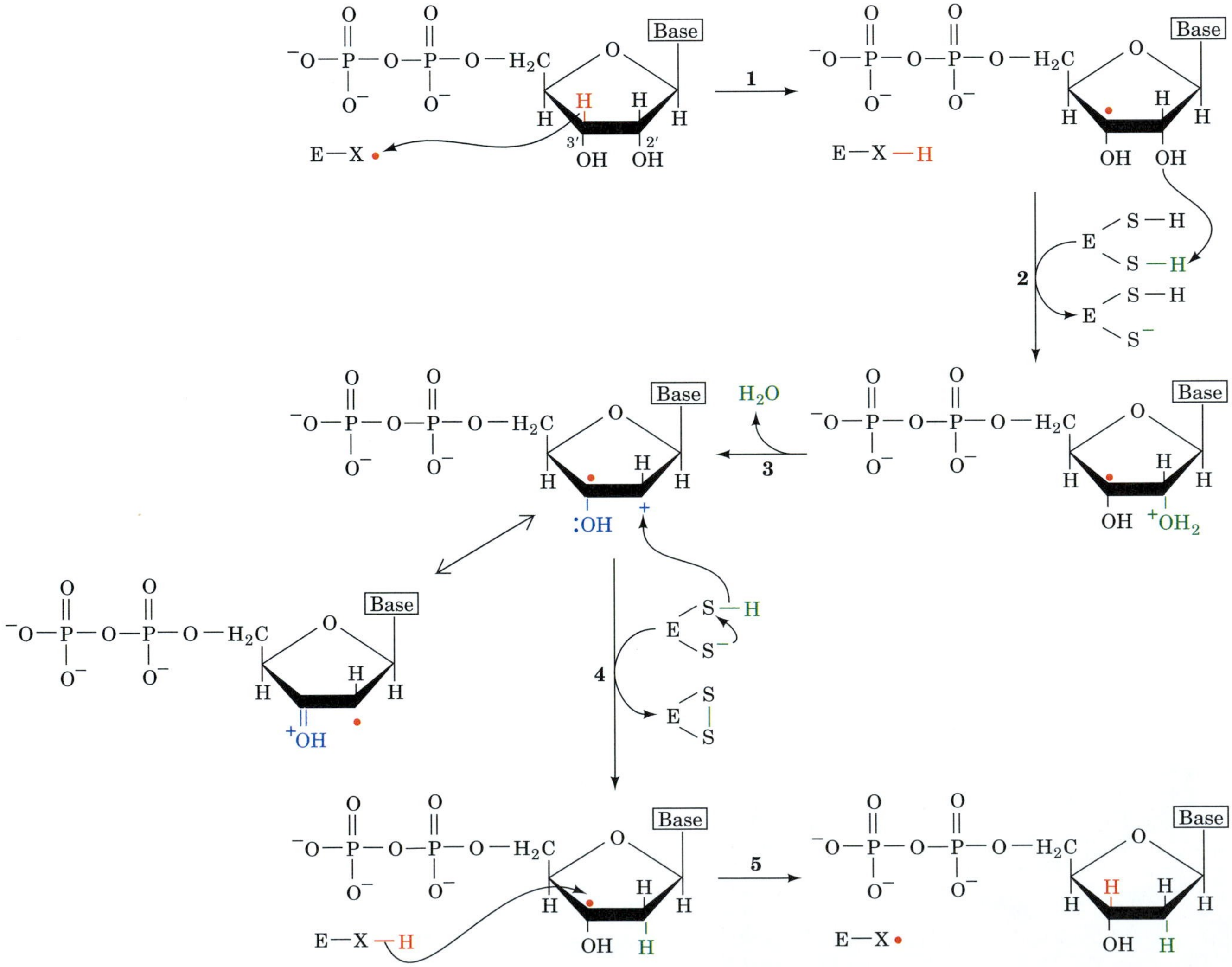

Figure 22-9.   The enzymatic mechanism of ribonucleotide reductase. The reaction occurs via a free radical–mediated process in which reducing equivalents are supplied by the formation of an enzyme disulfide bond. [After Stubbe, J.A., *J. Biol. Chem.* **265**, 5330 (1990).]

The Tyr 122 radical in R2 is too far away (>10 Å) from the enzyme's catalytic site to abstract an electron directly from the substrate. Evidently, the protein mediates electron transfer from this tyrosyl radical to another group that is closer to the substrate, probably the thiyl radical (—S·) form of Cys 439 in R1 (represented as X· in Fig. 22-9). Two other R1 Cys residues probably form the redox-active sulfhydryl pair that directly reduces the substrate. The resulting disulfide bond is reduced via disulfide interchange with yet two other Cys residues, which are positioned to accept electrons from external reducing agents to regenerate the active enzyme. Thus, each subunit of R1 contains at least five Cys residues that chemically participate in nucleotide reduction.

Thioredoxin and Glutaredoxin Reduce Ribonucleotide Reductase

The final step in the ribonucleotide reductase catalytic cycle is reduction of the enzyme's newly formed disulfide bond to re-form its redox-active sulfhydryl pair. One of the enzyme's physiological reducing agents is **thioredoxin,** a ubiquitous monomeric 108-residue protein with a pair of neighboring Cys residues (and which also participates in regulating the Calvin cycle; Section 18-3B). Thioredoxin reduces oxidized ribonucleotide reductase via disulfide interchange.

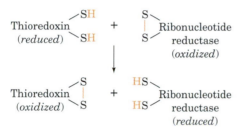

The X-ray structure of thioredoxin (Fig. 22-10) reveals that its redox-active disulfide group is located on a molecular protrusion, making this protein the only known example of a "male" enzyme.

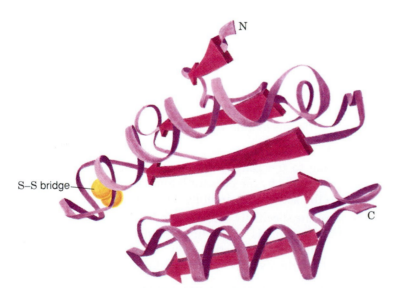

Figure 22-10.   The X-ray structure of *E. coli* thioredoxin in its oxidized (disulfide) state. The two yellow balls in the protrusion on the left represent the protein's redox-active disulfide group. [After a drawing by B. Furugren *in* Holmgren, A., Söderberg, B.-O., Eklund, H., and Brändén, C.-I., *Proc. Natl. Acad. Sci.* **72**, 2307 (1975).]

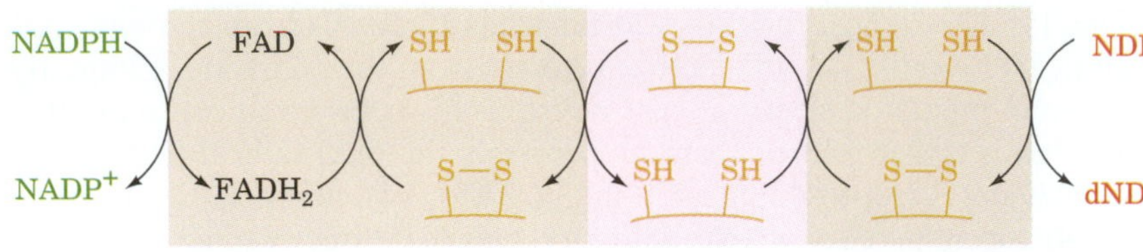

Figure 22-11. **An electron-transfer pathway for nucleoside diphosphate (NDP) reduction.** NADPH provides the reducing equivalents for this process through the intermediacy of thioredoxin reductase, thioredoxin, and ribonucleotide reductase.

Oxidized thioredoxin is, in turn, reduced in a reaction mediated by **thioredoxin reductase,** which contains redox-active thiol groups and an FAD prosthetic group. This enzyme is a homolog of glutathione reductase (Box 14-3) and catalyzes a similar reaction: the NADPH-mediated reduction of a substrate disulfide bond. NADPH therefore serves as the terminal reducing agent in the ribonucleotide reductase–catalyzed reduction of NDPs to dNDPs (Fig. 22-11).

The existence of a viable *E. coli* mutant devoid of thioredoxin indicates that this protein is not the only substance capable of reducing oxidized ribonucleotide reductase *in vivo*. This observation led to the discovery of **glutaredoxin,** a disulfide-containing 85-residue protein that can also reduce ribonucleotide reductase via disulfide interchange (mutants devoid of both thioredoxin and glutaredoxin are nonviable). The relative importance of thioredoxin and glutaredoxin in the reduction of ribonucleoside diphosphates remains to be established.

Ribonucleotide Reductase Is Regulated by a Complex Feedback Network

The synthesis of the four dNTPs in the amounts required for DNA synthesis is accomplished through feedback control. Maintaining the proper intracellular ratios of dNTPs is essential for normal growth. Indeed, *a deficiency of any dNTP is lethal, whereas an excess is mutagenic because the probability that a given dNTP will be erroneously incorporated into a growing DNA strand increases with its concentration relative to those of the other dNTPs.*

The activities of both *E. coli* and mammalian ribonucleotide reductases are remarkably responsive to the levels of nucleotides in the cell. The R1 subunit of *E. coli* ribonucleotide reductase has two independent allosteric sites (Fig. 22-8*a*):

1. ATP binding to the **activity site** activates the enzyme toward substrates determined by the effector bound at the **specificity site.** dATP binding to the activity site inhibits the enzyme toward all substrates.

2. At the substrate specificity site, ATP or dATP binding stimulates CDP and UDP reduction, dTTP binding stimulates GDP reduction but inhibits CDP and UDP reduction, and dGTP binding stimulates ADP reduction but inhibits CDP, UDP, and GDP reduction.

In the absence of any of these effectors, ribonucleotide reductase is inactive.

These allosteric effects suggest the following sequence of events for the reduction of ribonucleotides (Fig. 22-12). In the presence of a mixture of dNDPs, ribonucleotide reductase commences dNDP production by the

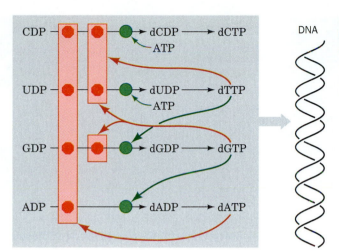

Figure 22-12. **The control networks for regulating deoxyribonucleotide biosynthesis.** Green dots and arrows represent activation; red octagons and arrows represent inhibition. [After Thelender, L. and Reichard, P., *Annu. Rev. Biochem.* **48**, 153 (1978).] ✳ See the Animated Figures.

ATP-stimulated reduction of CDP and UDP. The resulting dUDP is converted to dTTP (as described in the following section), which inhibits further CDP and UDP reduction but stimulates dGDP production. dGDP, after its phosphorylation to dGTP, inhibits the reduction of CDP, UDP, and GDP but stimulates production of dADP and thus dATP. As the dATP accumulates, it binds to the activity site of ribonucleotide reductase, thereby inhibiting all NDP reduction unless the ATP level is sufficiently high to displace the dATP. Although this scheme is no doubt an oversimplified description of a dynamic process, it accounts for the observed ability of cells to synthesize deoxynucleotides in the amounts of each required for DNA synthesis.

dNTPs Are Produced by Phosphorylation of dNDPs

The final step in the production of all dNTPs is the phosphorylation of the corresponding dNDPs:

$$dNDP + ATP \rightleftharpoons dNTP + ADP$$

This reaction is catalyzed by nucleoside diphosphate kinase, the same enzyme that phosphorylates NDPs (Section 22-1B). As before, the reaction is written with ATP as the phosphoryl donor, although any NTP or dNTP can function in this capacity.

B. Origin of Thymine

The dTMP component of DNA is synthesized by methylation of dUMP. The dUMP is generated through the hydrolysis of dUTP by **dUTP diphosphohydrolase (dUTPase):**

$$dUTP + H_2O \longrightarrow dUMP + PP_i$$

dTMP, once it is formed, is phosphorylated to form dTTP. The apparent reason for the energetically wasteful process of dephosphorylating dUTP and rephosphorylating dTMP is that cells must minimize their concentration of dUTP in order to prevent incorporation of uracil into their DNA (the enzyme system that synthesizes DNA from dNTPs does not efficiently discriminate between dUTP and dTTP).

Human dUTPase is a homotrimer of 141-residue subunits. Its X-ray structure, determined by John Tainer, reveals the basis for this enzyme's exquisite specificity for dUTP. Each subunit binds dUTP in a snug-fitting

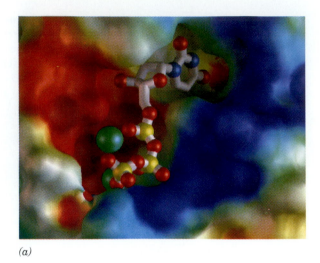

(a)

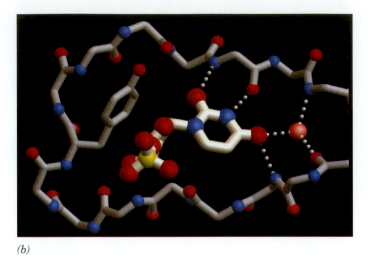

(b)

Figure 22-13.   The X-ray structure of human dUTPase.
(*a*) The active site region of dUTPase in complex with dUTP. The protein is represented by its molecular surface colored according to its electrostatic potential (negative, red; positive, blue; and near neutral, white). The dUTP is shown in ball-and-stick form with its N, O, and P atoms blue, red, and yellow. Mg^{2+} ions that have been modeled into the structure are represented by green spheres. Note the complementary fit of the uracil ring into its binding pocket, particularly the close contacts that discriminate against a methyl group on C5 of the pyrimidine ring and a 2′-OH group

on the ribose ring. (*b*) The binding site of dUMP, showing the hydrogen bonding system responsible for the enzyme's specific binding of a uracil ring. The dUMP and the polypeptide backbone binding it are shown in ball-and-stick form with atoms colored as in *a*; hydrogen bonds are indicated by white dotted lines; and a conserved water molecule is represented by a pink sphere. The side chain of a conserved Tyr is tightly packed against the ribose ring so as to discriminate against the presence of a 2′-OH group. [Courtesy of John Tainer and Clifford Mol, The Scripps Research Institute, La Jolla, California.]

cavity that sterically excludes thymine's C5 methyl group via the side chains of conserved residues (Fig. 22-13*a*). The enzyme differentiates uracil from the similarly shaped cytosine via a set of hydrogen bonds that in part mimic adenine's base pairing interactions (Fig. 22-13*b*). The 2′-OH group of ribose is likewise sterically excluded by the side chain of a conserved Tyr.

Thymidylate Synthase

Thymidylate (dTMP) is synthesized from dUMP by **thymidylate synthase** with N^5,N^{10}-methylenetetrahydrofolate (N^5,N^{10}-methylene-THF) as the methyl donor:

dUMP + N^5,N^{10}-**Methylene-tetrahydrofolate** → **dTMP** + **Dihydrofolate**

$$R = \text{—} \bigcirc \text{—} \overset{\overset{O}{\|}}{C} \text{—} \left(\overset{H}{N} \text{—} \overset{\overset{COO^-}{|}}{CH} \text{—} CH_2 \text{—} CH_2 \text{—} \overset{\overset{O}{\|}}{C} \right)_n \text{—} O^- ; \quad n = 1\text{–}6$$

Note that the transferred methylene group (in which the carbon has the oxidation state of formaldehyde) is reduced to a methyl group (which has the oxidation state of methanol) at the expense of the oxidation of the THF cofactor to **dihydrofolate (DHF).**

Thymidylate synthase, a highly conserved 70-kD dimeric protein, follows a mechanistic scheme proposed by Daniel Santi (Fig. 22-14):

1. An enzyme nucleophile, identified as the thiolate group of Cys 146, attacks C6 of dUMP to form a covalent adduct.

2. C5 of the resulting enolate ion attacks the CH_2 group of the iminium cation in equilibrium with N^5,N^{10}-methylene-THF to form an enzyme–dUMP–THF ternary covalent complex.

3. An enzyme base abstracts the acidic proton at the C5 position of the enzyme-bound dUMP, forming an exocyclic methylene group and eliminating the THF cofactor. The abstracted proton subsequently exchanges with solvent.

4. The redox change occurs via the migration of the C6-H atom of THF as a hydride ion to the exocyclic methylene group, converting it to a methyl group and yielding DHF. This reduction promotes the displacement of the Cys thiolate group from the intermediate to release product, dTMP, and re-form the active enzyme.

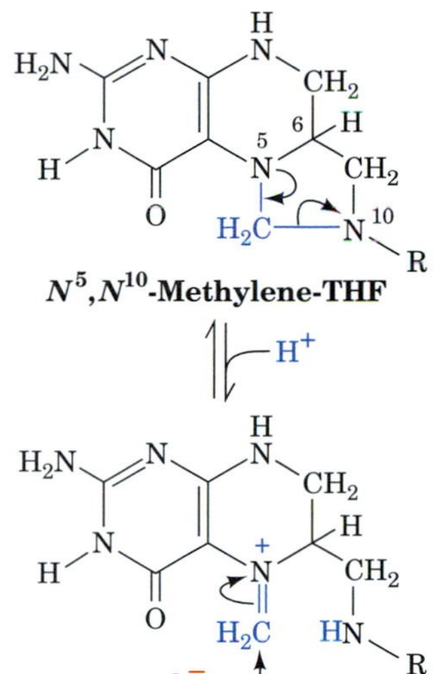

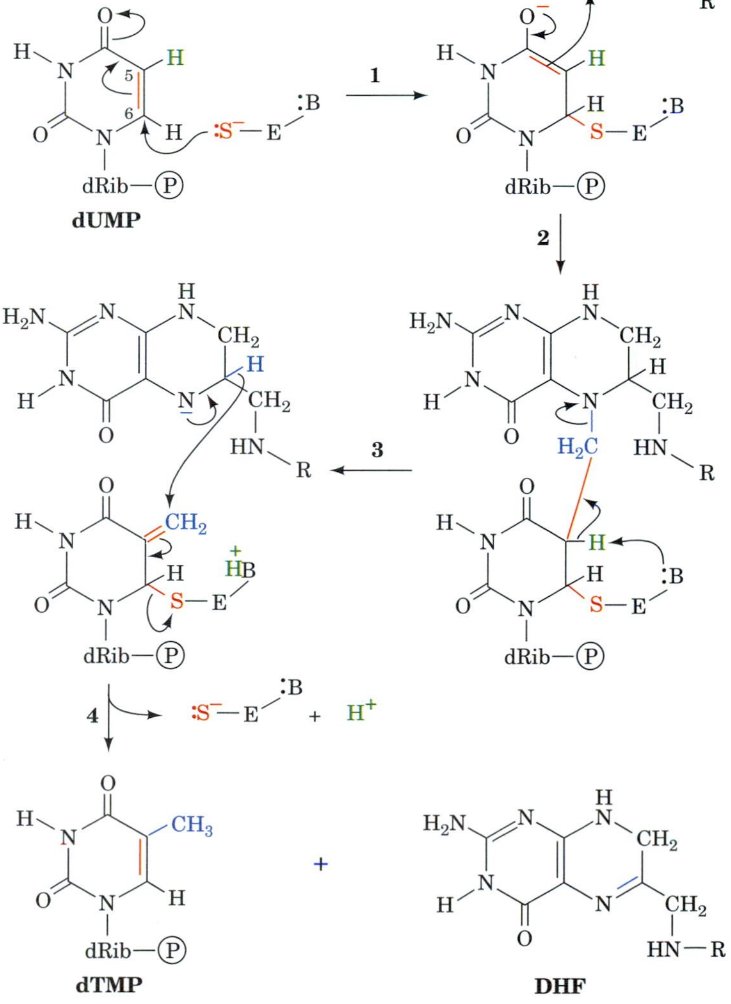

Figure 22-14.   The catalytic mechanism of thymidylate synthase. The methyl group is supplied by N^5,N^{10}-methylene-THF, which is concomitantly oxidized to dihydrofolate.

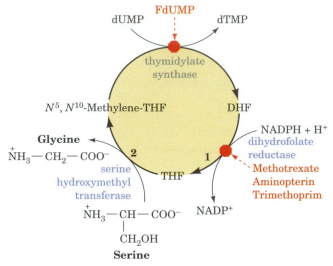

Figure 22-15.   Regeneration of N^5,N^{10}-methylenetetrahydrofolate. The DHF product of the thymidylate synthase reaction is converted back to N^5,N^{10}-methylene-THF by the sequential actions of (**1**) dihydrofolate reductase and (**2**) serine hydroxymethyltransferase. The sites of action of some inhibitors are indicated by red octagons. Thymidylate synthase is inhibited by FdUMP, whereas dihydrofolate reductase is inhibited by the antifolates methotrexate, aminopterin, and trimethoprim (see Box 22-1).

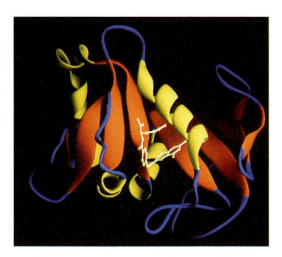

Figure 22-16.   A ribbon diagram of human dihydrofolate reductase in complex with folate. The helices of this monomeric enzyme are drawn in yellow, the β sheets in orange, and the other polypeptide segments in blue. [Courtesy of Jay F. Davies II and Joseph Kraut, University of California at San Diego.] ● **See the Interactive Exercises.**

Tetrahydrofolate Is Regenerated in Two Reactions

The thymidylate synthase reaction is biochemically unique in that it oxidizes THF to DHF; no other enzymatic reaction employing a THF cofactor alters this coenzyme's net oxidation state. The DHF product of the thymidylate synthase reaction is recycled back to N^5,N^{10}-methylene-THF through two sequential reactions (Fig. 22-15):

1. DHF is reduced to THF by NADPH as catalyzed by **dihydrofolate reductase (DHFR;** Fig. 22-16). Although in most organisms DHFR is a monomeric, monofunctional enzyme, in protozoa and some plants DHFR and thymidylate synthase occur on the same polypeptide chain to form a bifunctional enzyme that has been shown to channel DHF from its thymidylate synthase to its DHFR active sites.

2. Serine hydroxymethyltransferase (Section 20-4A) transfers the hydroxymethyl group of serine to THF yielding N^5,N^{10}-methylene-THF and glycine.

Inhibition of thymidylate synthase or DHFR blocks dTMP synthesis and is therefore the basis of cancer chemotherapies (see Box 22-1).

4. NUCLEOTIDE DEGRADATION

Most foodstuffs, being of cellular origin, contain nucleic acids. Dietary nucleic acids survive the acid medium of the stomach; they are degraded to their component nucleotides, mainly in the intestine, by pancreatic nucleases and intestinal phosphodiesterases. The ionic nucleotides, which cannot pass through cell membranes, are then hydrolyzed to nucleosides by a variety of group-specific nucleotidases and nonspecific phosphatases. Nu-

BIOCHEMISTRY IN HEALTH AND DISEASE

Inhibition of Thymidylate Synthesis in Cancer Therapy

dTMP synthesis is a critical process for rapidly proliferating cells, such as cancer cells, which require a steady supply of dTMP for DNA synthesis. Interruption of dTMP synthesis can therefore kill these cells. Most normal mammalian cells, which grow slowly if at all, require less dTMP and so are less sensitive to agents that inhibit thymidylate synthase or dihydrofolate reductase (notable exceptions are the bone marrow cells that constitute the blood-forming tissue and much of the immune system, the intestinal mucosa, and hair follicles).

5-Fluorodeoxyuridylate (FdUMP)

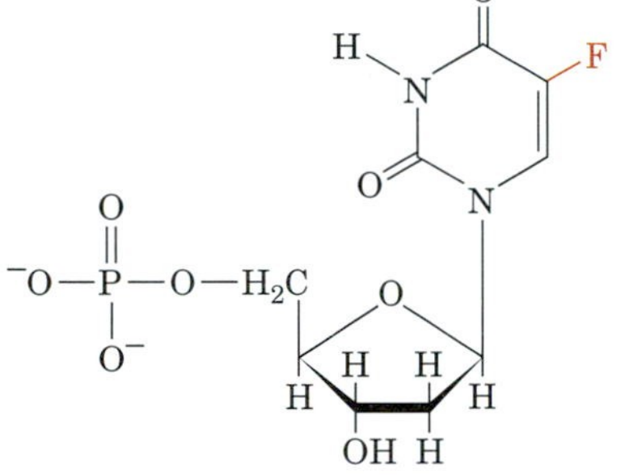

5-Fluorodeoxyuridylate (FdUMP)

is an irreversible inhibitor of thymidylate synthase. This substance, like dUMP, binds to the enzyme (an F atom is not much larger than an H atom) and undergoes the first two steps of the normal enzymatic reaction (Fig. 22-14). In Step 3, however, the enzyme cannot abstract the F atom as F^+ (F is the most electronegative element) so that the enzyme is frozen in an enzyme–FdUMP–THF ternary covalent complex.

Enzyme inhibitors such as FdUMP, which inactivate an enzyme only after undergoing part or all of its normal catalytic reaction, are called **mechanism-based inhibitors** (alternatively, **suicide substrates** because they cause the enzyme

to "commit suicide"). Because of their extremely high specificity, mechanism-based inhibitors are among the most useful therapeutic agents.

Inhibition of DHFR blocks dTMP synthesis as well as all other THF-dependent biological reactions, because the THF converted to DHF by the thymidylate synthase reaction cannot be regenerated. **Methotrexate (amethopterin), aminopterin**, and **trimethoprim**

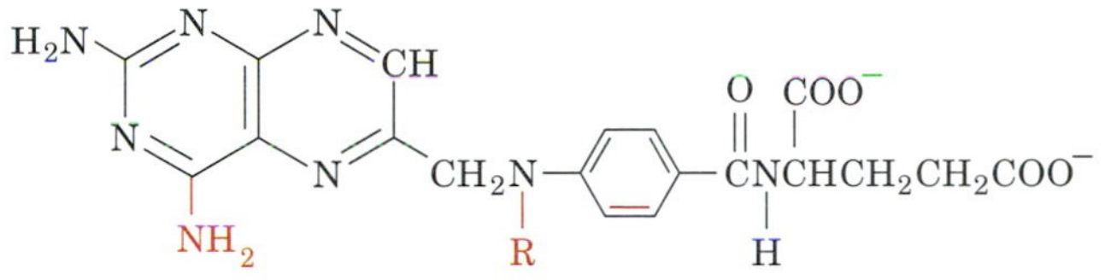

R = H   **Aminopterin**
R = CH$_3$ **Methotrexate (amethopterin)**

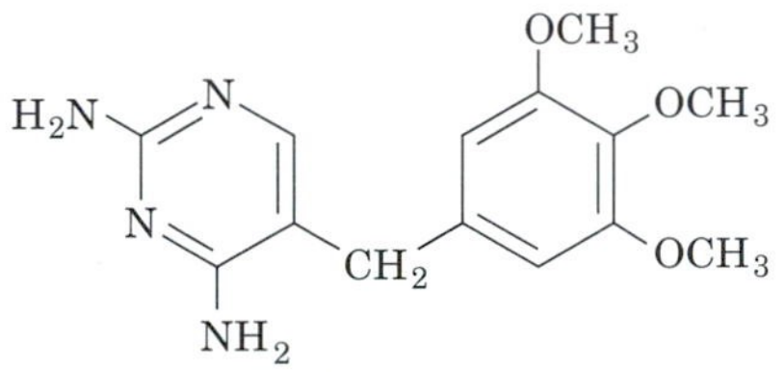

Trimethoprim

are DHF analogs that competitively although nearly irreversibly bind to DHFR with an ~1000-fold greater affinity than does DHF. These **antifolates** (substances that interfere with the action of folate cofactors) are effective anticancer agents, particularly against childhood leukemias. In fact, a successful chemotherapeutic strategy is to treat a cancer victim with a lethal dose of methotrexate and some hours later "rescue" the patient (but hopefully not the cancer) by administering massive doses of 5-formyl-THF and/or thymidine. Trimethoprim, which was discovered by George Hitchings and Gertrude Elion, binds much more tightly to bacterial DHFRs than to those of mammals and is therefore a clinically useful antibacterial agent.

---

cleosides may be directly absorbed by the intestinal mucosa or further degraded to free bases and ribose or ribose-1-phosphate through the action of **nucleosidases** and **nucleoside phosphorylases:**

$$\text{Nucleoside} + H_2O \xrightarrow{\text{nucleosidase}} \text{base} + \text{ribose}$$

$$\text{Nucleoside} + P_i \xrightarrow[\text{phosphorylase}]{\text{nucleoside}} \text{base} + \text{ribose-1-P}$$

Radioactive labeling experiments have demonstrated that only a small fraction of the bases of ingested nucleic acids are incorporated into tissue

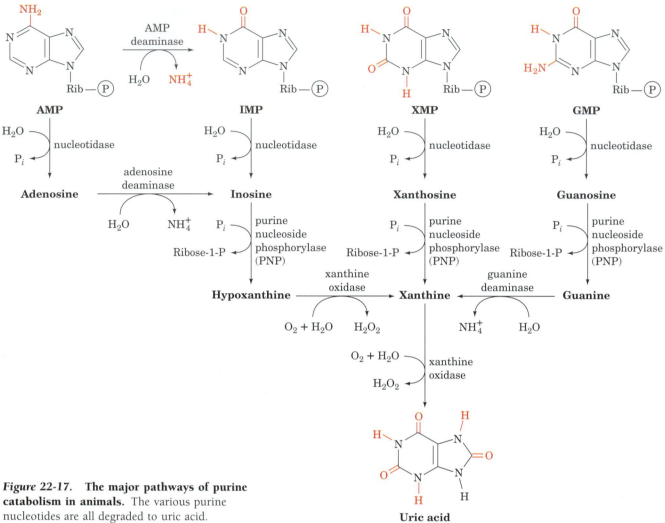

Figure 22-17.   The major pathways of purine catabolism in animals. The various purine nucleotides are all degraded to uric acid.

Uric acid

nucleic acids. Evidently, *the de novo pathways of nucleotide biosynthesis largely satisfy an organism's need for nucleotides.* Consequently, ingested bases are mostly degraded and excreted. Cellular nucleic acids are also subject to degradation as part of the continual turnover of nearly all cellular components. In this section, we outline these catabolic pathways and discuss the consequences of several of their inherited defects.

A. Catabolism of Purines

The major pathways of purine nucleotide and deoxynucleotide catabolism in animals are diagrammed in Fig. 22-17. The pathways in other organisms differ somewhat, but all the pathways lead to uric acid. Of course, the pathway intermediates may be directed to purine nucleotide synthesis via salvage reactions. In addition, ribose-1-phosphate, a product of the reaction catalyzed by **purine nucleoside phosphorylase (PNP;** Fig. 22-18), is a precursor of PRPP.

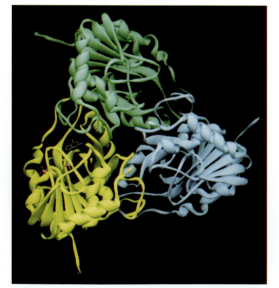

Figure 22-18.   The X-ray structure of human erythrocyte purine nucleoside phosphorylase. Each identical subunit is differently colored. The yellow subunit is shown in complex with a guanine molecule and two phosphate ions. [Courtesy of Mike Carson, University of Alabama at Birmingham; X-ray structure determined by Stephen Ealick and Charles Bugg, University of Alabama at Birmingham.] ● See the Interactive Exercises.

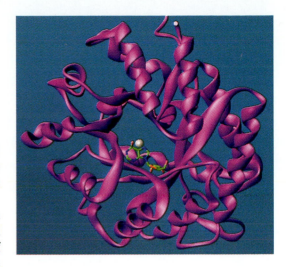

Figure 22-19. A ribbon diagram of murine adenosine deaminase. The enzyme is viewed approximately down the axis of its α/β barrel from the N-terminal ends of its β strands. A transition state analog, **6-hydroxyl-1,6-dihydropurine ribonucleoside (HDPR)**, is shown in skeletal form with its C, N, and O atoms green, blue, and red. The enzyme-bound Zn^{2+} ion, which is coordinated by HDPR's 6-hydroxyl group, is represented by a silver sphere. [Based on an X-ray structure by Florante Quiocho, Baylor College of Medicine.] ● **See the Interactive Exercises.**

Adenosine and deoxyadenosine are not degraded by mammalian PNP. Rather, adenine nucleosides and nucleotides are deaminated by **adenosine deaminase (ADA)** and **AMP deaminase** to their corresponding inosine derivatives, which can then be further degraded.

ADA is an eight-stranded α/β barrel (Fig. 22-19) with its active site in a pocket at the C-terminal end of the β barrel, as in all known α/β barrel enzymes. A catalytically essential zinc ion is bound in the deepest part of the active site pocket. Mutations that affect the active site of ADA cause immune system disorders (see Box 22-2).

The Purine Nucleotide Cycle

The deamination of AMP to IMP, when combined with the synthesis of AMP from IMP (Fig. 22-3, *left*), has the net effect of deaminating aspar-

---

Box 22-2

BIOCHEMISTRY IN HEALTH AND DISEASE

Severe Combined Immunodeficiency Disease

Abnormalities in purine nucleoside metabolism arising from rare genetic defects in adenosine deaminase selectively kill lymphocytes (a type of white blood cell). Since lymphocytes mediate much of the immune response, ADA deficiency results in **severe combined immunodeficiency disease (SCID)**. Without special protective measures, this disease is invariably fatal in infancy because of overwhelming infection. The mutations in all eight known ADA variants obtained from SCID patients appear to structurally perturb the active site of ADA.

Biochemical considerations provide a plausible explanation of SCID's etiology (causes). In the absence of active ADA, deoxyadenosine is phosphorylated to yield levels of dATP that are 50-fold greater than normal. This high concentration of dATP inhibits ribonucleotide reductase (Section 22-3A), thereby preventing the synthesis of the other dNTPs, choking off DNA synthesis and thus cell proliferation. The tissue-specific effect of ADA deficiency on the immune system can be explained by the observation that lymphoid tissue is particularly active in deoxyadenosine phosphorylation.

Immune system function in SCID patients can be boosted to a limited extent by injecting normal ADA to which several molecules of the biologically inert **polyethylene glycol (PEG)**

have been covalently linked. Without this modification, the liver clears ADA from the circulation within minutes, whereas PEG–ADA remains in the blood for 1–2 weeks. Evidently, the PEG masks the ADA from the receptors that otherwise filter it from the blood. Nevertheless, a more efficient treatment for ADA deficiency may be gene therapy (Section 3-5D). In this approach, lymphocytes are extracted from the blood of an ADA-deficient child and grown in the laboratory. Genetic engineering techniques are used to insert a normal ADA gene into the cells, which are then returned to the child. In one case, ADA-producing cells have persisted in the child for several years.

Other immune system defects result from a deficiency of purine nucleoside phosphorylase. This enzyme deficiency kills the so-called *T* lymphocytes but not the *B* lymphocytes and therefore causes an immunodeficiency syndrome of lesser severity than SCID (the *T* and *B* lymphocytes mediate different aspects of the immune response). This observation suggests that the selective inhibition of PNP may suppress the excess *T* lymphocyte activity associated with such autoimmune diseases as rheumatoid arthritis, psoriasis, and insulin-dependent diabetes and impede the growth of cancers such as *T* cell lymphomas and leukemias.

$$HO-(CH_2-CH_2O)_n-H$$

Polyethylene glycol

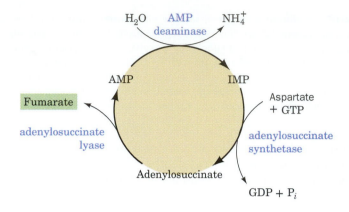

Net: H_2O + Aspartate + GTP $\longrightarrow$ NH_4^+ + GDP + P_i + fumarate

Figure 22-20. **The purine nucleotide cycle.** This pathway functions in muscle to prime the citric acid cycle by generating fumarate.

tate to yield fumarate (Fig. 22-20). John Lowenstein demonstrated that this **purine nucleotide cycle** has an important metabolic role in skeletal muscle. An increase in muscle activity requires an increase in the activity of the citric acid cycle. This process usually occurs through the generation of additional citric acid cycle intermediates (Section 16-5B). Muscles, however, lack most of the enzymes that catalyze these anaplerotic (filling up) reactions in other tissues. Instead, muscle replenishes its citric acid cycle intermediates with fumarate generated in the purine nucleotide cycle.

The importance of the purine nucleotide cycle in muscle metabolism is indicated by the observation that the activities of the three enzymes involved are all severalfold higher in muscle than in other tissues. In fact, individuals with an inherited deficiency in muscle AMP deaminase **(myoadenylate deaminase deficiency)** are easily fatigued and usually suffer from cramps after exercise.

Xanthine Oxidase Is a Mini-Electron-Transport System

Xanthine oxidase converts hypoxanthine (the base of IMP) to xanthine, and xanthine to uric acid (Fig. 22-17, *bottom*). The reaction product is an enol (which has a pK of 5.4; hence the name uric *acid*). The enol tautomerizes to the more stable keto form.

Hypoxanthine   **Xanthine**   **Uric acid (enol tautomer)**   **Uric acid (keto tautomer)**

$pK = 5.4$

Urate

In mammals, xanthine oxidase occurs almost exclusively in the liver and the small intestinal mucosa. It is a dimeric protein of identical 130-kD subunits, each of which contains an entire "zoo" of electron-transfer agents: an FAD, a Mo complex that cycles between its Mo(VI) and Mo(IV) oxidation states, and two different Fe–S clusters. The final electron acceptor is O_2, which is converted to H_2O_2, a potentially harmful oxidizing agent (Section 17-5B) that is subsequently converted to H_2O and O_2 by catalase.

B. Fate of Uric Acid

In humans and other primates, the final product of purine degradation is uric acid, which is excreted in the urine. The same is true of birds, terrestrial reptiles, and many insects, but these organisms, which do not excrete urea, also catabolize their excess amino acid nitrogen to uric acid via purine biosynthesis. This complicated system of nitrogen excretion has a straightforward function: *It conserves water.* Uric acid is only sparingly soluble in water, so its excretion as a paste of uric acid crystals is accompanied by very little water. In contrast, the excretion of an equivalent amount of the much more water-soluble urea osmotically sequesters a significant amount of water.

In all other organisms, uric acid is further processed before excretion (Fig. 22-21). Mammals other than primates oxidize it to their excretory product, **allantoin,** in a reaction catalyzed by the Cu-containing enzyme **urate oxidase.** A further degradation product, **allantoic acid,** is excreted by teleost (bony) fish. Cartilaginous fish and amphibia further degrade allantoic acid to urea prior to excretion. Finally, marine invertebrates decompose urea to NH_4^+.

Gout Is Caused by an Excess of Uric Acid

Gout is a disease characterized by elevated levels of uric acid in body fluids. Its most common manifestation is excruciatingly painful arthritic joint inflammation of sudden onset, most often of the big toe (Fig. 22-22), caused by deposition of nearly insoluble crystals of sodium urate. Sodium urate and/or uric acid may also precipitate in the kidneys and ureters as stones, resulting in renal damage and urinary tract obstruction.

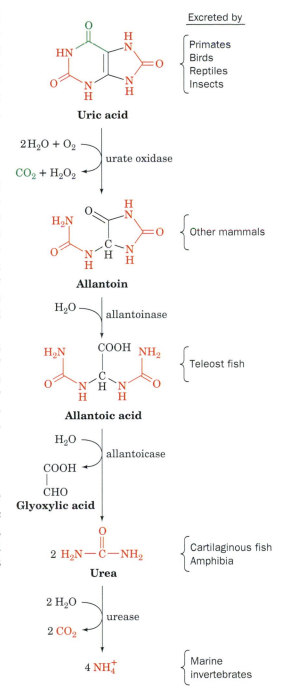

Figure 22-21.   The degradation of uric acid to ammonia. The process is arrested at different stages in the indicated species, and the resulting nitrogen-containing product is excreted.

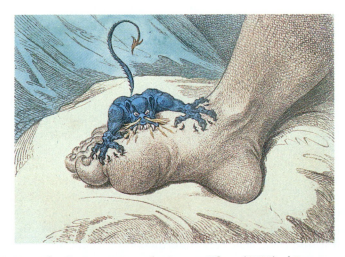

Figure 22-22.   *The Gout,* a cartoon by James Gilroy (1799). [Courtesy of Yale University Medical Historical Library.]

Gout, which affects ~3 per 1000 persons, predominantly males, has been traditionally, although inaccurately, associated with overindulgent eating and drinking. The probable origin of this association is that in previous centuries, when wine was often contaminated with lead during its manufacture and storage, heavy drinking resulted in chronic lead poisoning that, among other things, decreases the kidney's ability to excrete uric acid.

The most prevalent cause of gout is impaired uric acid excretion (although usually for reasons other than lead poisoning). Gout may also result from a number of metabolic insufficiencies, most of which are not well characterized. One well-understood cause is HGPRT deficiency (Lesch–Nyhan syndrome in severe cases), which leads to excessive uric acid production through PRPP accumulation (Section 22-1D).

Gout can be treated by administering the xanthine oxidase inhibitor **allopurinol,** a hypoxanthine analog with interchanged N7 and C8 positions.

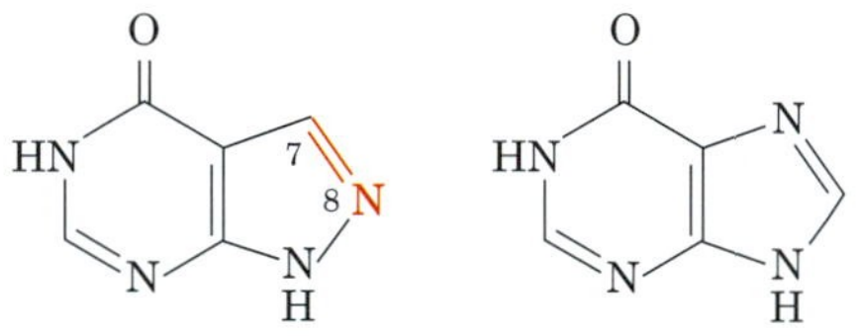

Allopurinol   **Hypoxanthine**

Xanthine oxidase hydroxylates allopurinol, as it does hypoxanthine, yielding **alloxanthine,**

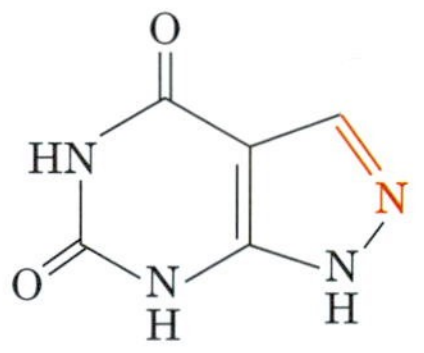

Alloxanthine

which remains tightly bound to the reduced form of the enzyme, thereby inactivating it. Allopurinol consequently alleviates the symptoms of gout by decreasing the rate of uric acid production while increasing the levels of the more soluble hypoxanthine and xanthine. Although allopurinol controls the gouty symptoms of Lesch–Nyhan syndrome, it has no effect on its neurological symptoms.

C. Catabolism of Pyrimidines

Animal cells degrade pyrimidine nucleotides to their component bases (Fig. 22-23, *top*). These reactions, like those of purine nucleotides, occur through dephosphorylation, deamination, and glycosidic bond cleavages. The resulting uracil and thymine are then broken down in the liver through reduction (Fig. 22-23, *middle*) rather than by oxidation as occurs in purine catabolism. The end products of pyrimidine catabolism, **β-alanine** and **β-aminoisobutyrate,** are amino acids and are metabolized as such. They are converted, through transamination and activation reactions, to malonyl-CoA and methylmalonyl-CoA (Fig. 22-23, *bottom left*). Malonyl-CoA is a precursor of fatty acid synthesis (Fig. 19-22), and methylmalonyl-CoA is converted to the citric acid cycle intermediate succinyl-CoA (Fig. 19-13). Thus, *to a limited extent, catabolism of pyrimidine nucleotides contributes to the energy metabolism of the cell.*

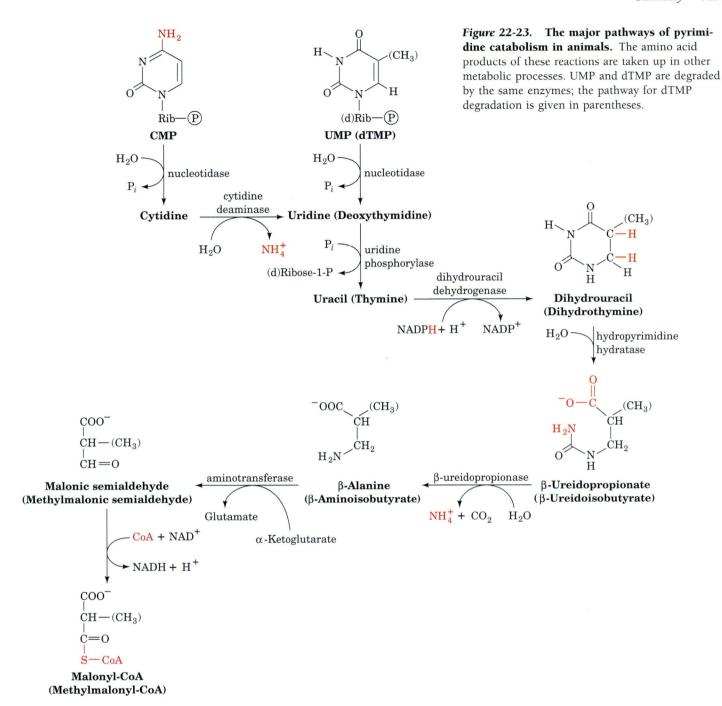

Figure 22-23.   The major pathways of pyrimidine catabolism in animals. The amino acid products of these reactions are taken up in other metabolic processes. UMP and dTMP are degraded by the same enzymes; the pathway for dTMP degradation is given in parentheses.

S U M M A R Y

1. The purine nucleotide IMP is synthesized in 11 steps from ribose-5-phosphate, aspartate, fumarate, glutamine, glycine, and HCO_3^-. Purine nucleotide synthesis is regulated at its first and second steps.

2. IMP is the precursor of AMP and GMP, which are phosphorylated to produce the corresponding di- and triphosphates.

3. The pyrimidine nucleotide UMP is synthesized from 5-phosphoribosyl pyrophosphate, aspartate, glutamine, and HCO_3^-

in six reactions. UMP is converted to UTP and CTP by phosphorylation and amination.

4. Pyrimidine nucleotide synthesis is regulated in bacteria at the ATCase step and in animals at the step catalyzed by carbamoyl phosphate synthetase II.

5. Deoxyribonucleoside diphosphates are synthesized from the corresponding NDP by the action of ribonucleotide reductase, which contains a binuclear Fe(III) prosthetic group, a

tyrosyl radical, and several redox-active sulfhydryl groups. Enzyme activity is regenerated through disulfide interchange with thioredoxin or glutaredoxin.

6. Ribonucleotide reductase is regulated by allosteric effectors, which ensure that deoxynucleotides are synthesized in the amounts required for DNA synthesis.

7. dTMP is synthesized from dUMP by thymidylate synthase. The dihydrofolate reduced in this reaction is converted back to tetrahydrofolate by dihydrofolate reductase (DHFR).

8. Purine nucleotides are degraded by nucleosidases and purine

nucleoside phosphorylase (PNP). Adenine nucleotides are deaminated by adenosine deaminase and AMP deaminase. The synthesis and degradation of AMP in the purine nucleotide cycle yield the citric acid cycle intermediate fumarate in muscles. Xanthine oxidase catalyzes the oxidation of hypoxanthine to xanthine and of xanthine to uric acid.

9. In humans, the ultimate product of purine degradation is uric acid, which is excreted. Other organisms degrade urate further.

10. Pyrimidines are broken down to intermediates of fatty acid metabolism.

REFERENCES

Carreras, C.W. and Santi, D.V., The catalytic mechanism and structure of thymidylate synthase, *Annu. Rev. Biochem.* **64,** 721–762 (1995).

Jordan, A. and Reichard, P., Ribonucleotide reductases, *Annu. Rev. Biochem.* **67,** 71–98 (1998).

Mao, S.S., Holler, T.P., Yu, G.X., Bollinger, J.M., Jr., Booker, S., Johnston, M.I., and Stubbe, J., A model for the role of multiple cysteine residues involved in ribonucleotide reduction: Amazing and still confusing, *Biochemistry* **31,** 9733–9743 (1992).

Nordlund, P. and Eklund, H., Structure and function of the *Escherichia coli* ribonucleotide reductase protein R2, *J. Mol.*

Biol. **232,** 123–164 (1993); *and* Uhlin, U. and Eklund, H., Structure of ribonucleotide reductase protein R1, *Nature* **370,** 533–539 (1994).

Scriver, C.R., Beaudet, A.L., Sly, W.S., Valle, D., Stanbury, J.B., Wyngaarden, J.B., and Frederickson, D.S. (Eds.), *The Metabolic and Molecular Bases of Inherited Disease* (7th ed.), Chapters 49–55, McGraw–Hill (1995). [These chapters describe normal and abnormal pathways of nucleotide metabolism.]

Smith, J.L., Enzymes of nucleotide synthesis, *Curr. Opin. Struct. Biol.* **5,** 752–757 (1995).

Zalkin, H. and Dixon, J.E., *De novo* purine nucleotide biosynthesis, *Prog. Nucleic Acid Res. Mol. Biol.* **42,** 259–285 (1992).

KEY TERMS

PRPP	feedforward activation	purine nucleotide cycle	mechanism-based inhibitor
sulfonamides	salvage pathway		

STUDY EXERCISES

1. Review the nomenclature of bases, nucleosides, and nucleotides.

2. Compare the pathways of purine and pyrimidine nucleotide synthesis with respect to (a) precursors, (b) energy cost, (c) acquisition of the ribose moiety, and (d) number of enzymatic steps.

3. How do PRPP levels influence purine and pyrimidine nucleotide synthesis?

4. How are folate cofactors involved in nucleotide metabolism?

5. Why are antifolates effective drugs?

6. Describe the metabolic defects of Lesch–Nyhan syndrome, orotic aciduria, SCID, and gout.

7. How does the cell balance the production of (a) purine and pyrimidine nucleotides and (b) ribonucleotides and deoxynucleotides?

8. What compounds are produced by the degradation of purines and pyrimidines?

PROBLEMS

1. Calculate the cost, in ATP equivalents, of synthesizing *de novo* (a) IMP, (b) AMP, and (c) CTP. Assume all substrates (e.g., ribose-5-phosphate and glutamine) and cofactors are available.

2. Certain glutamine analogs irreversibly inactivate enzymes that bind glutamine. Identify the nucleotide biosynthetic intermediates that accumulate in the presence of these compounds.

3. Rats are given cytidine that is ^{14}C-labeled at both its base and ribose components. Their DNA is then extracted and degraded with nucleases. Describe the labeling pattern of the recovered deoxycytidylate residues if deoxycytidylate production in the cell followed a pathway in which (a) intact CDP is reduced to dCDP, and (b) CDP is broken down to cytosine and ribose before reduction.

4. Explain why hydroxyurea,

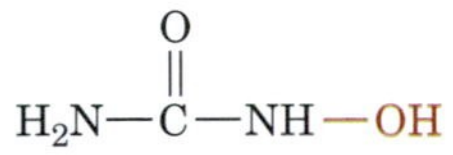

$$H_2N-\overset{\overset{\displaystyle O}{\|}}{C}-NH-OH$$

Hydroxyurea

which destroys tyrosyl radicals, is useful as an antitumor agent.

5. Why is deoxyadenosine toxic to mammalian cells?

6. Why do individuals who are undergoing chemotherapy with FdUMP or methotrexate temporarily go bald?

7. Normal cells die in a nutrient medium containing thymidine and methotrexate, whereas mutant cells defective in thymidylate synthase survive and grow. Explain.

8. Explain why methotrexate inhibits the synthesis of histidine and methionine.

9. Explain whether or not the following are mechanism-based inhibitors: (a) trimethoprim with bacterial dihydrofolate reductase, and (b) allopurinol with xanthine oxidase.

10. Why does von Gierke's glycogen storage disease (Box 15-2) cause symptoms of gout?

V

GENE EXPRESSION AND REPLICATION

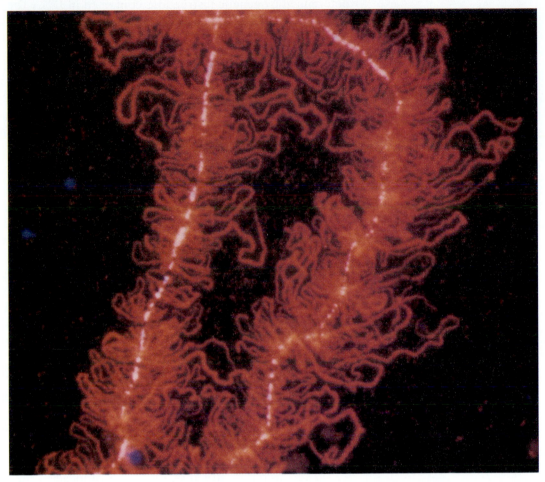

*The information encoded in DNA can be "read" by proteins that bind to the DNA. Even when the DNA molecule is arranged in loops to increase its accessibility to proteins, the proteins must be able to recognize specific nucleotide sequences. The looped DNA segment above is part of a "lampbrush" chromosome from a newt oocyte. [From Roth, M.B., and Gall, J.G., J. Cell Biol. **105**, 1049 (1987), by copyright permission of Rockefeller University Press.]*

NUCLEIC ACID STRUCTURE

In every organism, the ultimate source of biological information is nucleic acid. The shapes and activities of individual cells are, to a large extent, determined by genetic instructions contained in DNA (or RNA, in some viruses). According to the central dogma of molecular biology (Section 3-3B), sequences of nucleotide bases in DNA encode the amino acid sequences of proteins. Many of the cell's proteins are enzymes that carry out the metabolic processes we have discussed in Chapters 14–22. Other proteins have a structural or regulatory role or participate in maintaining and transmitting genetic information.

Two kinds of nucleic acids, DNA and RNA, store information and make it available to the cell. The structures of these molecules must be consistent with the following:

1. Genetic information must be stored in a form that is manageable in size and stable over a long period.

2. Genetic information must be decoded—often many times—in order to be used. **Transcription** is the process by which nucleotide sequences in DNA are copied onto RNA so that they can direct protein synthesis, a process known as **translation.**

3. Information contained in DNA or RNA must be accessible to proteins and other nucleic acids. These agents must recognize nucleic acids (in many cases, in a sequence-specific fashion) and bind to them in a way that alters their function.

4. The progeny of an organism must be equipped with the same set of instructions as in the parent. Thus, DNA is **replicated** (an exact copy made) so that each daughter cell receives the same information.

As we shall see, many cellular components are required to execute all the functions of nucleic acids. Yet nucleic acids are hardly inert "read-only" entities. RNA in particular, owing to its single-stranded nature, is a dynamic molecule that provides structural scaffolding as well as catalytic proficiency in a number of processes that decode genetic information. In this chapter, we shall focus on the structural properties of nucleic acids, including their interactions with proteins, that allow them to carry out their duties. In subsequent chapters, we shall examine the processes of replication (Chapter 24), transcription (Chapter 25), and translation (Chapter 26).

1. THE DNA HELIX

We begin our discussion of nucleic acid structure by examining the various forms of DNA, with an eye toward understanding how this molecule safeguards genetic information while leaving it accessible.

A. *The Geometry of DNA*

DNA is a two-stranded polymer of deoxynucleotides linked by phosphodiester bonds (Figs. 3-6 and 3-11). The biologically most common form of DNA is known as **B-DNA,** which has the structural features first described by James Watson and Francis Crick (Section 3-2B):

1. The two antiparallel polynucleotide strands wind in a right-handed manner around a common axis to produce an ~20-Å-diameter double helix.

2. The planes of the nucleotide bases, which form hydrogen-bonded pairs, are nearly perpendicular to the helix axis. In B-DNA, the bases

○ See Guided Exploration 21:

DNA structures.

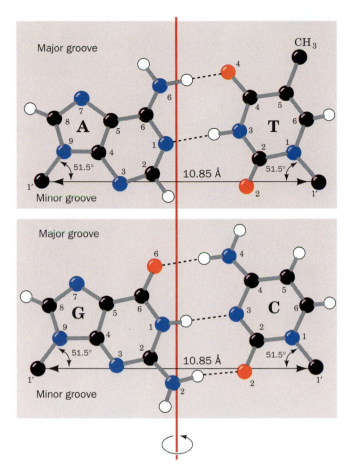

Figure 23-1.   The Watson–Crick base pairs. The line joining the C1′ atoms is the same length in both A · T and G · C base pairs and makes equal angles with the glycosidic bonds to the bases. This gives DNA a series of pseudo-two-fold symmetry axes that pass through the center of each base pair (*red line*) and are perpendicular to the helix axis. [After Arnott, S., Dover, S.D., and Wonacott, A.J., *Acta Cryst.* **B25**, 2196 (1969).] ⬤ **See Kinemage Exercise 17-2.**

occupy the core of the helix while the sugar–phosphate backbones wind around the outside, forming the major and minor grooves. Only the edges of the base pairs are exposed to solvent.

3. Each base pair has approximately the same width (Fig. 23-1), which accounts for the near-perfect symmetry of the DNA molecule, regardless of base composition. A · T and G · C base pairs are interchangeable: *They can replace each other in the double helix without altering the positions of the sugar–phosphate backbones' C1′ atoms.* Likewise, the partners of a Watson–Crick base pair can be switched (i.e., by changing a G · C to a C · G or an A · T to a T · A). In contrast, any other combination of bases would significantly distort the double helix.

4. The "ideal" B-DNA helix has 10 base pairs (bp) per turn (a helical twist of 36° per bp) and, since the aromatic bases have van der Waals thicknesses of 3.4 Å and are partially stacked on each other, the helix has a pitch (rise per turn) of 34 Å.

Double-helical DNA can assume several distinct structures depending on the solvent composition and base sequence. The major structural variants

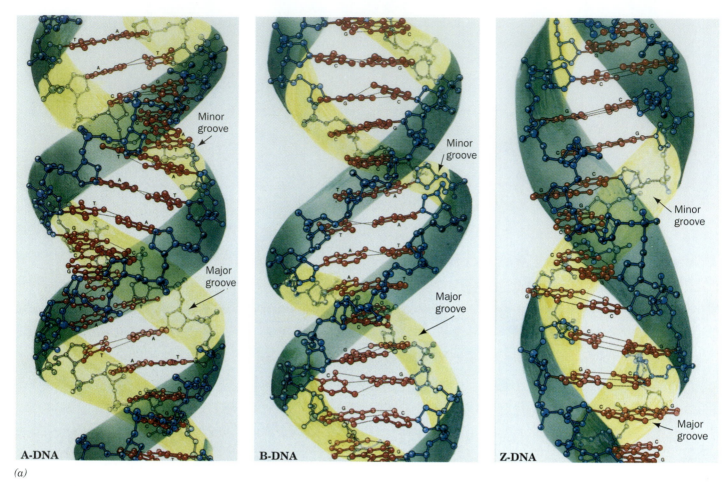

(a)

Figure 23-2.   Structures of A-, B-, and Z-DNA. (*a*) View perpendicular to the helix axis. The sugar–phosphate backbones are outlined by a blue-green ribbon and the bases are red. (*b*) (*Opposite*) Space-filling models with C, N, O, and P atoms colored white, blue, red, and green, respectively. H atoms have been omitted for clarity. (*c*) (*Opposite*) View down the helix axis. The ribose ring O atoms are red and the nearest base pair is white. [Drawings copyrighted © by Irving Geis.] ● **See Kinemage Exercises 17-1, 17-4, 17-5, and 17-6.**

Table 23-1.   **Key to Structure.** **Structural Features of Ideal A-, B-, and Z-DNA**

	A	B	Z
Helical sense	Right handed	Right handed	Left handed
Diameter	~26 Å	~20 Å	~18 Å
Base pairs per helical turn	11	10	12 (6 dimers)
Helical twist per base pair	33°	36°	60° (per dimer)
Helix pitch (rise per turn)	28 Å	34 Å	45 Å
Helix rise per base pair	2.6 Å	3.4 Å	3.7 Å
Base tilt normal to the helix axis	20°	6°	7°
Major groove	Narrow and deep	Wide and deep	Flat
Minor groove	Wide and shallow	Narrow and deep	Narrow and deep
Sugar pucker	C3′-*endo*	C2′-*endo*	C2′-*endo* for pyrimidines; C3′-*endo* for purines
Glycosidic bond	Anti	Anti	Anti for pyrimidines; syn for purines

of DNA are **A-DNA** and **Z-DNA.** The geometries of these molecules are summarized in Table 23-1 and Fig. 23-2.

A-DNA's Base Pairs Are Inclined to the Helix Axis

Under dehydrating conditions, B-DNA undergoes a reversible conformational change to A-DNA, which forms a wider and flatter right-handed

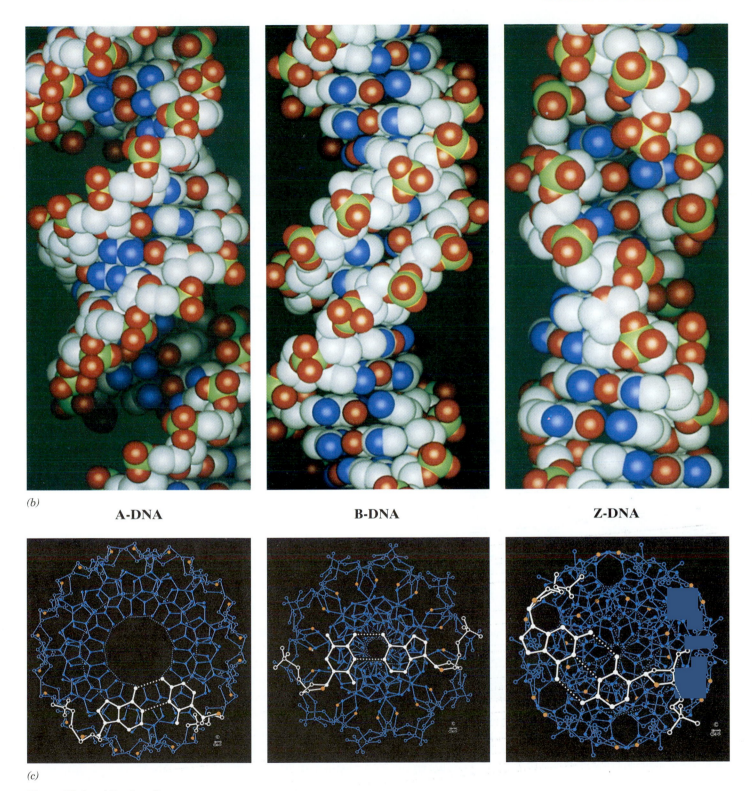

A-DNA   **B-DNA**   **Z-DNA**

(b)

(c)

Figure 23-2.   (*Continued*)

helix than does B-DNA. A-DNA has 11 bp per turn and a pitch of 28 Å, which gives it an axial hole (Fig. 23-2c, *left*). A-DNA's most striking feature, however, is that the planes of its base pairs are tilted 20° with respect to the helix axis. Since the axis does not pass through its base pairs, A-DNA has a deep major groove and a very shallow minor groove; it can be described as a flat ribbon wound around a 6-Å-diameter cylindrical hole.

Z-DNA Forms a Left-Handed Helix

Occasionally, a familiar system exhibits quite unexpected properties. Over 25 years after the discovery of the Watson–Crick DNA structure, the crystal structure determination of d(CGCGCG) by Andrew Wang and Alexander Rich revealed, quite surprisingly, a left-handed double helix. This structure, which was dubbed Z-DNA, has 12 Watson–Crick base pairs per turn, a pitch of 45 Å, a deep minor groove, and no discernible major groove. Z-DNA therefore resembles a left-handed drill bit in appearance (Fig. 23-2, *right*).

Fiber diffraction and NMR studies have shown that complementary polynucleotides with alternating purines and pyrimidines, such as poly d(GC)·poly d(GC) or poly d(AC)·poly d(GT), assume the Z conformation at high salt concentrations. The salt stabilizes Z-DNA relative to B-DNA by reducing the electrostatic repulsions between closest approaching phosphate groups on opposite strands (which are 8 Å apart in Z-DNA and 12 Å apart in B-DNA).

Is Z-DNA biologically significant? Rich has proposed that the reversible conversion of specific segments of B-DNA to Z-DNA under appropriate circumstances acts as a kind of switch in regulating gene expression. Yet the *in vivo* existence of Z-DNA has been difficult to prove. A major problem is demonstrating that a particular probe for detecting Z-DNA, for example, a Z-DNA-specific antibody, does not in itself cause what would otherwise be B-DNA to assume the Z conformation—a kind of biological uncertainty principle (i.e., the act of measurement inevitably disturbs the system being measured).

RNA Can Form an A Helix

Double-stranded RNA is the genetic material of certain viruses, but it is synthesized only as a single strand. Nevertheless, single-stranded RNA can fold back on itself so that complementary sequences base-pair to form double-stranded stems with single-stranded loops (Fig. 3-12). Double-stranded RNA is unable to assume a B-DNA-like conformation because of steric clashes involving its 2′-OH groups. Rather, it usually assumes a conformation resembling A-DNA (Fig. 23-2), with ~10.6 bp per helical turn, a pitch of ~26.5 Å, and base pairs inclined to the helix axis by ~16°.

Hybrid double helices, which consist of one strand each of RNA and DNA, apparently also have an A-DNA-like conformation (Fig. 23-3). Short stretches of RNA–DNA hybrid helices occur during the initiation of DNA replication by small segments of RNA (Section 24-1) and during the transcription of RNA on DNA templates (Section 25-1C).

B. Flexibility of DNA

The structurally distinct A, B, and Z forms of DNA are not thought to freely interconvert *in vivo*. Rather, the transition from one form to another requires unusual physical conditions (e.g., low humidity) or the influence of DNA-binding proteins. In addition, real DNA molecules deviate from the ideal structures described in the preceding section. X-Ray structures of B-DNA segments reveal that *individual residues significantly depart from the average conformation in a sequence-dependent manner*. For example, the helical twist per base pair may range from 28° to 42°. Each base pair can also deviate from its ideal conformation by rolling or twisting like the

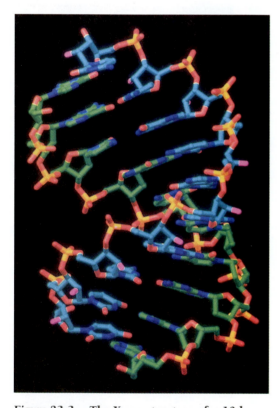

Figure 23-3. **The X-ray structure of a 10-bp RNA–DNA hybrid.** The complex consists of the RNA 5′-UUCGGGCGCC-3′ that is base paired to its DNA complement so as to form a largely A-type double helix, although its DNA strand has B-like conformational deviations. The structure is shown in stick form with RNA C atoms cyan, DNA C atoms green, N atoms blue, O atoms red except for the RNA O2′ atoms which are magenta, and P atoms gold. [Based on an X-ray structure by Barry Finzel, Pharmacia & Upjohn, Inc., Kalamazoo, Michigan.] ● See the Interactive Exercises.

blade of a propeller. Such conformational variations appear to be important for the sequence-specific recognition of DNA by the proteins that process genetic information.

DNA molecules, which are 20 Å thick and many times as long, are not perfectly rigid rods. In fact, it is imperative that these molecules be somewhat flexible so that they can be packaged in cells. DNA helices can adopt different degrees of curvature ranging from gentle arcs to sharp bends. The more severe distortions from linearity, as we shall see, generally occur in response to the binding of specific proteins.

The Conformational Flexibility of DNA Is Limited

The conformation of a nucleotide unit, as Fig. 23-4 indicates, is specified by the six torsion angles of the sugar–phosphate backbone and the torsion angle describing the orientation of the base around the glycosidic bond (the bond joining C1′ to the base). It would seem that these seven degrees of freedom per nucleotide would render polynucleotides highly flexible. Yet these torsion angles are subject to a variety of internal constraints that greatly restrict their rotational freedom.

The rotation of a base around its glycosidic bond (angle χ) is greatly hindered. Purine residues have two sterically permissible orientations known as the **syn** (Greek: with) and **anti** (Greek: against) conformations (Fig. 23-5). Only the anti conformation of pyrimidines is stable, because, in the syn conformation, the sugar residue sterically interferes with the pyrimidine's C2 substituent. *In most double-helical nucleic acids, all bases are in the anti conformation.* The exception is Z-DNA (Section 23-1A), in which the alternating pyrimidine and purine residues are anti and syn, respectively (this is one reason why the repeating unit of Z-DNA is considered to be a dinucleotide).

The flexibility of the ribose ring itself is also limited. The vertex angles of a regular pentagon are 108°, a value quite close to the tetrahedral angle (109.5°), so one might expect the ribofuranose ring to be nearly flat. However, the ring substituents are eclipsed when the ring is planar. To relieve this crowding, which occurs even between hydrogen atoms, the ring puckers; that is, it becomes slightly nonplanar. In the great majority of known nucleoside and nucleotide crystal structures, four of the ring atoms are coplanar to within a few hundredths of an angstrom and the remaining atom is out of this plane by several tenths of an angstrom. The out-of-plane

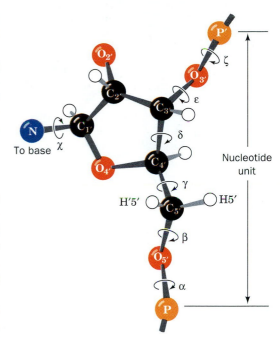

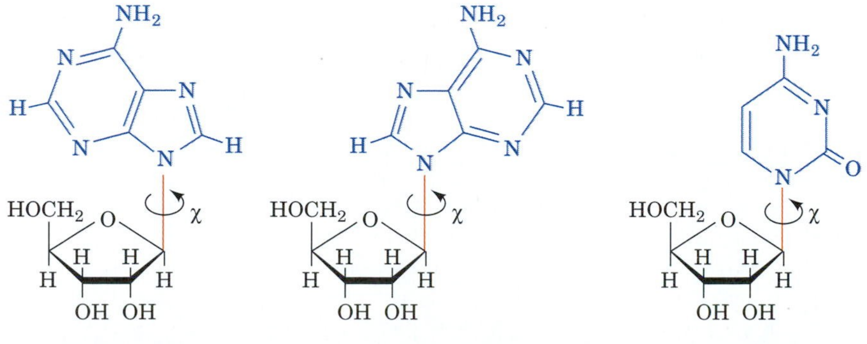

Figure 23-4.   The seven torsion angles that determine the conformation of a nucleotide unit.

syn-Adenosine   **anti-Adenosine**   **anti-Cytidine**

Figure 23-5.   **The sterically allowed orientations of purine and pyrimidine bases with respect to their attached ribose units.** In B-DNA, the nucleotide residues all have the anti conformation.

(a)

(b)

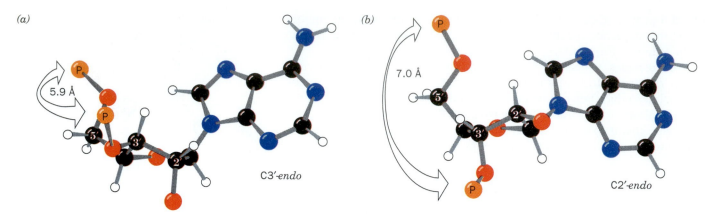

C3'-endo

7.0 Å

C2'-endo

Figure 23-6.   Nucleotide sugar conformations.
(a) The C3'-endo conformation (C3' is displaced to the same side of the ring as C5'), which occurs in A-RNA. (b) The C2'-endo conformation, which occurs in B-DNA. The distances between adjacent P atoms in the sugar–phosphate backbone are indicated. [After Saenger, W., *Principles of Nucleic Acid Structure*, p. 237, Springer-Verlag (1983).]

🔵 See Kinemage Exercise 17-3.

atom is almost always C2' or C3' (Fig. 23-6). The two most common ribose conformations are known as **C3'-endo** and **C2'-endo**; *"endo"* (Greek: *endon,* within) indicates that the displaced atom is on the same side of the ring as C5' (displacement toward the opposite side of C5' is called the **exo** conformation; Greek: *exo, out of*).

The ribose pucker is conformationally important in nucleic acids because it governs the relative orientations of the phosphate substituents to each ribose residue. In fact, B-DNA has the C2'-endo conformation, whereas A-DNA is C3'-endo. In Z-DNA, the purine nucleotides are all C3'-endo and the pyrimidine nucleotides are C2'-endo.

The Sugar–Phosphate Backbone Is Conformationally Constrained

Finally, if the torsion angles of the sugar–phosphate chain (angles α to ζ in Fig. 23-4) were completely free to rotate, there could probably be no stable nucleic acid structure. However, these angles are actually quite restricted. This is because of noncovalent interactions between the ribose ring and the phosphate groups and, in polynucleotides, steric interference between residues. The overall result is that *the sugar–phosphate chains of the double helix are stiff, yet their sugar–phosphate conformational angles are reasonably strain-free.*

🔵 See Guided Exploration 22:

DNA Supercoiling.

C. Supercoiled DNA

The chromosomes of many viruses and bacteria are circular molecules of duplex DNA. In electron micrographs (e.g., Fig. 23-7), some of these molecules have a peculiar twisted appearance, a phenomenon known as

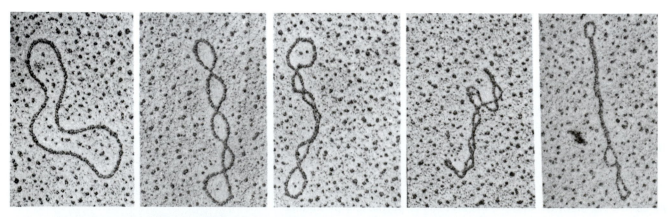

Figure 23-7.   Electron micrographs of circular duplex DNAs.
Their conformations vary from no supercoiling (*left*) to tightly supercoiled (*right*). [Electron micrographs by Laurien Polder. From

Kornberg, A. and Baker, T.A., *DNA Replication* (2nd ed.), p. 36, Freeman (1992). Used by permission.]

supercoiling or **superhelicity.** Supercoiled DNA molecules are more compact than "relaxed" molecules with the same number of nucleotides. This has important consequences for packaging DNA in cells (Section 23-5) and for the unwinding events that occur as part of DNA replication and transcription.

Superhelix Topology Can Be Simply Described

Consider a double-helical DNA molecule in which both strands are covalently joined to form a circular duplex molecule. A geometric property of such an assembly is that *its number of coils cannot be altered without first cleaving at least one of its polynucleotide strands.* You can easily demonstrate this with a buckled belt in which each edge of the belt represents a strand of DNA. The number of times the belt is twisted before it is buckled cannot be changed without unbuckling the belt (cutting a polynucleotide strand).

This phenomenon is mathematically expressed

$$L = T + W \qquad [23\text{-}1]$$

in which:

1. *L,* the **linking number,** is the number of times that one DNA strand winds around the other. This integer quantity is most easily counted when the molecule is made to lie flat on a plane. The linking number cannot be changed by twisting or distorting the molecule, as long as both its polynucleotide strands remain covalently intact.

2. *T,* the **twist,** is the number of complete revolutions that one polynucleotide strand makes around the duplex axis. By convention, *T* is positive for right-handed duplex turns so that, for B-DNA, the twist is normally the number of base pairs divided by 10.4 (the observed number of base pairs per turn of the B-DNA double helix in aqueous solution).

3. *W,* the **writhing number,** is the number of turns that the duplex axis makes around the superhelix axis. *It is a measure of the DNA's superhelicity.* The difference between writhing and twisting is illustrated in Fig. 23-8. When a circular DNA is constrained to lie in a plane, *W* = 0.

The two DNA conformations diagrammed on the right of Fig. 23-9 are topologically equivalent; that is, they have the same linking number, *L,* but differ in their twists and writhing numbers (topology is the study of the

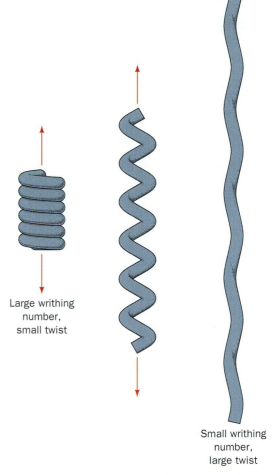

Large writhing
number,
small twist

Small writhing
number,
large twist

Figure 23-8. **The difference between writhing and twist as demonstrated by a coiled telephone cord.** In its relaxed state (*left*), the cord is in a helical form that has a large writhing number and a small twist. As the coil is pulled out (*center*) until it is nearly straight (*right*), its writhing number becomes small and its twist becomes large.

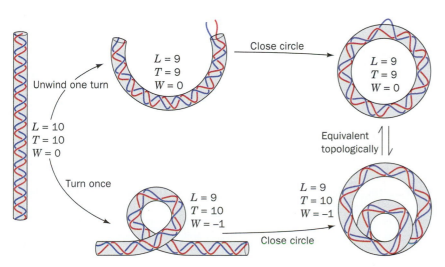

Unwind one turn

$L = 9$
$T = 9$
$W = 0$

Close circle

$L = 9$
$T = 9$
$W = 0$

$L = 10$
$T = 10$
$W = 0$

Turn once

$L = 9$
$T = 10$
$W = -1$

Equivalent
topologically

$L = 9$
$T = 10$
$W = -1$

Close circle

Figure 23-9. **Two ways of introducing one supercoil into a DNA that has 10 duplex turns.** The two closed circular forms shown (*right*) are topologically equivalent; that is, they are interconvertible without breaking any covalent bonds. The linking number, *L,* twist, *T,* and writhing number, *W,* are indicated for each form. Strictly speaking, the linking number is defined only for a covalently closed circle.

geometric properties of objects that are unaltered by deformation but not by cutting).

Since L is a constant in an intact duplex DNA circle, for every new double-helical twist, ΔT, there must be an equal and opposite superhelical twist; that is, $\Delta W = -\Delta T$. For example, a closed circular DNA without supercoils (Fig. 23-9, *upper right*) can be converted to a negatively supercoiled conformation (Fig. 23-9, *lower right*) by winding the duplex helix the same number of positive (right-handed) turns.

Supercoiled DNA Is Relaxed by Nicking One Strand

Supercoiled DNA may be converted to **relaxed circles** (as appears in the leftmost panel of Fig. 23-7) by treatment with **pancreatic DNase I,** an **endonuclease** (an enzyme that cleaves phosphodiester bonds within a polynucleotide strand), which cleaves only one strand of a duplex DNA. *One single-strand nick is sufficient to relax a supercoiled DNA.* This is because the sugar–phosphate chain opposite the nick is free to swivel about its backbone bonds (Fig. 23-4) so as to change the molecule's linking number and thereby alter its superhelicity. Supercoiling builds up elastic strain in a DNA circle, much as it does in a rubber band. This is why the relaxed state of a DNA circle is not supercoiled.

Naturally Occurring DNA Circles Are Underwound

The linking numbers of natural DNA circles are less than those of their corresponding relaxed circles; that is, they are underwound. However, because DNA tends to adopt an overall conformation that maintains its normal twist of 1 turn/10.4 bp, the molecule is negatively supercoiled ($W < 0$; Fig. 23-10, *left*). If the duplex is unwound (if T decreases), then W increases (L must remain constant). At first, this reduces the superhelicity of an underwound circle. However, with continued unwinding, the value of W passes through zero (a relaxed circle; Fig. 23-10, *center*) then becomes positive, yielding a positively coiled superhelix (Fig. 23-10, *right*).

Topoisomerases Control DNA Supercoiling

DNA functions normally only if it is in the proper topological state. In such basic biological processes as replication and transcription, complementary polynucleotide strands separate. The negative supercoiling of naturally occurring DNAs in both prokaryotes and eukaryotes promotes such

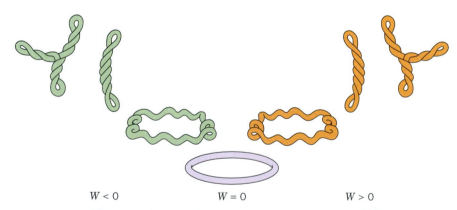

$$W < 0 \qquad\qquad W = 0 \qquad\qquad W > 0$$

Figure 23-10. **Progressive unwinding of a negatively supercoiled DNA molecule.** As the negatively supercoiled circle ($W < 0$) is unwound (without breaking covalent bonds), W approaches 0. Further unwinding forces the DNA to supercoil in the opposite direction, yielding a positively coiled superhelix ($W > 0$).

separations since it tends to unwind the duplex helix (an increase in W must be accompanied by a decrease in T). *If DNA lacks the proper superhelical tension, the above vital processes cannot occur.*

The supercoiling of DNA is controlled by a remarkable group of enzymes known as **topoisomerases.** They are so named because they alter the topological state (linking number) of circular DNA but not its covalent structure. There are two classes of topoisomerases in prokaryotes and eukaryotes:

1. **Type I topoisomerases** act by creating transient single-strand breaks in DNA.
2. **Type II topoisomerases** act by making transient double-strand breaks in DNA.

Type I Topoisomerases Incrementally Relax Supercoiled DNA

Type I topoisomerases are also known as **nicking–closing enzymes.** The prokaryotic enzymes catalyze the relaxation of negative supercoils in DNA by increasing its linking number in increments of one turn. Exposing a negatively supercoiled DNA to nicking–closing enzyme sequentially increases its linking number until the supercoil is entirely relaxed. Eukaryotic Type I topoisomerases can relax both negatively and positively supercoiled DNA.

A clue to the mechanism of action of this enzyme was provided by the observation that it reversibly **catenates** (interlinks) single-stranded circles (Fig. 23-11*a*). Apparently, the enzyme cuts a single strand, passes a single-strand loop through the resulting gap, and then reseals the break (Fig. 23-11*b*), thereby twisting double-helical DNA by one turn.

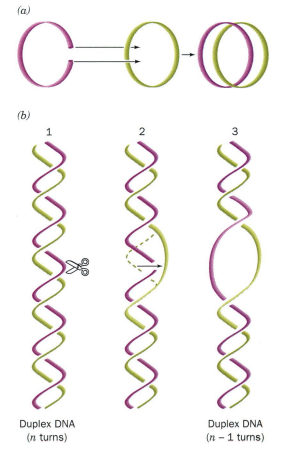

(a)

(b)

1   2   3

Duplex DNA   Duplex DNA
(n turns)   ($n - 1$ turns)

Figure 23-11. Type I topoisomerase action. By cutting a single-stranded DNA, passing a loop of it through the break, and then resealing the break, Type I topoisomerases can (*a*) catenate two single-stranded circles or (*b*) unwind duplex DNA by one turn.

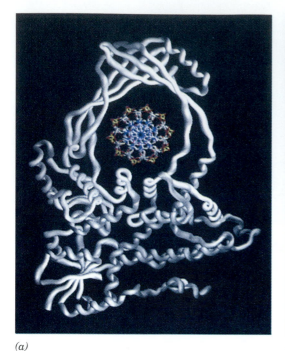

(a)

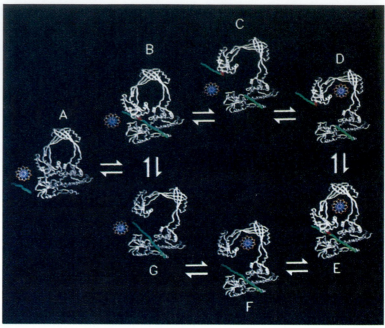

(b)

Figure 23-12. E. coli DNA topoisomerase I. (*a*) The X-ray
structure of the 67-kD N-terminal fragment of topoisomerase I.
The protein surrounds a hole large enough to enclose B-DNA,
modeled here as a double helix viewed down its axis. (*b*) The
proposed mechanism of topoisomerase I. The protein (A) binds a
single-stranded DNA (*green tube*) near the catalytic tyrosine (*red
dot;* B). The single-stranded DNA is then cleaved (C), the protein
opens up to permit a duplex DNA (or a single-stranded DNA) to
pass through the break (D), and the two segments of single-
stranded DNA are rejoined (E). The protein again opens up (F),
which allows the duplex DNA to escape (G). The complex can
then return to state B to initiate another strand passage reaction
or it can decompose to its component macromolecules (A).
Although this model shows duplex DNA passing through a transiently cleaved single strand, the passing segment could just as
well be single-stranded DNA. [Courtesy of Alfonso Mondragón,
Northwestern University.]

The X-ray structure of the 67-kD N-terminal fragment of *E. coli* topoisomerase I shows that the protein encloses a 27.5-Å-diameter hole that
can accommodate a single- or double-stranded DNA (Fig. 23-12*a*). The
hole's inner surface is lined with positively charged residues (18 Arg and
Lys but only 9 Asp and Glu residues), as expected for a DNA-binding surface. This bound DNA is passed through a break in a single strand of DNA
(Fig. 23-12*b*). During catalysis, an active site Tyr residue forms a phosphotyrosine diester linkage with one end of the broken DNA strand,

and the process is reversed when the break is resealed. Some topoisomerases form a 5′-phosphotyrosine intermediate (shown above); others form a 3′-phosphotyrosine intermediate. *Both types of covalent enzyme–DNA adducts conserve the free energy of the cleaved phosphodiester bond so that no free energy input is needed to reseal the nick in the DNA.*

Type II Topoisomerases Hydrolyze ATP

Type II topoisomerases are multimeric enzymes that require ATP hydrolysis to complete a reaction cycle in which two DNA strands are cleaved, duplex DNA is passed through the break, and the break is resealed. Both prokaryotic and eukaryotic Type II enzymes relax negative and positive supercoils, but only the prokaryotic enzyme (also known as **DNA gyrase**) can introduce negative supercoils. The negative supercoils in eukaryotic chromosomes result primarily from its packaging in nucleosomes (Section 23-5B) rather than from topoisomerase action.

Type II topoisomerases superficially resemble the Type I enzymes: They have a pair of catalytic Tyr residues, which form transient covalent intermediates with the 5′ ends of duplex DNA. They thereby mediate the cleavage of the two DNA strands at staggered sites to produce four-base "sticky ends." Stephen Harrison and James Wang have determined the X-ray structure of a large homodimeric fragment of yeast Type II topoisomerase that can cleave duplex DNA but cannot transport it through the break because it lacks the intact protein's ATPase domain. The heart-shaped dimer has a triangular hole that is 55 Å wide at its base and 60 Å high (Fig. 23-13), far larger than the diameter of B-DNA.

A proposed mechanism for Type II topoisomerases, based on this X-ray structure, is diagrammed in Fig. 23-14. The DNA to be cleaved first binds to the enzyme and is clamped in place. In the presence of ATP, the bound DNA is cleaved and a second duplex DNA is passed through the opening into the central hole of the protein. The cleaved DNA is then resealed and the transported DNA exits the complex at a point opposite its point of entry. ATP hydrolysis to ADP + P$_i$ prepares the enzyme for additional catalytic cycles.

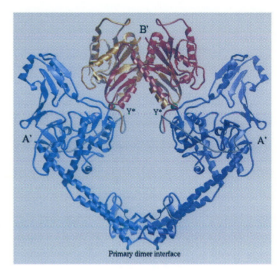

***Figure 23-13.*   The X-ray structure of yeast Type II topoisomerase** (residues 410–1202 of the 1429-residue subunit). The homodimeric protein is shown in ribbon form with its two subfragments A′ light blue and dark blue, and its two subfragments B′ red and gold. The active site Tyr residues are marked by green spheres and labeled Y*. [Courtesy of James Berger, Stephen Harrison, and James Wang, Harvard University.] ● See the Interactive Exercises.

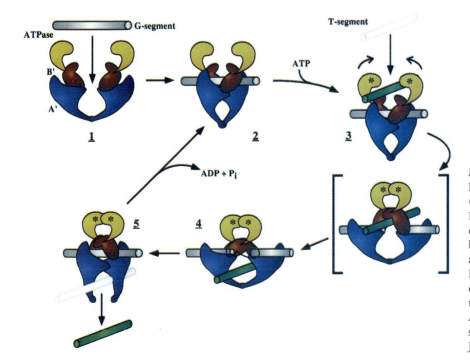

***Figure 23-14.*   Proposed mechanism for Type II topoisomerase.** (1) An unliganded enzyme (*yellow, red, and blue subfragments*) binds duplex DNA (the G-segment; *gray*), and a conformational change (2) clamps the DNA in place. (3) ATP binding (represented by asterisks) promotes cleavage of the G-segment DNA so that another duplex DNA (the T-segment; *green*) can pass into the central hole of the enzyme (4). The G-segment is resealed (5), and the T-segment exits the enzyme. ATP hydrolysis and release regenerate the starting state (2). [Courtesy of Stephen Harrison and James Wang, Harvard University.]

The importance of topoisomerases in maintaining DNA in its proper topological state is indicated by the fact that many antibiotics and chemotherapeutic agents are inhibitors of topoisomerases (see Box 23-1).

Box 23-1

BIOCHEMISTRY IN HEALTH AND DISEASE

Inhibitors of Type II Topoisomerases as Antibiotics and Anticancer Chemotherapeutic Agents

Type II topoisomerases are inhibited by a variety of compounds. For example, **ciprofloxacin** and **novobiocin** specifically inhibit DNA gyrase but not eukaryotic topoisomerase II and are therefore antibiotics. In fact, ciprofloxacin is the most efficacious oral antibiotic presently in clinical use (novobiocin's adverse side effects and the rapid generation of bacterial resistance to its presence have resulted in the discontinuation of its use in the treatment of human infections). A number of substances, including **doxorubicin** and **etoposide**, inhibit eukaryotic Type II topoisomerases and are therefore widely used in cancer chemotherapy.

Different Type II topoisomerase inhibitors act in one of two ways. Many of these agents, including novobiocin, inhibit their target enzyme's ATPase activity. They therefore kill cells by blocking topoisomerase activity, which results in the

arrest of DNA replication and RNA transcription. However, other substances, including ciprofloxacin, doxorubicin, and etoposide, enhance the rate at which their target Type II topoisomerases cleave double-stranded DNA and/or reduce the rate at which these breaks are resealed. Consequently, these substances induce higher than normal levels of transient protein-bridged breaks in the DNA of treated cells. The protein bridges are easily ruptured by the passage of the replication and transcription machinery, thereby rendering the breaks permanent. Although all cells have extensive enzymatic machinery to repair damaged DNA (Section 24-5), a sufficiently high level of DNA damage results in cell death. Consequently, since rapidly replicating cells such as cancer cells have elevated levels of Type II topoisomerases, they are far more likely to incur lethal DNA damage through the inhibition of their Type II topoisomerases than are slow-growing or quiescent cells.

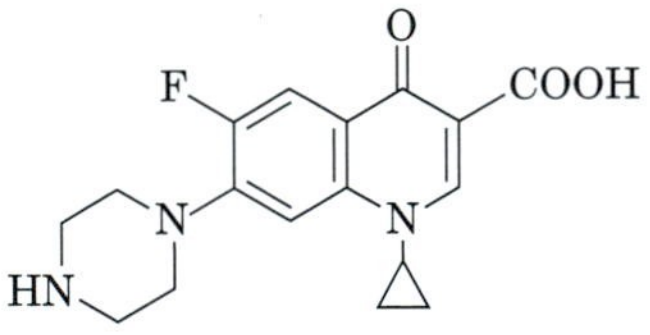

Ciprofloxacin

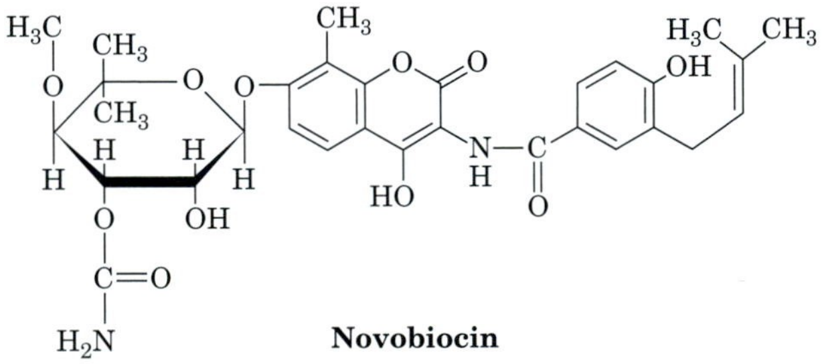

Novobiocin

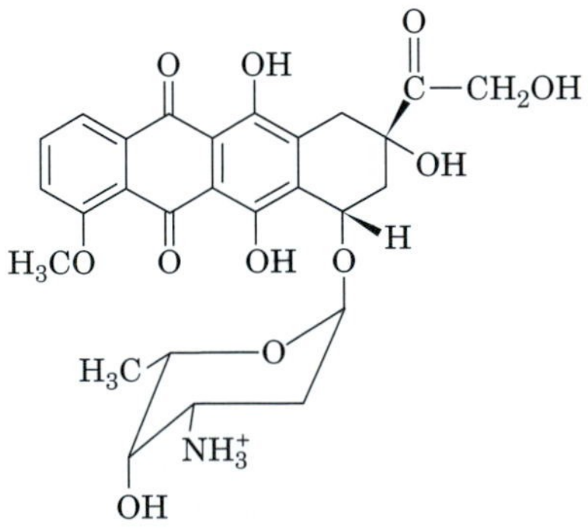

Doxorubicin

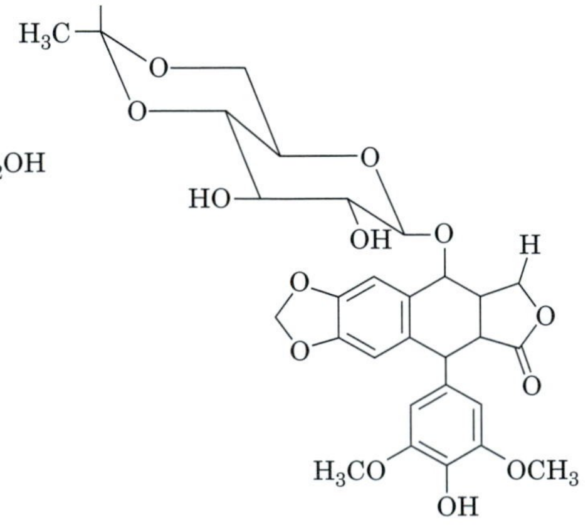

Etoposide

2. FORCES STABILIZING NUCLEIC ACID STRUCTURES

DNA does not exhibit the structural complexity of proteins because it has only a limited repertoire of secondary structures and no comparable tertiary or quaternary structures. This is perhaps to be expected since the 20 amino acid residues of proteins have a far greater range of chemical and physical properties than do the four DNA bases. Nevertheless, many RNAs have well-defined tertiary structures. In this section, we examine the forces that give rise to the structures of nucleic acids.

A. Denaturation and Renaturation

When a solution of duplex DNA is heated above a characteristic temperature, its native structure collapses and its two complementary strands separate and assume a random conformation (Fig. 23-15). This denaturation process is accompanied by a qualitative change in the DNA's physical properties. For example, the characteristic high viscosity of native DNA solutions, which arises from the resistance to deformation of its rigid and rod-like duplex molecules, drastically decreases when the DNA decomposes to the conformationally flexible single chains. Likewise, DNA's ultraviolet absorbance, which is almost entirely due to its aromatic bases, increases by ~40% on denaturation (Fig. 23-16) as a consequence of the disruption of the electronic interactions among neighboring bases.

Monitoring the changes in absorbance at a single wavelength (usually 260 nm) as the temperature increases reveals that the increase in absorbance occurs over a narrow temperature range (Fig. 23-17). This indicates that *the denaturation of DNA is a cooperative phenomenon in which the collapse of one part of the structure destabilizes the remainder.* In anal-

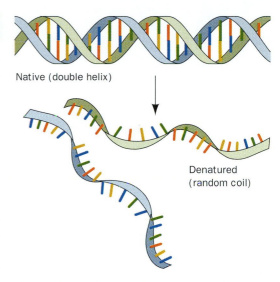

Figure 23-15.   A schematic representation of DNA denaturation.

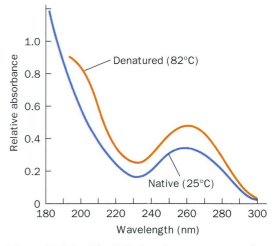

Figure 23-16.   The UV absorbance spectra of native and heat-denatured *E. coli* DNA. Note that denaturation does not change the general shape of the absorbance curve but only increases its intensity. [After Voet, D., Gratzer, W.B., Cox, R.A., and Doty, P., *Biopolymers* **1**, 205 (1963).] ✳ See the Animated Figures.

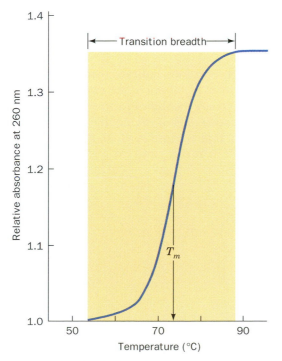

Figure 23-17.   An example of a DNA melting curve. The relative absorbance is the ratio of the absorbance (customarily measured at 260 nm) at the indicated temperature to that at 25°C. The melting temperature, T_m, is the temperature at which half of the maximum absorbance increase is attained. ✳ See the Animated Figures.

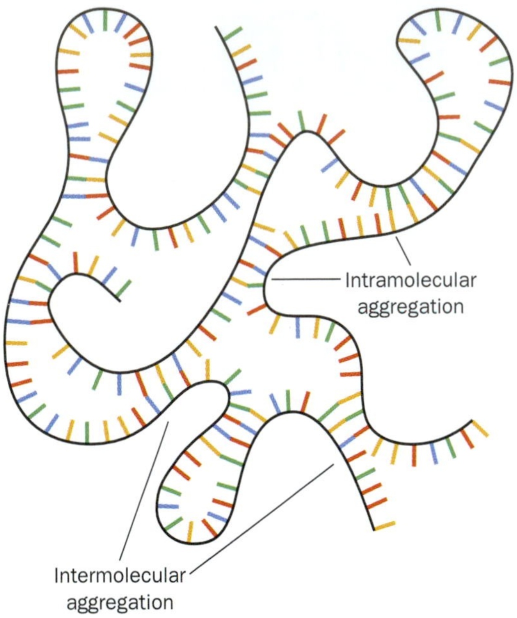

Figure 23-18.   **Partially renatured DNA.** This schematic representation shows the imperfectly base-paired structures assumed by DNA that has been heat denatured and then rapidly cooled. Note that both intramolecular and intermolecular aggregation can occur.

ogy with the melting of a solid, Fig. 23-17 is referred to as a **melting curve,** and the temperature at its midpoint is defined as the **melting temperature,** T_m.

The stability of the DNA double helix, and hence its T_m, depends on several factors, including the nature of the solvent, the identities and concentrations of the ions in solution, and the pH. T_m also increases linearly with the mole fraction of $G \cdot C$ base pairs, although this is not, as one might expect, entirely because $G \cdot C$ base pairs contain one more hydrogen bond than $A \cdot T$ base pairs (see below).

Denatured DNA Can Be Renatured

If a solution of denatured DNA is rapidly cooled below its T_m, the resulting DNA will be only partially base paired (Fig. 23-18) because the complementary strands will not have had sufficient time to find each other before the randomly base-paired structure becomes effectively "frozen in." If, however, the temperature is maintained ~25°C below the T_m, enough thermal energy is available for short base-paired regions to rearrange by melting and re-forming. Under such **annealing** conditions, as Julius Marmur discovered in 1960, denatured DNA eventually completely renatures. Likewise, complementary strands of RNA and DNA, in a process known as **hybridization,** form RNA–DNA hybrid double helices that are only slightly less stable than the corresponding DNA double helices.

B. Base Pairing

Base pairing is apparently a "glue" that holds together double-stranded nucleic acids. Only Watson–Crick pairs occur in the crystal structures of self-complementary oligonucleotides. However, other hydrogen-bonded base pairs with reasonable geometries are known (Fig. 23-19). For example, when monomeric adenine and thymine derivatives are cocrystallized, the $A \cdot T$ base pairs that form invariably have adenine N7 as the hydrogen bond acceptor (**Hoogsteen** geometry; Fig. 23-19*b*) rather than N1 (Watson–Crick geometry; Fig. 23-1).

These and other observations indicate that *Watson–Crick geometry is the most stable mode of base pairing in the double helix, even though non-Watson–Crick base pairs are theoretically possible.* Initially, it was believed that the geometrical constraints of the double helix precluded other types of base pairing. Recall that because $A \cdot T$, $T \cdot A$, $G \cdot C$, and $C \cdot G$ base pairs

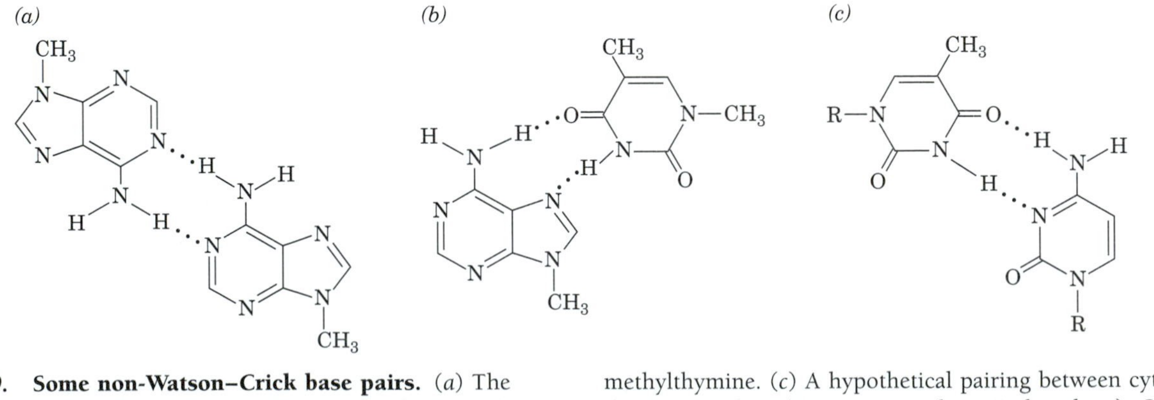

Figure 23-19.   **Some non-Watson–Crick base pairs.** (*a*) The pairing of adenine residues in the crystal structure of 9-methyladenine. (*b*) Hoogsteen pairing between adenine and thymine residues in the crystal structure of 9-methyladenine · 1-methylthymine. (*c*) A hypothetical pairing between cytosine and thymine residues (R represents ribose-5-phosphate). Compare these base pairs to those shown in Fig. 23-1.

are geometrically similar, they can be interchanged without altering the conformation of the sugar–phosphate chains. However, experimental measurements show that a major reason that other base pairs do not appear in the double helix is that *Watson–Crick base pairs have an intrinsic stability that non-Watson–Crick base pairs lack; that is, the bases in a Watson–Crick pair have a higher mutual affinity than those in a non-Watson–Crick pair.* Nevertheless, as we shall see, the double-helical segments of many RNAs contain unusual base pairs, such as $G \cdot U$, that help stabilize their tertiary structures.

Hydrogen Bonds Contribute Little to the Stabilities of Nucleic Acid Structures

It is clear that hydrogen bonding is required for the specificity of base pairing in DNA. Yet, as is also true for proteins (Section 6-4A), *hydrogen bonding contributes little to the stability of nucleic acid structures.* This is because, on denaturation, the hydrogen bonds between the base pairs of a native nucleic acid are replaced by energetically equivalent hydrogen bonds between the bases and water. Other types of forces must therefore play an important role in stabilizing nucleic acid structures.

C. Base Stacking and Hydrophobic Interactions

Purines and pyrimidines tend to form extended stacks of planar parallel molecules. This has been observed in the structures of nucleic acids (Fig. 23-2) and in the structures of crystallized nucleic acid bases (e.g., Fig. 23-20). These **stacking interactions** are a form of van der Waals interaction (Section 2-1A). Interactions between stacked G and C bases are greater than those between stacked A and T bases (Table 23-2), which largely accounts for the greater thermal stability of DNAs with a high G + C content. Note also that differing sets of base pairs in a stack have different stacking energies. Thus, the stacking energy of a double helix is sequence-dependent.

Single-stranded polynucleotides such as poly(A) also exhibit stacking interactions, as shown by the increase in UV absorbance with temperature (i.e., the stacked bases "melt" apart at high temperatures, which causes

Table 23-2.   Stacking Energies for the Ten Possible Dimers in B-DNA

Stacked dimer	Stacking energy ($kJ \cdot mol^{-1}$)
$C \cdot G$ $G \cdot C$	-61.0
$C \cdot G$ $A \cdot T$	-44.0
$C \cdot G$ $T \cdot A$	-41.0
$G \cdot C$ $C \cdot G$	-40.5
$G \cdot C$ $G \cdot C$	-34.6
$G \cdot C$ $A \cdot T$	-28.4
$T \cdot A$ $A \cdot T$	-27.5
$G \cdot C$ $T \cdot A$	-27.5
$A \cdot T$ $A \cdot T$	-22.5
$A \cdot T$ $T \cdot A$	-16.0

Source: Ornstein, R.L., Rein, R., Breen, D.L., and MacElroy, R.D., *Biopolymers* **17**, 2356 (1978).

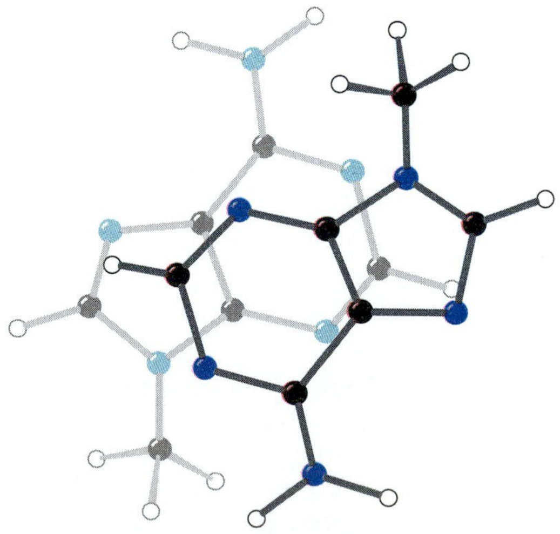

Figure 23-20.   The stacking of adenine rings in the crystal structure of 9-methyladenine. The partial overlap of the rings is typical of the association between bases in crystal structures and in double-helical nucleic acids. [After Stewart, R.F. and Jensen, L.H., *J. Chem. Phys.* **40**, 2071 (1964).]

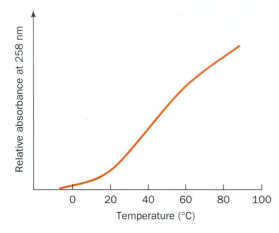

Figure 23-21.   **Melting curve for poly(A).** The broad shape of the curve indicates noncooperative conformational changes. Compare this figure with Fig. 23-17. [After Leng, M. and Felsenfeld, G., *J. Mol. Biol.* **15**, 457 (1966).]

their absorbance to increase; Fig. 23-21). This process is not highly cooperative, as is indicated by the broad melting curve.

Stacking Interactions in Aqueous Solution Result from Hydrophobic Forces

One might assume that hydrophobic interactions in nucleic acids are qualitatively similar to those that stabilize protein structures. This is not the case. Recall that folded proteins are stabilized primarily by the increase in entropy of the solvent water molecules (the hydrophobic effect; Section 6-4A); in other words, protein folding is enthalpically opposed and entropically driven. In contrast, thermodynamic measurements reveal that the base stacking in nucleic acids is enthalpically driven and entropically opposed, although the theoretical basis for this observation is not understood. This difference between hydrophobic forces in proteins and in nucleic acids may reflect the fact that nucleic acid bases are much more polar than most protein side chains (e.g., adenine versus the side chain of Phe or Leu). Whatever their origin, hydrophobic forces are of central importance in determining nucleic acid structures, as is clear from the denaturing effect of adding nonpolar solvents to aqueous solutions of DNA.

D. Ionic Interactions

Any theory of the stability of nucleic acid structures must take into account the electrostatic interactions of their charged phosphate groups. For example, the melting temperature of duplex DNA increases with the Na^+ concentration because these ions electrostatically shield the anionic phosphate groups from each other. Other monovalent cations such as Li^+ and K^+ have similar nonspecific interactions with phosphate groups. Divalent cations, such as Mg^{2+}, Mn^{2+}, and Co^{2+}, in contrast, specifically bind to phosphate groups, so *they are far more effective shielding agents for nucleic acids than are monovalent cations.* For example, an Mg^{2+} ion has an influence on the DNA double helix comparable to that of 100 to 1000 Na^+ ions. Indeed, enzymes that mediate reactions with nucleic acids or nucleotides usually require Mg^{2+} for activity. Mg^{2+} ions also play an essential role in stabilizing the complex structures assumed by many RNAs.

E. RNA Structure

The structure of RNA is stabilized by the same forces that stabilize DNA, and its conformational flexibility is limited by many of the same features that limit DNA conformation. In fact, RNA may be even more rigid than DNA owing to the presence of a greater number of water molecules that form hydrogen bonds to the 2'-OH groups of RNA. Even so, RNA comes in a greater variety of shapes and sizes than DNA and, in some cases, has catalytic activity. These attributes have led to the hypothesis that RNA carried out many of the basic activities of early life, before DNA or proteins had evolved (see Box 23-2).

Ribosomal RNAs Contain Double-Stranded Segments

Bacterial ribosomes, which are two-thirds RNA and one-third protein, contain three highly conserved RNA molecules (Section 26-3). The smallest of these is the 120-residue **5S RNA** (so named because it sediments in the ultracentrifuge at the rate of 5 Svedbergs; Section 5-2E). In Figure 23-22a, the *E. coli* 5S RNA nucleotide sequence is arranged to show the polynucleotide's predicted secondary structure, that is, the antiparallel segments that are highly base paired. The X-ray structure of its 61-nt fragment I (Fig. 23-22b), which is obtained by the mild nuclease digestion of 5S RNA, was determined by Peter Moore and Thomas Steitz. Fragment I assumes an ~94-Å-long A-DNA-like helix, although it has severe distortions from this geometry in the 10-bp segment which encompasses much of its so-called loop E (which, as the structure reveals, is not a loop), and which contains 9 non-Watson–Crick base pairs (red bases in Fig. 23-22; the two A·U base pairs here have Hoogsteen geometry).

(a)

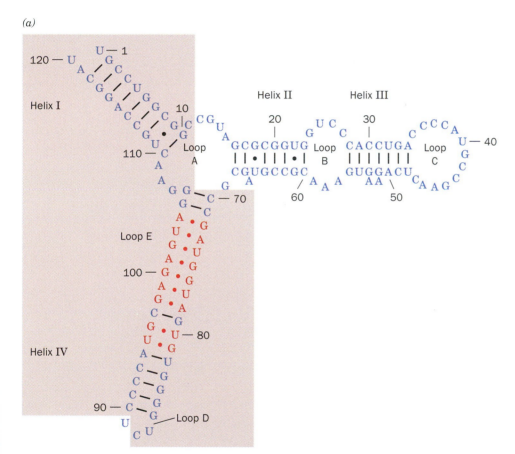

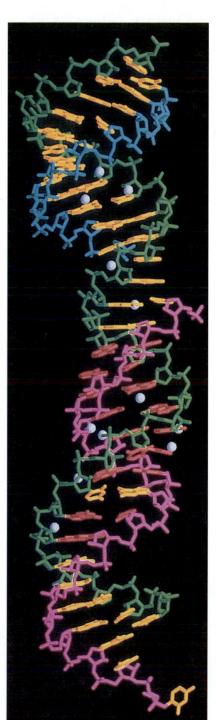

Figure 23-22. *E. coli* 5S RNA. (a) Sequence and predicted secondary structure. Dashes between bases indicate Watson–Crick base pairing and dots indicate other modes of base pairing. [After Noller, H.F., *Annu. Rev. Biochem.* **53**, 134 (1984).] (b) The X-ray structure of its fragment I (the shaded area of a). The sugar–phosphate chains of residues 1 to 11, 70 to 88, and 90 to 120 are cyan, magenta, and green, respectively. The bases are yellow except for the 9 non-Watson–Crick base pairs in the loop E region, which are red in both a and b. Bound Mg^{2+} ions, which are liganded by phosphate oxygens and base groups, are represented by light blue spheres. [Based on an X-ray structure by Peter Moore and Thomas Steitz, Yale University.] ● See the Interactive Exercises.

(b)

Box 23-2
BIOCHEMISTRY IN CONTEXT

Why Did DNA Evolve?

RNA molecules can, in principle, fulfill all the cellular functions performed by proteins and DNA. For example, RNA provides structural scaffolding for ribosomes and other subcellular particles, and RNA enzymes (ribozymes) catalyze a variety of chemical reactions *in vivo* and *in vitro*. In addition, RNA can serve as a template for the synthesis of a complementary polynucleotide strand, an essential criterion for the transmission of genetic information. Indeed, numerous viruses carry either single-stranded or double-stranded RNA as their genetic material.

Even assuming a preeminent position of RNA in primordial life forms, the ascendancy of proteins as the catalytic powerhouses can readily be explained by the greater chemical versatility of the 20 amino acid side chains compared to that of the four chemically quite similar nucleic acid bases. However, a different explanation is required to rationalize why DNA co-opted the function of safeguarding and transmitting genetic information. Simply put, DNA is more suitable for this function because it is more stable than RNA. RNA is highly susceptible to base-catalyzed hydrolysis by the reaction mechanism shown at right.

The base-induced deprotonation of the 2′-OH group facilitates its nucleophilic attack on the adjacent phosphorus atom, thereby cleaving the RNA backbone. The resulting 2′,3′-cyclic phosphate group subsequently hydrolyzes to produce a 2′- or 3′-nucleotide product (the RNase A–catalyzed hydrolysis of RNA follows a nearly identical reaction sequence but generates only 3′-nucleotides; Section 11-3A). DNA is not susceptible to degradation under alkaline conditions because it lacks the 2′-OH group required for the nucleophilic displacement involved in phosphodiester bond cleavage. This greater chemical stability of DNA makes it more suitable as a long-term repository of genetic information.

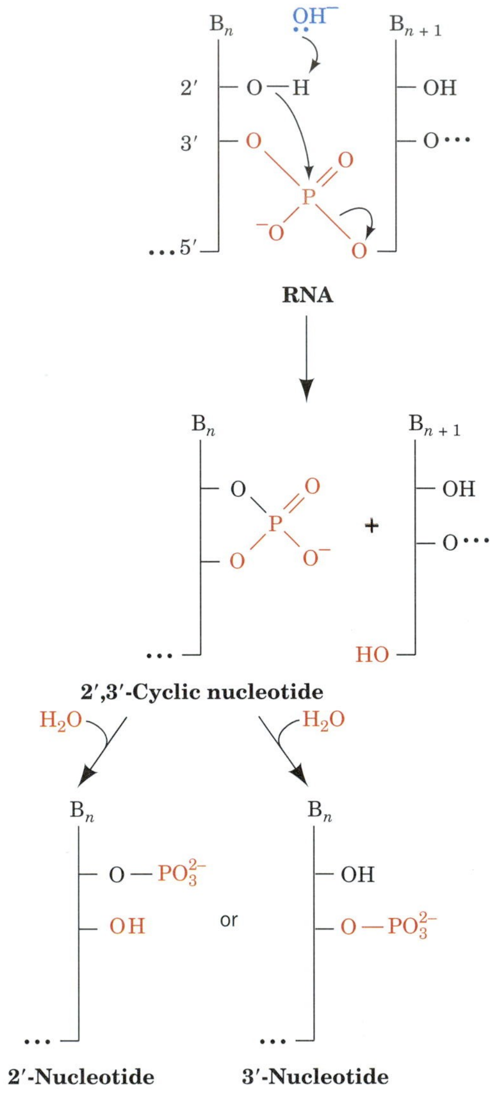

The tertiary structures of the other *E. coli* ribosomal RNA molecules have been studied by X-ray crystallography. Even in the absence of diffraction data, the secondary structures of these RNAs can be analyzed by computer. For example, Harry Noller examined the sequence of the 1542-nucleotide *E. coli* **16S RNA,** which yields many plausible but exclusive secondary structures. However, a comparison of the sequences of 16S RNAs from several prokaryotes, under the assumption that their structures have been evolutionarily conserved, led to the proposed flowerlike secondary structure in Fig. 23-23. This four-domain structure, which is 46% base paired, includes many stems that are short (<8 bp) and imperfectly base paired. The nucleotides that occupy the loops are presumably free to interact with ribosomal proteins or with other unpaired nucleotides in the fully folded conformation of the 16S RNA.

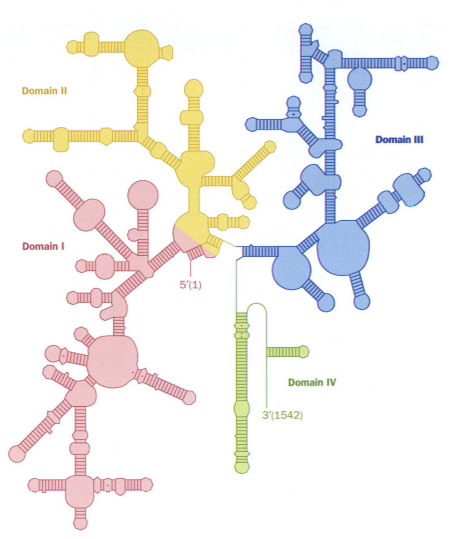

Figure 23-23.   The proposed secondary structure of the 1542-nucleotide *E. coli* 16S rRNA. The structure is based on the comparisons of sequences from different species under the assumption that secondary structure is evolutionarily conserved. The flowerlike series of stems and loops forms four domains. [After Gutell, R.R., Weiser, B., Woese, C.R., and Noller, H.F., *Prog. Nucleic Acid Res. Mol. Biol.* **32**, 183 (1985).]

Large RNA molecules presumably fold in stages, as do multidomain proteins (Section 6-4C). RNA folding is almost certainly a cooperative process, with the rapid formation of short duplex regions preceding the collapse of the structure into its mature conformation. The complete folding process may take several minutes, it may involve relatively stable intermediates, and it may require the assistance of proteins.

Transfer RNA Molecules Are Stabilized by Stacking Interactions

The three-dimensional structure of the yeast transfer RNA that forms a covalent complex with Phe **(tRNA^Phe)** was elucidated in 1974 by Alexander Rich in collaboration with Sung-Hou Kim and, in a different crystal form, by Aaron Klug. The molecule is compact and L-shaped, with each

Figure 23-24. **The X-ray structure of yeast tRNA^Phe.** The 76-nucleotide RNA is shown in stick form with C green, N blue, O red, and P gold. Note that both arms of the L-shaped molecule consist of double-helical stems. [After an X-ray structure by Alexander Rich and Sung-Hou Kim, MIT.]

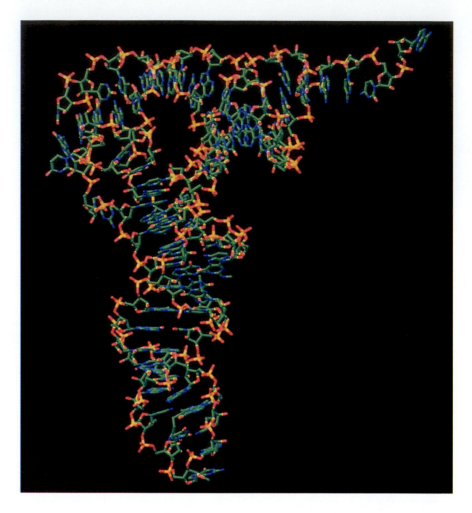

leg of the L being ~60 Å long (Fig. 23-24). The structural complexity of yeast tRNA^Phe is reminiscent of that of a protein. Although only 42 of its 76 bases occur in double-helical stems, 71 of them participate in stacking associations.

tRNA structures are characterized by the presence of covalently modified bases and unusual base pairs, including hydrogen-bonding associations involving three bases. These tertiary interactions contribute to the compact structure of the tRNAs (tRNA structure is discussed in greater detail in Section 26-2A).

The Hammerhead Ribozyme Has Catalytic Activity

Another complex RNA molecule with a known tertiary structure is the **hammerhead ribozyme** that occurs in the RNA of certain plant viruses. Ribozymes, as this name implies, are RNAs that exhibit enzymelike catalytic activity. The hammerhead ribozyme cleaves a specific RNA, in much the same way as does RNase A (Section 11-3A), during the posttranscriptional processing of RNA (Section 25-3). Hammerhead ribozymes have three duplex stems and a conserved core of two nonhelical segments (Fig. 23-25a).

The X-ray structure of a hammerhead ribozyme (Fig. 23-25b), determined by Aaron Klug, reveals that it indeed forms three A-type helical segments, although its overall shape more closely resembles a wishbone than a hammerhead. The nucleotides in the helical stems form normal Watson–Crick base pairs.

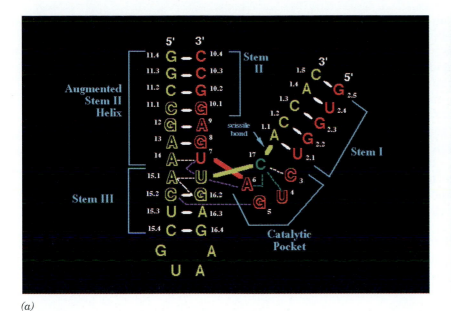

(a)

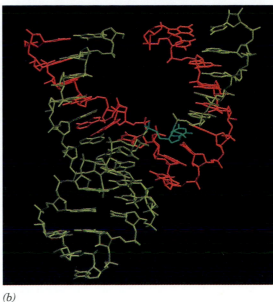

(b)

Figure 23-25.   The X-ray structure of a hammerhead ribozyme. (*a*) The sequence and schematic structural representation of the ribozyme drawn with its 16-nucleotide enzyme strand in red, its 25-nucleotide substrate strand in yellow, and its cleavage site base (C 17) in green. Essential nucleotides are indicated by shadowed letters, and the universal numbering scheme is provided. Watson–Crick as well as two G · A Hoogsteen base pairing interactions are denoted by white ovals, single hydrogen bonds between non-Watson–Crick bases are shown as white dashed lines, single hydrogen bonds between bases and backbone riboses are indicated by pink dashed lines, and aromatic interactions between C 17 and the catalytic pocket are represented by dashed green lines. (*b*) A stick model of the ribozyme colored as in *a*. [Courtesy of Aaron Klug, Laboratory of Molecular Biology, Cambridge, England.] ● See the Interactive Exercises.

As we shall see, ribozymes mediate a number of essential processes in the expression of genetic information *in vivo*. The catalytic versatility of synthetic RNAs is even greater. Through oligonucleotide synthesis or molecular cloning techniques (Section 3-5), a large library of RNA molecules can be prepared and screened for activity such as ligand binding and catalysis. NMR studies of RNAs that bind other molecules with high affinity ($K \approx 10^{-5}$ M) reveal that these RNAs, which are known as **aptamers,** assume compact conformations so that various functional groups are brought together to form a binding pocket. Studies of catalytic RNA molecules suggest that there is no limit to the types of reactions that RNA can catalyze, including phosphoryl-group transfer, isomerization of C—C bonds, and hydrolytic reactions. These results imply that the structures of ribozymes can support specific binding of substrates, proximity and orientation effects, and transition state stabilization—all hallmarks of protein enzymes (Section 11-3).

3. FRACTIONATION OF NUCLEIC ACIDS

In Section 5-2 we considered the most common procedures for isolating and characterizing proteins. Most of these methods, often with some modifications, are also used to fractionate nucleic acids according to size, composition, and sequence. There are also many techniques that apply only to nucleic acids. In this section, we outline some of the most useful nucleic acid fractionation procedures.

Nucleic acids in cells are invariably associated with proteins. Once cells have been broken open, their nucleic acids are usually deproteinized. This can be accomplished by shaking the protein–nucleic acid mixture with a phenol solution so that the protein precipitates and can be removed by centrifugation. Alternatively, the protein can be dissociated from the nucleic acids by detergents, guanidinium chloride, or high salt concentration, or it can be enzymatically degraded by proteases. In all cases, the nucleic acids, a mixture of RNA and DNA, can then be isolated by precipitation with ethanol. The RNA can be recovered from such precipitates by treating them with pancreatic DNase to eliminate the DNA. Conversely, the DNA can be freed of RNA by treatment with RNase.

In all these and subsequent manipulations, the nucleic acids must be protected from degradation by nucleases that occur both in the experimental material and on human hands. Nucleases can be inhibited by chelating agents such as EDTA, which sequester the divalent metal ions they require for activity. Laboratory glassware can also be autoclaved to heat denature the nucleases. Nevertheless, nucleic acids are generally easier to handle than proteins because most lack a complex tertiary structure and are therefore relatively tolerant of extreme conditions.

A. Chromatography

Many of the chromatographic techniques that are used to separate proteins (Section 5-2C) also apply to nucleic acids. However, **hydroxyapatite,** a form of calcium phosphate [$Ca_5(PO_4)_3OH$], is particularly useful in the chromatographic purification and fractionation of DNA. Double-stranded DNA binds to hydroxyapatite more tightly than do most other molecules. Consequently, DNA can be rapidly isolated by passing a cell lysate through a hydroxyapatite column, washing the column with a phosphate buffer of concentration low enough to release only the RNA and protein, and then eluting the DNA with a concentrated phosphate solution.

Affinity chromatography is used to isolate specific nucleic acids. For example, most eukaryotic messenger RNAs (mRNAs) have a poly(A) sequence at their 3′ ends (Section 25-3A). They can be isolated on agarose or cellulose to which poly(U) is covalently attached. The poly(A) sequences specifically bind to the complementary poly(U) in high salt and at low temperature and can later be released by altering these conditions.

B. Electrophoresis

Nucleic acids of a given type can be separated by gel electrophoresis (Section 3-4B) because their electrophoretic mobilities in such gels vary inversely with their molecular masses. DNAs of more than a few thousand base pairs cannot penetrate even a weakly cross-linked polyacrylamide gel and so must be separated in agarose gels. However, conventional gel electrophoresis is still limited to DNAs of <100,000 bp, because larger DNA molecules tend to worm their way through the agarose at a rate independent of their size. Charles Cantor and Cassandra Smith overcame this limitation by developing **pulsed-field gel electrophoresis (PFGE),** which can resolve DNAs of up to 10 million bp (6.6 million kD). In the simplest type of PFGE apparatus, the polarity of the electrodes is periodically reversed, with the duration of each pulse varying from 0.1 to 1000 s, depending on the sizes of the DNAs being separated. With each change in polarity, the migrating DNA must reorient to the new electrical field before it can resume movement. A DNA molecule may actually migrate backward for part of the time. Because small molecules reorient more quickly than large molecules, different-sized DNA molecules gradually separate (Fig. 23-26). In other variations of PFGE, electrodes are arranged in a circular pattern around the gel, or the gel rotates, so that the electrical field changes by less than 180° with each pulse. Forcing the DNA to reorient in this manner improves the resolving power of PFGE, but the DNA tends to move in a curved or zig-zag path, which may make interpretation of the results more difficult.

Intercalation Agents Stain Duplex DNA

The various DNA bands in a gel can be stained by planar aromatic cations such as **ethidium ion, proflavin,** and **acridine orange.**

Ethidium

Proflavin

Acridine orange

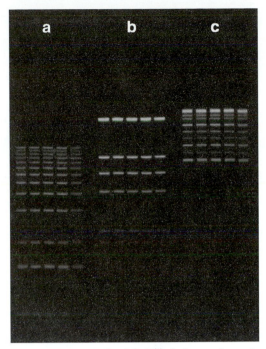

Figure 23-26. Pulsed-field gel electrophoresis. The three samples contain DNA fragments ranging from 0.5 to 48.5 kb. The smallest (fastest migrating) molecules are at the bottom. [Courtesy of Hoefer Scientific Instruments.]

These dyes bind to double-stranded DNA by **intercalation** (slipping in between the stacked bases; Fig. 23-27), where they exhibit a fluorescence under UV light that is far more intense than that of the free dye. As little as 50 ng of DNA can be detected in a gel by staining it with ethidium bromide. Single-stranded DNA and RNA also stimulate the fluorescence of ethidium but to a lesser extent than does duplex DNA.

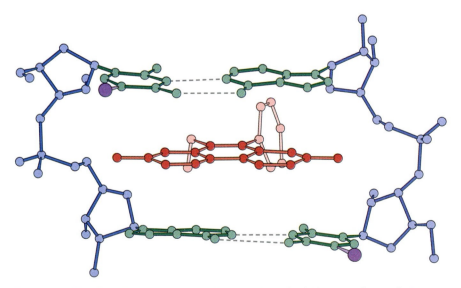

Figure 23-27. The X-ray structure of a complex of ethidium with 5-iodo-UpA. Ethidium (*red*) intercalates between the base pairs of the double-helically paired dinucleotide and thereby provides a model of the binding of ethidium to duplex DNA. [After Tsai, C.-C., Jain, S.C., and Sobell, H.M., *Proc. Natl. Acad. Sci.* **72**, 629 (1975).]

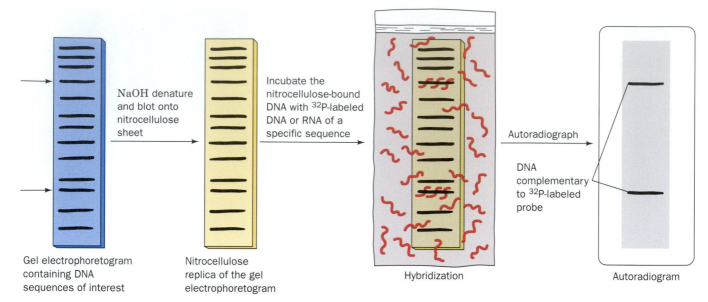

Gel electrophoretogram containing DNA sequences of interest

Nitrocellulose replica of the gel electrophoretogram

Hybridization

Autoradiogram

NaOH denature and blot onto nitrocellulose sheet

Incubate the nitrocellulose-bound DNA with ^{32}P-labeled DNA or RNA of a specific sequence

Autoradiograph

DNA complementary to ^{32}P-labeled probe

Figure 23-28.   **The detection of DNAs containing specific base sequences by Southern blotting.**

Southern Blotting Identifies DNAs with Specific Sequences

DNA with a specific base sequence can be identified through a procedure developed by Edwin Southern known as **Southern blotting** (Fig. 23-28). This procedure takes advantage of the ability of nitrocellulose to tenaciously bind single-stranded but not duplex DNA. Following gel electrophoresis of double-stranded DNA, the gel is soaked in 0.5 M NaOH to convert the DNA to its single-stranded form. The gel is then overlaid by a sheet of nitrocellulose paper. Molecules in the gel are forced through the nitrocellulose by drawing out the liquid with a stack of absorbent towels compressed against the far side of the nitrocellulose, or by using an electrophoretic process **(electroblotting).** The single-stranded DNA binds to the nitrocellulose at the same position it had in the gel. After drying at 80°C, which permanently fixes the DNA in place, the nitrocellulose sheet is moistened with a minimal quantity of solution containing a ^{32}P-labeled single-stranded DNA or RNA probe that is complementary in sequence to the DNA of interest. The moistened nitrocellulose is held at a suitable renaturation temperature for several hours to permit the probe to hybridize to its target sequence(s), washed to remove the unbound radioactive probe, dried, and then autoradiographed by placing it for a time over a sheet of X-ray film. The positions of the molecules that are complementary to the radioactive probe are indicated by a blackening of the developed film.

Specific RNA sequences can be detected through a variation of the Southern blot, punningly named a **Northern blot,** in which the RNA is immobilized on nitrocellulose paper and probed with a complementary radioactive RNA or DNA. A specific protein can be analogously detected in an **immunoblot** or **Western blot** through the use of antibodies directed against the protein in a procedure similar to that used in an ELISA (Fig. 5-3).

C. *Ultracentrifugation*

Equilibrium density gradient ultracentrifugation (Section 5-2E) in CsCl is one of the most commonly used DNA separation procedures. The buoy-

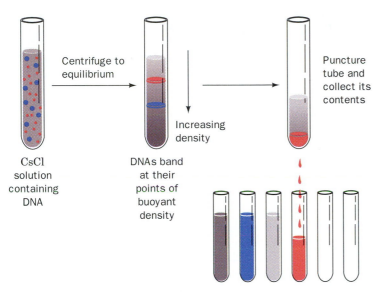 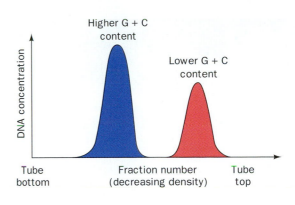

Figure 23-29. **The separation of DNAs by equilibrium density gradient ultracentrifugation in CsCl solution.** An initially 8 M CsCl solution forms a density gradient that varies linearly from ~1.80 g·cm^{-3} at the bottom of the centrifuge tube to ~1.55 g·cm^{-3} at the top. The sedimentation rates of the DNAs depend on base composition. The amount of DNA in each fraction is estimated from its UV absorbance, usually at 260 nm. ✳ See the Animated Figures.

ant density of double-stranded Cs$^+$DNA depends on its base composition, with DNAs of higher G + C content having a greater density (Fig. 23-29).

Single-stranded DNA is ~0.015 g·cm^{-3} denser than the corresponding double-stranded DNA, so the two can be separated by equilibrium density gradient ultracentrifugation. Circular DNA can also be separated from linear DNA in this manner.

RNA is too dense to band in CsCl but does so in Cs$_2$SO$_4$ solutions. RNA–DNA hybrids band in CsCl but at a higher density than the corresponding duplex DNA. RNA can also be fractionated by zonal ultracentrifugation through a sucrose gradient. RNAs are separated by this technique largely on the basis of their size.

4. DNA–PROTEIN INTERACTIONS

The accessibility of genetic information depends on the ability of proteins to recognize and interact with DNA in a manner that allows the encoded information to be copied as DNA (in replication) or as RNA (in transcription). Even the most basic steps of these processes require many proteins that interact with each other and with the nucleic acids. In addition, organisms regulate the expression of most genes, which requires yet other proteins that act as repressors or activators of transcription.

Many proteins bind DNA nonspecifically, that is, without regard to the sequence of nucleotides. For example, histones (which are involved in packaging DNA; Section 23-5A) and certain DNA replication proteins (which must potentially interact with all the sequences of an organism's genome) bind to DNA primarily through interactions between protein functional groups and the sugar–phosphate backbone of DNA. Proteins that recognize specific DNA sequences presumably also bind nonspecifically but loosely to DNA so that they can scan the polynucleotide chain for their target sequences before binding specifically and tightly. Sequence-specific DNA–protein interactions must be extremely precise so that the proteins

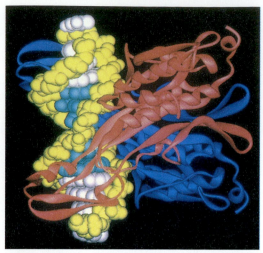

(a)

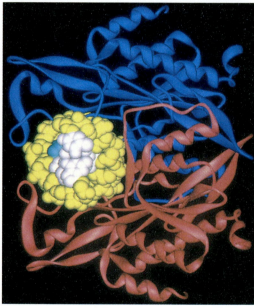

(b)

Figure 23-30. **The X-ray structure of *Eco*RI endonuclease in complex with DNA.** The segment of duplex DNA has the self-complementary sequence TCGC-<u>GAATTC</u>GCG (12 bp with an overhanging T at both 5′ ends; the enzyme's 6-bp target sequence is underlined) and is drawn as a space-filling model with its sugar–phosphate chains yellow, its recognition sequence bases cyan, and its other bases white. The protein is drawn in ribbon form with its two identical subunits red and blue. The complex is shown with (*a*) its DNA helix axis vertical and (*b*) in end view. The complex's two-fold axis is horizontal and the DNA's major groove faces right (toward the protein) in both views. [Based on an X-ray structure by John Rosenberg, University of Pittsburgh.] ● **See Kinemage Exercise 18-1.**

can exert their effects at sites selected from among—in the human genome—billions of base pairs. How do such proteins interact with their target sites on DNA?

Sequence-specific DNA-binding proteins generally do not disrupt the base pairs of the duplex DNA to which they bind. They do, however, *discriminate among the four base pairs (A·T, T·A, G·C, and C·G) according to the functional groups of the base pairs that project into DNA's major and minor grooves.* As illustrated in Fig. 23-2, these groups are more exposed in the major groove of B-DNA than in its narrower minor groove. Moreover, the major groove contains more sequence-specific functional groups than does the minor groove (Fig. 23-1). DNA–protein binding interactions primarily take the form of hydrogen bonds, although these may be indirect, involving intervening water molecules. Ionic interactions with backbone phosphate groups also occur. The complementarity of the binding partners in many cases is augmented by an "induced fit" phenomenon in which the protein and the nucleic acid change their conformations for greater stability. In some cases, these changes allow the binding proteins to interact with other proteins or alter the accessibility of the DNA to other molecules.

The forces that stabilize DNA–protein interactions must necessarily derive from those that stabilize proteins (Section 6-4A) and nucleic acids (Section 23-2). But because DNA stability is imperfectly understood, the interactions that underlie DNA–protein complexes are also somewhat murky. There is no doubt, however, that these forces are substantial. Most DNA–protein complexes have dissociation constants ranging from 10^{-9} to 10^{-12} M^{-1}; these values represent binding that is 10^3 to 10^7 times stronger than nonspecific binding. Below, we examine some examples of specific DNA–protein binding. We shall encounter many more examples of nucleic acid–protein interactions in subsequent chapters.

A. Restriction Endonucleases

Type II restriction endonucleases rid bacterial cells of foreign DNA by cleaving the DNA at specific sites that have not yet been methylated by the host's modification methylase (Section 3-4A). Restriction enzymes recognize palindromic DNA sequences of ~4 to 8 base pairs with such remarkable specificity that a single base change can reduce their activity by a millionfold. This degree of specificity is necessary to prevent accidental cleavage of other sites in a DNA sequence.

The X-ray structure of *Eco*RI endonuclease in complex with a segment of B-DNA containing the enzyme's recognition sequence was determined by John Rosenberg. The DNA binds in the symmetric cleft between the two identical 276-residue subunits of the dimeric enzyme (Fig. 23-30), thereby accounting for the DNA's palindromic recognition sequence. Pro-

Figure 23-31.   **The X-ray structure of *Eco*RV endonuclease in complex with DNA.** The 10-bp segment of DNA has the self-complementary sequence GGGATATCCC (the enzyme's 6-bp target sequence is underlined). The DNA and protein are colored as in Fig. 23-30. (*a*) The complex as viewed along its two-fold axis, facing the DNA's major groove. The two symmetry-related protein loops that overlie the major groove (composed of residues 182 to 186) are the only parts of the enzyme that make base-specific contacts with the DNA. (*b*) The complex as viewed from the right in *a* (the DNA's major groove faces left). Note how the protein kinks the DNA toward its major groove. [Based on an X-ray structure by Fritz Winkler, Hoffman-LaRoche Ltd., Switzerland.] ● See Kinemage Exercise 18-2.

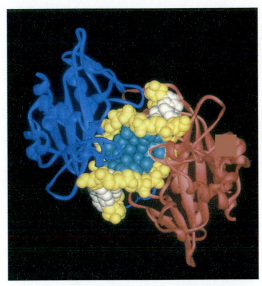

(a)

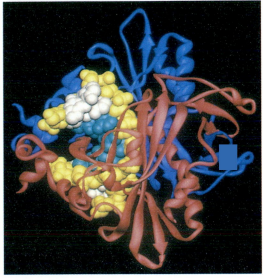

(b)

tein binding causes the dihedral angle between the recognition sequence's central two base pairs to open up by ~50° toward the minor groove. These base pairs thereby become unstacked, but the DNA remains nearly straight due to compensating bends at the adjacent base pairs. Nevertheless, this unwinds the DNA by 28° and widens the major groove by 3.5 Å at the recognition site. The N-terminal ends of a pair of parallel helices from each protein subunit are inserted into the widened major groove, where they participate in a hydrogen-bonded network with the bases of the recognition sequence. The phosphodiester cleavage points are located two bases from the center of the palindrome (Table 3-3).

Certain other restriction endonucleases, including *Eco*RV (Fig. 23-31), also induce kinks in the DNA, in some cases by opening up the minor groove and compressing the major groove. However, *complementary hydrogen bonding between the nucleotide bases and protein side chain and backbone groups, rather than the DNA distortion per se, is the primary prerequisite for the formation of a sequence-specific endonuclease–DNA complex.* For example, in the *Bam*HI–DNA complex, every potential hydrogen bond donor and acceptor in the major groove of the recognition site takes part in direct or water-mediated hydrogen bonds with the protein. No other DNA sequence could support this degree of complementarity with *Bam*HI.

B. Prokaryotic Transcriptional Control Motifs

In prokaryotes, the expression of many genes is governed at least in part by **repressors,** proteins that bind at or near the gene so as to prevent its transcription (Section 27-2). These repressors often contain ~20-residue polypeptide segments that form a **helix–turn–helix (HTH)** motif consisting of two α helices that cross at an angle of ~120°. HTH motifs, which are apparently evolutionarily related, occur as components of domains that otherwise have widely varying structures, although all of them bind DNA. Note that HTH motifs are structurally stable only when they are components of larger proteins.

Like restriction endonuclease recognition sites, the sequences to which repressors bind (called **operators**) exhibit palindromic symmetry or nearly so. Typically, the repressors are dimeric, although their interactions with DNA may not be perfectly symmetrical. Binding involves interactions between the amino acid side chains extending from the second helix of the HTH motif (the "recognition" helix) and the bases and sugar–phosphate chains of the DNA.

The X-ray structure of a dimeric repressor from **bacteriophage 434** bound to its 20-bp recognition sequence has been determined by Stephen Harrison. The **434 repressor** associates with the DNA in a two-fold symmetric manner with a recognition helix from each subunit bound in suc-

● See Guided Exploration 23:

Transcription Factor–DNA Interactions (HTH Motif, Zinc-Fingers, Leucine Zipper).

(a)   (b)   (c)

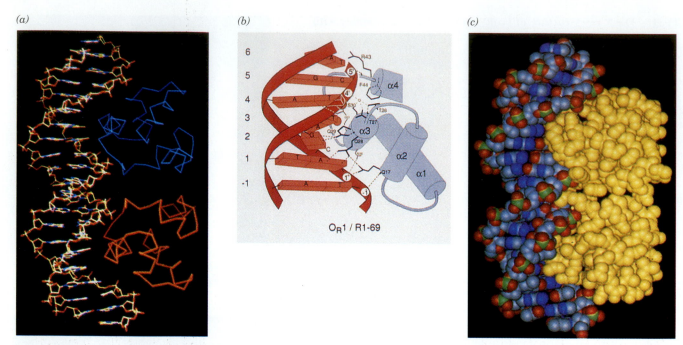

OR1 / R1-69

Figure 23-32.   The X-ray structure of a portion of the 434 phage repressor in complex with its target DNA. One strand of the 20-bp DNA (*left*) has the sequence d(TATACAA-GAAAGTTTGTACT). (*a*) A skeletal model showing the DNA and the first 63 residues of the protein's two identical subunits (*blue and red,* C_α backbone only). (*b*) A schematic drawing indicating how the helix–turn–helix motif, which encompasses helices α2 and α3, interacts with the DNA. Residues are identified by single-letter codes. Short bars emanating from the polypeptide chain represent peptide NH groups, dashed lines represent hydrogen bonds, and numbered circles represent DNA phosphates. The small circle is a water molecule. (*c*) A space-filling model corresponding to *a*. All the protein's non-H atoms are drawn in yellow. [Courtesy of Aneel Aggarwal, John Anderson, and Stephen Harrison, Harvard University.] ● **See the Interactive Exercises and Kinemage Exercise 19-1.**

cessive turns of the DNA's major groove (Fig. 23-32). The repressor closely conforms to the DNA surface and interacts with its paired bases and sugar–phosphate chains through elaborate networks of hydrogen bonds, salt bridges, and van der Waals contacts. In the repressor–DNA complex, the DNA bends around the protein in an arc of radius ~65 Å, which compresses the minor groove by ~2.5 Å near its center (between the two protein monomers) and widens it by ~2.5 Å toward its ends.

The *E. coli trp* Repressor Binds DNA Indirectly

The *E. coli **trp* repressor** regulates the transcription of genes required for tryptophan biosynthesis (Section 27-2C). Paul Sigler has determined the X-ray structure of this protein in complex with a DNA containing an 18-bp palindrome (of single-strand sequence TGTACTAGTTAACTAGTAC, where the *trp* repressor's target sequence is underlined) that closely resembles the *trp* operator. The homodimeric repressor protein also has HTH motifs whose recognition helices bind, as expected, in successive major grooves of the DNA, each in contact with half of the operator sequence (ACTAGT; Fig. 23-33). There are numerous hydrogen bonding contacts between the *trp* repressor and the DNA's nonesterified phosphate oxygens. Astoundingly, however, there are no direct hydrogen bonds or nonpolar contacts that can explain the repressor's specificity for its operator. Rather, all but one of the side chain–base hydrogen-bonding interactions are mediated by bridging water molecules. In addition, the operator contains several base pairs that are not in contact with the repressor but whose mutation nevertheless greatly decreases repressor binding affinity. This suggests that *the operator assumes a sequence-specific conformation that makes favorable contacts with the repressor.* Other DNA sequences could conceivably assume the same conformation but at too high an energy cost to form a stable complex with the repressor. This phenomenon, in which a protein senses the base sequence of DNA through the DNA's backbone conformation and/or flexibility, is referred to as **indirect readout.** This finding puts to rest the notion that proteins recognize nucleic acid sequences exclusively via particular sets of pairings between amino acid side chains and nucleotide bases analogous to Watson–Crick base pairing.

The *met* Repressor Binds DNA via a Two-Stranded β Sheet

Simon Phillips first determined the X-ray structure of the *E. coli met* **repressor** in the absence of DNA (the *met* repressor regulates the transcription of genes involved in methionine biosynthesis). The homodimeric protein lacks an HTH motif, but model-building studies suggested that the repressor might bind to its palindromic target DNA via a symmetry-related pair of protruding α helices, reminiscent of the way the recognition helices of HTH motifs interact with DNA. However, the subsequently determined X-ray structure of the *met* repressor–operator complex showed that the protein actually binds its target DNA sequence through a pair of symmetrically related β strands (located on the opposite side of the protein from the protruding α helices) that form a two-stranded antiparallel β sheet that inserts into the DNA's major groove (Fig. 23-34). The β strands make sequence-specific contacts with the DNA via hydrogen bonding and, probably, indirect readout. This result indicates that the conclusions of even what appear to be straightforward model-building studies should be viewed with skepticism. In the case of the *met* repressor–operator complex, the model-building study favored the incorrect model because it could not take into

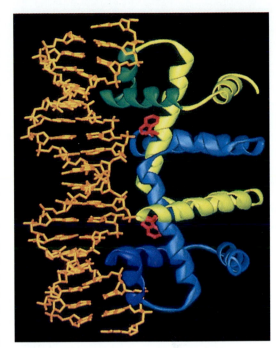

Figure 23-33. The X-ray structure of an *E. coli trp* repressor–operator complex. The molecular two-fold axis is horizontal and in the plane of the paper. The protein's two identical subunits are shown in ribbon form in green and blue with the HTH motifs more deeply colored. The 18-bp self-complementary DNA is yellow. *trp* repressor binds its operator only when L-tryptophan (*red*) is also bound. Note that the protein's recognition helices bind, as expected, in successive major grooves of the DNA but extend approximately perpendicular to the DNA helix axis, whereas those of 434 phage repressor are nearly parallel to the major grooves of its bound DNA (Fig. 23-32). [Based on an X-ray structure by Paul Sigler, Yale University.] ● See the Interactive Exercises.

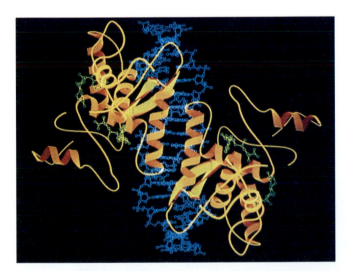

Figure 23-34. The X-ray structure of the *E. coli met* repressor–operator complex. The 104-residue repressor subunits are shown in gold. The self-complementary 19-bp DNA is shown as a blue ball-and-stick model. The methionine derivative *S*-adenosylmethionine, shown in green, must be bound to the repressor for it to bind DNA. Note that the DNA has four bound repressor subunits: Pairs of subunits form symmetric dimers in which each subunit donates one strand of the two-stranded β sheet that is inserted in the DNA's major groove (*upper left and lower right*). Two such dimers pair across the complex's two-fold axis via their antiparallel N-terminal helices, which contact each other over the DNA's minor groove. [Courtesy of Simon Phillips, University of Leeds, Leeds, U.K.] ● See the Interactive Exercises.

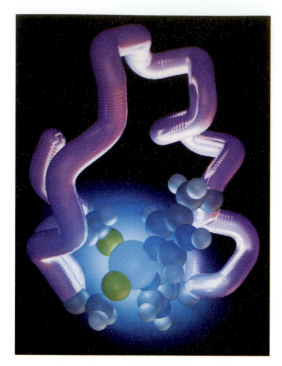

***Figure 23-35.   The NMR structure of a zinc finger from the *Xenopus* protein Xfin.** The Zn^{2+} ion together with the atoms of its His and Cys ligands are represented as spheres with Zn light blue, C gray, N blue, S yellow, and H white. [Courtesy of Michael Pique, The Scripps Research Institute, La Jolla, California. Based on an NMR structure by Peter E. Wright, The Scripps Research Institute.]*

account the small conformational adjustments that both the protein and the DNA make on binding one another.

C. Eukaryotic Transcription Factors

In eukaryotes, genes are selectively expressed in different cell types; this requires more complicated regulatory machinery than in prokaryotes. Prokaryotic repressors of known structure either contain an HTH motif or resemble the *met* repressor. However, eukaryotic DNA-binding proteins employ a much wider variety of structural motifs to bind DNA. A number of proteins known as **transcription factors** (Section 25-2C) promote the transcription of genes by binding to DNA sequences at or near those genes. In this section, we describe a variety of DNA-binding motifs in eukaryotic transcription factors.

Zinc Finger DNA-Binding Motifs

The first of the predominantly eukaryotic DNA-binding motifs, the **zinc finger,** was discovered by Aaron Klug in **transcription factor IIIA (TFIIIA)** from *Xenopus laevis* (an African clawed toad). The 344-residue TFIIIA contains nine similar, tandemly repeated, ~30-residue modules, each of which contains two invariant Cys residues and two invariant His residues. Each of these units binds a Zn^{2+} ion, which is tetrahedrally liganded by these Cys and His residues (Fig. 23-35). In some zinc fingers, the two Zn^{2+}-liganding His residues are replaced by two additional Cys residues, whereas others have six Cys residues liganding two Zn^{2+} ions. Indeed, structural diversity is a hallmark of zinc finger proteins. In all cases, however, the Zn^{2+} ions appear to knit together relatively small globular domains, thereby eliminating the need for much larger hydrophobic protein cores (Section 6-4A).

The **Cys$_2$–His$_2$ zinc finger** (Figs. 6-35 and 23-35) contains a two-stranded antiparallel β sheet and one α helix. Three of these motifs are incorporated into a 72-residue segment of the mouse protein **Zif268,** whose X-ray structure in complex with a target DNA was elucidated by Carl Pabo (Fig. 23-36). The three zinc fingers are arranged as separate domains in a C-shaped structure that fits snugly into the DNA's major groove. Each zinc finger interacts in a conformationally identical manner with successive 3-bp segments of the DNA, predominantly through hydrogen bonds between the zinc finger's α helix and one strand of the DNA. Each zinc finger specifically hydrogen bonds to two bases in the major groove. Interestingly, five of these six associations involve Arg–guanine pairs. In addition

***Figure 23-36.   The X-ray structure of a three–zinc finger segment of Zif268 in complex with a 10-bp DNA.** The protein and DNA (with a single nucleotide overhang at each end) are shown in stick form, with superimposed cylinders and ribbons marking the protein's α helices and β sheets. Finger 1 is orange, finger 2 is yellow, finger 3 is pink, the DNA is blue, and the Zn^{2+} ions are represented by cyan spheres. Note how the N-terminal (*lower*) end of each zinc finger's helix extends into the DNA's major groove to contact three base pairs. [Courtesy of Carl Pabo, MIT.]*

🔵 See the Interactive Exercises.

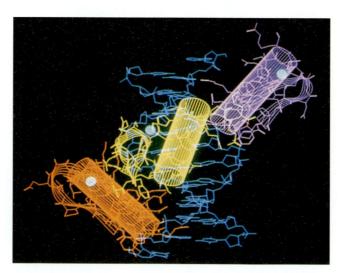

to these sequence-specific interactions, each zinc finger hydrogen bonds with the DNA's phosphate groups via conserved Arg and His residues.

The Cys_2–His_2 zinc finger broadly resembles the prokaryotic HTH motif as well as most other DNA-binding motifs we shall encounter (including other types of zinc fingers). All these DNA-binding motifs provide a platform for inserting an α helix into the major groove of B-DNA. However, the Cys_2–His_2 zinc fingers, unlike other DNA-binding motifs, occur as modules that each contact successive DNA segments (some transcription factors contain up to 37 zinc fingers). *Such a modular system can recognize extended asymmetric base sequences.*

A binuclear **Cys_6 zinc finger** mediates the DNA binding of the yeast protein **GAL4,** a transcriptional activator of several genes that encode galactose-metabolizing enzymes. The N-terminal portion of this 881-residue protein includes six Cys residues that collectively bind two Zn^{2+} ions (Fig. 23-37*a*). Each Zn^{2+} ion is tetrahedrally coordinated by four Cys

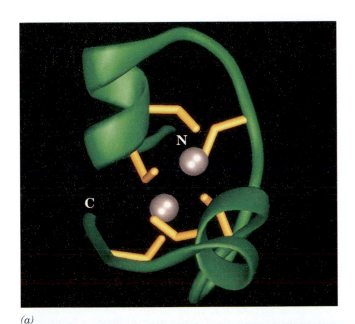

(a)

Figure 23-37. **The X-ray structure of the GAL4 DNA-binding domain in complex with DNA.** (*a*) A ribbon model of the protein's zinc finger domain (residues 8–40) with its six Cys side chains in stick form (*yellow*) and its Zn^{2+} ions shown as silver spheres. Compare this structure with Fig. 6-35 or 23-35. (*b*) The complex of the dimeric GAL4 protein with a palindromic 19-bp DNA (except for the central base pair) containing the protein's binding sequence. The structure is shown in tube form with the DNA red, the protein backbone cyan, and the Zn^{2+} represented by yellow spheres. The views are along the complex's two-fold axis (*left*) and turned 90° with the two-fold axis horizontal (*right*). Note how the C-terminal end of each subunit's N-terminal helix extends into the DNA's major groove. [Part *b* courtesy of and Part *a* based on an X-ray structure by Ronen Mamorstein and Stephen Harrison, Harvard University.] ● See the **Interactive Exercises.**

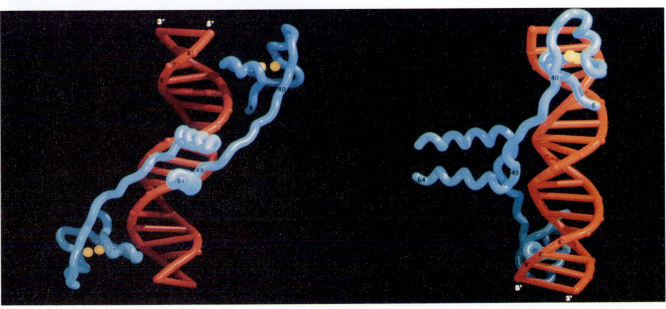

(b)

residues, with two of these residues ligating both metal ions. GAL4 binds to its 17-bp target DNA as a symmetric dimer (Fig. 23-37*b*), although in the absence of DNA it is a monomer. Each subunit includes a compact zinc finger (residues 8–40) that binds DNA, an extended linker (residues 41–49), and an α helix (residues 50–64) that assists in GAL4 dimerization. The N-terminal helix of the zinc finger inserts into the DNA's major groove, making sequence-specific contacts with a highly conserved CCG sequence at each end of the recognition sequence. The bound DNA retains its B conformation.

The dimerization helices of GAL4 (center of Fig. 23-37*b*) are positioned over the minor groove of the DNA. The linkers connecting these helices to the zinc fingers wrap around the DNA, largely following its minor groove. The two symmetrically related DNA-binding zinc fingers thereby approach the major groove from opposite sides of the DNA, ~1.5 helical turns apart, rather than from the same side the DNA, ~1 turn apart, as do HTH motifs. The resulting relatively open structure could permit other proteins to bind simultaneously to the DNA.

Transcription Factors with Leucine Zippers

Segments of certain eukaryotic transcription factors, such as the yeast protein **GCN4,** contain a Leu at every seventh position. We have already seen that α helices with the seven-residue pseudorepeating sequence (*a-b-c-d-e-f-g*)$_n$, in which the *a* and *d* residues are hydrophobic, have a hydrophobic strip along one side that allows them to dimerize as a coiled coil (e.g., α-keratin; Section 6-1C). Steven McKnight suggested that DNA-binding proteins containing such **heptad repeats** also form coiled coils in which the Leu side chains interdigitate, much like the teeth of a zipper (Fig. 23-38). In fact, *these* **leucine zippers** *mediate the dimerization of certain DNA-binding proteins but are not themselves DNA-binding motifs.*

The X-ray structure of the 33-residue polypeptide corresponding to the leucine zipper of GCN4 was determined by Peter Kim and Thomas Alber. The first 30 residues, which contain ~3.6 heptad repeats (Fig. 23-38*a*), coil

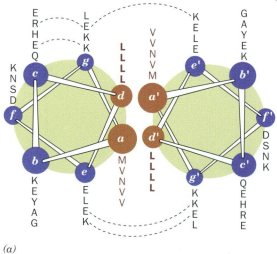

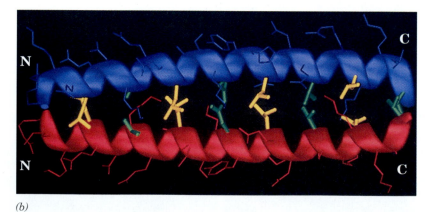

(a) *(b)*

Figure 23-38. The GCN4 leucine zipper motif. (*a*) A helical wheel representation of the motif's two helices as viewed from their N-termini. The sequences of residues at each position are indicated by the adjacent column of one-letter codes. Residues that form ion pairs in the crystal structure are connected by dashed lines. Note that all residues at positions *d* and *d'* are Leu (L), those at positions *a* and *a'* are mostly Val (V), and those at other positions are mostly polar. [After O'Shea, E.K., Klemm, J.D., Kim, P.S., and Alber, T., *Science* **254,** 540 (1991).] (*b*) The X-ray structure, in side view, in which the helices are shown in ribbon form. Side chains are shown in stick form with the contacting Leu residues at positions *d* and *d'* yellow and residues at positions *a* and *a'* green. [Based on an X-ray structure by Peter Kim, MIT, and Tom Alber, University of Utah School of Medicine.]

🔵 See Kinemage Exercise 4-1.

Figure 23-39. The X-ray structure of a portion of GCN4 in complex with its target DNA. The DNA (*red*), shown in stick form, consists of a 19-bp segment with a single nucleotide overhang at each end and contains the protein's palindromic (except for the central base pair) 7-bp target sequence. The two identical GCN4 subunits, shown in ribbon form, each contain a continuous 52-residue α helix. At their C-terminal ends (*yellow*), the two subunits associate in a parallel coiled coil (a leucine zipper), and at their basic regions (*green*), they smoothly diverge to each engage the DNA in its major groove at the target sequence. The N-terminal ends are white. [Based on an X-ray structure by Stephen Harrison, Harvard University.]
● See the Interactive Exercises.

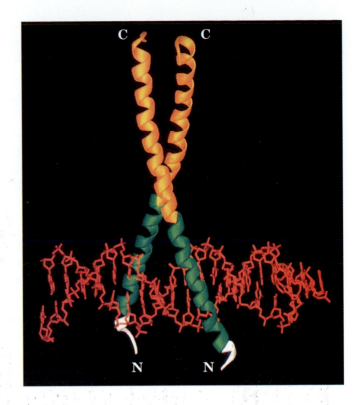

into an ~8-turn α helix that dimerizes as McKnight predicted to form ~1/4 turn of a parallel left-handed coiled coil (Fig. 23-38b). The dimer can be envisioned as a twisted ladder whose sides consist of the helix backbones and whose rungs are formed by the interacting hydrophobic side chains. The conserved Leu residues in the heptad position *d*, which corresponds to every second rung, are not interdigitated as McKnight originally suggested, but instead make side-to-side contacts. The alternate rungs are likewise formed by the *a* residues of the heptad repeat (which are mostly Val). These contacts form an extensive hydrophobic interface between the two helices.

In many leucine zipper–containing proteins, a DNA-binding region that is rich in basic residues is immediately N-terminal to the leucine zipper. For example, in GCN4, the C-terminal 56 residues form an extended α helix. The last 25 residues of two such helices associate as a leucine zipper. The N-terminal portions of the helices smoothly diverge to bind in the major grooves on opposite sides of the DNA, thereby clamping the DNA in a sort of scissors grip (Fig. 23-39). The basic residues in these portions of GCN4 make numerous contacts with both the bases and the phosphate oxygens of the DNA target sequence without distorting its conformation.

Other leucine zipper proteins contain a basic region that forms a **helix–loop–helix (HLH)** motif reminiscent of the helix–turn–helix motif of some prokaryotic DNA-binding proteins. The N-terminal portion of the first helix of the HLH motif binds in the major groove of its target DNA. The C-terminal helix of the HLH motif is continuous with a leucine zipper helix and likewise forms a coiled coil. The dimeric protein thus grips the DNA on either side, as shown in Fig. 23-40 for the protein **Max.**

Figure 23-40. The X-ray structure of Max binding to DNA. Residues 22–113 of the Max dimer form a complex with a 22-bp DNA containing the protein's palindromic 6-bp target sequence. The DNA (*red*) is shown in stick form, and the homodimeric protein is shown in ribbon form. The protein's N-terminal basic region (*green*) forms an α helix that engages its target sequence in the DNA's major groove and then merges smoothly with the H1 helix (*yellow*) of the helix–loop–helix motif. Following the loop (*magenta*), the protein's two H2 helices (*purple*) of the HLH motif merge smoothly with the leucine zipper helices (*cyan*) to form a parallel coiled coil. The protein's N- and C-terminal ends are white. [Based on an X-ray structure by Stephen Burley, The Rockefeller University.] ● See the Interactive Exercises.

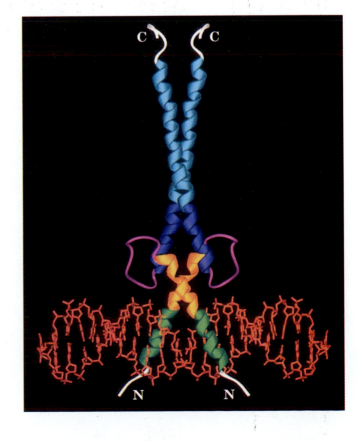

Each basic region contacts specific bases of the DNA as well as phosphate groups. Side chains of both the loop and the N-terminal end of the H2 helix also contact DNA phosphate groups.

5. EUKARYOTIC CHROMOSOME STRUCTURE

DNA molecules are generally enormous (Fig. 23-41). Although each base pair of B-DNA contributes only ~3.4 Å to its **contour length** (the end-to-end length of a stretched-out native molecule), a DNA molecule containing 166 kb (e.g., T2 bacteriophage DNA) is 55 μm long. The 23 chromosomes of the 3 billion-bp human genome have a total contour length of almost 1 m. One of the enduring questions of molecular biology is how such vast quantities of genetic information can be scanned and decoded in a reasonable time while stored in a small portion of the cell's volume.

The elongated shape of duplex DNA (its diameter is only 20 Å) and its relative stiffness make it susceptible to mechanical damage when outside the protective environment of the cell. For example, a *Drosophila* chromosome, if expanded by a factor of 500,000, would have the shape and some of the mechanical properties of a 6-km-long strand of uncooked spaghetti. In fact, the **shear degradation** of DNA by stirring, shaking, or pipetting a DNA solution is a standard laboratory method for preparing DNA fragments.

Prokaryotic genomes typically comprise a single circular DNA molecule. However, most eukaryotes condense and package their genome in several **chromosomes.** Each chromosome is a complex of a single linear DNA molecule and protein, a material known as **chromatin,** and is a dynamic entity whose appearance varies dramatically with the stage of the **cell cycle** (the general sequence of events that occur in the lifetime of a eukaryotic cell).

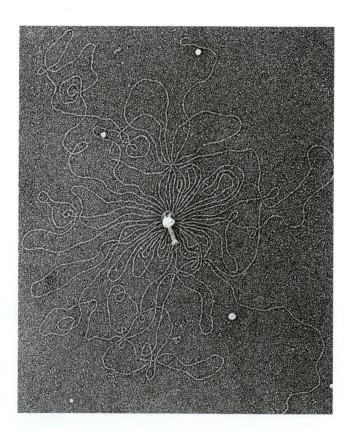

Figure 23-41. **An electron micrograph of a T2 bacteriophage and its DNA.** The phage has been osmotically lysed in distilled water so that its DNA has spilled out. The DNA is rendered visible in the electron microscope by first coating it with denatured cytochrome *c* to fatten it to ~200 Å in diameter and then shadowing the resultant preparation with platinum. [From Kleinschmidt, A.K., Lang, D., Jacherts, D., and Zahn, R.K., *Biochim. Biophys. Acta* **61**, 861 (1962).]

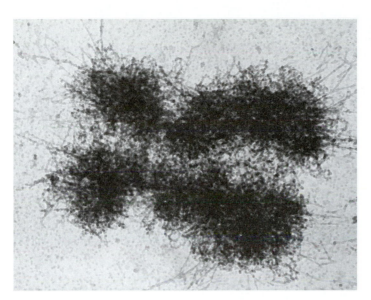

Figure 23-42. **An electron micrograph of a human metaphase chromosome.** [Courtesy of Gunther Bahr, Armed Forces Institute of Pathology.]

For example, chromosomes assume their most condensed forms (Fig. 23-42) only during the metaphase stage of cell division. During the remainder of the cell cycle, when the DNA is transcribed and replicated, the chromosomes of most cells become so highly dispersed that they cannot be distinguished. Yet the DNA of these chromosomes is nevertheless still compacted relative to its free B-helix form. Human chromosomes have contour lengths between 1.6 and 8.2 cm but in their most condensed state are only 1.3 to 10 μm long. In this section, we examine how DNA is packaged in cells to achieve this degree of condensation.

A. Histones

Chromatin is about one-half protein by mass, and most of this protein consists of **histones.** To understand how DNA is packaged, we must first examine these proteins. The five major classes of histones, **H1, H2A, H2B, H3,** and **H4,** all have a large proportion of positively charged residues (Arg and Lys; Table 23-3). These proteins can therefore bind DNA's negatively charged phosphate groups through electrostatic interactions.

The amino acid sequences of histones H2A, H2B, H3, and H4 are remarkably conserved. For example, histones H4 from cows and peas, species that diverged 1.2 billion years ago, differ by only two conservative residue changes, which makes this protein among the most evolutionarily conserved proteins known (Section 5-4A). *Such evolutionary stability implies that the histones have critical functions to which their structures are so well tuned that they are all but intolerant to change.* The fifth histone, H1, is more variable than the other histones; we shall see below that it also has a somewhat different role.

Histones are subject to posttranslational modifications that include methylation, acetylation, and phosphorylation of specific Arg, His, Lys, Ser, and Thr residues. *These modifications, many of which are reversible, all decrease the histones' positive charges, thereby significantly altering histone–DNA interactions.* Despite the histones' great evolutionary stability, their degree of modification varies enormously with the species, the tissue, and the stage of the cell cycle. As we shall see (Section 27-3A), modification of histones has been linked to transcriptional activation. The modifications probably create new protein binding sites on the histones themselves or on the DNA with which they associate.

Table 23-3. **Calf Thymus Histones**

Histone	Number of Residues	Mass (kD)	% Arg	% Lys
H1	215	23.0	1	29
H2A	129	14.0	9	11
H2B	125	13.8	6	16
H3	135	15.3	13	10
H4	102	11.3	14	11

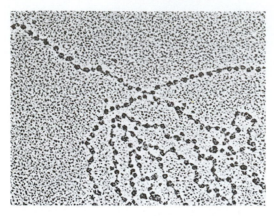

Figure 23-43.   An electron micrograph of *D. melanogaster* chromatin showing strings of closely spaced nucleosomes. [Courtesy of Oscar L. Miller, Jr. and Steven McKnight, University of Virginia.]

● See Guided Exploration 24:

Nucleosome Structure.

B. Nucleosomes

The first level of chromatin organization was pointed out by Roger Kornberg in 1974 from several lines of evidence:

1. Chromatin contains roughly equal numbers of histones H2A, H2B, H3, and H4, and no more than half that number of histone H1.

2. Electron micrographs of chromatin preparations at low ionic strength reveal ~100-Å-diameter particles connected by thin strands of apparently naked DNA, rather like beads on a string (Fig. 23-43).

3. Brief digestion of chromatin by **micrococcal nuclease** (which hydrolyzes double-stranded DNA) cleaves the DNA only between the above particles; apparently the particles protect the DNA closely associated with them from nuclease digestion. Gel electrophoresis indicates that each particle contains ~200 bp of DNA.

Kornberg proposed that the chromatin particles, which are called **nucleosomes,** consist of the octamer $(H2A)_2(H2B)_2(H3)_2(H4)_2$ in association with ~200 bp of DNA. The fifth histone, H1, was postulated to be associated in some other manner with the nucleosome (see below).

DNA Coils around a Histone Octamer to Form the Nucleosome Core Particle

Micrococcal nuclease initially degrades chromatin to single nucleosomes in complex with histone H1. Further digestion trims away additional DNA, releasing histone H1. This leaves the so-called **nucleosome core particle,** which consists of a 146-bp strand of DNA associated with the histone octamer. A segment of **linker DNA,** the DNA that is removed by the nuclease, joins neighboring nucleosomes. Its length varies between 8 and 114 bp among organisms and tissues, although it is usually ~55 bp.

The X-ray structure of the nucleosome core particle, determined by Timothy Richmond, reveals a nearly two-fold symmetric complex in which B-DNA is wrapped around the outside of the histone octamer in 1.65 turns of a left-handed superhelix (Fig. 23-44a). Despite having only weak sequence similarity, all four types of histones share a similar fold in which a long central helix is flanked on each side by a loop and a shorter helix (Fig. 23-44b). Pairs of histones interdigitate in a sort of "molecular handshake" to form the crescent-shaped heterodimers H2A–H2B and H3–H4, each of which binds 2.5 turns of duplex DNA that curves around it in a 140° bend. The H3–H4 pairs interact, via a bundle of four helices from the two H3 histones, to form an $(H3–H4)_2$ tetramer with which each H2A–H2B pair interacts, via a similar four-helix bundle between H2A and H4, to form the histone octamer (Fig. 23-44b).

The protein binds the DNA primarily via its sugar–phosphate backbone through hydrogen bonds, salt bridges, and helix dipoles (their positive N-terminal ends), all interacting with phosphate oxygens, as well as through hydrophobic interactions with the deoxyribose rings. In addition, an Arg side chain is inserted into the DNA's minor groove at each of the 14 positions at which it faces the histone octamer. The DNA superhelix has a radius of 42 Å and a pitch of 24 Å. However, the DNA does not follow a uniform superhelical path but, rather, is bent fairly sharply at several locations due to outward bulges of the histone core. Moreover, the DNA double helix exhibits considerable conformational variation along its length such that its twist, for example, varies from 9.4 to 10.9 bp/turn with an average value of 10.2 bp/turn (versus 10.4 bp/turn for DNA in solution).

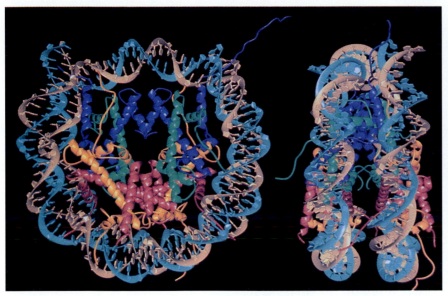

(a)

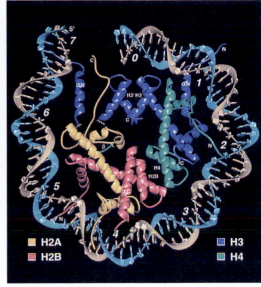

(b)

■ H2A		■ H3
■ H2B		■ H4

Figure 23-44.   The X-ray structure of the nucleosome core particle. (*a*) The entire core particle as viewed (*left*) along its superhelical axis and (*right*) rotated 90° about the vertical axis. The proteins of the histone octamer are drawn in ribbon form with H3 blue, H4 green, H2A yellow, and H2B red. The sugar–phosphate backbones of the 146-bp DNA are drawn as brown and turquoise ribbons whose attached bases are represented by polygons of the same color. In both views, the pseudo-two-fold axis is vertical and passes through the DNA center at the top.

(*b*) The top half of the nucleosome core particle as viewed in *a*, *left*, and identically colored. The numbers 0 through 7 arranged about the inside of the 73-bp DNA superhelix mark the positions of sequential double-helical turns. Those histones that are drawn in their entirety are primarily associated with this DNA segment, whereas only fragments of H3 and H2B from the other half of the particle are shown. The two four-helix bundles shown are labeled H3′ H3 and H2B H4. [Courtesy of Timothy Richmond, Eidgenössische Technische Hochschule, Switzerland.]

Linker Histones Bring Nucleosomes Together

In the micrococcal nuclease digestion of chromatin fibers, the ~200-bp DNA is first degraded to 166 bp. Then there is a pause before histone H1 is released and the DNA is further shortened to 146 bp. Since the 146-bp DNA of the core particle makes 1.65 superhelical turns, the 166-bp intermediate should make nearly two full superhelical turns, which would bring its two ends close together. Klug has proposed that histone H1 binds to nucleosomal DNA at this point, where the DNA segments enter and leave the core particle (Fig. 23-45). Chromatin fibers containing H1 have closely

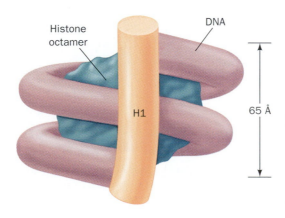

Histone octamer

DNA

H1

65 Å

Figure 23-45.   A model of histone H1 binding to the DNA of the 166-bp nucleosome. The DNA's two complete superhelical turns enable H1 to bind to the DNA's two ends and its middle. Here the histone octamer is represented by the central spheroid and the H1 molecule is represented by the yellow cylinder.

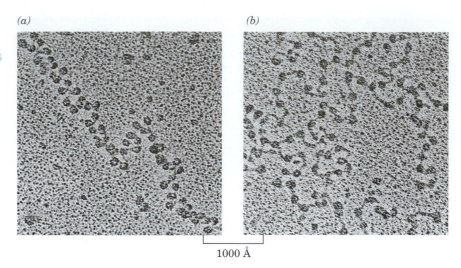

(a)   *(b)*

1000 Å

Figure 23-46.   **Electron micrographs of chromatin.** Both H1-containing chromatin (*a*) and H1-depleted chromatin (*b*) are in 5 to 15 mM salt. [Courtesy of Fritz Thoma, Eidgenössische Technische Hochschule, Switzerland.]

spaced nucleosomes (Fig. 23-46*a*), consistent with DNA entering and exiting the nucleosome on the same side. In H1-depleted chromatin, the entry and exit points tend to occur on opposite sides of the nucleosome, producing a more dispersed arrangement (Fig. 23-46*b*). Evidently, these linker histones have a relatively active role in condensing chromatin fibers and regulating the access of other proteins to the DNA.

C. Higher Levels of Chromatin Organization

Winding the DNA helix around a nucleosome reduces its contour length seven-fold: The 560-Å length of 166 bp is compressed to an ~80-Å-high supercoil. At physiological salt concentrations, chromatin condenses further by folding in a zigzag fashion to form a filament with a diameter of ~300 Å (Fig. 23-47).

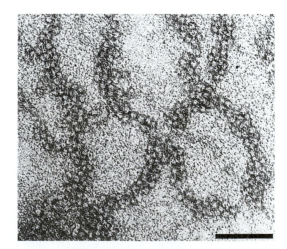

Figure 23-47.   **An electron micrograph of chromatin filaments.** Note that the 300-Å filaments are two to three nucleosomes across. The bar represents 1000 Å. [Courtesy of Jerome B. Rattner, University of Calgary, Canada.]

Model-building studies suggest that this filament is a solenoid with ~6 nucleosomes per turn and a pitch of 110 Å (the diameter of a nucleosome; Fig. 23-48). The solenoid is thought to be stabilized by interactions between histone H1 molecules in adjacent nucleosomes. Experimental evidence suggests that *the solenoid model may only be approximated in vivo, where 300-Å filaments have an unorganized, irregular structure owing to variations in the length of the linker DNA between nucleosome core particles.* For example, the addition of just one base pair to the linker DNA would cause an ~36° change in relative rotation between the nucleosomes at either end. Nonhistone DNA-binding proteins may also interfere with the regular packing of nucleosomes. One implication of this irregular condensed structure is that the accessibility of a particular DNA sequence may not depend on its position within a filament of uniform structure but may be a function of the degree of compaction of the chromatin, which may in turn vary in a sequence-dependent fashion.

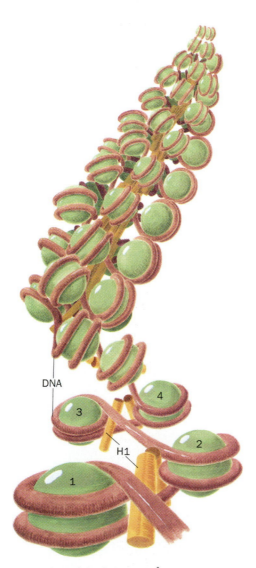

Figure 23-48.   A proposed model of the 300-Å chromatin filament. The zigzag pattern of nucleosomes (1, 2, 3, 4) closes up to form a solenoid with ~6 nucleosomes per turn. In this model, the H1 molecules (*yellow*) run along the center of the solenoid.

In the X-ray structure of the nucleosome, the N-terminal tails of histones H2B and H3 pass between the gyres (turns) of the DNA superhelix (Fig. 23-44*a*). The N-terminal segments of these tails are highly basic and contain the histones' acetylation sites. It is therefore likely that these tails stabilize the formation of chromatin filaments through their interactions with neighboring nucleosomes. Moreover, in the X-ray structure of the nucleosome, an H4's positively charged N-terminal tail binds, in an extended conformation, to an intensely negatively charged region on an exposed face of an H2A–H2B dimer from a neighboring nucleosome. Acetylation of H4's N-terminal tail reduces its positive charge and would therefore weaken this interaction. Since histone tails contain numerous acetylation sites, the disruption of higher order chromatin structure is likely to be controlled, at least in part, by the acetylation of certain residues but not others.

Loops of DNA Are Attached to a Scaffold

Histone-depleted metaphase chromosomes exhibit a central fibrous protein "scaffold" surrounded by an extensive halo of DNA (Fig. 23-49*a*). The strands of DNA that can be followed form loops that enter and exit the scaffold at nearly the same point (Fig. 23-49*b*). Most of these loops are 15 to 30 μm long (which corresponds to 45–90 kb), so when condensed as 300-Å filaments, they would be ~0.6 μm long. Electron micrographs of chromosomes in cross section, such as Fig. 23-50*a*, strongly suggest that the chromatin fibers of metaphase chromosomes are radially arranged. If the observed loops correspond to these radial fibers, they would each con-

(a)

(b)

Figure 23-49.   **Electron micrographs of a histone-depleted metaphase human chromosome.** (*a*) The central protein matrix (scaffold) anchors the surrounding DNA. (*b*) Higher magnification reveals that the DNA is attached to the scaffold in loops. [Courtesy of Ulrich Laemmli, University of Geneva, Switzerland.]

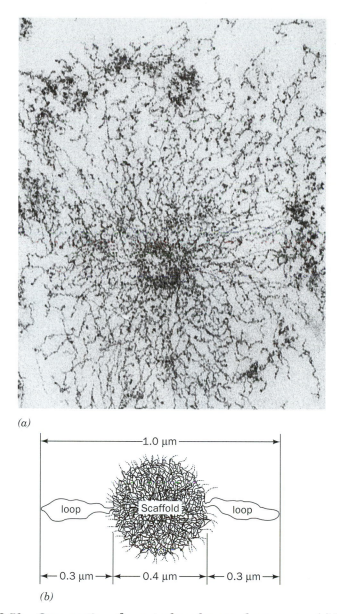

(a)

(b)

Figure 23-50.   Cross section of a metaphase human chromosome. (*a*) Electron micrograph. Note the mass of chromatin fibers radially projecting from the central scaffold. [Courtesy of Ulrich Laemmli, University of Geneva, Switzerland.] (*b*) Interpretive diagram indicating how the 0.3-μm-long radial loops are thought to combine with the 0.4-μm-wide scaffold to form the 1.0-μm-diameter metaphase chromosome.

tribute 0.3 μm to the diameter of the chromosome (a fiber must double back on itself to form a loop). Taking into account the 0.4-μm width of the scaffold, this model predicts the diameter of the metaphase chromosome to be 1.0 μm, in agreement with observations (Fig. 23-50*b*). A typical human chromosome, which contains ~140 million bp, would therefore have ~2000 of these ~70-kb radial loops. The 0.4-μm-diameter scaffold of such a chromosome has sufficient surface area along its 6-μm length to bind this number of radial loops. The radial loop model therefore accounts for DNA's observed packing ratio in metaphase chromosomes.

Almost nothing is known about how the 300-Å filaments organize to form radial loops or about how metaphase chromosomes and the far more dispersed chromosomes of nondividing cells interconvert. Certainly, non-

Box 23-3

BIOCHEMISTRY IN FOCUS

Packaging Viral Nucleic Acids

Viruses are parasites that consist of a nucleic acid molecule encased by a protein **capsid** and, in some cases, an additional lipid bilayer and glycoprotein envelope. Since most viruses contain only a few genes, they depend on the metabolic apparatus of their host cells for replication (and hence viruses are not considered to be alive). The small size of the viral "genome" limits the number of proteins that it can encode. Therefore, its capsid must be built from one or a few kinds of protein subunits that are often arranged in a symmetrical fashion. In the **helical viruses**, the coat protein subunits associate to form helical tubes, whereas in the **spherical viruses**, coat proteins aggregate as closed polyhedral shells. In both cases, the viral nucleic acid occupies the capsid's core.

One of the best-studied viruses is **tobacco mosaic virus** (**TMV**), a rod-shaped particle ~3000 Å long and 180 Å in diameter. The coat consists of ~2130 identical copies of a 158-residue protein. Its assembly begins when 34 of these wedge-shaped proteins spontaneously form a shallow protohelix (the yellow structure in Part *a* below). The viral genome, an ~6400-nucleotide RNA molecule (*red*), forms a hairpin loop that inserts into the protohelix's central cavity so that three nucleotides bind to each protein subunit (*b*). As additional protohelices associate with the growing viral

particle, the RNA is pulled up into the central cavity to form a coil with a radius of ~40 Å (*c*). The model below, of a portion of TMV, illustrates the helical arrangement of its coat protein subunits and the RNA (exposed at the top of the viral helix).

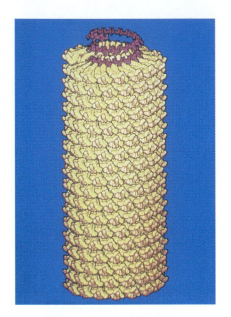

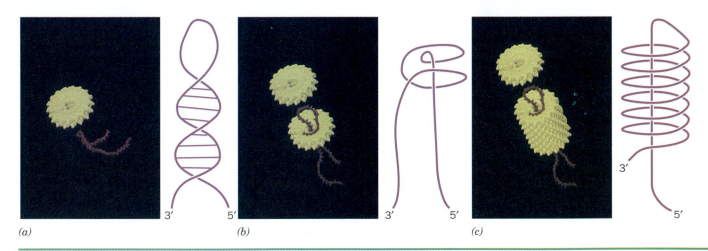

(*a*)   (*b*)   (*c*)

histone proteins, which constitute ~10% of the chromosomal proteins, must be involved in these processes.

In prokaryotes, DNA is also packaged through its association with highly basic proteins that functionally resemble histones. Nucleosomelike particles condense to form large loops that are attached to a protein scaffold, yielding a relatively compact chromosome. The demands of packaging nucleic acid molecules in small volumes are particularly acute in viruses, which are essentially DNA or RNA molecules surrounded by protein coats (see Box 23-3).

Spherical viruses have icosahedral capsids, which require a minimum of 60 protein subunits. A symmetrical arrangement of protein subunits on each triangular face of the icosahedron permits the capsid to contain more than 60 proteins. For example, the 175-Å-diameter **tomato bushy stunt virus** (**TBSV**) has 180 identical coat proteins arranged, as schematically illustrated, around a single ~4800-nucleotide RNA molecule.

Satellite tobacco mosaic virus (**STMV**), perhaps the smallest known virus (it can replicate only in cells that are coinfected with TMV), is a spherical virus with a diameter of 172 Å. It consists of 60 identical protein subunits that enclose an RNA molecule of only 1059 nucleotides. This RNA (*yellow*), which nestles just below the surface of the viral capsid (*blue*) is extensively base paired.

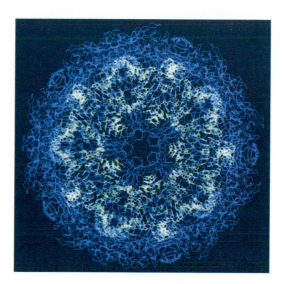

The capsid proteins of TBSV actually form two layers, and the RNA is sandwiched in between them. The volume constraints imposed by this arrangement require that the RNA be tightly packed. The negative charges of the RNA phosphate groups are presumably neutralized by the numerous positively charged Arg and Lys residues of the capsid proteins.

[Image of TMV assembly courtesy of Hong Wang and Gerald Stubbs, Vanderbilt University. TMV model courtesy of Gerald Stubbs and Keiichi Namba, Vanderbilt University; and Donald Caspar, Brandeis University. Drawing of TBSV copyrighted © by Irving Geis. X-Ray structure of STMV courtesy of Alexander McPherson, University of California at Riverside.]

SUMMARY

1. The most common form of DNA is B-DNA, which is a right-handed double helix containing A · T and G · C base pairs of similar geometry. The A-DNA helix, which also occurs in double-stranded RNA, is wider and flatter than the B-DNA helix. The left-handed Z-DNA helix may occur in sequences of alternating purines and pyrimidines.

2. The flexibility of nucleotides in nucleic acids is constrained by the allowed rotation angles around the glycosidic bond, the puckering of the ribose ring, and the torsion angles of the sugar–phosphate backbone.

3. Naturally occurring DNA is negatively supercoiled (underwound). Topoisomerases relax supercoils by cleaving one or both strands of the DNA, passing the DNA through the break, and resealing the broken strand(s).

4. Nucleic acids can be denatured by increasing the temperature above their T_m and renatured by lowering the temperature to ~25°C below their T_m.

5. The structures of nucleic acids are stabilized by Watson–Crick base pairing, by hydrophobic interactions between stacked base pairs, and by divalent cations that shield adjacent phosphate groups.

6. RNA molecules assume a variety of structures containing double-stranded stems and single-stranded loops. Some RNAs have catalytic activity.

7. Nucleic acids are fractionated by methods similar to those used to fractionate proteins, including solubilization, affinity chromatography, electrophoresis, and ultracentrifugation.

8. Sequence-specific DNA-binding proteins interact primarily with bases in the major groove and with phosphate groups through direct and indirect hydrogen bonds, van der Waals interactions, and ionic interactions. The conformations of both the protein and the DNA may change on binding.

9. Common structural motifs in DNA-binding proteins include the helix–turn–helix motif in prokaryotic repressors, and zinc fingers and leucine zippers in eukaryotic transcription factors.

10. The DNA of eukaryotic chromatin winds around histone octamers to form nucleosome core particles that further condense in the presence of linker histones. Additional condensation is accomplished by folding chromatin into 300-Å-diameter filaments, which are then attached in loops to a protein scaffold to form a condensed (metaphase) chromosome.

REFERENCES

Nucleic Acid Structure

Bates, A.D. and Maxwell, A., *DNA Topology,* IRL Press (1993).

Berger, J.M., Gamblin, S.J., Harrison, S.C., and Wang, J.C., Structure and mechanism of DNA topoisomerase II, *Nature* **379,** 225–232 (1996).

Chu, G., Pulsed-field gel electrophoresis: theory and practice, *Methods* **1,** 129–142 (1990).

Dickerson, R.E., DNA structures from A to Z, *Methods Enzymol.* **211,** 67–111 (1992).

Hagerman, P.J. and Amiri, K.M.A., Hammering away at RNA global structure, *Curr. Opin. Struct. Biol.* **6,** 317–321 (1996). [A brief review of developments in tRNA and hammerhead ribozyme structure.]

Herbert, A. and Rich, A., The biology of left-handed Z-DNA, *J. Biol. Chem.* **271,** 11595–11598 (1996).

Jaeger, L., The new world of ribozymes, *Curr. Opin. Struct. Biol.* **7,** 324–335 (1997). [Summarizes the catalytic versatility of RNA, with examples of naturally occurring and artificial ribozymes.]

Joshua-Tor, L. and Sussman, J.L., The coming of age of DNA crystallography, *Curr. Opin. Struct. Biol.* **3,** 323–335 (1993).

Patikoglou, G. and Burley, S.K., Eukaryotic transcription factor-DNA complexes, *Annu. Rev. Biophys. Biomol. Struct.* **26,** 289–325 (1997).

Rossman, M.G. and Johnson, J.E., Icosahedral RNA virus structure, *Annu. Rev. Biochem.* **58,** 533–573 (1989).

Sharma, A. and Mondragón, A., DNA topoisomerase, *Curr. Opin. Struct. Biol.* **5,** 39–47 (1995).

Sinden, R.R., *DNA Structure and Function,* Academic Press (1994). [Reviews the structure and conformations of DNA with simple and clear diagrams.]

Snustad, D.P., Simmons, M.J., and Jenkins, J.B., *Principles of Genetics,* Wiley (1997). [This and other textbooks review DNA structure and function.]

Wahl, M.C. and Sundaralingam, M., New crystal structures of nucleic acids and their complexes, *Curr. Opin. Struct. Biol.* **5,** 282–295 (1995).

Chromatin

Gruss, C. and Knippers, R., Structure of replicating chromatin, *Prog. Nucleic Acid Res. Mol. Biol.* **52,** 337–365 (1996).

Luger, K., Mäder, A.W., Richmond, R.K., Sargent, D.F., and Richmond, T.J., Crystal structure of the nucleosome core particle at 2.8 Å resolution, *Nature* **389,** 251–260 (1997).

van Holde, K. and Zlatonova, J., Chromatin higher order structure: chasing a mirage? *J. Biol. Chem.* **270,** 8373–8376 (1995). [Argues that chromatin structure is fundamentally irregular.]

Widom, J., Structure, dynamics, and function of chromatin in vitro, *Annu. Rev. Biophys. Biomol. Struct.* **27,** 285–327 (1998).

Wu, C., Chromatin remodeling and the control of gene expression, *J. Biol. Chem.* **272,** 28171–28174 (1997).

DNA–Protein Interactions

Aggarwal, A.K., Structure and function of restriction endonucleases, *Curr. Opin. Struct. Biol.* **5,** 11–19 (1995).

Berg, J.M. and Shi, Y., The galvanization of biology: a growing appreciation of the roles of zinc, *Science* **271,** 1081–1085 (1996). [Summarizes different types of zinc fingers and discusses why zinc is suitable for stabilizing small protein domains.]

Choo, Y. and Klug, A., Physical basis of a protein–DNA recognition code, *Curr. Opin. Struct. Biol.* **7,** 117–125 (1997).

Harrison, S.C., A structural taxonomy of DNA-binding domains, *Nature* **353,** 715–719 (1991). [A succinct review of common structural motifs in DNA-binding proteins.]

Shakked, Z., Guzikevich-Guerstein, G., Frolow, F., Rabinovich, D., Joachimiak, A., and Sigler, P.B., Determinants of repressor/operator recognition from the structure of the *trp* operator binding site, *Nature* **368,** 469–473 (1994).

Somers, W.S. and Phillips, S.E.V., Crystal structure of the *met* repressor–operator complex at 2.8 Å resolution reveals DNA recognition by β-strands, *Nature* **359,** 387–393 (1992).

KEY TERMS

syn conformation	supercoiling	topoisomerase	anneal
anti conformation	linking number	catenate	hybridize
C2'-endo	twist	melting curve	Watson–Crick base pair
C3'-endo	writhing number	T_m	Hoogsteen base pair

stacking interactions
ribozyme
aptamer
pulsed-field gel elec-
trophoresis
intercalation agent
Southern blotting

repressor
operator
HTH motif
indirect readout
transcription factor
zinc finger
heptad repeat

leucine zipper
HLH motif
contour length
shear degradation
chromosome
chromatin
histone

linker histone
nucleosome
nucleosome core particle
linker DNA

STUDY EXERCISES

1. Summarize the differences between A-, B-, and Z-DNA.

2. How do the structures of RNA and DNA differ?

3. How do Type I and Type II topoisomerases alter DNA topology?

4. Explain the molecular events of nucleic acid denaturation and renaturation.

5. Describe the forces that stabilize nucleic acid structure.

6. Describe the types of interactions between nucleic acids and proteins.

7. How does the DNA binding of a zinc finger–containing transcription factor differ from that of a prokaryotic repressor?

8. Describe how DNA is packaged in eukaryotic cells.

PROBLEMS

1. Amino acid residues in proteins are each specified by three contiguous bases. What is the contour length of a segment of B-DNA that encodes a 50-kD protein? Calculate the contour length for the same gene if it assumed an A-DNA conformation.

2. Unusual bases and non-Watson–Crick base pairs frequently appear in tRNA molecules. (a) Which base is most likely to pair with hypoxanthine (Section 22-1)? Draw this base pair. (b) Draw the structure of a G·U base pair.

3. The degree of supercoiling of a circular DNA molecule can be assessed by using an ultracentrifuge to measure its sedimentation velocity relative to the corresponding relaxed circular DNA molecule. (a) Does the supercoiled circle sediment faster or slower than the relaxed circle? (b) Can an ultracentrifuge be used to distinguish between negatively and positively supercoiled molecules? Explain.

4. You have discovered an enzyme secreted by a particularly virulent bacterium that cleaves the C2′—C3′ bond in the deoxyribose residues of duplex DNA. What is the effect of this enzyme on supercoiled DNA?

5. A closed circular duplex DNA has a 100-bp segment of alternating C and G residues. On transfer to a high salt solution, this segment undergoes a transition from the B conformation to the Z conformation. What is the change in its linking number, writhing number, and twist?

6. Compare the melting temperature of a 1-kb segment of DNA containing 20% A residues to that of a 1-kb segment containing 30% A residues under the same conditions.

7. How is the melting curve of duplex DNA affected by (a) decreasing the ionic strength of the solution and (b) adding a small amount of ethanol?

8. *E. coli* ribosomes contain three RNA molecules named for their sedimentation behavior: 5S, 18S, and 28S RNAs. Draw a diagram of a centrifuge tube showing the approximate positions of the three RNA species following ultracentrifugation in a sucrose density gradient.

9. What is the probability that the palindromic symmetry of the *trp* repressor target DNA sequence (Section 23-4B) is merely accidental?

10. For a linear B-DNA molecule of 50,000 kb, calculate (a) the contour length, (b) the length of the DNA as packaged in nucleosomes with linker histones present, and (c) the length of the DNA in a 300-Å filament.

11. Mouse genomic DNA is treated with a restriction endonuclease and electrophoresed in an agarose gel. A radioactive probe made from the human gene *rxr-1* is used to perform a Southern blot. The experiment was repeated three times. Explain the results of these repeated experiments:

Experiment 1. The autoradiogram shows a large smudge at a position corresponding to the top of the lane in which the mouse DNA was electrophoresed.

Experiment 2. The autoradiogram shows a smudge over the entire lane containing the mouse DNA.

Experiment 3. The autoradiogram shows three bands of varying intensity.

[Problem by Bruce Wightman, Muhlenberg College.]

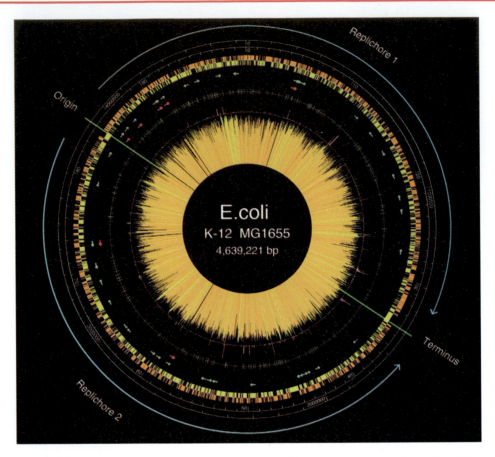

The E. coli *chromosome contains thousands of genes, here arranged by type in concentric circles. The entire chromosome, a single DNA molecule, must be replicated rapidly and accurately. Errors that do occur can often be repaired to ensure the faithful transmission of genetic information. [Courtesy of Frederick Blattner, University of Wisconsin.]*

DNA REPLICATION, REPAIR, AND RECOMBINATION

Watson and Crick's seminal paper describing the DNA double helix ended with the statement: "It has not escaped our notice that the specific pairing we have postulated immediately suggests a possible copying mechanism for the genetic material." As they predicted, when DNA replicates, each polynucleotide strand acts as a template for the formation of a complementary strand through base pairing interactions. The two strands of the parent molecule must therefore separate so that a complementary daughter strand can be enzymatically synthesized on the surface of each parent strand. This results in two molecules of duplex DNA, each consisting of one polynucleotide strand from the parent molecule and a newly synthesized complementary strand (Fig. 3-14). Such a mode of replication is termed **semiconservative** (in **conservative** replication, the parental DNA would remain intact and both strands of the daughter duplex would be newly synthesized).

The semiconservative nature of DNA replication was elegantly demonstrated in 1958 by Matthew Meselson and Franklin Stahl. The density of DNA was increased by labeling it with ^{15}N, a heavy isotope of nitrogen (^{14}N is the naturally abundant isotope). This was accomplished by growing *E. coli* in a medium that contained $^{15}NH_4Cl$ as its only nitrogen source. The labeled bacteria were then abruptly transferred to an ^{14}N-containing medium and the density of their DNA was monitored over several generations by equilibrium density gradient ultracentrifugation (Section 5-2E).

Meselson and Stahl found that after one generation (one doubling of the cell population), all the DNA had a density exactly halfway between the densities of fully ^{15}N-labeled DNA and unlabeled DNA. This DNA must therefore contain equal amounts of ^{14}N and ^{15}N as is expected after one generation of semiconservative replication (Fig. 24-1). Conservative DNA replication, in contrast, would preserve the fully ^{15}N-labeled parental strand and generate an equal amount of unlabeled DNA. After two generations, half of the DNA molecules were unlabeled and the remainder were ^{14}N–^{15}N hybrids. In succeeding generations, the amount of unlabeled DNA increased relative to the amount of hybrid DNA although the hybrid never totally disappeared. This is in accord with semiconservative replication but at odds with conservative replication, in which hybrid DNA never forms.

The details of DNA replication, including the unwinding of the parental strands and the assembly of complementary strands from nucleoside triphosphates, have emerged gradually since 1958. DNA replication is far more complex than the overall chemistry of this process might suggest, in large part because replication must be extremely accurate in order to preserve the integrity of the genome from generation to generation. In this chapter, we examine the protein assemblies that mediate DNA replication in prokaryotes and eukaryotes. We also discuss the mechanisms for ensuring fidelity during replication and for correcting polymerization errors and other types of DNA damage.

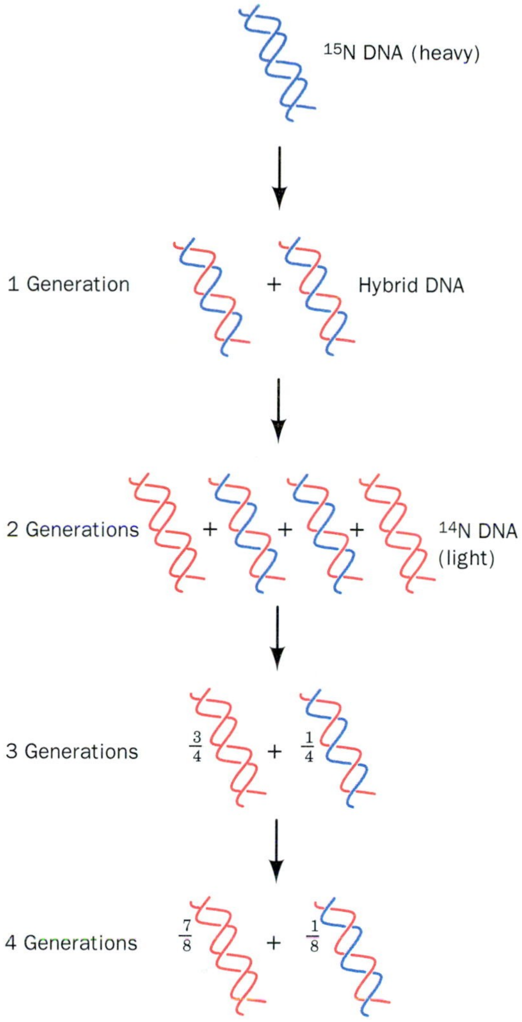

Figure 24-1. **The Meselson and Stahl experiment.** Parental (^{15}N-labeled or "heavy") DNA is replicated semiconservatively so that in the first generation, DNA molecules contain one parental strand (*blue*) and one newly synthesized strand (*red*). In succeeding generations, the proportion of ^{14}N-labeled ("light") strands increases, but hybrid molecules containing one heavy and one light strand persist. ✳ See the Animated Figures.

1. OVERVIEW OF DNA REPLICATION

DNA is replicated by enzymes known as **DNA-directed DNA polymerases** or simply **DNA polymerases.** These enzymes use single-stranded DNA as templates on which to catalyze the synthesis of the complementary strand

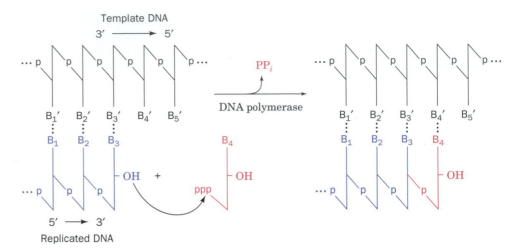

Figure 24-2.   Action of DNA polymerases. These enzymes assemble incoming deoxynucleoside triphosphates on single-stranded DNA templates such that the growing strand is elongated in its $5' \to 3'$ direction.

from the appropriate deoxynucleoside triphosphates (Fig. 24-2). The reaction occurs through the nucleophilic attack of the growing DNA chain's $3'$-OH group on the α-phosphoryl of an incoming nucleoside triphosphate. The otherwise reversible reaction is driven by the subsequent hydrolysis of the eliminated PP_i. The incoming nucleotides are selected by their ability to form Watson–Crick base pairs with the template DNA so that the newly synthesized DNA strand forms a double helix with the template strand. Nearly all known DNA polymerases can add a nucleotide only to the free $3'$-OH group of a base-paired polynucleotide so that *DNA chains are extended only in the $5' \to 3'$ direction.*

DNA Replication Occurs at Replication Forks

John Cairns observed DNA replication through the autoradiography of chromosomes from *E. coli* grown in a medium containing [³H]thymidine. These circular chromosomes contain replication "eyes" or "bubbles" (Fig. 24-3), called **θ structures** (after their resemblance to the Greek letter theta), that form when the two parental strands of DNA separate to allow the synthesis of their complementary daughter strands. DNA replication involving θ structures is known as **θ replication.**

A branch point in a replication eye at which DNA synthesis occurs is called a **replication fork.** A replication bubble may contain one or two replication forks (**unidirectional** or **bidirectional replication**). Autoradiographic

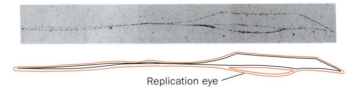

Replication eye

Figure 24-3.   An autoradiogram and its interpretive drawing of a replicating *E. coli* chromosome. The bacterium had been grown in a medium containing [³H]thymidine, thereby labeling the subsequently synthesized DNA so that it appears as a line of dark grains in the photographic emulsion (*red lines in the drawing*). The size of the replication eye indicates that this circular chromosome is about one-sixth duplicated. [Courtesy of John Cairns, Cold Spring Harbor Laboratory.]

(a)

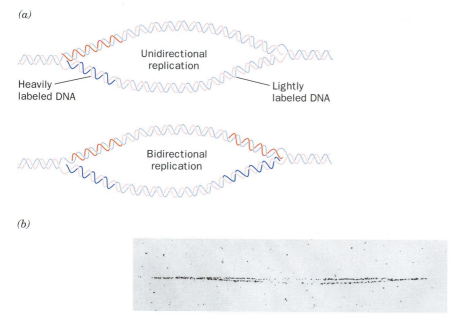

Figure 24-4. **The autoradiographic differentia-tion of unidirectional and bidirectional θ repli-cation of DNA.** (*a*) An organism is grown for several generations in a medium that is lightly la-beled with [³H]thymidine so that all of its DNA will be visible in an autoradiogram. A large amount of [³H]thymidine is then added to the medium for a few seconds before the DNA is iso-lated (**pulse labeling**) in order to heavily label bases near the replication fork(s). Unidirectional DNA replication will exhibit only one heavily labeled branch point, whereas bidirectional DNA replication will exhibit two such branch points. (*b*) An autoradiogram of *E. coli* DNA demonstrat-ing that it is bidirectionally replicated. [Courtesy of David M. Prescott and P.L. Kuempel, University of Colorado.]

(b)

studies have demonstrated that θ replication is almost always bidirectional (Fig. 24-4). In other words, *DNA synthesis proceeds in both directions from the point where replication is initiated.*

Replication Is Semidiscontinuous

The low-resolution images provided by autoradiograms such as Figs. 24-3 and 24-4*b* suggest that duplex DNA's two antiparallel strands are si-multaneously replicated at an advancing replication fork. Yet, since DNA polymerases extend DNA strands only in the 5′ → 3′ direction, how can they copy the parent strand that extends in the 5′ → 3′ direction past the replication fork? This question was answered in 1968 by Reiji Okazaki through the following experiment. If a growing *E. coli* culture is pulse la-beled for 30 s with [³H]thymidine, much of the radioactive and hence newly synthesized DNA consists of 1000- to 2000-nucleotide **(nt)** fragments (in eukaryotes, these so-called **Okazaki fragments** are 100–200 nt long). When the cells are transferred to an unlabeled medium after the [³H]thymidine pulse, the size of the labeled fragments increases over time. *The Okazaki fragments must therefore become covalently in-corporated into larger DNA molecules.*

Okazaki interpreted his experimental results in terms of the **semidiscontinuous replication** model (Fig. 24-5). The two par-ent strands are replicated in different ways. The newly syn-thesized DNA strand that extends 5′ → 3′ in the direction of replication fork movement, the **leading strand,** is continuously synthesized in its 5′ → 3′ direction as the replication fork ad-vances. The other new strand, the **lagging strand,** is also syn-thesized in its 5′ → 3′ direction. However, it can only be made discontinu-ously, as Okazaki fragments, as single-stranded parental DNA becomes newly exposed at the replication fork. The Okazaki fragments are later co-valently joined together by the enzyme **DNA ligase.**

DNA Synthesis Extends RNA Primers

Given that DNA polymerases require a free 3′-OH group to extend a DNA chain, how is DNA synthesis initiated? Careful analysis of Okazaki

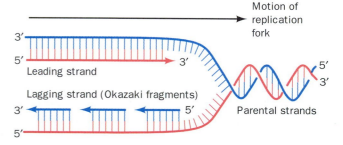

Figure 24-5. **Semidiscontinuous DNA replica-tion.** Both daughter strands (*leading strand red, lagging strand blue*) are synthesized in their 5′ → 3′ direction. The leading strand is synthe-sized continuously, whereas the lagging strand is synthesized discontinuously.

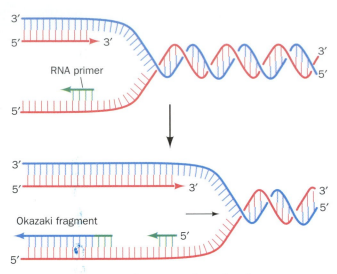

Figure 24-6. Priming of DNA synthesis by short RNA segments.

fragments revealed that their 5' ends consist of RNA segments of 1 to 60 nt (a length that is species dependent) that are complementary to the template DNA chain (Fig. 24-6). In *E. coli,* these **RNA primers** are synthesized by the enzyme **primase.** Multiple priming events are required for lagging strand synthesis, but only one priming event is required to initiate synthesis of the leading strand. Mature DNA, however, does not contain RNA; *the RNA primers are eventually replaced with DNA.*

2. PROKARYOTIC DNA REPLICATION

DNA replication involves a great variety of enzymes in addition to those mentioned above. Many of the required enzymes were first isolated from prokaryotes and are therefore better understood than their eukaryotic counterparts. Accordingly, we begin with a detailed consideration of prokaryotic DNA replication. Replication in eukaryotes is discussed in Section 24-3.

A. DNA Polymerases

In 1957, Arthur Kornberg discovered an *E. coli* enzyme that catalyzes the synthesis of DNA, based on its ability to incorporate the radioactive label from [^{14}C]thymidine triphosphate into DNA. This enzyme, which is now known as **DNA polymerase I** or **Pol I,** consists of a single 928-residue polypeptide. Pol I is said to be a **processive** enzyme because *it catalyzes a series of successive nucleotide polymerization steps, typically 20 or more, without releasing the single-stranded template.*

Pol I Has Exonuclease Activity

In addition to its polymerase activity, Pol I has two independent hydrolytic activities that occupy separate active sites: a $3' \rightarrow 5'$ exonuclease and a $5' \rightarrow 3'$ exonuclease. *The $3' \rightarrow 5'$ exonuclease activity allows Pol I to edit its mistakes.* If Pol I erroneously incorporates a mispaired nucleotide at the end of a growing DNA chain, the polymerase activity is inhibited and the $3' \rightarrow 5'$ exonuclease hydrolytically excises the offending nucleotide (Fig. 24-7). The polymerase activity then resumes DNA replication. This

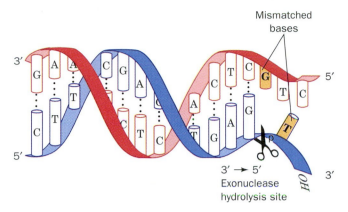

Figure 24-7.   The 3′ → 5′ exonuclease function of DNA polymerase I. This enzymatic activity excises mispaired nucleotides from the 3′ end of the growing DNA strand (*blue*).

proofreading mechanism explains the high fidelity of DNA replication by Pol I.

The Pol I 5′ → 3′ exonuclease binds to duplex DNA at single-strand nicks (breaks). It cleaves the nicked DNA strand in a base-paired region beyond the nick to excise the DNA as either mononucleotides or oligonucleotides of up to 10 residues (Fig. 24-8).

Although Pol I was the first of the *E. coli* DNA polymerases to be discovered, it is not *E. coli*'s primary replicase. Rather, its most important (and only essential) function is in lagging strand synthesis, in which it removes the RNA primers and replaces them with DNA. This process involves the 5′ → 3′ exonuclease and polymerase activities of Pol I working in concert to excise the ribonucleotides on the 5′ end of the single-strand nick between the new and old (previously synthesized) Okazaki fragments and to replace them with deoxynucleotides that are appended to the 3′ end of the old Okazaki fragment (Fig. 24-9). The nick is thereby translated (moved) toward the DNA strand's 3′ end, a process known as **nick translation.** When the RNA has been entirely excised, the nick is sealed by the action of DNA ligase (Section 24-2C), thereby linking the new and old Okazaki fragments.

Biochemists use nick translation to prepare radioactive DNA. Double-stranded DNA is nicked in only a few places by treating it with small amounts of pancreatic **DNase I.** Radioactively labeled dNTPs are then added and Pol I translates the nicks, thereby replacing unlabeled deoxynucleotides with labeled deoxynucleotides.

Pol I also functions in the repair of damaged DNA. As we discuss in Section 24-5, damaged DNA is detected by a variety of DNA repair systems, many of which endonucleolytically cleave the damaged DNA on the 5′ side of the lesion. Pol I's 5′ → 3′ exonuclease activity then excises the damaged DNA while its polymerase activity fills in the resulting single-strand gap in the same way it replaces the RNA primers of Okazaki fragments. Thus, Pol I has indispensable roles in *E. coli* DNA replication and

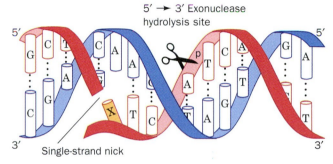

Figure 24-8.   The 5′ → 3′ exonuclease function of DNA polymerase I. This enzymatic activity excises up to 10 nucleotides from the 5′ end of a single-strand nick. The nucleotide immediately past the nick (X) may or may not be paired.

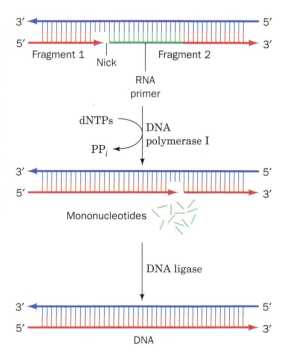

Figure 24-9.   The replacement of RNA primers by DNA in lagging strand synthesis. The RNA primer at the 5′ end of a newly synthesized Okazaki fragment (*Fragment 2*) is excised through the action of Pol I's 5′ → 3′ exonuclease function and replaced through its polymerase function, which adds deoxynucleotides to the 3′ end of the previously synthesized Okazaki fragment (*Fragment 1*). This translates the nick originally at the 5′ end of the RNA to the position that was occupied by its 3′ end (nick translation). The nick is sealed by the action of DNA ligase.

repair although it is not, as was first supposed, responsible for the bulk of DNA synthesis.

Structure of the Klenow Fragment

E. coli DNA polymerase I, a protein whose three enzymatic activities occupy three separate active sites, can be proteolytically cleaved to a large or **"Klenow" fragment** (residues 324–928), which contains both the polymerase and the $3' \rightarrow 5'$ exonuclease activities, and a small fragment (residues 1–323), which contains the $5' \rightarrow 3'$ exonuclease activity. The X-ray structure of the Klenow fragment, determined by Thomas Steitz, reveals that it consists of two domains (Fig. 24-10). The smaller domain (residues 324–517) contains the $3' \rightarrow 5'$ exonuclease site. The larger domain (residues 521–928; helix G and beyond in Fig. 24-10*b*) contains the polymerase active site at the bottom of a prominent cleft, a surprisingly large distance (~25 Å) from the $3' \rightarrow 5'$ exonuclease site. The cleft, which is lined with positively charged residues, has the appropriate size and shape (~22 Å × ~30 Å) to bind a B-DNA molecule in a manner resembling a right hand grasping a rod: The "thumb" consists of helices H–I; the "fingers" are helices L–P; and the remainder of the larger domain, the "palm," includes a six-stranded antiparallel β sheet that forms the floor of the cleft. The active sites of all DNA and RNA polymerases of known structure are located at the bottoms of similarly shaped clefts.

Steitz cocrystallized the Klenow fragment with a short DNA "template" strand and a complementary "primer" strand. The protein contacts only the phosphate backbone of the duplex DNA, consistent with Pol I's lack of sequence specificity in binding DNA. The separation of the polymerase and $3' \rightarrow 5'$ exonuclease active sites suggests that the bound DNA undergoes a large conformational shift in shuttling between these sites.

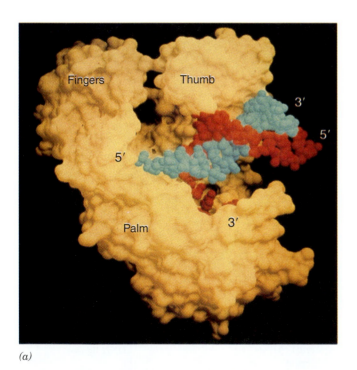

(a)

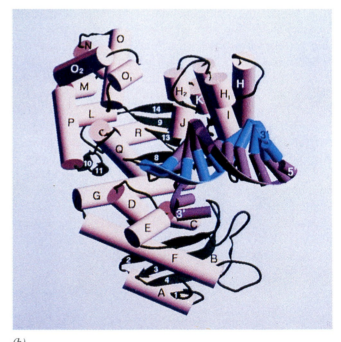

(b)

Figure 24-10.   The X-ray structure of *E. coli* DNA polymerase I Klenow fragment in complex with a double-helical DNA.
(*a*) The solvent-accessible surface of the Klenow fragment with a 12-nt DNA template strand in cyan and a 14-nt primer strand in red. (*b*) A tube-and-arrow representation of the complex in the same orientation as *a* in which the template strand is blue and the primer strand is magenta. [Courtesy of Thomas Steitz, Yale University.] ● **See the Interactive Exercises.**

Table 24-1.   **Properties of *E. coli* DNA Polymerases**

	Pol I	Pol II	Pol III
Mass (kD)	103	90	130
Molecules/cell	400	?	10–20
Turnover number[a]	600	30	9000
Structural gene	polA	polB	polC
Conditionally lethal mutant	+	−	+
Polymerization: 5′ → 3′	+	+	+
Exonuclease: 3′ → 5′	+	+	+
Exonuclease: 5′ → 3′	+	−	−

[a]Nucleotides polymerized $\min^{-1} \cdot molecule^{-1}$ at 37°C.

Source: Kornberg, A. and Baker, T.A., *DNA Replication* (2nd ed.), p. 167, Freeman (1992).

DNA Polymerase III Is *E. coli's* DNA Replicase

The discovery of normally growing *E. coli* mutants that have very little (but not entirely absent) Pol I activity stimulated the search for additional DNA polymerizing activities. This effort was rewarded by the discovery of two more enzymes, designated, in the order they were discovered, **DNA polymerase II (Pol II)** and **DNA polymerase III (Pol III).** The properties of these enzymes are compared with those of Pol I in Table 24-1. Pol II and Pol III had not previously been detected because their combined activities in the assays used are normally <5% that of Pol I. Pol II participates in DNA repair; mutant cells lacking Pol II can therefore grow normally. The absence of Pol III, however, is lethal, demonstrating that it is *E. coli's* DNA replicase.

The catalytic core of Pol III consists of three subunits: α (the product of the *polC* gene in *E. coli*), which contains the complex's DNA polymerase activity; ε, its 3′ → 5′ exonuclease; and θ. At least seven other types of subunits (Table 24-2) make up a labile multisubunit enzyme known as the **Pol III holoenzyme.** The catalytic properties of the Pol III core resemble those of Pol I except that Pol III lacks 5′ → 3′ exonuclease activity on double-stranded DNA. Thus, *Pol III can synthesize a DNA strand complementary to a single-stranded template and can edit the polymerization reaction to increase replication fidelity, but it cannot catalyze nick translation.*

Table 24-2.   **Components of DNA Polymerase III Holoenzyme**

Subunit	Mass (kD)	Structural Gene
α	130	polC (dnaE)
ε	27.5	dnaQ
θ	10	holE
τ	71	dnaX[a]
γ	45.4	dnaX[a]
δ	35	holA
δ′	33	holB
χ	15	holC
ψ	12	holD
β	40.6	dnaN

[a]The γ and τ subunits are encoded by the same gene sequence; the γ subunit corresponds to the N-terminal end of the τ subunit.

Sources: Kornberg, A. and Baker, T.A., *DNA Replication* (2nd ed.), p. 169, Freeman (1992); *and* Baker, T.A. and Wickner, S.H., *Annu. Rev. Genet.* **26**, 450 (1992).

B. Initiation of Replication

Replication of the 4.6×10^6-bp *E. coli* chromosome is initiated at a 245-bp region known as the ***oriC*** locus. Elements of this sequence are highly conserved among gram-negative bacteria. A complex of up to 30 subunits of the 52-kD **DnaA protein** binds to *oriC* and causes a segment of the DNA to melt open. The melting process requires ATP hydrolysis and is probably facilitated by the AT-rich nature of the DNA segment.

The hexameric **DnaB protein,** a so-called **helicase,** further unwinds the DNA strands in both directions in an ATP-dependent manner. This unwinding produces positive supercoils, which are offset to some extent by naturally occurring negative supercoils. However, continued unwinding at the replication fork requires the action of DNA gyrase (a Type II topoisomerase; Section 23-1C) to generate additional negative supercoils.

The separated DNA strands behind the advancing helicase are prevented from reannealing by the binding of **single-strand binding protein (SSB).** SSB, a tetrameric protein, coats single-stranded DNA, thereby maintaining it in an unpaired state. DNA must be stripped of SSB before it can be replicated.

> ● See Guided Exploration 25:
>
> **The Replication of DNA in *E. coli*.**

Figure 24-11. **The X-ray structure of SSB complexed with DNA.** The DNA-binding core of bacteriophage T4 SSB (gp32) is shown as its solvent-accessible surface colored according to its electrostatic potential (*blue,* most positive; *red,* most negative; and *white,* neutral). The single-stranded DNA, which is represented by (dT)$_4$ (the bound hexadeoxynucleotide's two remaining nucleotides are not visible in the X-ray structure), is shown in space-filling form with C yellow, O red, N green, and P black. The protein binds to the DNA via electrostatic interactions with the phosphate backbone, whereas the bases contact hydrophobic pockets on the protein surface. [Courtesy of Thomas Steitz, Yale University.]

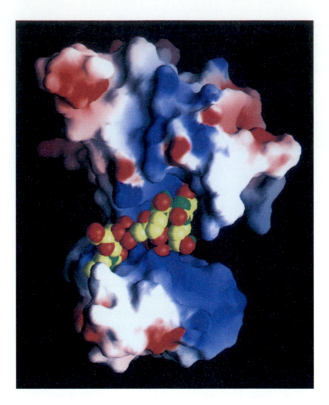

The X-ray structure of the DNA-binding core of the bacteriophage T4–encoded SSB known as **gp32** (gp for *gene product*), in its complex with a hexadeoxynucleotide, reveals that the protein has a deep positively charged cleft in which the DNA binds (Fig. 24-11). The central portion of this cleft is only 15 Å wide, which accounts for gp32's low affinity for duplex DNA (B-DNA has a diameter of ~20 Å).

The Primosome

All DNA synthesis, both of leading and lagging strands, requires the prior synthesis of an RNA primer. Primer synthesis in *E. coli* is mediated by an ~600-kD protein assembly known as a **primosome** (Table 24-3). The complex includes a helicase (DnaB) and primase **(DnaG).** The primosome is propelled in the $5' \rightarrow 3'$ direction along the DNA template for the lagging strand (i.e., toward the replication fork) by **PriA-** and DnaB-catalyzed ATP hydrolysis. This motion, which displaces the SSB in its path, is opposite in direction to that of template reading during DNA chain synthesis. Consequently, the primosome reverses its migration momentarily to allow primase to synthesize an RNA primer in the $5' \rightarrow 3'$ direction (Fig. 24-6).

The primosome is required to initiate each Okazaki fragment. The single RNA segment that primes the synthesis of the leading strand can be synthesized, at least *in vitro,* by either primase or **RNA polymerase** (the enzyme that synthesizes RNA transcripts from a DNA template; Section 25-1), but its rate of synthesis is greatly enhanced when both enzymes are present.

Table 24-3. **Proteins of the Primosome**[a]

Protein	Subunit Structure	Subunit Mass (kD)
PriA	Monomer	76
PriB	Dimer	11.5
PriC	Monomer	23
DnaT	Trimer	22
DnaB	Hexamer	50
DnaC	Monomer	29
Primase (DnaG)	Monomer	60

[a]The complex of all primosome proteins but primase is known as the preprimosome.

Source: Kornberg, A. and Baker, T.A., *DNA Replication* (2nd ed.), pp. 286–288, Freeman (1992).

C. Synthesis of the Leading and Lagging Strands

The Pol III holoenzyme catalyzes the synthesis of both the leading and lagging strands. This occurs in a single multiprotein particle, the **replisome,** which contains two Pol III enzymes. *In order for the replisome to move as*

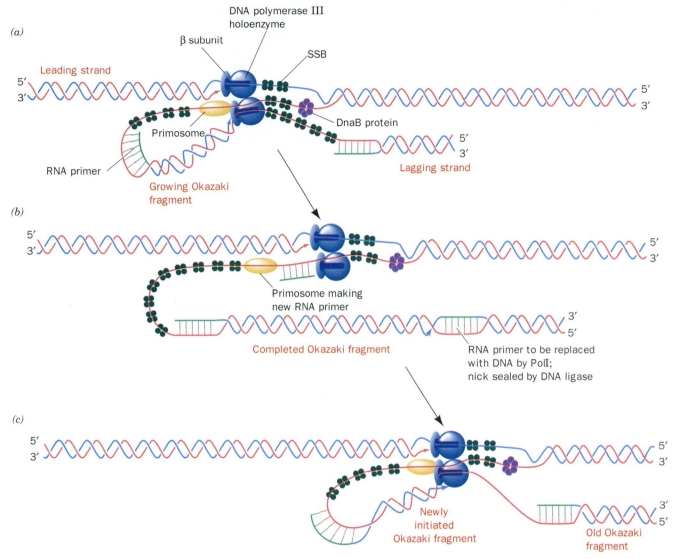

Figure 24-12. *Key to Function.* **The replication of *E. coli* DNA.** (*a*) The replisome, which contains two DNA polymerase III holoenzymes, synthesizes both the leading and the lagging strands. The lagging strand template must loop around to permit the holoenzyme to extend the primed lagging strand. (*b*) The holoenzyme releases the lagging strand template when it encoun- ters the previously synthesized Okazaki fragment. This may signal the primosome to initiate synthesis of a lagging strand RNA primer. (*c*) The holoenzyme rebinds the lagging strand template and extends the RNA primer to form a new Okazaki fragment. Note that in this model, leading strand synthesis is always ahead of lagging strand synthesis.

a single unit in the 5' → 3' direction along the leading strand, the lagging strand template must loop around (Fig. 24-12). After completing the syn- thesis of an Okazaki fragment, the lagging strand holoenzyme relocates to a new primer near the replication fork and resumes synthesis. *The result of this process is a continuous leading strand and a series of RNA-primed Okazaki fragments separated by single-strand nicks.* The RNA primers are replaced with DNA through Pol I–catalyzed nick translation, and the nicks in the lagging strand are sealed through the action of DNA ligase.

The DNA Ligase Reaction Is Activated by NAD⁺ or ATP

The free energy for the DNA ligase reaction is obtained, in a species-dependent manner, through the coupled hydrolysis of either NAD^+ to **nicotinamide mononucleotide (NMN⁺)** + AMP, or ATP to PP_i + AMP.

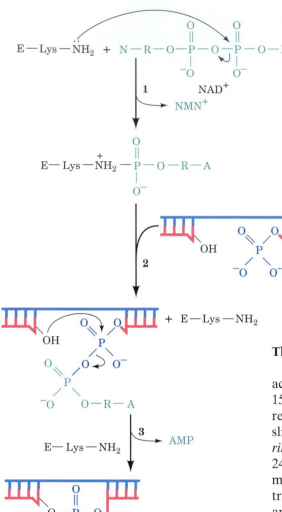

Figure 24-13.   The reactions catalyzed by E. coli DNA ligase. In eukaryotic and T4 ligases, NAD$^+$ is replaced by ATP so that PP$_i$ rather than NMN$^+$ is eliminated in the first reaction step. Here A, R, and N represent the adenine, ribose, and nicotinamide residues, respectively.

E. coli DNA ligase, a 77-kD monomer that uses NAD$^+$, catalyzes a three-step reaction (Fig. 24-13):

1. The adenyl group of NAD$^+$ is transferred to the ε-amino group of an enzyme Lys residue to form an unusual phosphoamide adduct.
2. The adenyl group of this activated enzyme is transferred to the 5′-phosphoryl terminus of the nick to form an adenylated DNA. Here, AMP is linked to the 5′-nucleotide via a pyrophosphate rather than the usual phosphodiester bond.
3. DNA ligase catalyzes the formation of a phosphodiester bond by attack of the 3′-OH on the 5′-phosphoryl group, thereby sealing the nick and releasing AMP.

ATP-requiring DNA ligases, such as those of eukaryotes, release PP$_i$ in the first step of the reaction rather than NMN$^+$. The DNA ligase from the bacteriophage T4 is notable because it can link together two duplex DNAs that lack complementary single-stranded ends **(blunt end ligation)** in a reaction that is a boon to genetic engineering (Section 3-5).

The β Subunit of Pol III Promotes Processivity

Many of the subunits listed in Table 24-2 modulate Pol III's polymerase activity. For example, the Pol III core has a processivity of only 10 to 15 residues. However, the β subunit increases Pol III's processivity to >5000 residues. The observation that a β subunit bound to a cut circular DNA slides to the break and falls off suggests that *the β subunit forms a closed ring or clamp around the DNA*. The X-ray structure of the β subunit (Fig. 24-14), determined by John Kuriyan, reveals that it is a dimer of C-shaped monomers that form an ~80-Å-diameter donut-shaped structure. The central ~35-Å-diameter hole is larger than the 20- and 26-Å diameters of B- and A-DNAs (the hybrid helices containing RNA primers and DNA have an A-DNA-like conformation; Section 23-1A). Each β monomer forms three domains of similar structure so that the dimeric ring is a pseudo-symmetrical six-pointed star. Electrostatic calculations indicate that the interior surface of the ring is positively charged, whereas its outer surface is negatively charged.

Model-building studies in which a B-DNA helix is threaded through the central hole (Fig. 24-14) indicate that the protein's α helices span the major and minor grooves of the DNA rather than entering into them as do, for example, the recognition helices of helix–turn–helix motifs (Section 23-4B). It appears that the β subunit is designed to minimize its associations with DNA. This presumably permits the protein to freely slide along the DNA helix.

E. coli DNA is replicated at a rate of ~1000 nt/s. Thus, in lagging strand synthesis, the DNA polymerase holoenzyme must be reloaded onto the template strand every 1 to 2 s (Okazaki fragments are 1000–2000 nt long). This requires that the β clamp, which promotes Pol III's processivity, be rapidly unclamped and reclamped. The **γ complex** of the Pol III holoenzyme (subunit composition γ$_2$δδ′χψ) is also known as the **clamp loader** because it opens the β clamp to load it onto the DNA template in an ATP-dependent manner. Once the β clamp has been loaded onto the DNA, the Pol III core binds to the β clamp more tightly than does the γ complex, thereby displacing it and permitting processive DNA replication to proceed. When the polymerase encounters the previously synthesized Okazaki

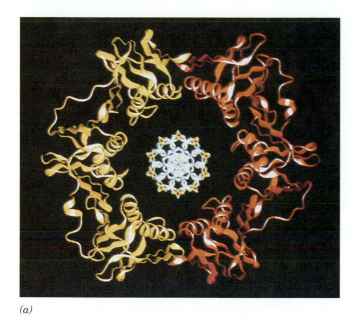

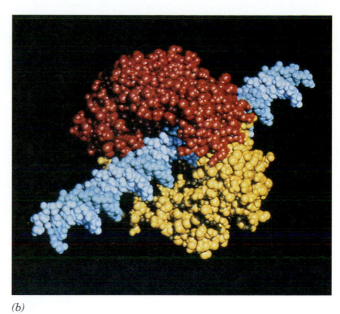

(a)   *(b)*

Figure 24-14.   The X-ray structure of the β subunit of *E. coli* Pol III holoenzyme. (*a*) A ribbon drawing showing the two monomeric units of the dimeric protein in yellow and red as viewed along the dimer's twofold axis. A stick model of B-DNA is placed with its helix axis coincident with the protein dimer's twofold axis. (*b*) A space-filling model of the protein, colored as in *a*, in a hypothetical complex with B-DNA (*cyan*). [Courtesy of John Kuriyan, The Rockefeller University.]

fragment, the Pol III core releases the DNA and loses its affinity for the β clamp, thereby allowing the γ complex access to the β clamp, which is unloaded from the DNA. The γ subunit and the Pol III core can then rapidly initiate the synthesis of a new Okazaki fragment because they are held in the vicinity of the lagging strand template through their linkage to the Pol III core engaged in leading strand synthesis.

D. Termination of Replication

The *E. coli* replication terminus is a large (350-kb) region flanked by seven nearly identical nonpalindromic ~23-bp terminator sites, **TerE, TerD,** and **TerA** on one side and **TerG, TerF, TerB,** and **TerC** on the other (Fig. 24-15; note that *oriC* is directly opposite the termination region on the *E. coli* chromosome). A replication fork, traveling counterclockwise as drawn in Fig. 24-15 passes through *TerG, TerF, TerB,* and *TerC* but stops on encountering either *TerA, TerD,* or *TerE* (*TerD* and *TerE* are presumably backup sites for *TerA*). Similarly, a clockwise-traveling replication fork passes *TerE, TerD,* and *TerA* but halts at *TerC* or, failing that, *TerB, TerF,* or *TerG*. Thus, these termination sites are polar; they act as one-way valves that allow replication forks to enter the terminus region but not to leave it. *This arrangement guarantees that the two replication forks generated by bidirectional initiation at oriC will meet in the replication terminus even if one of them arrives there well ahead of its counterpart.*

The arrest of replication fork motion at *Ter* sites requires the action of **Tus protein,** a 309-residue monomer that is the product of the *tus* gene (for *t*erminator *u*tilization *s*ubstance). Tus protein specifically binds to a *Ter* site, where it prevents strand displacement by DnaB helicase, thereby arresting replication fork advancement.

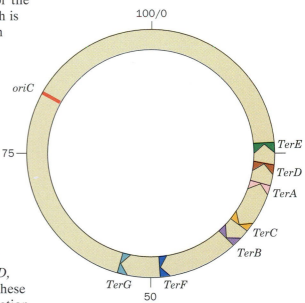

Figure 24-15.   A map of the *E. coli* chromosome showing the positions of the *Ter* sites. The *TerC, TerB, TerF,* and *TerG* sites, in combination with Tus protein, allow a counterclockwise-moving replisome to pass but not a clockwise-moving replisome. The opposite is true of the *TerA, TerD,* and *TerE* sites. Consequently, two replication forks that initiate bidirectional DNA replication at *oriC* will meet between the oppositely facing *Ter* sites.

Figure 24-16. **The X-ray structure of *E. coli* Tus in complex with a 15-bp *Ter*-containing DNA.** The protein is shown in ribbon form with its N-and C-terminal domains colored green and blue. The DNA is represented in stick form with its bases yellow and its sugar-phosphate backbone gold. [From Kamada, K., Horiuchi, T., Ohsumi, K., Shimamoto, N., and Morikawa, K., *Nature* 383, 599 (1996). Used with permission.] ● See the Interactive Exercises.

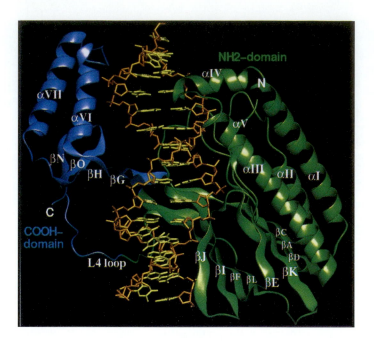

The X-ray structure of Tus in complex with a 15-bp *Ter* fragment (Fig. 24-16) reveals that the protein forms a deep positively charged cleft in which the DNA binds. A 5-bp segment of the DNA near the side of Tus that permits the passage of the replication fork is deformed and underwound relative to canonical (normal) DNA: Its major groove is deeper and its minor groove is significantly expanded. Protein side chains at the bottom of the cleft penetrate the DNA's deepened major groove to make sequence-specific contacts such that the protein cannot release the bound DNA without a large conformational change. Nevertheless, the mechanism through which Tus prevents replication fork advancement from one side of a *Ter* site but not the other remains unclear.

The final step in *E. coli* DNA replication is the topological unlinking of the catenated parental DNA strands, thereby separating the two replication products. This reaction is probably catalyzed by one or more topoisomerases.

E. Fidelity of Replication

Since a single polypeptide as small as the Pol I Klenow fragment can replicate DNA by itself, why does *E. coli* maintain a battery of >20 intricately coordinated proteins to replicate its chromosome? The answer apparently is *to ensure the nearly perfect fidelity of DNA replication required to accurately transmit genetic information.*

The rates of reversion of mutant *E. coli* or T4 phages to the wild type indicates that only one mispairing occurs per 10^8 to 10^{10} base pairs replicated. This corresponds to ~1 error per 1000 bacteria per generation. Such high replication accuracy arises from four sources:

1. Cells maintain balanced levels of dNTPs through the mechanisms discussed in Sections 22-1C and 22-2C. This is important because a dNTP present at aberrantly high levels is more likely to be misincorporated and, conversely, one present at low levels is more likely to be replaced by one of the dNTPs present at higher levels.

2. The polymerase reaction itself has extraordinary fidelity because it occurs in two stages. First, the incoming dNTP base pairs with the

template while the enzyme is in an open, catalytically inactive conformation. Polymerization occurs only after the polymerase has closed around the newly formed base pair, which properly positions the catalytic residues (induced fit). *The protein conformational change constitutes a double check for correct Watson–Crick base pairing between the dNTP and the template.*

3. The $3' \rightarrow 5'$ exonuclease functions of Pol I and Pol III detect and eliminate the occasional errors made by their polymerase functions.

4. A remarkable set of enzyme systems in all cells repairs residual errors in the newly synthesized DNA as well as any damage that may occur after its synthesis through chemical and/or physical insults. We discuss these DNA repair systems in Section 24-5.

In addition, the inability of a DNA polymerase to initiate chain elongation without a primer increases DNA replication fidelity. The first few nucleotides of a chain are those most likely to be mispaired because of the cooperative nature of base pairing interactions (Section 23-2). The use of RNA primers eliminates this source of error since the RNA is eventually replaced by DNA under conditions that permit more accurate base pairing.

3. EUKARYOTIC DNA REPLICATION

Eukaryotic and prokaryotic DNA replication mechanisms are remarkably similar, although the eukaryotic system is vastly more complex. Other modes of DNA replication occur in mitochondria and in certain viruses and bacteria (see Box 24-1). In this section, we consider some of the proteins of eukaryotic DNA replication and the challenges of replicating the ends of linear chromosomes.

A. Eukaryotic DNA Polymerases

Animal cells contain at least five distinct DNA polymerases, designated, in the order of their discovery, **DNA polymerases α, β, γ, δ, and ε** (Table 24-4). Their functions were largely elucidated by their different responses to inhibitors.

DNA polymerase α, which occurs only in the cell nucleus, participates in the replication of chromosomal DNA. This multisubunit protein (four types of subunits in *Drosophila;* five in rat liver), like all DNA polymerases, replicates DNA by extending a primer in the $5' \rightarrow 3'$ direction under the direction of a single-stranded DNA template. DNA polymerase α has a tightly associated primase activity. However, it lacks exonuclease activity, so the DNA it replicates must be proofread by some other means.

Table 24-4. Properties of Animal DNA Polymerases

	α	β	γ	δ	ε
Location	Nucleus	Nucleus	Mitochondrion	Nucleus	Nucleus
Mass (kD)	>250	36–38	160–300	170	256
Inhibitors:					
Aphidicolin	Strong	None	None	Strong	Strong
Dideoxy NTPs	None	Strong	Strong	Weak	Weak
N-Ethylmaleimide	Strong	None	Strong	Strong	Strong

Source: Kornberg, A. and Baker, T.A., *DNA Replication* (2nd ed.), p. 199, Freeman (1992).

Box 24-1

BIOCHEMISTRY IN FOCUS

Other Modes of DNA Replication

Whereas many viral and bacterial genomes are replicated by the leading strand–lagging strand mechanism that occurs in *E. coli* and eukaryotes, small circular chromosomes can also be replicated in other ways. For example, some bacteriophages contain a single-stranded circular DNA, the (+) strand. This serves as a template for the synthesis of the complementary (−) strand by a mechanism that resembles conventional leading strand synthesis to form a circular duplex DNA known as the **replicative form** (*left*). Additional copies of the (+) strand may then be synthesized via the **rolling-circle** or **σ replication** mode (so called because of the resemblance of the replicating structure to the Greek lowercase letter sigma). This process initiates at a single-strand break in the (+) strand and uses the (−) strand as a template (*right*). As the new (+) strand is synthesized, it displaces the existing (+) strand. Multiple rounds of rolling-circle replication can generate a large number of tandemly linked (+) strands that are later separated by an endonuclease for packaging into individual phage particles.

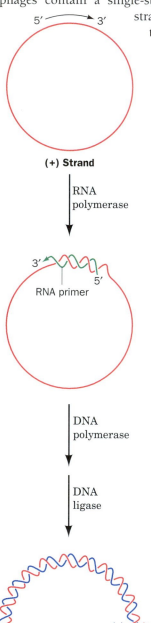

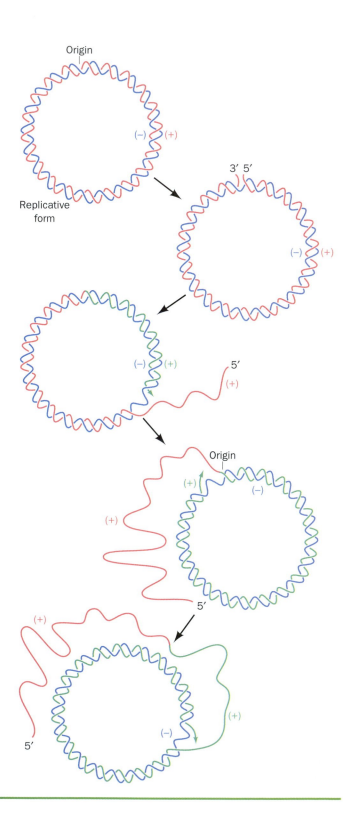

The replication of the 5386-nt single-stranded DNA of **bacteriophage ϕX174** begins with the synthesis of the (−) strand to produce a double-stranded replicative form. This process requires the binding of a nearly 600-kD primosome. The primosome–DNA complex, which is characterized by one or two associated small DNA loops (*below*), remains intact for (+) strand synthesis.

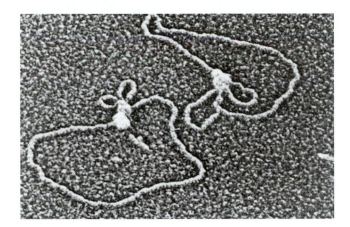

[Electron micrograph courtesy of Jack Griffith, University of North Carolina.]

Mitochondrial DNA is replicated by a process in which leading strand synthesis precedes lagging strand synthesis. The leading strand therefore displaces the lagging strand template to form a **displacement** or **D loop** (*right*). The 15-kb circular mitochondrial chromosome of mammals normally contains a single 500- to 600-nt D loop that undergoes frequent cycles of degradation and resynthesis. During replication, the D loop is extended. When the D loop has reached a point ~2/3 of the way around the chromosome, the lagging strand origin is exposed and its synthesis proceeds in the opposite direction around the chromosome. Lagging strand synthesis is therefore only ~1/3 complete when leading strand synthesis terminates.

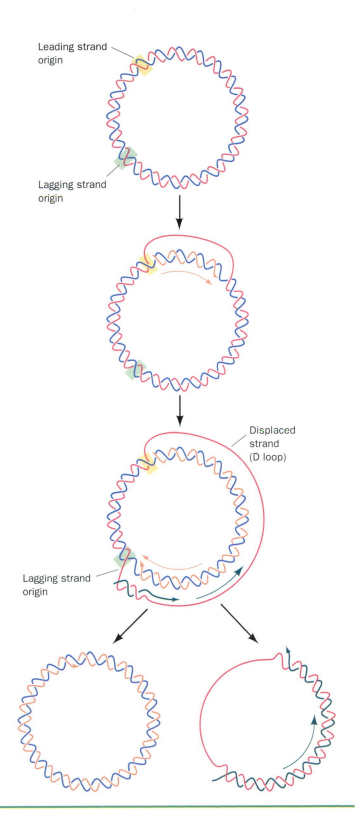

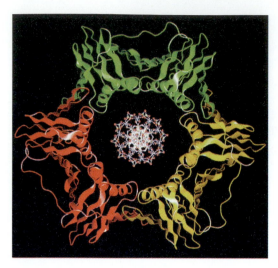

Figure 24-17.   The X-ray structure of PCNA.
The three protein monomers (*red, green, and yellow*) form a three-fold symmetric ring structure. A model of duplex B-DNA (viewed along its helix axis) has been placed in the center of the PCNA ring. Compare this structure to that of the β subunit of the *E. coli* Pol III holoenzyme (Fig. 24-14). [Courtesy of John Kuriyan, The Rockefeller University.] ● See the Interactive Exercises.

DNA polymerase δ, a nuclear enzyme with inhibitor sensitivities similar to those of DNA polymerase α (Table 24-4), lacks an associated primase but exhibits a proofreading 3′ → 5′ exonuclease activity. Moreover, whereas DNA polymerase α exhibits only moderate processivity (~100 nt), DNA polymerase δ is essentially infinitely processive (it can replicate the entire length of a template) but only when it is in complex with a protein known as **proliferating cell nuclear antigen (PCNA).** It is therefore likely that *a complex between DNA polymerase δ and PCNA is the eukaryotic leading strand replicase (which requires high processivity but only occasional need of a primer), whereas DNA polymerase α is the lagging strand replicase (which requires frequent priming but a processivity of only 100–200 nt).*

The amino acid sequence of the 261-residue PCNA aligns only weakly with that of the 366-residue β subunit of *E. coli* Pol III holoenzyme (Section 24-2C). However, the crystal structure of PCNA suggests that it forms a trimeric ring around DNA almost identical in structure to the dimeric β clamp (Fig. 24-17).

DNA polymerase ε, which superficially resembles DNA polymerase δ, differs from it in being highly processive in the absence of PCNA. DNA polymerase ε also has a 3′ → 5′ exonuclease activity that degrades single-stranded DNA to six- or seven-residue oligonucleotides rather than to mononucleotides, as does the exonuclease activity of DNA polymerase δ. DNA polymerase ε is required to repair UV-induced DNA lesions, but it may also be necessary for DNA replication *in vivo.*

DNA polymerase β is remarkable for its small size (335 residues in rat). The biological function of this nuclear enzyme is unknown, but it may participate in DNA repair. The X-ray structure of a stable proteolytic fragment (residues 85–335) of rat DNA polymerase β (Fig. 24-18), by Zdenek Hostomsky, reveals that this protein has the same U shape as other polymerases, including the Klenow fragment (Fig. 24-10), although it lacks strict structural homology.

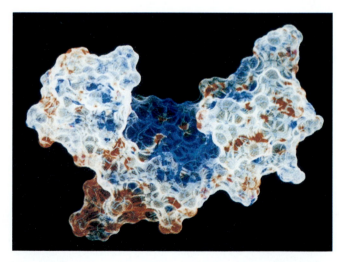

Figure 24-18.   The X-ray structure of the catalytic domain of rat DNA polymerase β. The protein, which has thumb, fingers, and palm subdomains resembling those of other DNA polymerases, is oriented with its N-terminal fingers subdomain at the top left as represented by its solvent-accessible surface. The surface is colored according to charge, with red negative, blue positive, and white neutral. The strong positive charge of the putative DNA-binding cleft no doubt facilitates the binding of the polyanionic DNA. [Courtesy of Zdenek Hostomsky, Agouron Pharmaceuticals, La Jolla, California.]

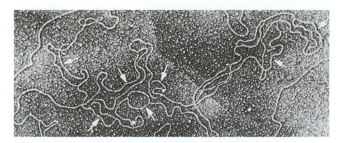

Figure 24-19. **An electron micrograph of a fragment of replicating *Drosophila* DNA.** The arrows indicate the multiple replication eyes. [From Kreigstein, H.J. and Hogness, D.S., *Proc. Natl. Acad. Sci.* **71**, 136 (1974).]

DNA polymerase γ occurs exclusively in the mitochondrion, where it presumably replicates the mitochondrial genome. Chloroplasts contain a similar enzyme. An additional member of the polymerase family of proteins is the viral enzyme **reverse transcriptase,** an RNA-directed DNA polymerase (see Box 24-2).

B. Initiation of Eukaryotic DNA Replication

DNA polymerase α synthesizes DNA at the rate of ~50 nt/s (~20 times slower than prokaryotic DNA polymerases). Since a eukaryotic chromosome typically contains 60 times more DNA than does a prokaryotic chromosome, its bidirectional replication from a single origin, as in prokaryotes, would require ~1 month. Electron micrographs such as Fig. 24-19, however, show that *eukaryotic chromosomes contain multiple origins,* one every 3 to 300 kb, depending on both the species and the tissue, so replication is complete in a few hours.

Cytological observations indicate that the various chromosomal regions are not all replicated simultaneously. Rather, clusters of 20 to 80 adjacent **replicons** (replicating units; DNA segments that are each served by a replication origin) are activated at once. The mechanisms for sequentially activating different replicons and for preventing the further replication of those that have already been replicated are not well understood.

In yeast cells, the initiation of DNA replication occurs at **autonomously replicating sequences (ARS),** which are conserved 11-bp sequences adjacent to easily unwound DNA. As in prokaryotes, a helicase is required to prepare the DNA for replication. Unlike prokaryotic DNA, however, eukaryotic DNA is packaged in nucleosomes (Section 23-5B). Some alteration of this structure is probably necessary for initiation, but once replication is under way, nucleosomes do not seem to impede the progress of DNA polymerases. Experiments with labeled histones indicate that nucleosomes just ahead of the replication fork disassemble and the freed histones immediately reassociate with the emerging daughter duplexes. The parental histones randomly associate with the leading and lagging duplexes. DNA replication (which occurs in the nucleus) is coordinated with histone protein synthesis in the cytosol so that new histones are available in the required amounts.

C. Telomeres and Telomerase

The ends of linear chromosomes present a problem for the replication machinery. Specifically, *DNA polymerase cannot synthesize the extreme 5' end of the lagging strand* (Fig. 24-20). Even if an RNA primer were paired with the 3' end of the DNA template, it could not be replaced with DNA (recall that DNA polymerase operates only in the 5' → 3' direction; it can

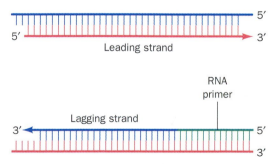

Figure 24-20. Replication of a blunt-ended chromosome. Leading strand synthesis can proceed to the end of the chromosome (*top*). However, DNA polymerase cannot synthesize the extreme 5' end of the lagging strand because it can only extend an RNA primer that is paired with the 3' end of a template strand (*bottom*). Removal of the primer and degradation of the remaining single-stranded extension would cause the chromosome to shorten with each round of replication.

only extend an existing primer, and the primer must be bound to its complementary strand). Consequently, *in the absence of a mechanism for completing the lagging strand, DNA molecules would be shortened at both ends by the length of an RNA primer with each round of replication.* This would eventually lead to the loss of essential genetic information at the ends of the chromosomes.

The ends of eukaryotic chromosomes, the **telomeres** (Greek: *telos,* end) have an unusual structure. Telomeric DNA consists of 1000 or more tandem repeats of a short G-rich sequence (TTGGGG in the protozoan *Tetrahymena* and TTAGGG in humans) on the 3'-ending strand of each chromosome end. Moreover, this strand has a 12- to 16-nt single-strand overhang. *This 3' extension can serve as a template for the primer that initiates the final Okazaki fragment of the lagging strand.*

Box 24-2 BIOCHEMISTRY IN FOCUS

Reverse Transcriptase

Reverse transcriptase (**RT**) is an essential enzyme of **retroviruses**, which are RNA-containing eukaryotic viruses such as **human immunodeficiency virus** (**HIV,** the causative agent of AIDS). RT, which was independently discovered in 1970 by Howard Temin and David Baltimore, synthesizes DNA in the 5' → 3' direction from an RNA template. Although the activity of this enzyme was initially considered antithetical to the central dogma of molecular biology (Section 3-3B), there is no thermodynamic prohibition to the RT reaction (in fact, under certain conditions, Pol I can copy RNA templates). RT catalyzes the first step in the conversion of the virus' single-stranded RNA genome to a double-stranded DNA.

After the virus enters a cell, its RT uses the viral RNA as a template to synthesize a complementary DNA strand, yielding an RNA–DNA hybrid helix. The DNA synthesis is primed by a host cell tRNA whose 3' end unfolds to base pair with a complementary segment of viral RNA. The viral RNA strand is then nucleolytically degraded by an **RNase H** (an RNase activity that hydrolyzes the RNA of an RNA–DNA hybrid helix). Finally, the DNA strand acts as a template for the synthesis of its complementary DNA, yielding double-stranded DNA that is then integrated into a host cell chromosome.

RT has been a particularly useful tool in genetic engineering because it can transcribe mRNAs to complementary strands of DNA (**cDNA,** *right*). cDNAs can be used, for ex-

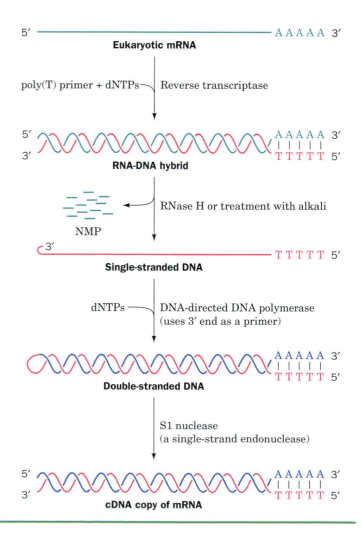

Elizabeth Blackburn has shown that telomeric DNA is synthesized and maintained by an enzyme named **telomerase.** *Tetrahymena* telomerase, for example, adds tandem repeats of the telomeric sequence TTGGGG to the 3′ end of any G-rich telomeric oligonucleotide independently of any exogenously added template. A clue as to how this occurs came from the discovery that telomerases are **ribonucleoproteins** (complexes of protein and RNA) whose RNA components contain a segment that is complementary to the repeating telomeric sequence. This RNA apparently acts as a template for a reaction in which nucleotides are added to the 3′ end of the DNA. Telomerase thus functions similarly to reverse transcriptase (Box 24-2); in fact, its protein component is homologous to reverse transcriptase. Telomerase repeatedly translocates to the new 3′ end of the DNA strand, thereby adding multiple telomeric sequences to the DNA

ample, to express eukaryotic structural genes in *E. coli* (Section 3-5D). Since *E. coli* lacks the machinery to splice out introns (Section 25-3A), the use of genomic DNA to express a eukaryotic structural gene in *E. coli* would require the prior excision of its introns—a technically difficult feat.

HIV-1 reverse transcriptase is a dimeric protein whose subunits are synthesized as identical 66-kD polypeptides, known as **p66,** that each contain a polymerase domain and an RNase H domain. However, the RNase H domain of one of the two subunits is proteolytically excised, thereby yielding a 51-kD polypeptide named **p51.** Thus, RT is a dimer of p66 and p51.

The X-ray structure of HIV-1 RT shows that the two subunits have different structures, although each has a fingers, palm, and thumb domain as well as a "connection" domain. The RNase H domain of the p66 subunit follows the connection domain. The p66 and p51 subunits are not related by twofold molecular symmetry (a rare but not unprecedented phenomenon) but instead associate in a sort of head-to-tail arrangement. Consequently, RT has only one polymerase active site.

Reverse transcriptase lacks a proofreading exonuclease function and hence is highly error prone. Indeed, it is HIV's capacity to rapidly evolve, even within a single host, that presents a major obstacle to the development of an anti-HIV vaccine. This rapid rate of mutation is also the main contributor to the ability of HIV to quickly develop resistance to drugs that inhibit virally encoded enzymes, including RT. ● See the Interactive Exercises.

[Structure of RT courtesy of Thomas Steitz, Yale University.]

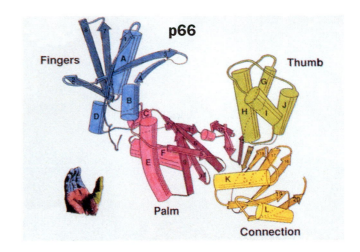

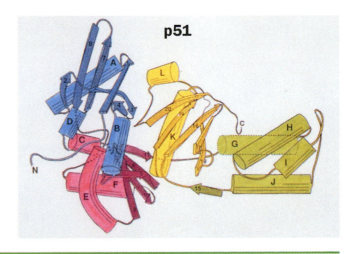

Figure 24-21. **The proposed mechanism for the synthesis of telomeric DNA by *Tetrahymena* telomerase.** The telomere's 5'-ending strand is later extended by normal lagging strand synthesis. [After Greider, C.W. and Blackburn, E.H., *Nature* 337, 336 (1989).]

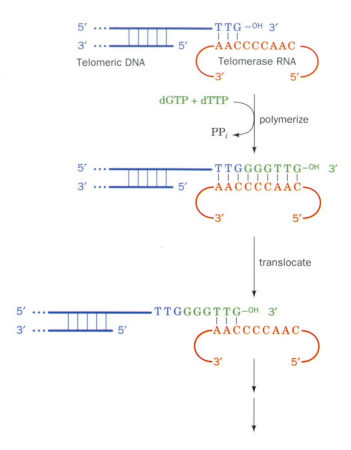

(Fig. 24-21). The DNA strand complementary to the telomeric G-rich strand is apparently synthesized by the normal cellular machinery for lagging strand synthesis, which necessarily leaves a 3' overhang on the G-rich strand.

Guanine-rich polynucleotides are notorious for their propensity to aggregate via Hoogsteen-type base pairing (Section 23-2B). This tendency allows the G-rich overhanging strands of telomeres to fold back on themselves to form hairpins, two of which assemble in a four-stranded structure (Fig. 24-22*a*). G residues from each strand hydrogen bond in an arrangement known as a **G-quartet** (Fig. 24-22*b*). The NMR structure in solution of the telomerelike dodecamer $d(G_4T_4G_4)$ differs from its X-ray crystal structure although the reasons for this are poorly understood. Formation of the telomeric quadruplex, whatever its structure, may regulate telomerase activity. The absence of telomerase, which allows the gradual truncation of chromosomes with each round of DNA replication, may contribute to the normal senescence of cells. Conversely, enhanced telomerase activity may permit the uncontrolled replication and cell growth that occurs in cancer (see Box 24-3).

4. MUTATION

The fidelity of DNA replication carried out by DNA polymerases and their attendant proofreading functions is essential for the accurate transmission of genetic information during cell division. Yet errors in polymerization occasionally occur and, if not corrected, may alter the nucleotide sequences of genes. DNA can also be altered by environmental

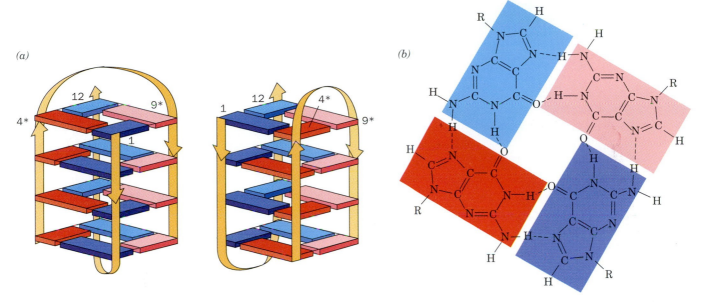

Figure 24-22. The structure of the telomeric oligonucleotide d(GGGGTTTTGGGG). (*a*) Diagram of the NMR solution structure (*left*) and X-ray crystal structure (*right*) in which the strand directions are indicated by arrows. The nucleotides are numbered 1 to 12 in one strand and 1* to 12* in the other strand. Guanine residues

G1 to G4 are represented by blue rectangles, G9 to G12 are cyan, G1* to G4* are red, and G9* to G12* are pink. (*b*) The base pairing interactions in a G-quartet. [After Schultze, P., Smith, F.W., and Feigon, J., *Structure* **2**, 227 (1994).]

Box 24-3

BIOCHEMISTRY IN HEALTH AND DISEASE

Telomerase, Aging, and Cancer

Without telomerase, a chromosome would become progressively shorter with each cell division, and the cell's descendants would eventually die from the loss of essential genes. Indeed, otherwise immortal *Tetrahymena* cultures with mutationally impaired telomerase resemble senescent mammalian cells before dying off. Since the somatic cells of multicellular organisms, in fact, lack telomerase activity, this suggests that the loss of telomerase function in somatic cells may contribute to the aging process.

Somatic mammalian cells grown in culture can divide only a limited number of times (20–60) before reaching senescence and dying. Yet there is only a weak correlation between the proliferative capacity of a cell culture and the age of its donor. There is, however, a strong correlation between the initial telomere length in a cell culture and its proliferative capacity. Cells that initially have relatively short telomeres undergo significantly fewer doublings than cells with longer telomeres. Moreover, fibroblasts from individuals with **progeria** (a rare disease characterized by rapid and premature aging resulting in childhood death) have short telomeres, an observation that is consistent with their known reduced proliferative capacity in culture. In contrast, sperm (which are essentially immortal) have telomeres that do not vary in

length with donor age, which indicates that telomerase is active during germ-cell growth. Likewise, those few cells in culture that become immortal (capable of unlimited proliferation) exhibit an active telomerase and a telomere of stable length, as do the cells of unicellular eukaryotes (which are also immortal). It therefore appears that telomere loss is a significant cause of cellular senescence and hence aging.

What selective advantage might multicellular organisms gain by eliminating the telomerase activity in their somatic cells? An intriguing possibility is that cellular senescence is a mechanism that protects multicellular organisms from cancer. Indeed, cancer cells, which are immortal and grow uncontrollably, contain active telomerase. For example, the enzyme is active in ovarian cancer cells but not in normal ovarian tissue. Interestingly, the telomeres in some tumors are quite short, possibly because the cells express telomerase only after they have begun to replicate uncontrollably (and have therefore "used up" much of their telomeres). Because telomerase apparently stabilizes even short telomeres, inhibition of telomerase should eventually lead to the complete loss of telomeres and hence cessation of cell division. This hypothesis makes telomerase inhibitors an attractive target for antitumor drug development.

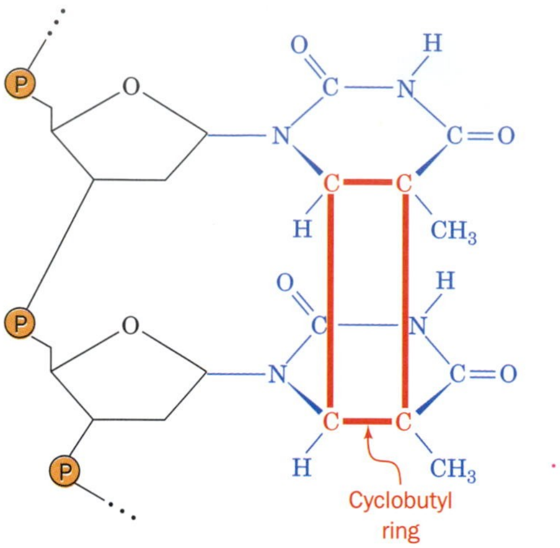

Figure 24-23.   The cyclobutylthymine dimer.
The dimer forms on UV irradiation of two adjacent thymine residues on a DNA strand. The ~1.6-Å-long covalent bonds joining the thymine rings (*red*) are much shorter than the normal 3.4-Å spacing between stacked rings in B-DNA, thereby locally distorting the DNA.

factors such as alkylating agents and ionizing radiation. For example, UV radiation (200–300 nm) promotes the formation of a cyclobutyl ring between adjacent thymine residues on the same DNA strand to form an intrastrand **thymine dimer** (Fig. 24-23). Similar cytosine and thymine–cytosine dimers also form but less frequently. Such **pyrimidine dimers** locally distort DNA's base-paired structure, interfering with transcription and replication.

In many cases, damaged DNA can be repaired, as discussed below (Section 24-5). Severe lesions, however, may be irreversible, leading to the loss of genetic information and, often, cell death. Even when damaged DNA can be mended, the restoration may be imperfect, producing a **mutation,** a heritable alteration of genetic information. In multicellular organisms, genetic changes are usually notable only when they occur in germ-line cells so that the change is passed on to all the cells of the organism's offspring. Damage to the DNA of a somatic cell, in contrast, rarely has an effect beyond that cell unless the mutation contributes to a malignant transformation (cancer).

A. Chemical Mutagenesis

The DNA damage produced by **chemical mutagens,** substances that induce mutations, falls into two major classes:

1. **Point mutations,** in which one base pair replaces another. These are subclassified as:
    (a) **Transitions,** in which one purine (or pyrimidine) is replaced by another.
    (b) **Transversions,** in which a purine is replaced by a pyrimidine or vice versa.
2. **Insertion/deletion mutations,** in which one or more nucleotide pairs are inserted in or deleted from DNA.

Point Mutations Are Generated by Altered Bases

Point mutations can result from the treatment of an organism with base analogs or substances that chemically alter bases. For example, the base analog **5-bromouracil (5BU)** sterically resembles thymine (5-methyluracil) but, through the influence of its electronegative Br atom, frequently assumes a tautomeric form that base pairs with guanine instead of adenine (Fig. 24-24). Consequently, when 5BU is incorporated into DNA in place of thymine, it may induce an $A \cdot T \rightarrow G \cdot C$ transition in subsequent rounds

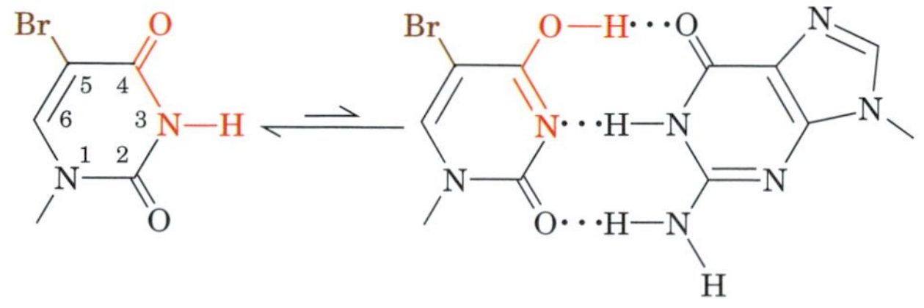

5-Bromouracil (5BU)   **5BU**
(keto tautomer)   **(enol tautomer)**   **Guanine**

Figure 24-24.   5-Bromouracil. The keto form of 5BU (*left*) is its most common tautomer. However, 5BU frequently assumes the enol form (*right*), which base pairs with guanine.

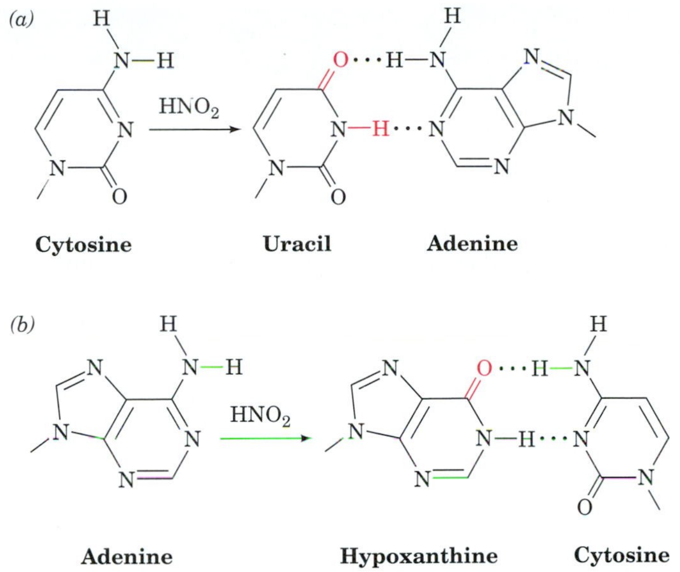

Figure 24-25. **Oxidative deamination by nitrous acid.** (*a*) Cytosine is converted to uracil, which base pairs with adenine. (*b*) Adenine is converted to hypoxanthine, a guanine derivative (it lacks guanine's 2-amino group) that base pairs with cytosine.

of DNA replication. Occasionally, 5BU is incorporated into DNA in place of cytosine, which instead generates a $G \cdot C \rightarrow A \cdot T$ transition.

In aqueous solutions, nitrous acid (HNO_2) oxidatively deaminates aromatic primary amines, so it can convert cytosine to uracil (Fig. 24-25*a*) and adenine to the guaninelike hypoxanthine (which forms two of guanine's three hydrogen bonds with cytosine; Fig. 24-25*b*). Hence, treatment of DNA with nitrous acid results in both $A \cdot T \rightarrow G \cdot C$ and $G \cdot C \rightarrow A \cdot T$ transitions. Despite its potential mutagenic activity, nitrite (the conjugate base of nitrous acid) is used as a preservative in prepared meats such as frankfurters because it also prevents the growth of *Clostridium botulinum,* the organism that causes botulism.

Alkylating agents such as dimethyl sulfate, **nitrogen mustard,** and **ethylnitrosourea**

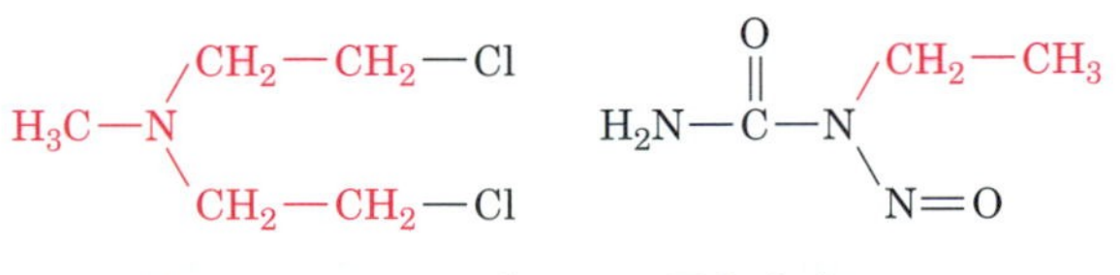

Nitrogen mustard   **Ethylnitrosourea**

often generate transversions. The alkylation of the N7 position of a purine nucleotide renders its glycosidic bond susceptible to hydrolysis, leading to loss of the base. The resulting gap in the sequence is filled in by an error-prone enzymatic repair system. Transversions arise when the missing purine is replaced by a pyrimidine. Even in the absence of alkylating agents, the glycosidic bonds of an estimated 10,000 purine nucleotides in each human cell spontaneously hydrolyze every day.

Nucleotide bases can also be modified by methylation. For example, the metabolic compound *S*-adenosylmethionine (Section 20-4D) occasionally nonenzymatically methylates a base to form derivatives such as 3-methyladenine and 7-methylguanine residues. Methylation reactions, however, frequently have normal physiological functions (see Box 24-4).

Box 24-4

BIOCHEMISTRY IN FOCUS

DNA Methylation

Not all DNA modifications are detrimental. For example, the A and C residues of DNA may be methylated, in a species-specific pattern, to form N^6-methyladenine (m^6A), N^4-methylcytosine (m^4C), and 5-methylcytosine (m^5C) residues, respectively.

N^6-Methyladenine (m^6A)
residue

5-Methylcytosine (m^5C)
residue

N^4-Methylcytosine (m^4C)
residue

These methyl groups project into B-DNA's major groove, where they can interact with DNA-binding proteins. In most cells, only a few percent of the susceptible bases are methylated, although this figure rises to >30% of the C residues in some plants. 5-Methylcytosine is the only methylated base in vertebrates.

Bacterial DNAs are methylated at their own particular restriction sites, thereby preventing the corresponding restriction endonuclease from degrading the DNA (Section 3-4A). **Methyltransferases** also modify other residues in a sequence-specific manner.

Methylation in prokaryotes functions most conspicuously as a marker of parental DNA in the repair of mismatched base pairs. Any replicational mispairing that has eluded the editing functions of Pol I and Pol III may still be corrected by a process known as **mismatch repair**. However, if this system is to correct errors rather than perpetuate them, it must distinguish the parental DNA, which has the correct base, from the daughter strand, which has the incorrect although normal base. The observation that *E. coli* with a deficient methyltransferase have higher mutation rates than wild-type bacteria suggests how this distinction is made. A newly replicated daughter strand is undermethylated compared to the parental strand because DNA methylation lags behind DNA synthesis. Experiments with model DNAs have demonstrated that the mismatch repair system corrects a mismatch by resynthesizing the unmethylated strand between the mismatch and a point (up to ~1000 bp away) opposite a methylated site on the parent strand.

In eukaryotes, the pattern of DNA methylation varies with the species, the tissue, and the position along a chromosome. Experimental evidence suggests that DNA methylation switches off eukaryotic gene expression, particularly when it occurs in the control regions upstream of a gene's transcribed sequence. For example, globin genes are less methylated in erythroid cells than they are in nonerythroid cells. In fact, the methylation pattern of a parental DNA strand directs the methylation of its daughter strand, so that the "inheritance" of a methylation pattern in a cell line permits all the cells to have the same differentiated phenotype. Variations in methylation may be responsible for **genomic imprinting** in mammals, the phenomenon in which certain maternal and paternal genes are differentially expressed in the offspring.

Insertion/Deletion Mutations Are Generated by Intercalating Agents

Insertion/deletion mutations may arise from the treatment of DNA with intercalating agents such as acridine orange or proflavin (Section 23-3B). The distance between two consecutive base pairs is roughly doubled by the intercalation of such a molecule between them. The replication of this distorted DNA occasionally results in the insertion or deletion of one or more nucleotides in the newly synthesized polynucleotide. (Insertions and deletions of large segments generally arise from aberrant crossover events; Section 24-6A.)

All Mutations Are Random

The bulk of the scientific data regarding mutagenesis is that mutations, whether the result of polymerase errors, spontaneous modification, or chemical damage to DNA, occur at random. This paradigm has been challenged by John Cairns, who demonstrated that bacteria unable to digest lactose preferentially acquired the mutations they needed in order to use lactose when it was the only nutrient available. This observation, which suggests that bacteria can "direct" mutations that benefit them, more likely reflects a nonspecific adaptive response in which the overall rate of mutation—useful as well as nonuseful—increases when the cells are under metabolic stress. *The hypermutable state appears to reflect the activation of error-prone DNA repair and recombination systems that are relatively inactive in normally growing cells.*

B. Carcinogens

Not all alterations to DNA have phenotypic consequences. For example, mutations in noncoding segments of DNA are often invisible. Similarly, the redundancy of the genetic code (more than one trinucleotide may specify a particular amino acid; Section 26-1C) can mask point mutations. Even when a protein's amino acid sequence is altered, its function may be preserved if the substitution is conservative (Section 5-4A) or occurs on a surface loop. Nevertheless, even a single point mutation, if appropriately located, can irreversibly alter cellular metabolism, for example, by causing cancer. As many as 80% of human cancers may be caused by **carcinogens** that damage DNA or interfere with its replication or repair. Consequently, many mutagens are also carcinogens.

There are presently over 60,000 man-made chemicals of commercial importance, and ~1000 new ones are introduced every year. The standard animal tests for carcinogenesis, exposing rats or mice to high levels of the suspected carcinogen and checking for cancer, are expensive and require ~3 years to complete. Thus, relatively few substances have been tested in this manner. Likewise, epidemiological studies in humans are costly, time-consuming, and often inconclusive.

Bruce Ames has devised a rapid and effective bacterial assay for carcinogenicity that is based on the high correlation between carcinogenesis and mutagenesis. He constructed special tester strains of the bacterium *Salmonella typhimurium* that are *his$^-$* (cannot synthesize histidine and therefore cannot grow in its absence). Mutagenesis in these strains is indicated by their reversion to the *his$^+$* phenotype.

In the **Ames test,** ~10^9 tester strain bacteria are spread on a culture plate that lacks histidine. A mutagen placed in the culture medium causes some of these *his$^-$* bacteria to become *his$^+$*, which is detected by their growth into visible colonies after 2 days at 37°C (Fig. 24-26). The mutagenicity of a substance is scored as the number of such colonies minus the few spontaneously revertant colonies that occur in the absence of the mutagen.

Many noncarcinogens are converted to carcinogens in the liver or in other tissues via a variety of detoxification reactions. A small amount of rat liver homogenate is therefore included in the Ames test medium in order to approximate the effects of mammalian metabolism.

About 80% of the compounds determined to be carcinogens in whole-animal experiments are also mutagenic by the Ames test. Dose–response curves, which are generated by testing a given compound at a number of concentrations, are almost always linear, indicating that *there is no threshold concentration for mutagenesis.* Several compounds to which humans have been extensively exposed that were found to be mutagenic by the Ames test were later found to be carcinogenic in animal tests. These include **tris(2,3-dibromopropyl)phosphate,** which was used as a flame retardant in children's sleepwear in the mid-1970s and can be absorbed through

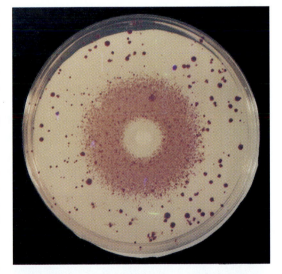

Figure 24-26.   The Ames test for mutagenesis. A filter paper disk containing a mutagen, in this case the alkylating agent ethyl methanesulfonate, is centered on a culture plate containing *his$^-$* tester strains of *Salmonella typhimurium* in a medium that lacks histidine. A dense halo of revertant bacterial colonies appears around the disk from which the mutagen diffused. The larger colonies distributed around the culture plate are spontaneous revertants. The bacteria near the disk have been killed by the toxic mutagen's high concentration. [Courtesy of Raymond Devoret, Institut Curie, Orsay, France.]

the skin; and **furylfuramide,** which was used in Japan in the 1960s and 1970s as an antibacterial additive in many prepared foods (and which had passed two animal tests before it was found to be mutagenic).

5. DNA REPAIR

DNA damage must be repaired to maintain the integrity of genetic information. The biological importance of DNA repair is indicated by the great variety of repair mechanisms in even a simple organism such as *E. coli*. These systems include enzymes that simply reverse the chemical modification of nucleotide bases as well as more complicated multienzyme systems that depend on the inherent redundancy of the information in duplex DNA to restore the damaged molecule.

A. Direct Reversal of Damage

Several enzymes recognize modified nucleotide bases in DNA and restore them to their original state. For example, the alkylated base O^6-methylguanine

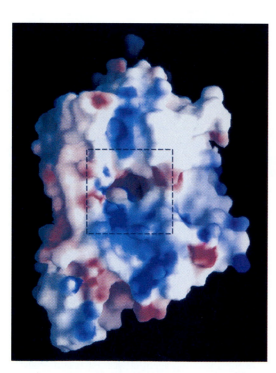

O^6-**Methylguanine residue**

frequently causes the incorporation of thymine instead of guanine during DNA replication. The alkylated base is recognized by O^6-**methylguanine–DNA methyltransferase,** which directly transfers the offending methyl group to one of its own Cys residues. This reaction inactivates the protein, which therefore cannot be strictly classified as an enzyme.

Some forms of UV-damaged DNA can also be repaired in a single step. Pyrimidine dimers may be restored to their monomeric forms by **photoreactivation** catalyzed by light-absorbing enzymes known as **DNA photolyases.** These 55- to 65-kD monomeric enzymes are found in many prokaryotes and eukaryotes but not in humans. Photolyases contain two prosthetic groups: a light-absorbing cofactor and $FADH^-$. In the *E. coli* enzyme, the cofactor N^5,N^{10}-methenyltetrahydrofolate (Section 20-4D) absorbs UV–visible light (300–500 nm) and transfers the excitation energy to the $FADH^-$, which then transfers an electron to the pyrimidine dimer, thereby splitting it. The resulting pyrimidine anion reduces the $FADH\cdot$ to regenerate the enzyme.

The *E. coli* DNA photolyase binds single-stranded or double-stranded DNA without regard to base sequence. Its X-ray structure reveals its DNA-binding site to be a positively charged flat surface with a hole that has a size and polarity complementary to that of a pyrimidine dimer (Fig. 24-27). In order to bind in this site and contact the isoalloxazine ring of the $FADH^-$ for electron transfer, the pyrimidine dimer must flip out of the double helix. This conformational change, which may be a general feature of base-modifying enzymatic reactions, is probably facilitated by the relatively weak base pairing of the pyrimidine dimer and the distortion it imposes on the double helix.

Figure 24-27. **The X-ray structure of *E. coli* DNA photolyase.** The solvent-accessible surface of the enzyme is shaded according to its electrostatic potential (*blue*, most positive; *red*, most negative; and *white*, neutral). The dashed lines enclose the hole in the protein surface where the pyrimidine dimer is thought to bind. [Courtesy of Johann Deisenhofer, University of Texas Southwestern Medical Center.]

B. Nucleotide Excision Repair

Pyrimidine dimers may also be mended by **nucleotide excision repair (NER)**, in which an oligonucleotide containing the lesion is excised from the DNA and the resulting single-strand gap is filled in. In *E. coli*, pyrimidine dimers are recognized by a multisubunit enzyme, the product of the *uvrA, uvrB,* and *uvrC* genes. This **UvrABC endonuclease,** in an ATP-dependent reaction, cleaves the dimer-containing DNA strand at the seventh and fourth phosphodiester bonds on the dimer's 5' and 3' sides, respectively (Fig. 24-28). The excised oligonucleotide is replaced through the action of a DNA polymerase, probably Pol I, followed by DNA ligase. UvrABC endonuclease also excises other types of DNA lesions in which the bases are displaced from their normal positions or contain bulky substituents. Apparently, *UvrABC endonuclease recognizes helix distortion rather than any particular group.*

Eukaryotic NER requires as many as 16 polypeptides and removes 24–32 nucleotides from a damaged DNA strand. Defective NER is associated with two human diseases. The rare inherited condition **xeroderma pigmentosum (XP;** Greek: *xeros,* dry + *derma,* skin) is mainly characterized by the inability of skin cells to repair UV-induced DNA lesions. Individuals with this autosomal recessive disease are extremely sensitive to sunlight. During infancy they develop marked skin changes such as dryness, excessive freckling, and keratoses (a type of skin tumor), together with eye damage, such as opacification and ulceration of the cornea. Moreover, they develop often fatal skin cancers at a 2000-fold greater rate than normal. Curiously, many individuals with XP also have a bewildering variety of seemingly unrelated symptoms including progressive neurological degeneration and developmental deficits. XP results from defects in any of seven gene products that are apparently involved in repairing UV-damaged DNA.

Cockayne syndrome (CS), an inherited disease also associated with defective NER, arises from defects in some of the same genes that are defective in XP as well as in two additional genes. Individuals with CS are hypersensitive to UV radiation and exhibit stunted growth as well as neurological dysfunction due to neuron demyelination but, intriguingly, have a normal incidence of skin cancer.

Glycosylases Remove Altered Bases

DNA containing a damaged base may be restored to its native state through a form of excision repair mediated by a variety of **DNA glycosylases.** Each of these enzymes cleaves the glycosidic bond of a specific type of altered nucleotide (Fig. 24-29), thereby leaving a deoxyribose residue in the backbone. Such **apurinic** or **apyrimidinic (AP)** sites are also generated under normal physiological conditions by the spontaneous hydrolysis of a glycosidic bond (Section 24-4A). The deoxyribose residue is then cleaved on one side by an **AP endonuclease,** the deoxyribose and several adjacent residues are removed by the exonuclease activity of DNA polymerase or some other cellular exonuclease, and the gap is filled in and sealed by DNA polymerase and DNA ligase.

Uracil, which does not normally occur in DNA, is excised by **uracil *N*-glycosylase** and replaced by C through nucleotide excision repair. This process is essential for the integrity of genetic information (see Box 24-5).

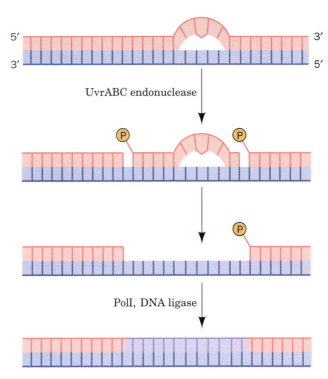

Figure 24-28. The mechanism of nucleotide excision repair (NER) of pyrimidine dimers.

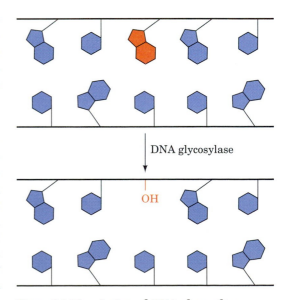

Figure 24-29. Action of DNA glycosylases. These enzymes hydrolyze the glycosidic bond of their corresponding altered base (*red*) to yield an AP site.

Box 24-5

BIOCHEMISTRY IN CONTEXT

Why Doesn't DNA Contain Uracil?

Three of the deoxynucleotide bases in DNA (adenine, guanine, and cytosine) also occur as ribonucleotide bases in RNA. The fourth deoxynucleotide base, thymine, is synthesized—at considerable metabolic effort (Section 22-3B)—from uracil, which occurs in RNA. Since uracil and thymine have identical base-pairing properties, why do cells bother to synthesize thymine at all?

This enigma was solved by the discovery of cytosine's penchant for conversion to uracil by deamination, either spontaneously or by reaction with nitrites (Section 24-4A). If U were a normal DNA base, the deamination of C would be highly mutagenic because there would be no indication of whether the resulting mismatched $G \cdot U$ base pair had initially been $G \cdot C$ or $A \cdot U$. Since T is DNA's normal base, however, any U in DNA is almost certainly a deaminated C and can be removed by uracil *N*-glycosylase.

Uracil *N*-glycosylase also has an important function in DNA replication. dUTP, an intermediate in dTTP synthesis, is present in all cells in small amounts. DNA polymerases do not discriminate well between dUTP and dTTP, both of which can base pair with a template A. Consequently, newly synthesized DNA contains an occasional U. These U's are rapidly replaced by T through excision repair.

C. The SOS Response and Recombination Repair

Agents that damage DNA induce a complex system of cellular changes in *E. coli,* known as the **SOS response.** Cells undergoing the SOS response cease dividing and increase their capacity to repair damaged DNA. **LexA,** a repressor, and **RecA,** a DNA-binding protein, regulate the activity of this system. During normal growth, LexA represses SOS gene expression. However, when DNA is damaged (and cannot fully replicate), the resulting single strands bind to RecA, and the resulting $DNA \cdot RecA$ complex activates LexA to cleave and thereby inactivate itself. The SOS genes, which include *recA* and *lexA* as well as the excision repair genes *uvrA* and *uvrB* (Section 24-5B), are thereby expressed. On repair of the DNA, the $DNA \cdot RecA$ complex is no longer present, so the newly synthesized LexA again represses the expression of the SOS genes.

SOS repair is error prone and therefore mutagenic. This is because the SOS repair system replaces the bases at a DNA lesion even when there is no information as to which bases were originally present. The SOS repair system is therefore a testimonial to the proposition that survival with a chance of loss of function (and the possible gain of a new one) is advantageous, in the Darwinian sense, over death.

Recombination Repair Resembles Genetic Recombination

RecA protein also mediates **genetic recombination,** the exchange of polynucleotide strands between homologous DNA segments (recombination is described in detail in the following section). *E. coli* cells with a mutant *recA* gene are deficient in both genetic recombination and **recombination repair,** which is also called **postreplication repair.** This system may act when damaged DNA undergoes replication before it has been repaired. For example, the replication machinery halts when it encounters a pyrimidine dimer and resumes polymerization at some point past the dimer site. The resulting daughter strand has a gap opposite the pyrimidine dimer (Fig. 24-30). This genetic lesion cannot be eliminated by excision repair, which requires an intact complementary strand. Yet such an intact strand occurs in the sister duplex that was formed at the same replication fork. *The lesion can therefore be corrected by exchanging the corresponding segments of sister DNA strands.* This places the gapped DNA segment opposite an undamaged strand, so that the gap can be filled in according to a template

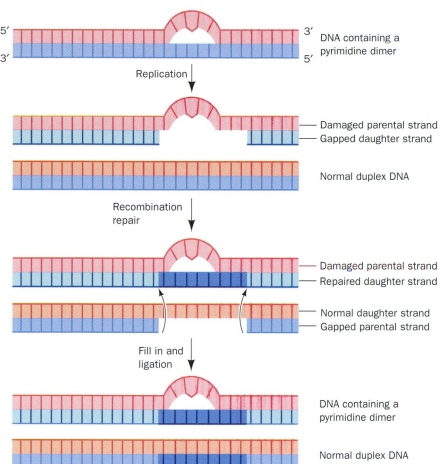

5′ ——————————— 3′　DNA containing a
3′ ——————————— 5′　pyrimidine dimer

↓ Replication

—— Damaged parental strand
—— Gapped daughter strand

Normal duplex DNA

↓ Recombination repair

—— Damaged parental strand
—— Repaired daughter strand

—— Normal daughter strand
—— Gapped parental strand

↓ Fill in and ligation

DNA containing a pyrimidine dimer

Normal duplex DNA

Figure 24-30. **Recombination repair.** This system allows the gap in a newly synthesized DNA strand opposite a damage site to be filled by the corresponding segment from its sister duplex.

and then sealed. The pyrimidine dimer, which is now also associated with an intact complementary strand, can then be eliminated by NER or photoreactivated by photolyase.

6. RECOMBINATION

Over the years, genetic studies have shown that genes are not immutable. In higher organisms, pairs of genes may exchange by crossing-over when homologous chromosomes are aligned (Fig. 24-31). Bacteria, which do not contain duplicate chromosomes, also have an elaborate mechanism for recombining genetic information. In addition, foreign DNA can be installed in a host's chromosome through recombination. In this section, we examine the molecular events of recombination and discuss the biochemistry of **transposons,** which are mobile genetic elements.

Figure 24-31. **Crossing-over.** (*a*) An electron micrograph and (*b*) an interpretive drawing of two homologous pairs of chromatids during meiosis in the grasshopper *Chorthippus parallelus*. Nonsister chromatids (*different colors*) may recombine at any of the points where they cross over. [Courtesy of Bernard John, The Australian National University.]

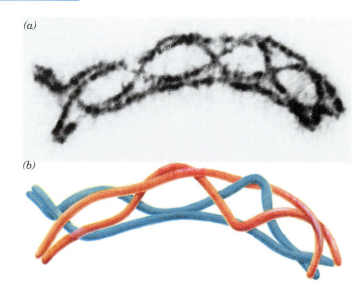

(a)

(b)

A. The Mechanism of General Recombination

General recombination occurs between DNA segments with extensive homology; **site-specific recombination** occurs between two short, specific DNA sequences. The prototypical model for general recombination (Fig. 24-32) was proposed by Robin Holliday in 1964. The corresponding strands of two aligned homologous DNA duplexes are nicked, and the nicked strands cross over to pair with the nearly complementary strands on the homologous duplex, thereby forming a segment of **heterologous DNA,** after which the nicks are sealed (Fig. 24-32*a–e*). The crossover point is a four-stranded structure known as a **Holliday junction** (Fig. 24-33). The crossover point can move in either direction in a process known as **branch migration** (Fig. 24-32*e,f*).

The Holliday junction can be "resolved" into two duplex DNAs in two equally probable ways (Fig. 24-32*g–l*):

1. Cleavage of the strands that did not cross over exchanges the ends of the original duplexes to form, after nick sealing, the traditional recombinant DNA molecule (right branch of Fig. 24-32*j–l*).

2. Cleavage of the strands that crossed over exchanges a pair of homologous single-stranded segments (left branch of Fig. 24-32*j–l*).

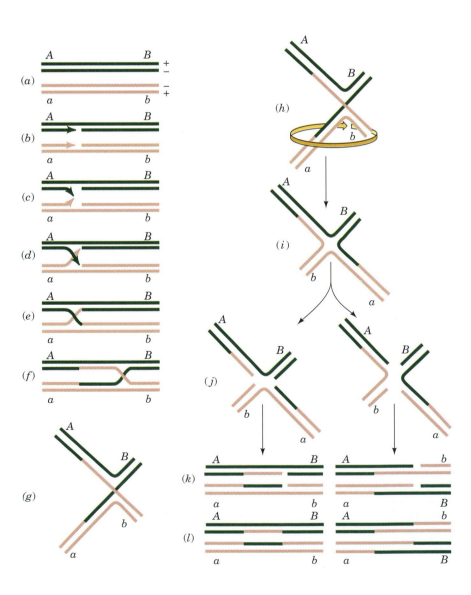

Figure 24-32.   *Key to Function.* The Holliday model of general recombination between homologous DNA duplexes. ✳ See the Animated Figures.

(a)

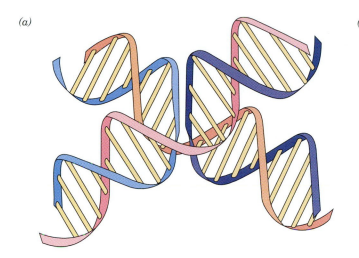

(b)

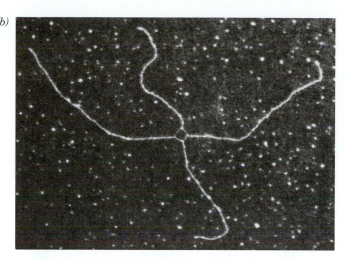

Figure 24-33. **The Holliday junction.** (*a*) A proposed model of its most energetically favorable conformation, which does not violate stereochemical principles. [After Murchie, A.I.H., Clegg, R.M., von Kitzing, E., Duckett, D.R., Diekman, S., and Lilley,

D.M.J., *Nature* **341**, 765 (1989).] (*b*) Electron micrograph of recombining DNA. Note the thinner single-stranded connections in the crossover region. [Courtesy of Huntington Potter, Harvard Medical School, and David Dressler, Oxford University.]

RecA Promotes Recombination in *E. coli*

The 352-residue RecA protein polymerizes on single-stranded DNA or duplex DNA that contains a single-strand gap. Electron microscopy reveals that RecA filaments bound to DNA form a right-handed helix with ~6.2 RecA monomers per turn (Fig. 24-34). Three nucleotides bind to each

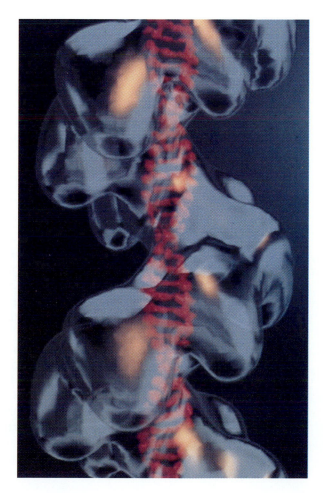

Figure 24-34. **Model of a RecA–DNA complex.** In this electron microscopy–based image, the transparent surface represents an *E. coli* RecA filament. The extended and untwisted duplex DNA (*red*) has been modeled into this image. [Courtesy of Edward Egelman, University of Minnesota Medical School.]

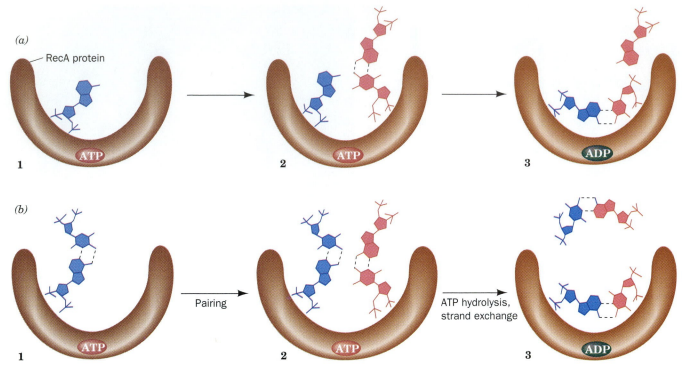

Figure 24-35.   Proposed models for RecA-mediated pairing and strand exchange. (*a*) Exchange involving a single-stranded and a duplex DNA. (*b*) Exchange involving two duplex DNAs. (**1**) A single-stranded DNA (duplex in *b*) binds to RecA to form an initiation complex. (**2**) Duplex DNA binds to the initiation complex so as to transiently form a three-stranded helix (four-stranded in *b*). (**3**) RecA rotates the bases of the aligned homologous strands to effect strand exchange in an ATP-driven process. [After West, S.C., *Annu. Rev. Biochem.* **61**, 618 (1992).]

RecA monomer so that the DNA assumes an extended conformation with ~18.6 nucleotides per turn (B-DNA has 10.4 bp per turn). The X-ray structure of RecA confirms that the central cavity of the RecA filament has a diameter of 25 Å, large enough to accommodate DNA.

How does RecA mediate DNA strand exchange between single-stranded and duplex DNA? On encountering a duplex DNA with a strand that is complementary to its bound single-stranded DNA, RecA partially unwinds the duplex and, in a reaction driven by ATP hydrolysis, exchanges the single-stranded DNA with the corresponding strand on the duplex. ATP hydrolysis is thought to drive the rearrangement of a three-stranded DNA intermediate bound to the protein (Fig. 24-35*a*).

RecA-catalyzed exchange of two duplex DNA segments proceeds similarly (Fig. 24-35*b*). Homologous DNA segments are paired in advance of strand exchange, so regions of triple- or quadruple-stranded DNA presumably form in the cavity of the RecA helix (Fig. 24-36). Eukaryotes contain proteins that are homologous to *E. coli* RecA and that apparently function in a similar ATP-dependent manner to mediate DNA recombination.

Recombination Requires Several Protein Activities

Numerous proteins in addition to RecA participate in *E. coli* recombination. The single-strand nicks to which RecA binds are made by the **RecBCD protein,** the 330-kD product of the SOS genes ***recB, recC,*** and ***recD.*** RecBCD unwinds duplex DNA in an ATP-driven process and nicks the resulting single-stranded DNA. Topoisomerases are necessary to maintain the level of supercoiling required by the recombination process. SSB also participates by maintaining DNA in its single-stranded state and

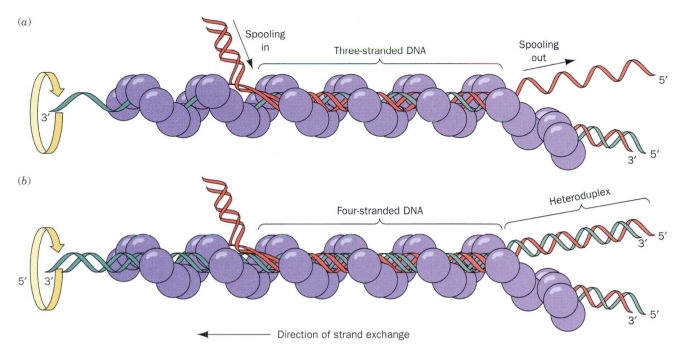

(a)

Spooling in

Three-stranded DNA

Spooling out

5'

3'

3' 5'

(b)

Four-stranded DNA

Heteroduplex

5' 3'

3' 5'

3' 5'

← Direction of strand exchange

Figure 24-36. Models for three- and four-stranded DNA helices in RecA-mediated strand exchange. (*a*) Three-stranded helix. (*b*) Four-stranded helix. The rotation of the RecA filament (*purple*) around its helix axis causes duplex DNA to be "spooled in" to the filament, right to left as drawn. Deproteinization of the structure in *b* and disruption of its four-stranded DNA would yield a Holliday junction (Fig. 24-33). [After West, S.C., *Annu. Rev. Biochem.* **61**, 617 (1992).]

by modulating RecA function. **RuvB protein** (37 kD), a DNA-dependent ATPase, is apparently the motor that drives branch migration although, by itself, it binds only weakly to DNA. **RuvA protein** (22 kD) binds specifically to both Holliday junctions and RuvB, thereby targeting RuvB to DNA. **RuvC protein** (19 kD) is the nuclease that cuts apart the two recombinant duplexes (Fig. 24-32*j*). DNA ligase is, of course, required to ligate the DNA strands at the resolved crossover junctions (Fig. 24-32*l*).

The X-ray structure of *E. coli* RuvA protein, determined by David Rice, reveals that this homotetramer of 209-residue subunits has fourfold rotational symmetry (Fig. 24-37*a*). RuvA protein has an appearance reminiscent of a four-petaled flower: It is rather flat (80 × 80 × 45 Å) with one face concave and the other convex. The concave face, which has four grooves emanating from four centrally located negatively charged projections or "pins" toward the corners of a square, is otherwise positively charged and contains numerous conserved residues, whereas the convex

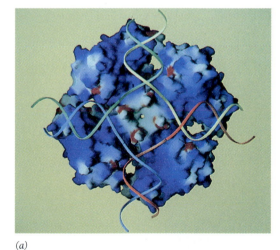

(a)

Figure 24-37. Structural basis of RuvA action. (*a*) The X-ray structure of *E. coli* RuvA as viewed down the tetramer's four-fold axis of rotation. The protein is represented by its solvent-accessible surface colored according to its electrostatic potential (*blue,* most positive; *red,* most negative; and *white,* neutral). Note the small negatively charged region around each central pin. Holliday junction DNA, here shown as red, cyan, green, and yellow ribbons, has been modeled into the structure. (*b*) Diagram of RuvA and RuvB in complex with Holliday junction DNA during branch migration. RuvA and RuvB are shown in gray outline with RuvA viewed along its fourfold axis as in Part *a,* and the two RuvB hexameric rings in cross section. The DNA (*red and blue ribbons*) is shown passing through the centers of the proteins. [Courtesy of David Rice, University of Sheffield and Robert Lloyd, University of Nottingham, U.K.]

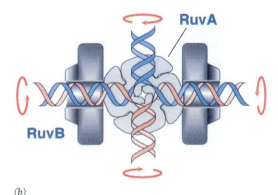

RuvA

RuvB

(b)

face is negatively charged and is poorly conserved. A model has therefore been proposed in which the duplex portions of the Holliday junction are bound in the grooves of RuvA's convex, positively charged face, with the single-stranded DNA segments that form the four-way junction surrounding RuvA's central pins (Fig. 24-37*a*). The repulsive forces between these negatively charged pins and the Holliday junction's anionic phosphate groups probably facilitate the separation of the single-stranded DNA segments.

Electron micrographs of the RuvA · RuvB · Holliday junction complex suggest that two RuvB hexamers contact opposite (180° apart) corners of the RuvA tetramer, where each wraps about a DNA arm of the RuvA-bound Holliday junction (Fig. 24-37*b*). RuvB-mediated ATP hydrolysis in this complex presumably counter-rotates these double-helical DNA stems in a manner that drives branch migration (Fig. 24-32*e,f*).

B. Transposition

In the early 1950s, Barbara McClintock reported that the variegated pigmentation pattern of maize (Indian corn) kernels results from the action of genetic elements that can move within the maize genome. This proposal was resoundingly ignored because it was contrary to the then-held orthodoxy that chromosomes consist of genes linked in fixed order. Another 20 years were to pass before evidence of mobile genetic elements was found in another organism, *E. coli.*

Transposons Move Genes between Unrelated Sites

It is now known that **transposable elements,** or **transposons,** are common in both prokaryotes and eukaryotes, where they influence the variation of phenotypic expression over the short term and evolutionary development over the long term. Each transposon codes for the enzymes that insert it into the recipient DNA. This process differs from general recombination in that it requires no homology between donor and recipient DNA and occurs at a rate of only one event in every 10^4 to 10^7 cell divisions.

Transposons with three levels of complexity have been characterized:

1. The simplest transposons are named **insertion sequences** or **IS elements.** They are normal constituents of bacterial chromosomes and **plasmids** (autonomously replicating circular DNA molecules that usually consist of several thousand base pairs). For example, a common *E. coli* strain has eight copies of **IS1** and five copies of **IS2.** IS elements generally consist of <2000 bp, comprising a so-called **transposase** gene, and in some cases, a regulatory gene, flanked by short inverted (having opposite orientation) terminal repeats. An inserted IS element is flanked by a directly (having the same orientation) repeated segment of host DNA (Fig. 24-38). This suggests that *an IS element is inserted in the host DNA at a staggered cut that is later filled in* (Fig. 24-39). The length of this target sequence (most commonly 5–9 bp), but not its sequence, is characteristic of the IS element.

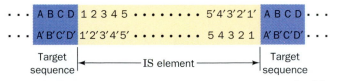

Figure 24-38. Structure of IS elements. These and other transposons have inverted terminal repeats (*numerals*) and are flanked by direct repeats of host DNA sequences (*letters*).

2. *More complex transposons carry genes not involved in the transposition process,* such as antibiotic resistance genes. For example, **Tn3** (Fig. 24-40) consists of 4957 bp and has inverted terminal repeats of 38 bp each. The central region of Tn3 codes for three proteins: (1) a 1015-residue transposase named **TnpA;** (2) a 185-residue protein known as **TnpR,** which represses the expression of both *tnpA* and *tnpR* and mediates the site-specific recombination reaction necessary for trans-

Staggered cut
within host DNA

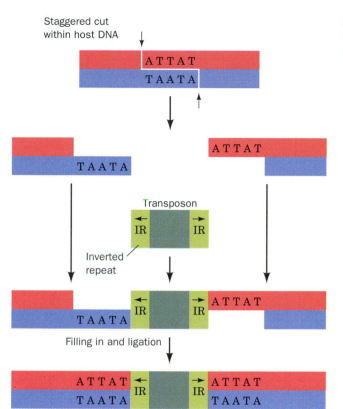

**Figure 24-39.   A model for transposon inser-
tion.** A staggered cut that is later filled in generates direct repeats of the target sequence.

position (see below); and (3) a **β-lactamase** that inactivates ampicillin (Box 8-2). The site-specific recombination occurs in an AT-rich region, the **internal resolution
site,** between *tnpA* and *tnpR*.

3. The so-called **composite transposons** (Fig. 24-41) consist of a gene-containing central region flanked by two
identical or nearly identical IS-like modules that have
either the same or an inverted relative orientation.
Composite transposons apparently arose by the association of two originally independent IS elements. Experiments demonstrate that *composite transposons can
transpose any sequence of DNA in their central region.*

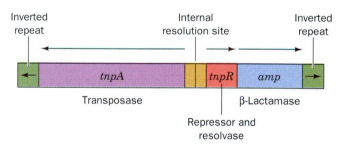

Figure 24-40. A map of transposon Tn3.

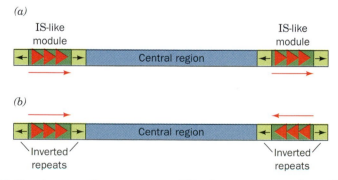

Figure 24-41.   A composite transposon. This element consists of two identical or
nearly identical IS-like modules (*green*) flanking a central region carrying various
genes. The IS-like modules may have either (*a*) direct or (*b*) inverted relative orientations.

A Proposed Transposition Mechanism

Transposons do not simply jump from point to point within a genome, as their name implies. Instead, *the mechanism of transposition involves the replication of the transposon*. A model for the movement of a transposon between two plasmids consists of the following steps (Fig. 24-42):

1. A pair of staggered single-strand cuts (such as in Fig. 24-39) is made at the target sequence of the recipient plasmid. Similarly, single-strand cuts are made on opposite strands on either side of the transposon.

2. Each of the transposon's free ends is ligated to a protruding single strand at the insertion site. This forms a replication fork at each end of the transposon.

3. The transposon is replicated, thereby yielding a **cointegrate** (the fusion of the two plasmids). Such cointegrates have been isolated.

4. Through a site-specific crossover between the internal resolution sites of the two transposons, the cointegrate is resolved into two separate plasmids, each of which contains the transposon. This recombination process is catalyzed by a transposon-coded **resolvase** (TnpR in Tn3) rather than by RecA.

Transposition Is Responsible for Much Genetic Rearrangement

In addition to mediating their own insertion into DNA, *transposons promote inversions, deletions, and rearrangements of the host DNA*. Inversions can occur when the host DNA contains two copies of a transposon in inverted orientation. The recombination of these transposons inverts the region between them (Fig. 24-43a). If, instead, the two transposons have the same orientation, recombination deletes the segment between them (Fig. 24-43b). The deletion of a chromosomal segment in this manner, followed by its integration into the chromosome at a different site by a separate recombination event, results in chromosomal rearrangement.

Transposons can be considered nature's genetic engineering "tools." For example, the rapid evolution, since antibiotics came into common use, of plasmids that confer resistance to several antibiotics (Section 3-5A) has resulted from the accumulation of the corresponding antibiotic-resistance transposons in these plasmids. Transposon-mediated rearrangements may also have been responsible for forming new proteins by linking two formerly independent gene segments. Moreover, transposons can apparently mediate the transfer of genetic information between unrelated species.

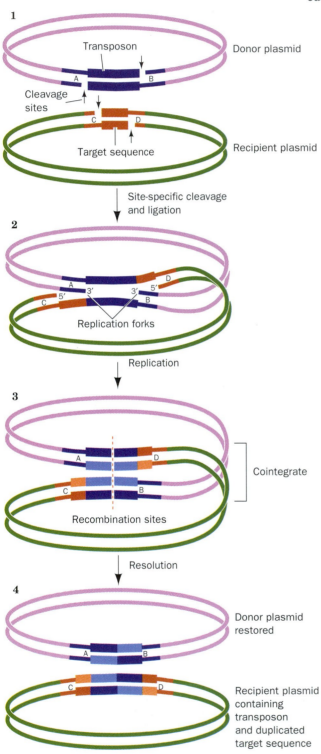

Figure 24-42. **A model for transposition involving the intermediacy of a cointegrate.** More lightly shaded bars represent newly synthesized DNA. [After Shapiro, J.A., *Proc. Natl. Acad. Sci.* 76, 1934 (1979).]

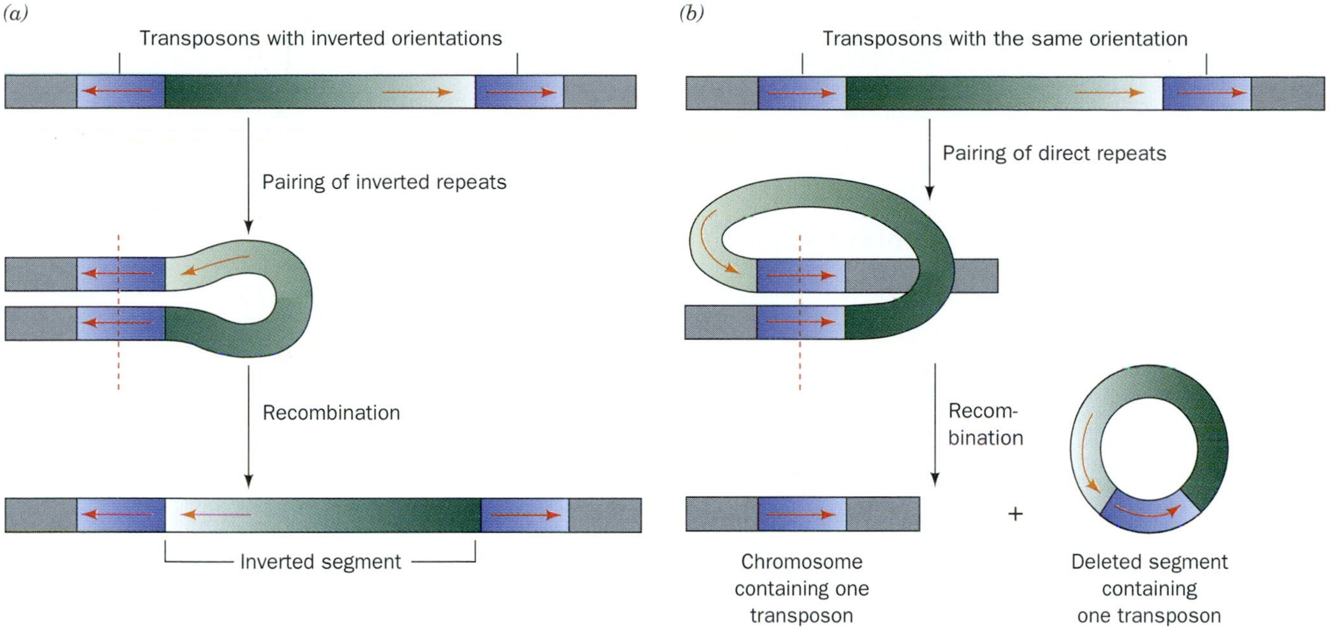

Figure 24-43. **Chromosomal rearrangement via recombination.** (*a*) The inversion of a DNA segment between two identical transposons with inverted orientations. (*b*) The deletion of a DNA segment between two identical transposons with the same orientation.

Eukaryotic Transposons Resemble Retroviruses

Transposons occur in such distantly related eukaryotes as yeast, maize, and fruit flies. In fact, ~3% of the *Drosophila* genome consists of transposons at various sites.

Retroviruses copy their single-stranded RNA genomes to double-stranded DNA (Box 24-2), which randomly inserts into the host genome before transcriptions can take place. The similar sequences of many eukaryotic transposons and retroviral genomes (and their dissimilarity to bacterial transposons) suggest that these transposons are degenerate retroviruses. They are therefore called **retrotransposons.** Transposition of a retrotransposon begins with its transcription to RNA, followed by synthesis of DNA from the RNA template by reverse transcriptase. As in retroviral infection, the newly synthesized DNA then inserts randomly into the host genome.

S U M M A R Y

1. DNA is replicated semiconservatively through the action of DNA polymerases that use the separated parental strands as templates for the synthesis of complementary daughter strands.

2. Replication is semidiscontinuous: The leading strand is syn-

thesized continuously while the lagging strand is synthesized as RNA-primed Okazaki fragments that are later joined.

3. *E. coli* DNA polymerase I has $3' \rightarrow 5'$ and $5' \rightarrow 3'$ exonuclease activities in addition to its $5' \rightarrow 3'$ polymerase activity. This enables it to carry out its physiological functions of

excising RNA primers to replace them with DNA and participating in the repair of damaged DNA.

4. Pol III, which consists of 10 different subunits, is *E. coli's* DNA replicase. Its β subunit, which functions as a sliding clamp, holds Pol III onto the DNA it is replicating and thereby increases Pol III's processivity. Two Pol III units combine to form a replisome.

5. To initiate replication, parental strands are first melted apart at a specific region and further unwound by a helicase. SSB prevents the single strands from reannealing. A primase-containing primosome synthesizes an RNA primer.

6. Because DNA polymerases operate only in the $5' \rightarrow 3'$ direction, the lagging strand template must loop back to the replisome. In *E. coli,* Pol III-mediated replication proceeds bidirectionally from the replication origin, *oriC,* until the two oppositely moving replication forks meet between oppositely facing *Ter* sequences.

7. The high fidelity of DNA replication is achieved by the regulation of dNTP levels, by the requirement for RNA priming, by $3' \rightarrow 5'$ proofreading, and by DNA repair mechanisms.

8. Eukaryotes contain at least five DNA polymerases, each of which appears to have the same handlike shape as the Klenow fragment of *E. coli* DNA Pol I. In eukaryotes, replication has multiple origins organized in replicons and proceeds through nucleosomes.

9. In order to replicate the $5'$ end of the lagging strand,

eukaryotic chromosomes end with repeated telomeric sequences added by the ribonucleoprotein telomerase. The $3'$ extension at the end of each chromosome serves as a template for primer synthesis.

10. Mutations in nucleotide sequences arise spontaneously, from replication errors, or through the action of chemical mutagens. Many compounds that are mutagenic in the Ames test are also carcinogenic.

11. Some forms of DNA damage (e.g., alkylated bases and pyrimidine dimers) may be reversed in a single step. In nucleotide excision repair, an oligonucleotide containing the lesion is removed and replaced. The SOS response in *E. coli* invokes error-prone repair. Recombination repair can restore damaged DNA that has already been replicated.

12. General recombination, in which strands of homologous DNA segments are exchanged, involves a crossover structure (Holliday junction) that can be resolved in two ways. *E. coli* RecA polymerizes on DNA to promote ATP-dependent strand exchange via a three- or four-stranded intermediate. Recombination also requires proteins to unwind DNA, produce nicks, maintain proper supercoiling, drive branch migration, resolve the Holliday junction, and ligate the DNA segments.

13. Transposons are genetic elements that move within a genome by a mechanism involving their replication. Transposons mediate rearrangement of the host DNA.

REFERENCES

DNA Replication

Arnold, E., Ding, J., Hughes, S.H., and Hostomsky, Z., Structures of DNA and RNA polymerases and their interactions with nucleic acid substrates, *Curr. Opin. Struct. Biol.* **5,** 27–38 (1995).

Brautigan, C.H. and Steitz, T.A., Structural and functional insights provided by crystal structures of DNA polymerases and their substrate complexes, *Curr. Opin. Struct. Biol.* **8,** 54–56 (1998).

DePamphilis, M.L., Origins of DNA replication in metazoan chromosomes, *J. Biol. Chem.* **268,** 1–4 (1993).

Herendeen, D.R. and Kelly, T.J., DNA Polymerase III: Running rings around the fork, *Cell* **84,** 5–8 (1996).

Kelman, Z. and O'Donnell, M., DNA polymerase III holoenzyme: Structure and function of a chromosome replicating machine, *Annu. Rev. Biochem.* **64,** 171–200 (1995).

Kong, X.-P., Onrust, R., O'Donnell, M., and Kuriyan, J., Three-dimensional structure of the β subunit of E. coli DNA polymerase III holoenzyme: A sliding DNA clamp, *Cell* **69,** 425–437 (1992).

Kornberg, A. and Baker, T.A., *DNA Replication* (2nd ed.), Freeman (1992). [A compendium of information about DNA replication.]

Sousa, R., Structural and mechanistic relationships between nucleic acid polymerases, *Trends Biochem. Sci.* **21,** 186–190 (1996).

Waga, S. and Stillman, B., The DNA replication fork in eukaryotic cells, *Annu. Rev. Biochem.* **67,** 721–751 (1998).

DNA Damage and Repair

Lindahl, T., Instability and decay of the primary structure of DNA, *Nature* **362,** 709–715 (1993). [Discusses spontaneous damage to DNA.]

Modrich, P. and Lahue, R., Mismatch repair in replication fidelity, genetic recombination and cancer biology, *Annu. Rev. Biochem.* **65,** 101–133 (1996).

Sancar, A., DNA excision repair, *Annu. Rev. Biochem.* **65,** 43–81 (1996).

Shore, D., Telomerase and telomere-binding protein: controlling the end game, *Trends Biochem. Sci.* **22,** 233–235 (1997).

Siegfried, Z. and Cedar, H., DNA methylation: A molecular lock, *Curr. Biol.* **7,** R305–R307 (1997).

Tanaka, K. and Wood, R.D., Xeroderma pigmentosum and nucleotide excision repair of DNA, *Trends Biochem. Sci.* **19,** 83–86 (1994).

Verdine, G.L., The flip side of DNA methylation, *Cell* **76,** 197–200 (1994).

Verdine, G.L. and Bruner, S.G., How do DNA repair proteins locate damaged bases in the genome, *Chem. Biol.* **4,** 329–334 (1997).

Wood, R.D., DNA repair in eukaryotes, *Annu. Rev. Biochem.* **65,** 135–167 (1996).

Recombination and Transposition

Egelman, E.H., What do X-ray crystallographic and electron microscopic structural studies of RecA protein tell us about recombination? *Curr. Opin. Struct. Biol.* **3,** 189–197 (1993).

Grindley, N.D.F. and Leschziner, A.E., DNA transpositions: from a black box to a color monitor, *Cell* **83,** 1063–1066 (1995). [Describes the mechanism of transposition and the structure and function of transposases.]

Ohtsubo, E. and Sekine, Y., Bacterial insertion sequences, *Curr. Topics Microbiol. Immunol.* **204,** 1–26 (1995).

Rice, D.W., Rafferty, J.B., Artymiuk, P.J., and Lloyd, R.G., Insights into the mechanisms of homologous recombination from the structure of RuvA, *Curr. Opin. Struct. Biol.* **7,** 798–803 (1997).

Shinagawa, H. and Iwasaki, H., Processing the Holliday junction in homologous recombination, *Trends Biochem. Sci.* **21,** 107–111 (1996).

KEY TERMS

θ structure	primosome	transversion	Holliday junction
replication fork	primase	insertion/deletion mutation	branch migration
Okazaki fragment	replisome	carcinogen	transposon
pulse-labeling	blunt end ligation	Ames test	IS element
leading strand	replicon	photoreactivation	inverted repeat
lagging strand	telomere	nucleotide excision repair	direct repeat
ligase	telomerase	(NER)	internal resolution site
primer	ribonucleoprotein	DNA glycosylase	composite transposon
nick translation	G-quartet	AP site	plasmid
processivity	mutagen	SOS response	cointegrate
proofreading	pyrimidine dimer	general recombination	retrovirus
helicase	point mutation	recombination repair	retrotransposon
SSB	transition	heterologous DNA	

STUDY EXERCISES

1. Explain how DNA replication is semiconservative, bidirectional, and semidiscontinuous.

2. Summarize the role of RNA in replicating the lagging strand and the ends of linear chromosomes.

3. Describe the functions of the three catalytic activities of *E. coli* DNA Pol I.

4. What is the purpose of the β clamp and PCNA?

5. Which reactions required for DNA replication are driven by ATP hydrolysis? What drives the non-ATP-dependent reactions?

6. What are the sources of high DNA replication fidelity?

7. How does DNA replication differ in eukaryotes and prokaryotes?

8. Describe the structure and function of telomeres.

9. List some of the ways mutations can arise.

10. Distinguish the general features of direct DNA repair and NER.

11. Describe the steps in general recombination.

12. What protein activities are required to support recombination?

13. How do transposons mediate genetic rearrangements?

PROBLEMS

1. Approximately how many Okazaki fragments are synthesized in the replication of the *E. coli* chromosome?

2. Explain why a DNA polymerase that could synthesize DNA in the $3' \rightarrow 5'$ direction would have a selective disadvantage even if it had $5' \rightarrow 3'$ proofreading activity.

3. Why are there no Pol I mutants that completely lack $5' \rightarrow 3'$ exonuclease activity?

4. Why is it advantageous to use only the Klenow fragment, rather than intact *E. coli* Pol I, in DNA sequencing reactions (Section 3-4C)?

5. Explain why DNA gyrase is required for efficient unwinding of DNA by helicase at the replication fork.

6. Why is the observed mutation rate of *E. coli* 10^{-8} to 10^{-10} per base pair replicated, even though the error rates of Pol I and Pol III are 10^{-6} to 10^{-7} per base pair replicated?

7. Why can't linear duplex DNAs, such as that of bacteriophage T7, be fully replicated by just *E. coli*-encoded proteins?

8. **Hydroxylamine** (NH_2OH) converts cytosine to the compound shown below.

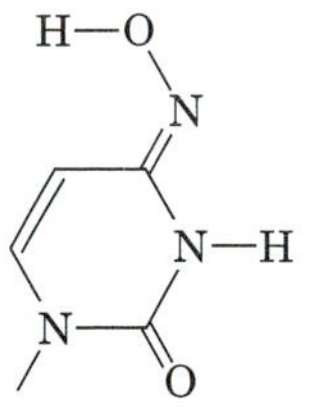

With which base does this modified cytosine residue pair? Does this generate a transition or a transversion mutation?

9. Certain sites in the *E. coli* chromosome are known as **hot spots** because they have unusually high rates of point mutations. Many of these sites contain a 5-methylcytosine residue. Explain the existence of such hot spots.

10. Predict whether loss of the following *E. coli* genes would be lethal or not: (a) *dnaB*, (b) *polA*, (c) *ssb*, (d) *recA*.

11. In *E. coli*, all newly synthesized DNA appears to be fragmented (an observation that could be interpreted to mean that the leading strand as well as the lagging strand is synthesized discontinuously). However, in *E. coli* mutants that are defective in uracil *N*-glycosylase, only about half the newly synthesized DNA is fragmented. Explain.

CHAPTER 25

A cell's transcription machinery reads discrete portions of the DNA template. The information contained in the DNA, as in the Rosetta stone shown here, must first be decoded, in this case to direct the synthesis of the cell's proteins.
[Photo courtesy of The British Museum. Reproduced with permission.]

TRANSCRIPTION AND RNA PROCESSING

DNA is confined almost exclusively to the nucleus of eukaryotic cells, as shown by microscopists in the 1930s. By the 1950s, the site of protein synthesis was identified by the demonstration that radioactively labeled amino acids that had been incorporated into proteins were associated with cytosolic RNA–protein complexes called ribosomes. Thus, *protein synthesis is not immediately directed by DNA because, at least in eukaryotes, DNA and ribosomes are never in contact.* The intermediary between DNA and the protein biosynthesis machinery, as outlined in Francis Crick's central dogma of molecular biology (Section 3-3B), is RNA.

Cells contain three major types of RNA: **ribosomal RNA (rRNA),** which constitutes two-thirds of the ribosomal mass; **transfer RNA (tRNA),** a set of small compact molecules that deliver amino acids to the ribosomes for assembly into proteins; and **messenger RNA (mRNA),** whose nucleotide sequences direct protein synthesis. In addition, a host of other small RNA species play various roles in the processing of newly transcribed RNA molecules. All types of RNA can be shown to hybridize with complementary sequences on DNA from the same organism. Thus, *all cellular RNAs are transcribed from DNA templates.*

The transcription of DNA to RNA is carried out by **RNA polymerases** that operate as multisubunit complexes, as do the DNA polymerases that catalyze DNA replication. In this chapter, we examine the catalytic properties of RNA polymerases and discuss how these proteins—unlike DNA polymerases—are targeted to specific genes. We shall also see how newly synthesized RNA is processed to become fully functional.

1. RNA POLYMERASE

RNA polymerase, the enzyme responsible for the DNA-directed synthesis of RNA, was discovered independently in 1960 by Samuel Weiss and Jerard Hurwitz. The enzyme couples together the ribonucleoside triphosphates (NTPs) ATP, CTP, GTP, and UTP on DNA templates in a reaction that is driven by the release and subsequent hydrolysis of PP_i:

$$(\text{RNA})_{n \text{ residues}} + \text{NTP} \rightleftharpoons (\text{RNA})_{n+1 \text{ residues}} + PP_i$$

$$\downarrow H_2O$$

$$2\,P_i$$

All cells contain RNA polymerase. In bacteria, one enzyme synthesizes all of the cell's RNA except the short RNA primers employed in DNA replication (Section 24-2B). Eukaryotic cells contain four or five RNA polymerases that each synthesize a different class of RNA. We shall first consider the *E. coli* enzyme because it is the best characterized RNA polymerase.

A. Enzyme Structure

The *E. coli* RNA polymerase **holoenzyme** is an ~449-kD protein with subunit composition $\alpha_2\beta\beta'\sigma$. Once RNA synthesis has been initiated, however, the σ subunit (also called **σ factor**) dissociates from the **core enzyme** $\alpha_2\beta\beta'$, which carries out the actual polymerization process. RNA polymerase is large enough to be clearly visible in electron micrographs (Fig. 25-1). Its β' subunit contains two Zn^{2+} ions, which are thought to participate in catalysis. The active enzyme also requires Mg^{2+}.

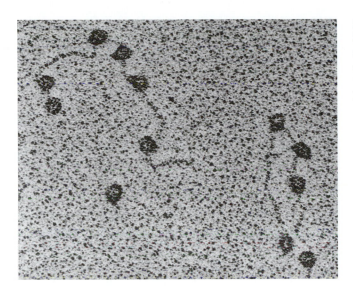

Figure 25-1. An electron micrograph of *E. coli* RNA polymerase holoenzyme. This soluble enzyme, one of the largest known, is attached to various promoter sites on bacteriophage T7 DNA. [From Williams, R.C., *Proc. Natl. Acad. Sci.* **74**, 2313 (1977).]

The low-resolution structure of *E. coli* RNA polymerase (Fig. 25-2*a*) was determined by Roger Kornberg via electron crystallography (Section 10-1A). The enzyme's most striking feature is a thumblike projection that flanks a cylindrical channel ~25 Å in diameter and 55 Å long. This channel has the dimensions appropriate to bind ~16 bp of B-DNA. RNA polymerase shares the overall "hand" shape of DNA polymerase (Section 24-2A) and reverse transcriptase (Box 24-2). X-Ray studies of the considerably smaller (99 kD) bacteriophage T7 RNA polymerase (Fig. 25-2*b*) suggest that during RNA chain elongation, the thumb closes over the DNA-binding channel.

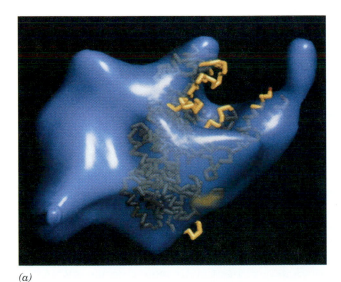

(*a*)

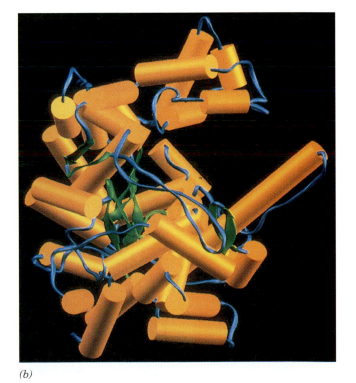

(*b*)

Figure 25-2. RNA polymerase structures. (*a*) The structure of *E. coli* RNA polymerase at ~27-Å resolution as determined by electron crystallography. The irregularly shaped enzyme (*blue*) is ~100 × 100 × 160 Å in size and is viewed along its ~25-Å-wide and 55-Å-long cylindrical channel. Note the thumblike projection (*right*) flanking this putative DNA-binding channel. The similarly oriented C$_\alpha$ backbone of DNA polymerase I Klenow fragment (Section 24-2A) is superimposed in yellow. [Courtesy of Roger Kornberg, Stanford University.] (*b*) The X-ray structure of T7 RNA polymerase oriented so as to match the low-resolution image of *E. coli* RNA polymerase shown in *a*. Helices are orange cylinders, β sheets are green arrows, and other segments of the protein are blue. Portions of the monomeric protein are not visible in the electron density map, and hence the polypeptide chain appears to be discontinuous. [Based on an X-ray structure by Rui Sousa and Bi-Cheng Wang, University of Pittsburgh.]

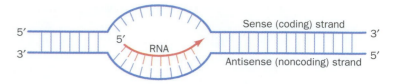

Figure 25-3.   Sense and antisense DNA strands. The template strand of duplex DNA is known as its antisense or noncoding strand. Its sense or coding strand has the same nucleotide sequence and orientation as the transcribed RNA.

B. *Template Binding*

RNA synthesis is normally initiated only at specific sites on the DNA template. In contrast to replication, which requires that both strands of the chromosome be entirely copied, the regulated expression of genetic information involves much smaller, single-strand portions of the genome. The DNA strand that serves as a template during transcription is known as the **antisense** or **noncoding strand** since its sequence is complementary to that of the RNA. The other DNA strand, which has the same sequence (except for the replacement of U with T) as the transcribed RNA, is known as the **sense** or **coding strand** (Fig. 25-3). The two strands of DNA in an organism's chromosome can therefore contain different sets of genes.

Keep in mind that "gene" is a relatively loose term that refers to sequences that encode polypeptides, as well as those that correspond to the sequences of rRNA, tRNA, and other RNA species. Furthermore, a gene typically includes sequences that participate in initiating and terminating transcription (and translation) that are not actually transcribed (or translated). The expression of many genes also depends on regulatory sequences that do not directly flank the coding regions but may be located a considerable distance away.

Most protein-coding genes (called **structural genes**) in eukaryotes are transcribed individually. In prokaryotic genomes, however, genes are frequently arranged in tandem along a single strand so that they can be transcribed together. These genetic units, called **operons,** typically contain genes with related functions. For example, *E. coli's* three different rRNA genes occur in single operons (Section 25-3B). The *E. coli **lac** operon* whose expression is described in detail in Section 27-2A, contains three genes encoding proteins involved in lactose metabolism as well as sequences that control their transcription (Fig. 25-4). Other operons contain genes encoding proteins required for biosynthetic pathways, for example, the *trp* **operon,** whose six **gene products** (proteins are often referred to as gene products) catalyze tryptophan synthesis. An operon is transcribed as a single unit, giving rise to a **polycistronic mRNA** that directs the more-or-less simultaneous synthesis of each of the encoded polypeptides (the term **cistron** is a synonym for gene). In contrast, eukaryotic structural genes, which are not part of operons, give rise to only **monocistronic mRNAs.**

RNA Polymerase Holoenzyme Binds to Promoters

How does RNA polymerase recognize the correct DNA strand and initiate RNA synthesis at the beginning of a gene (or operon)? *RNA polymerase binds to its initiation sites through base sequences known as **promoters** that are recognized by the corresponding σ factor.* The existence of promoters was first revealed through mutations that enhanced or diminished the transcription rates of certain genes. Promoters consist of ~40-bp

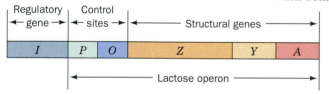

Figure 25-4.   The E. coli lac operon. This DNA includes genes encoding the proteins mediating lactose metabolism and the genetic sites that control their expression. The *Z, Y,* and *A* genes, respectively, specify the proteins β-**galactosidase,** **galactoside permease,** and **thiogalactoside transacetylase.** The closely linked regulator gene, *I,* which is not part of the *lac* operon, encodes a repressor that inhibits transcription of the *lac* operon.

sequences that are located on the 5′ side of the transcription start site. By convention, the sequence of this DNA is represented by its sense (non-template) strand so that it will have the same sequence and directionality as the transcribed RNA. A base pair in a promoter region is assigned a negative or positive number that indicates its position, upstream or downstream in the direction of RNA polymerase travel, from the first nucleotide that is transcribed to RNA; this start site is +1 and there is no 0. Because RNA is synthesized in the 5′ → 3′ direction (Section 25-1C), the promoter is said to lie upstream of the RNA's starting nucleotide.

The holoenzyme forms tight complexes with promoters (dissociation constant $K \approx 10^{-14}$ M). This tight binding can be demonstrated by showing that the holoenzyme protects the bound DNA segments from digestion *in vitro* by the endonuclease DNase I. Sequence determinations of the protected regions from numerous *E. coli* genes have identified the "consensus" sequence of *E. coli* promoters (Fig. 25-5). Their most conserved sequence is a hexamer centered at about the −10 position (sometimes called the **Pribnow box,** after David Pribnow, who described it in 1975). It has a consensus sequence of TATAAT in which the leading TA and the final T are highly conserved. Upstream sequences around position −35 also have a region of sequence similarity, TTGACA. The initiating (+1) nucleotide, which is nearly always A or G, is centered in a poorly conserved CAT or CGT sequence. Most promoter sequences vary considerably from the consensus sequence (Fig. 25-5). Nevertheless, a mutation in one of the partially conserved regions can greatly increase or decrease a promoter's initiation efficiency. This is because *the rates at which E. coli genes are transcribed vary directly with the rates that their promoters form stable initiation complexes with the holoenzyme.*

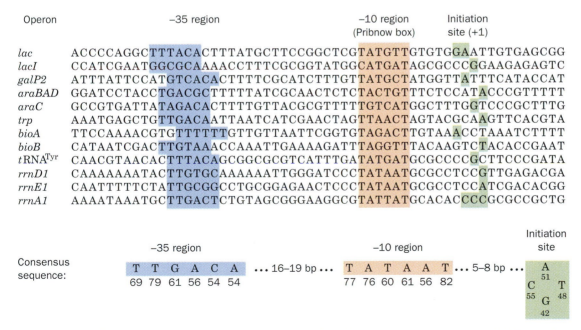

Figure 25-5. The sense (coding) strand sequences of selected *E. coli* promoters. A 6-bp region centered around the −10 position (*red shading*) and a 6-bp sequence around the −35 region (*blue shading*) are both conserved. The transcription initiation sites (+1), which in most promoters occur at a single purine nucleotide, are shaded in green. The bottom row shows the consensus sequence of 298 *E. coli* promoters with the number below each base indicating its percentage occurrence. [After Rosenberg, M. and Court, D., *Annu. Rev. Genet.* **13**, 321–323 (1979). Consensus sequence from Lisser, D. and Margalit, H., *Nucleic Acids Res.* **21**, 1512 (1993).]

Initiation Requires the Formation of an Open Complex

The promoter regions in contact with the RNA polymerase holoenzyme have been identified by a procedure named **footprinting.** In this procedure, DNA is incubated with a protein to which it binds and is then treated with an alkylation agent such as dimethyl sulfate (DMS). This results in alkylation of the DNA's bases followed by backbone cleavage at the alkylated positions. However, those portions of the DNA that bind proteins are protected from alkylation and hence cleavage. The resulting pattern of protection is called the protein's footprint.

The footprint of RNA polymerase holoenzyme indicates that it contacts the promoter primarily at its -10 and -35 regions. In some genes, additional upstream sequences may also influence RNA polymerase binding to DNA.

DMS methylates G residues at N7 and A residues at N3 in both double- and single-stranded DNA. DMS also methylates N1 of A and N3 of C, but only if these latter positions are not involved in base-pairing interactions. The pattern of DMS methylation therefore reveals whether the DNA is single or double stranded. Footprinting studies indicate that holoenzyme binding "melts" (separates) ~ 11 bp of DNA (from -9 to $+2$). The resulting **open complex** is analogous to the region of unwound DNA at the replication origin (Section 24-2B).

Core enzyme, which does not specifically bind promoters, tightly binds duplex DNA (the complex's dissociation constant is $K \approx 5 \times 10^{-12}$ M and its half-life is ~ 60 min). *Holoenzyme, in contrast, binds to nonpromoter DNA comparatively loosely* ($K \approx 10^{-7}$ M and a half-life of >1 s). Evidently, the σ subunit allows holoenzyme to move rapidly along a DNA strand in search of the σ subunit's corresponding promoter. Once transcription has been initiated and the σ subunit jettisoned, the tight binding of core enzyme to DNA apparently stabilizes the ternary enzyme–DNA–RNA complex.

Gene Expression Is Controlled by Different σ Factors

Because different σ factors recognize different promoters, *a cell's complement of σ factors determines which genes are transcribed.* Development and differentiation, which involve the temporally ordered expression of sets of genes, can be orchestrated through a "cascade" of σ factors. For example, infection of *Bacillus subtilis* by **bacteriophage SPO1** requires the expression of different sets of phage genes at different times. The first set, known as the **early genes,** are transcribed using the bacterial σ factor. One of the early phage gene products is a σ subunit known as σ^{gp28}, which displaces the host σ factor and thereby permits the RNA polymerase to recognize only the phage **middle gene** promoters. The phage middle genes, in turn, specify $\sigma^{gp33/34}$, which promotes transcription of only phage **late genes.**

The various σ factors in a bacterial cell are not necessarily used in a sequential manner. For example, σ factors in *E. coli* that differ from its primary σ factor (which is named σ^{70} because its molecular mass is 70 kD) control the transcription of coordinately expressed groups of special-purpose genes whose promoters are quite different from those recognized by σ^{70}.

C. Chain Elongation

Because RNA synthesis, like DNA synthesis, proceeds in the $5' \rightarrow 3'$ direction (Fig. 25-6), the growing RNA molecule has a 5'-triphosphate group. Mature RNA molecules, as we shall see, may also be chemically modified at one or both ends.

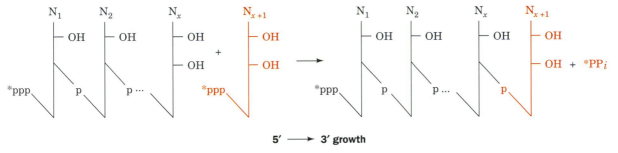

5′ ⟶ 3′ growth

Figure 25-6. 5′ → 3′ RNA chain growth. Nucleotides are added to the 3′ end of the growing RNA chain via attack of the 3′-OH group on the incoming nucleoside triphosphate. A radioactive label at the γ position of an NTP (indicated by an asterisk) is retained in the initial nucleotide of the RNA (the 5′ end) but is lost as PP_i during the polymerization of subsequent nucleotides.

A portion of the double-stranded DNA template remains opened up at the point of RNA synthesis. This allows the antisense strand to be transcribed onto its complementary RNA strand. The RNA chain transiently forms a short length of RNA–DNA hybrid duplex. The unpaired "bubble" of DNA in the open initiation complex apparently travels along the DNA with the RNA polymerase (Fig. 25-7).

As DNA's helical turns are pushed ahead of the advancing transcription bubble, they become more tightly wound (more positively supercoiled) while the DNA behind the bubble becomes equivalently unwound (more negatively supercoiled). This scenario is supported by the observation that the transcription of plasmids in *E. coli* induces their positive supercoiling in gyrase mutants (which cannot relax positive supercoils; Section 23-1C) and their negative supercoiling in topoisomerase I mutants (which cannot relax negative supercoils).

RNA Polymerase Is Processive

Once the open complex has been formed, transcription proceeds without dissociation of the enzyme from the template. Processivity is accomplished without an obvious clamplike structure (e.g., the β clamp of *E. coli* DNA polymerase III; Fig. 24-14), although the RNA polymerase thumb quite probably wraps around the DNA template to minimize dissociation

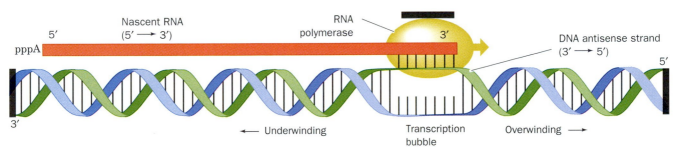

Figure 25-7. DNA supercoiling during transcription. In the region being transcribed, the DNA double helix is unwound by about a turn to permit the DNA's antisense strand to form a short segment of DNA–RNA hybrid double helix. As the RNA polymerase advances along the DNA template (here to the right), the DNA unwinds ahead of the RNA's growing 3′ end and rewinds behind it, thereby stripping the newly synthesized RNA from the template strand. Because the ends of the DNA as well as the RNA polymerase are apparently prevented from rotating by attachments within the cell (*black bars*), the DNA becomes overwound ahead of the advancing transcription bubble and unwound behind it (consider the consequences of placing your finger between the twisted DNA strands in this model and pushing toward the right). [After Futcher, B., *Trends Genet.* **4**, 272 (1988).]

Figure 25-8.   A model explaining the processivity of RNA polymerase. The enzyme grasps the DNA on either side of the transcription bubble.

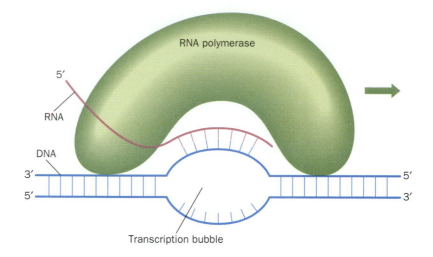

Transcription bubble

of the enzyme from the template. Processivity may also result from RNA polymerase binding to the DNA with high affinity but low specificity at points both ahead of and behind the transcription bubble (Fig. 25-8). The multiple attachment sites may explain why a transcription complex does not completely dissociate from a DNA template even when transcription is interrupted by the DNA replication machinery proceeding along the same strand of DNA (see Box 25-1).

Transcription Is Rapid

The *in vivo* rate of transcription is 20 to 50 nucleotides per second at 37°C (but still several times slower than the DNA replication rate; Section 24-2C). The error frequency in RNA synthesis is one wrong base incorporated for every $\sim 10^4$ transcribed. This frequency, which is 10^4 to 10^6 times higher than that for DNA synthesis, is tolerable because of the repeated transcription of most genes, because the genetic code contains numerous

---

Box 25-1

BIOCHEMISTRY IN FOCUS

Collisions between DNA Polymerase and RNA Polymerase

In rapidly proliferating bacterial cells, DNA synthesis is likely to occur even as genes are being transcribed. The DNA replication machinery moves along the circular chromosome at a rate many times faster than the movement of the transcription machinery. Collisions between DNA polymerase and RNA polymerase seem unavoidable. What happens when the two enzyme complexes collide? Using *in vitro* model systems, Bruce Alberts has shown that when both enzymes are moving in the same direction, the replication fork passes the RNA polymerase without displacing it and leaving it fully competent to resume RNA chain elongation.

When the replication fork collides head-on with a transcription complex, however, the replication fork pauses briefly before moving past the RNA polymerase. Surprisingly,

this causes the RNA polymerase to switch its template strand. The growing RNA chain dissociates from the original template DNA strand and hybridizes with the newly synthesized daughter DNA strand of the same sequence before RNA elongation resumes.

Head-on collisions are disadvantageous because (1) replication slows when DNA polymerase pauses and (2) dissociation of the RNA polymerase during the jump from one template strand to the other could abort the transcription process. Indeed, in many bacterial and phage genomes, the most heavily transcribed genes are oriented so that replication and transcription complexes move in the same direction. It remains to be seen whether a similar arrangement holds in eukaryotic genomes, which contain multiple replication origins and genes that are much larger than are prokaryotic genes.

synonyms (Section 26-1C), and because amino acid substitutions in proteins are often functionally innocuous.

Once an RNA polymerase molecule has initiated transcription and moved away from the promoter, another RNA polymerase can follow suit. The synthesis of RNAs that are needed in large quantities, rRNAs, for example, is initiated as often as is sterically possible, about once per second. This gives rise to an arrowhead appearance of the transcribed DNA (Fig. 25-9). mRNAs encoding proteins are generally synthesized at less frequent intervals, and there is enormous variation in the amounts of different polypeptides produced. For example, an *E. coli* cell may contain 10,000 copies of a ribosomal protein, whereas a regulatory protein may be present in only a few copies per cell. Many enzymes, particularly those involved in basic cellular "housekeeping" functions, are synthesized at a more or less constant rate; they are called **constitutive enzymes.** Other enzymes, termed **inducible enzymes,** are synthesized at rates that vary with the cell's circumstances. To a large extent, the regulation of gene expression relies on mechanisms that govern the rate of transcription, as we shall see in Section 27-2. The products of transcription also vary in their stability. Ribosomal RNA turns over much more slowly than mRNA, which in prokaryotes is rapidly synthesized and rapidly degraded (sometimes so fast that the 5' end of an mRNA is degraded before its 3' end has been synthesized).

D. Chain Termination

Electron micrographs such as Fig. 25-9 suggest that DNA contains specific sites at which transcription is terminated. The transcription termination sequences of many *E. coli* genes share two common features (Fig. 25-10a):

1. A series of 4 to 10 consecutive A·T base pairs, with the A's on the template strand. The transcribed RNA is terminated in or just past this sequence.

2. A G + C-rich region with a palindromic sequence that immediately precedes the series of A·T's.

The RNA transcript of this region can therefore form a self-complementary "hairpin" structure that is terminated by several U residues (Fig. 25-10b).

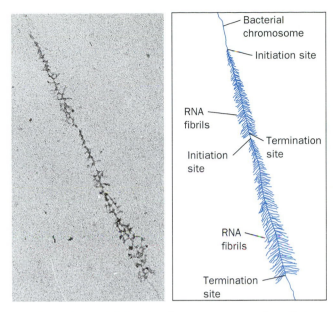

Figure 25-9. **An electron micrograph and its interpretive drawing of two contiguous *E. coli* ribosomal genes undergoing transcription.** The "arrowhead" structures result from the increasing lengths of the nascent RNA chains as the RNA polymerase molecules synthesizing them move from an initiation site on the DNA to the following termination site. [Courtesy of Oscar L. Miller, Jr. and Barbara A. Hamkalo, University of Virginia.]

(a)

<pre>
                    G · C   A · T
               rich region rich region
5' ··· NNAAGCGCCGNNNNCCGGCGCTTTTTTTNNN ··· 3' DNA
3' ··· NNTTCGCGGCNNNNGGCCGCGAAAAAANNN ··· 5' template

5' ··· NNAAGCGCCGNNNNCCGGCGCUUUUUU—OH 3'   RNA
                                            transcript
</pre>

(b)

<pre>
                        N
                      N   N
                      | |
                      N   C
                       \ /
                   G · C
                   C · G
                   C · G
                   G · C
                   C · G
                   G · C
                   A · U
                   A · U
          ···NNNN   UUUU—OH 3'
</pre>

Figure 25-10. **A hypothetical strong (efficient) *E. coli* terminator.** The base sequence was deduced from the sequences of several transcripts. (*a*) The DNA sequence together with its corresponding RNA. The A·T-rich and G·C-rich sequences are shown in blue and red, respectively. The twofold symmetry axis (green symbol) relates the flanking shaded segments that form an inverted repeat. (*b*) The RNA hairpin structure and poly(U) tail that trigger transcription termination. [After Pribnow, D., *in* Goldberger, R.F. (Ed.), *Biological Regulation and Development*, Vol. 1, p. 253, Plenum Press (1979).]

The structural stability of an RNA transcript at its terminator's G + C-rich hairpin and the weak base pairing of its oligo(U) tail to template DNA appear to be important factors in ensuring proper chain termination. The formation of the G + C-rich hairpin causes RNA polymerase to pause for several seconds at the termination site. This probably induces a conformational change in the RNA polymerase that permits the nontemplate DNA strand to displace the weakly bound oligo(U) tail from the template strand, thereby terminating transcription.

Despite the foregoing, experiments by Michael Chamberlin indicate that the RNA-terminator hairpin and U-rich 3′ tail do not function independently of their upstream and downstream flanking regions. Indeed, terminators that lack a U-rich segment can be highly efficient when joined to the appropriate sequence immediately downstream from the termination site. Termination efficiency also varies with the concentrations of nucleoside triphosphates, with the level of supercoiling in the DNA template, with changes in the salt concentration, and with the sequence of the terminator. These results suggest that *termination is a complex multistep process.*

Termination Often Requires Rho Factor

The termination sequences described above induce the spontaneous termination of transcription. Other termination sites, however, lack any obvious similarities and are unable to form strong hairpins; *they require a protein known as **rho factor** to terminate transcription.* Rho factor, a hexamer of identical 419-residue subunits, enhances the termination efficiency of spontaneously terminating transcripts and induces the termination of nonspontaneously terminating transcripts.

Several key observations have led to a model of rho-dependent termination:

1. Rho factor is an enzyme that catalyzes the unwinding of RNA–DNA and RNA–RNA double helices. This process is powered by the hydrolysis of nucleoside triphosphates to nucleoside diphosphates + P_i with little preference for the identity of the base.

2. Genetic manipulations indicate that rho-dependent termination requires a specific recognition sequence on the nascent (still being synthesized) RNA upstream of the termination site.

These observations suggest that rho factor attaches to the RNA at its recognition sequence and then migrates along the RNA in the $5′ \rightarrow 3′$ direction until it encounters an RNA polymerase paused at the termination site (without the pause, rho might not be able to overtake the RNA polymerase). There, rho unwinds the RNA–DNA duplex at the transcription bubble, thereby releasing the RNA transcript. Rho-terminated transcripts have 3′ ends that typically vary over a range of ∼50 nt. This suggests that rho gradually pries the RNA away from its template DNA rather than liberating the RNA at a specific point.

2. TRANSCRIPTION IN EUKARYOTES

Although the fundamental principles of transcription are similar in prokaryotes and eukaryotes, *eukaryotic transcription is distinguished by having multiple RNA polymerases and by relatively complicated control sequences.* The eukaryotic transcription machinery is also more complicated, requiring up to 50 proteins to recognize the control sequences and initiate transcription.

A. Eukaryotic RNA Polymerases

Eukaryotic nuclei contain three distinct types of RNA polymerase that differ in the RNAs they synthesize:

1. **RNA polymerase I,** which is located in the **nucleoli** (dark-staining nuclear bodies where ribosomes are assembled; Fig. 1-8 and Section 26-3), synthesizes the precursors of most rRNAs.

2. **RNA polymerase II,** which occurs in the nucleoplasm, synthesizes the mRNA precursors.

3. **RNA polymerase III,** which also occurs in the nucleoplasm, synthesizes the precursors of 5S rRNA, the tRNAs, and a variety of other small nuclear and cytosolic RNAs.

In addition to these nuclear enzymes, eukaryotic cells contain separate mitochondrial and (in plants) chloroplast RNA polymerases. The essential function of RNA polymerases in all cells makes them attractive targets for antibiotics and other drugs (see Box 25-2).

Eukaryotic RNA polymerases, whose molecular masses vary between 500 and 700 kD, are characterized by subunit compositions of Byzantine complexity. Each type of enzyme contains two nonidentical "large" (>100 kD) subunits and an array of up to 12 different "small" (<50 kD) subunits. For example, yeast RNA polymerase II, which has a molecular mass of ~550 kD, contains 12 subunits. The three largest subunits are homologs of the prokaryotic RNA polymerase subunits α, β, and β′ and are therefore thought to constitute the structural and functional core of the enzyme.

The low-resolution structures of yeast RNA polymerases I and II have been determined by electron crystallography (Fig. 25-11). Both proteins have irregular shapes that resemble each other and that of *E. coli* RNA polymerase (Fig. 25-2a) in that their most prominent feature is a thumb flanking an ~25-Å-wide channel that curves around the protein's surface and presumably binds B-DNA. In the RNA polymerase II structure (Fig. 25-11b), a channel wide enough to accommodate a single strand of RNA branches off from the DNA-binding channel in a nearly perpendicular direction.

B. Eukaryotic Promoters

Eukaryotic promoters are more complex and diverse than prokaryotic promoters. Interestingly, many of the elements required to initiate eukaryotic gene transcription have variable positions and orientations relative to their corresponding transcribed sequences.

RNA Polymerase I Promoters

Both prokaryotic and eukaryotic genomes contain multiple copies of their rRNA genes in order to meet the enormous demand for these rRNAs (which comprise, e.g., 80% of an *E. coli* cell's RNA content). Since the numerous rRNA genes in a given eukaryotic cell have essentially identical sequences, its RNA polymerase I recognizes only one promoter. Yet in contrast to RNA polymerase II and III promoters, RNA polymerase I promoters are species specific; that is, an RNA polymerase I recognizes only its own promoter and those of closely related species.

RNA polymerase I promoters were identified by determining how the transcription rate of an rRNA gene is affected by a series of increasingly

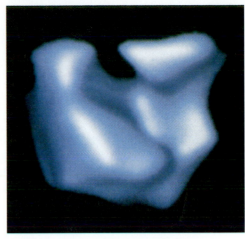

(a)

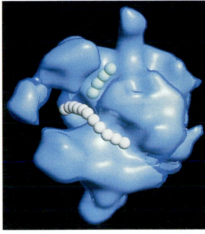

(b)

Figure 25-11. The structures of eukaryotic RNA polymerases as determined by electron crystallography. (a) Yeast RNA polymerase I at ~30-Å resolution. The irregularly shaped protein is around 150 × 110 × 100 Å in size. The 30-Å-wide and 100-Å-long groove that extends from the upper middle to the lower right of the image is the putative DNA-binding site. [Courtesy of Patrick Schultz and Pierre Oudet, Laboratoire de Genétique Moléculaire des Eucaryotes, Strasbourg, France.] (b) Yeast RNA polymerase II at ~16-Å resolution. The enzyme is about 140 × 136 × 110 Å in size. The beads are 8 Å in diameter and are placed every 6.8 Å. The chain of pink beads marks the putative DNA-binding channel (which is ~25 Å wide), and the chain of light green beads marks the putative RNA-binding channel (which is 12–15 Å wide) that branches off of the DNA-binding channel. [Courtesy of Roger Kornberg, Stanford University.]

Inhibitors of Transcription

A wide variety of compounds inhibit transcription in prokaryotes and eukaryotes. These agents are therefore toxic to susceptible organisms; that is, they function as antibiotics. Such compounds are also useful research tools since they arrest the transcription process at well-defined points.

Two related antibiotics, **rifamycin B**, which is produced by *Streptomyces mediterranei*, and its semisynthetic derivative **rifampicin,**

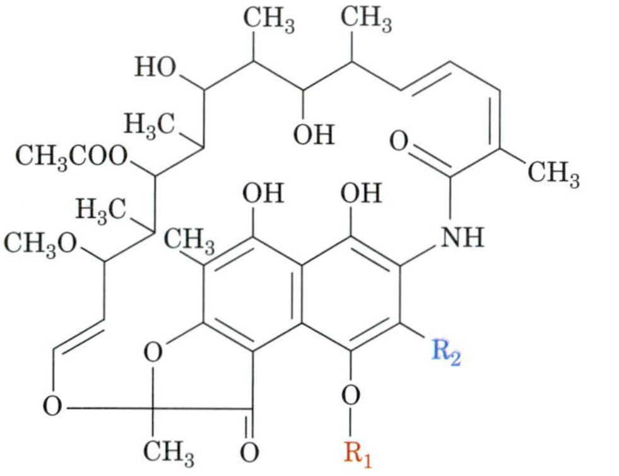

Rifamycin B R_1 = CH_2COO^-; R_2 = H

Rifampicin R_1 = H; R_2 = CH=$\overset{+}{N}$ N—CH$_3$

specifically inhibit transcription by prokaryotic but not eukaryotic RNA polymerases. The selectivity and high potency of rifampicin (2×10^{-8} M results in 50% inhibition of bacterial RNA polymerase) makes it a medically useful bactericidal agent. Rifamycins inhibit neither the binding of RNA polymerase to the promoter nor the formation of the first phosphodiester bonds, but they prevent further chain elongation. The inactivated RNA polymerase remains bound to the promoter, thereby blocking initiation by uninhibited enzyme. Once RNA chain initiation has occurred, however, rifamycins have no effect on the subsequent elongation process. The rifamycins can therefore be used in the laboratory to dissect transcriptional initiation and elongation.

The bacterial compound **actinomycin D,**

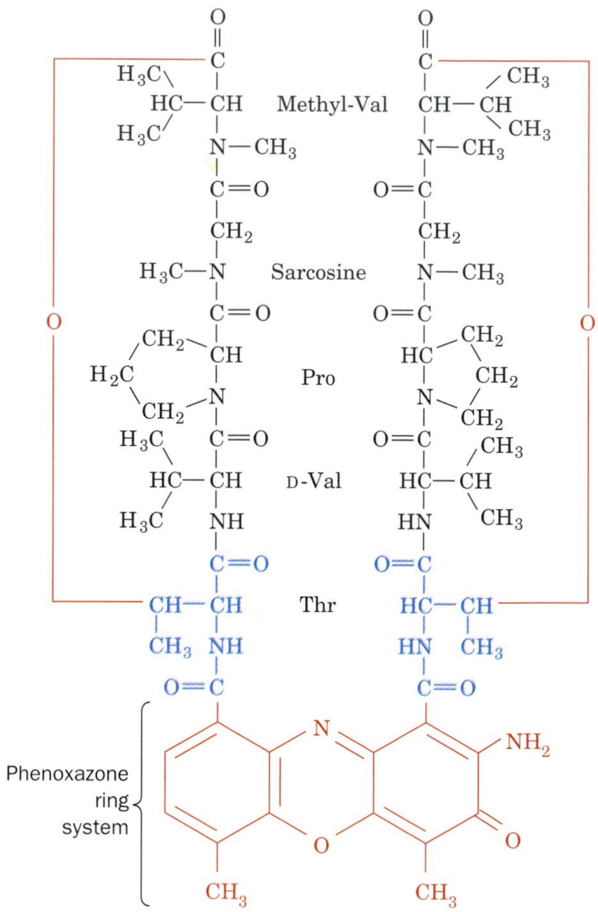

Actinomycin D

a useful antineoplastic (anticancer) agent produced by *Streptomyces antibioticus,* tightly binds to duplex DNA and, in doing so, strongly inhibits both transcription and DNA replication, presumably by interfering with the passage of RNA and DNA polymerases. The X-ray structure of actinomycin D in complex with an 8-bp DNA is shown at right as two vertically stacked complexes in space-filling form with the DNA's sugar–phosphate backbone yellow, its bases white, and

longer deletions approaching its transcription start site from either its upstream or its downstream side. Such studies have indicated, for example, that mammalian RNA polymerase I requires a so-called **core promoter element,** which spans positions −31 to +6 and hence overlaps the transcribed region. However, efficient transcription also requires an **upstream promoter element,** which is located between residues −187 and −107.

the actinomycins colored according to atom type (C green, N blue, and O red).

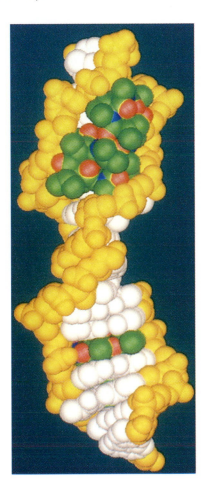

The actinomycin's phenoxazone ring system (which is visible in the lower molecule) intercalates between the DNA's base pairs, thereby unwinding the DNA helix by 23° and separating the neighboring base pairs by 7.0 Å. Actinomycin's chemically identical cyclic **depsipeptides** (which are seen in the upper molecule; depsipeptides have both peptide bonds and ester linkages) extend in opposite directions from the intercalation site along the minor groove of the DNA. Other intercalation agents, including ethidium and proflavin (Section 23-3B), also inhibit nucleic acid synthesis, presumably by similar mechanisms.

The poisonous mushroom *Amanita phalloides* (death cap), which is responsible for the majority of fatal mushroom poisonings in Europe, contains several types of toxic substances, including a series of unusual bicyclic octapeptides known as **amatoxins**. **α-Amanitin,**

α-Amanitin

which is representative of the amatoxins, forms a tight 1:1 complex with RNA polymerase II ($K = 10^{-8}$ M) and a looser one with RNA polymerase III ($K = 10^{-6}$ M) so as to specifically block their elongation steps. RNA polymerase I as well as mitochondrial, chloroplast, and prokaryotic RNA polymerases are insensitive to α-amanitin.

Despite the amatoxins' high toxicity (5–6 mg, contained in ~40 g of fresh mushrooms, is sufficient to kill a human adult), they act slowly. Death, usually from liver dysfunction, occurs no earlier than several days after mushroom ingestion (and after recovery from the effects of other mushroom toxins). This, in part, reflects the slow turnover rate of eukaryotic mRNAs and proteins.

[Structure of actinomycin D–DNA complex based on an X-ray structure by Fusao Takusagawa, University of Kansas.]

RNA Polymerase II Control Sequences

The promoters recognized by RNA polymerase II exhibit considerable diversity. The structural genes expressed in all tissues (the housekeeping genes, which are thought to be constitutively transcribed) have one or more copies of the sequence GGGCGG or its complement (the **GC box**) located upstream from their transcription start sites. The analysis of deletion and

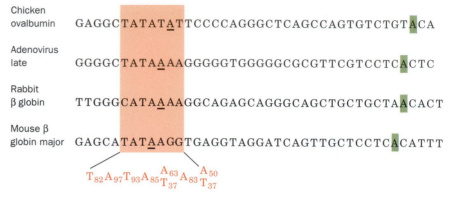

Figure 25-12. **The promoter sequences of selected eukaryotic structural genes.**
The homologous segment, the TATA box, is shaded in red with the base at position
−27 underlined. The initial nucleotide to be transcribed (+1) is shaded in green.
The bottom row indicates the consensus sequence of several such promoters, with
the subscripts indicating the percentage occurrence of the corresponding base.
[After Gannon, F., O'Hare, K., Perrin, F., Le Pennec, J.P., Benoist, C., Cochet, M.,
Breathnach, R., Royal, A., Garapin, A., Cami, B., and Chambon, P., *Nature* **278**, 433
(1978).]

point mutations indicates that *GC boxes function analogously to prokaryotic promoters.* On the other hand, structural genes that are selectively
expressed in one or a few cell types often lack these GC-rich sequences.
Instead, *they contain a conserved AT-rich sequence located 25 to 30
bp upstream from their transcription start sites* (Fig. 25-12). Note that this
so-called **TATA box** resembles the −10 region of a prokaryotic promoter
(TATAAT), although it differs in its location relative to the transcription
start site (−27 versus −10). The functions of the two promoter elements
are not strictly analogous, however, since the deletion of the TATA box
does not necessarily eliminate transcription. Rather, TATA box deletion or
mutation generates heterogeneities in the transcription start site, thereby
indicating that the TATA box participates in selecting this site.

The gene region extending between about −50 and −110 also contains
promoter elements. For instance, many eukaryotic structural genes have a
conserved consensus sequence of CCAAT (the **CCAAT box**) located between about −70 and −90 whose alteration greatly reduces the gene's transcription rate. Evidently, *the promoter sequences upstream of the TATA box
constitute DNA-binding sites for RNA polymerase II and other proteins
involved in transcription initiation.*

In addition, sequences located hundreds or even thousands of bases from
the transcribed sequence may act as **enhancers** or **silencers** of gene transcription. These regions may alter DNA's local conformation in a way that
promotes or interferes with RNA polymerase II binding. For example,
some enhancers contain a segment of alternating purines and pyrimidines,
just the type of sequence most likely to form Z-DNA (Section 23-1A). Alternatively, enhancers may be sites that lack the histones that normally coat
eukaryotic DNA and likely block RNA polymerase II binding. Regulatory
proteins that act as **activators** or **repressors** may bind to enhancers and silencers so as to influence RNA polymerase binding to promoters (some of
these mechanisms for regulating transcription are described in Section
27-3B). Some enhancers have the intriguing property that they need not
have a fixed position or orientation relative to their corresponding transcribed sequence.

RNA Polymerase III Promoters

The promoters of some genes transcribed by RNA polymerase III are located entirely within the genes' transcribed regions. Deletions of base sequences that start from outside one or the other end of the transcribed portion of these genes prevent transcription only if they extend into the segment between nucleotides +40 and +80. This portion of the gene is effective as a promoter because it contains the binding site for a transcription factor that stimulates upstream binding of RNA polymerase III. The promoters of other RNA polymerase III-transcribed genes may lie partially or even entirely upstream of their start sites.

C. Transcription Factors

Differentiated eukaryotic cells possess a remarkable capacity for the selective expression of specific genes. The synthesis rates of a particular protein in two cells of the same organism may differ by as much as a factor of 10^9. Thus, for example, reticulocytes (immature red blood cells) synthesize large amounts of hemoglobin but no detectable insulin, whereas the pancreatic β cells produce large quantities of insulin but no hemoglobin. In contrast, prokaryotic systems generally exhibit no more than a thousand-fold range in their transcription rates so that at least a few copies of all the proteins they encode are present in any cell. Nevertheless, *the basic mechanism for initiating transcription of structural genes in eukaryotes resembles that in prokaryotes: Protein factors bind selectively to the promoter regions of DNA.* In eukaryotes, a set of at least six **general transcription factors (GTFs;** Table 25-1) operate as a formal equivalent of a prokaryotic σ factor. The structures of some eukaryotic transcription factors are described in Section 23-4C.

The six GTFs, which are highly conserved from yeast to humans, are required for the synthesis of all mRNAs, even those with strong promoters. These factors allow a low (basal) level of transcription that can be augmented by the participation of other gene-specific factors. The GTFs, whose names include TF (for transcription factor) followed by the Roman numeral II to indicate that they are involved in transcription by RNA polymerase II, combine with the enzyme and promoter DNA in an ordered pathway.

Formation of a **preinitiation complex (PIC),** the multiprotein assembly capable of initiating transcription, begins when **TATA-binding protein (TBP),** a component of **TFIID,** binds to the TATA box of a promoter. TBP, whose X-ray structure was independently determined by Roger Kornberg and Stephen Burley, is a saddle-shaped molecule that consists of two struc-

Table 25-1. Human Transcription Factors Required for Basal Transcription by RNA Polymerase II

Factor	Number of Subunits	Mass (kD)
TFIID		
TBP	1	38
TAFs	12	15–250
TFIIA	3	12, 19, 35
TFIIB	1	35
TFIIE	2	34, 57
TFIIF	2	30, 74
TFIIH	9	35–89

Source: Roeder, R.G., *Trends Biochem. Sci.* **21,** 329 (1996).

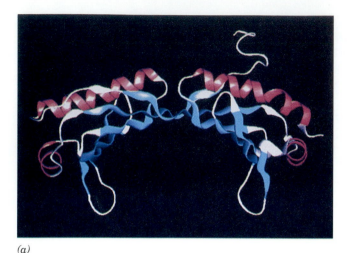

(a)

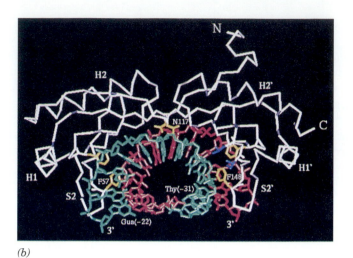

(b)

Figure 25-13.   The X-ray structure of TATA-binding protein (TBP) from the plant *Arabidopsis thaliana*. (*a*) A ribbon diagram of the protein in the absence of DNA, in which α helices are red, β strands are blue, and the remainder of the polypeptide backbone is white. The protein's pseudo-twofold axis of symmetry is vertical. Note that the protein seems to be precisely the proper size and shape to sit astride a 20-Å-diameter cylinder of undistorted B-DNA. This, however, is not what happens. (*b*) TBP in complex with a 14-bp TATA box–containing segment of DNA. The protein is represented by its C_α backbone (*white*); together with the side chains of Phe residues 57, 74, 148, and 165 (*yellow*), which induce sharp kinks in the DNA; Asn residues 27

and 117 (*also yellow*), which make hydrogen bonds in the minor groove; and Ile 52 and Leu 163 (*blue*), which are implicated in specific DNA recognition. The DNA is drawn in stick form with the sense and antisense strands in green and red, respectively. B-DNA enters its binding site with the 5′ end of the sense strand on the right and exits on the left with its helix axis nearly perpendicular to the page. Between the kinks, which are located at each end of the TATA box, the DNA is partially unwound with the protein's eight-stranded β sheet inserted into the DNA's greatly widened minor groove. [Courtesy of Stephen Burley, The Rockefeller University.] ● See the Interactive Exercises.

turally similar domains arranged with pseudo-twofold symmetry (Fig. 25-13*a*). TBP's overall structure suggests that it could fit snugly astride an undistorted B-DNA helix. However, the X-ray structures of TBP–DNA complexes, independently determined by Burley and Paul Sigler, reveal a quite different interaction. The DNA indeed binds to the concave surface of TBP but with its duplex axis nearly perpendicular rather than parallel to the saddle's cylindrical axis (Fig. 25-13*b*). The bound DNA is so sharply bent (kinked) in its interaction with TBP that it has a cranklike shape. The TBP, which undergoes little conformational change on binding the DNA, does so via hydrogen bonding and van der Waals interactions. The kinked and partially unwound DNA is stabilized by a wedge of two Phe side chains on each side of the saddle structure that pry apart the two base pairs flanking each kink from their minor groove sides. TBP is the only GTF that binds a specific DNA sequence and hence identifies the transcription start site. The bent conformation of DNA creates a stage for the assembly of other proteins into the preinitiation complex.

The other GTFs required for basal transcription assemble as is shown in Fig. 25-14. The preinitiation complex requires, at a minimum, TBP, **TFIIB, TFIIE, TFIIF,** and **TFIIH. TFIIA** probably stabilizes the TBP–TATA box interaction and hence may not be needed at strong promoters. The remaining components of TFIID (other than TBP), which are known as **TBP-associated factors (TAFs),** are probably activators that augment the rate of transcription and are not strictly required for the preinitiation complex. *The order of protein binding is critical for the recruitment of GTFs and RNA polymerase.* For example, the TBP component of TFIID, TFIIB, and TFIIA must bind to DNA before RNA polymerase II binds.

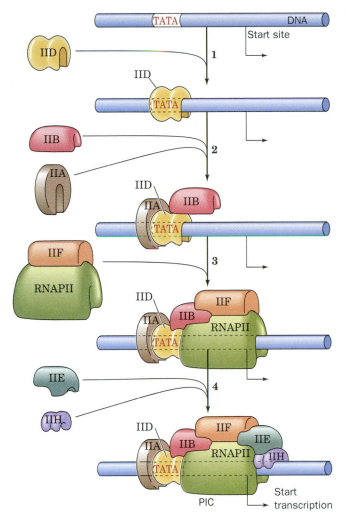

Figure 25-14. *Key to Function.* **The assembly of the preinitiation complex (PIC) on a TATA box–containing promoter.** (1) The TBP component of TFIID binds to the TATA box. (2) TFIIA and TFIIB then bind. (3) TFIIF binds to RNA polymerase II and escorts it to the complex. (4) Finally, TFIIE and TFIIH are sequentially recruited, thereby completing the PIC. [After Zawel, L. and Reinberg, D., *Curr. Opin. Cell Biol.* 4, 490 (1992).]

An X-ray structure–based model of the TBP–TFIIA–TFIIB–DNA complex (Fig. 25-15) reveals that even with three proteins bound just upstream of the transcription start site, there is still ample room for the additional protein–DNA and protein–protein interactions that regulate the frequency of RNA polymerase II recruitment to a gene's promoter.

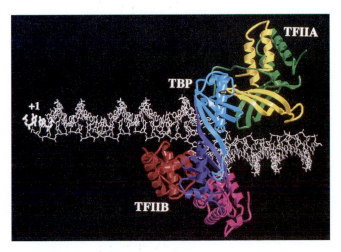

Figure 25-15. **Model of the TFIIA–TBP–TFIIB–promoter complex.** The arrangement of proteins (*ribbons*) and DNA (*white stick model*) was constructed from the independently determined X-ray crystal structures of TFIIA–TBP–promoter and TFIIB–TBP–promoter complexes. In the model, the DNA has been extended in both directions beyond the TATA box. The pseudosymmetrical TBP (*cyan and purple*), viewed from above, induces two sharp kinks in the DNA, giving it a cranklike shape. Both TFIIA (*yellow and green*) and TFIIB (*magenta and red*) interact with TBP, but each interacts with independent sites on the DNA. The transcription start site (+1) is at the left. [Courtesy of Stephen Burley, The Rockefeller University.]

Once the preinitiation complex has formed, the ATP-dependent helicase activity of TFIIH stimulates formation of the open complex. TFIIH also contains kinase activity. The reversible phosphorylation of the C-terminal region of the largest subunit of RNA polymerase II correlates with the point at which the transcription machinery moves away from ("clears") the promoter and enters the chain elongation phase. TFIIF appears to remain bound to the transcription machinery during elongation. The other GTFs may dissociate at some point after transcription initiation. The TFIID that remains bound to the promoter facilitates the reassembly of the preinitiation complex and hence promotes repeated transcription of the gene.

Certain RNA polymerase II promoters lack TATA boxes. These promoters, nevertheless, require many of the same GTFs that initiate transcription from TATA box–containing promoters, including, most surprisingly, TBP. How TBP binds to TATA-less promoters is unknown, although it seems unlikely that it does so in the same way it binds to a TATA box. The scheme outlined in Fig. 25-14 should therefore be taken as a flexible framework for transcription initiation in eukaryotes, with the exact protein requirements depending on the nature of the promoter and the presence of additional protein factors. Gene transcription by RNA polymerases I and III requires different sets of transcription factors, as might be expected from their different promoter organizations (Section 25-2B), although TBP appears to be essential in all cases (making TBP the only universal transcription factor). Unraveling the molecular events of transcription initiation in eukaryotes is difficult because many transcription factors are present in only a few copies per cell and may not be required *in vitro* even though their absence is ultimately lethal *in vivo*.

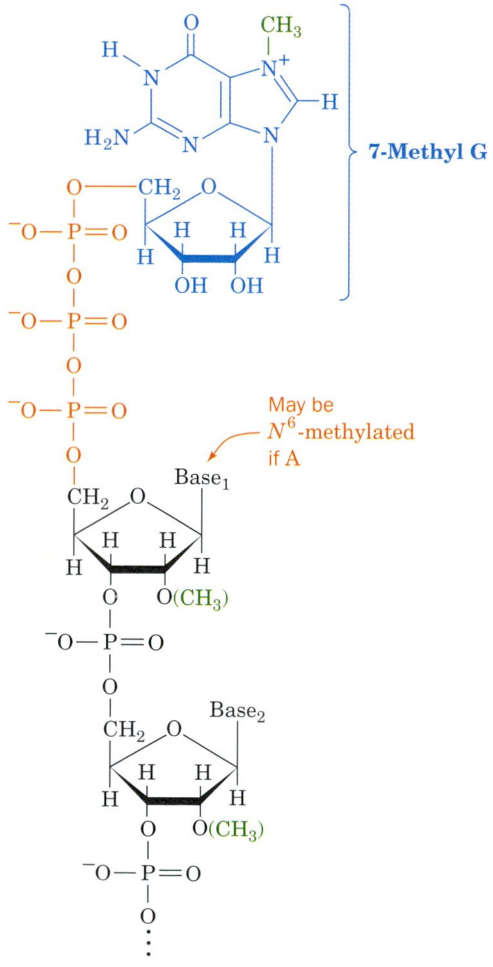

3. POSTTRANSCRIPTIONAL PROCESSING

The immediate products of transcription, the **primary transcripts,** are not necessarily functional. In order to acquire biological activity, many of them must be specifically altered: (1) by the exo- and endonucleolytic removal of polynucleotide segments; (2) by appending nucleotide sequences to their 3' and 5' ends; and/or (3) by the modification of specific nucleotide residues. The three major classes of RNA—mRNA, rRNA, and tRNA—are altered in different ways in prokaryotes and in eukaryotes. In this section, we shall outline these **posttranscriptional modification** processes.

A. Messenger RNA Processing

In prokaryotes, most primary mRNA transcripts are translated without further modification. Indeed, protein synthesis usually begins before transcription is complete (Section 26-4). In eukaryotes, however, mRNAs are synthesized in the cell nucleus, whereas translation occurs in the cytosol. Eukaryotic mRNA transcripts can therefore undergo extensive posttranscriptional processing while still in the nucleus.

Eukaryotic mRNAs Are Capped

Eukaryotic mRNAs have a cap structure consisting of a 7-methylguanosine residue joined to the transcript's initial (5') nucleotide via a 5'−5' triphosphate bridge (Fig. 25-16). The cap, which a specific **guanylyltransferase** adds to the growing transcript before it is >20 nt long, identifies the eukaryotic translation start site (Section 26-4A). The cap may be modified by methylation at the first and second nucleotides of the transcript. Capped mRNAs are resistant to 5'-exonucleolytic degradation.

Figure 25-16.   The structure of the 5' cap of eukaryotic mRNAs. The first two nucleotides of the transcript may be $O^{2'}$-methylated. If the first nucleotide is A, it may be N^6-methylated.

Eukaryotic mRNAs Have Poly(A) Tails

Eukaryotic RNA transcripts have heterogeneous 3' sequences, largely because the transcription termination process is imprecise. *Mature eukaryotic mRNAs, however, have well-defined 3' ends terminating in poly(A) tails of 20 to 50 nt.* The poly(A) tails are enzymatically appended to the primary transcripts in two reactions:

1. A transcript is cleaved 15 to 25 nucleotides past a highly conserved AAUAAA sequence and less than 50 nucleotides before a less-conserved U-rich or GU-rich sequence. The precision of this cleavage reaction has apparently eliminated the need for accurate transcription termination.

2. The poly(A) tail is subsequently generated from ATP through the stepwise action of **poly(A) polymerase.** This enzyme is activated by **cleavage and polyadenylation specificity factor (CPSF)** when this latter protein recognizes the AAUAAA sequence. Once the poly(A) tail has grown to ~10 residues, CPSF probably disengages from its recognition site in a manner reminiscent of the way σ factor is released from the transcription initiation site during the elongation of prokaryotic RNA.

Poly(A) polymerase is part of a 500- to 1000-kD complex that also contains the proteins required for mRNA cleavage. Consequently, the cleaved transcript is polyadenylated before it can dissociate and be digested by cellular nucleases. The maximum length of the poly(A) tail (~250 nt) may be determined by the stoichiometric binding of multiple copies of **poly(A) binding protein (PABP).**

In vitro studies indicate that a poly(A) tail is not required for mRNA translation. Rather, the observation that an mRNA's poly(A) tail shortens as it ages in the cytosol suggests that poly(A) tails have a protective role. In fact, the only mature mRNAs that lack poly(A) tails, those of histones (which are required in large quantities only during the relatively short period in a cell's lifetime when DNA is being replicated), have cytosolic lifetimes of <30 min versus hours or days for most other mRNAs. The binding of PABP to the tail may render the entire transcript resistant to nuclease digestion.

Eukaryotic Genes Consist of Alternating Exons and Introns

The most striking difference between eukaryotic and prokaryotic structural genes is that *the coding sequences of most eukaryotic genes are interspersed with unexpressed regions.* The primary transcripts, also called **pre-mRNAs** or **heterogeneous nuclear RNAs (hnRNAs),** are variable in length and are much larger (~2000 to >20,000 nt) than expected from the known sizes of eukaryotic proteins. Rapid labeling experiments demonstrated that little of the hnRNA is ever transported to the cytosol; most of it is quickly degraded in the nucleus. Yet the hnRNA's 5' caps and 3' tails eventually appear in cytosolic mRNAs. The straightforward explanation of these observations, that pre-mRNAs are processed by the excision of internal sequences, seemed so bizarre that it came as a great surprise in 1977 when Phillip Sharp and Richard Roberts independently demonstrated that this is actually the case. Thus, *pre-mRNAs are processed by the excision of non-expressed intervening sequences (introns), following which the flanking expressed sequences (exons) are joined, or spliced, together.*

A pre-mRNA typically contains around eight introns whose aggregate length averages 4–10 times that of its exons. This situation is graphically

Figure 25-17.   The chicken ovalbumin gene and its mRNA. The electron micrograph and its interpretive drawing show the hybridization of the antisense (template) strand of the chicken ovalbumin gene and its corresponding mRNA. The complementary segments of the DNA (*purple line in drawing*) and mRNA (*red line*) have annealed to reveal the exon positions (L, 1–7). The looped-out segments (I–VII), which have no complementary sequences in the mRNA, are the introns. [From Chambon, P., *Sci. Am.* **244**(5), 61 (1981).]

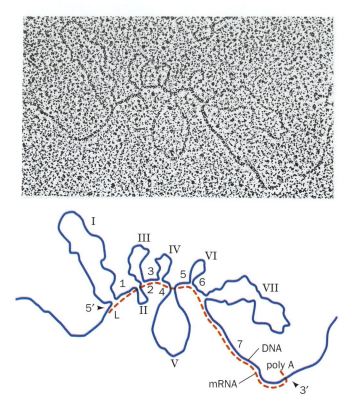

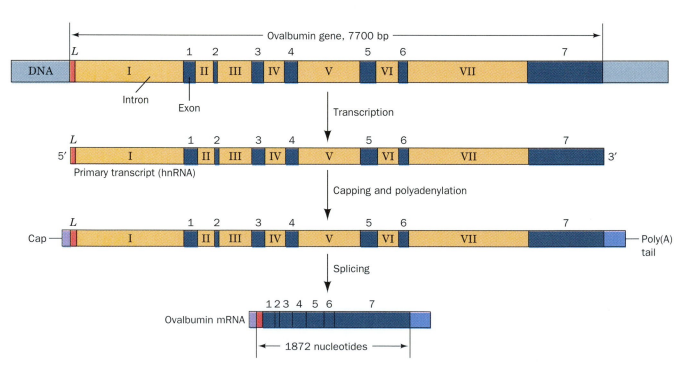

Figure 25-18.   The sequence of steps in the production of mature eukaryotic mRNA. This example shows the chicken ovalbumin gene. Following transcription, the primary transcript is capped and polyadenylated. The introns are then excised and the exons spliced together to form the mature mRNA.

illustrated in Fig. 25-17 (*opposite*), which is an electron micrograph of chicken **ovalbumin** mRNA hybridized to the antisense (template) strand of the ovalbumin gene (ovalbumin is the major protein component of egg white). The lengths of introns in vertebrate genes range from ~65 to ~200,000 nt. In fact, the introns from corresponding genes in two vertebrate species also vary extensively in both length and sequence (but rarely in number and positions).

The production of a translation-competent eukaryotic mRNA begins with the transcription of the entire gene, including its introns (Fig. 25-18; *opposite*). The resulting pre-mRNA is then capped and polyadenylated. Splicing generates the mature mRNA. Before it can be translated into protein, however, the mRNA must be transported to the cytosol, where the ribosomes are located (see Box 25-3).

Box 25-3

BIOCHEMISTRY IN FOCUS

Nucleocytoplasmic Transport

In eukaryotes, many of the RNAs synthesized in the nucleus perform their functions in the cytosol. In some cases [small nuclear RNAs (snRNAs), for example] the RNA is translocated to the cytosol for processing before re-entering the nucleus (in the case of snRNAs, to form spliceosomes). The nuclear membrane, which consists of two lipid bilayers, presents a formidable barrier to the movement of macromolecules in and out of the nucleus. RNA molecules as well as proteins exit and enter the nucleus via **nuclear pore complexes (NPCs)**, which number in the thousands in a typical mammalian nucleus. Rapidly proliferating cells contain even more NPCs.

Electron micrographs reveal that the NPC is shaped roughly like a cylinder with eightfold rotational symmetry, as indicated in the following cutaway interpretive drawing.

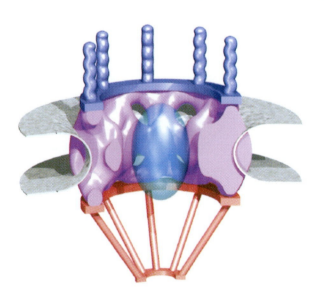

The cytoplasmic face is at the top, and the nuclear face is at the bottom. The NPC may contain as many as a thousand proteins of 50–100 kinds. The 9-nm inner diameter of the NPC permits the free diffusion of solutes smaller than 40–60 kD. Larger particles with diameters of up to 28 nm are transported in an energy-dependent manner through the ~100-nm length of the NPC.

Proteins to be imported into the nucleus contain basic sequences that are recognized by the transport machinery. Interestingly, histones, which are small enough to diffuse through the NPC on their own, also enter the nucleus via a mediated route. The signals for RNA translocation through the NPC are not well understood but are associated with RNA-binding proteins. Each species of RNA has a different recognition system. For example, rRNAs move as **preribosomal particles** containing both rRNA and ribosomal proteins. mRNA transport may depend on proteins such as the **cap-binding complex** or other proteins that bind to mRNA to form **heterogeneous nuclear ribonucleoproteins (hnRNPs)**.

Carrier proteins known as **importins** or **transportins** are essential for nucleocytoplasmic transport, as is a GTPase known as **Ran**. The NPC does not contain a recognizable "motor" protein such as myosin, but the GTPase activity of Ran (which resembles that of the G-proteins involved in signal transduction pathways; Section 21-3B) suggests that cargo moves through the NPC via a one-way, ratcheting mechanism that depends on protein conformational changes. After transport, the RNA or protein cargo and its carrier protein must dissociate so that the carrier can be recycled. The signals that govern the directionality of transport (particularly for substances such as snRNAs that move both in and out of the nucleus) have yet to be elucidated.

[Figure from Nigg, E.A., *Nature* **386**, 780 (1997). Used with permission.]

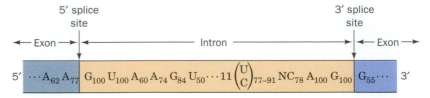

Figure 25-19.   The consensus sequences at the exon–intron junctions of eukaryotic pre-mRNAs. The subscripts indicate the percentage of pre-mRNAs in which the specified base(s) occurs. Note that the 3′ splice site is preceded by a tract of 11 predominantly pyrimidine nucleotides. [Based on data from Padgett, R.A., Grabowski, P.J., Konarska, M.M., Seiler, S.S., and Sharp, P.A., *Annu. Rev. Biochem.* **55**, 1123 (1986).]

Exons Are Spliced in a Two-Stage Reaction

Sequence comparisons of exon–intron junctions from a diverse group of eukaryotes indicate that they have a high degree of homology (Fig. 25-19), with most of them having an invariant GU at the intron's 5′ boundary and an invariant AG at its 3′ boundary. *These sequences are necessary and sufficient to define a splice junction.* The splicing reaction occurs via two transesterification reactions (Fig. 25-20):

1. A 2′,5′-phosphodiester bond forms between an intron adenosine residue and the intron's 5′-terminal phosphate group. The 5′ exon is thereby released and the intron assumes a **lariat structure** (so called because of its shape). The adenosine at the lariat branch point is typically located in a conserved sequence 20 to 50 residues upstream of the 3′ splice site. Mutations that change this branch point A residue abolish splicing at that site.

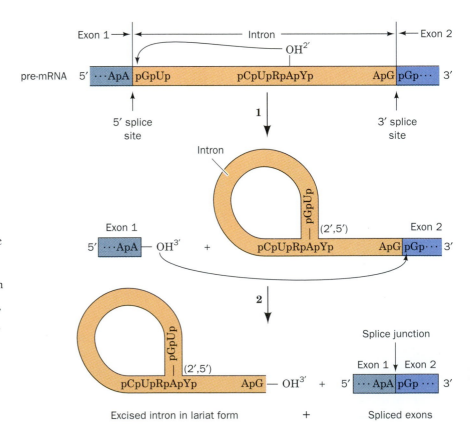

Figure 25-20.   *Key to Function.* The splicing reaction. Two transesterification reactions splice together the exons of eukaryotic pre-mRNAs. The exons and introns are drawn in blue and orange, and R and Y represent purine and pyrimidine residues. (**1**) The 2′-OH group of a specific intron A residue nucleophilically attacks the 5′-phosphate at the 5′ intron boundary to displace the 3′ end of the 5′ exon, thereby yielding a 2′,5′-phosphodiester bond and thus forming a lariat structure. (**2**) The liberated 3′-OH group attacks the 5′-phosphate of the 5′-terminal residue of the 3′ exon, forming a 3′,5′-phosphodiester bond, thereby displacing the intron in lariat form and splicing the two exons together.

2. The free 3'-OH group of the 5' exon displaces the 3' end of the intron, forming a phosphodiester bond with the 5'-terminal phosphate of the 3' exon and yielding the spliced product. The intron is thereby eliminated in its lariat form and, *in vivo,* is rapidly degraded. Mutations that alter the conserved AG at the 3' splice junction block this second step, although they do not interfere with lariat formation.

Note that splicing proceeds without free energy input; its transesterification reactions preserve the free energy of each cleaved phosphodiester bond. A metal ion is probably required for the first catalytic step.

Splicing Is Mediated by snRNPs

How are splice junctions recognized and how are the two exons to be joined brought together in the splicing process? Part of the answer to this question was established by Joan Steitz going on the assumption that one nucleic acid is best recognized by another. The eukaryotic nucleus contains numerous copies of highly conserved 60- to 300-nt RNAs called **small nuclear RNAs (snRNAs),** which form protein complexes termed **small nuclear ribonucleoproteins (snRNPs;** pronounced "snurps"). Steitz noted that the 5' end of one of these snRNAs, **U1-snRNA,** is partially complementary to the consensus sequence of 5' splice junctions. Apparently, *U1-snRNP recognizes the 5' splice junction.* Other snRNPs that participate in splicing are **U2-snRNP, U4–U6-snRNP** (in which the **U4-** and **U6-snRNAs** associate via base pairing), and **U5-snRNP.**

Splicing takes place in a 50S to 60S particle dubbed the **spliceosome.** The spliceosome brings together a pre-mRNA, the snRNPs, and a variety of pre-mRNA binding proteins. The binding of these proteins to immature mRNAs probably directs their processing and contributes to the heterogeneity of hnRNA. A rare class of introns whose splice sites are marked by the sequences AT and AC (rather than GU and AG) are excised through the action of a spliceosome that contains, except for U5-snRNA, a different set of snRNAs. Nevertheless, splice-site recognition and transesterification in both classes of introns appear to proceed by identical mechanisms.

A simplistic interpretation of Fig. 25-20 suggests that any 5' splice site could be joined with any following 3' splice site, thereby eliminating all the intervening exons together with the introns joining them. This does not normally occur. Rather, introns are individually excised in more or less 5' → 3' order. The mechanism for the proper splicing of mammalian genes remains somewhat mysterious since the sequences at splice sites tend not to be highly conserved and the introns are often many times larger than their flanking exons. In fact, splicing pathways do occur in which certain exons are deleted or alternative exons are incorporated into the mature mRNA. Alternative splicing can thus generate more than one protein product from a given gene. This mechanism is particularly important for the differential expression of genes in different tissues (Section 27-3D). The biological significance of introns is discussed in Box 25-4.

mRNA May Be Edited

Additional posttranscriptional modifications of eukaryotic mRNAs include the methylation of certain A residues. In a few cases, mRNAs undergo **editing,** which involves base changes, deletions, or insertions. In the most extreme examples of this phenomenon, which occur in the trypanosomes and related protozoa, several hundred U residues may be added and removed to produce a translatable mRNA. Not surprisingly, the mRNA editing machinery includes RNAs known as **guide RNAs (gRNAs)** that base-pair with the immature mRNA to direct its alteration.

Box 25-4

BIOCHEMISTRY IN CONTEXT

Why Do Introns Interrupt Structural Genes?

Introns in structural genes are most abundant in higher eukaryotes, less so in simple eukaryotes such as yeast, and rare in prokaryotes. Are introns merely examples of "selfish" DNA, that is, nucleotide sequences that persist only because there is little selective pressure to eliminate them? This hypothesis is consistent with the different lifestyles of prokaryotes and eukaryotes: That of prokaryotes places a premium on simplicity and miniaturization and hence would be expected to selectively eliminate introns. Yet the complexity and highly conserved nature of the splicing machinery argues that there is (or was) some selective advantage to so-called split genes comprising exons separated by introns.

Two competing theories attempt to describe the origins—and therefore the functions—of split genes. In the **introns-early theory**, exons are the descendants of "minigenes" and introns are descendants of the spacers between them. Modern proteins were assembled from sets of exons, and the leftover sequences, the introns, have persisted in higher organisms but have been eliminated from the genomes of bacteria. Support for this theory comes from observations that in some structural genes, exons correspond to well-defined protein structural elements. Most exons are too small to encode entire globular protein domains but specify segments corresponding to secondary structural motifs. This is seen, for example, in the enzyme pyruvate kinase:

The structure on the left shows a subunit of pyruvate kinase from cat muscle, with each of its 10 exons colored differently. The exploded view on the right is made by separating the protein's structural segments at its exon boundaries. Note that each of the pyruvate kinase gene's 10 exons encodes a discrete element of protein secondary structure, with most of the introns marking positions at which the polypeptide chain makes a reverse turn. This suggests that the pyruvate kinase gene was assembled by combining smaller protein-coding units.

In some proteins, the evidence for modular assembly is particularly striking. For example, many of the protein modules that occur in fibronectin and other vertebrate proteins (Fig. 5-21) are encoded by discrete exons. However, this pattern is more prevalent in recently evolved proteins than in ancient proteins (those present in both prokaryotes and eukaryotes). Furthermore, the analysis of ancient genes reveals no significant tendency for introns to avoid interrupting secondary structural elements.

A second theory of intron origins explains these inconsistencies. In the **introns-late theory**, split genes arose from the insertion of introns into uninterrupted primordial genes. This theory proposes that these ancient genes evolved without the participation of introns and hence bacterial structural genes lack introns because they never had them to begin with. What, then, is the origin of introns? One possibility is that

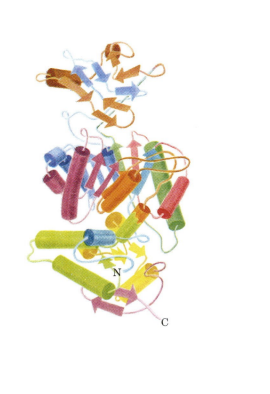

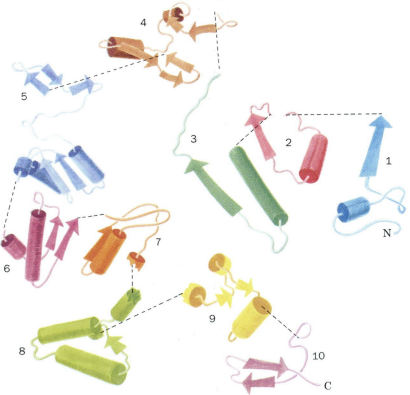

introns are the remnants of transposons (Section 24-6B) that inserted by chance into sequences corresponding to structural genes. There is evidence that some introns encode reverse transcriptase–like proteins or endonucleases, which could explain their potential for mobility. Of course, transposonlike elements may appear in introns simply because there is little selective pressure to eliminate them from these relatively nonessential segments of the genome.

Another possibility is that structural gene introns are the products of reverse splicing carried out by the self-splicing elements that form the introns of rRNA genes. In fact, some *Tetrahymena* introns are observed to undergo reverse splicing reactions *in vitro*. However, this scenario does not account for the elaborate splicing machinery present in eukaryotic nuclei.

Emerging information on genome organization may ultimately provide additional clues indicating the origins and functions of split genes. Already, portions of the human genome that were once thought to be useless or "junk" DNA are being assigned roles in gene organization and regulation. It is possible that in large genomes, introns have evolved to become indispensible elements for the proper storage and expression of genetic information.

[Figure after Lonberg, N. and Gilbert, W., *Cell* **40**, 84 (1985).]

B. Ribosomal RNA Processing

E. coli has three types of rRNAs, the **5S, 16S,** and **23S rRNAs.** These are specified by seven operons, each of which contains one nearly identical copy each of the three rRNA genes. The polycistronic primary transcripts of these operons are >5500 nt long and contain, in addition to the rRNAs, transcripts for as many as four tRNAs (Fig. 25-21). The steps in processing these primary transcripts to mature rRNAs were elucidated with the aid of mutants defective in one or more of the processing enzymes.

The initial processing, which yields products known as **pre-rRNAs,** commences while the primary transcript is still being synthesized. It consists of specific endonucleolytic cleavages by **RNase III, RNase P, RNase E,** and **RNase F** at the sites indicated in Fig. 25-21. The cleavage sites are probably recognized on the basis of their secondary structures.

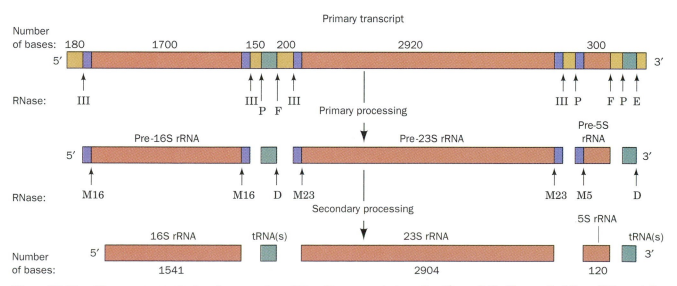

Figure 25-21. The posttranscriptional processing of *E. coli* rRNA. The transcriptional map is shown approximately to scale. The labeled arrows indicate the positions of the various nucleolytic cuts and the nucleases that generate them. [After Apiron, D., Ghora, B.K., Plantz, G., Misra, T.K., and Gegenheimer, P., *in* Söll, D., Abelson, J.N., and Schimmel, P.R. (Eds.), *Transfer RNA: Biological Aspects,* p. 148, Cold Spring Harbor Laboratory (1980).]

The 5' and 3' ends of the pre-rRNAs are trimmed away in secondary processing steps through the action of **RNases D, M16, M23,** and **M5** to produce the mature rRNAs. These final cleavages occur only after the pre-rRNAs become associated with ribosomal proteins. During ribosomal assembly (Section 26-3), specific rRNA residues are methylated, which may help protect them from inappropriate nuclease digestion.

Eukaryotic rRNA Processing

The eukaryotic genome typically has several hundred tandemly repeated copies of rRNA genes. These genes are transcribed and processed in the nucleolus. The primary eukaryotic rRNA transcript is an ~7500-nt **45S RNA** that contains the **18S, 5.8S,** and **28S rRNAs** separated by spacer sequences (Fig. 25-22). In the first stage of its processing, 45S RNA is specifically methylated at ~110 sites that occur mostly in its rRNA sequences.

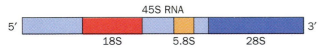
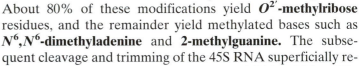

Figure 25-22.   The organization of the 45S primary transcript of eukaryotic rRNA.

About 80% of these modifications yield $O^{2'}$**-methylribose** residues, and the remainder yield methylated bases such as N^6, N^6**-dimethyladenine** and **2-methylguanine.** The subsequent cleavage and trimming of the 45S RNA superficially resembles that of prokaryotic rRNAs. In fact, enzymes exhibiting RNase III– and RNase P–like activities occur in eukaryotes. Eukaryotic ribosomes contain four different rRNAs (Section 26-3). The fourth type, 5S eukaryotic rRNA, is separately processed in a manner resembling that of tRNA (Section 25-3C).

Some Eukaryotic rRNAs Are Self-splicing

Only a few eukaryotic rRNA genes contain introns. Nevertheless, Thomas Cech's study of how such genes are spliced in the ciliated protozoan *Tetrahymena thermophila* led, in 1982, to an astonishing discovery: *RNA can act as an enzyme.* When the isolated pre-rRNA of this organism is incubated with guanosine or a free guanine nucleotide (GMP, GDP, or GTP), but in the absence of protein, its single 413-nt intron excises itself and splices together its flanking exons; that is, *the pre-rRNA is self-splicing.* The three-step reaction sequence of this process (Fig. 25-23) resembles that of mRNA splicing:

1. The 3'-OH group of the guanosine attacks the intron's 5' end, displacing the 3'-OH group of the 5' exon and thereby forming a new phosphodiester linkage with the 5' end of the intron.
2. The 3'-terminal OH group of the newly liberated 5' exon attacks the 5'-phosphate of the 3' exon, forming a new phosphodiester bond, thereby splicing together the two exons and displacing the intron.
3. The 3'-terminal OH group of the intron attacks a phosphate of the nucleotide 15 residues from the intron's end, displacing the 5'-terminal fragment and yielding the 3'-terminal fragment in cyclic form.

The self-splicing process consists of a series of transesterifications and therefore does not require free energy input. Self-splicing RNAs that react as shown in Fig. 25-23 are known as **group I introns** and occur in the nuclei, mitochondria, and chloroplasts of diverse eukaryotes (but not vertebrates), and in some bacteria. **Group II introns,** which occur in the mitochondria and chloroplasts of fungi and plants, react via a lariat intermediate and do not require an external nucleotide.

Although Cech's discovery of an RNA catalyst (ribozyme) was initially surprising, *there is no fundamental reason that an RNA, or any other macromolecule, cannot have catalytic activity.* In order to be an efficient catalyst,

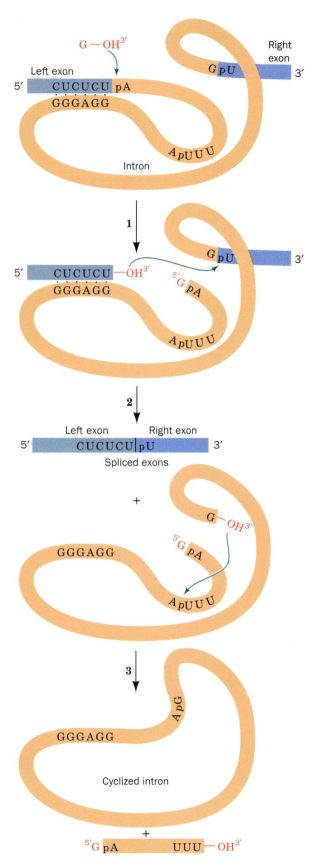

Figure 25-23.   **The sequence of reactions in the self-splicing of *Tetrahymena* pre-rRNA.** (1) The 3′-OH group of a guanine nucleotide attacks the intron's 5′-terminal phosphate to form a phosphodiester bond, displacing the 5′ exon. (2) The newly generated 3′-OH group of the 5′ exon attacks the 5′-terminal phosphate of the 3′ exon, thereby splicing the two exons and displacing the intron. (3) The 3′-OH group of the intron attacks the phosphate of the nucleotide that is 15 residues from the 5′ end so as to cyclize the intron and displace its 5′-terminal fragment. Throughout this process, the RNA maintains a folded, internally hydrogen-bonded conformation that permits the precise excision of the intron.

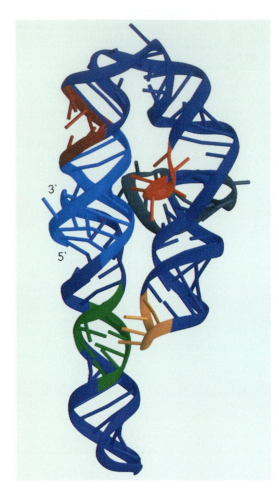

a macromolecule must be able to assume a stable structure, and as we have seen in Section 23-2E, many RNAs (e.g., the hammerhead ribozyme; Fig. 23-25) do just that.

The X-ray structure of a 160-nt portion of the *Tetrahymena* self-splicing intron was determined by Jennifer Doudna. This RNA, the largest of known structure, folds autonomously into a hairpin with most of its bases stacked in helices (Fig. 25-24). Comparative sequence analyses and model building had implicated short, conserved oligonucleotide sequences in stabilizing the intron's tertiary structure (e.g., the regions colored green, red, and orange in Fig. 25-24). In fact, structural motifs as small as two consecutive A residues appear to be stabilizing elements in the *Tetrahymena* intron. Extensive hydrogen bonding and Mg^{2+} ions coordinated by the oxygens of two or more phosphate groups contribute to a structure whose interior is densely packed and solvent inaccessible, much like the interior of a protein enzyme. Throughout this structure, the defining characteristic of RNA, its 2'-OH group, is both a donor and acceptor of hydrogen bonds to phosphates, bases, and other 2'-OH groups, thereby explaining why single-stranded DNA seems unable to form the compact structures required to generate efficient active sites.

The snRNA components of spliceosomes presumably evolved from primordial self-splicing RNAs, and the protein components of snRNPs serve mainly to fine-tune ribozymal structure and function. Similarly, the RNA components of ribosomes almost certainly have catalytic functions in addition to the structural and recognition roles traditionally attributed to them (Section 26-3). All these RNA activities are consistent with the hypothesis that RNAs were the original biological catalysts in precellular times (the so-called RNA world).

C. Transfer RNA Processing

A tRNA molecule, as discussed in Section 23-2E, consists of ~80 nucleotides, many of which are chemically modified, that assume a cloverleaf-shaped secondary structure with four base-paired stems (Fig. 25-25). The *E. coli* chromosome contains ~60 tRNA genes. Some of them are components of rRNA operons; the others are distributed, often in clusters, throughout the chromosome. The primary tRNA transcripts, which contain as many as five identical tRNA species, have extra nucleotides at the 3' and 5' ends of each tRNA sequence. The excision and trimming of these tRNA sequences resembles *E. coli* rRNA processing (Fig. 25-21) in that the two processes employ some of the same nucleases.

E. coli RNase P, which processes rRNA and generates the 5' ends of tRNAs, is a particularly interesting enzyme because its catalytic activity requires a 377-nt RNA (~125 kD; the protein component of RNase P has only 119 residues and a mass of ~14 kD). This RNA was first proposed to recognize the substrate RNA and thereby guide the protein subunit, which was presumed to be the actual nuclease, to the cleavage site. However, Sidney Altman has shown that *the RNA component of RNase P is, in fact, the enzyme's catalytic subunit by demonstrating that free RNase P RNA*

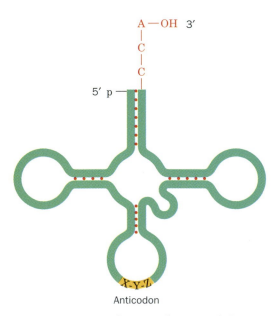

Figure 25-25.   **A schematic diagram of the tRNA cloverleaf secondary structure.** Each dot indicates a base pair in the hydrogen-bonded stems. The position of the **anticodon** (the 3-nt sequence that binds to mRNA during translation) and the 3'-terminal —CCA are indicated.

catalyzes the cleavage of substrate RNA at high salt concentrations. RNase P protein, which is basic, evidently functions to electrostatically reduce the repulsions between the polyanionic ribozyme and its substrate RNAs. RNase P activity occurs in eukaryotes (nuclei, mitochondria, and chloroplasts) as well as in prokaryotes.

Many Eukaryotic Pre-tRNAs Have Introns

Eukaryotic genomes contain from several hundred to several thousand tRNA genes. Many eukaryotic primary tRNA transcripts, for example, yeast **tRNA^{Tyr}** (Fig. 25-26), contain a small intron as well as extra nucleotides at their 5' and 3' ends. tRNA processing therefore includes nucleolytic removal of these extra nucleotides. The sequence of three nucleotides, CCA, the site where amino acids become attached (Section 26-2B), is lacking in the immature tRNA transcript. This trinucleotide is appended by the enzyme **tRNA nucleotidyltransferase,** which sequentially adds two C's and an A to tRNA using CTP and ATP as substrates (prokaryotic tRNA primary transcripts include the CCA sequence).

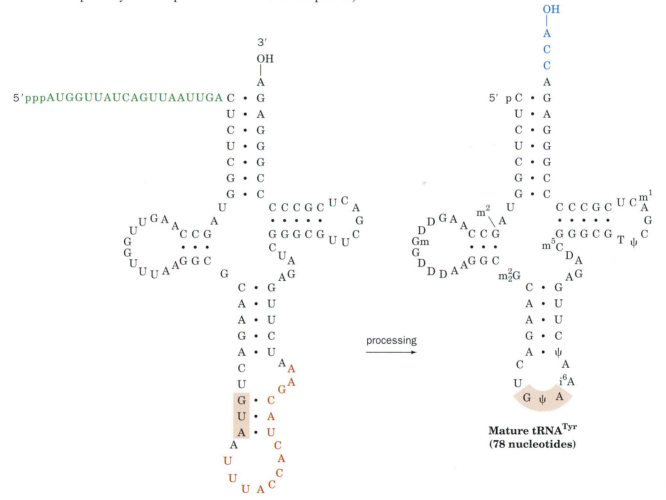

**tRNA^{Tyr} primary transcript
(108 nucleotides)**

Mature tRNA^{Tyr}
(78 nucleotides)

Figure 25-26. **The posttranscriptional processing of yeast tRNA^{Tyr}.** A 14-nt intervening sequence (*red*) and a 19-nt 5'-terminal sequence (*green*) are excised from the primary transcript, a —CCA (*blue*) is enzymatically appended to its 3' end, and several of its nucleotides are modified to form the mature tRNA (m^{2}G, N^2-methylguanosine; Gm, 2'-methylguanosine; m$_2^2$G, N^2,N^2-dimethylguanosine; ψ, pseudouridine; i^{6}A, N^6-isopentenyladenosine; m^{5}C, 5-methylcytosine; m^{1}A, 1-methyladenosine; see Fig. 26-5). The anticodon is shaded. [After DeRobertis, E.M. and Olsen, M.V., *Nature* **278**, 142 (1989).]

SUMMARY

1. RNA polymerase synthesizes a polynucleotide chain from ribonucleoside triphosphates using a single strand (the antisense or noncoding strand) of DNA as a template.

2. The σ factor of *E. coli* RNA polymerase holoenzyme recognizes and binds to a promoter sequence to position the enzyme to initiate transcription.

3. RNA synthesis requires the formation of an open complex. The transcription bubble travels along the DNA as the RNA chain is elongated by the processive activity of RNA polymerase.

4. In *E. coli*, RNA synthesis is terminated in response to specific secondary structural elements in the transcript and may require the action of rho factor.

5. Eukaryotes contain three nuclear RNA polymerases that synthesize different types of RNAs.

6. Eukaryotic promoters are diverse: They vary in position relative to the transcription start site and may consist of multiple upstream sequences. Enhancers and silencers form the binding sites for regulatory proteins that function as activators and repressors of transcription.

7. Basal transcription of eukaryotic genes requires that general transcription factors and RNA polymerase assemble in a prescribed order to form a preinitiation complex.

8. The primary transcripts of most eukaryotic structural genes are posttranscriptionally modified by the addition of a 5′ cap and a 3′ poly(A) tail. mRNAs that contain introns undergo splicing, in which the introns are excised and the flanking exons are joined together via two transesterification reactions mediated by an snRNA-containing spliceosome.

9. The processing of pre-rRNAs includes nucleolytic cleavage and methylation. Some eukaryotic rRNA transcripts undergo splicing catalyzed by the intron itself.

10. tRNA transcripts may be processed by the addition, removal, and modification of nucleotides.

REFERENCES

Transcription

Burley, S.K., The TATA box binding protein, *Curr. Opin. Struct. Biol.* **6,** 69–75 (1996). [Reviews TBP structure and the events of transcription initiation.]

Darst, S.A., Edwards, A.M., Kubalek, E.W., and Kornberg, R.D., Three-dimensional structure of yeast RNA polymerase II at 16 Å resolution, *Cell* **66,** 121–128 (1991).

Greenblatt, J., RNA polymerase II holoenzyme and transcriptional regulation, *Curr. Biol.* **9,** 310–319 (1997).

Liu, B. and Alberts, B.M., Head-on collision between a DNA replication apparatus and RNA polymerase transcription complex, *Science* **267,** 1131–1137 (1995).

Nudler, E., Avetissova, E., Markovtsov, V., and Goldfarb, A., Transcription processivity: protein-DNA interactions holding together the elongation complex, *Science* **273,** 211–217 (1996).

Richardson, J.P., Structural organization of transcription termination factor rho, *J. Biol. Chem.* **271,** 1251–1254 (1996).

Roeder, R.G., The role of general initiation factors in transcription by RNA polymerase II, *Trends Biochem. Sci.* **21,** 327–335 (1996).

Tan, S. and Richmond, T.J., Eukaryotic transcription factors, *Curr. Opin. Struct. Biol.* **8,** 41–48 (1998).

Tijan, R., Molecular machines that control genes, *Sci. Am.* **272**(2), 54–61 (1995). [Describes the identification of the components of eukaryotic transcription factors.]

RNA Processing

Apiron, D. and Miczak, A., RNA processing in prokaryotic cells, *BioEssays* **15,** 113–119 (1993).

Cate, J.H., Gooding, A.R., Podell, E., Zhou, K., Golden, B.L., Kundrot, C.E., Cech, T.R., and Doudna, J.A., Crystal structure of a group I ribozyme domain: principles of RNA packing, *Science* **273,** 1678–1690 (1996), *and* Doudna, J.A. and Cate, J.H., RNA structure: crystal clear? *Curr. Opin. Struct. Biol.* **7,** 310–316 (1997).

Cech, T.R., Self-splicing of group I introns, *Annu. Rev. Biochem.* **59,** 543–568 (1990).

Frank, D.N. and Pace, N.R., Ribonuclease P: Unity and diversity in a tRNA processing ribozyme, *Annu. Rev. Biochem.* **67,** 153–180 (1998).

Nigg, E.A., Nucleocytoplasmic transport: signals, mechanism and regulation, *Nature* **386,** 779–787 (1997).

Sharp, P.A., Split genes and RNA splicing, *Cell* **77,** 805–815 (1994). [An overview of mRNA splicing by a Nobel prize winner.]

Stoltzfus, A., Spencer, D.F., Zuker, M., Logsdon, J.M., Jr., and Doolittle, W.F., Testing the exon theory of genes: the evidence from protein structure, *Science* **265,** 202–207 (1994).

Symons, R.H., Ribozymes, *Curr. Opin. Struct. Biol.* **4,** 322–330 (1994). [Reviews the naturally occurring ribozymes, including group I introns and RNase P.]

Wahle, E. and Keller, W., The biochemistry of polyadenylation, *Trends Biochem. Sci.* **21,** 247–250 (1996).

KEY TERMS

mRNA	holoenzyme	antisense strand	noncoding strand
rRNA	core enzyme	sense strand	structural gene
tRNA	σ factor	coding strand	gene product

operon	inducible enzyme	primary transcript	snRNA
cistron	rho factor	posttranscriptional modifica-	snRNP
polycistronic RNA	nucleolus	tion	spliceosome
monocistronic RNA	GC box	cap structure	RNA editing
promoter	CCAAT box	poly(A) tail	gRNA
Pribnow box	general transcription factor	hnRNA	group I intron
footprinting	preinitiation complex	intron	group II intron
open complex	TBP	exon	
constitutive enzyme	TAFs	splicing	

STUDY EXERCISES

1. Why is it difficult to precisely define the term "gene"?

2. Compare DNA and RNA polymerases with respect to overall structure, substrates, mechanism of action, error rate, and template specificity.

3. What are the advantages and disadvantages of arranging genes in operons?

4. What is a consensus sequence?

5. What is the significance of the different DNA-binding properties of prokaryotic RNA polymerase core enzyme and holoenzyme?

6. What are the functions of the three eukaryotic RNA polymerases?

7. Describe the assembly of the eukaryotic preinitiation complex.

8. Summarize the posttranscriptional modification of eukaryotic mRNA, rRNA, and tRNA.

9. Compare the mechanism of splicing in mRNA processing and in group I introns.

PROBLEMS

1. The antibiotic **cordycepin** inhibits bacterial RNA synthesis.

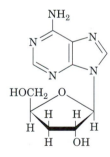

Cordycepin

(a) Of which nucleoside is cordycepin a derivative?

(b) Explain cordycepin's mechanism of action.

2. Indicate the −10 region, the −35 region, and the initiating nucleotide on the sense strand of the *E. coli* tRNA^{Tyr} promoter shown below.

```
5' CAACGTAACACTTTACAGCGGCGCGTCATTTGATATGATGCGCCCCGCTTCCCGATA 3'
3' GTTGCATTGTGAAATGTCGCCGCGCAGTAAACTATACTACGCGGGGCGAAGGGCTAT 5'
```

3. Design a six-residue nucleic acid probe that would hybridize with the greatest number of *E. coli* gene promoters.

4. Explain why inserting 5 bp of DNA at the −50 position of a eukaryotic gene decreases the rate of RNA polymerase II transcription initiation to a greater extent than inserting 10 bp at the same site.

5. Why does promoter efficiency tend to decrease with the number of G·C base pairs in the −10 region of a prokaryotic gene?

6. A eukaryotic ribosome contains 4 different rRNA molecules and ~82 different proteins. Why does a cell contain many more copies of the rRNA genes than the ribosomal protein genes?

7. Design an oligonucleotide-based affinity chromatography system for purifying mature mRNAs from eukaryotic cell lysates.

8. Explain why the $O^{2'}$-methylation of ribose residues protects rRNA from RNases.

9. Would you expect spliceosome-catalyzed intron removal to be reversible in a highly purified *in vitro* system and *in vivo*? Explain.

10. Infection with certain viruses inhibits snRNA processing in eukaryotic cells. Explain why this favors the expression of viral genes in the host cell.

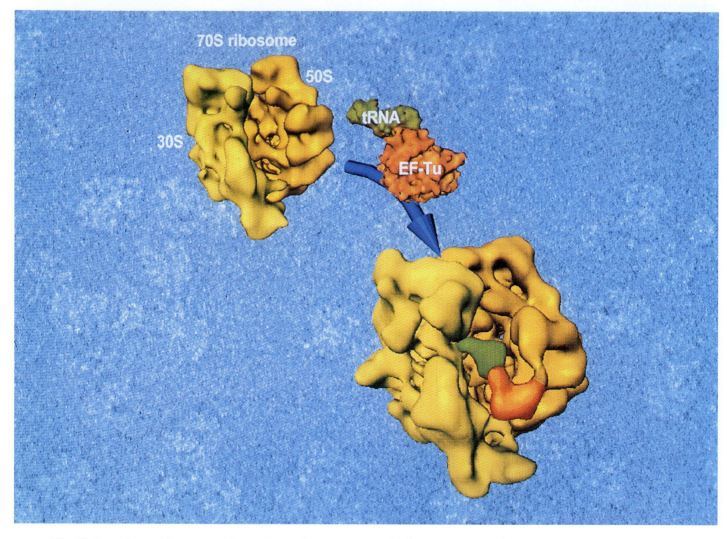

All cellular polypeptides are synthesized on ribosomes. How do the components of this macromolecular machine—
proteins, rRNAs, tRNAs, and mRNA—interact in such a complex process?
[Courtesy of Holgar Stark, Imperial College, London, U.K.]

TRANSLATION

How is the genetic information encoded in DNA decoded? In the preceding chapter, we saw how the base sequence of DNA is transcribed into that of RNA. In this chapter, we shall consider the remainder of the decoding process by examining how the base sequences of RNAs are translated into the amino acid sequences of proteins. This second part of the central dogma of molecular biology (DNA → RNA → protein) shares a number of features with both DNA replication and transcription, the other major events of nucleic acid metabolism. First, all three processes are carried out by large, complicated protein-containing macromolecular machines whose proper functioning depends on a variety of specific and nonspecific protein–nucleic acid interactions. Accessory factors may also be required for the initiation, elongation, and termination phases of these processes. Furthermore, translation, like replication and transcription, must be executed with accuracy. Although translation involves base pairing between complementary nucleotides, it is amino acids, rather than nucleotides, that are ultimately joined to generate a polymeric product. This process, like replication and transcription, is endergonic and requires the cleavage of "high-energy" phosphoanhydride bonds.

An understanding of translation requires not only a knowledge of the macromolecules that participate in polypeptide synthesis, but an appreciation for the mechanisms that produce a chain of linked amino acids in the exact order specified by its corresponding mRNA. Accordingly, we begin this chapter by examining the **genetic code,** the correspondence between nucleic acid sequences and polypeptide sequences. Next, we consider in turn the structures and properties of tRNAs and ribosomes. Finally, we examine how the translational machinery operates as a coordinated whole.

1. THE GENETIC CODE

One of the most fascinating puzzles in molecular biology is how a sequence of nucleotides composed of only four types of residues can specify the sequence of up to 20 types of amino acids in a polypeptide chain. Clearly, a one-to-one correspondence between nucleotides and amino acids is not possible. *A group of several bases, termed a* **codon,** *is necessary to specify a single amino acid.* A triplet code, that is, one with 3 bases per codon, is more than sufficient to specify all the amino acids since there are $4^3 = 64$ different triplets of bases. A doublet code with 2 bases per codon ($4^2 = 16$ possible doublets) would be inadequate. The triplet code allows many amino acids to be specified by more than one codon. Such a code, in a term borrowed from mathematics, is said to be **degenerate.**

How does the polypeptide-synthesizing apparatus group DNA's continuous sequences of bases into codons? For example, the code might be overlapping; that is, in the sequence

ABCDEFGHIJ ···

ABC might code for one amino acid, BCD for a second, CDE for a third, etc. Alternatively, the code might be nonoverlapping, so that ABC specifies one amino acid, DEF a second, GHI a third, etc. In fact, as we shall see, *the genetic code is a nonoverlapping, degenerate, triplet code.* A number of elegant experiments, some of which are outlined below, revealed the nature of the genetic code.

A. Codons Are Triplets That Are Read Sequentially

In genetic experiments on bacteriophage T4, Francis Crick and Sydney Brenner found that a mutation that resulted in the deletion of a nucleotide could abolish the function of a specific gene. However, a second mutation, in which a nucleotide was inserted at a different but nearby position, could restore gene function. These two mutations are said to be **suppressors** of one another; that is, they cancel each other's mutant properties. On the basis of these experiments, Crick and Brenner concluded that *the genetic code is read in a sequential manner starting from a fixed point in the gene.* The insertion or deletion of a nucleotide shifts the **reading frame** (grouping) in which succeeding nucleotides are read as codons. Insertions or deletions of nucleotides are therefore known as **frameshift mutations.**

In further experiments, Crick and Brenner found that whereas two closely spaced deletions or two closely spaced insertions could not suppress each other (restore gene function), three closely spaced deletions or insertions could do so. These observations clearly established that *the genetic code is a triplet code.*

The foregoing principles are illustrated by the following analogy. Consider a sentence (gene) in which the words (codons) each consist of three letters (bases):

THE   BIG   RED   FOX   ATE   THE   EGG

Here the spaces separating the words have no physical significance; they are present only to indicate the reading frame. The deletion of the fourth letter, which shifts the reading frame, changes the sentence to

THE   IGR   EDF   OXA   TET   HEE   GG

so that all words past the point of deletion are unintelligible (specify the wrong amino acids). An insertion of any letter, however, say an X in the ninth position,

THE   IGR   EDX   FOX   ATE   THE   EGG

restores the original reading frame. Consequently, only the words between the two changes (mutations) are altered. As in this example, such a sentence might still be intelligible (the gene could still specify a functional protein), particularly if the changes are close together. Two deletions or two insertions, no matter how close together, would not suppress each other but just shift the reading frame. However, three insertions, say X, Y, and Z in the 5th, 8th, and 12th positions, respectively, would change the sentence to

THE   BXI   GYR   EDZ   FOX   ATE   THE   EGG

which, after the third insertion, restores the original reading frame. The same would be true of three deletions. As before, if all three changes were close together, the sentence might still retain its meaning. Like this textual analogy, *the genetic code has no internal punctuation to indicate the reading frame; instead, the nucleotide sequence is read sequentially, triplet by triplet.*

Since any nucleotide sequence may have three reading frames, it is possible, at least in principle, for a polynucleotide to encode two or even three different polypeptides. In fact, some single-stranded DNA bacteriophages (which presumably must make maximal use of their small complement of DNA) contain completely overlapping genes that have different reading frames. A similar form of coding economy is exhibited by bacteria, in which the ribosomal initiation sequence of one gene in a polycistronic mRNA often overlaps the end of the preceding gene.

B. Deciphering the Genetic Code

In order to understand how the genetic code dictionary was elucidated, we must first review how proteins are synthesized. An mRNA does not directly recognize amino acids. Rather, *it specifically binds molecules of tRNA that each carry a corresponding amino acid* (Fig. 26-1). Each tRNA contains a trinucleotide sequence, its **anticodon,** that is complementary to an mRNA codon specifying the tRNA's amino acid. During translation, amino acids carried by tRNAs are joined together according to the order in which the tRNA anticodons bind to the mRNA codons at the ribosome (Fig. 3-16).

The genetic code could, in principle, be determined by simply comparing the base sequence of an mRNA with the amino acid sequence of the polypeptide it specifies. In the 1960s, however, techniques for isolating and sequencing mRNAs had not yet been developed. Moreover, techniques for synthesizing RNAs were quite rudimentary. They utilized **polynucleotide phosphorylase,** an enzyme from *Azotobacter vinelandii* that links together nucleotides without the use of a template.

$$(RNA)_n + NDP \rightleftharpoons (RNA)_{n+1} + P_i$$

Thus, the NDPs are linked together at random so that the base composition of the product RNA reflects that of the reactant NDP mixture. The elucidation of the genetic code therefore proved to be a difficult task, even with the development of cell-free translation systems.

E. coli cells that have been gently broken open and centrifuged to remove cell walls and membranes yield an extract containing DNA, mRNA, ribosomes, enzymes, and other cell constituents necessary for protein synthesis. When fortified with ATP, GTP, and amino acids, this system synthesizes small amounts of protein. A cell-free translation system, of course, produces proteins specified by the cell's DNA. Adding DNase halts protein synthesis after a few minutes because the system can no longer synthesize mRNA, and the mRNA originally present is rapidly degraded. At this point, purified mRNA or synthetic mRNA can be added to the system and the resulting polypeptide products can be subsequently recovered.

In 1961, Marshall Nirenberg and Heinrich Matthaei added the synthetic polyribonucleotide poly(U) to a cell-free translation system containing isotopically labeled amino acids and recovered a labeled poly(Phe) polypeptide. They concluded that *UUU must be the codon that specifies Phe*. Similar experiments with poly(A) and poly(C) yielded poly(Lys) and poly(Pro), respectively, thereby identifying AAA as a codon for Lys and CCC as a codon for Pro.

The Genetic Code Was Elucidated through Triplet Binding Assays and the Use of Polyribonucleotides with Known Sequences

In the absence of GTP, which is necessary for protein synthesis, trinucleotides but not dinucleotides are almost as effective as mRNAs in promoting the ribosomal binding of specific tRNAs. This phenomenon, which Nirenberg and Philip Leder discovered in 1964, permitted the various codons to be identified by a simple binding assay. Ribosomes, together with their bound tRNAs, are retained by a nitrocellulose filter but free tRNA is not. The bound tRNA was identified by using charged tRNA mixtures in which only one of the pendant amino acid residues was radioactively labeled. For instance, it was found that UUU stimulates the ribosomal binding of only Phe tRNA. Likewise, UUG, UGU, and GUU stimulate the binding of Leu, Cys, and Val tRNAs, respectively. Hence UUG, UGU, and GUU must be codons that specify Leu, Cys, and Val, respectively. In this

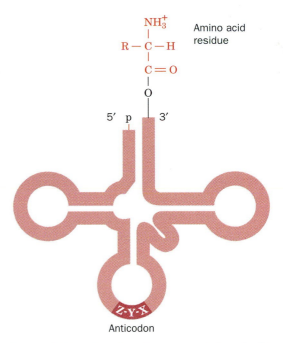

Figure 26-1. **Transfer RNA in its "cloverleaf" form.** Its covalently linked amino acid residue is at the top, and its anticodon (a trinucleotide segment that base pairs with the complementary mRNA codon during translation) is at the bottom.

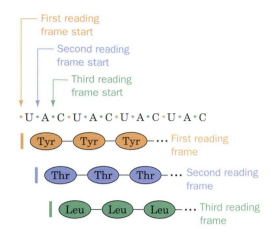

Figure 26-2. **The three potential reading frames of an mRNA.** Each reading frame would yield a different polypeptide.

way, the amino acids specified by some 50 codons were identified. For the remaining codons, the binding assay was either negative (no tRNA bound) or ambiguous.

The genetic code dictionary was completed and previous results confirmed through H. Gobind Khorana's chemical synthesis of polynucleotides with specified repeating sequences (at the time, an extremely laborious process). In a cell-free translation system, UCUCUCUC ···, for example, is read

$$\text{UCU CUC UCU CUC UCU C} \cdots$$

so that it specifies a polypeptide chain of two alternating amino acid residues. This particular mRNA stimulated the production of

$$\text{Ser—Leu—Ser—Leu—Ser—Leu} \cdots$$

since UCU codes for Ser and CUC codes for Leu.

Alternating sequences of three nucleotides, such as poly(UAC), specify three different homopolypeptides because ribosomes may initiate polypeptide synthesis on these synthetic mRNAs in any of the three possible reading frames (Fig. 26-2). Analyses of the polypeptides specified by various alternating sequences of two and three nucleotides confirmed the identity of many codons and filled out missing portions of the genetic code.

C. The Nature of the Genetic Code

The genetic code dictionary is presented in Table 26-1. Examination of this table indicates that the genetic code has several remarkable features:

1. *The code is highly degenerate.* Three amino acids—Arg, Leu, and Ser—are each specified by six different codons, and most of the rest are specified by either four, three, or two codons. Only Met and Trp, two of the least common amino acids in proteins (Table 4-1), are represented by a single codon. Codons that specify the same amino acid are termed **synonyms.**

2. *The arrangement of the code table is nonrandom.* Most synonyms occupy the same box in Table 26-1; that is, they differ only in their third nucleotide. XYU and XYC always specify the same amino acid; XYA and XYG do so in all but two cases. Moreover, changes in the first codon position tend to specify similar (if not the same) amino acids, whereas codons with second position pyrimidines encode mostly hydrophobic amino acids (tan in Table 26-1), and those with second position purines encode mostly polar amino acids (blue, red, and purple in Table 26-1). These observations suggest a nonrandom origin of the genetic code and indicate that the code evolved so as to minimize the deleterious effects of mutations (see Box 26-1).

3. *UAG, UAA, and UGA are **stop codons.*** These three codons (also known as **nonsense codons**) do not specify amino acids but signal the ribosome to terminate polypeptide chain elongation. UAG, UAA, and UGA are often referred to as **amber, ochre,** and **opal codons** (these names are the result of a laboratory joke: The German word for *amber* is Bernstein, the name of an individual who helped discover amber mutations, which change some other codon to UAG; *ochre* and *opal* are puns on *amber*).

4. *AUG and GUG are chain initiation codons.* The codons AUG and, less frequently, GUG specify the starting point for polypeptide chain

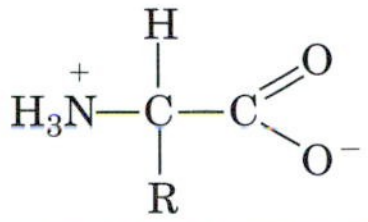

Table 26-1. *Key to Function.* **The "Standard" Genetic Code**[a]

First position (5′ end)	Second position				Third position (3′ end)
	U	**C**	**A**	**G**	
U	UUU, UUC — Phe; UUA, UUG — Leu	UCU, UCC, UCA, UCG — Ser	UAU, UAC — Tyr; UAA, UAG — STOP	UGU, UGC — Cys; UGA — STOP; UGG — Trp	U / C / A / G
C	CUU, CUC, CUA, CUG — Leu	CCU, CCC, CCA, CCG — Pro	CAU, CAC — His; CAA, CAG — Gln	CGU, CGC, CGA, CGG — Arg	U / C / A / G
A	AUU, AUC, AUA — Ile; AUG — Met[b]	ACU, ACC, ACA, ACG — Thr	AAU, AAC — Asn; AAA, AAG — Lys	AGU, AGC — Ser; AGA, AGG — Arg	U / C / A / G
G	GUU, GUC, GUA, GUG — Val	GCU, GCC, GCA, GCG — Ala	GAU, GAC — Asp; GAA, GAG — Glu	GGU, GGC, GGA, GGG — Gly	U / C / A / G

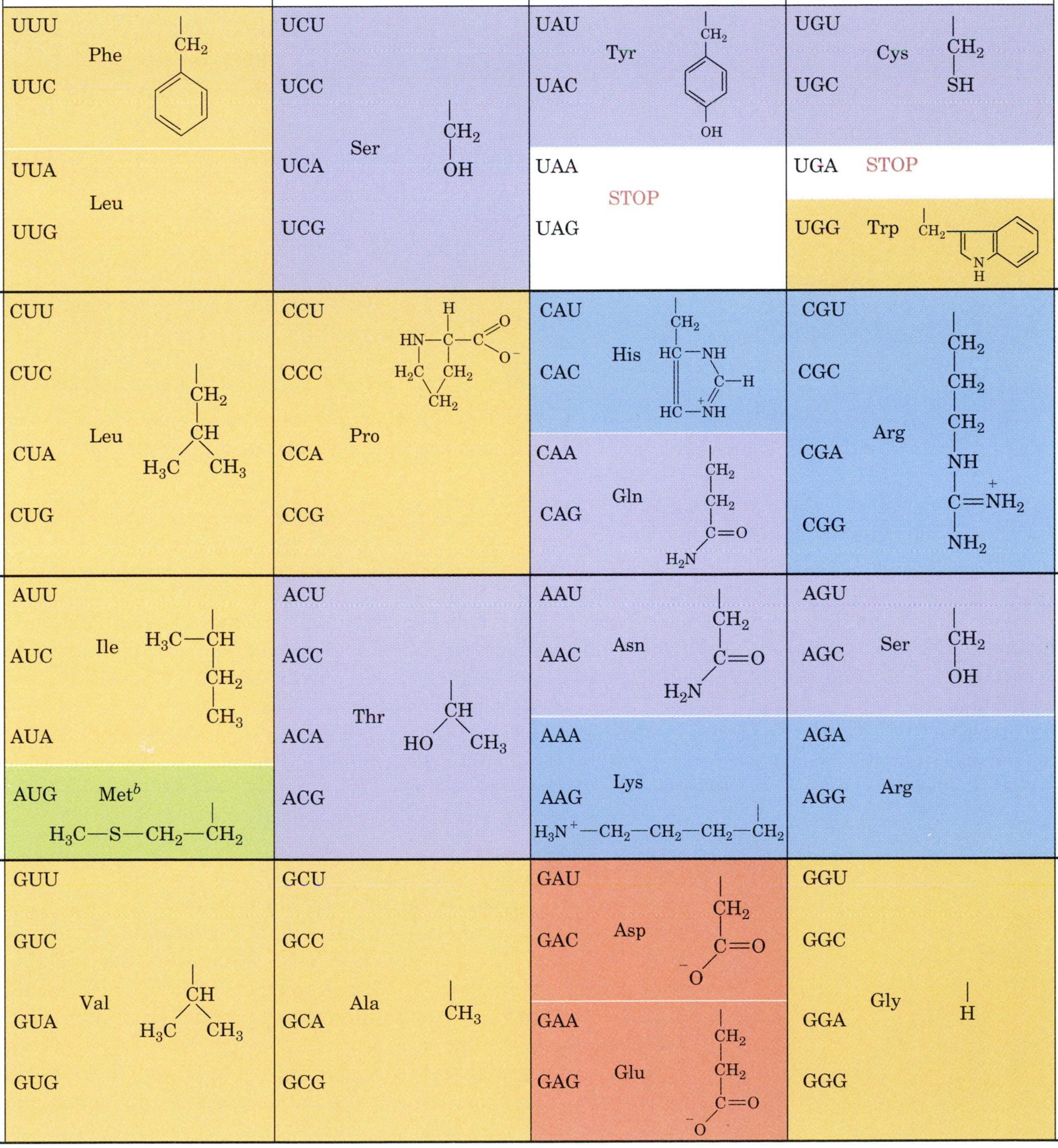

[a] Nonpolar amino acid residues are tan, basic residues are blue, acidic residues are red, and polar uncharged residues are purple.

[b] AUG forms part of the initiation signal as well as coding for internal Met residues.

synthesis. However, they also specify the amino acids Met and Val, respectively, at internal positions in polypeptide chains. We shall see in Section 26-4A how ribosomes differentiate the two types of codons.

Box 26-1

BIOCHEMISTRY IN CONTEXT

Evolution of the Genetic Code

Because of the degeneracy of the genetic code, a point mutation in the third position of a codon seldom alters the specified amino acid. For example, a GUU → GUA transition still codes for Val and is therefore said to be phenotypically silent. Other point mutations, even at the first or second codon positions, producing AUU (Ile) or GCU (Ala), for instance, result in the substitution of a chemically similar amino acid and may have minimal impact on the encoded protein's overall structure or function. This built-in protection against mutation may be more than accidental.

In the 1960s, the perceived universality of the genetic code led Francis Crick to propose the "frozen accident" theory, which holds that codons were allocated to different amino acids entirely by chance. Once assigned, the meaning of a codon could not change because of the high probability of disrupting the structure of the encoded protein. Thus, once established, the genetic code was thought to have ceased evolving.

However, the distribution of codons as presented in Table 26-1 suggests an alternative evolutionary history of the genetic code, in which a few simple codons corresponding to a handful of amino acids gradually became more complex. One scenario begins with an RNA-based world containing only A and U nucleotides. Uracil is almost certainly a primordial base since the pyrimidine biosynthetic pathway yields uracil nucleotides before cytosine or thymine nucleotides (Section 22-2). Adenine would have been required as uracil's complement.

Assuming that a triplet-based genetic code was established at the outset (and it is difficult to envision how any other arrangement could have given rise to the present-day triplet code), the two bases could have coded for $2^3 = 8$ amino acids. In fact, the contemporary genetic code assigns these all-U/A codons to six amino acids and a stop signal:

$$
\begin{array}{ll}
\text{UUU} = \text{Phe} & \text{AAA} = \text{Lys} \\
\text{UUA} = \text{Leu} & \text{AAU} = \text{Asn} \\
\text{UAU} = \text{Tyr} & \text{AUA} = \text{Ile} \\
\text{AUU} = \text{Ile} & \text{UAA} = \text{Stop}
\end{array}
$$

The AUA codon may well have originally specified the initiating Met (now encoded by AUG), reducing the possible total to seven amino acids.

When G and C appeared in evolving life forms, these nucleotides were incorporated into RNA. Codons containing three or four types of bases could specify additional amino acids, but because of selective pressure against introducing disruptive mutations into proteins, the level of codon redundancy increased. An inspection of Table 26-1 shows that triplet codons made entirely of G and C specify only four different amino acids, which is half the theoretical maximum of eight:

$$
\begin{array}{l}
\text{GGG, GGC} = \text{Gly} \\
\text{GCG, GCC} = \text{Ala} \\
\text{CGG, CGC} = \text{Arg} \\
\text{CCC, CCG} = \text{Pro}
\end{array}
$$

This nonrandom allocation of codons to amino acids argues against a completely random origin for the genetic code. The gradual introduction of two new bases (G and C) to a primitive genetic code based on only U and A must have allowed greater information capacity (i.e., coding for 20 amino acids) while minimizing the rate of deleterious substitutions.

The "Standard" Genetic Code Is Not Universal

For many years, it was thought that the "standard" genetic code (that given in Table 26-1) was universal. This assumption was based in part on the observation that one kind of organism (e.g., *E. coli*) can accurately translate the genes for quite different organisms (e.g., humans). Indeed, this phenomenon is the basis of genetic engineering. DNA studies in 1981 nevertheless revealed that *the genetic codes of certain mitochondria are variants of the "standard" genetic code.* For example, in mammalian mitochondria, AUA, as well as the standard AUG, is a Met/initiation codon; UGA specifies Trp rather than "Stop"; and AGA and AGG are "Stop" rather than Arg. Apparently, mitochondria, which contain their own genes and protein synthesizing systems, are not subject to the same evolutionary constraints as are nuclear genomes. An alternate genetic code also appears to have evolved in ciliated protozoa, which branched off very early in eukaryotic evolution. Thus, the "standard" genetic code, although very widely utilized, is not universal.

2. TRANSFER RNA AND ITS AMINOACYLATION

● See Guided Exploration 26:

The Structure of tRNA.

Cells must translate the language of RNA base sequences into the language of polypeptides. Yet nucleic acids do not specifically bind amino acids. In 1955, Francis Crick hypothesized that translation occurs through the mediation of "adaptor" molecules, which carry a specific enzymatically appended amino acid and recognize the corresponding nucleic acid codon (Fig. 26-3). Subsequent experimental work indicated that small soluble RNAs, now known as tRNAs, act as the adaptors.

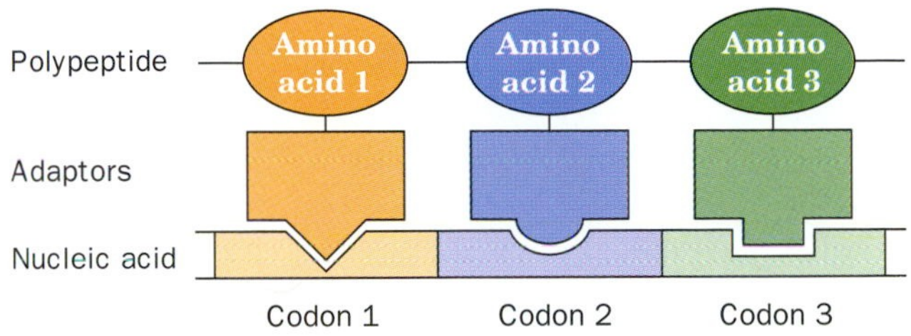

A. tRNA Structure

In 1965, after a 7-year effort, Robert Holley reported the first known base sequence of a biologically significant nucleic acid, that of the 76-residue yeast **alanine tRNA (tRNA^{Ala})**. Currently, the base sequences of thousands of tRNAs from hundreds of organisms and organelles are known (most from their DNA sequences). They vary in length from 60 to 95 nucleotides (18–28 kD) although most have ~76 nucleotides.

Almost all known tRNAs can be schematically arranged in the so-called cloverleaf secondary structure (Fig. 26-4). Starting from the 5′ end, they have the following common features:

1. A 5′-terminal phosphate group.
2. A 7-bp stem that includes the 5′-terminal nucleotide and that may contain non-Watson–Crick base pairs such as G · U. This assembly is known as the **acceptor** or **amino acid stem** because the amino acid residue carried by the tRNA is appended to its 3′-terminal OH group.
3. A 3- or 4-bp stem ending in a 5-to 7-nt loop that frequently contains the modified base **dihydrouridine (D)**. This stem and loop are therefore collectively termed the **D arm**.
4. A 5-bp stem ending in a loop that contains the anticodon. These features are known as the **anticodon arm**.
5. A 5-bp stem ending in a loop that usually contains the sequence TψC (where ψ is the symbol for **pseudouridine**). This assembly is called the **TψC** or **T arm**.
6. A 3′ CCA sequence with a free 3′-OH group. The —CCA may be genetically specified or enzymatically appended to the immature tRNA, depending on the species (Section 25-3C).

tRNAs have 15 invariant positions (always have the same base) and 8 **semivariant** positions (only a purine or only a pyrimidine) that occur mostly in the loop regions. The purine on the 3′ side of the anticodon is invariably modified. The site of greatest variability among the known tRNAs occurs in the so-called **variable arm**. It has from 3 to 21 nucleotides and may have a stem consisting of up to 7 bp.

tRNAs Have Numerous Modified Bases

One of the most striking characteristics of tRNAs is their large proportion, up to 25%, of posttranscriptionally modified bases. Nearly 80 such bases, found at >60 different tRNA positions, have been characterized. A few of them, together with their standard abbreviations, are indicated in

Figure 26-3. The adaptor hypothesis. This hypothesis postulates that the genetic code is read by molecules that recognize a particular codon and carry the corresponding amino acid.

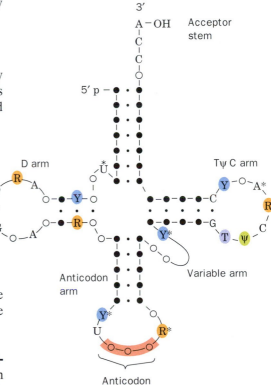

Figure 26-4. The cloverleaf secondary structure of tRNA. Filled circles connected by dots represent Watson–Crick base pairs, and open circles indicate bases involved in non-Watson–Crick base pairing. Invariant positions are indicated: R and Y represent invariant purines and pyrimidines, and ψ represents pseudouracil. The starred nucleotides are often modified. The D and variable arms contain different numbers of nucleotides in the various tRNAs.

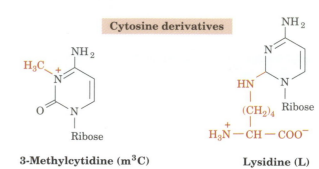

Pseudouridine (ψ)   Dihydrouridine (D)

3-Methylcytidine (m³C)   Lysidine (L)

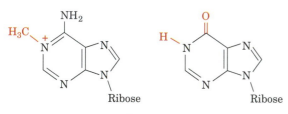

Adenine derivatives

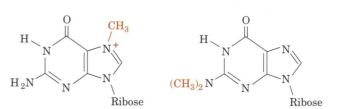

Guanine derivatives

1-Methyladenosine (m¹A)   Inosine (I)

N^7-Methylguanosine (m⁷G) N^2,N^2-Dimethylguanosine (m₂²G)

Figure 26-5. A few of the modified nucleosides that occur in tRNAs. Note that although inosine chemically resembles guanosine, it is biochemically derived from adenosine. Nucleosides may also be methylated at their ribose 2′ positions to form residues symbolized, for instance, by Cm, Gm, and Um. ● See Kinemage Exercise 20-2.

(a)

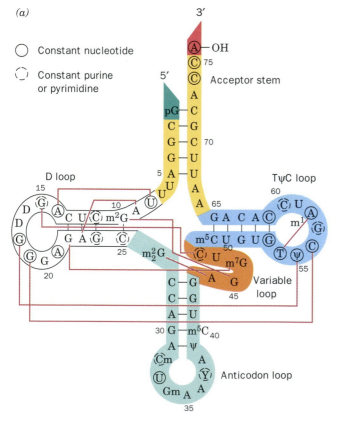

○ Constant nucleotide

◌ Constant purine or pyrimidine

(b)

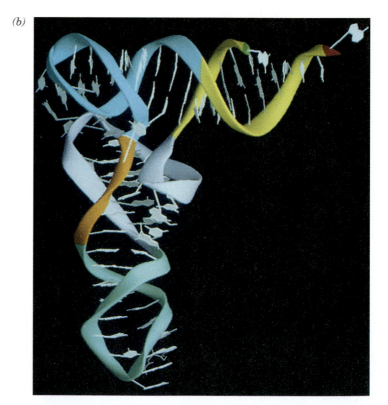

Figure 26-6. *Key to Structure.* The structure of yeast tRNA^Phe. (a) The base sequence drawn in cloverleaf form. Tertiary base pairing interactions are represented by thin red lines connecting the participating bases. Bases that are conserved or semiconserved in all tRNAs are circled by solid and dashed lines, respectively. The 5′ terminus is colored bright green, the acceptor stem is yellow, the D arm is white, the anticodon arm is light green, the variable arm is orange, the TψC arm is cyan, and the 3′ terminus is red. (b) The X-ray structure drawn to show how its base-paired stems are arranged to form the L-shaped molecule. The sugar–phosphate backbone is represented by a ribbon with the same color scheme as in a. [Courtesy of Mike Carson, University of Alabama at Birmingham.] ● See Kinemage Exercise 20-1.

Fig. 26-5. None of these modifications are essential for maintaining a tRNA's structural integrity or for its proper binding to the ribosome. However, base modifications may help promote attachment of the proper amino acid to the acceptor stem or strengthen codon–anticodon interactions.

tRNA Has a Complex Tertiary Structure

As described in Section 23-2E, tRNA molecules have an L-shape in which the acceptor and T stems form one leg and the D and anticodon stems form the other (Fig. 26-6). Each leg of the L is ~60 Å long, and the anticodon and amino acid acceptor sites are at opposite ends of the molecule, some 76 Å apart. The narrow 20- to 25-Å width of tRNA is essential to its biological function: During protein synthesis, two tRNA molecules must simultaneously bind in close proximity at adjacent codons on mRNA (Section 26-4B).

A tRNA's complex tertiary structure is maintained by extensive stacking interactions and base pairing within and between the helical stems. The structure of tRNA^{Phe}, for example, includes nine cross-links, all but one of which are non-Watson–Crick associations (Fig. 26-7). Moreover, most of the bases involved in these interactions are either invariant or semi-invariant, con-

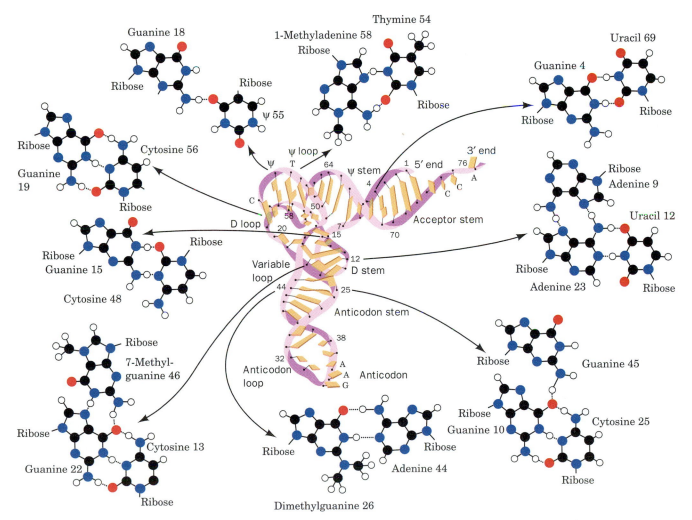

Figure 26-7.   The nine tertiary base pairing interactions in yeast tRNA^{Phe}. Note that all but one involve non-Watson–Crick pairs and that they are all located near the corner of the L. [After Kim, S.H., *in* Schimmel, P.R., Söll, D., and Abelson, J.N. (Eds.),

Transfer RNA: Structure, Properties and Recognition, p. 87, Cold Spring Harbor Laboratory (1979). Drawing of tRNA copyrighted © by Irving Geis.] ● **See Kinemage Exercise 20-3.**

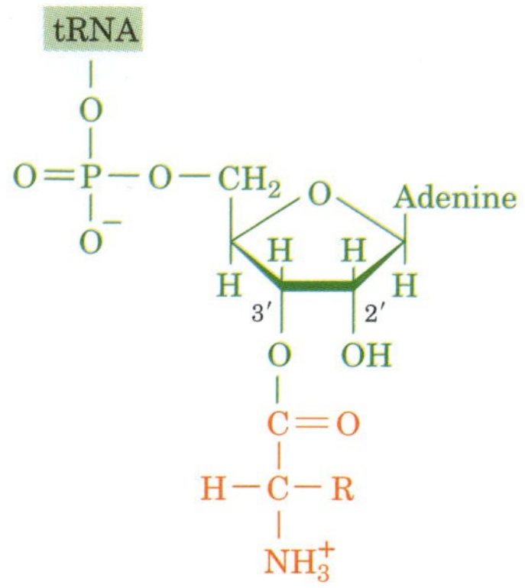

Aminoacyl–tRNA

Figure 26-8. **An aminoacyl–tRNA.** The amino acid residue is esterified to the tRNA's 3′-terminal nucleotide at either its 3′-OH group, as shown here, or its 2′-OH group.

> ● **See Guided Exploration 27:**
>
> The Structures of Aminacyl–tRNA Synthetases and Their Interactions with tRNAs.

sistent with the notion that *all tRNAs have similar conformations.* The structure is also stabilized by several unusual hydrogen bonds between bases and with phosphate groups or the 2′-OH groups of ribose residues.

The compact structure of yeast tRNA^{Phe} renders most of its bases inaccessible to solvent. The most notable exceptions are the anticodon bases and those of the amino acid–bearing —CCA terminus. Both of these groupings must be accessible in order to carry out their biological functions.

B. Aminoacyl–tRNA Synthetases

Accurate translation requires two equally important recognition steps:

1. The correct amino acid must be selected for covalent attachment to a tRNA by an **aminoacyl–tRNA synthetase** (discussed below).
2. The correct **aminoacyl–tRNA** must pair with an mRNA codon at the ribosome (discussed in Section 26-4D).

An amino acid–specific aminoacyl–tRNA synthetase (**aaRS**) appends an amino acid to the 3′-terminal ribose residue of its cognate tRNA to form an aminoacyl–tRNA (Fig. 26-8). Aminoacylation occurs in two sequential reactions that are catalyzed by a single enzyme.

1. The amino acid is first "activated" by its reaction with ATP to form an **aminoacyl–adenylate,**

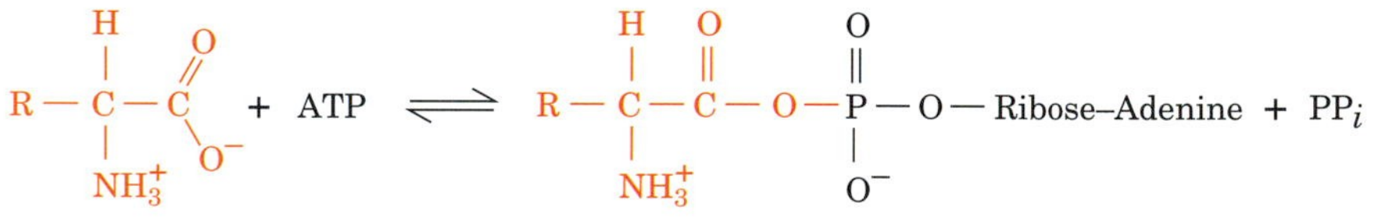

Amino acid   **Aminoacyl–adenylate (aminoacyl–AMP)**

which, with all but three aaRSs, can occur in the absence of tRNA. Indeed, this intermediate can be isolated, although it normally remains tightly bound to the enzyme.

2. This mixed anhydride then reacts with tRNA to form the aminoacyl–tRNA:

$$\text{Aminoacyl–AMP} + \text{tRNA} \rightleftharpoons \text{aminoacyl–tRNA} + \text{AMP}$$

Some aaRSs append an amino acid exclusively to the terminal 2′-OH group of their cognate tRNAs, and others do so at the 3′-OH group. This selectivity was established with the use of chemically modified tRNAs that lack either the 2′- or 3′-OH group of their 3′-terminal ribose residue. The use of these derivatives was necessary because, in solution, the aminoacyl group rapidly equilibrates between the 2′ and 3′ positions.

The overall aminoacylation reaction

$$\text{Amino acid} + \text{tRNA} + \text{ATP} \longrightarrow \text{aminoacyl–tRNA} + \text{AMP} + \text{PP}_i$$

is driven to completion by the hydrolysis of the PP$_i$ generated in the first reaction step. The aminoacyl–tRNA product is a "high-energy" compound (Section 13-2A); for this reason, the amino acid is said to be "activated" and the tRNA is said to be "charged." Amino acid activation resembles fatty acid activation (Section 19-2A); the major difference is that tRNA is the acyl acceptor in amino acid activation whereas CoA performs this function in fatty acid activation.

There Are Two Classes of Aminoacyl–tRNA Synthetases

Cells must have at least one aaRS for each of the 20 amino acids. The similarity of the reaction catalyzed by these enzymes and the structural similarities among tRNAs suggest that all aaRSs evolved from a common ancestor and should therefore be structurally related. This is not the case. In fact, *the aaRSs form a diverse group of enzymes with different sizes and quaternary structures and little sequence similarity.* Nevertheless, these enzymes can be grouped into two classes that each have the same 10 members in all organisms (Table 26-2). **Class I** and **Class II aminoacyl–tRNA synthetases** differ in several ways:

1. **Structural motifs.** The Class I enzymes share two homologous polypeptide segments that have the consensus sequences His-Ile-Gly-His (HIGH) and Lys-Met-Ser-Lys-Ser (KMSKS). Both of these segments are components of a dinucleotide-binding fold (Rossmann fold, which is also present in many NAD^+- and ATP-binding proteins; Section 6-2B). The Class II synthetases lack the foregoing sequences but have three other sequences in common. These sequences occur in a motif found only in Class II enzymes that consists of a seven-stranded antiparallel β sheet with three flanking helices and that forms the core of their catalytic domains.

2. **Anticodon recognition.** Many Class I aaRSs must recognize the anticodon to aminoacylate their cognate tRNAs. In contrast, several Class II enzymes do not interact with their bound tRNA's anticodon.

3. **Site of aminoacylation.** All Class I enzymes aminoacylate their bound tRNA's 3'-terminal 2'-OH group, whereas Class II enzymes charge the 3'-OH group. Nevertheless, an aminoacyl group attached at the 2' position rapidly equilibrates between the 2' and 3' positions (it must be at the 3' position to take part in protein synthesis).

4. **Amino acid specificity.** The amino acids for which the Class I synthetases are specific tend to be larger and more hydrophobic than those for Class II synthetases.

Table 26-2. **Classification of *E. coli* Aminoacyl–tRNA Synthetases**

	Amino Acid	Quaternary Structure	Number of Residues
Class I	Arg	α	577
	Cys	α	461
	Gln	α	551
	Glu	α	471
	Ile	α	939
	Leu	α	860
	Met	α_2	676
	Trp	α_2	325
	Tyr	α_2	424
	Val	α	951
Class II	Ala	α_4	875
	Asn	α_2	467
	Asp	α_2	590
	Gly	$\alpha_2\beta_2$	303/689
	His	α_2	424
	Lys	α_2	505
	Pro	α_2	572
	Phe	$\alpha_2\beta_2$	327/795
	Ser	α_2	430
	Thr	α_2	642

Source: Carter, C.W., Jr., *Annu. Rev. Biochem.* **62**, 717 (1993).

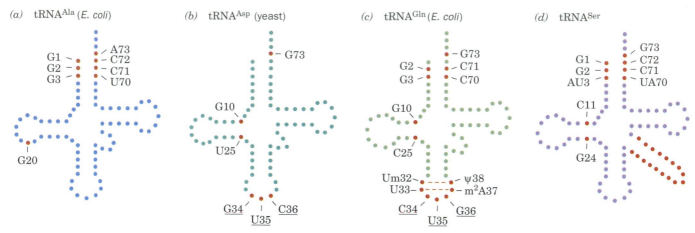

Figure 26-9.   Major identity elements in four tRNAs. Each base in the tRNA is represented by a filled circle. Red circles indicate positions that have been shown to be identity elements for the recognition of the tRNA by its cognate aminoacyl–tRNA synthetase. The anticodon bases that are identity elements are underlined.

Aminoacyl–tRNA Synthetases Recognize Unique Structural Features of tRNA

How does an aaRS recognize a tRNA so that it can be charged with the proper amino acid? First, all tRNAs have similar structures, so the features that differentiate them must be subtle variations in sequence or local structure. On the other hand, since the genetic code is degenerate, more than one tRNA may carry a given amino acid. These so-called **isoaccepting tRNAs** must be recognized by their cognate aaRSs. Finally, the tRNA must be charged with only the amino acid that corresponds to its anticodon, and not any of the 19 other amino acids. Clues to the specificity of synthetase–tRNA interactions have been gleaned from studies using tRNA fragments, mutationally altered tRNAs, chemical cross-linking agents, computerized sequence comparisons, and X-ray crystallography. The most common synthetase contact sites on tRNA occur on the inner (concave) face of the L. Other than that, there appears to be little regularity in how the various tRNAs are recognized by their cognate synthetases.

The elements in four tRNAs that are recognized by different aaRSs are shown in Fig. 26-9. In all of these cases, *the acceptor stem is critical for the enzyme–tRNA interaction, as are other sites, not necessarily including the anticodon.* A single base or base pair may constitute an identity element. When the experimentally determined identity elements for a variety of tRNAs are mapped onto a three-dimensional model of a tRNA molecule, they cluster in the acceptor stem and the anticodon loop (Fig. 26-10). These sites lie at opposite ends of the molecule, so that aaRSs that interact with both of them must have a size and structure adequate to bind both legs of the L-shaped tRNA.

The X-ray structures of synthetase–tRNA complexes reveal extensive contacts between the protein and the inside face of the tRNA L. The structure of *E. coli* **glutaminyl–tRNA synthetase (GlnRS)** in complex with **tRNA^Gln** and ATP (Fig. 26-11), determined by Thomas Steitz, was the first to be elucidated. GlnRS, a 553-residue monomeric Class I enzyme, has an elongated shape so that it binds the anticodon near one end of the protein and the acceptor stem near the other. Genetic and biochemical data indicate that the identity elements for tRNA^Gln include all seven bases of the anticodon loop (Fig. 26-9c). The bases of the anticodon itself are unstacked and splay outward so as to bind in separate recognition pockets of GlnRS.

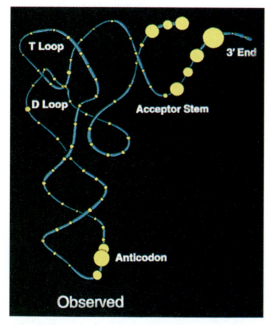

Figure 26-10.   The experimentally observed identity elements of tRNAs. The tRNA backbone is blue, and each of its nucleotides is represented by a yellow circle whose diameter is proportional to the fraction of the 20 tRNA acceptor types for which the nucleotide is an observed determinant. [Courtesy of William H. McClain, University of Wisconsin.]

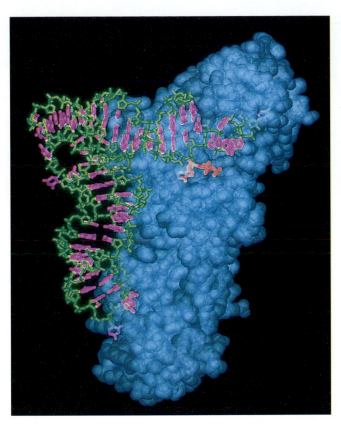

Figure 26-11. The X-ray structure of E. coli GlnRS · tRNA^Gln · ATP. The tRNA and ATP are shown in skeletal form with the tRNA sugar–phosphate backbone green, its bases magenta, and the ATP red. The protein is represented by a translucent cyan space-filling model that reveals the buried portions of the tRNA and ATP. Note that both the 3' end of the tRNA (*top right*) and its anticodon bases (*bottom*) are inserted into deep pockets in the protein. [Based on an X-ray structure by Thomas Steitz, Yale University.] ● See Kinemage Exercise 21-1.

The 3' end of tRNA^Gln plunges deeply into a protein pocket that also binds the enzyme's ATP and glutamine substrates.

Yeast **AspRS,** a Class II enzyme, is an α_2 dimer of 557-residue subunits. Its X-ray structure in complex with **tRNA^Asp**, determined by Dino Moras, reveals that the protein symmetrically binds two tRNA molecules by contacting them principally at their acceptor stem and anticodon regions (Fig. 26-12). The anticodon arm of tRNA^Asp is bent by as much as 20 Å toward the inside of the L relative to that in the X-ray structure of uncomplexed

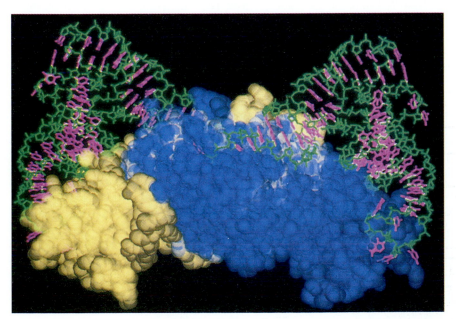

Figure 26-12. The X-ray structure of yeast AspRS · tRNA^Asp · ATP. The homodimeric enzyme with its two symmetrically bound tRNAs is viewed with its twofold axis approximately vertical. The tRNAs are shown in skeletal form with their sugar–phosphate backbones green and their bases magenta. The two protein subunits are represented by translucent yellow and blue space-filling models that reveal buried portions of the tRNAs. [Based on an X-ray structure by Dino Moras, Institut de Biologie Moléculaire et Cellulaire du CNRS, Strasbourg, France.]

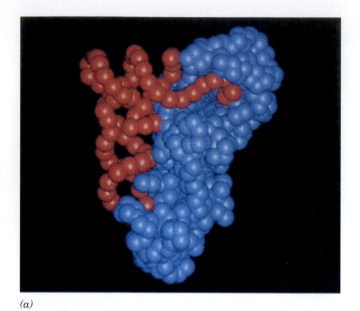

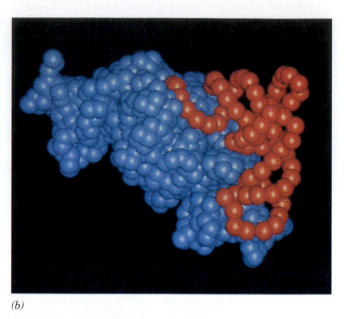

(a)

(b)

Figure 26-13.   Comparison of the binding of GlnRS and AspRS to their cognate tRNAs. The proteins and tRNAs are represented by blue and red spheres marking the C$_\alpha$ and P atom positions. Note how GlnRS (*a*), a Class I synthetase, binds tRNA^{Gln} from the minor groove side of its acceptor stem so as to bend its

3′ end into a hairpin conformation. In contrast, AspRS (*b*), a Class II synthetase, binds tRNA^{Asp} from the major groove side of its acceptor stem so that its 3′ end continues its helical path on entering the active site. [Courtesy of Dino Moras, Institut de Biologie Moléculaire et Cellulaire du CNRS, Strasbourg, France.]

tRNA^{Asp}, and its anticodon bases are unstacked. The hinge point for this bend is a G30 · U40 base pair in the anticodon stem (nearly all other species of tRNA contain a Watson–Crick base pair at this point). The anticodon bases of tRNA^{Gln} are also unstacked in contacting GlnRS but with a backbone conformation that differs from that in tRNA^{Asp}. Evidently, the conformation of a tRNA in complex with its cognate synthetase is dictated more by its interactions with the protein (induced fit) than by its sequence. This is perhaps one reason why *the members of each set of isoaccepting tRNAs in a cell are recognized by a single aaRS.*

The different modes of tRNA binding by GlnRS and AspRS are highlighted in Fig. 26-13. Although both tRNAs approach their synthetases along the inside of the L shapes, tRNA^{Gln} does so from the direction of the minor groove of its acceptor stem, whereas tRNA^{Asp} does so from the direction of its major groove. The 3′ end of tRNA^{Asp} thereby continues its helical track as it plunges into AspRS's catalytic site, whereas the 3′ end of tRNA^{Gln} bends backward into a hairpin turn as it enters its active site. These structural differences account for the observation that Class I and Class II enzymes aminoacylate different OH groups on the 3′ terminal ribose of tRNA (Fig. 26-14).

Proofreading Enhances the Fidelity of Amino Acid Attachment to tRNA

The charging of a tRNA with its cognate amino acid is a remarkably accurate process. Experimental measurements indicate, for example, that at equal concentrations of isoleucine and valine, IleRS transfers ~50,000 isoleucines to tRNA^{Ile} for every valine it so transfers. This high degree of accuracy is surprising because valine, which differs from isoleucine only by the lack of a single methylene group, should fit easily into the isoleucine-binding site of IleRS. Consideration of the additional ~12 kJ · mol^{-1} binding free energy that a methylene group is estimated to provide suggests that isoleucyl–tRNA synthetase should discriminate between isoleucine and valine by no more than a factor of ~100.

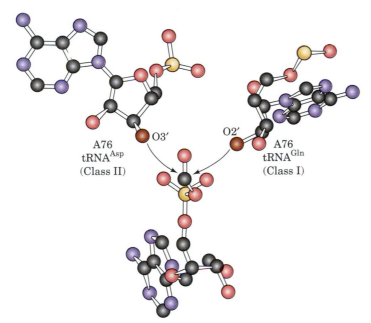

Aminoacyl–AMP

Figure 26-14. **A comparison of the stereo-
chemistries of aminoacylation by Class I and
Class II synthetases.** For GlnRS (Class I), the
enzyme-bound tRNA's 3′-terminal adenosine
residue (A76) is positioned to the right of the
aminoacyl–AMP (whose aminoacyl group is
represented here by only its carbonyl group),
whereas that for AspRS (Class II) is positioned to
the left of this group. Note that only atoms O3′
of tRNA^Gln and O2′ of tRNA^Asp are suitably
positioned to attack this group and thereby
transfer the aminoacyl residue to the tRNA. [After
Cavarelli, J., Eriani, G., Rees, B., Ruff, M., Boeglin,
M., Mitschler, A., Martin, F., Gangloff, J., Thierry,
J.-C., and Moras, D., *EMBO J.* **13**, 335 (1994).]

Paul Berg resolved this apparent paradox by demonstrating that, in the presence of tRNA^Ile, IleRS catalyzes the quantitative hydrolysis of valine–adenylate, the intermediate of the aminoacylation reaction, to valine + AMP rather than forming Val–tRNA^Ile. Thus, *isoleucyl–tRNA synthetase subjects aminoacyl–adenylates to a proofreading or editing step*, a process reminiscent of that carried out by DNA polymerase I (Section 24-2A). Proofreading occurs at a separate active site that presumably binds Val but excludes the larger Ile. The enzyme's overall selectivity is the product of the selectivities of its adenylation and proofreading steps, thereby contributing to the high fidelity of translation. Many other synthetases discriminate against noncognate amino acids in a similar fashion. Note that editing occurs at the expense of ATP hydrolysis, the thermodynamic price of high fidelity (increased order).

C. Codon–Anticodon Interactions

In protein synthesis, the proper tRNA is selected only through codon–anticodon interactions; the aminoacyl group does not participate in this process (this is one reason why accurate aminoacylation is critical for protein synthesis). The three nucleotides of an mRNA codon pair with the three nucleotides of a complementary tRNA anticodon in an antiparallel fashion. One might naively guess that each of the 61 codons specifying an amino acid would be read by a different tRNA. Yet even though most cells contain numerous groups of isoaccepting tRNAs, *many tRNAs bind to two or three of the codons specifying their cognate amino acids*. For example, yeast tRNA^Phe, which has the anticodon GmAA (where Gm indicates G with a 2′-methyl group), recognizes the codons UUC and UUU,

```
              3′              5′ 3′              5′
Anticodon:   —A—A—Gm—          —A—A—Gm—
             :   :   :          :   :   :
          5′ :   :   : 3′  5′ :   :   : 3′
Codon:      —U—U—C—          —U—U—U—
```

and yeast tRNA^Ala, which has the anticodon IGC (where I is inosine), recognizes the codons GCU, GCC, and GCA.

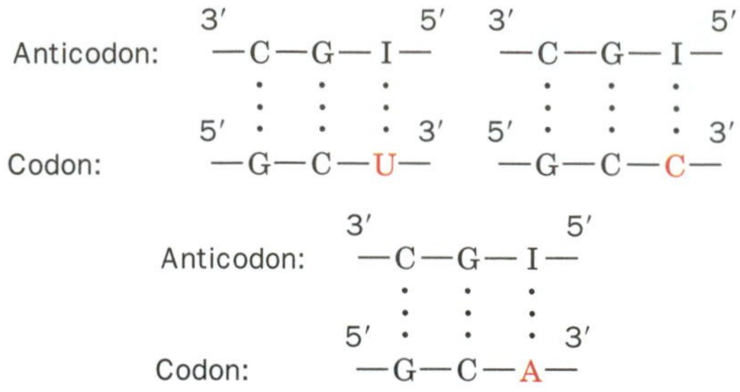

$$\begin{array}{ccc}
& 3' & 5' \quad 3' & 5' \\
\text{Anticodon:} & -\text{C}-\text{G}-\text{I}- & -\text{C}-\text{G}-\text{I}- \\
& \vdots \quad \vdots \quad \vdots & \vdots \quad \vdots \quad \vdots \\
& 5' \quad \vdots \quad \vdots \quad \vdots \quad 3' & 5' \quad \vdots \quad \vdots \quad \vdots \quad 3' \\
\text{Codon:} & -\text{G}-\text{C}-\text{U}- & -\text{G}-\text{C}-\text{C}-
\end{array}$$

$$\begin{array}{cc}
& 3' \quad\quad 5' \\
\text{Anticodon:} & -\text{C}-\text{G}-\text{I}- \\
& \vdots \quad \vdots \quad \vdots \\
& 5' \quad \vdots \quad \vdots \quad \vdots \quad 3' \\
\text{Codon:} & -\text{G}-\text{C}-\text{A}-
\end{array}$$

It therefore seems that non-Watson–Crick base pairing can occur at the third codon–anticodon position (the anticodon's first position is defined as its 3' nucleotide), the site of most codon degeneracy (Table 26-1). Note that the third (5') anticodon position commonly contains a modified base such as Gm or I.

The Wobble Hypothesis Accounts for Codon Degeneracy

By combining structural insight with logical deduction, Crick proposed the **wobble hypothesis** to explain how a tRNA can recognize several degenerate codons. He assumed that *the first two codon–anticodon pairings have normal Watson–Crick geometry and that there could be a small amount of play or "wobble" in the third anticodon position to allow limited conformational adjustments in its pairing geometry.* This permits the formation of several non-Watson–Crick pairs such as U·G and I·A (Fig. 26-15). The allowed pairings for the third codon–anticodon position are listed in Table 26-3. An anticodon with C or A in its third position can potentially pair only with its Watson–Crick complementary codon (although, in fact, there is no known instance of a tRNA with an A in its third anticodon position). If U or G occupies the third anticodon position, two codons can potentially be recognized. I at the third anticodon position can pair with U, C, or A.

A consideration of the various wobble pairings indicates that at least 31 tRNAs are required to translate all 61 coding triplets of the genetic code (there are 32 tRNAs in the minimal set because translation initiation requires a separate tRNA; Section 26-4A). Most cells have >32 tRNAs, some of which have identical anticodons. In fact, mammalian cells have >150 tRNAs. Some organisms contain a unique tRNA, **tRNA^{Sec},** that is charged with **selenocysteine (Sec),** a constituent of several enzymes (see Box 26-2).

Frequently Used Codons Are Complementary to the Most Abundant tRNA Species

The analysis of the base sequences of several highly expressed structural genes of baker's yeast, *Saccharomyces cerevisiae,* has revealed a remarkable bias in their codon usage. Only 25 of the 61 coding triplets are com-

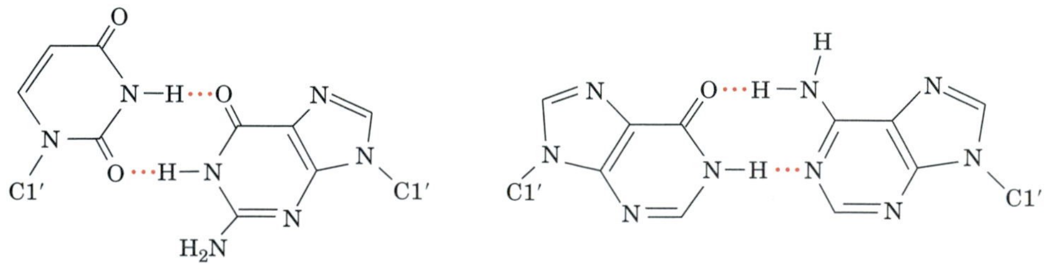

Table 26-3. **Allowed Wobble Pairing Combinations in the Third Codon–Anticodon Position**

5'-Anticodon Base	3'-Codon Base
C	G
A	U
U	A or G
G	U or C
I	U, C, or A

Figure 26-15. **U·G and I·A wobble pairs.** Both have been observed in X-ray structures.

Box 26-2
BIOCHEMISTRY IN FOCUS

Selenocysteine

Selenium, an essential trace element, is a component of several enzymes in both prokaryotes and eukaryotes. These **selenoproteins** contain selenocysteine (Sec) residues

$$\begin{array}{c} | \\ NH \\ | \\ CH-CH_2-Se-H \\ | \\ C=O \\ | \end{array}$$

**The selenocysteine
residue**

that are thought to participate in redox reactions such as that catalyzed by the mammalian selenoprotein **glutathione peroxidase** (Box 14-3).

Selenocysteine is not the result of a posttranslational modification but is incorporated into proteins as they are synthesized by the ribosome. Incorporation of this "21st amino acid" requires some variation in the standard translation machinery, including a unique tRNA and a UGA stop codon that is reinterpreted as a Sec codon.

tRNA^{Sec} is initially charged with serine in a reaction catalyzed by the same SerRS that charges tRNA^{Ser}. The resulting Ser–tRNA^{Sec} is enzymatically selenylated to produce selenocysteinyl–tRNA^{Sec}. Although tRNA^{Sec} must resemble tRNA^{Ser} enough to interact with the same SerRS, its acceptor stem has 8 bp (rather than 7), its D arm has a 6-bp stem and a 4-base loop (rather than a 4-bp stem and a 7- to 8-base loop), its TψC stem has 4 bp rather than 5, and its anticodon, UCA, recognizes a UGA stop codon rather than a Ser codon. In addition, several of the invariant residues of other tRNAs are altered in tRNA^{Sec}. These changes explain why tRNA^{Sec} is not recognized by the machinery that normally adds tRNA-bound aminoacyl groups to an elongating polypeptide chain at the ribosome. Instead, a dedicated protein (a special elongation factor) is required to deliver Sec–tRNA^{Sec} to the ribosome. This protein allows the UGA codon to be read as "Sec" rather than "Stop," provided that the local structure of the mRNA supports the readthrough of the stop codon. An mRNA element in the 3′ untranslated region determines whether UGA is read as a stop codon or as a Sec codon.

---

monly used. *The preferred codons are those that are most nearly complementary, in the Watson–Crick sense, to the anticodons in the most abundant species in each set of isoaccepting tRNAs.* A similar phenomenon occurs in *E. coli*, although several of its 22 preferred codons differ from those in yeast. The degree with which the preferred codons occur in a given gene is strongly correlated, in both organisms, with the gene's level of expression. This probably permits the mRNAs of proteins that are required in high abundance to be rapidly and smoothly translated.

3. RIBOSOMES

Ribosomes, small organelles that were once thought to be artifacts of cell disruption, were identified as the site of protein synthesis in 1955 by Paul Zamecnik, who demonstrated that ^{14}C-labeled amino acids are transiently associated with ribosomes before they appear in free proteins. The ribosome is both enormous (2520 kD in *E. coli* and 4220 kD in mammals) and complex (ribosomes contain several large RNA molecules and dozens of different proteins). This complexity is necessary for the ribosome to carry out the following functions:

1. The ribosome binds mRNA such that its codons can be read with high fidelity.
2. The ribosome includes specific binding sites for tRNA molecules.
3. The ribosome mediates the interactions of nonribosomal protein factors that promote polypeptide chain initiation, elongation, and termination.

4. The ribosome catalyzes peptide bond formation.

5. The ribosome undergoes movement so that it can translate sequential codons.

Structure of the Prokaryotic Ribosome

The *E. coli* ribosome has a sedimentation coefficient of 70S and, as James Watson discovered, is composed of two unequal subunits (Table 26-4). The small (30S) subunit consists of a **16S rRNA** molecule and 21 different polypeptides, whereas the large (50S) subunit contains a **5S** and a **23S rRNA** together with 31 different polypeptides. The up to 20,000 ribosomes in an *E. coli* cell account for ~80% of its RNA content and 10% of its protein.

The structure of the ribosome first came into focus through electron microscopy (Fig. 26-16*a*) and later through **cryoelectron microscopy** (Fig. 26-16*b*), a technique in which the sample is embedded in vitreous (glasslike) ice and hence retains its native shape to a greater extent than in conventional electron microscopy. The fine structure of the ribosome has recently been determined by X-ray crystallography (Fig. 26-16*c*), a landmark achievement due to the enormous size of the particle. Even at relatively low resolution, *the ribosome clearly has a highly irregular shape, about 250 Å across, with numerous lobes and bulges as well as channels and tunnels.* The X-ray structures of isolated large and small ribosomal subunits reveal additional details about how ribosomal structure relates to the various steps of protein synthesis. The structures also reveal considerable information about RNA–protein and RNA–RNA interactions.

Most of the structure of the 30S subunit from the heat-tolerant bacterium *Thermus thermophilus* has been reconstructed at a resolution of 3 Å by V. Ramakrishnan. This structure includes nearly all of the 16S rRNA and all but 5% of the structures of 20 proteins. The secondary structure of the 16S rRNA is shown in Fig. 26-17*a*. Its four domains include approximately 50 double-stranded helices connected by single-strand loops, many of which are also helical. The RNA helices assemble end to end and side by side to form an asymmetric mass with clearly delineated domains (Fig. 26-17*b*). Many of the RNA helices interact through their wide and shallow minor grooves (recall that RNA adopts an A-form helix, as described in Section 23-1A). These interactions are facilitated by nonstandard base pairs, which further widen the grooves.

The proteins of the 30S subunit associate primarily with the exterior of the ribosome (Fig. 26-17*c*). None are buried entirely within the RNA, and

Table 26-4.   **Components of *E. coli* Ribosomes**

	Ribosome	Small Subunit	Large Subunit
Sedimentation coefficient	70S	30S	50S
Mass (kD)	2520	930	1590
RNA			
Major		16S, 1542 nucleotides	23S, 2904 nucleotides
Minor			5S, 120 nucleotides
RNA mass (kD)	1664	560	1104
Proportion of mass	66%	60%	70%
Proteins		21 polypeptides	31 polypeptides
Protein mass (kD)	857	370	487
Proportion of mass	34%	40%	30%

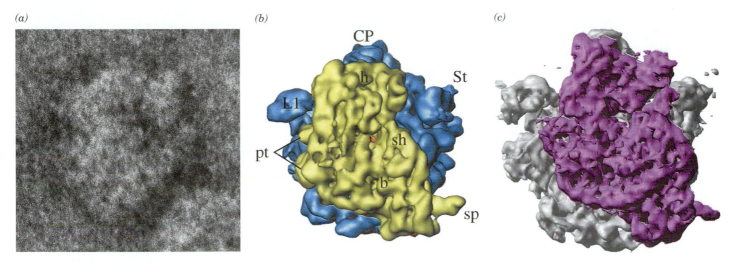

Figure 26-16. Views of the ribosome. (*a*) A conventional electron micrograph of an *E. coli* ribosome. [Courtesy of James Lake, UCLA.] (*b*) Density map of the *E. coli* ribosome based on cryoelectron microscopy. This map, which has a resolution of 11.5 Å, reveals the architecture of the large (*cyan*) and small (*yellow*) subunits of the ribosome. [Courtesy of Joachim Frank, State University of New York at Albany.] (*c*) X-Ray crystallographic structure of a ribosome from *Thermus thermophilus* at a resolution of 7.5 Å. [Courtesy of Harry Noller, University of California, Santa Cruz.]

very few impinge on the interface with the 50S subunit. Interestingly, the structure of the 16S rRNA largely determines the shape of the 30S subunit (compare Figs. 26-17*b* and *c*).

In contrast to the multidomain 30S subunit, the 50S subunit appears to be somewhat monolithic. The structure of the large subunit from the halophilic (salt-loving) bacterium *Haloarcula marismortui* has been determined to a

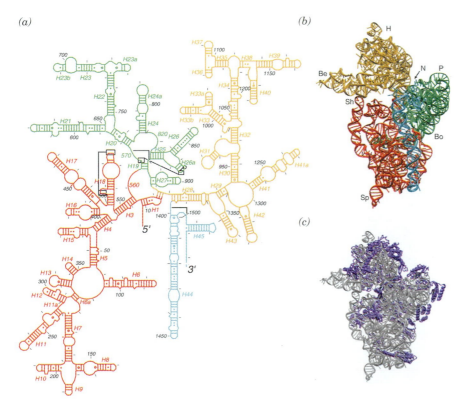

Figure 26-17. Structure of the 16S rRNA and the 30S subunit from *T. thermophilus*. (*a*) Secondary structure diagram of the 16S rRNA, with the four domains in different colors. Compare this structure to the proposed secondary structure of the *E. coli* 16S rRNA shown in Fig. 23-23. (*b*) Tertiary structure of the 16S rRNA with the same coloring of domains. (*c*) Structure of the 30S subunit. The 16S rRNA is shown in gray, and the 20 proteins are shown in purple. [Courtesy of V. Ramakrishnan, MRC, Cambridge.]

(a)

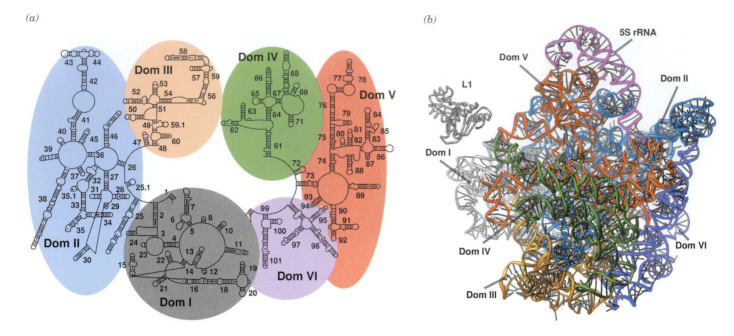

(b)

Figure 26-18.   Structure of the 23S rRNA from the large ribosomal subunit of *H. marismortui*. (*a*) Secondary structure diagram, with the six domains in different colors. (*b*) Tertiary structure of the rRNA with the same coloring of domains. The 5S rRNA (*magenta*) is included. The ribosomal protein L1 is shown for purposes of orientation. [Courtesy of Thomas Steitz and Peter Moore, Yale University.]

resolution of 2.4 Å by Thomas Steitz and Peter Moore. The structure includes 2833 of the 3045 nucleotides and 27 of the 31 proteins in the 50S subunit. The large 23S rRNA has a secondary structure that suggests six domains, but when fully folded, the domains interlock and, along with the small 5S rRNA, form a single mass of tightly packed helices (Fig. 26-18). As in the 30S subunit, ribosomal proteins are arranged mostly on the outer surface of the 50S subunit, leaving the subunit interface clear (Fig. 26-19).

The structures of many ribosomal proteins have been independently determined. By convention, proteins from the small and large subunits are designated with the prefixes S and L, respectively, followed by a number. All the proteins occur just once in the ribosome structure, except for one protein whose four copies associate with the large subunit. Several ribosomal proteins contain homologous structural motifs that consist of a three-stranded antiparallel β sheet with two strands connected by an α helix (Fig. 26-20). This structural motif has also been observed or implicated, through sequence homology, in other RNA-binding proteins and has therefore been

Figure 26-19.   Two views of the 50S subunit from *H. marismortui*. The rRNA is shown in gray and the proteins are shown in gold. (*a*) This view matches that of Fig. 26-18*b*. The central protein-free area forms the interface with the 30S subunit. (*b*) This view represents the back side of the structure shown in part (*a*) and shows how proteins are distributed over the outer surface of the 50S subunit. [Courtesy of Thomas Steitz and Peter Moore, Yale University.]

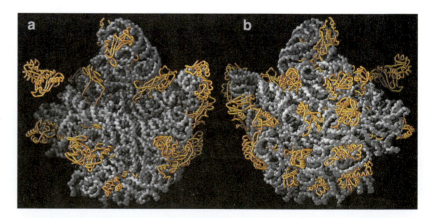

dubbed the **RNA recognition motif (RRM).** All these proteins presumably evolved from an ancient RNA-binding protein. Surprisingly, the X-ray structures of the ribosomal subunits indicate that the RRM-containing proteins interact with RNA in different ways.

The most notable feature of the mostly globular ribosomal proteins is that more than half of them include a long narrow extension consisting of an α helix, a β hairpin, or a single strand with no defined secondary structure. These polypeptide tails interact with multiple sites on the rRNA and are rich in basic residues that neutralize the charge repulsion of the RNA backbone phosphate groups. These proteins appear to function as mortar between RNA helices. This model is consistent with theories that the protein-synthesizing ribosome was originally made entirely of RNA molecules and that proteins were added later to stabilize it and assist with its assembly.

Ribosomes Have Multiple Binding Sites

Biochemical studies, X-ray structures, and model building have shown that up to three tRNA molecules at a time bind to the ribosome. The three binding sites are known as the **A** or **aminoacyl site** (it accommodates the incoming aminoacyl–tRNA), the **P** or **peptidyl site** (it accommodates the **peptidyl–tRNA,** the tRNA to which the growing peptide chain is attached), and the **E** or **exit site** (it accommodates a deacylated tRNA that is about to exit the ribosome; Fig. 26-21). The tRNA binding sites span both the ribosomal subunits. The anticodon arms of the tRNAs lie within the 30S subunit, and the acceptor stems project into the 50S subunit.

In order for the ribosome to assemble a polypeptide according to the information in an mRNA, the ribosome must bind an mRNA molecule in such a way that adjacent codons align with the anticodons of the tRNAs in the A and P sites. The mRNA apparently binds in a channel on the 30S subunit that is too narrow to admit more than a single strand of RNA, thereby preventing any interstrand base pairing from interfering with translation.

Correct alignment of codons and anticodons may elicit a conformational change in the 30S subunit that sends a signal to the 50S subunit, which catalyzes peptide bond formation. The small subunit and the tRNA molecules must then move so that the ribosome can translate the next codon. The

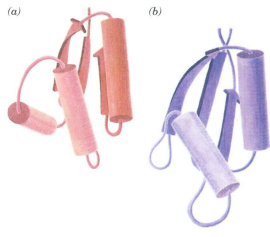

(a)   *(b)*

Figure 26-20. Models of two ribosomal proteins. (*a*) The 74-residue C-terminal fragment of the *E. coli* ribosomal protein known as L7/L12. (*b*) *Bacillus stearothermophilus* protein L30 (61 residues). The RNA recognition motifs in each protein are shaded more darkly. [After Leijonmarck, M., Appelt, K., Badger, J., Liljas, A., Wilson, K.S., and White, S.W., *Proteins* **3**, 244 (1988).]

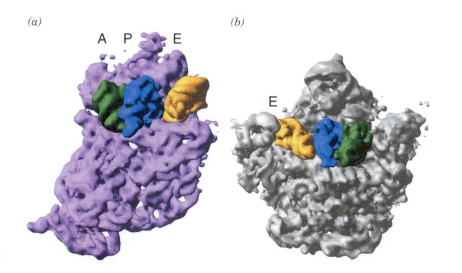

(a)   *(b)*

Figure 26-21. Ribosomal tRNA binding sites. In this model, the two subunits of the ribosome have been separated to expose the subunit interfaces. The 30S subunit is in purple on the left, and the 50S subunit is in gray on the right. The tRNA in the A site is green, the tRNA in the P site is blue, and the tRNA in the E site is gold. [Courtesy of Harry Noller, University of California, Santa Cruz.]

Table 26-5. **Components of Rat Liver Cytoplasmic Ribosomes**

	Ribosome	*Small Subunit*	*Large Subunit*
Sedimentation coefficient	80S	40S	60S
Mass (kD)	4220	1400	2820
RNA			
Major		18S, 1874 nucleotides	28S, 4718 nucleotides
Minor			5.8S, 160 nucleotides
			5S, 120 nucleotides
RNA mass (kD)	2520	700	1820
Proportion of mass	60%	50%	65%
Proteins		33 polypeptides	49 polypeptides
Protein mass (kD)	1700	700	1000
Proportion of mass	40%	50%	35%

multidomain structure of the 30S subunit probably gives the ribosome the flexibility it needs to complete this process.

The growing polypeptide fits into a tunnel on the 50S subunit that extends from the P site to the outer ribosomal surface (the tunnel exit is approximately in the center of the 50S subunit as shown in Fig. 26-19b). The ~100-Å-long tunnel is lined with mostly hydrophilic residues and has an average diameter of ~15 Å. This is barely large enough for an α helix, so significant protein folding probably cannot occur until the polypeptide exits the ribosome.

Eukaryotic Ribosomes Are Larger and More Complex Than Prokaryotic Ribosomes

Although eukaryotic and prokaryotic ribosomes resemble each other in both structure and function, they differ in nearly all details, including mass and subunit composition (Table 26-5; compare with Table 26-4). The small (40S) subunit of the rat liver cytoplasmic ribosome, the best-characterized eukaryotic ribosome, consists of 33 unique polypeptides and an **18S rRNA.** Its large (60S) subunit contains 49 different polypeptides and three rRNAs of 28S, 5.8S, and 5S.

Eukaryotic 18S and **28S rRNAs** are similar in secondary structure to prokaryotic 16S and 23S rRNAs. The **5.8S rRNA,** which occurs in the large eukaryotic subunit in a base-paired complex with 28S rRNA, is homologous in sequence to the 5′ end of prokaryotic 23S rRNA. Apparently, 5.8S rRNA arose through mutations that altered rRNA posttranscriptional processing to produce a fourth rRNA.

Ribosome Assembly

Experiments *in vitro* have shown that *E. coli* ribosomal subunits can self-assemble from mixtures of their numerous macromolecular components. This process begins with the binding of certain proteins to the rRNA molecules, which become a scaffold for binding other proteins. Self-assembly almost certainly occurs *in vivo,* but in a cell, the expression of rRNA and ribosomal protein genes must be coordinated so that the appropriate quantities of rRNA and proteins are available.

Cells regulate their ribosome content to match the rate at which they synthesize proteins under the prevailing growth conditions. In *E. coli,* for example, the rate of rRNA synthesis is tied to the rate of protein synthesis via the **stringent response.** This regulatory mechanism acts when a short-age of any species of charged tRNA (usually a result of "stringent" or poor growth conditions) limits the rate of protein synthesis. A ribosome that binds an uncharged tRNA apparently signals the shortage of the amino acid by stimulating a protein known as **stringent factor** to catalyze the synthesis of **ppGpp:**

$$\text{ATP} + \text{GDP} \rightleftharpoons \text{AMP} + \text{ppGpp}$$

This unusual nucleotide acts as an intracellular messenger to reduce, by 10- to 20-fold, the rate of transcription of rRNA and tRNA genes. In addition, ppGpp stimulates the transcription of genes involved in amino acid biosynthesis and represses the transcription of genes involved in such metabolic processes as DNA replication and carbohydrate biosynthesis. The cell is thereby prepared to withstand nutritional deprivation. The ability of ppGpp to activate the transcription of some genes while repressing that of others suggests that it acts by altering RNA polymerase promoter specificity.

4. POLYPEPTIDE SYNTHESIS

In order to appreciate the manner in which the ribosome orchestrates the translation of mRNA to synthesize polypeptides, it is helpful to assimilate the following points:

1. *Polypeptide synthesis proceeds from the N-terminus to the C-terminus;* that is, a **peptidyl transferase** activity appends an incoming amino acid to a growing polypeptide's C-terminus. This was shown to be the case in 1961 by Howard Dintzis, who exposed reticulocytes (immature red blood cells) that were actively synthesizing hemoglobin to [3H]-labeled leucine for less time than it takes to synthesize an entire polypeptide. The extent to which the tryptic peptides from the soluble (completed) hemoglobin molecules were labeled increased with their proximity to the C-terminus (Fig. 26-22), thereby indicating that incoming amino acids are appended to the growing polypeptide's C-terminus.

2. *Chain elongation occurs by the linkage of the growing polypeptide to the incoming tRNA's amino acid residue.* If the growing polypeptide

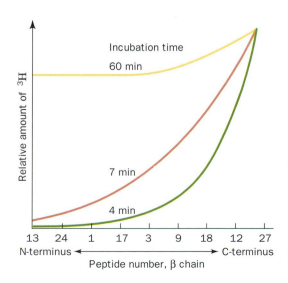

Figure 26-22. **Demonstration that polypeptide synthesis proceeds from the N- to the C-terminus.** Rabbit reticulocytes were incubated with [3H]leucine for the indicated times. The curves show the distribution of [3H]Leu among the tryptic peptides from the β subunit of soluble rabbit hemoglobin. The numbers on the horizontal axis are peptide identifiers, arranged from the N-terminus to the C-terminus. [After Dintzis, H.M., *Proc. Natl. Acad. Sci.* **47**, 255 (1961).]

is released from the ribosome by treatment with high salt concentrations, its C-terminal residue is esterified to a tRNA molecule as a peptidyl–tRNA.

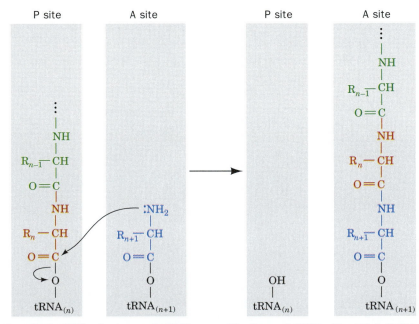

Peptidyl–tRNA

The nascent polypeptide must therefore grow by being transferred from the peptidyl–tRNA in the P site to the incoming aminoacyl–tRNA in the A site to form a peptidyl–tRNA with one more residue

Figure 26-23.   The ribosomal peptidyl transferase reaction forming a peptide bond. The amino group of the aminoacyl–tRNA in the A site nucleophilically displaces the tRNA of the peptidyl–tRNA ester in the P site, thereby forming a new peptide bond and transferring the nascent polypeptide to the A-site tRNA.

(Fig. 26-23). After the peptide bond has formed, the new peptidyl–tRNA, which now occupies the A site, is translocated to the P site so that a new aminoacyl–tRNA can enter the A site. The uncharged tRNA in the P site moves to the E site before it dissociates from the ribosome (the details of chain elongation are described in Section 26-4B).

3. *Ribosomes read mRNA in the 5′ → 3′ direction.* This was shown through the use of a cell-free protein-synthesizing system in which the mRNA was poly(A) with a 3′-terminal C:

$$5' \ A—A—A—A— \ \cdots \ —A—A—A—C \ 3'$$

Such a system synthesizes a poly(Lys) that has a C-terminal Asn:

$$H_3N^+–Lys—Lys—Lys— \ \cdots \ —Lys—Lys—Asn–COO^-$$

Together with the knowledge that AAA and AAC code for Lys and Asn (Table 26-1) and the polarity of peptide synthesis, this establishes that the mRNA is read in the 5′ → 3′ direction. Because mRNA is also synthesized in the 5′ → 3′ direction, prokaryotic ribosomes can commence translation as soon as a nascent mRNA emerges from RNA polymerase (Fig. 26-24). This, however, is not possible in eukaryotes because the nuclear membrane separates the site of tran-

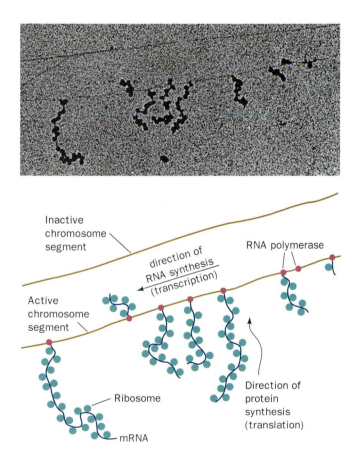

Figure 26-24. The simultaneous transcription and translation of an E. coli gene. The electron micrograph and its interpretive drawing show RNA polymerase molecules transcribing the DNA from right to left while ribosomes are translating the nascent RNAs (mostly from bottom to top). [Courtesy of Oscar L. Miller, Jr., and Barbara A. Hamkalo, University of Virginia.]

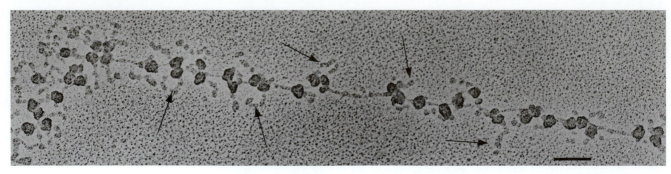

Figure 26-25. Electron micrograph of polysomes from silk gland cells of the silkworm Bombyx mori. The 3′ end of the mRNA is on the left. Arrows point to the silk fibroin polypeptides. The bar represents 0.1 μm. [Courtesy of Oscar L. Miller, Jr. and Steven L. McKnight, University of Virginia.]

scription (the nucleus) from the site of translation (the cytosol). Since most prokaryotic mRNAs are enzymatically degraded within 1 to 3 min of their synthesis (which eliminates the wasteful synthesis of unneeded proteins after a change in conditions), the 5′ ends of some mRNAs are degraded before their 3′ ends have been synthesized.

4. *Active translation occurs on polysomes.* In both prokaryotes and eukaryotes, multiple ribosomes can bind to a single mRNA transcript, giving rise to a beads-on-a-string structure called a **polyribosome** (**polysome;** Fig. 26-25). Individual ribosomes are separated by gaps of 50 to 150 Å. Polysomes arise because once an active ribosome has cleared its initiation site on mRNA, a second ribosome can initiate translation at that site.

A. Chain Initiation

The first indication of how ribosomes initiate polypeptide synthesis was the observation that almost half of the *E. coli* proteins begin with the otherwise uncommon amino acid Met. In fact, the tRNA that initiates translation is a peculiar form of Met–tRNA^{Met} in which the Met residue is *N*-formylated:

$$\text{HC}\overset{\displaystyle \text{O}}{\underset{\displaystyle \|}{}}-\text{NH}-\text{CH}-\overset{\displaystyle \text{O}}{\underset{\displaystyle \|}{\text{C}}}-\text{O}-\text{tRNA}_f^{Met}$$

with side chain $-\text{CH}_2-\text{CH}_2-\text{S}-\text{CH}_3$

***N*-Formylmethionine–tRNA$_f^{Met}$**
(fMet–tRNA$_f^{Met}$)

Because the ***N*-formylmethionine** residue **(fMet)** already has an amide bond, it can only be the N-terminal residue of a polypeptide. *E. coli* proteins are posttranslationally modified by deformylation of their fMet residue and, in many proteins, by the subsequent removal of the resulting N-terminal Met. This processing usually occurs on the nascent polypeptide, which accounts for the observation that mature *E. coli* proteins all lack fMet.

● See Guided Exploration 28:

Translational Initiation.

The tRNA that recognizes the initiation codon, **tRNA$_f^{Met}$**, differs from the tRNA that carries internal Met residues, **tRNA$_m^{Met}$**, although they both recognize the same AUG codon. Presumably, the conformations of these tRNAs are different enough to permit them to be distinguished in the reactions of chain initiation and elongation.

In *E. coli*, uncharged tRNA$_f^{Met}$ is aminoacylated with Met by the same MetRS that charges tRNA$_m^{Met}$. The resulting Met–tRNA$_f^{Met}$ is specifically *N*-formylated to yield fMet–tRNA$_f^{Met}$ by a transformylase that employs N^{10}-formyltetrahydrofolate (Section 20-4D) as its formyl donor. This transformylase does not recognize Met–tRNA$_m^{Met}$.

Base Pairing between mRNA and the 16S rRNA Helps Select the Translation Initiation Site

AUG codes for internal Met residues as well as the initiating Met residue of a polypeptide. Moreover, mRNAs usually contain many AUGs (and GUGs) in different reading frames. Clearly, *a translation initiation site must be specified by more than just an initiation codon.*

In *E. coli*, the 16S rRNA contains a pyrimidine-rich sequence at its 3′ end. This sequence, as John Shine and Lynn Dalgarno pointed out in 1974, is partially complementary to a purine-rich tract of 3 to 10 nucleotides, the **Shine–Dalgarno sequence,** that is centered ~10 nucleotides upstream from the start codon of nearly all known prokaryotic mRNAs (Fig. 26-26). *Base pairing interactions between an mRNA's Shine–Dalgarno sequence and the 16S rRNA apparently permit the ribosome to select the proper initiation codon.*

Initiation Requires Soluble Protein Factors

Translation initiation in *E. coli* is a complex process in which the two ribosomal subunits and fMet–tRNA$_f^{Met}$ assemble on a properly aligned mRNA to form a complex that can commence chain elongation. This

Initiation
codon

araB	– U U U G G A U G G A G U G A A A C G A U G G C G A U U –
galE	– A G C C U A A U G G A G C G A A U U A U G A G A G U U –
lacI	– C A A U U C A G G G U G G U G A U U G U G A A A C C A –
lacZ	– U U C A C A C A G G A A A C A G C U A U G A C C A U G –
Q β phage replicase	– U A A C U A A G G A U G A A A U G C A U G U C U A A G –
φX174 phage A protein	– A A U C U U G G A G G C U U U U U U A U G G U U C G U –
R17 phage coat protein	– U C A A C C G G G G G U U U G A A G C A U G G C U U C U –
Ribosomal S12	– A A A A C C A G G A G C U A U U U A A U G G C A A C A –
Ribosomal L10	– C U A C C A G G A G C A A A G C U A A U G G C U U U A –
trpE	– C A A A A U U A G A G A A U A A C A A U G C A A A C A –
trp leader	– G U A A A A A G G G U A U C G A C A A U G A A A G C A –

3′ end of 16S rRNA	3′ $_{HO}$A U U C C U C C A C U A G – 5′

Figure 26-26.   Some translation initiation sequences recognized by *E. coli* ribosomes. The RNAs are aligned according to their initiation codons (*blue shading*). Their Shine–Dalgarno sequences (*red shading*) are complementary, counting G·U pairs, to a portion of the 16S rRNA's 3′ end (*below*). [After Steitz, J.A., in Chambliss, G., Craven, G.R., Davies, J., Davis, K., Kahan, L., and Nomura, M. (Eds.), *Ribosomes. Structure, Function and Genetics*, pp. 481–482, University Park Press (1979).]

Table 26-6.   **The Soluble Protein Factors of *E. coli* Protein Synthesis**

Factor	Mass (kD)	Function
Initiation Factors		
IF-1	9	Assists IF-3 binding
IF-2	97	Binds initiator tRNA and GTP
IF-3	22	Releases 30S subunit from inactive ribosome and aids mRNA binding
Elongation Factors		
EF-Tu	43	Binds aminoacyl–tRNA and GTP
EF-Ts	74	Displaces GDP from EF-Tu
EF-G	77	Promotes translocation by binding GTP to the ribosome
Release Factors		
RF-1	36	Recognizes UAA and UAG Stop codons
RF-2	38	Recognizes UAA and UGA Stop codons
RF-3	46	Binds GTP and stimulates RF-1 and RF-2 binding

process also requires **initiation factors** that are not permanently associated with the ribosome, designated **IF-1, IF-2,** and **IF-3** in *E. coli* (Table 26-6).

Translation initiation in *E. coli* occurs in three stages (Fig. 26-27):

1. On completing a cycle of polypeptide synthesis, the 30S and 50S subunits remain associated as an inactive 70S ribosome. IF-3 binds to the 30S subunit so as to promote the dissociation of this complex. IF-1 increases this dissociation rate, perhaps by assisting in the binding of IF-3.

2. mRNA and IF-2 in a ternary complex with GTP and fMet–tRNA$_f^{Met}$ subsequently bind to the 30S subunit in either order. Since the ternary complex containing fMet–tRNA$_f^{Met}$ can bind to the ribosome before mRNA, fMet–tRNA$_f^{Met}$ binding must not be mediated by a codon–anticodon interaction; it is the only tRNA–ribosome association that does not require one, although this interaction helps binds fMet–tRNA$_f^{Met}$ to the ribosome. IF-3 also functions in this stage of the initiation process by helping bind the 30S subunit to the mRNA.

3. In a process that is preceded by IF-3 release, the 50S subunit joins the 30S initiation complex in a manner that stimulates IF-2 to hydrolyze its bound GTP to GDP + P$_i$. This irreversible reaction conformationally rearranges the 30S subunit and releases IF-1 and IF-2 for participation in further initiation reactions.

Initiation results in the formation of an fMet–tRNA$_f^{Met}$ · mRNA · ribosome complex in which the fMet–tRNA$_f^{Met}$ occupies the ribosome's P site while its A site is poised to accept an incoming aminoacyl–tRNA. Note that tRNA$_f^{Met}$ is the only tRNA that directly enters the P site. All other tRNAs must first enter the A site during chain elongation.

Eukaryotic Initiation Resembles That of Prokaryotes

Eukaryotic initiation resembles the overall prokaryotic process but differs from it in detail. Eukaryotes have a far more extensive "zoo" of initiation factors (designated eIF-*n;* "e" for eukaryotic) than do prokaryotes. One of these, **eIF-2,** participates in the second step of initiation when it forms a complex with GTP and the initiator tRNA and binds to the 40S ribosomal subunit along with mRNA. The initiator tRNA is **Met–tRNA$_i^{Met}$**

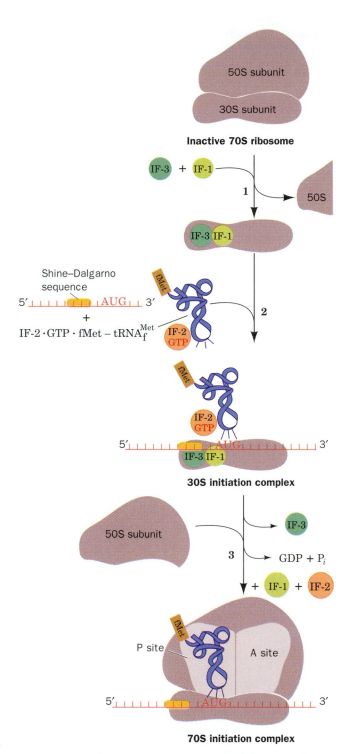

Figure 26-27. **The translation initiation pathway in** ***E. coli* ribosomes.**

(here the subscript "i" indicates the initiator tRNA^{Met}). Its appended Met residue is not formylated as in prokaryotes. Nevertheless, both species of initiator tRNAs are readily interchangeable *in vitro*.

Eukaryotic mRNAs lack the complementary sequences to bind 18S rRNA in the Shine–Dalgarno manner. Rather, *translation of eukaryotic mRNAs, which are invariably monocistronic, almost always starts at their first AUG.* Since **eIF-2** is a **cap-binding protein,** the 40S subunit may bind at or near the eukaryotic mRNA's 5′ cap (Section 25-3A) and migrate downstream until it encounters the first AUG. This hypothesis explains the greatly reduced translation initiation rates of improperly capped mRNAs.

B. Chain Elongation

Ribosomes elongate polypeptide chains in a three-stage reaction cycle (Fig. 26-28). This process, which occurs at a rate of up to 40 residues per second, requires several nonribosomal proteins known as **elongation factors** (Table 26-6).

1. Aminoacyl–tRNA Binding

In the "binding" stage of the *E. coli* elongation cycle, a binary complex of GTP and the elongation factor **EF-Tu** combines with an aminoacyl–tRNA. The resulting ternary complex binds to the ribosome. Binding of the aminoacyl–tRNA in a codon–anticodon complex to the ribosomal A site is accompanied by the hydrolysis of GTP to GDP so that EF-Tu·GDP and P_i are released. The EF-Tu·GTP complex is regenerated when GDP

● See Guided Exploration 29:

Translational Elongation.

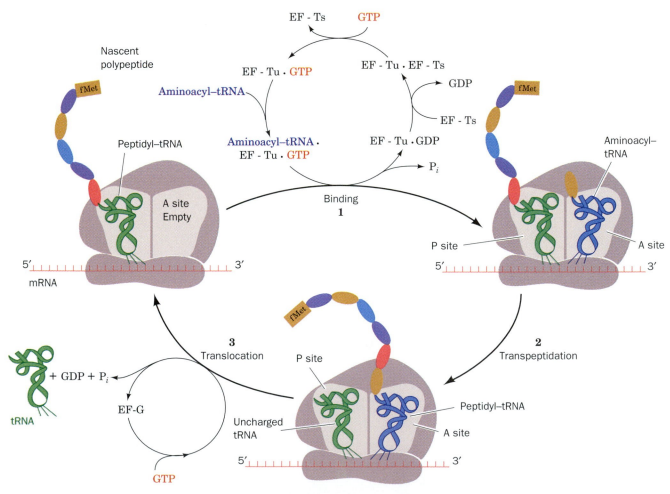

Figure 26-28. *Key to Function.* **The elongation cycle in *E. coli* ribosomes.** The E site is not shown. Eukaryotic elongation follows a similar cycle, but EF-Tu and EF-Ts are replaced by a single multisubunit protein, eEF-1, and EF-G is replaced by eEF-2.

is displaced from EF-Tu · GDP by the elongation factor **EF-Ts,** which, in turn, is displaced by GTP.

Aminoacyl–tRNAs can bind to the ribosomal A site without EF-Tu but at a rate too slow to support cell growth. The importance of EF-Tu is indicated by the fact that it is the most abundant *E. coli* protein; it is present in ~100,000 copies per cell (>5% of the cell's protein), which is approximately the number of tRNA molecules in the cell. Consequently, *the cell's entire complement of aminoacyl–tRNAs is essentially sequestered by EF-Tu.*

Morten Kjeldgaard and Jens Nyborg determined the X-ray structures of bacterial EF-Tu in complex with GDP and in complex with the slowly hydrolyzing GTP analog **guanosine-5′-(β,γ-imido)triphosphate (GDPNP).**

Guanosine-5′-(β,γ-imido)triphosphate
(GDPNP)

EF-Tu folds into three distinct domains that are connected by flexible peptides. The N-terminal domain 1, which binds GTP/GDP and catalyzes GTP hydrolysis, structurally resembles other known GTP-binding proteins. Comparison of the GDPNP and GDP complexes (Fig. 26-29) indicates that, on hydrolyzing GTP, EF-Tu undergoes a major structural reorganization in which domain 1 changes its orientation with respect to domains 2 and 3 by ~91°.

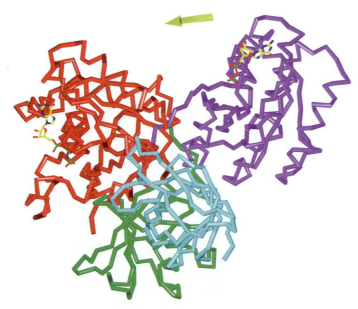

Figure 26-29. **The superposition of the X-ray structures of ribosomal elongation factor EF-Tu in its complexes with GDP and GDPNP.** The protein is represented by its C_α backbone with domain 1, its GTP-binding domain, magenta in the GDP complex and red in the GDPNP complex. Domains 2 and 3, which have the same orientation in both complexes, are green and cyan. The bound GDP and GDPNP are shown in skeletal form with C yellow, N blue, O red, and P green. [Courtesy of Morten Kjeldgaard and Jens Nyborg, Aarhus University, Århus, Denmark.]

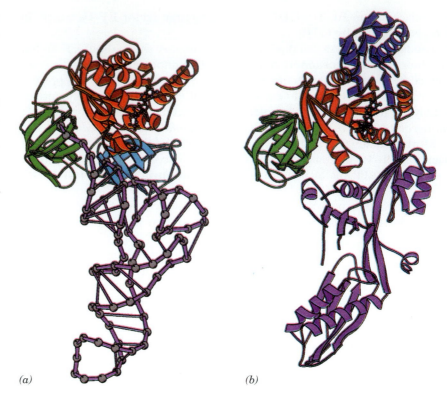

Figure 26-30.   **A comparison of the ternary structures of EF-Tu·tRNA and EF-G.** (*a*) In EF-Tu, domain 1 is red, domain 2 is green, and domain 3 is blue. The tRNA backbone of yeast tRNA^{Phe} is shown in purple. (*b*) In EF-G, domain 1 is red, domain 2 is green, an extra domain is dark blue, and domains 3, 4, and 5 are purple. A 25-residue segment of EF-G's domain 1 is not visible in its X-ray structure and domain 3 is poorly defined, thereby accounting for the several polypeptide segments seen in this model. Courtesy of Jens Nyborg, Aarhus University, Århus, Denmark.] ● See the Interactive Exercises.

(*a*)

(*b*)

The X-ray structure of the Phe–tRNA^{Phe}·EF-Tu·GDPNP ternary complex (Fig. 26-30*a*) reveals that the two macromolecules associate to form a corkscrew-shaped complex in which the EF-Tu and the tRNA's acceptor stem form a knoblike handle and the tRNA's anticodon helix forms the screw. The conformations of the two macromolecules closely resemble those seen in the X-ray structures of EF-Tu·GDPNP (Fig. 26-29) and un-complexed tRNA^{Phe} (Fig. 26-6*b*). The macromolecules appear to associate rather tenuously: The 3'-CCA–Phe segment of the Phe–tRNA^{Phe} binds in the cleft between domains 1 and 2 of EF-Tu·GDPNP (red and green in Fig. 26-30*a*); the 5'-phosphate of the tRNA binds in a depression at the junction of EF-Tu's three domains; and one side of the TψC stem of the tRNA makes contacts with the exposed main chain and side chains of EF-Tu domain 3 (blue in Fig. 26-30*a*). Presumably, all species of elongator tRNAs make these contacts with EF-Tu.

The X-ray structure of the ternary complex suggests why EF-Tu binds all aminoacylated tRNAs except initiator tRNAs, but never uncharged tRNAs. Evidently, the tight association of the aminoacyl group with EF-Tu greatly increases the affinity of EF-Tu for the otherwise loosely bound tRNA. What would be the first base pair in elongator tRNAs (e.g., G1·C72 in tRNA^{Phe}; Fig. 26-6*a*) is mismatched (C·A) in tRNA$_f^{Met}$, and hence this initiator tRNA has a 3' overhang of 5 nt versus 4 nt for elongator tRNAs. This, together with the formyl group on its appended fMet residue, apparently prevents fMet–tRNA$_f^{Met}$ from binding to EF-Tu, thereby explaining why the initiator tRNA never translates internal AUG codons.

2. Transpeptidation

In the second stage of the elongation cycle shown in Fig. 26-28, a peptide bond is formed through the nucleophilic displacement of the P-site tRNA by the amino group of the 3'-linked aminoacyl–tRNA in the A site (Fig. 26-23). This reaction requires that the tRNAs in the A and P sites be arranged so that their acceptor stems are close together. The anticodons of the two tRNAs must also be arranged in side-by-side fashion so that they can bind to consecutive codons in mRNA. X-Ray crystallographic models of ribosomes with bound tRNAs, such as the one shown in Fig. 26-21, indicate that this is the case.

The nascent polypeptide chain is lengthened at its C-terminus by one residue and transferred to the A-site tRNA, a process called **transpeptidation.** The reaction occurs without the need of activating cofactors such as ATP because the ester linkage between the nascent polypeptide and the P-site tRNA is a "high-energy" bond.

Peptide bond formation is catalyzed by the 23S rRNA of the large ribosomal subunit. Considerable evidence had suggested that the peptidyl transferase was indeed a ribozyme, and the identification of the active site was confirmed by X-ray analysis of the ribosome with a bound analog of the reaction intermediate. The ribosome's active site lies in a highly conserved area of the 23S rRNA and, most significantly, the nearest protein side chain is ~18 Å away from the newly formed peptide bond—too far to be involved in catalysis.

Catalytic activity appears to be centered on an invariant adenine residue whose N3 atom is positioned ~3 Å from the new peptide bond. Normally, adenine is too acidic ($pK < 3.5$) to enhance the nucleophilicity of the amino group of the amino acid attached to the tRNA. In the ribosome, the pK of the adenine is perturbed to ~7.6, possibly due to the influence of a buried phosphate group that is hydrogen bonded to the adenine via an intervening guanine residue. This possible hydrogen-bonded relay system is similar to the catalytic triad of serine proteases (see Section 11-5), which carries out the reverse reaction, peptide bond hydrolysis. According to one model of ribosome action, N3 of adenine acts as a general base to abstract a proton from the amino group of the aminoacyl–tRNA as it attacks the carbonyl carbon of the peptidyl–tRNA (see reaction at right). The protonated adenine then stabilizes the tetrahedral reaction intermediate before it donates the proton to the oxyanion leaving group of the P-site tRNA.

3. Translocation

Formation of a peptide bond causes the acceptor end of the new peptidyl–tRNA to shift from the A site to the P site while its anticodon end remains in the A site (Fig. 26-31). The acceptor end of the tRNA formerly in the P site moves to the E site while its anticodon end remains in the P site. This structure is resolved in the final stage of the elongation cycle, a process known as **translocation,** in which the peptidyl–tRNA, together with its bound RNA, moves entirely to the P site (thereby vacating the A site) and the uncharged tRNA moves fully into the E site. This prepares the ribosome for the next elongation cycle. The maintenance of the peptidyl–tRNA's codon–anticodon association acts as a place-keeper that permits the ribosome to translocate by exactly three nucleotides along the mRNA so as to preserve the reading frame. Nevertheless, instances of "frameshifting" during translation often occur at homopolymeric segments

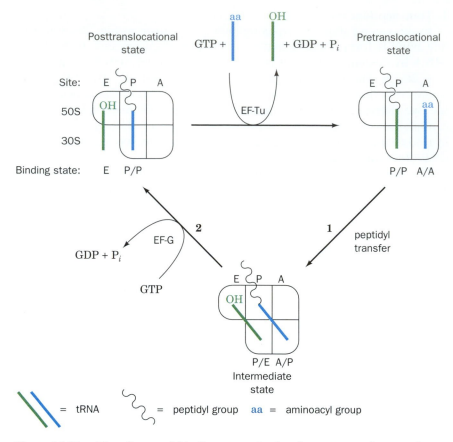

Figure 26-31.   **The ribosomal binding states in the elongation cycle.** Note how this scheme elaborates the classic elongation cycle diagrammed in Fig. 26-28. [In part after Moazed, D. and Noller, H.F., *Nature* **342**, 147 (1989).]

of mRNA, for example, poly(U), where the codon–anticodon pairing can easily slip forward or backward by one nucleotide.

The translocation process requires an elongation factor, **EF-G** in *E. coli*, that binds to the ribosome together with GTP and is released only on hydrolysis of the GTP to GDP + P$_i$ (Fig. 26-28). EF-G release is prerequisite for beginning the next elongation cycle because the ribosomal binding sites of EF-G and EF-Tu partially or completely overlap and hence their ribosomal binding is mutually exclusive. GTP hydrolysis, which precedes the translocation, provides the free energy for tRNA movement.

The X-ray structure of EF-G (Fig. 26-30*b*), determined by Thomas Steitz and Peter Moore, reveals that this protein has an elongated shape comprising five domains, the first two of which are arranged similarly to the first two domains of the EF-Tu · GDPNP complex (Fig. 26-29). Most intriguingly, the remaining three domains of EF-G have a conformation reminiscent of the shape of tRNA bound to EF-Tu (Fig. 26-30*a*). Such molecular mimicry raises the possibility that EF-G drives translocation not just by inducing a conformational change in the ribosome but by actively displacing the peptidyl–tRNA from the A site. The structural similarity between EF-G domains 3 to 5 and tRNA also suggests that in the earliest cells, whose functions were based on RNA, proteins might have evolved by mimicking shapes already used successfully by RNA.

The binding of tRNAs to the A and E sites of the ribosome exhibits negative allosteric cooperativity. At the end of an elongation cycle, the E site binds the newly deacylated tRNA with high affinity, whereas the now empty A site has low affinity for aminoacyl–tRNA. The arrival of a new aminoacyl–tRNA · EF-Tu · GTP complex induces the ribosome to undergo a conformational change that converts the A site to a high-affinity state and the E site to a low-affinity state. Thus, *the alternating activities of the GTP-hydrolyzing proteins EF-Tu and EF-G keep the ribosome cycling unidirectionally through the transpeptidation and translocation stages of translation.*

The Eukaryotic Elongation Cycle Resembles That of Prokaryotes

Elongation in eukaryotes closely resembles that in prokaryotes. In eukaryotes, the functions of EF-Tu and EF-Ts are assumed by two different subunits of the eukaryotic elongation factor **eEF-1.** Likewise, **eEF-2** functions in a manner analogous to prokaryotic EF-G. However, the corresponding eukaryotic and prokaryotic elongation factors are not interchangeable. eEF-2 is a specific target for inactivation by diphtheria toxin (see Box 26-3).

Box 26-3

BIOCHEMISTRY IN HEALTH AND DISEASE

Diphtheria Toxin Inactivates eEF-2

Diphtheria is a disease resulting from bacterial infection by *Cornyebacterium diphtheriae* that harbor the bacteriophage **cornyephage β**. Diphtheria was a leading cause of childhood death until the late 1920s, when immunization became prevalent. Although the bacterial infection is usually confined to the upper respiratory tract, the bacteria secrete a phage-encoded protein, known as **diphtheria toxin**, that is responsible for the disease's lethal effects. Diphtheria toxin specifically inactivates the eukaryotic elongation factor eEF-2.

The pathogenic effects of diphtheria are prevented, as was discovered in the 1880s, by immunization with **toxoid,** formaldehyde-inactivated toxin. Individuals who have contracted diphtheria are treated with antitoxin from horse serum, which binds to and thereby inactivates diphtheria toxin, as well as with antibiotics to combat the bacterial infection.

Diphtheria toxin is a monomeric 535-residue protein that is readily cleaved by trypsinlike enzymes, yielding two fragments, A and B, that remain linked by a disulfide bond. The B fragment binds to a specific receptor on the plasma membrane of susceptible cells, whereupon it facilitates the A fragment's entry into the cytosol via receptor-mediated endocytosis. The intracellular reducing environment then cleaves the disulfide bond linking the A and B fragments.

In the cytosol, the A fragment catalyzes the **ADP-ribosylation** of eEF-2 by NAD^+,

$$\text{eEF-2} + NAD^+$$
$$(active)$$

↓ diphtheria toxin

$$\text{ADP-ribosyl-eEF-2} + \text{Nicotinamide} + H^+$$
$$(inactive)$$

thereby inactivating the elongation factor (this reaction is similar to that catalyzed by cholera toxin; Box 21-1). Since the A fragment acts catalytically, one molecule is sufficient to ADP-ribosylate all of a cell's eEF-2, which halts protein synthesis and kills the cell. Only a few micrograms of diphtheria toxin are therefore sufficient to kill an unimmunized individual.

Diphtheria toxin specifically ADP-ribosylates a modified His residue on eEF-2 known as **diphthamide:**

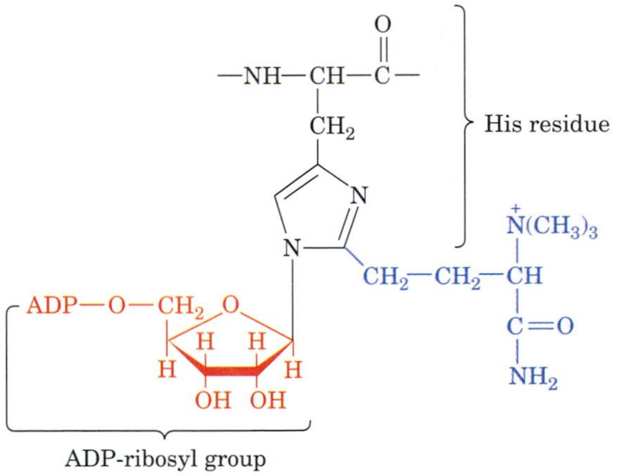

ADP-Ribosylated diphthamide

Diphthamide occurs only in eEF-2, which accounts for the specificity of diphtheria toxin. Since diphthamide occurs in all eukaryotic eEF-2's, it probably is essential to eEF-2 activity. Yet certain mutated cultured animal cells lacking the enzymes that posttranslationally modify His to diphthamide synthesize proteins normally. Perhaps the diphthamide residue has a control function.

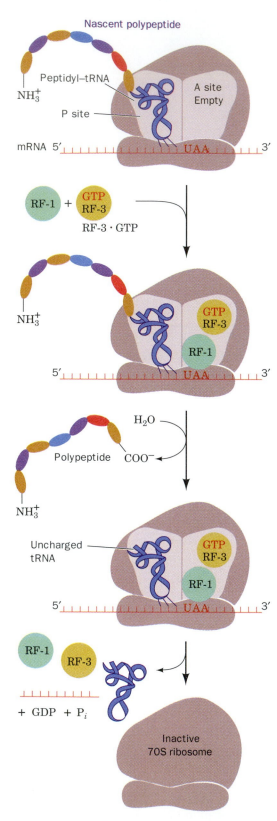

Figure 26-32.   The translation termination pathway in *E. coli* ribosomes. RF-1 recognizes the termination codons UAA and UAG, whereas RF-2 recognizes UAA and UGA.

C. Chain Termination

Polypeptide synthesis under the direction of synthetic mRNAs such as poly(U) results in a peptidyl–tRNA "stuck" in the ribosome. However, the translation of natural mRNAs, which contain the termination codons UAA, UGA, or UAG, yields free polypeptides. In *E. coli*, the termination codons, which normally have no corresponding tRNAs, are recognized by protein **release factors** (Table 26-6): **RF-1** recognizes UAA and UAG, whereas **RF-2** recognizes UAA and UGA. A third release factor, **RF-3,** a GTP-binding protein, in complex with GTP, stimulates the ribosomal binding of RF-1 and RF-2 (Fig. 26-32). In eukaryotes, a single release factor, **eRF,** recognizes all three termination codons.

The binding of a release factor—rather than an aminoacyl–tRNA—to a termination codon induces the ribosomal peptidyl transferase to transfer the peptidyl group to water. The reaction products are a free polypeptide and an uncharged tRNA, which dissociate from the ribosome. The release factors are expelled with the concomitant hydrolysis of the RF-3–bound GTP to GDP + P_i. Eukaryotic eRF, which also binds GTP, is similarly released from the ribosome by GTP hydrolysis. The resulting inactive ribosome must then release its bound mRNA before it can undertake a new round of polypeptide synthesis.

Nonsense Suppressors Prevent Termination

A mutation that converts an aminoacyl-coding ("sense") codon to a stop codon is known as a **nonsense mutation** and leads to the premature termination of translation. An organism with such a mutation may be "rescued" by a second mutation in a tRNA gene that causes the tRNA to recognize a nonsense codon. This **nonsense suppressor** tRNA carries the same amino acid as its wild-type progenitor and appends it to the growing polypeptide at the stop codon, thereby preventing chain termination. For example, the *E. coli* **amber suppressor** known as *su*3 is a tRNA^{Tyr} whose anticodon has mutated from the wild-type GUA (which reads the Tyr codons UAU and UAC) to CUA (which recognizes the amber stop codon UAG). An *su*3$^+$ *E. coli* cell with an otherwise lethal amber mutation in a gene coding for an essential protein would be viable if the replacement of the wild-type amino acid residue by Tyr does not inactivate the protein.

How do cells tolerate a mutation that both eliminates a normal tRNA and prevents the termination of polypeptide synthesis? They survive because the mutated tRNA is usually a minor member of a set of isoaccepting tRNAs and because nonsense suppressor tRNAs must compete with release factors for binding to stop codons. Consequently, many suppressor-rescued mutants grow more slowly than wild-type cells.

GTP Hydrolysis Speeds up Ribosomal Processes

What is the role of the GTP hydrolysis reactions mediated by the various GTP-binding factors (IF-2, EF-Tu, EF-G, and RF-3 in *E. coli*) and which are essential for normal ribosomal function? Translocation occurs in the absence of GTP, albeit extremely slowly, so that the free energy of the transpeptidation reaction is sufficient to drive the entire translational process. Moreover, none of the GTP hydrolysis reactions yields a "high-energy" covalent intermediate as does, say, ATP hydrolysis in numerous biosynthetic reactions. It is therefore thought that the ribosomal binding of a GTP-binding factor in complex with GTP allosterically causes ribosomal components to change their conformations in a way that facilitates a particular process such as translocation (e.g., Fig. 26-29). This conformational change also catalyzes GTP hydrolysis, which, in turn, permits the ribosome to relax to its initial conformation with the concomitant release of products including GDP + P$_i$. *The high rate and irreversibility of the GTP hydrolysis reaction therefore ensures that the various complex ribosomal processes to which it is coupled—initiation, elongation, and termination— will themselves be fast and irreversible.* GTP hydrolysis also facilitates translational accuracy (see below).

D. Translational Accuracy

Ribosomes translate mRNAs with remarkable fidelity; they incorporate wrong amino acids into polypeptides at a rate of only $\sim10^{-4}$ per residue (some antibiotics act by increasing the rate of amino acid misincorporation; see Box 26-4). Aminoacyl–tRNAs are selected by the ribosome only according to their anticodon. Yet the binding energy loss arising from a single base mismatch in a codon–anticodon interaction is estimated to be ~12 kJ·mol^{-1}, which cannot account for a ribosomal decoding accuracy of less than $\sim10^{-2}$ errors per codon. Evidently, the ribosome has some sort of proofreading mechanism that increases its overall decoding accuracy.

Aminoacyl–tRNA synthetases (Section 26-2B) improve their fidelity by a selective binding mechanism. However, no evidence has been found for such a mechanism in ribosomes. Rather, most evidence suggests that ribosomes exclude improper codon–anticodon interactions by a **kinetic proofreading** mechanism. As proposed by John Hopfield, this process operates as follows (Fig. 26-33):

1. The initial binding reaction discriminates between cognate (specified) and noncognate (unspecified) tRNAs according to their codon–anticodon binding energies.

2. Next, the GTP bound to EF-Tu irreversibly hydrolyzes, yielding an activated intermediate (indicated by the asterisk in Fig. 26-33). This complex can then react in one of two ways:

   (a) EF-Tu·GDP can dissociate from the ribosome with rate constant k_3, thereby committing the ribosome to form a peptide bond.

   (b) Alternatively, the aminoacyl–tRNA can dissociate from the ribosome with rate constant k_4, thereby aborting the elongation step. EF-Tu·GDP subsequently dissociates from the ribosome, permitting it to reinitiate the elongation step.

If the ratio of k_4/k_3 is greater for noncognate than for cognate tRNAs, then a second screening will have occurred. It is thought that the physical basis of this second screening is that k_3 is independent of the tRNA's identity, whereas k_4 is larger for a relatively weakly bound noncognate tRNA than it is for a cognate tRNA. Thus, a noncognate tRNA usually dissociates from the ribosome before EF-Tu·GDP does, whereas a cognate tRNA usually remains bound. The activated intermediate is essential for this process because otherwise its tRNA dissociation step (that characterized by k_4) would be identical to that of the initial recognition step (that characterized by k_{-1}). GTP hydrolysis is therefore the price of kinetic proofreading.

Proteins May Be Posttranslationally Modified

Even before a polypeptide has been completely synthesized, it begins to assume its mature conformation. Protein folding begins after the first 30–40 residues have been linked together, the point at which the polypeptide

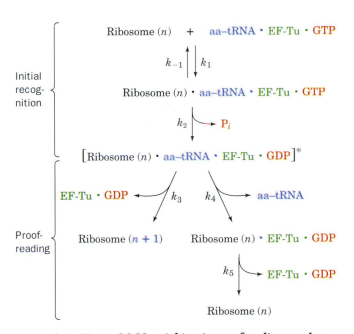

Figure 26-33. **A kinetic proofreading mechanism for selecting a correct codon–anticodon interaction.** The initial recognition reaction screens the aminoacyl–tRNA (aa–tRNA) for the correct codon–anticodon interaction. The resulting complex converts, in a GTP hydrolysis-driven process, to a "high-energy" intermediate (*) that either releases EF-Tu·GDP preparatory to forming a peptide bond or releases aminoacyl–tRNA before EF-Tu·GDP is released. If k_4/k_3 is greater for a codon–anticodon mismatch than it is for a match, then these latter steps constitute a proofreading mechanism for proper tRNA binding.

The Effects of Antibiotics on Protein Synthesis

The majority of known antibiotics, including a great variety of medically useful substances, block translation. This situation is presumably a consequence of the translation machinery's enormous complexity, which makes it vulnerable to disruption in many ways. Antibiotics have also been useful in analyzing ribosomal mechanisms because the blockade of a specific function often permits its biochemical dissection into its component steps. For example, the ribosomal elongation cycle was originally characterized through the use of the antibiotic **puromycin**, which resembles the 3′ end of Tyr–tRNA.

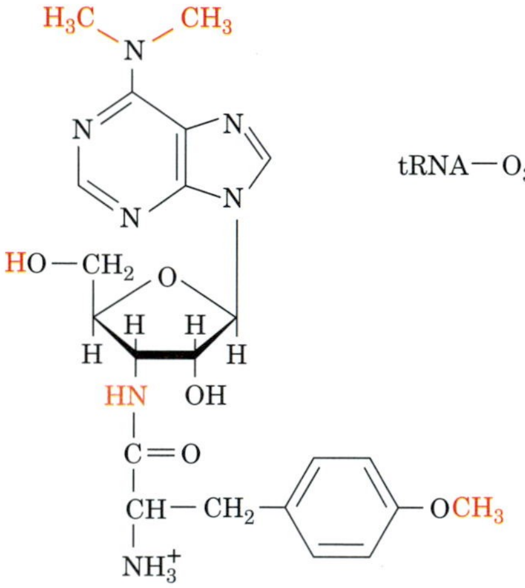

Puromycin binds to the ribosomal A site without the need of elongation factors. The transpeptidation reaction yields a peptidyl–puromycin in which puromycin's "amino acid residue" is linked to its "tRNA" via an amide rather than an ester bond. The ribosome therefore cannot catalyze further transpeptidation, and polypeptide synthesis is aborted.

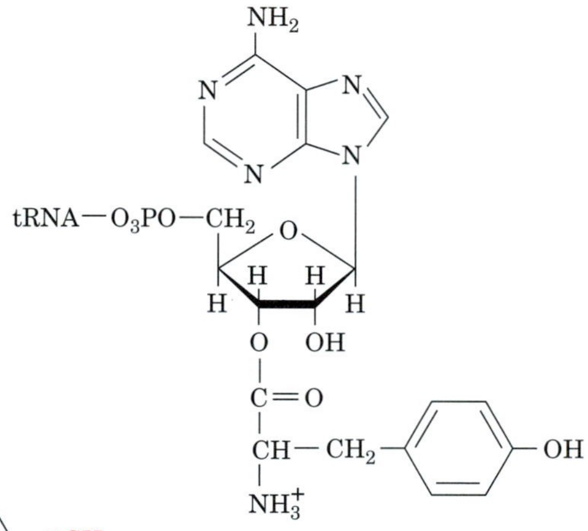

Streptomycin

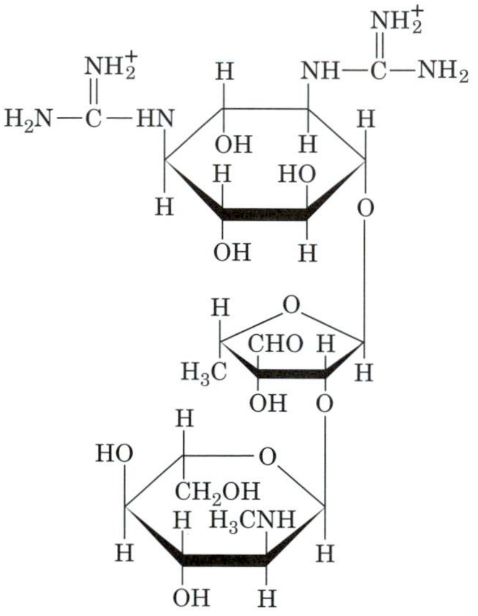

Puromycin

Tyrosyl–tRNA

Streptomycin

is a medically important member of a family of antibiotics known as **aminoglycosides** that inhibit prokaryotic ribosomes in a variety of ways. At low concentrations, streptomycin induces the ribosome to characteristically misread

chain begins to emerge from the ribosome. Molecular chaperones (Section 6-4C) may bind to newly synthesized polypeptides to facilitate their folding and association with other subunits. In addition, many proteins undergo **posttranslational modifications** that include the removal and/or derivatization of specific residues. For example, the initiating Met or fMet residue is often excised. Limited proteolysis is also required to activate proteins that are synthesized as inactive precursors, called **proproteins.** The zymogens of serine proteases (Section 11-5D) are activated in this fashion. Proteins that are translocated into the endoplasmic reticulum for export from the cell typically contain a signal peptide that must be cleaved off (Section 10-2D). Proteins bearing a signal peptide are known as **preproteins,** or **preproproteins** if they undergo additional proteolysis during their maturation.

Other common posttranslational modifications are the hydroxylation of Pro and Lys residues in collagen (Section 6-1C), phosphorylation, and glycosylation (Sections 8-3C and 15-5). Over 150 different types of side chain modifications are known; these involve all side chains except those of Ala, Gly, Ile, Leu, Met, and Val. The functions of such modifications are varied and in many cases remain enigmatic.

mRNA: One pyrimidine may be mistaken for the other in the first and second codon positions, and either pyrimidine may be mistaken for adenine in the first position. This inhibits the growth of susceptible cells but does not kill them. At higher concentrations, however, streptomycin prevents proper chain initiation and thereby causes cell death.

Chloramphenicol,

Chloramphenicol

the first of the "broad-spectrum" antibiotics, inhibits the peptidyl transferase activity of prokaryotic ribosomes. However, its clinical uses are limited to severe infections because of its toxic side effects, which are caused in part by the chloramphenicol sensitivity of mitochondrial ribosomes. Chloramphenicol appears to bind to the large subunit near the A site, since chloramphenicol competes for binding with puromycin and the 3' end of aminoacyl–tRNAs but not with peptidyl–tRNAs.

Tetracycline

Tetracycline

and its derivatives are broad-spectrum antibiotics that bind to the small subunit of prokaryotic ribosomes and inhibit aminoacyl–tRNA binding. Tetracycline also blocks the stringent response (Section 26-3) by inhibiting ppGpp synthesis. This indicates that deacylated tRNA must bind to the A site in order to induce the production of stringent factor.

Tetracycline-resistant bacterial strains have become quite common, thereby precipitating a serious clinical problem. Most often resistance is conferred by a decrease in bacterial cell membrane permeability to tetracycline rather than any alteration of ribosomal components that would overcome the inhibitory effect on translation.

SUMMARY

1. The genetic code, by which nucleic acid sequences are translated into amino acid sequences, is composed of three-nucleotide codons that do not overlap and are read sequentially by the protein-synthesizing machinery.

2. The standard genetic code of 64 codons includes numerous synonyms, three stop codons, and initiation codons.

3. All tRNAs have numerous chemically modified bases and a similar cloverleaf secondary structure comprising an acceptor stem, D arm, TψC arm, anticodon arm, and variable arm. The three-dimensional structures of tRNAs are likewise similar and are maintained by stacking interactions and non-Watson–Crick hydrogen-bonded cross-links.

4. Aminoacyl–tRNA synthetases (aaRSs) catalyze the ATP-dependent attachment of an amino acid to the appropriate tRNA. Class I and Class II aaRSs differ in their structures, their manner of tRNA recognition, and their site of aminoacylation. The acceptor stem and anticodon loop are common identity elements for tRNA–aaRS interactions. The fidelity of aminoacylation is enhanced by proofreading.

5. Wobble pairing between mRNA codons and tRNA anticodons at the third position accounts for much of the degeneracy of the genetic code. Within a species, the preferred codons correspond to the most abundant species in each set of isoaccepting tRNAs.

6. Ribosomes are large complexes of several RNA molecules and numerous proteins. They catalyze peptide bond formation between an incoming aminoacyl–tRNA and a tRNA to which the growing polypeptide is attached (the peptidyl–tRNA) according to the codon sequence of a bound mRNA. Ribosomal proteins and RNA self-assemble to form two unequal subunits of highly irregular structure, the small subunit (30S in *E. coli* and 40S in rat liver) and the large subunit (50S in *E. coli* and 60S in rat liver) which, in turn, associate to form the ribosome (70S in *E. coli* and 80S in rat liver).

7. Polypeptides are synthesized from the N-terminus to the C-terminus. Multiple ribosomes attached to a single mRNA constitute a polysome. In prokaryotes, transcription and translation occur simultaneously.

8. Translation is divided into initiation, elongation, and termination phases, each of which requires specific translation factors whose action is driven by GTP hydrolysis. Translation initiation requires an initiating tRNA charged with fMet (in prokaryotes) or Met (in eukaryotes). The ribosomal subunits assemble with the initiating tRNA, an initiation factor in complex with GTP, and mRNA. In prokaryotes, the mRNA's initiation site is selected through the binding of a Shine–Dalgarno sequence on the mRNA to complementary sequences on the 16S rRNA.

9. During polypeptide chain elongation, an aminoacyl–tRNA

in complex with a GTP-binding elongation factor enters the ribosome's A site. The amino group attacks the peptidyl group at its ester attachment site on the tRNA in the ribosome's P site in a transpeptidation reaction catalyzed by rRNA. The new peptidyl–tRNA is translocated from the A site to the P site, and the deacylated tRNA is translocated from the P to the previously vacated E site in preparation for another round of elongation.

10. Translation termination requires release factors that recognize stop codons.

11. The accuracy of protein synthesis is enhanced by kinetic proofreading at the codon–anticodon binding step of translation. The polypeptide product of translation may be modified by proteolysis or derivatization.

REFERENCES

The Genetic Code and tRNA

Arnaz, J.G. and Moras, D., Structural and functional considerations of the aminoacylation reaction, *Trends Biochem. Sci.* **22**, 211–216 (1997).

Björk, G.R., Ericson, J.U., Gustafsson, C.E.D., Hagervall, T.G., Jösson, Y.H., and Wikström, P.M., Transfer RNA modification, *Annu. Rev. Biochem.* **56**, 263–287 (1987).

Cusack, S., Aminoacyl-tRNA synthetases, *Curr. Opin. Struct. Biol.* **7**, 881–889 (1997). [Summarizes the features of 14 known aaRS structures.]

Fox, T.D., Natural variation in the genetic code. *Annu. Rev. Genet.* **21**, 67–91 (1987).

Jiminez-Sanchez, A., On the origin and evolution of the genetic code, *J. Mol. Evol.* **41**, 712–716 (1995).

Judson, J.F., *The Eighth Day of Creation,* Part II, Simon & Schuster (1979). [A fascinating historical narrative on the elucidation of the genetic code.]

Saks, M.E., Sampson, J.R., and Abelson, J.N., The transfer identity problem: a search for rules, *Science* **263**, 191–197 (1994). [Discusses the determinants of specificity in aminoacyl–tRNA synthetases.]

Stadtman, T.C., Selenocysteine, *Annu. Rev. Biochem.* **65**, 83–100 (1996).

Ribosome Structure and Function

Abel, K. and Jurnak, F., A complex profile of protein elongation: translating chemical energy into molecular movement, *Structure* **4**, 229–238 (1996).

Agrawal, R.K., Penczek, P., Grassucci, R.A., Li, Y., Leith, A., Nierhaus, K.H., and Frank, J., Direct visualization of A-, P-, and E-site transfer RNAs in the *Escherichia coli* ribosome, *Science* **271**, 1000–1002 (1996).

Ban, N., Freeborn, B., Nissen, P., Penczek, P., Grassucci, R.A., Sweet, R., Frank, J., Moore, P.B., and Steitz, T.A., The 9 Å resolution X-ray crystallographic map of the large ribosomal subunit, *Cell* **93**, 1105–1115 (1998).

Burgess, S.M. and Guthrie, C., Beat the clock: paradigms for NTPases in the maintenance of biological fidelity, *Trends Biochem. Sci.* **18**, 381–390 (1993). [Discusses kinetic proofreading.]

Clark, B.F.C. and Nyborg, J., The ternary complex of EF-Tu and its role in protein biosynthesis, *Curr. Opin. Struct. Biol.* **7**, 110–116 (1997).

Frank, J., The ribosome at higher resolution—the donut takes shape, *Curr. Opin. Struct. Biol.* **7**, 266–272 (1997).

Green, R. and Noller, H.F., Ribosomes and translation, *Annu. Rev. Biochem.* **66**, 679–716 (1997).

Gualerzi, C.O. and Pon, C.L., Initiation of mRNA translation in prokaryotes, *Biochemistry* **29**, 5881–5889 (1990).

Moore, P.B., The three-dimensional structure of the ribosome and its components, *Annu. Rev. Biophys. Biomol. Struct.* **27**, 35–58 (1998).

Stansfield, I., Jones, K.M., and Tuite, M.F., The end is in sight: terminating translation in eukaryotes, *Trends Biochem. Sci.* **201**, 489–491 (1995).

Woodson, S.A. and Leontis, N.B., Structure and dynamics of ribosomal RNA, *Curr. Opin. Struct. Biol.* **8**, 254–300 (1998).

KEY TERMS

genetic code
degeneracy
suppressor
reading frame
frameshift mutation
codon
anticodon
synonym
stop codon
acceptor stem
D arm

anticodon arm
TψC arm
variable arm
aaRS
aminoacyl–tRNA
isoaccepting tRNA
wobble hypothesis
peptidyl–tRNA
cryoelectron microscopy
affinity labeling
RRM

stringent response
peptidyl transferase
A site
P site
E site
polysome
fMet
Shine–Dalgarno sequence
initiation factor
elongation factor

release factor
transpeptidation
translocation
nonsense suppressor
kinetic proofreading
posttranslational modification
proprotein
preprotein
preproprotein

STUDY EXERCISES

1. List the overall similarities and differences among replication, transcription, and translation.

2. Explain why codons must consist of at least three nucleotides.

3. Describe the major structural features of tRNA.

4. Summarize the differences between Class I and Class II aminoacyl–tRNA synthetases.

5. Describe the proofreading mechanisms that enhance translation fidelity.

6. Describe the functions of the three tRNA-binding sites in the ribosome.

7. Summarize the roles of initiation, elongation, and release factors in translation.

8. Which aspects of ribosomal function require free energy input?

PROBLEMS

1. Could a single nucleotide deletion restore the function of a protein-coding gene interrupted by the insertion of a 4-nt sequence? Explain.

2. List all possible codons present in a ribonucleotide polymer containing U and G in random sequence. Which amino acids are encoded by this RNA?

3. Which amino acids are specified by codons that can be changed to an amber codon by a single point mutation?

4. Explain why the translation of a given mRNA can be inhibited by a segment of its complementary sequence, a so-called **antisense RNA.**

5. TyrRS selects its target amino acid by forming a specific hydrogen bond with the Tyr —OH group. Why is a separate proofreading function not required to prevent the formation of Phe–tRNA^{Tyr}?

6. Explain the significance of the observation that peptides such as fMet-Leu-Phe "activate" the phagocytic (particle-engulfing) functions of mammalian leukocytes (white blood cells).

7. How many types of macromolecules must be minimally contained in a cell-free protein-synthesizing system from *E. coli*? Count each type of ribosomal component as a different macromolecule.

8. Explain why prokaryotic ribosomes can translate a circularized mRNA molecule, whereas eukaryotic ribosomes normally cannot, even in the presence of the required cofactors.

9. Calculate the energy required, in ATP equivalents, to synthesize a 100-residue protein from free amino acids in *E. coli* (assume no ribosomal proofreading occurs).

10. Design an mRNA with the necessary prokaryotic control sites that codes for the octapeptide Lys-Pro-Ala-Gly-Thr-Glu-Asn-Ser.

11. A double-stranded fragment of viral DNA, one of whose strands is shown below, encodes two peptides, called *vir-1* and *vir-2*. Adding this double-stranded DNA fragment to an *in vitro* transcription and translation system yields peptides of 10 residues (*vir-1*) and 5 residues (*vir-2*).

    AGATCGGATGCTCAACTATATGTGATTAACAGAG-
    CATGCGGCATAAATCT

    (a) Identify the DNA sequence that encodes each peptide.

    (b) Determine the amino acid sequence of each peptide.

    (c) In a mutant viral strain, the T at position 23 has been replaced with G. Determine the amino acid sequences of the two peptides encoded by the mutant virus.

    [Problem by Bruce Wightman, Muhlenberg College.]

12. Shown below is the sequence of the sense strand of a mammalian gene. Determine the sequences of the mature RNA and the encoded protein. Assume that transcription initiates approximately 25 bp downstream of the TATAAT sequence, that each 5′ splice site has the sequence AG/GUAAGU, and that each 3′ splice site has the sequence AGG/N, where N stands for any of the four RNA bases and the / marks the location of the splice.

    TATAATACGCGCAATACAATCTACAGCTTCGCGTA
    AATCGTAGGTAAGTTGTAATAAATATAAGTGAGT
    ATGATAGGGCTTTGGACCGATAGATGCGACCCTG
    GAGGTAAGTATAGATAATTAAGCACAGGCATGCA
    GGGATATCCTCCAAATAGGTAAGTAACCTTACGG
    TCAATTAATTAGGCAGTAGATGAATAAACGATAT
    CGATCGGTTAGGTAAGTCTGAT

    [Problem by Bruce Wightman, Muhlenberg College.]

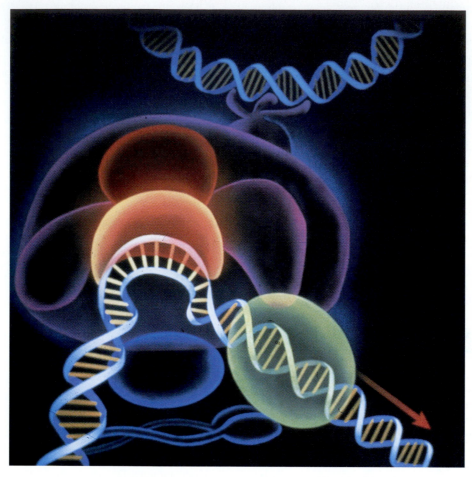

The controlled expression of genetic information requires a multitude of factors that interact specifically and nonspecifically with DNA. This painting schematically diagrams transcription factors in the preinitiation complex whose formation in eukaryotes must precede the transcription of DNA to mRNA. [Figure copyrighted © by Irving Geis.]

REGULATION OF GENE EXPRESSION

The faithful replication of DNA ensures that all the descendants of a single cell contain virtually identical sets of genetic instructions. Yet individual cells may differ—sometimes dramatically—from their progenitors, depending on how those instructions are read. The **expression** of genetic information in a given cell or organism, that is, the synthesis of RNA and proteins specified by the DNA sequence, is neither random nor fully preprogrammed. Rather, *the information in an organism's genome must be tapped in an orderly fashion during development and yet must also be available to direct the organism's responses to changes in internal or external conditions.*

Much of the mystery surrounding the regulation of gene expression has to do with how genetic information is organized and how it can be located and accessed on an appropriate time scale. The complexity of the mechanisms for transcribing and translating genes suggests the potential for even more complicated systems for enhancing or inhibiting these processes. Indeed, we have already seen numerous examples of how accessory protein factors influence the rates or specificities of transcription and translation. In this chapter, we consider some additional aspects of gene expression by examining a variety of strategies used by prokaryotes and eukaryotes to control how genetic information specifies cell structures and metabolic functions.

1. GENOME ORGANIZATION

Genomics, the study of organisms' genomes, was established as a discipline with the advent of techniques for rapidly sequencing enormous tracts of DNA, such as make up the chromosomes of living things. Studies that once relied on the hybridization of oligonucleotide probes to identify genes can now be conducted by comparing nucleotide sequences deposited in databases (Section 5-3D). In fact, a computerized database is the only practical format for storing the vast amount of information provided by a whole-genome sequence. Even a simplified diagram of the genome of the relatively simple bacterium *Helicobacter pylori* reveals a bewildering array of genes (Fig. 27-1). Nevertheless, the ability to sequence and map the entire

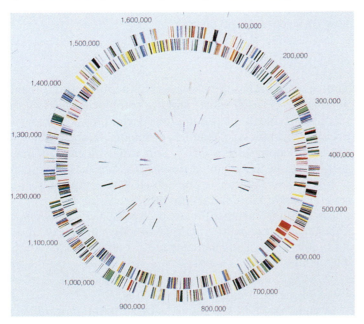

Figure 27-1. A map of the 1670-kb *H. pylori* circular chromosome. The 1590 predicted protein-coding sequences (91% of the genome) are indicated in different colors in the outermost ring, which represents the (+) strand of the DNA, and in the second ring, which represents the (−) strand. The third and fourth rings indicate intervening sequence elements and other repeating sequences (2.3% of the genome). The fifth and sixth rings show the genes for tRNAs, rRNAs, and other small RNAs (0.7% of the genome). Intergenic sequences account for 6% of the genome. [Courtesy of J. Craig Venter, The Institute for Genomic Research, Rockville, Maryland.]

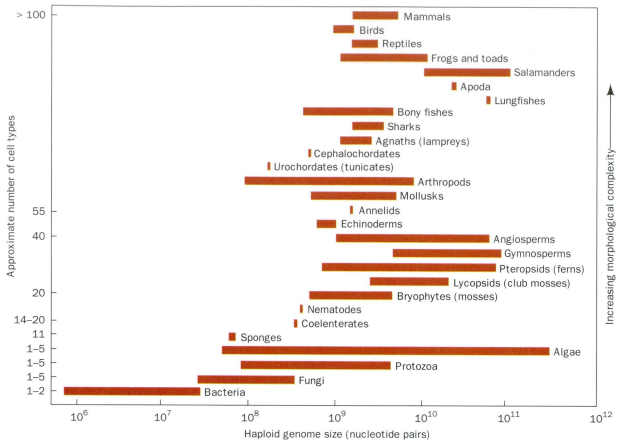

Figure 27-2. **The range of haploid genome DNA contents in various categories of organisms.** The morphological complexity of the organisms, as estimated from their number of different cell types, increases from bottom to top. Haploid genome size roughly correlates with complexity; exceptions are described by the C-value paradox. [After Raff, R.A. and Kaufman, T.C., *Embryos, Genes, and Evolution*, p. 314, Macmillan (1983).]

genome of an organism makes it possible to draw far-reaching conclusions about how many genes an organism contains, how they are organized, and how they serve the organism.

A. Gene Number

The rough correlation between the quantity of an organism's unique genetic material (its **C value**) and the complexity of its morphology and metabolism (Table 3-4) has numerous exceptions, known as the **C-value paradox.** For example, the genomes of lungfishes are 10 to 15 times larger than those of mammals (Fig. 27-2). Some algae have genomes 10 times larger still. We know that much of this "extra" DNA is unexpressed, but its function is largely a matter of conjecture. The complete genomic sequences of several prokaryotes and eukaryotes, including data from some large complicated genomes, nevertheless indicate that *the apparent number of genes, like the overall quantity of DNA, roughly parallels the organism's complexity* (Table 27-1). Thus, humans have ~40,000 genes compared to *E. coli*'s 4288 genes.

What Is the Minimum Number of Genes?

A comparison of the genomes of *Haemophilus influenzae* (1743 genes) and *Mycoplasma genitalium* (470 genes), both sequenced by J. Craig Ven-

Table 27-1. **Estimated Number of Protein-Coding Genes in Various Organisms**

Organism	Estimated Number of Genes
Prokaryotes	
Mycoplasma genitalium	470
Helicobacter pylori	1590
Methanococcus jannaschii	1738
Haemophilus influenzae	1743
Escherichia coli	4288
Eukaryotes	
Saccharomyces cerevisiae	5885
Caenorhabditis elegans	19,000
Zea mays (corn)	30,000
Homo sapiens	40,000

ter, reveals that they have 233 genes in common. These organisms, which are human parasites, lack some genes needed for central biochemical pathways (they rely on their host to supply certain intermediates). "Adding" these genes back and "subtracting" the genes necessary for parasitism gives an estimate of 250 for the minimum number of genes required to sustain cellular life. Since many of these genes appear to be related, the ancestral gene set was presumably even smaller. Of course, evolution, which is constrained by its history, does not necessarily follow the principle of greatest possible efficiency.

Identifying Gene Function

The yeast genome, consists of 12,100 kb of DNA allocated among 16 chromosomes and contains what appear to be 5885 protein-coding genes. The genome is relatively compact, with almost 70% of the total sequence consisting of **open reading frames (ORFs),** potential protein-coding sequences that are not interrupted by stop codons and that exhibit the same codon-usage preference as other genes in the organism. Computer analysis of the ORFs allows classification of about half of the yeast proteins on the basis of their sequence similarity to proteins of known function. Such homology searches reveal, for example, that yeast devotes at least 11% of its proteins to general metabolism, 3% to energy production, and 7% to transcription. These numbers are almost certainly underestimates, since not all the ORFs can be assigned a function unequivocally. In fact, some 2000 genes have no homologs of known function. Elucidating the functions of these so-called **orphan genes** represents a significant challenge to molecular biologists.

A similar situation exists for the 4639-kb *E. coli* genome, whose sequence was determined by Frederick Blattner. Even though *E. coli* has been intensively studied for the past 50 years and hence is the best characterized organism, 30 to 40% of its genes have no known function (Fig. 27-3). Some of these orphan genes may yet be identified as familiar components of biochemical or regulatory pathways. Recall that amino acid substitutions that obliterate sequence homology without significantly altering protein structure (Section 6-2C) may mask functional similarities. The development of more sophisticated database-searching tools that can identify protein structural motifs will no doubt improve the rate of identification of newly sequenced genes.

Some orphan genes may represent novel proteins whose specialized functions have yet to be detected under standard laboratory conditions. In the future, it may be possible to examine an organism's genetic inventory and "subtract" the genes for core metabolic processes, leaving a set of genes that determine the organism's unique metabolic capabilities. Genes determining bacterial virulence, which are often absent in nonpathogenic strains of the same species, could be identified in this manner. Such information might lead to new therapies based on peculiarities of the organism's metabolism.

Cataloging an organism's genes may also confirm suspected evolutionary relationships. The complete genome of the archaeote *Methanococcus jannaschii,* which contains 1738 protein-coding genes, shows that this organism has, as expected, the enzymes necessary for fixing nitrogen and for reducing CO_2 with H_2 to produce methane. Much of its translation machinery resembles that of both bacteria and eukaryotes, but, surprisingly, its transcription and replication proteins more clearly resemble their eukaryotic counterparts. The examination of several aspects of cellular function indicates that the anabolic genes of *M. jannaschii* reflect an ancient metabolic world shared by bacteria and archaeotes. In contrast, its cellular informa-

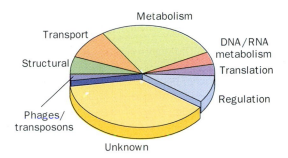

Figure 27-3. **Functions of the genes in the *E. coli* genome.** Approximately 40% of the probable 4288 protein-coding genes had previously been identified. Whole-genome sequencing reveals that another 40% of the genes have no identifiable function on the basis of sequence similarity to other genes of known function. [Based on data from Blattner, F.R., *et al., Science* **277,** 1458 (1997).]

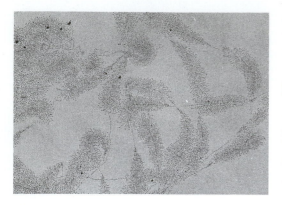

Figure 27-4.   An electron micrograph of tandem arrays of actively transcribing 18S, 5.8S, and 28S rRNA genes from the newt *Notophthalmus viridescens*. The axial fibers are DNA. The fibrillar "Christmas tree" matrices, which consist of newly synthesized RNA strands in complex with proteins, outline each transcriptional unit. Note that the longest ribonucleoprotein branches are only ~10% of the length of their corresponding DNA. Apparently, the RNA strands are compacted through secondary structure interactions and/or protein associations. The matrix-free segments of DNA are the nontranscribed spacers. [Courtesy of Oscar L. Miller, Jr., and Barbara R. Beatty, University of Virginia.]

tion-processing systems demonstrate the common ancestry of eukaryotes and archaeotes. A whole-genome comparison thus provides more information than gene-by-gene comparisons.

B. Gene Clusters

Figure 27-1, in which genes are color-coded by their putative functions, suggests that genes are randomly distributed in the genome. This appears to be the case in all the sequenced genomes; in other words, no discrete portion of the genome contains relatively more genes or genes related to a particular function than any other portion of the genome. However, closer examination shows that some protein-coding genes and other chromosomal elements exhibit a certain degree of organization. Prokaryotic genomes, for example, often contain operons, in which genes with related functions (e.g., encoding the proteins involved in a particular metabolic pathway) occur close together, sometimes in the same order in which their encoded proteins act in a metabolic reaction sequence.

Gene clusters also occur in both prokaryotes and eukaryotes. Although most genes occur only once in an organism's haploid genome, genes such as those for rRNA and tRNA, whose products are required in relatively large amounts, may occur in multiple copies. As we saw in Section 25-3B, large rRNA transcripts are cleaved to yield the mature rRNA molecules. Furthermore, the transcribed blocks of 18S, 5.8S, and 28S eukaryotic rRNA genes are arranged in tandem repeats that are separated by nontranscribed spacers (Fig. 27-4). The rRNA genes, which may be distributed among several chromosomes, vary in haploid number from less than 50 to over 10,000, depending on the species. Humans, for example, have 50 to 200 blocks of rRNA genes spread over 5 chromosomes. tRNA genes are similarly reiterated and clustered.

Protein-coding genes almost never occur in multiple copies, presumably because the repeated translation of a few mRNA transcripts provides adequate amounts of most proteins. One exception is histone proteins, which are required in large amounts during the short period when eukaryotic DNA synthesis occurs. Not only are histone genes reiterated (up to ~100 times in *Drosophila*), they often occur as sets of each of the five different histone genes separated by nontranscribed sequences (Fig. 27-5). The gene order and the direction of transcription in these quintets is preserved over large evolutionary distances. The spacer sequences vary widely among species and, to a limited extent, among the repeating quintets within a genome. In birds and mammals, which contain 10 to 20 copies of each of the five histone genes, the genes occur in clusters but in no particular order.

Figure 27-5.   The organization and lengths of the histone gene cluster repeating units in a variety of organisms. Coding regions are indicated in color, and spacers are gray. The arrows denote the direction of transcription (the top three organisms are distantly related sea urchins).

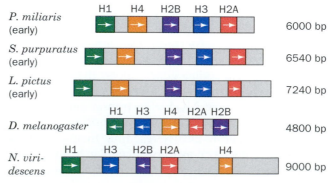

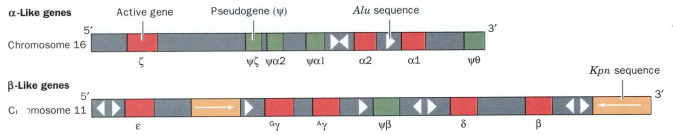

Figure 27-6. The organization of human globin genes. Red boxes represent active genes; green boxes represent pseudogenes; yellow boxes represent repetitive *Kpn* sequences, with the arrows indicating their relative orientations; and triangles represent repeating *Alu* sequences in their relative orientations. [After Karlsson, S. and Nienhuis, A.W., *Annu. Rev. Biochem.* **54**, 1074 (1985).]

Other gene clusters contain genes of similar but not identical sequence. For example, human globin genes are arranged in two clusters on separate chromosomes (Fig. 27-6). The various genes are transcribed at different developmental stages. Adult hemoglobin is an $\alpha_2\beta_2$ tetramer, whereas the first hemoglobin made by the human embryo is a $\zeta_2\varepsilon_2$ tetramer in which ζ and ε are α- and β-like subunits, respectively (Fig. 27-7). At approximately 8 weeks after conception, fetal hemoglobin containing α and γ subunits appears. The γ subunit is gradually supplanted by β starting a few weeks before birth. Adult human blood normally contains ~97% $\alpha_2\beta_2$ hemoglobin, 2% $\alpha_2\delta_2$ (in which δ is a β variant), and 1% $\alpha_2\gamma_2$.

The **α-globin gene cluster** (Fig. 27-6, *top*), which spans 28 kb, contains three functional genes: the embryonic ζ gene and two slightly different α genes, $\alpha1$ and $\alpha2$, which encode identical polypeptides. The α cluster also contains four pseudogenes (nontranscribed relics of ancient gene duplications): $\psi\zeta$, $\psi\alpha2$, $\psi\alpha1$, and $\psi\theta$. The **β-globin gene cluster** (Fig. 27-6, *bottom*), which spans >60 kb, contains five functional genes: the embryonic ε gene, the fetal genes $^G\gamma$ and $^A\gamma$ (duplicated genes encoding polypeptides that differ only by having either Gly or Ala at position 136), and the adult genes δ and β. The β-globin cluster also contains one pseudogene, $\psi\beta$. Both α and β gene clusters also include copies of the **repetitive DNA sequences** known as **Kpn** and **Alu sequences** (Section 27-1C).

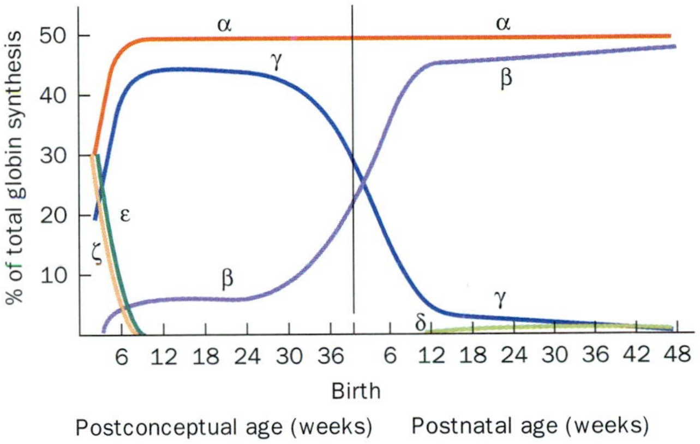

C. Nontranscribed DNA

Approximately 11% of the *E. coli* genome consists of nontranscribed regions. These sequences could more properly be called "nontranscribable," meaning that they do not contain ORFs (many genes that are transcribed only at specific times in an organism's life cycle or under specific environmental conditions may appear to be "nontranscribed" at other times). Nontranscribed regions include the regulatory sequences that separate individual genes and sites that govern the origin and termination of replication. In addition, bacterial genomes typically contain insertion sequences (Section 24-6B) and the remnants of integrated bacteriophages (Section 27-2D).

Eukaryotic Genomes Contain Repetitive Sequences

The small sizes of prokaryotic genomes probably exert selective pressure against the accumulation of useless DNA. Eukaryotic genomes, however, which are usually much larger than prokaryotic genomes, apparently are not subject to the same selective forces. About 30% of the yeast genome, for ex-

Figure 27-7. The progression of human globin chain synthesis with fetal development. Note that any red blood cell contains only one type each of α- and β-like subunits. [After Weatherall, D.J. and Clegg, J.B., *The Thalassaemia Syndromes* (3rd ed.), p. 64, Blackwell Scientific Publications (1981).]

ample, consists of nontranscribed sequences. The proportion is far greater in higher eukaryotes (as much as 98% in humans). Much of this DNA consists of repetitive sequences. High concentrations of repetitive DNA are located at the **centromeres** of eukaryotic chromosomes, the regions attached to the microtubular spindle during mitosis. These sequences may help align chromosomes and facilitate recombination. The telomeres also are composed of repeating DNA sequences (Section 24-3C). At least a dozen human diseases result from excessively reiterated trinucleotide sequences (see Box 27-1).

Highly repetitive sequences (present at $>10^6$ copies per haploid genome) are clusters of nearly identical sequences up to 10 bp long that are tandemly repeated thousands of times. These DNA regions are also known as **satellite DNA** because their distinctive base compositions cause them to form bands separate from the main DNA band in density gradient ultracentrifugation of fragmented DNA (Fig. 27-8). The three satellites of *Drosophila* DNA are closely related heptanucleotide repeats:

5′—ACAAACT—3′	5′—ATAAACT—3′	5′—ACAAATT—3′
3′—TGTTTGA—5′	3′—TATTTGA—5′	3′—TGTTTAA—5′
Satellite I	**Satellite II**	**Satellite III**

Trinucleotide Repeat Diseases

Several human neurological diseases are associated with repeated trinucleotides in certain genes. These trinucleotide repeats exhibit an unusual genetic instability: Above a threshold of about 35 to 50 copies (100–150 bp), the repeats tend to expand, by an unknown mechanism, with successive generations. Because the overall length of the repeat typically correlates with the age of onset of the disease, descendants of an individual with a trinucleotide repeat disease tend to be more severely affected and at an earlier age. The disease is therefore said to exhibit **genetic anticipation**.

Some types of trinucleotide repeat diseases are caused by massive expansion (usually to hundreds of copies) of a trinucleotide in the noncoding region of a gene, for example, in a region upstream of the transcription start site, in a 5′ or 3′ untranslated region (**UTR**), or in an intron (see Table). These expansions generally affect gene expression. For example, **myotonic dystrophy** results from aberrant expression of a protein kinase. The severity of the symptoms, progressive muscle weakness and wasting, correlate with the number of CTG repeats (>2000 in some cases).

In **fragile X syndrome**, the most common cause of mental retardation after Down's syndrome, trinucleotide expansion ranging from hundreds to thousands of copies promotes the breakage of the tip of the X chromosome's long arm, which is connected to the rest of the chromosome by a slender thread. Like many trinucleotide repeat diseases, fragile X syndrome exhibits non-Mendelian inheritance: Between 20% and 50% of males bearing the fragile X mutation are asymp-

tomatic. Their daughters are likewise asymptomatic, but these daughters' children may have the syndrome. Evidently, the fragile X defect is somehow activated by passage through a female.

Other trinucleotide repeat diseases result from the moderate expansion of a CAG triplet, which codes for glutamine, in the protein-coding region of a gene. Presumably, these expansions yield nonfunctional proteins that kill cells, particularly in the nervous system. Such a loss of neurons occurs in Huntington's disease, a devastating condition characterized by progressively disordered movements (**chorea**), cognitive decline, and emotional disturbances. This genetically dominant and invariably fatal disease has an age of onset of ~40 years and may follow a 10- to 20-year course. The repeated CAG sequences in the relevant gene, which codes for a 3145-residue polypeptide called **Huntingtin**, are normally present in 11 to 34 copies but increase to between 37 and 876 copies in affected individuals.

Some Diseases Associated with Trinucleotide Repeats

Disease	Repeat	Site of Repeat
Fragile X syndrome	CGG or CCG	5′ UTR
Myotonic dystrophy	CTG	Upstream region, 3′ UTR
Friedrich's ataxia	GAA	Intron
Spinobulbar muscular atrophy	CAG	Exon
Huntington's disease	CAG	Exon

Moderately repetitive sequences ($<10^6$ copies per haploid genome) occur in segments of 100 to several thousand base pairs that are interspersed with larger blocks of unique DNA. The best characterized of these sequences belong to the *Alu* family (so named because most of its ~300-bp segments contain a cleavage site for the restriction endonuclease *Alu*I; Table 3-3). The human genome contains 300,000 to 500,000 widely distributed *Alu* sequences that are, on average, 80 to 90% homologous with their consensus sequence. The *Alu* family is the most prominent moderately repetitive DNA in many organisms, but vertebrate genomes generally contain several different varieties of moderately repetitive DNA (e.g., the *Kpn* sequences in Fig. 27-6).

No function has been unequivocally assigned to moderately repetitive DNA, which therefore has been termed **selfish** or **junk DNA**. This DNA apparently is a molecular parasite that, over many generations, has disseminated itself throughout the genome through transposition (Section 24-6B). The theory of natural selection predicts that the increased metabolic burden imposed by the replication of an otherwise harmless selfish DNA would eventually lead to its elimination. Yet for slowly growing eukaryotes, the relative disadvantage of replicating an additional 100 bp of selfish DNA in an ~1-billion-bp genome would be so slight that its rate of elimination would be balanced by its rate of propagation.

As much as half of the 2-billion-bp maize (corn) genome may comprise transposable elements, many of which are inserted within other transposable elements. This freeloading DNA may contribute to the remarkably large sizes of plant genomes, which range from 10^8 bp to 10^{11} bp. The non-mobile remnants of transposons also make up a large portion of mammalian genomes. *These unexpressed sequences are subject to little selective pressure and accumulate mutations at a greater rate than do expressed sequences.* The resulting variations can be used to trace evolutionary relationships (see Box 27-2).

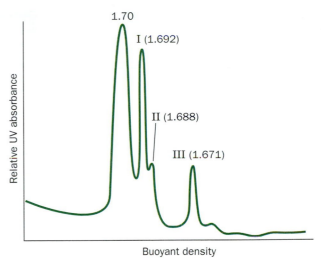

Figure 27-8. The buoyant density pattern of *Drosophila virilus* DNA centrifuged to equilibrium in neutral CsCl. Three prominent bands of satellite DNA (with densities of 1.692, 1.688, and 1.671 g·cm^{-3}) are present, in addition to the main DNA band (density of 1.70 g·cm^{-3}). [After Gall, J.G., Cohen, E.H., and Atherton, D.D., *Cold Spring Harbor Symp. Quant. Biol.* **38**, 417 (1973).]

Box 27-2
BIOCHEMISTRY IN CONTEXT

Inferring Genealogy from DNA Sequences

Mutations in DNA sequences can create or obliterate restriction sites, giving rise to different sized fragments when the DNA is digested by prokaryotic restriction endonucleases (Section 3-4A and Box 3-1). The genealogy of several human populations has been inferred from such restriction fragment length polymorphisms in five segments of their β-globin gene clusters. This study has led to the construction of a "family tree" in which the length of the horizontal axis indicates the genetic distance between the populations and hence the times between their divergence.

British
Italian
Cypriot
Indian
Melanesian
Polynesian
Thai
African

This tree suggests that Eurasian populations are much more closely related to each other than they are to African populations. Fossil evidence indicates that anatomically modern humans arose in Africa about 100,000 years ago and rapidly spread throughout that continent. This family tree therefore suggests that all Eurasian populations are descended from a surprisingly small "founder population" (perhaps only a few hundred individuals) that left Africa ~50,000 years ago.

[Figure after Wainscoat, J.S., Hill, A.V.S., Boyce, A.L., Flint, J., Hernandez, M., Thein, S.L., Old, J.M., Lynch, J.R., Falusi, A.G., Weatherall, D.J., and Clegg, J.B., *Nature* **319**, 493 (1986).]

2. REGULATION OF PROKARYOTIC GENE EXPRESSION

A complete genome sequence reveals the metabolic capabilities of an organism, but *gene sequences alone do not necessarily indicate when or where the encoded molecules are produced.* Various approaches have been used to identify transcriptionally active genes in certain cell types at particular developmental stages. This information may ultimately provide a more complete picture of the factors that control the pathogenicity of bacteria, for example, or the predispositions of humans to diseases.

The route from gene sequence to fully functional gene product offers many potential points for regulation, but *in prokaryotes, gene expression is almost entirely controlled at the level of transcription.* This is perhaps because prokaryotic mRNAs have lifetimes of only a few minutes, so translational control is unnecessary. In this section, we examine a few well-documented examples of transcriptional control in prokaryotes. In the following section, we shall consider how eukaryotic cells regulate gene expression.

● See Guided Exploration 30:

The Regulation of Gene Expression by the *lac* Repressor System.

A. *The lac Repressor*

Bacteria adapt to their environments by producing enzymes that metabolize certain nutrients only when those substances are available. For example, *E. coli* cells grown in the absence of lactose are initially unable to metabolize this disaccharide. To do so they require two proteins: **β-galactosidase,** which catalyzes the hydrolysis of lactose to its component monosaccharides:

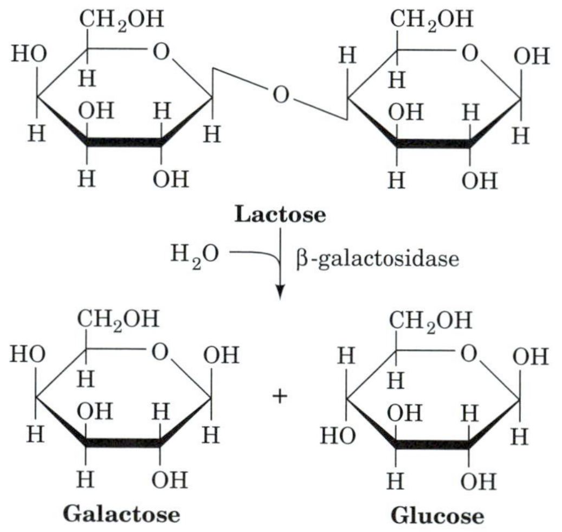

Lactose

H_2O — β-galactosidase

Galactose   **Glucose**

and **galactoside permease** (also known as **lactose permease**), which transports lactose into the cell. Cells grown in the absence of lactose contain only a few molecules of these proteins. Yet a few minutes after lactose is introduced into their medium, the cells increase the rate at which they synthesize these proteins by ~1000-fold and maintain this pace until lactose is no longer available. *This ability to produce a series of proteins only when the substances they metabolize are present permits the bacteria to adapt to their environment without the debilitating need to continuously synthesize large quantities of otherwise unnecessary enzymes.*

Lactose or one of its metabolic products must somehow act as an **inducer** to trigger the synthesis of the above proteins. The physiological inducer of the lactose system, the lactose isomer **1,6-allolactose,**

1,6-Allolactose

arises from lactose's occasional transglycosylation by β-galactosidase. Most *in vitro* studies of lactose metabolism use **isopropylthiogalactoside (IPTG),**

Isopropylthiogalactoside (IPTG)

a synthetic inducer that structurally resembles allolactose but is not degraded by β-galactosidase. Natural and synthetic inducers also stimulate the synthesis of **thiogalactoside transacetylase,** an enzyme whose physiological role is unknown.

The genes specifying β-galactosidase, galactoside permease, and thiogalactoside transacetylase, designated *Z, Y,* and *A,* respectively, are contiguously arranged in the *lac* operon (Fig. 25-4). All three structural genes are translated from a single mRNA transcript. A nearby gene, *I,* encodes the *lac* **repressor,** a protein that inhibits the synthesis of the three *lac* proteins.

lac Repressor Recognizes Operator Sequences

The target of the *lac* repressor is a region of the *lac* operon known as its **operator,** which lies near the beginning of the β-galactosidase gene. *In the absence of inducer, lac repressor specifically binds to the operator to prevent the transcription of mRNA (Fig. 27-9a). On binding inducer, the repressor*

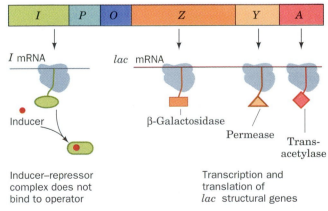

Figure 27-9.   *Key to Function.* The expression of the *lac* operon. (*a*) In the absence of inducer, the repressor (the product of the *I* gene) binds to the operator (*O*), thereby preventing transcription of the *lac* operon from the promoter (*P*). (*b*) On binding inducer, the repressor dissociates from the operator, which permits the transcription and subsequent translation of the *lac* structural genes (*Z, Y,* and *A,* which respectively encode β-galactosidase, lactose permease, and thiogalactoside transacetylase).

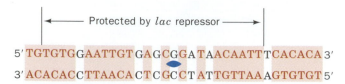

5′ TGTGTGGAATTGTGAGCGGATAACAATTTCACACA 3′
3′ ACACACCTTAACACTCGCCTATTGTTAAAGTGTGT 5′

Figure 27-10.   The base sequence of the *lac* operator O_1. Its symmetry-related regions, which comprise 28 of its 35 bp, are shaded in red.

dissociates from the operator, thereby permitting the transcription and subsequent translation of the lac enzymes (Fig. 27-9b).

The *lac* operator actually contains three operator sequences to which *lac* repressor binds with high affinity, known as O_1, O_2, and O_3. O_1, the primary repressor-binding site, was identified through its protection by *lac* repressor from nuclease digestion. The 26-bp protected sequence lies within a nearly twofold symmetrical sequence of 35 bp (Fig. 27-10). O_1 overlaps the transcription start site of the *lacZ* gene. The operator sequence O_2 is centered 401 bp downstream, fully within the *lacZ* gene, and O_3 is centered 93 bp upstream of O_1, at the end of the *lacI* gene. Genetic engineering experiments show that all three operator sequences must be present for maximum repression *in vivo*.

The observed rate constant for the binding of *lac* repressor to *lac* operator is $k \approx 10^{10}$ M$^{-1} \cdot$ s^{-1}. This "on" rate is much greater than that calculated for the diffusion-controlled process in solution: $k \approx 10^7$ M$^{-1} \cdot$ s^{-1} for molecules the size of *lac* repressor. Since it is impossible for a reaction to proceed faster than its diffusion-controlled rate, *lac* repressor must not encounter operator from solution in a random three-dimensional search. Rather, it appears that *lac repressor finds its operator by nonspecifically binding to DNA and sliding along it in a far more efficient one-dimensional search.*

lac Repressor Binds Two DNA Segments Simultaneously

The isolation of *lac* repressor, by Beno Müller-Hill and Walter Gilbert in 1966, was exceedingly difficult because the repressor constitutes only ~0.002% of the protein in wild-type *E. coli*. Molecular cloning techniques (Section 3-5) later made it possible to produce large quantities of *lac* repressor. Nevertheless, it was not until 1996 that Ponzy Lu and Mitchell Lewis reported the complete three-dimensional structure of this protein. Each 360-residue subunit of the repressor homotetramer has four functional units (Fig. 27-11):

1. An N-terminal "headpiece," which contains a helix–turn–helix (HTH) motif that resembles those in other prokaryotic DNA-binding proteins (Section 23-4B) and which specifically binds operator DNA sequences.

2. A linker, which contains a short α helix that acts as a hinge and also binds DNA. In the absence of DNA, the hinge helices of the *lac* repressor tetramer are disordered, allowing the headpieces to move freely.

3. A two-domain core, which binds inducers such as IPTG.

4. A C-terminal α helix, which is required for the quaternary structure of *lac* repressor. In the tetramer, all four C-terminal helices associate. Surprisingly, the repressor homotetramer does not exhibit the D_2 symmetry of nearly all homotetrameric proteins of known structure (three mutually perpendicular twofold axes; Section 6-3) but is instead V-shaped (with only twofold symmetry) and is therefore best considered to be a dimer of dimers (Fig. 27-12).

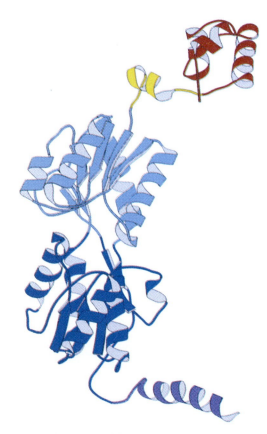

Figure 27-11.   Ribbon diagram of the *lac* repressor monomer. The DNA-binding domain containing its helix–turn–helix motif is red; the DNA-binding domain hinge helix is yellow; the inducer-binding core is light and dark blue; and the tetramerization helix is purple. [Courtesy of Ponzy Lu and Mitchell Lewis, University of Pennsylvania.]

The X-ray structure of a complex between *lac* repressor and a 21-bp synthetic DNA containing a high-affinity binding sequence reveals that each repressor tetramer binds two DNA segments (Fig. 27-12). The repressor's HTH motif fits snugly into the major groove, bending the DNA so that it has a radius of curvature of 60 Å. The two bound DNA segments are laterally separated by ~25 Å.

IPTG binds to the repressor core at the interface between the two domains colored light and dark blue in Fig. 27-11. This binding induces a conformational change in the repressor dimer that is communicated through the hinge helices to the headpieces. This causes the two DNA-binding domains in each dimer to separate by ~3.5 Å so that they can no longer simultaneously bind DNA, thereby causing the repressor to dissociate from the DNA.

The *lac* repressor is an allosteric protein: IPTG binding to one subunit alters the DNA-binding activity of its dimeric partner (but not of the other dimer in the repressor tetramer). Since the allosteric transition occurs within the dimer, why does full repressor activity require a tetramer? Model-building studies provide a plausible answer to this puzzle. *A lac repressor tetramer simultaneously binds to two operators so that it brings them together, forming a loop of DNA either 93 or 401 bp long, depending on whether the repressor binds O_1 and O_3 or O_1 and O_2.* The formation of a stable looped structure may require additional DNA-binding proteins; one candidate is **CAP** (Section 27-2B), a DNA-binding protein that binds to the DNA between O_1 and O_3 (Fig. 27-13). In the model shown in Fig. 27-13, the *lac* promoter is part of the looped DNA.

It was widely assumed for years that the *lac* repressor simply physically obstructs the binding of RNA polymerase to the *lac* promoter. However, experiments have demonstrated that RNA polymerase can bind to the promoter in the presence of the repressor but cannot properly initiate transcription. Dissociation of the repressor in response to an inducer would allow unimpeded transcription. If RNA polymerase were already bound to the *lac* promoter, transcription could begin immediately. Nevertheless, in the model shown in Fig. 27-13, the contact surface for RNA polymerase is on the inside of the DNA loop, which presumably would preclude the binding of RNA polymerase. Further studies are needed to resolve this apparent contradiction.

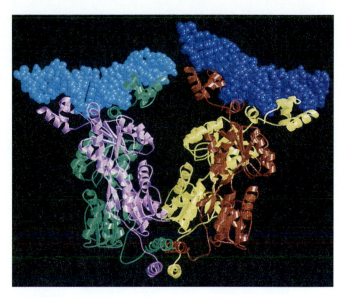

Figure 27-12. **The X-ray structure of *lac* repressor tetramer bound to two 21-bp segments of DNA.** The monomeric units of the repressor, shown in ribbon form, are green, pink, yellow, and red, and the DNA, shown in space-filling form, is blue. [Courtesy of Ponzy Lu and Mitchell Lewis, University of Pennsylvania.]

B. Catabolite Repression: An Example of Gene Activation

Glucose is *E. coli*'s metabolic fuel of choice; adequate amounts of glucose prevent the full expression of genes specifying proteins involved in the fermentation of numerous other catabolites, including lactose, arabinose, and galactose, even when they are present in high concentrations. This phenomenon, which is known as **catabolite repression,** prevents the wasteful duplication of energy-producing enzyme systems. Catabolite repression is overcome in the absence of glu-

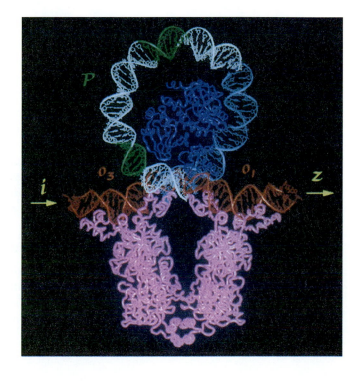

Figure 27-13. **A model of the 93-bp loop formed when the *lac* repressor tetramer binds to O_1 and O_3.** The proteins are represented by their C_α backbones, and the DNA is shown in skeletal form with its sugar–phosphate backbones traced by helical ribbons. The model was constructed from the X-ray structure of *lac* repressor (*pink*) in complex with two 21-bp operator DNA segments (*red*) and the X-ray structure of CAP (*blue*) in complex with its 30-bp target DNA (*cyan*). The remainder of the DNA loop was generated by applying a smooth curvature to B-DNA (*light blue*) with the −10 and −35 regions of the *lac* promoter highlighted in green. [Courtesy of Ponzy Lu and Mitchell Lewis, University of Pennsylvania.]

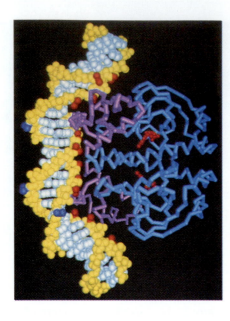

Figure 27-14.   The X-ray structure of the CAP–cAMP dimer in complex with DNA. The protein, viewed with its twofold axis of symmetry horizontal, is represented by its C_α backbone, with its N-terminal cAMP-binding domain blue and its C-terminal DNA-binding domain magenta. The 30-bp self-complementary DNA is shown in space-filling form with its sugar–phosphate backbone yellow and its bases white. The DNA phosphates whose ethylation interferes with CAP binding are red. Those in the complex that are hypersensitive to DNase I are blue (these phosphates bridge the CAP-induced kinks where the minor groove has been dramatically widened). The bound cAMPs are shown in ball-and-stick form in red. [Courtesy of Thomas Steitz, Yale University.] ● See the Interactive Exercises.

cose by a cAMP-dependent mechanism. cAMP levels are low in the presence of glucose but rise when glucose becomes scarce.

CAP–cAMP Complex Stimulates the Transcription of Catabolite Repressed Operons

Certain *E. coli* mutants, in which the absence of glucose does not relieve catabolite repression, are missing a cAMP-binding protein that is synonymously named **catabolite gene activator protein (CAP)** or **cAMP receptor protein (CRP).** CAP is a dimeric protein of identical 210-residue subunits that undergoes a large conformational change on binding cAMP. The CAP–cAMP complex, but not CAP alone, binds to the promoter region of the *lac* operon (among others) and stimulates transcription in the absence of repressor. CAP is therefore a **positive regulator** (it turns transcription on), in contrast to *lac* repressor, which is a **negative regulator** (it turns transcription off).

How does the CAP–cAMP complex operate? The *lac* operon has a weak (low-efficiency) promoter. One possibility is that CAP–cAMP binding enhances the ability of RNA polymerase to transcribe the lac operon by inducing a conformational change in the promoter DNA. In fact, the X-ray structure of CAP–cAMP in complex with a 30-bp segment of DNA, whose sequence resembles the CAP-binding site, reveals that the CAP dimer binds in successive major grooves of the DNA via its two HTH motifs so as to bend the DNA by ~90° around the protein dimer (Fig. 27-14). The bend arises from two ~45° kinks in the DNA between the fifth and sixth bases out from the complex's two-fold axis in both directions and results in a closing of the major groove and in an enormous widening of the minor groove at each kink. The distorted DNA may be a more efficient substrate for transcription initiation than linear DNA. A second possibility is that CAP–cAMP stimulates transcription initiation through direct contact with RNA polymerase. Indeed, the CAP–cAMP binding site on the *lac* operon overlaps the *lac* promoter.

The model for *lac* repressor binding (Fig. 27-13) paradoxically includes CAP, which is an activator. This dual binding may be a mechanism for conserving cellular energy in the absence of both glucose and lactose. If lactose became available, *lac* repressor would dissociate, and CAP–cAMP would be poised to promote transcription of the *lac* operon.

C. Attenuation

The *E. coli* **trp operon** encodes five polypeptides (which form three enzymes) that mediate the synthesis of tryptophan from chorismate (Section 20-5B). These five *trp* operon genes (*A–E*; Fig. 27-15) are coordinately expressed under the control of the *trp* repressor, which binds L-tryptophan to form a complex that specifically binds to the *trp* operator to reduce the rate

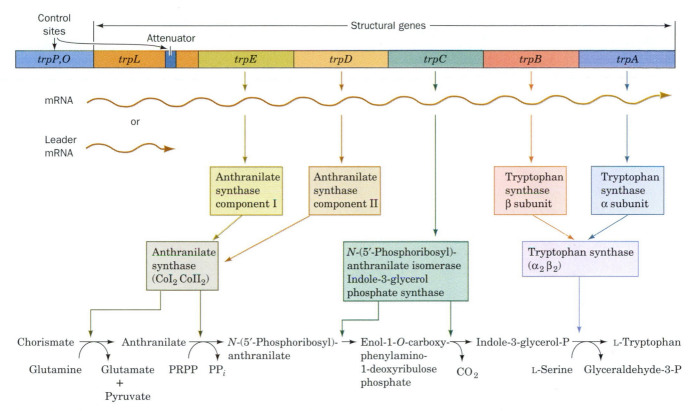

Figure 27-15. **A genetic map of the *E. coli trp* operon indicating the enzymes it specifies and the reactions they catalyze.** The gene product of *trpC* catalyzes two sequential reactions in the synthesis of tryptophan. [After Yanofsky, C., *J. Am. Med. Assoc.* **218**, 1027 (1971).]

of *trp* operon transcription 70-fold (Section 23-4B). In this system, tryptophan acts as a **corepressor**; its presence prevents what would be superfluous tryptophan biosynthesis.

The *trp* repressor–operator system was at first thought to fully account for the regulation of tryptophan biosynthesis in *E. coli*. However, the discovery of *trp* deletion mutants located downstream from the operator (*trpO*) that increase *trp* operon expression six-fold indicated the existence of an additional transcriptional control element. This element is located ~30 to 60 nucleotides upstream of the structural gene *trpE* in a 162-nucleotide **leader sequence** (*trpL;* Fig. 27-15).

When tryptophan is scarce, the entire 6720-nucleotide polycistronic *trp* mRNA, including the *trpL* sequence, is synthesized. When the encoded enzymes begin synthesizing tryptophan, the rate of *trp* operon transcription decreases as tryptophan binds to *trp* repressor. Of the *trp* mRNA that is transcribed, however, an increasing proportion consists of only a 140-nucleotide segment corresponding to the 5' end of *trpL. The availability of tryptophan therefore results in the premature termination of trp operon transcription.* The control element responsible for this effect is consequently termed an **attenuator.**

The *trp* Attenuator's Transcription Terminator Is Masked when Tryptophan Is Scarce

The attenuator transcript contains four complementary segments that can form one of two sets of mutually exclusive base-paired hairpins (Fig.

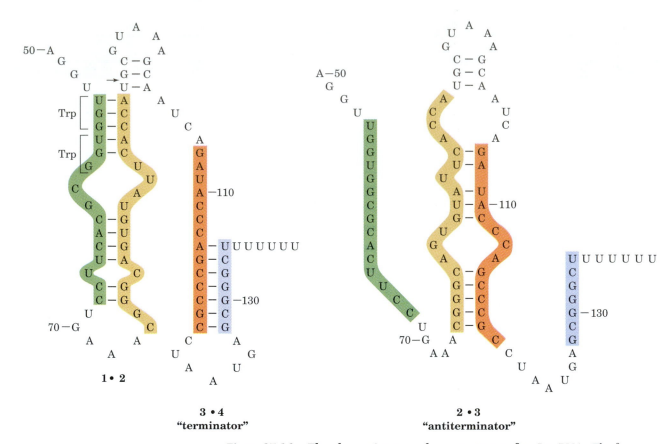

Figure 27-16.   The alternative secondary structures of *trpL* mRNA. The formation of the base-paired 2·3 (antiterminator) hairpin (*right*) precludes formation of the 1·2 and 3·4 (terminator) hairpins (*left*) and vice versa. Attenuation results in the premature termination of transcription immediately after nucleotide 140 when the 3·4 hairpin is present. The arrow indicates the mRNA site past which RNA polymerase pauses until approached by an active ribosome. [After Fisher, R.F. and Yanofsky, C., *J. Biol. Chem.* **258**, 8147 (1983).]

27-16). Segments 3 and 4 together with the succeeding residues constitute a transcription terminator (Section 25-1D): a G + C-rich hairpin followed by several sequential U's (compare with Fig. 25-10). Transcription rarely proceeds beyond this termination site unless tryptophan is in short supply.

How does a shortage of tryptophan cause transcription to proceed past this terminator? A section of the leader sequence, which includes segment 1 of the attenuator, is translated to form a 14-residue polypeptide that contains two consecutive Trp residues (Fig. 27-16, *left*). The position of this particularly rare dipeptide (~1% of the residues in *E. coli* proteins are Trp) provided an important clue to the mechanism of attenuation, as proposed by Charles Yanofsky (Fig. 27-17): An RNA polymerase that has escaped repression initiates *trp* operon transcription. Soon after the ribosomal initiation site of the *trpL* sequence has been transcribed, a ribosome attaches to it and begins translating the leader peptide. When tryptophan is abundant (i.e., there is a plentiful supply of tryptophanyl–tRNA[Trp]), the ribosome follows closely behind the transcribing RNA polymerase. Indeed, RNA polymerase pauses past position 92 of the transcript and continues transcribing only on the approach of a ribosome, thereby ensuring the close coupling of transcription and translation. The progress of the ribosome prevents the formation of the 2·3 hairpin and permits the formation of the

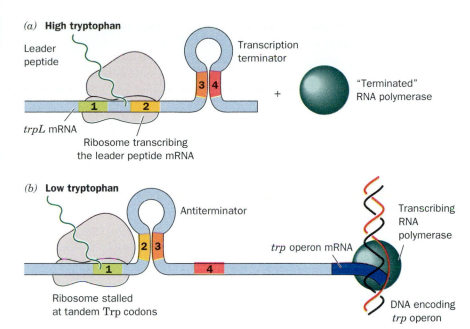

(a) **High tryptophan**

Leader
peptide

Transcription
terminator

trpL mRNA

Ribosome transcribing
the leader peptide mRNA

+

"Terminated"
RNA polymerase

(b) **Low tryptophan**

Antiterminator

trp operon mRNA

Transcribing
RNA
polymerase

Ribosome stalled
at tandem Trp codons

DNA encoding
trp operon

Figure 27-17. Attenuation in the *trp* operon. (*a*) When Trp–tRNA^{Trp} is abundant, the ribosome translates *trpL* mRNA. The presence of the ribosome on segment 2 prevents the formation of the base-paired 2·3 hairpin. The 3·4 hairpin, an essential component of the transcriptional terminator, can then form, thus aborting transcription. (*b*) When Trp–tRNA^{Trp} is scarce, the ribosome stalls on the tandem Trp codons of segment 1. This situation permits the formation of the 2·3 hairpin, which precludes the formation of the 3·4 hairpin. RNA polymerase therefore transcribes through this unformed terminator and continues transcribing the *trp* operon.

3·4 hairpin, which terminates transcription (Fig. 27-17*a*). When tryptophan is scarce, however, the ribosome stalls at the tandem UGG codons (which specify Trp) because of the lack of Trp–tRNA^{Trp}. As transcription continues, the newly synthesized segments 2 and 3 form a hairpin because the stalled ribosome prevents the otherwise competitive formation of the 1·2 hairpin (Fig. 27-17*b*). The 3·4 hairpin does not form, allowing transcription to proceed past this region and into the remainder of the *trp* operon. Thus, attenuation regulates *trp* operon transcription according to the tryptophan supply.

The leader peptides of the five other amino acid–biosynthesizing operons known to be regulated by attenuation are all rich in their corresponding amino acid residues. For example, the *E. coli* **his operon,** which specifies enzymes synthesizing histidine, has seven tandem His residues in its 16-residue leader peptide, whereas the **ilv operon,** which specifies enzymes participating in isoleucine, leucine, and valine biosynthesis, has five Ile, three Leu, and six Val residues in its 32-residue leader peptide. The leader transcripts of these operons resemble that of the *trp* operon in their capacity to form two alternative secondary structures, one of which contains a trailing transcription terminator.

D. Bacteriophage λ

One of the best characterized regulatory systems in molecular biology controls the life cycle of the *E. coli* **bacteriophage λ** (Fig. 3-28). Soon after entering the host, the 48,502-bp linear phage DNA, which has complementary single-stranded ends of 12 nucleotides (called cohesive ends or *cos* **sites**), circularizes and is covalently closed. At this stage, the virus has a "choice" of two alternative life styles (Fig. 27-18):

1. It can follow the **lytic** mode in which the phage is replicated by the host such that, after 45 min at 37°C, the host lyses to release ~100 progeny phage particles.

2. The phage may take up the **lysogenic** life cycle, in which its DNA is inserted at a specific site in the host chromosome such that the phage DNA passively replicates with the host DNA. The phage is then de-

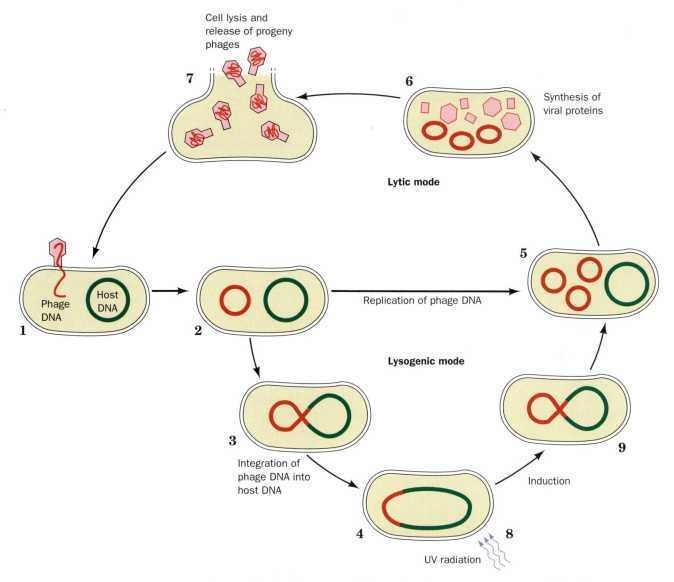

Figure 27-18. The λ phage life cycle. The infection of the *E. coli* host begins when the virus specifically adsorbs to the cell and injects its DNA (**1**). The linear DNA then circularizes (**2**) and commences directing the infection process. In the lysogenic mode, the phage DNA is stably integrated at a specific site in the host chromosome (**3** and **4**) so that it is passively replicated with the bacterial cell. Alternatively, the phage may take up the lytic mode in which the DNA directs its own replication (**5**) and the synthesis of viral proteins (**6**), resulting in lysis of the host cell and release of ~100 progeny phages (**7**). DNA damage caused, for example, by UV radiation (**8**) induces excision of the prophage DNA from the lysogenic bacterial chromosome (**9**) and causes the phage to take up the lytic mode.

scribed as a **prophage** and the host is called a **lysogen.** Under appropriate conditions, even after many bacterial generations, the phage DNA can be excised from the host DNA to initiate a lytic cycle in a process known as **induction.**

The advantage of lysogeny is clear: A parasite that can form a stable association with its host has a better chance of long-term survival than one that invariably destroys its host. However, when the host DNA is damaged, the phage enters the lytic mode and thereby escapes its doomed host (a phenomenon that has been described as the "lifeboat" response). The choice of life cycle depends on a complex system that regulates the transcription of phage genes.

The Lytic Pathway

The bacteriophage λ genome encodes ~50 gene products and contains numerous control sites (Fig. 27-19). In the lytic replication of phage λ, as in love and war, proper timing is essential. This is because the DNA must be replicated in sufficient quantity before it is made unavailable by packaging into phage particles and because packaging must be completed before the host cell is enzymatically lysed. The transcription of the λ genome, which is carried out by host RNA polymerase, is controlled in both the lytic and the lysogenic programs by the regulatory genes that are shaded in red in Fig. 27-19.

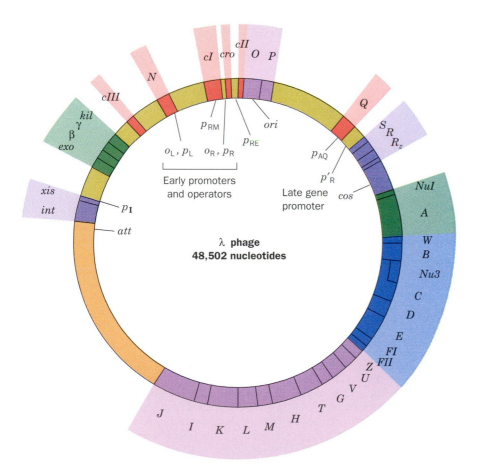

Figure 27-19. **A genetic map of bacteriophage λ.** Structural genes are indicated outside the circle, and control sites are indicated inside the circle. The genes encoding regulatory proteins are shaded red.

The lytic transcriptional process has three phases (Fig. 27-20):

1. In the first phase, **early transcription,** RNA polymerase begins transcribing "leftward" from the promoter p_L and "rightward" from the promoters p_R and p'_R (Fig. 27-20*a*). The leftward transcript, L1, which terminates at the termination site t_{L1}, encodes the *N* gene product, **gp*N*** (gp for *gene product*). Rightward transcription from p_R terminates with ~50% efficiency at t_{R1} to yield transcript R1, and otherwise at t_{R2} to yield transcript R2. R1 encodes only **Cro protein** (the product of the *cro* gene; see below), whereas R2 additionally encodes cII (the *cII* gene product; see below), **gp*O*,** and **gp*P*** (which participate in λ DNA synthesis). Rightward transcription from p'_R, which terminates at t'_R, yields a short transcript, R4, that specifies no protein.

2. The second phase, **delayed early transcription,** commences as soon as a significant quantity of gp*N* has accumulated. This protein acts as a **transcriptional antiterminator** at termination sites t_{L1}, t_{R1}, and t_{R2}, thereby yielding the extended transcripts L2 and R3 (Fig. 27-20*b*). L2 encodes proteins required for the excision of λ DNA from the *E. coli* chromosome (the products of the *xis* and *int* genes), whereas R3

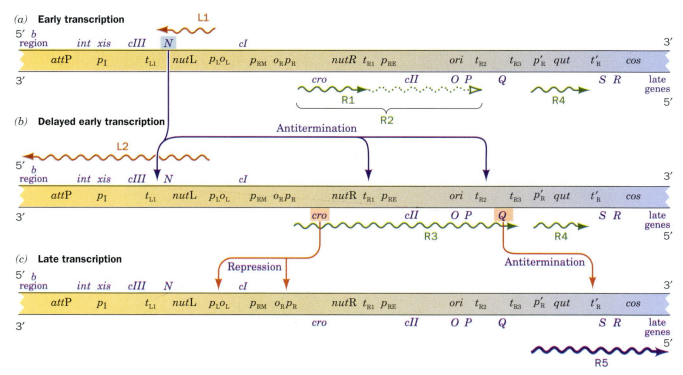

Figure 27-20.   *Key to Function.* Gene expression in the lytic pathway of phage λ. Genes specifying proteins that are transcribed to the "left" and "right" are shown above and below the phage chromosome. Control sites are indicated between the DNA strands. The genetic map is not drawn to scale, and not all of the genes or control sites are indicated. Transcripts are represented by wiggly arrows pointing in the direction of mRNA elongation; the actions of regulatory proteins are denoted by arrows pointing from each regulatory protein to the site(s) it controls. The lytic pathway has three transcriptional phases: (*a*) early transcription, (*b*) delayed early transcription, and (*c*) late transcription. Gene expression in each of the latter two phases is regulated by proteins synthesized in the preceding phase as is explained in the text. [After Arber, W., in Hendrix, R.W., Roberts, J.W., Stahl, F.W., and Weisberg, R.A. (Eds.), *Lambda II,* p. 389, Cold Spring Harbor Laboratory (1983).]

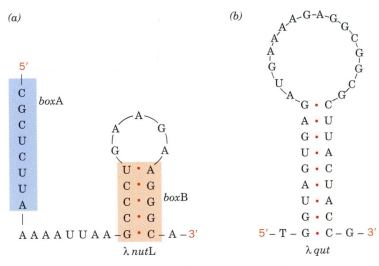

Figure 27-21. The RNA sequences of phage λ control sites. (*a*) *nut*L. This site closely resembles *nut*R. (*b*) *qut*. Each of these control sites is thought to form a base-paired hairpin.

encodes the transcriptional antiterminator **gpQ** as well as gpO, gpP, and Cro protein. Eventually, Cro protein, a repressor, accumulates and represses transcription from p_L and p_R (*cro* stands for *c*ontrol of *re*pressor and *o*ther things).

3. In the final phase, **late transcription,** gpQ has accumulated to the extent that it prevents transcriptional termination at t'_R (Fig. 27-20c). The proteins encoded by the resulting transcript, R5, include those that form the phage's capsid and those that catalyze host cell lysis.

This three-phase process, in which gene expression in the second and third phases are each regulated by proteins synthesized in the preceding phase, ensures the efficient production of new phage particles. These begin to appear ~22 min postinfection.

The antitermination effects of gpN depend on the formation of small base-paired hairpins on the RNA transcripts, at **nut** (*N* utilization) sites to the left of p_L (*nut*L) and to the right of p_R (*nut*R). These sites have similar sequences consisting of two elements called *box*A and *box*B (Fig. 27-21a). Antitermination also requires several *E. coli* proteins that form a complex with RNA polymerase and gpN at the *nut* sites on the mRNA. The complex travels with RNA polymerase during elongation and prevents it from pausing at the termination sites t_{L1} and t_{R1} (which are rho-dependent; Section 25-1D), and t_{R2} (which is rho-independent).

gpQ binds to a **qut** (*Q* utilization) site downstream of p'_R. *qut*, like *nut*, forms an RNA hairpin (Fig. 27-21b), but gpQ binds to DNA (rather than to RNA) and together with a host protein induces RNA polymerase to transcribe through the t'_R termination site to produce a lengthy mRNA transcript, R5 (Fig. 27-20c).

The Lysogenic Pathway

Lysogeny is established by the integration of the viral DNA into the host chromosome accompanied by the shutdown of all lytic gene expression. With phage λ, integration takes place through a **site-specific recombination** process that differs from general recombination (Section 24-6A) in that it occurs only between the chromosomal sites designated *att*P on the phage

***Figure 27-22.* Site-specific recombination in phage λ.** This schematic diagram shows (1) the circularization of the linear phage DNA through base pairing between its complementary ends to form the *cos* site and (2) the integration/excision of this DNA into/from the *E. coli* chromosome through site-specific recombination between the phage *att*P and host *att*B sites. The darker colored regions in the *att* sites represent the homologous 15-bp crossover sequences (O), whereas the lighter colored regions symbolize the unique sequences of bacterial (B and B′) and phage (P and P′) origin. This process requires the participation of **integration host factor (IHF)**, a histonelike *E. coli* protein that specifically binds to *cos* sites so as to induce a sharp bend in the DNA. [After Landy, A. and Weisberg, R.A., *in* Hendrix, R.W., Roberts, J.W., Stahl, F.W., and Weisberg, R.A. (Eds.), *Lambda II*, p. 212, Cold Spring Harbor Laboratory (1983).]

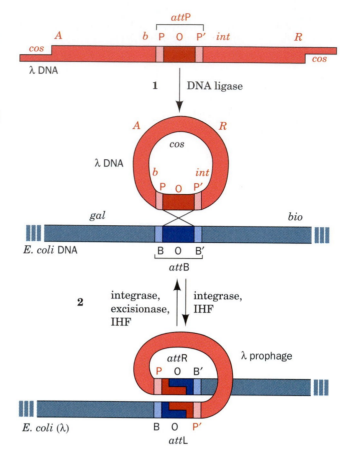

and *att*B on the bacterial host (Fig. 27-22). These two *att*achment sites have a 15-bp identity (Fig. 27-23, *top*) so that they can be represented as having the sequences POP′ for *att*P and BOB′ for *att*B, where O represents the common sequence. Phage integration occurs through a process that yields the inserted phage chromosome flanked by the sequences BOP′ (the *att*L site) on the "left" and POB′ (the *att*R site) on the "right." The nature of the crossover site was determined through the use of ^{32}P-labeled bacterial DNA and unlabeled phage DNA: The crossover site occurs at a unique position on each strand that is displaced with respect to its complementary strand so as to form a staggered recombination joint (Fig. 27-23, *bottom*).

Phage integration is mediated by **integrase,** the product of the *int* gene, which acts as a Type I topoisomerase: It nicks and reseals one strand of double-helical DNA (Section 23-1C). Integrase also functions in concert with **excisionase** (the *xis* gene product) in removing the prophage from the host DNA in the lytic pathway.

The establishment of lysogeny is triggered by high concentrations of cII. This protein stimulates transcription from the promoters p_I (I for *i*ntegrase) and p_{RE} (RE for *r*epressor *e*stablishment; Fig. 27-20). Transcription from p_I, which is located within the *xis* gene, yields integrase but not excisionase. λ DNA is consequently integrated into the host chromosome. The reason why a high cII concentration is required to establish lysogeny is that this early gene product can stimulate transcription from p_I and p_{RE} only when it is in oligomeric form. Consequently, when the ratio of infecting phages to bacteria is high (and cII is synthesized at a high rate), lysogeny is favored, thereby preventing the large number of phages from eliminating their host cells.

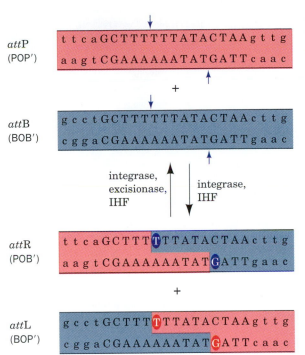

attP
(POP′)
t t c a G C T T T T T T A T A C T A A g t t g
a a g t C G A A A A A A T A T G A T T c a a c
+
attB
(BOB′)
g c c t G C T T T T T T A T A C T A A c t t g
c g g a C G A A A A A A T A T G A T T g a a c
integrase,
excisionase,
IHF
integrase,
IHF
attR
(POB′)
t t c a G C T T T T T T A T A C T A A c t t g
a a g t C G A A A A A A T A T G A T T g a a c
+
attL
(BOP′)
g c c t G C T T T T T T A T A C T A A g t t g
c g g a C G A A A A A A T A T G A T T c a a c

Figure 27-23. **The site-specific recombination process that inserts/excises phage λ DNA into/from the *E. coli* chromosome.** Exchange occurs between the phage *att*P site (*red*) and the bacterial *att*B site (*blue*), and the prophage *att*L and *att*R sites. The strand breaks occur at the approximate positions indicated by the short blue arrows. The sources of the more darkly shaded bases in *att*R and *att*L are uncertain. The uppercase letters represent bases common to the phage and bacterial DNAs, whereas lowercase letters symbolize bases in the flanking B, B′, P, and P′ sites.

The transcript initiated from p_{RE} includes the ***cI*** gene, which encodes the **λ** or **cI repressor.** This protein, like Cro protein, blocks transcription from p_L and p_R, thereby shutting down the synthesis of the early gene products necessary for lytic growth. λ repressor also stimulates the transcription of its own gene from p_{RM} (RM for *r*epressor *m*aintenance). The continued synthesis of the λ repressor maintains lysogeny from generation to generation.

The λ Genetic Switch

The difference between the Cro protein (which represses all λ mRNA synthesis) and λ repressor (which stimulates transcription of its own gene while repressing all other λ mRNA synthesis) is the basis for a genetic switch that determines whether phage λ follows a lytic or lysogenic program. The switch mechanism, which was largely elucidated by Mark Ptashne, is finely tuned to tightly repress lytic growth and yet remain poised to turn it on efficiently. Thus, under normal conditions, lysogens spontaneously induce only about once per 10^5 cell divisions, but transient exposure to inducing conditions triggers lytic growth in almost every cell of a lysogenic culture.

The action of the λ switch is centered on the operator $\boldsymbol{o_R}$ (Fig. 27-20) which is located between p_{RM} and p_R and consists of three subsites designated $\boldsymbol{o_{R1}}$, $\boldsymbol{o_{R2}}$, and $\boldsymbol{o_{R3}}$ (Fig. 27-24). Each subsite consists of a homologous 17-bp segment that has approximately palindromic symmetry.

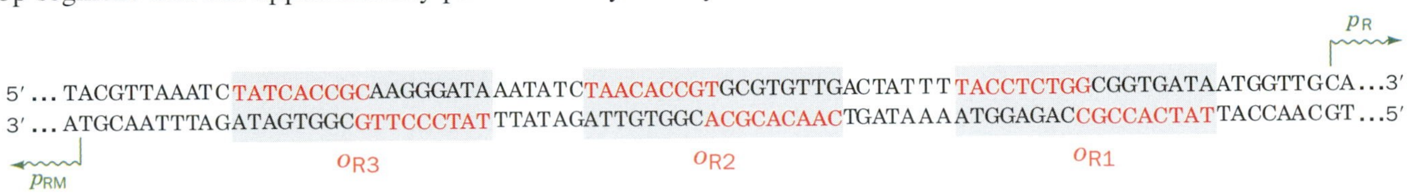

pR
5′... TACGTTAAATCTATCACCGCAAGGGATAAATATCTAACACCGTGCGTGTTGACTATTTTACCTCTGGCGGTGATAATGGTTGCA...3′
3′... ATGCAATTTAGATAGTGGCGTTCCCTATTTATAGATTGTGGCACGCACAACTGATAAAATGGAGACCGCCACTATTACCAACGT...5′
pRM
oR3
oR2
oR1

Figure 27-24. **The base sequence of the o_R region of the phage λ chromosome.** The operator consists of three homologous 17-bp subsites separated by short AT-rich spacers. Each subsite has approximate palindromic symmetry as is demonstrated by the comparison of the two sets of red letters in each subsite. The wiggly arrows mark the transcription start sites and directions at the indicated promoters.

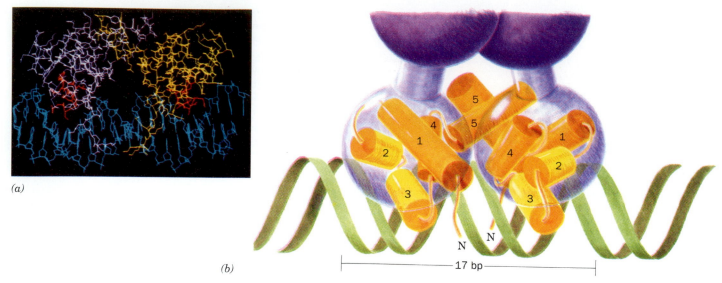

(a)

(b)

Figure 27-25.   The X-ray structure of a dimer of λ repressor N-terminal domains in complex with B-DNA. (*a*) Stick-form representation of the complex in which the DNA is blue, the two repressor N-terminal domains are yellow and white, and the recognition helices of the HTH motif are red. Note that the protein's N-terminal arms wrap around the DNA. [Courtesy of Carl Pabo, The Johns Hopkins University.] (*b*) An interpretive drawing indicating how contacts between the repressor's C-terminal domains (*upper lobes; not part of the X-ray structure) maintain the intact protein's dimeric character. The λ repressor binds to the 17-bp operator subsites as symmetric dimers with the N-terminal domain of each subunit specifically binding to a half-subsite. Note how the α3 recognition helices of the symmetry-related α2–α3 HTH units (*light yellow*) fit into successive turns of the DNA's major groove. [After Ptashne, M., *A Genetic Switch* (2nd ed.), p. 38, Cell Press & Blackwell Scientific Publications (1992).] ● See the Interactive Exercises.

λ repressor binds to DNA as a homodimer so that its two-fold symmetry matches that of the operator subsites. The monomer's 236-residue polypeptide chain is folded into two roughly equal-sized domains connected by an ~30-residue segment that is readily cleaved by proteolytic enzymes. The isolated N-terminal domains retain their ability to bind specifically to operators (although with only half of the binding energy of the intact repressor) but cannot dimerize. The C-terminal domains can still dimerize but lack the capacity to bind DNA. Evidently, *λ repressor's N-terminal domain binds operator, whereas its C-terminal domain provides the contacts for dimer formation.*

Although the λ repressor has not been crystallized, its N-terminal domain comprising residues 1 to 92, as excised by treatment with the papaya protease **papain,** does crystallize. The X-ray structure of this protein, both alone and in complex with a 20-bp DNA containing an operator subsite sequence, was determined by Carl Pabo. The N-terminal domain crystallizes as a symmetric dimer even though it does not dimerize in solution. Each subunit contains an N-terminal arm and five α helices (Fig. 27-25), two of which, α2 and α3, form a helix–turn–helix (HTH) motif, much like those in other prokaryotic repressors of known structure (Section 23-4B). The α3 helix, the recognition helix, protrudes from the protein surface such that the two α3 helices of the dimeric protein fit into successive grooves of the operator DNA.

Cro protein also forms homodimers. Its 66-residue subunit forms a single domain that contains its operator recognition sites as well as its dimerization contacts. The X-ray structure of Cro protein in complex with its operator DNA (Fig. 27-26), determined by Brian Matthews, reveals that this dimer, like the λ repressor, interacts with DNA via HTH motifs and, in addition, induces the DNA to bend around the protein by 40°.

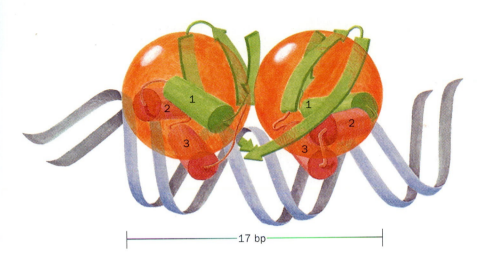

├──────── 17 bp ────────┤

Figure 27-26. **The X-ray structure of the Cro protein dimer in complex with B-DNA.** Note that the λ repressor (Fig. 27-25), although otherwise dissimilar, contains HTH units that also bind in successive turns of the DNA's major groove. [After Ptashne, M., *A Genetic Switch* (2nd ed.), p. 40, Cell Press & Blackwell Scientific Publications (1992).] ● See the Interactive Exercises.

Chemical and nuclease protection experiments have indicated that the λ repressor has the following order of intrinsic affinities for the subsites of o_R (Fig. 27-27):

$$o_{R1} > o_{R2} > o_{R3}$$

Despite this order, o_{R1} and o_{R2} are filled nearly together. This is because λ repressor bound at o_{R1} cooperatively binds repressor at o_{R2} through associations between their C-terminal domains (Fig. 27-27c). o_{R1} and o_{R2} are therefore both occupied at low λ repressor concentrations, whereas o_{R3} becomes occupied only at higher repressor concentrations.

The binding of λ repressor to o_R, as previously mentioned, abolishes transcription from p_R and stimulates transcription from p_{RM}. This stimulation results from the repressor at o_{R2} directly contacting RNA polymerase, thereby helping the polymerase bind to p_{RM} (Fig. 27-27c). Conse-

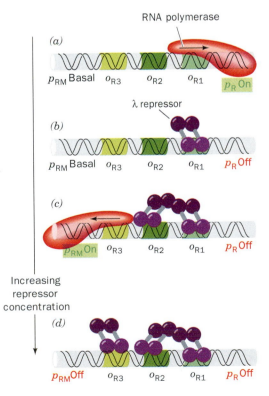

Figure 27-27. **The binding of λ repressor to the three subsites of o_R.** (a) In the absence of repressor, RNA polymerase initiates transcription at a high level from p_R (*right*) and at a basal level from p_{RM}. (b) The repressor has ~10 times higher affinity for o_{R1} than it has for o_{R2} or o_{R3}. Repressor dimer therefore first binds to o_{R1} so as to block transcription from p_R. (c) A second repressor dimer binds to o_{R2} at only slightly higher repressor concentrations due to specific binding between the C-terminal domains of neighboring repressors. In doing so, it stimulates RNA polymerase to initiate transcription from p_{RM} at a high level (*left*). (d) At high repressor concentrations, repressor binds to o_{R3} so as to block transcription from p_{RM}. [After Ptashne, M., *A Genetic Switch* (2nd ed.), p. 23, Cell Press & Blackwell Scientific Publications (1992).]

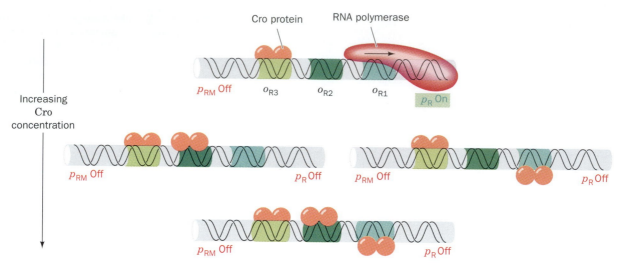

Figure 27-28.   The binding of Cro protein to the three o_R subsites. o_{R3} binds Cro ~10 times more tightly than does o_{R1} or o_{R2}. Cro dimer therefore first binds to o_{R3}. A second dimer then binds to either o_{R1} or o_{R2} and in each case blocks transcription from p_R.

At high Cro concentrations, all three operator subsites are occupied. Compare this binding sequence with that of λ repressor (Fig. 27-27). [After Ptashne, M., *A Genetic Switch* (2nd ed.), p. 27, Cell Press & Blackwell Scientific Publications (1992).]

quently, *λ repressor prevents the synthesis of all phage gene products but itself.* At very high concentrations of λ repressor, however, transcription from p_{RM} is also repressed (Fig. 27-27*d*). This maintains the repressor concentration within reasonable limits.

Cro protein binds to the subsites of o_R in an order opposite to that of λ repressor (Fig. 27-28):

$$o_{R3} > o_{R2} \approx o_{R1}$$

This binding is noncooperative. Cro protein binding to o_{R3} abolishes transcription from p_{RM} and thereby prevents the synthesis of λ repressor. Additional Cro binding to o_{R2} and/or o_{R1} turns off transcription from p_R. *Lytic versus lysogenic growth therefore depends on which repressor occupies the o_R subsites.*

DNA damage that induces *E. coli*'s SOS response (Section 24-5C) leads to RecA activation. This protein stimulates the self-cleavage of the λ repressor between its N- and C-terminal domains, thereby abolishing the repressor's ability to bind cooperatively to o_{R2}. The diminished binding of λ repressor to o_R allows the early genes, including *cro*, to be transcribed. As Cro accumulates, it first binds to o_{R3} so as to block even basal levels of λ repressor synthesis. Since there is no mechanism for selectively inactivating Cro, the phage irreversibly enters the lytic mode: The λ switch, once thrown, cannot be reset. The prophage is subsequently excised by the newly synthesized integrase and excisionase, new phage particles are synthesized, and the damaged host is lysed, yielding phage particles prepared to infect a new host (Fig. 27-18).

3. REGULATION OF EUKARYOTIC GENE EXPRESSION

The general principles that govern the expression of prokaryotic genes apply also to eukaryotic genes: *The expression of specific genes may be actively inhibited or stimulated through the effects of proteins that bind to DNA or RNA.* As in prokaryotes, the majority of known regulatory mechanisms act at the level of gene transcription, but unlike prokaryotic control systems, the eukaryotic mechanisms must contend with much larger amounts of DNA that is packaged in seemingly inaccessible structures. In this section, we describe a few of the strategies whereby eukaryotic cells manage their genetic information.

A. Chromatin Structure and Gene Expression

The vast majority of DNA in multicellular organisms is not transcribed. This includes the large portion of the genome that does not encode protein or RNA, as well as the genes whose expression is inappropriate for a particular cell type. Although nearly all the cells in an organism contain identical sets of DNA, genes are expressed in a highly tissue-specific manner. For example, most pancreatic cells synthesize and secrete digestive enzymes, but the pancreatic islet cells synthesize insulin or glucagon instead (Section 21-3A).

Nonexpressed DNA is typically highly condensed in a form known as **heterochromatin.** An extreme example of this is the complete inactivation of one of the two X chromosomes in female mammals, which is visible as a darkly staining **Barr body** in the cell nucleus (Fig. 27-29). Transcriptionally active DNA (known as **euchromatin**) is less condensed, presumably to provide access to the transcription machinery. In fact, Harold Weintraub demonstrated that transcriptionally active chromatin is more susceptible to digestion by pancreatic DNase I than transcriptionally inactive chromatin. Yet nuclease sensitivity apparently reflects a gene's potential for transcription rather than transcription itself.

The banding pattern of **polytene chromosomes** found in certain secretory cells of dipteran (two-winged) flies reflects the selective condensation of nontranscribed DNA (Fig. 27-30). Polytene chromosomes result from the multiple replications of chromosomes such that the replicas remain attached to each other and in register. The light bands of such chromosomes correspond to specific genes; when undergoing transcription, these bands are further decondensed to form **chromosome puffs** (Fig. 27-31). In *Drosophila,* these puffs reproducibly form and regress as part of the nor-

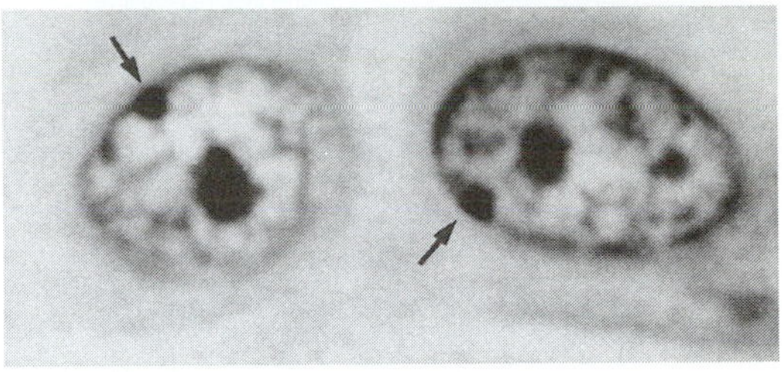

Figure 27-29. Barr body. The dark area at the periphery of each stained nucleus (*arrows*) is the Barr body, a condensed X chromosome. [George Wilder/Visuals Unlimited.]

Figure 27-30. An electron micrograph of a segment of a polytene chromosome from *D. melanogaster.* Note that its interband regions consist of chromatin fibers that are more or less parallel to the long axis of the chromosome, whereas its bands, which contain ~95% of the chromosome's DNA, are much more highly condensed. [Courtesy of Gary Burkholder, University of Saskatchewan, Canada.]

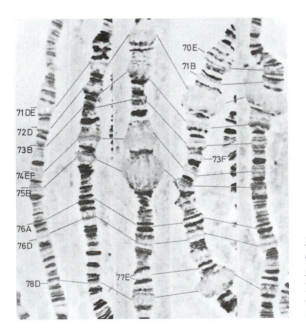

Figure 27-31. Formation and regression of chromosome puffs. This series of photomicrographs shows puffs (*lines*) in a *D. melanogaster* polytene chromosome over a 22-h period of larval development. [Courtesy of Michael Ashburner, Cambridge University, UK.]

mal larval development program and in response to physiological stimuli such as hormones and heat.

Histone Acetylation Influences Nucleosome Structure

Since nucleosomes bind DNA tightly and quite stably, how does RNA polymerase and its associated transcription factors get access to a gene's promoter in order to initiate transcription? A likely answer is that *histone proteins are acetylated and deacetylated in a controlled fashion to alter the structure of chromatin.* The ε-amino groups of Lys residues near the N-termini of histones H3 and H4 are acetylated by cytoplasmic **histone acetylases;** this modification appears to be necessary for the subsequent assembly of the nucleosome's histone octamer core (Section 23-5B). In addition, a set of nuclear histone acetylases can modify specific Lys residues in all four core histones.

As we discussed in Section 23-5C, acetylation neutralizes the positively charged Lys residues that extend out from the nucleosome interior (see Fig. 23-44*a*), thereby weakening their affinity for neighboring nucleosomes and, possibly, for negatively charged DNA. This disruption of nucleosome/chromatin structure is thought to "open up" the nucleosome for transcription. The link between histone acetylation and transcriptional competence is provided by the observation that the nuclear histone acetylases associate with transcription factors. In fact, one of the TAF components of TFIID (Section 25-2C) is a histone acetylating enzyme. A family of **histone deacetylases,** which remove the acetyl groups, apparently reverse the transcription activating effects of histone acetylation. The "silencing" of genes through histone deacetylation may facilitate the recruitment of RNA polymerase to other genes by limiting the amount of chromatin that must be scanned by transcription factors. In mammals, transcriptionally silent DNA is also typically more highly methylated than is transcribed DNA.

Histone acetylases and deacetylases appear to occur in all eukaryotes, although the sites of modification vary by species. The different acetylases also differ in which histone Lys residues they modify; this may affect which transcription factors are recruited to that site. Alternatively, the presence of a transcription factor may determine to what extent acetylation opens up the nucleosome to transcription.

It is not known whether targeted histone acetylation merely renders DNA more accessible to the transcription machinery or completely displaces the entire nucleosome. The first possibility is more likely, given the observation that, in model systems, actively transcribing RNA polymerase dislodges the nucleosome but only to a point 40 to 95 bp upstream of its original site. Consideration of DNA topology also makes this scenario plausible. A moving transcription bubble generates positive supercoils in the DNA ahead of it and negative supercoils behind it (Fig. 25-7). However, nucleosomal DNA is wound around its histone core in a left-handed toroidal coil and is therefore negatively supercoiled. Consequently, an advancing RNA polymerase molecule should destabilize the nucleosome ahead of it while facilitating nucleosome assembly in its wake, precisely what is observed.

B. Control of Transcription in Eukaryotes

As described in Section 25-2, the process of initiating the transcription of eukaryotic genes requires many different proteins and binding sites on DNA. *Activators and repressors of transcription may bind some distance from the promoters at sites known as* **enhancers** *and* **silencers.** For example, William Rutter has linked the 5′-flanking sequences of either the insulin or the chymotrypsin gene to the sequence encoding **chloramphenicol acetyl-transferase (CAT),** an easily assayed enzyme not normally present in eukaryotic cells. A plasmid containing the insulin gene recombinant elicits expression of the CAT gene only when introduced into cultured cells that normally produce insulin. Likewise, the chymotrypsin recombinants are active only in chymotrypsin-producing cells. Dissection of the insulin control sequence indicates that the enhancer lies between positions -103 and -333, and, in insulin-producing cells only, it stimulates the transcription of the CAT gene with little regard to its position and orientation relative to the promoter.

Enhancers and silencers are recognized by proteins, such as TAFs (Section 25-2C), that act in addition to the general transcription factors that are essential to transcribe all genes. How do these "upstream transcription factors" stimulate (or inhibit) transcription? Evidently, when these proteins bind to their target DNA sites, they somehow activate (or repress) transcription initiation. The proteins may bind cooperatively to each other and/or to the polymerase in a manner resembling the binding of the λ repressor tetramer and RNA polymerase to the operator o_R of phage λ (Fig. 27-27c). Molecular cloning experiments indicate that many enhancers consist of modules whose individual deletion reduces but does not eliminate enhancer activity. Such complex arrangements presumably permit transcriptional control systems to respond to a variety of stimuli in a graded manner.

The transcription factors themselves typically consist of a DNA-binding domain and a domain that activates the transcription machinery. Genetically engineered hybrid proteins containing the DNA-binding activity of one transcription factor and the transcription-activating function of another activate the same genes as the first transcription factor. Indeed, it makes little difference whether the activation domain is placed on the N-terminal side of the DNA-binding domain or on its C-terminal side. This geometric permissiveness in binding is extended to the enhancer DNA sequence itself: *Transcription factors are largely insensitive to the orientations of their enhancers relative to the transcription start site.*

The foregoing observations suggest a model for transcription activation in which transcription factors interact with enhancer elements together with the preinitiation complex [PIC; the complex of general transcription factors (GTFs) assembled on the TATA box that recruits RNA polymerase to the transcription start site so as to support a basal level of transcription; Fig. 25-14]. This causes the DNA to loop out (Fig. 27-32). The rate of tran-

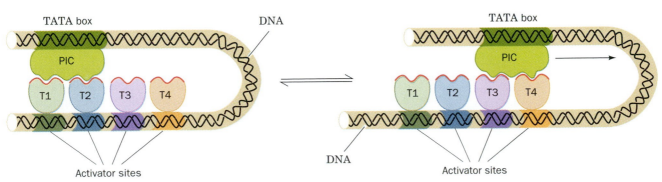

Figure 27-32. A model for the action of transcription factors. Here, four transcription factors, T1, T2, T3, and T4, are bound to their target DNA sequences and simultaneously, in groups of two via relatively nonspecific interactions (*red*), to the preinitiation complex (PIC; Section 25-2C) bound at the TATA box. The tran-

scription factor–PIC interaction is probably rather weak, but it is conjectured that continual dissociation and reassociation maintains this interaction so that RNA polymerase can attach to the PIC and initiate transcription. [After Ptashne, M., *Nature* **335**, 687 (1988).]

scription of a given gene tends to increase with the number of upstream transcription factors because opportunities for multisite binding increase the chances of recruiting the PIC and RNA polymerase and hence successfully completing transcription initiation.

Extracellular Signals Influence Gene Transcription

Different cell types express cell-specific proteins as part of their internally programmed development and in response to extracellular signals. We saw in Section 21-3C how ligand binding to a cell-surface receptor can stimulate a kinase cascade leading to the activation of transcription factors in the nucleus. At least two other types of signaling pathways transmit extracellular messages to the transcription machinery.

Cytoplasmic proteins known as **signal transducers and activators of transcription (STATs)** are activated in response to ligand binding to a variety of cell-surface receptors in organisms as diverse as *Dictyostelium* (slime molds) and mammals. STAT proteins contain a DNA-binding domain, an SH2 domain, and an SH3 domain. SH2 and SH3 domains, as we have seen, are common elements of proteins that participate in receptor tyrosine kinase signaling pathways (Section 21-3C). STAT proteins are activated by ligand-activated tyrosine kinases via their phosphorylation of a single STAT Tyr residue. The phosphorylated STAT then dimerizes by reciprocal SH2–phosphotyrosine interactions and enters the nucleus, where it binds DNA so as to stimulate the transcription of specific genes. For example, **STAT5A** binds to the promoter for the **β-casein** gene (β-casein is a major protein component of milk) in mammary gland cells stimulated by the hormone **prolactin.** Gene "knockout" mice (Sections 3-5D and 13-4B) that lack STAT5A are unable to lactate. DNA-bound STATs interact with other transcription activators, raising the possibility of regulating gene expression through synergistic actions with other transcription-activating pathways. The duration of STAT-induced transcription depends on the rate at which the protein is dephosphorylated.

Steroid hormones, which mediate a wide variety of physiological processes, likewise act by altering gene expression. These hormones, which are nonpolar molecules, readily pass through the plasma membranes of cells in order to bind to their cognate receptors in the cytosol. The steroid–receptor complexes, in turn, enter the nucleus, where they bind to enhancers known as **hormone response elements** so as to induce, or in some cases repress, transcription of the associated genes. Thus, eukaryotic steroid receptors are inducible transcription factors whose actions resemble those of prokaryotic transcription regulators such as the *E. coli* CAP–cAMP complex (Section 27-2B). The **glucocorticoid receptor** (glucocorticoids are steroids that affect carbohydrate metabolism; Section 9-1E) is representative of a large family of proteins that have a common organization including domains for ligand binding, DNA binding, and transcriptional regulation. The DNA-binding domain of the glucocorticoid receptor contains two sets of four Cys residues that each coordinate a Zn^{2+} ion. The X-ray structure of this zinc finger domain bound to DNA shows that the receptor forms a homodimer in which each domain contacts half of a symmetrical hormone response element (Fig. 27-33).

Cyclin-Dependent Kinases Regulate Cell Growth

The **cell cycle,** the general sequence of events that occurs during the lifetime of a eukaryotic cell, is divided into four distinct phases (Fig. 27-34):

1. Mitosis and cell division occur during the relatively brief **M phase** (for mitosis).

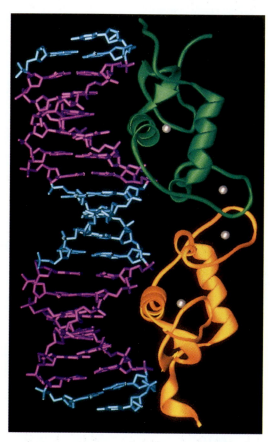

Figure 27-33. **The X-ray structure of the dimeric glucocorticoid receptor DNA-binding domain in complex with DNA.** The 18-bp DNA has a single nucleotide overhang at each of its 5' ends and contains two symmetrical 6-bp glucocorticoid response element half-sites (*magenta*) separated by a 4-bp spacer (*cyan*). The protein subunits are represented by green and gold ribbons, and the Zn^{2+} ions are shown as silver spheres. Note how the glucocorticoid receptor's two N-terminal helices are inserted into successive major grooves in a manner reminiscent of the recognition helices of prokaryotic HTH domains. [Based on an X-ray structure by Paul Sigler, Yale University.] ● **See the Interactive Exercises.**

2. This is followed by the **G₁ phase** (for gap), which covers the longest part of the cell cycle.

3. G₁ gives way to the **S phase** (for synthesis), which, in contrast to events in prokaryotes, *is the only period in the cell cycle when DNA is synthesized.*

4. During the relatively short **G₂ phase,** the now tetraploid cell prepares for mitosis. It then enters M phase once again and thereby commences a new round of the cell cycle.

The cell cycle for cells in culture typically occupies a 16- to 24-h period. In contrast, cell cycle times for the different types of cells in a multicellular organism may vary from as little as 8 h to > 100 days. Most of this variation occurs in the G₁ phase. Moreover, many terminally differentiated cells, such as neurons or muscle cells, never divide; they assume a quiescent state known as the **G₀ phase.**

Progression through the cell cycle depends on the expression of certain proteins at appropriate times. Both basal and activated transcription are affected, largely through the phosphorylation of various components of the transcription apparatus. The first clues to the regulation of the cell cycle came from studies of marine invertebrate embryos, in which a class of proteins named **cyclins** accumulate steadily throughout the cell cycle and then abruptly disappear during mitosis. Cyclins, which have been discovered in many eukaryotes, combine with a 34-kD protein to form an active Ser/Thr protein kinase, **cyclin-dependent protein kinase (CDK),** whose full activity may also require phosphorylation.

The cell cycle in humans is governed by several CDKs. They do so by phosphorylating nuclear proteins, among them histone H1, several oncogene proteins (see below), and proteins involved in nuclear disassembly and cytoskeletal rearrangement. This presumably initiates a cascade of cellular events that culminates in mitosis.

The X-ray structure of one of these CDKs, human **CDK2,** has been determined in several conformational states: (1) in its complex with ATP; (2) in its complex with ATP and the C-terminal fragment of **cyclin A** (residues 173–432); and (3) in its complex with ATPγS (a slowly hydrolyzable ATP analog) and the C-terminal fragment of cyclin A in which Thr 160 of CDK2 is phosphorylated (a covalent modification required for full catalytic activity of the CDK2–cyclin A complex; Fig. 27-35). In the CDK2–ATP com-

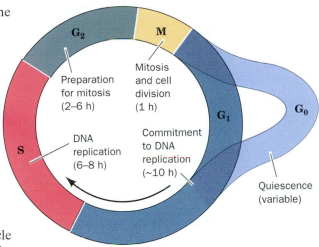

Figure 27-34.   The eukaryotic cell cycle. Cells may enter a quiescent phase (G₀) rather than continuing about the cycle.

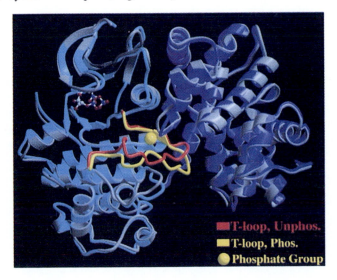

Figure 27-35.   The X-ray structure of the CDK2–cyclin A–ATPγS complex in which CDK2 is phosphorylated at its Thr 160. Cyclin A is purple, and CDK2 is cyan with its T-loop yellow and its phosphate group represented by a yellow ball. This structure is superimposed on that of the unphosphorylated complex, which is gray with its T-loop red. [Courtesy of Nikola Pavletich, Memorial Sloan-Kettering Cancer Center, New York.] ● See the Interactive Exercises.

plex, the so-called T-loop (residues 152–170) blocks the entrance to the catalytic cleft, whereas in the cyclin A–CDK2–ATP complex, the T-loop assumes a conformation similar to that of the catalytically active cAMP-dependent protein kinase (cAPK; Fig. 15-14). However, in the X-ray structure of the cyclin A–CDK2–ATPγS complex in which CDK2's Thr 160 is phosphorylated, the phosphate group fits snugly into a positively charged pocket that forms in part on cyclin binding, thereby moving the T-loop by as much as 7 Å. This further reorganizes CDK2 such that it resembles cAPK even more closely and results in additional CDK2–cyclin A contacts. Thus, although cyclin A is the primary activator of CDK2, the phosphate group at CDK2's Thr 160 functions as a major reorganizing center in the cyclin A–CDK2 complex.

Some Cancers Result from Loss of Tumor Suppressors

One of the targets of certain CDKs is the **retinoblastoma protein (Rb),** which inhibits cell growth by inhibiting the function of the transcription factor **E2F.** E2F activates genes that promote cell proliferation, including the cyclins. Phosphorylation of Rb by CDK reverses Rb's ability to restrain cell growth. Rb is known as a **tumor suppressor** because its loss may lead to **retinoblastoma** (a cancer of the developing retina) and other tumors. The activity of E2F is also directly inhibited through phosphorylation by some CDKs.

The transcription factor and tumor suppressor **p53,** unlike E2F, is activated by CDK-catalyzed phosphorylation. Phosphorylated p53 binds to specific genes and acts as a powerful transcription activator. Mutations in the p53 gene are present in ~50% of human cancers. Many of the mutant p53's that are implicated in cancer have lost their sequence-specific DNA-binding properties (Fig. 27-36). p53 normally acts as a "molecular policeman" in monitoring genome integrity. When the genome is damaged, p53 accumulates through an unknown mechanism, thereby activating the transcription of a gene, *Pic1.* The 21-kD product of *Pic1* binds to and inhibits various CDKs and consequently arrests the cell cycle in order to allow time for DNA repair. If DNA repair is unsuccessful, p53 may trigger cell suicide, a process named **apoptosis,** which prevents the proliferation of the genetically damaged and hence cancer-prone cell. Thus, the inactivation of p53 permits these cancerous cells to proliferate.

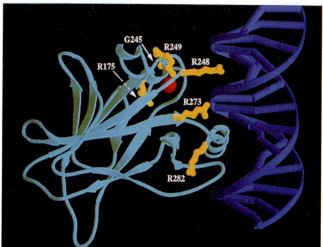

Figure 27-36.   The X-ray structure of the DNA-binding domain of human p53 in complex with its target DNA. The protein is shown in ribbon form (*cyan*) and the DNA in ladder form with its bases represented by cylinders (*blue*). A tetrahedrally liganded Zn^{2+} ion is shown as a red sphere, and the side chains of the six most frequently mutated side chains in human tumors are shown in stick form (*yellow*) and identified with their one-letter codes. [Courtesy of Nikola Pavletich, Memorial Sloan-Kettering Cancer Center, New York.] ● See the Interactive Exercises.

C. Somatic Recombination and Antibody Diversity

Genetic recombination in germline cells is a fundamental feature of reproduction in multicellular organisms. Genetic rearrangement also occurs in certain other types of cells to generate new sets of genetic information. For example, **somatic recombination** (Greek: *soma,* body) is responsible for the expression of an enormous variety of immunoglobulin (antibody) gene products. As a result, the immune system can potentially recognize an almost limitless number of different antigens. The required antibody diversity also results from an accelerated rate of mutation during the development of antibody-producing *B* cells (Section 7-4D).

κ Light Chain Genes Are Assembled from Multiple Gene Segments

One of the two types of immunoglobulin light chains, the **κ chain,** is encoded by four exons (Fig. 27-37):

1. A **leader** or L_κ **segment,** which encodes a 17- to 20-residue hy-

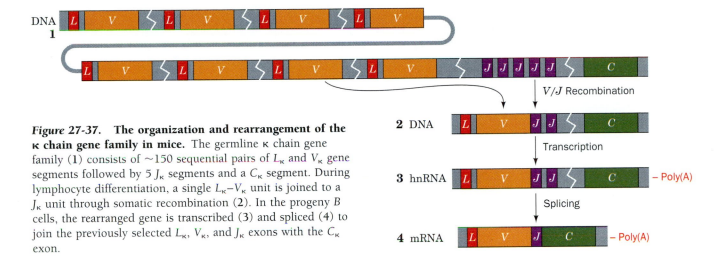

Figure 27-37. The organization and rearrangement of the κ chain gene family in mice. The germline κ chain gene family (1) consists of ~150 sequential pairs of $L_κ$ and $V_κ$ gene segments followed by 5 $J_κ$ segments and a $C_κ$ segment. During lymphocyte differentiation, a single $L_κ$–$V_κ$ unit is joined to a $J_κ$ unit through somatic recombination (2). In the progeny B cells, the rearranged gene is transcribed (3) and spliced (4) to join the previously selected $L_κ$, $V_κ$, and $J_κ$ exons with the $C_κ$ exon.

drophobic signal peptide. This polypeptide directs newly synthesized κ chains to the endoplasmic reticulum and is then excised (Section 10-2D).

2. A **$V_κ$ segment,** which encodes the first 95 residues of the κ chain's 108-residue variable region.

3. A **joining** or **$J_κ$ segment,** which encodes the variable region's remaining 13 residues.

4. The **$C_κ$ segment,** which encodes the κ chain's constant region.

In embryonic tissues (which do not make antibodies), these exons occur in clusters. The κ chain gene family is made of ~150 $L_κ$ and $V_κ$ segments, separated by introns, with the $L_κ$–$V_κ$ units separated from each other by ~7-kb spacers. This sequence of exon pairs is followed, well downstream, by 5 $J_κ$ segments at intervals of ~300 bp, a 2.4-kb spacer, and a single $C_κ$ segment.

The assembly of a κ chain mRNA is a complex process involving both somatic recombination and selective gene splicing (Section 27-3D). The first step of this process, which occurs in B cell progenitor cells, is an intrachromosomal recombination that joins an $L_κ$–$V_κ$ unit to a $J_κ$ segment and deletes the intervening sequences (Fig. 27-37). Then, in later cell generations, the entire modified gene is transcribed and selectively spliced so as to join the $L_κ$–$V_κ$–$J_κ$ unit to the $C_κ$ segment. The $L_κ$ and $V_κ$ segments are also spliced together in this step, yielding an mRNA that encodes one of each of the four elements of a κ chain gene.

Recombinational Flexibility Contributes to Antibody Diversity

The joining of 1 of 150 $V_κ$ segments to 1 of 5 $J_κ$ segments can generate only 150 × 5 = 750 different κ chains, far less than the number observed. However, studies of many joining events involving the same $V_κ$ and $J_κ$ segments revealed that *the V/J recombination site is not precisely defined; these two gene segments can join at different crossover points* (Fig. 27-38). Consequently, the amino acids specified by the codons in the vicinity of the V/J recombination site depend on what part of the sequence is supplied by the germline $V_κ$ segment and what part is supplied by the germline $J_κ$ segment. Assuming that this recombinational flexibility increases the possible κ chain diversity 10-fold, the expected number of possible different κ chains is increased to 150 × 5 × 10 = 7500.

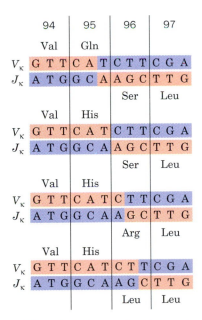

Figure 27-38. Variation at the $V_κ$/$J_κ$ joint. The crossover point at which the $V_κ$ and $J_κ$ sequences somatically recombine varies by several nucleotides, thereby giving rise to different nucleotide sequences (*brown bands*) in the active κ gene. For example, as indicated here, amino acid 96, which occurs in the κ chain's third hypervariable region, can be Ser, Arg, or Leu.

Germline heavy chain DNA

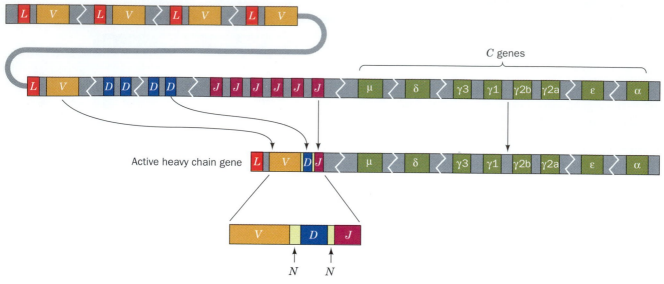

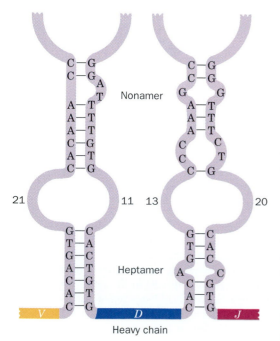

Figure 27-39. **The organization and rearrangement of the heavy chain gene family in humans.** This gene family consists of ~250 sequential pairs of L_H and V_H gene segments followed by ~10 D segments, 6 J_H segments, and 8 C_H segments (one for each class or subclass of heavy chains; Table 7-2). During lymphocyte differentiation, an L_H–V_H unit is joined to a D segment and a J_H segment. In this process, the D segment becomes flanked by short stretches of random sequence called N regions. In the B cell and its progeny, transcription and splicing joins the L_H–V_H–N–D–N–J_H unit to one of the 8 C_H gene segments.

The other type of immunoglobulin light chain, the **λ chain,** is similarly encoded by a gene family containing L_λ, V_λ, J_λ, and C_λ segments whose recombination likewise yields a large number of possible polypeptides.

Heavy Chain Genes Are Also Assembled from Sets of Gene Segments

Heavy chain genes are assembled in much the same way as are light chain genes but with the additional inclusion of an ~13-bp **diversity** or **D** segment between their V_H and J_H segments. The human heavy chain gene family consists of clusters of ~250 different L_H–V_H units, ~10 D segments, 6 J_H segments, and 8 C_H segments (Fig. 27-39). Germline V_H, D, and J_H segments are joined in a particular order (D is joined to J_H before V_H is joined to DJ_H), and the joining sites are subject to the same recombination flexibility as are light chain V/J sites.

Highly conserved sequences on each side of the V, D, and J gene segments can form stem-and-loop secondary structures that act as recombination signals (Fig. 27-40). The structural similarities and functional interchangeabilities of these sites suggest that all $V(D)J$ joining reactions are

Figure 27-40. **Stem-and-loop recombination sites in the germline heavy chain gene family.** These structures mediate somatic recombination between the V_H and D segments (*left*) and between the D and J_H segments (*right*). The κ chain V/J recombination signal consists of a similar heptamer–nonamer stem-and-loop structure. The recombination system's requirement for both the 20/21- and the 11/13-bp spacers prevents it from inadvertently skipping the D segment by directly joining the V_H and J_H segments.

catalyzed by an evolutionarily conserved ***V(D)J recombinase*** system. Indeed, David Baltimore discovered two proteins, **RAG1** and **RAG2,** that work in concert to recognize the cleavage site and participate in the required double-stranded cleavage of the DNA as well as in its subsequent joining.

Assuming that recombinational flexibility contributes a factor of 100 toward heavy chain diversity, somatic recombination can generate some $250 \times 10 \times 6 \times 100 = 1.5 \times 10^6$ different heavy chains. Then, taking into account κ chain diversity (and neglecting that of λ chains), there can be as many as $7500 \times 1.5 \times 10^6 = 11$ billion different types of immunoglobulins formed by somatic recombination among ~400 different gene segments.

Somatic Mutation Is a Further Source of Antibody Diversity

Despite the enormous antibody diversity generated by somatic recombination, immunoglobulins are subject to even more variations through **somatic mutations** of two types:

1. During V_H/D and D/J_H joining, a few nucleotides may be added or removed from the recombination joints. The added nucleotides, which form so-called ***N* regions,** yield *NDN* units of up to 30 bp that encode enormously variable heavy chain segments of up to 10 amino acid residues (Fig. 27-39). The *N* regions appear to arise through the action of **terminal deoxynucleotidyl transferase,** a template-independent DNA polymerase present in the *B* cell progenitors that make the heavy chain joints. This enzyme is probably absent in later cell generations when the light chain joints are formed.

2. The variable regions of both heavy and light chains are more diverse than is expected on the basis of comparisons of their amino acid sequences with their corresponding germline nucleotide sequences. Indeed, these regions mutate at rates of up to 10^{-3} base changes per nucleotide per cell generation, rates that are at least a millionfold higher than the rates of spontaneous mutation in other genes. *B* cells and/or their progenitors apparently possess enzymes that mediate this **somatic hypermutation** of immunoglobulin gene segments.

These somatic mutation processes increase the possible number of different antibodies that humans can produce by many orders of magnitude beyond the 11 billion estimated from somatic recombination alone. The diversity arising from both recombination and mutation thereby permits an individual organism to cope, in a kind of Darwinian struggle, with the rapid mutation rates of pathogenic microorganisms.

D. Posttranscriptional and Translational Control

Virtually every stage of a protein's existence from its transcription to its posttranslational modification and ultimate demise through proteolysis offers opportunities for regulation. Indeed, the expression of most genes is probably controlled by a variety of processes. In this section, we examine a few of these regulatory mechanisms.

Alternative mRNA Splicing Yields Multiple Proteins from a Single Gene

The expression of numerous cellular genes is modulated by the selection of alternative splice sites. Thus, genes containing multiple exons may give rise to transcripts containing mutually exclusive exons. In effect, certain exons in one type of cell may be introns in another. For example, a

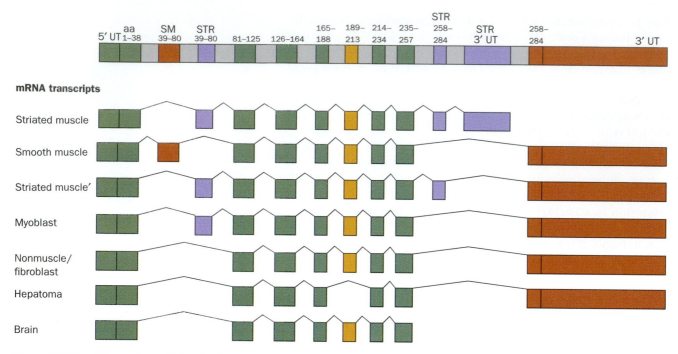

Figure 27-41.   Alternative splicing in the rat α-tropomyosin gene. Seven alternative splicing pathways give rise to cell-specific α-tropomyosin variants. The thin kinked lines indicate the positions occupied by the introns before they are spliced out to form the mature mRNAs. Tissue-specific exons are indicated together with the amino acid (aa) residues they encode: "constitutive" exons (those expressed in all tissues) are blue; those expressed only in smooth muscle (SM) are maroon; those expressed only in striated muscle (STR) are violet; and those variably expressed are yellow. Note that the smooth and striated muscle exons encoding amino acid residues 39 to 80 are mutually exclusive, and, likewise, there are alternative 3′ untranslated (UT) exons. [After Breitbart, R.E., Andreadis, A., and Nadal-Ginard, B., *Annu. Rev. Biochem.* **56**, 481 (1987).]

single rat gene encodes seven tissue-specific variants of the muscle protein **α-tropomyosin** through the selection of alternative splice sites (Fig. 27-41).

Alternative splicing arrangements can potentially take many forms, not all of which generate functional proteins. For example, failure to remove an intron from a primary transcript may yield nonfunctional mRNA. A splicing event that occurs in one tissue and not another therefore contributes to the tissue-specific synthesis of a gene product. Exon skipping, which likewise may yield nonfunctional gene products, is also observed.

mRNAs Are Degraded at Different Rates

The range of mRNA stability in eukaryotic cells, measured in half-lives, varies from a few minutes to many hours or days. The mRNA molecules themselves appear to contain elements that dictate their decay rates. These elements include the poly(A) tail, the 5′ cap, and sequences that are located within the coding region.

Deadenylation by the progressive action of exonucleases appears to be a prerequisite for mRNA decay. When the residual poly(A) tail is less than 10 nt and no longer capable of interacting with poly(A) binding protein (Section 25-3A), the mRNA is rapidly degraded by endo- or exonucleases. In some cases, deadenylation converts the mRNA to a substrate for a **decapping enzyme** that removes the 5′ cap, thereby rendering the mRNA susceptible to 5′ → 3′ exonucleolytic degradation. It is not known how events at the 3′ end of the molecule influence decapping at the 5′ end, but secondary structure in the 3′ untranslated region of the mRNA may participate in recruiting ribonucleases.

In vitro experiments suggest that the action of translation somehow shortens the lifetimes of some mRNAs, although the mechanisms underlying this phenomenon are not understood. One possibility is that the ribosome itself is equipped with machinery for destroying mRNAs.

Translational Control

In some cells, altering the rates of mRNA production or degradation do not provide the necessary level of control. For example, the early embryonic development of sea urchins, insects, and frogs depends on the rapid translation of mRNA that has been stockpiled in the oocyte. The mRNA is stored in inactive form in association with proteins but on fertilization becomes available for translation. This permits embryogenesis to commence immediately, without waiting for mRNAs to be synthesized.

Globin synthesis in reticulocytes (immature red blood cells) also proceeds rapidly, but only if heme is available. The inhibition of globin synthesis occurs at the level of translation initiation. In the absence of heme, reticulocytes accumulate a protein, **heme-controlled repressor (HCR),** that phosphorylates a specific Ser residue, Ser 51, on the α subunit of eIF-2 (the initiation factor that delivers GTP and Met–tRNA$_i^{Met}$ to the ribosome; Section 26-4A). HCR is generated, in the absence of heme, from a preexisting proinhibitor by a poorly characterized process.

Phosphorylated eIF-2 participates in translation initiation in much the same way as unphosphorylated eIF-2, but it is not regenerated normally. At the completion of the initiation process, unmodified eIF-2 exchanges its bound GDP for GTP in a reaction mediated by another initiation factor, **eIF-2B:**

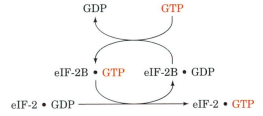

Phosphorylated eIF-2 forms a much tighter complex with eIF-2B than does unphosphorylated eIF-2. This sequesters eIF-2B (Fig. 27-42), which is present in lesser amounts than is eIF-2, thereby preventing regeneration of the

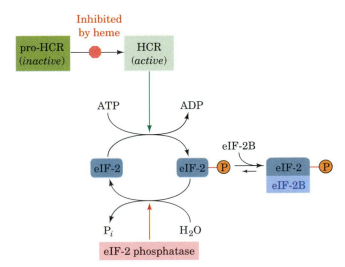

Figure 27-42.   A model for heme-controlled protein synthesis in reticulocytes.

eIF-2 · GTP required for translation. The phosphorylated eIF-2 molecules are reactivated through the action of **eIF-2 phosphatase,** which is unaffected by heme.

E. The Molecular Basis of Development

Perhaps the most awe-inspiring event in biology is the growth and development of a fertilized ovum to form an extensively differentiated multicellular organism. No outside instruction is required to do so; *fertilized ova contain all the information necessary to form complex multicellular organisms such as human beings.* Much of what we know about the molecular basis of cell differentiation is based on studies of the fruit fly *Drosophila melanogaster.* We therefore begin this section with a synopsis of *Drosophila* embryogenesis.

Drosophila Development

Almost immediately after the *Drosophila* egg (Fig. 27-43*a*) is laid (which, rather than the earlier fertilization, triggers development), it commences a series of rapid, synchronized nuclear divisions, one every 6 to 10 min. Here, the nuclear division process is not accompanied by the formation of new cell membranes; the nuclei continue sharing their common cytoplasm to form a so-called **syncytium** (Fig. 27-43*b*). After the 8th round of nuclear division, the ~256 nuclei begin to migrate toward the cortex (outer layer) of the egg where, by around the 11th nuclear division, they have formed a single layer surrounding a yolk-rich core (Fig. 27-43*c*; the germ-cell progenitors, the pole cells, are set aside after the 9th division). At this stage, the mitotic cycle time begins to lengthen while the nuclear genes, which have heretofore been fully engaged in DNA replication, become transcriptionally active (a freshly laid egg contains an enormous store of mRNA that has been contributed by the developing oocyte's surrounding "nurse" cells). In the 14th nuclear division cycle, which lasts ~60 min, the egg's plasma membrane invaginates around each of the ~6000 nuclei to yield a cellular monolayer called a **blastoderm** (Fig. 27-43*d*). At this point, after ~2.5 h of development, transcriptional activity reaches its maximum and mitotic synchrony is lost.

During the next few hours, the embryo undergoes **gastrulation** (migration of cells to form a triple-layered structure) and organogenesis. A striking aspect of this remarkable process, in *Drosophila* as well as in higher animals, is the division of the embryo into a series of segments corresponding to the adult organism's organization (Fig. 27-43*e*). The *Drosophila* embryo has at least three segments that eventually merge to form its head (Md, Mx, and Lb for mandibulary, maxillary, and labial), three thoracic segments (T1–T3), and eight abdominal segments (A1–A8). As development continues, the embryo elongates and several of its abdominal segments fold over its thoracic segments (Fig. 27-43*f*). At this stage, the segments become subdivided into anterior (forward) and posterior (rear) compartments. The embryo then shortens and unfolds to form a larva that hatches 1 day after beginning development (Fig. 27-43*g*). Over the next 5 days, the larva feeds, grows, molts twice, pupates, and commences metamorphosis to form an adult (Fig. 27-43*h*). In this latter process, the larval epidermis is almost entirely replaced by the outgrowth of apparently undifferentiated patches of larval epithelium known as **imaginal disks** that are committed to their developmental fates as early as the blastoderm stage. These structures, which maintain the larva's segmental boundaries, form the adult's legs, wings, antennae, eyes, etc. About 10 days after commencing development, the adult emerges and, within a few hours, initiates a new reproductive cycle.

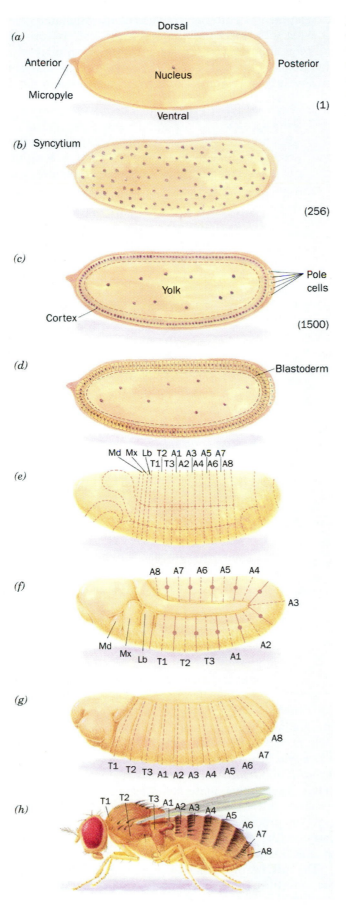

Figure 27-43.   Development in *Drosophila*. The various stages are explained in the text. Note that the embryos and newly hatched larva are all the same size, ~0.5 mm long. The adult is, of course, much larger. The approximate numbers of cells in the early stages of development are given in parentheses.

Developmental Patterns Are Genetically Mediated

What is the mechanism of embryonic pattern formation? Much of what we know about this process stems from genetic analyses of a series of bizarre mutations in three classes of *Drosophila* genes that normally specify progressively finer regions of cellular specialization in the developing embryo:

1. *Maternal-effect genes, which define the embryo's polarity,* that is, its anteroposterior (head to tail) and dorsoventral (back to belly) axes. Mutations of these genes globally alter the embryonic body pattern, producing, for example, nonviable embryos with two anterior or two posterior ends pointing in opposite directions.

2. *Segmentation genes, which specify the correct number and polarity of embryonic body segments.* These are subclassified as follows:

   (a) **Gap genes,** the first of a developing embryo's to be transcribed, are so named because their mutations result in gaps in the embryo's segmentation pattern. Embryos with defective **hunchback (hb)** genes, for example, lack mouthparts and thorax structures.

   (b) **Pair-rule genes** specify the division of the embryo's broad gap domains into segments. These genes are so named because their mutations usually delete portions of every second segment.

   (c) **Segment polarity genes** specify the polarities of the developing segments. Thus, homozygous **engrailed (en)** mutants lack the posterior compartment of each segment.

3. *Homeotic selector genes, which specify segmental identity;* their mutations transform one body part into another. For instance, **Antennapedia (Antp,** antenna-foot) mutants have legs in place of antennae (Fig. 27-44a), whereas the mutations **bithorax (bx), anteriorbithorax (abx),** and **postbithorax (pbx)** each transform sections of halteres (vestigial wings that function as balancers), which normally occur only on segment T3, to the corresponding sections of wings, which normally occur only on segment T2 (Fig. 27-44b).

*The properties of maternal-effect gene mutants suggest that maternal-effect genes specify substances known as **morphogens** whose distributions in the egg cytoplasm define the future embryo's spatial coordinate system.* Indeed, immunofluorescence studies by Christiane Nüsslein-Volhard have demonstrated that the product of the **bicoid (bcd)** gene is distributed in a gradient that decreases toward the posterior end of the normal embryo (Fig. 27-45a), whereas embryos with *bcd*-deficient mothers lack this gradient. The gradient arises through the secretion, by ovarian nurse cells, of *bcd* mRNA into the anterior end of the oocyte during oogenesis. The **nanos** gene mRNA is similarly deposited near the egg's posterior pole. The *bcd* and *nanos* gene products regulate the expression of specific gap genes. Some other maternal-effect genes specify proteins that trap the localized mRNAs in their area of deposition. This explains why early embryos produced by females homozygous for maternal-effect mutations can often be "rescued" by the injection of cytoplasm, or sometimes just the mRNA, from early wild-type embryos.

The mRNA of the gap gene *hunchback* (*hb*) is deposited uniformly in the unfertilized egg (Fig. 27-45a). However, **Bicoid protein** activates the transcription of the embryonic *hb* gene, whereas **Nanos protein** inhibits the

(a)

(b)

Figure 27-44.   Developmental mutants of Drosophila. (a) Head of an adult that is homozygous for the homeotic *Antennapedia* mutation. Absence of the *Antp* gene product causes the imaginal disks that normally form antennae to develop as the legs that normally occur only on segment T2. [Courtesy of Walter Gehring, University of Basel, Switzerland.] (b) A four-winged *Drosophila* (it normally has two wings) that results from the presence of three mutations in the bithorax complex. These mutations cause the normally haltere-bearing segment T3 to develop as if it were the wing-bearing segment T2. [Courtesy of Edward B. Lewis, Caltech.]

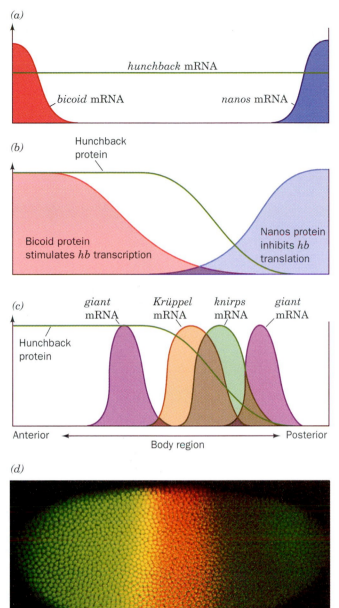

(a)

hunchback mRNA

bicoid mRNA

nanos mRNA

(b)

Hunchback
protein

Bicoid protein
stimulates *hb* transcription

Nanos protein
inhibits *hb*
translation

(c)

giant
mRNA

Krüppel
mRNA

knirps
mRNA

giant
mRNA

Hunchback
protein

Anterior

Posterior

Body region

(d)

Figure 27-45.   The formation and effects of the Hunchback protein gradient in *Drosophila* embryos. (*a*) The unfertilized egg contains maternally supplied *bicoid* and *nanos* mRNAs placed at its anterior and posterior poles, together with a uniform distribution of *hunchback* mRNA. (*b*) On fertilization, the three mRNAs are translated. Bicoid and Nanos proteins are not bound in place as are their mRNAs and hence their gradients are broader than those of the mRNAs. Bicoid protein stimulates the transcription of *hunchback* mRNA whereas Nanos protein inhibits its translation, resulting in a gradient of Hunchback protein that decreases nonlinearly from anterior to posterior. (*c*) Specific concentrations of Hunchback protein induce the transcription of the *giant*, *Krüppel*, and *knirps* genes. The gradient of Hunchback protein thereby specifies the positions at which these latter mRNAs are synthesized. (*d*) A photomicrograph of a *Drosophila* embryo (*anterior end left*) that has been immunofluorescently stained for both Hunchback (*green*) and Krüppel proteins (*red*). The region where these proteins overlap is yellow. [Parts *a*, *b*, and *c* after Gilbert, S.F., *Developmental Biology* (5th ed.), pp. 550 and 565, Sinauer Associates (1997); Part *d* courtesy of Jim Langeland, Stephen Paddock, and Sean Carroll, University of Wisconsin-Madison.]

translation of *hb* mRNA. Consequently, **Hunchback protein** becomes distributed in a gradient that decreases from anterior to posterior (Fig. 27-45*b*). Footprinting studies have demonstrated that Bicoid protein binds to five homologous sites (consensus sequence TCTAATCCC) in the *hb* gene's upstream promoter region.

Hunchback protein controls the expression of several other gap genes (Fig. 27-45*c*,*d*): High levels of Hunchback protein induce ***giant*** expression; ***Krüppel*** (German: cripple) is expressed where the level of Hunchback protein begins to decline; ***knirps*** (German: pigmy) is expressed at even lower levels of Hunchback protein; and *giant* is again activated in regions where Hunchback protein is undetectable. These patterns of gene expression are stabilized and maintained by additional interactions. For example, **Krüppel protein** binds to the promoters of the *hb* gene, which it activates, and the *knirps* gene, which it represses. Conversely, **Knirps protein** represses the *Krüppel* gene. This mutual repression is thought to be responsible for the sharp boundaries between the various gap domains.

***Figure 27-46. Drosophila* embryos stained for pair-rule genes.** The Fushi tarazu protein (Ftz) is brown, and the Eve protein is gray. These proteins are each expressed in seven stripes. [Courtesy of Peter Lawrence, MRC Laboratory of Molecular Biology, Cambridge, U.K.]

Pair-rule genes are expressed in sets of seven stripes, each just a few nuclei wide, along the early embryo's anterior–posterior axis (Fig. 27-46). The gap gene products directly control three **primary pair-rule genes:** *hairy,* ***even-skipped* (*eve*),** and **runt.** The promoters of most primary pair-rule genes consist of a series of modules, each of which contains a particular arrangement of activating and inhibitory binding sites for the various gap gene proteins. As a result, the expression of a pair-rule gene reflects the combination of gap gene proteins present, giving rise to a "zebra stripe" pattern. As with the gap genes, the patterns of expression of the primary pair-rule genes become stabilized through interactions among themselves. The primary pair-rule gene products also induce or inhibit the expression of five **secondary pair-rule genes** including *fushi tarazu* (*ftz;* Japanese for not enough segments). Thus, as Walter Gehring demonstrated, *ftz* transcripts first appear in the nuclei lining the cortical cytoplasm during the embryo's 10th nuclear division cycle. By the 14th division cycle, when the cellular blastoderm forms, *ftz* is expressed in a pattern of seven belts around the blastoderm, each 3 or 4 cells wide (Fig. 27-46).

The expression of eight known segment polarity genes is initiated by pair-rule gene products. For example, by the 13th nuclear division cycle, as Thomas Kornberg demonstrated, *engrailed* (*en*) transcripts become detectable but are more or less evenly distributed throughout the embryonic cortex. However, since *en* is preferentially expressed in nuclei containing high concentrations of either Eve or Ftz proteins, by the 14th cycle they form a striking pattern of 14 stripes around the blastoderm (half the spacing of *ftz* expression). The *en* gene product thereby induces the posterior half of each segment to develop in a different fashion from its anterior half.

Homeotic Genes Direct Development of Individual Body Parts

The structural components of developmentally analogous body parts, say, *Drosophila* antennae and legs, are nearly identical; only their organizations differ (Fig. 27-47). *Consequently, developmental genes must control the pattern of structural gene expression rather than simply turning these genes on or off.*

The *Drosophila* homeotic selector genes map into two large gene families: the **bithorax complex (*BX-C*),** which controls differentiation in the thoracic and abdominal segments, and the **antennapedia complex (*ANT-C*),** which primarily affects head and thoracic segments. *Homozygous mutations in BX-C cause one or more segments to develop as if they were more anterior segments* (e.g., segment T3 develops as if it were segment T2; Fig. 27-44b). The entire deletion of *BX-C* causes all segments posterior to T2 to resemble T2; apparently T2 is the developmental "ground state" of the more distal segments. The evolution of homeotic gene families, it is thought, permitted arthropods (the phylum containing insects) to arise from the more primitive annelids (segmented worms) in which all segments are nearly alike.

Detailed genetic analysis of *BX-C* led Edward B. Lewis to formulate a model for segmental differentiation (Fig. 27-48): *BX-C* contains at least one gene for each segment from T3 to A8 (numbered 0 to 8 in Fig. 27-48). Starting with segment T3, progressively more posterior segments express successively more *BX-C* genes until, in segment A8, all of these genes are expressed. Such a pattern of gene expression may result from a gradient in the concentration of a *BX-C* repressor that decreases from the anterior to the posterior end. The developmental fate of a segment is thereby determined by its position in the embryo.

In characterizing the *Antennapedia* (*Antp*) gene, Gehring and Matthew Scott independently discovered that *Antp* cDNA hybridizes to both the *Antp* and the *ftz* genes, indicating that *these genes share a common base se-*

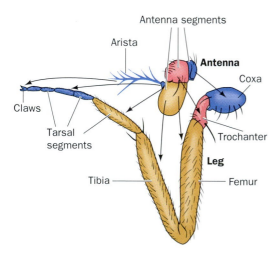

Figure 27-47.   The correspondence between *Drosophila* antennae and legs. [After Postlethwait, J.H. and Schneiderman, H.A., *Dev. Biol.* 25, 622 (1971).]

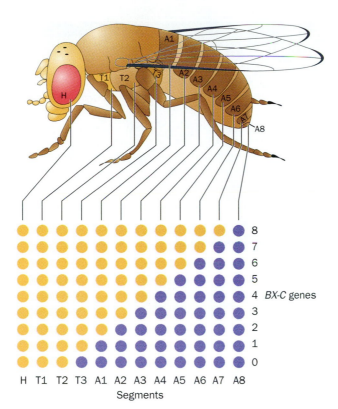

Figure 27-48. A model for the differentiation of embryonic segments in *Drosophila*. Segments T2, T3, and A1–A8, as the lower drawing indicates, are each characterized by a unique combination of active (*purple circles*) and inactive (*yellow circles*) BX-C genes. These genes, here numbered 0 to 8, are thought to be sequentially activated from anterior to posterior in the embryo so that segment T2, the developmentally most primitive segment, has no active *BX-C* genes, while in segment A8 all of them are active. [After Ingham, P., *Trends Genet.* **1**, 113 (1985).]

quence. Subsequent experiments revealed that a similar sequence, called a **homeodomain** or **homeobox,** occurs in many *Drosophila* homeotic genes. These sequences, which are 70 to 90% homologous to one another, encode even more homologous 60-residue polypeptide segments.

Further hybridization studies using homeodomain probes led to the truly astonishing finding that *homeodomains are also present in the genomes of many animals.* Homeodomain-containing genes have collectively become known as ***Hox* genes.** In vertebrates, they are organized in four clusters of 9 to 11 genes, each located on a separate chromosome and spanning more than 100 kb. In contrast, *Drosophila,* as we saw, has two *Hox* clusters, whereas nematodes (roundworms), which are evolutionarily more primitive than insects, have only one *Hox* cluster. The various *Hox* clusters, as well as their component genes, almost certainly arose through a series of gene duplications.

Hox Genes Encode Transcription Factors

Some *Hox* genes are remarkably homologous; for example, the homeodomains of the *Drosophila Antp* gene and the frog ***MM3* gene** encode polypeptides that have 59 of their 60 amino acids in common. Since vertebrates and invertebrates diverged over 600 million years ago, this strongly suggests that the product of the homeodomain has an essential function.

The polypeptide encoded by the homeodomain of the *Drosophila engrailed* gene specifically binds to the DNA sequences just upstream from the transcription start sites of both the *en* and the *ftz* genes. Moreover, fusing the *ftz* gene's upstream sequence to other genes imposes *ftz*'s pattern of stripes (Fig. 27-46) on the expression of these genes in *Drosophila* embryos. *These observations suggest that homeodomain-containing genes encode transcription factors that regulate the expression of other genes.*

Thomas Kornberg and Carl Pabo determined the X-ray structure of the 61-residue homeodomain from the *Drosophila* Engrailed protein in

***Figure 27-49.   The X-ray structure of the En-
grailed protein homeodomain in complex with
its target DNA.*** The protein is shown in ribbon
form (*green*) with its recognition helix (helix 3;
yellow) bound in the DNA's major groove. The
N-terminus (*red*) binds in the minor groove. The
DNA is shown in stick form (*blue*) with the base
pairs containing its TAAT subsite highlighted in
magenta. [Based on an X-ray structure by Carl
Pabo, The Johns Hopkins University.] ● **See the
Interactive Exercises.**

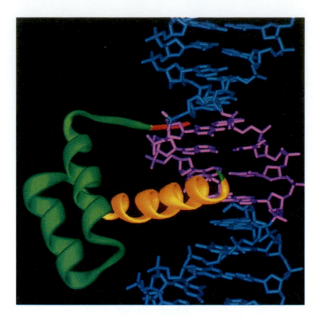

complex with a 21-bp DNA (Fig. 27-49). The homeodomain consists largely
of three α helices, the last two of which form an HTH motif that is closely
superimposable on the HTH motifs of prokaryotic repressors such as the
λ repressor (Fig. 27-25). However, the interaction of helix 3, the HTH motif's recognition helix, with the DNA's major groove differs in the two proteins. In the λ repressor complex, for example, the N-terminal end of the
recognition helix inserts into the major groove, whereas in the homeodomain complex, the C-terminal end of the helix, which is longer than that
of the λ repressor, inserts into the major groove.

Vertebrate *Hox* genes, like those of *Drosophila,* are expressed in specific patterns and at particular stages during embryogenesis. That the *Hox*
genes directly specify the identities and fates of embryonic cells was shown,
for example, by the following experiment. Mouse embryos were made
transgenic for the *Hox-1.1* gene that had been placed under the control of
a promoter that is active throughout the body even though *Hox-1.1* is normally expressed only below the neck. The resulting mice had severe cra-
niofacial abnormalities such as a cleft palate and an extra vertebra and an
intervertebral disk at the base of the skull. Some also had an extra pair of
ribs in the neck region. Thus, the altered expression of the *Hox-1.1* gene
induced a homeotic mutation, that is, a change in the development pattern,
analogous to those observed in *Drosophila* (Fig. 27-44).

Homozygotic mice whose *Hox-3.1* coding sequence has been replaced
with that of *lacZ* are born alive but usually die within a few days. They exhibit skeletal deformities in their trunk regions in which several skeletal
segments are transformed into the likenesses of more anterior segments.
The pattern of β-galactosidase activity, as colorimetrically detected through
the use of a substrate analog whose hydrolysis products are blue (Fig.
27-50), indicates that *Hox-3.1* deletion modifies the properties but not the
positions of the embryonic cells that normally express *Hox-3.1*.

***Figure 27-50.   The pattern of expression of the Hox-3.1 gene in a 12.5-day-old
mouse embryo.*** The protein-coding portion of the *Hox-3.1* gene was replaced by the
lacZ gene. The regions of this transgenic embryo in which *Hox-3.1* is expressed are
revealed by soaking the embryo in a buffer containing a substance that turns blue
when hydrolyzed by the *lacZ* gene product, β-galactosidase. [Courtesy of Phillipe
Brûlet, Collège de France and the Institut Pasteur, France.]

SUMMARY

1. The complete sequence of an organism's genome makes it possible to identify the functions of its component genes, and therefore its metabolic capabilities, on the basis of homology between its open reading frames and genes of known function.

2. Certain genes are found in clusters, for example, those in bacterial operons, rRNA and tRNA genes, and eukaryotic histone genes. The human globin gene clusters contain genes expressed at different developmental stages.

3. Prokaryotic genomes contain small amounts of nontranscribed DNA, which include control regions for replication and transcription. The genomes of higher eukaryotes contain much larger proportions of nontranscribed DNA in the form of repetitive sequences and the remnants of transposons.

4. Prokaryotic gene expression is controlled at the level of transcription. Regulation of the *lac* operon is mediated by the binding of the *lac* repressor to its operator sequences. This binding, which prevents transcription of the operon, is reversed by the binding to the *lac* repressor of an inducer whose presence signals the availability of lactose, the substrate for the *lac* operon–encoded enzymes.

5. In catabolite repression, a complex of CAP and cAMP, which signals the scarcity of glucose, binds to DNA to stimulate the transcription of genes encoding proteins that participate in the metabolism of other sugars.

6. Attenuation is a mechanism whereby the translation-dependent formation of alternate mRNA secondary structures in an operon's leader sequence determines whether transcription proceeds or terminates.

7. The lytic growth of bacteriophage λ requires the sequential expression of the antiterminators gp*N* and gp*Q* as well as Cro protein, which represses transcription of λ repressor. Lysogenic growth requires λ repressor, which prevents the expression of all λ genes except its own. The switch from lysogenic to lytic growth depends on whether Cro protein or λ repressor occupies their mutual operator sequences.

8. Transcriptionally active eukaryotic DNA is less condensed than "silent" DNA. The reversible acetylation of histones is thought to regulate the accessibility of nucleosomes to the transcription machinery.

9. "Upstream" transcription factors bind to enhancer or silencer sequences and interact with the preinitiation complex (PIC) to regulate gene expression. These factors include STATs and steroid hormone receptors, both of which respond to extracellular signals, and tumor suppressors, which are in turn regulated by cyclin-dependent protein kinases.

10. Antibody diversity results from somatic recombination involving the rearrangement of clustered gene sequences encoding segments of the immunoglobulin light and heavy chains. Diversity is augmented by imprecise *V/D/J* joining and by somatic hypermutation.

11. Additional control of eukaryotic gene expression is effected by alternative splicing, variable rates of mRNA degradation, and regulation of translation initiation.

12. The development of the *Drosophila* embryo is controlled by maternal-effect genes, which define the embryo's polarity; gap, pair-rule, and segment polarity genes, which specify the number and polarity of embryonic body segments; and homeotic selector genes (*Hox* genes), which encode transcription factors that regulate the expression of genes and therefore govern cell differentiation. *Hox* genes similarly regulate vertebrate development.

REFERENCES

Genomics

Lander, E.S., The new genomics: global views of biology, *Science* **274**, 536–539 (1996). [An exposition of the goals of genome sequencing, focusing on the human genome.]

Pearson, C.E. and Sinden, R.R., Trinucleotide repeat DNA structures: dynamic mutations from dynamic DNA, *Curr. Opin. Struct. Biol.* **8**, 321–330 (1998).

Strauss, E.J. and Falkow, S., Microbial pathogenesis: genomics and beyond, *Science* **276**, 707–712 (1997). [Discusses some practical outcomes of studying microbial genomes.]

Prokaryotic Gene Expression

Kolb, A., Busby, S., Buc, H., Garges, S., and Adhya, S., Transcriptional regulation by cAMP and its receptor protein, *Annu. Rev. Biochem.* **62**, 749–795 (1993).

Lewis, M., Chang, G., Horton, N.C., Kercher, M.A., Pace, H.C., Schumacher, M.A., Brennan, R.G., and Lu, P., Crystal structure of the lactose operon repressor and its complexes with DNA and inducer, *Science* **271**, 1247–1254 (1996).

Matthews, K.S. and Nichols, J., Lactose repressor protein: Functional properties and structure, *Prog. Nucl. Acid Res. Mol. Biol.* **58**, 127–164 (1998).

Ptashne, M., *A Genetic Switch* (2nd ed.), Chapters 1–4, Cell Press & Blackwell Scientific Publications (1992). [A detailed description of the bacteriophage λ genetic switch.]

Yanofsky, C., Transcription attenuation, *J. Biol. Chem.* **263**, 609–612 (1988). [A general discussion of attenuation.]

Eukaryotic Gene Expression

Adams, C.C. and Workman, J.L., Nucleosome displacement in transcription, *Cell* **72**, 305–308 (1993).

Beato, M., Herrlich, P., and Schütz, G., Steroid hormone receptors: many actors in search of a plot, *Cell* **83**, 851–857 (1995). [Reviews actions of steroid hormone receptors and their interactions with transcription factors.]

Beelman, C.A. and Parker, R., Degradation of mRNA in eukaryotes, *Cell* **81,** 179–183 (1995).

Cho, Y., Gorina, S., Jeffrey, P.D., and Pavletich, N.P., Crystal structure of a p53 tumor suppressor–DNA complex: understanding tumorigenic mutations, *Science* **265,** 346–355 (1994).

Darnell, J.E., Jr., STATs and gene regulation, *Science* **277,** 1630–1635 (1997).

Dynlacht, B.D., Regulation of transcription by proteins that control the cell cycle, *Nature* **389,** 149–152 (1997).

Gao, C.Y. and Zelenka, P.S., Cyclins, cyclin-dependent kinases and differentiation, *BioEssays* **19,** 307–314 (1997).

Gilbert, S.F., *Developmental Biology* (5th ed.), Chapter 14, Sinauer Associates (1997).

Grunstein, M., Histone acetylation in chromatin structure and transcription, *Nature* **389,** 349–352 (1997).

Lewis, S.M. and Wu, G.E., The origins of V(D)J recombination, *Cell* **88,** 159–162 (1997).

Mitas, M., Trinucleotide repeats associated with human disease, *Nucl. Acids Res.* **25,** 2245–2253 (1997).

Nüsslein-Volhard, C., Gradients that organize embryo development, *Sci. Am.* **272**(7), 54–61 (1996).

Ptashne, M. and Gann, A., Transcriptional activation by recruitment, *Nature* **386,** 569–577 (1997). [A broad review of how regulatory factors and the transcription machinery cooperate in gene expression.]

Schatz, D.G., Oettinger, M.A., and Schlissel, M.S., V(D)J recombination: molecular biology and regulation, *Annu. Rev. Immunol.* **10,** 359–383 (1992).

Tijan, R. and Maniatis, T., Transcriptional activation: a complex puzzle with few easy pieces, *Cell* **77,** 5–8 (1994). [An overview of eukaryotic gene regulation through modular binding of transcription factors.]

Wolberger, C., Homeodomain interactions, *Curr. Opin. Struct. Biol.* **6,** 62–68 (1996).

KEY TERMS

gene expression	operator	euchromatin	blastoderm
genomics	catabolite repression	Barr body	gastrulation
C value	positive regulator	polytene chromosome	imaginal disk
C-value paradox	negative regulator	chromosome puff	maternal-effect gene
ORF	corepressor	enhancer	segmentation gene
gene cluster	leader sequence	silencer	gap gene
centromere	attenuation	STAT	pair-rule gene
highly repetitive sequences	lytic mode	hormone response element	segment polarity gene
satellite DNA	lysogenic mode	tumor suppressor	morphogen
moderately repetitive sequences	prophage	apoptosis	homeotic selector gene
selfish DNA	lysogen	somatic recombination	homeodomain
repressor	induction	somatic mutation	*Hox* gene
inducer	site-specific recombination	deadenylation	
	heterochromatin	syncytium	

STUDY EXERCISES

1. List some of the elements responsible for the large sizes of eukaryotic genomes relative to prokaryotic genomes.

2. List several reasons why the functions of some ORFs cannot be determined.

3. Why are rRNA and tRNA genes, but not protein-coding genes, generally found in clusters?

4. Describe the regulation of the *lac* operon by *lac* repressor and CAP.

5. How does attenuation regulate gene expression?

6. Describe the genetic switch mechanism in bacteriophage λ.

7. How does histone modification influence eukaryotic transcription initiation?

8. Why is the spacing between enhancers and promoters variable?

9. Describe how antibody diversity is generated.

10. Describe embryogenesis in *Drosophila*.

PROBLEMS

1. DNA isolated from an organism can be sheared into fragments of uniform size (~300 kb), heated to separate the strands, and then cooled to allow complementary strands to reanneal. The renaturation process can be followed over time. Explain why the renaturation of *E. coli* DNA is a monophasic process whereas the renaturation of human DNA is biphasic (an initial rapid phase followed by a slower phase).

2. Explain why the organization of genes in operons facilitates the assignment of functions to previously unidentified ORFs in a bacterial genome.

3. Explain why (a) inactivation of the O_1 sequence of the *lac* operator almost completely abolishes repression of the *lac* operon; (b) inactivation of O_2 or O_3 causes only a twofold

loss in repression; and (c) inactivation of both O_2 and O_3 reduces repression ~70-fold.

4. Why do *E. coli* cells with a defective *lacZ* gene fail to show galactoside permease activity after the addition of lactose in the absence of glucose?

5. Describe the probable genetic defect that abolishes the sensitivity of the *lac* operon to the absence of glucose when other metabolic operons continue to be sensitive to the absence of glucose.

6. Why can't eukaryotic transcription be regulated by attenuation?

7. Predict the effect of deleting the leader peptide sequence on regulation of the *trp* operon.

8. Predict the effect of the following gene deletions on the bacteriophage λ life cycle: (a) *N*; (b) *cII*; (c) *cro*.

9. Is it possible for a transcription enhancer to be located within the protein-coding sequence of a gene? Explain.

10. Explain why natural selection has favored the instability of RNA.

11. In *Drosophila*, an *esc*⁻ homozygote develops normally unless its mother is also an *esc*⁻ homozygote. Explain.

APPENDICES

APPENDICES

GLOSSARY

Numbers and Greek letters are alphabetized as if they were spelled out.

ab initio. From the beginning.

ABO blood group antigens. The oligosaccharide components of glycolipids on the surfaces of erythrocytes and other cells.

Absolute configuration. The spatial arrangement of chemical groups around a chiral center.

Abzyme. An antibody that can catalyze a chemical reaction, much like an enzyme.

Accessory pigment. A molecule in the photosynthetic system that absorbs light at wavelengths other than those absorbed by chlorophyll.

Acid. A substance that can donate a proton.

Acid–base catalysis. A catalytic mechanism in which partial proton transfer from an acid and/or partial proton abstraction by a base lowers the free energy of a reaction's transition state.

Acidic solution. A solution whose pH is less than 7.0 ($[H^+] > 10^{-7}$ M).

Acidosis. A pathological condition in which the pH of the blood drops below its normal value of 7.4.

Active site. The region of an enzyme in which catalysis takes place.

Active transport. The transmembrane movement of a substance from low to high concentrations by a protein that couples this endergonic transport to an exergonic process such as ATP hydrolysis.

Acyl group. A portion of a molecule with the formula —COR, where R is an alkyl group.

Adenylate cyclase system. A signal transduction pathway in which a hormone binding to a cell-surface receptor activates a G-protein that in turn stimulates adenylate cyclase to synthesize the second messenger 3′,5′-cyclic AMP (cAMP) from ATP.

Adenylylation. Addition of an adenylyl (AMP) group.

Adipose tissue. Fat cells; distributed throughout an animal's body.

Aerobe. An organism that uses O_2 as an oxidizing agent for nutrient breakdown.

Affinity chromatography. A procedure in which a molecule is separated from a mixture of other molecules by its ability to bind specifically to an immobilized ligand.

Affinity labeling. A technique in which a labeled substrate analog reacts irreversibly with, and can thereby be used to identify, a group in an enzyme's active site.

Agarose. Linear carbohydrate polymers, made by red algae, that form a loose mesh.

Agarose gel electrophoresis. See gel electrophoresis.

Alcoholic fermentation. A metabolic pathway that synthesizes ethanol from pyruvate through decarboxylation and reduction.

Alditol. A sugar produced by reduction of an aldose or ketose to a polyhydroxy alcohol.

Aldonic acid. A sugar produced by oxidation of an aldose aldehyde group to a carboxylic acid group.

Aldose. A sugar whose carbonyl group is an aldehyde.

Alkalosis. A pathological condition in which the pH of the blood rises above its normal value of 7.4.

Allosteric effector. A small molecule whose binding to a protein affects the function of another site on the protein.

Allosteric interaction. The binding of ligand at one site in a macromolecule that affects the binding of other ligands at other sites in the molecule. See also cooperative binding.

α-amino acid. See amino acid.

α anomer. See anomers.

α/β barrel. A β barrel in which successive parallel β strands are connected by α helices such that a barrel of α helices surrounds the β barrel.

α-carbon. The carbon atom of an amino acid to which the amino and carboxylic acid groups are attached.

α cell. A pancreatic islet cell that secretes the hormone glucagon in response to low blood glucose levels.

α helix. A regular secondary structure of polypeptides, with 3.6 residues per right-handed turn, a pitch of 5.4 Å, and hydrogen bonds between each backbone N—H group and the backbone C=O group that is four residues earlier.

Alzheimer's disease. A neurodegenerative disease characterized by the precipitation of β amyloid protein in the brain.

Ames test. A method for assessing the mutagenicity of a compound from its ability to cause genetically defective strains of bacteria to revert to normal growth.

Amido group. A portion of a molecule with the formula —CONH—.

Amino acid. A compound consisting of a carbon atom to which are attached a primary amino group, a carboxylic acid group, a side chain (R group), and an H atom. Also called an α-amino acid.

Amino acid composition. A tally of the number and type of amino acids in a polypeptide.

Amino group. A portion of a molecule with the formula —NH₂, —NHR, or —NR₂, where R is an alkyl group. Amino groups are usually protonated at physiological pH.

Amino sugar. A sugar in which one or more OH groups are replaced by an amino group, which is often acetylated.

Amino terminus. The end of a polypeptide that has a free amino group. Also called the N-terminus.

Aminopeptidase. An enzyme that catalyzes the hydrolytic excision of a polypeptide's N-terminal residue.

Amphibolic. A term to describe a metabolic process that can be either catabolic or anabolic.

Amphipathic substance. See amphiphilic substance.

Amphiphilic substance. A substance that contains both polar and nonpolar regions and is therefore both hydrophilic and hydrophobic. Also called an amphipathic substance.

Amyloid deposit. An accumulation of certain types of insoluble protein aggregates in tissues, e.g., in the brain in Alzheimer's disease.

Anabolism. The reactions by which biomolecules are synthesized from simpler components.

Anaerobe. An organism that does not use O_2 as an oxidizing agent for nutrient breakdown. An obligate anaerobe cannot grow in the presence of O_2, whereas a facultative anaerobe can grow in the presence or absence of O_2.

Anaplerotic reaction. A reaction that replenishes the intermediates of a metabolic pathway.

Anion exchange. A chromatographic procedure in which anionic molecules bind to a cationic matrix.

Anneal. To maintain conditions that allow loose base pairing between complementary single polynucleotide strands so that properly paired double-stranded segments form.

Annular lipids. Membrane lipids that surround a membrane protein in a specific manner.

Anomeric carbon. The carbonyl carbon of a monosaccharide, which becomes a chiral center when the sugar cyclizes to a hemiacetal or hemiketal.

Anomers. Sugars that differ only in the configuration around the anomeric carbon. In the α anomer, the OH substituent of the anomeric carbon is on the opposite side of the ring from the CH_2OH group at the chiral center that designates the D or L configuration. In the β anomer, the OH substituent is on the same side.

Antibody. A protein produced by an animal's immune system in response to the introduction of a foreign substance (an antigen); it contains at least two pairs of identical heavy and light chains. Also called an immunoglobulin (Ig).

Anticodon. The sequence of three nucleotides in tRNA that recognizes an mRNA codon through complementary base pairing.

Antigen. A substance that elicits an immune response (production of antibodies) when introduced into an animal; it is specifically recognized by an antibody.

Antioxidant. A substance that destroys an oxidative free radical such as $O_2^- \cdot$ or $OH \cdot$.

Antiparallel. Running in opposite directions.

Antiport. The simultaneous transmembrane movement of two molecules in opposite directions.

Antisense strand. The DNA strand that serves as a template for transcription; it is complementary to the RNA. Also called the noncoding strand.

Antiterminator. A protein that blocks the termination of transcription.

AP site. An apurinic or apyrimidinic site; the deoxyribose residue remaining after the removal of a base from a DNA strand.

Apoenzyme. An enzyme that is inactive due to the absence of a cofactor.

Apolipoprotein. The protein component of a lipoprotein. Also called an apoprotein.

Apoprotein. A protein without the prosthetic group or metal ion that renders it fully functional. See also apoenzyme and apolipoprotein.

Aptamer. A nucleic acid whose conformation allows it to bind a particular ligand with high specificity and high affinity.

Archaea. One of the two major groups of prokaryotes (the other is eubacteria). Also known as archaebacteria.

Archaebacteria. See archaea.

Assay. A laboratory technique for detecting, and in some cases quantifying, a macromolecule or its activity.

Asymmetric center. See chiral center.

Atherosclerosis. A disease characterized by the formation of cholesterol-containing fibrous plaques in the walls of blood vessels, leading to loss of elasticity and blockage of blood flow.

ATP mass action ratio. The ratio $[ATP]/[ADP][P_i]$, which influences the rate of electron transport and oxidative phosphorylation.

ATPase. An enzyme that catalyzes the hydrolysis of ATP to $ADP + P_i$.

Attenuation. A mechanism in prokaryotes for regulating gene expression in which the availability of an amino acid determines whether an operon (consisting of genes for the enzymes that synthesize the amino acid) is transcribed.

Attenuator. A prokaryotic control element that governs transcription of an operon according to the availability of an amino acid synthesized by the proteins encoded by the operon.

Autocatalytic reaction. A reaction in which a product molecule can act as a catalyst for the same reaction; the reactant molecule therefore appears to catalyze its own reaction.

Autoimmune disease. A disease in which the immune system has lost some of its self-tolerance and produces antibodies against certain self-antigens.

Autolysis. An autocatalytic process in which a molecule catalyzes its own degradation.

Autophosphorylation. The kinase-catalyzed phosphorylation of itself or an identical molecule.

Autoradiography. A process in which X-ray film records the positions of radioactive entities, such as proteins or nucleic acids, that have been immobilized in a matrix such as a nitrocellulose membrane or an electrophoretic gel.

Autotroph. An organism that can synthesize all its cellular components from simple molecules using the energy from sunlight (photoautotroph) or from the oxidation of inorganic compounds (chemolithotroph).

Axial substituent. A group that extends perpendicularly from the plane of the ring to which it is bonded. See also equatorial substituent.

Backbone. The atoms that form the repeating linkages between successive residues of a polymeric molecule, exclusive of the side chains. Also called the main chain.

Bacteria. The organisms comprising the two major groups of prokaryotes, the archaea and the eubacteria.

Bacteriophage. A virus specific for bacteria. Also known as a phage.

Base. (1) A substance that can accept a proton. (2) A purine or pyrimidine component of a nucleoside, nucleotide, or nucleic acid.

Base pair. The specific hydrogen-bonded association between nucleic acid bases. The standard base pairs are $A \cdot T$ and $G \cdot C$.

Basic solution. A solution whose pH is greater than 7.0 ($[H^+] < 10^{-7}M$).

Beriberi. A disease caused by a deficiency of thiamine (vitamin B_1), which is a precursor of the cofactor thiamine pyrophosphate.

β anomer. See anomers.

β barrel. A protein motif consisting of a β sheet rolled into a cylinder.

β bend. See reverse turn.

β bulge. An irregularity in a β sheet resulting from an extra residue that is not hydrogen bonded to a neighboring chain.

β cell. A pancreatic islet cell that secretes the hormone insulin in response to high blood glucose levels.

β hairpin. A protein motif in which two antiparallel β strands are connected by a reverse turn.

β oxidation. A series of enzyme-catalyzed reactions in which fatty acids are progressively degraded by the removal of two-chain units as acetyl-CoA.

β sheet. A regular secondary structure in which extended polypeptide chains form interstrand hydrogen bonds. In parallel β sheets, the polypeptide chains all run in the same direction; in antiparallel β sheets, neighboring chains run in opposite directions.

Bilayer. An ordered, two-layered arrangement of amphiphilic molecules in which polar segments are oriented toward the two solvent-exposed surfaces and the nonpolar segments associate in the center.

Bile acid. An amphiphilic cholesterol derivative that acts as a detergent to solubilize lipids for digestion and absorption.

Binding change mechanism. The mechanism whereby the subunits of ATP synthase adopt three successive conformations to convert $ADP + P_i$ to ATP as driven by the dissipation of the transmembrane proton gradient.

Biochemical standard state. A set of conditions including unit activity of the species of interest, a temperature of 25°C, a pressure of 1 atm, and a pH of 7.0.

Blunt ends. The fully base-paired ends of a DNA fragment that has been cleaved by a restriction endonuclease that cuts the DNA strands at opposing sites.

Bohr effect. The increase in O_2 binding affinity of hemoglobin in response to an increase in pH.

Bovine spongiform encephalopathy. An invariably fatal neurodegenerative disease in cattle resulting from prion infection; it is similar to scrapie in sheep and Creutzfeldt–Jakob disease in humans. Also known as mad cow disease.

bp. Base pair, the unit of length used for DNA molecules. Thousands of base pairs (kilobase pairs); abbreviated kb.

Buffer. A solution of a weak acid and its conjugate base in approximately equal quantities. Such a solution resists changes in pH on the addition of acid or base.

Buffering capacity. The ability of a buffer solution to resist pH changes on addition of acid or base. Buffers are most useful when the pH is within one unit of its component acid's pK.

C_4 cycle. A photosynthetic process, which occurs only in certain plants, in which CO_2 is first concentrated by incorporating it into oxaloacetate (a C_4 compound).

Cahn–Ingold–Prelog system (*RS* system). A system for unambiguously describing the configurations of molecules with one or more asymmetric centers by assigning a priority ranking to the substituent groups of each asymmetric center.

Calvin cycle. The sequence of photosynthetic dark reactions in which ribulose-5-phosphate is carboxylated, converted to three-carbon carbohydrate precursors, and regenerated. Also called the reductive pentose phosphate cycle.

CAM. See crassulacean acid metabolism.

Cap. A 7-methylguanosine residue that is posttranscriptionally appended to the 5′ end of a eukaryotic mRNA.

Capillary electrophoresis (CE). An electrophoretic procedure carried out in small diameter capillary tubes.

Carbamate. The product of a reaction between CO_2 and an amino group: —NH—COO⁻.

Carbohydrate. A compound with the formula $(C \cdot H_2O)_n$ where $n \geq 3$. Also called a saccharide.

Carbonyl group. A portion of a molecule with the formula $\overset{\diagdown}{\underset{\diagup}{C}} = O$.

Carboxyl group. A portion of a molecule with the formula —COOH. Carboxyl groups are usually ionized at physiological pH.

Carboxyl terminus. The end of a polypeptide that has a free carboxylate group. Also called the C-terminus.

Carboxypeptidase. An enzyme that catalyzes the hydrolytic excision of a polypeptide's C-terminal residue.

Carcinogen. An agent that damages DNA so as to induce a mutation that leads to uncontrolled cell proliferation (cancer).

Catabolism. The degradative metabolic reactions in which nutrients and cell constituents are broken down for energy and raw materials.

Catabolite repression. A phenomenon in which the presence of glucose prevents the expression of genes involved in the metabolism of other fuels.

Catalyst. A substance that promotes a chemical reaction without undergoing permanent change. A catalyst increases the rate at which a reaction approaches equilibrium but does not affect the free energy change of the reaction.

Catalytic triad. The hydrogen-bonded Ser, His, and Asp residues that participate in catalysis in serine proteases.

Catenate. To interlink circular DNA molecules like the links of a chain.

Cation exchange. A chromatographic procedure in which cationic molecules bind to an anionic matrix.

CCAAT box. A eukaryotic promoter element with the consensus sequence CCAAT that is located 70 to 90 nucleotides upstream from the transcription start site.

CE. See capillary electrophoresis.

Cell cycle. The sequence of events between eukaryotic cell divisions; it includes mitosis and cell division (M phase), a gap stage (G_1 phase), a period of DNA synthesis (S phase), and a second gap stage (G_2 phase) before the next M phase.

Cellular immunity. Immunity mediated by *T* lymphocytes (*T* cells).

Central dogma of molecular biology. The paradigm that DNA directs its own replication as well as its transcription to RNA, which is then translated into a polypeptide. The flow of information is from DNA to RNA to protein.

Centromere. The eukaryotic chromosomal region that attaches to the mitotic spindle during cell division; it contains high concentrations of repetitive DNA.

Ceramide. A sphingosine derivative with an acyl group attached to its amino group.

Cerebroside. A ceramide with a sugar residue as a head group.

Chain-termination procedure. A technique for determining the nucleotide sequence of a DNA using dideoxy nucleotides so as to yield a collection of daughter strands of all different lengths. Also called the dideoxy method.

Channeling. The transfer of an intermediate product from one enzyme active site to another in such a way that the intermediate remains in contact with the protein.

Chaotropic agent. A substance that increases the solubility of nonpolar substances in water and thereby tends to denature proteins.

Chaperonins. Large cagelike molecular chaperones comprising Hsp60 and Hsp10 proteins (GroEL and GroES in *E. coli*) and which provide a protected microenvironment for proteins to fold.

Chargaff's rules. The observation, first made by Erwin Chargaff, that DNA has equal numbers of adenine and thymine residues and equal numbers of guanine and cytosine residues.

Chemiosmotic theory. The postulate that the free energy of electron transport is conserved in the formation of a transmembrane proton gradient. The electrochemical potential of this gradient is used to drive ATP synthesis.

Chemolithotroph. An autotrophic organism that obtains energy from the oxidation of inorganic compounds.

Chimera. See recombinant.

Chiral center. An atom whose substituents are arranged such that it is not superimposable on its mirror image. Also called an asymmetric center.

Chirality. The property of being asymmetric. A chiral molecule cannot be superimposed on its mirror image.

Chloroplasts. The plant organelles in which photosynthesis takes place.

Chromatin. The complex of DNA, RNA, and protein that comprises the eukaryotic chromosomes.

Chromatography. A technique for separating the components of a mixture of molecules on the basis of their partition between a mobile solvent phase and a porous matrix (stationary phase).

Chromosome. The complex of protein, RNA, and a single DNA molecule that comprises some or all of an organism's genome.

Chromosome puff. A portion of a polytene chromosome that has been decondensed to allow gene transcription.

Chylomicrons. Lipoprotein particles that transport dietary triacylglycerols and cholesterol from the intestines to the tissues.

Cis peptide. A conformation in which successive C_α atoms are on the same side of the peptide bond.

Citric acid cycle. A set of eight enzymatic reactions, arranged in a cycle, in which free energy in the form of ATP, NADH, and $FADH_2$ is recovered from the oxidation of the acetyl group of acetyl-CoA to CO_2. Also called the Krebs cycle and the tricarboxylic acid (TCA) cycle.

Clone. A collection of identical cells derived from a single ancestor.

Cloning. The production of an exact copy of a DNA segment or the organism that harbors it.

Cloning vector. A DNA molecule such as in a plasmid, virus, or artificial chromosome that can accommodate a segment of foreign DNA for cloning.

Closed system. A thermodynamic system that cannot exchange matter with its surroundings.

Coated pit. The protein-coated site on a cell surface where receptor–ligand complexes are endocytosed.

Coated vesicle. A membranous intracellular transport vesicle that is enclosed in a cagelike framework of the protein clathrin.

Coding strand. See sense strand.

Codon. The sequence of three nucleotides in DNA or RNA that specifies a single amino acid.

Coenzyme. A small organic molecule that is required for the catalytic activity of an enzyme. A coenzyme may be either a cosubstrate or a prosthetic group.

Cofactor. A small organic molecule (coenzyme) or metal ion that is required for the catalytic activity of an enzyme.

Coiled coil. An arrangement of polypeptide chains in which two α helices wind around each other, as in α keratin.

Cointegrate. The product of the fusion of two plasmids, which occurs as an intermediate in transposition.

Colligative property. A physical property, such as freezing point depression or osmotic pressure, that depends on the concentration of a dissolved substance rather than on its chemical nature.

Colony hybridization. A procedure in which DNA from multiple cell colonies is transferred to a membrane or filter and incubated with a DNA or RNA probe to test for the presence of a desired DNA fragment in the cell colonies. Also called *in situ* hybridization.

Compartmentation. The division of a cell into smaller functionally discrete systems.

Competitive inhibition. A form of enzyme inhibition in which a substance competes with the substrate for binding to the enzyme active site and thereby appears to increase K_M.

Condensation reaction. The formation of a covalent bond between two molecules, during which the elements of water are lost; the reverse of hydrolysis.

Conjugate acid. The compound that forms when a base accepts a proton.

Conjugate base. The compound that forms when an acid donates a proton.

Conjugate redox pair. An electron donor and acceptor that form a half-reaction.

Conservative replication. A hypothetical mode of DNA duplication in which the parental molecule remains intact and both strands of the daughter duplex are newly synthesized.

Conservative substitution. A change of an amino acid residue in a protein to one with similar properties, e.g., Leu to Ile or Asp to Glu.

Constant region. The C-terminal portion of an antibody (immunoglobulin) subunit, which does not exhibit the high sequence variability of the antigen-recognizing (variable) region of the antibody.

Constitutive enzyme. An enzyme that is synthesized at a more or less steady rate and that is required for basic cell function. Also called a housekeeping enzyme. See also inducible enzyme.

Contact inhibition. The inhibition of proliferation in cultured animal cells when the cells touch each other.

Contour length. The end-to-end length of a stretched-out polymer molecule.

Contour map. A map containing lines (contours) that trace positions of equal value of some property of the map (e.g., height above sea level, electron density).

Convergent evolution. The independent development of similar characteristics in unrelated species or proteins.

Cooperative binding. A situation in which the binding of a ligand at one site on a macromolecule affects the affinity of other sites for the same ligand. Both negative and positive cooperativity occur. See also allosteric interactions.

Corepressor. A substance that acts together with a protein repressor to block gene transcription.

Cori cycle. A metabolic pathway in which lactate produced by glycolysis in the muscles is transported via the bloodstream to the liver, where it is used for gluconeogenesis. The resulting glucose returns to the muscles.

Cosubstrate. A coenzyme that is only transiently associated with an enzyme so that it functions as a substrate.

Coupled enzymatic reaction. A technique in which the activity of an enzyme is measured by the ability of a second enzyme to use the product of the first enzymatic reaction to produce a detectable product.

Covalent catalysis. A catalytic mechanism in which the transient formation of a covalent bond between the catalyst and a reactant lowers the free energy of a reaction's transition state.

Crassulacean acid metabolism (CAM). A variation of the C_4 photosynthetic cycle in which CO_2 is temporarily stored as malate.

Creutzfeldt–Jakob disease. A human neurodegenerative disease similar to bovine spongiform encephalopathy.

Cristae. The invaginations of the inner mitochondrial membrane.

Cross talk. The interactions of different signal-transduction pathways through activation of the same signaling components, generation of a common second messenger, or similar patterns of target protein phosphorylation.

Cryoelectron microscopy. A technique in electron microscopy in which a sample is rapidly frozen to very low temperatures so that it retains its native shape to a greater extent than in conventional electron microscopy.

C-terminus. See carboxyl terminus.

Curved arrow convention. A notation for indicating the movement of an electron pair in a chemical reaction by drawing a curved arrow emanating from the electrons and pointing to the electron-deficient center that attracts the electron pair.

C-value paradox. The exceptions to the rule that the quantity of an organism's unique genetic material (its C value) correlates with the complexity of its morphology and metabolism.

Cyanosis. A bluish skin color indicating the presence of deoxyhemoglobin in the arterial blood.

Cyclic symmetry. A type of symmetry in which the asymmetric units of a symmetric object are related by a single axis of rotation.

Cyclin. A member of a family of proteins that participate in regulating the stages of the cell cycle and whose concentrations vary dramatically over the course of the cell cycle.

Cytochrome. A redox-active protein that carries electrons via a prosthetic Fe-containing heme group.

Cytoplasm. The entire contents of a cell excluding the nucleus.

Cytoskeleton. The network of intracellular fibers that gives a cell its shape and structural rigidity.

Cytosol. The contents of a cell (cytoplasm) excluding its nucleus and other membrane-bounded organelles.

D. Dalton, a unit of molecular mass; 1/12th the mass of a ^{12}C atom.

Dark reactions. The portion of photosynthesis in which NADPH and ATP produced by the light reactions are used to incorporate CO_2 into carbohydrates.

Darwinian evolution. See evolution.

ddNTP. An abbreviation for any dideoxynucleoside triphosphate.

Deamination. The hydrolytic removal of an amino group.

Degenerate code. A code in which more than one "word" encodes the same entity.

$\Delta\Psi$. See membrane potential.

Denature. To disrupt the native conformation of a polymer such that it becomes fully unfolded and does not retain significant secondary structure.

Deoxy sugar. A saccharide produced by replacement of an OH group by H.

Deoxynucleotide. See deoxyribonucleotide.

Deoxyribonucleic acid. See DNA.

Deoxyribonucleotide. A nucleotide in which the pentose is 2'-deoxyribose. Also known as a deoxynucleotide.

Desensitization. A cell's or organism's adaptation to a long-term stimulus through a reduced response to the stimulus.

Dextrorotatory. Rotating the plane of plane-polarized light clockwise from the point of view of the observer; the opposite of levorotatory.

Diabetes mellitus. A disease in which the pancreas does not secrete sufficient insulin (also called type I, insulin-dependent, or juvenile-onset diabetes) or in which the body has insufficient response to circulating insulin (type II, non-insulin-dependent, or maturity-onset diabetes). Diabetes is characterized by elevated levels of glucose in the blood.

Dialysis. A procedure in which solvent molecules and solutes smaller than the pores in a semipermeable membrane freely exchange with the bulk medium, while larger solutes are retained, thereby changing the solution in which the larger molecules are dissolved.

Diazatroph. A bacterium that can fix nitrogen.

Dideoxy method. See chain-termination procedure.

Differential labeling. Treatment of a macromolecule with a labeling reagent in the presence and absence of a molecule to which it is thought to bind, in order to identify those portions of the macromolecule that are shielded by the bound molecule.

Diffraction pattern. The record of the destructive and constructive interference of radiation scattered from an object. In X-ray crystallography, this takes the form of a series of discrete spots resulting from a collimated beam of X-rays scattering from a single crystal.

Diffusion. The transport of molecules throught their random movement.

Diffusion-controlled limit. The theoretical maximum rate of an enzymatic reaction in solution, about 10^8 to 10^9 $M^{-1} \cdot s^{-1}$.

Dihedral angle. See torsion angle.

Dihedral symmetry. A type of symmetry in which the asymmetric units are related by a 2-fold rotational axis that intersects another rotation axis at a right angle.

Dimer. An assembly consisting of two monomeric units (protomers).

Dinucleotide binding fold. A protein structural motif consisting of two $\beta\alpha\beta\alpha\beta$ units, which binds a dinucleotide such as NAD^+. Also called a Rossmann fold.

Diphosphoryl (pyrophosphoryl) group. Two phosphoryl groups linked by a phosphoanhydride bond $(-O_3P-O-PO_3-)^{2-}$.

Diploid. Having two equivalent sets of chromosomes.

Dipolar ion. See zwitterion.

Disaccharide. A carbohydrate consisting of two monosaccharides linked by a glycosidic bond.

Dissociation constant (K). The ratio of the products of the concentrations of the dissociated species to those of their parent compounds at equilibrium.

Disulfide bond. A covalent —S—S— linkage.

DNA. Deoxyribonucleic acid. A polymer of deoxynucleotides whose sequence of bases encodes genetic information in all living cells.

dNTP. A deoxyribonucleoside triphosphate.

Domain. A group of one to a few polypeptide segments comprised of around 100–200 polypeptide residues that fold into a globular unit.

Double-displacement reaction. A reaction in which a substrate binds and a product is released in the first stage, and another substrate binds and another product is released in the second stage.

Double-reciprocal plot. See Lineweaver–Burk plot.

$\mathscr{E}$. See reduction potential.

Edman degradation. A procedure for the stepwise removal and identification of the N-terminal residues of a polypeptide.

EF hand. A widespread helix–loop–helix structural motif that forms a Ca^{2+}-binding site.

Eicosanoids. C_{20} compounds derived from the C_{20} fatty acid arachidonic acid and which act as local mediators. Prostaglandins, prostacyclins, thromboxanes, and leukotrienes are eicosanoids.

Electrogenic transport. The transmembrane movement of a charged substance in a way that generates a charge difference across the membrane.

Electromotive force (emf). $\Delta\mathscr{E}$. See also reduction potential.

Electron crystallography. A technique for determining molecular structure, in which the electron beam of an electron microscope is used to elicit diffraction from a two-dimensional crystal of the molecules of interest.

Electron-transport chain. A series of membrane-associated electron carriers that pass electrons from reduced coenzymes (NADH and $FADH_2$) to molecular oxygen so as to recover free energy for the synthesis of ATP.

Electrophile. A group that contains an unfilled valence electron shell, or contains an electronegative atom. An electrophile (electron-lover) reacts readily with a nucleophile (nucleus-lover).

Electrophoresis. See gel electrophoresis.

Electrostatic catalysis. A catalytic mechanism in which the distribution of charges about the catalytic site lowers the free energy of a reaction's transition state.

Elementary reaction. A simple one-step chemical process, several of which may occur in sequence in a chemical reaction.

ELISA. See enzyme-linked immunosorbent assay.

Elongation factor. A protein that interacts with tRNA and/or the ribosome during polypeptide synthesis.

Eluant. The solution used to wash a chromatographic column.

Elution. The process of dislodging a molecule that has bound to a chromatographic matrix.

emf. Electromotive force. See reduction potential.

Enantiomers. Molecules that are nonsuperimposable mirror images of one another. Enantiomers are a type of stereoisomer.

Endergonic process. A process that has an overall positive free energy change (a nonspontaneous process).

Endocrine gland. A tissue in higher animals that synthesizes and releases hormones into the bloodstream.

Endoglycosidase. An enzyme that catalyzes the hydrolysis of the glycosidic bonds between two monosaccharide units within a polysaccharide.

Endonuclease. An enzyme that catalyzes the hydrolysis of the phosphodiester bonds between two nucleotide residues within a polynucleotide strand.

Endopeptidase. An enzyme that catalyzes the hydrolysis of a peptide bond within a polypeptide chain.

Endoplasmic reticulum (ER). A labyrinthine membranous organelle in eukaryotic cells in which membrane lipids are synthesized and some proteins undergo posttranslational modification.

Endosome. A membrane-bounded vesicle that receives materials that the cell ingests via receptor-mediated endocytosis and passes them to lysosomes for degradation.

Energy coupling. The conservation of the free energy of electron transport in a form that can be used to synthesize ATP from ADP + P_i.

"Energy-rich" compound. See "High-energy" compound.

Enhancer. A eukaryotic DNA sequence located some distance from the transcription start site, where an activator of transcription may bind.

Enthalpy (H). A thermodynamic quantity, $H = U + PV$, that is equivalent to the heat absorbed at constant pressure (q_P).

Entropy (S). A measure of the degree of randomness or disorder of a system. It is defined as $S = k_B \ln W$, where k_B is the Boltzmann constant and W is the number of equivalent ways the system can be arranged in its particular state.

Enzyme. A biological catalyst. Most enzymes are proteins; a few are RNA.

Enzyme-linked immunosorbent assay (ELISA). A technique in which a molecule is detected, and in some cases quantified, by its ability to bind an antibody to which an enzyme with an easily detected reaction product is attached.

Epimers. Sugars that differ only by the configuration at one C atom (excluding the anomeric carbon).

Equatorial substituent. A group that extends largely in the plane of the ring to which it is bonded. See also axial substituent.

Equilibrium. The point in a process at which the forward and reverse rates are exactly balanced so that it undergoes no net change.

Equilibrium constant (K_{eq}). The ratio, at equilibrium, of the product of the concentrations of reaction products to that of its reactants. K_{eq} is related to $\Delta G°$ for the reaction: $\Delta G° = -RT \ln K_{eq}$. Usually abbreviated K.

Equilibrium density gradient centrifugation. A technique in which a mixture of molecules is subjected to ultracentrifugation in a concentrated solution containing a dense, fast-diffusing substance such as CsCl that forms a density gradient as the centrifuge spins. Centrifugation is continued to equilibrium, thereby separating the mixture according to the densities of its component molecules.

ER. See endoplasmic reticulum.

Erythrocyte. A red blood cell, which functions to transport O_2 to the tissues. It is essentially a membranous sack of hemoglobin.

Erythrocyte ghost. Membranous particles derived from erythrocytes, which retain their original shape but are devoid of cytoplasm.

Essential amino acid. An amino acid that an animal cannot synthesize and must therefore obtain in its diet.

Essential fatty acid. A fatty acid that an animal cannot synthesize and must therefore obtain in its diet.

Ester group. A portion of a molecule with the formula —COOR, where R is an alkyl group.

Ether. A molecule with the formula ROR′, where R and R′ are alkyl groups.

Eubacteria. One of the two major groups of prokaryotes (the other is archaea).

Euchromatin. The transcriptionally active, relatively uncondensed chromatin in a eukaryotic cell.

Eukarya. See eukaryote.

Eukaryote. An organism consisting of a cell (or cells) whose genetic material is contained in a membrane-bounded nucleus.

Evolution. The gradual alteration of an organism or one of its components as a result of genetic changes that are passed from parent to offspring.

Exciton transfer. A mode of decay of an energetically excited molecule, in which electronic energy is transferred to a nearby unexcited molecule. Also known as resonance energy transfer.

Exergonic process. A process that has an overall negative free energy change (a spontaneous process).

Exoglycosidase. An enzyme that catalyzes the hydrolytic excision of a monosaccharide unit from the end of a polysaccharide.

Exon. A portion of a gene that appears in both the primary and mature mRNA transcripts. Also called an expressed sequence.

Exonuclease. An enzyme that catalyzes the hydrolytic excision of a nucleotide residue from the end of a polynucleotide strand.

Exopeptidase. An enzyme that catalyzes the hydrolytic excision of an amino acid residue from one end of a polypeptide chain.

Expression vector. A plasmid containing the transcription and translation control sequences required for the production of a foreign DNA gene product (RNA or protein) in a host cell.

Extrinsic protein. See peripheral protein.

Fab fragment. A proteolytic fragment of an antibody molecule that contains the antigen binding site. See also Fc fragment.

Facilitated diffusion. See passive-mediated transport.

Familial hypercholesterolemia. See hypercholesterolemia.

Fat. A mixture of triacylglycerols that is solid at room temperature.

Fatty acid. A carboxylic acid with a long-chain hydrocarbon side group.

Fc fragment. A proteolytic fragment of an antibody molecule that contains the C-terminal portions of its two heavy chains. See also Fab fragment.

Feedback inhibition. The inhibition of an early step in a reaction sequence by the product of a later step.

Feed-forward activation. The activation of a later step in a reaction sequence by the product of an earlier step.

Fermentation. An anaerobic catabolic process.

Fe–S cluster. See iron–sulfur cluster.

Fibrous protein. A protein characterized by a stiff, elongated conformation, that tends to form fibers.

First-order reaction. A reaction whose rate is proportional to the concentration of a single reactant.

Fischer convention. A system for describing the configurations of chiral molecules by relating their structures to that of D- or L-glyceraldehyde.

Fischer projection. A graphical convention for specifying molecular configuration in which horizontal lines represent bonds that extend above the plane of the paper and vertical lines represent bonds that extend below the plane of the paper.

5′ end. The terminus of a polynucleotide whose C5′ is not esterified to another nucleotide residue.

Flipase. An enzyme that catalyzes the translocation of a membrane lipid across a lipid bilayer (a flip-flop).

Flip-flop. See transverse diffusion.

Fluid mosaic model. A model of biological membranes in which integral membrane proteins float and diffuse laterally in a fluid lipid bilayer.

Fluorescence. A mode of decay of an excited molecule, in which electronic energy is emitted in the form of a photon.

Fluorescence photobleaching recovery. A technique for assessing the diffusion of membrane components from the rate at which the fluorescently labeled component moves into an area previously bleached by a pulse of laser light.

Fluorophore. A fluorescent group.

Flux. (1) The rate of flow of metabolites through a metabolic pathway. (2) The rate of transport per unit area.

Footprinting. A procedure in which the DNA sequence to which a protein binds is identified by determining which bases are protected by the protein from chemical or enzymatic modification.

Fractional saturation (Y). The fraction of a protein's ligand-binding sites that are occupied by ligand.

Fractionation procedure. A laboratory technique for separating the components of a mixture of molecules through differences in their chemical and physical properties.

Frameshift mutation. An insertion or deletion of nucleotides in DNA that alters the sequential reading (the frame) of sets of three nucleotides (codons) during translation.

Free energy (G). A thermodynamic quantity, $G = H - TS$, whose change at constant pressure is indicative of the spontaneity of a process. For spontaneous processes, $\Delta G < 0$, whereas for a process at equilibrium, $\Delta G = 0$. Also called Gibbs free energy.

Free energy of activation ($\Delta G^{\ddagger}$). The free energy of the transition state minus the free energies of the reactants in a chemical reaction.

Free radical. A molecule with an unpaired electron.

Freeze-etching. An elaboration of freeze-fracture electron microscopy in which additional membrane surfaces are exposed by subliming ice away.

Freeze-fracture electron microscopy. A technique used to visualize the interior portions of a membrane, in which a frozen membrane is shattered such that it splits between the leaflets of the bilayer.

Functional group. A portion of a molecule that participates in interactions with other substances. Common functional groups in biochemistry are acyl, amido, amino, carbonyl, carboxyl, diphosphoryl (pyrophosphoryl), ester, ether, hydroxyl, imino, phosphoryl, and sulfhydryl groups.

Furanose. A sugar with a five-membered ring.

Futile cycle. See substrate cycle.

G. See free energy.

$G^{\ddagger}$. The free energy of the transition state. See also free energy of activation.

G-protein. A guanine nucleotide–binding protein involved in signal transduction that is inactive when it binds GDP and active when its binds GTP. The GTPase activity of the G-protein limits its own activity. Heterotrimeric G-proteins consist of three subunits, which dissociate to form G_{α} (to which GTP binds) and $G_{\beta\gamma}$ components on activation.

Ganglioside. A ceramide whose head group is an oligosaccharide containing at least one sialic acid residue.

Gap genes. See segmentation genes.

Gastrulation. The stage of embryonic development in which cells migrate to form a triple-layered structure.

Gel electrophoresis. A procedure in which macromolecules are separated on the basis of charge or size by their differential migration through a gel-like matrix under the influence of an applied electric field. In polyacrylamide gel electrophoresis (PAGE), the matrix is cross-linked polyacrylamide. Agarose gels are used to separate molecules of very large masses, such as DNAs. See also SDS-PAGE and pulsed-field gel electrophoresis.

Gel filtration chromatography. A procedure in which macromolecules are separated on the basis of their size and shape. Also called size exclusion or molecular sieve chromatography.

Gene. A unique sequence of nucleotides that encodes a polypeptide or RNA; it may include nontranscribed and nontranslated sequences that have regulatory functions.

Gene duplication. An event, such as aberrant crossover, that gives rise to two copies of a gene on the same chromosome, each of which can then evolve independently.

Gene expression. The decoding, via transcription and translation, of the information contained in a gene to yield a functional RNA or protein product.

Gene knockout. A genetic engineering process that deletes or inactivates a specific gene in an animal.

Gene product. The RNA or protein that is encoded by a gene and that is the end point of the gene's expression through transcription and translation.

Gene therapy. The transfer of genetic material to the cells of an individual in order to produce a therapeutic effect.

General acid catalysis. A catalytic mechanism in which partial proton transfer from an acid lowers the free energy of a reaction's transition state.

General base catalysis. A catalytic mechanism in which partial proton abstraction by a base lowers the free energy of a reaction's transition state.

General transcription factor (GTF). One of a set of eukaryotic proteins that are required for the synthesis of all mRNAs.

Genetic code. The correspondence between the sequence of nucleotides in a nucleic acid and the sequence of amino acids in a polypeptide; a series of three nucleotides (a codon) specifies an amino acid.

Genetic engineering. See recombinant DNA technology.

Genetic recombination. See recombination.

Genome. The complete set of genetic instructions in an organism.

Genomic imprinting. The differential expression of maternal and paternal genes in an offspring, which results from different patterns of methylation in these genes.

Genomic library. A set of cloned DNA fragments representing an organism's entire genome.

Genomics. The study of the size, organization, and gene content of organisms' genomes.

Genotype. An organism's genetic characteristics.

Gibbs free energy. See free energy.

Globin. The polypeptide component of myoglobin and hemoglobin.

Globoside. A ceramide whose head group is a neutral oligosaccharide.

Globular protein. A water-soluble protein characterized by a compact, highly folded structure.

Glucogenic amino acid. An amino acid whose degradation yields a gluconeogenic precursor. See also ketogenic amino acid.

Gluconeogenesis. The synthesis of glucose from noncarbohydrate precursors.

Glucose–alanine cycle. A metabolic pathway in which pyruvate produced by glycolysis in the muscles is converted to alanine and transported to the liver, where it is converted back to pyruvate for gluconeogenesis. The resulting glucose returns to the muscles.

Glycan. See polysaccharide.

Glycerophospholipid. An amphiphilic lipid in which two fatty acyl groups are attached to a glycerol-3-phosphate whose phosphate group is linked to a polar group.

Glycoconjugate. A molecule, such as a glycolipid or glycoprotein, that contains covalently linked carbohydrate.

Glycoforms. Glycoproteins that differ only in the sequence, location, and number of covalently attached carbohydrates.

Glycogen. An $\alpha(1\rightarrow6)$ branched polymer of $\alpha(1\rightarrow4)$-linked glucose residues that serves as a glucose storage molecule in animals.

Glycogen storage disease. An inherited disorder of glycogen metabolism affecting the size and structure of glycogen molecules or their mobilization in the muscle and/or liver.

Glycogenolysis. The enzymatic degradation of glycogen to glucose-6-phosphate.

Glycolipid. A lipid to which carbohydrate is covalently attached.

Glycolysis. The 10-reaction pathway by which glucose is broken down to 2 pyruvate with the concomitant production of 2 ATP and the reduction of 2 NAD^+ to 2 NADH.

Glycoprotein. A protein to which carbohydrate is covalently attached.

Glycosaminoglycan. An unbranched polysaccharide consisting of alternating residues of uronic acid and hexosamine.

Glycosidic bond. The covalent linkage (acetal or ketal) between the anomeric carbon of a saccharide and an alcohol (*O*-glycosidic bond) or an amine (*N*-glycosidic bond). Glycosidic bonds link the monosaccharide residues of a polysaccharide.

Glycosylation. The attachment of carbohydrate chains to a protein through *N*- or *O*-glycosidic linkages.

Glyoxylate pathway. A variation of the citric acid cycle in plants that allows acetyl-CoA to be converted quantitatively to gluconeogenic precursors.

Glyoxysome. A membrane-bounded plant organelle in which the reactions of the glyoxylate cycle take place. It is a specialized type of peroxisome.

Golgi apparatus. A eukaryotic organelle consisting of a set of flattened membranous sacs in which newly synthesized proteins and lipids are modified.

Gout. A disease characterized by elevated levels of uric acid, usually the result of impaired uric acid excretion. Its most common manifestation is painful arthritic joint inflammation caused by the deposition of sodium urate.

Gram-negative bacterium. A bacterium that does not take up Gram stain, indicating that its cell wall is surrounded by a complex outer membrane that excludes Gram stain.

Gram-positive bacterium. A bacterium that takes up Gram stain, indicating that its outermost layer is a cell wall.

Grana (*sing.* granum). The stacked disks of the thylakoid in a chloroplast.

Growth factor. A protein hormone that stimulates the proliferation and differentiation of its target cells.

GTF. See general transcription factor.

H. See enthalpy.

Half-reaction. The single oxidation or reduction process, involving an electron donor and its conjugate electron acceptor, that occurs in electrical cells but requires direct contact with another such reaction to form a complete oxidation–reduction reaction. Also called a redox couple.

Half-life. See half-time.

Half-time ($t_{1/2}$). The time required for half the reactant initially present to undergo reaction. Also called half-life.

Halophile. An organism that thrives in (and may require) high salinity.

Haploid. Having one set of chromosomes.

Haworth projection. A representation of a sugar ring in which ring bonds that project in front of the plane of the paper are drawn as heavy lines and ring bonds that project behind the plane of the paper are drawn as light lines.

Heat shock protein (Hsp). See molecular chaperone.

Helicase. An enzyme that unwinds DNA, producing positive supercoils.

Helix cap. A protein structural element in which the side chain of a residue preceding or succeeding a helix folds back to form a hydrogen bond with the backbone of one of the helix's four terminal residues.

Helix–turn–helix (HTH) motif. An ~20-residue protein motif that forms two α helices that cross at an angle of ~120°. This motif occurs in numerous prokaryotic DNA-binding proteins.

Hemiacetal. The product of the reaction between an alcohol and the carbonyl group of an aldehyde.

Hemiketal. The product of the reaction between an alcohol and the carbonyl group of a ketone.

Hemolytic anemia. Loss of red blood cells through their lysis (destruction) in the bloodstream.

Henderson–Hasselbalch equation. The mathematical expression of the relationship between the pH of a solution of a weak acid and its pK: pH = pK + log ([A$^-$]/[HA]).

Heterochromatin. Highly condensed, nonexpressed eukaryotic DNA.

Heterogeneous nuclear RNA (hnRNA). Eukaryotic mRNA primary transcripts whose introns have not yet been excised. Also called pre-mRNA.

Heterologous DNA. A segment of DNA consisting of imperfectly complementary strands during recombination.

Heterolytic cleavage. Cleavage of a bond in which one of two chemically bonded atoms acquires both of the electrons that formed the bond.

Heteropolysaccharide. A polysaccharide consisting of more than one type of monosaccharide.

Heterotrimeric G-protein. See G-protein.

Heterotroph. An organism that obtains free energy from the oxidation of organic compounds produced by other organisms.

Heterozygous. Having one each of two gene variants.

Hexose monophosphate shunt. See pentose phosphate pathway.

"High-energy" compound. A substance whose degradation is highly exergonic (yields at least as much free energy as is required to synthesize ATP from ADP + P$_i$; ≥30.5 kJ·mol^{-1} under standard biochemical conditions). Also called an "energy-rich" compound.

High-performance liquid chromatography (HPLC). An automated chromatographic procedure for fractionating molecules using precisely fabricated matrix materials and pressurized flows of precisely mixed solvents.

Hill constant. The exponent in the Hill equation. It provides a measure of the degree of cooperative binding of a ligand to a molecule.

Hill equation. A mathematical expression for the degree of saturation of ligand binding to a molecule as a function of the ligand concentration.

Histones. Highly conserved basic proteins that form a core to which DNA is bound in a nucleosome.

HIV. Human immunodeficiency virus, the causative agent of acquired immunodeficiency syndrome (AIDS).

hnRNA. See heterogeneous nuclear RNA.

Holliday junction. The four-stranded structure that forms as an intermediate in DNA recombination.

Holoenzyme. A catalytically active enzyme–cofactor complex.

Homeobox. See homeodomain.

Homeodomain. An ~60-amino acid DNA-binding motif common to many genes that specify the identities and fates of embryonic cells; such genes encode transcription factors. Also called a homeobox.

Homeostasis. The maintenance of a steady state in an organism.

Homeotic selector genes. Insect genes that specify the identities of body segments.

Homolactic fermentation. The reduction of pyruvate to lactate with the concomitant oxidation of NADH to NAD$^+$.

Homology. The evolutionary relatedness of two entities.

Homology modeling. A technique in which the three-dimensional structure of a polypeptide is deduced from the structure of a polypeptide with a similar (homologous) sequence.

Homolytic cleavage. Cleavage of a bond in which each participating atom acquires one of the electrons that formed the bond.

Homopolysaccharide. A polysaccharide consisting of one type of monosaccharide unit.

Homozygous. Having two identical copies of a particular gene.

Hormone. A substance (e.g., a peptide or a steroid) that is secreted by one tissue into the bloodstream and which induces a physiological response (e.g., growth and metabolism) in other tissues.

Hormone response element. A DNA sequence to which a steroid hormone–receptor complex binds so as to enhance or repress the transcription of an associated gene.

Hot spot. A DNA site that has an unusually high rate of point mutations.

Housekeeping enzyme. See constitutive enzyme.

HPLC. See high-performance liquid chromatography.

Hsp. Heat shock protein. See molecular chaperone.

HTH motif. See helix–turn–helix motif.

Humoral immunity. Immunity mediated by antibodies (immunoglobulins) produced by *B* lymphocytes (*B* cells).

Hybridization. The formation of double-stranded segments of complementary DNA and/or RNA sequences.

Hybridoma. The cell clones produced by the fusion of an antibody-producing lymphocyte and an immortal myeloma cell. These are the cells that produce monoclonal antibodies.

Hydration. The molecular state of being surrounded by and interacting with several layers of solvent water molecules, that is, solvated by water.

Hydrogen bond. A mainly electrostatic interaction between a weakly acidic donor group such as O—H or N—H and a weakly basic acceptor atom such as O or N.

Hydrolase. An enzyme that catalyzes a hydrolytic reaction.

Hydrolysis. The cleavage of a covalent bond accomplished by adding the elements of water; the reverse of a condensation.

Hydronium ion. A proton associated with a water molecule, H_3O^+.

Hydropathy. A measure of the combined hydrophobicity and hydrophilicity of an amino acid residue; it is indicative of the likelihood of finding that residue in a protein interior.

Hydrophilic substance. A substance whose high polarity allows it to readily interact with water molecules and thereby dissolve in water.

Hydrophobic collapse. A driving force in protein folding, resulting from the tendency of hydrophobic residues to avoid contact with water and hence form the protein core.

Hydrophobic effect. The tendency of water to minimize its contacts with nonpolar substances, thereby inducing the substances to aggregate.

Hydrophobic interaction chromatography. A procedure in which molecules are selectively retained on a nonpolar matrix by virtue of their hydrophobicity.

Hydrophobic substance. A substance whose nonpolar nature reduces its ability to be solvated by water molecules. Hydrophobic substances tend to be soluble in nonpolar solvents but not in water.

Hydroxyl group. A portion of a molecule with the formula —OH.

Hyperammonemia. Elevated levels of ammonia in the blood, a toxic situation.

Hypercholesterolemia. High levels of cholesterol in the blood, a risk factor for heart disease. Familial hypercholesterolemia usually results from an inherited defect in the LDL receptor.

Hyperglycemia. Elevated levels of glucose in the blood.

Hypervariable residue. An amino acid residue occupying a position in a protein that is occupied by many different residues among evolutionarily related proteins. The opposite of a hypervariable residue is an invariant residue.

Hypoxia. A condition in which the oxygen level in the blood is lower than normal.

I-cell disease. A hereditary deficiency in a lysosomal hydrolase that leads to the accumulation of glycosaminoglycan and glycolipid inclusions in the lysosomes.

Ig. Immunoglobulin. See antibody.

Imaginal disk. A patch of apparently undifferentiated but developmentally committed cells in an insect larva that ultimately gives rise to a specific external structure in the adult.

Imino group. A portion of a molecule with the formula $\diagdown$C=NH.

Immunoaffinity chromatography. A procedure in which a molecule is separated from a mixture of other molecules by its ability to bind specifically to an immobilized antibody.

Immunoassay. A procedure for detecting, and in some cases quantifying the activity of, a macromolecule by using an antibody or mixture of antibodies that reacts specifically with that substance.

Immunoblot. A technique in which a molecule immobilized on a membrane filter can be detected through its ability to bind to an antibody directed against it. A Western blot is an immunoblot to detect an immobilized protein after electrophoresis.

Immunofluorescence microscopy. A technique in microscopy in which a fluorescence-tagged antibody is used to reveal the presence of the antigen to which it binds.

Immunoglobulin (Ig). See antibody.

Immunoglobulin fold. A disulfide-linked domain consisting of a sandwich of a three-stranded and a four-stranded antiparallel β sheet that occurs in antibody molecules.

in situ. In place.

in situ **hybridization.** See colony hybridization.

in vitro. In the laboratory (literally, in glass).

in vivo. In a living organism.

Inactivator. An inhibitor that binds irreversibly to an enzyme.

Indirect readout. The ability of a DNA-binding protein to detect its target base sequence through the sequence-dependent conformation and/or flexibility of its DNA backbone rather than through direct interaction with its bases.

Induced fit. An interaction between a protein and its ligand, which induces a conformational change in the protein that increases the protein's affinity for the ligand.

Inducer. A substance that facilitates gene expression.

Inducible enzyme. An enzyme that is synthesized only when required by the cell. See also constitutive enzyme.

Inhibitor. A substance that reduces an enzyme's activity by affecting its substrate binding or turnover number.

Initiation factor. A protein that interacts with mRNA and/or the ribosome and which is required to initiate translation.

Inorganic compound. A compound that does not contain the element carbon.

Insertion sequence. A simple transposon that is flanked by short inverted repeats. Also called an IS element.

Insulin resistance. The inability of cells to respond to insulin by increasing their glucose uptake.

Integral protein. A membrane protein that is embedded in the lipid bilayer and can be separated from it only by treatment with agents that disrupt membranes. Also called an intrinsic protein.

Intercalation agent. A planar aromatic cation that slips in between the stacked bases of a polynucleotide.

Interconvertible enzyme. A protein that undergoes phosphorylation and dephosphorylation so as to modulate its activity.

Interfacial activation. The increase in activity when a lipid-specific enzyme contacts the lipid–water interface.

Intermembrane space. The compartment between the inner and outer mitochondrial membranes. Because of the porosity of the outer membrane, the intermembrane space is equivalent to the cytosol in its small-molecule composition.

Internal conversion. A mode of decay of an excited molecule, in which electronic energy is converted to heat (the kinetic energy of molecular motion).

Intervening sequence. See intron.

Intrinsic protein. See integral protein.

Intron. A portion of a gene that is transcribed but excised prior to translation. Also called an intervening sequence.

Invariant residue. A residue in a protein that is the same in all evolutionarily related proteins. The opposite of an invariant residue is a hypervariable residue.

Ion exchange chromatography. A fractionation procedure in which molecules are selectively retained by a matrix bearing oppositely charged groups.

Ion pair. An electrostatic interaction between two ionic groups of opposite charge. In proteins, it is also called a salt bridge.

Ionophore. An organic molecule, often an antibiotic, that increases the permeability of a membrane to a particular ion.

Iron–sulfur cluster. A prosthetic group consisting most commonly of equal numbers of iron and sulfur ions (i.e., [2Fe–2S] and [4Fe–4S]) and that usually participates in oxidation–reduction reactions.

IS element. See insertion sequence.

Isoaccepting tRNA. A tRNA that carries the same amino acid as another tRNA but has a different anticodon.

Isoelectric point (p*I*). The pH at which a molecule has no net charge.

Isolated system. A thermodynamic system that cannot exchange matter or energy with its surroundings.

Isomerase. An enzyme that catalyzes an isomerization reaction.

Isopeptide bond. An amide linkage between an α-carboxylate group of an amino acid and the ε-amino group of Lys, or between the α-amino group of an amino acid and the β- or γ-carboxylate group of Asp or Glu.

Isoschizomers. Two restriction endonucleases that cleave at the same nucleotide sequence.

Isozymes. Enzymes that catalyze the same reaction but are encoded by different genes.

Jaundice. A yellowing of the skin and whites of the eyes as a result of the deposition of the heme degradation product bilirubin in these tissues. It is a symptom of liver dysfunction, bile-duct obstruction, or a high rate of red cell destruction.

Junk DNA. See selfish DNA.

K. See dissociation constant and equilibrium constant.

k. See rate constant.

kb. Kilobase pair; 1 kb = 1000 base pairs (bp).

k_{cat}. The catalytic constant for an enzymatic reaction, equivalent to the ratio of the maximal velocity (V_{max}) and the enzyme concentration ($[E]_T$). Also called a turnover number.

kD. Kilodaltons; 1000 daltons (D).

K_{eq}. See equilibrium constant.

Ketogenesis. The synthesis of ketone bodies from acetyl-CoA.

Ketogenic amino acid. An amino acid whose degradation in animals yields compounds that can be converted to fatty acids or ketone bodies, but not to glucose. See also glucogenic amino acid.

Ketone bodies. Acetoacetate, D-β-hydroxybutyrate, and acetone; these compounds are produced from acetyl-CoA by the liver for use as metabolic fuels in peripheral tissues when glucose is unavailable.

Ketose. A sugar whose carbonyl group is a ketone.

Ketosis. A pathological condition in which ketone bodies are produced in excess of their utilization.

Kinase. An enzyme that transfers a phosphoryl group between ATP and another molecule.

Kinetic proofreading. The process by which a noncognate tRNA dissociates from the ribosome at a rate faster than the dissociation of EF-Tu·GDP, thereby promoting translational accuracy.

K_M. See Michaelis constant.

Krebs cycle. See citric acid cycle.

K_w. The ionization constant of water; equal to 10^{-14} M^2.

L. See linking number.

Lactose intolerance. The inability to digest the disaccharide lactose due to a deficiency of the enzyme β-galactosidase (lactase).

Lagging strand. The DNA strand whose parent (template) strand extends 5′→3′ in the direction of travel of the replication fork. This strand is synthesized as a series of discontinuous fragments that are later joined.

Lateral diffusion. The movement of a lipid within one leaflet of a bilayer.

Leader peptide. See signal peptide.

Leader sequence. (1) A sequence of nucleotides upstream of the first gene in an operon, whose translation governs transcription of the entire operon. See attenuation. (2) An N-terminal sequence of a polypeptide, often not part of the mature protein, that is responsible for its passage into or through a membrane.

Leading strand. The DNA strand whose parent (template) strand extends 3′→5′ in the direction of travel of the replication fork. This strand is synthesized continuously.

Lectin. a protein that binds to a specific saccharide.

Lesch–Nyhan syndrome. A genetic disease caused by the deficiency of hypoxanthine–guanine phosphoribosyltransferase, an enzyme required for purine salvage reactions. Affected individuals produce excessive uric acid and exhibit neurological abnormalities.

Leucine zipper. A protein structural motif in which two α helices, each with a hydrophobic strip along one side, associate as a coiled coil. This motif, which has a Leu at nearly every seventh residue, mediates the association of many types of DNA-binding proteins.

Leukocyte. White blood cell.

Levorotatory. Rotating the plane of polarized light counterclockwise from the point of view of the observer; the opposite of dextrorotatory.

Ligand. (1) A small molecule that binds to a larger molecule. (2) A molecule or ion bound to a metal ion.

Ligase. An enzyme that catalyzes bond formation coupled with the hydrolysis of ATP.

Ligation. The joining together of two molecules such as two DNA segments.

Light reactions. The portion of photosynthesis in which specialized pigment molecules capture light energy and are thereby oxidized. Electrons are transferred to generate NADPH and a transmembrane proton gradient that drives ATP synthesis.

Limited proteolysis. A technique in which a polypeptide is incompletely digested by proteases.

Lineweaver–Burk plot. A rearrangement of the Michaelis–Menten equation that permits the determination of K_M and V_{max} from a linear plot. Also called a double-reciprocal plot.

Linker DNA. The ~55-bp segment of DNA that links nucleosome core particles in chromatin.

Linking number (*L*). The number of times that one strand of a covalently closed circular double-stranded DNA winds around the other; it cannot be changed without breaking covalent bonds.

Lipid. Any member of a broad class of biological molecules that are largely or wholly hydrophobic and therefore tend to be insoluble in water but soluble in organic solvents such as hexane.

Lipid bilayer. See bilayer.

Lipid-linked protein. A protein that is anchored to a biological membrane via a covalently attached lipid such as a farnesyl, geranylgeranyl, myristoyl, palmitoyl, or glycosylphosphatidylinositol group.

Lipoprotein. A globular particle consisting of a nonpolar lipid core surrounded by an amphiphilic coat of protein, phospholipid, and cholesterol. Lipoproteins transport lipids between tissues via the bloodstream.

Liposome. A vesicle bounded by a single lipid bilayer.

London dispersion forces. The weak attractive forces between electrically neutral molecules in close proximity, which arise from electrostatic interactions among their fluctuating dipoles.

Lung surfactant. The amphipathic protein and lipid mixture that prevents collapse of the lung alveoli (microscopic air spaces) on the expiration of air.

Lyase. An enzyme that catalyzes the elimination of a group to form a double bond.

Lymphocyte. An immune system cell.

Lysogen. A bacterium that harbors bacteriophage DNA inserted in its chromosome.

Lysogenic growth. The portion of a bacteriophage life cycle during which its DNA, which is inserted into the host chromosome, is passively replicated with the host DNA.

Lysosome. A membrane-bounded organelle in a eukaryotic cell that contains a battery of hydrolytic enzymes and which functions to digest ingested material and to recycle cell components.

Lytic growth. The portion of a bacteriophage life cycle during which its DNA is expressed and new phage particles are produced, resulting in lysis of the host cell.

Main chain. See backbone.

Major groove. The groove on a DNA double helix onto which the glycosidic bonds of a base pair form an angle of $>180°$. In B-DNA, this groove is wider than the minor groove.

Malaria. A mosquito-borne disease caused by the protozoan *Plasmodium falciparum*, which resides in red blood cells during much of its life cycle.

Malignant tumor. A mass of cells that proliferates uncontrollably; a cancer.

Maternal-effect genes. Insect genes whose mRNA or protein products are deposited by the mother in the ovum and which define the polarity of the embryonic body.

Matrix. The gel-like solution of enzymes, substrates, cofactors, and ions in the interior of the mitochondrion.

Mechanism-based inhibitor. A molecule that chemically inactivates an enzyme only after undergoing part or all of its normal catalytic reaction. Also called a suicide substrate.

Mediated transport. The transmembrane movement of a substance through the action of a specific carrier protein; the opposite of nonmediated transport.

Melting temperature (*T*$_m$). The midpoint temperature of the melting curve for the thermal denaturation of a macromolecule.

Membrane potential ($\Delta\Psi$). The electrical potential difference across a membrane.

Memory *B* cell. A type of *B* cell that can recognize its corresponding antigen and rapidly proliferate to produce specific antibodies weeks to years after this antigen was first encountered.

Mercaptan. A compound containing an —SH group.

Messenger RNA (mRNA). A ribonucleic acid whose sequence is complementary to that of a protein-coding gene in DNA. In the ribosome, mRNA directs the polymerization of amino acids to form a polypeptide with the corresponding sequence.

Metabolic fuel. A molecule that can be oxidized to provide free energy for an organism.

Metabolism. The total of all degradative and biosynthetic cellular reactions.

Metabolite. A reactant, intermediate, or product of a metabolic reaction.

Metabolon. The loose association of the enzymes of a metabolic pathway in one region of the cytoplasm or organelle.

Metal ion catalysis. A catalytic mechanism that requires the presence of a metal ion to lower the free energy of a reaction's transition state.

Metal-activated enzyme. An enzyme that loosely binds a metal ion, typically Na^+, K^+, Mg^{2+}, or Ca^{2+}.

Metalloenzyme. An enzyme that contains a tightly bound metal ion cofactor, typically a transition metal ion such as Fe^{2+}, Zn^{2+}, or Mn^{2+}.

Methanogen. An organism that produces CH_4.

Micelle. A globular aggregate of amphiphilic molecules in aqueous solution that are oriented such that polar segments form the surface of the aggregate and the nonpolar segments form a core that is out of contact with the solvent.

Michaelis constant (*K*$_M$). For an enzyme that follows the Michaelis–Menten model, $K_M = (k_{-1} + k_2)/k_1$; K_M is equal to the substrate concentration at which the reaction velocity is half-maximal.

Michaelis–Menten equation. A mathematical expression that describes the activity of an enzyme in terms of the substrate concentration ([S]), the enzyme's maximal velocity (V_{max}), and its Michaelis constant (K_M): $v_o = V_{max}[S]/(K_M + [S])$.

Microfilament. A 70-Å-diameter cytoskeletal element composed of actin.

Microheterogeneity. The variability in carbohydrate composition in glycoproteins.

Minor groove. The groove on a DNA double helix onto which the glycosidic bonds of a base pair form an angle of $<180°$. In B-DNA, this groove is narrower than the major groove.

Mitochondria (*sing.* mitochondrion). The double-membrane-enveloped eukaryotic organelles in which aerobic metabolic reactions occur, including those of the citric acid cycle, fatty acid oxidation, and oxidative phosphorylation.

Mixed inhibition. A form of enzyme inhibition in which an inhibitor binds to both the enzyme and the enzyme–substrate complex and thereby differently affects K_M and V_{max}. Also called noncompetitive inhibition.

Modification methylase. A bacterial enzyme that methylates a specific sequence of DNA as part of a restriction–modification system.

Molecular chaperone. A protein that binds to unfolded or misfolded proteins so as to promote normal folding and the formation of native quaternary structure. Also known as a heat shock protein (Hsp). See also chaperonins.

Molecular cloning. See recombinant DNA technology.

Molecular sieve chromatography. See gel filtration chromatography.

Molecular weight. See M_r.

Molecularity. The number of molecules that participate in an elementary chemical reaction.

Molten globule. A collapsed but conformationally mobile intermediate in protein folding that has much of the native protein's secondary structure but little of its tertiary structure.

Monocistronic mRNA. The RNA transcript of a single gene.

Monoclonal antibody. A single type of antibody molecule produced by a clone of hybridoma cells, which are derived by the fusion of a myeloma cell with a lymphocyte producing that antibody.

Monomer. (1) A structural unit from which a polymer is built up. (2) A single subunit or protomer of a multisubunit protein.

Monoprotic acid. An acid that can donate only one proton.

Monosaccharide. A carbohydrate consisting of a single saccharide (sugar).

Morphogen. A substance whose distribution in an embryo directs, in part, the embryo's developmental pattern.

Motif. A common grouping of secondary structural elements. Also called a supersecondary structure.

M_r. Relative molecular mass. A dimensionless quantity that is defined as the ratio of the mass of a particle to 1/12th the mass of a ^{12}C atom. Also known as molecular weight.

mRNA. See messenger RNA.

Multienzyme complex. A group of noncovalently associated enzymes that catalyze two or more sequential steps in a metabolic pathway.

Multiple myeloma. A disease in which a cancerous B cell proliferates and produces massive quantities of a single antibody known as a myeloma protein.

Mutagen. An agent that induces a mutation in an organism.

Mutase. An enzyme that catalyzes the transfer of a functional group from one position to another on a molecule.

Mutation. A heritable alteration in an organism's genetic material.

MWC model of allosterism. See symmetry model of allosterism.

Myocardial infarction. The death of heart tissue caused by the loss of blood supply (a heart attack).

Myofibril. The bundle of fibers that are arranged in register in striated muscle cells.

Native structure. The fully folded conformation of a macromolecule.

Natural selection. The evolutionary process by which the continued existence of a replicating entity depends on its ability to survive and reproduce under the existing conditions.

NDP. A ribonucleoside diphosphate.

Near-equilibrium reaction. A reaction whose ΔG value is close to zero, so that it can operate in either direction depending on the substrate and product concentrations.

N-end rule. The correlation between the identity of a polypeptide's N-terminal residue and its half-life in the cell.

NER. See nucleotide excision repair.

Nernst equation. An expression of the relationship between reduction potential difference ($\Delta \mathscr{E}$) and the concentrations of the electron donors and acceptors (A, B): $\Delta \mathscr{E} = \Delta \mathscr{E}° - RT/n\mathscr{F} \ln ([A_{red}][B_{ox}]/[A_{ox}][B_{red}])$.

Neurotransmitter. A substance released by a nerve cell to alter the activity of another nerve cell.

Neutral drift. Evolutionary changes that become fixed at random rather than through natural selection.

Neutral solution. A solution whose pH is equal to 7.0 ($[H^+] = 10^{-7}$ M).

N-glycosidic bond. See glycosidic bond.

Nick translation. The progressive movement of a single-strand break (nick) in duplex DNA through the coordinated actions of a $5' \rightarrow 3'$ exonuclease function that removes residues from the $5'$ side of the break and a polymerase function that adds residues to the $3'$ side.

Nitrogen fixation. The process by which atmospheric N_2 is converted to a biologically useful form such as NH_3.

N-linked oligosaccharide. An oligosaccharide linked via a glycosidic bond to the amide group of a protein Asn residue.

NMR. See nuclear magnetic resonance.

Noncoding strand. See antisense strand.

Noncompetitive inhibition. (1) A synonym for mixed inhibition. (2) A special case of mixed inhibition in which the inhibitor binds the enzyme and enzyme–substrate complex with equal affinities ($K_I = K_I'$), thereby reducing the apparent value of V_{max} but leaving K_M unchanged.

Noncooperative binding. A situation in which binding of a ligand to a macromolecule does not affect the affinities of other binding sites on the same molecule.

Nonessential amino acid. An amino acid that an organism can synthesize from common intermediates.

Nonmediated transport. The transmembrane movement of a substance through simple diffusion; the opposite of mediated transport.

Nonrepetitive structure. A segment of a polymer in which the backbone has an ordered arrangement that is not characterized by a repeating conformation.

Nonsense codon. See stop codon.

Nonsense mutation. A mutation that converts a codon that specifies an amino acid to a stop codon, thereby causing the premature termination of translation.

Nonsense suppressor tRNA. A mutated tRNA that recognizes a stop codon so that its attached aminoacyl group is appended to the growing polypeptide chain; it mitigates the effect of a nonsense mutation in a structural gene.

Nonshivering thermogenesis. The production of heat via fuel oxidation without the synthesis of ATP and without shivering or other muscle movement.

Northern blotting. A procedure for identifying an RNA containing a particular base sequence through its ability to hybridize with a complementary single-stranded segment of DNA or RNA. See also Southern blotting.

N-terminus. See amino terminus.

nt. Nucleotide.

NTP. A ribonucleoside triphosphate.

Nuclear magnetic resonance (NMR). A spectroscopic method for characterizing atomic and molecular properties on the basis of the signals emitted by radiofrequency-excited atomic nuclei in a magnetic field. It can be used to determine the three-dimensional molecular structure of a protein or nucleic acid.

Nuclease. An enzyme that degrades nucleic acids.

Nucleic acid. A polymer of nucleotide residues. The major nucleic acids are deoxyribonucleic acid (DNA) and ribonucleic acid (RNA). Also known as a polynucleotide.

Nucleolus (pl. nucleoli). The dark-staining region of the eukaryotic nucleus, where ribosomes are assembled.

Nucleophile. A group that contains unshared electron pairs that readily reacts with an electron-deficient group (electrophile). A nucleophile (nucleus-lover) reacts with an electrophile (electron-lover).

Nucleoside. A compound consisting of a nitrogenous base and a five-carbon sugar (ribose or deoxyribose) in *N*-glycosidic linkage.

Nucleosome. The complex of a histone octamer and ~200 bp of DNA that represents the lowest level of DNA organization in the eukaryotic chromosome.

Nucleosome core particle. The complex of histones and ~146 bp of DNA that forms a compact disk-shaped particle in which the DNA is wound in ~2 helical turns around the outside of the histone octamer.

Nucleotide. A compound consisting of a nucleoside esterified to one or more phosphate groups. Nucleotides are the monomeric units of nucleic acids.

Nucleotide excision repair (NER). A multistep process in which a portion of DNA containing a lesion is excised and replaced by normal DNA.

Nucleus. The membrane-enveloped organelle in which the eukaryotic cell's genetic material is located.

O-glycosidic bond. See glycosidic bond.

O-linked oligosaccharide. An oligosaccharide linked via a glycosidic bond to the hydroxyl group of a protein Ser or Thr side chain.

Oil. A mixture of triacylglycerols that is liquid at room temperature.

Oil body. A 1-μm-diameter, lipoproteinlike particle that stores triacylglycerols in plant cells.

Okazaki fragments. The short segments of DNA formed in the discontinuous lagging-strand synthesis of DNA.

Oligomer. (1) A short polymer consisting of a few linked monomer units. (2) A protein consisting of a few protomers (subunits).

Oligosaccharide. A polymeric carbohydrate containing a few monosaccharide residues.

Ω loop. A loop of 6 to 16 polypeptide residues that occurs on the surface of a protein and which forms a compact globular entity.

Oncogene. A mutant version of a normal gene (a proto-oncogene), which may be acquired through viral infection; it interferes with the mechanisms that normally control cell growth and differentiation and thereby contributes to uncontrolled proliferation (cancer).

Open reading frame (ORF). A portion of the genome that potentially codes for a protein; this stretch of nucleotides contains no stop codons, is flanked by the proper control sequences, and exhibits the same codon-usage preference as other genes in the organism.

Open system. A thermodynamic system that can exchange matter and energy with its surroundings.

Operator. A DNA sequence at or near the transcription start of a gene, to which a repressor binds so as to control transcription of the gene.

Operon. A prokaryotic genetic unit that consists of several genes with related functions that are transcribed as a single mRNA molecule.

Optical activity. The ability of a molecule to rotate the plane of polarized light.

Ordered mechanism. A Sequential reaction with a compulsory order of substrate addition to the enzyme.

ORF. See open reading frame.

Organelle. A differentiated structure within a eukaryotic cell, such as a mitochondrion, ribosome, or lysosome, that performs specific functions.

Organic compound. A compound that contains the element carbon.

Osmosis. The movement of solvent across a semipermeable membrane from a region of low solute concentration to a region of high solute concentration.

Osmotic pressure. The pressure that must be applied to a solution containing a high concentration of solute to prevent the net flow of solvent across a semipermeable membrane (a membrane that is permeable to solvent but not to solute) separating it from a solution with a lower concentration of solute. The osmotic pressure of a 1 M solution of any solute separated from solvent by a semipermeable membrane is 22.4 atm.

Overproducer. A genetically engineered organism that produces massive quantities of a foreign DNA gene product.

Oxidative phosphorylation. The process by which the free energy obtained from the oxidation of metabolic fuels is used to generate ATP from ADP + P_i.

Oxidizing agent. A substance that can accept electrons, thereby becoming reduced.

Oxidoreductase. An enzyme that catalyzes an oxidation–reduction reaction.

Oxygen debt. The postexertion continued elevation in O_2 consumption that is required to replenish the ATP consumed by the liver during operation of the Cori cycle.

P/O ratio. The ratio of the number of molecules of ATP synthesized from ADP + P_i to the number of atoms of oxygen reduced.

p_{50}. For a gaseous ligand, the ligand concentration, in units of torr, at which a binding protein such as hemoglobin is half-saturated with ligand.

PAGE. Polyacrylamide gel electrophoresis. See gel electrophoresis.

Pair-rule genes. See segmentation genes.

Paper chromatography. A technique for separating molecules on the basis of their rate of movement with solvent on a paper matrix.

Partial oxygen pressure (pO_2). The concentration of gaseous O_2 in units of torr.

Passive-mediated transport. The thermodynamically spontaneous carrier-mediated transmembrane movement of a substance from high to low concentration. Also called facilitated diffusion.

Pasteur effect. The greatly increased sugar consumption of yeast grown under anaerobic conditions compared to that of yeast grown under aerobic conditions.

Pathogen. A disease-causing microorganism.

PCR. See polymerase chain reaction.

Pellagra. The human disease resulting from a deficiency of the vitamin niacin (nicotinic acid), a precursor of the nicotinamide-containing cofactors NAD^+ and $NADP^+$.

Pentose phosphate pathway. A pathway for glucose degradation that yields ribose-5-phosphate and NADPH. Also called the hexose monophosphate shunt.

Peptidase. An enzyme that hydrolyzes peptide bonds. Also called a protease.

Peptide. A polypeptide of less than about 40 residues.

Peptide bond. An amide linkage between the α-amino group of one amino acid and the α-carboxylate group of another. Peptide bonds link the amino acid residues in a polypeptide.

Peptide group. The planar —CO—NH— group that encompasses the peptide bond between amino acid residues in a polypeptide.

Peptidoglycans. The cross-linked bag-shaped macromolecules consisting of polysaccharide and polypeptide chains that form bacterial cell walls.

Peripheral protein. A protein that is weakly associated with the surface of a biological membrane. Also called an extrinsic protein.

Periplasmic compartment. The space between the cell wall and the outer membrane of gram-negative bacteria.

Peroxisome. A eukaryotic organelle with specialized oxidative functions.

Perutz mechanism. A model for the cooperative binding of oxygen to hemoglobin, in which O_2 binding induces the protein to shift conformation from the deoxy (T state) to the oxy (R state).

PFGE. See pulsed-field gel electrophoresis.

pH. A quantity used to express the acidity of a solution, equivalent to $-\log [H^+]$.

Phage. See bacteriophage.

Phenotype. An organism's physical characteristics.

ϕ (phi). The torsion angle that describes the rotational position around the C_α—N bond in a peptide group.

Phosphatase. An enzyme that hydrolyzes phosphoryl ester groups. See also protein phosphatase.

Phosphodiester bond. The linkage in which a phosphate group is esterified to two alcohol groups, e.g., the phosphate groups that join the adjacent nucleoside residues in a polynucleotide.

Phosphoinositide pathway. A signal-transduction pathway in which hormone binding to a cell-surface receptor induces phospholipase C to catalyze the hydrolysis of phosphatidylinositol-4,5-bisphosphate (PIP_2), which yields inositol-1,4,5-trisphosphate (IP_3) and 1,2-diacylglycerol (DG), both of which are second messengers.

Phospholipase. An enzyme that hydrolyzes one or more bonds in a glycerophospholipid.

Phosphorolysis. The cleavage of a chemical bond by the substitution of a phosphate group rather than water.

Phosphoryl group. A portion of a molecule with the formula —PO_3H_2.

Phosphoryl group-transfer potential. A measure of the tendency of a phosphorylated compound to transfer its phosphoryl group to water; the opposite of its free energy of hydrolysis.

Photoautotroph. An autotrophic organism that obtains energy from sunlight.

Photon. A packet of light energy. See also Planck's law.

Photooxidation. A mode of decay of an excited molecule, in which oxidation occurs through the transfer of an electron to an acceptor molecule.

Photophosphorylation. The synthesis of ATP from ADP + P_i coupled to the dissipation of a proton gradient that has been generated through light-driven electron transport.

Photoreactivation. The conversion of pyrimidine dimers, a form of DNA damage, to monomers using light energy.

Photorespiration. The consumption of O_2 and evolution of CO_2 by plants (a dissipation of the products of photosynthesis) resulting from the competition between O_2 and CO_2 for binding to ribulose bisphosphate carboxylase.

Photosynthesis. The reduction of CO_2 to $(CH_2O)_n$ in plants and bacteria as driven by light energy.

Phylogenetic tree. A reconstruction of the probable path of evolution of a set of organisms, often based on sequence variations in homologous proteins and nucleic acids; a sort of family tree.

Phylogeny. The study of the evolutionary relationships among organisms.

pI. See isoelectric point.

PIC. See preinitiation complex.

Ping Pong reaction. A group-transfer reaction in which one or more products are released before all substrates have bound to the enzyme.

Pitch. The distance a helix rises along its axis per turn; 5.4 Å for an α helix, 34 Å for B-DNA.

pK. A quantity used to express the tendency for an acid to donate a proton (dissociate); equal to $-\log K$, where K is the acid's dissociation constant. Also known as pK_a.

Planck's law. An expression for the energy (E) of a photon: $E = hc/\lambda$, where c is the speed of light, λ is its wavelength, and h is Planck's constant (6.626×10^{-34} J·s).

Plaque. (1) A region of lysed cells on a "lawn" of cultured bacteria, which indicates the presence of infectious bacteriophage. (2) A deposit of insoluble material in an animal's tissues.

Plasmalogen. A glycerophospholipid in which the C1 substituent is attached via an ether rather than an ester linkage.

Plasmid. A small circular DNA molecule that autonomously replicates in a bacterial or yeast cell. Plasmids are often modified for use as cloning vectors.

pmf. See protonmotive force.

pO_2. See partial oxygen pressure.

Point mutation. The substitution of one base for another in DNA. Point mutations may arise from mispairing during DNA replication or from chemical alterations of existing bases.

Polarimeter. A device that measures the optical rotation of a solution. It can be used to determine the optical activity of a substance.

Poly(A) tail. The sequence of adenylate residues that is posttranslationally appended to the 3′ end of eukaryotic mRNAs.

Polyacrylamide gel electrophoresis (PAGE). See gel electrophoresis.

Polycistronic mRNA. The RNA transcript of a bacterial operon. It encodes several polypeptides.

Polycythemia. A condition characterized by an increased number of erythrocytes.

Polyelectrolyte. A macromolecule that bears multiple charged groups.

Polymer. A molecule consisting of numerous smaller units that are linked together in an organized manner. Polymers may be linear or branched and may contain one or more kinds of structural units (monomers).

Polymerase. An enzyme that catalyzes the addition of nucleotide residues to a polynucleotide through nucleophilic attack of the chain's 3′-OH group on the α-phosphoryl group of the incoming nucleoside triphosphate. DNA- and RNA-directed polymerases require a template molecule with which the incoming nucleotide must base pair.

Polymerase chain reaction (PCR). A procedure for amplifying a segment of DNA by repeated rounds of replication centered between primers that hybridize with the two ends of the DNA segment of interest.

Polynucleotide. See nucleic acid.

Polypeptide. A polymer consisting of amino acid residues linked in linear fashion by peptide bonds.

Polyprotic acid. A substance with more than one proton that can be donated. Polyprotic acids have multiple ionization states.

Polyribosome. An mRNA transcript bearing multiple ribosomes in the process of carrying out translation. Also called a polysome.

Polysaccharide. A polymeric carbohydrate containing multiple monosaccharide residues. Also called a glycan.

Polysome. See polyribosome.

Polytene chromosome. A chromosome resulting from multiple rounds of DNA replication such that the daughter molecules remain associated and in register.

Polyunsaturated fatty acid. A fatty acid that contains more than one double bond in its hydrocarbon chain.

Porphyrias. Genetic defects in heme biosynthesis that result in the accumulation of porphyrin and/or its precursors.

Postreplication repair. See recombination repair.

Posttranscriptional modification. The removal or addition of nucleotide residues or their modification following the synthesis of RNA.

Posttranslational modification. The removal or derivatization of amino acid residues following their incorporation into a polypeptide.

Prebiotic era. The period of time between the formation of the earth ~4.6 billion years ago and the appearance of living organisms at least 3.5 billion years ago.

Precursor. The entity that gives rise, through a process such as evolution or chemical reaction, to another entity.

Preinitiation complex (PIC). The assembly of transcription factors bound to DNA that renders the DNA available for transcription by RNA polymerase.

pre-mRNA. See heterogeneous nuclear RNA.

pre-rRNA. An immature rRNA transcript.

pre-tRNA. An immature tRNA transcript.

Preproprotein. A protein bearing both a signal peptide (preprotein) and a propeptide (proprotein).

Preprotein. A protein bearing a signal peptide that is cleaved off following the translocation of the protein through the endoplasmic reticulum membrane.

Pribnow box. The prokaryotic promoter element with the consensus sequence TATAAT that is centered at the −10 position relative to the transcription start site.

Primary active transport. Transmembrane transport that is driven by the exergonic hydrolysis of ATP.

Primary structure. The sequence of residues in a polymer.

Primary transcript. The immediate product of transcription, which may be modified before becoming fully functional.

Primer. An oligonucleotide that serves as a starting point for additional polymerization reactions catalyzed by DNA polymerase to form a polynucleotide. A primer base-pairs with a segment of a template polynucleotide so as to form a short double-stranded segment that can then be extended through template-directed polymerization.

Primosome. The protein complex that initiates synthesis of an Okazaki fragment in discontinuous DNA synthesis.

Prion. A protein whose misfolding causes it to aggregate and produce the neurodegenerative symptoms of bovine spongiform encephalopathy and related diseases. Misfolded prions induce properly folded prions to misfold and thereby act as infectious agents.

Probe. A labeled single-stranded DNA or RNA segment that can hybridize with a DNA or RNA of interest in a screening procedure.

Processive enzyme. An enzyme that catalyzes many rounds of a polymerization reaction without dissociating from the growing polymer.

Prochirality. A property of a nonchiral molecule such that it contains a group whose substitution by another group yields a chiral molecule.

Proenzyme. An inactive precursor of an enzyme.

Prokaryote. A unicellular organism that lacks a membrane-bounded nucleus. All bacteria are prokaryotes.

Promoter. The DNA sequence at which RNA polymerase binds to initiate transcription.

Proofreading. An additional catalytic activity of an enzyme, which acts to correct errors made by the primary enzymatic activity.

Prophage. A bacteriophage that is stably inserted in the host genome.

Propeptide. A polypeptide segment of an immature protein that must be proteolytically excised to activate the protein.

Proprotein. The inactive precursor of a protein that, to become fully active, must undergo limited proteolysis to excise its propeptide.

Prostaglandin. See eicosanoids.

Prosthetic group. A cofactor that is permanently (often covalently) associated with an enzyme.

Protease. See peptidase.

Proteasome. A multiprotein complex with a hollow cylindrical core in which cellular proteins are degraded to peptides (recycled) in an ATP-dependent process.

Protein. A macromolecule that consists of one or more polypeptide chains.

Protein kinase. An enzyme that catalyzes the transfer of a phosphoryl group from ATP to the OH group of a protein Ser, Thr, or Tyr residue.

Protein module. A sequence motif of ~40–100 residues that may occur in unrelated proteins or as multiple arrays within one protein.

Protein phosphatase. An enzyme that catalyzes the hydrolytic excision of phosphoryl groups from proteins.

Proteoglycan. An extracellular aggregate of protein and glycosaminoglycans.

Proto-oncogene. The normal cellular analog of an oncogene; the mutation of a proto-oncogene may yield an oncogene that contributes to uncontrolled cell proliferation (cancer).

Protomer. One of two or more identical units of an oligomeric protein. A protomer may consist of one or more polypeptide chains.

Proton jumping. The rapid movement of a proton among hydrogen-bonded water molecules. Proton jumping is largely responsible for the rapid rate at which hydronium and hydroxyl ions appear to move through an aqueous solution.

Protonmotive force (pmf). The free energy of the electrochemical proton gradient that forms during electron transport.

Proximity effect. A catalytic mechanism in which a reaction's free energy of activation is reduced by the bringing together of its reacting groups.

Pseudo-first-order reaction. A bimolecular reaction whose rate appears to be proportional to the concentration of only a single reactant.

Pseudogene. An unexpressed sequence of DNA that is apparently the defective remnant of a duplicated gene.

ψ (psi). The torsion angle that describes the rotational position around the C_α—C bond in a peptide group.

Pulse labeling. A technique for tracing metabolic fates, in which cells or a reacting system are exposed briefly to high levels of a labeled compound.

Pulsed-field gel electrophoresis (PFGE). An electrophoretic procedure in which electrodes arrayed around the periphery of an agarose slab gel are sequentially pulsed so that DNA molecules must continually reorient, thereby allowing very large molecules to be separated by size.

Purine nucleotide cycle. The conversion of aspartate to fumarate, which replenishes citric acid cycle intermediates, through the deamination of AMP to IMP.

Purines. Derivatives of the compound purine, a planar, aromatic, heterocyclic compound. Adenine and guanine, two of the nitrogenous bases of nucleotides, are purines.

Pyranose. A sugar with a six-membered ring.

Pyrimidines. Derivatives of the compound pyrimidine, a planar, aromatic, heterocyclic compound. Cytosine, uracil, and thymine, three of the nitrogenous bases of nucleotides, are pyrimidines.

Pyrophosphoryl group. See diphosphoryl group.

q. The thermodynamic term for heat absorbed.

q_P. The thermodynamic term for heat absorbed at constant pressure.

Q cycle. The cyclic flow of electrons accompanied by the transport of protons, involving a stable semiquinone intermediate of CoQ in Complex III of mitochondrial electron transport and in photosynthetic electron transport.

Quantum (*pl.* quanta). A packet of energy. See also photon.

Quantum yield. The ratio of molecules reacted to photons absorbed.

Quaternary structure. The spatial arrangement of a macromolecule's individual subunits.

R group. A symbol for a variable portion of an organic molecule, such as the side chain of an amino acid.

R state. One of two conformations of an allosteric protein; the other is the T state.

Racemic mixture. A sample of a compound in which both enantiomers are present in equal amounts.

Radioimmunoassay (RIA). A technique for measuring the concentration of a molecule on the basis of its ability to block the binding of a small amount of the radioactively labeled molecule to its corresponding antibody.

Radionuclide. A radioactive isotope.

Ramachandran diagram. A plot of ϕ and ψ values that indicates the sterically allowed conformations of a polypeptide.

Random coil. A totally disordered and rapidly fluctuating polymer conformation.

Random mechanism. A Sequential reaction without a compulsory order of substrate addition to the enzyme.

Rate constant (k). The proportionality constant between the velocity of a chemical reaction and the concentration(s) of the reactant(s).

Rate enhancement. The ratio of the rates of a catalyzed to an uncatalyzed chemical reaction.

Rate equation. A mathematical expression for the time-dependent progress of a reaction as a function of reactant concentration.

Rate-determining step. The step with the highest transition state free energy in a multistep reaction; the slowest step.

Reaction coordinate. The path of minimum free energy for the progress of a reaction.

Reaction order. The sum of the exponents of the concentration terms that appear in a reaction's rate equation.

Reading frame. The grouping of nucleotides in sets of three whose sequence corresponds to a polypeptide sequence.

Receptor. A binding protein that is specific for its ligand and elicits a discrete biochemical effect when its ligand is bound.

Receptor tyrosine kinase. A hormone receptor whose intracellular domain is activated, as a result of hormone binding, to phosphorylate tyrosine residues on other proteins and/or on other subunits of the same receptor.

Receptor-mediated endocytosis. A process in which an extracellular ligand binds to a specific cell-surface receptor and the resulting receptor–ligand complex is engulfed by the cell.

Recombinant. A DNA molecule constructed by combining DNA from different sources. Also called a chimera.

Recombinant DNA technology. The isolation, amplification, and modification of specific DNA sequences. Also called molecular cloning or genetic engineering.

Recombination. The exchange of polynucleotide strands between separate DNA segments. General recombination occurs between DNA segments with extensive homology, whereas site-specific recombination occurs between two short, specific DNA sequences.

Recombination repair. A mechanism for repairing damaged DNA, in which recombination exchanges a portion of a damaged strand for a homologous segment that can then serve as a template for the replacement of the damaged bases. Also called postreplication repair.

Redox center. A group that can undergo an oxidation–reduction reaction.

Redox couple. See half-reaction.

Reducing agent. A substance that can donate electrons, thereby becoming oxidized.

Reducing equivalent. A term used to describe the electrons that are transferred from one molecule to another during a redox reaction.

Reducing sugar. A saccharide bearing an anomeric carbon that has not formed a glycosidic bond and can therefore be oxidized by mild oxidizing agents.

Reduction potential ($\mathscr{E}$). A measure of the tendency of a substance to gain electrons.

Reductive pentose phosphate cycle. See Calvin cycle.

Regular secondary structure. A segment of a polymer in which the backbone adopts a regularly repeating conformation.

Release factor. A protein that recognizes a stop codon and thereby helps induce ribosomes to terminate polypeptide synthesis.

Renaturation. The refolding of a denatured macromolecule so as to regain its native conformation.

Repetitive DNA. Stretches of DNA of up to several thousand bases that occur in multiple copies in an organism's genome; they are often arranged in tandem.

Replica plating. The transfer of yeast colonies, bacterial colonies, or phage plaques from a culture plate to another culture plate, a membrane, or a filter in a manner that preserves the distribution of the cells on the original plate.

Replication. The process of making an identical copy of a DNA molecule. During DNA replication, the parental polynucleotide strands separate so that each can direct the synthesis of a complementary daughter strand, resulting in two complete DNA double helices.

Replication fork. The branch point in a replicating DNA molecule at which the two strands of the parental molecule are separated and serve as templates for the synthesis of the daughter strands.

Replicon. A unit of eukaryotic DNA that is replicated from one replication origin.

Replisome. The DNA polymerase–containing protein assembly that catalyzes the synthesis of both the leading and lagging strands of DNA at the replication fork.

Repressor. A protein that binds at or near a gene so as to prevent its transcription.

RER. See rough endoplasmic reticulum.

Residue. A term for a monomeric unit of a polymer.

Resonance energy transfer. See exciton transfer.

Respiratory distress syndrome. Difficulty in breathing in prematurely born infants, caused by alveolar collapse resulting from insufficient synthesis of lung surfactant.

Restriction endonuclease. A bacterial enzyme that cleaves a specific DNA sequence as part of the restriction–modification system.

Restriction fragment length polymorphism (RFLP). Inherited differences in DNA sequences (polymorphisms) among members of the same species leading to variation in the sites of cleavage of the DNA by particular restriction endonucleases and hence to DNA fragments of different lengths in digests of these DNAs by these restriction endonucleases.

Restriction map. A diagram of a DNA molecule, showing the sites recognized by restriction endonucleases, constructed by analysis of the fragments generated by digestion of the DNA with those endonucleases.

Restriction–modification system. A matched pair of bacterial enzymes that recognize a specific DNA sequence: a modification methylase that methylates bases in that sequence, and a restriction endonuclease that cleaves the DNA if it has not been methylated in that sequence. It is a defensive system that eliminates foreign (e.g., viral) DNA.

Reticulocyte. An immature red blood cell, which actively synthesizes hemoglobin.

Retrovirus. A virus whose genetic material is RNA that must be reverse-transcribed to double-stranded DNA during host cell infection.

Reverse transcriptase. A DNA polymerase that uses RNA as its template.

Reverse turn. A polypeptide conformation in which the chain makes an abrupt reversal in direction; usually consisting of four successive residues. Also called a β bend.

RFLP. See restriction fragment length polymorphism.

RIA. See radioimmunoassay.

Ribonucleic acid. See RNA.

Ribonucleotide. A nucleotide in which the pentose is ribose.

Ribosomal RNA (rRNA). The RNA molecules that constitute the bulk of the ribosome, the site of polypeptide synthesis. rRNA provides structural scaffolding for the ribosome and apparently catalyzes peptide bond formation.

Ribosome. The organelle that synthesizes polypeptides under the direction of mRNA. It consists of around two-thirds RNA and one-third protein.

Ribozyme. An RNA molecule that has catalytic activity.

Rickets. The vitamin D-deficiency disease in children that is characterized by stunted growth and deformed bones.

Rigor mortis. The stiffening of muscles after death.

RNA. Ribonucleic acid. A polymer of ribonucleotides. The major forms of RNA include messenger RNA (mRNA), transfer RNA (tRNA), and ribosomal RNA (rRNA).

RNA editing. The posttranscriptional insertion, deletion, or alteration of bases in mRNA.

Rossmann fold. See dinucleotide binding fold.

Rough endoplasmic reticulum (RER). That portion of the endoplasmic reticulum associated with ribosomes; it is the site of synthesis of membrane proteins and proteins destined for secretion or residence in certain organelles.

rRNA. See ribosomal RNA.

RS system. See Cahn–Ingold–Prelog system.

S. See entropy.

S. Svedberg, a unit for the sedimentation coefficient, equivalent to 10^{-13} s.

Saccharide. See carbohydrate.

Salt bridge. See ion pair.

Salting in. The increase in solubility of a protein (or other molecule) with increasing (low) salt concentration.

Salting out. The decrease in solubility of a protein (or other molecule) with increasing (high) salt concentration.

Sarcomere. The repeating unit of a myofibril, consisting of thin and thick filaments that slide past each other during muscle contraction.

Satellite DNA. DNA regions that consist of highly repetitive DNA segments; their distinctive base composition causes them to form bands known as satellites that are separate from that of other DNA in density gradient ultracentrifugation.

Saturated fatty acid. A fatty acid that does not contain any double bonds in its hydrocarbon chain.

Saturation. The state in which all of a macromolecule's ligand-binding sites are occupied by ligand.

Schiff base. An imine that forms between an amine and an aldehyde or ketone.

SCID. See severe combined immunodeficiency disease.

Scrapie. See bovine spongiform encephalopathy.

Screening. A technique for identifying clones that contain a desired gene.

Scurvy. A disease caused by vitamin C (ascorbic acid) deficiency, which results in inadequate cross-linking of collagen chains in connective tissue.

SDS-PAGE. Polyacrylamide gel electrophoresis (PAGE) in the presence of the detergent sodium dodecyl sulfate (SDS), which denatures and imparts a uniform charge density to polypeptides and thereby permits them to be fractionated on the basis of size rather than inherent charge.

Second messenger. An intracellular ion or molecule that acts as a signal for an extracellular event such as ligand binding to a cell-surface receptor.

Secondary active transport. Transmembrane transport that is driven by the energy stored in an electrochemical gradient, which itself is generated utilizing the free energy of ATP hydrolysis or electron transport.

Secondary structure. The local spatial arrangement of a polymer's backbone atoms without regard to the conformations of its substituent side chains. α helices and β sheets are common secondary structural elements of proteins.

Second-order reaction. A reaction whose rate is proportional to the square of the concentration of one reactant or to the product of the concentrations of two reactants.

Segment polarity genes. See segmentation genes.

Segmentation genes. Insect genes that specify the correct number and polarity of body segments. Gap genes, pair-rule genes, and segment polarity genes are all segmentation genes.

Selectable marker. A gene whose product has an activity, such as antibiotic resistance, such that, under the appropriate conditions, cells harboring the gene can be distinguished from those that lack the gene.

Selfish DNA. Genomic DNA that has no apparent function. Also called junk DNA.

Semiconservative replication. The natural mode of DNA duplication in which each new duplex molecule contains one strand from the parent molecule and one newly synthesized strand.

Semidiscontinuous replication. The mode of DNA replication in which one strand is replicated as a continuous polynucleotide strand (the leading strand) while the other is replicated as a series of discontinuous fragments (Okazaki fragments) that are later joined (the lagging strand).

Sense strand. The DNA strand complementary to the strand that is transcribed; it has the same base sequence (except for the replacement of U with T) as the synthesized RNA. Also called the coding strand.

Sequential model of allosterism. A model for allosteric behavior in which the subunits of an oligomeric protein change conformation in a stepwise manner as the number of bound ligands increases.

Sequential reaction. A reaction in which all substrates must combine with the enzyme before a reaction can occur; it can proceed by an Ordered or Random mechanism.

Serine protease. A peptide-hydrolyzing enzyme that has a reactive Ser residue in its active site.

Severe combined immunodeficiency disease (SCID). An inherited disease that greatly impairs the immune system. One such defect is a deficiency of the enzyme adenosine deaminase.

Shine–Dalgarno sequence. A purine-rich sequence ~10 nucleotides upstream from the start codon of many prokaryotic mRNAs that is partially complementary to the $3'$ end of the 16S rRNA. This sequence helps position the ribosome to initiate translation.

Shotgun cloning. The cloning of an organism's genome in the form of a set of random fragments.

Sickle-cell anemia. An inherited disease in which erythrocytes are deformed and damaged by the presence of a mutant hemoglobin (Glu 6β→Val) that in its deoxy form polymerizes into fibers.

σ factor. A component of the bacterial RNA polymerase holenzyme, which recognizes a gene's promoter and is released once chain initiation has occurred.

Signal hypothesis. The description of a pathway for the recognition, transmembrane movement, and proteolytic processing of secreted and transmembrane proteins that is directed by signal peptides.

Signal peptide. A short peptide sequence that determines the cellular location of a protein, for example, the N-terminal sequence of 13 to 36 residues that mediates the recognition of nascent secretory and transmembrane proteins by the apparatus that translocates the protein into the endoplasmic reticulum. This leader peptide is subsequently cleaved away by signal peptidase.

Signal transduction. The transmittal of an extracellular signal to the cell interior by the binding of a ligand to a cell-surface receptor so as to elicit a cellular response through the activation of a sequence of intracellular events that often include the generation of second messengers.

Silencer. A DNA sequence, some distance from the transcription start site, where a repressor of transcription may bind.

Single-displacement reaction. A reaction in which a group is transferred from one molecule to another in a concerted fashion (with no intermediates).

Site-directed mutagenesis. A technique in which a cloned gene is mutated in a specific manner.

Site-specific recombination. See recombination.

Size exclusion chromatography. See gel filtration chromatography.

Small nuclear ribonucleoprotein (snRNP). A complex of protein and small nuclear RNA that participates in mRNA splicing.

Small nuclear RNA (snRNA). Highly conserved 60- to 300-nt RNAs that participate in mRNA splicing.

snRNA. See small nuclear RNA.

snRNP. See small nuclear ribonucleoprotein.

Soap. A salt of a long-chain fatty acid, which contains a polar head group and a long hydrophobic tail.

Solvation. The state of being surrounded by several layers of ordered solvent molecules. Hydration is solvation by water.

Somatic recombination. Genetic rearrangement that occurs in cells other than germline cells.

Sonication. Irradiation with high frequency sound waves. Such treatment is used to disrupt cells and subcellular membranous structures.

SOS response. A bacterial system that recognizes damaged DNA, halts its replication, and repairs the damage, although in an error-prone fashion.

Southern blotting. A procedure for identifying a DNA base sequence after electrophoresis, through its ability to hybridize with a complementary single-stranded segment of labeled DNA or RNA. See also Northern blotting.

Spherocytosis. A hereditary abnormality in the erythrocyte cytoskeleton that renders the cells rigid and spheroidal and which causes hemolytic anemia.

Sphingolipid. A derivative of the C_{18} amino alcohol sphingosine. Sphingolipids include the ceramides, cerebrosides, and gangliosides. Sphingolipids with phosphate head groups are called sphingophospholipids.

Spliceosome. A 50S to 60S particle containing proteins, snRNPs, and pre-mRNA; it carries out the splicing reactions whereby the pre-mRNA is converted to a mature mRNA.

Splicing. The usually ribonucleoprotein-catalyzed process by which introns are removed and exons are joined to produce a mature transcript. Some RNAs are self-splicing.

Spontaneous process. A thermodynamic process that occurs without the input of free energy from outside the system. Spontaneity is independent of the rate of a process.

Stacking interactions. The stabilizing van der Waals interactions between successive (stacked) bases and base pairs in a polynucleotide.

Standard state. A set of conditions including unit activity of the species of interest, a temperature of 25°C, a pressure of 1 atm, and a pH of 0.0. See also biochemical standard state.

Starch. A mixture of linear and branched glucose polymers that serve as the principal energy reserves of plants.

State function. Quantities such as energy, enthalpy, entropy, and free energy, whose values depend only on the current state of the system, not on how they reached that state.

Steady state. A set of conditions in an open system under which the formation and degradation of individual components are balanced such that the system does not change over time.

Stem-and-loop. A secondary structural element in a single-stranded nucleic acid, in which two complementary segments form a base-paired stem whose strands are connected by a loop of unpaired bases.

Stereoisomers. Chiral molecules with different configurations about at least one of their asymmetric centers but which are otherwise identical.

Steroid. Any of numerous naturally occurring lipids composed of four fused rings; many are hormones that are derived from cholesterol.

Sterol. An alcohol derivative of a steroid.

Sticky end. The single-stranded extension of a DNA fragment that has been cleaved at a specific sequence (often by a restriction endonuclease) in a staggered cut such that this single-stranded extension is complementary to those of similarly cleaved DNAs.

Stop codon. A sequence of three nucleotides that does not specify an amino acid but instead causes the termination of translation. Also called a nonsense codon.

Striated muscle. The voluntary or skeletal muscles, which have a striped microscopic appearance.

Stringent response. The sweeping metabolic shutdown in bacterial metabolism accompanying poor growth conditions. The response, mediated by ppGpp, includes a decrease in rRNA synthesis.

Stroma. The concentrated solution of enzymes, small molecules, and ions in the interior of a chloroplast; the site of carbohydrate synthesis.

Stromal lamellae. The membranous assemblies that connect grana in a chloroplast.

Strong acid. An acid that is essentially completely ionized in aqueous solution. A strong acid has a dissociation constant much greater than unity ($pK < 0$).

Structural gene. A gene that encodes a protein.

Substrate. A reactant in an enzymatic reaction.

Substrate cycle. Two opposing metabolic reactions that function together to provide a control point for regulating metabolic flux. Also called a futile cycle.

Substrate-level phosphorylation. The direct transfer of a phosphoryl group to ADP to generate ATP.

Subunit. One of several polypeptide chains that make up a protein.

Sugar. A simple mono- or disaccharide.

Suicide substrate. See mechanism-based inhibitor.

Sulfhydryl group. A portion of a molecule with the formula —SH.

Supercoiling. The topological state of covalently closed circular double-helical DNA that arises through the over- or underwinding of the double helix and which gives the DNA circle a peculiar twisted appearance. Also called superhelicity.

Superhelicity. See supercoiling.

Supersecondary structure. See motif.

Suppressor mutation. A mutation that cancels the effect of another mutation.

Surface labeling. A technique in which a lipid-insoluble protein-labeling reagent is used to identify the portion of a membrane protein that is exposed to solvent.

Symbiosis. A mutually dependent relationship between two organisms.

Symmetry model of allosterism. A model for allosteric behavior in which all the subunits of an oligomeric protein are constrained to change conformation in a concerted manner so as to maintain the symmetry of the oligomer. Also called the MWC model of allosterism.

Symport. The simultaneous transmembrane movement of two molecules in the same direction.

Syncytium. A single cell containing multiple nuclei that results from repeated nuclear division without formation of new plasma membranes.

Synonymous codon. A codon that specifies the same amino acid as another.

T state. One of two conformations of an allosteric protein; the other is the R state.

T. See twist.

$t_{1/2}$. See half-time.

T→R transition. A shift in conformation of an allosteric protein induced by ligand binding.

TATA box. A eukaryotic promoter element with the consensus sequence TATAAAA located 10 to 27 nucleotides upstream from the transcription start site.

Tautomers. Isomers that differ only in the positions of their hydrogen atoms.

Taxonomy. The study of biological classification.

Tay-Sachs disease. A fatal sphingolipid storage disease caused by a deficiency of hexosaminidase A, the enzyme that breaks down ganglioside G_{M2}.

TCA cycle. Tricarboxylic acid cycle. See citric acid cycle.

Telomerase. An RNA-containing DNA polymerase that, using the RNA as a template, catalyzes the repeated addition of a specific G-rich sequence to the 3′ end of a eukaryotic DNA molecule to form a telomere.

Telomere. The end of a linear eukaryotic chromosome, which consists of tandem repeats of a short G-rich sequence on the 3′-ending strand and its complementary sequence on the 5′-ending strand.

Tertiary structure. The entire three-dimensional structure of a single-chain polymer, including that of its side chains.

Tetramer. An assembly consisting of four monomeric units.

Thermodynamics. The study of the relationships among various forms of energy.

Thermophile. An organism that thrives at high temperatures.

Thick filament. The sarcomere element that is composed primarily of several hundred myosin molecules.

Thin filament. The sarcomere element that is composed primarily of actin, along with tropomyosin and troponin.

Threading. A technique for determining the three-dimensional structure of a polypeptide by computationally determining its stability when it is fitted to a known polypeptide structure.

3′ end. The terminus of a polynucleotide whose C3′ is not esterified to another nucleotide residue.

Thylakoid. The innermost compartment in chloroplasts, which is formed by invaginations of the chloroplast's inner membrane. The thylakoid membrane is the site of the light reactions of photosynthesis.

Titration curve. The graphic presentation of the relationship between the pH of an acid- or base-containing solution and the degree of proton dissociation (roughly equivalent to the amount of strong base or strong acid that has been added to the solution).

T_m. See melting temperature.

Topoisomerase. An enzyme that alters DNA supercoiling by catalyzing breaks in one or both strands, passing DNA through the break, and resealing the break.

Topology. The study of the geometric properties of an object that are not altered by deformations such as bending and stretching.

Torsion angle. The dihedral angle described by the bonds between four successive atoms. The torsion angles ϕ and ψ indicate the backbone conformation of a peptide group in a polypeptide.

Trans peptide. A conformation in which successive C_α atoms are on opposite sides of the peptide bond.

Transamination. The transfer of an amino group from an amino acid to an α-keto acid to yield a new α-keto acid and a new amino acid.

Transcription. The process by which RNA is synthesized using a DNA template, thereby transferring genetic information from the DNA to the RNA. Transcription is catalyzed by RNA polymerase as facilitated by numerous other proteins.

Transcription factor. A protein that promotes the transcription of a gene by binding to DNA sequences at or near the gene or by interacting with other proteins that do so.

Transfer RNA (tRNA). The small L-shaped RNAs that deliver specific amino acids, which have been covalently linked to the tRNA's 3′ ends, to ribosomes according to the sequence of a bound mRNA. The proper tRNA is selected through the complementary base pairing of its three-nucleotide anticodon with the mRNA's codon, and its amino acid group is transferred to the growing polypeptide.

Transferase. A enzyme that catalyzes the transfer of a functional group from one molecule to another.

Transformation. The permanent alteration of a bacterial cell's genetic message through the introduction of foreign DNA.

Transgene. A foreign gene that is stably expressed in a host organism.

Transgenic organism. An organism that stably expresses a foreign gene (transgene).

Transition state. A molecular assembly at the point of maximal free energy in the reaction coordinate diagram of a chemical reaction.

Transition state analog. A stable substance that geometrically and electronically resembles the transition state of a reaction.

Transition. A mutation in which one purine (or pyrimidine) replaces another.

Transition temperature. The temperature at which a lipid bilayer shifts from a gel-like solid to a more fluid liquid crystal form.

Translation. The process of transforming the information contained in the nucleotide sequence of an RNA to the corresponding amino acid sequence of a polypeptide as specified by the genetic code. Translation is catalyzed by ribosomes and requires the additional participation of messenger RNA, transfer RNA, and a variety of protein factors.

Transmembrane protein. An integral protein that completely spans the membrane.

Transpeptidation. The ribosomal process in which a tRNA-bound nascent polypeptide is transferred to a tRNA-bound aminoacyl group so as to form a new peptide bond, thereby lengthening the polypeptide by one residue at its C-terminus.

Transposable element. See transposon.

Transposition. The movement (copying) of genetic material from one part of the genome to another or, in some cases, from one organism to another.

Transposon. A genetic unit that can move (be copied) from one position to another in a genome; some transposons carry genes. Also called a transposable element.

Transverse diffusion. The movement of a lipid from one leaflet of a bilayer to the other. Also called flip-flop.

Transversion. A mutation in which a purine is replaced by a pyrimidine or vice versa.

Treadmilling. The addition of monomeric units to one end of a linear aggregate, such as an actin filament, and their removal from the opposite end such that the length of the aggregate remains unchanged.

Triacylglycerol. A lipid in which three fatty acids are esterified by a glycerol backbone. Also called a triglyceride.

Tricarboxylic acid (TCA) cycle. See citric acid cycle.

Triglyceride. See triacylglycerol.

Trimer. An assembly consisting of three monomeric units.

tRNA. See transfer RNA.

Tumor suppressor. A protein whose loss or inactivation may lead to cancer.

Turnover number. See k_{cat}.

Twist (T). The number of complete revolutions that one strand of a covalently closed circular double-helical DNA makes around the duplex axis. It is positive for right-handed superhelical coils and negative for left-handed superhelical coils.

U. The thermodynamic symbol for energy.

Ubiquitination. The covalent attachment of ubiquitin to a eukaryotic intracellular protein, which, in most cases, marks it for degradation by a proteasome.

Ultracentrifugation. A procedure that subjects macromolecules to a strong centrifugal force (in an ultracentrifuge), thereby separating them by size and/or density and providing a method for determining their mass and subunit structure.

Uncompetitive inhibition. A form of enzyme inhibition in which an inhibitor binds to the enzyme–substrate complex and thereby decreases its apparent K_M and its apparent V_{max} by the same factor.

Uncoupler. A substance that allows the proton gradient across a membrane to dissipate without ATP synthesis so that electron transport proceeds without oxidative phosphorylation.

Uniport. The transmembrane movement of a single molecule.

Unsaturated fatty acid. A fatty acid that contains at least one double bond in its hydrocarbon chain.

Urea cycle. A catalytic cycle in which amino groups donated by ammonia and aspartate combine with a carbon atom from HCO_3^- to form urea for excretion and which provides the route for the elimination of nitrogen from protein degradation.

Uridylylation. Addition of a uridylyl (UMP) group.

Uronic acid. A sugar produced by oxidation of an aldose primary alcohol group to a carboxylic acid group.

Vacuole. An intracellular vesicle for storing water or other molecules.

van der Waals distance. The distance of closest approach between two nonbonded atoms.

van der Waals forces. The noncovalent associations between molecules that arise from the electrostatic interactions among permanent and/or induced dipoles.

Variable region. The N-terminal portions of an antibody molecule, where antigen binding occurs and which is characterized by high sequence variability.

Variant. A naturally occurring mutant form.

Vector. See cloning vector.

Vesicle. A fluid-filled sac enclosed by a membrane.

Virulence. The disease-evoking power of a microorganism.

Virus. A nonliving entity that co-opts the metabolism of a host cell to reproduce.

Vitamin. A metabolically required substance that cannot be synthesized by an animal and must therefore be obtained from the diet.

V_{max}. Maximal velocity of an enzymatic reaction.

v_o. Initial velocity of an enzymatic reaction.

W. See writhing number.

w. The thermodynamic term for the work done by a system on its surroundings.

Water of hydration. The shell of relatively immobile water molecules that surrounds and interacts with (solvates) a dissolved molecule.

Weak acid. An acid that is only partially ionized in aqueous solution. A weak acid has a dissociation constant less than unity (p$K > 0$).

Western blot. See immunoblot.

Wild type. The naturally occurring version of an organism or gene.

Wobble pairing. The permissive tRNA–mRNA pairing at the third anticodon position that includes non-Watson–Crick base pairs. This allows many tRNAs to recognize two or three different (degenerate) codons.

Writhing number (W). The number of turns that the duplex axis of a covalently closed circular double-helical DNA makes around the superhelix axis. It is a measure of the DNA's superhelicity.

X-Ray crystallography. A method for determining three-dimensional molecular structures from the diffraction pattern produced by exposing a crystal of a molecule to a beam of X-rays.

Y. See fractional saturation.

YAC. See yeast artificial chromosome.

Yeast artificial chromosome (YAC). A linear DNA molecule that contains the chromosomal structures required for normal replication and segregation in a yeast cell. YACs are commonly used as cloning vectors.

Ylid. A molecule with opposite charges on adjacent atoms.

Zinc finger. A protein structural motif, often involved in DNA binding, that consists of 25 to 60 residues that include His and/or Cys residues to which one or two Zn^{2+} ions are tetrahedrally coordinated.

Zonal ultracentrifugation. A preparative technique in which a mixture of molecules is applied to the surface of a preformed density gradient before ultracentrifugation.

Z-scheme. A Z-shaped diagram indicating the sequence of events and their reduction potentials in the two-center photosynthetic electron-transport system of plants and cyanobacteria.

Zwitterion. A compound bearing oppositely charged groups. Also called a dipolar ion.

Zymogen. The inactive precursor (proenzyme) of a proteolytic enzyme.

SOLUTIONS TO PROBLEMS

CHAPTER 1

1.

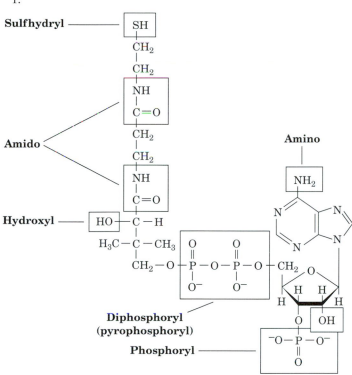

Sulfhydryl — SH

Amido

Amino

Hydroxyl — HO

Diphosphoryl (pyrophosphoryl)

Phosphoryl

2. The cell membrane must be semipermeable so that the cell can retain essential compounds while allowing nutrients to enter and wastes to exit.

3. Concentration = (number of moles)/(volume)
Volume = $(4/3)\pi r^3 = (4/3)\pi(5 \times 10^{-7} \text{ m})^3$
$= 5.24 \times 10^{-19} \text{ m}^3 = 5.24 \times 10^{-16} \text{ L}$
Moles of protein = (2 molecules)/
$(6.022 \times 10^{23} \text{ molecules} \cdot \text{mol}^{-1}) = 3.32 \times 10^{-24} \text{ mol}$
Concentration = $(3.32 \times 10^{-24} \text{ mol})/(5.24 \times 10^{-16} \text{ L})$
$= 6.3 \times 10^{-9} \text{ M} = 6.3 \text{ nM}$

4. Number of molecules = (molar conc.)(volume)
$(6.022 \times 10^{23} \text{ molecules} \cdot \text{mol}^{-1})$
$= (10^{-3} \text{ mol} \cdot \text{L}^{-1})(5.24 \times 10^{-16} \text{ L})$
$(6.022 \times 10^{23} \text{ molecules} \cdot \text{mol}^{-1})$
$= 3.2 \times 10^5$ molecules

5. (a) Liquid water; (b) ice has less entropy at the lower temperature.

6. (a) Decreases; (b) increases; (c) increases; (d) no change.

7. (a) $T = 273 + 10 = 283$ K
$\Delta G = \Delta H - T\Delta S$
$\Delta G = 15 \text{ kJ} - (283 \text{ K})(0.05 \text{ kJ} \cdot \text{K}^{-1})$
$= 15 - 14.15 \text{ kJ} = 0.85 \text{ kJ}$

ΔG is greater than zero, so the reaction is not spontaneous.
(b) $T = 273 + 80 = 353$ K
$\Delta G = \Delta H - T\Delta S$
$\Delta G = 15 \text{ kJ} - (353 \text{ K})(0.05 \text{ kJ} \cdot \text{K}^{-1}) =$
$15 - 17.65 \text{ kJ} = -2.65 \text{ kJ}$
ΔG is less than zero, so the reaction is spontaneous.

8. $K_{eq} = e^{-\Delta G^{\circ\prime}/RT} = e^{-(-20,900 \text{ J} \cdot \text{mol}^{-1})/(8.314 \text{ J} \cdot \text{K}^{-1} \text{mol}^{-1})(298 \text{ K})}$
$= 4.6 \times 10^3$

9. $\Delta G^{\circ\prime} = -RT \ln K_{eq} = -RT \ln [C][D]/[A][B]$
$= -(8.314 \text{ J} \cdot \text{K}^{-1} \cdot \text{mol}^{-1})(298 \text{ K}) \ln (3)(5)/(10)(15)$
$= 5700 \text{ J} \cdot \text{mol}^{-1} = 5.7 \text{ kJ} \cdot \text{mol}^{-1}$
Since $\Delta G^{\circ\prime}$ is positive, the reaction is endergonic under standard conditions.

10. From Eq. 1-17, $K_{eq} = [G6P]/[G1P] = e^{-\Delta G^{\circ\prime}/RT}$
$[G6P]/[G1P] = e^{-(-7100 \text{ J} \cdot \text{mol}^{-1})/(8.314 \text{ J} \cdot \text{K}^{-1} \cdot \text{mol}^{-1})(298 \text{ K})}$
$[G6P]/[G1P] = 17.6$
$[G1P]/[G6P] = 0.057$

CHAPTER 2

1. (a) Donors: NH1, NH_2 at C2, NH9; acceptors: N3, O at C6, N7. (b) Donors: NH1, NH at C4; acceptors: O at C2, N3. (c) Donors: NH_3^+ group, OH group; acceptors: COO^- group, OH group.

2. (a) Water, (b) water, (c) micelle.

3. Water molecules move from inside the dialysis bag to the surrounding seawater by osmosis. Ions from the seawater diffuse into the dialysis bag. At equilibrium, the compositions of the solutions inside and outside the dialysis bag are identical. If the membrane were solute-impermeable, essentially all the water would leave the dialysis bag.

4. Because the bacterium and the fish have identical shapes (length/diameter = 6 for each), the ratio of their surface-to-volume ratios is the same as the inverse ratio of their lengths or diameters (e.g., 3 μm versus 30 cm). Thus, the bacterium has 100,000 times more surface area per unit volume than the fish.

5. (a) COO^-
          CH
          ‖
          HC
          COO^-

(b) COO^-
       H—C—H
       NH_3^+

(c) COO^-
       H—C—H
       NH_2

(d) COO^-
       H—C—CH_2—COO^-
       NH_3^+

6. (a) pH 4, NH_4^+; pH 8, NH_4^+; pH 11, NH_3.
(b) pH 4, $H_2PO_4^-$; pH 8, HPO_4^{2-}; pH 11, HPO_4^{2-}.

7. (a) $(0.01 \text{ L})(5 \text{ mol} \cdot \text{L}^{-1} \text{ NaOH})/(1 \text{ L}) =$

$$0.05 \text{ M NaOH} \equiv 0.05 \text{ M OH}^-$$

$[\text{H}^+] = K_w/[\text{OH}^-] = (10^{-14} \text{ M}^2)/(0.05 \text{ M}) = 2 \times 10^{-13} \text{ M}$
$\text{pH} = -\log[\text{H}^+] = -\log (2 \times 10^{-13}) = 12.7$

(b) $(0.02 \text{ L})(5 \text{ mol} \cdot \text{L}^{-1} \text{ HCl})/(1 \text{ L}) =$

$$0.1 \text{ M HCl} \equiv 0.1 \text{ M H}^+$$

Since the contribution of $0.01 \text{ L} \times 100 \text{ mM}/(1\text{L}) = 1 \text{ mM}$
glycine is insignificant in the presence of 0.1 M HCl,
$\text{pH} = -\log[\text{H}^+] = -\log (0.1) = 1.0$

(c) $\text{pH} = \text{p}K + \log ([\text{acetate}]/[\text{acetic acid}])$
$[\text{acetate}] = (5 \text{ g})(1 \text{ mol}/82 \text{ g})/(1 \text{ L}) = 0.061 \text{ M}$
$[\text{acetic acid}] = (0.01 \text{ L})(2 \text{ mol} \cdot \text{L}^{-1})/(1 \text{ L}) = 0.02 \text{ M}$
$\text{pH} = 4.76 + \log (0.061/0.02) = 4.76 + 0.48 = 5.24$

8. Let HA = sodium succinate and A^- = disodium succinate.
$[\text{A}^-] + [\text{HA}] = 0.05 \text{ M}$, so $[\text{A}^-] = 0.05 \text{ M} - [\text{HA}]$
From Eq. 2-9, $\log ([\text{A}^-]/[\text{HA}] = \text{pH} - \text{p}K =$

$$6.0 - 5.64 = 0.36$$

$[\text{A}^-]/[\text{HA}] = \text{antilog } 0.36 = 2.29$
$(0.05 \text{ M} - [\text{HA}])/[\text{HA}] = 2.29$
$[\text{HA}] = 0.015 \text{ M}$
$[\text{A}^-] = 0.05 \text{ M} - 0.015 \text{ M} = 0.035 \text{ M}$
grams of sodium succinate =

$$(0.015 \text{ mol} \cdot \text{L}^{-1})(140 \text{ g} \cdot \text{mol}^{-1})/(1 \text{ L}) = 2.1 \text{ g}$$

grams of disodium succinate =

$$(0.035 \text{ mol} \cdot \text{L}^{-1})(162 \text{ g} \cdot \text{mol}^{-1})/(1 \text{ L}) = 5.7 \text{ g}$$

9. At pH 4, essentially all the phosphoric acid is in the
H_2PO_4^- form, and at pH 9, essentially all is in the HPO_4^{2-}
form (Fig. 2-16). Therefore, the concentration of OH^- required is equivalent to the concentration of the acid:
$(0.1 \text{ mol} \cdot \text{L}^{-1} \text{ phosphoric acid})(0.1 \text{ L}) = 0.01 \text{ mol}$
NaOH required $= (0.01 \text{ mol})(1 \text{ L}/5 \text{ mol NaOH}) =$

$$0.002 \text{ L} = 2 \text{ mL}$$

10. (a) Succinic acid, (b) ammonia, (c) HEPES.

CHAPTER 3

1. (a) Yes; (b) no; (c) no; (d) yes.

2. Since the haploid genome contains 21% G, it must contain
21% C (because G = C) and 58% A + T (or 29% A and
29% T, because A = T). Each cell is diploid, containing
90,000 kb or 9×10^7 bases. Therefore,
$A = T = (0.29)(9 \times 10^7) = 2.61 \times 10^7$ bases
$C = G = (0.21)(9 \times 10^7) = 1.89 \times 10^7$ bases

3. *(a)*

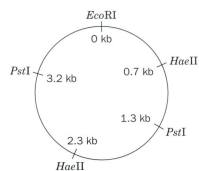

(b)

4. The number of possible sequences of four different nucleotides is 4^n where n is the number of nucleotides in the
sequence. Therefore, (a) $4^1 = 4$, (b) $4^2 = 16$, (c) $4^3 = 64$,
and (d) $4^4 = 256$.

5. (a) *Alu*I, *Eco*RV, *Hae*III, *Pvu*II; (b) *Hpa*II and *Msp*I;
(c) *Bam*HI and *Bgl*II; *Hpa*II and *Taq*I; *Sal*I and *Xho*I.

6. Let the *Eco*RI site be set at 0 (or 4) kb.

*Eco*RI
0 kb
*Hae*II
0.7 kb
*Pst*I 3.2 kb
1.3 kb
2.3 kb
*Pst*I
*Hae*II

OR

*Eco*RI
0 kb
*Hae*II
3.3 kb
*Pst*I
0.8 kb
2.7 kb
*Pst*I
1.7 kb
*Hae*II

7. (a) Newly synthesized chains would be terminated less frequently, so the bands representing truncated fragments on
the sequencing gel would appear faint.
(b) Chain termination would occur more frequently, so
longer fragments would be less abundant.
(c) The amount of DNA synthesis would decrease and the
resulting gel bands would appear faint.
(d) No effect.

8. The *C. elegans* genome contains 100,000 kb, so
$f = 5/100,000 = 5 \times 10^{-5}$. Using Eq. 3-2,
$N = \log (1 - P)/\log (1 - f)$
$N = \log (1 - 0.99)/\log (1 - 5 \times 10^{-5})$
$N = -2/(-2.17 \times 10^{-5}) = 9.21 \times 10^4$

9. Nontransformed bacteria would not grow in the presence
of ampicillin or tetracycline. Clones transformed with the
plasmid only would be resistant to both antibiotics. Clones
containing the plasmid with the insert would be sensitive to
ampicillin and resistant to tetracycline.

10. (a) Only single DNA strands of variable length extending
from the remaining primer would be obtained. The number of these strands would increase linearly with the number of cycles rather than geometrically.
(b) PCR would yield a mixture of DNA segments whose
lengths correspond to the distance between the position of
the primer with a single binding site and the various sites
where the multispecific primer binds.

(c) The first cycle of PCR would yield only the new strand that is complementary to the intact DNA strand, since DNA synthesis cannot proceed when the template is broken. However, since the new strand has the same sequence as the broken strand, PCR can proceed normally from the second cycle on.

(d) DNA synthesis would terminate at the breaks in the first cycle of PCR.

11. ATAGGCATAGGC and CTGACCAGCGCC.

CHAPTER 4

1. Ser and Thr; Val, Leu, and Ile; Asn and Gln; Asp and Glu.

2. Hydrogen bond donors: α-amino group, amide nitrogen. Hydrogen bond acceptors: α-carboxylate group, amide carbonyl.

3.

4. The first residue can be one of five residues, the second one of the remaining four, etc.
$$N = 5 \times 4 \times 3 \times 2 \times 1 = 120$$

5. (a) +1; (b) 0; (c) −1; (d) −2.

6. (a) $pI = (2.35 + 9.87)/2 = 6.11$
   (b) $pI = (6.04 + 9.33)/2 = 7.68$
   (c) $pI = (2.10 + 4.07)/2 = 3.08$

7.

8.

9. (2S,3S)-isoleucine

10. (a) Serine (*N*-acetylserine); (b) lysine (5-hydroxylysine); (c) methionine (*N*-formylmethionine).

11.

(a) The pK's of the ionizable side chains (Table 4-1) are 3.90 (Asp) and 10.54 (Lys); assume that the terminal Lys carboxyl group has a pK of 3.5 and the terminal Ala amino group has a pK of 8.0 (Section 4-1D). The pI is approximately midway between the pK's of the two ionizations involving the neutral species (the pK of Asp and the N-terminal pK):

$$pI \approx \tfrac{1}{2}(3.90 + 8.0) \approx 5.95$$

(b) The net charge at pH 7.0 is 0 (as drawn above).

12. At position A8, duck insulin has a Glu residue, whereas human insulin has a Thr residue. Since Glu is negatively charged at physiological pH and Thr is neutral, human insulin has a higher pI than duck insulin. (The other amino acids that differ between the proteins do not affect the pI because they are uncharged.)

CHAPTER 5

1. (a) Leu, His, Arg. (b) Lys, Val, Glu.

2. Peptide B, because it contains more Trp and other aromatic residues.

3. The protein has an elongated shape, so it behaves like a larger protein during gel filtration. The mass determined by SDS-PAGE is more accurate since the mobility of a denatured SDS-coated protein depends only on its size.

4. The protein contains two 60-kD polypeptides and two 40-kD polypeptides. Each 40-kD chain is disulfide bonded to a 60-kD chain. The 100-kD units associate noncovalently to form a protein with a molecular mass of 200 kD.

5. The protein aggregates at the higher salt concentration.

6. Dansyl chloride reacts with primary amino groups, including the ε-amino group of Lys residues.

7. (a) Gly; (b) Thr; (c) none (the N-terminal amino group is acetylated and hence unreactive with Edman's reagent)

8. Asp-Met-Leu-Phe-Met-Arg-Ala-Tyr-Gly-Asn

9.

10. Oxidative cleavage of disulfide bonds by performic acid also oxidizes Met residues, thereby rendering them impervious to the CNBr cleavage reaction. Since this is one of the mainstays of protein sequence analysis, oxidative cleavage is rarely used.

11. Because protein 1 has a greater proportion of hydrophobic residues (Ala, Ile, Pro, Val) than do proteins 2 and 3, hydrophobic interaction chromatography could be used to isolate it.

12. Arg-Ile-Pro-Lys-Cys-Arg-Lys-Phe-Gln-Gln-Ala-Gln-His-Leu-Arg-Ala-Cys-Gln-Gln-Trp-Leu-His-Lys-Gln-Ala-Asn-Gln-Ser-Gly-Gly-Gly-Pro-Ser

CHAPTER 6

1.

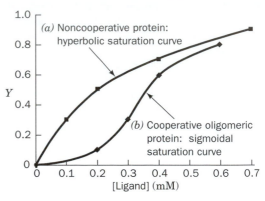

2. (a) 3.6_{13}; (b) steeper

3. (100 residues)(1 α-helical turn/3.6 residues)
   (5.1 Å/keratin turn) = 142 Å

4. Yes

5. (a) Gln; (b) Ser; (c) Ile; (d) Cys.

6. βαβαβααβαββββαββαββαββββαββββββα

7. (a) C_4 and D_2; (b) C_6 and D_3.

8. No

9. Hydrophobic effects, van der Waals interactions, and hydrogen bonds are destroyed during denaturation. Covalent cross-links are retained.

10. At physiological pH, the positively charged Lys side chains repel each other. Increasing the pH above the pK (> 10.5) would neutralize the side chains and allow an α helix to form.

11. The molecular mass of O_2 is 32 D. Hence the ratio of the masses of hemoglobin and 4 O_2, which is equal to the ratio of their volumes, is $65,000/(4 \times 32) = 508$. The 70 kg office worker has a volume of 70 kg $\times$ 1 cm^3/g $\times$ (1000 g/kg) $\times$ (1 m/100 cm)^3 = 0.070 m^3. Hence the ratio of the volumes of the office and the office worker is $(4 \times 4 \times 3)/0.070 = 686$. These ratios are similar in magnitude, which you may not have expected.

12. Peptide c is most likely to form an α helix with its three charged residues (Lys, Glu, and Arg) aligned on one face of the helix. Peptide a has adjacent basic residues (Arg and Lys), which would destabilize a helix. Peptide b contains Gly and Pro, both of which are helix-breaking (Table 6-1). The presence of Gly and Pro would also inhibit the formation of β strands, so peptide b is least likely to form a β strand.

CHAPTER 7

1.

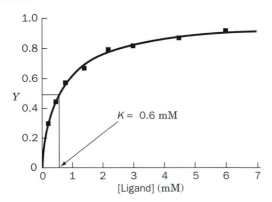

$K = 0.6$ mM

2. b describes sigmoidal binding to an oligomeric protein and hence represents cooperative binding.

(a) Noncooperative protein: hyperbolic saturation curve

(b) Cooperative oligomeric protein: sigmoidal saturation curve

[Ligand] (mM)

3. (a) Lower; (b) higher. The Asp 99β → His mutation of hemoglobin Yakima disrupts a hydrogen bond at the $α_1 - β_2$ interface of the T state (Fig. 7-10a) causing the $T \rightleftharpoons R$ equilibrium to shift toward R state (lower p_{50}). The Asn 102β → Thr of hemoglobin Kansas causes the opposite shift in the $T \rightleftharpoons R$ equilibrium by abolishing an R-state hydrogen bond (Fig. 7-10b).

4. The increased BPG helps the remaining erythrocytes deliver O_2 to tissues. However, BPG stabilizes the T conformation of hemoglobin, so it promotes sickling and therefore aggravates the disease.

5. As the crocodile remains underwater without breathing, its metabolism generates CO_2 and hence the HCO_3^- content of its blood increases. The HCO_3^- preferentially binds to the crocodile's deoxyhemoglobin, which allosterically prompts the hemoglobin to assume the deoxy conformation and thus release its O_2. This helps the crocodile stay underwater long enough to drown its prey.

6. Because many myosin heads bind along a thin filament where it overlaps a thick filament, and because the myosin molecules do not execute their power strokes simultaneously, the thick and thin filaments can move past each other by more than 60 Å in the interval between power strokes of an individual myosin molecule.

7. In the absence of ATP, each myosin head is in its low energy configuration and cannot release its bound actin molecule. Consequently, thick and thin filaments form a rigid, cross-linked array.

8. (a) 150–200 kD; (b) 150–200 kD; (c) ~23 kD and 53–75 kD.

9. The loops are on the surface of the domain, so they can tolerate more amino acid substitutions. Amino acid changes in the β sheets would be more likely to destabilize the domain.

10. The antigenic site in the native protein usually consists of several peptide segments that are no longer contiguous when the tertiary structure of the protein is disrupted.

11. (a) Fab fragments are monovalent and therefore cannot cross-link antigens to produce a precipitate. (b) A small antigen has only one antigenic site and therefore cannot bind more than one antibody to produce a precipitate. (c) When antibody is in great excess, most antibodies that

are bound to antigen bind only one per immunoglobulin molecule. When antigen is in excess, most immunoglobulins bind to two independent antigens.

CHAPTER 8

1. (a) 4; (b) 8; (c) 16.

2. (a) and (d)

3.

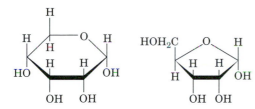

4. (a) Yes; (b) no (its symmetric halves are superimposable); (c) no.

5.

$$
\begin{array}{c}
\text{O} \diagdown \text{H} \\
\text{C} \\
\text{HO}-\text{C}-\text{H} \\
\text{H}-\text{C}-\text{OH} \\
\text{H}-\text{C}-\text{OH} \\
\text{HO}-\text{C}-\text{H} \\
\text{CH}_3
\end{array}
$$

L-Fucose

L-Fucose is the deoxy form of L-galactose.

6. α-D-glucose-$(1 \rightarrow 1)$-α-D-glucose or α-D-glucose-$(1 \rightarrow 1)$-β-D-glucose

7. 19

8. One

9. Amylose (it has only one nonreducing end from which glucose can be mobilized).

10. −200

CHAPTER 9

1. *trans*-Oleic acid has a higher melting point because, in the solid state, its hydrocarbon chains pack together more tightly than those of *cis*-oleic acid.

2. Of the $4 \times 4 = 16$ pairs of fatty acid residues at C1 and C3, only 10 are unique because a molecule with different substituents at C1 and C3 is identical to the molecule with the reverse substitution order. However, C2 may have any of the four substituents for a total of $4 \times 10 = 40$ different triacylglycerols.

3. The triacylglycerol containing the stearic acid residues yields more energy since it is fully reduced.

4. (a) Palmitic acid and 2-oleoyl-3-phosphatidylserine; (b) oleic acid and 1-palmitoyl-3-phosphatidylserine; (c) phosphoserine and 1-palmitoyl-2-oleoyl-glycerol; (d) serine and 1-palmitoyl-2-oleoyl-phosphatidic acid.

5. No; the two acyl chains of the "head group" are buried in the bilayer interior, leaving a head group of diphosphoglycerol.

6. Steroid hormones, which are hydrophobic, can diffuse through the cell membrane to reach their receptors.

7. Eicosanoids synthesized from arachidonic acid are necessary for intercellular communication. Cultured cells do not need such communication and therefore do not require linoleic acid.

8. Triacylglycerols lack a polar head group, so they do not orient themselves in a bilayer with their acyl chains inward and their glycerol moiety toward the surface.

9. The large oligosaccharide head groups of gangliosides would prevent efficient packing of the lipids in a bilayer.

10. (a) Saturated; (b) long-chain. By increasing the proportion of saturated and long-chain fatty acids, which have higher melting points, the bacteria can maintain constant membrane fluidity at the higher temperature.

CHAPTER 10

1. (a) (1 turn/5.4 Å)(30 Å) = 5.6 turns
   (b) (3.6 residues/turn)(5.6 turns) = 20 residues
   (c) The additional residues form a helix, which partially satisfies backbone hydrogen bonding requirements, when the lipid head groups do not offer hydrogen bonding partners.

2. No. Although the β strand could span the bilayer, a single strand would be unstable because its backbone could not form hydrogen bonds.

3. (a) Inner; (b) outer. See Fig. 10-17.

4. (a) Both the intra- and extracellular portions will be labeled. (b) Only the extracellular portion will be labeled. (c) Only the intracellular portion will be labeled.

5. The mutant signal peptidase would cleave many preproteins within their signal peptides, which often contain Leu-Leu sequences. This would not affect translocation into the ER, since signal peptidase acts after the signal peptide enters the ER lumen. Proteins lacking the Leu-Leu sequence would retain their signal peptides. These proteins, and those with abnormally cleaved signal sequences, would be more likely to fold abnormally and therefore function abnormally.

6. (a) Nonmediated; (b) mediated; (c) nonmediated; (d) mediated.

7. (a) $\Delta \bar{G} = RT \ln ([Na^+]_{in}/[Na^+]_{out})$
   $= (8.314 \text{ J} \cdot \text{K}^{-1} \cdot \text{mol}^{-1})(310 \text{ K})$
   $(\ln [10 \text{ mM}/150 \text{ mM}])$
   $= (8.314)(310)(-2.71) \text{ J} \cdot \text{mol}^{-1}$
   $= -6980 \text{ J} \cdot \text{mol}^{-1} = -7.0 \text{ kJ} \cdot \text{mol}^{-1}$
   (b) $\Delta \bar{G} = RT \ln ([Na^+]_{in}/[Na^+]_{out}) + Z_A \mathcal{F} \Delta \Psi$
   $= -6980 + (1)(96,485 \text{ C} \cdot \text{mol}^{-1})(-0.06 \text{ J} \cdot \text{C}^{-1})$
   $= -6980 \text{ J} \cdot \text{mol}^{-1} - 5790 \text{ J} \cdot \text{mol}^{-1}$
   $= -12,770 \text{ J} \cdot \text{mol}^{-1} = -12.8 \text{ kJ} \cdot \text{mol}^{-1}$

8. (a) K^+ transport ceases because the ionophore–K^+ complex cannot diffuse through the membrane when the lipids are immobilized in a gel-like state. (b) No. An unblocked N-terminus would be a protonated amino group, which would repel another gramicidin A N-terminus rather than hydrogen bond to it, and which is less likely to be buried in the hydrophobic interior of a lipid bilayer.

9. The number of ions to be transported is
   $(10 \text{ mM})(100 \text{ μm}^3)(N)$
   $= (0.01 \text{ mol} \cdot \text{L}^{-1})(10^{-13} \text{ L})(6.02 \times 10^{23} \text{ ions} \cdot \text{mol}^{-1})$
   $= 6.02 \times 10^8 \text{ ions}$
   Since there are 100 ionophores, each must transport 6.02×10^6 ions. The time required is
   $(6.02 \times 10^6 \text{ ions})(1 \text{ s}/10^4 \text{ ions}) = 602 \text{ s} = 10 \text{ min}$

10. In the absence of ATP, Na^+ extrusion by the $(Na^+–K^+)$– ATPase would cease, so no glucose could enter the cell by Na^+–glucose symport. Any glucose in the cell would then exit via the passive-mediated glucose transporter, and the cellular [glucose] would decrease until it matched the extracellular [glucose] (of course, the cell would probably osmotically burst before this could occur).

11. The hyperbolic curve for glucose transport into pericytes indicates a protein-mediated sodium-dependent process. The transport protein has binding sites for sodium ions. At low $[Na^+]$, glucose transport is directly proportional to $[Na^+]$. However, at high $[Na^+]$, all Na^+-binding sites on the transport protein are occupied, and thus glucose transport reaches a maximum velocity. Glucose transport into endothelial cells is not sodium-dependent and occurs at a high rate whether or not Na^+ is present. There is not enough information in the figure to determine whether glucose transport into endothelial cells is protein-mediated.

12. Rapid transport at pH 5 indicates the involvement of hydrogen ions. One possibility is that lactate is transported in symport with H^+. At pH 5, $[Ca^{2+}]$ is 1000-fold greater than at pH 8, which would explain the greater rate of transport at pH 5. A second possibility is that lactate is transported in the protonated form and at pH 5, a greater percentage of lactate is protonated. Yet another possibility is that amino acid side chains in the transport protein that are involved in lactate transport are protonated at pH 5 and not at pH 8.

CHAPTER 11

1. (a) isomerase (alanine racemase); (b) lyase (pyruvate decarboxylase); (c) oxidoreductase (lactate dehydrogenase); (d) ligase (glutamine synthetase).

2. The tighter S binds to the enzyme, the greater the value of $\Delta G_E^{\ddagger}$. As $\Delta G_E^{\ddagger}$ approaches $\Delta G_N^{\ddagger}$, the rate of the enzyme-catalyzed reaction approaches the rate of the nonenzymatic reaction.

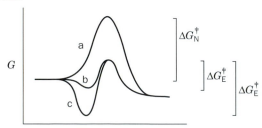

Reaction coordinate

3. Glu has a pK of ~4 and, in its ionized form, acts as a base catalyst. Lys has a pK of ~10 and, in its protonated form, acts as an acid catalyst.

4. The active form of the enzyme contains the thiolate anion. The increased pK would increase the nucleophilicity of the thiolate and thereby increase the rate of the reaction catalyzed by the active form of the enzyme. However, at physiological pH, there would be less of the active form of the enzyme and therefore the overall rate would be decreased.

5. DNA lacks the 2'-OH group required for the formation of the 2',3'-cyclic reaction intermediate.

6. Two such analogs are

Furan-2-carboxylate   **Thiophene-2-carboxylate**

Both of these molecules are planar, particularly at the C atom to which the carboxylate is bonded, as is true of the transition state for the proline racemase reaction.

7. The preferential binding of the transition state to an enzyme is an important (often the most important) part of an enzyme's catalytic mechanism. Hence, the substrate binding site is the catalytic site.

8. The lysozyme active site is arranged to cleave oligosaccharides between the fourth and fifth residues. Moreover, since the lysozyme active site can bind at least six monosaccharide units, $(NAG)_6$ would be more tightly bound to the enzyme than $(NAG)_4$, and this additional binding free energy would be applied to distorting the D ring to its half-chair conformation, thereby facilitating the reaction.

9. Asp 101 and Arg 114 form hydrogen bonds with the substrate molecule (Fig. 11-17). Ala cannot form these hydrogen bonds, so the substituted enzyme is less active.

10.
Tosyl-L-alanine chloromethylketone   or

Tosyl-L-valine chloromethylketone

11. The observation that subtilisin and chymotrypsin are genetically unrelated indicates that their active site geometries arose by convergent evolution. Assuming that evolution has optimized the catalytic efficiencies of these enzymes and that there is only one optimal arrangement of catalytic groups, any similarities between the active sites of subtilisin and chymotrypsin must be of catalytic significance. Conversely, any differences are unlikely to be catalytically important.

12. (a) Little or no effect; (b) catalysis would be much slower because the mutation disrupts the function of the catalytic triad.

13. If the soybean trypsin inhibitor were not removed from tofu, it would inhibit the trypsin in the intestine. At best, this would reduce the nutritional value of the meal by rendering its protein indigestible. It might very well also lead to intestinal upset.

14. The peptide bond that is cleaved to activate both trypsinogen and chymotrypsinogen is preceded by a cationic residue for which trypsin but not chymotrypsin is specific. Thus, trypsin, but not chymotrypsin, can activate both of these zymogens.

CHAPTER 12

1. From Eq. 12-7, $[A] = [A]_o e^{-kt}$. Since $t_{1/2} = 0.693/k$, $k = 0.693/14$ d $= 0.05$ d^{-1}. (a) 7 μmol; (b) 5 μmol; (c) 3.5 μmol; (d) 0.3 μmol.

2. For Reaction A, only a plot of 1/[reactant] versus t gives a straight line, so the reaction is second order. The slope, k, is 0.15 mM$^{-1} \cdot$s^{-1}. For Reaction B, only a plot of ln [reactant] versus t gives a straight line, so the reaction is first order. The negative of the slope, k, is 0.17 s^{-1}.

Time (s)	Reaction A 1/[reactant] (mM^{-1})	Reaction B ln [reactant]
0	0.16	1.69
1	0.32	1.53
2	0.48	1.36
3	0.62	1.16
4	0.78	0.99
5	0.91	0.83

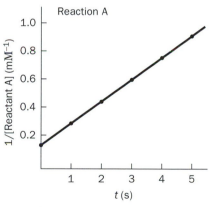

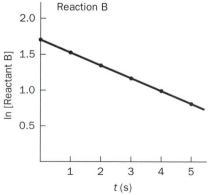

3. $v_o = V_{max}[S]/(K_M + [S])$
$v_o/V_{max} = [S]/(K_M + [S])$
$0.95 = [S]/(K_M + [S])$
$[S] = 0.95K_M + 0.95 [S]$
$[S] = (0.95/0.05)K_M = 19K_M$

4. Acetylcholinesterase, carbonic anhydrase, catalase, and fumarase.

5. Set A corresponds to $[S] > K_M$, and set B corresponds to $[S] < K_M$. Ideally, a single data set should include [S] values that are both larger and smaller than K_M.

Set A [S] (mM)	v_o (μM·s^{-1})	Set B [S] (mM)	v_o (μM·s^{-1})
2	0.42	0.12	0.17
1	0.38	0.10	0.15
0.67	0.34	0.08	0.13
0.50	0.32	0.07	0.11

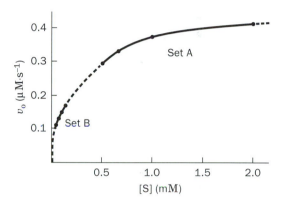

6. Construct a Lineweaver-Burk plot.

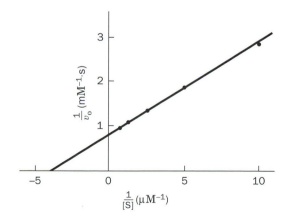

$K_M = -1/x$-intercept $= -1/(-4$ μM$^{-1}) = 0.25$ μM
$V_{max} = 1/y$-intercept $= 1/(0.8$ mM$^{-1} \cdot$s$) = 1.25$ mM·s^{-1}

7. (a) and (b) A* will appear only if the reaction follows a Ping Pong mechanism, since only a double-displacement reaction can exchange an isotope from P back to A in the absence of B. Hence, in a reaction that has a sequential mechanism, A will not become isotopically labeled.

8. The lines of the double reciprocal plots intersect to the left of the $1/v_o$ axis (on the $1/[S]$ axis). Hence, inhibition is mixed (with $\alpha = \alpha'$).

[S]	1/[S]	1/v_o	1/v_o with I
1	1.00	0.7692	1.2500
2	0.50	0.5000	0.8333
4	0.25	0.3571	0.5882
8	0.125	0.2778	0.4545
12	0.083	0.2500	0.4167

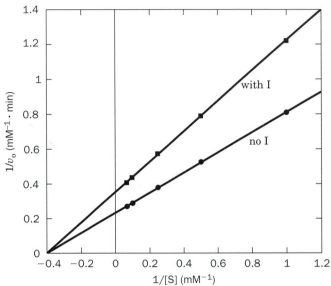

9. From Eq. 12-33, α is 3.
$\alpha = 3 = 1 + [I]/K_I = 1 + 5\ \text{mM}/K_I$
$K_I = 2.5\ \text{mM}$

10. Enzyme Y is more efficient at low [S]; enzyme X is more efficient at high [S].

11.

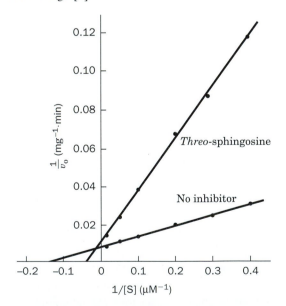

(a) K_M is derived from the x-intercept ($= -1/K_M$). In the absence of inhibitor, $K_M = 1/0.14\ \mu\text{M}^{-1} = 7\ \mu\text{M}$. In the

presence of inhibitor, $K_M = 1/0.05\ \mu\text{M}^{-1} = 25\ \mu\text{M}$. $V_{\max}$ is derived from the y-intercept ($= 1/V_{\max}$). In the absence of inhibitor, $V_{\max} = 1/0.008\ \text{mg}^{-1}\cdot\text{min} = 125\ \text{mg}\cdot\text{min}^{-1}$. In the presence of inhibitor, $V_{\max} = 1/0.01\ \text{mg}^{-1}\cdot\text{min} = 100\ \text{mg}\cdot\text{min}^{-1}$.

(b) The lines in the double-reciprocal plots intersect very close to the $1/v_o$ axis. Hence, *threo*-sphingosine is most likely a competitive inhibitor. Competitive inhibition is likely also because of the structural similarity between the inhibitor and the substrate, which allows them to compete for binding to the enzyme active site.

CHAPTER 13

1. The theoretical maximum yield of ATP is equivalent to ($\Delta G^{\circ\prime}$ for fuel oxidation) $\div$ ($\Delta G^{\circ\prime}$ for ATP synthesis).
   (a) $(-2850\ \text{kJ}\cdot\text{mol}^{-1})/(30.5\ \text{kJ}\cdot\text{mol}^{-1}) \approx 93$ ATP
   (b) $(-9781\ \text{kJ}\cdot\text{mol}^{-1})/(30.5\ \text{kJ}\cdot\text{mol}^{-1}) \approx 320$ ATP

2. At pH 6, the phosphate groups are more ionized than they are at pH 5, which increases their electrostatic repulsion and therefore increases the magnitude of ΔG for hydrolysis (makes it more negative).

3. The exergonic hydrolysis of PP$_i$ by pyrophosphatase ($\Delta G^{\circ\prime} = -33.5\ \text{kJ}\cdot\text{mol}^{-1}$) drives fatty acid activation.

4. Calculating ΔG for the reaction ATP + creatine $\rightleftharpoons$ phosphocreatine + ADP, using Eq. 13–1:

$$\Delta G = \Delta G^{\circ\prime} + RT \ln\left(\frac{[\text{phosphocreatine}][\text{ADP}]}{[\text{creatine}][\text{ATP}]}\right)$$

$$= 12.6\ \text{kJ}\cdot\text{mol}^{-1} + (8.3145\ \text{J}\cdot\text{K}^{-1}\cdot\text{mol}^{-1})(298\ \text{K})$$

$$\ln\left(\frac{(2.5\ \text{mM})(0.15\ \text{mM})}{(1\ \text{mM})(4\ \text{mM})}\right)$$

$$= 12.6\ \text{kJ}\cdot\text{mol}^{-1} - 5.9\ \text{kJ}\cdot\text{mol}^{-1} = 6.7\ \text{kJ}\cdot\text{mol}^{-1}$$

Since $\Delta G > 0$, the reaction will proceed in the opposite direction as written above, that is, in the direction of ATP synthesis.

5. Using the data in Table 13-2, we calculate $\Delta G^{\circ\prime}$ for the adenylate kinase reaction.

	$\Delta G^{\circ\prime}$
ATP + H$_2$O $\longrightarrow$ AMP + PP$_i$	$-32.2\ \text{kJ}\cdot\text{mol}^{-1}$
2 ADP + 2 P$_i$ $\longrightarrow$ 2 ATP + 2 H$_2$O	2×30.5 $= 61.0\ \text{kJ}\cdot\text{mol}^{-1}$
PP$_i$ + H$_2$O $\longrightarrow$ + 2 P$_i$	$-33.5\ \text{kJ}\cdot\text{mol}^{-1}$
2 ADP $\longrightarrow$ ATP + AMP	$\Delta G^{\circ\prime}\ -4.7\ \text{kJ}\cdot\text{mol}^{-1}$

Since ΔG for a reaction at equilibrium is zero, Eq. 13-1 becomes $\Delta G^{\circ\prime} = -RT \ln K_{\text{eq}}$ so that $K_{\text{eq}} = e^{-\Delta G^{\circ\prime}/RT}$.

$$K_{\text{eq}} = \frac{[\text{ATP}][\text{AMP}]}{[\text{ADP}]^2} = e^{-\Delta G^{\circ\prime}/RT}$$

$$[\text{AMP}] = \frac{(5 \times 10^{-4}\ \text{M})^2}{(5 \times 10^{-3}\ \text{M})} \times e^{-(-4700\ \text{J}\cdot\text{mol}^{-1})/(8.3145\ \text{J}\cdot\text{K}^{-1}\cdot\text{mol}^{-1})(298\ \text{K})}$$

$$[\text{AMP}] = 3.3 \times 10^{-4}\ \text{M} = 0.33\ \text{mM}$$

6. The more positive the reduction potential, the greater the oxidizing power. From Table13-3,

Compound	$\mathscr{E}^{\circ\prime}$ (V)
Acetoacetate	−0.346
NAD^+	−0.315
Pyruvate	−0.185
Cytochrome b (Fe^{3+})	0.077
SO_4^{2-}	0.480

7. The balanced equation is

2 cyto c(Fe^{3+}) + ubiquinol $\longrightarrow$
$\qquad$ 2 cyto c(Fe^{2+}) + ubiquinone + 2 H^+

Using the data in Table 13-3,

$\Delta\mathscr{E}^{\circ\prime} = \mathscr{E}^{\circ\prime}_{(e^-\text{ acceptor})} - \mathscr{E}^{\circ\prime}_{(e^-\text{ donor})}$ = 0.235 V − 0.045 V
$\qquad$ = 0.190 V

$\Delta G^{\circ\prime} = -n\mathscr{F}\Delta\mathscr{E}^{\circ\prime} = -(2)(96{,}485\text{ J}\cdot\text{V}^{-1}\cdot\text{mol}^{-1})(0.190\text{ V})$
$\qquad\qquad = -36.7\text{ kJ}\cdot\text{mol}^{-1}$

8. Using the data in Table 13-3:

(a) $\Delta\mathscr{E}^{\circ\prime} = \mathscr{E}^{\circ\prime}_{(e^-\text{ acceptor})} - \mathscr{E}^{\circ\prime}_{(e^-\text{ donor})} = \mathscr{E}^{\circ\prime}_{\text{(fumarate)}} - \mathscr{E}^{\circ\prime}_{(NAD^+)} = 0.031\text{ V} - (-0.315\text{ V}) = 0.346\text{ V}$. Because $\Delta\mathscr{E}^{\circ\prime} > 0$, $\Delta G^{\circ\prime} < 0$ and the reaction will spontaneously proceed as written.

(b) $\Delta\mathscr{E}^{\circ\prime} = \mathscr{E}^{\circ\prime}_{(e^-\text{ acceptor})} - \mathscr{E}^{\circ\prime}_{(e^-\text{ donor})} = \mathscr{E}^{\circ\prime}_{\text{(cyto }b)} - \mathscr{E}^{\circ\prime}_{\text{(cyto }a)} = 0.077\text{ V} - (0.290\text{ V}) = -0.213\text{ V}$. Because $\Delta\mathscr{E}^{\circ\prime} < 0$, $\Delta G^{\circ\prime} > 0$ and the reaction will spontaneously proceed in the opposite direction from that written.

9. Using the data in Table 13-3, for the oxidation of free $FADH_2$ ($\mathscr{E}^{\circ\prime} = -0.219$ V) by ubiquinone ($\mathscr{E}^{\circ\prime} = 0.045$ V),

$\Delta\mathscr{E}^{\circ\prime} = \mathscr{E}^{\circ\prime}_{\text{(ubiquinone)}} - \mathscr{E}^{\circ\prime}_{(FADH_2)}$
$\qquad = (0.045\text{ V}) - (-0.219\text{ V}) = 0.264\text{ V}$

$\Delta G^{\circ\prime} = -n\mathscr{F}\Delta\mathscr{E}^{\circ\prime} = -(2)(96{,}485\text{ J}\cdot\text{V}^{-1}\cdot\text{mol}^{-1})$
$\qquad\qquad (0.264\text{ V}) = -50.9\text{ kJ}\cdot\text{mol}^{-1}$

This is more than enough free energy to drive the synthesis of ATP from ADP + P_i ($\Delta G^{\circ\prime} = +30.5$ kJ $\cdot$ mol^{-1}; Table 13-2).

10. $Z \xrightarrow{B} W \xrightarrow{C} Y \xrightarrow{A} X$

CHAPTER 14

1. C1 of DHAP and C1 of GAP are achiral but become chiral in FBP (as C3 and C4). There are four stereoisomeric products that differ in configuration at C3 and C4: fructose-1,6-bisphosphate, psicose-1,6-bisphosphate, tagatose-1,6-bisphosphate, and sorbose-1,6-bisphosphate (see Fig. 8-2).

2. (a) Glucose + 2 NAD^+ + 2 ADP + 2 P_i
$\qquad \longrightarrow$ 2 pyruvate + 2 NADH + 2 ATP + 2 H_2O

(b) Glucose + 2 NAD^+ + 2 ADP + 2 AsO_4^{2-}
$\qquad \longrightarrow$ 2 pyruvate + 2 NADH + 2 ADP—AsO_3^{2-} + 2 H_2O

$\underline{2\text{ ADP—}AsO_3^{2-} + 2\text{ }H_2O \longrightarrow 2\text{ ADP} + AsO_4^{3-}}$

Overall: Glucose + 2 NAD^+
$\qquad\qquad \longrightarrow$ 2 pyruvate + 2 NADH

(c) Arsenate is a poison because it uncouples ATP generation from glycolysis. Consequently, glycolytic energy generation cannot occur.

3.
ribulose-5-phosphate isomerase reaction:	ribulose-5-phosphate epimerase reaction:

1,2-Enediolate intermediate $\qquad$ **2,3-Enediolate intermediate**

4. (a) ΔG values differ from $\Delta G^{\circ\prime}$ values because $\Delta G = \Delta G^{\circ\prime} + RT$ ln [Products]/[Reactants] and cellular reactants and products are not in their standard states.

(b) Yes. The same *in vivo* conditions that decrease the magnitude of ΔG relative to $\Delta G^{\circ\prime}$ may also decrease ΔG for ATP synthesis.

5. When [GAP] = 10^{-4} M, [DHAP] = 5.5×10^{-4} M. According to Eq. 1-17, $K = e^{-\Delta G^{\circ\prime}/RT}$

$\dfrac{[GAP][DHAP]}{[FBP]} = e^{-(22{,}800\text{ J}\cdot\text{mol}^{-1})/(8.3145\text{ J}\cdot\text{K}\cdot\text{mol}^{-1})(310\text{ K})}$

$\dfrac{(10^{-4})(5.5 \times 10^{-4})}{[FBP]} = 1.4 \times 10^{-4}$

[FBP] = 3.8×10^{-4} M
[FBP]/[GAP] = $(3.8 \times 10^{-4}\text{ M})/(10^{-4}\text{ M})$ = 3.8

6. For the coupled reaction

Pyruvate + NADH + H^+ $\longrightarrow$ lactate + NAD^+

$\Delta\mathscr{E}^{\circ\prime} = (-0.185\text{ V}) - (-0.315\text{ V}) = 0.130$ V. According to Eq. 13-7,

$$\Delta\mathscr{E} = \Delta\mathscr{E}^{\circ\prime} - \frac{RT}{n\mathscr{F}} \ln\left(\frac{[\text{lactate}][NAD^+]}{[\text{pyruvate}][NADH]}\right)$$

and $\Delta G = -n\mathscr{F}\Delta\mathscr{E}$ (Eq. 13-6). Since two electrons are transferred in the above reaction, $n = 2$.

(a) $\Delta\mathscr{E} = 0.130\text{ V} - \dfrac{RT}{n\mathscr{F}} \ln(1) = 0.130$ V

$\Delta G = -(2)(96{,}485\text{ J}\cdot\text{V}^{-1}\cdot\text{mol}^{-1})(0.130\text{ V})$
$\qquad\qquad = -25.1\text{ kJ}\cdot\text{mol}^{-1}$

(b) $RT/n\mathscr{F} = (8.3145\text{ J}\cdot\text{K}^{-1}\cdot\text{mol}^{-1})(298\text{ K})/$
$\qquad\qquad (2)(96{,}485\text{ J}\cdot\text{V}^{-1}\cdot\text{mol}^{-1}) = 0.01284$ V

$\Delta\mathscr{E} = 0.130\text{ V} - 0.01284\text{ V}\ln(160 \times 160)$
$\Delta\mathscr{E} = 0.130\text{ V} - 0.130\text{ V} = 0$
$\Delta G = 0$

(c) $\Delta\mathscr{E} = 0.130\text{ V} - 0.01284\text{ V}\ln(1000 \times 1000)$
$\Delta\mathscr{E} = 0.130\text{ V} - 0.177\text{ V} = -0.047$ V
$\Delta G = -(2)(96{,}485\text{ J}\cdot\text{V}^{-1}\cdot\text{mol}^{-1})(-0.047\text{ V})$
$\qquad\qquad = 9.1\text{ kJ}\cdot\text{mol}^{-1}$

7. Pyruvate kinase regulation is important for controlling the flux of metabolites such as fructose (in liver), which enter glycolysis after the PFK step.

8. The three glucose molecules that proceed through glycolysis yield six ATP. The bypass through the pentose phosphate pathway results in a yield of five ATP.

9. The label will appear at C1 and C3 of F6P (see Fig. 14-29).

10. Even when the flux of glucose through glycolysis and hence the citric acid cycle is blocked, glucose can be oxidized by the pentose phosphate pathway, with the generation of CO_2.

11. The liver enzyme is far more sensitive than the brain enzyme to the three activators. It is possible that liver PFK-1 is subject to a greater degree of regulation than brain PFK-1. Fuel must be supplied to the brain continuously and thus glycolysis is always active, but the liver has a wide variety of physiological roles and is more likely to regulate cellular pathways.

CHAPTER 15

1. (a) The equation for glycolysis is

$$\text{Glucose} + 2\,NAD^+ + 2\,ADP + 2\,P_i \longrightarrow$$
$$2\,\text{pyruvate} + 2\,NADH + 4\,H^+ + 2\,ATP + 2\,H_2O$$

The equation for gluconeogenesis is

$$2\,\text{Pyruvate} + 2\,NADH + 4\,H^+ + 4\,ATP + 2\,GTP + 6\,H_2O$$
$$\longrightarrow \text{glucose} + 2\,NAD^+ + 4\,ADP + 2\,GDP + 6\,P_i$$

For the two processes operating sequentially,

$$2\,ATP + 2\,GTP + 4\,H_2O \longrightarrow 2\,ADP + 2\,GDP + 4\,P_i$$

(b) The equation for catabolism of 6 G6P by the pentose phosphate pathway is

$$6\,G6P + 12\,NADP^+ + 6\,H_2O \longrightarrow$$
$$6\,Ru5P + 12\,NADPH + 12\,H^+ + 6\,CO_2$$

Ru5P can be converted to G6P by transaldolase, transketolase, and gluconeogenesis:

$$6\,Ru5P + H_2O \longrightarrow 5\,G6P + P_i$$

The net equation is therefore

$$G6P + 12\,NADP^+ + 7\,H_2O \longrightarrow$$
$$12\,NADPH + 12\,H^+ + 6\,CO_2 + P_i$$

2. Phosphoglucokinase activity generates G1,6P, which is necessary to "prime" phosphoglucomutase that has become dephosphorylated and thereby inactivated through the loss of its G1,6P reaction intermediate.

3. The overall free energy change for debranching is

Breaking $\alpha(1 \rightarrow 4)$ bond	$\Delta G^{\circ\prime} = -15.5\ \text{kJ} \cdot \text{mol}^{-1}$
Forming $\alpha(1 \rightarrow 4)$ bond	$+15.5\ \text{kJ} \cdot \text{mol}^{-1}$
Hydrolyzing $\alpha(1 \rightarrow 6)$ bond	$-7.1\ \text{kJ} \cdot \text{mol}^{-1}$
Total	$\Delta G^{\circ\prime} = -7.1\ \text{kJ} \cdot \text{mol}^{-1}$

The overall free energy change for branching is

Breaking $\alpha(1 \rightarrow 4)$ bond	$\Delta G^{\circ\prime} = -15.5\ \text{kJ} \cdot \text{mol}^{-1}$
Forming $\alpha(1 \rightarrow 6)$ bond	$+7.1\ \text{kJ} \cdot \text{mol}^{-1}$
Total	$\Delta G^{\circ\prime} = -8.4\ \text{kJ} \cdot \text{mol}^{-1}$

The sum of the two reactions of branching has $\Delta G < 0$, but debranching would be endergonic ($\Delta G > 0$) without the additional step hydrolyzing the $\alpha(1 \rightarrow 6)$ bond to form glucose.

4. A glycogen molecule with 28 tiers would represent the most efficient arrangement for storing glucose, and its outermost tier would contain considerably more glucose residues than a glycogen molecule with only 12 tiers. However, densely packed glucose residues would be inaccessible to phosphorylase. In fact, such a dense glycogen molecule could not be synthesized because glycogen synthase and branching enzyme would have no room to operate (see Box 15-1 for a discussion of glycogen structure).

5. In the course of glucose catabolism, a detour through glycogen synthesis and glycogen breakdown begins and ends with G6P. The energy cost of this detour is one ATP equivalent, consumed in the UDP–glucose pyrophosphorylase step. The overall energy lost is therefore 1/38 or ~3%.

6. (a) Circulating [glucose] is high because cells do not respond to the insulin signal to take up glucose.
(b) Insulin is unable to activate phosphoprotein phosphatase-1 in muscle, so glycogen synthesis is not stimulated. Moreover, glycogen synthesis is much reduced by the lack of available glucose in the cell.

7. Epinephrine binding to its receptor stimulates production of cAMP and therefore promotes phosphorylation catalyzed by cAPK. When cAMP phosphodiesterase is inhibited, [cAMP] remains high and thereby prolongs the effects of epinephrine.

8. A defect in G6P transport would have the symptoms of glucose-6-phosphatase deficiency: accumulation of glycogen and hypoglycemia.

9. The conversion of circulating glucose to lactate in the muscle generates 2 ATP. If muscle glycogen could be mobilized, the energy yield would be 3 ATP, since phosphorolysis of glycogen bypasses the hexokinase-catalyzed step that consumes ATP in the first stage of glycolysis.

10. The deficiency is in branching enzyme (Type IV glycogen storage disease). The high ratio of G1P to glucose indicates abnormally long chains of $\alpha(1 \rightarrow 4)$-linked residues with few $\alpha(1 \rightarrow 6)$-linked branch points (the normal ratio is ~10).

CHAPTER 16

1. (a) Because citric acid cycle intermediates such as citrate and succinyl-CoA are precursors for the biosynthesis of other compounds, anaerobes must be able to synthesize them.
(b) These organisms do not need a complete citric acid cycle, which would yield reduced coenzymes that must be reoxidized.

2. Citrate must be cleaved to generate an acetyl group and oxaloacetate. The oxaloacetate can then be converted to succinate to complete the cycle.

3. (a) The labeled carbon becomes C4 of the succinyl moiety of succinyl-CoA. Because succinate is symmetrical, the label appears at C1 and C4 of succinate. When the resulting oxaloacetate begins the second round, the labeled carbons appear as $^{14}CO_2$ in the isocitrate dehydrogenase and the α-ketoglutarate dehydrogenase reactions (see Fig. 16-2).
(b) The labeled carbon becomes C3 of the succinyl moiety of succinyl-CoA and hence appears at C2 and C3 of succinate, fumarate, malate, and oxaloacetate. Neither C2 nor C3 of oxaloacetate is released as CO_2 in the second round of the cycle. However, the ^{14}C label appears at C1 and C2 of the succinyl moiety of succinyl-CoA in the second round and therefore appears at all four positions of the resulting oxaloacetate. Thus, in the third round, ^{14}C is released as $^{14}CO_2$.

4.

$$\begin{array}{c} H_3C \\ \searrow \\ CH-\overset{\overset{\displaystyle O}{\displaystyle \|}}{C}-S-CoA \\ \nearrow \\ H_3C \end{array}$$

5. NAD^+ ($\mathscr{E}^{\circ\prime} = -0.315$ V) does not have a high enough reduction potential to support oxidation of succinate to fumarate ($\mathscr{E}^{\circ\prime} = +0.031$ V); that is, the succinate dehydrogenase reaction has insufficient free energy to reduce NAD^+. Enzyme-bound FAD ($\mathscr{E}^{\circ\prime} \approx 0$) is more suitable for oxidizing succinate.

6. Competitive inhibition can be overcome by adding more substrate, in this case succinate. Oxaloacetate overcomes malonate inhibition because it is converted to succinate by the reactions of the citric acid cycle.

7. To synthesize citrate, pyruvate must be converted to oxaloacetate by pyruvate carboxylase:

   Pyruvate + CO_2 + ATP + H_2O ⟶
    oxaloacetate + ATP + P_i

   A second pyruvate is converted to acetyl-CoA by pyruvate dehydrogenase:

   Pyruvate + CoASH + NAD^+ ⟶
    acetyl-CoA + CO_2 + NADH

   The acetyl-CoA then combines with oxaloacetate to produce citrate:

   Oxaloacetate + acetyl-CoA ⟶ citrate + CoASH

   The net reaction is

   2 Pyruvate + ATP + NAD^+ + H_2O ⟶
    citrate + ADP + P_i + NADH

8. For the reaction isocitrate + NAD^+ ⇌ α-ketoglutarate + NADH + CO_2 + H^+, we assume $[H^+]$ and $[CO_2] = 1$. According to Eq. 13-1,

$$\Delta G = \Delta G^{\circ\prime} + RT \ln \left(\frac{[\text{NADH}][\alpha\text{-ketoglutarate}]}{[\text{NAD}^+][\text{isocitrate}]} \right)$$

$$= -21 \text{ kJ} \cdot \text{mol}^{-1} + (8.3145 \text{ J} \cdot \text{K} \cdot \text{mol}^{-1})$$

$$(298 \text{ K}) \ln \left[\frac{(1)(0.1)}{(8)(0.02)} \right]$$

$$= -21 \text{ kJ} \cdot \text{mol}^{-1} - 1.17 \text{ kJ} \cdot \text{mol}^{-1} = -22.17 \text{ kJ} \cdot \text{mol}^{-1}$$

With such a large negative free energy of reaction under physiological conditions, isocitrate dehydrogenase is likely to be a metabolic control point.

9. This observation is consistent with the metabolon hypothesis. Close association between enzymes in a metabolic pathway may allow intermediates to pass directly from one active site to another without entering the bulk solution. Thus, the orientation and asymmetric labeling pattern of a symmetric molecule may be preserved to some extent.

10. Animals cannot carry out the net synthesis of glucose from acetyl-CoA (to which acetate is converted). However, ^{14}C-labeled acetyl-CoA enters the citric acid cycle and is converted to oxaloacetate. Some of this oxaloacetate may be converted to glucose through gluconeogenesis and subsequently taken up by muscle and incorporated into glycogen.

CHAPTER 17

1. Mitochondria with more cristae have more surface area and therefore more proteins for electron transport and oxidative phosphorylation. Tissues with a high demand for ATP synthesis (such as heart) contain mitochondria with more cristae than tissues with lower demand for oxidative phosphorylation (such as liver).

2. When NADH participates in the glycerophosphate shuttle, the electrons of NADH flow to FAD and then to CoQ, bypassing Complex I. Thus, 2 ATP are synthesized per NADH. Three ATP are produced when NADH participates in the malate–aspartate shuttle.

3. The relevant half-reactions (Table 13-3) are

   FAD + 2 H^+ + $2e^-$ ⇌ $FADH_2$ $\Delta\mathscr{E}^{\circ\prime} = -0.219$ V
   $\frac{1}{2} O_2$ + 2 H^+ + 2 e^- ⇌ H_2O $\Delta\mathscr{E}^{\circ\prime} = 0.815$ V

   Since the O_2/H_2O half-reaction has the more positive $\Delta\mathscr{E}^{\circ\prime}$, the FAD half-reaction is reversed and the overall reaction is

   $\frac{1}{2} O_2$ + $FADH_2$ ⇌ H_2O + FAD
    $\Delta\mathscr{E}^{\circ\prime} = 0.815$ V $- (-0.219$ V$) = 1.034$ V

   Since $\Delta G^{\circ\prime} = -n\mathscr{F}\Delta\mathscr{E}^{\circ\prime}$,

   $$\Delta G^{\circ\prime} = -(2)(96,485 \text{ kJ} \cdot \text{V}^{-1} \cdot \text{mol}^{-1})(1.034 \text{ V})$$
   $$= -200 \text{ kJ} \cdot \text{mol}^{-1}$$

   The maximum number of ATP that could be synthesized is therefore

   $200 \text{ kJ} \cdot \text{mol}^{-1}/30.5 \text{ kJ} \cdot \text{mol}^{-1}$
    = 6.6 mol ATP/mol $FADH_2$ oxidized by O_2.

4.

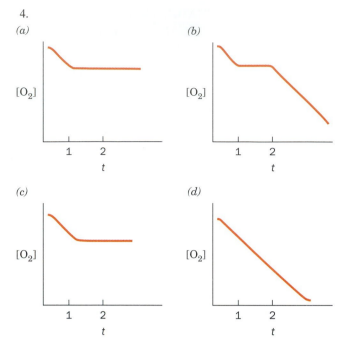

(a) O_2 consumption ceases because amytal blocks electron transport in Complex I.
(b) Electrons from succinate bypass the amytal block by entering the electron-transport chain at Complex II and thereby restoring electron transport through Complexes III and IV.
(c) CN^- blocks electron transport in Complex IV, after the point of entry of succinate.
(d) Oligomycin blocks oxidative phosphorylation and hence O_2 consumption. DNP uncouples electron transport from oxidative phosphorylation and thereby permits O_2 consumption to resume.

5. $\mathscr{E}$ may differ from $\mathscr{E}°'$, depending on the redox center's microenvironment and the concentrations of reactants and products. In addition, the tight coupling between successive electron transfers within a complex may "pull" electrons so that the overall process is spontaneous.

6. (a)

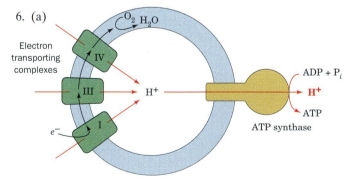

(b) An increase in external pH (decrease in $[H^+]$) increases the electrochemical potential across the mitochondrial membrane and therefore leads to an increase in ATP synthesis.

7. DNP and related compounds dissipate the proton gradient required for ATP synthesis. The dissipation of this gradient decreases the rate of synthesis of ATP, decreasing the ATP mass action ratio. Decreasing this ratio relieves the inhibi-

tion of the electron transport chain, causing an increase in metabolic rate.

8. Hormones stimulate the release of fatty acids from stored triacylglycerols, which activates UCP and also provides the fuel whose oxidation yields electrons for the heat-generating electron-transfer process.

9. The switch to aerobic metabolism allows ATP to be produced by oxidative phosphorylation. The phosphorylation of ADP increases the [ATP]/[ADP] ratio, which then increases the [NADH]/[NAD$^+$] ratio because a high ATP mass action ratio slows electron transport. The increases in [ATP] and [NADH] inhibit their target enzymes in glycolysis and the citric acid cycle (Fig. 17-23) and thereby slow these processes.

10. Glucose is shunted through the pentose phosphate pathway to provide NADPH, whose electrons are required to reduce O_2 to $O_2^-\cdot$.

CHAPTER 18

1. The color of the seawater indicates that the photosynthetic pigments of the algae absorb colors of visible light other than red.

2. The label appears as $^{18}O_2$:

$$H_2^{18}O + CO_2 \xrightarrow{\text{light}} (CH_2O) + {}^{18}O_2$$

3. (a) The energy per photon is $E = hc/\lambda$, so the energy per mole of photons is
$$\begin{aligned} E &= Nhc/\lambda \\ &= (6.022 \times 10^{23}\ \text{mol}^{-1})(6.626 \times 10^{-34}\ \text{J}\cdot\text{s}) \\ &\qquad (2.998 \times 10^8\ \text{m}\cdot\text{s}^{-1})/(7 \times 10^{-7}\ \text{m}) \\ &= 1.71 \times 10^5\ \text{J}\cdot\text{mol}^{-1} \\ &= 171\ \text{kJ}\cdot\text{mol}^{-1} \end{aligned}$$
(b) $(171\ \text{kJ}\cdot\text{mol}^{-1})/(30.5\ \text{kJ}\cdot\text{mol}^{-1}) = 5.6$
Five moles of ATP could theoretically be synthesized (at least under standard biochemical conditions).

4. (a) The relevant half-reactions are (Table 13-3):

$$O_2 + 4\,e^- + 4\,H^+ \longrightarrow 2\,H_2O \qquad \Delta\mathscr{E}°' = 0.815\ \text{V}$$
$$NADP^+ + H^+ + 2\,e^- \longrightarrow NADPH \qquad \Delta\mathscr{E}°' = -0.320\ \text{V}$$

The overall reaction is

$$2\,NADP^+ + 2\,H_2O \longrightarrow 2\,NADPH + O_2 + 2\,H^+$$
$$\Delta\mathscr{E}°' = -0.320\ \text{V} - (0.815\ \text{V}) = -1.135\ \text{V}$$

$$\begin{aligned} \Delta G°' &= -n\mathscr{F}\Delta\mathscr{E}°' \\ &= -(4)(96{,}485\ \text{J}\cdot\text{V}^{-1}\cdot\text{mol}^{-1})(-1.135\ \text{V}) \\ &= 438\ \text{kJ}\cdot\text{mol}^{-1} \end{aligned}$$
(b) One mole of photons of red light ($\lambda = 700$ nm) has an energy of 171 kJ. Therefore, $438/171 = 2.6$ moles of photons are theoretically required to drive the oxidation of H_2O by $NADP^+$ to form one mole of O_2.
(c) The energy of a mole of photons of UV light ($\lambda = 220$ nm) is
$$\begin{aligned} E &= Nhc/\lambda \\ &= (6.022 \times 10^{23}\ \text{mol}^{-1})(6.626 \times 10^{-34}\ \text{J}\cdot\text{s}) \\ &\qquad (2.998 \times 10^8\ \text{m}\cdot\text{s}^{-1})/(2.2 \times 10^{-7}\ \text{m}) \\ &= 544\ \text{kJ}\cdot\text{mol}^{-1} \end{aligned}$$

The number of moles of 220-nm photons required to produce one mole of O_2 is $438/544 = 0.8$.

5. Both systems mediate cyclic electron flows. The photooxidized bacterial reaction center passes electrons through a series of electron carriers so that the electrons return to the reaction center ($P870^+$) and restore it to its original state. During cyclic electron flow in PSI, electrons from photooxidized P700 are transferred to cytochrome b_6f and, via plastoquinone, back to $P700^+$. In both cases, there is no net change in redox state of the reaction center, but light-driven electron movements are accompanied by the transmembrane movement of protons.

6. When cyclic electron flow occurs, photoactivation of PSI drives electron transport independently of the flow of electrons derived from water. Thus, the oxidation of H_2O by PSII is not linked to the number of photons consumed by PSI.

7. An increase in $[O_2]$ increases the oxygenase activity of RuBP carboxylase–oxygenase and therefore lowers the efficiency of CO_2 fixation.

8. After the light is turned off, ATP and NADPH levels fall as these substances are used up in the Calvin cycle without being replaced by the light reactions. 3PG builds up because it cannot pass through the phosphoglycerate kinase reaction in the absence of ATP. The RuBP level drops because it is consumed by the RuBP carboxylase reaction (which requires neither ATP nor NADPH) and its replenishment is blocked by the lack of ATP for the phosphoribulokinase reaction.

9. The net synthesis of 2 GAP from 6 CO_2 in the initial stage of the Calvin cycle (Fig. 18-20) consumes 18 ATP and 12 NADPH (equivalent to 36 ATP). The conversion of 2 GAP to glucose-6-phosphate (G6P) by gluconeogenesis does not require energy input (Section 15-4B), nor does the isomerization of G6P to glucose-1-phosphate (G1P). The activation of G1P to its nucleotide derivative consumes 2 ATP equivalents (Section 15-5), but ADP is released when the glucose residue is incorporated into starch. These steps represent an overall energy investment of $18 + 36 + 1 = 55$ ATP.

   Starch breakdown by phosphorolysis yields G1P, whose subsequent degradation by glycolysis yields 3 ATP, 2 NADH (equivalent to 6 ATP), and 2 pyruvate. Complete oxidation of 2 pyruvate to 6 CO_2 by the pyruvate dehydrogenase reaction and the citric acid cycle (Section 16-1) yields 8 NADH (equivalent to 24 ATP), 2 $FADH_2$ (equivalent to 4 ATP), and 2 GTP (equivalent to 2 ATP). The overall ATP yield is therefore $3 + 6 + 24 + 4 + 2 = 39$ ATP.

   The ratio of energy spent to energy recovered is $55/39 = 1.4$.

10. These plants store CO_2 by CAM. At night, CO_2 reacts with PEP to form malate. By morning, so much malate (malic acid) has accumulated that the leaves have a sour taste. During the day, the malate is converted to pyruvate + CO_2. The leaves therefore become less acidic and hence tasteless. Late in the day, when all the malate is consumed, the leaves become slightly basic, that is, bitter.

CHAPTER 19

1. A defect in carnitine palmitoyl transferase II prevents normal transport of activated fatty acids into the mitochondria for β oxidation. Tissues such as muscle that use fatty acids as metabolic fuels therefore cannot generate ATP as needed. The problem is more severe during a fast because other fuels, such as dietary glucose, are not readily available.

2. The first three steps of β oxidation resemble the reactions that convert succinate to oxaloacetate (Section 16-3 F–H).

Succinate → **Fumarate**

L-Malate → **Oxaloacetate**

3. (a) Six cycles are required.

   (b) 3 acetyl-CoA, 3 propionyl-CoA, and 1 methylpropionyl-CoA.

4. There are not as many usable nutritional calories per gram in unsaturated fatty acids as there are in saturated fatty acids. This is because oxidation of fatty acids containing double bonds yields fewer reduced coenzymes whose oxidation drives the synthesis of ATP. In the oxidation of fatty acids with a double bond at an odd-numbered carbon, the enoyl-CoA isomerase reaction bypasses the acyl-CoA dehydrogenase reaction and therefore does not generate $FADH_2$ (equivalent to 2 ATP). A double bond at an even-numbered carbon must be reduced by NADPH (equivalent to the loss of 3 ATP).

5. 3-Ketoacyl-CoA transferase is required to convert ketone bodies to acetyl-CoA. If the liver contained this enzyme, it would be unable to supply ketone bodies as fuels for other tissues.

6. Palmitate (C_{16}) synthesis requires 14 NADPH. The transport of 8 acetyl-CoA to the cytosol by the tricarboxylate transport system supplies 8 NADPH (Fig. 19-20), which represents $8/14 \times 100 = 57\%$ of the required NADPH.

7. The label does not appear in palmitate because $^{14}CO_2$ is released in Reaction 3 of fatty acid synthesis (Fig. 19-22).

8. This fatty acid (**linolenate**) cannot be synthesized by animals because it contains a double bond closer than 6 carbons from its noncarboxylate end.

9. The synthesis of stearate (18:0) from mitochondrial acetyl-CoA requires 9 ATP to transport 9 acetyl-CoA from the mitochondria to the cytosol. Seven rounds of fatty acid synthesis consume 7 ATP (in the acetyl-CoA carboxylase reaction) and 14 NADPH (equivalent to 42 ATP). Elongation of palmitate to stearate requires 1 NADH and 1 NADPH (equivalent to 6 ATP). The energy cost is therefore $9 + 7 + 42 + 6 = 64$ ATP.

(a) The degradation of stearate to 9 acetyl-CoA consumes 2 ATP (in the acyl-CoA synthetase reaction) but generates, in eight rounds of β oxidation, 8 FADH$_2$ (equivalent to 16 ATP) and 8 NADH (equivalent to 24 ATP). Thus, the energy yield is $16 + 24 - 2 = 38$ ATP. This represents only about half of the energy consumed in synthesizing stearate (38 ATP versus 64 ATP).

(b) The complete oxidation of the 9 acetyl-CoA to CO$_2$ by the citric acid cycle yields an additional 9 GTP (equivalent to 9 ATP), 27 NADH (equivalent to 81 ATP), and 9 FADH$_2$ (equivalent to 18 ATP) for a total of $38 + 9 + 81 + 18 = 146$ ATP. Thus, more than twice the energy investment of synthesizing stearate is recovered (146 ATP versus 64 ATP).

10. (a)

$$H_3C - \overset{O}{\overset{||}{\underset{14}{C}}} - CH_2 - \overset{O}{\overset{||}{\underset{14}{C}}} - O^-$$

Acetoacetate

See Fig. 19-17.

(b)

$$\begin{array}{c} OH \\ \underset{14}{|} \\ CH - (CH_2)_{14} - CH_3 \\ | \\ H_2N - C - H \\ | \\ CH_2OH \end{array}$$

Sphinganine

See Fig. 19-31.

CHAPTER 20

1. Proteasome-dependent proteolysis requires ATP to activate ubiquitin in the first step of linking ubiquitin to the target protein (Fig. 20-2) and for unfolding the protein as it enters the proteasome.

2. The urea cycle transforms excess nitrogen from protein breakdown to an excretable form, urea. In a deficiency of a urea cycle enzyme, the preceding urea cycle intermediates may build up to a toxic level. A low-protein diet minimizes the amount of nitrogen that enters the urea cycle and therefore reduces the concentrations of the toxic intermediates.

3. A high concentration of ammonia drives the glutamate dehydrogenase reaction in reverse: α-ketoglutarate + NH$_4^+$ + NAD(P)H → glutamate + NAD(P)$^+$. The citric acid cycle slows as α-ketoglutarate, an intermediate of the cycle, is consumed.

4. Since the three reactions converting tiglyl-CoA to acetyl-CoA and propionyl-CoA are analogous to those of fatty acid oxidation (β oxidation; Fig. 19-9), the reactions are

$$CH_3 - CH = \overset{CH_3}{\underset{|}{C}} - \overset{O}{\overset{||}{C}} - SCoA$$

Tiglyl-CoA

H$_2$O ↘ (a hydratase)

$$CH_3 - \overset{H}{\underset{OH}{\overset{|}{C}}} - \overset{CH_3}{\underset{|}{CH}} - \overset{O}{\overset{||}{C}} - SCoA$$

NAD$^+$ ↘ (a dehydrogenase)
NADH ↙

$$CH_3 - \overset{O}{\overset{||}{C}} - \overset{CH_3}{\underset{|}{CH}} - \overset{O}{\overset{||}{C}} - SCoA$$

CoASH ↘ (a thiolase)

$$CH_3 - \overset{O}{\overset{||}{C}} - SCoA \quad + \quad CH_3 - CH_2 - \overset{O}{\overset{||}{C}} - SCoA$$

Acetyl-CoA   **Propionyl-CoA**

5.

bond to be cleaved

6. (a) Ala, Arg, Asn, Asp, Cys, Gln, Glu, Gly, His, Met, Pro, Ser, and Val
(b) Leu and Lys
(c) Ile, Phe, Thr, Trp, and Tyr

7. Tryptophan can be considered a member of this group since one of its degradation products is alanine, which is converted to pyruvate by transamination.

8. In the absence of uridylyl-removing enzyme, adenylyltransferase·P$_{II}$ will be fully uridylylated, since there is no mechanism for removing the uridylyl groups once they are attached. Uridylylated adenylyltransferase·P$_{II}$ adenylylates glutamine synthetase, which activates it. Hence, the defective *E. coli* cells will have a hyperactive glutamine synthetase and thus a higher than normal glutamine concentration. Reactions requiring glutamine will therefore be

accelerated, thereby depleting glutamate and the citric acid cycle intermediate α-ketoglutarate. Consequently, biosynthetic reactions requiring transamination, as well as energy metabolism, will be suppressed.

9. Since only plants and microorganisms synthesize aromatic amino acids, herbicides that inhibit these pathways do not affect amino acid metabolism in animals.

10. The pigment coloring skin and hair is melanin, which is synthesized from tyrosine. When tyrosine is in short supply, as when dietary protein is not available, melanin cannot be synthesized in normal amounts, and the skin and hair become depigmented.

and adipose tissue to synthesize glycogen, fat, and protein from the excess nutrients while inhibiting the breakdown of these metabolic fuels. Hence, ingesting glucose before a race will gear the runner's metabolism for resting rather than for running.

9. During starvation, the synthesis of glucose from liver oxaloacetate depletes the supply of citric acid cycle intermediates and thus decreases the ability of the liver to metabolize acetyl-CoA via the citric acid cycle.

10. The leptin produced by the normal mouse will enter the circulation of the *ob/ob* mouse, resulting in its decreased appetite and weight.

CHAPTER 21

1. Hyperinsulinemia would result in a decrease in blood glucose. The decrease in [glucose] for the brain would cause loss of brain function (leading to coma and death).

2. ATP generating pathways such as glycolysis and fatty acid oxidation require an initial investment of ATP (the hexokinase and phosphofructokinase steps of glycolysis and the acyl-CoA synthetase activation step that precedes β oxidation). This "priming" cannot occur when ATP has been exhausted.

3. The fraction of receptors occupied by ligand is $\dfrac{[RL]}{[R] + [RL]}$.

   Since $[R] + [RL] = 0.3\ \mu M$ and $[L] + [RL] = 2\ \mu M$,
   $[R] = 0.3\ \mu M - [RL]$ and $[L] = 2\ \mu M - [RL]$

   Therefore, $K = \dfrac{[R][L]}{[RL]} = 1 \times 10^{-6}\ M$

   $$= \dfrac{(0.3\ \mu M - [RL])(2\ \mu M - [RL])}{[RL]}$$

   $[RL]$ (in units of μM) $= 0.6 - 2.3[RL] + [RL]^2$
   or $[RL]^2 - 3.3[RL] + 0.6 = 0$
   Using the quadratic equation and taking the negative square root, $[RL] = 0.19\ \mu M$. (The positive square root yields $[RL] = 3.11\ \mu M$, which is greater than the total receptor concentration, an impossible result.)

   Thus, $\dfrac{[RL]}{[R]_{total}} \times 100 = \dfrac{0.19\ \mu M}{0.3\ \mu M} \times 100 = 63\%$

4. Because the GTP analog cannot be hydrolyzed, G_α remains active. Analog binding to G_s therefore increases cAMP production. Analog binding to G_i decreases cAMP production.

5. No. Although the diacylglycerol second messengers are identical, phosphatidylethanolamine does not generate an IP_3 second messenger that triggers the release of Ca^{2+}, which in turn alters protein kinase C activity.

6. In the presence of the viral protein, the cell would undergo more cycles of cell division in response to the growth factor.

7. Insulin promotes the uptake of glucose via the increase in GLUT4 receptors on the adipocyte surface. A source of glucose is necessary to supply the glycerol-3-phosphate backbone of triacylglycerols.

8. Ingesting glucose while in the resting state causes the pancreas to release insulin. This stimulates the liver, muscle,

CHAPTER 22

1. (a) 7 ATP; (b) 8 ATP; (c) 7 ATP.

2. PRPP and FGAR accumulate because they are substrates of Reactions 2 and 5 in the IMP biosynthetic pathway (Fig. 22-1). XMP also accumulates because the GMP synthase reaction is blocked (Fig. 22-3). Although glutamine is a substrate of carbamoyl phosphate synthetase II (the first enzyme of UMP synthesis; Fig. 22-5), the other substrates of this enzyme do not accumulate.

3. (a) The recovered deoxycytidylate would be equally labeled in its base and ribose components (i.e., the same labeling pattern as in the original cytidine). (b) The recovered deoxycytidylate would be unequally labeled in its base and ribose components because the separated ^{14}C-cytosine and ^{14}C-ribose would mix with the different-sized pools of unlabeled cellular cytosine and ribose before recombining as the deoxycytidylate that becomes incorporated into DNA. [This experiment established that deoxyribonucleotides, in fact, are synthesized from their corresponding ribonucleotides (alternative a).]

4. Hydroxyurea destroys the tyrosyl radical that is essential for the activity of ribonucleotide reductase. Tumor cells are generally fast-growing and cannot survive without this enzyme, which supplies dNTPs for nucleic acid synthesis. In contrast, most normal cells grow slowly, if at all, and hence have less need for nucleic acid synthesis.

5. Deoxyadenosine inhibits ribonucleotide reductase, thereby preventing the synthesis of the deoxynucleotides required for DNA synthesis.

6. FdUMP and methotrexate kill rapidly proliferating cells, such as cancer cells and those of hair follicles. Consequently, hair falls out.

7. The mutant cells grow because the medium contains the thymidine they are unable to make. Normal cells, however, continue to synthesize their own thymidine and thereby convert their limited supply of THF to DHF. The methotrexate inhibits dihydrofolate reductase, so THF cannot be regenerated. Without a supply of THF for the synthesis of nucleotides and amino acids, the cells die.

8. The synthesis of histidine and methionine requires THF. The cell's THF is converted to DHF by the thymidylate synthase reaction, but in the presence of methotrexate, THF cannot be regenerated.

9. (a) Trimethoprim binds to bacterial dihydrofolate reductase but does not permanently inactivate the enzyme. Therefore, it is not a mechanism-based inhibitor.
(b) Allopurinol is oxidized by xanthine oxidase to a product that irreversibly binds to the enzyme. It is therefore a mechanism-based inhibitor of xanthine oxidase.

10. In von Gierke's disease (glucose-6-phosphatase deficiency), glucose-6-phosphate accumulates in liver cells, thereby stimulating the pentose phosphate pathway. The resulting increase in ribose-5-phosphate production boosts the concentration of PRPP, which in turn stimulates purine biosynthesis. High levels of uric acid derived from the breakdown of these excess purines causes gout.

CHAPTER 23

1. Since amino acids have an average molecular mass of ~110 D, the 50-kD protein contains 50,000 D ÷ 110 D/ residue = ~455 residues. These residues are encoded by 455 × 3 = 1365 nucleotides. In B-DNA, the rise per base pair is 3.4 Å, so the contour length of 1365 bp is 3.4 Å/bp × 1365 bp = 4641 Å, or 0.46 μm. In A-DNA, the contour length would be 1365 bp × 2.6 Å/bp = 3549 Å, or 0.35 μm.

2. (a) Hypoxanthine, which lacks guanine's 2-amino-group, pairs with cytosine in much the same way as does guanine.

C   **Hypoxanthine**

(b)

U   **G**

3. (a) The supercoiled molecule is more compact than the relaxed circle and therefore sediments more rapidly. (b) Both overwound and underwound DNA molecules are supercoiled (see Fig. 23-10), so the ultracentrifuge-based measurements cannot distinguish between them.

4. The enzyme has no effect on the supercoiling of DNA since cleaving the C2′—C3′ bond of ribose does not sever the sugar–phosphate chain of DNA.

5. In the B-DNA to Z-DNA transition, a right-handed helix with one turn per 10.4 base pairs converts to a left-handed helix with one turn per 12 base pairs. Since a right-handed duplex helix has a positive twist, the twist decreases:

$$\Delta T = -\frac{100}{10.4} + \frac{-100}{12} = -17.9 \text{ turns}$$

The linking number must remain constant ($\Delta L = 0$) since no covalent bonds are broken. Hence, the change in writhing number is $\Delta W = -\Delta T = 17.9$ turns.

6. The segment with 20% A residues (i.e., 40% A·T base pairs) contains 60% G·C base pairs and therefore melts at a higher temperature than a segment with 30% A residues (i.e., 40% G·C base pairs).

7. (a) Its T_m decreases because the charges on the phosphate groups are less shielded from each other at lower ionic strength and hence repel each other more strongly, thereby destabilizing the double helix. (b) The nonpolar solvent diminishes the hydrophobic forces that stabilize double-stranded DNA and hence lowers the T_m.

8. The largest (and therefore the heaviest) RNA forms a band closest to the bottom of the centrifuge tube; the smallest (lightest) RNA forms a band near the top of the tube.

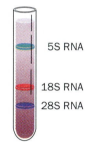

9. The target half-site consists of 6 base pairs. Since there are 4 possible base pairs (A·T, T·A, G·C, and C·G), the probability that any two base pairs are randomly related by symmetry is 1/4. Hence, the probability of finding all 6 pairs of base pairs by random chance is $(1/4)^6 = 2.4 \times 10^{-4}$.

10. (a) The contour length is 5×10^7 bp × 3.4 Å = 1.7×10^8 Å = 17 mm.
(b) A nucleosome, which binds ~200 bp, compresses the DNA to an 80-Å-high supercoil. The length of the DNA is therefore (80 Å/200 bp) × (5×10^7 bp) = 2×10^7 Å = 2 mm.
(c) The 300-Å filament contains 6 nucleosomes per turn and has a pitch of 110 Å. Therefore, the length of the DNA is (110 Å/6 nucleosomes) × (1 nucleosome/200 bp) × (5×10^7 bp) = 4.6×10^6 Å = 0.46 mm.

11. Experiment 1. The restriction enzyme failed to digest the genomic DNA, leaving the DNA too large to enter the gel during electrophoresis.

Experiment 2. The hybridization conditions were too "relaxed," resulting in nonspecific hybridization of the probe to all the DNA fragments. This problem could be corrected by boiling the blot to remove the probe and repeating the hybridization at a higher temperature and/or lower salt concentration.

Experiment 3. The probe hybridized with three different mouse genes. The different intensity of each band reflects the relatedness of the sequences. The most intense band is most similar to the human *rxr-1* gene, whereas the least intense band is least similar to the *rxr-1* gene.

(a) 3′ → 5′ Polymerase

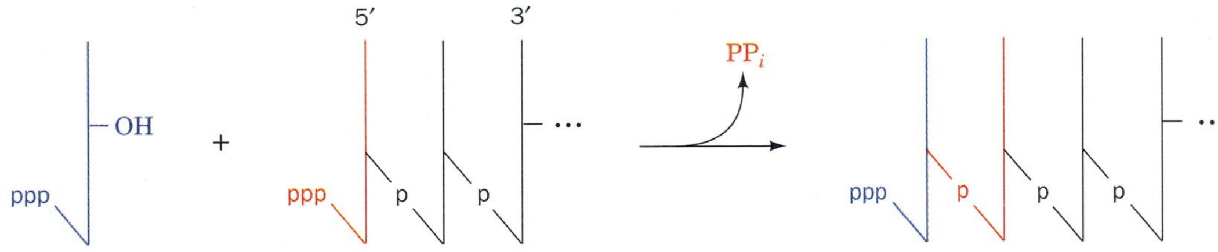

(b) 5′ → 3′ Exonuclease

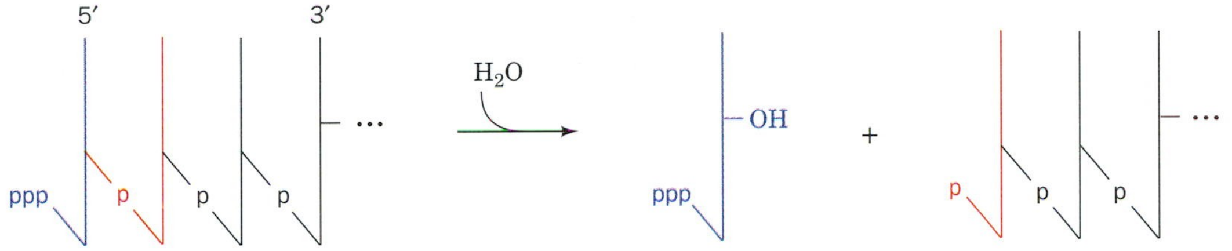

CHAPTER 24

1. Okazaki fragments are 1000 to 2000 nt long, and the *E. coli* chromosome contains 4.6×10^6 bp. Therefore, *E. coli* chromosomal replication requires 2300 to 4600 Okazaki fragments.

2. As indicated in Fig. *a* (*above*), nucleotides would be added to a polynucleotide strand by attack of the 3′-OH of the incoming nucleotide on the 5′ triphosphate group of the growing strand with the elimination of PP_i. The hydrolytic removal of a mispaired nucleotide by the 5′ → 3′ exonuclease activity (Fig. *b, above*) would leave only an OH group or monophosphate group at the 5′ end of the DNA chain. This would require an additional activation step before further chain elongation could commence.

3. The 5′ → 3′ exonuclease activity is essential for DNA replication because it removes RNA primers and replaces them with DNA. Absence of this activity would be lethal.

4. The Klenow fragment, which lacks 5′ → 3′ exonuclease activity, is used to ensure that all the replicated DNA chains have the same 5′ terminus, a necessity if sequence is assigned according to fragment length.

5. DNA gyrase adds negative supercoils to relieve the positive supercoiling that helicase-catalyzed unwinding produces ahead of the replication fork.

6. Mismatch repair and other repair systems correct the errors missed by the proofreading functions of DNA polymerases.

7. The *E. coli* replication system can fully replicate only circular DNAs. Bacteria do not have a mechanism (e.g., telomerase-catalyzed extension of telomeres) for replicating the extreme 3′ ends of linear template strands.

8. The cytosine derivative base pairs with adenine, generating a $C \cdot G \rightarrow T \cdot A$ transition.

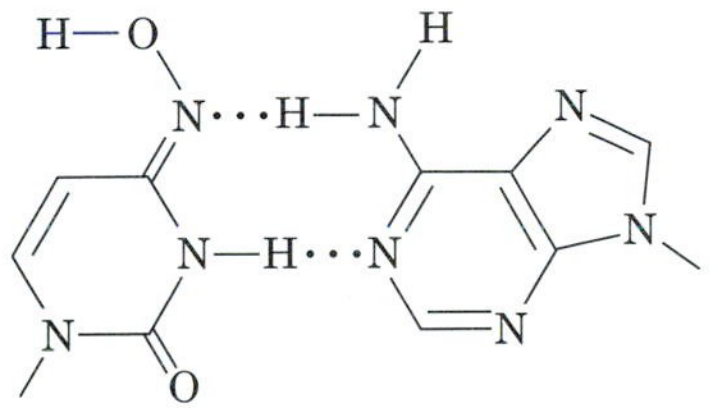

Adenine

9. When 5-methylcytosine residues deaminate, they form thymine residues.

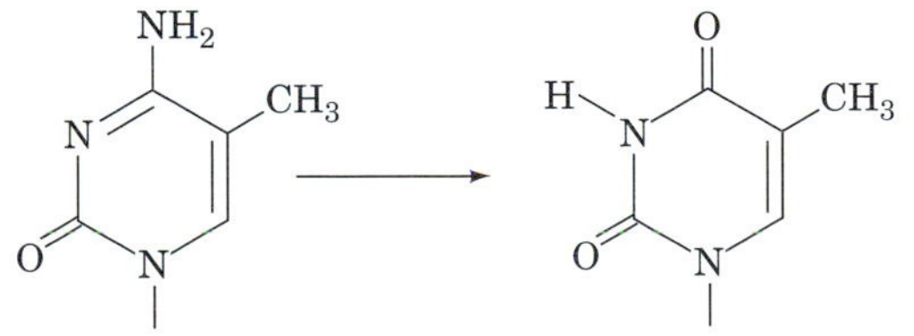

5-Methyl-C   **T**

Since thymine is a normal DNA base, the repair systems cannot determine whether such a T or its opposing G is the mutated base. Consequently, only about half of the deaminated 5-methylcytosines are correctly repaired.

10. (a) *dnaB* encodes the helicase DnaB. Loss of DnaB, which unwinds DNA for replication, would be lethal.
(b) *polA* encodes Pol I. Loss of Pol I would prevent the excision of RNA primers and would therefore be lethal.
(c) *ssb* encodes single-strand binding protein (SSB), which prevents reannealing of separated single strands. Loss of SSB would be lethal.

(d) *recA* encodes RecA protein, which mediates general recombination and the SOS response. Loss of RecA would be harmful but not necessarily lethal.

11. *E. coli* contains a low concentration of dUTP, which DNA polymerase incorporates into DNA in place of dTTP. The resulting uracil bases are rapidly excised by uracil-*N*-glycosylase followed by nucleotide excision repair (NER), which temporarily causes a break in the DNA chain. DNA that is isolated before DNA polymerase I and DNA ligase can complete the repair process would be fragmented. However, in the absence of a functional uracil-*N*-glycosylase, the inappropriate uracil residues would remain in place, and hence the leading strand DNA would be free of breaks. The lagging strand, being synthesized discontinuously, would still contain breaks, although fewer than otherwise.

CHAPTER 25

1. (a) Cordycepin is the 3′-deoxy analog of adenosine.
   (b) Because it lacks a 3′-OH group, the cordycepin incorporated into a growing RNA chain cannot support further chain elongation in the 5′ → 3′ direction.

2. The top strand is the sense strand.

5′ CAACGTAACACTTTACAGCGGCGCGTCATTTGATATGATGCGCCCCGCTTCCCGATA 3′

−35 region	−10 region	start point

Its TATGAT segment differs by only one base from the TATAAT consensus sequence of the promoter's −10 sequence; its TTTACA sequence differs by only one base from the TTGACA consensus sequence of the promoter's −35 sequence and is appropriately located ~25 nt to the 5′ side of the −10 sequence; and the initiating G nucleotide is the only purine that is located ~10 nt downstream of the −10 sequence.

3. The probe should have a sequence complementary to the consensus sequence of the 6-nt −10 promoter sequence: 5′-ATTATA-3′.

4. Promoter elements for RNA polymerase II include sequences at −27 (the TATA box) and between −50 and −100. The insertion of 10 bp would separate the promoter elements by the distance of one turn of the DNA helix, thereby diminishing the binding of proteins required for transcription initiation. However, the protein-binding sites would still be on the same side of the helix. Inserting 5 bp (half of a helical turn) would move the protein-binding sites to opposite sides of the helix, making it even more difficult to initiate transcription.

5. G · C base pairs are more stable than A · T base pairs. Hence, the more G · C base pairs that the promoter contains, the more difficult it is to form the open complex during transcription initiation.

6. Transcription of an rRNA gene yields a single rRNA molecule that is incorporated into a ribosome. In contrast, transcription of a ribosomal protein gene yields an mRNA that can be translated many times to produce many copies of its corresponding protein. The greater number of rRNA genes relative to ribosomal protein genes helps ensure the bal-

anced synthesis of rRNA and proteins necessary for ribosome assembly.

7. The cell lysates can be applied a column containing a matrix with immobilized poly(dT). The poly(A) tails of processed mRNAs will bind to the poly(dT) while other cellular components are washed away. The mRNAs can be eluted by decreasing the salt concentration to destabilize the A · T base pairs.

8. The mechanism of RNase hydrolysis requires a free 2′-OH group to form a 2′,3′-cyclic phosphate intermediate (Section 11-3A). Nucleotide residues lacking a 2′-OH group would therefore be resistant to RNase-catalyzed hydrolysis.

9. The mRNA splicing reaction, which requires no free energy input and results in no loss of phosphodiester bonds, is theoretically reversible *in vitro*. However, the degradation of the excised intron makes the reaction irreversible in the cell.

10. Inhibition of snRNA processing interferes with mRNA splicing. As a result, host mRNA cannot be translated, so the host ribosomes will synthesize only viral proteins.

CHAPTER 26

1. A 4-nt insertion would add one codon and shift the gene's reading frame by one nucleotide. The proper reading frame could be restored by deleting a nucleotide. Gene function, however, would not be restored if (1) the 4-nt insertion interrupted the codon for a functionally critical amino acid; (2) the 4-nt insertion created a codon for a structure-breaking amino acid; (3) the 4-nt insertion introduced a stop codon early in the gene; or (4) the 1-nt deletion occurred far from the 4-nt insertion so that even though the reading frame was restored, a long stretch of frameshifted codons separated the insertion and deletion points.

2. The possible codons are UUU, UUG, UGU, GUU, UGG, GUG, GGU, and GGG. The encoded amino acids are Phe, Leu, Cys, Val, Trp, and Gly (see Table 26-1).

3. An amber mutation results from any of the point mutations XAG, UXG, or UAX → UAG. The XAG codons specify Gln, Lys, and Glu; the UXG codons specify Leu, Ser, and Trp; and the UAX codons that are not stop codons both specify Tyr. Hence some of the codons specifying these amino acids can undergo a point mutation to UAG.

4. Ribosomes cannot translate double-stranded RNA, so the base pairing of a complementary antisense RNA to an mRNA prevents its translation.

5. Phe lacks the —OH group that occurs in Tyr and therefore binds too weakly to TyrRS to participate in the aminoacylation reaction to a significant extent.

6. Only newly synthesized bacterial polypeptides have fMet at their N-terminus. Consequently, the appearance of fMet in a mammalian system signifies the presence of invading bacteria. Leukocytes that recognize the fMet residue can therefore combat these bacteria through phagocytosis.

7. The ribosome contains 52 proteins and 3 RNAs. A minimum of 31 noninitiating tRNAs and their 20 cognate aminoacyl–tRNA synthetases are required. tRNA$_f^{Met}$ is

also required (it is charged by the aminoacyl–tRNA synthetase for Met). Initiation requires 3 factors: IF-1, IF-2, and IF-3. Elongation requires 3 factors: EF-Tu, EF-Ts, and EF-G. Termination requires 3 factors: RF-1, RF-2, and RF-3. mRNA is also required. Thus, the total number of different macromolecules is

$$52 + 3 + 31 + 20 + 1 + 3 + 3 + 3 + 1 = 117$$

8. Prokaryotic ribosomes can select an initiation codon located anywhere on the mRNA molecule as long as it lies just downstream of a Shine–Dalgarno sequence. In contrast, eukaryotic ribosomes usually select the AUG closest to the 5′ end of the mRNA. Eukaryotic ribosomes therefore cannot recognize a translation initiation site on a circular mRNA.

9. Assuming the leading fMet residue is cleared from the mature polypeptide as it often is:

   Initiation requires 1 GTP (1 ATP equivalent).

   100 cycles of elongation require 100 GTP (100 ATP) for EF-Tu action, and 100 GTP (100 ATP) for EF-G action.

   Termination requires 1 GTP (1 ATP).

   The synthesis each aminoacyl-tRNA occurs via the pyrophosphate cleavage of ATP and hence requires 2 ATP equivalents. A total of 101 aminoacyl-tRNAs are required: the leading Met-tRNA$_f^{Met}$ + 100 others.

   Thus the number of ATP equivalents = $1 + 100 \times 2 + 1 + 101 \times 2 = 404$ (if the leading fMet remained on the mature polypeptide, this number would be 400 since one less aminoacyl-tRNA and one less elongation step would be required to form the 100-residue polypeptide).

10.

	Start	Lys	Pro	Ala
5′-AGGAGCUX$_{-4}$	^A_GUG	AA^A_G	CCX	GCX-

Shine–Dalgarno sequence.
3–10 base pairs with G · U's allowed

Gly	Thr	Glu	Asn	Ser	STOP
GGX	ACX	GA^A_G	AA^U_C	UCX	UAA
				or	UAG - 3′
				AG^U_C	UGA

11. (a) Each ORF begins with an initiation codon (ATG) and ends with a stop codon (TGA):
    ATGCTCAACTATATGTGA encodes *vir-2* and
    ATGCCGCATGCTCTGTTAATCACATATAGTTGA on the complementary strand encodes *vir-1*.
    (b) *vir-1*: MPHALLITYS; *vir-2*: MLNYM.
    (c) *vir-1*: MPHALLIPYS; *vir-2*: MLNYMGLTEHAA.

12. There are four exons (the underlined bases)

    TATAATACGCGCAATACAATCTACAGCTTC<u>GCGTA</u>
    <u>AATCGTAGGT</u>AAGTTGTAATAAATATAAGTGAGT
    <u>ATGATAGGGCTTTGGACCGATAGATGCGACCCTG</u>
    <u>GAGGT</u>AAGTATAGATAATTAAGCACAGG<u>CATGCA</u>
    <u>GGGATATCCTCCAAATAGGT</u>AAGTAACCTTACGG
    TCAATTAATTAGG<u>CAGTAGATGAATAAACGATAT</u>
    <u>CGATCGGTTAGGT</u>AAGTCTGAT

The mature mRNA, which has a 5′ cap and a 3′ poly(A) tail, therefore has the sequence

GCGUAAAUCGUAGGCUUUGGACCGAUAG**AUG**
CGACCCUGGAGCAUGCAGGGAUAUCCUCCAAA
UAGCAGUAGA**UGA**AUAAACGAUAUCGAUCGG
UUAGGU

The initiation codon and termination codon are shown in boldface. The encoded protein has the sequence

MRPWSMQGYPPNSSR

CHAPTER 27

1. Virtually all the DNA sequences in *E. coli* are present as single copies, so the renaturation of *E. coli* DNA is a straightforward process of each fragment reassociating with its complementary strand. In contrast, the human genome contains many repetitive DNA sequences. The many DNA fragments containing these sequences find each other to form double-stranded regions (renature) much faster than the single-copy DNA sequences that are also present, giving rise to a biphasic renaturation curve.

2. Because genes encoding proteins with related functions often occur in operons, the identification of one or several genes in an operon may suggest functions for the remaining genes in that operon.

3. (a) O_1 is the primary repressor-binding site, so *lac* repressor cannot stably bind to the operator in its absence and repression cannot occur.
   (b) Both O_2 and O_3 are secondary repressor-binding sequences. If one is absent, the other can still function, resulting in only a small loss of repressor effectiveness.
   (c) In the absence of both O_2 and O_3, the repressor can bind only to O_1, which partially interferes with transcription but does not repress transcription as fully as when a DNA loop forms through the cooperative binding of *lac* repressor to O_1 and another operator sequence.

4. In the absence of β-galactosidase (the product of the *lacZ* gene), lactose is not converted to the inducer allolactose. Consequently, *lac* enzymes, including galactoside permease, are not synthesized.

5. Since operons other than the *lac* operon maintain their sensitivity to the absence of glucose, the defect is probably not in the gene that encodes CAP. Instead, the defect is probably located in the portion of the *lac* operon that binds CAP–cAMP.

6. In eukaryotes, transcription takes place in the nucleus and translation occurs in the cytoplasm. Hence, in eukaryotes, ribosomes are never in contact with nascent mRNAs, an essential aspect of the attenuation mechanism in prokaryotes.

7. Deletion of the leader peptide sequence from *trpL* would eliminate sequence 1 of the attenuator. Consequently, the 2·3 hairpin rather than the 3·4 terminator hairpin would form. Transcription would therefore continue into the remainder of the *trp* operon, which would then be regulated solely by *trp* repressor.

8. (a) gp*N* is an antiterminator; without it, the phage would be unable to transcribe the genes encoding proteins re-

quired for phage replication. Consequently, only lysogenic growth would be possible.

(b) cII is required for lysogeny. The absence of *cII* would force the phage to follow the lytic pathway.

(c) *cro* encodes a protein that represses the transcription of all λ phage genes including the λ repressor. In the absence of *cro*, λ repressor is continuously produced, allowing only lysogenic growth.

9. A sequence located downstream of the gene's promoter (i.e., within the coding region) could regulate gene expression if it were recognized by the appropriate transcription factor such that the resulting DNA–protein complex successfully recruited RNA polymerase to the promoter.

10. The susceptibility of RNA to degradation *in vivo* makes it possible to regulate gene expression by adjusting the rate of mRNA degradation. If mRNA were very stable, it might continue to direct translation even when the cell no longer needed the encoded protein.

11. The *esc* gene is apparently a maternal-effect gene. Thus, the proper distribution of the *esc* gene product in the fertilized egg, which is maternally specified, is sufficient to permit normal embryonic development regardless of the embryo's genotype.

INDEX

Page numbers in **bold face** refer to a major discussion of the entry. F after a page number refers to a figure. T after a page number refers to a table. Positional and configurational designations in chemical names (e.g , 3-, α-, *N*-, *p*-, *trans*-, D-) are ignored in alphabetizing. Numbers and Greek letters are otherwise alphabetized as if they were spelled out.

TABLE OF CONTENTS FOR THE ANIMATED FIGURES

TABLE OF CONTENTS FOR THE GUIDED EXPLORATIONS

TABLE OF CONTENTS FOR THE KINEMAGES

EXERCISE 11 (E11_ATCs.kin): Aspartate Transcarbamoylase (ATCase) – Allosteric Interactions

Kinemage 1: Quaternary structure of ATCase and its allosteric conformational changes (Section 12-3; Fig. 12-12)

Kinemage 2: Conformational changes caused by the binding of substrate (Fig. 12-13)

EXERCISE 12 (E12_TIM.kin): Triose Phosphate Isomerase (TIM) – Catalytic Mechanism

Kinemage 1: The eight-stranded alpha/beta barrel (Section 6-2B; Figs. 6-27c and 6-28d)

Kinemage 2: TIM active site (Section 14-2E; Fig. 14-6)

EXERCISE 13 (E13_PFK.kin): Phosphofructokinase (PFK) – Allosteric Interactions

Kinemage 1: Conformational changes in a dimeric unit of PFK (Section 14-4A; Fig. 14-22 and 14-24)

Kinemage 2: The major conformational changes in a subunit of PFK (Fig. 14-24)

EXERCISE 14 (E14_Phos.kin): Glycogen Phosphorylase – Allosteric Interactions

Kinemage 1: Structure of glycogen phosphorylase *a* (Section 15-1A; Fig. 15-3)

Kinemage 2: Conformational differences between T and R states of glycogen phosphorylase b (Fig. 15-5)

Kinemage 3: The dimer interface on activation of glycogen phosphorylase

EXERCISE 15 (E15_cAPK.kin): cAMP-Dependent Protein Kinase (cAPK)

Kinemage 1: The catalytic subunit of cAPK (Section 15-3B; Fig. 15-14)

EXERCISE 16 (E16_CaM.kin): Calmodulin (CaM)

Kinemage 1: The structure of CaM highlighting its EF-hand motifs (Section15-3B; Figs. 15-16 and 15-17)

Kinemage 2: The structure of CaM in complex with its target peptide (Fig. 15-18)

EXERCISE 17 (E17_DNA2.kin): DNA – Structures of A, B, and Z forms

Kinemage 1: Comparison of the structures of A-, B-, and Z-DNAs (Section 23-1; Fig. 23-2)

Kinemage 2: The Watson–Crick base pairs (Section 23-1A; Fig. 23-1)

Kinemage 3: Sugar pucker, 3'-*exo* and 3'-*endo* (Section 23-1B; Fig 23-6)

Kinemage 4: B-DNA (Section 23-1; Fig. 23-2)

Kinemage 5: A-DNA (Section 23-1; Fig. 23-2)

Kinemage 6: Z-DNA (Section 23-1; Fig. 23-2)

EXERCISE 18 (E18_EcoR.kin): *Eco*RI and *Eco*RV Restriction Endonucleases – Complexes with Target DNAs

Kinemage 1: *Eco*RI–DNA (Fig. 23-30)

Kinemage 2: *Eco*RV–DNA (Fig. 23-31)

EXERCISE 19 (E19_Rprs.kin): 434 Repressor /DNA Interactions

Kinemage 1: 434 Repressor – DNA interactions (Fig. 23-33)

EXERCISE 20 (E20_tRNA.kin): tRNA^{Phe} – Structural Interactions

Kinemage 1: Overview of functional domains (Fig. 26-6)

Kinemage 2: Structural features (Section 26-2A)

Kinemage 3: Tertiary base-pairing interactions (Fig. 26-7)

EXERCISE 21 (E21_GnRS.kin): Glutaminyl–tRNA Synthetase (GlnRS) – Complex with tRNA^{Gln} and ATP

Kinemage 1: The overall complex (Fig. 26-11)

TABLE OF CONTENTS FOR THE INTERACTIVE EXERCISES

One- and Three-Letter Symbols for the Amino Acids[a]

A	Ala	Alanine
B	Asx	Asparagine or aspartic acid
C	Cys	Cysteine
D	Asp	Aspartic acid
E	Glu	Glutamic acid
F	Phe	Phenylalanine
G	Gly	Glycine
H	His	Histidine
I	Ile	Isoleucine
K	Lys	Lysine
L	Leu	Leucine
M	Met	Methionine
N	Asn	Asparagine
P	Pro	Proline
Q	Gln	Glutamine
R	Arg	Arginine
S	Ser	Serine
T	Thr	Threonine
V	Val	Valine
W	Trp	Tryptophan
Y	Tyr	Tyrosine
Z	Glx	Glutamine or glutamic acid

[a]The one-letter symbol for an undetermined or nonstandard amino acid is X.

Thermodynamic Constants and Conversion Factors

Joule (J)

$1\ J = 1\ kg \cdot m^2 \cdot s^{-2}$ $1\ J = 1\ C \cdot V$ (coulomb volt)

$1\ J = 1\ N \cdot m$ (newton meter)

Calorie (cal)

1 cal heats 1 g of H_2O from 14.5 to 15.5°C

$1\ cal = 4.184\ J$

Large calorie (Cal)

$1\ Cal = 1\ kcal$ $1\ Cal = 4184\ J$

Avogadro's number (N)

$N = 6.0221 \times 10^{23}$ molecules $\cdot mol^{-1}$

Coulomb (C)

$1\ C = 6.241 \times 10^{18}$ electron charges

Faraday ($\mathcal{F}$)

$1\ \mathcal{F} = N$ electron charges

$1\ \mathcal{F} = 96,485\ C \cdot mol^{-1} = 96,485\ J \cdot V^{-1} \cdot mol^{-1}$

Kelvin temperature scale (K)

0 K = absolute zero   273.15 K = 0°C

Boltzmann constant (k_B)

$k_B = 1.3807 \times 10^{-23}\ J \cdot K^{-1}$

Gas constant (R)

$R = Nk_B$ $R = 1.9872\ cal \cdot K^{-1} \cdot mol^{-1}$

$R = 8.3145\ J \cdot K^{-1} \cdot mol^{-1}$ $R = 0.08206\ L \cdot atm \cdot K^{-1} \cdot mol^{-1}$

The Standard Genetic Code

First Position (5′ end)	Second Position				Third Position (3′ end)
	U	C	A	G	
U	UUU Phe	UCU Ser	UAU Tyr	UGU Cys	U
	UUC Phe	UCC Ser	UAC Tyr	UGC Cys	C
	UUA Leu	UCA Ser	UAA Stop	UGA Stop	A
	UUG Leu	UCG Ser	UAG Stop	UGG Trp	G
C	CUU Leu	CCU Pro	CAU His	CGU Arg	U
	CUC Leu	CCC Pro	CAC His	CGC Arg	C
	CUA Leu	CCA Pro	CAA Gln	CGA Arg	A
	CUG Leu	CCG Pro	CAG Gln	CGG Arg	G
A	AUU Ile	ACU Thr	AAU Asn	AGU Ser	U
	AUC Ile	ACC Thr	AAC Asn	AGC Ser	C
	AUA Ile	ACA Thr	AAA Lys	AGA Arg	A
	AUG Met[a]	ACG Thr	AAG Lys	AGG Arg	G
G	GUU Val	GCU Ala	GAU Asp	GGU Gly	U
	GUC Val	GCC Ala	GAC Asp	GGC Gly	C
	GUA Val	GCA Ala	GAA Glu	GGA Gly	A
	GUG Val	GCG Ala	GAG Glu	GGG Gly	G

[a]AUG forms part of the initiation signal as well as coding for internal Met residues.

Some Common Biochemical Abbreviations

A	adenine		ETF	electron-transfer flavoprotein
aaRS	aminoacyl–tRNA synthetase		F1P	fructose-1-phosphate
ACAT	acyl-CoA:cholesterol acyltransferase		F2,6P	fructose-2,6-bisphosphate
ACP	acyl-carrier protein		F6P	fructose-6-phosphate
ADA	adenosine deaminase		FAD	flavin adenine dinucleotide, oxidized form
ADP	adenosine diphosphate		FADH·	flavin adenine dinucleotide, radical form
AIDS	acquired immunodeficiency syndrome		$FADH_2$	flavin adenine dinucleotide, reduced form
ALA	δ-aminolevulinic acid		FBP	fructose-1,6-bisphosphate
AMP	adenosine monophosphate		FBPase	fructose-1,6-bisphosphatase
ATCase	aspartate transcarbamoylase		Fd	ferredoxin
ATP	adenosine triphosphate		FH	familial hypercholesterolemia
BChl	bacteriochlorophyll		fMet	N-formylmethionine
bp	base pair		FMN	flavin mononucleotide
BPG	D-2,3-bisphosphoglycerate		G	guanine
BPheo	bacteriopheophytin		G1P	glucose-1-phosphate
BPTI	bovine pancreatic trypsin inhibitor		G6P	glucose-6-phosphate
C	cytosine		G6PD	glucose-6-phosphate dehydrogenase
CaM	calmodulin		GABA	γ-aminobutyric acid
CAM	crassulacean acid metabolism		Gal	galactose
cAMP	cyclic AMP		GalNAc	N-acetylgalactosamine
CAP	catabolite gene activator protein		GAP	glyceraldehyde-3-phosphate
cAPK	cAMP-dependent protein kinase		GAPDH	glyceraldehyde-3-phosphate dehydrogenase
CDK	cyclin-dependent protein kinase		GDP	guanosine diphosphate
cDNA	complementary DNA		Glc	glucose
CDP	cytidine diphosphate		GlcNAc	N-acetylglucosamine
CE	capillary electrophoresis		GMP	guanosine monophosphate
Chl	chlorophyll		GPI	glycosylphosphatidylinositol
CM	carboxymethyl		GSH	glutathione
CMP	cytidine monophosphate		GSSH	glutathione disulfide
CoA or			GTF	general transcription factor
CoASH	coenzyme A		GTP	guanosine triphosphate
CoQ	coenzyme Q (ubiquinone)		Hb	hemoglobin
CPS	carbamoyl phosphate synthetase		HDL	high density lipoprotein
CTP	cytidine triphosphate		HIV	human immunodeficiency virus
D	dalton		HMG-CoA	β-hydroxy-β-methylglutaryl-CoA
d	deoxy		hnRNA	heterogeneous nuclear RNA
DCCD	dicyclohexylcarbodiimide		HPLC	high-performance liquid chromatography
dd	dideoxy		Hsp	heat shock protein
ddNTP	2′,3′-dideoxynucleoside triphosphate		HTH	helix–turn–helix
DEAE	diethylaminoethyl		Hyl	5-hydroxylysine
DG	1,2-diacylglycerol		Hyp	4-hydroxyproline
DHAP	dihydroxyacetone phosphate		IDL	intermediate density lipoprotein
DHF	dihydrofolate		IF	initiation factor
DHFR	dihydrofolate reductase		IgG	immunoglobulin G
DNA	deoxyribonucleic acid		IMP	inosine monophosphate
DNP	2,4-dinitrophenol		IP_3	inositol-1,4,5-trisphosphate
dNTP	2′-deoxynucleoside triphosphate		IPTG	isopropylthiogalactoside
E4P	erythrose-4-phosphate		IS	insertion sequence
EF	elongation factor		ISP	iron–sulfur protein
ELISA	enzyme-linked immunosorbent assay		kb	kilobase pair
emf	electromotive force		kD	kilodalton
ER	endoplasmic reticulum		K_M	Michaelis constant

(table continued on following page)